# Calculus for Scientists and Engineers
## An Analytical Approach

## Other Books of Interest

Advanced Methods of Mathematical Physics
*R.S. Kaushal and D. Parashar*

Aspects of Combinatorics and Combinatorial Number Theory (1-84265-049-1)
*S.D. Adhikari*

An Elementary Course in Partial Differential Equations (81-7319-170-0)
*T. Amarnath*

Engineering Mathematics (1-84265-086-6)
*R.K. Jain and S.R.K. Iyengar*

A First Course in Algebraic Topology (1-84265-003-3)
*B.K. Lahiri*

A First Course in Mathematics Analysis (81-7319-064-X)
*D. Somasundaram and B. Choudhary*

Foundations of Complex Analysis (81-7319-040-2)
*S. Ponnusamy*

Foundations of Functional Analysis (1-84265-079-3)
*S. Ponnusamy*

Function Spaces and Applications (1-84265-002-5)
*D.E. Edmunds et al*

Functional Analysis: Selected Topics (81-7319-199-9)
*Pawan K. Jain*

Fundamentals of Approximation Theory (1-84265-016-5)
*Hrushikesh N. Mhaskar, Devidas V. Pai*

Fuzzy Topology (1-84265-098-X)
*N. Palaniappan*

An Introduction to Commutative Algebra and Number Theory (81-7319-304-5)
*S.D. Adhikari*

An Introduction to Measure and Integration (81-7319-120-4)
*Inder K. Rana*

Introduction to rings and Modules (81-7319-037-2)
*C. Musili*

Linear Algebra (1-84265-080-7)
*Vivek Sahai and Vikas Bist*

Metric Spaces Corrected Edition (81-85198-99-3)
*P.K. Jain and K. Ahmad*

Nonlinear Functional Analysis (81-7319-230-8)
*R. Akerkar*

Nonlinearities in Complex Systems (81-7319-182-4)
*Sanjay Puri and Sushanta Dattagupta*

Ordinary Differential Equations (1-84265-069-6)
*D. Somasundaram*

Partial Differential Equations for Engineers and Scientists (1-84265-028-9)
*J.N. Sharma and Kehar Singh*

Sequence Spaces and Applications (81-7319-239-1)
*P.K. Jain and E. Malkowsky*

Topics in Sobolev Spaces and Applications (1-84265-094-7)
*D. Bahuguna, V. Raghavendra and B.V. Rathish Kumar*

Topological Algebras (81-7319-282-0)
*V.K. Balachandran*

Wavelets and Allied Topics (1-84265-076-9)
*P.K. Jain*

# Calculus for Scientists and Engineers
## An Analytical Approach

**K.D. Joshi**

Alpha Science International Ltd.

Pangbourne England

Learning Resources
Centre

12295663

**K.D. Joshi**
Department of Mathematics
Indian Institute of Technology, Bombay
Mumbai-400 076, India

Alpha Science International Ltd.
P.O. Box 4067, Pangbourne RG8 8UT, UK

ISBN 1-84265-048-3

Printed in India.

Dedicated to the memory of

## Prof. M.S. Huzurbazar

*from whom I learnt
what mathematics really is*

# Preface

The importance of mathematics in the basic courses of any science and engineering program is too well recognised to need any comment. But an occasionally undesirable (albeit unintentionally so) outcome is that in such a program mathematics is often looked at only as a tool. As a result, the approach is practical rather than analytical. Put differently, the emphasis is more on the 'how's' and less on the 'why's' of mathematics. The latter are either omitted as too theoretical or are covered only perfunctorily, even though the brighter students are intellectually capable of understanding and appreciating them.

The present book, as its title indicates, lays more emphasis on the analytical approach to calculus. Each section is like an essay on a particular topic appearing in its title. But it also deals with some associated topics and questions that arise naturally. For example, the section titled 'Multitude of Tests for Convergence of Series' not only gives the standard tests for convergence of series, but also answers why so many tests are needed in the first place. In addition, a conscious effort is made to give a strong motivation for the abstract concepts covered and thereby help a beginner overcome most of the aversion he has for abstraction. For example, instead of merely giving the $\epsilon$-$\delta$ definition of a limit, a whole section is devoted to explain how the definition evolved and why it is a most natural formulation of the basic concept of a limit. This will help a student get over the clumsiness of the definition, which can be quite repulsive otherwise.

The first two chapters are devoted to the various facets, especially as they appear to a beginner, of the nature of mathematics in general and of calculus in particular. The next four chapters deal systematically with the standard topics encountered in a first course in collegiate mathematics. There is a large number of exercises of various levels of difficulties. There are very few exercises meant solely to give computational drill. Nearly all exercises require some thinking. Answers to the exercises are provided at the end of the book.

The book can be used as a regular textbook for a first course in engineering mathematics. Most students in such a course already have a working knowledge of the elementary techniques of calculus. The present book will complement it by giving them deeper insights and thereby increasing their level of maturity. Such familiarity is, however, not a strict prerequisite. The book can also be used by students with litle or no background of calculus, because the treatment is self-contained. (The occasional references in the initial chapters to more advanced topics are of an incidental nature and are usually explained later on at appropriate places.) Even though the emphasis is more on the 'Why's' than on the 'how's' of mathematics, the latter are not entirely ignored. There is adequate exposure to the techniques through the exercises. The answers to these exercises will enable the sincere student to acquire the how's as well.

Yet another class of students who may find the book useful is those who want to pursue mathematics as their major subject. Such students often have some difficulty in the transition from calculus to analysis. The treatment given here will enable them get over the rough edges.

Even for those who want mathematics strictly as a tool, I believe there is a need to know at least some of the 'why's. With easy availability of software packages to do just about any mathematical task, the 'how's can now be handled by machines. Human brains should therefore concentrate more on the 'why's. Machines can add and subtract flawlessly and instantaneously. But there is still no machine which tells *why* we have to add and not subtract in a given problem.

This change of perspective has influenced both the contents and the style. Thus worked-out problems and diagrams are used only sparingly as compared with the traditional textbooks on the subject. These things do occur in the present book, but only to illustrate the underlying thoughts. Also missing are biographical references in the footnotes. Wherever a mathematician is mentioned by a name, it is because of his work. I personally believe this is an adequate tribute to him. Unless his time is directly relevant to the matter at hand, there is little point in giving the years of his birth and death, especially these days when such information (and much more) is easily available on many websites (e.g. http://www-history.mcs.st-andrews.ac.uk/history/index.html).

I am grateful to the Curriculum Development Programme at I.I.T. Bombay for funding the preparation of the manuscript of this book. The contents of a book of this type are so well-known that it is difficult to say which material I borrowed from which source. But I must mention the classic book (in two volumes) *Introduction to Calculus and Analysis* by R. Courant and F. John which has influenced me throughout right from the time I was a student. The present book is prompted by my long experience of teaching core mathematics courses to the B.Tech. students at I.I.T. Bombay. Many of the questions answered here were originally asked (in various forms) by the students. I express my gratitude to them and invite similar questions and comments from the readers.

K.D. Joshi

# SYMBOLS

## Standard Symbols

| symbol | meaning |
|---|---|

$\in$      belongs to (or 'belonging to')

$\exists$      there exists

$\forall$      for every (or 'for all')

$\subset$      contained in (and possibly equal to)

$\subsetneqq$      contained in but not equal to

$A \cup B$      union of (sets) $A$ and $B$

$A \cap B$      intersection of (sets) $A$ and $B$

$A - B$      complement of $B$ in $A$ ($B$ need not be a subset of $A$)

$|A|$      number of elements of (a finite set) $A$

$|a|$      absolute value of (a real number) $a$

$[a]$      the integral part of (a real number) $a$ (i.e. the greatest integer not exceeding $a$)

$n!$      '$n$ factorial' $(= 1.2.3.\ldots\ldots(n-1).n$ where $n \in I\!N)$

$\binom{n}{m}$      '$n$ choose $m$' $\left(= \dfrac{n(n-1)\ldots(n-m+1)}{m!}\right)$

$\emptyset$      the empty (or null) set

$I\!N$      the set of positive integers

$\mathbb{Z}$      the set of all integers

$Q$      the set of rational numbers

$I\!R$      the set of real numbers

$I\!R^n$      the set of ordered $n$-tuples of real numbers ($n \in I\!N$)

iff      if and only if

∎      end of the proof of a theorem. (Also used after the statement of a theorem if the proof is omitted or precedes the statement.)

Where single letter symbols are used to denote vectors, in the text they are put in bold face (e.g. **a**, **u**, **i** etc.). But in the hand-drawn diagrams they are denoted by putting an arrow on top (e.g. $\vec{a}, \vec{u}, \vec{i}$ etc.).

# Contents

# Chapter 1

# NATURE OF MATHEMATICS IN GENERAL

## 1.1    Reasons for Studying Mathematics

Frankly, for most persons, the major reason is that they have no choice! Mathematics is a required subject not only in school, but also in programmes leading to degrees in sciences and engineering. Nowadays even disciplines traditionally regarded as poles apart from mathematics (such as medicine and economics) require some parts of mathematics. And so the students and the researchers have to study mathematics as a tool, whether they like it or not.

There are, of course, those who study mathematics for various other reasons. In schools, there are always a few pupils in a class who like mathematics as a scoring subject. Even after the passion for marks wears out, some students continue to like mathematics for its precision and consequent certainty. Unlike the theories of science which are always subject to change (depending upon the discovery of new facts), there is a certain permanence to the theorems in mathematics which fascinates some persons. A mathematical result, once proved correctly will remain true forever. It may become useless or it may be completely subsumed by a stronger and a more general result. But it will never be false. (See Section 1.6 for more elaboration of this point.)

Then there are those who like mathematics as an intellectual sport. The initiation usually comes through mathematical puzzles. (See the exercises for a few examples of such puzzles.) Certain problems in schools (especially the riders in pure geometry) are challenges which trigger some of the ambitious students, occasionally leading to a healthy competition. In some cases, the love for pure geometry continues long after school days. Frequently problems in pure geometry can be done rather routinely if more advanced techniques such as co-ordinates, trigonometry or vectors are used. But it is quite a challenge

1

to solve them by methods of pure geometry only. (Again see the exercises for a few examples of such problems.) Some persons (including the laymen as well as professional mathematicians) spend hours and hours of their leisure time to meet these challenges. Number theory is yet another inexhaustible source of challenging problems. In fact some of these problems are so simple to state that even a school boy can understand them. (See Exercise (1.6) for an example of an unsolved problem.) Attempting to solve such problems has the same irresistible appeal to some persons as a treacherous peak has to a mountaineer.

There is also a view that mathematics is an art rather than a science. (See Section 1.7 for an elaboration of this view point.) The theorems of mathematics are like works of art such as paintings, sculptures, poems or melodies. They are created not because of any possible practical use, nor even so as to meet any challenge but purely for the aesthetic pleasures they give.

It must be admitted that those who enjoy mathematics as a sport or as an art are not very multitudinous. You don't have to be a singer yourself to get elated by a musical composition rendered by someone else. And when you watch a cricket match, you can fully share the excitement of a glorious sixer, even if you have never held a bat in your hand. In mathematics, on the other hand, it is difficult to appreciate it as a sport or as an art if you are merely a spectator.

Summing up, a vast majority of those who study mathematics does so for its applications. But there are others who study it for other reasons as well.

## EXERCISES

1.1   A fisherman wants to take a goat, a wolf and a pile of hay across a river. He has a boat in which he can accommodate only one of these at a time. He must not leave the wolf and the goat together unattended or the goat and the hay together unattended. Help him.

1.2   There are 11 genuine coins of equal weights and a fake coin of a different weight (it not being given whether the fake coin is lighter or heavier than a genuine one). Using a balance only thrice, detect the fake coin and also whether it is lighter or heavier than a genuine one.

1.3   Tablets, each weighting 10 grams are produced on a number of machines. One of the machines is faulty and produces tablets weighting 9 gms each. With only are weighing, detect the faulty machine.

1.4   $ABC$ is a triangle in which $AB = AC$. $D$ is the mid-point of $BC$ and $E$ is the foot of the perpendicular drawn from $D$ to $AC$. $F$ is the midpoint of $DE$. Show that $AF$ and $BE$ are perpendicular to each other.

1.5   $ABC$ and $A'B'C'$ are triangles in the same plane. If the perpendiculars from $A, B$ and $C$ to $B'C', C'A'$ and $A'B'$ respectively are concurrent, prove that so are the perpendiculars from $A', B'$ and $C'$ to $BC, CA$ and $AB$ respectively.

1.6 Prove that every even integer from 4 to 100 can be expressed as a sum of two primes (e.g. $10 = 7 + 3, 20 = 13 + 7$ or $17 + 3$). (One of the most famous conjectures in number theory, called **Goldbach's conjecture** states that every even integer $> 2$ is expressible as a sum of two primes, in at least one way. This deceptively simple statement has still eluded a proof, although it has been verified for a large number of even integers.)

## 1.2   The Intimidating Image of Mathematics

Probably no other subject invites as varied a reaction as mathematics. There are those (including some non-mathematicians) who simply love it. But the fact remains that many persons develop a phobia for mathematics often at a very young age. It is not uncommon to find students who are reasonably confident in other subjects but who dread mathematics. The framing of the curricula, the method of teaching and the attitude of the teacher are undoubtedly a few contributing factors, especially at young ages. There are, however, certain unique features of mathematics itself which serve to give it a formidable image.

The strongest among these is perhaps, the unrelenting precision demanded by mathematics. Even a single numerical slip can be costly in a mathematics problem and students who lose marks because of it can develop a grudge against mathematics. Even those who are good at calculating without mistakes often find it hard to handle the logical precision needed in mathematics because as laymen we rarely insist on such precision. For example, when we say "Those below 18 cannot vote", in real life we take it to mean that those who are 18 or above are eligible to vote. The more careful will add a rider "provided they are not disqualified for some other reason". But everybody will agree that the statement means that 18 is the voting age. But a mathematician will disagree. To him the statement is merely laying down a necessary condition for voting. And necessity should never be confused with sufficiency. (See Sec. 1.5 for further elaboration and Sec.1.8 for an exception.)

Another instance where the mathematical thinking differs from that in real life involves what is called quantification of truth. Every mathematical statement is either 'true' or 'false' and never both. There are no degrees of truth in between. Truth means complete and absolute truth, without qualification. There is no such thing as a statement which is 'almost true' or 'partially true' or 'having a shade of truth', although we commonly use such phrases in practice. The reason we use such expressions is that the statement involves, directly or indirectly, some quantity and the degree of truth of the statement is measured by how close the actual quantity is to the quantity implicit in the statement. For example, take a statement 'John is tall'. In this statement, some standard of tallness is implicit, which may, of course, depend upon the context. Let us suppose that this standard is a minimum height of 180 centimeters. Now if John's height is, say, 150, 160, 170, 175, 179, 180, 183, 186 and 190 centimeters then we would probably describe the statement, respectively, as grossly false, false, having a shade of truth, substantially true, almost true, true, quite true, very true and an understatement! From a mathamatical point of view the statement is false ('equally false') in the first five cases and true ('equally true') in the remaining four.

In a way, this feature of mathematics (which is also known as **bi-valued logic** or the **law of excluded middle**) makes mathematical statements easier to handle. But to a layman, it gives the impression that mathematics is a fastidious and uncompromising taskmaster. In particular, when it comes to proving the truth of some assertion, it has to be proved to the hilt in all cases. It is not enough to cite some examples where it is true. Such examples do have considerable illustrative value but they are no substitute for a mathematical proof. Nor is it enough to prove the assertion in 'almost all' cases and conveniently dismiss the remaining ones by saying "the exception proves the rule!". In mathematics, even one exception (or a **counterexample** as it is called) renders a statemnet about members of a class false (as false as millions of such exceptions would!) Phrases such as 'almost all' are occasionally used in mathematics too. But they occur with a precise, mathematical meaning and not the way we use them in real life. (See, for example, Section 3.2.)

The insistence on logical perfection also makes it necessary to define certain intuitively clear concepts in a rather clumsy manner. (See Section 1.8. for examples.) When this happens in the case of some of the most basic concepts such as limits, it confuses and alienates the students.

What makes things worse is that mathematics, unlike most other subjects, has a vertical growth. That is, the new builds on the old. In geography, if you don't know the geography of, say, Brazil, it hardly matters when you study the geography of Poland. But in mathematics, an earlier weakness can spoil your subsequent learning unless corrected immediately. In particular, mathematics does not allow you the luxury of putting off your studies till the night before the examination. And who would like that?

Abstraction is yet another feature of mathematics which makes it easier to dislike it. Hardly any instruments are needed in mathematics. So the digression and consequent relief which comes through the laboratory or field work is virtually non-existent in mathematics. Many mathematical results are not amenable to any experimental verification. And even in the case of those that are, experimentation is rarely encouraged because it is no substitute for a mathematical proof. (See Section 1.4 for more elaboration of this point.) To be sure, mathematics has many real life applications. But after the school level, there is usually many a gap between a result and its down-to-earth application. Even at the school level, not everything has a practical use. Take, for example, the nine-point circle in geometry. Unless a student is trained to appreciate it for its aesthetic value, he may dismiss it as unappealing and taxing.

The multifarious nature of mathematics can also be a source of frustration sometimes. Different aspects of mathematics need different faculties of the brain. It is not easy for everybody to be good at all of them. In schools, for example, many students who are a wiz at arithmetic and algebra, draw a blank in geometry. And those who love geometry and trigonometry problems, may fumble miserably in combinatorics and probability. An even more common example is that of a student who is good at computational problems but who cannot give logical arguments, especially those involving epsilons and deltas. A failure of this type can be more demoralising than a total failure everywhere, because it comes after raising false hopes.

Even those who are quite good at and experienced in mathematics occasionally find it hard to come to terms with the uncertainty of assessing the difficulty of some of its problems. You can spend years on a problem and make little progress. And then a sudden flash gives you the solution. This is, of course, a very thrilling experience. But what if the flash comes to someone else and not to you? If you are trying the problem as a hobby you can accept it is a part of the game. But nobody would like such uncertainty in his regular work. (See the exercises at the end for examples of deceptively simple problems.)

These are some of the factors which make mathematics intimidating. It is to be emphasised, however, that most of them can be overcome with a sincere and determined effort on the part of the students and a patient and helping attitude on the part of the instructor. When this is done, these very factors can, in fact, become assuring. For example, once a student learns how to handle the exactitude of mathematics, it becomes a great asset for him. If his answer is exact, he can rest assured that the examiner will have to mark it as correct, regardless of his own personal likings and dislikings. No other subject offers so much objectivity.

## EXERCISES

2.1  Suppose the points $P_1, \ldots, P_n$ ($n > 1$) are all in the same plane but not all on the same line. Prove that there is at least one line which passes through exactly two of these points. (This problem, proposed by the British mathematician Sylvester in 1893, remained unsolved for several decades. But it has an extremely short and elementary solution.)

2.2  Show how to bisect a given segment using only a compass (i.e. without a straight edge). (This problem, as well as its solution are old. But not many people seem to know it. Till you get the key idea, this problem can keep you busy for hours, perhaps days.)

## 1.3     Abstraction in Mathematics

The role of abstraction in mathemmatics has a lot to do with the identity of mathematics or, in simpler words, with what mathematics is all about. Our very first encounter with mathematics is usually through a problem like, "If Anil got 3 apples from his father and 2 from his mother, how many apples did he get in all?" The answer of course, is 5 apples. But obviously, the crux of the problem has little to do with apples. The underlying mathematics is simply to add the figures 3 and 2 and this would remain unchanged even if 'apples' are replaced by some other objects. Here the problem of adding 3 and 2 is the mathematical essence or **abstract** of the original problem which ostensibly deals with apples.

Although the example may seem too trivial, it illustrates the basic point quite well. Mathematics does not concern itself with what the numbers 3 and 2 stand for in real life nor with what their addition stands for, but rather with those features of these numbers and the operation of addition which are independent of any one particular interpretation. For example, in mathematics we study that $3+2 = 2+3$. This can be translated in terms of apples. But such a translation, even though it may be of some interest in a particular situation, is not really the business of mathematics.

Assigning numbers to various quantities involved in real life problems is the first step in abstraction. Depending on the nature of a problem, the numbers to be assigned may be more complicated than whole numbers. For example, in a problem dealing with the length of a piece of wire, we may have to assign fractions or even irrational numbers. (Occasionally, complex numbers come in handy, too, as, for example, in problems involving electric current in an inductive circuit.) In arithmetical problems, the numbers we assign are generally constants. However, in algebra and more advanced mathematics, they are mostly numerical variables, that is, variables taking numerical values.

Merely assigning numbers or numerical variables does not, of course, solve the problem. We have to subject these variables to various mathematical operations such as addition, multiplication, exponentiation, differentiation, integration and so on. Some of these operations correspond very naturally and directly to some real-life constructions or concepts. For example if $x$ and $y$ are the lengths of two (straight) pieces of wire, then $x+y$ is the length of their join while $xy$ is the area of the rectangle having these segments for two of its sides. In the case of operations like differentiation and integration, the correspondence with reality is far more subtle. But in the case of an operation like exponentiation, there is simply no real life equivalent. When $y$ is irrational, the expression $x^y$ has no easy physical interpretation. This is where mathematics really begins to appear abstract. Abstract does not mean practically useless. Even though exponentiation is abstract, calculations based on it often lead to the solutions of some real life problems. (See the exercises for one such problem.)

Abstracting the essence of a problem in terms of numbers, has been a function of mathematics for ages. The success of this approach depends upon the choice of the right mathematical 'model'. A model which is appropriate for one situation may not be so for another, even though there may be some superficial resemblance between the two. (A jovial illustration of this is that if three apples are hanging from a branch and one of them is shot down, then two will be left. Here subtraction is the right mathematical model. But if, instead, three birds are perched on a tree branch and one of them is shot down, then none will be left! Here straight subtraction fails because unlike the apples, the birds can leave the branch on their own.) The trouble is that mathematics itself does not tell you which is a right model for a particular real life problem. This has to be figured out by those who want to apply mathematics. Mathematics can teach you how to add but not *when* to add ! Content

Sometimes, when mathematics is applied to solve some real life problem, it so happens that the mathematical concepts or constructions available at that time are not adequate to solve that problem. In that case, new mathematical concepts and constructions have to be invented. So many developments in mathematics such as the Bessel functions owe their origins to some physical problems. Later on, these developments are studied for their own sake, i.e. regardless of any particular application. It is then very likely that they will appear abstract if no hint is given about their genesis.

Abstraction in terms of numbers is as old as mathematics. A parallel development was the geometric abstraction. Points, lines (or curves) and planes are idealised abstractions of a tiny dot, an object having only length, and a lamina having no thickness respectively. In real life, of course, such things do not exist. Still, the abstract concepts of geometry, along with a few axioms, led to a theory which had not only a certain intrinsic beauty but applications in mensuration problems. For many centuries, algebra and geometry developed mostly independently of each other. But in the seventeenth century, because of the introduction of Cartesian co-ordinates, geometry was completely subsumed by algebra. Although this was a significant achievement, it eliminated the need to draw diagrams (except as an aid to understanding) and took away wherever little visual appeal mathematics had, thereby makings mathematics appear still more abstract.

In the last hundred years or so, mathematics has undergone yet another abstraction. It is non-numerical and is far more pervasive than the numerical one. It has fundamentally changed the popular conception of mathematics as the science of numbers, by shifting its focus from numbers to sets. Admittedly, this kind of abstraction is more difficult to take than the traditional numerical one. But the gains make it worth the trouble. Instead of abstracting solely in terms of numbers, if we abstract in terms of sets with suitable additional structures, so many new avenues are opened, because the choice of the structure is ours. As a result, problems which look highly dis-similar may turn out to have the some essence. (See Section 1.9 for more elaboration of these points.)

Summing up, mathematics is abstract because the very business of mathematics is to strip a problem of its inconsequential details (which may be important in a particular application) and to look at the underlying core, rather like a surgeon who removes a patient's clothes (however elegant and expensive they may be) and operates on the naked body. Without abstraction, mathematics will lose its identity. It will reduce to physics, chemistry, biology, economics, commerce or whatever. But it will not be mathematics!

This is not to suggest that mathematics should always be taught in total disassociation with any particular real life problems. That would be a grave mistake both from the pedagogical and practical point of view. It would not only make mathematics appear dull but useless as well. A person who can rattle off formulas about the curvature of a plane curve but cannot find the point where a given road bends most abruptly is a social parasite.

## EXERCISES

3.1   The (continuous) rates of growth of the populations of two countries, $A$ and $B$, are 3% per annum and 4% per annum respectively. At the beginning of the year 1950, the population of $A$ was twice that of $B$, while at the beginning of 1980, it was 1.5 times that of $B$. Find when the two countries will have equal populations. (The solution requires the use of exponential functions, see Exercise (5.8.12). The problem illustrates how a totally abstract concept can have real life applications.)

### 1.4     Role of Deductive Logic in Mathematics

It is to be noted that deductive logic is not an exclusive prerogative of mathematics. Even in day-to-day life we resort to logical reasoning, sometimes to win an argument, sometimes to unravel a mystery or sometimes to predict the future. The degree of logical consistency expected of a scientist is, of course, much higher than of a layman. But, for a mathematician it is highest. To illustrate this point, suppose a layman, a physicist and a mathematician all start measuring the angles of hundreds of triangles of various shapes, find the sum in each case and keep a record. Suppose that the layman finds that with one or two exceptions, the sum in each case comes out to be 180 degrees. He will ignore the exceptions and say 'The sum of the three angles in a triangle is 180 degrees.' A physicist will be more cautious in dealing with the exceptional cases. He will examine them more carefully. If he finds that the sum in them is somewhere between 179 degrees to 181 degrees, say, then he will attribute the deviation to experimental errors. He will then state a law, 'The sum of the three angles of any triangle is 180 degrees'. He will then watch happily as the rest of the world puts his law to test and finds that it holds good in thousands of different cases, until somebody comes up with a triangle in which the law fails miserably. The physicist now has to withdraw his law altogether or else to replace it by some other law which holds good in all the cases tried. Even this new law may have to be modified at a later date. And this will continue without end.

A mathematician will be the fussiest of all. If there is even a single exception he will refrain from saying anything. Even when millions of triangles are tried without a single exception, he will not state it as a theorem that the sum of the three angles in *any* triangle is 180 degrees. The reason is that there are infinitely many different types of triangles. To generalise from a million to infinity is as baseless to a mathematician as to generalise from one to a million. He will at the most make a *conjecture* and say that there is a strong evidence suggesting that the conjecture is true. But that is not the same thing as proving a theorem. The only proof acceptable to a mathematician is the one which follows from earlier theorems by sheer logical implications. For example, in the present case, such a proof follows easily from the theorem that an external angle of a triangle is the sum of the other two internal angles.

The approach taken by the layman or the physicist is known as the **inductive** approach whereas the mathematician's approach is called the **deductive** approach. In the former, we make a few observations and generalise. In the latter, we deduce from something which is already proven. Of course, a question can be raised as to on what basis this supporting theorem is proved. The answer will be some other theorem. But then the same question can be asked about the other theorem. Eventually, a stage is reached where a certain statement cannot be proved from any other proved statements and must, therefore, be taken for granted to be true. Such a statement is known as an **axiom** or a **postulate**. Each branch of mathematics has its own postulates or axioms. For example, one of the axioms of geometry is that through two distinct points there passes exactly one line. The whole beautiful structure of gemostry is based on a few axioms such as this one. Every theorem in geometry can be ultimately deduced from these axioms.

Inductive logic is inevitable in real life as well as in the sciences, because we have to draw inferences about a whole class from a few sample cases, these being the only ones we can observe. The whole idea is to seek some order in the observed facts. Such order is expressed in terms of some simple laws (e.g. the law of gravity). Taking these laws as axioms, deductive logic is applied to predict the outcome in some subsequent event, that is, a case not observed when the law was framed. If the actual outcome tallies with the predicted one, then that constitutes an experimental proof of the law and thereby makes the law more acceptable. But, if an event occurs where the predicted and the actual outcomes differ, the latter always overrides the former. It is in this sense that the natural sciences are subservient to observed facts. Facts are supreme.

Because of its abstraction, mathematics is not tied down to any particular set of natural facts. So the question of experimental verification is really not relevant. The sole concern , instead, is the derivation of the various implications of the statements assumed as axioms. A mathematician is not even concerned with whether his axioms are true or not. All he says is what would happen if they are. And this he does strictly in a deductive manner. As a hypothetical (and admittedly over-simplified) example, a mathematician does not care whether all metals are electropositive nor whether all electropositive elements form halides, things which are vitally relevant in chemistry. All he does is to prove a 'theorem' to the effect that if all metals are electropositive and all electropositive elements form halides then all metals form halides. This hardly sounds very bright. But, all mathematical proofs consist of long chains of tiny bits of reasoning like this. Genius is not needed for individual bits but for combining them suitably. (See the exercises for an example of one relatively short but bright example of a proof.) A fitting analogy is that every single word in the sentence "To be or not to be is a question" is an ordinary, prosaic word. But the sentence itself is a masterpiece of the literary genius of Shakespear.

The fact that deductive logic is the only raw material of mathematics makes it unique in many respects, which will be elaborated in the answers to some of the questions to come. For example, its theorems are lifeless and, at the same time, immortal (Section 1.7)! From a pedagogical point of view, the discipline of deductive logic makes it necessary in mathematics to observe certain scruples which are considered unnecessary or too fussy in day-to-day life (Sections 1.5, 1.10 and 1.11).

A couple of points, however, deserve to be mentioned here. First, inductive logic has no place in a mathematical proof. (The so-called mathematical induction is, in reality, a case of logical deduction.) But this does not mean that inductive logic is totally useless in mathematics. Deductive logic is needed in proving a statement. But how does one think of that statement in the first place? Here the answer is often through inductive logic and experimentation. The ability to see some pattern in a few observed cases is a valuable asset even in mathematics. (Three such pattern recognition problems will be given in the exercises.)

In fact, historically, many mathematical results were left by their discoverers without giving complete deductive proofs. The insistence on such proofs is relatively recent. As great a mathematician as Euler has to his credit numerous results for none of which he gave a proof acceptable by modern standards. (One such result will be given in the exercises.) What is truly remarkable is that every single result of Euler was later established by a rigorous proof. That is, Euler's intuition never failed him. So, even though deductive logic is the only currency in mathematics, it would be quite wrong to regard mathematics as just an exercise in deductive logic. That would be as absurd as equating the Taj Mahal with the marble used in building it. Nor is a mathematical theorem valued simply because its proof is logically correct. Greatness of a theorem depends on many other qualities such as its depth, utility and beauty.

# EXERCISES

4.1 Prove that there are infinitely many prime numbers. [*Hint* : If $p_1, \ldots, p_n$ are prime then the number $p_1 p_2 \ldots p_n + 1$ has a prime factor different from each $p_i$. This argument, due to Euclid, is still considered as one of the best mathematical proofs.]

4.2 Some prime numbers can be expressed as sums of two perfect squares (e.g. $29 = 5^2 + 2^2$ , $37 = 6^2 + 1^2$) while some (e.g. 11, 23) cannot. Try more examples and come up with a guess as to which primes can be expressed as sums of two perfect squares. (Do not prove your guess. That may be very difficult.)

4.3 Take a planar map of countries and colour it so that no two countries sharing a common curve along their borders get the same colours. (It is okay to give the some colour if the borders have merely a few isolated points in common, e.g. if the two countries are non-adjacent sectors of a circle.) Try to use as few colours as possible. Do this exercise with other maps (even imaginary ones) and guess what is the smallest number of the colours that will suffice for every planar map. (If you can prove your guess, you are probably a genius.)

4.4 Suppose $n$ players participate in a knock-out tennis tournament. At each round, the players are paired off and a match is played between the players of each pair. (In case of an odd number of players at any round, one of them gets a bye.) The winners (and the player getting a bye, if any) enter the next round. This process continues till a champion is found. Experimenting with several values of $n$ (e.g. $n = 4, 10, 25, 100$ etc.) make a guess about the total number of matches played in the tournament. Then take some other values of $n$ and see if your guess works for them too.

4.5 Unlike your guesses in Exercises (4.2) and (4.3), the one in the last exercise is easy to prove (if correct!). Do this by induction on $n$. [*Hint* : Consider the two cases, $n$ even and $n$ odd separately. What you will need is the so-called second principle of mathematical induction. That is, in order to prove the assertion for $n = k$, you will need to know its truth not just for $n = k - 1$, but even for lower values of $n$.]

4.6 Prove your guess in Exercise (4.4), without induction, by a short, elegant argument.

4.7 Suppose $f(x) = a_0 + a_1 x + a_2 x^2 + \cdots + a_n x^n$ is a polynomial with non-zero roots, $\alpha_1, \ldots, \alpha_n$. Prove that

$$\frac{1}{\alpha_1} + \frac{1}{\alpha_2} + \cdots + \frac{1}{\alpha_n} = -\frac{a_1}{a_0}.$$

[*Hint* : Note that $f(x) = a_n(x - \alpha_1)(x - \alpha_2) \cdots (x - \alpha_n)$. Consider $f(1/x)$.]

4.8    Suppose $f(x) = 1 - \dfrac{x}{3!} + \dfrac{x^2}{5!} - \dfrac{x^3}{7!} + \dfrac{x^4}{9!} + \cdots + \dfrac{(-1)^n x^n}{(2n+1)!} + \cdots$. Show that the zeros of $f$ are of the form $(n\pi)^2$ where $n$ is a positive integer. [*Hint :* Clearly $f$ has no negative zeros. For $x > 0$, show that $f(x) = \dfrac{\sin y}{y}$ where $y^2 = x$. The power series expansions of the sine and the cosine functions will be taken up formally in Section 4.8]

4.9    Assuming that the result of Exercise (4.7) is valid for power series (i.e. for 'polynomials of infinite degree') and applying it to the function $f(x)$ in the last exercise, prove that $\displaystyle\sum_{n=1}^{\infty} \frac{1}{n^2} = \frac{\pi^2}{6}$. (An assumption like this is, of course, unwarranted and can lead to disasters. But Euler made it anyway and 'proved' this beautiful result. A logically rigorous proof can be given using theorems about Fourier series. But that is beyond our scope.)

## 1.5    Implication Statements

An **implication statement** is one in which an assertion is made that whenever a certain statement (say $p$) holds then so does some other statement (say $q$). Symbolically this is expressed by writing '$p{\rightarrow}q$' or '$p{\Rightarrow}q$'. A typical example of an implication statement is 'If two triangles are congruent then their areas are equal'. Here $p$ is the statement that the two (given) triangles are congruent and $q$ is the statement that they have equal areas. The statements $p$ and $q$ can be completely independent of each other. There need not be any semantic correlation between the two. For example, a statement like "If it rains then John is intelligent" is a perfectly good example of an implication statement. It, of course, sounds ridiculous and examples like this rarely figure in any serious study. The statements $p$ and $q$ are called respectively the **hypothesis** and the **conclusion** of the implication statement 'If $p$ then $q$' or '$p{\rightarrow}q$'.

Some statements are not ostensibly in the implication form but can be easily cast into it. For example the statement 'All rich men are intelligent' can be paraphrased as 'If a man is rich then he is intelligent'. Sometimes this becomes easier by introducing appropriate symbols. For example, the theorem 'The sum of the three angles of any triangle is 180 degrees' can be put as 'If $ABC$ is a triangle then $\angle A + \angle B + \angle C = 180$ degrees.' Even statements which have no obsensible hypothesis can be artificially cast as implication statements by adding a dummy hypothesis which is always true. For example, the statement 'The number $\pi$ is irrational' is equivalent to 'If $0 = 0$ then $\pi$ is irrational'. A far more satisfactory way, of course, would be to include the definition of $\pi$ or of rationality in the hypothesis. For example, "If $p, q$ are any two integers then the ratio $p/q$ can never equal the number $\pi$."

If '$p{\to}q$' is an implication statement, then '$q{\to}p$' is also an implication statement. The two are said to be **converses** of each other. Numerous examples can be given where an implication statement is true but its converse is false, or where, both are true but differ considerably in the difficulty of proofs (see the exercises). When both '$p{\to}q$' and '$q{\to}p$' hold we write '$p \leftrightarrow q$' or '$p \Leftrightarrow q$' and read it as "$p$ and $q$ imply each other" or "$p$ is (logically) equivalent to $q$" or "$p$ holds if and only if q does".

Although implication statements are common even in daily conversations, there are several reasons why they can be baffling in mathematics, if due care is not taken. The basic point to note is that implication statements (and more generally, any mathematical statements) should always be interpreted to mean exactly what they say. In practice we often attach an extra meaning to the statement. In the case of implication statements, especially, we often confuse them with their converses in real life. For example, when we say that if a person is below 18 then he cannot vote, we often take it to mean that if he is 18 or above then he can vote. In mathematics, however, the statement merely says that being 18 or above is a necessary condition for voting. It does not say that it is sufficient. If 18 were to be the voting age, the correct statement should have been "You can vote if and only if you are 18 or above". This, of course sounds very clumsy in real life.

The word 'unless' is also a cause of confusion. It is used when the hypothesis of an implication statement happens to be the negation of some statement. (Negations of statements will be studied in detail in Section 1.11.) The safest way to handle it is to replace it by 'if not'. For example, 'Unless it rains, the crops will die' simply means that if it does not rain, then the crops will die. In real life we take it to mean that the crops will be saved if it rains. In mathematics it never means so. Here again, the statement is silent as to what would happen if it rains. Even if the crops die despite raining, that does not render the implication statement false. The only way an implication statement '$p{\to}q$' becomes false is when $p$ holds and $q$ does not.

An interesting special case of this arises when the nature of the statement $p$ is such that it can never be true. For example, consider the statement, 'If a man has six legs then he is rich.' Since no man has six legs, a layman is most apt to declare this statement as false. But in mathematics it is true because the only way it would be false is if there is a six-legged man who is not rich. Since no such man exists, the statement is true. A truth of this kind is often called **vacuous truth**, because such statements can often be paraphrased in terms of properties of the members of a vacuous (i.e. empty) set. In the present case it is the set of all six-legged men. Note that the statements 'Every six-legged man is rich' and 'Every six-legged man is poor' are both true and still there is no contradiction because these statements are not the negations of each other. (See Section 1.11 for the correct negations of such statements.)

Is it not foolish to indulge in absurdities like this? Certainly yes. No mathematician consciously goes on proving results which are vacuously true. However, occasionally vacuous truth enters into a mathematical proof when we have to show that a certain condition holds for every member of some set arising in that proof. If this set happens to be empty then this condition is vacuously satisfied and we can proceed further. Such a situation is generally avoided in elementary courses. (However, see the comments after Theorem (2.6.1).)

Summing up, implication statements can be baffling to a beginner because of certain differences in their real-life and mathematical interpretations. However, they need not be so if care is taken to interpret them to mean exactly what they say. Here are a few paraphrases of the implication statement '$p \rightarrow q$'

(i)     $p$ implies (or leads to) $q$

(ii)    $q$ follows from (or is a logical consequence of) $p$

(iii)   whenever $p$ holds, $q$ holds

(iv)    whenever $q$ fails, $p$ fails

(v)     either $p$ fails or $q$ holds

(vi)    $p$ is a sufficient condition for $q$

(vii)   $q$ is a necessary condition for $p$

(viii)  $p$ is false unless $q$ holds.

(ix)    $p$ is true only if $q$ is true.

The double implication statement '$p \Leftrightarrow q$' also has many similar versions some of which were mentioned earlier. When $p$ and $q$ are properties of the same object and '$p \Leftrightarrow q$' holds, it is also customary to say that $p$ and $q$ are characterisations of each other. For example, a characterisation of a cyclic quadrilateral is that the sum of the opposite angles is 180 degrees.

## EXERCISES

5.1   If two triangles are congruent, then their areas are equal. Prove this statement and show by an example that its converse is false.

5.2   The opposite pairs of sides of a parallelogram are equal (in length). Cast this statement and its converse in the form of implication statements, and prove them. (Here you will find that the proof of the converse is obtained by essentially reversing the steps in the proof of the direct implication.)

5.3 Prove that the sides $AB$, $AC$ of a triangle $ABC$ are equal if and only if $\angle B = \angle C$, by drawing the angle bisector of the angle $A$. (Here both the direct and converse implications are true and are easy to prove. But the proofs are not quite the mirror images of each of other because the justifications for the congruency of triangles are different.)

5.4 Prove that a triangle $ABC$ is equilateral if and only if $\cos A + \cos B + \cos C = \frac{3}{2}$. (Here, the necessity of the condition is trivial but the sufficiency is not so easy to prove. This is, in fact, typical of most trignometrical characterisations of equilateral triangles. Try a few and you will soon be convinced.)

## 1.6    Mathematical Theorems: Lifeless and Immortal

It might seem a contradiction of terms to say that something can be lifeless and immortal at the same time. To illustrate what is meant, consider a plastic rose. If prepared by a skilled craftsman it will look exactly like a genuine rose and, if appropriately scented, may even attract insects. Of course it is lifeless. But age does not wither its beauty and, if care is taken to shield it from dust and direct sun, it will look fresh forever. So in this sense it is immortal.

The results in natural sciences are like natural flowers. They last as long as they are in conformity with observed facts. A mathematical theorem, however, it like a plastic rose. Although it might ostensibly deal with some particular natural facts, its allegiance is not *per se* to the facts but to making logical deductions from these facts. Even if these facts give way, the deductions remain intact. To illustrate the difference, suppose a physicist and a mathematician each makes some observations about the motion of the earth around the sun and that of the moon around the earth and then makes predictions about when the next solar eclipses will occur. In the case of the physicist, his theory will be of little value if the future solar eclipses do not occur at the predicted times. A mathematician, on the other hand, will not right away predict when the next solar eclipses will take place. Instead, he will state his theorem as "If the motion of the earth around the sun and that of the moon about the earth satisfy certain conditions then the next solar eclipses will take place on certain dates" and he will give a logical proof for this. If the actual eclipses do not follow his predictions, it would mean that his hypothesis about the motions of the earth and the moon is wrong. That does not mean that his theorem is false. We may at the most say it is useless. Maybe in future, some other solar system will be discovered in which some planet and its satellite satisfy the hypothesis. In that case the theorem will be useful again. But whether useful or not, it will remain as a true result forever. It is in this sense that a mathematical theorem is immortal and lifeless at the same time. It has no natural death because it has no natural life!

There is, however, a different kind of a death that can befall on a mathematical theorem. This happens when some logical mistake is detected in its proof. In that case, the theorem becomes invalid even if its conclusion may still continue to hold. Suppose, for instance, that in the example above, that the mathematician makes a mistake of reasoning or of calculation while predicting the dates of the solar eclipses. It may still happen that his predictions tally with the facts. But his theorem is invalidated unless he (or somebody else) patches up the hole in its proof.

How can a wrong proof give a correct conclusion? This can happen for a variety of reasons. Maybe the mistake is too minor to affect the truth of the conclusion. Or it may fortuitously cancel with some other mistake. A far more common situation is that the proof contains some statements without adequate justifications. There are many statements which are intuitively clear but whose rigorous proofs are from trivial. A well-known example is the **Jordan Curve Theorem** which says that every simple closed curve in a plane divides it into two regions (one bounded and the other unbounded) of which it is the common boundary. If the proof of a mathematical theorem is based on a result which is intuitively clear but which is not proved so far, then it is a wrong (or at least an incomplete) proof even though the conclusion reached may not be wrong. Occasionally it also happens that a proof (sometimes even a published one) contains a serious flaw, that is, a claim which is not only unjustified but downright false. Such a proof, of course, has to be discarded and the search for a correct proof continues. (See Exercise (4.9) for an example.)

Summing up, in natural sciences, what matters most is the conformity of the conclusion with observed facts, while in mathematics it is not the conclusion *per se* but the journey from the hypothesis to the conclusion that matters most.

## EXERCISES

6.1   Find the fallacy in the attempted proofs for the assertions given below. Where the assertions are correct, give a correct proof.

   (i)    **Assertion:**      Every cyclic quadrilateral is a rectangle.

          **Proof:**         Let $ABCD$ be a cyclic quadrilateral. Draw a circle with $AC$ as a diameter. Then since the quadrilateral is cyclic, $B$ lies on this circle. So $\angle ABC$, being an angle in a semi-circle, is a right angle. Similarly all other angles are right angles. Hence $ABCD$ is a rectangle.

(ii) **Assertion:**    $\lim\limits_{x \to \frac{\pi}{2}^-} (\sec x - \tan x) = 0.$

**Proof:**    $\lim\limits_{x \to \frac{\pi}{2}^-} \sec x = \infty = \lim\limits_{x \to \frac{\pi}{2}^-} \tan x.$ So the given limit is $\infty - \infty$, i.e. 0.

(iii) **Assertion:**    $\lim\limits_{n \to \infty} \dfrac{1}{n+1} + \dfrac{1}{n+2} + \cdots + \dfrac{1}{2n} == 0$

**Proof:**    As $n \to \infty$ each of the terms $\dfrac{1}{n+1} + \ldots, \dfrac{1}{2n}$ tends to 0. Hence their sum tends to 0.

(iv) **Assertion:**    If $f : \mathbb{R} \to \mathbb{R}$ is a function such that $f'(x) = 0$ for all $x \in \mathbb{R}$, then $f$ is a constant function.

**Proof:**    The hypothesis implies that the tangent to the graph of $f$ at every point is horizontal. So the graph must be a horizontal straight line. That is, $f$ is a constant.

(v) **Assertion:**    Every triangle is equilateral.

**Proof:**    Let $ABC$ be any triangle. Through each point, say $P$, on the side $AB$ draw a line parallel to $BC$ to cut the side $AC$ at a point, say $Q$. The points $P$ and $Q$ determine each other and so there is a one-to-one correspondence between the points on the side $AB$ with those on the side $AC$. Thus the two sides have the same number of points and hence the same length. So $AB \equiv AC$. Similarly $AB \equiv BC$. Hence $ABC$ is eqilateral.

## 1.7    Mathematics as an Art

The popular image of a mathematician (a bespectacled, stern-faced guy surrounded by a maze of horrible formulas) is so antithetic to that of an artist (an amorous, azure-eyed character absorbed in dulcect notes or strokes) that any suggestion that mathematics is an art would appear ridiculous. But, believe it or not, it is. Or more accurately, a good part of mathematics is an art. Its great theorems are works of art. And its medium is logical deduction.

A layman, even an educated one, is most apt to treat mathematics as a science. In fact, Gauss, one of the greatest mathematicians of all times (and unquestionably, the greatest mathematician of the nineteenth century) called mathematics as the queen of sciences. However, he probably used the term science in a very wide sense, as an intellectual activity amenable to reasoning. Sciences like physics, chemistry, biology etc. involve both experimentation and reasoning. Mathematics, by contrast, is pure reasoning. And so, it stands to reason that mathematics is the queen of sciences. (Gauss then went on to say that Number Theory is the queen of mathematics.)

Whether mathematics is an art or a science or both (or perhaps neither) obviously depends on what one understands by mathematics, by art and by science. The trouble is that these basic concepts have no universally accepted definitions. The line between arts and crafts, for example, has always been a thin one. Mathematics itself has undergone major changes in its constitution and function. So it is futile to try to prove conclusively that mathematics is an art. Instead, we shall indicate what it has in common with an art.

Like many other human endeavors, mathematics began as a useful activity. But gradually some parts of it were found to be aesthetically pleasing and later on some parts of it began to be developed more for their aesthetic appeal than for any possible use. As an analogy, the primary purpose of cooking is to make the food easier to digest. But the culinary art aims at preparing dishes which water your mouth even when you are not hungry! Similar remarks can be made about clothes, illustrative drawings and even fist fights (which later evolved to martial arts).

Mathematics, too, began as aid to solve real life problems. But even in ancient times certain elementary facts about numbers and geometric figures (such as the rule of three or the concurrence of the medians of a triangle) fascinated men. Discovery of such 'order in chaos ' is surely an artistic experience. It motivated some men to look for more, and more intricate, forms of order, even though they may have no immediate applications. (It is doubtful if the nine point circle of a triangle has found any practical use so far.)

The search for order in chaos is, of course, the most fundamental function of science. And in this sense mathematics resembles (and is often taken as) a science. There is, however, an important difference. The order which a scientist is after is an order in facts of nature. He has no freedom to choose these facts, much less to change them to suit his convenience. A mathematician, on the other hand, is concerned with order in things which are creations of his imagination. A number, for example, is only a concept. And so are the addition and multiplication of numbers. Even though numbers (and the operations on them) are highly useful in representing natural objects, they themselves have no natural existence. That every solid melts upon heating is a law of nature. But, commutativity of multiplication of numbers is not a law of nature. It is a consequence of the axioms about numbers. By changing these axioms, it is perfectly possible to have a number system where multiplication is not commutative. Such a system will probably be not as 'useful' as the usual number system. But, mathematically, that is no reason not to study it. In fact, if developed by a genius like Ramanujan, such a number system may turn out to be far more thrilling than the present one. (A development somewhat like this did take place in the case of geometry in the nineteenth century, when the Russian mathematician Lobachevsky studied geometries which were non-euclidean by dispensing with the so-called parallel postulate. In these geometries the sum of the three angles of a triangle may exceed 180 degrees. It is however, hard to say if the motivation was purely artistic.)

It is this freedom which distinguishes mathematics from sciences and in fact makes it more like an art. A photographer's goal is to make pictures which resemble the actual objects as closely as possible. An artist, on the other hand, has no such limitation. He can paint an object not the way it is but the way it *appears* to him. And in abstract masterpieces, we encounter shapes which do not even remotely resemble any real objects! Another analogy is the difference between a historian and a writer of historical fiction. History, like science, is a part of knowledge and has to be based strictly on what actually happened (or more precisely, on our understanding of what actually happened as gathered from the available evidence). Historical fiction, on the other hand, is a creative art and the author is free to invent characters and scenes in order to enhance the artistic impact.

There are, however, several respects in which mathematics differ from the more familiar arts. First the freedom which a mathematician has is restricted by the requirement that he has to observe the strict rules of deductive logic. The rigidity of these rules makes mathematics more like a sport than an art. Secondly, whatever freedom a mathematician has, he has to use it with discretion or else he will easily alienate himself. Unlike paintings, music and poetry, mathematics has no sensuous or emotional appeal and this already puts a severe restriction on the number of persons who can appreciate it as an art. The problem is further compounded by the fact that little conscious effort is made to train or help persons appreciate the artistic aspect of mathematics. The aesthetics of mathematics is yet in its infancy. Last but not least, the preparation and the time needed to really understand and appreciate a mathematical proof is often prohibitive. (It also takes a certain knowledgeability to appreciate highly cultured art forms such as classical music and abstract paintings. But its degree comes nowhere close to that needed in mathematics.)

Another interesting feature of mathematics is that nothing in it can be branded as permanently useless. When a dancer leaps, the exercise his feet get is incidental. In mathematics, however, there have been cases where certain theorems, originally proved purely as an intellectual pursuit, have later on found some down-to-earth applications. This is undoubtedly very romantic and does not happen frequently. It can hardly serve as a justification for mass-scale production of mathematical results with no artistic appeal and no hint of applicability to any real life problem.

## 1.8    Clumsy Mathematical Definitions

Let us first try to have a definition of the word 'definition' itself. Surely, a definition serves to describe the object in question. But the description may be so general that it may also apply to other objects as well. A more specific description will narrow this class further. Finally when the description is such that it is satisfied only by that type of an object, it becomes a definition. In short, a definition is a description which uniquely identifies[1] the object in question. For example when we define an equilateral triangle as a triangle all whose sides are equal in length, we mean not only that every equilateral triangle has this property but also that any triangle with this property is equilateral. On the other hand, "a quadrilateral having all sides equal" is a description, but not a definition of a square. It is a definition of a rhombus, which is a more general concept than a square. To define a square we would have to add the requirement that all angles also be equal in measure. Frequently, the same concept can be defined in several different ways. For example, an equilateral triangle can also be defined as one all whose angles are equal. It must, of course, be shown that the two definitions are equivalent.

It is also implicit in the very concept of a definition that the description which is to serve as a definition must not involve any unfamiliar terms. Otherwise the very purpose of a definition is lost. For example, to define a Monday as the day of the week immediately following a Sunday would be fine only if it is understood what is meant by a Sunday. The best thing is to have a definition of a Sunday. If we define a Sunday as a day immediately preceding a Monday, then we would be guilty of what is called a vicious cycle, for, in that case, in order to know what a Monday is we must know what a Sunday is and vice versa! We could of course define a Sunday as a day immediately following a Saturday. But then we must first know what a Saturday is.

The insistence that a definition may use only such terms which are previously familiar is structurally similar to the insistence that a mathematical proof may use only such results whose truth is known previously. The difficulty involved is also exactly similar. Just as in deductive logic, we ultimately come down to a few statements (called axioms) whose truth cannot be proved but simply has to be assumed, in the case of definition, we have to have a few terms which cannot be defined any further. Such terms are called **primitive terms**. Every branch of mathematics has its primitive terms. For example, among the primitive terms of classical geometry are a point, a line and the incidence relation (between a point and a line). Intuitively we may think of a point as a tiny dot, and a line as a set of points having only length but no breadth or thickness. But somebody

---

[1] Because of this, it is customary to use the word 'if' to mean 'if and only if' in a definition. Thus to say 'A triangle is called equilateral if all its sides are equal' means that it is called equilateral when *and only when* its three sides are equal. This is in sharp contrast with the warning given earlier about not confusing an implication statement with its converse. This usage is unfortunate but standard. Fortunately it appears only in definitions. The words 'defined', 'called' etc. make it clear that the statement is a definition and not just a description. It is also customary to stress this by underlining the words defined or by putting them in bold face or in italics. Hence no confusion need arise.

else may have an entirely different intuition. (For example think of a point as a lock, a line as a key and incidence of a point an a line to mean that the lock can be opened by the key.) It really does not matter at all.

Except for primitive terms, every term appearing in a mathematical definition must have been defined earlier. This insistence creates no problems for most definitions but renders some basic definitions very clumsy. For example, an injective function is defined as a function, say $f$, with the property that it takes distinct points to distinct points, or in other words, whenever $x$ and $y$ are points in the domain of $f$ and $x \neq y$ then $f(x) \neq f(y)$. This definition is very easy to understand. It is essentially in the nature of a glossary. Instead of having to say 'a function which takes distinct points to distinct points' it is convenient to have a short phrase and the definition merely coins such a phrase.

But what about the definition of a function, say, from a set $X$ to a set $Y$? In elementary texts, such a function is often defined as a rule or correspondence which assigns to each element of $X$, one and only one element of $Y$. This definition, followed by a couple of illustrative examples, generally suffices to give a clear idea of what a function is. But it has a flaw. It relies on the words 'rule' (or correspondence') and 'assigns' which have not been previously defined. One way out is to take these words as primitive terms. But this is hardly a solution. If these terms can be accepted as primitive, then why not take 'function' itself as a primitive term? If we go on doing this, there would be no need to define anything! Obviously, that would be chaotic.

Another way out is to paraphrase the definition in such a way that the terms 'rule' etc. are completely bypassed. To see how this is done, suppose $f : X \longrightarrow Y$ is a function in the intuitive sense given above. For each $x \in X$, we denote by $f(x)$ that unique element of $Y$ which $f$ assigns to $x$. Consider the ordered pair $(x, f(x))$. This is an element of the set $X \times Y$, often called the **cartesian product** of $X$ and $Y$, which consists of all ordered pairs of the form $(x, y)$ with $x \in X$ and $y \in Y$. Now let $G_f$ be the set of all ordered pairs we get this way. In the notation of set theory,

$$G_f = \{(x, y) : x \in X, \ y \in Y, \ y = f(x)\}.$$

$G_f$ is a subset of $X \times Y$. It is often called the **graph** of $f$ because that's what it is if $X, Y$ are each sets of real numbers. This set obviously depends on the function $f$. If we had a different function, say $g$ from $X$ to $Y$, its graph $G_g$ will be a different subset of $X \times Y$ than $G_f$. Note that each such graph has the property that for every $x \in X$, there is one and only one $y \in Y$ such that $(x, y)$ is in the graph. (Geometrically, every 'vertical' line meets the graph in exactly one point.) Conversely if $S$ is any subset of $X \times Y$ with this property, then $S$ is the graph of a (unique) function from $X$ to $Y$. A function and its graph are so intimately related to each other that knowing one is as good as knowing the other. *So why not identify the two?* If we do so, then a function[2] from $X$ to $Y$

---

[2]The sets $X$ and $Y$ are called, respectively the **domain** and the **codomain** of $f$. If $y_0 = f(x_0)$ then $y_0$ is called (the) **image** of $x_0$ and $x_0$ is called (a) **preimage** or (an) **inverse** image of $y_0$ under $f$. The **range** of $f$ is the set of those points of $Y$ which have at least one preimage each. (In the past, it was customary to call $Y$ as the range.)

can now be *defined* as a subset, say $f$, of $X \times Y$ having the property that for every $x \in X$, there is one and only one $y \in Y$ such that $(x, y) \in f$. As compared with the earlier definition of a function, this new one does appear clumsy. But it meets the requirement that it uses no loose or vague terms.

The trick adopted here (which let us call the **definition trick**) is applied in many other instances as well. It so happens sometimes that we want to define some concept, say, $A$ which is easy to understand but hard to define rigorously, i.e., without using previously undefined terms. However, there often is some other thing, say $B$, which is related to $A$ and which can be expressed in a rigorous manner. (In the example just given, $A$ is the concept of a function and $B$ is a subset of a certain kind.) If the relationship between $A$ and $B$ is such that each uniquely determines the other then $A$ can be defined as $B$.

Once this point is understood, it is easy to see why certain definitions look clumsy. For example, intuitively an (infinite) sequence is an (infinite) succession of terms. But the word 'succession ' is not defined previously. However every such succession determines a function from $I\!N$, the set of positive integers and vice versa. So a sequence is now *defined* as a function from $I\!N$. At the school level, the basic concept of an angle was formerly defined by a statement like "An angle is formed when two straight lines meet (or intersect) each other." This simple-minded definition is criticised today on the ground that it does not identify an angle. In other words, it tells us *when an angle is formed* but does not tell *what an angle is*. If the difficulty were merely grammatical (i.e., if the definitions must use only the verb 'to be' to link the object to be defined with its description) then we could dodge it by saying that an angle is what happens when two lines intersect. But this is too ambiguous because when two lines meet, so many things happen. One point gets distinguished as the point of intersection. Also each line gets divided into two parts called half-lines or rays. Unless we specify which of these objects we call an angle, our definition would be meaningless. Here the definition trick comes to our rescue. An angle is determined by the two rays along its arms. These rays are, in turn, determined by their union (unless they are collinear). So in modern texts it is customary to *define* an angle as a union of two (non-collinear) rays having the same initial point. (Other definitions of an angle are possible. Also the definition just given needs some modification to cover angles of 0 or 180 degrees. That does not concern us. We just want to illustrate how the definition trick works.) So here too we sacrifice lucidity for logical perfection.

The definition of a limit is yet another instance where insistence on logical perfection has entailed some clumsienesss. Because of its special importance, we shall discuss it in detail later (Section 2.3).

Summing up, some mathematical definitions appear clumsy because of the need to be precise. It is to be noted, however, that most definitions are in the form of a glossary and not clumsy. But those few that are, usually define the most basic concepts. And so, even though they are small in number, they can be intimidating to a beginner. Things would improve considerably if a conscious effort is made (at least at elementary levels) to unravel through the apparent clumsiness and explain how the underlying ideas are really very simple.

# EXERCISES

8.1 An **ordered pair** $(x, y)$ is defined as the set $\{\{x, y\}, x\}$. Explain why this definition is necessary to capture the essence of the concept. Consider especially the case $x = y$.

8.2 Give a rigorous definition of a finite sequence of length $n$ (where $n$ is a positive integer).

8.3 For any positive integer $n$, an **ordered $n$-tuple** can be defined as a finite sequence of length $n$. What is wrong if we define an ordered pair as an ordered 2-tuple?

8.4 An $m \times n$ **matrix** (where $m, n$ are positive integers) is defined popularly as a rectangular array with $m$ rows and $n$ columns. What is wrong with this definition? Give a rigorous definition of a matrix as a function defined an a suitable set.

8.5 Suppose $P$ is a point on a curve $C$. Can we define the tangent to $C$ at $P$ as (i) a line which touches $C$ at $P$, (ii) a line which intersects $C$ only at the point $P$, (iii) a line which intersects $C$ at two coincident points, both being equal to $P$ ? (For a correct definition of a tangent, see Section 2.2.)

## 1.9   Why Today's Mathematics Has Sets Rather Than Numbers at Its Nucleus

Till recently the popular image of mathematics was that it was the science of numbers. As explained in Section 1.3, the link of mathematics with real life was through numbers. Today, the picture has changed. Sets have taken over numbers as the focal point of mathematics. Even the elementary school books talk about sets.

Although mathematicians had proved many non-trivial results about real numbers, no rigorous definition of a real number was given till the middle of the nineteenth century. Intuitively the real numbers were taken to correspond to the points on an (infinite) straight line. This geometric approach had (and still has) its advantages. But it could not give a satisfactory treatment of such basic concepts as limits and continuity. The construction of rational numbers from natural numbers was a relatively simple matter. But there was a big jump in going from the rationals to reals. Several approaches were tried. The easiest to understand is due to Dedekind and is based on the simple idea that every real number is completely determined by the set of all rationals less than it. Of course not every subset of rationals arises this way. But it is easy to tell which ones do. So a real number can be defined as a certain set, namely as a subset of the set of rationals[3]. These sets are necessarily infinite sets and often defy our intuition based on finite sets. Not surprisingly, they were criticised as too abstract. But gradually they found acceptance. The number system could now be developed quite rigorously starting from the natural numbers which were so intuitive that they were considered as 'God given'. Their basic properties were taken as axioms (called the **Peano axioms**).

In a later development, it turned out that even the natural numbers can be defined as certain sets. We shall not give the construction except to remark that 0 corresponds to $\emptyset$, the empty set, 1 to the set $\{\emptyset\}$, 2 to the set $\{\emptyset, \{\emptyset\}\}$ and so on. The operations on natural numbers can also be defined in terms of certain simple operations on sets and then the Peano axioms can be proved as theorems starting from the axioms of set theory.

This means that the sets are more basic than numbers. Euclidean geometry can also be done entirely in terms of sets. Of course, from a pedagogical point of view, natural numbers and visual geometric figures still have their importance. But in principle, they are subsumed by sets. Sets and the derived concept of functions also serve to give precise definitions of many elusive terms (see Section 1.8 for example). Sets have become very convenient means of precise and yet concise expression of ideas. Figuratively we may say that sets are the alphabets of modern mathematics.

The real importance of sets is not just that they provide a more basic foundation of mathematics but also that they have tremendously increased the applicability and widened the horizons of mathematics. As remarked in Section 1.3, numbers (or numerical variables) have been used for ages to abstract the essence of a real-life problem. There are, however, many problems which are non-numerical in the sense that they involve no numbers, or even if they do, the numbers are not their crucial feature. The essence of many of these problems can be conveniently abstracted in terms of sets rather than in terms of numbers. Here is an example :

> Suppose at a dancing party every boy dances with at least one girl and no girl dances with every boy. Prove that there exist boys $b, b'$ and girls $g, g'$ so that $b$ dances with $g$ and $b'$ with $g'$, but neither $b$ dances with $g'$ nor $b'$ with $g$.

---

[3]Another method of constructioning the real numbers from the rational numbers is given in Appendix A. In that method, a real number is a certain set of sequences of rational numbers. So, either way, a real number is a certain set.

It is difficult to tackle this problem with numbers. But a set-theoretic formulation is easy. Let $B$ be the set of boys at the party. For each girl $g$, let $B_g$ be the set of those boys who dance with $g$. Each $B_g$ is a subset of $B$. The problem amounts to showing that under the given conditions, there exist girls $g, g'$ such that neither $B_g$ is subset of $B_{g'}$ nor $B_{g'}$ a subset of $B_g$. (See the Exercises for a solution and also for another example of a non-numerical problem.)

As in the case of numerical abstraction, when sets are used to abstract the mathematical essence of a problem, two apparently different problems may look similar. In fact, in the case of sets, the apparent dis-similarity of the original problems may be far more baffling than in the case of abstraction with numbers. Here is one such pair of problems:

(i) Given twenty points $P_1, \ldots, P_{20}$ in a plane, we want to draw mutually non-overlapping discs of equal radii, centred at these points. What is the bound on the radius of such discs?

(ii) 20 students appear for a multiple choice test with 10 questions, each of which has either a $T$ or an $F$ as an answer. Prove that no matter how they answer these questions, there will be at least two students whose answers to at least six questions will match with each other.

The first problem is, of course, very easy, the answer being $\frac{1}{2}d$ where $d$ is the minimum of the (euclidean) distances between various pairs of distinct points from $P_1, \ldots P_{20}$. It is far from obvious that this is also the key idea in the solution of the second problem, except that instead of the euclidean distance we have to work with another distance called the Hamming distance. (See Exercises for details.) But the point is that in order to see this similarity, one must first generalise the concept of the euclideam distance, by abstracting some of its more basic properties. This leads to what is called a metric. So the euclideam distance is a prime example of a metric. But the Hamming distance is another. In fact, there are many others. Consequently, a theorem proved for abstract metrics is applicable not only to the euclideam distance but also to the Hamming distance and to other situations.

The crucial step in applying mathematical reasoning to problems like this is to translate them in terms of suitable 'structures' on suitable sets. It is this freedom of choosing the set and the structure on it which makes sets far more adaptable and useful than mere numbers. We shall not go deeper into it, since for our purpose, the traditional numerical abstraction will suffice most of the time.

## EXERCISES

9.1 Solve the problem involving the dance party, assuming that the set $B$ of boys is finite. [*Hint* : Consider a girl who dances with the largest number of the boys. Why will this break down if the set $B$ is infinite?]

9.2 Suppose $B_1, \ldots, B_k$ are subsets of a (possibly infinite) set $B$ such that for all $i, j$ either $B_i \subset B_j$ or $B_j \subset B_i$. Prove that there is some $r, 1 \leq r \leq k$

such that $B_i \subset B_r$ for all $i = i, \ldots k$. Hence give a solution to the problem above when the set $B$ is infinite but the set $G$ of girls is finite.

9.3  Show by an example that the problem above is false if both $B$ and $G$ are infinite sets.

9.4  A box contains secret documents and only 5 persons are privileged to have access to them. As a further security measure, it is desired that when any three but no fewer of these 5 persons come together they should be able to open the box. Design a system of locks and keys which will achieve this. First translate the problem in terms of sets.

9.5  Let $X$ be the set of all binary sequences of length 10 (i.e. sequences of the form $(x_1, \ldots, x_{10})$ where each $x_i$ is either 1 or 0. Denote this sequence by $\vec{x}$.) For $\vec{x} = (x_1, \ldots, x_{10})$ and $\vec{y} = (y_1, \ldots, y_{10})$ in $X$ define their **Hamming distance** $d(\vec{x}, \vec{y})$ to be the number of those indices $i$ such that $x_i \neq y_i$, $1 \leq i \leq 10$. Prove that:

(i)  $d(\vec{x}, \vec{y}) \geq 0$ for all $\vec{x}, \vec{y} \in X$ with equality holding if and only if $\vec{x} = \vec{y}$.

(ii)  $d(\vec{x}, \vec{y}) = d(\vec{y}, \vec{x})$ for all $\vec{x}, \vec{y} \in X$

(iii)  $d(\vec{x}, \vec{z}) \leq d(\vec{x}, \vec{y}) + d(\vec{y}, \vec{z})$ for all $\vec{x}, \vec{y}, \vec{z} \in X$

(These properties, called respectively the **positivity, symmetry** and the **triangle inequality**, are the defining conditions of an abstract metric.)

9.6  In the last exercise, for $\vec{x} \in X$ show that the number of points $\vec{y}$ such that $d(\vec{x}, \vec{y}) \leq 2$ is 56. (The case $\vec{y} = \vec{x}$ is included.)

9.7  Using the last two exercises, solve the problem about 20 students appearing for a multiple choice test.

9.8  If $a$ is a real number, its **absolute value** (also called its **numerical value**, or **modulus**, or occasionally **magnitude**) is defined as $a$ if $a \geq 0$ and as $-a$ if $a < 0$. It is denoted by $|a|$. (Thus, $|3| = 3, |0| = 0$ and $|-3| = 3$.) Prove that (i) $|a| \geq 0$ for all $a$, (ii) $|-a| = |a|$ for all $a$ and, more generally, $|ab| = |a||b|$ for all $a, b$ and (iii) $|a + b| \leq |a| + |b|$ for all $a, b$. (The last property is called the **triangle inequality for absolute value** and is so frequently needed in calculus proofs that it is used often without an explicit mention. A less frequently needed version of it is $|\,|a| - |b|\,| \leq |a - b|$ for all $a, b$.)

9.9  Define the **(euclidean) distance** between two real numbers $x$ and $y$ as the real number $|x - y|$. Prove that the defining conditions of a metric, as given in Exercise (9.5) are satisfied. If $a < b$, prove geometrically that $|x - a| < |x - b|$ if and only if $x < \dfrac{a + b}{2}$. Similarly, show that if $\delta > 0$, then '$|x - c| < \delta$' is equivalent to '$c - \delta < x < c + \delta$'. (More generally we can define the euclidean distance between points of a higher dimensional euclidean space, as will be indicated in Section 4.8. The triangle inequality then corresponds to the familiar fact that the length of any side of a triangle cannot exceed the sum of the lengths of the other two sides, which justifies the name. See Exercise (4.8.16).)

9.10 For points, say, $(x_1, y_1)$ and $(x_2, y_2)$ of $I\!R^2$ define a new distance function $d$ by $d((x_1, y_1), (x_2, y_2)) = |x_1 - x_2| + |y_1 - y_2|$. Prove that this new distance function is a metric, i.e., it satisfies the conditions of positivity, symmetry and triangle inequality. Let $O = (0, 0)$ and $A = (3, 2)$. Let $D$ be the set of all points in $I\!R^2$ whose distance from $O$ is at most 5 and $S$ be the set of all points in $I\!R^2$ which are equidistant from $O$ and $A$ (w.r.t. the new distance function). Identify and sketch $D$ and $S$. (*Caution* : $D$ comes out to be a square and not a disc centred at $O$. Nor is $S$ a straight line as would have been the case with the usual euclidean distance function for $I\!R^2$. This exercise shows that many concepts in geometry are subject to the distance function and may change drastically if the distance function is changed.)

9.11 Repeat the last exercise for yet another distance function $d$ defined by $d((x_1, y_1), (x_2, y_2)) = \max\{|x_1 - x_2|, |y_1 - y_2|\}$.

## 1.10    The Order of Quantifiers in Mathematics

A **quantifier** is a phrase which gives quantitative information about those members of a class that have a certain property. Thus the italicised phrases in the statements ' *All* men are mortal', '*Hardly any* politician is honest', '*There is* a man eight feet tall' '*Three per cent* of the machines are faulty' and 'This bill is opposed by *some* highbrow socialites' are all quantifiers. Each of these statements involves a certain set, say $X$, and a certain subset, say $A$ of $X$ (e.g. in the first statement, $X$ is the set of all men and $A$ the subset of those men who are mortal). A quantifier tells us something about the size of $A$ (or the relative size of $A$ as in the case of the statement about machines). Adverbs like 'generally' 'usually' or 'seldom' can also be interpreted as quantifiers, although the quantitative information they convey is not exact.

In mathematics, the two 'extreme' quantifiers are most important. Given a subset $A$ of a set $X$, the statement that $A$ is non-empty (i.e. has at least one element in it) is expressed by saying that 'there exists $x \in X$ such that $x \in A$'. (In real life examples, a less clumsy diction is possible, e.g. in the statement above about a man eight feet tall.) This is often abbreviated as ' $\exists\, x \in X$ s.t. $x \in A$'. Note that here $x$ is a dummy variable. We are not specifying any particular element of $A$. Nor are we saying how many elements $A$ has. All we are saying is that there exists at least one element in $A$. For this reason, the quantifier $\exists$ is called an **existential quantifier**. It simply asserts the existence without necessarily giving any method for finding.

At the other extreme there is the **universal quantifier** which says that the subset $A$ equals the entire set $X$. This is expressed as, 'for every (or for all) $x \in X, x \in A$ and symbolically, ' $\forall\, x \in X, x \in A$'. The phrase 'for any' is confusing since it means 'for some' or 'for every' depending on the context. It is best to avoid it.

As with other statements in mathematics, those involving the universal quantifier are to be interpreted strictly. In particular, for such a statement to be true, it is *not* enough that it holds true in a vast majority of the cases, not even that it holds true in 'almost all' cases. Even one case where it fails (called a **counterexample**) makes the statement false. Thus even one immortal man renders the statement 'Every man is mortal' false (as false as millions of such men would!).

Statements involving only one quantifier are common even in real life and cause little confusion. A statement like 'Behind every great man there is a great woman' involves two quantifiers. If we elongate it further as 'Behind every great man there is a great woman such that whenever he runs into a problem she has a solution for it' we get a statement with four quantifiers. But such constructions are rarely used in day-to-day life as they appear very clumsy. In mathematics, however, statements involving two or more quantifiers are common. (The origin of this is probably in an attempt to give a precise definition of a limit. See Section 2.3.) Such statements need careful handling, especially when it comes to changing the order of the quantifiers as we now elaborate.

We have already seen two statements involving quantifiers. They were 'There is a man who is eight feet tall' and 'All men are mortal'. In mathematics, the statements involving quantifiers are rarely so simple. Moreover, a mathematician is apt to word them more clumsily. Thus, our statements would read 'There exists a man such that (or with the property that) he is eight feet tall' and 'For every man, it is the case that he is mortal'. One can even go further. Let $M$ denote the set (or the class) of all men. Then these statements may be written as ' $\exists\, x \in M$ such that $x$ is eight feel tall' and ' $\forall\, x \in M, x$ is mortal'. Here $x$ is a dummy variable taking values in the set $M$. We could have as well replaced it by any other symbol not previously used in the particular context (for example we cannot replace $x$ by $M$). In technical terms, the variable $x$ is said to be **bound** by the quantifiers. A sentence such as ' $x \in M, x$ is mortal' is meaningless. The expressions ' $\exists\, x \in M$' or ' $\forall\, x \in M$', so to speak, serve to 'introduce' the variable $x$. Strictly speaking, such introduction must be made before anything is said about the variable introduced. However, where confusion is not likely, it is customary to defer it a little. Thus, the present statements could be written ' $x$ is eight feet tall for some $x \in M$' and ' $x$ is mortal $\forall\, x \in M$' respectively.

The introduction of a variable bound by a quantifier cannot, however, be postponed at will. A quantifier cannot be shifted beyond another quantifier unless the variables governed by them are completely independent of each other. We illustrate this with examples. Consider the statement 'For every man there is a woman who loves him'. Letting $M, W$ denote, respectively, the sets of all men and women, we can write this statement as '$\forall x \in M$, $\exists y \in W$ such that $y$ loves $x$', or less clumsily as, '$\forall x \in M, y$ loves $x$ for some $y \in W$'. But we cannot write it as '$\exists y \in W$ such that $\forall x \in M, y$ loves $x$'. This latter statement would mean that there exists a woman who loves each and every man, something not asserted by the original statement. Here the order of quantifiers is crucial since the variable $y$ may depend on a particular value of $x$.

On the other hand, take the statement 'Every man loves every woman'. We can write it as '$\forall x \in M$ and $\forall y \in W, x$ loves $y$' or as '$\forall x \in M, x$ loves $y \forall y \in W$' or even as '$x$ loves $y \forall x \in M$ and $\forall y \in W$'. Here the variables $x, y$ are independent of each other and so no harm arises by interchanging the order of the quantifiers. Since both the quantifiers are of the same type, it is customary to write '$\forall x \in M, y \in W$' instead of '$\forall x \in M$ and $\forall y \in W$'. Equivalently, one could consider the cartesian product $M \times W$ and write '$\forall (x, y) \in M \times W$'.

As a last example take the statement 'A woman who loves a brave man does not love any other man'. Here we let $B$ be the set of all brave men and for each $x \in M$ we let $W_x$ be the set of all women who love $x$. Then the statement is '$\forall x \in B$, $\forall y \in W_x$ and $\forall z \in M - \{x\}, y$ does not love $z$'. Here the variations of $y$ and $z$ depend on $x$ and so the order of quantifiers cannot be changed.

A special word of caution is perhaps necessary for students who are simultaneously studying other subjects such as physics. In such subjects it is customary to write the main formula first and then proceed to explain the meanings of the symbols. (For example, $E = mc^2$ where $E$ is the energy, $m$ is the mass and $c$ the speed of light.) It is preferable to avoid this habit in mathematics. The explanatory or 'control' statements should always come in the proper order before the main assertion is made. Otherwise the statement may become meaningless or may convey a wrong meaning.

## EXERCISES

10.1 If a person has a friend who is an actor and also a friend who is a cricketer, does it mean that he has a friend who is both an actor and a cricketer? Explain the fallacy using existential quantifiers and sets.

10.2 For every positive real number $x$, there exists a positive real number $y$ such that $y < x$. Does this mean there exists a positive real number $y$ such that $y < x$ for every positive real number $x$? Explain the fallacy using quantifiers.

## 1.11    Positive Negation of a Statement

'Positive negation' may seem like a contradiction of terms. To understand what it means, let us first see what the negation of a statement, say $p$, means. This is simply a statement (often denoted by $p'$ or $-p$ or $\sim p$ or $\neg p$) to the effect that $p$ does not hold. For example if $p$ is the statement that 'John is intelligent' then $p'$ is the statement 'John is not intelligent' or 'It is not the case that John is intelligent'. Similarly if $p$ is 'Salt dissolves in water' then $p'$ is 'Salt does not dissolve in water'. Clearly $p'$ is true precisely when $p$ is false and vice versa. This is because, unlike in real life, in mathematics there is no quantification of truth. (See Section 1.2 for more elaboration of this point.) Every statement has only two possibilities, either 'true' or 'false' exactly one of which must hold. It then follows that the double negation (i.e. the negation of the negation) of a statement is logically equivalent to the original statement.

In mathematical statements, symbols are often used for brevity. The negation of such statements is expressed by putting a slash ($/$) over that symbol which incorporates the principal verb of the statement to be negated. Thus, '$x = y$' is negated as '$x \neq y$'. Similarly '$p \not\Rightarrow q$' (read as '$p$ does not imply $q$') is the negation of '$p \Rightarrow q$' (read as '$p$ implies $q$'). Note that '$x \neq y \Rightarrow x \in A$', '$x = y \not\Rightarrow x \in A$' and '$x = y \Rightarrow x \notin A$' are very different statements and only the middle one is the correct logical negation of '$x = y \Rightarrow x \in A$' because the principal verb in the original statement is 'implies', embodied in the symbol $\Rightarrow$.

Every statement can be negated mechanically by putting the phrase 'It is not the case that' before it. But this is not very informative. We look for a negation which will give some positive information. Such negations are called positive negations. They are often not possible. For example the statement that 'John is intelligent' cannot be negated without using the word 'not' somewhere. We may negate it as 'John is dumb'. But this presupposes that 'dumb' is the same as 'not intelligent ' and so the word 'not' is avoided only superficially. But things are better when the nature of the statement is such that when it fails, one of some other possibilities must hold. For example let $p$ be the statement '$x$ is less than $y$' or more compactly, '$x < y$'. (Here $x,y$ are some real numbers already defined.) The negation, $p'$, then is 'It is not the case that $x$ is less than $y$' or '$x$ is not less than $y$' or in symbols, '$x \not< y$'. But these get us nowhere. Now, one of the basic properties of real numbers (technically called the **law of trichotomy**) is that for any two real numbers $x$ and $y$, exactly one of the three possibilities holds: (i) $x < y$, (ii) $x = y$ and (iii) $x > y$. So we may also write the negation of $p$ by '$x$ is bigger than or equal to $y$' or in symbols '$x \geq y$'. This is a positive negation of $p$. It gives a positive information about the relationship between $x$ and $y$ and starting from it we may be able to conclude something.

The negations of implication statements and of statements involving quantifiers are especially important. From the remarks in Section 1.5, it is clear that the negation of '$p \to q$' is '$p'$ and $q$'. That is, the statement that $p$ holds and $q$ fails. For example, the negation of 'If it rains the streets are wet' is 'It rains and the streets are not wet'. In real life we often replace 'and' by 'but' or 'and still' in such sentences to emphasise that the expected conclusion ($q$) fails even though the hypothesis $(p)$ holds. In mathematics it is not necessary to do so.

Note, incidentally, that even if we agree that 'John is dumb' is the positive negation of 'John is intelligent', it does not follow that 'John is very dumb' is the negation of 'John is very intelligent.' The correct negation is 'John is not very intelligent'. In real life, we take this as a polite way of saying that John is dumb (in fact, very dumb). But in mathematics it is not so. The original statement refers to John's degree of intelligence. Just because he lacks a high degree of intelligence, it does not follow that he is at the other end. Maybe he is just average. Put differently, an antithesis should not be confused with a negation.

A statement which asserts that every member of a class has a certain property is false when there is at least one member of that class which does not have that property. Thus the negation of 'every man is mortal' is simply 'there exists a man who is not mortal '. A layman is likely to negate this as 'no man is mortal' or 'every man is immortal' or 'not every man is mortal' or 'every man is not mortal'. Of these the first two are incorrect, while the third one is correct but not very informative. The fourth one is correct but confusing because the scope of the word 'not' is ambiguous. If we club it with 'mortal' and replace 'not mortal' with 'immortal' (which is perfectly sensible) then the last sentence would read 'Every man is immortal' which is not the correct negation of the original statement. (In the same vein, a beginner often confuses a statement like '$x$ and $y$ are not both zero' with the statement '$x$ and $y$ are both non-zero'. The former means that *at least one* of $x$ and $y$ is non-zero and is also written symbolically as '$(x, y) \neq (0, 0)$'. It is not as strong as the latter.)

The easiest thing to keep in mind is that if a statement begins with the universal quantifier $\forall$ ('for every') then its negation will begin with the existential quantifier $\exists$ (there exists), and vice versa. Thus the negation of 'there exists $\epsilon > 0$ such that $x + \epsilon > y$' is 'For every $\epsilon > 0$, it is not the case that $x + \epsilon > y$,' or, in a positive form, 'For every $\epsilon > 0, x + \epsilon \leq y$'.

Sometimes a statement about a class ostensibly appears as an implication statement. Consider, for example, the statement 'If John is rich, he is intelligent.' Here the statement deals with a particular individual 'John' and the negation of the statement is simply 'John is rich but not intelligent'. But consider the statement, 'If (or whenever) a man is rich, he is intelligent'. This is a statement about a class (viz., the class of all men). Here the words 'for every man' are understood. Written fully, the statement would read, 'For every man, if he is rich, he is intelligent'. The correct negation of this statement is 'There exists a man who is rich but not intelligent.'

The positive negation of a complicated statement can be worked out step by step keeping in mind the general comments above and with a little practice, the process becomes almost mechanical. We illustrate this, by negating the following statement, say $p$.

"For every $\epsilon > 0$, there exists $\delta > 0$ such that whenever $0 < |x - c| < \delta$, $|f(x) - L| < \epsilon$."

(The reader will notice that $p$ is precisely the definition of a limit, but that is not relevant here.)

Let $p'$ denote the negation of $p$. The statement $p$ asserts that something is true for every positive $\epsilon$. This something, which we denote by $q$, is the statement "there exists $\delta > 0$ such that whenever $0 < |x - c| < \delta$, $|f(x) - L| < \epsilon$". So the original statement $p$ says that for every positive $\epsilon$, the statement $q$ holds. Hence its negation $p'$ is the statement that there exists some positive $\epsilon$ for which $q$ fails. We are thus reduced to having to negate the statement $q$. Now $q$ asserts the existence of a positive $\delta$ with a certain property, say $r$. To be precise, $r$ is the statement, "whenever $0 < |x - c| < \delta$, $|f(x) - L| < \epsilon$". To say that $q$ fails means that no matter which positive $\delta$ we take, the statement $r$ is false. Putting together our progress so far, $p'$ (i.e. the negation of the original statement) reads, "There exists $\epsilon > 0$, such that for every $\delta > 0$, $r$ fails i.e. $r'$ holds". So now we have to negate $r$. Although $r$ is ostensibly an implication statement, the words 'for every $x$' are understood at its beginning. So $r'$ would mean that there is some $x$ for which the implication statement fails, i.e. its hypothesis holds but conclusion fails. That is, there exists $x$ such that '$0 < |x - c| < \delta$' holds, but '$|f(x) - L| < \epsilon$' fails. The last one we can paraphrase positively by saying that '$|f(x) - L| \geq \epsilon$'. So $r'$ says "there exists $x$ such that $0 < |x - c| < \delta$ and $|f(x) - L| \geq \epsilon$." Adding this to our earlier work, $p'$, the negation of the statement $p$ above, now reads :

"There exists $\epsilon > 0$, such that, for every $\delta > 0$, there exists $x$ such that $0 < |x - c| < \delta$ and $|f(x) - L| \geq \epsilon$."

The exercises will provide some drill in negating statements in a positive way. The reason why positive negations are so important in mathematics is that frequently we prove an implication statement say $p \rightarrow q$ in an **indirect** manner. That is, we do not start by assuming that $p$ is true and then show that $q$ must also be true. (This approach is the **direct proof**.) Instead, we start by assuming that $q$ is false. Then with some deductive reasoning, we show that $p$ must also be false, thereby we get a **contradiction** or an absurd result. So the implication $p \rightarrow q$ must be true. This method of proof is called a proof by contradiction or *reductio-ad-absurdum*. In essence, instead of proving $p \rightarrow q$ we are proving its **contrapositive**, i.e. the implication $q' \rightarrow p'$ which is logically equivalent to $p \rightarrow q$.

This kind of a reasoning is not uncommon even in real life. When a defense lawyer argues that the accused did not physically assault the victim because at the time of the act, he (i.e. the accused) was at some other place, he is resorting to this reasoning. (In law such an argument is called an *alibi*.) In mathematics, however, it is needed more frequently. In that case when we attempt to show that $q'$ implies $p'$, merely knowing that $q$ fails will not be of much help. In order to proceed further, we need some positive information. That is, we need to express the negation $q'$ in a positive form. As an excellent example, in Section 3.1 we shall prove that if a function satisfies a certain condition then it has a limit. The proof will begin by negating the definition of a limit, as was done above in detail. (The school geometry, especially the earlier propositions, also provide several examples of indirect proofs. One such proposition is given in the exercises.)

## EXERCISES

11.1   Negate the following statements in a positive way. (You are allowed to use antonyms, i.e. words of opposite meanings such as 'poor' versus 'rich'.)

    (i)     No rich man is intelligent.

    (ii)    A man is rich only if he is intelligent.

    (iii)   John is rich and intelligent.

    (iv)   John is rich or intelligent. (This includes the possibility that John is both. In mathematics, 'or' is always used in this inclusive sense unless 'not both' is specifically stated.)

    (v)    John is either rich or intelligent but not both.

    (vi)   There exists a woman who loves every man.

    (vii)   For every man there exists a woman who loves him.

    (viii)  For every man there exists a woman such that whenever he asks her to dance she agrees.

    (ix)   For every $\epsilon > 0$, there exists a positive integer $m$ such that for all $n \geq m$ and for all $x \in S, |f_n(x) - f(x)| < \epsilon$.

    (x)    There exists $L$ such that for every $\epsilon > 0$ there exists $m$ such that for all $n \geq m, |a_n - L| < \epsilon$.

11.2   Assume that 'bad' is the antonym of 'good'. Is 'All happy men are bad' the correct negation of 'All happy men are good'? Is 'All bad men are happy' the correct negation of 'All good men are happy'? Is 'John is happy if and only if John is good' a correct negation of 'John is happy if and only if John is bad'? Is '30% of the men are bad' a correct logical negation of '30 per cent of the men are good'?

11.3   The statements 'John is good' and 'John is bad' can never hold together. But the statements 'Half the men are good' and 'Half the men are bad' are equivalent. Why is there no contradiction?

11.4   Assuming that in any triangle, the angle opposite a greater side is greater, prove the converse.

### 1.12   Why It Is Important to Know the Proofs and Not Just the Statements of Mathematical Theorems

As elaborated in Section 1.4, in mathematics a statement is never accepted without a deductive proof, no matter how evident it looks or how powerful experimental evidence there is in its favour. So the proof is the very heart of a theorem. Without it, a theorem has no existence. So from this point of view, one must know a proof if one is to know a theorem.

The question, however, is whether one must know the proof of *every* theorem which one applies, if the interest is more in applications to the sciences or engineering or to some other field. This is more of a practical issue. Admittedly some of the proofs are so lengthy and pendantic that little is gained by going through them. It is also true that because of the insistence on logical perfection in mathematics a lot of work often goes into rigorously establishing what is intuitively obvious even to a layman, possibly after a little experimentation. Take, for example, the Jordan Curve Theorem mentioned in Section 1.6. Even before it is stated, it is necessary to give clear-cut mathematical definitions of the intuitively clear concepts like a curve, a region and a boundary. Even though the theorem looks purely geometric, its proof requires a good dip into some other areas of mathematics such as algebra or combinatorics. Then again, there is the question of what degree of completeness one hopes to achieve. Even professional mathematicians are not in a position to give a complete proof of every theorem they use, starting from the axioms of set theory (which, as discussed in Section 1.9 is the basis of modern mathematics). Frequently they simply quote a result they need and give a reference where the proof may be found. For someone who wants to use mathematics only as a tool, it is tempting to think that the proofs matter little.

However, for someone who wants to apply mathematics intelligently, it pays to have some idea of the proof of the result he is using, just as it pays to know the basic mechanism of a car engine if you are driving your own car. Without the proof, the theorem may sometimes appear rather arbitrary or unappealing. And so, unless you use it frequently, you might forget it. A proof, however, consists of a logical sequence of deductions, each step smoothly leading to the next one. It is a natural flow of ideas and, as such, is often easier to remember than the mere theorem. It is somewhat like this. A theorem is rather like the name of a person. Unless you meet or hear of him frequently, you are likely to forget it. But if the name happens to have some meaning and this meaning aptly reflects some peculiar characteristic of that person, then this very fact will enable you to remember the name effortlessly. So, from a pedagogical point of view, it pays to know a proof.

There are, of course, other reasons too. Frequently, the proof does more than just proving the theorem. As a simple example, take the well-known proposition in geometry that any three non-collinear points, say $P, Q$, and $R$ lie on one and only one circle $C$. This, by itself, does not tell you how to construct the circle $C$. But the proof tells you how to locate the centre of $C$ (viz, as the point of intersection of the perpendicular bisectors of any two of the segments $PQ, QR$

and *RP*.) This information may be vitally needed in some problem (see, for example, Exercise (12.1)). It may also happen that the proof used to prove a particular result implies with a slight modification a more general or a possibly stronger result. For example, Euclid's proof in Exercise (4.1), can be easily modified (see the Exercises) to show that there are infinitely many primes of the form $4n - 1$ where $n$ is an integer. Merely knowing that there are infinitely many primes would not imply this. (Every prime, other than 2, is of the form $4n + 1$ or $4n - 1$. So, since there are infinitely many primes, it follows easily that either there are infinitely many primes of the form $4n + 1$ or of the form $4n - 1$. But that does not tell us which possibility holds. As a matter of fact, both are true. But the former is not so easy to prove.)

As a simple, but instructive example from calculus, it is well-known that if $0 < x < 1$, then $x^n \to 0$ as $n \to \infty$. But what about $nx^n$? Here we are in trouble because one of the factors (viz. $n$) tends to $\infty$ while the other (viz. $x^n$) tends to 0. In such cases nothing can be said in general about the behaviour of the product. If however, we look at the proof of the fact that $x^n \to 0$ as $n \to \infty$, then we can see that with a slight modification, it will also imply that $nx^n \to 0$ as $n \to \infty$. (We shall elaborate this more in Chapter 3, see Section 3.3.)

In order for a theorem to be valid, it is sufficient if it has just one proof. But many results can be proved in more than one ways. A comparative study of such proofs is often instructive. Some proofs are unusually short and elegant. (For example, Exercise (4.6).) Some proofs, on the other hand, are lengthy and involved. But sometimes they provide more information than a shorter proof. It also happens sometimes that out of the several proofs given, one easily generalizes to other situations. Occasionally, even a mere paraphrase of a proof shows how it will apply to more general situations. Examples to illustrate these statements will be given in the exercises.

By and large, it is to be expected that it will be more difficult to prove a stronger or a more general result. But an exception can arise in the case of a proof by mathematical induction. Suppose for every positive integer $n$, we want to prove some statement, say, $P(n)$. We begin by showing that $P(1)$ is true (which is usually easy). Then we show that for every postive integer $n$, the truth of $P(n)$ implies that of $P(n+1)$. Sometimes the nature of $P(n+1)$ may be such that the truth of $P(n)$ may not suffice to establish that of $P(n+1)$. But it may be possible to find another statement $Q(n)$ such that for every positive integer $n, Q(n) \Rightarrow P(n)$. If the nature of the statement $Q(n)$ is such that it is easy to prove the inductive step (viz., the implication statement $Q(n) \Rightarrow Q(n+1)$) then we have the apparently paradoxical situation in which it is easier to prove $Q(n)$ than $P(n)$ even though $Q(n)$ is stronger than $P(n)$. An example of this 'magic' of induction will be given in the exercises.

To appreciate such magic or the elegance of any proof, one must, of course, study the proof and not just the theorem. But the best way to really appreciate any proof is to try to prove the result yourself. If you succeed, you have a feeling of achievement. If you fail you have nothing to lose. In fact, in that case you know that the assertion is not all that trivial and so you can appreciate the proof better. If the proof turns out to be considerably shorter, simpler or trickier than you predicted, then you will remember that proof for a long time. You might be tempted to pose it as a challenging problem to a like-minded friend. After you gain some maturity, you will be in a position to sometimes improve upon a proof, or, give another proof. If you are keen-eyed, you may even be able to spot some flaw in the proof. It has happened a number of times in mathematics that a proof, even a published one, had to be withdrawn because a mistake was found in it. Although this is very unlikely to happen in the case of a proof of the theorems in calculus, which have well stood the test of time, the ability to quickly detect the fallacy in some piece of reasoning is a valuable asset in mathematics as well as in real life.

## EXERCISES

12.1    A **rational point** in the $x$-$y$ plane is a point whose $x$ and $y$ co-ordinates are both rational. Prove that if a circle of radius 1 contains three (or more) rational points, then it contains infinitely many such points..

12.2    Prove that every number of the form $4k - 1$ where $k$ is a positive integer has at least one prime divisor of the form $4n - 1$.

12.3    Using the last exercise, modify Euclid's argument to show that there are infinitely many primes of the form $4n - 1$.

12.4    Let $A_i = (x_i, y_i)$, $i = 1, 2, 3$ be the vertices of a triangle in the $x$-$y$ plane. Write down the equations of the three altitudes of the triangle $A_1 A_2 A_3$. Solve any two of them simultaneously and verify that the point of intersection lies on the third altitude also. Hence show that the three altitudes of a triangle are concurrent.

12.5    In the last exercise prove that the three equations of the altitudes have a common solution without solving them explicitly. Hence deduce that the three altitudes are concurrent. Compare the two proofs. Which is more elegant? Which is more informative?

12.6    Suppose two glasses contain 100 cc. each of milk and water respectively. Suppose 5 cc. of the liquid in the first glass is poured to that in the second, the mixture is stirred and 5 cc. of it is transferred back to the first glass. Let $\alpha$ and $\beta$ be respectively the percentages (by volume) of water in the first glass and of milk in the second glass. Calculate $\alpha$ and $\beta$ and determine which is bigger.

12.7    In the last exercise, show directly that $\alpha = \beta$ without calculating either of them. In fact, your argument should also work regardless of whether the mixture in the second glass was stirred or not.

12.8    In exercise (12.6), suppose the exchange of liquids (5 c.c.s at a time) continues back and forth. After how many such exchanges will there be more than 20 per cent water in the first glass? Will the reasoning in the last exercise give the answer here ?

12.9 The $n^{\text{th}}$ **Hilbert matrix** $H_n$ is defined as the $n \times n$ matrix whose $(i,j)^{\text{th}}$ entry is $\dfrac{1}{i+j+1}$. For example, $H_3$ is

$$\begin{bmatrix} 1 & 1/2 & 1/3 \\ 1/2 & 1/3 & 1/4 \\ 1/3 & 1/4 & 1/5 \end{bmatrix}$$

Try to prove by induction on $n$ that the determinant of the matrix $H_n$ is positive for every $n$. What difficulty do you run into?

12.10 A **Cauchy matrix** of order $n$ is an $n \times n$ matrix whose $(i,j)^{\text{th}}$ entry is of the form $\dfrac{1}{x_i + y_j}$, where $x_1, \ldots, x_n$ and $y_1, \ldots, y_n$ are some given real numbers. (Assume no $x_i + y_j$ is 0.) By induction an $n$, show that the determinant of $C_n$ equals

$$\prod_{1 \leq i < j \leq n} (x_j - x_i)(y_j - y_i) / \prod_{\substack{1 \leq i \leq n \\ 1 \leq j \leq n}} (x_i + y_j).$$

[*Hint* : Subtract the first column of $C_n$ from each of the remaining ones, take out factors common from the rows and the columns. Then subtract the first row from the remaining ones and do the same. What is left is a matrix whose determinant equals the determinant of another Cauchy matrix of order $n - 1$.]

12.11 Prove the result of Exercise (12.9) from that of the last one by expressing $H_n$ as a very special case of $C_n$. Thus it is easier to do a more general problem.

12.12 Let $C_1, C_2$ be concentric circles with the radius of $C_2$ being twice that of $C_1$. From a point $P$ on $C_2$, tangents $PA$ and $PB$ are drawn to $C_1$. Prove that the centroid of the triangle $PAB$ lies on $C_1$. Do the problem with methods from pure geometry. Then do it with co-ordinate geometry by taking the equations of $C_1, C_2$ as $x^2 + y^2 = 1$ and $x^2 + y^2 = 4$ respectively without loss of generality. Use the formulas for the chord of contact of a point w.r.t. a circle and for the mid-point of a chord. In this formulation show that the proof generalises to a similar result about two concentric ellipses with equations of the form $\dfrac{x^2}{a^2} + \dfrac{y^2}{b^2} = 1$ and $\dfrac{x^2}{a^2} + \dfrac{y^2}{b^2} = 4$.

### 1.13    How Impossible Things Are Possible in Mathematics

The question can be interpreted in several different ways. And depending on it, the answer would change.

If by impossible things, one means things like flying carpets and singing trees, then obviously they exist only in poetry and fairy tales. The only weak analogues mathematics can offer are things which defy our intuition which is based on our day-to-day experience of life. For example, we think of a 'curve' as a 'thin, one-dimensional' set of points, having length but no breadth. And this indeed is the case with the familiar curves such as circles, ellipses and parabolas. These curves are of course too specialised. The general concept of a curve may be captured as a locus, i.e. the path traced out by a continuously moving point. The notion of continuity has, of course, to be defined rigorously. This is not very hard to do and we shall in fact do it in Section 4.12. But it turns out that the standard definition of continuity is far more general than we expect. As a result, it is possible to have curves which occupy whole laminas, e.g. an entire disc or a square. Such space-filling curves (or **Peano curves** as they are called) do give us some shock. A few other examples are bounded regions with infinitely long boundaries (see Exercise (6.3.6) for an example) or their higher dimensional analogues, e.g. bounded solids with infinite surface areas.

The reason such impossible things exist in mathematics is that mathematical definitions are often made by abstracting some, but not all, of the properties of some familiar objects. Naturally, things so defined cannot meet all our expectations. For example, curves like circles and ellipses are not only continuous but also have continuously changing tangents. If this requirement is added to the definition of an abstract curve, then it is no longer possible to have space-filling curves. So, it is all a question of definitions.

Another possible interpretation of 'impossible becoming possible' is a tribute to an exceptional ability. The word 'impossible' did not exist in Napoleon's dictionary. In Indian mythology, the king Bhageerath, through his relentless endeavour, achieved the impossible task of changing the course of Ganga. (The Greek mythology also has a similar feat attributed to Hercules.) So here, 'impossible' means 'impossible for ordinary mortals'. Impossibility of this kind is subject to the prevailing conditions. Climbing the Everest or running a mile in less than four minutes were considered impossible at one time because of the invariable failures of even the most determined efforts. But today these feats stand achieved. Mathematics, too, is not lagging behind in such accomplishments. Two best recent illustrations are the **Four Colour Theorem** and the **Fermat's Last Theorem**. The former says that every planar map of countries can be coloured with four colours so that no two adjacent countries get the same colour (cf. Exercise (4.3)), while the latter says that for an integer $n > 2$, it is impossible to find positive integers $x, y$ and $z$ such that $x^n + y^n = z^n$. Rigorous proofs of these results were achieved in 1976 and 1994, respectively, after more than a century each of attempts.

In fact, in the twentieth century the phenomenol advances in science and technology have rendered possible many things which could exist only in fairy tales. Man can now fly and is on the way to be cloned. One really wonders whether anything can now be called as permanently impossible. In other words, is it possible to have things which are impossible not only today but will remain so forever, no matter how assiduously they are pursued? This is where mathematics differs sharply from the natural sciences and takes us back to the quesiton in the title. As discussed in Section 1.6, the natural sciences are subservient to observed facts. Their basic laws are subject to change and so nothing can be ruled out as impossible forever, no matter how difficult or ridiculous it appears to be now. In mathematics, on the other hand, the allegiance is to deductive logic and so its results are immortal. Every such result automatically implies an impossibility result. For example, the theorem that the three altitudes of any triangle are concurrent is equivalent to saying that it is impossible to have a triangle whose altitudes are not concurrent. We can say this with absolute certainty, far more so than we can say it is impossible to have a flying carpet. The Fermat's Last Theorem just mentioned is also an impossibility result.

Things become more subtle when the problem asks for the possibility or otherwise of a *method* to do something that is already known to exist. A classic example of such a problem is the **angle trisection problem**. There is an easy geometric construction for bisecting any given angle. It is but natural to think that there is a similar construciton, using a compass and a straight edge only, for an angle which is exactly one third of a given angle. Such an angle definitely exists. The problem is not its existence, but giving an *exact* construction for it. Such a construction eluded mathematicians for centuries. But, of course, that does not mean it could not exist. Finally, it was proved that it is *impossible* to have such a construction. The proof ultimately reduces to showing that a certain cubic polynomial can have no rational root and is not particularly difficult. Two other famous results of this type are that it is impossible to square a circle (i.,e. draw a square whose area equals that of a given circle) and to double a cube (i.e., to draw the side of a cube whose volume is twice that of a given cube). In both cases, the desired constructions must use only a compass and a straight edge. (The edge is not marked with a scale.)

It may be argued that both these insruments are highly primitive and that with modern advances in technology or with some consummate skill not known to man so far, it may be possible in future to trisect an angle using only a compass and a straight edge (or a ruler as it is sometimes called). How can we permanently rule out something dealing with physical objects? For example, our experience today is that gold does not dissolve in water. But a technology may come up to divide gold particles so finely that it may indeed dissolve in water.

The answer depends on what one understands by a ruler and a compass. To a layman they are physical objects. But to a mathematician, a ruler is any device which permits one to draw an entire straight line as soon as two distinct points on it are known. Similarly a compass is an instrument to draw a circle with a given radius and a given centre. Any other use of a ruler or a compass is illegitimate. The mechanism of a compass and possible improvements in it are of no concern to a mathematician. In effect, the mathematician is *defining* a ruler and a compass (perhaps rather narrowly). And once these definitions are agreed upon, the impossibility of trisecting an angle is an indelible fact. So once again we see that it is all a quesiton of definitions.

Another celebrated impossibility result deals with solving polynomial equations. It is trivial to solve a first degree equation, say, $ax + b = 0$ (with $a \neq 0$). For the quadratic equation $ax^2 + bx + c = 0$ (with $a \neq 0$ again), there is the well-known formula $\dfrac{-b \pm \sqrt{b^2 - 4ac}}{2a}$ for its roots. The expression $\sqrt{b^2 - 4ac}$ is called a **radical** (more precisely, a radical of order 2). It is natural to expect that for a polynomial of higher degree there are similar formulas involving radicals (possibly of higher orders) which express the roots of the polynomial in terms of its coefficients. For polynomials of degree 3 or 4 such formulas indeed exist (although they are more complicated than the quadratic formula and not as well-known). But the search for such formulas for a fifth (or higher) degree polynomial proved futile. It is not that such polynomials do not have roots. In fact, there is a well known result called the **Fundamental Theorem of Algebra** which says that every polynomial of degree $n$ has $n$ roots (some of which may be complex and some may coincide with each other.) So the existence of roots is no problem. The question is whether they can be expressed by a formula involving radicals. Intense search for such a formula continued till a young mathematicain Abel proved that *no such formula exists*. Put differently, this means that it is impossible to solve a general polynomial equation (of degree 5 or mroe) by radicals. No matter who attempts it and how assidously that person works, there is something inherently impossible about it, something like making two parallel lines meet.

Such impossibility results are a landmark in the history of human thought. They also serve to underscore the unique position which mathematics has. In every field including mathematics, impossible things are possible in the sense that what looked impossible at one time may become possible some time later. But it is only in mathematics that you can prove the permanent impossibility of something because in mathematics once you make your definitions, everything is sealed. So mathematics stands out as the only subject where it is possible to have truly impossible things. Or, by a juggling of words, in mathematics and only in mathematics impossible things are possible!

# Chapter 2

# NATURE OF CALCULUS IN GENERAL

## 2.1. Reasons for Studying Calculus

As with mathematics in general, the most practical reason from a student's point of view for studying calculus is to get through the courses! More seriously, however, calculus is an indispensable tool for almost anything ranging from calculating the areas and volumes of various figures, finding how much interest you earn in a bank if your money is growing continuously, designing a safe curving road on a mountain surface, figuring out the most economical way to do some task and so on. The pre-calculus tools such as arithmetic, classical algebra, geometry and trignometry are undoubtedly still useful. In fact the layman often finds them adequate. Indeed, in many walks of life you can get by with mere arithmetic. But these tools begin to show their limitations when the problems get more subtle. We illustrate this with several simple examples.

Suppose a driver, while he is driving a car, is suddenly asked how fast he is moving at that particular moment. He will simply read the speedometer and answer, say, at 30 k.p.h. But ask him what he means by his (i.e. the car's) speed at that moment. He knows the answer but most likely he won't be able to tell you. If he knows a little arithmetic, he will tell you that it is the distance the car will go in one hour if it continues to run at that same speed. This is correct, but not of much use, because the car might not run for one full hour. If the driver is intelligent, he may be able to get over this difficulty. "Keep on driving for just a minute and multiply the distance you cover by 60" will be his answer. In fact this time interval can be made even smaller, say, a fraction of a second. We of course have to assume that we have instruments to measure the time and even the small distances accurately. This is indeed an important practical problem. But let us not worry about it here. Even if we assume that we have such instruments, there is a far more nagging difficulty we still have to face. What guarantee is there that the speed will remain constant throughout the time interval we have chosen? If the speed can fluctuate within an hour, then

41

theoretically it can fluctuate even within a fraction of a second. In symbols, suppose $t_0$ is the time at which we want the instantaneous speed. If the car runs for a time, say[1] $\Delta t$, starting from $t_0$, and covers a distance $\Delta s$ in this time interval, then if the speed remains constant throughout, it must equal $\dfrac{\Delta s}{\Delta t}$. Otherwise this ratio will only give what is called the **average speed** over the time interval from $t_0$ to $t_0 + \Delta t$. No matter how small $\Delta t$ is, the difficulty will always remain. We cannot set $\Delta t$ equal to 0, because then $\Delta s$ would also be 0 and we would be dealing with a meaningless ratio $\dfrac{0}{0}$. So this is not a solution either. For the solution we have to turn to calculus. It is based on the simple idea that if the time interval is sufficiently short then the average speed over it is a sufficiently good approximation to the instantaneous speed.

What is involved in this example is the concept of rate of change. The speed is the rate of change of the distance w.r.t. the time. More generally whenever one quantity, say $y$, depends upon (or, technically, is a function of) some other quantity, say $x$, then we can talk of the rate of change of $y$ w.r.t. $X$. For example, the coefficient of linear expension of a metal is the rate of change of its length (w.r.t. temperature) per unit length. In real life we can only measure the average rate of change. The instantaneous (or momentary) rate of change can be understood intuitively. Its exact definition and properties are studied in calculus. Since rates of various kinds are important in many fields besides physics (e.g. the rate of a reaction in chemistry, the rate of blood flow in medicine and the rate of inflation in economics), it is hardly surprising that the applications of calculus are so wide-spread.

In the example above, the problem was to find a certain rate. Let us now consider a problem where the rate of growth is given and we have to find the actual growth. In school we learn that even if the rate of interest is the same, compound interest is more than simple interest and this difference grows the more frequently it is compounded. Let us assume a bank gives interest at the rate of 10 per cent per annum. If a depositor invests 1 rupee, then at the end of the year the total will be $1 + \frac{1}{10}$. However, if compounding is done every six months, i.e. twice a year, then at the end of the first half year, the total will be $\left(1 + \frac{1}{20}\right)$. During the second half this would earn an interest equal to $\frac{1}{20}\left(1 + \frac{1}{20}\right)$. So the total at the end of the first year will be $(1 + \frac{1}{20})^2$. By a similar reasoning, if the interest is compounded thrice a year (at equal rests, of course), then the total will be $(1 + \frac{1}{30})^3$ which is slightly bigger than $(1 + \frac{1}{20})^2$. (With a principal of just one rupee, the difference is, in fact, too small to be entered into the passbook. But let us ignore this and take the miser's view that every part of a rupee, however small, matters.) More generally, if the compounding is done $n$ times annually, then a principal of 1 rupee will yield a total of $(1 + \frac{1}{10n})^n$ rupees

---

[1]It is customary in calculus, and more generally in mathematics to denote small changes in a variable by putting a $\Delta$ (or occasionally a $\delta$) in front of the symbol for that variable. Thus, if $t$ and $s$ denote time and distance respectively, then $\Delta t$ and $\Delta s$ are the changes in time and distance respectively. A beginner is advised to bear in mind that $\Delta t$ is a single symbol and not the product of the two symbols $\Delta$ and $t$. Similar remarks apply for many other notations such as $dx$, $dy$, $d^2x$, $dt^2$ to be encountered later.

at the end of the first year.

So far we have used nothing more than arithmetic. With a little algebra, it is not hard to show (cf. Exercise (1.1)) that the expression $(1 + \frac{1}{10n})^n$ grows as $n$ grows. (This is intuitively obvious anyway since the more frequently we compound, the more interest we get.) To give this fact a practical touch, suppose two competing banks $A$ and $B$, constrained by law not to increase the rate of interest beyond 10 p.c.p.a., try, instead to woo the depositors by compounding more and more frequently. Where will this 'interest war' end? The answer is when one of the banks conceives the bright idea of what is called **continuous compounding.** In the compoundings we considered above, no matter how frequently we compound, money grows only at the rests, but not in between two successive rests. In continuous componding, on the other hand, the money keeps growing every moment. So, if a depositor leaves the bank 10 minutes after depositing his money, by that time it has already grown. And moreover, it has grown a little more in the second half of these ten minutes than during the first! As with instanstaneous speed, the concept of continuous compounding is easy to understand intuitively. But without calculus it cannot be handled satisfactorily. (By the way, in this problem, if the interest is continuously compounded at the rate of 10 p.c.p.a. then at the end of one year, the total will be approximately 1.1051709. With monthly and daily compoundings the totals will, respectively, be 1.1047131 and 1.1051558, both approximately.)

The essence of continuous compounding is that the rate of growth is proportional to the quantity existing at that time. This phenomenon is fairly common in nature. Bacterial colonies, and to some extent, human populations (cf. Exercise (1.3.1)) obey the law of continuous growth. The growth can be negative, in which case it is called a decay. Radioactive decay is a good example of continuous decay. So we have another reason why calculus is so important and all-pervasive. (Continuous growth will be studied in more detail in Section 5.8).

The difference between the pre-calculus mathematics and calculus is brought about most vividly when it comes to finding the areas of plane figures. Starting from the formula of the area of a rectangle (viz. the product of its length and breadth) we can easily derive the well-known formula for the area of a triangle, viz $\frac{1}{2} \times$ base $\times$ height (see Figure 2.1.1.(a)). In trigonometry we study a few other formulas for the area of a triangle while co-ordinate geometry expresses it in terms of the co-ordinates of the verties of the triangle. All that these derivations require is some algebraic manipulations.

How do we find the areas of other figures? If the figure is a quadrilateral or more generally a polygon, then we can cut it into a finite number of triangles as shown in Fig.2.1.1 (b). (This can be done in many different ways. If the polygon has some special features such as regularity, a particular dissection into triangles may stand out as being the easiest to work with.) We can determine the sides of all these triangles using trigonometry or co-ordinate geometry. We then find their areas and simply add to get the area of the polygon. The computations many be lengthy, but theoretically we get the exact answer in a finite amount of time.

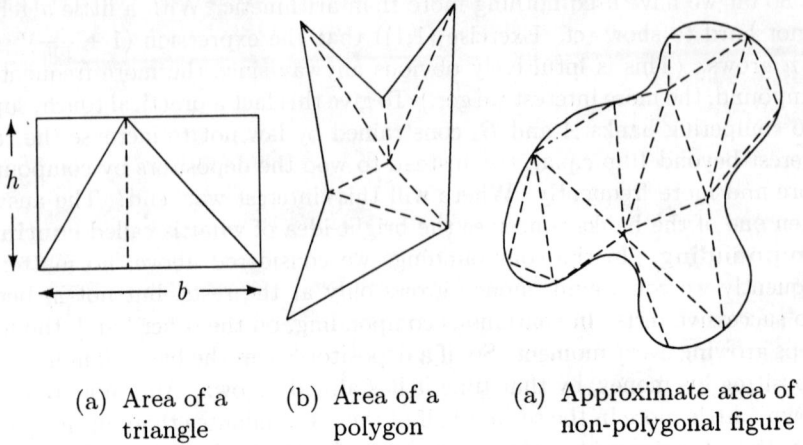

| (a) Area of a triangle | (b) Area of a polygon | (a) Approximate area of non-polygonal figure |

Figure 2.1.1 : Areas of Plane Figures

This method breaks down for non-polygonal figures. They cannot be subdivided into a finite number of triangles. By taking sufficiently small triangles, we may be able to cover with only finitely many triangles, most of the figure (see Fig.2.1.1.(c)) and so we can get an approximate area of the figure. To get the exact asnwer we need calculus. (The well-known high school formula for the area of a circle also requires calculus. That is why it is given without proof in schools. There is, however, an ingeneous way to make the foumula plausible without calculus. It will be indicated in Exercise (1.4).)

All the three examples so far were of the type where calculus gave an exact answer while other methods gave only an approximate answer. It can be argued that in practice we have to settle for an approximate answer anyway, becaue whatever be the nature of the calculating agency (whether humans or machines), it can calculate correctly only upto a certain number of decimals. This is a very relevant point but it does not render calculus dispensable. Even in practice, when we have a method to give only an approximate answer to a problem, we need to have some estimate, or an upper bound on the error, i.e. the difference between the exact and the approximate answer. (The error, by the way, is generally not known exactly, because if it is, then it is as good as knowing the exact answer! So what we look for is some assurance that we are within a certain degree of accuracy.) The methods of calculus generally give us such estimates. For example, in the problem of continuous compounding discussed above, calculus tells us how many times a year we should compound if the two totals are to agree upto, say, the fifth place of decimals.

Secondly for some real life problems there is no natural method even to give an approximate answer. Suppose, for example, we want to design open metallic boxes (i.e. those with no tops) to hold a fixed volume. If the metal is costly, we would naturally like to choose the dimensions of the box so as to minimise its surface area. There is no method short of trial and error to give even an approximate answer here. But with calculus the problem is easy. If the box is to have a volume $V$ c.c.s, then for the most economical box, the length and breadth are $(2V)^{1/3}$ c.m.s. each while the height is $\left(\dfrac{V}{4}\right)^{1/3}$ c.m.s. In other words, the base should be a square and the height should be half the side of the square. Here claculus is not needed for expressing the surface area or the volume of the box but for a very different purpose. There are infinitely many possible boxes with a given volume. So we cannot select the one with the least surface area by straight comparison, the way we find out who is the shortest student in a class. The problem here is that of **optimisation**. We are given a set $S$ and a real valued function, say $f$, on $S$. We have to find a point of $S$ at which the function $f$ attains its maximum or minimum, depending upon our interest. (In the present problem, $S$ is the set of all boxes having a given volume $V$ and the function $f$ assigns to each such box its surface area.) When the set $S$ is finite, we can do this merely by comparing the values $f$ attains at every point of $S$. This method fails when the set $S$ is infinite. There is no golden method for optimisation over an infinite set. Occasionally, we can get the answer without calculus. For example, even though there are infinitely many points on a straight line, say $L$, the one among them which is closest to a given point $P$ (not on $L$) can be obtained simply by drawing a perpendicular from $P$ to the line $L$. This requires only some simple geometry. A few other examples where optimisation over an infinite set is possible without calculus willl be indicated in the exercises. Interesting as these examples are, they form a rather specialised class, often based on a clever use of certain inequalities such as the **A.M.-G.M. inequality** (which asserts that the geometric mean of positive numbers can never exceed their arithmetic mean). Moreover, they can also be done with calculus methods which are less tricky (see Exercises (1.7) and (4.10.16) together). In fact, the A.M.-G.M. inequality can itself be proved using calculus (see Exercise (4.10.15)). On the other hand, there are many optimisation problems, which cannot be solved by elementary methods, but which can be solved using calculus. We thus have yet another answer why calculus is so important.

The four examples given in this section are only indicative. It is hardly possible to give even a glimpse of all major applications of calculus in one section. The ones we have given are not the most powerful. The reason for their choice is that they will be referred to in the rest of the book. In fact, the first three of them will serve as a motivation for the concept of a limit, which lies at the very heart of calculus.

# EXERCISES

**1.1**  Prove that $\left(1+\dfrac{1}{10n}\right)^{n} < \left(1+\dfrac{1}{10(n+1)}\right)^{n+1}$ for every positive integer $n$. Thus the more frequently we compound, the more is the total. [*Hint* : Expand by binomial theorem and compare term-by-term. The expression on the right has one more term than that on the left.]

**1.2**  A subset $S$ of a plane is called **convex** if for every two points of $S$, the entire line segment joining them is contained in $S$. Figure 2.1.2 shows a few convex and non-convex figures.

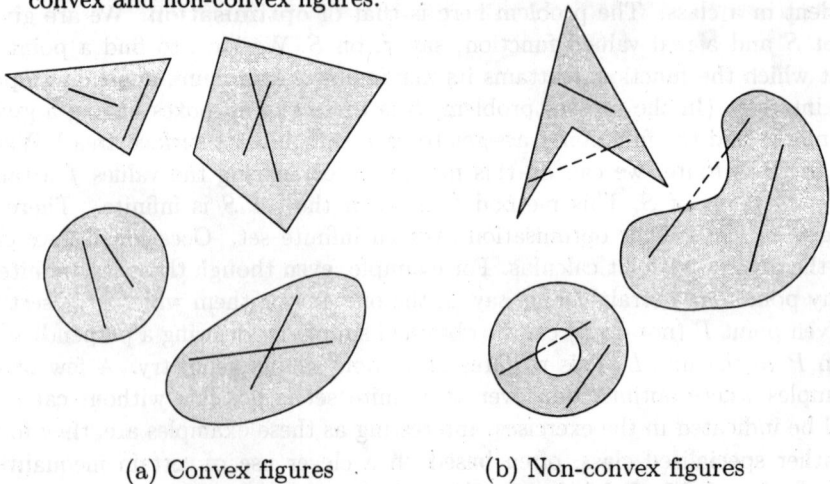

     (a) Convex figures           (b) Non-convex figures

Figure 2.1.2    Convex and Non-convex Figures

If $A_i = (x_i, y_i), i = 1, 2, \ldots, n$ are the vertices of a convex $n$-gon and $P = (a, b)$ is any point inside it, show that its area is

$$\frac{1}{2}\sum_{i=1}^{n}(x_i - a)(y_{i+1} - b) - (x_{i+1} - a)(y_i - b),$$

where we assume that $x_{n+1} = x_1$ and $y_{n+1} = y_1$. Show by an example that this result may not hold if the polygon is not convex.

**1.3**  Suppose a regular $n$-gon $P_n$ is inscribed in a circle of radius $r$. Prove that its area is $\dfrac{1}{2}nr^2 \sin\left(\dfrac{2\pi}{n}\right)$. Show that area of $P_n$ is less than that of $P_{2n}$.

(In fact, area of $P_n$ < area of $P_m$ for all $m > n$. This is intuitively clear since $P_m$ captures more of the area of the circle. A rigorous proof can be given using the fact that for small values of $\theta, \theta < \tan\theta$. But the obvious proof of this fact itself pre-supposes the formula for the area of a sector of a circle. However, a proof based only on the concept of the arc length will be given in Section 4.7. See Exercise (4.7.9).)

1.4 Suppose a circle is divided into $n$ equal sectors. Cut one of these sectors into two equal halves and rearrange those two halves and the remaining sectors so as to form a figure shown below in Fig.2.1.3(b).

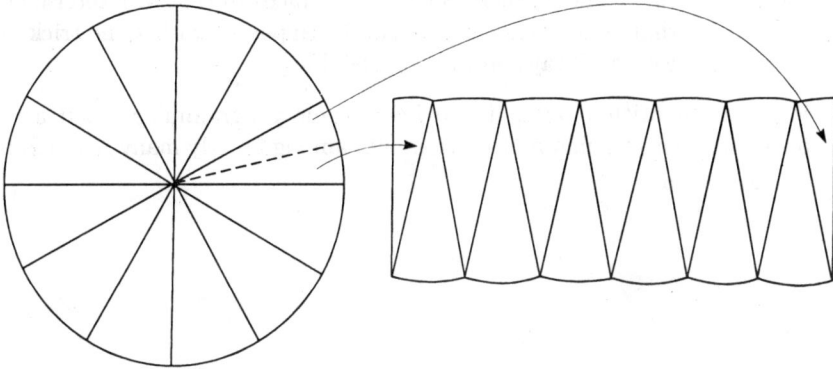

(a) Circle divided into sectors      (b) Almost rectangular figure

Figure 2.1.3 : Area of a Circle.

This figure is almost like a rectangle, and resembles a rectangle the more closely the larger is $n$. What is the area of this figure? Hence derive a formula for the area of a circle.

1.5 Find the points where the following functions (in which $a, b, c$ are some constants) attain their maximum and minimum values on the set of real numbers.

(i) $f(x) = ax^2 + bx + c$ (distinguish the cases $a > 0$ and $a < 0$)

(ii) $g(x) = a \cos x + b \sin x$

Vary the problem by letting $S$ be some subset of $I\!R$, e.g. a closed interval.

1.6 In any triangle $ABC$ prvoe that the maximum value of $\cos A + \cos B + \cos C$ is $\dfrac{3}{2}$ and that it occurs if $ABC$ is equilateral.

1.7 Prove that in any acute angled triangle $ABC$, the minimum value of $\tan A + \tan B + \tan C$ is $3\sqrt{3}$. When is it attained? [*Hint* : Use the identity $\tan A + \tan B + \tan C = \tan A \tan B \tan C$ and the A.M.-G.M. inequality.]

1.8 In the problem about boxes containing a given volume, suppose that the boxes have tops also. Prove, using the G.M.-H.M. inequality, that among all such boxes, the cube has the least surface area.

1.9 Prove that among all boxes (with tops) having a fixed surface area, the cube has the maximum volume. [This and the last problem are said to be **duals** of each other. A solution to either implies a solution to the other.]

1.10 Prove that the problem of minimising the surface area of a box (without top) so that it holds a given volume can also be done without calculus, using the G.M.-H.M. inequality. [*Hint*: If $x, y, z$ are the dimensions of the box, $z$ being the height, put $x = 2u, y = 2v$ and work in terms of $u, v, z$ instead of $x, y, z$. The problem effectively reduces to that in Exercise (1.8). This shows that sometimes calculus can be bypassed with some trick. But such tricks are not always obvious or possible.]

1.11 Write an algorithm (and, if you know some programming, then a computer program) for finding the maximum among finitely many real numbers $x_1, x_2, \ldots, x_n$.

## 2.2.  Limiting Process - the Heart of Calculus

In the last section, we encountered three problems, where an approximate answer could be obtained without calculus, viz. the problem of finding the instantaneous speed, the problem of finding the total under continuously compounded interest and the problem of finding the area of a plane figure. Moreover, in each case, we saw how the approximations can be improved so as to get closer to the exact answer. Specifically, we saw that the average speed $\dfrac{\Delta s}{\Delta t}$ is very close to the instantaneous speed if $\Delta t$ is sufficiently small, the total with compounding $n$ times a year is very close to that with continuous compounding if the integer $n$ is large and finally, that the sum of the areas of the triangles fitted in a figure (without overlapping) is very close to the area of that figure if the triangles are sufficiently fine.

What exactly is the relationship between this behaviour of appropximate answers and the exact answers? A layman will say that the approximate answer tends to (or approaches) the exact answer. Specifically he will say, for example, that $\dfrac{\Delta s}{\Delta t}$ tends to the instantaneous speed as $\Delta t$ gets smaller and smaller or that the total $(1 + \frac{1}{10n})^n$ will approach the total on continuous compounding as $n$ becomes larger and larger. A mathematician uses the word 'limit' to express the same idea. Thus, he says, "The limit of $\dfrac{\Delta s}{\Delta t}$ as $\Delta t$ tends to 0 is the instantaneous speed". In symbols, he writes " $\lim\limits_{\Delta t \to 0} \dfrac{\Delta s}{\Delta t}$ = instantaneous speed". Similarly, he writes " $\lim\limits_{n \to \infty} (1 + \dfrac{1}{10n})^n$ = total on continuous compounding" and reads it as "the limit of $(1 + \dfrac{1}{10n})^n$ as $n$ tends to infinity is the total on continuous compounding". The area of a plane figure is similarly the limit of the sum of the areas of the triangles fitted in it. (The symbol $\infty$ is read as 'infinity'. We shall say more about it in Section 2.4.) (Instead of $\lim\limits_{\Delta t \to 0}$ it is also customary to write $\lim_{\Delta t \to 0}$ especially in running text.)

The limits involved in these three examples represent the three major types of limits studied in calculus. The first, $\lim\limits_{\Delta t \to 0} \dfrac{\Delta s}{\Delta t}$, is a special case of what is called a **derivative**. More generally, any rate of change is a derivative. The second limit, viz., $\lim\limits_{n \to \infty} \left(1 + \dfrac{1}{10n}\right)^n$ is the **limit of a sequence**. It is conceptually different from the earlier limit in that while the variable $\Delta t$ approaches 0 through all possible positive values, the variable $n$ jumps to $\infty$ only through integral values. (Technically we say that $\Delta t$ is a **continuous variable** while $n$ is a **discrete variable**.) The third limit, viz., the area of a figure, is an interesting limit. As the triangles into which the figure is cut become finer and finer, their individual areas tend to 0. But the number of triangles becomes larger and larger. So we have a case of a sum in which each summand tends to 0 but the number of summands tends to infinity. This makes the limit intriguing. Limits of this type are called **integrals**.

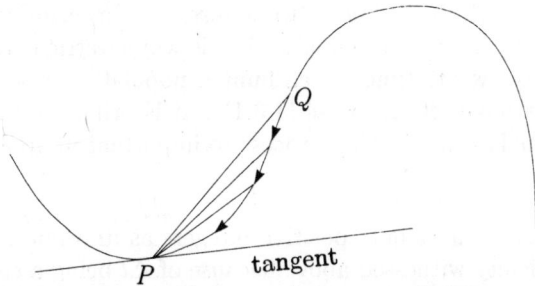

Figure 2.2.1 : Tangent to a Curve

Almost all basic concepts in calculus are based on limits of one of these types (or their extensions). It is therefore, no exaggeration, to say that the limiting process (in one form or the other) is the heart of calculus. Take away the limits and calculus will reduce to arithmetic, algebra, geometry or trigonometry. By the way, the concept of tangency in geometry, although intuitively clear, requires the limiting process for a rigorous definition. For a particular curve, we may be able to define it differently. For example, if $P$ is a point on a circle with centre $O$ then we may define the tangent at $P$ as the line through $P$ perpendicular to

*OP*. But, for a general curve, a tangent at a point $P$ has to be defined as the limiting position of a chord (or rather, a secant) $PQ$ as the point $Q$ approaches $P$ along the curve (see Figure 2.2.1).

So far, we have not *defined* a limit. This we shall do in the next section. The definition of limit has been one of the most elusive ones in mathematics. Let us see what are the difficulties involved by taking the first two of our exmaples of limits. When we consider $\lim_{\Delta t \to 0} \dfrac{\Delta s}{\Delta t}$, we have in mind values of $\Delta t$ which are very close to 0 but not 0 itself. If we put $\Delta t = 0$ then we get a meaningless expression $\dfrac{0}{0}$. On the other hand, suppose we put $\Delta t$ equal to some very small poistive real number, say $\Delta t = 10^{-6}$. Let the corresponding value of $\Delta s$ be, say, $4.2 \times 10^{-5}$. Then $\dfrac{\Delta s}{\Delta t} = 42$. If $\Delta t$ is in hours and $\Delta s$ in kilometers, this would be 42 k.p.h. But this will be the average speed and not the instantaneous speed. A lower value of $\Delta t$, say, $2.21 \times 10^{-9}$ will give a different answer, but that will not be exact either. The difficulty is that *there is no such thing as the smallest positive real number*. Given any positive real number $x$, we can always find another one (e.g. $\dfrac{x}{2}$) which is smaller than that. If time were a discrete variable, that is, if time existed only in whole multiples of some basic indivisible unit (which may be extremely small, such as a nanosecond), then it would have made sense to be "as close to 0 as possible without actually hitting it", because in that case the smallest $\Delta t$ could be is this one unit of time. The instantaneous speed would then be simply a matter of arithmetic. But time is not a discrete variable. It is a continuous variable. In fact the presumption of classical physics is that all physical variables such as length, mass, energy etc. are continuous. Even those variables which are known to be discrete (for example the electrical charge) are often treated as continuous. That is why electric current is defined as the rate of charge w.r.t. time. Even human population is sometimes taken as a continuous variable (cf. Exercise (1.3.1)). It is primarily this assumption of physics which makes the limiting process so important in applications.

In the second limit that we considered, viz. $\lim_{n \to \infty} (1 + \dfrac{1}{10n})^n$, the variable $n$ is a discrete one. It takes only positive integers as its values. So we do not encounter the difficulty witnessed above because of $\Delta t$ being a continuous variable. But we have a different type of a problem here. Even though we read '$n \to \infty$' as '$n$ approaches or tends to infinity', it does not signify getting very close, or closer and closer. In fact even a very large number like $10^{100}$ is as far away from $\infty$ as 1 is. So here limit has a different interpreation. Once again, putting $n = \infty$ in $(1 + \dfrac{1}{10n})^n$ would give $1^\infty$ which is meaningless. To see what $\lim_{n \to \infty} (1 + \dfrac{1}{10n})^n$ really means, let us study the behaviour of the expression $(1 + \dfrac{1}{10n})^n$ as $n$ changes. From Exercise (1.1), we know that as $n$ increases so does $\left(1 + \dfrac{1}{10}\right)^n$. But after a certain stage this increase becomes negligible.

As a result, if $m$ is a sufficiently large integer, then the infinitely many terms

$$\left(1 + \frac{1}{10(m+1)}\right)^{m+1}, \left(1 + \frac{1}{10(m+2)}\right)^{m+2}, \cdots, \left(1 + \frac{1}{10(m+k)}\right)^{m+k}, \cdots,$$

although all distinct, cluster around some real number. And this real number is the limit of $(1 + \frac{1}{10n})^n$ as $n$ tends to infinitly. This description, of course, lacks mathematical precision. That will be given in the coming sections.

Although the two limits we considered are of different kinds, there is something common to them. The concept of the infinite is involved in both. The second limit, viz., $\lim\limits_{n\to\infty} (1 + \frac{1}{10n})^n$ involves it directly. The first limit involves it in a reciprocal form. As $\Delta t$ approaches $0$ through positive values, $\frac{1}{\Delta t}$ tends to infinity. If we call $\frac{1}{\Delta t}$ as $y$, then $\lim\limits_{\Delta t \to 0} \frac{\Delta s}{\Delta t}$ is the same as $\lim\limits_{y \to \infty} y\Delta s$. (Note however that unlike $n, y$ tends to infinity through all real and not just the integral values.) Infinity is a very baffling concept. We can approach it only in imagination but not in real life. The artifice of a limit is a way man has learnt to deal with the infinite. It is undoubtedly a landmark in the history of human thought. Those who understand and use limits correctly are a class above the others. The limiting concept separates the men from the boys.

## EXERCISES

2.1  Go back to Exercise (1.12.6). If the exchange of 5 c.c.'s of liquid mixtures goes on indefinitely, what would happen intuitively? Express your answer in the form of a limit of a sequence. (Do not prove it.)

2.2  A white ant eats half a piece of a paper on the first day. Thereafter everyday it eats half of what was left on the previous day. When will it consume the whole paper? Express the answer as a limit of a sequence.

2.3  Exress the coefficient of linear expansion of a metal in terms of a derivative. (The coefficient of linear expansion of a metal is the rate per unit length of the change of length of a wire of that metal w.r.t. the temperature and varies from metal to metal. Note the distinction between the 'rate' and the 'rate per unit length'. Similarly the coefficient of laminar expansion is the rate per unit area of the change of area of a lamina (i. e. a thin surface) w.r.t. the temperature. The coefficient of cubical expansion involves volume instead of area.)

2.4  A well-known result in school geometry is that the angle between a chord $PQ$ of a circle and the tangent to the circle at $P$ equals the angle in the opposite arc $PQ$ (see Figure 2.2.2). Prove this first by regarding the tangent at $P$ as the line perpendicular to the radius through $P$. Then prove it by regarding the tangent at $P$ as the limiting position of the secant $PR$ as $R$ tends to $P$ along the arc $QP$.

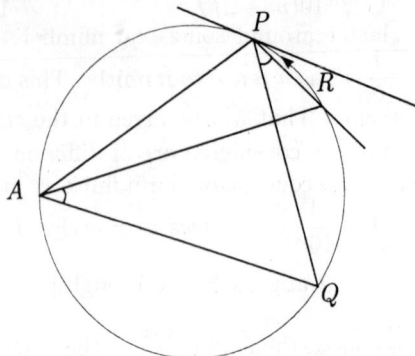

Figure 2.2.2 : Angle between a Chord and a Tangent

2.5 If $(x_1, y_1)$ is a point on the circle whose equation in cartesian co-ordinates is $x^2 + y^2 = a^2$, then the equation of the tangent to it at $(x_1, y_1)$ is $xx_1 + yy_1 = a^2$. Give two proofs of this result, using the two interpretations of the tangent as in the last exercise. Using the second of these methods, obtain equations to tangents to ellipses and hyperbolas in the standard form.

2.6 There is an alternate, tricky way to find the equation of the tangent at a point $P = (x_1, y_1)$ on an ellipse $\dfrac{x^2}{a^2} + \dfrac{y^2}{b^2} = 1$. Let $u = \dfrac{x}{a}$ and $v = \dfrac{y}{b}$. Under this transformation, the ellipse in the $x$-$y$ plane is transformed into the circle $u^2 + v^2 = 1$ in the $u$-$v$ plane. Show that under this transformation, straight lines are mapped to straight lines (but the angle betweeen them may not be preserved). Let $Q = \left(\dfrac{x_1}{a}, \dfrac{y_1}{b}\right)$ be the point in the $u$-$v$ plane, which corresponds to $P$. Then the tangent at $P$ is mapped to the tangent at $Q$. Using this find the equation of the tangent to the ellipse at $P$.

2.7 Suppose we want to find how much water falls over a country $C$ as rain. What will be the answer if the rainfall is the same all over $C$? In other cases, express the answer as the limit of a certain sum in which each summand tends to 0 but the number of summands tends to infinity.

2.8 Do a similar problem for obtaining the mass of a solid whose denisty is not uniform.

## 2.3  Why the Definition of a Limit Is So Clumsy

Limiting concept is the very heart of calculus. So it is only fair that the study of calculus should begin with a clear-cut definition of a limit. Most texts on calculus do, in fact, give it at the start. But a beginner generally finds it extremely clumsy. The reason is that even though the concept of a limit is intuitively clear, attempts to define it in simple terms proved to be futile. (The reason for this will be explained below.) As a result, for a long time, mathematicians kept on obtaining important results in calculus, without having a precise definition of a limit. The first such definition (which is followed today) was given in 1872, long after the deaths of such giants as Euler, Leibnitz, Lagrange, Cauchy, Bessel and even Gauss. Their work did not suffer because of lack of a rigorous definition of a limit. Still, by the middle of the nineteenth century, a need for such a definition was felt because of the absurdities that result by recklessly manipulating infinite series. As a simple example, take the identity:

$$f(x) \quad \frac{1}{1-x} = 1 + x + x^2 + x^3 + \cdots$$

This can be 'proved' by multiplying both the sides by $(1-x)$. On the right hand side, all terms except 1 cancel out in pairs and so the identity holds whenever $1 - x \neq 0$, i.e., $x \neq 1$. But what if we put $x = 2$? Then we get the absurd result that

$$1 + 2 + 4 + 8 + 16 + \cdots = -1.$$

So it is necessary to give a precise meaning, first of all, to the infinite sum appearing in the identity. The most natural thing is to take the sum, say $S_n$, of the first $n$ terms of it and then take the limit of $S_n$ as $n \to \infty$. This necessitates the definition of a limit of a sequence.

Let us now begin to define a limit. Limits involving infinity require a little different (although substantially similar) handling. So let us confine ourselves to the situation where we have a real-valued function $f(x)$ of a real variable $x$, which is defined in a neighbourhood of a point, say $c$, but not necessarily at $c$. We want to give a precise meaning to the statement "As $x$ tends to (or approaches) $c$, $f(x)$ tends to (a limit) $L$". (Written more compactly as "$\lim_{x \to c} f(x) = L$".)

(In the case of instantaneous speed, the variable was $\Delta t$ instead of $x$, the point $c$ was 0, and $f(\Delta t)$ was the number $\dfrac{\Delta s}{\Delta t}$, where $\Delta s$ was the distance travelled in the time interval from $t_0$ (the moment at which we want the speed) to $t_0 + \Delta t$. Note that $\Delta t$ could be negative, in which case $\Delta s$ would also be negative. But $\Delta t$ cannot be 0.)

A layman is most likely to define the statement "$x$ tends to $c$" as "the difference between $x$ and $c$ is very small but not zero" or in symbols, "$|x - c|$ is positive but very small". The trouble is to decide how small is "very small". In practice, depending upon the context, some standard of smallness is generally understood. But such imprecise meaning cannot be allowed in a mathematical statement. Saying "very very small" will not solve the problem either. In fact no matter how many very's you put before 'small' the difficulty will remain. There is no such thing as a real number greater than 0 but smaller then every positive real number you can think of. In the last centuries, an attempt was made to imagine or to hypothecate such numbers. They were called **infinitesimals.** The expression "$x$ tends to $c$" was then taken to mean that $|x - c|$ is infinitesimally small. And so the definition of a limit was given as, "If $x$ is infinitesimally close to $c$ then $f(x)$ is infinitesimally close to $L$". Continuity of a function $f(x)$ at a point meant that an infinitesimal change in the value of $x$ would change the value of $f(x)$ infinitesimally. That is, $f(x)$ is continuous at $c$ if whenever $\triangle x$ is infinitesimally small, so is $f(c + \triangle x) - f(c)$. The ratio of these two infinitesmimally small changes i.e., $\dfrac{f(c + \triangle x) - f(c)}{\triangle x}$ was taken as the definition of $f'(c)$, i.e., the derivative of $f$ at $c$.

Thus, once you swallowed the concept of an infinitesimal, the rest was smooth sailing. Even today, many persons, especially the non-mathematicians find it very convenient to think and talk in terms of infinitesimals (or in terms of **'elemental quantities'** as they are sometimes called). The fact that infinitesimals have no real existence but have to be imagined hardly bothers them. After all, such hypothecations are not new to mathematics. The complex number $i$, for example, is introduced as an imaginary square-root of the real number $-1$. It is easy to show that $-1$ has no square root within the real number system. But the complex number system, which is an extension of the real number system, does contain such a square root (in fact two of them). Similarly, it was thought, that it should be possible to have an extension of the real number system in which there is room for infinitesimals, that is, elements which are bigger than 0 but smaller than every positive real number.

Unfortunately, the attempt to formalise such a system which would accommodate all real numbers and also infinitesimals were not met with success till recently (till 1966 to be precise, almost a century after the first rigorous definition of a limit was given). Today we do have such a system called the system of **hyperreal numbers.** But its construction is far too involved to be presentable in an elementary course in calculus[2]. Moreover, to apply such a system, we would first have to extend our function $f$, which has been given to be defined only on a neighbourhood of the real number $c$ (except possibly at $c$ itself). It $x$ is real number different from $c$, then $x$ can never be infinitesimally close to $c$. In order that $|x - c|$ be an infinitesimal, $x$ must be a hyperreal number. And so, before we can talk of whether $f(x)$ is infinitesimally close to $L$ or not, we must first define $f$ on all hyperreal numbers in a neighbourhood of $c$. This we can do only if we know what a hyperreal number is. But as we just said, they are not easy to define.

Summing up, the attempt to define '$x$ approaches $c$' by saying '$|x - c|$ is infinitesimally small', although theoretically successful by now, is still a long way from finding room in an elementary calculus course. Its only advantage, perhaps, is that those who use infinitesimals without defining them rigorously can now

---

[2]See Appendix B for an outline of the construction.

have some satisfaction that even if they do not know their rigorous definition, somebody else does. This is not an entirely bad idea, because in elementary courses we do use certain terms without formal definitions and many theorems without formal proofs. But to have to do so for the most pivotal concept in the study is rather undesirable.

But then what is the way out? As we just saw, there is no way to give a mathematically precise meaning to the expression '$x$ tends to $c$. The same, of course, holds for the expression '$f(x)$ tends to $L$'. How can then we possibly give a precise meaning to the statement 'As $x$ tends to $c$, $f(x)$ tends to $L$.'? Do we just give up and take a limit as a primitive term? That is basically what people did till a mathematician called Weierstrass and his disciple Heine came up with an extremely ingenious way out of this dilemma. It is rather baffling and so, in order to understand it clearly, let us first take a linguistic digression and look at idioms like 'a black sheep' or 'a big fish in a small pond'. To understand their literal meanings, you of course have to know the meaning of each and every word appearing in them and to put these meanings together. But this is not the case with their figurative meanings. When a person is described as a 'black sheep' in his family, in order to know what is meant, you have to know the phrase 'black sheep' as a whole. It is of little use to know 'black' as a colour and 'sheep' as a certain mammal. The person referred to may not be black at all, and even if he is, he is definitely not a sheep! So here 'a black sheep' does not mean 'a sheep which is black'. To avoid confusion, it would probably be better to write 'blacksheep' as a single word like 'blacksmith'. But that would take away the poetry in it. Similarly to define or to understand the figurative meaning of a sentence like, 'As a ship starts to sink, the mice on it start leaving it' it is neither necessary not sufficient to know the separate meaning of its two clauses, viz, 'As a ship starts to sink' and 'The mice on it start leaving it'. The sentence as a whole has a meaning quite independent of the literal meanings of its clauses.

Now, going back to our main theme, recall that our original task is to give a precise mathematical meaning to the sentence 'As $x$ tends to $c$, $f(x)$ tends to $L$.' We saw that it is futile to try to define the statements '$x$ tends to $c$' or '$f(x)$ tends to $L$'. But the situation it not entirely hopeless. In view of the examples given above, we may still be able to define the statement 'As $x$ tends to $c$, $f(x)$ tends to $L$' in its entirety, that is, without defining its two clauses separately. In effect, we shall be cancelling the difficulties involved in defining the two clauses individually. Instead of giving the the final definition right away, we shall develop it gradually through a series of intermediate statements, which become progressively precise. We begin with the layman's understanding:

"$|f(x) - L|$ is small, whenever $|x - c|$ is small (but not 0)".

As a first step, let us paraphrase this so as to bring out the fact that the smallness of $|f(x) - L|$ depends on that of $|x - c|$. So,

"$|f(x) - L|$ can be made small by taking $|x - c|$ small (but not 0)"

This statement is still vague since it does not specify how small is small. The standard of smallness can, in fact, vary. The next statement makes allowance for such variation.

"$|f(x) - L|$ can be made as small as desired (or even smaller)
by taking $|x - c|$ sufficiently small (but not 0)"

This essentially is the definition of a limit. The subsequent improvement does little more than introducing appropriate symbols. The catch here is that the standard of smallness of $|f(x) - L|$ has to be specified first. Only after knowing this standard, can we decide what degree of smallness of $|x - c|$ will enable us to meet this standard. The standard method of specifying the standard of smallness is, of course, to specify an upper bound on the size. This naturally leads us into the following:

"For every $\epsilon > 0$, there exists $\delta > 0$ such that whenever
$$0 < |x - c| < \delta, \quad |f(x) - L| < \epsilon.\text{"}$$

This is the final (or "polished") definition of a limit (or rather one form of limits). Once again we emphasise that $\epsilon$ has to be specified first. Only then can an appropriate $\delta$ be found. A $\delta$ which works for a particular value of $\epsilon$, will obviously work for any larger value of $\epsilon$, but in general not for a smaller value. Unless the function is a constant equal to $L$ (in some neighbourhood of $c$ ),there is no single $\delta$ which will work for every $\epsilon$. Thus $\delta$ very much depends on $\epsilon$. There is however, no restriction on $\epsilon$ other than that it be positive. It could be as small as we like. To emphasise these facts to a beginner they are sometimes incorporated right into the definition of a limit. For example, after the phrase "For every $\epsilon > 0$", the words "however small" or "no matter how small" are added. But they are really redundant. Similarly, to indicate that $\delta$ depends on $\epsilon$, we say "there exists $\delta > 0$, depending on $\epsilon$, such that ......" or write "there exits $\delta = \delta(\epsilon) > 0$, such that ...". (Of course $\delta$ also depends on the function $f$ and the point $c$. But since these are already fixed, this need not be emphasised in the definition.)

It is easy to interpret the definition of a limit geometrically. But first we need to introduce some basic terms and notations. An **interval** is a portion of the real line stretching from a point $a$ to a point $b$ (say). Here $a, b$ are any two real numbers, called the end-points of the interval. (We tacitly assume $a < b$ as otherwise we get a degenerate interval.) Depending upon the inclusion or the exclusion of the end-points, the notations and the names (and also some of the properties) change. If neither of the two end-points is included, we get the set $\{x \in \mathbb{R} : a < x < b\}$ called an **open interval** and denoted by $(a, b)$. (This is likely to be confused with the ordered pair $(a, b)$. For this reason, some authors use $]a, b[$. But $(a, b)$ is more common.) If both the end-points are included, we get the set $\{x \in \mathbb{R} : a \le x \le b\}$ called a **closed interval** and denoted by $[a, b]$. We can also consider the sets $\{x \in \mathbb{R} : a \le x < b\}$ and $\{x : a < x \le b\}$ denoted, respectively, by $[a, b)$ and $(a, b]$. They are called **semi-open intervals**. In all cases, the real numbers $\dfrac{a+b}{2}$ and $b-a$ are the **centre** (or the mid-point) and the **length** of the interval. Every point of an interval (of whatever type) other than an end-point is called an **interior point** of it. Clearly an interval is open if and only if every point of it is an interior point.

The open intervals are especially important in the study of limits (and in calculus in general) because as seen from Exercise (1.9.9), the statement '$|x-c| < \delta$ is equivalent to saying that $x$ lies in the open interval of length $2\delta$ centered at $c$. This interval is often called the $\delta$-**neighbourhood** of the point $c$. The definition of '$\lim\limits_{x \to c} f(x) = L$' can be paraphrased by saying that for every $\epsilon$-neighbourhood of $L$, there is some $\delta$-neighbourhood of $c$, every point (except possibly the centre) of which is mapped by the function $f$ into the given $\epsilon$-neighbourhood of $L$. We show this in Fig. 2.3.1(a). More vividly, we draw the graph of the function $f$. The points of the form $(x, y)$ for which $L - \epsilon < y < L + \epsilon$ constitute a horizontal strip of width $2\epsilon$ with centreline $y = L$ as shown in Fig. 2.3.1(b). Then, given in any such strip, however narrow it be, we can find some $\delta$-neighbourhood of $c$ such that the portion of the graph of $f$ lying over it (again, except possibly its centre) is completely contained within this strip. Naturally, the narrower this strip, the smaller the $\delta$ we would need.

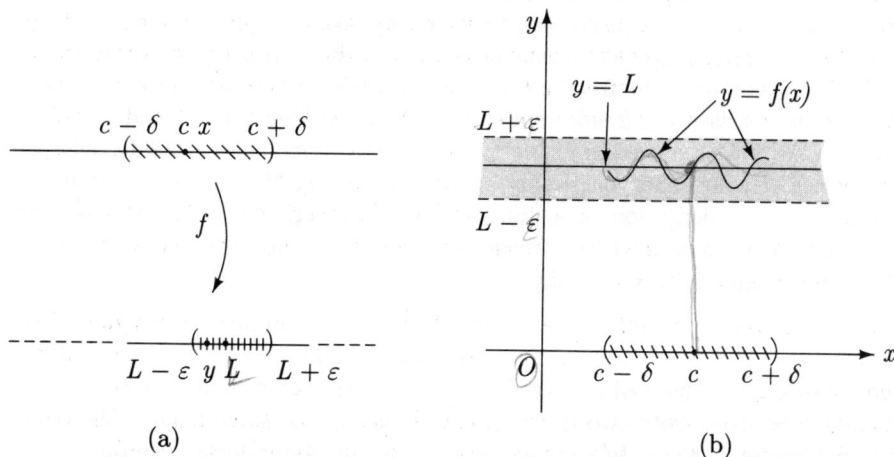

(a)   (b)

Figure 2.3.1 : Geometric Interpretation of Definition of Limit

To make the definition of a limit still more lively, think of a card game between two players $A$ and $B$. Each player has with him an infinite pack of cards, one for each real number. The player $A$ is making the statement, "As $x$ tends to $c$, $f(x)$ tends to $L$." The player $B$ is challenging it. $B$ begins the game by playing a card which bears a positive real number $\epsilon$ (of his choice). $A$ looks at that card, ponders a bit and plays a card which bears a positive number $\delta$. Now it is $B$'s turn again. He plays a card which is marked with a number $x$. The rules of the game require that this number must satisfy $0 < |x - c| < \delta$.

Now an umpire looks at $x$, calculates $f(x)$ and compares $|f(x) - L|$ with $\epsilon$. If every time $|f(x) - L|$ is less than $\epsilon$, $A$ wins.

Besides giving a lighter touch to the definition of a limit, this analogy also helps a beginner correctly handle statements about limits. When you are given some data and asked to prove a statement of the from "As $x \to c, f(x) \to L$", your position is that of the player $A$. You do not have the freedom to choose $\epsilon > 0$. Your opponent $B$ has that freedom and you should be prepared for any choice he makes. So your proof must begin with a statement like "Let $\epsilon > 0$ be given (arbitrarily)." Your task is now to use the given data and to come up with a $\delta$ which is suitable for this $\epsilon$. Typically your $\delta$ will be something like $\epsilon, \dfrac{\epsilon}{2}, \epsilon^2, \min\left\{\dfrac{\epsilon}{2}, M\right\}$, or $\min\left\{\epsilon^2, \dfrac{\epsilon}{K}\right\}$ where $M, K$ are some (positive) numbers given in the data or obtainable from it. It is important to know that you are generally *not* required to find the best (i.e. the largest) $\delta$ that will work for a given $\epsilon$. All you need to find is *some* positive $\delta$ which will work. Strictly speaking, you are not even required to define your $\delta$ by an explicit formula. It is enough to prove merely that it exists. In some problems, for example, your answer may come out to be $\delta = \min\{\delta_1, \delta_2\}$ where $\delta_1$ and $\delta_2$ are some positive numbers whose existence is obtainable from the data, without being able to know their exact expressions in terms of $\epsilon$.

On the other hand, when a statement like "As $x \to c, f(x) \to L$" is given to you as a part of the hypothesis, your position is that of the player $B$, except that, instead of challenging that statement, you are trying to use it to derive some result. Now you have the freedom to choose a positive real number $\epsilon$. Naturally you have to exercise this freedom judiciously. You are then guaranteed a positive real number $\delta$ with a certain property. (In practice, of course, nobody tells you $\delta$ in terms of your $\epsilon$. So you simply proceed by saying "Let $\delta > 0$ be a number as given by the definition of a limit" without bothering how this $\delta$ has been obtained.) You now have the freedom to choose $x$ and, once again, a clever choice will take you to your goal.

It is important to realise that $\epsilon$ and $\delta$ are just dummy symbols denoting positive real numbers. They are very standard. But we could as well replace them with any other symbols we like (taking care, of course, not to use the same symbol to denote two different quantities at the same time). We could even interchange them. In fact, in many proofs involving limits, we often have to choose as our $\epsilon$, some $\delta$ which comes from some other limit. A beginner should be alert lest he get addicted to a particular notation.

As a typical illustration of these remarks, we prove some basic results about limits. They are needed so frequently, that an explicit mention is rarely made. The proofs may appear rather abstract but the underlying ideas are very simple and natural. Note the frequent use of the properties of the absolute value, and especially, the triangle inequality (see Exercises (1.9.8) and (1.9.9)). This is to be expected because the definition of a limit is crucially based on the smallness of certain absolute values and the triangle inequality implies, in particular, that if two numbers are small (in absolute value) then so is their sum.

**3.1 Theorem** : Suppose $\lim\limits_{x \to c} f(x) = L$ and $\lim\limits_{x \to c} g(x) = M$. Then
(i) $\lim\limits_{x \to c} (f(x) + g(x)) = L + M$   (ii) $\lim\limits_{x \to c} (f(x)g(x)) = LM$
(iii) $\lim\limits_{x \to c} (kf(x)) = kL$ where $k$ is any constant
(iv) $\lim\limits_{x \to c} (f(x) - g(x)) = L - M$, and finally
(v) if $M \neq 0$, then $\lim\limits_{x \to c} \dfrac{f(x)}{g(x)} = \dfrac{L}{M}$

**Proof:** In every assertion, we have to prove that some limit equals some number. So we must begin with 'Let $\epsilon > 0$ be given' in each case. In (i), we want to make $|f(x) + g(x) - (L + M)|$ less than $\epsilon$. By triangle inequality,

$$\begin{aligned} |f(x) + g(x) - (L + M)| &= |(f(x) - L) + (g(x) - M)| \\ &\leq |f(x) - L| + |g(x) = -M| \end{aligned} \tag{1}$$

So if we could make each of $|f(x) - L|$ and $|g(x) - M|$ less than $\epsilon/2$ we would be through. For this, we use the data. Since $f(x) \to L$ as $x \to c$, taking $\dfrac{\epsilon}{2}$ (instead of $\epsilon$) in the definition of a limit, we get that there exists some $\delta_1 > 0$ such that

$$|f(x) - L| < \frac{\epsilon}{2} \quad \text{whenever} \quad 0 < |x - c| < \delta_1. \tag{2}$$

Similarly, from $\lim\limits_{x \to c} g(x) = M$, we get some $\delta_2 > 0$ such that

$$|f(x) - L| < \frac{\epsilon}{2} \quad \text{whenever} \quad 0 < |x - c| < \delta_2. \tag{3}$$

Now let $\delta = \min\{\delta_1, \delta_2\}$. Then $\delta > 0$. Also when $0 < |x - c| < \delta$, both (2) and (3) are applicable and from (1), we get that

$$|f(x) + g(x) - (L + M)| < \frac{\epsilon}{2} + \frac{\epsilon}{2} = \epsilon \quad \text{whenever} \quad 0 < |x - c| < \delta. \tag{4}$$

The proof of (ii) is conceptually similar but a bit more complicated, because $f(x)g(x) - LM$ does not factor as $(f(x) - L)(g(x) - M)$. But if we add and subtract $f(x)M$, we get

$$\begin{aligned} |f(x)g(x) - LM| &= |f(x)(g(x) - M) + M(f(x) - L)| \\ &\leq |f(x)|\,|g(x) - M| + |M|\,|f(x) - L| \end{aligned} \tag{5}$$

Here to make each term less than $\epsilon/2$ requires more work than for (1), because of the presence of the coefficients $|f(x)|$ and $|M|$. First, by a result which we relegate to the exercises (Exercise (3.5)), there exists $R$ ($> 0$) and $\delta_0$ such that

$$|f(x)| \leq R \quad \text{whenever} \quad 0 < |x - c| < \delta_0 \tag{6}$$

Now, putting $\dfrac{\epsilon}{2R}$ in the definition of $\lim\limits_{x \to c} g(x) = M$, we get $\delta_1 > 0$ such that

$$|g(x) - M| < \frac{\epsilon}{2R} \quad \text{whenever} \quad 0 < |x - c| < \delta_1 \tag{7}$$

Similarly from $\lim_{x \to c} f(x) = L$, we get $\delta_2 > 0$ such that

$$|f(x) - L| < \frac{\epsilon}{2|M|} \quad \text{whenever} \quad 0 < |x - c| < \delta_2 \tag{8}$$

(Here we are tacitly assuming $M \neq 0$. If $M = 0$, the second term in (5) drops out and the argument becomes even simpler. Or we can play it safe by taking $|M| + 1$ instead of $|M|$ in (8).)

Now let $\delta = \min\{\delta_0, \delta_1, \delta_2\}$. Then from (6), (7), (8) and (5) we have

$$|f(x)g(x) - LM| < R\frac{\epsilon}{2R} + |M|\frac{\epsilon}{2|M|} = \epsilon, \quad \text{whenever} \quad 0 < |x - c| < \delta$$

which proves (ii).     (iii) now follows as a special case by taking $g(x) = k$, a constant function. (A direct argument for (iii) is recommended as an exercise.) (iv) can be proved the same way as (i). A slicker way is to write $f(x) - g(x)$ as $f(x) + (-1)g(x)$ and apply (i) and (iii) with $k = -1$.

Before proving (v) , note that we are only given that $M \neq 0$ and not that $g(x) \neq 0$. The function $g$ may, in fact vanish at some points. We claim that this does not happen if $x$ is sufficiently close to $c$. Since $M \neq 0$, $|M| > 0$. Taking $\epsilon = \frac{1}{2}|M|$ in the definition of $\lim_{x \to c} g(x) = M$, there exists $\delta_0 > 0$ such that $0 < |x - c| < \delta_0$ implies $|g(x) - M| < \frac{1}{2}|M|$, which in particular means $g(x) \neq 0$, for otherwise, $|g(x) - M| = |0 - M| = |M| > \frac{1}{2}|M|$, a contradiction. Since $g(x) \neq 0$ whenever $0 < |x - c| < \delta_0$, the quotient function $\dfrac{f(x)}{g(x)}$ is defined for such values of $x$ and it makes sense to talk of its limit as $x \to c$. We have to show it equals $\dfrac{L}{M}$. It suffices to show that $\lim_{x \to c} \dfrac{1}{g(x)} = \dfrac{1}{M}$, for one can then apply (ii) to the product $f(x) \cdot \dfrac{1}{g(x)}$. Now once again, we consider $\left|\dfrac{1}{g(x)} - \dfrac{1}{M}\right| = \dfrac{|g(x) - M|}{|g(x)|\,|M|}$. Just now we proved that if $0 < |x - c| < \delta_0$, then $|g(x) - M| < \frac{1}{2}|M|$, i.e., $g(x)\epsilon(M - \frac{1}{2}|M|, M + \frac{1}{2}|M|)$ which equals $(\frac{1}{2}M, \frac{3}{2}M)$ if $M > 0$ and $(-\frac{3}{2}|M|, -\frac{1}{2}|M|)$ if $M < 0$. In either case, whenever $0 < |x - c| < \delta_0$, $|g(x)| > \frac{1}{2}|M|$ and hence $\left|\dfrac{1}{g(x)} - \dfrac{1}{M}\right| < \dfrac{|g(x) - M|}{\frac{1}{2}|M|^2}$. Now given $\epsilon > 0$, there exists $\delta_1 > 0$ such that $|g(x) - M| < \dfrac{\epsilon|M|^2}{2}$ whenever $0 < |x - c| < \delta_1$. Let $\delta = \min\{\delta_0, \delta_1\}$. Then, $0 < |x - c| < \delta$ implies $\left|\dfrac{1}{g(x)} - \dfrac{1}{M}\right| < \epsilon$. So $\dfrac{1}{g(x)} \to \dfrac{1}{M}$ as $x \to c$. As noted earlier, this complates the proof of (v). ∎

Another point to note about limits is that throughout the discussion of $\lim_{x \to c} f(x)$, there is absolutely no reference to $f(c)$, that is, the value of the function $f$ at the point $c$. The function $f$ need not even be defined at $c$, and even if it is, $f(c)$ need not equal $L$. If this happens then the function $f$ is said to be continuous at $c$. Most of the familiar functions (such as the polynomials, exponential functions, the sines and cosines) are continuous everywhere. Functions which occur "in nature" are also generally continuous. (When we turn off a switch we think that the current in a wire suddenly goes off. But in reality it falls off gradually over an extremely short time interval.) So, a beginner is likely to get the feeling that $\lim_{x \to c} f(x)$ is the same as $f(c)$. A misconception like this cuts at the very root of calculus and no effort is too great to avoid it. Perhaps a social analogy will help. Think of $c$ and its neighbouring points as houses on a road. Suppose the function $f$ gives the income (or some other attribute such as education, life style etc.) of the occupant of a house. Then $f(c)$ is the actual income etc. of the occupant of the house $c$ while $\lim_{x \to c} f(x)$ is the impression we form about it from the incomes of his neighbours. Such impressions can often be wrong. A particular house in a very posh locality may be occupied by a relatively poor person. In fact, to stretch the simile further, it may not be occupied at all (which corresponds to the case where $\lim_{x \to c} f(x)$ exists but $f(c)$ is not defined).

Another possible misconception about limits is to interpret the definition to mean something like "As $|x - c|$ gets smaller and smaller, so does $|f(x) - L|$." If this is taken to mean only that the smaller $\delta$'s will work for smaller $\epsilon$'s, it is not wrong. But if it is taken literally, it would mean that whenever $x_1, x_2$ are such that $0 < |x_1 - c| < |x_2 - c|$, then we must have $|f(x_1) - L| \leq |f(x_2) - L|$. And this interpretation is wrong as can be shown by simple examples (e.g. consider $\lim_{x \to 0} x \sin \frac{1}{x}$). (see Exercise (3.4)). Going back to the comments we made before the definition of a limit, we defined the sentence 'As $x$ tends to $c$, $f(x)$ tends to $L$' in its entirety, without defining the clauses '$x$ tends to $c$' and '$f(x)$ tends to $L$' separately. But what if we have to use these clauses by themselves, perhaps just one at a time? It turns out that the need for doing so never arises. In informal writing we do sometimes use such unpaired phrases, like "Now let $x$ tend to $c$" especially when we want to use some results about limits. Take, for example, the so-called **Sandwich Theorem** which asserts that if $f, g, h$ are three functions which satisfy $f(x) \leq g(x) \leq h(x)$ for all $x$ in some neighbourhood of a point $c$ (except perhaps at $x = c$), and $\lim_{x \to c} f(x) = \lim_{x \to c} h(x)$ then $\lim_{x \to c} g(x)$ also exists and equals this common value. (For a proof, see Exercise (3.4).) As a typical application of this theorem, suppose we have shown by some geometric reasoning that for all sufficiently small $|x|(x \neq 0)$,

$$\cos x \leq \frac{\sin x}{x} \leq 1.$$

It is then very customary to say "By letting $x \to 0$ and applying the Sandwich Theorem, we get $\lim_{x \to 0} \frac{\sin x}{x} = 1$. Here the phrase 'letting $x \to 0$' is only a convenient language. Strictly speaking, it is meaningless. But everybody understands what is meant. If we are very fussy, we can simply dispense with it and say "since $\lim\limits_{x \to 0} \cos x = 1 = \lim\limits_{x \to 0} 1$, it follows by the Sandwich Theorem that $\lim\limits_{x \to 0} \dfrac{\sin x}{x} = 1$".

Summing up, the clumsiness of the definition of a limit is due primarily to the elusive nature of the concept of an infinitesimally small number. In order to circumvent this difficulty, the definition had to take the most unusual step of defining a sentence in its entirety without defining its clauses separately. In doing so, quantifiers have to be introduced and their proper order is extremely vital. Predictably, such a definition invites a lot of controversy especially from the pedagogical point of view. The definition of a limit has been criticised for its stodgy formalism. Because of the symbols $\epsilon$ and $\delta$ appearing in it (and hence in the proofs of so many results pertaining to limits), the term $\epsilon$-$\delta$ **mathematics** is a colloquial euphemism for that fussy, ritualistic part of mathematics which lays heavy emphasis on formal perfection with total disregard for intuition.

What nobody disputes, however, is the precision of the definition. Even after a century, in essence, it is still the only definition of a limit. In fact it set the trend for the kind of rigour the mathematics to come must have. Its major achievement is that it put an end to the futile search for infinitesimals that was going or for more than a century. It showed that this most basic concept in calculus can be defined without having to extend the real number system. A little clumsiness (which is not hard to get rid of with patient handling) is a small price to pay for such a historic achievement.

## EXERCISES

3.1   In each of the following, a function $f(x)$, and numbers $c, L$ are given. In each case, for a given $\epsilon > 0$, find $\delta > 0$ such that $|f(x) - L| < \epsilon$ whenever $0 < |x - c| < \delta$.

  (i)  $f(x) = x, c = 2, L = 2$
  (ii) $f(x) = 3x + 5, c = 2, L = 11$
  (iii) $f(x) = x^3, c = 2, L = 8$
  (iv) $f(x) = x^3 - 2x^2 + 5x - 4,\ c = 2, L = 6$

3.2   In each of the examples of the last exercise change the value of $L$ and then show that as $x \to c, f(x)$ does not tend to $L$, by showing that in each case the negation of the definition of the limit holds. [*Hint* : The negation of the definition was already given in Section 1.11.]

3.3   Prove that $\lim\limits_{x \to 0} \sin \dfrac{1}{x}$ does not exist by showing that for every $L$, there exists $\epsilon > 0$ such that no $\delta$ can be found. [*Hint* : The crucial point is that there are arbitrarily small values of $x$ for which $\sin \dfrac{1}{x} = 1$ and also arbitrarily small values of $x$ for which $\sin \dfrac{1}{x} = -1$. Choose $\epsilon$ so that the interval $(L - \epsilon, L + \epsilon)$ excludes either 1 or $-1$.]

3.4 Prove the Sandwich Theorem and deduce from it that $\lim\limits_{x \to 0} x \sin \dfrac{1}{x} = 0$. Show, however, that every neighbourhood of $x$ contains points $x_1, x_2$ such that $0 < x_1 < x_2$ but $|x_1 \sin \dfrac{1}{x_1} - 0| > |x_2 \sin \dfrac{1}{x_2} - 0|$. So it is not correct to interpret the definition of $\lim\limits_{x \to c} f(x) = L$ literally to mean that as $|x - c|$ gets smaller and smaller, so does $|f(x) - L|$.

3.5 Show that if $\lim\limits_{x \to c} f(x)$ exists then $f$ is bounded in some sufficiently small neighbourhood of $c$ (excluding the point $c$, of course). That is, show that there exists some $\delta > 0$ and some $M \geq 0$ such that $|f(x)| \leq M$ whenever $0 < |x - c| < \delta$.

3.6 A **syllogism** is a logical argument in which given the truth of two implication statements of the form $p \Rightarrow q$ and $q \Rightarrow r$, one concludes $p \Rightarrow r$. (This is, of course, sheer common sense. But in formal logic, even this needs a proof.) For example, if we are given that "As it rains, the streets get wet" and "As the streets get wet, accidents happen" then we conclude that "As it rains, accidents happen." Here we are using syllogism. Suppose now we are given two functions $f, g$ and three numbers $c, L$ and $M$. Suppose further that we are given that

      (i)   as $x$ tends to $c, f(x)$ tends to $L$

and   (ii)  as $y$ tends to $L, g(y)$ tends to $M$.

From these two statements, can we conclude by syllogism that as $x$ tends to $c, h(x)$ tends to $M$, where $h = g \circ f$ is the composite function, i.e., $h(x) = g(f(x))$?

3.7 In the last exercise suppose further that $g(L) = M$. Then give a rigorous proof of the assertion, using the definition of a limit. [*Hint* : Start with a given $\epsilon > 0$. First find a $\delta > 0$ as given by (ii) above. Now use this $\delta$ as your $\epsilon$ in (i) and get some $\delta_1$.] Why is the additional condition $g(L) = M$ needed?

3.8 If $\lim\limits_{x \to c} f(x) = L > 0$, prove that $\lim\limits_{x \to c} \sqrt{f(x)} = \sqrt{L}$.

3.9 If $\lim\limits_{x \to c} = L$, prove that $\lim\limits_{x \to c} |f(x)| = |L|$. Show by an example that the converse is true for $L = 0$ but not in general.

## 2.4    Limits Involving Infinity

The concept of infinity is fascinating from an emotional and a philoophical point of view. In mathematics it appears in a peculiar way which can be confusing to a beginner. And so anything involving infinity needs a careful handling. Basically, the reason is that $\infty$ (and also $-\infty$) behave like ordinary real numbers in certain respects but not in all respects. If we indiscriminately apply the familiar rules about real numbers to $\infty$, we easily run into disasters. For example, $1 + \infty = 2 + \infty$, each being equal to $\infty$. But if we cancel $\infty$ from this equation, we get the absurd result $1 = 2$.

Exactly, what do the symbols $\infty$ and $-\infty$ stand for? There is no unanimous answer to this. Euler freely regarded $\infty$ as the reciprocal of the number 0. He would have no hesitation in writing equations like

$$\ln \infty = 1 + \frac{1}{2} + \frac{1}{3} + \cdots + \frac{1}{n} + \cdots$$

(obtained by putting $x = -1$ in the power series expansion of $\ln \frac{1}{1+x}$, viz.

$$\ln \frac{1}{1+x} = -\ln(1+x) = -x + \frac{x^2}{2} - \frac{x^3}{3} + \cdots + (-1)^n \frac{x^n}{n} + \cdots$$

which will be obtained in Exercise (5.7.12)). From a modern view-point, this equation about $\ln \infty$ would mean merely that the series $\sum_{n=1}^{\infty} \frac{1}{n}$ is divergent. But it can also be given a more subtle meaning. If we interpret $\ln \infty$ as $\lim_{n \to \infty} \ln n$, then, since the right hand side of this equation is $\lim_{n \to \infty} H_n$ where $H_n = 1 + \frac{1}{2} + \cdots + \frac{1}{n}$, the equation above can be taken to suggest that as $n \to \infty, \ln n$ and $H_n$ are nearly equal. This indeed is the case as was proved by Euler himself. (For more elaboration of this, see Section 3.6)

So here we have a case where the use of $\infty$ as a number $\left( = \frac{1}{0} \right)$, in the hands of a genius like Euler, conveys something remarkable. There are, however, many absurdities which arise from an equation like $0 \cdot \infty = 1$. For, in that case we have, $1 = 0 \cdot \infty = (0 + 0) \cdot \infty = 0 \cdot \infty + 0 \cdot \infty = 1 + 1 = 2$.

It follows that if we want to extend the real number system by adding $\infty$ and $-\infty$ to it, then these two extra members will have to be given a special treatment. For example, $\infty + \infty = \infty, \infty \cdot \infty = \infty$ but $0 \cdot \infty$ and $\infty + (-\infty)$ have to be left undefined. Every real number is (strictly) less than $\infty$ and (strictly) bigger than $-\infty$. Hence for any real number $a$, the open interval $(a, \infty)$ is simply the set $\{x \in I\!R : a < x\}$ since the other condition (viz. $x < \infty$) is redundant. Similarly $[a, \infty) = \{x \in I\!R : a \le x\}$, $(-\infty, a) = \{x \in I\!R : x < a\}$ and $(-\infty, a] = \{x \in I\!R : x \le a\}$. The interval $(-\infty, \infty)$ is simply the entire real line. Expressions like $(a, \infty]$ or $[-\infty, a]$ are meaningless because every interval is a subset of $I\!R$ and hence cannot contain either $\infty$ or $-\infty$.

These notations illustrate the convenience gained by introducing the symbols $\infty$ and $-\infty$. There are four kinds of open intervals (i) those that are bounded at both the ends, i.e. of the form $(a, b)$ or $\{x \in \mathbb{R} : a < x < b\}$ where $a, b$ are real numbers, (ii) those which are bounded below but not above, i.e., of the form $\{x \in \mathbb{R} : a < x\}$ for some $a \in \mathbb{R}$, (iii) those which are bounded above but not below and finally (iv) those which are bounded neither above nor below. Of the last type, there is only one viz. the entire real line. Because of the symbols $\infty$ and $-\infty$ we can denote an open interval of *any* of these four types by the symbol $(a, b)$ where $a < b$. Here $a, b$ are allowed to assume the values $-\infty$ and $\infty$ respectively. Frequently we have to state a theorem which is valid for all open intervals, whether bounded or not. The notation just introduced covers all possibilities and hence saves a lot of space. Without the symbols $\infty$ and $-\infty$, we would have to list and consider all the four cases separately.

At the same time, these notations also suggest that neither $\infty$ nor $-\infty$ is indispensable. Without them, we could still write $(a, \infty)$ as $\{x \in \mathbb{R} : a < x\}$ etc.

Although this is a rather trivial example, it illustrates the modern view of $\infty$ and $-\infty$, viz., that they are merely symbols. They are convenient but not indispensable. *They are not real numbers.* Every time they occur, their occurrence can be circumvented. In effect, $\infty$ and $-\infty$ are non-entities. We know absolutely nothing about them. Every time they appear in some sentence or notation, that sentence or notation will be defined in its entirety. Lest this appear paradoxical, take a real-life analogy. When you ask a person some question and he replies "My boss knows", it means that the person has a boss and he (i.e., the boss) knows the answer to the question. But if he says "God knows", it does not mean that there is some such thing as God and He knows the answer to the question. In fact, the person who says "God knows" may himself be a total atheist! His statement is merely another way of saying that he does not know and moreover, no person of his knowledge knows the answer to the question you asked. Similarly when we say "$\lim_{x \to c} f(x) = \infty$", this statement is qualitatively different from the statement "$\lim_{x \to c} f(x) = L$", where $L$ is a real number. We defined the latter in the last section. You cannot get the definition of the former merely by replacing $L$ throughout by $\infty$, just as you do not get the meaning of "God knows" from from the meaning of "My boss knows" by replacing "My boss" by "God". The sentence 'As $x$ tends to $c$, $f(x)$ tends to $\infty$' has to be given a separate definition of its own, as we now proceed to do.

Just as the statement "$f(x)$ tends to $L$" intuitively means that $|f(x) - L|$ is very small, but has no mathematically precise meaning, the same is the case with the statement "$f(x)$ tends to $\infty$". Intuitively it means $f(x)$ is very large. Ideally, it should mean $f(x)$ is larger than every real number $R$ (no matter how large). But this would be a non-starter. There simply does not exist a number which would be bigger than every real number just as there does not exist a positive number which is smaller than every positive real number. In the last section we remarked that if we enlarge the real number system to what is called the system of hyperreal numbers, then in it there is some room for infinitesimally small numbers, i.e. positive (hyperreal) numbers which are smaller than every positive (real) number. If we take the reciprocals of such infinitesimally small numbers, we do get infinitely large (hyperreal) numbers, that is, numbers bigger than every real number. So if we are working in the system of hyperreal numbers, the statement 'As $x \to c, f(x) \to \infty$' can be defined as 'whenever $|x - c|$ is an infinitesimal, so is $\dfrac{1}{f(x)}$'. (It is possible that when Euler treated $\infty$ as $\dfrac{1}{0}$, what he had in mind was not literally the reciporcal of the real number 0 (which obviously does not exist), but the reciprocal of some infinitesimal.)

However, as remarked in the last section, the construction of hyperreal numbers is far too complicated to be accessible to a beginner. So, let us define "$\lim_{x \to c} f(x) = \infty$" staying strictly within real numbers. Here we adopt the same trick as we did in the last section. We define the entire statement 'As $x$ tends to $c, f(x)$ tends to $L$' without defining its clauses separately. The standard of largeness has to be specified by giving some (large) number $R$. We skip the intermediate statements and write down only the final definition, putting into the parentheses words which are really redundant, but added for emphasis.

> "For every real number $R$ (however large), there exists $\delta > 0$ such that whenever $0 < |x - c| < \delta, f(x) > R$."

In an entirely analogous manner, we define "$\lim_{x \to c} f(x) = -\infty$." Or alternately we can define it to mean that "$\lim_{x \to c} -f(x) = \infty$". It is easy to show that the two definitions are equivalent. (All that is needed is that $a < b$ if and only if $-a > -b$.)

Familiar theorems about limits of sums, differences, products and quotients of limits do not in general hold true for infinite limits. For example, if $\lim_{x \to c} f(x) = \infty = \lim_{x \to c} g(x)$ then nothing can be said about $\lim_{x \to c} (f(x) - g(x))$ in general. This limit need not even exist. And even when it does, it need not be 0. It could be a (finite) real number. Or it could be $\infty$ or $-\infty$. Similarly, if $\lim_{x \to c} f(x) = 0$ and $\lim_{x \to c} g(x) = \infty$, then nothing can be said about $\lim_{x \to c} f(x)g(x)$. (Examples of all cases will be given in the exercises.)

Because of anomalies like this, most authors nowadays prefer to say that $\lim_{x \to c} f(x)$ does not exist if $\lim_{x \to c} f(x) = \infty$ or $\lim_{x \to c} f(x) = -\infty$. If we allow such limits, then we have to attach riders to the theorems about limits so as to exclude the anomalous situations. For example the theorem "The limit of a sum of two functions is the sum of the limits of the two functions" has to be qualified by a proviso, "provided the two limits are not of the form $\infty$ and $-\infty$." To say that such limits do not exist means that they do not exist as real numbers. But in doing so, we are definitely withholding some information. For example the statement '$\lim_{x \to 0} \dfrac{1}{x^2}$ does not exist' is quite correct in the sense that there is no real number $L$ such that $\lim_{x \to 0} \dfrac{1}{x^2} = L$. But to say that $\lim_{x \to 0} \dfrac{1}{x^2} = \infty$ definitely gives some positive information about the values of $\dfrac{1}{x^2}$ for small $x$.

Compare this with $\lim\limits_{x \to 0} \sin\left(\dfrac{1}{x}\right)$. This limit not only does not exist as a (finite)

real number (see Exercise (3.3)), but it is also false that $\lim\limits_{x \to 0} \sin\left(\dfrac{1}{x}\right) = \infty$

or $\lim\limits_{x \to 0} \sin\left(\dfrac{1}{x}\right) = -\infty$. So the statement "$\lim\limits_{x \to c} f(x)$ does not exist" is not so
informative in today's sense than in the past when $\infty$ and $-\infty$ were allowed as
possible values of a limit.

So far we considered limits where the function $f(x)$ tended to $\infty$ or to $-\infty$.
We can also consider limits where the variable $x$, instead of tending to a (finite)
real number tends to $\infty$ or to $-\infty$. That is, we can consider $\lim\limits_{x \to \infty} f(x)$ and
$\lim\limits_{x \to -\infty} f(x)$. Again, depending upon whether these limits are finite or infinite,
the definitions would be different. We give only two of these definitions and
leave the formulation of the others as exercises. If $L$ is a real number, we say
$\lim\limits_{x \to \infty} f(x) = L$ if for every $\epsilon > 0$, there exists $R$ such that for every $x > R$, we
have $|f(x) - L| < \epsilon$. And we say $\lim\limits_{x \to -\infty} f(x) = \infty$ if for every $R$ (however large)
there exists $M$ such that whenever $x < M$, $f(x) > R$. Theorems about limits
of the type $\lim\limits_{x \to c} f(x)$ (where $c$ is a real number) continue to hold for limits of
the type $\lim\limits_{x \to \infty} f(x)$ or $\lim\limits_{x \to -\infty} f(x)$. The proofs require slight modifications. A
very suggestive and convenient way to handle the two types of limits together
is to use the concept of a neighbourhood. An open interval centred at a point
$c$ is called a **neighbourhood** of the point $c$. (It is not really vital that $c$ be
at the centre. But generally this is taken to be the case.) If we remove the
point $c$ from a neighbourhood of $c$, it is called a **deleted neighbourhood**
of $c$. Thus a typical deleted neighbourhood of a real number $c$ is of the form
$(c - \delta, c + \delta) - \{c\}$. Equivalently, it is the union of the two intervals $(c - \delta, c)$
and $(c, c + \delta)$ (see Figure 2.4.1(a)). Deleted neighbourhoods of $\infty$ are intervals
of the form $(R, \infty)$ for some real number $R$. Similarly those of $-\infty$ are of the
form $(-\infty, M)$ for some real number $M$ (see Figure 2.4.1(b) and (c)).

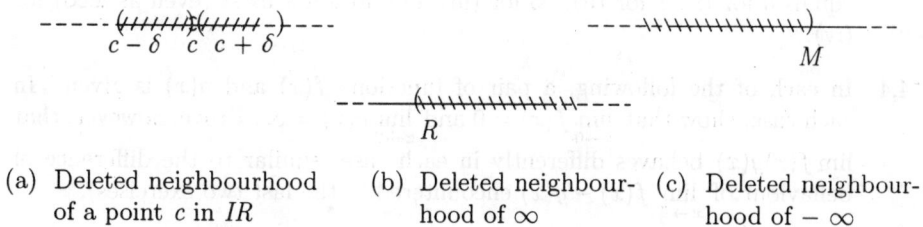

(a) Deleted neighbourhood    (b) Deleted neighbour-  (c) Deleted neighbour-
    of a point $c$ in $I\!R$        hood of $\infty$         hood of $-\infty$

Figure 2.4.1 Deleted Neighbourhood

We can now unify the three definitions of $\lim\limits_{x \to c} f(x) = L$ into a single one,
regadless of whether $c$ is a real number, or $c = \infty$ or $c = -\infty$. All we have

to say is that for every $\epsilon > 0$, there exists a deleted neighbourhood of $c$ such that $|f(x) - L| < \epsilon$ whenever $x$ is in this deleted neighbourhood. We can even go one step further and allow $L$ to be $\infty$ or $-\infty$. Thus all possible nine interpreatations of $\lim_{x \to c} f(x) = L$ can be combined together in a single statement, viz., for every neighboudhood $V$ of $L$, there exists a deleted neighbourhood $U$ of $c$ such that $f(x) \in V$ whenever $x \in U$. Note that unlike for a real number, a neighbourhood of $\infty$ or of $-\infty$ is already a deleted neighbourhood, because, a neighbourhood contains only real numbers and neither $\infty$ nor $-\infty$ is a real number. Figuratively, we may say that $\infty$ and $-\infty$ have got real numbers as close neighbours. But they are themselves not real numbers. This, in a nutshell, is the reason why limits involving infinity require a slightly different handling.

# EXERCISES

4.1 Défine "$\lim_{x \to \infty} f(x) = -\infty$" directly and also as "$\lim_{x \to \infty} -f(x) = \infty$" and show that the two definitions are equivalent. Similarly, show that "$\lim_{x \to -\infty} f(x) = L$" can also be defined as "$\lim_{x \to \infty} f(-x) = L$".

4.2 In each of the following, a pair of functions $f(x)$ and $g(x)$ is given. In each case show that $\lim_{x \to \frac{\pi}{2}} f(x) = \infty$ and $\lim_{x \to \frac{\pi}{2}} g(x) = \infty$. (You may assume here that $\sin h \to 0$ as $h \to 0$. This will be proved later, along with many other properties of the trigonometric functions; see Sections 4.7 and 4.8.)

   (i)   $f(x) = |\sec x|$, $g(x) = |\tan x|$ (cf. also Exercise (1.6.1), part (ii))

   (ii)  $f(x) = \sec^2 x$, $g(x) = \tan^2 x$

   (iii) $f(x) = \sec^4 x$, $g(x) = \tan^4 x$

   (iv)  $f(x) = \tan|x| + \sin\left(\dfrac{1}{x - \frac{\pi}{2}}\right)$, $g(x) = \tan|x|$.

4.3 In the last exercise, let $h(x) = f(x) - g(x)$ in each case. Prove that $\lim_{x \to \frac{\pi}{2}} h(x)$ equals 0 for (i), 1 for (ii), $\infty$ for (iii) and does not exist (even as $\pm\infty$) for (iv).

4.4 In each of the following, a pair of functions $f(x)$ and $g(x)$ is given. In each case, show that $\lim_{x \to 0} f(x) = 0$ and $\lim_{x \to 0} g(x) = \infty$. Prove, however, that $\lim f(x)g(x)$ behaves differently in each case (similar to the difference of behaviour of $\lim_{x \to \frac{\pi}{2}} f(x) - g(x)$ encountered in the last two exercises).

   (i) $f(x) = x^4$, $g(x) = \dfrac{1}{x^2}$    (ii) $f(x) = 2x^2$, $g(x) = \dfrac{1}{x^2}$

   (iii) $f(x) = x^2$, $g(x) = \dfrac{1}{x^4}$    (iv) $f(x) = x^2 \sin\left(\dfrac{1}{x}\right)$, $g(x) = \dfrac{1}{x^2}$

4.5   Take the standard proof of the Sandwich Theorem (see Exercise (3.4)). How will you modify it for limits of the form $\lim\limits_{x \to \infty} f(x)$ or $\lim\limits_{x \to -\infty} f(x)$? Similarly modify the proofs of the other standard results about limits (e.g. limit of a sum of two functions equals the sum of their limits).

4.6   Prove that the intersection of two deleted neighbourhoods of a real number $c$ is also a deleted neighbourhood of $c$. Show that this is also true if $c$ is replaced by $\infty$ or by $-\infty$. Using this fact and the unified definition of $\lim\limits_{x \to c} f(x)$ (given in the last paragraph of the text), give a unified proof of the Sandwich Theorem which is applicable even to the cases $c = \infty$ or $c = -\infty$. Give similar unified proofs of other results about limits.

## 2.5   Restricted Limits

When we consider $\lim\limits_{x \to c} f(x)$ (where $c$ can be a real number or $\pm \infty$ as well), it is implicit that the function $f$ is defined at all points in some deleted neighbourhood of $c$. (As emphasised in Section 2.3, it is immaterial whether $f(c)$ is itself defined or not. But at all nearby points, $f$ has to be defined.) Many times this condition is not satisfied. For example, the function $f(x) = \sqrt{x}$ is defined only for $x \geq 0$. So, if we have to consider $\lim\limits_{x \to 0} f(x)$, we run into a difficulty because $f$ is not defined in any deleted neighbourhood of $0$. Another important example is that of a sequence. By definition, a **sequence** is a function whose domain is $I\!N$, the set of all positive numbers. For a sequence whose $n^{\text{th}}$ term, say $a_n$, is $\dfrac{n}{\sqrt{n^2 + 1}}$, the corresponding function $f$ is $f(n) = \dfrac{n}{\sqrt{n^2 + 1}}$. Here the variable $n$ takes only integral values. In this case we could of course as well define $f(x) = \dfrac{x}{\sqrt{x^2 + 1}}$ for all real $x$. In other words, we are extending the domain from $I\!N$ to the entire $I\!R$. But sometimes there is no natural way of doing so. For example if the sequence is $b_n = \dfrac{2^n}{n!}$, how do we define $x!$ if $x$ is not an integer? We also encounter situations where, even though a function $f$ is defined on an entire deleted neighbourhood of a point $c$, the formulas defining

$f(x)$ are different for different values of $x$. As two typical examples, consider the functions $f, g$ defined by

$$f(x) = \begin{cases} x & \text{if } x \leq 1 \\ x^2 + 1 & \text{if } x > 1 \end{cases} \tag{1}$$

$$g(x) = \begin{cases} x & \text{if } x \text{ is rational} \\ 1 - x & \text{if } x \text{ is irrational} \end{cases} \tag{2}$$

In such cases, while considering $\lim_{x \to c} f(x)$ it is convenient to let $x$ approach $c$ through some restricted values. This leads to the concept of a **restricted limit**. Specifically, let $S$ be a subset of $\mathbb{R}$. It does not matter whether $c$ is itself in $S$ or not. We define the **limit** of $f(x)$ as $x$ tends to $c$ **through** $S$, written as $\lim_{\substack{x \to c \\ x \in S}} f(x)$ in much the same way as $\lim_{x \to c} f(x)$, except that we have to add the restriction ' $x \in S$'. That is, we say $\lim_{\substack{x \to c \\ x \in S}} f(x) = L$ (where $L$ is a real number) if for every $\epsilon > 0$, there exists $\delta > 0$ such that for all $x \in S, |f(x) - L| < \epsilon$ whenever $0 < |x - c| < \delta$. (We leave as an exercise the modification that is necessary if $c$ is $\infty$ or $-\infty$. Or we can word the definition in terms of a deleted neighbourhood so that it will apply equally well to all cases.)

Note that $\lim_{\substack{x \to c \\ x \in S}} f(x)$ depends not only on the function $f$ and the point $c$ but on the subset $S$ as well. For example, in (2) above let $S$ be the set of rational numbers. Then $\lim_{\substack{x \to 2 \\ x \in S}} g(x) = \lim_{\substack{x \to 2 \\ x \in S}} x = 2$ as is easy to show. If, however, we take $S$ to be the set of all irrational numbers then $\lim_{\substack{x \to 2 \\ x \in S}} g(x)$ comes out to be $-1$.

Restricted limits are also called relative limits sometimes. If $\lim_{\substack{x \to c \\ x \in S}} f(x) = L$, we say $L$ is the **limit** of $f(x)$ **relative** to $S$ as $x$ tends to $c$.

An important special case of restricted limits arises by taking $S$ to be the set $\{x \in \mathbb{R} : x > c\}$, i.e., $S = (c, \infty)$ or verbally, $S$ is the set of real numbers to the right of $c$ or above $c$. In this case, $\lim_{\substack{x \to c \\ x \in S}} f(x)$ is often denoted by $\lim_{x \to c+} f(x)$ and called the **limit** of $f(x)$ as $x$ tends to $c$ **plus** or as $x$ tends to $c$ **from the right** or **from abvoe**, or sometimes the **right-handed limit** for short. Similarly we get the **left-handed limit** or the **limit from below**, $\lim_{x \to c-} f(x)$, by taking $S = (-\infty, c) = \{x \in \mathbb{R} : x < c\}$.

It is trivial to check that if $\lim_{x \to c} f(x) = L$, then $\lim_{\substack{x \to c \\ x \in S}} f(x)$ also exists and equals $L$ for every subset $S$ of $\mathbb{R}$. In particular, in this case $\lim_{\substack{x \to c \\ x \in S_1}} f(x) = \lim_{\substack{x \to c \\ x \in S_2}} f(x)$ for any two subsets $S_1$ and $S_2$ of $\mathbb{R}$. It follows that if we can find two subsets $S_1$ and $S_2$ for which the restricted limits are not equal then $\lim_{x \to c} f(x)$ cannot exist. In fact this is a very standard way of showing $\lim_{x \to c} f(x)$ does not exist.

For example, in (2) above, let $S_1$ and $S_2$ be the sets of all rational and irrational numbers respectively. As we already saw, $\lim\limits_{\substack{x \to c \\ x \in S_1}} g(x) = 2 \neq -1 = \lim\limits_{\substack{x \to c \\ x \in S_2}} g(x)$.

So we conclude that $\lim\limits_{x \to 2} g(x)$ does not exist. In (1), $\lim\limits_{x \to 1^+} f(x) = 2$ while $\lim\limits_{x \to 1^-} f(x) = -1$ and so $\lim\limits_{x \to 1} f(x)$ does not exist. (For $c \neq 1$, however, $\lim\limits_{x \to c} f(x)$ exists and equals $c$ if $c < 1$ and $c^2 + 1$ if $c > 1$ as is easy to check.) Another example will be given in the Exercises.

The logic adopted here may be illustrated with a real-life analogy. Suppose in some community, there is a law that once a man is engaged to a woman, he can go out only with that woman. Suppose we want to spy on a man Mr. $X$ in that community but we have no time to do so every day. We then make a spot check. Suppose on all Mondays we see him go out with a woman, say $A$. This does not prove that $X$ is engaged to $A$. It only shows that if at all he is engaged, then he is engaged to $A$. Suppose, however, that we also find him date another woman $B$ on all Thursdays. Then it proves, not only that $X$ is not engaged either to $A$ or to $B$, it also proves that $X$ is not engaged to any woman, i.e., $\cdot X$'s fiance does not yet exist. A limit is like watching the behaviour of the man on all days, while a restricted limit is like a spot check. Tacitly we are assuming here that fiances and limits, if at all they exist, are unique. The question of uniqueness of limits will be taken up later (Section 2.7).

What if we find two subsets $S_1$ and $S_2$ of $\mathbb{R}$ such that $\lim\limits_{\substack{x \to c \\ x \in S_1}} f(x)$ and $\lim\limits_{\substack{x \to c \\ x \in S_2}} f(x)$ are equal? Does it mean that $\lim\limits_{x \to \infty} f(x)$ exists? The answer depends on how the subsets $S_1, S_2$ are related. If, for example, $S_1 \subset S_2$, then the restricted limit of $f(x)$ through $S_1$ gives us no more information than the restricted limit of $f(x)$ through $S_2$. If however, the subsets $S_1$ and $S_2$ together cover some deleted neighbourhood of $c$ then the things are better as we now show.

**5.1 Thoerem:** If $S_1, S_2$ are subsets of $R$ such that $S_1 \cup S_2$ contains some deleted neighbourhood of $c$, and $\lim\limits_{\substack{x \to c \\ x \in S_1}} f(x)$ and $\lim\limits_{\substack{x \to c \\ x \in S_2}} f(x)$ both exist and are equal then $\lim\limits_{x \to c} f(x)$ also exists and equals this common value.

**Proof:** Just for the sake of variation we give the proof here for the case $c = \infty$ and leave as an exercise the other two cases. From the hypothesis, there exists $R$ such that $(R, \infty) \subset S_1 \cup S_2$. Let $L = \lim\limits_{\substack{x \to \infty \\ x \in S_1}} f(x) = \lim\limits_{\substack{x \to \infty \\ x \in S_2}} f(x)$. We have to show $\lim\limits_{x \to \infty} f(x) = L$. So let $\epsilon > 0$ be given. Since $\lim\limits_{\substack{x \to \infty \\ x \in S_1}} = L$, there exists $R_1$ such that for all $x \in S_1, |f(x) - L| < \epsilon$ whenever $x > R_1$. Similarly, there exists $R_2$ such that for all $x \in S_2, |f(x) - L| < \epsilon$ whenever $x > R_2$. Now let $R' = \max\{R, R_1, R_2\}$ and suppose $x > R'$. Then $x > R$ and hence by our assumption $x \in S_1$ or $x \in S_2$. In the first case, $x > R_1$ (since $R' \geq R_1$) gives $|f(x) - L| < \epsilon$. Similarly, in the second case $|f(x) - L| < \epsilon$. So $|f(x) - L| < \epsilon$ for all $x > R'$. Hence $\lim\limits_{x \to \infty} f(x) = L$. ∎

As a corollary we get,

**5.2 Corollary:** If $c \in \mathbb{R}$, then $\lim\limits_{x \to c} f(x)$ exists if and only if $\lim\limits_{x \to c^+} f(x)$ and $\lim\limits_{x \to c^-} f(x)$ both exist and are equal.

**Proof:** As noted before, $\lim\limits_{x \to c^+} f(x)$ and $\lim\limits_{c \to c^-} f(x)$ are simply $\lim\limits_{\substack{x \to c \\ x \in S_1}} f(x)$ and $\lim\limits_{\substack{x \to c \\ x \in S_2}} f(x)$ respectively where $S_1 = (c, \infty)$ and $S_2 = (-\infty, c)$. Then $S_1 \cup S_2$ contains a deleted neighbourhood (in fact *every* deleted neighbourhood of $c$). So if $\lim\limits_{x \to c^+} f(x)$ and $\lim\limits_{x \to c^-} f(x)$ both exist and are equal then by the last theorem $\lim\limits_{x \to c} f(x)$ exists. The other way implication is trivial. ∎

Some authors *define* $\lim\limits_{x \to c} f(x)$ by first defining the two restricted limits $\lim\limits_{x \to c^+} f(x)$ and $\lim\limits_{x \to c^-} f(x)$ separately and then requiring that these two be equal. Although in view of the corollary above the two approaches are equivalent, this practice is not very desirable for several reasons. First, it gives undue importance to a particular kind of restricted limits. Secondly it is prone to give a wrong impression that in order for $x$ to tend to $c$, $x$ must approach $c$ only along a ray ending at $c$. This may make a beginner believe that a similar result is true for functions of several variables. But that is not the case as can be shown by simple counter-examples. (We shall not systematically study functions of two or more real variables. Nevertheless we mention one such example here. For $(x, y) \neq (0, 0)$ let $f(x, y) = \dfrac{x^2 y}{x^4 + y^2}$. Then $f(x, y) \to 0$ as $(x, y)$ approaches $(0, 0)$ along any straight line, say along the line $y = mx$, for it is easily seen that $f(x, mx) = \dfrac{mx^3}{x^4 + m^2 x^2} = \dfrac{mx}{x^2 + m^2}$ tends to 0 as $x$ tends to 0 for every $m$. But if $y = x^2$, then $f(x, y) = 1/2$. Hence there are points arbitrarily close to $(0, 0)$ at which $f$ has value $1/2$. So $f(x, y)$ does not tend to 0 as $(x, y) \to (0, 0)$. In fact, $\lim\limits_{(x,y) \to (0,0)} f(x, y)$ does not exist.)

As remarked earlier, the limit of a sequence is a special case of a restricted limit, where the variable $x$ tends to $\infty$ through integer values. To avoid having to write $\lim\limits_{\substack{x \to \infty \\ x \in \mathbb{N}}} f(x)$, the variable $x$ is often replaced by a symbol like $n, m, r$ etc. (these symbols being generally used to denote integer variables) and instead of $f(n)$ one writes $a_n$ or whatever be the $n^{\text{th}}$ term of the sequence. The limit of the sequence $\{a_n\}$ can then be written simply as $\lim\limits_{n \to \infty} a_n$. For example, let $a_n$ be the total which a principal of one rupee will yield at the end of the first year if interest is paid at the rate of 10 per cent per annum and compounded $n$ times a year (at equal intervals). As we saw in Section 2.1, $a_n = (1 + \dfrac{1}{10n})^n$. Moreover $\lim\limits_{n \to \infty} a_n$ will give the total one would get under continuous compounding. Note that here if we replace $n$ by a real variable $x$, then the expression $\left(1 + \dfrac{1}{10x}\right)^x$ can be given a meaning and it turns out that as $x \to \infty$, it approaches the same limit. But when $x$ is not an integer (or at least a rational number), the expression $(1 + \dfrac{1}{10x})^x$ has a rather complicated definition (see Section 5.7). When $n$ is a positive integer, however, $\left(1 + \dfrac{1}{10n}\right)^n$ has a very natural meaning. So here, the restricted limit $\lim\limits_{\substack{x \to \infty \\ x \in I\!N}} \left(1 + \dfrac{1}{10x}\right)^x$ is far more natural than $\lim\limits_{x \to \infty} \left(1 + \dfrac{1}{10x}\right)^x$.

Sometimes, in the case of sequences, instead of letting the variable $n$ tend to $\infty$ through all integral values, it may be convenient to put some more restriction on $n$, for example, that $n$ approach $\infty$ only through even values or odd values. We then have a case of a doubly restricted limit. For example, suppose $a_n = (-1)^n$. Then $a_n = 1$ for even $n$ while $a_n = -1$ for odd $n$. So $\lim\limits_{\substack{n \to \infty \\ n \text{ even}}} a_n = 1$ and $\lim\limits_{\substack{n \to \infty \\ n \text{ odd}}} a_n = -1$. As these two restricted limits are not equal, it follows easily that $\lim\limits_{n \to \infty} (-1)^n$ does not exist. Note, incidentally, that $(-1)^x$ has no meaning if $x$ is just a real number.

Summing up, restricted limits are important because by choosing the set $S$ suitably they are easier to calculate than the unrestricted limits. And the information gathered can throw some light on the existence and the value (if any) of the unrestricted limit.

# EXERCISES

5.1 Show that $\lim\limits_{x \to 0} \sin \dfrac{1}{x}$ does not exist by showing that $\sin \dfrac{1}{x} \to 1$ as $x \to 0$ through values of the form $\dfrac{1}{2n\pi + \frac{\pi}{2}}$ where $n \in I\!N$, while $\sin \dfrac{1}{x} \to 0$ as $x$ tends to 0 through values of the form $\dfrac{1}{n\pi}$, where $n \in I\!N$. Compare this proof with that in Exercise (3.3).

5.2 Prove Theorem (5.1) in the other two cases, that is, where $c \in I\!R$ and $c = -\infty$.

5.3 Extend Theorem (5.1) to the case where instead of two subsets we have a finite number of subsets, say, $S_1, S_2, \ldots, S_k$ whose union contains some deleted neighbourhood of $c$. [*Hint*: Either adapt the argument given in the proof or apply induction on $k$.]

5.4 As an application of the result of last exercise prove that $\lim\limits_{x \to 1} f(x)$ exists where the function $f$ is defined by

$$f(x) = \begin{cases} x & \text{if} & x < 1 \text{ and } x \text{ is rational} \\ 1 & \text{if} & x < 1 \text{ and } x \text{ is irrational} \\ x^2 & \text{if} & x > 1 \text{ and } x \text{ is rational} \\ 2 - x & \text{if} & x > 1 \text{ and } x \text{ is irrational} \end{cases}$$

[*Hint* : Consider 4 suitably chosen sets $S_1, S_2, S_3, S_4$ and take restricted limits as $x \to 1$ through each].

5.5   Let $S_0$ be the set of all real numbers other than positive integers. That is, $S_0$ is the set $I\!R - I\!N$. For every positive integer $n$, let $S_n = S_0 \cup \{n\}$, that is, $S_n$ is obtained by adding just one more element, viz., $n$, to $S_0$. Prove that $I\!R$ equals the union of infinitely many sets $S_1, S_2, S_3, \ldots$, but not the union of any finite number of them. (In case you don't know how to take the union of an infinite family of sets, in the present case it is defined as the set $\{x \in I\!R : \text{there is some } k \in I\!N \text{ such that } x \in S_k\}$. It is denoted by symbols like $\bigcup_{n \in I\!N} S_n$ or, similar to the sigma notation for an infinite series, by $\bigcup_{n=1}^{\infty} S_n$.)

5.6   Let $f(x) = \begin{cases} 1 & \text{if } x \text{ is a positive integer} \\ 0 & \text{otherwise} \end{cases}$

Let $S_1, S_2, \ldots$ be the sets of the last exercise. Prove that for every $k \in I\!N$, $\lim_{\substack{x \to \infty \\ x \in S_k}} f(x) = 0$. Show, however, that $\lim_{x \to \infty} f(x)$ does not exist. (This example, which is admittedly contrived, shows that although Theorem (5.1) can be extended to the case of finitely many subsets as in Exercise (5.3), it cannot be extended to infinitely many subsets.)

5.7   Analyse the failure of the attempted extension of Theorem (5.1) in the last exercise. Show that it ultimately boils down to the fact that while the intersection of two, and hence any finite number of deleted neighbourhoods is again a deleted neighbourhood, the intersection of infinitely many deleted neighbourhoods need not be a deleted neighbourhood; it may even be the empty set.

5.8   Show that $\lim_{x \to 0^+} \sqrt{x} = 0$.

5.9   Show that $\lim_{x \to 0^+} \dfrac{1}{x} = \infty$ while $\lim_{x \to 0^-} \dfrac{1}{x} = -\infty$.

## 2.6 Existence and Uniqueness of Limits

Since so many concepts in mathematics (e.g., the instantaneous speed, the arc length or area) are defined as limits of some kind or the other, the question of existence and uniqueness of limits acquires tremendous importance. If you invite some new male acquaintence of yours to your place, then before asking him to bring along his wife too, you better ensure that he is married and monogamous. If he is a confirmed bachelor or a divorcee, you might embarass him. And if he has two or more wives, he might embarass you by asking 'which one?'! A similar embarassment can occur while dealing with the limits. For example, we saw in Sections 2.1 and 2.2 that $\lim_{n \to \infty} (1 + \frac{1}{10n})^n$ is the total you will get at the end of the year if you put in one rupee and interest is given at the rate of 10 p.c. p.a. compounded continuously. If this limit does not exist or exists but is not unique, there will be a terrible row between you and the banker.

Fortunately, the question of uniqueness of limits turns out to be very easy. Limits of all kinds are unique. By way of illustration, we prove that limits of the form $\lim_{x \to \infty} f(x)$ are unique. Essentially the same argument applies to other limits.

**6.1 Theorem:** Suppose $\lim_{x \to c} f(x) = L$ and also that $\lim_{x \to c} f(x) = L'$. Then $L = L'$.

**Proof:** First assume both $L$ and $L'$ are real numbers. If $L \neq L'$ then let $\epsilon = \frac{1}{2}|L - L'|$. Then $\epsilon > 0$. By definition of a limit, there exist $\delta_1, \delta_2 > 0$ such that

$$\text{(i)} \quad 0 < |x - c| < \delta_1 \Rightarrow |f(x) - L| < \epsilon$$

and (ii) $\quad 0 < |x - c| < \delta_2 \Rightarrow |f(x) - L'| < \epsilon$

Now let $x$ be a point which is so close to $c$ that $0 < |x - c| < \delta_1$ and $0 < |x - c| < \delta_2$ are both true. Then we have $|f(x) - L| < \epsilon$ and also $|f(x) - L'| < \epsilon$. But then using the well-known triangle inequality property of absolute values, we have

$$
\begin{aligned}
|L - L'| &= |L - f(x) + f(x) - L'| \\
&\leq |L - f(x)| + |f(x) - L'| \\
&= |f(x) - L| + |f(x) - L'| \\
&< \epsilon + \epsilon = 2\epsilon = |L - L'|
\end{aligned}
$$

So we get $|L - L'| < |L - L'|$ a contradiction. Thus we must have $L = L'$. (Pictorially, the conditions $|f(x) - L| < \epsilon$ and $|f(x) - L'| < \epsilon$ are equivalent to saying that $f(x)$ lies, respectively, in the open intervals $(L - \epsilon, L + \epsilon)$ and $(L' - \epsilon + L' + \epsilon)$. If $L < L'$ then $L + \epsilon = L' - \epsilon$. If $L > L'$ then $L - \epsilon = L' + \epsilon$. In either case these two open intervals are mutually disjoint and so $f(x)$ can never lie in both of them. The case $L < L'$ is shown in Figure 2.6.1.)

We still have to consider the case where one or both of $L, L'$ is infinite. We do the case where $L$ is finite and $L' = \infty$ and leave the others as exercise. This case is in fact even simpler than the one above. We take $\epsilon$ arbitrarily, say $\epsilon = 1$ and get $\delta_1$ such that $0 < |x - c| < \delta_1$ implies $L - 1 < f(x) < L + 1$. Now, taking $R = L + 1$ in the definition of $\lim_{x \to c} f(x) = \infty$, we get $\delta_2 > 0$ such that $f(x) > L+1$ whenever $0 < |x-c| < \delta_2$. Any $x$ such that $0 < |x-c| < \min\{\delta_1, \delta_2\}$ will give us a contradiction because $f(x) < L+1$ on one hand and $f(x) > L+1$ on the other. ∎

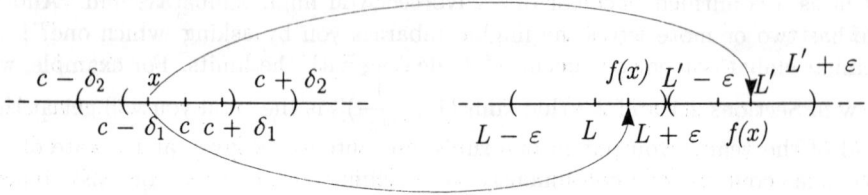

Figrue 2.6.1 : Uniqueness of Limits

In the proof above it was tacitly assumed that we could find $x$ such that $x$ was very close to $c$ and $x$ was in the domain of $f$ (i.e. $f(x)$ was defined). If we were considering relative (i.e., restricted) limits then we would further have to restrict $x$ to belong to some subset $S$ of $\mathbb{R}$. And depending upon how this set is chosen there may not exist such $x$. In other words, it may happen that some deleted neighbourhood of $c$ is disjoint from $S$. (An extreme example of this occurs when $S = \emptyset$, the empty set.) In such a case, we cannot guarantee the uniqueness of the restricted limit $\lim_{\substack{x \to c \\ x \in S}} f(x)$. For, here the condition in the definition of $\lim_{\substack{x \to c \\ x \in S}} = L$ is satisfied vacuously no matter what the real number $L$ is. To avoid trivial exceptions like this, we assume henceforth that whenever we consider a limit relative to a set $S$, the set $S$ intersects every deleted neighbourhood of the point approached.

Thus, we have nothing to worry as far as uniqueness of limits is concerned. If at all limits exist, they are unique. The existence of limits, however, cannot be disposed off so lightly. In fact, not all limits exist. For example, $\lim_{x \to 0} \sin\left(\dfrac{1}{x}\right)$ does not exist (see Exercises (3.3) and (5.1)). Similarly we saw at the end of the last section that the sequence $a_n = (-1)^n$ has no limit as $n \to \infty$. This is, of course, not very surprising, because, $\sin(\frac{1}{x})$ assumes different values as $x$ tends to 0 through different sets of values (see again Exercise (5.1).) Similarly, $(-1)^n$ is 1 for even $n$ and $-1$ for odd $n$ and since $n$ can approach $\infty$ through even as well as through odd values, $(-1)^n$ is not clustering around any single real number $L$ for all sufficiently large $n$.

So these are examples where we really do not expect the limit to exist. The behaviour of $\sin(\frac{1}{x})$ as $x \to 0$ or that of $(-1)^n$ as $n \to \infty$ is rather like that of a girl who flirts with different men on different days of the week. We really do not expect such a girl should get married. But to stretch the social analogy, what about a girl who thinks of just one man, adores him secretly, tries to get close to him but is too shy to actually hit upon him? We all agree that such a girl *should* get married to that man (and, of course, live happily everafter). It would be a heart-breaking tragedy if her idol turns out to be a mere illusion.

A sequence like $a_n = (1 + \dfrac{1}{10n})^n$ is a mathematical analogue of such a nice girl. As we saw in Section 2.1, $a_n$ is the total you get at the end of one year if you invest one rupee and interest at the rate of 10 p.c.p.a. is compounded $n$ times a year. It is intuitively clear and also easy to prove (see Exercise (1.1)) that $a_n$ increases as $n$ increases. But, for large $n$ the difference between $a_{n+1}$ and $a_n$ is very small. If you calculate (with a calculator, of course!) $a_n$ for large values of $n$, say $n = 100$ or $n = 1000$ etc. you will realise that as $n \to \infty$, $a_n$ is definitely heading to something. Depending upon how many places of decimals your calculator can handle, $a_n$ will eventually look like a constant on your calculator, although in reality it is not. Even without calculating, there is another way to see that $a_n$ *should* tend to some limit as $n \to \infty$. Although $a_n$ grows as $n$ grows, it is not difficult to show (see Exercise (6.3)) that for no value of $n$, can $a_n$ exceed the number $\frac{21}{19}$. So this number is like an upper bound which $a_n$ can never cross. Still, $a_n$ is getting closer and closer to $\frac{21}{19}$. It does not follow, of course, that $a_n$ will tend to $\frac{21}{19}$ as $n \to \infty$ because conceivably, we can find a smaller upper bound for $a_n$. What if we could find the *smallest* upper bound for $a_n$ and call it $L$? Then it is intuitively clear that $a_n$ should tend to $L$ as $n$ tends to $\infty$. Because if all the $a_n$'s fall short of $L$ by a margin of some positive $\epsilon$ (say) or more, then $L - \epsilon$ will be a new upper bound for $a_n$, which is even smaller than $L$. (See Figure (2.6.2) for a graphic description.)

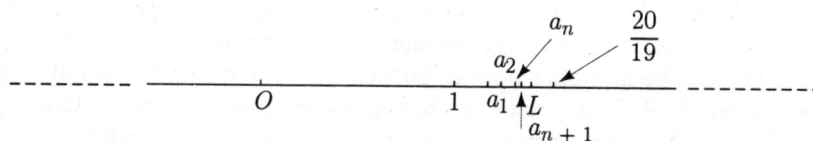

Figure 2.6.2 : A limit which should exist

Thus we see that whether $\lim\limits_{n \to \infty} a_n$ exists or not will depend on whether there exists a smallest upper bound for the set of values which this sequence assumes. It truns out that in the present case the smallest upper bound does exist. But it is not rational, even though every term of the sequence $(1 + \dfrac{1}{10n})^n$ is rational. In other words, if we were living entirely in the world of ratinoal numbers then we would have a sequence which should tend to a limit but doesn't. If you look at rational numbers, they might appear packed as sardines in a tin. Between any two rational numbers, there are infinitely many rational numbers, e.g. between 2.5 and 2.7 we have 2.6, 2.61, 2.611, 2.6111, ... and many others. So it may come as a surprise that there are lots of 'holes' in the rational number system. In fact every irrational number (e.g. $\pi$ or $\sqrt{2}$) can be considered as a hole in the rational number system. By taking the decimal expansions of irrational numbers we get plenty of examples of rational sequences which should have limits but don't within the world of rational numbers, (e.g. the sequence 3, 3.1, 3.14, 3.141, 3.1415, ... converges to $\pi$ in the real number system, but in the rational number system it has no limit.) Because of holes like these, the rational number system is incomplete in some sense.

What about the real number system? Certainly, the holes which correspond to the irrational numbers are patched up. But are there any other holes? Put differently, are there any limits which should exist but don't? If there are any, should we extend the real number system still further to patch these holes?

These are very basic questions. Fortunately, they have a very satisfactory answer. It turns out that the real number system has no holes in it. It is therefore already complete and needs no extension. (It may need extensions for other purposes, e.g. for solving an equation like $x^2 + 1 = 0$, but not for the purpose of ensuring the existence of limits.) Completeness of the real number system is, by far, its most crucial property as far as calculus is concerned. There is hardly a significant result of calculus which does not depend at least indirectly on completness.

We are, of course yet to define completeness formally. (Describing it as absence of holes is suggestive but not mathematically precise.) There are many equivalent formulations of completeness. In the discussion above of the behaviour of the sequence $a_n = (1 + \dfrac{1}{10n})^n$ we saw that this sequence would have a limit if there is a smallest upper bound for the set of its terms. The existence of such smallest or least upper bounds is, in fact, a very convenient formulation of completeness. But first we need some terminology. A subset $A$ of real numbers is said to be **bounded above** if there exists some real number $x$ such that for every $a \in A, a \leq x$. Any such number $x$ is called an **upper bound** of $A$. Note that $x$ may or not be an element of $A$. Obviously any number bigger than $x$ will also be an upper bound. So we try to look for smaller and smaller upper bounds of $A$. Ideally we look for the **least upper bound** (or **l.u.b.** for short) of $A$. As the name suggests, this is a real number, say $y$, which is first of all an upper bound for $A$, and secondly, is smallest among all upper bounds of $A$; that is, if $x$ is any other upper bound of $A$ then we must have $y < x$. It is trivial to show that the l.u.b., if it exists, is unique. That is, no set can have more than one least upper bounds. If $y$ is the least upper bound of a set $A$ then $y$ may or may not belong to $A$. If it does, then clearly it is the largest element of the set $A$, also called the **maximum** element of $A$. For example, if $A$ is the set of all rational numbers $\leq 2$, then 2 is the maximum element of $A$. If we let $A$ be the set of all rationals $< 2$, then $A$ has no maximum element, but 2 is still its least upper bound. To indicate its resemblance with a maximum but at the same time to distinguish from it, an l.u.b. is also called a **supremum.**

Now suppose $A$ is a non-empty subset of $\mathbb{R}$, which is bounded above. Then by definition, $A$ has at least one upper bound. But it is not obvious that $A$ has a supremum. Just as the set of positive real numbers has no least element, it is conceivable that the set of all upper boundes of $A$ has no least element. In fact, if instead of $\mathbb{R}$, we live entirely in $\mathbb{Q}$, the set of all rational numbers, then it is easy to construct many non-empty sets $A$ (contained in $\mathbb{Q}$) which are bounded above but have no supremum (in $Q$). The set $\left\{(1 + \dfrac{1}{10n})^n : n = 1, 2, \ldots\right\}$ is actually one such set. But proving that $A$ has no supremum in $Q$ is not so easy. A simpler example will be given in the exercises (Exercise (6.6)). (We could, of ocourse, take any irrational number, say $\pi$, and let $A$ be the set $\{x : x \in Q, x < \pi\}$. Then $A$ would have no supremum in $Q$. But such a construction begins with something which is outside $Q$. We want to construct $A$ in a manner which will make sense to anyone who knows nothing beyond rational numbers.)

With the real number system, such anomalies do not arise. This property of real numbers, which is the cornerstone of all major theorems of calculus deserves a special display.

**6.2 Axiom:** (The least upper bound axiom): The real number system is **complete**, that is, every non-empty subset of $\mathbb{R}$ which is bounded above has a least upper bound (i.e. a supremum) in $\mathbb{R}$.

The reason this property is called an axiom is that in elementary courses in calculus, it is customary to take a real number as a primitive term and some of their properties as axioms. In more advanced courses it is possible to construct real numbers starting from natural numbers (which, too, can be constructed, if desired, as certain sets as mentioned in Section 1.9). Although we shall not go that far, in Appendix A we shall show how the real numbers can be constructed from rational numbers. In that case, completeness of the real number system can be proved as a theorem, instead of assuming it as an axiom.

We already had an application of completeness in showing that the sequence $(1 + \dfrac{1}{10n})^n$ has a limit. Many basic theorems about sequences and continuous functions are consequences of completeness of $\mathbb{R}$. Some of them are, in fact, logically equivalent to the completeness of $\mathbb{R}$, in that if we assume their truth (as an axiom) then we can deduce completeness as a consequence. We shall study a few equivalent verions of completeness later. Here we mention only one which is obtained by reversing the order. That is, we define a **lower bound** for a subset $A$ of $\mathbb{R}$ to be a real number $x$ such that for every $a \in A, a \geq x$. The **greatest lower bound** or **g.l.b.** or **infimum** of $A$ is defined as a lower bound of $A$ which is bigger than every other lower bound of $A$. We then have the following simple but useful result.

**6.3 Theroem:** The following two statements are equivalent:

(i) $\mathbb{R}$ is complete, i.e. every non-empty subset of $\mathbb{R}$ which is bounded above has a supremum.

(ii) every non-empty subset of $\mathbb{R}$ which is bounded below has an infimum (in $\mathbb{R}$).

**Proof:** Because of the perfect symmetry of the two statements it suffices to prove either one of the two implications. We show that (i) implies (ii). Let $A$

be a subset of $\mathbb{R}$ such that $A \neq \phi$ and $A$ is bounded below. Let $y$ be any lower bound of $A$. We let $B$ be the set $\{x \in \mathbb{R} : -x \in A\}$. The sets $A$ and $B$ are mirror image of each other through the origin, i.e., the point 0. So it is easy to see (cf. Fig. 2.6.3) that $-y$ is an upper bound for $B$. Also $B \neq \phi$ since $A \neq \phi$. Hence by (i), $B$ has a supremum say $z$ in $\mathbb{R}$. Then $-z$ will be an infimum of $A$. Thus (ii) holds. ∎

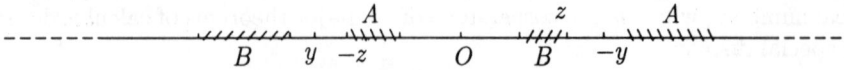

Figure 2.6.3 : Another Version of Completeness of $\mathbb{R}$.

Completenes of $\mathbb{R}$ ensures that the limits which should exist indeed exist. We shall see many instances of this. Perhaps the most immmediate application of this sort is the following result. The argument is a straight generalisation of the one we used for showing that $\lim\limits_{n \to \infty} (1 + \dfrac{1}{10n})^n$ exists and so we omit the proof. To understand the statement of the theorem, a sequence $\{a_n\}$ is said to be **monotonically increasing** if $a_{n+1} \geq a_n$ for all $n$. (Note that we are not insisting that strict inequality, i.e. $a_{n+1} > a_n$ should hold. If this is the case for every $n \in \mathbb{N}$, then the sequence $\{a_n\}$ is called **strictly monotonically increasing**. We caution the reader that some authors prefer to use 'steadily increasing' instead of 'monotonically increasing', the latter term being reserved for what we call 'strictly monotonically increasing'. ) Monotonically decreasing and strictly monotonically decreasing sequences are defined analogously. A monotonically increasing sequence is evidently bounded below (in fact, the first term $a_1$ is itself a lower bound). But it need not be bounded above (e.g. let $a_n = n^2$ for all $n$). If, however, it is bounded above then we have:

**6.4 Theorem:** If $\{a_n\}$ is a monotonically increasing sequence which is bounded above then it is convergent; i.e., $\lim\limits_{n \to \infty} a_n$ exists (as a finite real number). In fact, this limit equals the supremum of the set $\{a_n : n \in \mathbb{N}\}$. Similarly, every monotonically decreasing sequence which is bounded below is convergent. ∎

As we shall see later (Q. 3.1), many other limits can be expressed in terms of limits of sequences. So the theorem above (and a few other similar theorems about existence of limtis of sequences) actually imply the existence of many other limits. Summing up, it is because of the completeness of the real number system that the limtis exist. As noted at the beginning, unqiueness of limits is in general not a problem at all. Figuratively, if limiting process is the heart of calculus, then completeness of $\mathbb{R}$ is what makes the heart beat!

Although the preceding theorem is generally sufficient in applications, sometimes it is convenient to have a result which is basically of the same spirit as the last theorem (and can, in fact be derived from it) but has a slightly different setting. A sequence is, by definition, a function of a discrete variable. But the concept of monotonicity makes sense for a real valued function defined on any interval as well. Suppose $f : S \longrightarrow \mathbb{R}$ is a function. Then we say $f$ is monotonically increasing on $S$ if whenever $x, y \in S$ and $x < y$, we have $f(x) \leq f(y)$. (Once again, strict inequality is not required. If it holds always, the function is called strictly monotonically increasing.) The function $f(x) = x^2$ is monotonically increasing on the interval $[0, \infty)$ and monotonically decreasing on $(-\infty, 0]$. (On the entire real line it is neither.)

We now have the following result whose proof is essentially a repetition of the proof of the last theorem.

**6.5 Theorem:** Suppose a function $f : (a, b) \longrightarrow \mathbb{R}$ is monotonically increasing. If $f$ is bounded above, then the left handed limit $\lim\limits_{x \to b^-} f(x)$ exists. Similarly, if $f$ is bounded below, then the right handed limit $\lim\limits_{x \to a^+} f(x)$ exists. The same holds if $f$ is monotonically decreasing.

**Proof:** For the first assertion, we are given that the set $\{f(x) : a < x < b\}$ is bounded above. Obviously it is non-empty. So, by completeness, it has a supremum, say $L$. We claim $\lim\limits_{x \to b^-} f(x) = L$. Let $\epsilon > 0$ be given. Then since $L$ is the least upper bound of this set, $L - \epsilon$ is not an upper bound for it and so there exists $c \in (a, b)$ such that $f(c) > L - \epsilon$. Let $\delta = b - c$. Then $\delta > 0$. Also, whenever $x \in (b - \delta, b)$ we have, by monotonicity of $f$, $f(c) \leq f(x)$ and hence $L - \epsilon < f(x)$. Since $f(x) \leq L$ holds anyway ($L$ being an upper bound), we get $L - \epsilon < f(x) < L + \epsilon$ whenever $b - \delta < x < b$ as desired. The proofs of the other assertions are similar. $\blacksquare$

## EXERCISES

6.1 Complete the proof of Theorem (6.1) by considering the reamining cases, e.g. when $L = -\infty$ and $L' = \infty$. Also extend the theroem to limtis of the form $\lim\limits_{x \to \infty} f(x)$ and $\lim\limits_{x \to -\infty} f(x)$. (Paraphrasing the proof in terms of deleted neighbourhoods will make this easy).

6.2 Let $a_n = (1 + \dfrac{1}{10n})^n$. Expanding by the binomial theorem, show that

$$a_n \leq 1 + \frac{1}{10} + \frac{1}{2.10^2} + \frac{1}{3!10^3} + \cdots + \frac{1}{n!10^n}$$

and hence further that

$$a_n < 1 + \frac{1}{10} + \frac{1}{2.10^2} + \frac{1}{2^2.10^3} + \cdots + \frac{1}{2^{n-1}10^n}, \text{ for } n \geq 3.$$

6.3   Using the last exercise and the formula for the sum of a geometric progression, show that $a_n < \dfrac{21}{19}$ for all $n$.

6.4   Let $x = \dfrac{p}{q}$ and $y = \dfrac{p+2q}{p+q}$, where $p$ and $q$ are positive integers.

Prove that :            (i) $x^2$ can never be 2
                                    (ii) if $x^2 > 2$ then $y^2 < 2$.
                                    (iii) if $x^2 < 2$ then $y^2 > 2$.

6.5   Let $x, y$ be as in the last exercise and let $z = \dfrac{3p + 4q}{2p + 3q}$.

Prove that      (i)  if $x^2 < 2$ then $z^2 < 2$ and $x < z$
   and      (ii) if $x^2 > 2$ then $z^2 > 2$ and $z < x$.

[*Hint* : $z$ is obtained from $y$ exactly the same way $y$ is obtained from $x$.]

6.6   Let $A = \{x \in Q : x > 0, x^2 < 2\}$. Using the last two exercises show that $A$ is a non-empty subset of $Q$ which is bounded above but has no supremum in $Q$.

6.7   Starting from $p = 1, q = 1$ and applying the construction in Exercise (6.4) repeatedly, generate a sequence of rationals which converges to $\sqrt{2}$ (in the real number system). (The first few terms of the sequence are $1, \frac{3}{2}, \frac{7}{5}, \frac{17}{12}$.) Is this sequence monotonically increasing / decreasing?

6.8   If a subset $A$ of $I\!R$ has a maximum element show that it is unique and also that it is its supremum. Give an example of a non-empty subset which has a supremum but no maximum. Show that such a set can never be finite. Prove similar results for minimum and infimum.

6.9   Let $A$ be a non-empty subset of $I\!R$ which is bounded above. Let $B$ be the set of all upper bounds of $A$. Show that $B$ is non-empty and bounded below. Show further that the supremum of $A$ is the same as the infimum of $B$. What can be said about $A \cap B$?

6.10  Using the last exercise give an alternate proof of Theorem (6.3) which does not depend on the idea of taking reflection.

6.11  Prove Theorem (6.4).

6.12  Prove that a number $y$ is the supremum of a subset $A$ of $I\!R$ if and only if (i) $x \leq y$ for all $x \in A$ and (ii) for every $\epsilon > 0$, there exists $x \in A$ such that $x > y - \epsilon$.

6.13 Suppose $y$ is the supremum of a non-empty subset $A$ of $\mathbb{R}$. Prove that there exists a monotonically increasing sequence $\{x_n\}$ such that (i) $x_n \in A$ for all $n$ and (ii) $x_n \to y$ as $n \to \infty$. Prove a similar result about the infimum of a set. (These results are sort of converses to the assertions in Theorem (6.4).)

## 2.7    Why Calculus Is Difficult

A more appropriate question to consider would be 'why calculus appears difficult especially to a beginner?' For, like many other subjects, the difficulty of calculus is more apparent than real and can be overcome with a few simple tricks and a determined effort.

Probably the major reason calculus appears so difficult is that it is different in many respects from the kind of mathematics one generally studies prior to it. The latter typically includes arithmetic, algebra, geometry and trigonometry. As noted earlier, these topics do not involve limits. Limiting process is inherent in certain geometric concepts such as tangency and areas but usually they are studied as intuitively clear concetps, without formal definitions. Calculus, on the other hand, begins with the concept of a limit. The formal definition of a limit is far too clumsy (see Section 2.3) for a beginner. Sentences of the type "For every ....... there exists ..... such that ......." almost never occur in the mathematics studied till then (or anywhere else for that matter). An incorrect order of quantifiers is one of the most common mistakes at this stage. Many students, even when they have the right ideas, are unable to express them properly because they have not mastered the peculiar phraseology. It is almost like trying to express oneself in a foreign language.

With limits, the difficulty is not merely that of having to master a peculiar diction. The very concept of a limit has certain clashes with our intuition as laymen. A layman's intuition is based on a discrete, finitistic understanding of of the world around him. In such a set-up, there can be no limiting process. As mentioned in Section 2.2, a limiting process can exist only when there are quantities which are either arbitrarily small or arbitrarily large. Obviously the sets of values these quantities can assume are necessarily infinite. In real life we deal mostly with finite sets. As a result, we are spared of many difficulties which we have to face in calculus. For example, when we talk of the tallest student in the class, we never worry whether such a student exists or not. There could be more than one such students, of course. But we can never imagine a situation where there is none. In the case of an infinite set, on the other hand, the largest element need not exist, even when the set is bounded. For example, the set $A = \{x \in I\!R : x < 2\}$ has no maximum element. It does have 2 as its supremum. So we have to carefully distinguish between a supremum and a maximum, a distinction which is confusing to a beginner, because for a finite subset of $I\!R$, these two quantities are always equal.

The ground on which the game of calculus is played is the real number system. It stands to reason that before the game starts, the ground should be made familiar to the players. That is, a construction of real numbers from rational numbers (familiarity with which may be taken for granted), should form the first topic in calculus. Some authors do follow this approach. But it turns out to be time-consuming and often too involved for a beginner. So an alternate approach is taken in which the real numbers are treated as primitive terms and their basic properties are taken as axioms. Although this practice is logically unassailable, its main disadvantage is that it does not allow adequate time for developing any intuitive understanding of real numbers. A beginner tries to extend his intuitive understanding of natural numbers to the real numbers. And this attempt is very prone to mistakes as noted above. In particular, the need for completeness is not appreciated.

Even a student who grasps the need for completeness of $I\!R$, is nagged by yet another difficulty. The completeness asserts the existence of suprema (for non-empty sets bounded above) and, as a consequence, implies the existence of certain limits (see e.g. Theorem (6.4)). But it does not tell you how to find the supremum or the limit whose existence is established. And frequently there is ineed no easy way to actually identify it. This is in sharp contrast with real life and also with the kind of mathematics one generally studies before calculus. In real life, if somebody makes a statement like "There exists a man who is eight feet tall" it is generally expected that he give some evidence, either by actually producing such a man, or if not, his photograph or at least his whereabouts from which he can be traced. Otherwise the statement will be taken as of little value. In pre-calculus mathematics, too, whenever the existence of something is asserted, the proof generally gives a construction for it also. For example, the statement "Given three non-colliniear points, there exists a circle passing through them" is proved by actually showing which circle has this property. Naturally a beginner expects the same thing in calculus. For example, in Section 2.6 we showed that, as a consequence of completeness of real numbers, the sequence $a_n = (1 + \dfrac{1}{10n})^n$ has a limit as $n \to \infty$. But unless we express the limit in a tangible form, its mere existence is of little appeal to him. (It is not hard to show that this limit is $e^{1/10}$. But that is really not an answer, because the number $e$ itself is often defined as a certain limit of a very similar form, specifically, $e = \lim_{n\to\infty} (1 + \frac{1}{n})^n$. So, saying that $\lim_{n\to\infty} (1 + \frac{1}{10n})^n = e^{1/10}$ is really an evasive answer. It is almost like saying, "My daughter is married to my son-in-law".) Methods are, of course, available to find the approximate value of the limit of $(1 + \frac{1}{10n})^n$. But that is not likely to satisfy everybody. "If we have to settle anyway for an approximate value of $\lim_{n\to\infty} (1 + \frac{1}{10n})^n$, we might as well get it simply by evaluating $(1 + \frac{1}{10n})^n$ for a large value of $n$, say for $n = 1,000$. What good is completeness then ? Indeed, what good is limit and what good is calculus ?" is a likely reaction.

Questions of this kind have been disturbing some mathematicians for quite some time. They have developed their own school of mathematics called **constructivist mathematics**. As the name suggests, this kind of mathematics lays a heavy emphasis on giving an explicit construction for something instead of merely proving its existence. But it is yet to develop sufficiently so as to be in a position to provide an acceptable substitute to the present day calculus. At present, the most satisfactory approach is to first prove the existence of a solution to a problem using theoretical methods, such as the completeness of $\mathbb{R}$ and to put off the question of calculating the value of the answer till much later. The reason for this kind of a postponement is that frequently there is no method to give the exact value of the answer. But methods are generally available to calculate it to a desired degree of accuracy. These methods vary considerably in terms of their efficiency. Generally, more efficient methods require more theoretical background. A branch of mathematics, called **numerical analysis** is devoted to devising such methods. We shall not go into the details here. But just to illustrate, let us recall that the exact value of $\lim_{n\to\infty} (1 + \frac{1}{n})^n$ is $e$. If we put $n = 1000$ and calculate, we get an accuracy of only upto the second place of decimals. But there are other limits which are equal to $e$. For example, it can be shown (as we shall do in Section 5.7) that as $n \to \infty$, the sum $1 + \frac{1}{1!} + \frac{1}{2!} + \ldots + \frac{1}{n!}$ tends to $e$. Here the denominator $n!$ is very large even for relatively small values of $n$. So even by summing only the first 10 terms of this series (which can be done even by hand, i.e., without a calculator), we get an approximate value which is correct upto the seventh place of decimals (see Exercise (7.9)). The price we have to pay is the rather considerable theoretical work we have to do to show that $e$ is the limit of the sum $1 + \frac{1}{1!} + \ldots + \frac{1}{n!}$ as $n \to \infty$.

Thus, there is a very good justification for drawing a line between proving the existence of something first and worrying about its calculation later. This is difficult for a beginner to swallow, because in the pre-calculus mathematics, things are usually the other way. The question of existence is rarely considered and even when it arises, it is settled by an explicit construction. It is the various formulas and identities one establishes to calculate the answer that make the pre-calculus mathematics so very exciting. Take for example, the area of a triangle. Nobody worries what area means and whether every plane figure has an area. It is just taken for granted that the area of a rectangle is the product of its sides. One then proves that the area of a triangle is $\frac{1}{2} \times$ base $\times$ height. (cf. Figure(2.1.1) (a)). Then one goes on manipulating, using numerous identities in trigonometry and finally gets the well-known formula for the area of a triangle in terms of its sides $a, b, c$ (viz. $\sqrt{s(s-a)(s-b)(s-c)}$ where $s = \frac{a+b+c}{2}$). It is only when some such beautiful formula (look at its symmetry) is proved that a student feels that he has done some 'real' mathematics.

Unfortunately, calculus in its initial phases contains no such beautiful formulas. In the old days when infinite series and integrals were manipulated without much heed to rigour, the picture was different. One such unscrupulous proof was given in Exercise (1.4.9). As another example, it can be 'proved' that
$$\sum_{n=0}^{\infty} \frac{1}{n!} = e = \lim_{n \to \infty} (1 + \frac{1}{n})^n$$ as will be indicated in Exercise (7.1). In fact, as the subsequent exercises will reveal, once we accept the unscrupulous derivation of the result $e = \sum_{n=0}^{\infty} \frac{1}{n!}$, then we can get, as corollaries, a number of interesting results about the number $e$. The derivation of these corollaries is, by itself, not faulty. Still, they are considered as unacceptable because they have a shaky foundation. They have to be put off till the result $e = \sum_{n=0}^{\infty} \frac{1}{n!}$ is established by a rigorous proof. And the modern standards of rigour do not allow the charming, care-free proofs which prevailed in the past. While the need for such rigour cannot be disputed (in view of the disastrous consequences of reckless manipulations), it has to be admitted that insistence on it does give a dry appearance to calculus, especially in its initial stages. The rigorous definitions of even such familiar functions as logarithms, sines and cosines can be given only after a lot of spadework involving integrals and power series.

Sometimes proofs in calculus are based on techniques rarely used in precalculus mathematics and this makes the proofs appear strange to a beginner even though they are perfectly simple. Suppose, for example, that we want to prove the equality of two numbers, or expressions, say, $A$ and $B$. The most familiar method would be to start with $A$ (or with $B$ ) and to reduce it to the other or to show that both $A$ and $B$ can be reduced to some third expression, say $C$. Sometimes some novel methods such as showing that some expression having $A - B$ as a factor vanishes (with the other factor non-zero) are useful. In calculus proofs, we encounter a very different technique. Instead of showing directly that $A$ equals $B$, we show (separately) that $A \leq B$ and $B \leq A$ which would force $A = B$. Even the proofs of these inequalities are sometimes not direct. For example, instead of showing that $A \leq B$, we instead show that for every positive real number $\epsilon$, $A \leq B + \epsilon$. This implies that $A \leq B$ as otherwise, $A - B$ would be positive and by taking $\epsilon$ to be any positive number less than $A - B$ we would get $B + \epsilon < A$, a contradiction. It takes a while to appreciate such types of proofs just as it takes some time to swallow the first encounter with a proof based on *reductio-ad-absurdum*. Sometimes the equality of $A$ and $B$ is proved by showing that their differnce $|A - B|$ can be made arbitrarily small; that is, for every $\epsilon > 0$, we show that $|A - B| < \epsilon$ which forces $|A - B| = 0$, i.e., $A = B$. (The proof of Theorem (6.1) is of this spirit, except that there it is done indirectly.)

As a result, even a student who is good at and loves the pre-calculus mathematics, often finds the first course in calculus uninspiring because problems in it are not amenable to any of the skills he has mastered till then.

Thus we see that the root cause of most of the difficulties experienced by a beginner lies in the gap between the pre-calculus mathematics (and more generally, the layman's understanding of the world) on the one hand and calculus on the other. Any attempt to eliminate or at least to mitigate these difficulties must be aimed at reducing this gap. Good instructors and good authors are generally aware of this. But you, too, can supplement their efforts. One way to do this is to look for the similarities between the two. Whenever possible, illustrate with real life situations. For example, the distinction between a discrete and a continuous variable can be illustrated by thinking of a discrete variable as a mango and a continuous variable as a canful of mango juice. Such analogies are obviously not to be stretched too far. Some of them may even sound ridiculous to a mature person. But if they succeed in helping you develop a feeling for the relevant concept in calculus, they have already served their purpose. For example, if it appears strange to you that many times what matters more is whether a sequence has a limit or not than the actual value of the limit, think of a sequence as a person and its limit as the spouse. In real life, so many times what matters more is if someone is married and not who is the spouse.

Making calculus more akin to real life is one way to bring them closer. You can also try making day-to-day life more akin to calculus. Look at a tree and appreciate that its height is not a constant but is a function of time and has a positive derivative (however small it may be). But the second derivative of this function is usually negative. Ask yourself why. To conquer the clumsy phraseology encountered in calculus, try to use such phraseology in day-do-day life, if only for fun. For example tell your friend that you will go to the film show if and only if you are free. If you have not read even one line of Shakespeare, boast (vacuously) that you remember, word to word, each and every line you have read of him.

Another healthy habit which pays immensely is to draw appropriate diagrams. Many students feel that diagrams need to be drawn only when some geometric figures such as triangles, circles, cubes etc. or some physical objects such as poles, wheels etc. are involved. But this is not so. Diagrams can be used to illustrate even abstract concepts. For most human beings, such diagrams have a far greater appeal and a far more lasting impression than mere words, symbols or expressions. (Even the newspapers often illustrate the outcomes of statistical surveys of some population with 'pie cut' diagrams in which a disc representing the population under survey is divided into various sectors whose areas are proportional to the percentages of the categories they represent.)

Space considerations prevent us from illustrating many concepts or proofs with diagrams. You can make this up by drawing your own diagrams. Thus, whenever it is said ' $a < c < b$' or '$c \in (a,b)$' take a few seconds out to show this as :

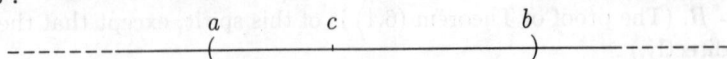

$$\text{-------}\underset{a}{(}\underline{\phantom{----}}\underset{c}{,}\underline{\phantom{--------}}\underset{b}{)}\text{-------}$$

on a piece of paper. An inequality of the form $|x-c| < \delta$ appears very frequently in calculus. Stated in this form, it does not carry much appeal. But interpret it geometrically by noting that $|x - c|$ is just the distance between $x$ and $c$. So the inequality says that $x$ lies in a $\delta$- neighbourhood of $c$, or equivalently, in the interval $(c - \delta, c + \delta)$, which can be pictured as

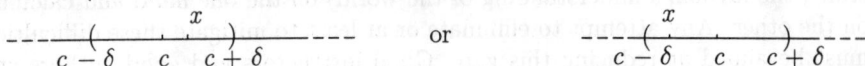

$$\text{---}\underset{c-\delta}{(}\underset{c}{\overset{x}{\phantom{--}}}\underset{c+\delta}{)}\text{---}\quad\text{or}\quad\text{---}\underset{c-\delta}{(}\underset{c}{\phantom{--}}\overset{x}{\underset{c+\delta}{)}}\text{---}$$

Similarly, whenever you encounter a sentence like '$x_n \to c$ as $n \to \infty$, where $\{x_n\}$ is an abstract sequence, take a piece of paper, draw a line, mark a point on it and label it $c$ and then mark off a few points to be labelled $x_1, x_2, \ldots x_n$ so as to indicate that $x_n$ approaches $c$ as $n$ tends to $\infty$. (Try to be as general as you can. For example, avoid taking all $x_n$'s on the same side of the point $c$ unless this is known to be the case.) You will wind up with a picture like this :

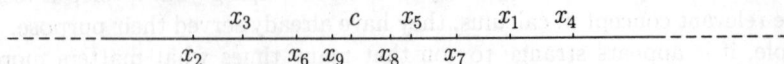

$$\underset{x_2}{\phantom{-----}}\overset{x_3}{\phantom{--}}\underset{x_6\ x_9}{\phantom{--}}\overset{c}{\phantom{--}}\underset{x_8}{\overset{x_5}{\phantom{--}}}\underset{x_7}{\phantom{--}}\overset{x_1\ x_4}{\phantom{--}}\text{--------}$$

It is not, however, so much this picture as the process of drawing it (and the fact that *you* have drawn it) that matters more. It is when the tip of your pen goes hopping towards the point marked $c$, that's when you really learn convergence!

Although in this book we shall mostly deal with sets which are subsets of the real line (and these too will be rather special, viz., intervals), in order to develop visual associations with (possibly abstract) sets, (and especially their subsets, their unions and intersections) it is a good idea to draw what are called **Venn diagrams**. In these diagrams, an ambient set is usually shown as a region in a plane[3] (often with an irregular shape, like that of an amoeba) and subsets of it are shown as subregions. Particular elements are shown by points. A function from one set to another is shown most vividly by an arrow from the domain set to the codomain set. When it is to be stressed that certain points or subsets of the domain set are mapped by the function to certain points or subsets of the codomain, the arrows are shown to originate at the appropriate place in the region which pictures the domain set and terminate at the appropriate place in the region representing the codomain set.

As an example, the Venn diagram in Figure 2.7.1 shows a function $f$ from a set $S$ to a set $T$ and another function $g$ from $T$ to some set $X$. The subsets

---

[3] A pie cut diagram mentioned above is also a Venn diagram. But it is used in a special context, the subsets in it usually being disjoint. Also it is drawn with greater precision than an ordinary Venn diagram in which the areas of the regions are not necessarily proportional to the sizes of the sets they represent.

$A$ and $B$ of $S$ are mutually disjoint. They contain points $x$ and $y$ respectively which are taken to the same point of $T$ under $f$. So the image sets $f(A)$ and $f(B)$ overlap. This shows very vividly that the function $f$ is not one-to-one. Also the range of $f$, i.e. the set $f(S)$, is shown as proper subset of $T$, thus making it clear without words that the function $f$ is not onto. The composite function $g \circ f$ is shown by a long arrow from $S$ to $X$. And finally, there is also a function $h$ from $X$ to itself. Such a function is shown by a circular arrow to stress that it has the same set for its domain and its codomain.

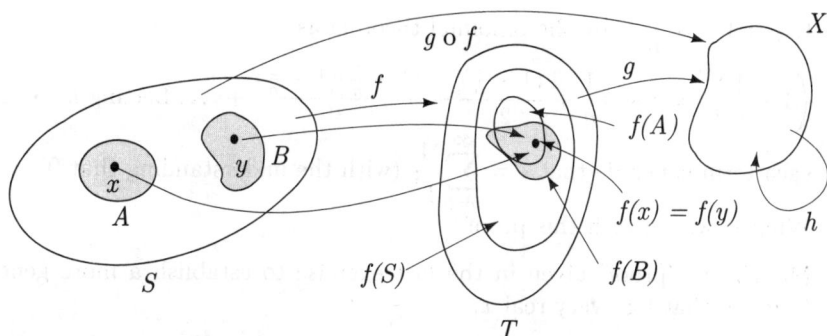

Figure 2.7.1 : Venn Diagram for Sets and Functions

Many concepts in mathematics such as tangents, slopes, concavity are geometric in origin. Modern mathematics takes pride in defining and studying such things exclusively in terms of numbers. The reason is partly that even though diagrams are a great help most of the time, they also often channelise your thinking unduly and thereby make you blind to some of the possibilities. For example, when we prove some geometric property of a triangle, we often draw only an acute angled triangle, even though sometimes a different argument is needed for an obtuse angled triangle. A proof based on co-ordinates, on the other hand, often applies equally well to all the cases. (A good illustration is the proof of concurrency of the three altitudes of a triangle, see Exercise (1.12.5).) A wrongly drawn diagram can, in fact, be misleading (see e.g. Exercise (4.8.11)). Therefore the crave for logical perfection shuns the use of diagrams as too intuition based. An undesirable side-effect is that the beginner finds the study very dry and unappealing. To make it lively, draw your own diagrams. They may not be a substitute for a formal proof. But they are definitely an invaluable help in understanding a proof and often an inspiration to come up with a proof. The time you spend in drawing them is never wasted.

Finally, have patience and faith in calculus. Calculus has evolved over a period of centuries and well stood the test of time. Some of its difficulties are easier to bear if you realise their inevitability and the underlying purpose. For example, clumsy sentences of the form "for every ......... there exists ..... such that ......" are necessary because without them a rigorous definition of a limit is impossible. Similarly, if the first course in calculus appears to contain only clumsy definitions, fussy proofs and contrived counter-examples but no interesting identities or formulas, then that is not because calculus lacks such formulas, but because a strong foundation is necessary to derive these formulas in a clean way. Unscrupulous short-cuts are sometimes possible but dangerous, like the medications given by a quack. A good physician, on the other hand, has to have undergone a thorough (and possibly boring) training in anatomy and physiology long before he starts treating patients.

## EXERCISES

7.1 Expand $(1 + \frac{1}{n})^n$ by the binomial theorem as

$$\left(1 + \frac{1}{n}\right)^n = 1 + n\frac{1}{n} + \frac{(1 - \frac{1}{n})}{2!} + \frac{(1 - \frac{1}{n})(1 - \frac{2}{n})}{3!} + \ldots. \text{ Letting } n \to \infty \text{ in}$$

each term conclude that $e = \sum_{n=0}^{\infty} \frac{1}{n!}$ (with the understanding that $0! = 1$).
What is wrong with this proof?

7.2 Modify the 'proof' given in the last exercise to establish a more general formula, that for every real $x$,

$$e^x = 1 + \frac{x}{1!} + \frac{x^2}{2!} + \ldots + \frac{x^n}{n!} + \ldots = \sum_{n=0}^{\infty} \frac{x^n}{n!}$$

7.3 For every positive integer $q$, show that $\sum_{n=q+1}^{\infty} \frac{1}{n!} < \frac{1}{q!q}$. [*Hint* : Consider the sum of the geometric progression

$$\frac{1}{q+1} + \frac{1}{(q+1)^2} + \frac{1}{(q+1)^3} + \ldots + \frac{1}{(q+1)^r}.$$

Show that this sum cannot exceed $\frac{1}{q}$, no matter how large $r$ is].

7.4 Using Exercises (7.1) and (7.3), show that the finite sum $\sum_{n=0}^{10} \frac{1}{n!}$ gives an approximate value of $e$ which is correct at least upto seven places of decimals.

7.5 Using the formula in Exercise (7.1), show that $e$ is irrational. [*Hint* : Suppose, if possible, that $e = \frac{p}{q}$ where $p, q$ are positive integers. Then $q!e$ is an integer. Now use Exercise (7.3) to get a contradiction.]

7.6 Show that $\lim_{n \to \infty} (1 + \frac{1}{10n})^n = e^{1/10}$ and hence is irrational even though $(1 + \frac{1}{10n})^n$ is rational for every positive integer $n$. (This fulfils a claim made in Section 2.6.)

# Chapter 3

# SEQUENCES AND SERIES

## 3.1 Why It Is Preferable to Deal with Limits of Sequences First than with Limits of Functions of a Real Variable

A sequence is, by definition, a function whose domain is $I\!N$, the set of all natural numbers. So sequences are special cases of functions and as elaborated in Section 2.5, the limits of sequences are a special case of restricted limits. For a variety of reasons, this special case has certain advantages both conceptual and practical. Throughout this section we shall denote the $n^{\text{th}}$ term of a sequence by symbols like $a_n, b_n$ etc.

Sequences arise more naturally from real life situations than functions of a (continuous) real variable. Take for example, the sequence $a_n = (1 + \frac{1}{10n})^n$. As we saw in Section 2.1, $a_n$ is the total that would accrue at the end of one year with a principal of 1 rupee, if the interest rate is 10 p.c.p.a. and is compounded $n$ times annually (at equally spaced rests). We can write $a_n$ as $f(n)$ where $f(x) = (1 + \frac{1}{10x})^x$. Although the expression $(1 + \frac{1}{10x})^x$ is mathematically well-defined for any positive real value of $x$, it will lose its real-life interpretation if $x$ is not a whole number. Surely you cannot compound an interest a non-integral number of items ! Whenever the variable $n$ denotes the number of times a certain process is carried out, it can only take integral lvalues. As another example, consider the situation in Exercise (1.12.8). Let $a_n$ be the percentage of water in the first glass after the exchange of liquids is carried out $n$ times. Here, even though the formula for $a_n$ may make sense even when $n$ is not an integer, it will have no meaning in terms of exchange of liquids.

The sequence $\{a_n\}$ in this example, is an example of a **recursively defined sequence.** In such a sequence $a_n$ is not given directly as a function of $n$. Instead, $a_n$ is expressed in terms of $a_{n-1}$, or sometimes, in terms of some other lower terms of the sequence. Any such relation is known as a **recurrence relation.** In the present problem, the recurrence relation (which is obtained by calculating how much water passes from one glass to the other and backward during the $n^{\text{th}}$ exchange) is

$$a_n \; = \; \frac{100}{21} + \frac{19}{21}a_{n-1} \tag{1}$$

Here the first term $a_1$ has to be calculated directly as $\dfrac{100}{21}$ (or we can apply (1) with $a_0 = 0$. In the definition of a sequence, it is often convenient to allow 0 in the domain, as in this case). We then use (1) again and again to get, successively, $a_2, a_3, a_4, \ldots$ and so on. This way we can find $a_n$ for any given $n$. But what we would like to have is an expression for $a_n$ from which we can get the value of $n$, just by plugging a given value of $n$ directly, i.e., without having to compute $a_{n-1}$ first. Such an expression is called a **closed form expression** for $a_n$ and obtaining it is known as **solving** the given **recurrence relation**. There is a systematic theory of solving recurrence relations. We shall not go into it. Still (1) can be solved by a somewhat *ad-hoc* method. We write down equations similar to (1) for $a_{n-1}, a_{n-2}, \ldots$, and finally for $a_1$. That is,

$$\left. \begin{aligned} a_n &= \frac{100}{21} + \frac{19}{21}a_{n-1} \\ a_{n-1} &= \frac{100}{21} + \frac{19}{21}a_{n-2} \\ &\;\;\vdots \\ a_2 &= \frac{200}{21} + \frac{19}{21}a_1 \\ a_1 &= \frac{100}{21} \end{aligned} \right\} \tag{2}$$

If we now multiply the first equation by 1, the second by $\frac{19}{21}$, the third by $\left(\frac{19}{21}\right)^2, \ldots$, and the last one by $\left(\frac{19}{21}\right)^{n-1}$ and add, we see that all the terms except $a_n$ on the left hand side cancel, giving

$$a_n = \frac{100}{21} + \frac{100}{21} \cdot \frac{19}{21} + \frac{100}{21} \cdot \left(\frac{19}{21}\right)^2 + \ldots + \frac{100}{21} \cdot \left(\frac{19}{21}\right)^{n-1}$$

which, using the formula for the sum of a geometric progression and a little simplification, gives

$$a_n \; = \; 50 \left(1 - \left(\frac{19}{21}\right)^n\right) \tag{3}$$

Let us now find $\lim\limits_{n \to \infty} a_n$. Intuitively, it is obvious that this limit should be 50, because as the exchange of the liquids taken from the mixtures goes on, we expect that both the mixtures will become equally watery (or equally milky). In Section 3.3 we shall prove that $(\frac{19}{21})^n \to 0$ as $n \to \infty$ and so we shall indeed get that $a_n \to 50$ as $n \to \infty$.

What if we were unable to solve (1) in the manner above ? Then we could have calculated $a_1, a_2, a_3, a_4$ by hand and probably we would be able to *guess* the formula (3). Of course, mere guess is no proof. But then we could *prove* (3) by induction on $n$. This is very easy to do and we leave it as an exercise. The point we want to make is something else. It is possible to find $\lim\limits_{n\to\infty} a_n$ directly from (1) even *without* solving it, i.e., without obtaining (3). We must first prove, however, that $\lim\limits_{n\to\infty} a_n$ exists. For this we apply Theorem (2.6.4).

It is intuitively obvious that the sequence $\{a_n\}$ is monotonically increasing, because it is clear that the mixture in the first glass (which had 100 c.c. of pure milk to begin with) will get more and more watery with each exchange of liquids. To prove mathematically, that $a_n \geq a_{n-1}$ from (1), is equivalent to showing that $a_{n-1} \leq 50$. This latter inequality can be easily shown by induction on $n$, again using (1). So the sequence $\{a_n\}$ is monotonically increasing. Also, it is bounded above since $a_n \leq 50$ for all $n$. So by completeness of the real line (in the form of Theorem (2.6.4)), $\lim\limits_{n\to\infty} a_n$ exists. Call it $L$.

We now take the limits of both the sides as $n \to \infty$. As $n \to \infty, a_n \to L$ and hence $a_{n-1}$ also tends to $L$. So (1) gives

$$L = \frac{100}{21} + \frac{19}{21}L$$

solving which we get $L = 50$, as expected.

Thus we were able to get the limit of a sequence from a recurrence relation for it, even without having an explicit closed form expression for its $n^{\text{th}}$ term. For functions of a real variable this method does not work very often. The analogue of recurrence relations for such functions is what is called a **functional equation**, i.e., an equation which relates the value of a function at a point with its values at some other points. For example, $\sin(\pi + x) = -\sin x$ is a functional equation for the sine function. But such functional equations are not very common for functions of a real variable. For sequences, that is, for functions of an integer variable $n$, on the other hand, recurrence relations are quite common. As a result, methods such as induction come in very handy as we saw in the example above. (The sequence of Exercise (2.6) is also recursively defined. Probably the most historically famous recursively defined sequence is the **Fibonacci sequence** which will be given in the exercises.)

Apart from their natural occurence, limits of the form $\lim\limits_{n\to\infty} a_n$ also have a pedagoical advantage over limits of the form $\lim\limits_{x\to c} f(x)$, where $c$ is a (finite) real number. As mentioned in Section 2.3, the latter limit often equals $f(c)$, even though conceptually they are quite different. Even after the distinction is stressed, a beginner continues to try to obtain $\lim\limits_{x\to c} f(x)$ by simply plugging $c$ into the formula for $f(x)$. This temptation is automatically precluded for limits of the form $\lim\limits_{n\to\infty} a_n$. The nature of the sequence $a_n$ is often such that it is obviously meaningless to put $n = \infty$ into the formula for $a_n$, even when such a formula is available.

The limit of a sequence is a case of a restricted limit. That is, if $a_n = f(n)$ and if the function $f$ is such that $f(x)$ makes sense for all (sufficiently large) real values of $x$, then $\lim_{n \to \infty} a_n$ is the restricted limit $\lim_{\substack{x \to \infty \\ x \in \mathbb{N}}} f(x)$. With a change of the variable, we can write $\lim_{x \to \infty} f(x)$ as $\lim_{y \to 0+} g(y)$ where $y = \dfrac{1}{x}$ and $g(y) = f(\frac{1}{y})$. Now if we let $S = \{\frac{1}{n} : n \in \mathbb{N}\}$ (i.e., $S$ is the set of numbers of the form $1, \frac{1}{2}, \frac{1}{3}, \frac{1}{4}, \ldots, \frac{1}{n}, \ldots)$, then $\lim_{n \to \infty} a_n$ is the restricted limit $\lim_{\substack{y \to 0 \\ y \in S}} g(y)$. More generally, for any real number $c$, we can view the limit of a sequence $a_n = f(n)$ as a special case of a restricted limit of the form $\lim_{y \to c} h(y)$ where we let $y$ approach $c$ through the values $c + 1, c + \frac{1}{2}, c + \frac{1}{3}, \ldots$, and the function $h(y)$ is defined as $f(\dfrac{1}{x - c})$.

It might therefore appear at the first sight that the limits of sequences are conceptually less general than limits of the form $\lim_{y \to c} h(y)$. Here the sequence $c + 1, c + \frac{1}{2}, \ldots, c + \frac{1}{n}, \ldots$ converges to $c$ and the image sequence $h(c + 1)$, $h(c + \frac{1}{2}), \ldots, h(c + \frac{1}{n}), \ldots$ is precisely the original sequence $a_1, a_2, \ldots, a_n, \ldots$. Suppose, however, that we have another sequence say $\{b_n\}$ converging to $c$ (not necessarily from the right). Then we may consider the image sequence $\{h(b_n)\}$ and its limits. The following theorem establishes an important relationship between the limits of such sequences and the limit of the function $h(y)$ as $y \to c$. (As recommended in Section 2.7, the reader is urged to illustrate the proof with his own diagram.)

**1.1 Theorem :** Let $h(y)$ be a function which is defined in some deleted neighbourhood $U$ of a point $c$. Then $\lim_{y \to c} h(y) = L$ if and only if for every sequence $\{b_n\}$ in $U$ (i.e, a sequence $b_n$ such that $b_n \in U$ for all $n$), which converges to $c$, the image sequence $h(b_n)$ converges to $L$.

**Proof:** First assume that $\lim_{y \to c} h(y) = L$. Let $\{b_n\}$ be a sequence in $U$ such that $\lim_{n \to \infty} b_n = c$. We have to show that $\lim_{n \to \infty} h(b_n) = L$. We do this in the most straightforward manner. Let $\epsilon > 0$ be given. Since $\lim_{y \to c} h(y) = c$, there exists some $\delta > 0$ such that, $|h(y) - L| < \epsilon$ whenever $0 < |y - c| < \delta$. (We assume $\delta$ is so small that the deleted neighbourhood $(c - \delta, c + \delta) - \{c\}$ is contained in the given deleted neigbhourhood $U$ of $c$.) Now, since $\lim_{n \to \infty} b_n = c$, and $\delta > 0$, there exists some positive integer $m$ such that for all $n \geq m, |b_n - c| < \delta$. Further, $b_n \neq c$ since $b_n$ is given to lie in $U$, which, being a deleted neighbourhood of $c$ does not contain the point $c$. So for all $n \geq m$, we have $0 < |b_n - c| < \delta$. But then, we have for all $n \geq m, |h(b_n) - b| < \epsilon$. Thus we have shown that $\lim_{n \to \infty} h(b_n) = L$, i.e., $h(b_n)$ converges to $L$.

The converse implication is not so straightforward. We are given that for every sequence $\{b_n\}$ in $U$ which converges to $c$, the sequence $\{h(b_n)\}$ converges to $L$. To apply this hpothesis, we should choose the sequence $\{b_n\}$ judiciously. But there is no way to tell *a priori* which choice will give us the result, viz., $\lim_{y \to c} h(y) = L$. So we proceed by the method of *reductio-ad-absurdum*. That is, we assume that $\lim_{y \to c} h(y) = L$ is false. Then its negation is true. From this we shall construct a sequence $\{b_n\}$ in $U$ converging to $c$ for which $\{h(b_n)\}$ does not converge to $L$. In order to do this, we need the negation in the positive form. We already obtained it in Section 1.11, except that the variable there was $x$ and the function was $f$. Replacing them with $y$ and $h$ respectively, we have, because of our assumption, that

"There exists $\epsilon > 0$ such that for every $\delta > 0$, there exists $y$ such that

$$0 < |y - c| < \delta \text{ and } |h(y) - L| \geq \epsilon \text{ "} \tag{4}$$

Note that here the choice of $\delta$ is ours. We first choose a $\delta = \delta_1$ so small that the detered neighbourhood $(c - \delta, c + \delta) - \{c\}$ is contained in $U$. Now by (4), there exists some $y$, which we call $b_1$ such that $0 < |b_1 - c| < \delta_1$ and $|h(b_1) - L| \geq \epsilon$. Next, we put $\delta = \dfrac{\delta_1}{2}$ and apply (4) again to get $b_2$ (which will, in general, be different from $b_1$ but may be equal to it by coincidence, it really does not matter) such that $0 < |b_2 - c| < \dfrac{\delta_1}{2}$ and $|h(b_2) = L| \geq \epsilon$.

We now go on taking $\delta = \dfrac{\delta_1}{n}$, for $n = 3, 4, 5, \ldots$ and everytime because of (4), we get some $b_n$ such that

$$0 < |b_n - c| < \frac{\delta_1}{n} \quad \text{and} \quad |h(b_n) - L| \geq \epsilon \quad \text{for all} \quad n. \tag{5}$$

Now let $n \to \infty$. Then $\dfrac{\delta_1}{n} \to 0$ since $\delta_1$ is a fixed real number. Here we are using the fact that $\dfrac{1}{n} \to 0$ as $n \to \infty$. This is generally taken too obvious. But it needs a proof and it will be given in Section 3.3. So, from the first part of (5), we get that the sequence $\{b_n\}$ in $U$ converges to $c$. But, by the second part, the image sequence $\{h(b_n)\}$ does not converge to $L$, contradicting the hypothesis. So $h(y)$ must tend to $L$ as $y$ tends to $c$. ∎

This theorem is of limited practical use. When in any particular situation, it comes to showing that $\lim_{y \to c} h(y) = L$, it turns out that whether this is done directly or by showing that $\lim_{n \to \infty} h(y_n) = L$ whenever $y_n \to c$, essentially the same amount of work is involved. But the theoretical significance of this theorem is profound. It shows that limits of sequences are no less general than limits of functions of a real variable. If one wants, one can completely dispense with limits of the form $\lim_{x \to c} f(x)$ and work exclusively with limits of various sequences.

Another important theoretical application of limits of sequences is that they provide a satisfactory means of handling infinite series. Finite sums of real numbers are probably the most ubiquitous mathematical concept in real life, ranging from a grocery shop to a nuclear laboratory. In forming such sums the order in which the terms are added makes no difference mathematically. For example, the sum $a_1 + a_2 + a_3$ also equals $a_2 + a_1 + a_3, a_1 + a_3 + a_2, a_3 + a_2 + a_1, a_2 + a_3 + a_1$ and $a_3 + a_1 + a_2$. In general, the sum of $n$ terms can be expressed in any of $n!$ different ways, one corresponding to each permutation of the terms. (For particular figures, there may be some practical advantage in adding them in a particular order. For example, instead of adding 437, 198 and 2 in the given order, it is better to add 198 and 2 first and then quickly get 637 as the answer. Those who can add rapidly often resort to such short-cuts).

Possibly as a result of man's never-ending greed and pursuit of perfection the idea of adding infinitely many terms has fascinated him for a long time. It is inherent in the decimal expansion of a real number. For example when a real number $x$ is written as, say,

$$b_3 b_2 b_1 \cdot a_1 a_2 a_3 \ldots a_n \ldots$$

what it really means it that

$$x = 100 b_3 + 10 b_2 + b_1 + \frac{a_1}{10} + \frac{a_2}{10^2} + \frac{a_3}{10^3} + \ldots + \frac{a_n}{10^n} + \ldots$$

The natural question is what meaning to assign to an infinite sum. The trouble arises because unlike a finite sum, the behaviour of an infinite sum may depend crucially on the order in which the terms are summed. For example consider the sum

$$1 - \frac{1}{2} + \frac{1}{3} - \frac{1}{4} + \frac{1}{5} - \frac{1}{6} + \ldots$$

If we evaluate this sum by adding one term at a time, then, as the grouping

$$\left(1 - \frac{1}{2}\right) + \left(\frac{1}{3} - \frac{1}{4}\right) + \left(\frac{1}{5} - \frac{1}{6}\right) + \ldots \tag{6}$$

shows, its value will be positive. But suppose we rearrange the terms as

$$-\frac{1}{2} - \frac{1}{4} - \frac{1}{6} - \frac{1}{8} + 1 - \frac{1}{10} - \frac{1}{12} - \frac{1}{14} - \frac{1}{16} - \frac{1}{18}$$
$$+ \frac{1}{3} - \frac{1}{20} - \frac{1}{22} - \frac{1}{24} - \frac{1}{26} - \frac{1}{28} + \frac{1}{5} - \frac{1}{30} \tag{7}$$

where, before each positive term we insert a block of sufficiently many negative terms, then by adding one term at a time the sum will never be positive. It follows that we cannot give any definite meaning to the sum of an infinite number of terms, without first ordering them in some definite manner. Any such ordering corresponds to a sequence of terms. So suppose we have an infinite sequence say $\{a_n\}$ of terms. We want to define $a_1 + a_2 + \ldots + a_n + \ldots$ or $\sum_{n=1}^{\infty} a_n$ for short. (Note, incidentally, that in the case of a finite summation of the form, say, $b_1 + b_2 + \ldots + b_m$ in the sigma notation $\sum_{n=1}^{m} b_n$ the upper index, viz. $m$, denotes the last value which the index variable $n$ assumes. In the case of an infinite sum $\sum_{n=1}^{\infty} a_n$, the variable $n$ never actually equals $\infty$. Perhaps $\sum_{1 \leq n < \infty}$ would be a better notation. But the notation $\sum_{n=1}^{\infty} a_n$ is quite standard[1].)

We go on adding one term at a time. Let $S_n$ denote the sum of the first $n$ terms, i.e., $S_n = a_1 + \ldots + a_n$ or $S_n = \sum_{k=1}^{n} a_k$. This sum is called the $n^{\text{th}}$ **partial sum** of the series $\sum_{n=1}^{\infty} a_n$. (Note by the way that in the notation $\sum_{n=1}^{\infty} a_n, n$ is just a dummy variable of summation. Do not confuse it with $S_n = \sum_{k=1}^{n} a_k$, where $n$ is not a dummy variable but $k$ is a dummy variable. If necessary, you may denote $\sum_{n=1}^{\infty} a_n$ by $\sum_{r=1}^{\infty} a_r$ or by $\sum_{k=1}^{\infty} a_k$ etc. It really does not matter, because the scope of a dummy variable of summation extends only to that summation. No harm arises in using it again to denote something else.)

Thus from the original sequence $\{a_n\}$ we get a new sequence of its partial sums, viz. $\{S_n\}$ or $S_1, S_2, S_3, \ldots, S_n, \ldots$ If we compare (6) and (7) we see that even though we are summing the same terms, the sequences $\{S_n\}$ that we get are entirely different. For (6), $S_1 = 1, S_2 = \frac{1}{2}, S_3 = \frac{5}{6}, S_4 = \frac{7}{12}, S_5 = \frac{47}{60}, \ldots$, while for (7), $S_1 = -\frac{1}{2}, S_2 = -\frac{3}{4}, S_3 = -\frac{11}{12}, S_4 = -\frac{25}{24}, S_5 = -\frac{1}{24}, \ldots$

We *define* $\sum_{n=1}^{\infty} a_n$ to be **convergent** if the sequence $\{S_n\}$ is convergent, i.e., if $\lim_{n \to \infty} S_n$ exists as a (finite) real number, say $L$. We then write $\sum_{n=1}^{\infty} a_n = L$ and say $L$ is the **sum** of $\sum_{n=1}^{\infty} a_n$. (Thus it will be seen that the same notation $\sum_{n=1}^{\infty} a_n$ is used for an infinite series and for its sum, when it exists. Sometimes we use a different symbol for the sum to avoid confusion.)

---

[1] As with the notation for a limit, in running text, it is common to denote $\sum_{n=1}^{\infty}$ by $\sum_{n=1}^{\infty}$, i.e. to write the expression '$n = 1$' as a subscript rather than directly below the summation sign $\Sigma$ and similarly to write $\infty$ as a superscript. Finte sums are also denoted more compactly as $\sum_{k=1}^{k=n}$ etc.

If the sequence $\{S_n\}$ is divergent (i.e., not convergent) then we say $\sum\limits_{n=1}^{\infty} a_n$ is divergent. It is now clear why the sums (6) and (7) are totally different. Later on we shall see that for series of positive terms such anomalies do not arise, i.e., any rearrangement of terms will give the same sum.

In the modern treatment of calculus, infinite series (and sequences) are somewhat de-emphasised in the very first course. In the classical approach, however, they played a central role. In fact many functions can be defined as certain series. For example, in old books it is common to find $e^x$ *defined* by

$$e^x = 1 + \frac{x}{1} + \frac{x^2}{21} + \frac{x^3}{31} + \cdots + \frac{x^n}{n!} + \ldots = \sum_{n=0}^{\infty} \frac{x^n}{n!} \tag{8}$$

and then the logarithm $\ln x$ is defined as the inverse function of this. Today, it is more common to define $\ln x$ by a certain integral and then to define $e^x$ as its inverse function (see Section 5.7). Still, infinite series retain their appeal. A few interesting problems involving them will appear as exercises in subsequent sections.

Summing up, it is preferable to study the limits of sequences first because they are more natural, easier to work with, less prone to mistakes, and conceptually as general as the other kinds of limits. Moreover, they are the gateway to infinite sums.

## EXERCISES

1.1   Vary Exercise (1.12.8) by assuming that the two glasses have unequal volumes of milk and water to begin with (say 100 cc. and 200 cc. respectively). Assume, however, that the liquid transferred everytime is the same viz. 5 cc. What are the analogues of (1) to (3)? What is $\lim\limits_{n\to\infty} a_n$?

1.2   Prove (3) by induction on $n$. Also using (1) and induction on $n$, prove that $a_n \leq 50$ for all $n$ and hence that the sequence $\{a_n\}$ is monotonically increasing.

1.3   Let $\{a_n\}$ be the sequence in Exercise (2.6.7), defined recursively by $a_n = \dfrac{p_n}{q_n}$ where $p_1 = 1 = q_1$ and for $n > 1, p_n = p_{n-1} + 2q_{n-1}$ and $q_n = p_{n-1} + q_{n-1}$. By induction on $n$ show that $p_n^2 - 2q_n^2 = (-1)^n$ for all $n$. Hence show that $a_n \to \sqrt{2}$ an $n \to \infty$.

1.4   The **Fibonacci sequence** $\{F_n\}$ is defined by $F_1 = 1, F_2 = 1$ and for $n > 2, F_n = F_{n-1} + F_{n-2}$. By induction on $n$, show that for all $n$,

$$F_n = \frac{1}{\sqrt{5}} \left[ \left( \frac{1 + \sqrt{5}}{2} \right)^n - \left( \frac{1 - \sqrt{5}}{2} \right)^n \right] \tag{9}$$

(This is obviously impossible to guess. But there are ways to arrive at this formula. One such method will be given as an application of power series, see Section 3.10.) The first few terms of the Fibonacci sequence, viz. $1, 1, 3, 5, 8, 13, 21, \ldots$ have a natural significance as the sizes of a population of certain organisms which produce one organism each every day after attaining maturity at the age of two days. Even more significant, if only historically, is the limit of the ratio $\dfrac{F_n}{F_{n-1}}$ as $n \to \infty$. This limit is popularly called the **golden ratio.** We shall calculate it later, Exercises (3.4) to (3.6).)

1.5   Prove theorem (1.1) when $c = \infty$ or $-\infty$.

1.6   In (7), show that from $\frac{1}{3}$ onwards, every positive term, except 1, is preceeded by exactly 5 negative terms. Specifically, show that for $n \geq 1$, $\dfrac{1}{2n+1}$ is preceeded by $-\dfrac{1}{10n} - \dfrac{1}{10n+2} - \dfrac{1}{10n+4} - \dfrac{1}{10n+6} - \dfrac{1}{10n+8}$, by showing that the (numerical) value of the sum of these 5 terms exceeds $\dfrac{1}{2n+1}$.

1.7   Show that neither of the series $\displaystyle\sum_{n=1}^{\infty} n, \sum_{n=1}^{\infty} (-1)^n$ is convergent but that the series $\displaystyle\sum_{n=1}^{\infty} \dfrac{1}{n(n+1)}$ is convergent.

1.8   If a series $\displaystyle\sum_{n=1}^{\infty} a_n$ is convergent, prove that $a_n \to 0$ as $n \to \infty$. (This gives a necessary condition for convergence of a series. But it is not sufficient as we shall see later).

1.9   Suppose the sequences $\{a_n\}, \{b_n\}$ convergence to $L, M$ respectively. If $a_n \leq b_n$ for all $n$, prove that $L \leq M$. Show by an example that even if we have strict inequality, i.e, $a_n < b_n$ for all $n$, we do not in general have $L < M$. [*Hint* : For the first part, proceed by contradiction, taking $\epsilon = (L - M)/2$ if $L > M$. The result may be expressed verbally by saying that under convergence, inequalities are preserved, but their strictness may be lost].

1.10  Can you deduce the Sandwich Theorem for sequences from the last exercise? How would you give a correct proof ?

### 3.2    Importance of Eventuality for Sequences and Series

Let us first see what eventuality means. In real life, 'eventually' means 'throughout, after some point of time'. For example, 'Eventually, he settled in India' means that the person talked about might have made many trips abroad, might have even thought of settling abroad or might have tried to settle in India but then abandoned the idea. But after all these wanderings, he finally settled in India and remained there till the end of his life.

When used in connection with sequences, eventuality has little to do with time. Still, the idea of passage of time is so inherent in the concept of a variable $n$ approaching $\infty$, that expressions like 'after $n$ is bigger than 1000' are used colloquially to mean 'for $n > 1000$'. Similarly, 'eventually or after some stage' means 'for all $n$ bigger than some value'.

It often happens that the terms of a sequence $\{a_n\}$ satisfy some condition not for every $n$, but for all $n$ after some stage. In such a case we say that the condition is satisfied eventually. For example, the sequence $a_n = \dfrac{n^3}{2^n}$ has terms $\dfrac{1}{2}, 2, \dfrac{27}{8}, 4, \dfrac{125}{32}, \dfrac{27}{8}, \dfrac{343}{128}, \ldots$. It is not monotonically decreasing. But after the fourth term, each term is less than its preceding term. So we say this sequence is **eventually monotonically decreasing**. Similarly two sequences $\{a_n\}$ and $\{b_n\}$ are said to be **eventually equal** if there is a positive integer $m$ such that for all $n \geq m, a_n = b_n$. Note that here we are not asserting that $m$ is the smallest integer with this property. All we are saying is what happens if $n \geq m$. We are saying nothing at all for $n < m$.

There is another interesting way to express eventuality. Suppose $a_n = b_n$ eventually. Then there exists some $m$ such that for all $n \geq m, a_n = b_n$. It follows that there are *at most* $m - 1$ values of $n$ (viz. $1, 2, \ldots, m - 1$) for which $a_n \neq b_n$. Conversely, if $a_n = b_n$ for all $n$ except for $n = n_1, n = n_2, \ldots, n = n_k$ (say) then we let $m = \max\{n_1, \ldots, n_k\} + 1$. Then we have $a_n = b_n$ for all $n \geq m$, i.e., $\{a_n\}$ and $\{b_n\}$ are eventually equal. So instead of saying that some condition holds eventually, we also say it holds for all except finitely many values of $n$. This expression is also rendered by saying that it holds for 'almost all' $n$. In Sections 1.2 and 1.11, we remarked that in mathematics there is no such thing as a statement being almost true. But here 'almost all' is used in a very precise mathematical sense, viz., 'with the possible expression of finitely many' and so there is nothing wrong in using it. It is in fact used commonly. Thus to say that a sequence $\{a_n\}$ is eventually monotonically increasing is equivalent to saying that $a_n \leq a_{n+1}$ for almost all $n$.

The importance of eventuality for sequences comes from the fact that as far as the convergence of a sequence is concerned, only its eventual behaviour matters. It does not matter at all how its first few terms behave, as long as the number of such 'bad' terms is finite. A precise version can be given as follows :

**2.1 Theorem:** Suppose two sequences $\{a_n\}$ and $\{b_n\}$ are eventually equal. Then if $\{a_n\}$ converges to $L$ so does $\{b_n\}$, and vice versa.

**Proof:** Let $\epsilon > 0$ be given. Then there exists $m_1$ such that for all $n \geq m_1$, $|a_n - L| < \epsilon$. Also by hypothesis, there exists $m_2$ such that for all $n \geq m_2, a_n = b_n$. It follows that if we let $m = \max\{m_1, m_2\}$, then for all $n \geq m, |b_n - L| < \epsilon$. So $b_n \to L$ as $n \to \infty$. Similarly if $b_n \to L$, then $a_m \to L$. ∎

Although this theorem is hardly profound, it enables us to extend almost every result about sequences. For example, Theorem (2.6.4) tells us that every monotonically increasing sequence which is bounded above is convergent. Now suppose $\{a_n\}$ is a sequence which is eventually monotonically increasing and bounded above. Then we cannot apply that theorem directly. But we construct another sequence as follows. We are given that there exists some $m$ such that for all $n \geq m, a_n \leq a_{n+1}$. We let $b_1 = b_2 = \ldots = b_m = a_m$ and for $n > m$, we let $b_n = a_n$. This new sequence $\{b_n\}$ is monotonically increasing. Also any upper bound for $\{a_n\}$ is also an upper bound for $\{b_n\}$. So Theorem (2.6.4) shows that $\{b_n\}$ is convergent. The theorem above now implies that $\{a_n\}$ is also convergent.

We leave it as an exercise to similarly extend other results about sequences (e.g. the Sandwich Theorem), where the hypothesis involves some condition on the $n^{th}$ term of a sequence. In fact such extensions are so routine to obtain that such theorems are often applied directly in situations where the condition on the $n^{th}$ term it satisfied only eventually. Another possible extension of theorems about sequences is, although trivial, sometimes useful. We defined a sequence as a function from the set $I\!N = \{1, 2, 3, \ldots\}$. It is often convenient to let the domain of a sequence be a set of the form $\{m, m+1, m+2, \ldots\}$ where $m$ is an integer (positive, negative or 0). In many series for example, it is convenient to start with the "$0^{\text{th}}$ term" instead of the "first" term. On the other hand if $a_n$ is a sequence of the form $\dfrac{n(n+3)}{(n-1)(n-3)}$, then $a_1$ and $a_3$ are not defined. So it is better to define this sequence only on $\{4, 5, 6, \ldots\}$. All theorems about sequences defined on $I\!N$ generalise without any difficulty to sequences defined on sets of the form $\{m, m+1, m+2, \ldots\}$. (Note that here $m$ can be a negative integer too. But we are not allowing the domain to be the entire set $Z\!\!\!Z$ of all integers. If we do, then convergence will need a substantially different treatment. Series of the form $\displaystyle\sum_{n=-\infty}^{\infty} a_n$ are indeed encountered sometimes in mathematics. But we shall not deal with them).

Let us now see the analogue of Theorem (2.1) to infinite series.

**2.2 Theorem :** Suppose $\{a_n\}$ and $\{b_n\}$ are eventually equal. Then the series $\displaystyle\sum_{n=1}^{\infty} a_n$ is convergent if and only if the series $\displaystyle\sum_{n=1}^{\infty} b_n$ is convergent. Their sums may be different. But if $\{b_n\}$ is obtained from $\{a_n\}$ by re-arranging a finite number of terms, then $\displaystyle\sum_{n=1}^{\infty} a_n = \sum_{n=1}^{\infty} b_n$.

**Proof:** Let $S_n$ and $T_n$ denote, respectively, the partial sums of the two series.

That is,

$$S_n = a_1 + a_2 + \ldots + a_n \quad \text{and} \quad T_n = b_1 + b_2 + \ldots + b_n.$$

We are given that there exists some $m$ such that $a_n = b_n$ for all $n \geq m$. It is clear then that for $n \geq m, T_n = S_n + D$ where $D = (b_1 - a_1) + (b_2 - a_2) + \ldots + (b_m - a_m)$. (In fact the last term is 0). By definition, $\sum_{n=1}^{\infty} a_n$ is convergent if the sequence $\{S_n\}$ is convergent. But then so is the sequence $\{S_n + D\}$. In fact we have

$$\sum_{n=1}^{\infty} b_n = \lim_{n \to \infty} T_n = \lim_{n \to \infty} (S_n + D) = \left( \lim_{n \to \infty} S_n \right) + D = \left( \sum_{n=1}^{\infty} a_n \right) + D.$$

So convergence of $\sum_{n=1}^{\infty} a_n$ implies that of $\sum_{n=1}^{\infty} b_n$ and vice versa. Since $D$ need not be 0, the sums may be different. But if $\{b_n\}$ is obtained merely by re-arranging the first $m - 1$ (say) terms of $\{a_n\}$, then $b_1, \ldots b_{m-1}$ are the same terms as $a_1, \ldots, a_{m-1}$, possibly in a different order and so

$$
\begin{aligned}
D &= (b_1 - a_1) + \cdots + (b_{m-1} - a_{m-1}) \\
&= b_1 + \cdots + b_{m-1} - (a_1 + a_2 + \ldots + a_{m-1}) \\
&= 0.
\end{aligned}
$$

and so $\sum_{n=1}^{\infty} a_n = \sum_{n=1}^{\infty} b_n.$ ∎

As with Theorem (2.1), Theorem (2.2) is more of a conveninet device than a profound fact. Later on we shall study several tests for convergece of series. Typically such tests assert that if the terms of a series satisfy a certain condition then the series is convergent. Theorem (2.2) allows us to apply such tests even for those series in which 'almost all' terms satisfy these conditions. For example, many tests are applicable only for series of non-negative terms. Because of Theorem (2.2), they are applicable even if a finite number of terms of the series is negative.

The difference between the two theorems deserves to be stressed. Suppose $\{a_n\}$ and $\{b_n\}$ are eventually equal. Then by Theorem (2.1), not only do these two sequences have the same *character* (i.e., either both are convergent or both divergent) but they also have equal limits (when they exist). Theorem (2.2) on the other hand says only that the series $\sum_{n=1}^{\infty} a_n$ and $\sum_{n=1}^{\infty} b_n$ have the same character, i.e., one of them is convergent if and only if the other is. It does not say that their sums are equal. This might seem to lessen the importance of Theorem (2.2), if one takes the view that it is useless to know merely that a series is convergent whithout knowing what its value is. It turns out, however, that the question of convergence of a given series arises far more frequently than that of knowing its exact sum. And, as mentioned above, Theorem (2.2) widens, albeit rather trivially, the applicability of various tests for convergence.

There is, in fact, an element of eventuality in the very concept of convergence of a sequence. Suppose $\{a_n\}$ converges to $L$. Let $\epsilon > 0$. Then $a_n$ lies eventually in the interval $(L - \epsilon, L + \epsilon)$. Let us see what effect this has on the averaging of terms. Given a sequence $\{a_n\}$, let us define a new sequence $\{b_n\}$ by $b_n = \frac{1}{n}(a_1 + a_2 + \cdots + a_n)$. Verbally, $b_n$ is the mean or the average of the first $n$ terms of the sequence. What will happen if the sequence $\{a_n\}$ were eventually constant? Then there exists $m \in I\!N$ and some $L$ such that $a_n = L$ for all $n \geq m$. This does not mean that $b_n$ will eventually equal $L$ because in forming $b_n$, all terms of $a_n$, right from the first term count. And the effect of the first $m - 1$ terms of $\{a_n\}$, which may differ widely from $L$ will be felt in $b_n$ for all $n$. Intuitively it is clear that as $n$ grows the effect of these 'irregular' terms will be less pronounced, because the number of 'regular' terms willl increase and the average will swing towards them. Thus we expect that $b_n \to L$ as $n \to \infty$.

A stronger result is, in fact, true.

**2.3 Theorem :** If $a_n \to L$ as $n \to \infty$ then $\dfrac{a_1 + a_2 + \cdots + a_n}{n} \to L$ as $n \to \infty$.

**Proof :** Intuitively the idea of the proof is the same as above. All except finitely many terms of $\{a_n\}$ are very close to $L$ and as the number of such terms grows, the average ought to be close to $L$. For a precise proof, let $\epsilon > 0$ be given. Then for all $n$,

$$
\left| \frac{a_1 + a_2 + \cdots + a_n}{n} - L \right| = \left| \frac{(a_1 - L) + (a_2 - L) + \cdots + (a_n - L)}{n} \right|
$$

$$
\leq \frac{1}{n}(|a_1 - L| + |a_2 - L| + \cdots + |a_n - L|). \quad (1)
$$

In the sum on the right in (1), all terms after some stage are small. The effect of the first few terms which aren't, can be subdued by taking $n$ large. To make this precise, first choose $k$ such that $|a_n - L| < \dfrac{\epsilon}{2}$ for all $n \geq k$. Now, $|a_1 - L| + \cdots + |a_k - L|$ is a fixed number and so there exists some $r$ such that for all $n \geq r$, $\dfrac{|a_1 - L| + \cdots + |a_k - L|}{n} < \dfrac{\epsilon}{2}$. (Here again we are using the fact that $\dfrac{1}{n} \to 0$ as $n \to \infty$. A proof will be given in the next section.)

Now let $m = \max\{k, r\}$. Then for all $n \geq m$, we have

$$
\frac{|a_1 - L| + \cdots + |a_n - L|}{n}
$$

$$
= \frac{|a_1 - L| + \cdots + |a_k - L|}{n} + \frac{|a_{k+1} - L| + \cdots + |a_n - L|}{n}
$$

$$
< \frac{\epsilon}{2} + \frac{\epsilon}{2} \frac{n - k}{n} < \epsilon.
$$

So by (1), $\dfrac{a_1 + a_2 + \cdots + a_n}{n} \to L$ as $n \to \infty$. ∎

# EXERCISES

2.1  Generalise the result of Exercise (1.9) to the case where $a_n \leq b_n$ for almost all $n$. Also generalise the Sandwich Theorem.

2.2  In the language of neighbourhood of $\infty$ introduced in Section 2.4, show that the expressions 'for almost all $n$' or 'eventually' mean the same thing as 'for all $n$ in some neighbourhood of $\infty$'. What would be the analogue of Theorem (2.1) for limits of the form $\lim\limits_{x \to c} f(x)$?

2.3  Show by an example that the converse of Theorem (2.3) is false.

## 3.3  Why the Sequences $\dfrac{1}{n}$, $x^n$ and $nx^n$ Converge to 0 when $|x| < 1$

That the first sequence viz., $\dfrac{1}{n}$ tends to 0 as $n \to \infty$ is often taken as too obvious to need a proof. After all, as $n$ gets larger and larger without bounds, its reciprocal $\dfrac{1}{n}$ should get closer and closer to 0. This argument would indeed be valid if we were living in the rational number system, $\mathbb{Q}$. For, in that case, the definition of convergence would require us to show that for every *rational* $\epsilon > 0$, there is some $m \in \mathbb{N}$ such that $\dfrac{1}{n} < \epsilon$ for all $n \geq m$. This is very easy. Since $\epsilon \in \mathbb{Q}$, it is of the form $\dfrac{p}{q}$ for some positive integers $p$ and $q$. All we need to do is to take $m = q + 1$. Then for all $n \geq m$, $\dfrac{1}{n} \leq \dfrac{1}{q+1} < \dfrac{1}{q} \leq \dfrac{p}{q} = \epsilon$. So $\left|\dfrac{1}{n} - 0\right| < \epsilon$ as desired.

But we are in the real number system $\mathbb{R}$. If we go to the construction of $\mathbb{R}$ from $\mathbb{Q}$ then it is easy to show that between any two real numbers there is at least one (and hence infinitely many) rational number. This is expressed by saying that $Q$ is **dense** in $I\!R$. This is a very useful property. Using it we can show that $\frac{1}{n} \to 0$ as $n \to \infty$. Given any (real) $\epsilon > 0$, all we need to do is to get a rational $\epsilon'$ such that $0 < \epsilon' < \epsilon$ and apply the argument above for $\epsilon'$.

However, as discussed in Section 2.6, the approach usually taken in calculus courses is to skip the construction of real numbers and instead to assume their basic properties as axioms, the most important such property being the completeness of $I\!R$ (see Axiom (2.6.3)). In that case we can no longer assume[2] that $Q$ is dense in $I\!R$. We shall have to prove it first and only then we can use it for showing that $\frac{1}{n} \to 0$ as $n \to \infty$. It turns out that it is easier to reverse this order. That is, we shall prove, directly from the completeness of $I\!R$, that $\frac{1}{n} \to 0$ as $n \to \infty$. And then, as a corollary it will follow that $Q$ is dense in $I\!R$.

**3.1 Theorem :** As $n \to \infty$, $\dfrac{1}{n} \to 0$.

**Proof:** Let a real $\epsilon > 0$ be given. Since $n \geq m > 0 \Rightarrow \dfrac{1}{n} \leq \dfrac{1}{m}$, it suffices to show that there is some $m \in I\!N$ such that $\dfrac{1}{m} < \epsilon$. If this is not the case, then we would have $\dfrac{1}{m} \geq \epsilon$ and hence $m \leq \dfrac{1}{\epsilon}$ for all $m \in I\!N$. This means that the set $I\!N$ is bounded above. Since it is also non-empty, by Axion (2.6.3), it has a least upper bound, say $L$ in $I\!R$. But then, the number $L - \dfrac{1}{2}$, being less than $L$, is not an upper bound of $I\!N$. So there exists some $k \in I\!N$ such that $k \geq L - \dfrac{1}{2}$. This gives $k + 1 \geq L + \dfrac{1}{2} > L$. But since $k \in I\!N, k + 1$ is also in $I\!N$. So some member of $I\!N$ exceeds $L$ contradicting that $L$ is an upper bound (in fact the least upper bound) of $I\!N$. So the assumption that $\frac{1}{m} \geq \epsilon$ for all $m \in I\!N$ was wrong. Hence there is some $m \in I\!N$ such that $\dfrac{1}{m} < \epsilon$. As noted before, this completes the proof that $\dfrac{1}{n} \to 0$ as $n \to \infty$. ∎

Note that the only property of $I\!N$ used in the proof above was that $x+1 \in I\!N$ whenever $x \in I\!N$. A subset of $I\!R$ which contains 1 and has this property is called an **inductive subset**, the name obviously coming from the method of mathematical induction. Clearly $I\!N$ itself is an inductive subset. But there are others, e.g., $Z\!\!\!Z$ (the set of all integers), $Q, Q^+$ (the set of all positive rationals) and indeed the whole $I\!R$ are all inductive subsets. It is clear that $I\!N$ is the smallest inductive subset. In fact, in a very formal approach to calculus where the real number system is taken as a primitive concept, the natural numbers are *defined* as the members of the smallest inductive subset of $I\!R$. Once this is done then we define $Z\!\!\!Z$ as $\{x \in I\!R : x \in I\!N$ or $x = 0$ or $-x \in I\!N\}$ and $Q$ as the

---

[2]In Appendix A, a construction of real numbers, starting from the rational numbers will be given and it will be proved, directly from the construction that $Q$ is dense in $I\!R$.

set $\{x \in I\!\!R : \text{there exist some } p, q \in Z\!\!\!Z \text{ such that } q \neq 0 \text{ and } x = p/q\}$.

We shall not follow this highly formal approach. The approach we (and most authors) follow is a hodge-podge between intuition and formalism. That is, we assume familiarity with $Q$. But thereafter, instead of actually constructing $I\!\!R$ from $Q$, we only state that a construction[3] is possible in which $I\!\!R$ will be complete. Taking completeness of $I\!\!R$ as an axiom we derive other properties. The thereom above, which is often called the **Archimedean property** of real numbers was an example. We now prove a highly useful corollary of it.

**3.2 Corollary :** The set of rationals is dense in $I\!\!R$, that is, given any two $a, b \in I\!\!R$, with $a < b$, there exists $x \in Q$ such that $a < x < b$.

**Proof:** Let $\epsilon = b - a$. Then $\epsilon > 0$. By the theorem above, there exists $m \in I\!\!N$ such that $\frac{1}{n} < \epsilon$ for all $n \geq m$. In particular $\frac{1}{m} < \epsilon$. We now cut the real line into pieces of length $\frac{1}{m}$, at 'node' points of the form $0, \pm\frac{1}{m}, \pm\frac{2}{m}, \ldots, \pm\frac{n}{m}, \ldots$. All these numbers are rational and any two consecutive ones among them differ by $\frac{1}{m}$. Since the interval $(a, b)$ has length greater than $\frac{1}{m}$, it is clear that it must contain at least one of the cut points, say $x$. But then $x$ is a desired rational. Let us now put this argument into a rigorous form. Instead of the interval $(a, b)$, we consider the interval $(ma, mb)$. We contend that this interval contains at least one integer. For this we let $S = \{z : z \geq mb, z \text{ is an integer}\}$. We leave it as an exercise (Exercise (3.3)) to show that $S$ has a smallest element, say $y$. But then $y - 1 \notin S$ and so $y - 1 < mb$. Also $y - 1 > ma$, for otherwise we would get

$$y - 1 \leq ma < mb \leq y$$

which would imply $1 = y - (y - 1) \geq mb - ma = m\epsilon$, contradicting that $\frac{1}{m} < \epsilon$. Thus we see that the integer $y - 1$ is in the interval $(ma, mb)$. Equivalently, the rational number $x = \frac{y - 1}{m}$ is in the interval $(a, b)$. ∎

As a corollary to this corollary, we get an interesting result.

**3.3 Corollary :** Every real number is the limit of some sequence of rational numbers.

**Proof:** Let $a \in I\!\!R$. We apply the Corollary above to intervals of the form $(a, a + \frac{1}{n})$, where $n = 1, 2, 3, \ldots$. Thus we get, for each $n \in I\!\!N$, some rational number $x_n$ such that

$$a < x_n < a + \frac{1}{n}.$$

Since $\lim_{n\to\infty} a = a$ and $\lim_{n\to\infty} a + \frac{1}{n} = \lim_{n\to\infty} a + \lim_{n\to\infty} \frac{1}{n} = a + 0 = a$, we get by the Sandwich Theorem (cf. Exercise (1.10)) for sequences that $\lim_{n\to\infty} x_n = a$. Thus the sequence $\{x_n\}$ of rationals converges to $a$. ∎

---

[3]See the footnote on the last page.

We shall see a few other application of the denseness of rationals later. We thus see that the seemingly trivial fact that $\lim_{n \to \infty} \frac{1}{n} = 0$ has a rather non-trivial proof and also non-trivial applications. We now turn to the other two limits in the title, viz. $\lim_{n \to \infty} x^n$ and $\lim_{n \to \infty} nx^n$. Here $x$ is a fixed real number. These limits depend on what value $x$ has. The following theorem tells the story of the first limit for postive $x$.

**3.4 Theroem :** Let $x > 0$. Then $\lim_{n \to \infty} x^n = 0, 1$ or $\infty$ according as $x < 1, x = 1$ or $x > 1$.

**Proof:** Case (i) : $0 < x < 1$. Here it is clear that the sequence $\{x^n\}$ is monotonically decreasing. Since all its terms are postive, it is bounded below. So by Theorem (2.6.4) it has a limit say $L$. We assert that $L = 0$. For this, we note that the sequence satisfies the recursive formula

$$x^n = xx^{n-1} \text{ for } n > 1.$$

Letting $n \to \infty$ on both sides, we get $L = xL$. Since $x \neq 1$, this implies $L = 0$.

Case (ii) $x = 1$. Here $x^n = 1$ for all $n$ and so $\lim_{n \to \infty} x^n = 1$.

Case (iii) Put $y = \frac{1}{x}$. Then $0 < y < 1$ and so $y^n \to 0$ as $n \to \infty$ by case (i) above. So $x^n = \frac{1}{y^n} \to \infty$ as $n \to \infty$.

In this proof, we proved (i) first using completeness of $\mathbb{R}$ and then derived (iii) from (i). We give another proof in which we prove (iii) first and then derive (i) from it using the same trick, viz., putting $y = \frac{1}{x}$.

Suppose $x > 1$. Then we can write $x = 1 + h$ where $h > 0$. (This simple trick, by the way, is very useful. A similar trick which is often conveneient in dealing with limits of the form $\lim_{x \to c} f(x)$ is to write $x$ as $c + h$ and to consider $\lim_{h \to 0} f(c + h)$.) So $x^n = (1+h)^n$, which, upon expansion by the binomial theorem gives,

$$x^n = (1 + h)^n = 1 + nh + \frac{n(n-1)}{2}h^2 + \ldots + nh^{n-1} + h^n \qquad (1)$$

In (1) every term on the right hand side is positive. So omitting all except the second term, we have

$$x^n \geq nh \text{ for all } n \geq 1. \qquad (2)$$

(In fact strict inequality holds. But $\geq$ is enough for our purpose.) Since $h > 0, nh \to \infty$ as $n \to \infty$ and so $x^n \to \infty$ as $n \to \infty$ ∎

When $x = 0, x^n = 0$ for all $n$ and so $\lim_{n \to \infty} x^n = 0$. If $x < 0$ then $|x^n - 0| = |x|^n$. If $-1 < x < 0$, then $0 < |x| < 1$ and so $|x|^n \to 0$ as $n \to \infty$ by the theorem above. But then $x^n \to 0$ as $n \to \infty$. Thus we see that $x^n \to 0$ as $n \to \infty$, whenever $-1 < x < 1$, i.e., whenever $|x| < 1$. This answers the second part of our main question. If $x = -1$, then $x^n = 1$ for even $n$ and $x^n = -1$ for odd $n$. So $\lim_{n \to \infty} x^n$ does not exist. If $x < -1$, then $x^n = (-x)^n$ if $n$ is even and $x^n = -(-x)^n$ if $n$ is odd. Here $(-x)^n \to \infty$ as $n \to \infty$. So $x^n \to \infty$ if $n \to \infty$ through even values and $x^n \to -\infty$ if $n \to \infty$ through odd values. So $\lim_{n \to \infty} x^n$ does not exist if $x < -1$.

The fact that $x^n \to 0$, for $|x| < 1$, is simple but quite useful. We shall see important applications of it later. As a rather trivial but immediate application, we go back to Exercise (1.12.8). If $a_n$ is the percentage of water in the first glass after $n$ exchanges of the liquids (5 cc.) at a time, then it is intuitively clear that $\lim_{n \to \infty} a_n = 50$ and we gave a proof of this in Section 3.1 based on a recurrence relation for $a_n$, without a closed form expression for $a_n$. Suppose, however, we do have the closed form expression for $a_n$ (which was obtained as (3) in Section 3.1), viz.

$$a_n = 50 \left( 1 - \left( \frac{19}{21} \right)^n \right) \tag{3}$$

Since $0 < \frac{19}{21} < 1$, we see from the theorem above that $\left( \frac{19}{21} \right)^n \to 0$ as $n \to 0$ and so from (3) we immediately get that $a_n \to 50$.

This proof has certain advantages over the earlier proof. Let us, for example, consider the question in Exercise (1.12.8). That is, we want to know after how many exchanges of liquids, will the first glass contain more than 20% water. This is equivalent to finding an $n$ such that $50 \left( 1 - \left( \frac{19}{21} \right)^n \right) > 20$ which, upon simplification, becomes $\left( \frac{19}{21} \right)^n < 0.6$. We can reduce this by taking logarithms of both the sides. That would give $n \geq 6$. So 6 is the least value of $n$ for which $a_n > 20$. Many times, however, we are not so fussy about finding the least $n$ which will work. All we want is *some* $n$ such that $a_n > 20$. Suppose we did not have access to logarithms. How could we find such an $n$? We could, of course, go on computing the powers $\frac{19}{21}, \left( \frac{19}{21} \right)^2, \left( \frac{19}{21} \right)^3, \ldots$ till we reach one that is less than 0.6. That would in fact give us the least $n$ (which is 6 in the present case). But this approach is not practicable if the least such $n$ were a big integer. (Try the problem with 40% instead of 20% water). So we look for another method. The first proof of the theorem above is not of much help here. It is based directly on completeness of $\mathbb{R}$. But the second proof is based on an estimation of $\left( \frac{19}{21} \right)^n$, or rather its reciprocal. In fact, taking $h = \frac{2}{19}$ in (2) gives

$$\left( \frac{21}{19} \right)^n > \frac{2n}{19} \tag{4}$$

We are looking for an integer $n$ such that $\left(\dfrac{19}{21}\right)^n < \dfrac{6}{10}$ or equivalently, $\left(\dfrac{21}{19}\right)^n$ $> \dfrac{10}{6}$. In view of (4), this would be the case if $\dfrac{2n}{19} > \dfrac{10}{6}$, i.e., if $n > \dfrac{95}{6}$. So $n = 16$ would work. This value is, of course, much higher than the least $n$ that would work, viz. 6. But see how effortlessly it was obtained. In fact, if we recall how (2) was obtained from (1), we get a slightly sharper inequality than (2) viz.

$$\left(1 + \frac{2}{19}\right)^n > 1 + \frac{2n}{19}$$

using which we get that even $n = 7$ would suffice. All this goes to substantiate our remarks in Section 1.12 that it pays to know the proof rather than just a result and moreover, when the same result can be established in more than one way, one of the proofs may be advantageous over others in some respects.

Probably the most important theoretical application of the fact that $x^n \to 0$ if $|x| < 1$ is that it tells us when an infinite **geometric series** $1 + x + x^2 + \ldots + x^n + \ldots$ is convergent. Recall that the convergence of a series means, by definition, the convergence of the sequence $S_n$, where $S_n$ is the sum of the first $n$ terms of that series. In the present case, using the formula for the sum of a geometric progression, we get, for $x \neq 1$

$$S_n = 1 + x + x^2 + \ldots + x^{n-1} = \frac{1 - x^n}{1 - x} \tag{5}$$

If $|x| < 1$, then the expression on the end has a limit, viz. $\dfrac{1 - 0}{1 - x}$, i.e., $\dfrac{1}{1 - x}$. More generally, if the first term of the series is $a$, so that the geometric series is $a + ax + ax^2 + \ldots + ax^n + \ldots$, then we get

**3.5 Theorem :** The geometric series $\displaystyle\sum_{n=0}^{\infty} ax^n$ converges to $\dfrac{a}{1 - x}$ if $|x| < 1$. ∎

For $x = 1$, (5) is meaningless,. However, in that case the sum $S_n$ is $n$ and so the series is divergent. If $x = -1$, then $S_n = 1$ if $n$ is odd while $S_n$ is 0 if $n$ is even. So in this case, too, the geometric series $\displaystyle\sum_{n=0}^{\infty} x^n$ diverges. If $|x| > 1$, then also this series diverges by Exercise (1.8) since $\lim\limits_{n \to \infty} x^n$ is not 0. So Theorem (3.5) can be improved to say that if $a \neq 0$, then $\displaystyle\sum_{n=0}^{\infty} ax^n$ is convergent if and only if $|x| < 1$. However, even as it is, it has a number of interesting application, a few of which will be given in the exercises.

The geometric series is one of the very few series whose convergence is decided by actually calculating its partial sum (i.e., the sum of the first $n$ terms) and then taking its limit. There are, however, many series which are comparable to the geometric series in some sense. That considerably widens the applicability of Theorem (3.5) as we shall see in Section 3.9.

We now also get an explanation for the absurd result we got in the answer to Q.2.3, because of an indiscriminate manipulation with infinite series, viz.

$$-1 = 1 + 2 + 4 + 8 + \dots$$

which was obtained by putting $x = 2$ in the formula for the sum of a geometric series. We now see that such a substitution is invalid becacause the series in question is not convergent for $x = 2$. If however, we give $x$ any value for which the series is convergent, say $x = \dfrac{1}{10}$, then we do get a valid result like

$$\frac{10}{9} = 1 + \frac{1}{10} + \frac{1}{100} + \frac{1}{1000} + \dots + \frac{1}{10^n} + \dots$$

which can be expressed in a more familiar form as the decimal expansion of the number $\dfrac{10}{9}$ viz., $1.1111\dots11\dots\dots$.

Let us now move to the third limit, viz. $\lim\limits_{n \to \infty} n x^n$ where $|x| < 1$. Let us first assume $x > 0$. So $0 < x < 1$. We cannot write this limit as the product $\left( \lim\limits_{n \to \infty} n \right) \left( \lim\limits_{n \to \infty} x^n \right)$ because that would be of the form $\infty \cdot 0$ which is meaningless. So the preceding theorem cannot be used directly here. But, once again, let us see if either of its two proofs would generalise. The first proof was based on the fact that the sequence $\{x^n\}$ is monotonically decreasing. Now the sequence $a_n = n x^n$ is not quite monotonically decreasing. If for example, $x = \dfrac{4}{5}$, then $a_1 = \dfrac{4}{5}$ and $a_2 = \dfrac{32}{25} > a_1$. But if we keep on computing the subsequent terms we see that from $a_5$ onwards they start decreasing again. Even though $n$ increases, $(\frac{4}{5})^n$ decreases and it decreases more rapidly so that the effect of increasing $n$ is offset. In other words although the sequence $\{n(\frac{4}{5})^n\}$ is not monotonically increasing right from the start, it is so after some stage, or 'eventually' monotonically increasing (cf. last section).

It is not hard to show that for every $0 < x < 1$, the sequence $a_n = n x^n$ is monotonically increasing after some stage, that is, there exists $m$ such that for all $n \geq m$, we have $a_{n+1} \leq a_n$. The 'stage' i.e. the integer $m$ could of course change as $x$ changes. If $x$ is very close to 1, a large $m$ will be needed. For a proof, let us compare $a_{n+1}$ and $a_n$ multiplicatively (as both are positive) and see when $\dfrac{a_{n+1}}{a_n} \leq 1$ holds. With a little calculation, this gives $\frac{1}{n} \leq \frac{1}{x} - 1$, i.e., $n \geq \dfrac{1}{\frac{1}{x} - 1}$. So we may take $m$ to be any integer greater than $\dfrac{1}{\frac{1}{x} - 1}$. Since $\{a_n\}$ is monotonically decreasing after some stage, and bounded below (by 0), by Theorem (2.6.4) (or rather its extension, as outlined in the last section), $\{a_n\}$ converges to some limit $L$. To find the value of $L$ we write

$$a_{n+1} = (n + 1)x^{n+1} = x n x^n + x^{n+1} = x a_n + x^{n+1}.$$

Letting $n \to \infty$ (and noting that $x^{n+1} \to 0$ by Theorem (2.3)) we get $L = xL + 0$, which yields $L = 0$ since $x \neq 1$.

The second proof of theorem (2.3) also generalises. Writing $\frac{1}{x}$ as $1 + h$ where $h > 0$, we get by the binomial theorem,

$$nx^n = n\frac{1}{(1+h)^n} = \frac{n}{1 + nh + \frac{n(n-1)}{2}h^2 + \ldots + nh^{n-1} + h^n} \tag{6}$$

All the terms in the denominator are positive. In the proof of theorem (2.3) we retained only the term $nh$ and dropped the others. That would not work now, because that would only imply that $nx^n < \dfrac{n}{nh} = h$ which is not strong enough to show that $nx^n \to 0$ as $n \to \infty$. However, if we retain the third term in the denominator of (6), then the situation is better, for then we have

$$nx^n < \frac{2n}{n(n-1)h^2} \qquad \text{for } n > 2$$

i.e., $\qquad 0 < nx^n < \dfrac{2}{(n-1)h^2} \qquad \text{for } n > 2 \tag{7}$

Since $h$ is a constant, $\dfrac{2}{(n-1)h^2} \to 0$ as $n \to \infty$ and so from (7) and the Sandwich Theorem we get that $nx^n \to 0$ as $n \to \infty$. If instead of the third term of the denominator of (6), we had taken only the fourth term, then we would have gotten the stronger result that $n^2 x^n \to 0$ as $n \to \infty$. Indeed, by a similar argument one can prove that for every positive integer $k, n^k x^n \to 0$ if $0 < x < 1$. If $-1 < x < 0$, then $|x^k x^n| = n^k |x|^n$ and so we get,

**3.6 Theorem :** For every $k \in I\!N$ and $x \in (-1, 1), n^k x^n \to 0$ as $n \to \infty$. ∎

A paraphrase of this result is noteworthy. Suppose $y > 1$. Then $y^n \to \infty$ as $n \to \infty$. Putting $x = \frac{1}{y}$, we get that $\dfrac{n^k}{y^n} \to 0$ as $n \to \infty$. This means that even though both the numerator and denominator tend to $\infty$ as $n \to \infty$, for sufficiently large $n$, the numerator becomes negligibly small as compared with the denominator. More generally if $a_n = \dfrac{b_n}{c_n}$ where both $b_n$ and $c_n$ tend to $\infty$ as $n \to \infty$, then we say $c_n$ **tends to $\infty$ more rapidly** than $b_n$ or that $c_n$ **dominates** $b_n$ if $a_n \to 0$. If $b_n$ and $c_n$ are polynomials of unequal degrees then it is easy to show (cf. Exercise (3.1)) that the polynomial of larger degree dominates the other. In the present case the contest is between $n^k$ and $y^n$ where $y > 1$. Both expressions are powers. In $n^k$, the exponent $k$ is fixed while the base $n$ tends to $\infty$. In $y^n$ it is the other way. The theorem above implies that $y^n$ dominates $n^k$, no matter how large $k$ is. Instead of $n^k$ we could take any polynomial expression in $n$. It is therefore customary to say that *exponential growth is stronger than algebraic growth*. A rather popular instance of this is to consider a deal between two persons $A$ and $B$, in which $A$ agrees to pay $B$ 1000 rupees on the first day, and thereafter, every day one thousand more rupees than on the previous day, while $B$ agrees to give $A$ one matchstick on the first day, and thereafter, every day twice as many matchsticks as on the previous day. Although this sounds like an incerdibly good bargain for $B$, it will not be long before he regrets it (see Exercise (3.13)). We shall return to this in Section 5.8.

# EXERCISES

**3.1**  Let $a_n = \dfrac{b_n}{c_n}$ where

$$b_n = p_0 + p_1 n + p_2 n^2 + \ldots + p_k n^k$$

and

$$c_n = q_0 + q_1 n + q_2 n^2 + \ldots + q_r n^r$$

for some integers $k, r$ and real numbers $p_0, \ldots, p_k, q_0, \ldots q_r$. (Such expressions are called **polynomials** in $n$. Generally it is assumed that $p_k \neq 0$ and $q_r \neq 0$. Then the integers $k$ and $r$ called their **degrees**). Prove that :

   (i)   if $k < r$ then $a_n \to 0$ as $n \to \infty$ (i.e., $c_n$ dominates $b_n$)

   (ii)  if $k = r$ then $a_n \to \dfrac{p_k}{q_r}$ as $n \to \infty$

   (iii) if $k > r$ then as $n \to \infty$, $a_n \to \infty$ if $\dfrac{p_k}{q_r} > 0$ and $a_n \to -\infty$ if $\dfrac{p_k}{q_r} < 0$.

   [*Hint* : Divide both the numerator and denominator by $n^r$.]

**3.2**  Let $S$ be a non-empty subset of $I\!N$. Prove that $S$ has a least element. [*Hint* : Prove by induction on $n$, the statement that every subset of $I\!N$ containing $n$ has a least element. Indeed, it can be shown that this property of $I\!N$, called the **well ordering** is equivalent to the principle of mathematical induction].

**3.3**  Let $x$ be any real number. Prove that there is a smallest integer greater than or equal to $x$. [*Hint* : Let $S = \{n \in Z\!\!\!Z : n \geq x\}$. First show that $S \neq \emptyset$, for otherwise $x$ will be an upper bound for $Z\!\!\!Z$ and a contradiction would come as in the proof of Theorem (3.1). If $x > 0$, then $S \subset I\!N$ and apply the last exercise. If $x < 0$, show that $S$ contains at most finitely many negative integers as otherwise $-x$ would be an upper bound for $I\!N$ and again there would be a contradiction].

**3.4**  In Exercise (1.4), a closed form expression for $F_n$, the $n^{th}$ Fibonacci number was given. Using it show that $\displaystyle \lim_{n \to \infty} \frac{F_n}{F_{n-1}} = \frac{1 + \sqrt{5}}{2}$. (As mentioned there, this limit is called the golden ratio).

**3.5**  Using only the recurrence relation and not the closed form expression for Fibonacci numbers, viz., $F_n = F_{n-1} + F_{n-2}$, prove that if $\displaystyle \lim_{n \to \infty} \frac{F_n}{F_{n-1}}$ exists then it must be $\dfrac{1 + \sqrt{5}}{2}$.

3.6 Let $a_n = \dfrac{F_{n+1}}{F_n}$ $(n \geq 1)$, where $F_n$ is the $n^{th}$ Fibonacci number. Let $L = \dfrac{1 + \sqrt{5}}{2}$. Prove that $a_n - L = \dfrac{L - a_{n-1}}{L a_{n-1}}$ and hence that the terms of the sequence $\{a_n\}$ are alternatively bigger and smaller than $L$. Show further that $|a_n - L| \leq \dfrac{|a_1 - L|}{L^{n-1}}$ and hence that $a_n \to L$ as $n \to \infty$. More specifically, show that the subsequence $a_2, a_4, a_6, a_8, \ldots$ is monotonically increasing while $a_1, a_3, a_5, a_7, \ldots$ is monotonically decreasing, and both converge to $L$, i.e., to $\dfrac{1 + \sqrt{5}}{2}$. (This and the last exercises illustrate a bit of cunning work. We just *assume* that $\lim\limits_{n \to \infty} a_n$ exists and find its value. We then give a rigorous proof that $a_n$ in fact converges to that number. Rather like a detective who gets a hunch as to who is the murderer and then sets up a trap to get a conclusive evidence. Subsequences will be taken up formally in Section 3.5.)

3.7 Suppose a rectangle has the property that when a square on its shorter side is cut off from it, the remaining rectangle is similar to the original one. Prove that the sides of the original rectangle must be in the golden ratio. (Such a rectangle is called a **golden rectangle**. It is said that among all rectangles other than a square, it has the most pleasing shape. Television screens are often golden rectangles).

3.8 Find at what time between 4.00 p.m. and 5.00 p.m. the hour and the minute hands of a clock coincide. Do the problem by simple arithmetic and also using infinite series. (For the latter, go on adding the time it takes the hour hand to reach the former position of the minute hand).

3.9 Two mathematicians A and B, 1 km. apart on a straight road start walking towards each other with constant speeds of 5 km./hour and 4 km./hour respectively. A fly, which flies at a constatnt speed of 20 km./hour is orignally on A's head. It flies towards B, and after touching B's head, immediately reverses its direction and flies towards A, then again towards B and so on. Find how much total distance the fly covers till A and B meet. Do the problem with geometric series.

3.10 Do the last exercise with plain arithmetic. (According to a popoular anecdote about the great mathematician von Neumann, at a social party he was asked this question. He answered it correctly in just a few seconds. The person who asked it said, somewhat defeatedly, "Well, I'm glad you got it the easy way. Many persons try to do this with infinite series." von Neumann was puzzled and said, "But that's exactly the way I did it.")

3.11 A game is played in which a coin is tossed by three players $A, B, C$ cyclically in that order. Whoever tosses a head first wins the game. If the coin has probability $p$ of showing a head, find the chances of $A, B, C$ of winning the game. Do the problem with as well as without infinite series.

3.12 Prove that for every positive integer $k$, the sequence $\{n^k x^n\}$ (where $0 < x < 1$) is monotonically decreasing after some stage.

3.13 Assuming that 1,000 matchsticks cost 1 rupee, prvoe that in the deal between the persons A and B given at the end of the section, B will be in the red within one month.

3.14 Suppose for every real number $x$ we define

$$e^x = 1 + x + \frac{x^2}{2!} + \frac{x^3}{3!} + \ldots = \sum_{n=0}^{\infty} \frac{x^n}{n!}$$

$$\cos x = 1 - \frac{x^2}{2!} + \frac{x^4}{4!} - \frac{x^6}{6!} + \ldots = \sum_{n=0}^{\infty} \frac{(-1)^n x^{2n}}{(2n)!}, \quad \text{and}$$

$$\sin x = x - \frac{x^3}{3!} + \frac{x^5}{5!} - \frac{x^7}{7!} + \ldots = \sum_{n=0}^{\infty} \frac{(-1)^n x^{2n+1}}{(2n+1)!}.$$

(i) Obtain expressions for the hyperbolic sines and cosines in terms of infinite series. (These functions are defined, respectively, as

$$\sinh(x) = \frac{e^x - e^{-x}}{2} \quad \text{and} \quad \cosh(x) = \frac{e^x + e^{-x}}{2}.)$$

(ii) Assuming that these definitions are valid even when $x$ is a complex number, obtain the well-known Euler's formula :

$$e^{i\theta} = \cos \theta + i \sin \theta.$$

Also obtain relationship between sines and hyperholic sines and between cosines and hyperholic cosines.

(iii) Assuming that these series can be differentiated term-by-term, show that $\frac{d}{dx}(e^x) = e^x$, $\frac{d}{dx}(\sin x) = \cos x$ and $\frac{d}{dx}(\cos x) = -\sin x$.
(Many other properties of these functions can be proved. This exercise, like Exercise (2.7.2) illustrates the charm of manipulations with infinite series. It is not difficult to show that the series above do converge for all real $x$, and in fact for all complex $x$ too. But considerably more work is needed to establish the validity of term-by-term differentiation. That is why this naive approach is generally not followed today.)

3.15 Prove that the set of irrationals is also dense in $\mathbb{R}$, i.e., gien any $a, b \in \mathbb{R}$ with $a < b$, prove that there exists an irrational number $x$, such that $a < x < b$. [*Hint* : If $\alpha$ is irrational and $q$ is a non-zero rational then $\alpha q$ is an irrational. Use this fact and Corollary (3.2).]

3.16 Strengthen Corollary (3.2) and the last exercise by showing that every open interval $(a, b)$ (with $a < b$) contains infinitely many rationals and infinitely many irrationals.

## 3.4 Why the Sequences $x^{1/n}$ (where $x > 0$) and $n^{1/n}$ Both Converge to 1

Before answering the question, let us see what the expression $x^{1/n}$ really means. It is called the $n^{\text{th}}$ **root** of $x$ and means the real numer whose $n^{th}$ power is $x$. The existence of such a number needs a proof. In fact in the rational number system it may not always exist. It is well-known for example, that there is no rational number whose square is 2. So $2^{1/2}$ (i.e., $\sqrt{2}$) does not exist in $Q$ even though there exist sequences of rational numbers converging to $\sqrt{2}$ (cf. Exercise (2.6.4) (i)). Historically, this was one of the earliest noted instances of incompleteness of the rational number system.

As is to be expected, the completeness of $I\!\!R$ will be vitally needed to establish the existence of $n^{\text{th}}$ roots. Using a certain property of continuous functions called Intermediate Value Property, the existence of $n^{\text{th}}$ roots can be proved quite easily (see Exercise (4.1.7)). The proof of that property, of course, needs completeness. Here we give a proof based directly on completeness.

**4.1 Theorem :** Let $x$ be a positive real number and $n$ a positive integer. Then there exists a unique positive real number $y$ such that $y^n = x$.

**Proof :** Before giving the proof we prove two lemmas. A **lemma** is an auxiliary result which is used in the course of the proof of some other, more important result, but which is usually not of much significance by itself. (This is not true of all the lemmas though. In fact some of the most important results in mathematics are called lemmas, because their discoveres used them originally as lemmas to prove something else.)

**4.2 Lemma :** If $y_1, y_2$ are non-negative then $y_1 < y_2$ if and only if $y_1^n < y_2^n$.

**Proof :** All we need to do is to write $y_2 = y_1 + h$ and expand $y_2^n$ by the binomial theorem. Then

$$y_2^n - y_1^n = ny_1^{n-1}h + \frac{n(n-1)}{2}y_1^{n-2}h^2 + \ldots + \frac{n(n-1)}{2}y_1^2 h^{n-2} + ny_1 h^{n-1} + h^n.$$

If $0 \leq y_1 < y_2$ then $h > 0$ and so every term on the right is non-negative and the last one is positive. Hence $y_2^n > y_1^n$. Conversely if $y_2^n > y_1^n$ then certainly $y_2 \neq y_1$. If $y_2 > y_1$ is false then we must have $0 \leq y_2 < y_1$. But then, by the earlier part, $y_2^n < y_1^n$, a contradiction. So $y_1 < y_2$. $\blacksquare$

**4.3 Lemma :** If $0 < y \leq 1$ and $y^n < x$ then there exists $h > 0$ such that $(y + h)^n < x$.

**Proof :** Expanding $(y + h)^n$ by the binomial theorem,

$$(y + h)^n = y^n + hE \tag{1}$$

where $E = ny^{n-1} + \frac{n(n-1)}{2}y^{n-2}h + \ldots + nyh^{n-2} + h^{n-1}$

If $0 < h \leq 1$, then, since we also have $0 < y \leq 1$, we get

$$0 < E \leq n + \frac{n(n-1)}{2} + \ldots + n + 1 = 2^n - 1 < 2^n. \tag{2}$$

(Actually it is not very important that the sum equals $2^n - 1$. All that matters is that it is a constant independent of $h$.)

Now choose $h$ so that $0 < h < \min\left\{1, \dfrac{x - y^n}{2^n}\right\}$. Then from (1) and (2),

$(y + h)^n < y^n + (x - y^n) = x$. ∎

We remark that both these lemmas will follow more easily if we use some results about continuity and derivatives. But, as said earlier, we are after a direct proof of Theorem (2.1). For this, assume first that $0 < x \leq 1$. Let $A = \{r \geq 0 : r^n < x\}$. Clearly $0 \in A$ and so A is non-empty. Also since $x \leq 1$, we see in view of Lemma (4.2) that no element of A can exceed 1. So A is bounded above. Hence by Axiom (2.6.3), A has a supremum $y$ in $\mathbb{R}$. We claim that $y^n = x$, by showing first that $y^n \leq x$ and then by showing that $y^n \geq x$.

By Exercise (2.6.13), there exists a sequence $\{r_m\}$ of elements of A such that $r_m \to y$ as $m \to \infty$. But then, $r_m^n \to y^n$ as $n \to \infty$. Now $r_m^n < x$ for every $m$. So by Exercise (1.9), $\lim_{m \to \infty} r_m^n \leq \lim_{m \to \infty} x$, i.e., $y^n \leq x$. For the other half, viz., $y^n \geq x$, suppose, instead that $y^n < x$. Then by Lemma (4.3), $(y + h)^n < x$ for some $h > 0$. But then $y + h$ will be an element of $A$, exceeding $y$ contradicting that $y$ is an upper bound for $A$. So $y^n = x$.

If $x \geq 1$, we apply the argument above to $\dfrac{1}{x}$ to get $y_1 > 0$ such that $y_1^n = \dfrac{1}{x}$. Now set $y = \dfrac{1}{y_1}$. Then $y^n = x$.

Thus we have completely proved the existence of a positive $y$ such that $y^n = x$. By Lemma (4.2), $y$ is also unique. ∎

The number $y$ given by the theorem above is called the (positive) $n^{\text{th}}$ **root** of $x$ and denoted by $x^{1/n}$ or by $\sqrt[n]{x}$. If $n$ is even, then $-y$ is also an $n^{th}$ root of $x$. But unless stated otherwise, by $x^{1/n}$ we mean only the positive $n^{th}$ root.

Let us now study the sequence $a_n = x^{1/n}$, where $x > 0$ is fixed. If $x = 1$, this is a constant sequence. For $x > 1$, it is monotonically decreasing while for $x < 1$ it is monotonically increasing. This is intuitively obvious but the proof requires a little argument. We omit it because we shall show directly that $x^{1/n} \to 1$ as $n \to \infty$. First assume $x > 1$. Then by Lemma (4.2) above, $a_n > 1$ for all $n$. So we can write $a_n = 1 + h_n$ where $h_n > 0$ for all $n$. The problem reduces to showing that $h_n \to 0$ as $n \to \infty$. The standard strategy for such things, as we know well by now, is to find an upper estimate for $h_n$ (i.e., a quantity not less than $h_n$) which tends to 0 and then to apply the Sandwich Theorem. We have $a_n^n = (1 + h_n)^n = x$. So by the binomial theorem,

$$x = 1 + nh_n + \frac{n(n-1)}{2}h_n^2 + \ldots + nh_n^{n-1} + h_n^n$$

$$\text{or} \quad x - 1 = nh_n + \frac{n(n-1)}{2}h_n^2 + \ldots + h_n^n \tag{3}$$

As every term on the right hand side is positive, if we drop all except the first term, we get

$$nh_n < x - 1, \quad \text{i.e.,} \quad 0 < h_n < \frac{x-1}{n}. \tag{4}$$

Since $x - 1$ is a fixed constant, $\frac{x-1}{n} \to 0$ as $n \to \infty$, and so by the Sandwich Theorem we get that $h_n \to 0$ as $n \to \infty$. So $a_n = 1 + h_n \to 1$ as $n \to \infty$, as was to be proved.

If $x = 1$, then $x^{1/n} = 1$ for all $n$ and so trivially $x^{1/n} \to 1$ as $n \to \infty$. The case $0 < x < 1$ still remains. In this case, $\frac{1}{x} > 1$ and so, $x^{1/n} = \left(\frac{1}{\frac{1}{x}}\right)^{1/n} = \frac{1}{(\frac{1}{x})^{1/n}}$ which tends to 1 as $n \to \infty$ (by the earlier part).

Thus we have completely proved :

**4.4 Theorem :** For $x > 0, x^{1/n} \to 1$ as $n \to \infty$. ∎

The argument for $n^{1/n}$ is very similar. If we calculate the terms we see that the sequence $a_n = n^{1/n}$ is eventually monotonically decreasing. But we do not need it. It is clear that for $n > 1, a_n > 1$ (by Lemma (4.2) again). So once again we write $a_n = 1 + h_n$ where $h_n > 0$ and try to show that $h_n \to 0$ by finding an upper estimate for $h_n$. This is done in the same manner as above, except that when we write

$$n - 1 = nh_n + \frac{n(n-1)}{2}h_n^2 + \ldots + h_n^n \tag{5}$$

it would not do to take the first term, for that would merely give us $h_n < \frac{n}{n-1}$ and $\frac{n}{n-1} \to 1$ as $n \to \infty$. However, if we retain only the second term on the right hand side of (5), we get

$$\frac{n(n-1)}{2}h_n^2 < n - 1$$

for $n > 1$ which gives

$$0 < h_n < \frac{2}{\sqrt{n}} \quad \text{for } n > 1 \tag{6}$$

So we would be through if we can show that $1/\sqrt{n} \to 0$ as $n \to \infty$. This we leave as an easy exercise. So, we have

**4.5 Theorem :** As $n \to \infty, n^{1/n} \to 1$. ∎

This result is interesting because here the base (viz. $n$) increases as $n \to \infty$ while the exponent (viz. $\frac{1}{n}$) decreases. So once again we have a sort of a race. In Section 3.3 we saw that the exponential growth is stronger than the algebraic growth. Theorem (4.5) can be generalised to say that for any polynomial $p(n)$

in $n$, whose leading coefficient is positive, $[p(n)]^{1/n} \to 1$ as $n \to \infty$. Verbally this means that the effect of the diminishing exponent is stronger than that of the base increasing algebraically.

It is tempting to prove the preceding two theorems using logarithms. The base of the logarithms does not really matter (as long as it is positive). If for example, we take natural logarithms (i.e., logarithms with base $e$) then for Theorem (4.4) we can argue that $\ln(x^n) = n \ln x$ which tends to $\infty$ or $-\infty$ depending upon whether $\ln x$ is positive or negative. Taking exponentials again, we get that $x^n \to \infty$ if $x > 1$ and $x^n \to 0$ if $x < 1$.

This approach is not really wrong. It is based on the assumption that the logarithm of a limit is the same as the limit of the logarithm. This indeed is the case. But it requires certain properties of the logarithm function (specifically its continuity and its behaviour in neighbourhoods of $\infty$ and of $-\infty$). To establish these properties we first need a precise definition of logarithms. The usual definition that $\ln x = y$ means $x = e^y$ is meaningless because we have not yet defined $e^y$ rigorously. As mentioned at the end of Section 3.1, it is more customary nowadays to define $\ln x$ first and *then* the exponential function. So a proof of Theorem (4.4) based on logarithms, although not quite wrong, is not acceptable at this stage. There is, however, nothing wrong in applying logarithms informally to *guess* certain limits and then to back up the guess by giving a proof without using logarithms. Or, sometimes, where we already know the limits, we can still take logarithms to conclude something interesting. Suppose we do this to the limit in Theorem (4.5). Then we get $\dfrac{\ln n}{n} \to \ln 1 = 0$ as $n \to \infty$. Here $n$ is an integer variable. But $\ln x$ is defined for all positive real $x$ and it can be shown that $\dfrac{\ln x}{x} \to 0$ as $x \to \infty$. In other words $\ln x \to \infty$ much slower than $x$ does. This fact is expressed by saying that logarithmic growth is weaker than algebraic growth. We shall return to this in Section 5.8.

## EXERCISES

4.1 Suppose $x \in \mathbb{R}$ and $n \in \mathbb{N}$. Prove that if $n$ is even then $x$ has precisely two $n^{th}$ roots (one positive and the other negative) if $x > 0$ and no $n^{th}$ root if $x < 0$. If $n$ is odd prove that every $x \in \mathbb{R}$ has exactly one $n^{th}$ root which is positive if $x > 0$ and negative if $x < 0$.

4.2 Let $x > 0$ be real and $y$ be a rational number. If $y = \dfrac{m}{n}$ where $m, n$ are integers and $n > 0$, we define $x^y$ to mean $(x^m)^{1/n}$. Prove that this is the same as $(x^{1/n})^m$. Prove further that if the same rational $y$ is expressed as $\dfrac{p}{q}$ where $p, q$ are integers with $q > 0$, then $x^{p/q} = x^{m/n}$. This allows us to unambiguously define powers with rational exponents. Extend all familiar laws of indices to such powers, e.g., $x^{y_1+y_2} = x^{y_1} x^{y_2}$ and $(x^y)^z = x^{yz}$. Extend Lemma (4.2) to positive rational exponents.

4.3 Prove that $\sqrt{n} \to \infty$ as $n \to \infty$. More generally for any rational $y > 0$, show that $n^y \to \infty$ as $n \to \infty$. What happens if $y < 0$?

4.4 Find $\lim_{n\to\infty} \sqrt{n+1} - \sqrt{n}$. [Hint : Rationalise.]

4.5 Suppose $p(n) = a_0 + a_1 n + \ldots + a_k n^k$ where $k$ is a positive integer, $a_0, \ldots, a_k$ are real numbers and $a_k \neq 0$. Prove that as $n \to \infty$, $p(n) \to \infty$ if $a_k > 0$ and $p(n) \to -\infty$ if $a_k < 0$. Prove, however, that $\dfrac{p(n)}{n^{k+1}} \to 0$ in either case. (cf. Exercise (3.1).)

4.6 In the last exercise assume $a_k > 0$. Prove that $[p(n)]^{1/n} \to 1$ as $n \to \infty$. (Note that $p(n)$ may be negative for some $n$. How do you overcome this difficulty ?)

4.7 Suppose $\{x_k\}$ is a sequence of real numbers converging to a positive real number $x$. Prove that the terms of $\{x_k\}$ are eventually positive. Prove further that for every positive integer $n, x_k^{1/n} \to x^{1/n}$ as $k \to \infty$ and hence also that $x_k^y \to x^y$ for every rational number $y$. [Hint : Use Lemma (4.2).]

## 3.5    Subsequences

The concept of a subsequence is not so fundamentally important as that of a sequence. Still, it is very easy to understand and convenient to apply. One of the formulations of the most vital property of the real number system, viz. its completeness, can be given in terms of subsequences and, as we shall see later, this formulation is very handy in proving certain properties of continuous functions.

A sequence is a function defined on $I\!N$. A subsequence of it is, basically, the restriction of this function to some infinite subset of $I\!N$. Formally, let $\{a_n\}$ be a sequence. Let $S$ be an infinite subset of $I\!N$. We arrange elements of $S$ in the ascending order, say $S = \{n_1, n_2, n_3, \ldots, n_k, \ldots\}$ where $n_1 < n_2 < n_3 < \ldots < n_k < n_{k+1} < \ldots$. Then we get a new sequence $a_{n_1}, a_{n_2}, \ldots a_{n_k}, a_{n_{k+1}}, \ldots$ which is a **subsequence** of $\{a_n\}$. Note that here the index variable is $k$ and not $n$. To stress this difference, sometimes the original sequence is denoted by $\{a_n\}_{n=1}^{\infty}$ and the subsequence by $\{a_{n_k}\}_{k=1}^{\infty}$. In the language of functions if $a_n = f(n)$ for $n \in I\!N$, where $f$ is a function defined on $I\!N$, then the subsequence $\{a_{n_k}\}_{k=1}^{\infty}$

is the composite function $f \circ g$ where $g : I\!N \to I\!N$ is the function defined by $g(k) = n_k$ for $k \in I\!N$. Note that there is absolutely no restriction on the integers $n_k$ other than they be strictly monotonically increasing. For example, it is not required that $n_1, n_2, n_3, \ldots$ form an arithmetic or geometric progression. In fact the reason subsequences are so convenient is that we can choose the subsequence to suit our needs. Note that a subsequence of a subsequence is a subsequence of the original sequence. Thus $a_2, a_4, a_6 \ldots, a_{2n}, \ldots$ is a subsequence of $\{a_n\}$ and $a_4, a_{14}, a_{24}, a_{34}, \ldots$ is a subsequence of this subsequence.

A subsequence being a sequence in its own right, we can talk of its convergence. Suppose $\{a_{n_k}\}_{k=1}^{\infty}$ is a subsequence. Then $\lim\limits_{k \to \infty} a_{n_k}$ is nothing but the restricted limit $\lim\limits_{\substack{n \to \infty \\ n \in S}} a_n$ where $S$ is the set $\{n_1, n_2, \ldots, n_k \ldots\}$. It follows that if a sequence is convergent to $L$ then every subsequence also converges to $L$. That the converse is false shown by $a_n = (-1)^n$. This sequence is not convergent. But the subsequence $\{a_{2n}\}_{n=1}^{\infty}$ (i.e., the subsequence $a_2, a_4, a_6, \ldots$) converges to 1, while $\{a_{2n-1}\}_{n=1}^{\infty}$ converges to $-1$. In the sequence $1, 2, \dfrac{1}{3}, 4, 5, \dfrac{1}{6}, 7, 8, \dfrac{1}{9}, 10, \ldots,$ one subsequence tends to $\infty$ while one subsequence tends to 0. In both cases the sequences themseles were not convergent but had certain subsequences which were convergent. We now prove a theorem, due to **Bolzano** and **Weierstrass** to the effect that every bounded sequence has this property. Here 'bounded', of course, means bounded from below as well as from above.

**5.1 Theorem :** Every bounded sequence has at least one convergent subsequence.

**Proof :** Let $\{x_n\}$ be a bounded sequence with lower and upper bounds $a$ and $b$ (say). Then the closed interval $[a, b]$ contains all the terms of the sequence, i.e., contains $x_n$ for all values of $n$. We call $a$ as $a_0$, $b$ as $b_0$ and let $c_0 = \dfrac{a_0 + b_0}{2}$. Then $c_0$ divides the interval $[a_0, b_0]$ into two subintervals $[a_0, c_0]$ and $[c_0, b_0]$ of equal lengths. Now at least one of these two subintervals must contain $x_n$ for infinitely many values of $n$, for otherwise their union would contain $x_n$ for only finitely many vaues of $n$, a contradiction. Note that we are not saying that infinitely many distinct terms of the sequence must lie either in $[a_0, c_0]$ or $[c_0, b_0]$. This may be false because the whole sequence $\{x_n\}$ may have only finitely many distinct terms, a constant sequence being an extreme example. All we are saying is that either there are infinitely many values of $n$ for which $x_n$ is in $[a_0, c_0]$ or there are infinitely many values of $n$ for which $x_n$ is in $[c_0, b_0]$. In the notations of sets, either the set $\{n \in I\!N : x_n \in [a_0, c_0]\}$ is infinite or the set $\{n \in I\!N : x_n \in [c_0, b_0]\}$ is infinite. And this is true because the union of these two sets is the entire $I\!N$, which is infinite. If the first possibility holds, we rename $a_0$ as $a_1$ and $c_0$ as $b_1$. If the first possibility does not hold then the second must hold and we set $a_1 = c_0$ and $b_1 = b_0$. In either case, we have found

an interval $[a_1, b_1]$ such that

$$\left.\begin{array}{ll} \text{(i)} & a_0 \le a_1 < b_1 \le b_0, \text{ i.e. }, [a_1, b_1] \subset [a_0, b_0] \\ \text{(ii)} & b_1 - a_1 = \dfrac{b_0 - a_0}{2} \\ \text{(iii)} & [a_1, b_1] \text{ contains } x_n \text{ for infinitely many values of } n. \end{array}\right\} \quad (1)$$

We now repeat this argument for the interval $[a_1, b_1]$ instead of $[a_0, b_0]$. That is, we bisect it at $c_1 = \dfrac{a_1 + b_1}{2}$. Then either $[a_1, c_1]$ contains $x_n$ for infintely many values of $n$ or $[c_1, b_1]$ does (as otherwise, their union $[a_1, b_1]$ would contain $x_n$ for only finitely many values of $n$). If the first possibility holds we take $[a_2, b_2]$ to be $[a_1, c_1]$, otherwise we take it as $[c_1, b_1]$. Continuing this process of successive bisection *ad infinitum* we get an infinite sequence of closed intervals $[a_0, b_0], [a_1, b_1], [a_2, b_2], \ldots, [a_n, b_n], \ldots$ such that

$$\left.\begin{array}{ll} \text{(i)} & [a_0, b_0] \supset [a_1, b_1] \supset \ldots \supset [a_k, b_k] \supset [a_{k+1}, b_{k+1}] \supset \ldots \\ \text{(ii)} & b_k - a_k = \dfrac{b_0 - a_0}{2^k} \text{ for } k = 0, 1, 2 \ldots \\ \text{(iii)} & \text{for every } k = 0, 1, 2, \ldots [a_k, b_k] \text{ contains } x_n \\ & \text{for infinitely values of } n. \end{array}\right\} \quad (2)$$

(Figure 3.5.1 shows a situation where $a_1 = a_2 = a_3 = c_0, b_1 = b_0, b_2 = c_1, b_3 = c_2, a_4 = c_3$ and $b_4 = b_3$.)

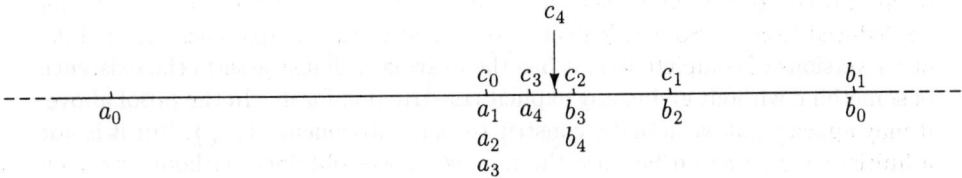

Figure 3.5.1 : Proof of Bolzano-Weisrstrass Theorem.

The proof of (2) is by induction on $k$ and left to the reader. The property (i) is expressed by saying that we have a sequence of **nested intervals.**

By (2)(i), the sequence $\{a_k\}_{k=0}^\infty$ is monotonically increasing. It is also bounded above (by $b_0$ for example). So by Theorem (2.6.3) there is some $L$ such that $a_k \to L$ as $k \to \infty$. Similarly, $\{b_k\}$ is monotonically decreasing, bounded below and so there is some $M$ such that $b_k \to M$ as $k \to \infty$. If we let $k \to \infty$ in 2(ii), and use Theorem (3.4), we see that $a_k - b_k \to 0$ and so $L = M$.

We assert that some subsequence of $\{x_n\}$ converges to $L$. For this we first claim that for every $\epsilon > 0$, the interval $(L - \epsilon, L + \epsilon)$ contains $x_n$ for infinitely many values of $n$. Indeed given any such $\epsilon$, since both $\{a_k\}$ and $\{b_k\}$ converge to $L, a_k$ and $b_k$ are in $(L - \epsilon, L + \epsilon)$ for all sufficiently large $k$. Fix any such $k$.

Then $[a_k, b_k] \subset (L - \epsilon, L + \epsilon)$. But $[a_k, b_k]$ contains $x_n$ for infinitely many values of $n$, by 2(iii). So the same is true of $(L - \epsilon, L + \epsilon)$.

It is now easy to finish the proof. Put $\epsilon = 1$. Then there exists some $n_1$ such that $x_{n_1} \in (L - 1, L + 1)$, i.e., $|x_{n_1} - L| < 1$. Now put $\epsilon = \frac{1}{2}$. Since $x_n \in (L - \frac{1}{2}, L + \frac{1}{2})$ for infinitely many values of $n$, there exists $n_2 > n_1$ such that $x_{n_2} \in (L - \frac{1}{2}, L + \frac{1}{2})$, i.e., $|x_{n_2} - L| < \frac{1}{2}$. Now put $\epsilon = \frac{1}{3}$. Continuing in this manner, we get integers $n_1, n_2, n_3, \ldots, n_k, n_{k+1}, \ldots$ such that

$$n_1 < n_2 < \ldots < n_k < n_{k+1} < \ldots$$

and

$$|x_{n_k} - L| < \frac{1}{k} \text{ for } k = 1, 2, 3, \ldots$$

Since $\dfrac{1}{k} \to 0$ as $k \to \infty$ (by Theorem (3.1)), we see that the subsequence $\{x_{n_k}\}_{k=1}^{\infty}$ of the given sequence $\{x_n\}$ converges to $L$. ∎

We have given the proof a little more elaborately than is usual. The underlying idea is very simple. We go on dividing a closed, bounded interval into two halves and choose one of them at each stage. The sequence of nested subintervals got this way has at least one (and, in fact, only one) point in it by completeness of $\mathbb{R}$. Every neighbourhood of this point contains almost all intervals in our construction. And this usually implies the desired result. This technique, known as **successive bisection** is used in the proofs of many basic theorems asserting the existence of something.

We remark that this theorem was derived from Axiom (2.6.2) (through Theorem (2.6.4)). But it can be shown (cf. Exercise (5.10)) that Axiom (2.6.2) can be deduced from it. So it is logically equivalent to the completeness of $\mathbb{R}$. Like other versions of completeness, it has the drawback that it asserts the existence of something without giving any explicit construction for it. (In the proof above, it may appear that we actually constructed the subsequence $\{x_{n_k}\}$. But it is not a finitistic construction because the number $L$ was obtained without any such construction). When it actually comes to finding a convergent subsequence of a particular sequence, some other methods, or even inspection, may be used.

But the theoretical importance of the Bolzano-Weierstrass theorem is profound. As we shall see later, it implies certain interesting properties of continuous functions. For the time being, we use it to prove another basic result about sequences, which is often taken as an alternate definition of completeness. First we need a definition.

**5.2 Definition :** A sequence $\{a_n\}$ is said to be a **Cauchy sequence** if for every $\epsilon > 0$, there exists some $p \in \mathbb{N}$ such that for all $m \geq p$ and $n \geq p$ (written often as $m, n \geq p$), $|a_m - a_n| < \epsilon$.

This definition resembles that of a convergent sequence. But in the case of a convergent sequence, all its terms after some stage are close to some number $L$ (viz., the limit of that sequence). Here, there is no number $L$. Instead, we are saying that the terms of $\{x_n\}$ are *mutually* close after some stage. Since being close to something in common implies mutual closeness (possibly of a less intimate degree), it is intuitively obvious that every convergent sequence is a Cauchy sequence as we indeed prove.

**5.3 Theorem :** Every convergent sequence is a Cauchy sequence.

**Proof :** Suppose a sequence $\{a_n\}$ converges to $L$. Let $\epsilon > 0$ be given. There eixsts some $p \in I\!N$ such that for all $n \geq p, |a_n - L| < \dfrac{\epsilon}{2}$. Now for $m, n \geq p$ we have both $|a_n - L| < \dfrac{\epsilon}{2}$ and $|a_m - L| < \dfrac{\epsilon}{2}$ and so,

$$
\begin{aligned}
|a_m - a_n| &= |(a_m - L) + (L - a_n)| \\
&\leq |a_m - L| + |a_n - L| \\
&< \frac{\epsilon}{2} + \frac{\epsilon}{2} = \epsilon.
\end{aligned}
$$

Thus $\{a_n\}$ is a Cauchy sequence. ∎

The natural question now is whether the converse is true, that is whether every Cauchy sequence is convergent. In the rational number system the answer is in general in the negative. For example, it is not hard to show that the sequence in Exercise (2.6.7) is a Cauchy sequence. But it is not convergent. (In fact in view of Corollary (3.3), we can get lots of counter-examples. All we have to do is to construct a sequence converging to an irrational number.) A non-convergent Cauchy sequence is an example of a sequence which should converge but does not. As is to be expected, with the real number system the situation is better because of completeness.

**5.4 Theorem :** A sequence of real numbers is convergent if and only if it is a Cauchy sequence.

**Proof :** The direct implication was already proved. For the converse, suppose $\{a_n\}$ is a Cauchy sequence. We first contend that it is bounded. For this apply Definition (5.2) with $\epsilon = 1$. Then there is some $p \in I\!N$ such that for all $m, n \geq p, |a_m - a_n| < 1$. In particular, $|a_n - a_p| < 1$ for all $n \geq p$, i.e.,

$$a_p - 1 < a_n < a_p + 1 \quad \text{for all } n \geq p.$$

Now $a_p - 1$ and $a_p + 1$ are some fixed real numbers. Let $a = \min\{a_1, \ldots, a_{p-1}, a_p - 1\}$ and $b = \max\{a_1, \ldots a_{p-1}, a_p + 1\}$. Then for *all* $n \in I\!N, a \leq a_n \leq b$ and so $\{a_n\}$ is bounded. We now apply Theorem (5.2) to get a subsequence $\{a_{n_k}\}_{k=1}^{\infty}$ and a real number $L$ such that $a_{n_k} \to L$ as $k \to \infty$. We contend that $a_n \to L$ as $n \to \infty$, which would finish the proof.

Let $\epsilon > 0$ be given. As $\{a_n\}$ is a Cauchy sequence, there exists some $p \in I\!N$ such that

$$|a_m - a_n| < \frac{\epsilon}{2} \text{ for all } m, n \geq p \tag{3}$$

Also since $a_{n_k} \to L$ as $k \to \infty$, there exists $q \in \mathbb{N}$ such that

$$|a_{n_k} - L| < \frac{\epsilon}{2} \text{ for all } k \geq q \qquad (4)$$

We may assume $q$ is so large that $n_q \geq p$. (This is because $n_1 < n_2 < n_3 < \ldots$ which in particular implies $n_k \to \infty$ as $k \to \infty$.) Now in (3) we put $m = n_q$. Then from (3) and (4),

$$
\begin{aligned}
|a_n - L| &= |a_n - a_{n_q} + a_{n_q} - L| \\
&\leq |a_n - a_{n_q}| + |a_{n_q} - L| < \frac{\epsilon}{2} + \frac{\epsilon}{2} = \epsilon, \text{ for all } n \geq p.
\end{aligned}
$$

So $a_n \to L$ as $n \to \infty$, i.e., $\{a_n\}$ is convergent. ∎

As with Theorem (5.1), it can be shown that Theorem (5.4) is not just a consequence of completeness of $\mathbb{R}$, but, in fact, an equivalent statement of it (see Exercise (5.10)). There are several other formulations of completeness. But the one above (viz., the converse implication in Theorem (5.5)) is especially interesting because it is capable of a vast generalisation. In Section 1.9 (see also Exercises (1.9.5), (1.9.10) and (1.9.11)) we remarked that the concept of a euclidean distance can be generalised. When so done the generalised distance function is called a **metric**. The concepts of convergence and of a Cauchy sequence make sense even for an abstract metric space if we replace $|x_n - L|$ and $|x_m - x_n|$ by $d(x_n, L)$ and $d(x_m, x_n)$ respectively where $d$ is the (abstract) distance function. A metric space is called complete if every Cauchy sequence in it is convergent. The theory of complete metric spaces is one of the most fruitful ones. We shall not go into it. But we remark that the existence of solutions to certain differential equations is a consequence of the fact that a certain metric space (consisting of functions) is complete.

Inasmuch as the convergence of an infinite series is defined in terms of that of the sequence of its partial sums, the preceding theorem implies the following **Cauchy criterion** for series. The proof is left as an exercise.

**5.5 Theorem :** An infinite series $\displaystyle\sum_{n=1}^{\infty} a_n$ is convergent if and only if it has the property that for all $\epsilon > 0$, there exists $p \in \mathbb{N}$ such that for all $m, n$ with $n \geq m \geq p, |\displaystyle\sum_{k=m}^{n} a_k| < \epsilon$. ∎

We shall see applications of this theorem in Sections 3.6 and 3.7. As a little more 'practical' application of Theorem (5.5), we can define the power $x^y$ for any positive base $x$ and any real exponent $y$. In Section 3.4 we defined $x^y$ when $y$ was of the form $\frac{1}{n}$ for some $n \in \mathbb{N}$. In Exercise (4.2), it was indicated how $x^y$ is defined for all rational values of $y$. Now suppose $y$ is irrational. By Corollary (3.3), there exists a sequence $\{y_n\}$ of rationals converging to $y$. Then the sequence $\{x^{y_n}\}_{n=1}^{\infty}$ is well-defined and it is tempting to define $x^y$ as $\lim_{n \to \infty} x^{y_n}$.

It must, of course, be ensured first that the sequence $\{x^{y_n}\}$ is convergent. One way to do this is to show that it is a Cauchy sequence and then apply Theorem (5.4). Basically the idea is that since $\{y_n\}$ is a convergent and hence a Cauchy sequence, $|y_m - y_n|$ is small for sufficiently large $m, n$. It can then be shown that $|x^{y_m} - x^{y_n}|$ is also small. There is, however, another difficulty. What if there is another sequence, say, $\{z_n\}$ of rationals, also converging to $y$? In that case $x^y$ will be defined as $\lim_{n \to \infty} x^{z_n}$. If $x^y$ is to be well-defined, then we must show that whenever two sequences of rationals $\{y_n\}$ and $\{z_n\}$ have equal limits, the sequences $\{x^{y_n}\}$ and $\{x^{z_n}\}$ also have equal limits. This is possible, but a little messy to do. The alternate approach to define powers by first defining logarithms (see Section 5.7) turns out to be more convenient. Still, because of Theorem (5.4), if one wants, they can be defined in the manner indicated above. (In Exercise (5.7.3), this will be done for $x = e$. The same method works for any positive $x$.)

# EXERCISES

5.1  A number $L$ is called a **limit point** of a sequence $\{a_n\}$ if for every $\epsilon > 0$, and for every $m$, there exists $n \geq m$ such that $|a_n - L| < \epsilon$. Prove that the following statements are equivalent to each other:

(i)   $L$ is a limit point of $\{a_n\}$,

(ii)  every neighbourhood of $L$ contains $a_n$ for infinitely many values of $n$

(iii) there exists a subsequence of $\{a_n\}$ converging to $L$.

5.2  Imitating the argument in the proof of Theorem (5.1), show that every bounded sequence has at least one limit point. Then, using the last exercise, give another (although essentially the same) proof of Theorem (5.1). Prove also that Theorem (5.1) implies Theorem (2.6.4) and hence also Axiom (2.6.2).

5.3  What are the limit points of the sequence $1, 1, 2, 1, 2, 3, 1, 2, 3, 4, 1, 2, 3, 4, 5, 1, 2, \ldots$?

*5.4  Consider the sequence $1, \dfrac{1}{2}, 1, \dfrac{1}{3}, \dfrac{2}{3}, 1, \dfrac{1}{4}, \dfrac{2}{4}, \dfrac{3}{4}, 1, \dfrac{1}{5}, \dfrac{2}{5}, \dfrac{3}{5}, \dfrac{4}{5}, 1, \dfrac{1}{6}, \dfrac{2}{6}, \ldots$

Prove that every real number in $[0, 1]$ is a limit point of this sequence.

5.5  A subset $A$ of $\mathbb{R}$ is called **open** if for every $x \in A$, there exists some $r > 0$ such that the interval $(x - r, x + r)$ is contained in $A$. Prove that :

(i)   all open intervals, whether bounded or unbounded are open (as subsets of $\mathbb{R}$), the union of any number of open intervals is open, the empty set is also open,

(ii)  the intersection of any two (and hence any finite number of) open sets is open.

5.6   A subset $B$ of $\mathbb{R}$ is called **closed** if its component $\mathbb{R} - B$ is open. Prove
      that :

   (i)   all closed intervals, whether bounded or unbounded (e.g., $[0, \infty)$) are
         closed, the empty set is also closed (which incidentally, shows that
         unlike the real life usage of the terms 'open' and 'closed', the two pos-
         sibilities can co-exist)

   (ii)  the semi-open intervals $[a, b)$ and $(a, b]$ (where $a < b$) are neither open
         nor closed (another departure from real life, where 'closed' is generally
         the opposite of open)

   (iii) a subset $B$ is closed if and only if it has the property that for every
         convergent sequence $\{a_n\}$, if $a_n \in B$ for all $n$ then $\lim\limits_{n \to \infty} a_n \in B$, or in
         other words, a sequence in $B$ cannot converge to a point outside $B$. (To
         appreciate what is really involved here, take $B = (0, 1]$ and $a_n = \dfrac{1}{n}$ for
         $n \in \mathbb{N}$. Then $a_n \in B$ for all $n$, and $\{a_n\}$ is convergent (to 0) but $0 \notin B$.
         Here $B$ is not closed.)

   (iv)  the set of all limit points of any sequence $\{x_n\}$ is closed. (This set
         may possibly be empty as in the case of the sequence $a_n = n, n \in \mathbb{N}$.)
         [*Hint* : Use (iii) or show directly that its complement is open.]

5.7   Prove that evey non-empty closed, bounded subset has a least and a greatest
      element. [*Hint* : Use Exercise (2.6.13) and part (iii) of the last exercise.]
      In particular, it follows that every bounded sequence has a smallest and a
      largest limit point.

5.8   Prove that if a sequence $\{a_n\}$ converges to $L$, then there exists a monotonic
      subsequence of $\{a_n\}$ which converges to $L$. (Monotonic means one which is
      either monotonically increasing or monotonically decreasing.)

5.9   Prove that every sequence has a monotonic subsequence. [*Hint* : First
      dispose off the case of an unbounded sequence. For bounded sequences
      apply Theorem (5.1) and the last exercise.]

5.10  Using only the definitions and the fact that $\mathbb{N}$ has no upper bound in $\mathbb{R}$,
      prove that every monotonically increasing sequence which is bounded above
      must be a Cauchy sequence. [*Hint* : Proceed by contradiction.] Hence show
      that Theorem (5.4) implies Theorem (2.6.4) and hence also Axiom (2.6.2).

5.11  Prove Theorem (5.5).

5.12  Prove that a monotonically increasing sequence is convergent if and only if
      it contains some convergent subsequence.

## 3.6    **Divergence of** $\displaystyle\sum_{n=1}^{\infty} \frac{1}{n}$ **and Convergence of** $\displaystyle\sum_{n=1}^{\infty} \frac{1}{n^2}$

The series $\displaystyle\sum_{n=1}^{\infty} \frac{1}{n}$ is popularly known as the **harmonic series**, because its terms form a harmonic progression. Its $n^{\text{th}}$ partial sum, viz. $1 + \frac{1}{2} + \frac{1}{3} + \ldots + \frac{1}{n}$ is called the $n^{\text{th}}$ **harmonic number** and often denoted by $H_n$. Since the terms of the series are all positive, it is clear that $\{H_n\}_{n=1}^{\infty}$ is a monotonically increasing sequence. Since $H_{n+1} - H_n = \frac{1}{n+1}$ is very small for large $n$, it might appear that $H_n$ grows very slowly, and may in fact be bounded above, in which case, by Theorem (2.6.4) $H_n$ would have a (finite) limit. In fact, in Exercise (1.8), it was shown that a *necessary* condition for a series $\displaystyle\sum_{n=1}^{\infty} a_n$ to converge it that $a_n \to 0$ as $n \to \infty$. In real life we often confuse necessity and sufficiency (see the beginning of Section 1.2) and so a layman is apt to think that the series $\displaystyle\sum_{n=1}^{\infty} \frac{1}{n}$ is convergent. Another reason to believe so is that the analogous series $\displaystyle\sum_{n=1}^{\infty} \frac{1}{n^2}$ is convergent. In fact, in Exercise (1.4.9) we even calculated its sum as $\frac{\pi^2}{6}$, albeit somewhat shadily, so it stands to reason that $\displaystyle\sum_{n=1}^{\infty} \frac{1}{n}$ should also converge (and its value be related to $\pi$ in some way).

It turns out however, that $\displaystyle\sum_{n=1}^{\infty} \frac{1}{n}$ is divergent as we now show. The proof is a simple application of the Cauchy criterion for convergence of series :

**6.1 Theorem :** The harmonic series $\displaystyle\sum_{n=1}^{\infty} \frac{1}{n}$ is divergent.

**Proof :** By Theorem (5.5), $\displaystyle\sum_{n=1}^{\infty} \frac{1}{n}$ would be convergent if and only if, for every $\epsilon > 0$, there exists $p \in I\!\!N$ such that for all $n \geq m \geq p$, $\displaystyle\sum_{k=m}^{n} \frac{1}{k} < \epsilon$. (Note that we have omitted the absolute value since the terms are all positive.) We claim that for $\epsilon = \frac{1}{2}$, no such $p$ exists. For given any $p \in I\!\!N$, if we set $m = p+1$ and $n = 2p$ then each of the $p$ terms in the sum $\dfrac{1}{p+1} + \dfrac{1}{p+2} + \ldots + \dfrac{1}{2p-1} + \dfrac{1}{2p}$ is at least $\dfrac{1}{2p}$ and so $\displaystyle\sum_{k=m}^{n} \frac{1}{k} = \sum_{k=p+1}^{2p} \frac{1}{k} \geq p \cdot \frac{1}{2p} \geq \frac{1}{2}$. So the series is divergent.∎

Verbally, the harmonic series diverges because no matter how far you go out, there will always be some block of consecutive terms which adds up to at least $\frac{1}{2}$.

Even though the sequence $\{H_n\}$ of harmonic numbers does not turn out to bounded, the observation that $H_n$ grows very slowly with $n$ is not wrong. We shall in fact show later (Section 5.8) that $H_n$ is nearly equal to $\ln n$, where ln is the natural logarithm. As remarked at the end of Section 3.4, $\ln n \to \infty$ much slower than $n$.

It is instructive to see the divergence of the harmonic series by another method. If we group its terms (except the first one) as

$$1 + \left(\frac{1}{2}\right) + \left(\frac{1}{3} + \frac{1}{4}\right) + \left(\frac{1}{5} + \frac{1}{6} + \frac{1}{7} + \frac{1}{8}\right) + \left(\frac{1}{9} + \frac{1}{10} + \ldots + \frac{1}{16}\right) + \ldots \quad (1)$$

into blocks whose sizes are powers of 2 then the sum in each block is at least $\frac{1}{2}$. So the subsequence $\{H_{2^m}\}_{m=1}^{\infty}$ diverges to $\infty$. Since $\{H_n\}$ is monotonically increasing, $H_n$ also tends to $\infty$ as $n \to \infty$. On the other hand, if we group the terms of $\sum_{n=1}^{\infty} \frac{1}{n}$ as

$$(1) + \left(\frac{1}{2} + \frac{1}{3}\right) + \left(\frac{1}{4} + \frac{1}{5} + \frac{1}{6} + \frac{1}{7}\right) + \left(\frac{1}{8} + \frac{1}{9} + \ldots + \frac{1}{15}\right) + \ldots \quad (2)$$

then each block adds up to at most 1 and so, for $n = 2^m - 1$, we have $H_n \leq m$. This shows that $H_n$ grows very slowly. The smallest $n$ such that $H_n > 20$ is at least $2^{20} - 1$, i.e., more than a million.

This idea of grouping or 'condensing' the terms into suitable blocks of sizes which are powers of 2 can be generalised to prove the following result called **Cauchy's condensation test**.

**6.2 Theorem :** Let $\{a_n\}$ be a monotonically decreasing sequence of positive real numbers. Then the series $\sum_{n=1}^{\infty} a_n$ is convergent if and only if the series $\sum_{n=0}^{\infty} 2^n a_{2^n}$, i.e, the series $a_1 + 2a_2 + 4a_4 + 8a_8 + 16a_{16} + \ldots$ is convergent.

**Proof :** For notational simplicity write $b_n$ for $2^n a_{2^n}$. Let $S_n$ and $T_n$ denote the partial sums of $\sum_{n=1}^{\infty} a_n$ and $\sum_{n=0}^{\infty} b_n$. That is, $S_n = a_1 + \ldots + a_n$ and $T_n = b_0 + \ldots + b_n$. Then both the sequences $\{S_n\}$ and $\{T_n\}$ are monotonically increasing. By grouping the terms analogously to (1), we get that for every $n \geq 0$,

$$\begin{aligned} S_{2^n} &\geq a_1 + a_2 + 2a_4 + 4a_8 + \ldots + 2^{n-1} a_{2^n} \\ &= \frac{1}{2}a_1 + \frac{1}{2}(a_1 + 2a_2 + 4a_4 + 8a_8 + \ldots + 2^n a_{2^n}) \\ &= \frac{1}{2}a_1 + \frac{1}{2}T_n. \end{aligned} \quad (3)$$

On the other hand by grouping them analogously to (2), we get, for every $n \geq 1$.

$$S_{2^n-1} \leq a_1 + 2a_2 + 4a_4 + \ldots + 2^{n-1}a_{2^{n-1}} = T_{n-1} \tag{4}$$

Now suppose $\sum_{n=1}^{\infty} a_n$ is convergent. Let $L = \sum_{n=1}^{\infty} a_n$. Then $\{S_n\}$ is monotonically increasing and converges to $L$. In particular $S_k \leq L$ for all $k$. So by (3), for all $n \geq 0$,

$$T_n \leq 2S_{2^n} - a_1 \leq 2L - a_1$$

which shows that the sequence $\{T_n\}$ is bounded above. As it is also monotonically increasing, it is convergent by Theorem (2.6.4). Hence $\sum_{n=0}^{\infty} b_n$ i.e.,

$\sum_{n=0}^{\infty} 2^n a_{2^n}$ is convergent.

Conversely suppose $\sum_{n=0}^{\infty} b_n$ is convergent, then the sequence $\{T_n\}$ is bounded above. But then by (4), the subsequence $\{S_{2^n-1}\}_{n=1}^{\infty}$ of $\{S_n\}_{n=1}^{\infty}$ is also bounded above by some number, say $M$. We claim that $\{S_n\}$ itself is bounded above by $M$. For this, all we need to do is, given $n \in \mathbb{N}$, find some $r$ such that $n \leq 2^r - 1$. Then $S_n \leq S_{2^r-1} \leq M$. So $\{S_n\}$ is convergent, again by Theorem (2.6.4). This means $\sum_{n=1}^{\infty} a_n$ is convergent. ∎

We can now see very effortlessly that $\sum_{n=1}^{\infty} \frac{1}{n}$ is divergent. The condensed series $\sum_{n=0}^{\infty} 2^n \frac{1}{2^n}$ is simply the series $1 + 1 + 1 + \ldots$ and is obviously divergent.

As another application of the last theorem, we shall now show that the series $\sum_{n=1}^{\infty} \frac{1}{n^2}$ is convergent. The argument in Exercise (1.4.9) is based on unfounded manipulations. We can now prove, quite rigorously, that $\sum_{n=1}^{\infty} \frac{1}{n^2}$ is convergent.

(However, we still have no method for showing that its value is $\frac{\pi^2}{6}$.) In fact, we can prove a more general result.

**6.3 Theorem :** For every rational number $y > 1$, the series $\sum_{n=1}^{\infty} \frac{1}{n^y}$ is convergent.

**Proof :** By the last part of Exercise (4.2), the sequence $\{\frac{1}{n^y}\}$ is monotonically decreasing. Also all its terms are positive. So by the theorem above, to show the convergence of $\sum\limits_{n=1}^{\infty}\dfrac{1}{n^y}$, it suffices to show that the condensed series $\sum\limits_{n=0}^{\infty}2^n\dfrac{1}{(2^n)^y}$ is convergent. Now $(2^n)^y = (2^y)^n$ and so $2^n\dfrac{1}{(2^n)^y} = \left(\dfrac{2}{2^y}\right)^n = \left(\dfrac{1}{2^{y-1}}\right)^n$. So the series $\sum\limits_{n=0}^{\infty}2^n\dfrac{1}{(2^n)^y}$ is simply a geometric series with common ratio $\dfrac{1}{2^{y-1}}$. If $y > 1$, then $2^{y-1} > 1$ and $\dfrac{1}{2^{y-1}} < 1$. So by Theorem (3.5) the series $\sum\limits_{n=0}^{\infty}2^n\dfrac{1}{(2^n)^y}$ is convergent. $\blacksquare$

Note that the restriction that $y$ is rational appears merely because so far we have not defined powers with irrational exponents (see the comments at the end of the last section). When this is done, the theorem above holds for all $y > 1$. The sum of the series $\sum\limits_{n=1}^{\infty}\dfrac{1}{n^y}$ is denoted by $\zeta(y)$. This gives us a function $\zeta : (1, \infty) \to I\!R$, popularly known as the **Riemann zeta** function. The result of Exercise (1.4.9) can be stated as $\zeta(2) = \frac{\pi^2}{6}$. It can be shown by similar arguments that $\zeta(4) = \frac{\pi^4}{90}, \zeta(6) = \frac{\pi^6}{945}$ and more generally $\zeta(y)$ can be evaluated when $y$ is an even integer. For other values of $y$, there is no easy closed form expression for $\zeta(y)$. There is a huge literature on the Riemann zeta function and its applications.

The divergence of the harmonic sereis should serve as an eye-opener to those who go too much by intuition. Intuition is undoubtedly an invaluable asset. But when it clashes with logic, the latter always prevails in mathematics. Whenever this happens, initially there is some confusion, in fact a kind of shock. But subsequently, serveral advantages ensue. First, you learn to check your intuition, make it more guarded, more refined and as a consequence, your intuition becomes a sharper tool. Secondly, results which defy our intuition often open up new lines of thinking. Let us view the convergence of $\sum\limits_{n=1}^{\infty}\dfrac{1}{n^2}$ on the background of the divergence of the series $\sum\limits_{n=1}^{\infty}\dfrac{1}{n}$. The former is a sort of a 'sub-series' of the latter in the sense that every term in $\sum\limits_{n=1}^{\infty}\dfrac{1}{n^2}$ appears exactly once in $\sum\limits_{n=1}^{\infty}\dfrac{1}{n}$. The latter consists of reciprcals of all positive integers while the former of the reciprocals of only perfect squares, i.e., $1, 4, 9, 16, 25, 36, \ldots$. To say $\sum\limits_{n=1}^{\infty}\dfrac{1}{n^2}$ is convergent, says in some sense, that the set of perfect squares, although an infinite one, is rather thinly spread out. Its elements are not encountered very frequently. If instead of $\sum\limits_{n=1}^{\infty}\dfrac{1}{n^2}$, we take $\sum\limits_{n=1}^{\infty}\dfrac{1}{n^3}$ which has an even smaller value (as is easy to show), then we may say that this reflects the fact that the perfect cubes are even rarer than perfect squares. In general, given an infinite subset $S$ of $I\!N$, we write its elements in an ascending order say $n_1 < n_2 < n_3 \ldots$ and form the infinite series of their reciprocals, viz. $\sum\limits_{k=1}^{\infty} \dfrac{1}{n_k}$. If this series is divergent then we may say that the elements of the set $S$ occur rather frequently. For example, it is not hard to show that if $S$ consists of an arithmetic progression then this series $\sum\limits_{n \in S} \dfrac{1}{n}$ is divergent.

What if $S$ is the set of all primes, i.e., $S = \{2, 3, 5, 7, \ldots\}$? We already know (Exercise (1.4.1)) that $S$ is infinite. Write elements of $S$ as $p_1, p_2, p_3, \ldots,$ $p_k, \ldots$ in an ascending order. Then $p_k$ is called the $k^{\text{th}}$ prime. There is no easy formula to express $p_k$ directly in terms of $k$. Nor can much be said about the distribution of primes. There are arbitrarily large blocks of consecutive integers which contain no primes (Exercise (6.5)). On the other hand, sometimes even a very large prime is followed soon by another prime. So primes are probably not so rare. This is confirmed by the fact that the series of their reciprocals, i.e., $\sum\limits_{k=1}^{\infty} \dfrac{1}{p_k}$ is divergent. A proof will be indicated through the exercises.

## EXERCISES

6.1 Prove that for $n > 1$, $H_n$ is not an integer. [*Hint*: Let $2^m$ be the highest power of 2, not exceeding $n$. Consider the numerator and denominator of $2^{m-1} H_n$.]

6.2 If $y$ is a rational number with $y \leq 1$, prove that $\sum\limits_{n=1}^{\infty} \dfrac{1}{n^y}$ is divergent.

6.3 Suppose $S$ is an arithmetic progression, say, $S = \{a + d, 2a + d, \ldots,$ $ka + d, \ldots\}$ where $a, d$ are positive integers. Prove that the series $\sum\limits_{n \in S} \dfrac{1}{n}$, i.e., the series $\sum\limits_{k=1}^{\infty} \dfrac{1}{ka + d}$ is divergent.

6.4 Suppose a series $\sum\limits_{n=1}^{\infty} a_n$ is convergent. Prove that for evey $\epsilon > 0$, there exists $m$ such that for all $n \geq m$, $\left| \sum\limits_{k=n+1}^{\infty} a_k \right| < \epsilon$. (In other words, the 'remainder after the $n^{\text{th}}$ term' of a convergent series (also often called its tail-end) can be made as small as possible.)

6.5 For every positive integer $n$, show that none of the $n$ consecutive integers beginning with $(n + 1)! + 2$ is a prime.

6.6   Suppose $A$ is an infinite subset of primes for which $\sum\limits_{n \in A} \dfrac{1}{n}$ is convergent with

a sum equal to $L$ (say) where $L < 1$. Let $A_1 = A$ and $A_2 = \{p_1 p_2 : p_1 \in A, p_2 \in A\}$. That is, $A_2$ is the set of those integers which are the products of two (not necessarily distinct) elements of $A_1$. Prove that $\sum\limits_{n \in A_2} \dfrac{1}{n} \leq L^2$.

6.7   In the last exercise, more generally, show that for every positive integer

$m$, $\sum\limits_{n \in A_m} \dfrac{1}{n} \leq L^m$ where $A_m$ is the set of all integers which are the products of $m$ primes each coming from $A$.

6.8   Continuing, let $A_*$ be the set of all integers ($> 1$) whose prime factorisation contains only those primes which are in $A$. Show that $\sum\limits_{n \in A_*} \dfrac{1}{n} \leq \dfrac{L}{1-L}$.

(Note that every element of $A_*$ belongs to precisely one $A_m$. Strictly speaking, the manipulations on infinite series needed in this and the last two exercises require some justifications. These will be given in the next section.)

6.9   Let $\{p_1, \ldots, p_m\}$ be a finite set of primes and $A$ be the set of all primes other than $p_1, \ldots p_m$. Let $a = p_1 p_2 \ldots p_m$ and $S$ be the arithmetic progression, $\{a+1, 2a+1, 3a+1, \ldots, ka+1, \ldots\}$. Let $A_*$ be the set constructed from $A$ as in the last exercise. Show that $S \subset A_*$.

6.10  Prove that the series of reciprocals of the primes is divergent. [*Hint* : Suppose $\sum\limits_{n=1}^{\infty} \dfrac{1}{p_k}$ is convergent where $p_k$ is the $k^{\text{th}}$ prime. Apply Exercise (6.4) to

get an integer $m$ such that $\sum\limits_{k=m+1}^{\infty} \dfrac{1}{p_k} = L$ (say) $< 1$. Now let $A$ be the set of

all primes other than $p_1, \ldots, p_m$. Apply the last two exercises to show that

$\sum\limits_{n \in S} \dfrac{1}{n} \leq \sum\limits_{n \in A_*} \dfrac{1}{n} \leq \dfrac{L}{1-L} < \infty$ contradicting the result of Exercise (6.3).]

## 3.7 Series of Positive Terms

In Section 3.1, we saw that unlike a finite sum, the value of an infinite sum depends on the order in which the terms are added and hence that sequences are necessary to define them. We mentioned that for series of positive terms, such anomalies do not occur. As we shall see in this section, infinite series of postive terms can be manipulated much the same way as finite sums. (A few such manipulations were, in fact, needed in Exercises (6.7) to (6.9).) This is the primary reason why series of positive terms are easier to handle.

Although we shall prove the results for series of positive terms, everything here goes through, more generally, for series of non-negative terms. Even for series all whose terms are negative (or at least non-positive) there is little difficulty because all one has to do is to mutliply everything by $-1$ and apply the results in this session. When all except a finite number of terms of a series are all of the same sign, there is still no problem (cf. Section 3.2). It is only when the series contain infinitely many positive and infinitely many negative terms that trouble arises.

Throughout this section we deal with series of non-negative terms. Suppose $\sum_{n=1}^{\infty} a_n$ is such a series. Earlier we defined its value as $\lim_{n \to \infty} S_n$ where $S_n = a_1 + a_2 + \ldots + a_n$. Suppose by rearranging the terms of this series we obtain a new series $\sum_{n=1}^{\infty} b_n$. Formally, this means that there is a one-to-one correspondence between the $a_i$'s and $b_j$'s such that the corresponding terms are always equal. (In particular, each term occurs the same number of times in each summation.) To be still more precise, this means there is a bijection $\theta : I\!N \to I\!N$ (or a **permutation** as it is often called) such that for every $n \in I\!N, b_n = a_{\theta(n)}$. But the concept of rearrangement is clear enough to dispense with such a formal language.)

It is not immediately clear that $\sum_{n=1}^{\infty} a_n = \sum_{n=1}^{\infty} b_n$. We could prove this directly (Exercise (7.1)). Instead, we shall give an alternate meaning to $\sum_{n=1}^{\infty} a_n$ which is independent of any particular ordering of the terms. It would then follow as a corollary that a rearrangement of the terms does not change the value of the series.

The alternate meaning we want to give to $\sum_{n=1}^{\infty} a_n$ is not based on the concept of a limit, but on the concept of the supremum of a set. (These two concepts are, of course, interrelated in view of Theorem (2.6.4) and Exercise (2.6.13) which goes to show that our alternate interpretation is not going to be very surprising.) We cannot physically add an infinite number of terms. But surely we can add any finite number of them. So suppose $n_1, n_2, \ldots, n_k$ are some positive integers. Let us denote by A the subset $\{n_1, \ldots, n_k\}$. Let us denote the sum $a_{n_1} + \ldots + a_{n_k}$ by $\sigma(A)$ ($\sigma$ stands for summation). We can do this for every finite subset of $\mathbb{N}$. (If $A$ happens to be of the form $\{1, 2, \ldots, k\}$ for some $k \in \mathbb{N}$, then $\sigma(A)$ is precisely the $k^{\text{th}}$ partial sum of $\displaystyle\sum_{n=1}^{\infty} a_n$. So we may say that we are generalising the concept of a partial sum by taking the sum of any finite number of terms. If $A = \emptyset$, we set $\sigma(A) = 0$. Note that $\sigma(A)$ depends not only on the size of $A$, i.e., on the number of terms in it, but also on how big these terms are. But definitely, as $A$ grows, so does $\sigma(A)$. That is, if $A \subset B$ then $\sigma(A) \leq \sigma(B)$. So it is clear that if we let $S(\{a_n\})$ be the set of all such sums, i.e., $S(\{a_n\}) = \{\sigma(A) : A \subset \mathbb{N}, A \text{ finite}\}$ then $S(\{a_n\})$ can have no maximum element unless $a_n = 0$ for almost all $n$. For if $\sigma(B)$ were such a maximum element, where $B \subset \mathbb{N}$ is finite, we can enlarge $B$ by adding one more element, say, $k$ such that $a_k > 0$ to get a finite subset $C$ such that $\sigma(B) < \sigma(C)$. When the maximum element does not exist, the next best thing to look for is a supremum. The following theorem tells us the answer.

**7.1 Theorem :** Let $\{a_n\}_{n=1}^{\infty}$ be a sequence of positive terms. Let $S(\{a_n\})$ be the set $\{\sigma(A) : A \subset \mathbb{N}, A \text{ finite}\}$. Then $S(\{a_n\})$ is bounded above if and only if the series $\displaystyle\sum_{n=1}^{\infty} a_n$ is convergent. And when this happens, the sum of $\displaystyle\sum_{n=1}^{\infty} a_n$ is precisely the supremum of the set $S(\{a_n\})$.

**Proof :** As usual, let $S_n = a_1 + \ldots + a_n$. Then $S_n$ is nothing but $\sigma(\{1, 2, \ldots, n\})$ and so $S_n$ is a member of the set $S(\{a_n\})$. The sequence $\{S_n\}$ is monotonically increasing. If $\displaystyle\sum_{n=1}^{\infty} a_n$ is not convergent then this sequence is not bounded above, in which case a set which contains all its terms cannot be bounded above either. So $S(\{a_n\})$ is not bounded above.

Conversely, suppose $\displaystyle\sum_{n=1}^{\infty} a_n$ is convergent. Let $L = \displaystyle\sum_{n=1}^{\infty} a_n$. Then we claim first that $L$ is an upper bound for the set $S(\{a_n\})$. A typical member of $S(\{a_n\})$ is of the form $a_{n_1} + a_{n_2} + \ldots + a_{n_k}$ for some positive integers $n_1, \ldots, n_k$. Without loss of generality assume that $n_1 < n_2 < \ldots < n_k$. Now the partial sum $S_{n_k}$ contains all these terms and possibly some more. So $a_{n_1} + \ldots + a_{n_k} \leq S_{n_k}$. But the sequence $\{S_n\}$ is monotonically increasing and converges to $L$. So $S_n \leq L$ for all $n$. In particular $S_{n_k} \leq L$, whence $a_{n_1} + \ldots + a_{n_k} \leq L$. Thus $L$ is an upper bound for $S(\{a_n\})$. So $S(\{a_n\})$ is a bounded set.

It remains to prove that $L$ is the least upper bound of the set $S(\{a_n\})$. Let, if possible, $S(\{a_n\})$ have some upper bound of the form $L - \epsilon$ where $\epsilon > 0$. Then $S_n \to L$ implies there exists some $m$ such that $S_n \in (L - \epsilon, L + \epsilon)$ for all $n \geq m$. In particular, $S_m > L - \epsilon$. But $S_m = \sigma(\{1, 2, \ldots, m\})$ and so $S_m \in S(\{a_n\})$, contradicting that $L - \epsilon$ is an upper bound for $S(\{a_n\})$. So $L$ is the supremum of $S(\{a_n\})$. $\blacksquare$

As promised earlier, we have,

**7.2 Corollary :** Suppose the sequence $\{b_n\}$ (of non-negative terms) is obtained by rearranging the terms of the sequence $\{a_n\}$. Assume that $\sum_{n=1}^{\infty} a_n$ is convergent.

Then $\sum_{n=1}^{\infty} b_n$ is also convergent and $\sum_{n=1}^{\infty} b_n = \sum_{n=1}^{\infty} a_n$.

**Proof :** We merely note that the sets $S(\{a_n\})$ and $S(\{b_n\})$ are identical and hence their suprema are equal. The result follows from the theorem above. ∎

Interesting as this corllary is, the real significance of Theorem (7.1) is deeper. A sequence is, by definition, a function defined on $\mathbb{N}$. Let us denote the function by $f$ and instead of $a_n$ write $f(n)$. A typical member of the set $S(\{a_n\})$ constructed above is of the form $\sigma(A) = a_{n_1} + \ldots + a_{n_k}$ where $A = \{n_1, \ldots, n_k\}$ is a finite subset of $\mathbb{N}$. In our new notation $\sigma(A) = \sum_{i=1}^{k} f(n_i)$. Since this is a finite sum, the order of summation is immaterial and so we can also write it as $\sum_{n \in A} f(n)$. The theorem above says that if $f$ takes non-negative vaules and the series $\sum_{n=1}^{\infty} f(n)$ is convergent then its value is the supremum of the set $\{\sigma(A) : A \subset \mathbb{N}, A \text{ finite}\}$. In our new notation let us denote this set by $S(f)$ (rather than by $S(\{a_n\})$ as we did earlier) and its supremum by $\sum(f)$. The theorem above says that if $f$ takes only non-negative values then we could have defined $\sum_{n=1}^{\infty} f(n)$ as $\sum(f)$, i.e., as $\sup\{\sum_{n \in A} f(n) : A \subset \mathbb{N}, A \text{ finite}\}$. Note that this definition makes no reference to the order of the elements of the set $\mathbb{N}$. So it is called, quite appropriately, an **unordered summation**.

The advantage of unordered summation is that it can be easily generalised. Indeed if $X$ is any set whatsoever and $f$ is a function defined on $X$ which assumes only non-negative values then we can still consider the set $S(f)$ defined as $\left\{\sum_{x \in A} f(x) : A \subset X, A \text{ finite}\right\}$. If this set is bounded above then we say $f$ is **summable** over $X$ and the supremum of this set is denoted by $\sum(f)$ or by $\sum_{x \in X} f(x)$. More generally, for a subset $Y$ of $X$, we denote $\sum_{x \in Y} f(x)$ to be the supremum of the set $\left\{\sum_{x \in A} f(x) : A \subset Y, A \text{ finite}\right\}$. Clealry $\sum_{x \in Y} f(x) \le \sum_{x \in X} f(x)$ and so, summability of $f$ over $X$ implies its summability over every $Y \subset X$. But it may happen that even though $f$ is not summable over the entire set $X$, it is summable over some subset $Y$ of $X$. A trivial example is to take $Y$ to be a finite subset of $X$. As a non-trivial example, let $X = \mathbb{N}, Y = \{n^2 : n \in \mathbb{N}\}$ and let $f : X \to \mathbb{R}$ be the function $f(x) = \frac{1}{x}$. Then the divergence of the harmonic series (Theorem (6.1)) is equivalent to saying that $f$ is not summable over $X$. But $f$ is summable over $Y$ because the series $\sum_{n=1}^{\infty} \frac{1}{n^2}$ is convergent. In fact, if we use Exercise (1.4.9) then $\sum_{x \in Y} f(x) = \frac{\pi^2}{6}$. The result of Exercise (6.10) can be expressed by saying that the function $f(x) = \frac{1}{x}$ is not summable over the set of primes.

When a function $f$ is not summable over a set $X$, it means that the set
$$\left\{ \sum_{x \in A} f(x) : A \subset X, A \text{ finite} \right\} \text{ is not bounded above. In such a case we set}$$
$\sum_{x \in X} f(x)$ equal to $\infty$.

Even if the function $f$ assumes negative values, we could, of course, still define $\sum_{x \in A} f(x)$ as the supremum of the set $\{\sum_{x \in A} f(x) : A \subset X, A \text{ finite}\}$. But in doing so, we are, in effect, ignoring points where $f$ is negative since they contribute nothing towards this supremum. To see this, suppose $A$ is a finite subset of $X$. Write $A$ as $B \cup C$ where $B = \{x \in A : f(x) \geq 0\}$ and $C = \{x \in A : f(x) < 0\}$. Clearly, $\sigma(A) = \sum_{x \in A} f(x)) = \sigma(B) + \sigma(C) < \sigma(B)$ since $\sigma(C) < 0$ (assuming $C \neq \phi$). Since we are taking supremum, we might as well forget $\sigma(A)$ and consider only $\sigma(B)$. So $\sup\{\sigma(A) : A \subset X, A \text{ finite}\}$ will not be a satisfactory definition of $\sum_{x \in A} f(x)$. We shall take up this problem in the next section. Here we stick to the case of a function which takes only non-negative values.

A special case of unordered summation deserves explicit mention. Let $X = I\!N \times I\!N$, i.e., the cartesian product of $I\!N$ with itself. By defintion, $X$ consits of all ordered pairs of the form $(m, n)$ where $m, n$ are positive integers. Let $f$ be a function with $X$ as a domain. Then $f$ is called a **doubly infinite sequence** or a sequence with two parameters. For $m, n \in I\!N$, the value of $f$ at $(m, n)$ is denoted, instead of by the usual $f(m, n)$ (or more fussily, by $f((m, n))$), by symbols like $a_{(m,n)}$ which are abbreviated to $a_{m,n}$ or even to $a_{mn}$ where it is understood that $mn$ does not mean the prdoduct of the integers $m$ and $n$, but stands for the ordered pairs $(m, n)$. In Fig. (3.7.1) we picture the doubly infinite sequence $a_{mn} = \frac{1}{m + 2n}$ by writing its values at a few circled points of the domain.

There is no natural order on the set $I\!N \times I\!N$ and so the doubly infinite sum $\sum_{m=1}^{\infty} a_{m,n}$ cannot be defined in a manner which is a straightforward generalisation of the ordinary inifinite series (with only one parameter), viz. as the limit of the sequence of their partial sums. However, if $f$ takes only non-negative values then we can use unordered summation and define $\displaystyle\sum_{m=1}^{\infty} a_{m,n}$ as $\displaystyle\sum_{(m,n)\in I\!N \times I\!N} f(m,n)$,

the latter being defined as $\sup \left\{ \displaystyle\sum_{(m,n)\in A} a_{m,n} : A \subset I\!N \times I\!N, A \text{ finite} \right\}$.

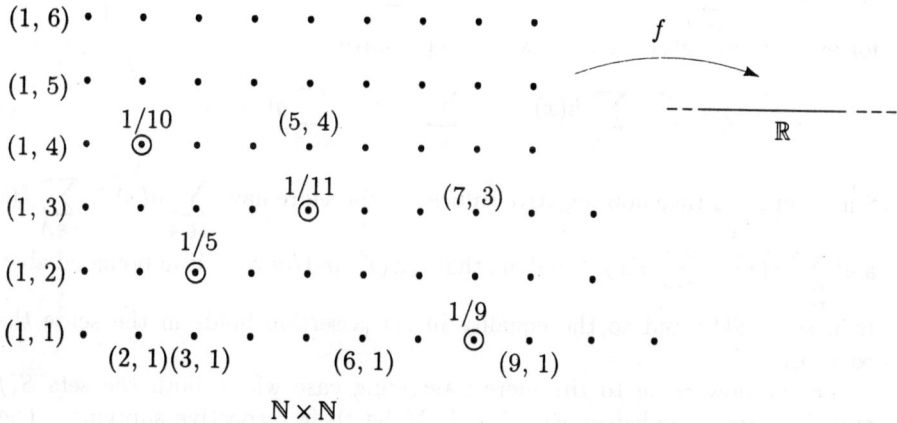

Figure 3.7.1 : A Doubly Infinite Sequence

Most of the theorems about infinite series generalise to unordered summations. But some care is necessary. Suppose, for example, that $f, g$ are two functions defined on the same set $X$. Then even if $f, g$ take only non-negative values, the same need not be true of $f - g$. So an equation like $\displaystyle\sum_{x\in X}(f - g)(x) = \sum_{x\in X} f(x) - \sum_{x\in X} g(x)$ may be meaningless. For $f + g$, however, there is no difficulty and we have the following result.

**7.3 Theorem :** Suppose $f, g$ are non-negative real valued functions on a set $X$. For $x \in X$, define $h(x) = f(x) + g(x)$. Then $\displaystyle\sum_{x\in X} h(x) = \sum_{x\in X} f(x) + \sum_{x\in X} g(x)$.

**Proof :** Let us first see what the theorem says. We are not assuming that $\displaystyle\sum_{x\in X} f(x)$ and $\displaystyle\sum_{x\in X} g(x)$ are finite, i.e., $f$ and $g$ are summable over $X$. One of these quantities could be $\infty$. In that case, the theorem asserts that $\displaystyle\sum_{x\in X} h(x)$ is also $\infty$ and the equality in the assertion holds in the sense that both the sides are infinite. If, on the other hand, both $f, g$ are summable then the theorem asserts that $h$ is also summable over $X$ and the equality holds as an equality of two real numbers. This will be typical of nearly all equalities about unordered summations. In the present theorem, we prove both the cases. But in the subsequent ones we shall prove only the case where both the sides of the equality are finite. The case of an equality of the form $\infty = \infty$ is usually easier and will be left as an exercise.

Let $S(f), S(g), S(h)$ be the sets whose suprema are, respectively, $\sum\limits_{x \in X} f(x)$, $\sum\limits_{x \in X} g(x)$ and $\sum\limits_{x \in X} h(x)$. That is, $S(f) = \left\{ \sum\limits_{x \in A} f(x) : A \subset X, A \text{ finite} \right\}$ etc. Now for every *finite* subset $A$ of $X$ we certainly have,

$$\sum_{x \in X} h(x) \;=\; \sum_{x \in A} f(x) + \sum_{x \in A} g(x). \tag{1}$$

Since both $f, g$ take non-negative values, we therefore have $\sum\limits_{x \in A} h(x) \geq \sum\limits_{x \in A} f(x)$ and $\sum\limits_{x \in A} h(x) \geq \sum\limits_{x \in A} g(x)$. It follows that if $S(f)$ and/or $S(g)$ is unbounded above then so is $S(h)$ and so the equality in the assertion holds in the sense that $\infty = \infty$.

Let us now come to the more interesting case where both the sets $S(f)$ and $S(g)$ are bounded above. Let $L, M$ be their respective suprema. Then $L = \sum\limits_{x \in X} f(x)$ and $M = \sum\limits_{x \in X} g(x)$. From (1) it is clear that $L + M$ is an upper bound for the set $S(h)$. To complete the proof we must show that it is, in fact, the least upper bound, i.e., the supremum of $S(h)$. In other words for every $\epsilon > 0$, we must show that $L + M - \epsilon$ is not an upper bound of the set $S(h)$. Write $L + M - \epsilon$ as $\left( L - \dfrac{\epsilon}{2} \right) + \left( M - \dfrac{\epsilon}{2} \right)$. Since $L - \dfrac{\epsilon}{2}$ is not an upper bound of the set $S(f)$, there exists a finite subset $A$ of $X$ such that

$$\sum_{x \in A} f(x) > L - \frac{\epsilon}{2} \tag{2}$$

Similarly, there exists a finite subset $B$ of $X$ such that

$$\sum_{x \in B} g(x) > M - \frac{\epsilon}{2} \tag{3}$$

The trouble is that the subsets $A$ and $B$ may be different. If they were the same, we could merely add (2) and (3) to get an element of $S(h)$ which exceeds $L + M - \epsilon$. To remedy the situation where $A \neq B$, we let $C = A \cup B$. Then $C$ is also a finite subset of $X$. Moreover since $f$ takes only non-negative values, (2) continues to hold if $A$ is replaced by the bigger set $C$. Similarly we can replace $B$ by $C$ in (3). Then adding, we get $\sum\limits_{x \in C} h(x) > L + M - \epsilon$ and so $L + M - \epsilon$ is not an upper bound of the set $S(h)$. As shown before, this completes the proof. ∎

We now prove a theorem to be called **decomposition theorem** whose truth is very obvious for finite summations. Suppose $X$ is a finite set and $f$ is a real-valued function on $X$. If $X = X_1 \cup X_2 \cup \ldots \cup X_k$ is a decomposition of the set $X$ into a finite number of mutually disjoint subsets $X_1, \ldots, X_k$ then it is clear that

$$\sum_{x \in X} f(x) = \sum_{x \in X_1} f(x) + \sum_{x \in X_2} f(x) + \ldots + \sum_{x \in X_k} f(x) = \sum_{i=1}^{k} \sum_{x \in X_i} f(x) \quad (4)$$

In effect, all we are doing is to group the terms to be added, add the terms in each group and finally add all these sums. In fact, here we do not need any restriction on $f$. If the set $X$ is infinite, and $f$ takes only non-negative values then also (4) holds (both sides possibly being infinite). We leave the proof as an exercise with the hint that for every finite subset $A$ of $X$, the sets $A \cap X_1, \ldots, A \cap X_k$ (some of which may be empty) give a decomposition of $A$.

What is not so obvious is that (4) remains valid even if we have a decomposition of $X$ into an *infinite* family of mutually disjoint subsets, say, $X_1, X_2, \ldots, X_k, \ldots$. Such decompositions arise, rather naturally, sometimes. For example, suppose $X = I\!\!N \times I\!\!N$, already pictured in Fig. 3.7.1. Then we can decompose $X$ either 'rowwise' (i.e., into subsets $I\!\!N \times \{1\}, I\!\!N \times \{2\}, \ldots$) or 'columnwise' or in many other ways. In Exercise (6.8), the set $A_*$ was decomposed into the sets $A_1, A_2, \ldots, A_k, \ldots$. The theorem we are about to prove provides the justification needed for that exercise.

**7.4 Theorem :** Suppose $X = X_1 \cup X_2 \cup \ldots \cup X_k \cup \ldots$ where for every $i \neq j, X_i \cap X_j = \emptyset$. Then for every non-negative real-valued function $f$

$$\sum_{x \in X} f(x) = \sum_{n=1}^{\infty} \sum_{x \in X_n} f(x). \quad (5)$$

**Proof :** As mentioned before, we prove (5) only in the more interesting case where both the sides are finite. So assume $f$ is summable over each $X_n$ and that the infinite series in (5) is convergent. For notational simplicity write $L_n$ for

$$\sup \left\{ \sum_{x \in A} f(x) : A \subset X_n, A \text{ finite} \right\} \text{ and let } L \text{ be the value of } \sum_{n=1}^{\infty} L_n. \text{ We have to}$$

show that $L$ is the supremum of the set $S(f) = \left\{ \sum_{x \in A} f(x) : A \subset X, A \text{ finite} \right\}$.

We first show that $L$ is an upper bound. For this let $A \subset X$ be a finite subset. Then $A$ intersects at most a finite number of the sets $X_1, \ldots, X_k, \ldots$. That is, there exist integers $n_1, n_2, \ldots, n_k$ (say) such that

$$A = (A \cap X_{n_1}) \cup (A \cap X_{n_2}) \cup \ldots \cup (A \cap X_{n_k}) \quad (6)$$

Since $A$ is finite, so is every set on the right. Therefore,

$$\sum_{x \in A} f(x) = \sum_{x \in A \cap X_{n_1}} f(x) + \ldots + \sum_{x \in A \cap X_{n_k}} f(x)$$

The terms on the right are at most $L_{n_1}, L_{n_2}, \ldots L_{n_k}$ respectively. Hence $\sum_{x \in A} f(x)$ is at most $L_{n_1} + L_{n_2} + \ldots + L_{n_k}$. But by Theorem (7.1) (applied to the series $\sum_{n=1}^{\infty} L_n$ of non-negative terms), $L_{n_1}, L_{n_2}, \ldots L_{n_k}$ is at most $L$. So $L$ is an upper bound for the set $S(f)$.

To finish the proof we have to show that for every $\epsilon > 0$, there is some finite subset $A$ of $X$ such that $\sum_{x \in A} f(x) > L - \epsilon$. This is the only non-trivial part of the proof. First, from the definition of convergence of the series $\sum_{n=1}^{\infty} L_n$, we get some $m \in I\!N$ such that

$$\sum_{i=1}^{m} L_i > L - \frac{\epsilon}{2} \tag{7}$$

Now, for each $i = 1, \ldots, m$, there exists a finite subset $A_i$ of $X_i$ such that

$$\sum_{x \in A_i} f(x) > L_i - \frac{\epsilon}{2m} \tag{8}$$

Let $A = A_1 \cup A_2 \cup \ldots \cup A_m$. Then $A$ is a finite subset of $X$ and

$$\sum_{x \in A} f(x) \;=\; \sum_{i=1}^{m} \sum_{x \in A_i} f(x) \tag{9}$$

Adding the inequalities in (8) and using (7), we get $\sum_{x \in A} f(x) > L - \epsilon$ as desired. So $f$ is summable over $X$ and (5) holds.

Conversely suppose that $f$ is summable over $X$. Write $L$ for $\sum_{x \in X} f(x)$. Then trivially $f$ is summable over each $X_n$. Let $L_n = \sum_{x \in X_n} f(x)$. We claim that the series $\sum_{n=1}^{\infty} L_n$ is convergent. Since this is a series of non-negative terms it suffices to show that the sequence of its partial sums is bounded. Now for each $n \in I\!N$, $L_1 + L_2 + \ldots + L_n$ equals $\sum_{x \in Y_n} f(x)$ where $Y_n = X_1 \cup X_2 \cup \ldots \cup X_n$. This follows from the remarks above about the validity of the decomposition theorem for a finite decomposition. But since $f$ takes only non-negative values, $\sum_{x \in Y_n} f(x) \le \sum_{x \in X} f(x) = L$. Putting it all together, $\sum_{i=1}^{n} L_i \le L$ for all $n$ and so the series $\sum_{n=1}^{\infty} L_n$ is convergent. Suppose it converges to $M$. Then by the part of the theorem proved above, $M$ must equal $\sum\limits_{x \in X} f(x)$. So $M = L$. The theorem is now completely established. ∎

As an interesting application of this theorem we have the following corollary about double summations. Suppose for every $i = 1, \ldots, m$ and $j = 1, \ldots, n$ we have a real number $a_{ij}$. Then we have

$$\sum_{i=1}^{n}\sum_{j=1}^{n} a_{ij} = \sum_{j=1}^{n}\sum_{i=1}^{m} a_{ij} \tag{10}$$

This is sometimes called the formula for the **change of order of summation,** because on the left hand side we are first summing over $j$ for a fixed $i$ and then adding these sums. On the right hand side we are first summing over $i$, for a fixed $j$. It can be shown by simple examples that the change of order of summation is not generally valid for doubly infinite summations (Exercise (7.8)). However, if the terms are non-negative then it is valid as we show.

**7.5 Corollary :** Suppose $\{a_{m,n}\}_{\substack{m=1 \\ n=1}}^{\infty}$ is a doubly infinite sequence of non-negative terms. Then

$$\sum_{m=1}^{\infty}\sum_{n=1}^{\infty} a_{m,n} = \sum_{n=1}^{\infty}\sum_{m=1}^{\infty} a_{m,n} \tag{11}$$

**Proof :** Let $X = I\!N \times I\!N$ and define $f : X \to I\!R$ by $f(m,n) = a_{m,n}$. If we let $X = X_1 \cup X_2 \cup \ldots \cup X_n \cup \ldots$ where $X_n$ is the '$n^{\text{th}}$ row' of $I\!N \times I\!N$, then for each $n$, $\sum\limits_{m=1}^{\infty} a_{m,n}$ is simply $\sum\limits_{x \in X_n} f(x)$ and the sum on the right of (11) is $\sum\limits_{n=1}^{\infty}\sum\limits_{x \in X_n} f(x)$. By the theorem above, this equals $\sum\limits_{x \in X} f(x)$. But if we decompose $X$ columnwise, then the left hand side of (11) also equals $\sum\limits_{x \in X} f(x)$. So (11) holds. ∎

The property expressed in this corollary is also known as **commutativity of two summations,** one with respect to $m$ and the other with respect to $n$. The name obviously comes from the well-known commutative laws for addition and multiplication, which simply say that the order you add two terms does not matter (i.e., $x + y = y + x$) and similarly for multiplication. More generally, two processes are said to commute with each other when the net result of performing them is the same no matter in which order they are performed. Many results in calculus can be stated succinctly in terms of commutativity of some process. For example, the simple result that the limit of a sum of two and hence any finite number of sequences is the sum of their limits can be paraphrased by saying that limits commute with finite summation.

Mere paraphrases do not, of course, prove a result. But they often enable you to see the similarity between two results, and sometimes to anticipate new results. For example, the corollary above says that under certain conditions, two limits (each of the form of an infinite sum) commute with each other. Later we shall encounter a few other results where certain other types of limits commute with each other. We only remark here that the validity of term-by-differentiation (cf. Exercise (3.14)) can be interpreted as commutativity of differentiation (which is a form of limits) with infinite summation (another form of limits.)

As another application of Theorem (7.4), we prove one more theorem about infinite series, whose truth is trivial for finite sums. Suppose $L = a_1 + a_2 + \ldots + a_m$ and $M = b_1 + \ldots b_n$. Then the product of these sums, i.e., $LM$, equals the sum of the $mn$ terms of the form $a_i b_j$. For infinite sums this is not true in general. But the situation is better if the terms are all non-negative.

**7.6 Theorem :** Suppose $\sum\limits_{n=1}^{\infty} a_n$ and $\sum\limits_{n=1}^{\infty} b_n$ are convergent series of non-negative terms. Then the doubly infinite series $\sum\limits_{\substack{m=1\\n=1}}^{\infty} a_m b_n$ is also convergent and its value

is $\left( \sum\limits_{m=1}^{\infty} a_m \right) \left( \sum\limits_{n=1}^{\infty} b_n \right)$.

**Proof :** Once again, let $X = \mathbb{N} \times \mathbb{N}$ and define $f : X \to \mathbb{R}$ by $f(m,n) = a_m b_n$. If we decompose $X$ rowwise as $X = X_1 \cup X_2 \cup \ldots \cup X_n \cup \ldots$ then

$$\sum_{x \in X_n} f(x) = \sum_{m=1}^{\infty} (a_m b_n) = b_n \sum_{m=1}^{\infty} a_m \text{ for each } n = 1, 2 \ldots . \text{ So the double sum}$$

$$\sum_{n=1}^{\infty} \sum_{x \in X_n} f(x) \text{ equals } \sum_{n=1}^{\infty} \left( b_n \sum_{m=1}^{\infty} a_m \right) \text{ which is simply } \left( \sum_{m=1}^{\infty} a_m \right) \left( \sum_{n=1}^{\infty} b_n \right)$$

since the sum $\sum\limits_{m=1}^{\infty} a_m$ is a common factor in each term. But by Theorem (7.4),

$$\sum_{n=1}^{\infty} \sum_{x \in X_n} f(x) \text{ equals } \sum_{x \in X} f(x), \text{ which in the present case is simply the doubly}$$

infinite sum $\sum\limits_{\substack{m=1\\n=1}}^{\infty} a_m b_n$. The result follows. $\blacksquare$

The theorem can be easily extended, by induction, to the product of a finite number of convergent series of non-negative series. We leave this extension as an exercise, but remark that it was needed in Exercise (6.7).

In summary, we see that series of non-negative terms are easier to handle because they obey more or less the same laws which hold for finite sums. This gives us considerable freedom in manipulating them, an excellent illustration of it being Exercise (6.7).

# EXERCISES

7.1  If $\sum\limits_{n=1}^{\infty} a_n$ is a series of non-negative terms and $\sum\limits_{n=1}^{\infty} b_n$ is obtained by rearranging its terms, prove directly that $\sum\limits_{n=1}^{\infty} a_n = \sum\limits_{n=1}^{\infty} b_n$. [*Hint* : Let $S_n, T_n$ be their partial sums. Show that for every $n$, there is some $m$ such that $T_n \leq S_m$ and vice versa.]

7.2  Prove (4) for the case where $X$ is infinite.

7.3  Let $X = \mathbb{N} \times \mathbb{N}$. For every positive integer $k$, let $Y_k = \{(m,n) \in X : m + n = k\}$. Prove that each $Y_k$ is a finite set with $k - 1$ elements. (In particular $Y_1 = \emptyset$.) Prove further that $X = Y_1 \cup Y_2 \cup \ldots \cup Y_k \cup \ldots$ is a decomposition of $X$ into mutually disjoint subsets. Interpret each $Y_k$ geometrically (i.e., indicate it in Fig. 3.7.1.).

7.4  The **Cauchy product** of two series $\sum\limits_{n=1}^{\infty} a_n$ and $\sum\limits_{n=1}^{\infty} b_n$ is defined as the series $\sum\limits_{n=1}^{\infty} c_n$ where $c_n = \sum\limits_{i=1}^{n-1} a_i b_{n-i}$, (set $c_1 = 0$.) Prove that if $\sum\limits_{n=1}^{\infty} a_n$ and $\sum\limits_{n=1}^{\infty} b_n$ are convergent series of non-negative terms then $\sum\limits_{n=1}^{\infty} c_n$ is convergent. Identify its sum.

*7.5  Extend Theorem (7.6) to the case of a product of any finite number of convergent series of non-negative terms.

7.6  Using Theorem (7.6), show that if $a_n \geq 0$ for all $n$ and $\sum\limits_{n=1}^{\infty} a_n$ is convergent then $\sum\limits_{n=1}^{\infty} a_n^2$ is also convergent. (A proof based directly on the Cauchy criterion for convergence is also possible.)

7.7  Suppose a non-negative real valued function $f$ is summable on a set $X$. For $X_1, X_2 \subset X$, prove that

$$\sum_{x \in X_1 \cup X_2} f(x) = \sum_{x \in X_1} f(x) + \sum_{x \in X_2} f(x) - \sum_{x \in X \cap X_2} f(x).$$

State a similar result for the case of three subsets $X_1, X_2, X_3$ of $X$.

7.8  Figure 3.7.2 shows the values of a doubly infinite sequence $\{a_{m,n}\}_{\substack{m=1 \\ n=1}}^{\infty}$.

| | | | | | | |
|---|---|---|---|---|---|---|
| 0 • | 0 • | 0 • | 0 • | $\overset{\bullet}{-1}$ | $\overset{\bullet}{1/2}$ | |
| 0 • | 0 • | 0 • | $\overset{\bullet}{-1}$ | $\overset{\bullet}{1/2}$ | $\overset{\bullet}{1/4}$ | |
| 0 • | 0 • | $\overset{\bullet}{-1}$ | $1/2$ | $\overset{\bullet}{1/4}$ | $\overset{\bullet}{1/8}$ | |
| 0 • | $\overset{\bullet}{-1}$ | $\overset{\bullet}{1/2}$ | $\overset{\bullet}{1/4}$ | $\overset{\bullet}{1/8}$ | $\overset{\bullet}{1/16}$ | |
| $\overset{\bullet}{-1}$ | $1/2$ | $1/4$ | $1/8$ | $1/16$ | $\overset{\bullet}{1/32}$ | $\overset{\bullet}{1/64}$ |

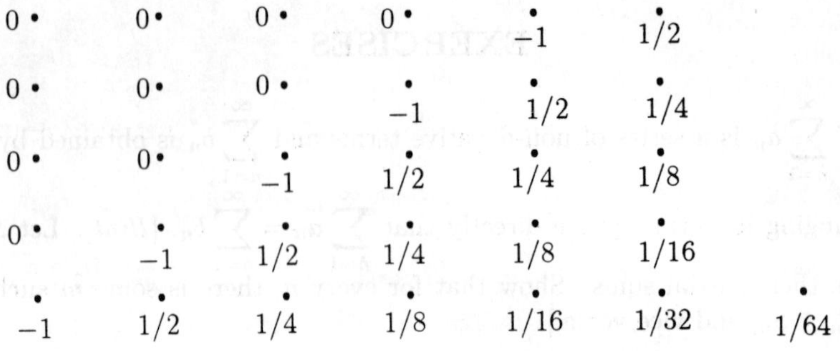

Fig. 3.7.2 : A Misbehaving Doubly Infinite Sequence

Write an explicit formula for $a_{m,n}$ in terms of $m$ and $n$. For this doubly infinite sequence show that $\sum_{n=1}^{\infty}\sum_{m=1}^{\infty} a_{m,n}$ and $\sum_{n=1}^{\infty}\sum_{m=1}^{\infty} a_{m,n}$ both exist but are not equal. (It is much easier to see this from the diagram, rather than from a formula for $a_{m,n}$.) In other words Corollary (7.5) may fail if the hypothesis about non-negativity is dropped.

7.9  For a positive integer $n$, let $a_n = \dfrac{1}{n+1} + \dfrac{1}{n+2} + \ldots + \dfrac{1}{2n-1} + \dfrac{1}{2n}$.
Prove that $a_n \geq \frac{1}{2}$ for all $n$ and hence $\lim_{n\to\infty} a_n$, if it exists, is $\geq \frac{1}{2}$. Can we find this limit by taking the limit of each term and adding, in view of commutativity of limits and finite summation? (For the actual value of the limit, see Exercise (5.7.20).)

## 3.8  Why an Absolutely Convergent Series is Convergent

Linguistically, the question may appear silly. The adverb 'absolutely' indicates certainty, no possibility of a mistake. And so it would be a contradiction of terms if we had a series which is absolutely convergent but not convergent.

This is, of course, no mathematical proof. If it were then we could prove anything merely by assigning suitable names! In mathematics, we have the freedom to define things. But after that, we are completely governed by the definitions and the rules of deductive logic. For example, in Exercise (5.5), an open interval is not an open set just because of its name. It needs a proof.

In the case of an absolutely convergent series, the word 'absolutely' refers to the absolute value or the modulus of a real number. We have already encountered this concept many times in the various definitions and proofs. Probably, the most important property of absolute values (which we have already used several times, e.g. in the proofs of Theorems (2.3.1),(5.3) and (5.4)) is the so-called triangle inequality (see Exercises (1.9.8) and (1.9.9)) which states that for any two real number $x_1$ and $x_2$,

$$|x_1 + x_2| \leq |x_1| + |x_2| \tag{1}$$

The triangle inequality can be generalised, by induction, to the sum of any finite real numbers. Thus, for any real numbers $x_1, x_2, \ldots, x_n$ we have

$$|\sum_{i=1}^{n} x_i| \leq \sum_{i=1}^{n} |x_i| \tag{2}$$

This fact, coupled with the Cauchy criterion for convergence of series (Theorem (5.5)) gives an easy proof that absolute convergence implies convergence. A series $\sum_{n=1}^{\infty} a_n$ is said to be **absolutely convergent** if the series of the absolute values of its terms, i.e. the series $\sum_{n=1}^{\infty} |a_n|$ is convergent.

**8.1 Theorem :** Every absolutely convergent series is convergent.

**Proof :** Assume $\sum_{n=1}^{\infty} |a_n|$ is convergent. We apply Theorem (5.5). So, let $\epsilon > 0$ be given. Then since $\sum_{n=1}^{\infty} |a_n|$ is convergent, by Theorem (5.5) there exists $p \in \mathbb{N}$ such that whenever $n \geq m \geq p$,

$$\sum_{k=m}^{n} |a_k| < \epsilon. \tag{3}$$

But by (2) (with a different notation), we have

$$\left| \sum_{k=m}^{n} a_k \right| \leq \sum_{k=m}^{n} |a_k| \tag{4}$$

From (3) and (4) we have

$$|\sum_{k=m}^{n} a_k| < \epsilon \text{ for all } n \geq m \geq p.$$

So by Theorem (5.5), the series $\sum_{n=1}^{\infty} a_n$ is convergent (Note, however, that in general the sums $\sum_{n=1}^{\infty} a_n$ and $\sum_{n=1}^{\infty} |a_n|$ are different. The relationship between them will be given in the exercises.) ∎

The preceding theorem may appear trivial because of its short proof. But it is not quite so. The proof appears short because the real hard work was done in the proofs of Theorems (5.4) and (5.5). As mentioned earlier, Theorem (5.4) is one of the versions of completeness of $\mathbb{R}$. So it is really the completeness which makes the preceding theorem true. A counter-example will be given in the exercises to show that the theorem is false in the rational number system.

The converse of the last Theorem is false. Probably the most standard counter-example is the series

$$1 - \frac{1}{2} + \frac{1}{3} - \frac{1}{4} + \cdots + (-1)^{n+1}\frac{1}{n} + \cdots \tag{5}$$

The convergence of this series follows from the well-known **alternating series test** (see Exercise (8.1)). That it is not absolutely convergent follows from Theorem (6.1). A series which is convergent but not absolutely convergent is sometimes called **conditionally convergent.**

In the last section we showed that series of non-negative terms have certain pleasant properties. It turns out that most of these properties are shared by absolutely convergent series. As a result, absolutely convergent series are almost as easy to handle as series of non-negative terms. In particular, the order of the terms is immaterial. For a conditionally convergent series, on the other hand, the order of the terms is vital as we shall see.

The transition from absolutely convergent series to series of non-negative terms depends on a simple trick. Suppose $a$ is a real number. Let $a^+ = \max\{a, 0\}$. That is, $a^+ = a$ if $a \geq 0$ and $0$ otherwise. Similarly, let $a^- = \max\{-a, 0\}$. That is $a^- = -a$ if $a \leq 0$ and $0$ otherwise. Note that $a^+$ and $a^-$ are both non-negative. They are called respectively the **positive** and the **negative parts** of $a$. Note that the negative part itself is not negative! Perhaps non-negative part and non-positive parts will be more appropriate terms. The following relations are easy to prove for every $a \in \mathbb{R}$.

$$a = a^+ - a^- \tag{6}$$

$$\text{and} \quad |a| = a^+ + a^- \tag{7}$$

Because of these simple formulas, it is easy to characterise absolutely convergent series.

**8.2 Theorem :** A series $\sum_{n=1}^{\infty} a_n$ is absolutely convergent if and ony if each of the two series $\sum_{n=1}^{\infty} a_n^+$ and $\sum_{n=1}^{\infty} a_n^-$ is convergent. Moreover, when this happens, its sum is $\displaystyle\sum_{k=1}^{\infty} a_n^+ - \sum_{n=1}^{\infty} a_n^-$.

**Proof** : Let $S_n, T_n, U_n$ denote the $n^{th}$ partial sums of $\displaystyle\sum_{n=1}^{\infty} |a_n|, \sum_{n=1}^{\infty} a_n^+$ and $\displaystyle\sum_{n=1}^{\infty} a_n^-$ respectively. Then by (7)

$$S_n = T_n + U_n \tag{8}$$

and hence $T_n \le S_n$ and $U_n \le S_n$. If $\displaystyle\sum_{n=1}^{\infty} |a_n|$ is convergent then $\{S_n\}$ is bounded above. But then so are $T_n$ and $U_n$. Moreover both of them are monotonically increasing. So by Theorem (2.6.4) they are convergent, i.e., $\displaystyle\sum_{n=1}^{\infty} a_n^+$ and $\displaystyle\sum_{n=1}^{\infty} a_n^-$ are both convergent. The converse follows directly from (8). Finally, for the last part, we merely write $a_n = a_n^+ - a_n^-$ by (6) and add. ∎

This simple theorem enables us to extend the results about series of non-negative terms to absolutely convergent series. By way of illustration we extend Corollary (7.2) and leave the extensions of other results as exercises.

**8.3 Corollary :** An absolutely convergent series remains so and has the same sum after any re-arrangement of terms.

**Proof :** Assume $\sum_{n=1}^{\infty} a_n^+$ and $\sum_{n=1}^{\infty} a_n^-$ are both onvergent and further

$$\sum_{n=1}^{\infty} a_n = \sum_{n=1}^{\infty} a_n^+ - \sum_{n=1}^{\infty} a_n^-.$$

Suppose $\{b_n\}_{n=1}^{\infty}$ is a re-arrangement of $\{a_n\}_{n=1}^{\infty}$. Then $\{b_n^+\}$ and $\{b_n^-\}$ are re-arrangements of $\{a_n^+\}$ and $\{a_n^-\}$ respectively. So by Corollary (7.2), $\displaystyle\sum_{n=1}^{\infty} b_n^+$ and $\displaystyle\sum_{n=1}^{\infty} b_n^-$ converge respectively to $\displaystyle\sum_{n=1}^{\infty} a_n^+$ and $\displaystyle\sum_{n=1}^{\infty} a_n^-$. But then the equality $b_n = b_n^+ - b_n^-$ for all $n$ implies that $\displaystyle\sum_{n=1}^{\infty} b_n$ converges and its sum is $\displaystyle\sum_{n=1}^{\infty} b_n^+ - \sum_{n=1}^{\infty} b_n^-$, which is the same as $\displaystyle\sum_{n=1}^{\infty} a_n^+ - \sum_{n=1}^{\infty} a_n^-$, i.e., as $\displaystyle\sum_{n=1}^{\infty} a_n$. ∎

What about a conditionally convergent series ? The following theorem gives an answer.

**8.4 Theorem :** If a series $\sum_{n=1}^{\infty}$ is conditionally convergent then the series $\sum_{n=1}^{\infty} a_n^+$

and $\sum_{n=1}^{\infty} a_n^-$ are both divergent.

**Proof :** Following the same notations as in the proof of Theorem (8.2), we get, using (6) instead of (7), that

$$S_n = T_n - U_n \tag{9}$$

for all $n \in \mathbb{N}$. We are given that $\sum_{n=1}^{\infty} a_n$ is convergent and so the sequence $\{S_n\}$ is convergent. So by (8), the convergence of either one of $\{T_n\}$ and $\{U_n\}$ implies that of the other (simply write $U_n = T_n - S_n$ and $T_n = S_n + U_n$). Therefore $\{T_n\}$ and $\{U_n\}$ are either both convergent or both divergent. The first possibility would imply by Theorem (8.2), that $\sum_{n=1}^{\infty} a_n$ is absolutely convergent, contrary to the hypothesis. So $\sum_{n=1}^{\infty} a_n^+$ and $\sum_{n=1}^{\infty} a_n^-$ are both divergent. ∎

For example, in the case of the series (3), the series of positive parts, i.e., $\sum_{n=1}^{\infty} a_n^+$ is ,

$$1 + 0 + \frac{1}{3} + 0 + \frac{1}{5} + 0 + \dots$$

or, dropping the zero terms,

$$1 + \frac{1}{3} + \frac{1}{5} + \dots + \frac{1}{2n+1} + \dots$$

and similarly, the series of negative parts is

$$\frac{1}{2} + \frac{1}{4} + \frac{1}{6} + \frac{1}{8} + \dots + \frac{1}{2n} + \dots$$

and both of these are divergent (cf. Exercise (6.3)).

The converse of this theorem is false as is shown by the series $\sum_{n=1}^{\infty} (-1)^n$. Here the series of positive as well as negative parts are both divergent but the series is not conditionally convergent because the $n^{th}$ term does not tend to 0, which is a necessary (although not sufficient) condition for convergence (Exercise (1.8)).

What if we are also given that $a_n \to 0$ as $n \to \infty$, in addition to the divergence of both $\sum_{n=1}^{\infty} a_n^+$ and $\sum_{n=1}^{\infty} a_n^-$? It turns out that in this case, the behaviour of the series $\sum_{n=1}^{\infty} a_n$ is subject, entirely, to the order of the terms. With one arrangement of the terms, it can be made to diverge to $\infty$ but with some other arrangement it will diverge to $-\infty$. This may seem paradoxical and illustrates how our intuition can mislead us sometimes, especially where infinite sets are involved.

We indicate here how, with a suitable rearrangement of the terms, we can make $\sum_{n=1}^{\infty} a_n$ diverge to $\infty$. Think of the positive terms of the series as an infinite collection of stones of weights equal to the terms. Similarly let the absolute values of the negative terms of the series correspond to stones in another pile. We ignore the zero terms. The total weight of each pile is infinite. Now we go on picking one stone at a time from the second pile and 'counter' it by taking a sufficient (but finite) number of stones from the first pile, so that the net weight is at least one unit. If we do this for every stone of the second pile then we get a re-arrangement of the terms of the series in which the partial sum can be made as large as we please.

To formalise this argument, let the positive terms of the series be $a_{n_1}, a_{n_2}, a_{n_3}, \ldots, a_{n_k}, \ldots$ where $n_1 < n_2 < n_3, \ldots$. Similarly let its negative terms be $a_{m_1}, a_{m_2}, a_{m_3}, \ldots$ where again, $m_1 < m_2 < m_3 < \ldots$. Then $\sum_{k=1}^{\infty} a_{n_k}$ and $\sum_{k=1}^{\infty} a_{m_k}$ diverge to $\infty$ and $-\infty$ respectively. Now take $a_{m_1}$. Since $\sum_{k=1}^{\infty} a_{n_k}$ diverges, there exists $k_1$ such that $\sum_{k=1}^{k_1} a_{n_k} \geq |a_{m_1}| + 1$. Now take $a_{m_2}$. Again divergence of $\sum_{k=1}^{\infty} a_{n_k}$ implies there exists $k_2 > k_1$ such that $\sum_{k=k_1+1}^{k_2} a_{n_k} \geq |a_{m_2}| + 1$. Continuing in this manner, we get an infinite series

$$a_{n_1} + a_{n_2} + \cdots + a_{n_{k_1}} + a_{m_1} + a_{n_{k_1+1}} + a_{n_{k_1+2}} + \cdots + a_{n_{k_2}} + a_{m_2} + a_{n_{k_2+1}} + \cdots$$

whose partial sum after $k_r + r$ terms is at least $r$. We can now insert the zero terms (if any) at will and get a desired re-arrangement of the series $\sum_{n=1}^{\infty} a_n$. Note that we have not used the hypothesis that $a_n \to 0$ as $n \to \infty$. If we do and are more careful in choosing the 'stones' from the two piles then it can be shown that for every real number $\alpha$, there exists a re-arrangement of $\sum_{n=1}^{\infty} a_n$, which converges to $\alpha$. We shall not go into a proof of this, because the result is of little actual use. But it has a profound negative significance. In a sense it says that the convergence of a conditionally convergent series and also its value are purely accidental. By re-arranging the terms the series may behave totally differently.

In the last section we generalised the concept of convergence of a series of non-negative terms to that of summability of a non-negative valued function on an arbitrary set $X$. Suppose now we have a function $f : X \to \mathbb{R}$ which takes both positive and negative values. How can we define the unordered sum $\sum_{x \in X} f(x)$? We saw in the last section why we cannot define it simply as

$\sup \left\{ \sum_{x \in A} f(x) : A \subset X, A \text{ finite} \right\}$. The correct approach is indicated by the last part of Theorem (8.2). Given a function $f : X \to \mathbb{R}$, we define two functions $f^+ : X \to \mathbb{R}$ and $f^- : X \to \mathbb{R}$ by

$$f^+(x) = [f(x)]^+ = \max\{f(x), 0\}$$
$$\text{and} \qquad f^-(x) = [f(x)]^- = \max\{-f(x), 0\}$$

respectively. Then $f^+$ and $f^-$ are both non-negative valued and so the unordered sums $\sum_{x \in X} f^+(x)$ and $\sum_{x \in X} f^-(x)$ make sense.

As $f = f^+ - f^-$, it is tempting to define $\sum_{x \in X} f(x)$ as $\sum_{x \in X} f^+(x) - \sum_{x \in X} f^-(x)$. But this would be meaningless if it is of the form $\infty - \infty$. When only one of $\sum_{x \in X} f^+(x)$ and $\sum_{x \in X} f^-(x)$ is $\infty$ and the other is a (finite) real number, $\sum_{x \in X} f(x)$ would come out to be $\infty$ or $-\infty$. It is only when both $\sum_{x \in X} f^+(x)$ and $\sum_{x \in X} f^-(x)$ are both finite that their difference would be finite. This leads to the following definition :

**8.5 Definition :** A real-valued function $f : X \to \mathbb{R}$ is said to be **summable** over $X$ if the functions $f^+$ and $f^-$ are both summable over $X$.

The following theorem is left as an exercise (Exercise (8.5)).

**8.6 Theorem :** A real-valued function $f : X \to \mathbb{R}$ is summable over $X$ if and only if the function $|f|$ is summable. ∎

This theorem shows that for unordered summations, summability is the same as 'absolute summability'. In particular, there is no such thing as conditional summability, analogus to the conditionally convergent series. This is to be expected, because as we saw above, a particular ordering of terms makes a vital difference in the case of a conditionally convergent series.

## EXERCISES

8.1   Suppose $\{a_n\}$ is a monotonically decreasing sequence converging to 0. Prove that the series $\sum_{n=1}^{\infty} (-1)^{n+1} a_n$ is convergent. If $S$ is its sum, prove also that $|S_n - S| \leq a_n$ for all $n$. [*Hint* : Prove that the subsequences $\{S_{2n}\}_{n=1}^{\infty}$ and $\{S_{2n+1}\}_{n=1}^{\infty}$ of $\{S_n\}$ are, respectively, monotonically increasing and decreasing. Then show that they have a common limit. This test is called **Leibnitz's test** or **alternating series test**.]

8.2　In contrast to theorem (8.1), show that for a *sequence* $\{a_n\}$, the implication is exactly the other way. That is, convergence of $\{a_n\}$ implies that of $\{|a_n|\}$ but the converse is false in general. [*Hint* : Use the fact that for any two real numbers $a$ and $b$, $\big||a| - |b|\big| \leq |a - b|$; see Exercise (1.9.8).] Prove, however, that if $|a_n| \to 0$ then $a_n \to 0$ as $n \to \infty$.

8.3　For a real number $a$, show that $a^+ = \frac{1}{2}(|a| + a)$ and $a^- = \frac{1}{2}(|a| - a)$.

8.4　Show that the series $\displaystyle\sum_{n=0}^{\infty} a_n$ where

$$a_n = \begin{cases} \dfrac{1}{m!} & \text{if } n = 2m, m = 0, 1, 2, \ldots \\[2mm] \dfrac{1}{m!} - \dfrac{1}{2^{m-1}} & \text{if } n = 2m+1, m = 0, 1, 2, \ldots \end{cases}$$

is absolutely convergent but not convergent in the rational number system, even though every term is rational. [*Hint* : Use the facts about the number $e$ given in the Exercises (2.7.1) to (2.7.5).]

8.5　Prove Theorem (8.6). [*Hint* : Observe that if $0 \leq g(x) \leq h(x)$ for all $x \in X$, then the summability of $h$ over $X$ implies that of $g$.]

8.6　Suppose $f : X \to I\!R$ is summable over $X$. Prove Theorem (7.4) for $f$.

8.7　Suppose $\{a_{m,n}\}_{m=1 \atop n=1}^{\infty}$ is a doubly infinite sequence which is summable over the set $I\!N \times I\!N$. Prove that Corollary (7.5) holds.

8.8　Let $a_n = b_n = \dfrac{(-1)^{n+1}}{\sqrt{n}}$. Prove that $\displaystyle\sum_{n=1}^{\infty} a_n$ and $\displaystyle\sum_{n=1}^{\infty} b_n$ are conditionally convergent but that their Cauchy product $\displaystyle\sum_{n=1}^{\infty} c_n$ (see Exercise (7.4)) is divergent by showing that $|c_n| \geq \dfrac{2(n-1)}{n}$ for all $n \geq 1$. [*Hint* : Use A.M. - G.M. inequality.]

8.9　If $\displaystyle\sum_{n=1}^{\infty} a_n$ and $\displaystyle\sum_{n=1}^{\infty} b_n$ are absolutely convergent, show that their Cauchy product $\displaystyle\sum_{n=1}^{\infty} c_n$ is convergent. (Actually, this also holds when one of the two series is absolutely convergent and the other is merely convergent. But the proof is more involved.) [*Hint* : Begin by showing that the function $f : I\!N \times I\!N \to I\!R$ defined by $f(i, j) = a_i b_j$ is summable over $I\!N \times I\!N$.]

### 3.9    Multitude of Tests for Convergence of Series

By definition, a series $\sum_{n=1}^{\infty} a_n$ is convergent if and only if the sequence $\{S_n\}$ of
its partial sums is convergent. So, theoretically, the convergence of series is just
a part of convergence of sequences. Still, there are very few tests for convergence
of sequences. Whether a given sequence is convergent or not is usually settled
directly in an *ad hoc* manner. For the convergence of series, on the other hand,
there are dozens of tests.

To explain this paradox, when a sequence $\{a_n\}$ is given, usually $a_n$ is given as
a closed form expression in $n$, for example $a_n = \dfrac{(-1)^n 2^n}{n!}$ or $a_n = \sqrt{n+1} - \sqrt{n}$

etc. In the case of a series $\sum_{n=1}^{\infty} a_n$, even though $a_n$ may be given as a closed form
expression, it is very rarely that from this expression we can obtain a closed
form expression for $S_n$. Consider for example the harmonic series $\sum_{n=1}^{\infty} \dfrac{1}{n}$ or the

series $\sum_{n=1}^{\infty} \dfrac{1}{n^2}$. There is no closed form expression for the partial sums of these
series. And so their convergence has to be decided by special methods given
in Section 3.6. (In Section 3.1, we remarked that we also encounter recursively
defined sequences. If the recursion relations are not easy to solve then for such
sequences we have the same difficulty in testing their convergence as in the case
of series. In fact the sequence of partial sums of a series $\sum_{n=1}^{\infty} a_n$ is a special case
of recursively defined sequences, the recurrence relation being $S_{n+1} = S_n + a_n$.)

Another limitation we have to live with is that even if we are somehow able
to show that a series is convergent, there is usually no easy method for finding
its sum (see the comments in Section 3.6 about series of the form $\sum_{n=1}^{\infty} \dfrac{1}{n^y}$ for

$y > 1$). Even the mere task of testing the convergence of a series can be a
challenge in itself, because there is no golden test which will apply in all cases.
There is the Cauchy criterion, of course, given in Theorem (5.5). But its utility
is limited again by the fact that very rarely we can explicitly evaluate the sum
$\sum_{k=m}^{n} a_k$.

Let us first see those very few cases where it *is* possible to obtain a closed
form expression for $S_n$. We already know this to be the case for geometric series
(Theorem (3.5)), i.e., series whose terms form a geometric progression. If the
$a_n$'s form an arithmetic progression, say with the first term $a$ and common
difference $d$, then also we have a formula for $S_n$. But in this case, since $a_n =
a + (n-1)d$, we see directly that if $d \neq 0$ then $a_n \nrightarrow 0$ and so this series is
not convergent by Exercise (1.8). On the other hand if $d = 0$, then $a_n = a$
for all $n$ and here, too, $a_n \nrightarrow 0$ except in the trivial case $a = 0$. So the closed

form expression for $S_n$ in the case of a series whose terms are in A.P., although available, is of little interest.

Another type of series $\sum_{n=1}^{\infty} a_n$ for which a closed form expression for $S_n$ is possible is what is called a **telescopic series**. The peculiar name comes from the fact that a telescope has a series of cylindrical shells which slide one inside the other, the outer diameter of a shell equalling the inner diameter of the shell it slides into. Typically, the $n^{th}$ term, $a_n$, of a telescopic series can be expressed as a difference of two (or more) not necessarily consecutive terms of some other sequence $\{b_n\}$ so that when we add $a_1, a_2, \ldots, a_n$ most of the $b_i$'s cancel out in pairs. Suppose, for example, that $a_n = \dfrac{1}{n(n+1)}$ (see Exercise (1.7)). Then $a_n = b_n - b_{n+1}$ where $b_n = \dfrac{1}{n}$. So $S_n = \sum_{k=1}^{n} a_k = (b_1 - b_2) + (b_2 - b_3) + \ldots + (b_n - b_{n+1}) = b_1 - b_{n+1} = 1 - \frac{1}{n+1}$. It is clear that $S_n \to 1$ as $n \to \infty$. And so $\sum_{n=1}^{\infty} a_n$ is convergent with sum 1.

Unfortunately, unlike the geometric series which arise in many interesting problems (cf. Exercises (3.8) to (3.11)), the telescopic series rarely arise in practice. So in effect, the geometric series are just about the only ones whose convergence is decided by explicitly calculating their partial sums. In all other cases, we try to get approximate rather than exact value of $S_n$. This approximation need not be very close, because the idea is not to calculate $S_n$ *per se* but only to conclude whether the sequence $\{S_n\}$ is convergent. Even a crude approximation can sometimes tell whether the series is convergent or not. This is what we did in the case of the harmonic series $\sum_{n=1}^{\infty} \dfrac{1}{n}$ in Section 3.6 when we showed that if $n$ is of the form $2^m$ then $S_n \geq \sum_{k=1}^{m} \dfrac{1}{2} = \dfrac{m}{2}$ and hence that $\sum_{n=1}^{\infty} \dfrac{1}{n}$ is divergent. Similarly, for the series $\sum_{n=1}^{\infty} \dfrac{1}{n^2}$, we showed, through Theorem (6.2), that when $n$ is of the form $2^m - 1$, $S_n \leq \sum_{k=1}^{m} \dfrac{1}{2^{k-1}}$. The convergence of the geometric series $\sum_{k=1}^{\infty} \dfrac{1}{2^{k-1}}$ then implied that $\sum_{n=1}^{\infty} \dfrac{1}{n^2}$ is convergent.

The strategy adopted in these examples is typical. We compare the given series with some known series. The following test known as the **comparison test** is the basic tool to relate the convergence or otherwise of one series to another.

**9.1 Theorem :** Let $\sum_{n=0}^{\infty} a_n$ and $\sum_{n=0}^{\infty} b_n$ be series of non-negative terms. Suppose

$a_n \le b_n$ eventually (i.e., for all $n$ after some stage). Then convergence of $\sum\limits_{n=1}^{\infty} b_n$

implies that of $\sum\limits_{n=1}^{\infty} a_n$. (And hence divergence of $\sum\limits_{n=1}^{\infty} a_n$ implies that of $\sum\limits_{n=1}^{\infty} b_n$.)

**Proof :** This follows easily from the Cauchy criterion, Theorem (5.5), because

for $n > m \ge r, |\sum\limits_{k=m}^{n} a_k| = \sum\limits_{k=m}^{n} a_k \le \sum\limits_{k=m}^{n} b_k = |\sum\limits_{k=m}^{n} b_k|$ where $r$ is chosen so that

$a_k \le b_k$ for all $k \ge r$. ∎

For example, it is easily seen that for $n \ge 2, n! \ge 2^n$ and so $\dfrac{1}{n!} \le (\dfrac{1}{2})^n$. Since

the series $\sum\limits_{n=0}^{\infty}(\dfrac{1}{2})^n$ is convergent, we get that $\sum\limits_{n=0}^{\infty} \dfrac{1}{n!}$ is convergent (cf. Exercise (2.7.1)).

**9.2 Corollary :** Suppose $\sum\limits_{n=1}^{\infty} a_n$ and $\sum\limits_{n=1}^{\infty} b_n$ are series of positive terms. Assume

$\dfrac{a_n}{b_n}$ tends to some finite non-zero limit $L$ as $n \to \infty$. Then convergence of either one of the two series implies that of the other.

**Proof :** Clearly $L \ge 0$ since $\dfrac{a_n}{b_n} > 0$ for all $n$. Since $L \ne 0$, we must have $L > 0$.

Choose $\epsilon > 0$ so small that $L - \epsilon > 0$. (For example $\epsilon = L/2$.) Then there exists $m$ such that for all $n \ge m, L - \epsilon < \dfrac{a_n}{b_n} < L + \epsilon$, and hence

$$(L - \epsilon)b_n \quad < \quad a_n < (L + \epsilon)b_n \tag{1}$$

Now suppose $\sum\limits_{n=1}^{\infty} a_n$ is convergent. Then from the first inequality in (1)

and the last theorem we get that $\sum\limits_{n=1}^{\infty}(L - \epsilon)b_n$ is convergent. But then so is

$\sum\limits_{n=1}^{\infty} b_n$ (since $L - \epsilon$ is a non-zero constant). Similarly, the second inequality in

(1) coupled with the theorem above, shows that the convergence of $\sum\limits_{n=1}^{\infty} b_n$ (and

hence of $\sum\limits_{n=1}^{\infty}(L + \epsilon)b_n$) implies that of $\sum\limits_{n=1}^{\infty} a_n$. ∎

This corollary is a very powerful tool if applied properly. Given a series $\sum\limits_{n=1}^{\infty} a_n$ of positive terms, we look for another, familiar series $\sum\limits_{n=1}^{\infty} b_n$ which is 'comparable' to $\displaystyle\sum_{n=1}^{\infty} a_n$ in the sense that $\dfrac{a_n}{b_n}$ approaches a non-zero limit and then

apply the Corollary above. The skill, of course, lies in choosing this series $\displaystyle\sum_{n=1}^{\infty} b_n$ correctly. There is no golden method to do this. But generally, one goes by the qualitative behaviour of the various parts of the expression for $a_n$. For example, multiplying or dividing by a non-zero constant does not change the behaviour of a series. Less trivially, if $a_n$ involvels a sum of two or more quantities, then one retains only the dominating among them (for large $n$) and drops the others. As

a concrete example consider $\displaystyle\sum_{n=1}^{\infty} \dfrac{3}{n(n+1)}$. Here the constant 3 has absolutely

no role to play and so, dropping it, we consider the series $\displaystyle\sum_{n=1}^{\infty} \dfrac{1}{n(n+1)}$. Earlier

we showed that this series is convergent, being a telescopic series. But that was more of a fortuitous coincidence. As a more systematic way, we note that the denominator of $a_n$ is $n^2 + n$. Here for large $n$, $n$ is insignificant as compared to $n^2$.

So dropping it, we set $b_n = \dfrac{1}{n^2}$. Then $\dfrac{a_n}{b_n} \to 3$ as $n \to \infty$. Since we already know

$\displaystyle\sum_{n=1}^{\infty} \dfrac{1}{n^2}$ is convergent we conclude $\displaystyle\sum_{n=1}^{\infty} a_n$ is convergent. (We could have turned

the tables around. Having already proved the convergence of $\displaystyle\sum_{n=1}^{\infty} \dfrac{1}{n(n+1)}$ as

a telescopic series, we can get that of $\displaystyle\sum_{n=1}^{\infty} \dfrac{1}{n^2}$, using the Corollary above. This

proof is simpler than that in Section 2.6.)

When we want to apply Theorem (9.1) (or Corollary (9.2)), to prove the convergence or divergence of a given series, we must, of course, have a ready supply of other series whose behaviour is known beforehand. The geometric series are of this type. By Theorem (3.5) we know precisely when they are convergent. Let us therefore see under what conditions would a given series

$\displaystyle\sum_{n=1}^{\infty} a_n$ (of positive terms) will look 'more or less like' a geometric series. If

$\displaystyle\sum_{n=1}^{\infty} a_n$ itself were a geometric series, then the ratio $\dfrac{a_{n+1}}{a_n}$ would be a constant,

say $L$. When this is not the case the next best thing is, of course, that the ratio $\dfrac{a_{n+1}}{a_n}$ is eventually constant. And when this does not hold either, the next best

thing is that the ratio $\dfrac{a_{n+1}}{a_n}$ tends to some limit $L$ as $n \to \infty$. In that case we

expect that the qualitative behaviour of the series $\displaystyle\sum_{n=1}^{\infty} a_n$ be the same as that of

a geometric series with common ratio $L$. This guess turns out to be more or less correct as the following theorem, named appropriately as the **ratio test** shows.

**9.3 Theorem :** Suppose $\sum\limits_{n=1}^{\infty} a_n$ is a series of positive terms for which $\dfrac{a_{n+1}}{a_n} \to L$

as $n \to \infty$. Then $\sum\limits_{n=1}^{\infty} a_n$ is convergent if $L < 1$ and divergent if $L > 1$.

**Proof :** Assume first that $L < 1$. Choose $\epsilon > 0$ so small that $L + \epsilon < 1$. Then there exists some $m$ such that for all $n \geq m$, $\dfrac{a_n}{a_{n-1}} < L + \epsilon$. Now we write, for $n \geq m$

$$a_n = a_m \frac{a_{m+1}}{a_m} \frac{a_{m+2}}{a_{m+1}} \cdots \frac{a_{n-1}}{a_{n-2}} \frac{a_n}{a_{n-1}} \tag{2}$$

As each of the ratios is less than $L + \epsilon$, we get, for all $n > m$

$$a_n < a_m (L + \epsilon)^{n-m}$$

or

$$a_n < K(L + \epsilon)^n \text{ where } K = a_m (L + \epsilon)^{-m} \tag{3}$$

Now $K$ is a fixed positive constant and the geometric series $\sum\limits_{n=1}^{\infty} K(L + \epsilon)^n$ is

convergent since $L + \epsilon < 1$. Hence by Theorem (9.1), $\sum\limits_{n=1}^{\infty} a_n$ is convergent.

A similar argument works when $L > 1$. We choose $\epsilon > 0$ so that $L - \epsilon > 1$ and find $m$ such that $\dfrac{a_n}{a_{n-1}} > L - \epsilon$ for all $n \geq m$. Then from (2), we get

$a_n > a_m (L - \epsilon)^{n-m}$ and comparison with the geometric series $\sum\limits_{n=1}^{\infty} K(L - \epsilon)^n$,

(where $K = a_m (L - \epsilon)^{-m}$), gives that $\sum\limits_{n=1}^{\infty} a_n$ diverges.

By a slight change of index, instead of the ratio $\dfrac{a_{n+1}}{a_n}$ we may consider the

ratio $\dfrac{a_n}{a_{n-1}}$ (defined for $n > 1$) which is sometimes a little more convenient in applying the preceding theorem.

As a simple, but not trivial, application of this theorem, consider the series $\sum\limits_{n=0}^{\infty} a_n$ where $a_n = \dfrac{x^n}{n!}$. If $x > 0$, then we can apply the theorem above. Here

$$\frac{a_n}{a_{n-1}} = \frac{x^n}{n!} \frac{(n-1)!}{x^{n-1}} = \frac{x}{n} \to 0 \text{ as } n \to \infty. \text{ So } L = 0 \text{ for all } x > 0. \text{ Hence for all}$$

$x > 0$ the series $\sum\limits_{n=0}^{\infty} \dfrac{x^n}{n!}$ is convergent. (Just a little latter we shall show that it is convergent for $x < 0$ too.) Recall that this series is sometimes taken as the definition of the exponential function $e^x$. (Exercise (3.14)).

Note that the theorem says nothing at all about the case $L = 1$. This is to be expected because consider the series $\sum_{n=1}^{\infty} \frac{1}{n}$ and $\sum_{n=1}^{\infty} \frac{1}{n^2}$. In both the cases $\frac{a_n}{a_{n-1}} \to 1$ as $n \to \infty$. But the former is divergent while the latter is convergent, as we saw in Section 3.6. So in general nothing can be said when $L = 1$. This fate is typical of many tests of convergence of series which depend upon the value of a certain limit, say L, associated with the terms of the series. The test guarantees the convergence or divergence of the series if $L$ lies in certain open intervals. But there is always some border point or a critical value of $L$ for which the test is inconclusive. When this happens more delicate tests are needed to settle the question of convergence. But even these sophisticated tests leave some grey areas.

As an example, we mention here a test, called **Raabe's test** without proof. (For a proof, see Exercise (9.13).) It is applicable when the ratio test fails.

**9.4 Theorem :** Suppose $a_n > 0$ for all $n$ and $\frac{a_n}{a_{n-1}} \to 1$ as $n \to \infty$. Assume $\lim_{n\to\infty} n \left( 1 - \frac{a_n}{a_{n-1}} \right)$ exists and equals $l$. Then $\sum_{n=1}^{\infty} a_n$ is convergent if $l > 1$ and divergent if $l < 1$. ∎

We can apply this test to the series $\sum_{n=1}^{\infty} \frac{1}{n^2}$. A little calculation shows that $\lim_{n\to\infty} n \left( 1 - \frac{(n-1)^2}{n^2} \right) = \lim_{n\to\infty} \frac{2n-1}{n} = 2 > 1$. So $\sum_{n=1}^{\infty} \frac{1}{n^2}$ is convergent, as we already know. Note, however, that even Raabe's test fails for $\sum_{n=1}^{\infty} \frac{1}{n}$, because $\lim_{n\to\infty} n \left( 1 - \frac{n-1}{n} \right) = 1$.

In the ratio test we assumed that each ratio $\frac{a_n}{a_{n-1}}$ (after some stage) was nearly $L$. As a result, $a_n$ was comparable to $L^n$ (except for some constant factor) which ultimately led to the proof. There are situations where the latter can hold without the former. Consider for example the series

$$\sum_{n=0}^{\infty} a_n = 1 + 2 + \frac{1}{4} + \frac{1}{2} + \frac{1}{16} + \frac{1}{8} + \frac{1}{64} + \frac{1}{32} + \ldots \qquad (4)$$

This is essentially a combination of two geometric series, each with common ratio $\frac{1}{4}$. Specifically, $a_{2m} = (\frac{1}{2})^{2m}$ and $a_{2m+1} = 2(\frac{1}{2})^{2m+1}$ for $m = 0, 1, 2, \ldots$. Here $\frac{a_n}{a_{n-1}} = \frac{1}{8}$ if $n$ is even and $\frac{a_n}{a_{n-1}} = 2$ if $n$ is odd. So $\lim_{n\to\infty} \frac{a_n}{a_{n-1}}$ does not exist. Still, $a_n = (\frac{1}{2})^n$ for even $n$ and $a_n = 2(\frac{1}{2})^n$ for odd $n$. In either case, the qualitative behaviour of $a_n$ is the same as that of $L^n$ where $L = \frac{1}{2}$. In fact, here $a_n^{1/n} \to \frac{1}{2}$ as $n \to \infty$, since $(\frac{1}{2})^{1/n} \to 1$ as $n \to \infty$ by Theorem (4.4).

Thus we are led to expect that there may be a test more general than the ratio test. This indeed is the case as we now show in the following theorem called the **root test**.

**9.5 Theorem :** Let $\displaystyle\sum_{n=1}^{\infty} a_n$ be a series of positive terms. Assume that $\displaystyle\lim_{n\to\infty} a_n^{1/n}$ exists and equals $L$. Then $\displaystyle\sum_{n=1}^{\infty} a_n$ is convergent if $L < 1$ said divergent if $L > 1$.

**Proof :** The proof is similar to, and in fact a little more direct, than that of the ratio test. If $L < 1$, then choose $\epsilon > 0$ so that $L + \epsilon < 1$ and get an $m$ such that for all $n \geq m$, $a_n^{1/n} < L + \epsilon$, which, by Lemma (4.2) implies $a_n < (L + \epsilon)^n$. So comparison with the geometric series $\displaystyle\sum_{n=1}^{\infty} (L + \epsilon)^n$ implies convergence of $\displaystyle\sum_{n=1}^{\infty} a_n$, by Theorem (9.1). The case $L > 1$ is left as an exercise. ∎

The root test is more general than the ratio test because it can be shown that if $\displaystyle\lim_{n\to\infty} \frac{a_{n+1}}{a_n}$ exists and equals $L$ then $\displaystyle\lim_{n\to\infty} a_n^{1/n}$ also exists and equals $L$. (In fact, a more general result is true; see Exercise(9.9).) So wherever the ratio test is applicable, so is the root test. But as the series in (4) shows, there are cases where the root test applies but the ratio test fails. This is not surprising because while the ratio test deals with the size of $a_n$ *relative to* that of $a_{n-1}$, the root test deals with the absolute size of $a_n$ which is obviously more relevant to convergence. Still, the ratio test has the advantage that it is generally much easier to apply. For example, using it we proved above the convergence of $\displaystyle\sum_{n=0}^{\infty} \frac{x^n}{n!}$. To do this by the root test we would have to find $\displaystyle\lim_{n\to\infty} \left(\frac{x^n}{n!}\right)^{1/n}$. Here $(x^n)^{1/n}$ is, of course, $x$. But $(n!)^{1/n}$ is not so easy to handle. Using logarithms (in the manner indicated at the end of the answer to Section 3.4) it can be shown that $(n!)^{1/n} \to \infty$ as $n \to \infty$. And so $\left(\frac{x^n}{n!}\right)^{1/n} \to 0$ as $n \to \infty$. Now perhaps you can appreciate how easy the ratio test was in this case.

As in the case of the ratio test, the root test is silent about the case $L = 1$ and the same examples as before, viz., $\displaystyle\sum_{n=1}^{\infty} \frac{1}{n}$ and $\displaystyle\sum_{n=1}^{\infty} \frac{1}{n^2}$ show that noth-ing can be concluded in this case, because $\left(\frac{1}{n}\right)^{1/n} = \frac{1}{n^{1/n}} \to \frac{1}{1} = 1$ and also $\left(\frac{1}{n^2}\right)^{1/n} = \left(\frac{1}{(n^{1/n})^2}\right) \to \frac{1}{1} = 1$ by Theorem (4.5).

Both the ratio and the root tests can be extended slightly to cover the cases where the limits in their hypotheses do not exist. In such cases one looks, instead, for certain limit points (see Exercise (5.1) for a definition) of the sequences $\left\{\dfrac{a_{n+1}}{a_n}\right\}$ and $\{a_n^{1/n}\}$. These extensions will be given as exercises.

Another important extension of both the tests deals with the case where the terms of the sequence $\{a_n\}$ are not necessarily positive. In this case we consider $\displaystyle\sum_{n=1}^{\infty} |a_n|$ instead of $\displaystyle\sum_{n=1}^{\infty} a_n$. For $\displaystyle\lim_{n\to\infty} \dfrac{|a_n|}{|a_{n-1}|}$ to make sense we must of course assume that $a_n$ is eventually non-zero. For $\displaystyle\lim_{n\to\infty} |a_n|^{1/n}$, no such restriction is necessary.

**9.6 Theorem :** If $\displaystyle\lim_{n\to\infty} \left|\dfrac{a_n}{a_{n-1}}\right|$ exists and is less than 1, then $\displaystyle\sum_{n=1}^{\infty} a_n$ is convergent. If $\displaystyle\lim_{n\to\infty} |a_n|^{1/n}$ exists and is less than 1 then also $\displaystyle\sum_{n=1}^{\infty} a_n$ is convergent.

**Proof :** All we need to do is to combine the first parts of the ratio and the root tests with Theorem (8.1), which says that every absolutely convergent series is convergent. ∎

This theorem is quite useful in proving the convergence of a series if that series also happens to be absolutely convergent. As an immediate application, we see that the series $\displaystyle\sum_{n=0}^{\infty} \dfrac{x^n}{n!}$ is convergent for all real $x$.

Note that nothing is said about the cases where the limit of $\left|\dfrac{a_n}{a_{n-1}}\right|$ or of $|a_n|^{1/n}$ exceeds 1. For, in these cases, even though we get that $\displaystyle\sum_{n=1}^{\infty} |a_n|$ is divergent, it tells us nothing about the convergence or divergence of $\sum_{n=1}^{\infty} a_n$. It could either be divergent or it could be conditionally convergent.

There are, in fact, very few general tests for conditional convergence. The alternating test (Exercise (8.1)) is perhaps the only elementary one. Fortunately, series arising in real life situations generally have positive terms. Even then, as mentioned earlier, there is no golden method applicable to all. But then that is precisely why the problem of deciding whether a given series is convergent becomes an interesting one. It is rather like finding out if a person is married or not without having a direct access to this information. First you look for such obvious signs as wedding rings. If they fail, (for example if the person comes from a community where wedding rings are not very customary), you look for more subtle signs such as the facial expressions. And if you are a Sherlock Holmes then even one sharp glance may tell you not only the marital status but the number of children as well !

## EXERCISES

9.1   Prove that a series of the form $\sum_{n=1}^{\infty} \dfrac{an+b}{n(n+1)(n+2)}$, where $a, b$ are constants is convergent and find its sum.

9.2   Suppose $\sum_{n=1}^{\infty} a_n$ and $\sum_{n=1}^{\infty} b_n$ are series of positive terms and $\dfrac{a_{n+1}}{a_n} \le \dfrac{b_{n+1}}{b_n}$ for all $n$. Prove that if $\sum_{n=1}^{\infty} b_n$ is convergent then so is $\sum_{n=1}^{\infty} a_n$.

9.3   Taking logarithms, show that $(n!)^{1/n} \to \infty$ as $n \to \infty$. [*Hint* : First prove an analogue of Theorem (2.3) for the case where $x_n \to \infty$ as $n \to \infty$.]

9.4   For a real number $x$, prove directly that $\dfrac{x^n}{n!} \to 0$ as $n \to \infty$. Also deduce this from the convergence of $\sum_{n=0}^{\infty} \dfrac{x^n}{n!}$. Compare the two proofs.

9.5   Let $r$ be a fixed integer $\ge 2$. For every sequence $\{a_n\}_{n=1}^{\infty}$ of integers with $0 \le a_n < r$ for every $n$, show that the series $\sum_{n=1}^{\infty} \dfrac{a_n}{r^n}$ is convergent. If $\{b_n\}$ is another such sequence, prove that $\sum_{n=1}^{\infty} \dfrac{a_n}{r^n} = \sum_{n=1}^{\infty} \dfrac{b_n}{r^n}$ if and only if there exists some $k \ge 1$ such that $a_i = b_i$ for $i = 1, \ldots, k-1$ and one of the following two possibilities hold:

(i)   $a_k = b_k + 1$, and for every $i > k, a_i = 0$ and $b_i = r-1$

(ii)  $b_k = a_k + 1$, and for every $i \ge k, b_i = 0$ and $a_i = r-1$.

Prove further that for every $x \in [0,1]$, there exists a sequence $\{a_n\}_{n=1}^{\infty}$ with $0 \le a_n < r$ for every $n$ such that $\sum_{n=1}^{\infty} \dfrac{a_n}{r^n} = x$. (Such a sequence is called an **$r$-adic expansion** of $x$, or expansion with **base** or **radix** $r$. The cases $r = 2, 3$ and $10$ are also known as **dyadic** (or **binary**), **triadic** (or **ternary**) and **decimal** expansions respectively.)

9.6   Prove that a real number is rational if and only if it has a decimal expansion which is either terminating or recurring after some stage. (For example, the decimal expansion of $\dfrac{115721}{24975}$ is $6.23507507507\ldots$ which is also denoted by $6.23\overline{507}$, indicating that the block $507$ repeats itself infinitely often.) Characterise those rationals which have terminating decimal expansions. Generalise to other bases.

*9.7 Prove that it is impossible to have a sequence $\{x_n\}$ of real numbers such that every real number in $[0,1]$ occurs in it at least once. [*Hint* : Assuming such a sequence exists, consider a real number $x$ whose decimal expansion differs from that of $x_n$ in the $n^{th}$ place of decimal.] This fact is expressed by saying that the set of real numbers is not **denumerable**. The argument given here is called a **diagonalisation** argument. By contrast, show that the set of rationals in $[0,1]$ is denumerable.

9.8 Let $L$ be the largest limit point of a sequence $\{a_n\}_{n=1}^{\infty}$. (If $\{a_n\}$ is bounded above, then $L$ is finite by Exercise (5.7). Otherwise take $L = \infty$.) $L$ is called the **limit superior** or **lim sup** of $\{a_n\}$ and is denoted by $\lim\sup_{n\to\infty} a_n$ (or by $\overline{\lim}_{n\to\infty} a_n$ ) Similarly define **limit inferior** or **lim inf** of $\{a_n\}$, denoted by $\liminf_{n\to\infty} a_n$ (or by $\underline{\lim}_{n\to\infty} a_n$ ) as the smallest limit point of $\{a_n\}$. Prove that:

(i) if $\lim\sup_{n\to\infty} = L < \infty$, then for every $\epsilon > 0$, there exists $m$ such that for all $n \geq m, a_n < L + \epsilon$. State and prove a similar statement for lim inf.

(ii) $\liminf_{n\to\infty} a_n \leq \lim\sup_{n\to\infty} a_n$ with equality holding if and only if $\{a_n\}$ is convergent.

9.9 If $\{a_n\}$ is a sequence of positive numbers, prove that

$$\liminf_{n\to\infty} \frac{a_{n+1}}{a_n} \leq \liminf_{n\to\infty} a_n^{1/n} \leq \lim\sup_{n\to\infty} a_n^{1/n} \leq \lim\sup_{n\to\infty} \frac{a_{n+1}}{a_n}$$

and show by examples that the inequalities may be strict. Note that this in particular implies that whenever $\lim_{n\to\infty} \frac{a_{n+1}}{a_n}$ exists, $\lim_{n\to\infty} a_n^{1/n}$ also exists and the two are equal.

9.10 Generalise the ratio test (Theorem (9.5)) by showing that $\sum_{n=1}^{\infty} a_n$ is convergent if $\overline{\lim}_{n\to\infty} \frac{a_{n+1}}{a_n} < 1$ and divergent if $\underline{\lim}_{n\to\infty} \frac{a_{n+1}}{a_n} > 1$.

9.11 Prove that the root test (Theorem (9.5)) has an even stronger generalisation by showing that $\sum_{n=1}^{\infty} a_n$ is convergent if $\overline{\lim}_{n\to\infty} a_n^{1/n} < 1$ and divergent if $\overline{\lim}_{n\to\infty} a_n^{1/n} > 1$.

9.12 Prove the results of Theorem (6.3) and Exercise (6.2) (for $y \neq 1$) using Raabe's test.

*9.13 Using the results of Theorem (6.3), Exercises (6.2) and (9.2) prove Raabe's test. [*Hint* : If $l > 1$, there exists a rational $y$ such that $l > y > 1$. Similarly, if $l < 1$, get a rational $y$ with $l < y < 1$. Take $b_n$ as $\frac{1}{n^y}$. Rather sophisticated arguments are needed to show that $\dfrac{a_n}{a_{n-1}} \le \dfrac{b_n}{b_{n-1}}$ in the first

case and $\dfrac{a_n}{a_{n-1}} \ge \dfrac{b_n}{b_{n-1}}$ in the second.]

9.14 Determine, using any of the tests studied so far, which of the following series are convergent. (Whenever functions like logarithms or trigonometric or exponential functions are involved, you may assume their standard properties without proofs.)

(i) $\displaystyle\sum_{n=2}^{\infty} \frac{1}{n \log n}$    (ii) $\displaystyle\sum_{n=2}^{\infty} \frac{(-1)^n}{n \log n}$    (iii) $\displaystyle\sum_{n=1}^{\infty} \frac{n^2}{2^n - n}$

(iv) $\displaystyle\sum_{n=1}^{\infty} \frac{\sin \frac{1}{n}}{n}$    (v) $\displaystyle\sum_{n=1}^{\infty} \frac{e^n}{n!}$    (vi) $\displaystyle\sum_{n=1}^{\infty} \frac{n^n}{n!}$

(vii) $\displaystyle\sum_{n=1}^{\infty} \frac{1}{n^{1+1/n}}$    (viii) $\displaystyle\sum_{n=1}^{\infty}(2^{1/n} - 1)^n$    (ix) $\displaystyle\sum_{n=1}^{\infty}(n^{1/n} - 1)^n$.

9.15 Prove that $\displaystyle\sum_{n=2}^{\infty} \log\left(1 - \frac{1}{n^2}\right) = -\log 2$ by first showing that the infinite prod-

uct $\displaystyle\prod_{n=2}^{\infty}\left(1 - \frac{1}{n^2}\right)$ equals $\dfrac{1}{2}$. (Infinite products are defined analogously to in-

finite sums. That is, given a sequence $\{a_n\}_{n=r}^{\infty}$ we define $\displaystyle\prod_{n=r}^{\infty} a_n$ as $\displaystyle\lim_{n\to\infty} P_n$,

where $P_n$ is the 'partial product' $\displaystyle\prod_{k=r}^{n} a_k$ (i.e., the product $a_r a_{r+1} \cdots a_n$) for $n \ge r$. Like infinite series, it is in general not easy to evaluate infinite products. The present problem is exceptional in that the partial product, because of cancellation of most of the factors, comes out to be unusually simple, very much like a telescopic series. Of course, one could show directly that the given series is telescopic. But it is a little easier to see this for the corresponding infinite product. Sometimes it is helpful to go the other way, i.e. to convert infinite products (with positive factors) to infinite series by taking logarithms. In both these conversions, we need not only the well-known property 'log $xy = \log x + \log y$' of logarithms, but also the continiuty of the logarithm function. This will be established later (see Exercise (4.1.22) and Section 5.8).

The case where the product converges to 0 (e.g. when $a_n = \frac{1}{n}$) is uninteresting. In all other cases of convergent products, say, $\displaystyle\prod_{n=k}^{\infty} a_n$, it can be shown (by an argument similar to that in Exercise (1.8)) that $a_n \to 1$ as $n \to \infty$. So it is customary to write $a_n$ as $1 + b_n$. It can then be shown (see Exercise (5.8.19)) that absolute convergence of the series $\displaystyle\sum_{n=k}^{\infty} b_n$ is a sufficient condition for the convergence of the product $\prod_{n=k}^{\infty}(1 + b_n)$. This fact, along with the divergence of the harmonic series, leads to another proof that the series of the reciprocals of the primes is divergent. (The proof in Exercise (6.10) is similar but by-passes infinite products.)

Historically, many beautiful formulas were obtained by manipulating the infinite products in the same carefree manner as infinite series. For example, Euler, who proved $\sum_{n=1}^{\infty} \dfrac{1}{n^2} = \dfrac{\pi^2}{6}$ by treating $\dfrac{\sin x}{x}$ as a polynomial of infinite degree in $x^2$ (see Exercise (1.4.9)), went on to 'factorise' this polynomial and get

$$\frac{\sin x}{x} = \left(1 - \frac{x^2}{\pi^2}\right)\left(1 - \frac{x^2}{4\pi^2}\right)\left(1 - \frac{x^2}{9\pi^2}\right)\cdots$$

which can now be proved rigorously using Fourier series.

We shall rarely need infinite products except for one well-known result called **Wallis formula** (see Sections 5.6 and 5.8) which says

$$\frac{\pi}{2} = \prod_{n=1}^{\infty} \frac{(2n)(2n)}{(2n-1)(2n+1)} = \frac{2.2}{1.3} \cdot \frac{4.4}{3.5} \cdot \frac{6.6}{5.7} \cdots \quad .)$$

## 3.10    Power Series

A trivial pun is to say that the power series are a very powerful tool in mathematics because there are literally infinitely many powers in them! Specifically, a **power series** is a series of the form $\sum_{n=0}^{\infty} a_n x^n$, where $a_0, a_1, \ldots, a_n, \ldots$ are some real numbers and $x$ is some real variable. More generally if $c$ is a fixed real number then a series of the form $\sum_{n=0}^{\infty} a_n (x - c)^n$ is called a power series centred at $c$. A simple change of variable converts the latter to the former and so we generally stick to power series centred at 0. The coefficients $a_n$ as well as the variable $x$ can in fact be complex. But here we confine only to the real case. By convention we set $x^0 = 1$, even when $x = 0$. So, written in the full form, $\sum_{n=0}^{\infty} a_n x^n$ is the series $a_0 + a_1 x + a_2 x^2 + \ldots + a_n x^n + \ldots$. If the $a_n$'s are eventually 0, then a power series reduces to a polynomial in $x$. So it is tempting to suppose that the power series would behave more or less like polynomials. Euler derived the beautiful formula $\sum_{n=1}^{\infty} \frac{1}{n^2} = \frac{\pi^2}{6}$ (cf. Exercise (1.4.9)) by assuming (without justification) that a certain formula for the sum of the reciprocals of the root of a polynomial (Exercise (1.4.7)) holds even for power series.

Such indiscriminate extensions of properties of polynomials to power series are not acceptable by modern standards of rigour. So a logical approach to power series is to regard the power series $\sum_{n=0}^{\infty} a_n x^n$ as an ordinary series of real numbers for each fixed (real) value of the variable $x$. Obviously, it is a different series for each different values of $x$ and its behaviour vis-a-vis convergence, would naturally differ for different values of $x$ and the variation would depend on the sequence of coefficients, viz. $\{a_n\}_{n=0}^{\infty}$. If each $a_n$ equals 1 (or some other non-zero constant) then Theorem (3.5) tells us that the power series $\sum_{n=0}^{\infty} a_n x^n$ is convergent for $x \in (-1, 1)$ and divergent for all other values of $x$. Conceivably with a more haphazard choice of the sequence $\{a_n\}_{n=0}^{\infty}$, we may be able to generate a power series which converges, say, for all rational values of $x$ and diverges for all irrational $x$. It turns out that this is not so. No matter what the sequence $\{a_n\}_{n=0}^{\infty}$, there is a remarkable regularity in the values of $x$ for which the series $\sum_{n=0}^{\infty} a_n x^n$ is convergent. Specifically, such values lie in an interval centred at 0. The following theorem, due to **Hadamard** identifies this interval in terms of the sequence $\{a_n\}_{n=0}^{\infty}$. We state the theorem in a form which is generally sufficient for applications.

**10.1 Theorem :** Suppose $|a_n|^{1/n} \to L$ as $n \to \infty$, where $L$ could be $\infty$. Let $R = \frac{1}{L}$ with the understanding that $\frac{1}{0} = \infty$ and $\frac{1}{\infty} = 0$. Then the power series $\sum_{n=0}^{\infty} a_n x^n$ is absolutely convergent for $|x| < R$ and divergent for $|x| > R$. The same conclusion holds if $\lim_{n \to \infty} |\frac{a_{n+1}}{a_n}| = L$.

**Proof :** The proof is an easy application of the root test (Theorem (9.5)). For each fixed $x \neq 0, |a_n x^n|^{1/n} = |a_n|^{1/n} |x| \to L|x|$. So $\sum_{n=0}^{\infty} a_n x^n$ is absolutely convergent and hence convergent if $L|x| < 1$, i.e., if $|x| < R$. If, on the other hand, $|x| > R$, then $L|x| > 1$ and it is not hard to show that in this case $a_n x^n \nrightarrow 0$ as $n \to \infty$. So $\sum_{n=0}^{\infty} a_n x^n$ is not convergent. The last assertion of the theorem follows from Exercise (9.9) ∎

The open interval $(-R, R)$ is called, quite appropriately, the **interval of convergence** of the power series $\sum_{n=0}^{\infty} a_n x^n$. $R$ is called the **radius of convergence**. Note that nothing is said about the convergence when $x = R$ or $-R$. Simple examples (see Exercise (10.2)) show that all possibilities can arise. For $x \in (-R, R)$, denote by $A(x)$ the sum $\sum_{n=0}^{\infty} a_n x^n$. Then $A$ is a real-valued function called the **sum function** of the series, defined on the domain $(-R, R)$. In Exercise (3.14), we considered the three power series, $\sum_{n=0}^{\infty} \frac{x^n}{n!}$, $\sum_{n=0}^{\infty} \frac{(-1)^n x^{2n}}{(2n)!}$ and $\sum_{n=0}^{\infty} \frac{(-1)^n x^{2n+1}}{(2n+1)!}$. Using the last part of the theorem above, it is easy to show that in each of these power series $L = 0$ and hence $R = \infty$. So these series define certain functions of $x$ for all real $x$. They are respectively denoted by $e^x, \cos x$ and $\sin x$. These definitions of trigonometric functions are purely analytical. They make no reference to the geometric interpretation of $x$ as a certain angle. It turns out that functions defined by power series have certain very nice properties. For example, they can be differentiated term-by-term, that is $A'(x) = \sum_{n=1}^{\infty} n a_n x^{n-1}$. The proof is not trivial (see Section 4.6). But once it is done, we see instantaneously that $\frac{d}{dx}(\sin x) = \cos x$ and $\frac{d}{dx}(\cos x) = -\sin x$ (see Section 4.8). All the familiar identities about trigonometric functions can be derived through various properties of power series. This approach to the trigonometric functions is the 'cleanest' approach. Nothing is left to geometric visualisation or to intuition.

Thus we see that power series can be used to define new functions. Sometimes, on the other hand, we already have a function, say, $f(x)$ and we want to study its behaviour in a neighbourhood of some point, say, $c$. If we can find a power series, say, $\sum_{n=0}^{\infty} a_n (x - c)^n$ centred at $c$, which converges to $f(x)$ for all $x$ in some open interval around $c$, then we can use its partial sum $S_n(x) = \sum_{k=0}^{n} a_k (x - c)^k$ as a very good polynomial approximation to $f(x)$ for large $n$. This method is frequently used when the function $f(x)$ comes out as the solution of some differential equation and we know it exists but cannot identify it in a closed form.

Because of absolute convergence, the power series are, within their intervals of convergence, amenable to the same algebraic manipulations as finite sums. These manipulations give rise to formulas for their sum functions. The following theorem gives the two basic formulas.

**10.2 Theorem :** Suppose $\sum\limits_{n=0}^{\infty} a_n x^n$ and $\sum\limits_{n=0}^{n} b_n x^n$ are power series with radii of convergence $R_1$ and $R_2$ respectively and $\alpha, \beta$ are some fixed real numbers. Let $R = \min\{R_1, R_2\}$. For each $n = 0, 1, 2, \ldots$ let .

$$
\begin{aligned}
c_n &= \alpha a_n + \beta b_n \\
\text{and} \quad d_n &= a_0 b_n + a_1 b_{n-1} + \ldots + a_{n-1} b_1 + a_n b_0.
\end{aligned}
$$

Then both $\sum\limits_{n=0}^{\infty} c_n x^n$ and $\sum\limits_{n=0}^{\infty} d_n x^n$ converge at least for all $x$ in $(-R, R)$. If, further, $A(x), B(x), C(x)$ and $D(x)$ are the sum functions of $\sum\limits_{n=0}^{\infty} a_n x^n, \sum\limits_{n=0}^{\infty} b_n x^n,$

$\sum\limits_{n=0}^{\infty} c_n x^n$ and $\sum\limits_{n=0}^{\infty} d_n x^n$ respectively, then for all $x \in (-R, R)$,

$$
C(x) = \alpha A(x) + \beta B(x) \quad \text{and} \quad D(x) = A(x)B(x).
$$

**Proof :** The first formula is an elementary property of convergent series. The second formula is essentially the Cauchy product (cf. Exercise (7.4)) of the two series $\sum\limits_{n=0}^{\infty} a_n x^n$ and $\sum\limits_{n=0}^{\infty} b_n x^n$ and holds true because both the series converge absolutely if $x \in (-R, R)$. (See Exercise (8.9).) Note, however, that unlike in that exercise where the series began witn $n = 1$, we are now starting the series with $n = 0$. This, in fact, is a little more convenient while dealing with Cauchy products.) ∎

If we apply the second part of this theorem to the power series $\sum\limits_{n=0}^{\infty} \dfrac{x^n}{n!}$ and $\sum\limits_{n=0}^{\infty} \dfrac{y^n}{n!}$ which converge, respectively, to $e^x$ and $e^y$, we can show $e^{x+y} = e^x e^y$. (Exercise (10.3)). Formulas for $\sin(x + y)$ and $\cos(x + y)$ can be obtained similarly.

We now give an application of power series which is somewhat of a cunning type. In Exercise (1.4) we defined the Fibonacci numbers by the recurrence relation

$$
F_n = F_{n-1} + F_{n-2} \quad (n > 2) \tag{1}
$$

with $F_1 = 1$ and $F_2 = 1$. If we set $F_0 = 0$, then (1) is valid for $n = 2$ as well.

We remarked that a closed form expression for $F_n$ is given by

$$F_n = \frac{1}{\sqrt{5}}\left[\left(\frac{1+\sqrt{5}}{2}\right)^n - \left(\frac{1-\sqrt{5}}{2}\right)^n\right] \text{ for } n \geq 0 \tag{2}$$

It is easy to *verify* (2) by induction on $n$ (cf. Exercise (1.4)). But it is obviously impossible to guess (2) just by calculating $F_n$ for a few values of $n$. We now show how power series can be used to guess (2). We consider the power series $\sum_{n=0}^{\infty} F_n x^n = 0 + x + x^2 + 2x^3 + 3x^4 + 5x^5 + 8x^6 + \ldots$ and let $F(x)$ be its sum function. We shall first identify $F(x)$ as a rather familiar function. Multiply both sides of (1) by $x^n$ and sum over from $n = 2$ to $\infty$. Then we get

$$\sum_{n=2}^{\infty} F_n x^n = x \sum_{n=2}^{\infty} F_{n-1} x^{n-1} + x^2 \sum_{n=2}^{\infty} F_{n-2} x^{n-2} \tag{3}$$

which, after a change of index variable in the series on the right becomes,

$$\sum_{n=2}^{\infty} F_n x^n = x \sum_{n=1}^{\infty} F_n x^n + x^2 \sum_{n=0}^{\infty} F_n x^n \tag{4}$$

Each of the power series involved here is almost the sum function $F(x)$ except possibly for a few missing terms. So, we get,

$$F(x) - F_0 - F_1 x = x(F(x) - F_0) + x^2 F(x). \tag{5}$$

Putting $F_0 = 0$ and $F_1 = 1$ and solving yields

$$F(x) = \frac{x}{1 - x - x^2} \tag{6}$$

We now expand (or 'develop') $F(x)$ as a power series in $x$. First we resolve $F(x)$ into partial fractions by factoring its denominator as a product of two linear factors, say, $(1 - x - x^2) = (1 - \alpha x)(1 - \beta x)$. A comparison of coefficients gives $\alpha$ and $\beta$ as roots of the quadratic $y^2 - y - 1 = 0$. We take $\alpha = \frac{1+\sqrt{5}}{2}$ and $\beta = \frac{1-\sqrt{5}}{2}$. A little calculation then gives

$$F(x) = \frac{x}{1 - x - x^2} = \frac{1}{\sqrt{5}}\left(\frac{1}{1 - \alpha x} - \frac{1}{1 - \beta x}\right) \tag{7}$$

Now $\frac{1}{1 - \alpha x}$ can be expanded as a geometric series, viz., $\sum_{n=0}^{\infty} \alpha^n x^n$. Similarly $\frac{1}{1 - \beta x} = \sum_{n=0}^{\infty} \beta^n x^n$. Using Theorem (10.2), we get from (7),

$$\sum_{n=0}^{\infty} F_n x^n = \sum_{n=0}^{\infty} \frac{1}{\sqrt{5}}(\alpha^n - \beta^n) x^n \tag{8}$$

It is tempting to equate the coefficients of like powers in (8) and thereby to conclude (2). But this requires some justification. First, it requires what is called the **uniqueness theorem for power series,** which asserts that if two power series define the same sum function in some interval of the form $(-R, R)$ where $R > 0$, then their coefficients are identical. We postpone the proof of this theorem to Section 4.6. To apply this theorem to (8), it is not difficult to show that the readius of convergence of the power series on the right is $\dfrac{1}{\alpha}$ (cf. Exercise (10.11)). But what about the power series $\sum\limits_{n=0}^{\infty} F_n x^n$? Unless we know how big $F_n$'s are, we cannot determine its radius of convergence. Still, using solely (1), it is easy to find an upper bound for $F_n$, using which it can be shown that $\sum\limits_{n=0}^{\infty} F_n x^n$ has a positive radius of convergence (cf. Exercise (10.5)). We can now apply the uniqueness theorem to get (2) from (8), legitimately.

But if our concern is merely to *guess* (2), then we need hardly worry about legitimacy. As long as the guess can be proved correct by mathematical induction, it is a perfectly valid form of proof. It is as if the work from (3) to (8) is done 'behind the curtain' and only the outcome, viz. (2) is shown to the public. This method of solving recurrence relations is used commonly and a few applications of it will be given as exercises. The interesting part is to observe that neither the recurrence relation (1), nor its solution (2) makes any reference to power series. The latter are used only as a powerful tool. More generally, if $\{a_n\}_{n=0}^{\infty}$ is any sequence of real numbers then the power series $\sum\limits_{n=0}^{\infty} a_n x^n$ is called its **generating function.** Thus (6) shows that the generating function of the Fibonacci sequence is $\dfrac{x}{1 - x - x^2}$. The strategy adopted above was to first get hold of the generating function and then to expand it to get a closed form expression for the $n^{\text{th}}$ term of the Fibonacci sequence. There are situations where the closed form expression for $a_n$ is fairly complicated (often involving some summation) but the generating function $\sum\limits_{n=0}^{\infty} a_n x^n$ has a succinct formula. In such cases, it is easier to remember the generating function and then to derive from it $a_n$ for a particular value of $n$ as and when needed.

## EXERCISES

10.1   Generalise theorem (10.1) by showing that it holds even when $L = \overline{\lim}_{n\to\infty} |a_n|^{1/n}$.

10.2   Prove that the power series $\sum\limits_{n=0}^{\infty} x^n, \sum\limits_{n=1}^{\infty} \dfrac{x^n}{n}$ and $\sum\limits_{n=1}^{\infty} \dfrac{x^n}{n^2}$ have the same interval of convergence, viz. $(-1, 1)$, but that the first power series converges at neither, the second at only one and the third at both of the end-points of this interval.

10.3 Considering the Cauchy product of the series $\sum_{n=0}^{\infty} \dfrac{x^n}{n!}$ and $\sum_{m=0}^{\infty} \dfrac{y^m}{m!}$, prove that $e^{x+y} = e^x e^y$. By a similar, but a little more elaborate calculation, establish the well-known identities for $\sin(x+y)$ and $\cos(x+y)$. Using the latter with $y = -x$, show that $\cos^2 x + \sin^2 x = 1$ for all $x$. [*Hint*: Use the binomial theorem.]

10.4 Find the radius of convergence of the following power series:

(i) $\sum_{n=0}^{\infty} r^n x^n (r \neq 0)$ (ii) $\sum_{n=0}^{\infty} (n^3 - 5n^2 + 7n - 2)x^n$ (iii) $\sum_{n=0}^{\infty} a_n x^n$ where $a_n$ is defined as $\dfrac{h(h-1)\ldots(h-n+1)}{n!}$, $h$ being a fixed real number.

10.5 By induction on $n$ and using only (1), show that the Fibonacci numbers $F_n$ satisfy $F_n \leq 2^n$ for all $n$. Hence show that the power series $\sum_{n=0}^{\infty} F_n x^n$ has a positive radius of convergence.

10.6 Suppose in a game, a fair coin is tossed at each round. The game is won if two consecutive tosses are heads. Let $p_n$ be the probability of winning the game at the end of the $n^{\text{th}}$ round (but not earlier). Prove that $p_0 = p_1 = 0, p_2 = \dfrac{1}{4}, p_3 = \dfrac{1}{8}, p_4 = \dfrac{1}{8}$ and in general,

$$p_n = \frac{1}{2}p_{n-1} + \frac{1}{4}p_{n-2} \text{ for } n \geq 3. \tag{9}$$

Show that the series $\sum_{n=0}^{\infty} p_n$ converges to 1. (Why is this intuitively obvious?) Let $P(x) = \sum_{n=0}^{\infty} p_n x^n$. Show that $P(x) = \dfrac{x^2}{4 - 2x - x^2}$. Hence obtain a closed form expression for $p_n$.

10.7 Assuming term-by-term differentiation of power series is valid, show that for $|x| < 1$, $\sum_{n=1}^{\infty} nx^n = \dfrac{x}{(1-x)^2}$.

10.8 Suppose that for the game in Exercise (10.6), each toss costs one rupee. Then $\sum_{n=0}^{\infty} np_n$ is what is called the **average** or **expected** cost of winning the game. For the game there show that this cost is 6. This can be obtained from the closed from expression for $p_n$, using the last exercise. Show, however, that it can also be obtained directly by multiplying both sides of (9) by $n$ and summing over from $n = 3$ to $\infty$. (The catch is that the convergence of $\sum_{n=0}^{\infty} np_n$ must be shown beforehand, as otherwise the equality would be of the form $\infty = \infty$.) Yet another method is to consider $P'(1)$ where $P(x)$ is as in Exercise (10.6).

10.9  Suppose for the game in Exercise (10.6), instead of two, three consecutive heads are necessary for winning. Show that this time the average cost of winning is 14 rupees. (There is a recurrence relation analogous to (9). But it is not easy to solve it. But the other methods given in the last exercise work. Problems like this have a practical application in deciding what reward a player should get if he is playing a game on a gambling machine.)

10.10 The generating function $\sum_{n=0}^{\infty} a_n x^n$ of a sequence $\{a_n\}_{n=0}^{\infty}$ is sometimes called its **ordinary generating function** so as to distinguish it from its **exponential generating function** defined as $\sum_{n=0}^{\infty} \frac{a_n}{n!} x^n$. Justify the name. Show that its radius of convergence is larger than that of the ordinary generating function. (Especially, consider the case $a_n = n!$ or $a_n = n^n$. Here the radius of convergence of the ordinary generating function is 0 and so theorems about power series cannot be applied. But the exponential generating functions behave better.)

10.11 If $\alpha$ is a real number, prove that the radius of convergence of $\sum_{n=0}^{\infty} \alpha^n x^n$ is $\frac{1}{|\alpha|}$. If $\alpha$ and $\beta$ are real numbers with $0 < |\beta| < |\alpha|$, prove that the radius of convergence of $\sum_{n=0}^{\infty} (\alpha^n + \beta^n) x^n$ and of $\sum_{n=0}^{\infty} (\alpha^n - \beta^n) x^n$ is $\frac{1}{|\alpha|}$.

10.12 Suppose we have an infinite supply of $k$ types of objects (where $k$ is a fixed positive integer). For a non-negative integer $n$, let $a_n$ be number of ways to choose $n$ out of these objects, repetitions being allowed freely. Prove that the (ordinary) generating function of the sequence $\{a_n\}_{n=0}^{\infty}$ equals $\frac{1}{(1-x)^k}$. (This expression can be expanded by the binomial theorem (for negative exponent, viz. $-k$, see Exercise (4.6.13)) to get a closed form expression for $a_n$ as $\binom{n+k-1}{n}$. This can, of course, be proved combinatorially. But this exercise shows how the ordinary generating functions can be applied to combinatorial problems. The exponential generating functions are more convenient in problems involving permutations, i.e. arrangements. But we shall not go into it.)

## 3.11   Strength of Uniform Convergence

To begin with, it is important to stress the basic conceptual difference between uniform convergence and pointwise convergence. Let $S$ be a set. Suppose for every positive integer $n$, $f_n$ is a real valued function defined on $S$. Then for every fixed $x$ in $S$, $\{f_n(x)\}_{n=1}^{\infty}$ is just a sequence of real numbers. In general, this sequence will converge for some values of $x$ in $S$ and diverge for other values. If $\{f_n(x)\}$ is convergent for every $x \in S$, we say $\{f_n\}$ **converges pointwise** in $S$. In this case, we can define the limit function of $\{f_n\}$ as the function $f : S \to \mathbb{R}$ defined by $f(x) = \lim_{n \to \infty} f_n(x)$ for $x \in S$. We already had examples of this situation. Suppose $\sum_{n=0}^{\infty} a_n x^n$ is a power series with radius of convergence $R$. Let $S$ be any subset of $(-R, R)$ and for $n \geq 1$, let $f_n(x)$ be the partial sum $\sum_{k=0}^{n} a_k x^k$. Then theorem (10.1) says that $\{f_n\}$ converges pointwise in $S$, and the limit function is precisely the sum function $\sum_{n=0}^{\infty} a_n x^n$.

In pointwise convergence, for each fixed $x \in S$, $f_n(x)$ is eventually close to $f(x)$. But the degree of eventuality may change as $x$ changes. In symbols, suppose $\epsilon > 0$ is given. Then for every fixed $x \in S$, there will exist some $N$ such that for all $n \geq N$, $|f_n(x) - f(x)| < \epsilon$. But this integer $N$ may depend on $x$. So it is better to write it as $N_x$ or $N(x)$. An integer larger than $N_x$ will obviously also work. But in general the smallest integer that will work for a particular value of $x$ may not work for some other value of $x$. For example, suppose $S$ is the closed unit interval $[0,1]$ and $f_n(x) = x^n$ for $n = 1, 2, 3, \ldots$ Then for every $x \in S$, $f_n(x) \to 0$ if $x \neq 1$ (Theorem (3.4)) while $f_n(1) \to 1$ as $n \to \infty$. So here the limit function $f$ is given by

$$f(x) = \begin{cases} 0 & \text{if } 0 \leq x < 1 \\ 1 & \text{if } x = 1. \end{cases}$$

Suppose $\epsilon = .001 = 10^{-3}$. For $x = 0$ and $1$, $f_n(x)$ is a constant sequence and so the least $N$ that will work is 1. For $0 < x < 1$, on the other hand, $N_x$ will have to be bigger than $\dfrac{\log_{10} \epsilon}{\log_{10} x}$, i.e. $\dfrac{-3}{\log_{10} x}$. (Note that $\log_{10} x < 0$.) For $x = \frac{1}{3}$, $N_x$ will have to be at least 7. But 7 will not work for $x = \frac{1}{2}$ since $|(\frac{1}{2})^7 - 0| = \frac{1}{128} > .001 = \epsilon$. For $x = \frac{1}{2}$, $N_x$ will have to be at least 10. In fact the closer $x$ gets to 1, $N_x$ will have to be higher and higher. There is no single $N$ which will work for every $x \in S$. If, however, we stop a little away from 1, things are better. That is, if $S = [0, 1 - \delta]$ where $\delta > 0$, then an $N$ that will work for the worst case, viz. for $x = 1 - \delta$, will obviously work for all $x < 1 - \delta$.

Let us look at this geometrically. Fig. 3.11.1 shows the graphs of the functions $f_1, f_2, \ldots$ and also the limit function $f$. We see that even though for each $x \in [0, 1)$, $f_n(x)$ finally 'descends' down to $f(x)$ (which is 0) this happens very rapidly for $x$ close to 0 and very slowly for $x$ close to 1. Given a strip of width $2\epsilon$ about the graph of the limit function $f$, none of the graphs of $f_n$'s lies completely in this strip. In other words, even though for every $x \in [0, 1)$ $|f_n(x) - f(x)| \to 0$ as $n \to \infty$; for every $n$, there are values of $x$ for which $|f_n(x) - f(x)|$ is very close to 1. Put differently, $\sup\limits_{x \in [0,1)} |f_n(x) - f(x)|$ (i.e., the supremum of the set $\{|f_n(x) - f(x)| : x \in [0, 1)\}$ is 1 for every $n$ and hence does not tend to 0 as $n \to \infty$. So the graph of $f_n$ as a whole, does not tend to the graph of $f$. However, if we fix some $\delta > 0$ and take only the portions of these graphs over $[0, 1 - \delta]$, then they do tend to the graph of $f$ over $[0, 1 - \delta]$, because, $\sup\limits_{x \in [0,1-\delta]} |f_n(x) - f(x)| = \sup\limits_{x \in [0,1-\delta]} x^n = (1 - \delta)^n$ and by Theorem (3.4), this does tend to 0 as $n \to \infty$, since $0 < 1 - \delta < 1$. (We assume $\delta < 1$, as otherwise $[0, 1 - \delta)$ is the empty set.)

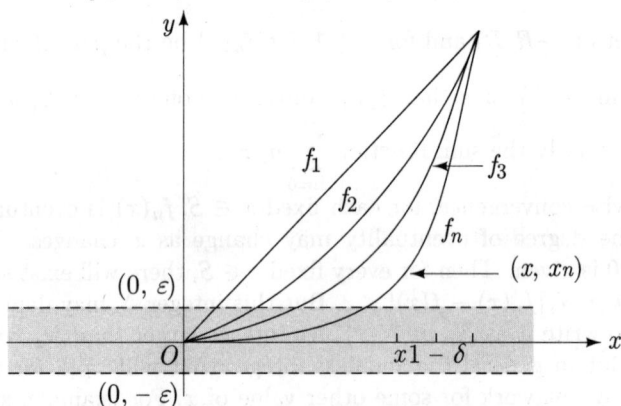

Figure 3.11.1 : Convergence of $\{x^n\}$ on $[0, 1]$

Summing up, in pointwise convergence, even though for every $x \in S$, the sequence $\{f_n(x)\}$ converges to $f(x)$ as $n \to \infty$, the rapidness of this converges can vary considerably as $x$ varies over the set $S$. There is no uniformity in it and so the entire graph of $f_n$ may not come close to the graph of $f$. The situation is rather like indisciplined pilgrims. Although each one eventually reaches his destination, he walks at this own pace. At any time, they may be so scattered, that it is really difficult to say that they are travelling as a coherent group. Compared this with a battalion of disciplined soldiers. Whey they march, their speeds are nearly uniform and the picture they create is that their battalion, as a whole, is a single object moving forward.

The concept of uniform convergence is intended to capture this sense of uniformity in marching. Let us once again suppose $\{f_n\}_{n=1}^{\infty}$ is a sequence of real-valued functions defined on a set $S$ and $f : S \to \mathbb{R}$ is another function. We want to define what is meant by saying that $f_n$ converges uniformly to $f$ on $S$. Ideally, this should mean that for every $n, |f_n(x) - f(x)|$ should be constant on $S$, i.e., independent of $x$. And if we call this constant $C_n$, then $C_n$ should tend to 0 as $n \to \infty$. It turns out that the requirement that $|f_n(x) - f(x)|$ should be independent of $x$ is far too stringent to be generally true. Nor is it really needed. It suffices to assume that, for each $n$, there is some constant $C_n$

(depending on $n$) such that for every $x \in S, |f_n(x) - f(x)| \leq C_n$ and further that $C_n \to 0$ as $n \to \infty$. The first requirement is equivalent to saying that $\sup_{x \in S} |f_n(x) - f(x)| \leq C_n$. In view of the Sandwich Theorem for sequences, $C_n \to 0$ implies that $\sup_{x \in S} |f_n(x) - f(x)| \to 0$ as $n \to \infty$. Keeping in mind that in the case of sequences it is only the eventual behaviour that matters (see Section 3.2) we are led to the following definition.

**11.1 Definition :** A sequence $\{f_n\}$ is said to **converge uniformly** on a set $S$ to a function $f$, if for almost all $n$, $\sup_{x \in S} |f_n(x) - f(x)|$ is finite and further $\sup_{x \in S} |f_n(x) - f(x)| \to 0$ as $n \to \infty$.

(Normally functions can be denoted either by symbols like $f_n, f$ or even by $f_n(x), f(x)$ etc. In the definition above, we have used $f_n \to f$ instead of the more conventional notatoin $f_n(x) \to f(x)$. There is, of course, nothing wrong in saying that $\{f_n(x)\}$ converges to $f(x)$ uniformly on $S$. But this is likely to be confused with the pointwise convergence. Sometimes uniform convergence is stressed by writing a double arrow, i.e., $f_n \rightrightarrows f$ on $S$, or $f_n(x) \rightrightarrows f(x)$ on $S$.)

The geometric formulation of this definition is that if we surround the graph of the function $f$ on $S$ by a strip of vertical height $2\epsilon$, where $\epsilon > 0$ is given, then there exists $N$ such that for all $n \geq N$, the graph of $f_n$ on $S$ lies completely within this strip (see Figure (3.11.2)).

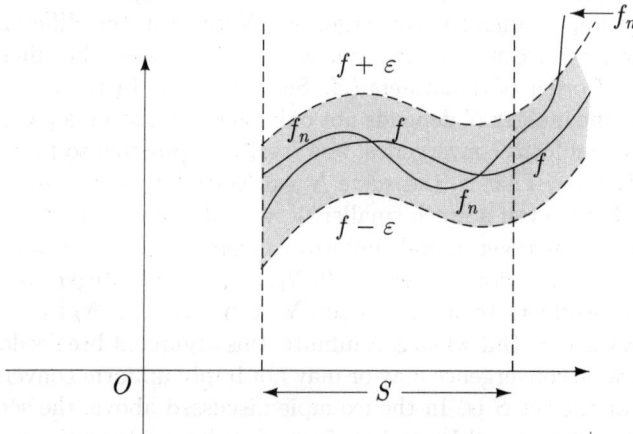

Figure 3.11.2 : Uniform Convergence

It is possible to paraphrase the definition above without using the concept of a supremum, as we now show.

**11.2 Theorem :** With the notation as above, $f_n \to f$ uniformly on $S$ if and only if for every $\epsilon > 0$, there exists $N$ such that for all $x \in S$ and for all $n \geq N, |f_n(x) - f(x)| < \epsilon$.

174 *Calculus for Scientists and Engineers*

**Proof :** Assume first $f_n \to f$ uniformly on $S$. Let $\epsilon > 0$ be given. Then since $\sup_{x \in S} |f_n(x) - f(x)| \to 0$ as $n \to \infty$, there exists $N$ such that for all $n \geq N$, $|\sup_{x \in S} |f_n(x) - f(x)| - 0| < \epsilon$. Since for every $x \in S$,

$$0 \leq |f_n(x) - f(x)| \leq \sup_{x \in S} |f_n(x) - f(x)|,$$

it follows that $|f_n(x) - f(x)| < \epsilon$, whenever $n \geq N$ and $x \in S$.

Conversely assume that the given condition holds. We first claim that $\sup_{x \in S} |f_n(x) - f(x)|$ is finite for almost all (i.e., all except finitely many) $n$. For this, set $\epsilon = 1$ in the condition. Then there exists some $N_0$ such that for all $n \geq N_0, |f_n(x) - f(x)| < 1$, for all $x \in S$. But then $\sup_{x \in S} |f_n(x) - f(x)| \leq 1$ for $n \geq N_0$, and hence is finite. We next prove that as $n \to \infty$, $\sup_{x \in S} |f_n(x) - f(x)| \to 0$. Let $\epsilon > 0$ be given. Applying the condition to $\frac{\epsilon}{2}$ (instead of $\epsilon$), we get some $N$ such that for all $n \geq N$, and for all $x \in S, |f_n(x) - f(x)| < \frac{\epsilon}{2}$. But in that case, $\sup_{x \in S} |f_n(x) - f(x)| \leq \frac{\epsilon}{2} < \epsilon$. So for all $n \geq N, |\sup_{x \in S} |f_n(x) - f(x)| - 0| < \epsilon$ whenever $n \geq N$. Hence $\lim_{n \to \infty} \sup_{x \in S} |f_n(x) - f(x)| = 0$, i.e., $f_n \to f$ uniformly on $S$. ∎

This theorem is hardly profound. The condition givn in it is often taken as the definition of uniform convergence. Note that the difference between pointwise convergence and uniform convergence is precisely the difference caused by a change of order of quantifiers (cf. Section 1.11). In the case of pointwise convergence, the integer $N$ depends not only on $\epsilon$ but also on a particular $x \in S$. In the case of uniform convergence, however, it is possible to find an $N$ which depends only on $\epsilon$. That is, the *same N* will work for *every* $x \in S$. (Of course, for a particular $x$, even a much smaller $N$ may also work.) If the set $S$ is finite then pointwise convergence and uniform convergence are the same, because if $S = \{x_1, \ldots, x_k\}$ and, for a given $\epsilon > 0, N_1, \ldots, N_k$ are integers which work for $x_1, \ldots, x_k$ respectively then we can set $N = \max\{N_1, \ldots N_k\}$ and this $N$ will work for every $x \in S$. But when $S$ is infinite, this argument breaks down. In such a case, pointwise convergence may or may not imply uniform convergence. It all depends what the set $S$ is. In the example discussed above, the sequence $\{x^n\}$ converges pointwise on $[0,1]$, and uniformly on intervals of the form $[0, 1 - \delta]$ where $\delta > 0$, but not on the interval $[0, 1)$ (and hence not on the superset $[0, 1]$ either).

We now turn to the main question, viz., why uniform convergence is such a strong concept. In the simile given above uniform convergence was compared with a march of an army battalion. The latter, by its discipline, has tremendous strength which may, at times, result in collapse of bridges, if the soldiers' steps are in perfect harmony ! In mathematics, too, uniform convergence is a powerhouse. Its strength comes primarily because it allows an interchange of orders of two limits (see the comments on Corollary (7.5) which also involves changing the order of certain limits). To appreciate what it is, let us once again take the example $f_n(x) = x^n$ discussed above. Let $S = [0, 1]$. For each $n$, $\lim_{x \to 1} x^n = 1$ and so $\lim_{n \to \infty} ( \lim_{x \to 1^-} x^n) = \lim_{n \to \infty} 1 = 1$. But if, for a fixed $x \in [0, 1)$, we take $\lim_{n \to \infty} x^n$, it is 0. So $\lim_{x \to 1^-} ( \lim_{n \to \infty} x^n) = \lim_{x \to 1^-} 0 = 0$. In other words, the two limiting processes, one obtained by letting $n \to \infty$ and the other by letting $x$ tend to 1 from below, do not commute. We already saw that $\{x^n\}$ does not converge uniformly on $[0, 1)$. It turns out that this is the real villain, because as the next theorem shows, the situation is better with uniform convergence.

**11.3 Theorem :** Suppose $f_n \to f$ uniformly on a set $S$. Let $c \in \mathbb{R}$. (It is not necessary that $c \in S$.) Suppose for each $n$, $\lim_{\substack{x \to c \\ x \in S}} f_n(x)$ is finite and equals $L_n$ (say). Assume further that $L_n \to L$ as $n \to \infty$. Then $f(x) \to L$ as $x \to c$ through $S$. That is, $\lim_{\substack{x \to c \\ x \in S}} ( \lim_{n \to \infty} f_n(x)) = \lim_{n \to \infty} (\lim_{\substack{x \to c \\ x \in S}} f_n(x))$.

**Proof :** Intitutively, the argument is that if $n$ is large then $f_n(x)$ is close to $f(x)$ for all $x \in S$. If $x$ happens to be close to $c$ then $f_n(x)$ is close to $L_n$. But $L_n$ is itself close to $L$. So $f(x)$ must be close to $L$.

To make this precise, let $\epsilon > 0$ be given. Then by uniform convergence, there exists $N_1$ such that

$$|f_n(x) - f(x)| < \frac{\epsilon}{3}, \text{ for all } n \geq N_1 \text{ and for all } x \in S \tag{1}$$

Also there exists $N_2$ such that for all $n \geq N_2$,

$$|L_n - L| < \frac{\epsilon}{3} \tag{2}$$

Let $N$ be any integer greater than or equal to both $N_1$ and $N_2$ (e.g., $N = \max\{N_1, N_2\}$). Since $\lim_{\substack{x \to c \\ x \in S}} f_N(x) = L_N$, there exists $\delta > 0$ such that

$$|f_N(x) - L_N| < \frac{\epsilon}{3} \text{ for all } x \in S, \text{ with } 0 < |x - c| < \delta \tag{3}$$

Applying (1) and (2) with $n = N$, and using (3) (and triangle inequality), we get that whenever $0 < |x - c| < \delta$ and $x \in S$,

$$
\begin{aligned}
|f(x) - L| &\leq |f(x) - f_N(x)| + |f_N(x) - L_N| + |L_N - L| \\
&< \frac{\epsilon}{3} + \frac{\epsilon}{3} + \frac{\epsilon}{3} = \epsilon.
\end{aligned}
$$

This shows $\lim_{\substack{x \to c \\ x \in S}} f(x) = L$ as desired. ∎

Since most of the important concepts in calculus are defined in terms of limits, it is not surprising that this theorem holds the secret of the power of uniform convergence. As we shall see later, it is because of this theorem (or something similar to it) that we can differentiate and integrate power series term by term.

In order for this theorem to be useful, it is necessary to be able to tell if a given sequence $\{f_n\}$ of functions converges uniformly or not on a given set $S$. There is no golden method for this. Generally one tries to see if for a given $n$, we can find some reasonable upper bound, say $C_n$ on the set $\{|f_n(x) - f(x)| : x \in S\}$. It is not necessary that $C_n$ should be the least upper bound (i.e., the supremum) of this set. Even a crude upper bound would do as long as it is not too extravagant. If $C_n$ could be shown to tend to 0 as $n$ tends to $\infty$, then $f_n$ would converge uniformly to $f$ on $S$. A desired upper bound is often obtained by the worst-case consideration as we did in the case of the sequence $\{x^n\}$ on the set $[0, 1 - \delta)$ $(\delta > 0)$, where, the worst case occurs at $x = 1 - \delta$. In more complicated problems, it may be necessary to split the given set suitably (see Exercise (11.1) (ii)).

When, on the other hand, we suspect (for example, because of our failure to find an upper bound such as $C_n$ ) that the sequence $f_n$ does *not* converge uniformly to its limit function $f$, then to prove this rigorously, we look for 'bad' points in the set $S$. That is, for every $n$, we try to find some $x_n \in S$ such that $|f_n(x_n) - f(x_n)|$ is at least some fixed positive number $\epsilon$ (of our choice). Thus, for example, in the case of the sequence $\{x_n\}$, if we take $\epsilon = 1/2$, then for every $n$, $|f_n(x_n) - 0| \geq \epsilon$ for $x_n = \dfrac{1}{2^{1/n}}$. So the sequence does not converge uniformly to the zero function on the set $[0, 1)$. (In some problems, it is easier to prove the absence of uniform convergence indirectly by showing that some consequence of uniform convergence fails, see, for example, Exercise (11.6)).

Just as the convergence of a series of numbers involves nothing new, conceptually, than the convergence of sequences, the same holds for series of functions. The pointwise as well as the uniform convergence of a series of functions are defined in terms of the respective convergences of the sequence of its partial sums. That is, we say a series $\displaystyle\sum_{n=1}^{\infty} f_n(x)$ is pointwise (or uniformly) convergent on a set $S$ if the sequence $\{S_n(x)\}_{n=1}^{\infty}$ is pointwise (respectively, uniformly) convergent on the set $S$, where $S_n(x) = \displaystyle\sum_{k=1}^{n} f_k(x)$, for $x \in S$. The following test, known as **Weierstrass $M$-test** is probably the most frequently used sufficiency criterion for uniform convergence of series.

**11.4 Theorem :** Suppose $\{f_n\}$ is a sequence of real-valued functions defined on a set $S$. Assume $\displaystyle\sum_{n=1}^{\infty} M_n$ is a convergent series of non-negative real numbers.

If for every $n \in I\!N$ and $x \in S, |f_n(x)| \leq M_n$, then the series $\displaystyle\sum_{n=1}^{\infty} f_n$ converges absolutely and uniformly on $S$.

**Proof :** By Theorem (9.1), $\sum\limits_{n=1}^{\infty} |f_n(x)|$ is convergent and hence by Theorem

(8.1), $\sum\limits_{n=1}^{\infty} f_n(x)$ is also convergent for every $x \in S$. Let $f(x) = \sum\limits_{n=1}^{\infty} f_n(x)$. We

have to show that $S_n \to f$ uniformly on $S$, where $S_n(x) = \sum\limits_{k=1}^{n} f_k(x)$. Clearly

$$|S_n(x) - f(x)| = |\sum\limits_{k=n+1}^{\infty} f_k(x)| = \lim\limits_{m \to \infty} |\sum\limits_{k=n+1}^{m} f_k(x)|. \text{ Since } |f_k(x)| \leq M_k \text{ for all}$$

$k \in S$, we have

$$| \sum\limits_{k=n+1}^{m} f_k(x) | \leq \sum\limits_{k=n+1}^{m} |f_k(x)| \leq \sum\limits_{k=n+1}^{m} M_k \tag{4}$$

Since $\sum\limits_{n=1}^{\infty} M_n$ is convergent, $\sum\limits_{k=n+1}^{m} M_k$ is bounded above by a bound inde-

pendent of $m$. So from (4), we see that $|S_n(x) - f(x)|$ is bounded above, for each
fixed $n$. Hence, $\sup\limits_{x \in S} |S_n(x) - f(x)|$ is finite for all $n$. To show that this supre-

mum tends to 0 as $n \to \infty$, we apply the Cauchy criterion (Theorem (5.5) with

a slight change of notation) to the series $\sum\limits_{n=1}^{\infty} M_n$. Let $\epsilon > 0$ be given. Then there

exists $p$ such that for all $n, m$ with $m \geq n \geq p$, $\sum\limits_{k=n+1}^{m} M_k < \dfrac{\epsilon}{2}$. Then, again by

(4), $\left| \sum\limits_{k=n+1}^{m} f_k(x) \right| < \dfrac{\epsilon}{2}$ for all $x \in S$. Hence, letting $m \to \infty$,

$$|S_n(x) - f(x)| = \lim\limits_{m \to \infty} \left| \sum\limits_{k=n+1}^{m} f_k(x) \right| \leq \dfrac{\epsilon}{2} \text{ for all } x \in S.$$

So $\sup\limits_{x \in S} |S_n(x) - f(x)| \leq \dfrac{\epsilon}{2} < \epsilon$, for all $n \geq p$. Hence $S_n$ converges uniformly to
$f$ on $S$. ∎

This theorem, whenever applicable, enables us to tap the pleasant consequences of both the absolute convergence and uniform convergence. Usually,
the functions $f_n(x)$ appearing in the series above are very simple functions with
nice properties. These properties, carry over without difficulty to fintie sums,
and hence in particular to the partial sums, viz. $S_n(x)$. Uniform convergence
then extends these properties to the sum function $f(x)$ while absolute convergence permits certain algebraic manipulations.

An excellent illustration is provided by the power series. Here the $n^{\text{th}}$ function $f_n(x)$ is of the form $a_n x^n$ where $a_n$ is some constant. (Note that we are

now allowing $n$ to be 0 as well with the understanding that $a_0 x^0$ is the constant function $a_0$.) In the last section (cf. Theorem (10.1)) we defined the interval of convergence $(-R, R)$ of a power series $\sum_{n=0}^{\infty} a_n x^n$. It is not true in general that the convergence is uniform over this entire interval (see Exercise (11.6)). But if we shrink it slightly, the situation is better. Again we prove the theorem only in an important special case and leave the general case as an exercise.

**11.5 Theorem :** Let $R = 1/L$ where $L = \lim_{n \to \infty} |a_n|^{1/n}$, with the understanding that $R = \infty$ if $L = 0$. Then for every $R'$ with $0 \le R' < R$, the power series $\sum_{n=0}^{\infty} a_n x^n$ converges uniformly on $[-R', R']$.

**Proof :** By Theorem (10.1), $\sum_{n=0}^{\infty} a_n R'^n$ is absolutely convergent. Hence $\sum_{n=0}^{\infty} M_n$ is convergent, where $M_n = |a_n R'^n|$. Also, for $x \in [-R, R], |a_n x^n| = |a_n||x|^n \le M_n$. Hence by the theorem above, $\sum_{n=0}^{\infty} a_n x^n$ converge uniformly (and also absolutely, which we know already anyway) on $[-R, R]$. ∎

The applicability of this theorem is wider than appears at first sight. Even though the convergence may not be uniform over $(-R, R)$, given any $c \in (-R, R)$, we can find $\delta > 0$ and $R'$ such that $(c - \delta, c + \delta) \subset [-R', R'] \subset (-R, R)$. (Specifically, we may take $\delta$ to be any positive number less than $R - |c|$ and $R' = |c| + \delta$. The reader is once again advised to draw suitable diagrams to see such things easily.) So by the theorem above, the convergence is uniform over $[-R', R']$ and hence on $(c - \delta, c + \delta)$. This is generally enough to study the behaviour of the sum function $f(x)$ at $c$. Many concepts in calculus are of a local nature. That is, they depend only on the values of a function in some neighbourhood of that point. For example, limit of $f(x)$ as $x$ tends to $c$ is a local concept. So when we want to prove that some local property of the sum function of $\sum_{n=0}^{\infty} a_n x^n$ holds true at every point in $(-R, R)$, we can apply the argument above. Later, we shall indeed do so to establish the validity of term-by-term differentiation of power series. In the last section we saw that the power series are a powerful tool. Figuratively, we may now say that uniform convergence is the fuel on which the power series work.

## EXERCISES

11.1    (i)    Prove that if $f_n \to f$ uniformly on a set $S$ then for every subset $T$ of $S, f_n \to f$ uniformly on $T$. (This is absolutely trivial, but its contrapositive statement is sometimes useful in showing the convergence is not uniform on $S$, by suitably identifying a subset $T$ on which it is eariser to show that it is not uniform.)

(ii) Let $T_1, T_2, \ldots, T_k$ be (not necessarily mutually disjoint) subsets of $S$ whose union is $S$. Prove that $f_n \rightrightarrows f$ on $S$ if and only if $f_n \rightrightarrows f$ on $T_i$ for every $i = 1, 2, \ldots, k$. Is this still true if, instead of $T_1, T_2, \ldots, T_k$, we have an infinite collection of subsets of $S$ whose union is $S$ ? (This result is useful when it is not easy to directly prove uniform convergence on the entire set $S$ but, for some suitably chosen (finite) decomposition of $S$, it is easy to prove it on each member of this decomposition. Sometimes the arguments needed to prove uniform convergence on $T_i$ are different for different $i$'s. The situation is analogous to that of a restricted limit; see Theorem(2.5.1) and Exercises (2.5.3) to (2.5.7).)

**11.2** Assume $f_n \to f$ uniformly on $S$ and that each $f_n$ is bounded on $S$ (i.e., there exists some $M_n$ such that $|f_n(x)| \le M_n$ for all $x \in S$). Prove that $f$ is bounded on $S$ and further that $\{f_n\}$ is **uniformly bounded** on $S$, i.e., there is some $M$, (independent of $n$), such that $|f_n(x)| \le M$ for all $x \in S$ and $n \in \mathbb{N}$.

**11.3** Prove that the result of the last exercise is not true in general for pointwise convergence. [*Hint* : Let $S = (0,1)$ and $f_n(x) = \dfrac{1}{x + \frac{1}{n}}$. See also Exercise (11.6) below.]

**11.4** Let $S = (0,1)$. Prove that each of the following sequences/series converges pointwise on (0,1). Identify the subintervals on whcih the convergence is uniform.

(i) $\dfrac{n}{nx+1}$    (ii) $\dfrac{x}{nx+1}$    (iii) $\dfrac{1}{nx+1}$

(iv) $\displaystyle\sum_{n=1}^{\infty} x^n(1-x)$    *(v) $\displaystyle\sum_{n=1}^{\infty}(-1)^n x^n(1-x)$.

**11.5** Suppose $\{f_n\}$ is a sequence of real-valued functions defined on $T$ such that (i) for every $x \in T$ we have $f_{n+1}(x) \le f_n(x)$ for all $n$ and (ii) $f_n(x) \to 0$ uniformly on $T$ as $n \to \infty$. Prove that the alternating series $\displaystyle\sum_{n=1}^{\infty}(-1)^{n+1} f_n(x)$ converges uniformly (but not necessarily absolutely) on $T$. [*Hint* : Model the proof after that of Theorem (11.4) except that instead of Theorem (9.1), use Leibnitz test for alternating series, Exercise (8.1).] Using this result give an easier solution to (v) of the last exercise.

**11.6** Prove that the power series $\displaystyle\sum_{n=0}^{\infty} x^n$ does not converge uniformly on $(-1,1)$, its interval of convergence. [*Hint* : Use Exercise (11.2). A direct argument is possible, but a little messy.]

11.7  Prove that the power series $\sum\limits_{n=1}^{\infty} \dfrac{x^n}{n^2}$ converges uniformly on its entire interval of convergence.

11.8  Prove that the power series $\sum\limits_{n=1}^{\infty} \dfrac{(-1)^n x^n}{n}$ has $(-1, 1)$ as its interval of convergence. Identify the subintervals on which the series converges uniformly.

11.9  Prove that Theorem (11.5) continues to hold if we replace $\lim\limits_{n\to\infty} |a_n|^{1/n}$ by $\overline{\lim}_{n\to\infty} |a_n|^{1/n}$.

11.10  Prove a uniform version of the Cauchy criterion for convergence, that is, suppose $\{f_n\}$ is a sequence of functions all defined on a set $S$. Prove that there exists a function $f$ on $S$ such that $f_n \to f$ uniformly on $S$ if and only if for every $\epsilon > 0$, there exists $p \in \mathbb{N}$ such that for all $m \geq n \geq p$ and for all $x \in S$, $|f_m(x) - f_n(x)| < \epsilon$.

11.11  Suppose $\{f_n\}$ and $\{g_n\}$ converge uniformly on $S$ to $f, g$ respectively. Prove that $\{f_n \pm g_n\}$ converges uniformly to $f \pm g$ on $S$. Show by an example that $\{f_n g_n\}$ and $\{f_n/g_n\}$ (even when defined) need not converge to $fg$ and $f/g$ uniformly on $S$. [*Hint* : The necessary counter-examples can be found from Exercise (11.4), with minor modifications.]

11.12  In the last exercise, prove that $\{f_n g_n\}$ converges unifromly to $fg$ on $S$ if $f$ and $g$ are bounded on $S$.

# Chapter 4

# CONTINUITY AND DERIVATIVES

## 4.1 Continuous Functions - Nature's Choice

Basically the reason continuous functions are so natural is that the natural functions are continuous! There is more to this than verbal juggling. Just as nature abhors vacuum, nature abhors a sudden or an instantaneous change. A puppy does not become a dog and a caterpillar does not turn into a butterfly instantaneously. Each goes through an infinite number of intermediate stages of growth. Even when we talk of things like sudden storms or earthquakes, it really does not mean that the objects in question (such as the air or the rocks) which were at rest till a particular moment suddenly acquired tremendous speeds the next moment. In fact, mathematically there is no such thing as the next moment, just as there is no such thing as the next real number. The change of speed occurs over an extremely short time interval. But within this interval, the speed changes continously and not suddenly from 0 to some positive value. The same thing happens when we turn an electrical switch on or off. The current in the wire appears to shoot up from 0 to some value (say $i_0$) instantaneously and then from $i_0$ to 0 again as in Fig. 4.1.1. (a). But in reality this change takes place over some time intervals (shown highly magnified in Fig. 4.1.1.(b)). The numbers $\delta_0$ and $\delta_1$ are so small that they can be taken to be 0 for many purposes. But in reality the current in a wire in a network is a continuous function of time. This, in fact, is the tenet on which the classical physics is based. All physical variables such as length, mass, energy, charge as well as the functions which express the relationships among these variables are assumed to be continuous. The quantum theory has challenged this premise by holding that certain physical quantities can exist only as whole multiples of some basic units. But these basic units are so small, that the theory based on treating these variables as continuous is not out-dated. In fact it is applicable even to situations which involve variables like population, which are most unquestionably discrete and not continuous (cf. Exercise (1.3.1)).

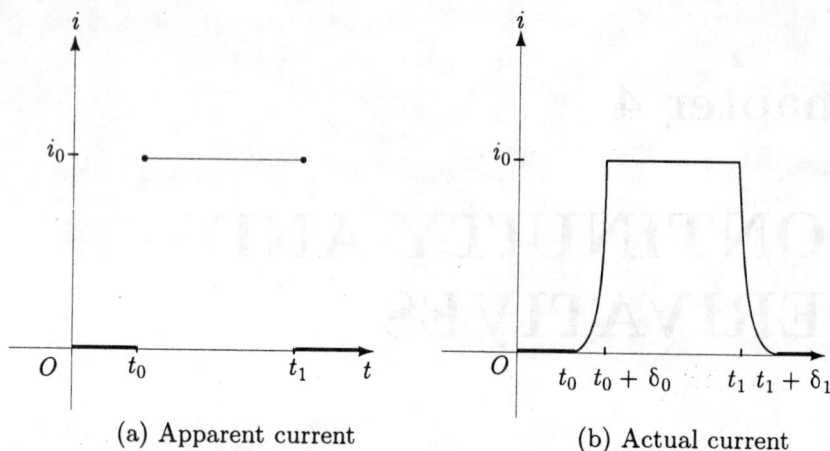

(a) Apparent current                    (b) Actual current

Figure 4.1.1 : Effects of a Switch on Network Current

Intuitively, continuity means the absence of a sudden jump or slump. In terms of graphs, it means the absence of a break. Thus continuity of a function $y = f(x)$ at a point $c$ in its domain means that as $x$ approaches $c$, $f(x)$ should approach $f(c)$. Put differently, if you go away from $c$ by a small distance, then the corresponding change in the values of $f$ should also be small. In the old days people used to say that an infinitesimally small change in the independent variable (viz. $x$) produces only an infinitesimal change in the dependent variable (viz. $y$). In symbols, continuity of $f$ at $c$ means whenever $h$ is infinitesimally small, so is the difference $f(c + h) - f(c)$.

As mentioned in Section 2.3, modern standards of mathematical rigour do not permit the use of the intuitively clear but hard to define term 'infinitesimal'. But the idea can be expressed quite succinctly in terms of limits, a concept for which we already have a perfectly rigorous definition.

**1.1 Definition :** A function $f$ is said to be **continuous** at a point $c$ if $\lim_{x \to c} f(x) = f(c)$.

It is inherent in this definition that $f$ is defined at $c$ and also near $c$. It was emphasised in Section 2.3 that $\lim_{x \to c} f(x)$ has nothing to do conceptually with $f(c)$. Either may exist without the other and even when both exist, they need not be equal. All possibilities can be illustrated by (rather artificial) examples. Continuity of $f(x)$ at $c$ puts an end to such pathologies. In order for $f$ to be continuous at $c$, both $\lim_{x \to c} f(x)$ and $f(c)$ must exist and equal each other. (As remarked in Section 2.3, the fact that most naturally occurring functions are continuous also sometimes serves to give a beginner the wrong impression that $\lim_{x \to c} f(x)$ is conceptually the same as $f(c)$.)

Since continuity is defined in terms of limits, it is not suprising that most of the elementary results about continuous functions follow from the corresponding properties of limits. Suppose for example, that both $f$ and $g$ are continuous at $c$. Then the functions $af + bg$ (where $a, b$ are some constants), $fg$ and $f/g$ (provided $g(c) \neq 0$) are continuous at $c$. A few other such results, which follow from some results about limits, will be given as exercises. But one of them deserves to be mentioned here because it will be used again and again.

**1.2 Theorem** : Suppose $f$ is defined in some neighbourhood $U$ of a point $c$. Then $f$ is continuous at $c$ if and only if for every sequence $\{x_n\}$ in $U$ converging to $c$, the sequence $\{f(x_n)\}$ converges to $f(c)$.

**Proof** : Merely apply Theorem (3.1.1) taking $L = f(c)$. ∎

This result is of a local nature in the sense that it deals with continuity of a function at a point. Such results usually follow easily from the corresponding results about the limits of a function at a point.

By contrast, there are certain properties of continuous functions which are global in nature. That is, they hold when the functions in question are continuous on some set, usually an interval, say $I$. So suppose $f : I \longrightarrow \mathbb{R}$ is a function. To say that $f$ is **continuous on the interval** $I$ means $f$ is continuous at every point $c$ of $I$. If $c$ is an interior point (i.e., not an end point of $I$) then we already know what this means. But if $I$ contains either of its end-points then a minor modification is necessary. Suppose for example, that $I$ is a closed interval of the form $[a, b]$ where $a < b$. Since $f$ may not be defined at points less than $a$, it is meaningless to talk of $\lim_{x \to a} f(x)$. However, in this case, we can talk of the right handed limit $\lim_{x \to a^+} f(x)$ (see Section 2.5). If this limit exists and equals $f(a)$ we say $f$ is **continuous from the right** at $a$. Similarly we define continuity from left. Whenever we say $f : [a, b] \longrightarrow \mathbb{R}$ is continuous, we mean $f$ is continous at every $c \in (a, b)$, continuous from the right at $a$ and from the left at $b$.

Functions which are continuous on closed and bounded intervals have certain remarkable properties, some of which are quite intuitive. The most well-known is the Intermediate Value Property. To illustrate what it means suppose that the temperature at some place at a particular time is, say, 15° and after 10 hours it is 20°. We do not know what the temperature was in between these two moments. It could have fluctuated up and down any number of times. Sometimes it could have gone higher than 20° or below 15°. But one thing we conclude, almost instinctively, is that at some moment in those ten hours, the temperature must have been exactly 17°.

Geometrically, if we plot the temperature (T) against time (t), as in Fig. 4.1.2, then we expect the graph to cross the horizontal line $T = 17$ at least once. For, if this does not happen, then intuitively some portion of the graph will lie entirely above this line, the rest would lie entirely below it and hence there would be a break in the graph, contradicting the continuity of the temperature function (which, in nature, is taken for granted). There is of course nothing special about the number 17 here. It is merely a value which is intermediate to (i.e. lies in between) the values of the temperature function at the two end-points, viz. 15 and 20. (Hence the name.)

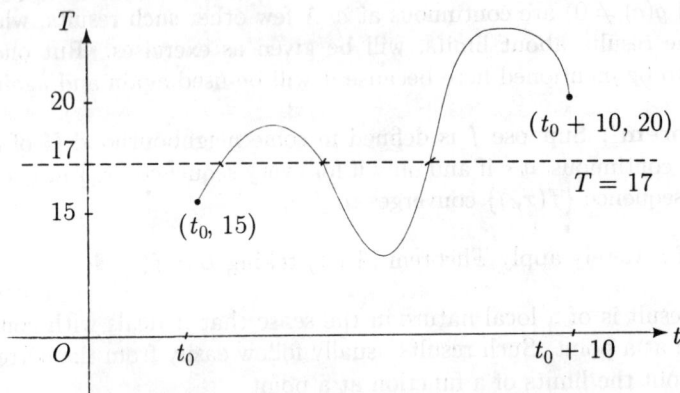

Figure 4.1.2 : Intermediate Value Property

We now prove the **Intermediate Value Property** (or **IVP** for short) rigorously. The proof is not trivial and makes a crucial use of completeness of the real number system. So IVP is typical of those theorems whose statements look obvious but whose rigorous proofs are far from trivial. (An even more telling example was the Jordan Curve Theorem mentioned in Section 1.6.)

**1.3 Theorem :** Suppose $f : [a, b] \longrightarrow \mathbb{R}$ is continuous. Let $m$ be a real number lying between $f(a)$ and $f(b)$. Then there exists some $c \in [a, b]$ such that $f(c) = m$.

**Proof :** Without loss of generality asusme $f(a) \leq f(b)$. (As otherwise replace $f$ by $-f$ which is also continuous). If $m = f(a)$ or $m = f(b)$ we take $c$ as $a$ or $b$ respectively. Hence assume $f(a) < m < f(b)$. We shall use the technique of successive bisection, encountered in the proof of Theorem (3.5.1). Call $a$ as $a_0$ and $b$ as $b_0$. Let $c_0 = \dfrac{a_0 + b_0}{2}$. If $f(c_0) = m$, we are done. Otherwise either $f(c_0) < m$ or $f(c_0) > m$. In the first case, call $c_0$ as $a_1$ and $b_0$ as $b_1$. In the second case, call $a_0$ as $a_1$ and $c_0$ as $b_1$. In either case we have $f(a_1) < m < f(b_1)$. We now subdivide $[a_1, b_1]$ into two equal subintervals and repeat the argument. We repeat this process if necessary. So either we shall reach some $c_r$ (viz. the mid-point of $[a_r, b_r]$) such that $f(c_r) = m$ or else we shall get an infinite sequence of nested intervals $[a_n, b_n]$ such that

$$
\left.
\begin{array}{rl}
\text{(i)} & [a_n, b_n] \subset [a_{n-1}, b_{n-1}] \text{ for every } n = 1, 2, \ldots \\[2mm]
\text{(ii)} & b_n - a_n = \dfrac{b_0 - a_0}{2^n} \text{ for all } n = 0, 1, 2, \ldots \\[2mm]
\text{and} \quad \text{(iii)} & f(a_n) < m < f(b_n) \text{ for all } n = 0, 1, 2, \ldots
\end{array}
\right\} \quad (1)
$$

Because of (i) and (ii), we are now exactly in the same situation as that encountered in the proof of Theorem (3.5.1). So, by the same reasoning used there, the sequences $\{a_n\}$ and $\{b_n\}$ converge to a common limit, say $c$. Note that for every $n, a_n \leq c \leq b_n$. By Exercise (3.1.9), this implies that $a \leq c \leq b$. We contend that $f(c) = m$. For this we use (iii).

Since $a_n \to c$ as $n \to \infty$, and $f$ is continuous at $c$, we have $f(a_n) \to f(c)$ as $n \to \infty$, by Theorem (1.2). But $f(a_n) < m$ for all $n$ and so by Exercise (3.1.9), $\lim_{n \to \infty} f(a_n) \leq m$. So $f(c) \leq m$. Similarly, using $m < f(b_n)$ for all $n$, we get $m \leq \lim_{n \to \infty} f(b_n) = f(c)$. Since $f(c) \leq m$ on one hand while $m \leq f(c)$ on the other, we must have $f(c) = m$ as was to be proved. ∎

The argument given here is very intuitive and akin to real-life experience. Suppose we are looking for a person $P$ who lives somewhere on a road stretchng from a point $A$ to a point $B$. If $P$ lives at either of these two end points, then there is no problem. Otherwise we reach the mid-point of the road and inquire. If $P$ does not live there we are told in which direction to go. We go half-way in that direction and inquire again. We go on repeating this process. Every time we are coming within half the distance from $P$ as at the last stage and that is why this method of searching for $P$ is called the **binary search**. In practice the search will terminate when we come within a striking distance of $P$. In the proof above, it could go on forever. Still, we have $a_n \leq c \leq b_n$ for all $n$ and condition (ii) in (1) shows that by taking $n$ sufficiently large we can come as close to the desired point $c$ as desired. So we can take $a_n$ (or $b_n$) as an approximation for $c$, if $n$ is large.

The Intermediate Value Property is often applied to show the existence of a zero of a continuous function $f$. All we have to do is to find two points $a, b$ (say) such that $f$ is continuous on $[a, b]$ and $f(a), f(b)$ are of opposite signs. The method of binary search can then be used to get an approximate location of a zero of $f$. If the function $f$ satisfies some stronger condition than continuity, then more efficient methods are available for locating a zero of $f$ as we shall see later (section 4.13).

A few interesting applications of IVP will be given in the exercises. We now prove another basic property of continuous functions which deals with their maximum and minimum values. Suppose $S$ is any set (not necessarily a subset of $I\!R$) and $f : S \longrightarrow I\!R$ is a function. We say that $f$ **attains (or has) a maximum** at a point $s_0$ of $S$ (or that $f(s_0)$ is a **maximum value** of $f$ on $S$) if for every $s \in S, f(s) \leq f(s_0)$. Similarly we define minimum value of $f$ on $S$. (In Section 2.1, we already had an instance of this where the set $S$ consisted of all boxes of a given volume $V$ and $f$ was the surface area function.) When the set $S$ is finite (and non-empty), the existence of a maximum value is obvious. Moreover, there is a simple method, based on straight comparison, to find it (see Exercise (2.1.11)). A similar algorithm is available for finding a minimum.

This simple-minded approach for finding maxima or minima (these words are, respectively, the plurals of 'maximum' and 'minimum') does not work when the set $S$ is infinite. There are various difficulties. First of all the function $f$ may not be bounded. Even when it is bounded it might not have a maximum. This may sound strange, but strange things do happen with infinite sets. A simple counter-example is to take $f(x) = x^2 + 3$ and $S = (0,1)$. Here $f$ has neither a maximum nor a minimum in $S$. (Note that the points 0 and 1 are not in $S$. If we include them, then $S$ becomes $[0,1]$ and $f$ does have a minimum (viz. 3) at 0 and a maximum (viz. 4) at 1. But, on $(0,1)$, $f$ takes values which are arbitrarily close to 3 but not the value 3 anywhere. So $f$ has an infimum but no minimum over $(0,1)$. Similarly $f$ has a supremum but no maximum over $(0,1)$. On $(0,1]$, $f$ has a maximum but no minimum.)

Let us now see how the function $f(x) = x^2 + 3$ behaves on the interval $[-1, 1]$. It is clear that $f$ assumes a minimum at $x = 0$ and maximum at 1 and also at $-1$. But suppose we change $f$ at 0 to say $f(0) = 3.5$, without changing it at any other point of $(-1, 1)$. This new function $f$ is discontinuous at 0 (see Figure 4.1.3). It has no minimum on $[-1, 1]$ even though it is bounded below and has 3 as its infimum. But there is no point $x_0$ in $[-1, 1]$ at which $f(x_0)$ actually equals 3. As for maximum, $f$ assumes its maximum at each of the two end-points viz. $-1$ and 1. If we throw these points out or keep them in but make the function discontinuous at them by setting, say, $f(-1) = 3.1$ and $f(1) = 3.2$, then $f$ would have no maximum either on $[-1, 1]$.

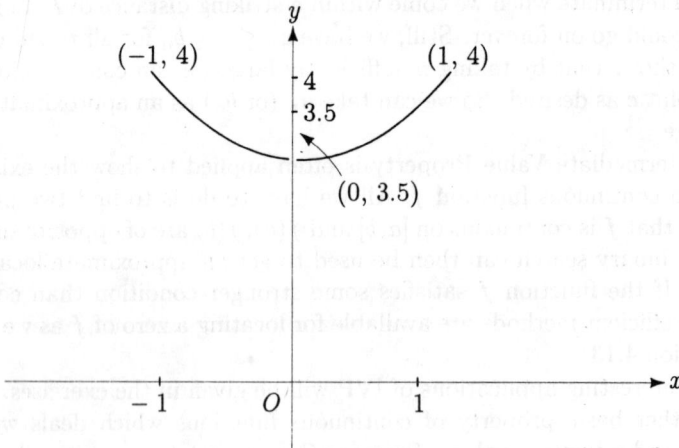

Figure 4.1.3 : A Discontinuous Function with no Minimum

Since a real valued function on an infinite set need not always have a maximum or a minimum (even when it is bounded), any theorem which guarantees the existence of maxima and minima under certain conditions is desirable. We the existence of maxima and minima under certain conditions is desirable. We prove one such result. In the examples given above, the failure of a function to have a maximum/minimum over an interval was due either to the discontinuity of the function or to the exclusion of the end-points from the interval. It turns out that if these defects are removed then the function does have a maximum and a minimum, as we now show. The proof again depends on the completeness of the real number system. We shall use completeness in the form of the Bolzano-Weierstrass theorem (Theorem (3.5.1)).

**1.4 Theorem** : Every continuous real-valued function on a closed and bounded interval is bounded and attains its bounds (i.e., has a maximum and a minimum).

**Proof** : Let $f : [a, b] \longrightarrow \mathbb{R}$ be continuous. We first show that $f$ is bounded above. If not, then for every $n \in \mathbb{N}$ there exists some $x_n \in [a, b]$ such that $f(x_n) \geq n$. By theorem (3.5.1), the sequence $\{x_n\}_{n=1}^{\infty}$ has a subsequence, say, $\{x_{n_k}\}_{k=1}^{\infty}$ which converges to some point, say, $c$. Since $a \leq x_{n_k} \leq b$ for all $k$, we have, by Exercise (3.1.9) that $a \leq \lim_{k \to \infty} x_{n_k} \leq b$, i.e., $a \leq c \leq b$. So $f$ is continuous at $c$, by assumption. Since $x_{n_k} \to c$ as $k \to \infty$, Theorem (1.2) tells us that $f(x_{n_k}) \to f(c)$ as $k \to \infty$. But, on the other hand, $f(x_{n_k}) \geq n_k$ for every $k$, implies that $f(x_{n_k}) \to \infty$ as $k \to \infty$, a contradiction. So $f$ is bounded above. By a similar reasoning (or by applying the reasoning above to $-f$) $f$ is bounded below. So $f$ is a bounded real-valued function on $[a, b]$. This means that the range of $f$, i.e., $f([a, b])$ is a bounded subset, say $T$, of $\mathbb{R}$. Let $M, m$ be respectively the supremum and the infimum of $T$ (which again exist by completeness, see Theorem (2.6.3)). We have to show that $M$, and $m$ are in $T$, i.e., there exist $\alpha, \beta \in$ in $[a, b]$ such that $f(\alpha) = m$ and $f(\beta) = M$. We do the first of these; the proof of the second is simlar.

Define $g : [a, b] \longrightarrow \mathbb{R}$ by $g(x) = f(x) - m$. Then $g(x) \geq 0$ for all $x \in [a, b]$. Further, for every $\epsilon > 0$, there exists $x \in [a, b]$ such that $g(x) < \epsilon$ since $m = \inf T$. Note also that $g(x)$ is continuous. Now if $g(x)$ is never 0 in $[a, b]$, then the function $h(x) = \dfrac{1}{g(x)}$ is also continuous on $[a, b]$. Since $g(x)$ takes arbitrarily small positive values on $[a, b]$, $h(x)$ takes arbitrarily large values. So $h$ is not bounded above, contradicting the first part of the proof (applied to the function $h$ instead of $f$). So, $g$ must vanish somewhere in $[a, b]$, i.e., there exists $\alpha \in [a, b]$ such that $g(\alpha) = 0$, which is equivalent to $f(\alpha) = m$. As remarked earlier, this completes the proof. ∎

Note that this theorem merely asserts the existence of a maximum and a minimum. It does not give a construction for them. As is to be expected, if $f$ satisfies some stronger condition that continuity, then it may be possible to *find* the points where a maximum (or a minimum) of $f$ is attained. One such result will be proved later in Section (4.11).

Because of the important properties of continuous functions, it is natural to inquire if there is a good source of continuous functions. As remarked earlier,

defined by mathematical formulas, in view of the elementary results about continuity, it is possible to construct new continuous functions from some elementary
ones, by taking finite sums, products and quotients (where defined). But things
are not so rosy when it comes to infinite sums. For example, let $f_1(x) = x$ and
for $n > 1$, let $f_n(x) = x^n - x^{n-1}$. Each $f_n$ is continuous. For each $n$, $\sum_{k=1}^{n} f_k(x)$ is
simply $x^n$ and is continuous. But if we take the infinite sum $\sum_{k=1}^{\infty} f_k(x)$, it is de-
fined only for $x \in (-1, 1)$ and equals $0$ for $|x| < 1$ and $1$ for $x = 1$. (See Theorem
(3.3.4)). So even though each $f_n(x)$ is continuous on $(-1, 1]$ (in fact continuous
on entire $\mathbb{R}$), the sum function $\sum_{n=1}^{\infty} f_n(x)$ is continuous only on $(-1, 1)$ but not
on $(-1, 1]$. If we recall (Section 3.11) that the sequence $\{x^n\}$ is not uniformly
convergent on $(-1, 1]$, we are led to believe that continuity of an infinite sum
of functions may hold when the convergence is uniform. This indeed is the
case as the following theorem, stated for sequences (rather than series) shows.
The proof, like that of theorem (1.2) is short because most of the hard work is
already done earlier.

**1.5 Theorem :** Suppose $\{f_n(x)\}_{n=1}^{\infty}$ is a sequence of real-valued functions,
each defined in some neighbourhood $N$ of a real number $c$. Suppose $f_n$ converges
uniformly to a function $f$ on $N$. Then if each $f_n$ is continuous at $c$, so is $f$.

**Proof :** Merely apply Theorem (3.11.3) with $L_n = f_n(c)$ and $L = f(c)$. We
then get, $\lim_{x \to c} f(x) = \lim_{x \to c} ( \lim_{n \to \infty} f_n(x)) = \lim_{n \to \infty} (\lim_{x \to c} f_n(x)) = \lim_{n \to \infty} (f_n(c)) = f(c)$.
So $f$ is continuous at $c$. ∎

Combining this theorem with Theorem (3.11.5) we get an important corollary.

**1.6 Corollary :** Every power series defines a function which is continuous in
its interval of convergence.

**Proof :** Let $(-R, R)$ be the interval of convergence of the power series $\sum_{n=0}^{\infty} a_n x^n$
(say) and let $A(x)$ be the sum function. For $n = 0, 1, 2, \ldots$, let $S_n(x) = \sum_{k=0}^{n} a_k x^k$.
Then each $S_n(x)$ is continuous on $(-R, R)$ (in fact continuous on entire $\mathbb{R}$). We
cannot apply the last theorem hastily, because the convergence of $S_n(x)$ to $A(x)$
need not be uniform, on the entire interval $(-R, R)$. However, for every $R'$ with
$0 \leq R' < R$, the convergence is uniform on $(-R', R')$ by Theorem (3.11.5). So
by Theorem (1.5) $A(x)$ is continuous at every point of $(-R', R')$. But given any
$c \in (-R, R)$ we can always find $R'$ such that $c \in (-R', R')$ and $0 \leq R' < R$.
Then $A(x)$ is continuous at $c$. Since this holds for every $c \in (-R, R)$, $A(x)$ is

Then $A(x)$ is continuous at $c$. Since this holds for every $c \in (-R, R), A(x)$ is continuous on $(-R, R)$. ∎

As an immediate consequence, we see that if the exponential, sine and cosine functions are defined by power series (cf. Exercise (3.3.14)) then they are continuous everywhere because the radius of convergence is $\infty$ in each case (Theorem (3.10.1)). The continuity of the exponential function and its basic property given in Exercise (3.10.3), viz. $e^{x+y} = e^x e^y$ enables us to define the natural logarithm as the inverse of the exponential function. We shall indicate this in Exercise (1.21). As mentioned earlier, nowadays it is more customary to define natural logarithms first (by means of integrals) and then the exponential function as the inverse of logarithms.

We conclude the section with a brief discussion of functions which are *not* continuous. It was mentioned at the beginning of the section that all 'naturally occuring' functions are continuous. It might, therefore, appear that the only role discontinuous functions have is as possible counter-examples to theorems about continuous functions (e.g. the function in Figure 4.1.3). But this is not quite the case for a variety of reasons.

First of all, in real life, we do not always deal with functions which are creations of mother nature. There are many arificial, man-made functions which enter our lives. Take, for example, the bank balance in our account. As a function of time, it is constant most of the time and then suddenly increases or decreases. (As mentioned in Chapter 2, continuous compounding of interest is possible but not practicable. And even if a bank allows it, what about the deposits and the withdrawls we make ?) The tax rates, bus fares and telephone charges are also discontinuous functions of income, distance and time respectively, for the simple reason that we measure money only in whole multiples of some unit, or in other words, money is a discrete variable. Because of the IVP, a non-constant, discrete valued function of a continuous variable is necessarily discontinuous.

Secondly, even when we are dealing with naturally occuring functions such as electric current in a wire, which are continuous in theory, it is sometimes advantageous to treat them as discontinuous. In other words, going back to Fig. 4.1.1, it is better to treat the apparent current as the actual one, because the change in the short time interval during which a switch becomes fully operative is so rapid that it is next to impossible, at least with ordinary instruments, to determine its exact value at points of this subinterval. To assume that the current changes instantaneously from 0 to $i_0$ and vice versa would make the current a discontinuous function. But for most practical purposes, this causes no harm and for many others the gain in simplicity outweighs the harm. True, there are a few situations where we need to assume that the current function is continuous. In such situations, methods are available to approximate the discontinuous function by a continuous function to a high degree of accuracy. Similar comments apply for other sudden changes such as the change in the speed of a cricket ball when it is hit by a bat.

Leaving this debate aside, let us turn to analysing discontinuous functions.

Suppose a function $f(x)$ is defined in a neighbourhood of a point $c$. If $f$ is not continuous at the point $c$, then we say that it is discontinuous at $c$ or that $c$ is a **point of discontinuity** (or simply a **discontinuity**) of $f$. In view of Definition (1.1), this can happen for either one of the two reasons :

(i) $\lim\limits_{x \to c} f(x)$ does not exist

(ii) $\lim\limits_{x \to c}$ exists but differs from $f(c)$.

(If $c$ is an end-point of the domain of $f$, then we replace $\lim\limits_{x \to c} f(x)$ by $\lim\limits_{x \to c^+} f(x)$ or by $\lim\limits_{x \to c^-} f(x)$ as the case may be.)

A discontinuity of type (ii) is called a **simple** or a **removable** discontinuity. The name is appropriate because in such cases the discontinuity can be removed (or 'cured') simply by redefining $f$ at just one point, viz. $c$. (Strictly speaking, this gives rise to a new function which is continuous ; the original function remains discontinuous. But it is a standard abuse of language to say that $f$ can be made continuous at $c$.) For example, suppose $f(x) = x^2$ for $x \neq 3$ and $f(3) = 10$. Then, $\lim\limits_{x \to 3} f(x) = \lim\limits_{x \to 3} x^2 = 9 \neq 10 = f(3)$. So $f$ is discontinuous at 3. But if we set $f(3) = 9$, then $f$ is continuous at 3. So 3 is a removable discontinuity of $f$. (Strictly speaking, one should say that the function g defined by $g(x) = x^2$ for $x \neq 3$ and $g(3) = 9$ is continuous at 3.) The function whose graph is shown in Fig. 4.1.3 also has a simple discontinuity at 0.

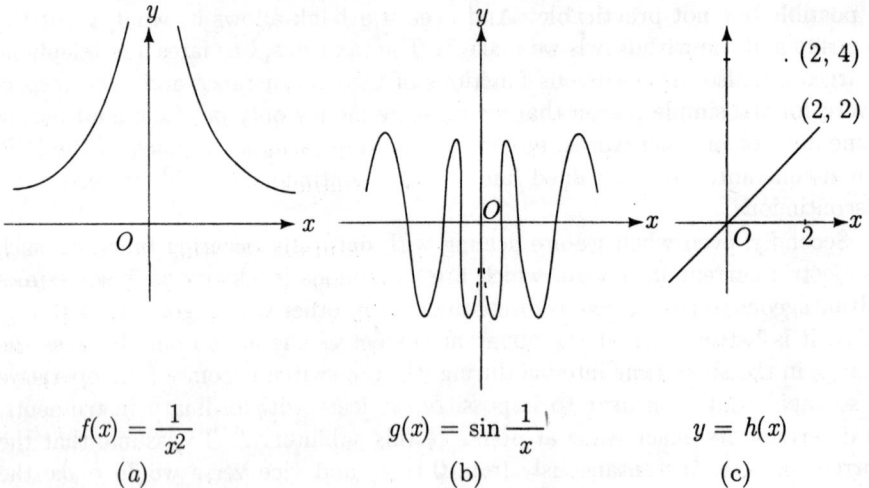

$$f(x) = \frac{1}{x^2}$$
$$(a)$$

$$g(x) = \sin\frac{1}{x}$$
$$(b)$$

$$y = h(x)$$
$$(c)$$

Figure 4.1.4 : Non-removable Discontinuities

Discontinuities of type (i) are more serious. For example, if $f(x) = \dfrac{1}{x^2}$ for $x \neq 0$ (see Fig. 4.1.4 (a) for a graph), then $f$ is discontinuous at 0 and will remain so no matter how we define $f(0)$. The function $g(x) = \sin\dfrac{1}{x}$, shown in Fig. 4.1.4 (b), also has a non-removable singularity a 0. Contrast this with the function $h(x)$ defined by

$$h(x) = \begin{cases} x & \text{if } x \leq 2 \\ x^2 & \text{if } x > 2 \end{cases}$$

whose graph is shown in Fig.4.1.4 (c).

Here, even though $\lim\limits_{x \to 2} h(x)$ does not exist, the situation is not as bad as that for the functions $f$ and $g$, because $\lim\limits_{x \to 2^+} h(x)$ and $\lim\limits_{x \to 2^-} h(x)$ both exist (and equal 2 and 4 respectively). Since $h(2) = 2$, we see that even though $h$ is discontinuous at 2, it is continuous from the left at 2. If we set $h(2) = 4, h$ would be right continuous at 2 (but not left continuous). (If we give $h(2)$ any value other than 2 and 4, then $h$ would be discontinuous from both the sides.) A discontinuity of this type (i.e. a discontinuity $c$ where both the right and the left handed limits of the function exist at $c$ but are unequal) is called a **jump discontinuity**. It is precisely these discontinuities which arise in real-life functions. For example, the apparent current in a wire shown in Fig. 4.1.1 (a) or the bank balance or the bus fares are functions having only jump discontinuities.

Next to the removable discontinuities, the jump discontinuities are the esasiest to tackle. The two can, in fact, be clubbed together. If $c$ is a removable discontinuity, then $\lim\limits_{x \to c^+}$ and $\lim\limits_{x \to c^-}$ both exist and are equal while if $c$ is a jump discontinuity, then also they exist but are unequal. A function which has got only finitely many discontinuities and each one of them is either a simple or a jump discontinuity is called a **piecewise continuous** function. To justify this name, suppose the function is $f : [a, b] \longrightarrow I\!\!R$ and the discontinuities are $t_1, t_2, \ldots, t_n$, where we may suppose that $t_1 < t_2 < \ldots < t_n$. For notational uniformity, call $a$ as $t_0$ and $b$ as $t_{n+1}$. Then for each $i = 1, 2, \ldots, n + 1, f$ is continuous on the open interval $(t_{i-1}, t_i)$. Also, $\lim\limits_{x \to t_{i-1}^+} f(x)$ and $\lim\limits_{x \to t_i^-} f(x)$ both exist. So if we set $f(t_i) = \lim\limits_{x \to t_i^-} f(x)$ and $f(t_{i-1}) = \lim\limits_{x \to t_{i-1}^+} f(x)$, then $f$ is continuous on the closed interval $[t - i - 1, t_i]$ and so enjoys all the good properties of a continuous function defined on a closed bounded interval. (Note, however, that for $i = 1, 2, \ldots, n$, the value we have to assign to $f(t_i)$ in order to make $f$ continuous on $[t_{i-1}, t_i]$ will differ from the value we have to assign to it to make $f$ continuous on $[t_i, t_{i+1}]$, unless $t_i$ is a simple discontinuity.) In other words, the function $f$ can be thought of as a continuous function on each of the *pieces* $[t_0, t_1], [t_1, t_2], \ldots, [t_n, t_{n+1}]$. (Geometrically, the graph of $f$ looks like it is 'broken' into a finite number of pieces each of which is the graph of a continuous function.)

Piecewise continuous functions are the next best thing to continuous functions. Although neither Theorem (1.3) nor Theorem (1.4) holds for them, it can be shown (see Exercise (1.26)) that every piecewise continuous function on a closed and bounded interval is bounded. (In other words, the first assertion in Theorem (1.4) holds for such functions.) Later on, we shall see more instances of how piecewise continuous functions share some of the pleasant properties of continuous functions. What is more important, the class of such functions is large enough to include many real-life functions, as we saw above. In a nutshell, if continuity is a very natural concept, then piecewise continuity is the most natural extension of it! Functions which are not even piecewise continuous can be very bizarre and have little other role to play in calculus than to serve as weird counter-examples.

## EXERCISES

1.1 Suppose $f$ is continuous at a point $c$ and $f(c) > 0$. Prove that there exists a neighbourhood $U$ of $c$ such that $f(x) > 0$ for all $x \in U$. Prove a similar result if $f(c) < 0$.

1.2 Prove that the set of zeros of a continuous function is closed (see Exercise (3.5.6) for a definition).

1.3 If $f, g$ are continuous at a point $c$, and $a, b$ are any constants, prove that the functions $af + bg$, $fg$ and $f/g$ are all continuous at $c$. (In the last one, it is assumed that $g(c) \neq 0$, which then implies $g(x) \neq 0$ in a neighbourhood of $c$ by Exercise (1.1) above.)

1.4 Prove that a continuous function which is 0 at every rational number is identically 0. [*Hint* : Use Corollary (3.3.3) and Theorem (1.2).] Deduce that two continuous functions which agree (i.e., take equal values) at evey rational agree everywhere.

1.5 Prove that the composite of two continuous functions is continuous. More specifically suppose $f : \mathbb{R} \longrightarrow \mathbb{R}$, $g : \mathbb{R} \longrightarrow \mathbb{R}$ are functions with $f$ continuous at some $c \in \mathbb{R}$ and $g$ continuous at the point $f(c)$. Prove that $g \circ f$ is continuous at $c$.

1.6 Prove that all constant functions and the identity function from $\mathbb{R}$ to $\mathbb{R}$ (i.e., the function which takes $x$ to $x$ for all $x \in \mathbb{R}$) are continuous. Hence using Exercise (1.3), prove that all polynomials are continuous functions. Prove also that the absolute value function $f(x) = |x|$ is continuous everywhere.

1.7 Using the Intermediate Value Property, prove that for every positive integer $n$, every positive real number has a positive $n^{\text{th}}$ root. (See the remark before Theorem (3.4.1).)

1.8 Prove that the function $f(x) = x^{1/n}$, where $n$ is a positive integer is continuous at every $x \geq 0$. Then prove this for functions of the form $g(x) = x^y$ where $y$ is a rational number. (See Exercise (3.4.2) for the definition of $x^y$.)

1.9 Prove that if $f : [a, b] \longrightarrow \mathbb{R}$ is continuous then the range of $f$, i.e., the set $f([a, b])$ is a closed and bounded interval, say, $[c, d]$. Show by an example that the end points $c, d$ need not be the images of $a, b$. Prove however, that if $f$ is also one-to-one then $f$ is either strictly monotonically increasing or strictly monotonically decreasing, with $f(a) = c, f(b) = d$ in the first case and $f(a) = d, f(b) = c$ in the second. In either case, prove that the inverse function $f^{-1} : [c, d] \longrightarrow [a, b]$ is continuous (The **inverse function** $f^{-1}$ associates to each $y \in [c, d]$, the unique $x \in [a, b]$ such that $f(x) = y$.)

1.10 Prove that the cubic polynomial $x^3 - 6x^2 + 3x + 1$ has three distinct roots. Prove also that every polynomial of an odd degree has at least one real root.

1.11 Suppose $S$ is a bounded solid. Prove that there is some horizontal plane which divides $S$ into two solids of equal volume. [*Hint* : Use Intermediate Value Property.]

1.12 Let $C$ be the ellipse $\dfrac{x^2}{a^2} + \dfrac{y^2}{b^2} = 1$ and $d$ be any number with $0 < d < 2b$. Let $L$ be a straight line. Prove that there exists a chord of length $d$ of $C$ which is parallel to $L$. (Do not actually find it. Simply prove its existence.)

1.13 Let $L_1$ and $L_2$ be two straight lines in a plane intersecting at an acute angle and $P$ be a point in that plane lying outside the acute angle. Prove that there exist points $A, C$ on $L_1$ and $B, D$ on $L_2$ such that :

 (a) $P, A, B$ are collinear and $PA : AB = 1 : 2$   and
 (b) $P, C, D$ are collinear and $CD$ is of unit length.

Can you give simple geometric constructions for these points?

1.14 Suppose $f : [a, b] \longrightarrow I\!R$ is continuous and $a \le f(x) \le b$ for all $x \in [a, b]$. Prove that there exists $c \in [a, b]$ such that $f(c) = c$. [*Hint* : Consider the function $g(x) = f(x) - x$.] A point $c$ like this is called a **fixed point** of $c$. How will you interpret the result geometrically? Does it hold if $[a, b]$ is replaced by $(a, b)$?

1.15 Give an alternate proof of the result of the last exercise as follows. If $f$ has no fixed point, then colour the points of $[a, b]$ by saying that $x$ is red if $f(x) > x$ and blue if $f(x) < x$. For every positive integer $n$, divide $[a, b]$ into $n$ equal parts by points $a = x_0 < x_1 < x_2 < \ldots < x_{n-1} < x_n$. (To be specific, $x_i = a + \dfrac{i}{n}(b - a)$, for $i = 0, 1, \ldots, n$.) Prove that for some $i = 0, 1, \ldots, n, x_{i-1}$ and $x_i$ are of different colours. So for every $n$, there exist points $y_n$ and $z_n$ in $[a,b]$ such that $|y_n - z_n| = \dfrac{b - a}{n}$ and $y_n$ is red, $z_n$ is blue. Now apply Bolzano-Weierstrass theorem to get a contradiction.

1.16 Suppose $f : [0, 1] \longrightarrow I\!R$ is continuous and $f(0) = f(1)$. Prove that there exists $c \in [0, \frac{1}{2}]$ such that $f(c) = f(c + \frac{1}{2})$.

1.17 Give an alternate proof of the first part of Theorem (1.4) using the method of successive bisection. [*Hint* : If $f$ is unbounded above on $[a, b]$ then it is unbounded above on at least one of the subintervals $[a, c]$ and $[c, b]$ where $c = \dfrac{a+b}{2}$].

1.18  Given a straight line $L$ and a point $P$ not on $L$, prove using Theorem (1.4) that there exists some point $Q$ on $L$ which is closest to $P$ (among all points on $L$). Why is there no farthest point ?

1.19  Given an ellipse $C$ and a point $P$ in a plane prove that there exist both closest and farthest points on $C$, from $P$. [*Hint* : Parametrise $C$.] Deduce that if $P$ is not on $C$, then there are at least two circles centred at $P$ which touch $C$.

*(1.20)  Given a point $P$ on an ellipse $C$, prove that there exist points $Q, R$ on $C$ such that the triangle $PQR$ is similar to a given triangle.

1.21  Suppose $e^x$ is defined as $\displaystyle\sum_{n=0}^{\infty} \frac{x^n}{n!}$ for $x \in \mathbb{R}$ as in Exercise (3.3.14). Prove that

(i) $e^x > 0$ for all $x \in \mathbb{R}$ and $e^x > 1$ for $x > 0$. [*Hint* : For $x < 0, e^{-x}e^x = e^0 = 1$ by Exercise (3.10.3)].

(ii) $e^x$ is strictly monotonically increasing, i.e., $e^x < e^y$ whenever $x < y$. [*Hint* : Write $y = x + h$ with $h > 0$ and use (i)]

(iii) $e^x \to \infty$ as $x \to \infty$ and $e^x \to 0$ as $x \to -\infty$.

(iv) For every $x > 0$, there exists a unique $y$ such that $e^y = x$. [*Hint* : Use IVP]. This unique real number $y$ is called the natural logarithm of $x$, denoted by $\ln x$.

1.22  Using the last exercise and Exercise (1.9), show that $\ln : (0, \infty) \longrightarrow \mathbb{R}$ is a continuous function which has the property that $\ln(xy) = \ln x + \ln y$ for all positive numbers $x, y$. Prove that $\ln x$ is a strictly monotonically increasing function.

1.23  Using logarithms, we can now define arbitrary powers. Let $x > 0$ and $y$ be any real number. We define $x^y$ to be the number $e^{y \ln x}$, i.e., the number $\displaystyle\sum_{n=0}^{\infty} \frac{(y \ln x)^n}{n!}$. Prove that if $y$ is a rational then this definition of $x^y$ coinicides with the earlier one (cf. Exercise (3.4.2)). [*Hint* : To show that two positive numbers are equal, it suffices to show that their logarithms are equal. Prove the assertion first for $y = 1$, then extend it by induction and Exercise (1.22) to the case $y \in \mathbb{N}$ and finally, again using Exercise (1.22) to the general case.]

1.24  Pinpoint precisely at what point the proof of the first part of Theorem (1.4) breaks down if we replace $[a, b]$ by an open interval $(a, b)$ or by a semi-open interval.

$$(iii) f(x) = \begin{cases} x \sin(1/x) & \text{if } x < 0 \\ 1 & \text{if } x = 0 \\ x & \text{if } 0 < x \le 3 \\ \dfrac{1}{x-3} & \text{if } x > 3 \end{cases}$$

$$(iv) f(x) = \begin{cases} 1 & \text{if } x \text{ is rational} \\ 0 & \text{if } x \text{ is irrational} \end{cases} \quad [\textit{Hint}: \text{Use Exercise (3.3.16).}]$$

1.26  Prove that a piecewise continuous function on a closed bounded interval is bounded but need not attain its bounds.

## 4.2    Uniform Continuity

In the last section we defined continuity of a function *at* a point. In this sense, it is a local concept. In the present section, we study unifrom continuity of a function *on a set*, which is a global concept. The distinction between continuity and uniform continuity is of the same spirit as that between pointwise and uniform convergence. So before defining uniform continuity, let us first recall what uniform convergence (Section 3.11) is. Suppose $\{f_n(x)\}_{n=1}^{\infty}$ converges to $f(x)$ pointwise on a set $S$. Let $\epsilon > 0$. Then for every $x$ in $S$ there is some $N$ (depending both on $\epsilon$ and $x$) such that for all $n \ge N, |f_n(x) - f(x)| < \epsilon$. In uniform convergence, on the other hand, there is some $N$, depending only on $\epsilon$, such that for all $x \in S$, and for all $n \ge N, |f_n(x) - f(x)| < \epsilon$. As we saw in Section 3.11 (and then in Theorem (1.5)), the existence of a uniform $N$ which works for every $x$ makes uniform convergence considerably stronger than pointwise convergence.

The distinction between uniform and ordinary continuity is of a similar type. Suppose $f : S \longrightarrow I\!R$ is continuous, where $S$ is an interval which may be bounded or unbounded and may or may not contain its end-points. If $S$ contains an end-point, then by continuity of $f$ at that point we mean the appropriate left or right continuity as the case may be. Now, continuity of $f$ on $S$ means simply the continuity of $f$ at each point $x$ of $S$. This means that for every $\epsilon > 0$, there exists some $\delta > 0$, which may depend both on $\epsilon$ and $x$ such that for every $y \in S, |f(x) - f(y)| < \epsilon$ whenever $|y - x| < \delta$.

A $\delta$ which works for one value of $x$ may not work for some other value of $x$. As a simple example, let $S$ be the open interval $(0,1)$ and let $f(x) = 1/x$. Then $f$ is continuous on $S$. Take $\epsilon = 1$. Let $x \in S$. We want a $\delta > 0$ such that

$$\frac{1}{x} - 1 < \frac{1}{y} < \frac{1}{x} + 1 \quad \text{for every} \quad y \in (x - \delta, x + \delta) \cap S \tag{1}$$

Let us give $x$ a particular value, say, $x = \dfrac{1}{100}$. Then the inequality in (1) will hold only when $\dfrac{1}{101} < y < \dfrac{1}{99}$. This interval $\left(\dfrac{1}{101}, \dfrac{1}{99}\right)$ is not centred at $\dfrac{1}{100}$. If $\left(\dfrac{1}{100} - \delta, \dfrac{1}{100} + \delta\right)$ is to be contained in $\left(\dfrac{1}{101}, \dfrac{1}{99}\right)$, then the largest that $\delta$ can be is min $\left\{\dfrac{1}{100} - \dfrac{1}{101}, \dfrac{1}{99} - \dfrac{1}{100}\right\}$, i.e., $\dfrac{1}{100 \times 101} = \dfrac{1}{10100}$. By a similar reasoning, for $x = \dfrac{1}{1000}$, the largest $\delta$ that will work (for $\epsilon = 1$) is $\dfrac{1}{1,001,000}$. So by taking $x$ closer and closer to 0 we see that there is no single $\delta$ that will work for every $x$ in $(0, 1)$.

As is to be expected, in uniform continuity, it is possible to find a $\delta$ which depends only on a given $\epsilon > 0$ (and, of course, on the function $f$) but not on $x$, i.e., which works for every $x \in S$. This leads to the following definition.

**2.1 Defintion :** Let $S$ be an interval. Then a function $f : S \longrightarrow I\!\!R$ is said to be **uniformly continuous** on $S$ if for every $\epsilon > 0$, there exists $\delta > 0$ such that for all $x, y \in S, |f(x) - f(y)| < \epsilon$ whenever $|x - y| < \delta$.

In the example given above, the function $f(x) = \dfrac{1}{x}$ is not uniformly continuous on the open interval $(0, 1)$. However, if we let $S = (t, 1]$, where $t > 0$, then it is easy to show that $f$ is uniformly continuous on $[t, 1]$ and hence also on $(t, 1]$, because given an $\epsilon > 0$, a $\delta$ which will work 'in the worst case', viz., at $t$, will also work for all $x > t$. So we see that in talking of uniform continuity, the set $S$ has as much role to play as the function $f$.

As another example, consider $f(x) = x^2$. It is not hard to show directly that $f$ is uniformly continuous on any bounded interval. This will also follow from a theorem we are about to prove. But $f$ is not uniformly continuous on an unbounded interval, say, $[0, \infty)$. When $x$ is large, even a small increase in its value changes $f(x)$ by a large margin, because $(x + h)^2 - x^2 = h(2x + h) > 2xh$ for $h > 0$. So if we take $\epsilon = 1$, then for a given $x > 0$, the largest $\delta$ that will work is at most $\dfrac{1}{2x}$. As $x \longrightarrow \infty$, no single $\delta$ will work for all $x$. Here too, if we let $S = [0, R]$, where $R$ is fixed, then $f$ is indeed uniformly continuous on $S$, because the $\delta$ that will work for the worst case, viz. for $x = R$ will work for all $x \in S$.

These examples suggest that while continuity does not always imply uniform continuity, the situation may be better if the interval were bounded and closed. This indeed turns out to be the case as we show now. The proof, once again crucially uses the completeness of the real number system in one form or the other. We prefer to give the proof using Bolzano-Weierstrass Theorem.

**2.2 Theorem :** Every continuous real-valued function on a closed and bounded interval is uniformly continuous (on that interval).

**Proof :** Let $f : [a, b] \longrightarrow \mathbb{R}$ be continuous. Suppose $f$ is not uniformly continuous on $[a, b]$. We then take the logical negation (Section 1.11) of the definition of uniform continuity and get that there exists some $\epsilon > 0$ such that for every $\delta > 0$, there exist $x, y \in [a, b]$ such that $|x - y| < \delta$ and $|f(x) - f(y)| \geq \epsilon$. In particular, for each $n \in \mathbb{N}$, by taking $\delta = 1/n$, we get two points, say $x_n$ and $y_n$ in $[a, b]$ such that

$$|x_n - y_n| < \frac{1}{n} \text{ and } |f(x_n) - f(y_n)| \geq \epsilon \text{ for all } n \in \mathbb{N} \tag{2}$$

Now the sequence $\{x_n\}_{n=1}^{\infty}$ is a bounded sequence and so has a subsequence, say, $\{x_{n_k}\}_{k=1}^{\infty}$ converging to some point, say $c$, in $[a, b]$. (The argument is an exact duplication of that at the beginning of the proof of Theorem (1.4)). So, by Theorem (1.2), $f(x_{n_k}) \to f(c)$. The first part of (2) implies

$$x_{n_k} - \frac{1}{n_k} < y_{n_k} < x_{n_k} + \frac{1}{n_k}$$

for all $k$. So by the Sandwich Theorem, $\{y_{n_k}\}$ also converges to $c$. But then again by Theorem (1.2), $f(y_{n_k}) \longrightarrow f(c)$. Thus we have that as $k \to \infty$, both $f(x_{n_k})$ and $f(y_{n_k})$ tend to the same limit, viz., $f(c)$. So there exist integers $K_1, K_2$ such that

$$|f(x_{n_k}) - f(c)| < \frac{\epsilon}{2} \text{ for all } k \geq K_1 \tag{3}$$

and

$$|f(y_{n_k}) - f(c)| < \frac{\epsilon}{2} \text{ for all } k \geq K_2 \tag{4}$$

Taking $k = \max\{K_1, K_2\}$, we get, from (3) and (4) and the triangle inequality that $|f(x_{n_k}) - f(y_{n_k})| < \frac{\epsilon}{2} + \frac{\epsilon}{2} = \epsilon$. But this contradicts the second part of (2) with $n = n_k$. This establishes the uniform continuity of $f$ on $[a, b]$. ∎

Thus, on closed, bounded intervals, continuity is as strong as uniform continuity. That this is not always so for other intervals we already know from the examples above. However, it is possible to tell which continuous functions on them are uniformly continuous. One such characterisation will be given in the exercises. The theorem above will be sufficient for our purpose and, as a matter of fact, we shall need it crucially only once. Still, the concept of uniform continuity deserves to be stressed. As discussed at the beginning of the last section, the intuitive meaning of continuity of a function $f$ is preservation of closeness. That is, whenever two points $x$ and $y$ in the domain of $f$ are close to each other, their images $f(x)$ and $f(y)$ are also close to each other. In pointwise continuity, the standard of closeness can vary as you move from one point to another in the domain and there may not be a common standard applicable at all the points. This is hard to swallow. In uniform continuity, on the other hand, the same standard of smallness applies all over the domain and this is consistent with our intuition. So, in this sense, it is the uniform continuity and not continuity which is a more natural concept.

## EXERCISES

2.1 Prove that if $f : S \longrightarrow I\!R$ is uniformly continuous and $T$ is a subinterval of $S$ then $f$ is uniformly continuous on $T$.

2.2 If $f : S \longrightarrow I\!R$ is uniformly continuous where $S$ is a bounded interval, show that $f$ is bounded on $S$. [*Hint* : Take $\delta$ corresponding to some fixed $\epsilon$, say $\epsilon = 1$. Cover $S$ by a finite number of subintervals of length less than $\delta$ each. Then $f$ is bounded on each of them.] This gives a somewhat simpler proof that $f(x) = \dfrac{1}{x}$ is not uniformly continuous on $(0, 1)$.

2.3 Prove that the converse of the result in the last exercise is not true by showing that $f(x) = \sin \dfrac{1}{x}$ is bounded but not uniformly continuous on $(0,1)$.

2.4 Suppose $f : S \longrightarrow I\!R$ is uniformly continuous. Prove that for every Cauchy sequence $\{x_n\}$ in $S$, $\{f(x_n)\}$ is also a Cauchy sequence. Use this fact and the Bolzano-Weierstrass theorem to give an alternate solution to Exercise (2.3).

2.5 Suppose $f : (a,b) \longrightarrow I\!R$ is uniformly continuous. Let $\{x_n\}$ be any sequence in $(a, b)$ which converges to $a$. Prove that $\{f(x_n)\}$ is a convergent sequence (in $I\!R$). Let $L = \lim\limits_{n\to\infty} f(x_n)$. Prove that $L$ is independent of $\{x_n\}$, that is, if $\{y_n\}$ is any other sequence in $(a, b)$ converging to $a$, then $\{f(y_n)\}$ also converges to $L$. Hence show that $\lim\limits_{x\to a^+} f(x)$ exists and equals $L$. Similarly show that $\lim\limits_{x\to b^-} f(x)$ exists.

2.6 Using the last exercise show that a function $f : (a,b) \longrightarrow I\!R$ is uniformly continuous if and only if there exists a continuous function $g : [a, b] \longrightarrow I\!R$ such that $g(x) = f(x)$ for all $x \in (a, b)$. (In such a situation we say that $f$ is a **restriction** of $g$ or that $g$ is an **extension** of $f$.)

2.7 A real-valued function $f$ on an interval $S$ is said to satisfy a **Lipschitz condition** if there exists a constant $M$ such that for all $x, y \in S, |f(x) - f(y)| \le M|x - y|$. Prove that such a function is uniformly continuous.

2.8 Prove that the function $f(x) = \sqrt{x}$ does not satisfy a Lipschitz condition on $[0,1]$, even though it is uniformly continuous. Prove that this function satisfies a Lipschitz condition on $[1, \infty)$ and hence that $f$ is uniformly continuous on $[1, \infty)$ and also on $[0, \infty)$.

2.9 Prove that for all $x, y$ with $0 \leq y \leq x, |\sqrt{x} - \sqrt{y}| \leq \sqrt{x - y}$. [*Hint* : Write $x$ as $y + h$.] Use this fact to give a direct proof that $f(x) = \sqrt{x}$ is uniformly continuous on $[0, \infty)$.

## 4.3    Differentiability - a Stronger Concept than Continuity

Mathematically, the title simply says that every differentiable function is continuous. This we shall prove in a moment. But it is the deeper inplications of this statement that will really concern us. As the name indicates, differentiability has to do with a difference of some kind. Specifically, suppose a variable $y$ is a function of some other variable $x$. In other words, $y$ depends on $x$ and hence $x, y$ are called the **independent** and the **dependent variables** respectively. Suppose $y = f(x)$. Let $c$ be a fixed value of $x$. Then the corresponding value of $y$ is $f(c)$. Suppose now we increase $c$ by an amount $\triangle x$ often called an **increment**. (The word is somewhat misleading, because $\triangle x$ can be negative in which case there is a decrease rather than an increase in the value of $x$.) The value of $f$ changes by $f(c + \triangle x) - f(c)$. Let us denote this by $\triangle y$ and call it the increment in $y$, caused by the increment $\triangle x$ in $x$. Continuity of $f$ at $c$ simply means that $\triangle y$ is also small. This is a qualitative statement. Differentiability of $f$ at $c$, on the other hand, is a quantitative statement. It tells us how big $\triangle y$ approximately is as compared to $\triangle x$. Specifically, it tells us the approximate value of the ratio $\triangle y / \triangle x$, often called the **incrementary ratio.** The limit of this incrementary ratio is, by definition, the **derivative** of $f$ at $c$, generally denoted by $f'(c)$. If this limit exists (as a finite real number) then $f$ is said to be **differentiable** at $c$.

Continuity of $f$ at $c$ merely tells us that $\triangle y$ is small whenever $\triangle x$ is small. Differentiability tells us that if $\triangle x$ is small then $\triangle y$ is approximately $f'(c)\triangle x$. In particular, $\triangle y$ is small if $\triangle x$ is small. So differentiability is intuitively stronger than continuity. Before giving a rigorous proof of this, let us be convinced of its truth with a real-life example. As was already discussed in Sections 2.1 and 2.2., if $x$ denotes time and $y$ the distance travelled by a car, then $f'(c)$ is the instantaneous speed at a point $c$ of time. Suppose the car is at a point $P$ on the road at, say, 9.00 a.m. Where will it be after ten seconds? Continuity of the distance function (w.r.t. time) merely gives the answer as "near $P$". This is little more informative than sheer common sense. But suppose we are also given that the speed of the car at 9:00 a.m. is 30 km./hour. Then we can say that 10 seconds later the car will be near the point $Q$ down the road at a distance $\dfrac{30 \times 1000}{60 \times 6}$ meters (approximately 83 meters) from $P$ and this is a lot more informative.

We now give a formal proof that differentiability implies continuity. We first prove a stronger result which characterises differentiability.

**3.1 Theorem :** Suppose $f$ is a real-valued function defined in a neighbourhood $U$ of a point $c \in \mathbb{R}$. Then $f$ is differentiable at $c$ if and only if there exist constants $A$ and $B$ and a function $g$ defined on $U$ such that :

$$\text{(i)} \quad f(x) = A + B(x - c) + g(x) \text{ for all } x \in U, \text{ (ii) } g(c) = 0$$

and $\quad$ (iii) $\quad \dfrac{g(x)}{x - c} \to 0$ as $x \to c$.

**Proof :** Suppose first that $f$ is differentiable at $c$, i.e, $f'(c)$ exists. We take $A = f(c)$ and $B = f'(c)$. Then in order to satisfy (i) we must define $g(x)$ as $f(x) - f(c) - f'(c)(x - c)$ for $x \in U$. Trivially (ii) holds and to complete the proof we have to show

$$\lim_{x \to c} \frac{f(x) - f(c) - f'(c)(x - c)}{x - c} = 0$$

which is clear since $\lim\limits_{x \to c} \dfrac{f(x) - f(c)}{x - c} = f'(c)$ by definition and

$\lim\limits_{x \to c} \dfrac{f'(c)(x - c)}{x - c} = \lim\limits_{x \to c} f'(c) = f'(c)$.

Conversely suppose the given condition holds. Then from (ii), $A = f(c)$. So using (i), $\lim\limits_{x \to c} \dfrac{f(x) - f(c)}{x - c} = \lim\limits_{x \to c} B + \dfrac{g(x)}{x - c} = B + 0 = B$ by (iii). Hence $f'(c)$ exists and equals $B$. ∎

**3.2 Corollary :** Every differentiable function is continuous.

$\quad$ **Proof :** With the notation of the theorem, $\lim\limits_{x \to c} g(x) = \lim\limits_{x \to c} \dfrac{g(x)}{x - c}(x - c)$ $= 0.0 = 0$. So by (i) with $A = f(c)$, we get

$$\lim_{x \to c} f(x) = \lim_{x \to c} f(c) + B \lim_{x \to c}(x - c) + \lim_{x \to c} g(x)$$
$$= f(c) + B.0 + 0 = f(c).$$

So $f$ is continuous at $c$. ∎

The converse of Corollary (3.2) is false. The most standard counterexample is the function $f(x) = |x|$, which is continuous everywhere (see Exercise (1.6)) and differentiable for all $x \neq 0$ (with $f'(x) = 1$ if $x > 0$ and $f'(x) = -1$ if $x < 0$ as can be easily seen simply from the graph). However at 0,

$$\lim_{\triangle x \to 0^+} \frac{f(0 + \triangle x) - f(0)}{\triangle x} = \lim_{\triangle x \to 0^+} \frac{|\triangle x| - 0}{\triangle x} = \lim_{\triangle x \to 0^+} \frac{\triangle x}{\triangle x} = \lim_{\triangle x \to 0^+} 1 = 1$$

while

$$\lim_{\triangle x \to 0^-} \frac{f(0 + \triangle x) - f(0)}{\triangle x} = \lim_{\triangle x \to 0^-} \frac{|\triangle x|}{\triangle x} = \lim_{\triangle x \to 0^-} \frac{-\triangle x}{\triangle x} = -1.$$

In other words, $f$ is differentiable from the right as well from the left at 0. But the right derivative at 0 (often denoted by $f'_+(0)$) equals 1 while the left derivative $f'_-(0)$ equals $-1$ and so $f'(0)$ does not exist. If we draw the graph of this function we see that it is an unbroken curve but has a 'kink' at $(0,0)$, and therefore fails to have a unique tangent there. By suitably modifying this function it is possible to construct a function which is continuous everywhere but not differentiable at any of a given finite set of points. In fact, by taking a uniformly convergent series of such functions one can even construct a continuous function which is not differentiable *anywhere* (Exercise (3.10)). The existence of such a function may come as a surprise. But an even greater surprise is in store. Let $C$ be the class of all continuous functions on some closed bounded interval, say $[0,1]$, and let $A$ be the subclass consisting of those functions in $C$ which are nowhere differentiable. The existence of such a function means that the class $A$ is nonempty. Still, since none of the familiar functions (like those constructed from polynomials, trigonometric functions etc.) behaves so miserably, we tend to think that elements of $A$ must be fairly rare, or in other words, the subclass $A$ is very small as compared to the class $C$. It turns out that the reality is precisely the opposite. In a certain sense, it is the complement of $A$ in $C$, (i.e., the class of continuous functions which are differentiable at at least one point) that is very small!

It will take us too far afield to tell precisely in what sense this apparently paradoxical statement holds. But perhaps an analogy will indicate what is meant. In arithmetic, we deal mostly with rational numbers and they dominate our day-to-day transactions. Even the most advanced digital computers can handle only rational numbers. So, even if we know that there are irrational numbers (e.g., $\sqrt{2}$ or $\sqrt{3}$, cf. Exercise (2.6.4)), we tend to think that they are exceptional. But in reality, there are lots of irrational numbers. In fact, it is the rationals that are in minority. True, there are infinitely many rationals, and in fact their set is dense (cf. Corollary (3.3.2)). Still, the set of rationals is small enough to be exhausted by enumeration, cf. Exercise (3.9.7). The set of irrationals, on the other hand cannot be so exhausted. So there is a difference between the infinitude of rationals and the infinitude of irrationals. The latter far outnumber the former. If we pick a real number 'at random', its probability of being a rational is 0. Put differently, in the world of real numbers, an irrational number is the rule while being a rational is an exception, even though we mostly deal with the latter.

Similarly, in the class of all continuous functions, being differentiable even at one point is a rarity, a pleasant coincidence, while being nowhere differentiable is the rule! In the last section it was mentioned that continuity is a natural concept because functions which occur in nature are continuous. Still, mere continuity is too shallow a concept to capture the essence of nature. Translated in real-life terms, a continuous, nowhere differentiable function is like a car which moves continuously but is so jerky that it has no definite speed at any point !

Leaving aside this philosphical discussion, we observe that differentiability must be a very stronger concept than continuity, in fact an 'exceptionally' strong concept in the sense above. It stands to reason that functions which are differentiable everywhere are still rarer and must be having some especially nice properties. We shall see ample evidence of this throughout this chapter.

For the time being, let us see how the existence of the derivative of a function even at one point reveals some important information about the behaviour of the function near that point. Going back to the notations in Theorem (3.1), if $f$ is differentiable at a point $c$, and we let

$$g(x) \;=\; f(x) - f(c) - f'(c)(x - c) \tag{1}$$

then $g(c) = 0$ and not only $g(x) \to 0$ as $x \to c$ but a much stronger statement is true. As $x \to c$, $x - c \to 0$. Theorem (3.1) says that $\dfrac{g(x)}{x - c} \to 0$ as $x \to 0$. In other words the numerator $g(x)$ is negligibly small as compared with the denominator $x - c$, which, itself is small.

It is instructive to paraphrase this. Define two functions $L_0(x)$ and $L_1(x)$ by

$$L_0(x) \;=\; f(c) \quad \text{for all } x \tag{2}$$
$$L_1(x) \;=\; f(c) + f'(c)(x - c) \text{ for all } x. \tag{3}$$

The graphs of both $L_0$ and $L_1$ are straight lines passing through the point $(c, f(c))$. But the first line is horizontal while the second has slope $f'(c)$. If we draw these graphs along with the graph of $y = f(x)$ together we get a figure like Fig. 4.3.1. The point $P = (c, f(c))$ lies on all three graphs. Let $Q = (x, f(x))$ be a piont on the graph of $y = f(x)$. As $x \to c$, the point $Q$ tends to $P$. For this we need only the continuity of $f$ at $c$. But differentiability tells us more. The slope of the chord $PQ$ is $\dfrac{f(x) - f(c)}{x - c}$ and as $x \to c$, this tends to $f'(c)$ by definition. So the limiting position of the secant line $PQ$ is the straight line of slope $f'(c)$ passing through $P$. But, by definition, the limiting position of the secant is the tangent. Thus we get the well-known geometric interpretation of the derivative as the slope of the tangent. Note that $y = L_1(x)$ is precisely the equation of the tangent at $P$.

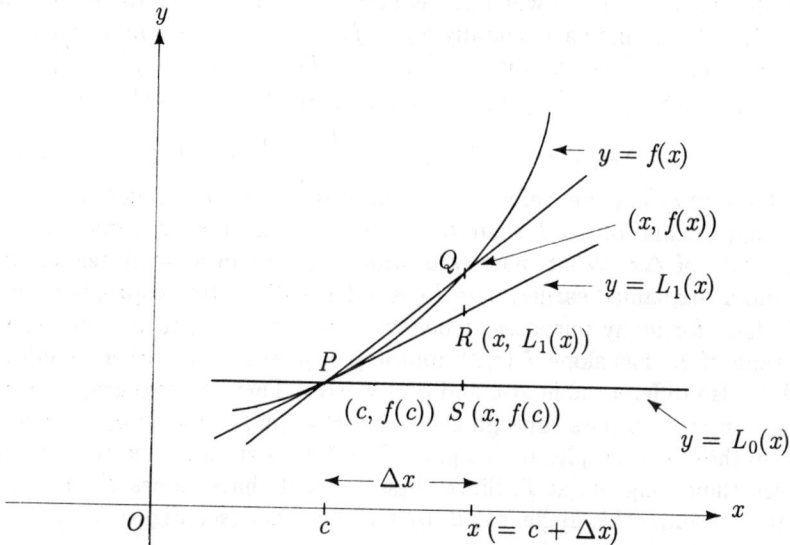

Figure 4.3.1 : First Order (= Tangent Line) Approximation

Let us now suppose $x = c + \triangle x$ where $\triangle x$ is small. Let the vertical line $x = c + \triangle x$ cut the three graphs at $Q, R, S$ respectively. Then the distance $QS$ represents the difference $|f(x) - L_0(x)|$ while the distance $QR$ represents the difference $|f(x) - L_1(x)|$ or, equivalently, $|g(x)|$ by (1). They are the errors if we approximate $f(x)$ by $L_0(x)$ and by $L_1(x)$ respectively. $L_0(x)$ is a constant function. It is sometimes called the **zeroth order approximation.** $L_1(x)$ is called the **first order** or the **tangent line** or the **linear** approximation. When we know nothing beyond the continuity of $f$ at $c$, the zeroth order approximation is the best we can do. But if we know $f$ is differentiable at $c$, then the first order approximation $L_1(x)$ can be defined and gives far more accurate estimation of $f(c + \triangle x)$ than $L_0(x)$ does, for small values of $\triangle x$. For even though both $|f(c + \triangle x) - f(c)|$ and $|f(c + \triangle x) - f(c) - f'(c)\triangle x|$ tend to 0 as $\triangle x \longrightarrow 0$, the second expression tends to 0 even after division by $|\triangle x|$. As a concrete example, suppose $f(x) = x^3 - 2x, c = 4$ and $\triangle x = .01$. Then $f(c) = 56, L_0(c + \triangle x) = 56$, $L_1(c + \triangle x) = 56 + 46\triangle x = 56.46$ while $f(c + \triangle x) = 56.461201$. So the first order approximation is correct upto two places of decimals while the zeroth order approximation has a much larger error (about 400 times that of the first order approxmation).

It is this approximability of a differentiable function by a linear function (i.e., a function whose graph is a straight line) which makes it easier to handle. Linear functions are of the form $ax + b$ where $a, b$ are some constants and are the easiest functions to study. (For functions of several variables, approximability by a linear function is taken as the very definition of differentiability. But the theory of such functions will not be covered in this book.)

The approximation of $f(c + \triangle x)$ by $L_1(c + \triangle x)$ i.e., by $f(c) + f'(c)\triangle x$ is, of course, not a good one when $\triangle x$ is not small. In fact for some large values of $\triangle x$, $L_0(c + \triangle x)$ may accidentally equal $f(c + \triangle x)$. For example, consider the function $f(x) = x^3 - 12x$ with $c = 1$. Here $L_0(c + \triangle x) = f(c) = -11$. The graph of the cubic $y = x^3 - 12x$ will again cut the line $y = -11$ at $x = \frac{-1 \pm \sqrt{45}}{2}$. So for $\triangle x = \frac{-1 + \sqrt{45}}{2} - 1$ and for $\triangle x = \frac{-1 - \sqrt{45}}{2} - 1$, $L_0(c + \triangle x)$ coincides with $f(c + \triangle x)$ and for values of $\triangle x$ near to these two values, $L_0$ will be a better approximation to $f$ than $L_1$ is. But we are not concerned with such large values of $\triangle x$. What matters is what happens in a small neighbourhood of $c$, and as explained earlier, here $L_1$ is a decidedly better approxiamtion than $L_0$. In fact, for many purposes, $f$ behaves exactly like $L_1$ near $c$. For example, the graph of $L_1$ has slope $f'(c)$. Suppose $y = g(x)$ is some other function of $x$ which is also differentiable at $c$ and $g(c) = f(c)$. Then the two graphs $y = f(x)$ and $y = g(x)$ both pass through the point $P = (c, f(c)) = (c, g(c))$. The angle between these two graphs at this piont $P$ of intersection is defined as the angle between their tangents at $P$. Since these tangents have slopes $f'(c)$ and $g'(c)$, we can determine this angle, say $\theta$, by trigonometry (see Figure 4.3.2).

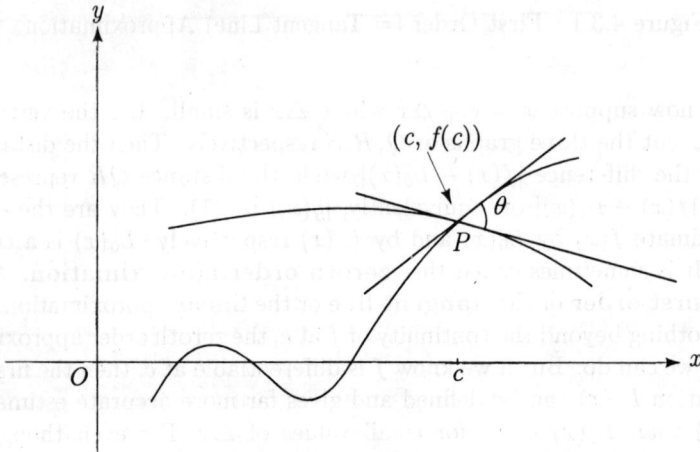

Figure 4.3.2 : Angle between two Graphs

So far we studied differentiability at a point. As with contunity, the differentiability of a function $f$ over an interval $S$ is defined as differentiability at each point of $S$ with appropriate right or left differentiability at the end points of $S$, should they belong to $S$. Differentiability of a function over an interval has many theoretically important consequences as we shall see later. Note that $f'$ itself is a well-defined function on $S$ in this situation. It is called the **derived function** of $f$. If this function $f'$ is continuous at every point of $S$, then $f$ is said to be **continuously differentiable** on $S$. Naturally such functions will be even more well-behaved than those which are merely differentiable on $S$. Thus we have a hierarchy of progressively stronger conditions : continuous, differentiable and continuously differentiable.

We can even go further and inquire if the desired function $f'$ is itself differentiable at a point $c \in S$. If it is, we say $f$ is **twice differentiable** at $c$, denote the derivative of $f'$ at $c$ by $f''(c)$ and call it the **second derivative** of $f$ at $c$. Note that the existence of $f''(c)$ presupposes that $f'(c)$ exists in a neighbourhood of $c$. If $f''(x)$ exists at every $x \in S$, we get a well-defined function $f''$ on $S$ and we can inquire if it is continuous, differentiable etc. So we can keep on defining derivatives of higher order. The $n^{th}$ derivative of $f$ at a point $c$ is denoted by $f^n(c)$, (and in case $n$ is small by putting $n$ dashes as superscripts e.g. $f''(c), f'''(c) = f^{(3)}(c)$ etc.). By convention, we define the zeroth derivative, $f^{(0)}$ of $f$ to be just the function $f$ itself. This is consistent because if you start with $f$ then you get $f^{(n)}$ by differentiating $f$ $n$ times, for every $n$, including $n = 0$.

The physical interpretation of the second derivative is that it is the rate of change of the rate of change. In the case of the distance as a function of time, the first derivative is the speed while the second derivative is the acceleration. Its geometric interpretation will be studied later (Section 4.10).

Just as the knowledge of the first derivative of a function $f$ at a point $c$ enables us to approximate $f(c + \Delta x)$ more accurately than mere continuity, it is to be expected that the knowledge of the higher and higher order derivatives at $c$ would give better and better approximations. This guess is quite correct and is given a precise expression in the following theorem whose proof is deferred because although the result itself is a straight analogue of the direct implication in Theorem (3.1), the proof does not follow directly from the definition. So we shall take it up in Section 4.9. (The peculiar notation $R_n(x)$ comes from the phrase "the remainder after the $n^{th}$ term", where the counting begins with $f(c)$ as the zeroth rather than as the first term.)

**3.3 Theorem :** If $f^{(n)}(c)$ exists, where $n \geq 1$, then we can write

$$f(x) = f(c) + f'(c)(x - c) + \frac{1}{2}f''(c)(x - c)^2 + \ldots + \frac{1}{n!}f^{(n)}(c)(x - c)^n + R_n(x)$$

where $R_n(c) = 0$ and $\frac{R_n(x)}{(x-c)^n} \longrightarrow 0$ as $x \longrightarrow c$. ∎

In other words, the function $f(c) + f'(c)(x - c) + \ldots + \frac{1}{n!}f^{(n)}(c)(x - c)^n$ can be taken as an approximation to $f(x)$, and the error $R_n(x)$ is so small that it tends to 0 more rapidly than $(x - c)^n$. This approximation is called the $n^{\text{th}}$ **order approximation** to $f$ at $c$. The tangent line approximation studied above correponds to the case $n = 1$. The approximation given by $n = 2$ is called a **quadratic approximation.** Although it is superior to the linear approximation, it turns out that the improvement is not very significant, unless $f'(c) = 0$. To see this, let $L_1(x) = f(c) + f'(c)(x - c)$ be the linear approximation and $Q(x) = f(c) + f'(c)(x - c) + \frac{1}{2}f''(c)(x - c)^2$ be the quadratic approximation of $f$ near $c$. The difference between them is $\frac{1}{2}f''(c)(x - c)^2$. If $f'(c) \neq 0$, then this difference is negligibly small as compared with $f'(c)(x-c)$, for small $|x-c|$, because $\lim\limits_{x\to c}\dfrac{\frac{1}{2}f''(c)(x-c)^2}{f'(c)(x-c)} = 0$. So, when $f'(c) \neq 0$, $f'(c)(x-c)$ is the dominating term and the improvement that results by taking $Q(x)$ instead of $L_1(x)$ is generally not very significant for many purposes. If, however, $f'(c) = 0$, then $\dfrac{1}{2}f''(c)(x-c)^2$ becomes the dominating term and is significant even for very small values of $|x-c|$. A real life analogy is that when the top boss in an office is present, his subordinates do not have much of a say. But when he is absent, his immediate subordinate rules the office. Extending the analogy, in the $n^{th}$ order approximation, the term $\dfrac{1}{n!}f^{(n)}(c)(x-c)^n$ is significant only when $f'(c) = f''(c) = \ldots = f^{(n-1)}(c) = 0$ and $f^{(n)}(c) \neq 0$.

It is probably for this reason that even though the existence of higher order derivatives is a stronger condition, derivatives beyond the second order are rarely encountered in applications. Indeed many times all that is needed is the first derivative (sometimes along with its continuity). Thus, there is a world of difference between continuity and the (first order) differentiability, but beyond that there is no marked difference.

## EXERCISES

3.1 If $f, g$ are both differentiable at a point $c$, and $h = af + bg$ where $a, b$ are some constants, prove that $h$ is differentiable at $c$ and $h'(c) = af'(c) + bg'(c)$. (This property is often expressed by saying that differentiation is a **linear process**. What do the special cases $a = 1, b = \pm 1$ signify ?)

3.2 With $f, g$ as above, let $h = fg$, i.e., $h(x) = f(x)g(x)$. Since limit of a product is the product of the limits and deriative is a special case of a limit, we should expect that $h'(c) = f'(c)g'(c)$. Show by an example that this is not so in general (e.g., let $f(x) = g(x) = x$ and $c = 1$). What goes wrong ? Show that the correct answer is $h'(c) = f(c)g'(c) + f'(c)g(c)$. (This is called the **product rule** for derivatives.) Similarly show that if $g(c) \neq 0$, then $f/g$ is differentiable at $c$ with derivative $\dfrac{g(c)f'(c)-g'(c)f(c)}{[g(c)]^2}$ (the **quotient rule for derivatives**).

3.3 Prove directly using the binomial theorem and also by induction that for every positive integer $n$, the derivative of the function $f(x) = x^n$ is $nx^{n-1}$. Then show that this holds for all $n \in \mathbb{Z}$, the set of integers.

3.4 Let $f$ be differentiable at $c$, with $f'(c) > 0$. Prove that $f$ is **strictly increasing at** $c$, i.e., there exists a neighbourhood, say, $U = (c-\delta, c+\delta)$ of $c$ such that for every $x \in U, f(x) > f(c)$ if $x > c$ and $f(x) < f(c)$ if $x < c$. Prove a similar result if $f'(c) < 0$.

3.5 Let $f(x) = x^3$ for all $x$ and define $g$ by

$$g(x) = \begin{cases} x & \text{if } x \leq 0 \\ 2x & \text{if } x > 0. \end{cases}$$

Prove that both $f, g$ are strictly increasing at $0$ in the sense of the last exercise but $f'(0) = 0$ while $g'(0)$ does not exist. (In other words, the converse of the result in the last exercise is false.)

3.6 Suppose $c_1, c_2, \ldots, c_n$ are (distinct) real numbers. Let $f(x) = \sum_{i=1}^{n} |x - c_i|$.

Prove that $f$ is continuous everywhere and differentiable everywhere except at $c_1, \ldots, c_n$.

3.7 For a real number $x$, let $\{x\}$ denote its **fractional part** defined as $x - [x]$, where $[x]$ (often called the **integral part** of $x$) is defined as the largest integer not exceeding $x$. Note that $\{x\}$ always lies in the interval $[0, 1)$. Now define $f : \mathbb{R} \longrightarrow \mathbb{R}$ by

$$f(x) = \begin{cases} \{x\} & \text{if } [x] \text{ is even} \\ 1 - \{x\} & \text{if } [x] \text{ is odd.} \end{cases}$$

Prove that $f(x + 2) = f(x)$ for all $x \in \mathbb{R}$. (It is customary to express this by saying that $f$ is **periodic** with $2$ as a period. For more on periodic functions, see Exercise (8.8).) Draw the graph of $f$. (It is called a **sawtooth curve** because of its shape.) Prove further that $f$ is continuous everywhere and differentiable everywhere except at points of $\mathbb{Z}$, the set of all integers.

3.8 Let $n$ be a positive integer and define $f_n(x) = f(4^n x)$ where $f$ is the function in the last exercise. The graph of $f_n$ is similar to that of $f$ except that the 'teeth' are $4^n$ times more condensed. Show that $f_n$ is continuous everywhere. Identify the set of points where it is non-differentiable. At all other points prove that $f_n'$ equals either $4^n$ or $-4^n$. (Geometrically, these are the slopes of the segments whose union is the graph.) In fact show that $|f_n(x) - f_n(y)| \leq 4^n |x - y|$ for all $x, y$.

3.9 For each $x$ show that the series $\sum_{n=0}^{\infty} (\frac{3}{4})^n f_n(x)$, with $f_n$ as above is convergent and defines a function $g(x)$ which is continuous everywhere. [*Hint*: Use Weierstrass $M$-test.]

*3.10 Prove that the function $g$ in the last exercise is not differentiable at any $x \in \mathbb{R}$. [*Hint*: Fix $x$. For every fixed $m \in \mathbb{N}$, let $\delta_m = \frac{1}{2}\frac{1}{4^m}$ if $0 \leq \{4^m x\} < \frac{1}{2}$ and $\delta_m = -\frac{1}{2}\frac{1}{4^m}$ if $\frac{1}{2} \leq \{4^m x\} < 1$. Show that for $n > m$, $f_n(x + \delta_m) = f_n(x)$ while $f_m(x + \delta_m) - f_m(x) = 4^m \delta_m$. So $g(x + \delta_m) - g(x) = 3^m \delta_m + \sum_{k=0}^{m-1} \left(\frac{3}{4}\right)^k [f_k(x + \delta_m) - f_k(x)]$. Hence show that $\left| \frac{g(x + \delta_m) - g(x)}{\delta_m} \right| \geq \frac{3^m}{2}$. But $m$ is arbitrary.]

3.11 Suppose $f$ is differentiable at $c$. Prove that there exists a function $\epsilon(x)$ such that $\epsilon(c) = 0$, $\lim_{x \to c} \epsilon(x) = 0$ and $f(x) = f(c) + f'(c)(x - c) + (x - c)\epsilon(x)$. (This is essentially a reformulation of the direct implication in Theorem (3.1) but is a little more convenient to apply, as will be seen in the next exercise).

3.12 Suppose $f$ is differentiable at a point $c$ and $g$ is differentiable at $g(c)$. Let $h = g \circ f$ be the composite function. Prove that $h$ is differentiable at $c$, and $h'(c) = g'(f(c))f'(c)$. (This is known as the **chain rule**.) Why is it intuitively obvious?

3.13 Let $f(x) = x^2$ if $x \geq 0$ and $f(x) = -x^2$ if $x < 0$. Prove that $f$ is differentiable everywhere but not twice differentiable at 0. Generalising, show that for every positive integer $n$, there exists a function $f$ which is $n^{\text{th}}$ order differentiable everywhere but not $(n + 1)^{\text{th}}$ order differentiable at some point.

3.14 Let $f_1(x) = x \sin \dfrac{1}{x}$, $f_2(x) = x^2 \sin \dfrac{1}{x}$ for $x \neq 0$ and let $f_1(0) = f_2(0) = 0$. Show that $f_1$ is not differentiable at 0. Prove also that $f_2$ is differentiable everywhere but $f_2'$ is not continuous at 0. (You may, for the time being, assume the elementary properties of the trigonometric functions, including the fact that the derivative of $\sin x$ is $\cos x$.)

3.15 Show that the function $f(x) = \sqrt{x}$ defined for $x \geq 0$ is differentiable for all $x > 0$, but not (right) differentiable at 0.

3.16 Let $C$ be the unit circle $\{(x, y) : x^2 + y^2 = 1\}$. Write the upper semi-circle in the form $y = f(x)$. Then verify that the tangent at every point on it (as found by taking derivatives) is indeed perpendicular to the radius. Do the same for points of the lower semi-circle. What happens at the points $(1, 0)$ and $(-1, 0)$?

3.17 Similarly verify that the formulas for the equations of the tangents at points of other conics, viz., ellipse, parabola and hyperbola are correct.

## 4.4 The Derivative Is not a Ratio but Behaves like a Ratio

If a variable $y$ is a function of a variable $x$, say $y = f(x)$, then the derivative of $f$, i.e., $f'$ is also popularly denoted by $\dfrac{dy}{dx}$ (read "dee wye by dee ex" or "dee wye upon dee ex"). Similarly the second derivative $f''$ is denoted by $\dfrac{d^2 y}{dx^2}$ (read "dee two wye by dee ex square") and in general $f^{(n)}$ is denoted by $\dfrac{d^n y}{dx^n}$. These notations, especially $\dfrac{dy}{dx}$, has certain advantages and disadvantages.

A minor disadvantage is that this is not a convenient notation for denoting the derivative at a point. $\dfrac{dy}{dx}$ is a function of $x$, obtained by differentiating $y$ w.r.t. $x$. For example, if $y = f(x) = x^3$ then $\dfrac{dy}{dx} = 3x^2$. But suppose we want the derivative at a particular point, say 4. In the functional notation, this is simply $f'(4)$ (which equals 48 in the present case). In the new notation, we would have to write it as $\left(\dfrac{dy}{dx}\right)_{x=4}$, which is rather clumsy.

This disadvantage is mostly notational. But there is also a conceptual disadvantage. The notation $\dfrac{dy}{dx}$ (also written sometimes as $dy/dx$) is very apt to be taken as the ratio of the two expressions, viz., $dy$ and $dx$. These quantities are often called the **differentials** of $y$ and $x$ respectively and the derivative $\dfrac{dy}{dx}$ is therefore also called sometimes the **differential quotient**. This notation is very suggestive in that $\dfrac{dy}{dx}$ is, by definition, the limit of the incrementary ratio, $\dfrac{\triangle y}{\triangle x}$. So it stands to reason that it should behave like a ratio.

And it certainly does in many respects. For example, we can add ratios with equal denominators by simply adding their numerators i.e., $\dfrac{a}{c} + \dfrac{b}{c} = \dfrac{a+b}{c}$. Now suppose $c = dx, a = dy$ and $b = dz$ where $y$ and $z$ are some function of $x$. Let $w = y + z$. Then $\dfrac{dw}{dx}$ is the derivative of $w$ w.r.t. $x$. By Exercise (3.1), $\dfrac{dw}{dx} = \dfrac{dy}{dx} + \dfrac{dz}{dx}$. It is customary to write this statement as $\dfrac{d(y+z)}{dx} = \dfrac{dy}{dx} + \dfrac{dz}{dx}$. In fact we can multiply both sides by $dx$ and write $d(y+z) = dy + dz$. A similar equality holds for $d(y - z)$. An advantage of these notations is that they hold regardless of which is the independent variable. That is to say, $y, z$ could be functions of some other variable, say, $t$ as well. Then $d(y + z) = dy + dz$ would mean $\dfrac{d(y+z)}{dt} = \dfrac{dy}{dt} + \dfrac{dz}{dt}$. Similarly we can paraphrase other rules for differentiation in the language of differentials. For example,

$$d(yz) = ydz + zdy \tag{1}$$

and

$$d(y/z) = \frac{zdy - ydz}{z^2} \tag{2}$$

(cf. Exercise (3.2)).

The property of derivatives which reduces to apparent triviality because of the new notation is the chain rule. Suppose $y$ is a function of $x$, say, $f(x)$ and $z$ is a function of $y$, say, $g(y)$. Then $z = h(x)$ where $h = g \circ f$. By Exercise (3.12), if $f, g$ are both differentiable, so is $h$ and moreover, $h'(x) = g'(f(x))f'(x)$. In the notation of differentials, this reads as

$$\frac{dz}{dx} = \frac{dz}{dy}\frac{dy}{dx} \qquad (3)$$

This in fact brings us to the crucial question. Is (3) really as obvious as it looks? If derivatives are ratios of differentials then certainly the answer is 'yes'. But in that case one would first have to carefully define what a differential is. Unfortunately there is no easy answer to this. In elementary textbooks, differentials are usually treated in one of two standard ways. The first is to define $dx$ as an infinitesimally small change in the value of $x$. Similarly $dy$ is an infinitesimally small change in $y$. When $y$ is a function of $x$, say $f(x)$, then these two differentials are related to each other by $\frac{dy}{dx} = f'(x)$. This is taken as the very definition of differentiability of $y$ w.r.t. $x$. Similarly if $z$ is a function of $y$ then $\frac{dz}{dy}$ is the derivative of $z$ w.r.t. $y$. Now to conclude (3), one would have to assume that differentials obey the same laws of algebra as ordinary real numbers do. This was an article of faith for many years. However, as discussed in Section 2.3, attempts to give a rigorous definition and treatment of infinitesimals have been successful only recently. Moreover, they need a certain degree of sophistication well beyond a first course in calculus.

Another approach that is tried sometimes, is to treat any one variable, say $x$, as the independent variable. No meaning is given to $dx$. It is treated as "just another variable". However, for every other variable $y$ which is a function of $x$, say $y = f(x)$, $dy$ is *defined* as $f'(x)dx$. It is now clear that $\frac{dy}{dx} = f'(x)$. This is really not a theorem, it is just a restatement of the definition of $dy$. Now suppose $z = g(y)$. Then $z = h(x)$ where $h$ is the composite function $g \circ f$. It is now clear that (3) holds. But the price we have to pay is that even though $\frac{dy}{dx}$ and $\frac{dz}{dx}$ are, by definition, the derivatives of $y$ and $z$, respectively, w.r.t. $x$, we cannot say immediately that $\frac{dz}{dy}$ is the derivative of $z$ w.r.t. $y$. This is because we have given a special status to the variable $x$. It is only when we differentiate w.r.t. $x$ that we can equate the derivative with a ratio of differentials. When we differentiate $z$ w.r.t. $y$, $\frac{dz}{dy}$ already has a meaning, because $dz = h'(x)dx$, $dy = f'(x)dx$ and so $\frac{dz}{dy} = \frac{h'(x)}{f'(x)}$. We must now *prove* that $\frac{h'(x)}{f'(x)}$ in fact equals the derivative of $z$ w.r.t. $y$, i.e., equals $g'(f(x))$. This, of course, means exactly that we have to prove the chain rule!

Summing up, there is no short cut to the chain rule. This is to be expected because it would be a miracle indeed if the conceptual work involved in anything could be bypassed merely by a convenient notation. Neither $dy$ nor $dx$ has any meaning of its own. The expression $\dfrac{dy}{dx}$, as a whole, has a definite meaning, viz., the derivative of $y$ w.r.t. $x$. But it is not the ratio of $dy$ and $dx$, even though it behaves like a ratio in many respects. Semantically, the situation is analogous to the meaning of the statement "As $x$ tends to $c$, $f(x)$ tends to $L$.". Here neither of the two clauses "As $x$ tends to $c$" and "$f(x)$ tends to $L$" has any meaning of its own. But the sentence as a whole has a meaning, given by the $\epsilon$-$\delta$ definition of a limit (see Section 2.3).

It is true, of course, that $\dfrac{dy}{dx}$ is the limit of $\dfrac{\Delta y}{\Delta x}$ as $\Delta x$ tends to 0. The expression $\dfrac{\Delta y}{\Delta x}$ is a genuine ratio and is amenable to all the rules applicable to ratios of real numbers. Since $\dfrac{dy}{dx}$ may be approximated by $\dfrac{\Delta y}{\Delta x}$ and the latter is a genuine ratio, it is not surprising that $\dfrac{dy}{dx}$ also behaves like a ratio in certain respects. For example, if we go through the proof of the chain rule, we see that in essence it is based on writing $\dfrac{\Delta z}{\Delta x}$ as $\dfrac{\Delta z}{\Delta y}\dfrac{\Delta y}{\Delta x}$ and then taking limits as $\Delta x \longrightarrow 0$. (The only thorny point is that there may be some non-zero values of $\Delta x$ for which $\Delta y$ may be 0. So the argument has to be modified so that we never divide by $\Delta y$. Or one can use the concept of restricted limit, see Exercise (4.1).)

To regard $\dfrac{dy}{dx}$ as a ratio simply because it is the limit of a ratio is as absurd as regarding a power series as a polynomial, just because it is the limit of a sequence of polynomials. Properties which are preserved under limits do carry over from polynomials to power series. But these are many properties which are not of this type. For example, every polynomial of an odd degree has at least one root (Exercise (1.10)). The power series $\displaystyle\sum_{n=0}^{\infty} \dfrac{x^n}{n!}$ (which represents the exponential function $e^x$) can be written as $\displaystyle\lim_{n \to \infty} P_n(x)$ where $P_n(x) = \displaystyle\sum_{k=0}^{2n+1} \dfrac{x^k}{k!}$. Then each $P_n$ has at least one root. But the limit function $e^x$ has no real root (Exercise (1.21)). So we cannot indiscriminately apply results about polynomials to power series. Although sometimes such unwarranted extensions lead to interesting results (e.g. Exercise (1.4.9)), they can also lead to disasters. (See, for example, the comments at the beginning of Section 2.3.)

Fortunately, with differentials, such disasters do not occur even if we treat them like ordinary real numbers, because it is possible, albeit after a lot of hard work, to give a rigourous definition of differentials. So manipulating differentials is like using some deep result (e.g., the Jordan Curve Theorem in Section 1.6) without proof. The convenience gained is worthwhile at least in informal writing. For example, a differential equation like $\dfrac{dy}{dx} = y^2$ is routinely written as $\dfrac{dy}{y^2} = dx$ (called **splitting the differentials**) and then solved by integrating both the sides. If one insists, in a formal writing this can be replaced by writing $\dfrac{1}{y^2}\dfrac{dy}{dx} = 1$ and then integrating both sides w.r.t. $x$, i.e., by writing $\displaystyle\int \dfrac{1}{y^2}\dfrac{dy}{dx}dx = \int 1\,dx$. The integral on the left can be shown to be $\displaystyle\int \dfrac{1}{y^2}dy$ as we shall see later. Thus we get exactly the same result whether we choose to split the differentials or not.

Those who prefer not to treat $\dfrac{dy}{dx}$ as a ratio of two quantities having their separate identities can always convert any meaningless formula involving differentials to a 'meaningful' formula involving derivatives. For example the equation

$$ds = \sqrt{(dx)^2 + (dy)^2 + (dz)^2} \tag{4}$$

(which is encountered when we study the arc length of a curve), can be written as

$$\frac{ds}{dt} = \sqrt{\left(\frac{dx}{dt}\right)^2 + \left(\frac{dy}{dt}\right)^2 + \left(\frac{dz}{dt}\right)^2} \tag{5}$$

where $t$ is any variable such that $x, y, z$ and $s$ are differentiable functions of $t$.

The notation $\dfrac{dy}{dx}$ makes it absolutely clear which variable is differentiated w.r.t. which. The notation $f'(x)$, on the other hand, deals more with what formula the derivative would have if we are given a formula for $y$ in terms of $x$. As a result, the former notation sometimes reveals something not brought out by the latter. Suppose for example $y$ is some distance and $x$ is time. Then the units of $\dfrac{\Delta y}{\Delta x}$ are length per unit time and the same would hold for $\dfrac{dy}{dx}$. In physical applications $x, y$ often denote some physical quantities. Then given their dimensions, the dimensions of the derivative are immediately revealed by $\dfrac{dy}{dx}$.

Another notation for derivatives is worth mentioning. Instead of $\dfrac{dy}{dx}$, the derivative of $y$ w.r.t. $x$ is also denoted by $\dfrac{d}{dx}(y)$. In this new notation the results about derivatives would translate as $\dfrac{d}{dx}(y+z) = \dfrac{d}{dx}(y) + \dfrac{d}{dx}(z)$, $\dfrac{d}{dx}(yz) = y\dfrac{d}{dx}(z) + \dfrac{d}{dx}(y)z$ and so on. In this new notation, $\dfrac{d}{dx}$ is treated as a single symbol. It is called the **differential operator** (w.r.t. $x$). When it is applied to a variable $y$ which is a function of $x$, the result is the derivative of $y$ w.r.t $x$. It is as if $\dfrac{d}{dx}$ is a machine. When an input $y$ is fed to it, it is processed and $\dfrac{dy}{dx}$ comes as the output.

When the variable $x$ is understood, we may denote $\dfrac{d}{dx}$ by the letter $D$. Thus $\dfrac{dy}{dx}$ becomes $Dy$. Or we may simply put an apostrophe ( ′, read 'dash' or 'prime') as a superscript and write $y'$ for $\dfrac{dy}{dx}$. This is called the $D$-notation or **opeartor notation for derivatives.** It is especially convenient when higher order derivatives are involved. For example, consider the second derivative of $y = f(x)$ w.r.t. $x$. The first derivative $f'(x)$ is simply $Dy$. Now $f''(x)$ is obtained by differentiating $f'(x)$ w.r.t. $x$. Hence $f''(x) = \dfrac{d}{dx}(\dfrac{dy}{dx}) = D(Dy)$. It is customary to write this as $D^2y$. Here the operator is $D^2$ and it consists of applying $D$ twice in succession. Similarly $D^3y$ is the third derivative of $y$ (w.r.t. $x$). If $\alpha$ is any constant then the ordinary product $\alpha y$ is the result of multiplying $y$ by $\alpha$. Here multiplication by $\alpha$ is an operator (but not a differential operator) and we denote it by $\alpha$ itself. Note that $\alpha$ commutes with $D$, i.e., $(\alpha D)y = (D\alpha)y$. Here the left hand side means $\alpha(Dy)$ i.e., $\alpha\dfrac{dy}{dx}$. That is, we first apply the differential operator to $y$, get $\dfrac{dy}{dx}$ and then apply $\alpha$ to it, that is, multiply it by $\alpha$. The right hand side $(D\alpha)y$, on the other hand means $D(\alpha y)$ i.e., $\dfrac{d}{dx}(\alpha y)$. Since $\alpha$ is a constant this equals $\alpha\dfrac{d}{dx}(y)$. We can also add operators by adding their actions. Thus, for example, $(D+3)y$ means $Dy + 3y$, i.e., $\dfrac{dy}{dx} + 3y$.

With this notation an expression like $\dfrac{d^3y}{dx^3} - 3\dfrac{d^2y}{dx^2} + 2\dfrac{dy}{dx} - 6y$ can be written as $(D^3 - 3D^2 + 2D - 6)y$. This may not seem much of an achievement. The real advantage of this notation becomes apparent while solving certain differential equations. It turns out that the solutions of the differential equation

$$\frac{d^3y}{dx^3} - 3\frac{d^2y}{dx^2} + 2\frac{dy}{dx} - 6y = 0$$

are related very intimately to the roots of the polynomial $D^3 - 3D^2 + 2D - 6$, in the variable $D$. But we shall not elaborate it here.

Summing up, if $y = f(x)$, then the derivative of $y$, w.r.t. $x$ can be denoted in various ways, viz., $f'(x)$, $\dfrac{dy}{dx}$, $y'$, $\dfrac{d}{dx}(y)$ and $Dy$. Each has certain advantages and so different notations are followed in different contexts so as to combine the advantages of all. Rather like a person who has a command over several languages and knows which one to use where to his advantage!

## EXERCISES

4.1 Give an alternate proof of the chain rule

$$\frac{dz}{dx} = \frac{dz}{dy}\frac{dy}{dx}$$

by writting $\dfrac{\Delta z}{\Delta x}$ as $\dfrac{\Delta z}{\Delta y}\dfrac{\Delta y}{\Delta x}$ if $\Delta y \neq 0$. [*Hint* : Fix $x_0$. Let $S$ be the set $\{\Delta x : f(x_0 + \Delta x) - f(x_0) \neq 0\}$. Then show $\displaystyle\lim_{\substack{\Delta x \to 0 \\ \Delta x \in S}} \dfrac{\Delta z}{\Delta x} = \dfrac{dz}{dy}\dfrac{dy}{dx}$. For $\Delta x \notin S, \Delta y = 0$ and hence $\Delta z = 0$. So write $\dfrac{\Delta z}{\Delta x}$ as $\cdot \left(\dfrac{dz}{dy}\right)_{f(x_0)} \dfrac{\Delta y}{\Delta x}$.]

4.2   A function $F$ defined on an interval $S$ is called an **anti-derivative** or a **primitive** of a function $f$ (also defined on $S$) if $F'(x) = f(x)$ for all $x \in S$. Prove that :

    (i)   Any constant function is an anti-derivative of the identically zero function on any $S$. (In particular we see that anti-derivatives are not unique.)

    (ii)   For an integer $n \neq 1, \dfrac{x^{n+1}}{n+1}$ is an anti-derivative of $x^n$ on $\mathbb{R}$ if $n > -1$ and on each of $(-\infty, 0)$ and $(0, \infty)$ if $n < -1$.

    (iii)   If $F, G$ are anti-derivatives of $f, g$ respectively on $S$ then for all constants $a, b$, $aF + bG$ is an anti-derivative of $af + bg$.

4.3   Let $f$ be a function defined at an interval $S$ (which may be bounded or unbounded, open or closed). By $\displaystyle\int f(x)dx$ we denote the set of all anti-derivatives of $f$ on $S$, i.e., $\displaystyle\int f(x)dx = \{F : F$ is a function on $S$ and $F'(x) = f(x)$ for all $x \in S\}$. In view of the last exercise, $\displaystyle\int f(x)dx$ in general consists of many functions. Translate parts (i) and (iii) of the last exercise in this notation.

$\left(\displaystyle\int f(x)dx\right.$ is, of course, what is more popularly called the **indefinite integral** of $f(x)$ w.r.t. $dx$. It is usually introduced after studying intergrals. But its definition is solely in terms of derivatives and conceptually has nothing to do with integration. So we introduce it here, except for the name. Note that here the entire symbol $\displaystyle\int dx$ is a single notation. $dx$ has no meaning of its own here. Also $x$ is just a dummy variable. That is, $\displaystyle\int f(x)dx$ is the same as $\displaystyle\int f(y)dy$ or $\displaystyle\int f(z)dz$ etc.)

4.4   Suppose $f, g$ are functions on $S$ having at least one anti-derivative each, i.e., the sets $\displaystyle\int f(x)dx$ and $\displaystyle\int g(x)dx$ are non-empty. (Later on we shall see that this is always the case if $f, g$ are continuous.) Let $h(x) = f(x) + g(x)$ for $x \in S$. Prove that the equation $\displaystyle\int f(x)dx + \displaystyle\int g(x)dx = \displaystyle\int h(x)dx$ is true in the sense that any member of $\displaystyle\int f(x)dx$ (i.e. any antiderivative of *f)* when added to any member of $\int g(x)dx$ gives a member of $\int h(x)dx$ (i.e., an antiderivative of $h$) and conversely every member of $\int h(x)dx$ can be expresssed as the sum of two functions one belonging to $\int f(x)dx$ and the other to $\int g(x)dx$. (This is a triviality to prove once you understand what it says.)

4.5 Suppose $f$ has an antiderivative over an interval $S$. Suppose $x = g(t)$ where $g$ is a function which maps an interval $T$ into the interval $S$ and that $g$ is differentiable over $S$. Prove that

$$\int f(x)dx = \int f(g(t))g'(t)dt = \int f(g(t))\frac{dx}{dt}dt$$

in the sense that every antiderivative of $f$ over $S$, will give an antiderivative of $f(g(t))g'(t)$ when $x$ is replaced by $g(t)$. (This is the **rule of subsitution** for indefinite integrals. It is nothing but the chain rule in disguise. To see this, it may be helpful to work out a few examples, e.g. $\int x\sin(x^2)dx$ with $x = \sqrt{t}$.)

4.6 Translate the product rule for derivatives (Exercise (3.2)) into a result about indefinite integrals. (The answer is known as **integration by parts.**)

4.7 Using the rule of subsitution solve the differential equation $\frac{dy}{dx} = y^2$ on $(0, \infty)$, that is, find a function $y = f(x)$ defined on $(0, \infty)$ such that for every $x \in (0, \infty)$, $f'(x) = [f(x)]^2$. (*Caution:* There are many such functions. Identifying all of them is known as the problem of finding the general solution of the differential equation. We shall not go into it.)

4.8 Suppose $f$ is continuous, one-to-one and maps an interval $[a, b]$ onto an interval $[c, d]$ (see Exercise (1.9)). Let $g : [c, d] \longrightarrow [a, b]$ be the inverse funciton of $f$. Suppose $f$ is differentiable at some $x_0 \in [a, b]$ and $f'(x_0) \neq 0$. Prove that $g$ is differentiable at $y_0$ and $g'(y_0) = \frac{1}{f'(x_0)}$. [*Hint :* If you rewrite this as $\frac{dx}{dy} = \frac{1}{dy/dx}$, the proof suggests itself.] As an application, find $\frac{d}{dx}\left(x^{1/n}\right)$ for $x > 0$, where $n$ is a positive integer. (The case $n = 2$ was done in Exercise (3.15).) Finally show that the result of Exercise (3.3) is true for $x > 0$ for any rational exponent $n$. What can be said about the relationship between the graphs of a function $y = f(x)$ and of its inverse function $y = g(x)$ drawn with the same axes? Illustrate with an example, say, $y = x^2$ and $y = \sqrt{x}$ for $x > 0$.

4.9 The **Leibnitz rule** for the derivatives of a product says that if $y = f(x)$ and $z = g(x)$ are $n$ times differentiable and $w = h(x) = f(x)g(x)$ then $h$ is $n$ times differentiable and $h^{(n)}(x) = \sum_{r=0}^{n} \binom{n}{r} f^{(r)}(x)g^{(n-r)}(x)$. Write this in the operator notation and prove it by idnuction on $n$. (The result as well as the proof closely resemble the binomial theorem. The resemblance is, of course, purely formal and not conceptual.)

4.10 Suppose $y = f(x)$ is a twice differentiable function of $x$. Express $D^2y^2$, $(Dy)^2$ and $DyDy$ in terms of $f$ (and its derivatives). For the function $f(x) = x^2$, show that they equal, respectively, $12x^2, 4x^2$ and $6x^2$. (In particular they are all different. While using operator notation, a little care is necessary to ensure that the familiar rules of algebra are not applied indiscriminately.)

## 4.5    Lagrange's Mean Value Theorem - a Basic Result

It is a peculiarity of mathematics that some of its most obvious results are also among the most non-trivial ones, a glaring example being the Jordan Curve Theorem, which we have mentioned several times without proof. Although not so profound, the Intermediate Value Property (Theorem (1.3)) of continous functions is another example. As we saw, the proof crucially uses completeness of the real number system. In fact, many results which depend directly on completeness are likely to appear as obvious. Completeness, in essence, means the absence of holes. Our picture of the real number system is an infinite straight line with no holes in it. As a result we, as laymen, are apt to take the completeness for granted and treat as obvious many results which, in a rigorous approach, require proofs.

Probably the most flagrant example of such a result is,

**5.1 Theorem :** A function whose derivative vanishes identically on an interval is constant. ∎

Many persons take this theorem as completely obvious. Partly, the reason is that its converse is indeed trivial. The derivative of a constant function is zero everywhere from the very definition because each incrementary ratio is zero. So those who confuse an implication statement with its converse would certainly treat this theorem as a triviality. Even those who realise the difference between a statement and its converse, often fail to appreciate the subtlety of the theorem above. Their logic is something like this. Geometrically, the derivative of a function $y = f(x)$ represents the slope of the tangent to the graph of the function. If $f'(x) = 0$ for all $x$ then the tangent at every point of the graph is horizontal and so the graph must be a horizontal straight line, i.e., the function must be a constant. Those who prefer not to rely on geometric intuition word this argument in terms of rates. The derivative $\dfrac{dy}{dx}$ represents the rate of change of $y$ w.r.t. $x$. If it is indentically zero then $y$ is neither increasing nor decreasing as $x$ increases. So $y$ must be constant, i.e., $f(x)$ is a constant function.

For a layman, these arguments are quite convincing and not easy to refute. Mathematically, too, these arguments have their value from the point of view of acceptablity of the result. They definitely make the theorem plausible. Still, they cannot serve as a mathematical proof because they are based on certain interpretations of a derivative. There is, of course, nothing wrong in such interpretations. In fact, mathematics would lose much of its use if its very basic concepts failed to have such lucid interpretations. Still, one must be careful not to let such interpretations dictate the underlying mathematics as otherwise it would be nothing short of attempting to prove something by an analogy. In the argument given above, for example, it was claimed that if the tangent to a curve at every point is horizontal, then the curve itself is a horizontal straight line. With our intuitive understanding of a tangent, this statement certainly looks obvious. But as we saw in Section 2.2, in a rigorous approach a tangent is *defined* as the limiting position of a secant. And that brings us very close to derivatives. In other words, tangency is a concept whose existence is subservient to that of a derivative. Therefore, any statement about tangency, no matter how obvious it looks, will have to be derived from some results about derivatives. In particular, the claim above, viz. that a curve whose tangent is horizontal is itself a horizontal line has to be established using some result about the derivatives. And that amounts to proving Theorem (5.1).

The argument based on the concept of a rate is also open to the same objection. We can define an average rate of change by simple arithmetic. But to define an instantaneous rate of change we need the concept of a derivative. So we cannot indiscriminately extend our intuition about average rate to an instantaneous rate of change. Anything we want to prove about the instantaneous rate of change has to be proved by first establishing the relevant result about derivatives. For example, the chain rule, which looks quite obvious when derivatives are interpreted as rates of change (cf. Exercise (3.12)), needs a proof, using strictly the definition of a derivative. Similarly, the claim that if the rate of change of $y$ w.r.t. $x$ is always 0 then $y$ is a constant function of $x$, no matter how obvious it looks, needs a rigourous mathematical proof. And that again brings us back to the need for a mathematical proof of Theorem (5.1).

Another evidence that Theorem (5.1) is not quite as trivial as it seems is that its proof will crucially need the completeness of the real number system. The analogous result is false if we were living in $Q$, the world of rational numbers. For example, define $f : Q \longrightarrow Q$ by $f(x) = 1$ if $x < \sqrt{2}$ and $f(x) = 2$ if $x > \sqrt{2}$. (In case you don't like the mention of the irrational number $\sqrt{2}$, the definition can be worked without making a direct reference to it, see Exercise (5.1)). This function is differentiable at every point $c$ of $Q$. (Note that now that we are living in $Q$, the definition of differentiability at $c$ means $\dfrac{f(c + \Delta x) - f(c)}{\Delta x}$ tends to a rational limit as $\Delta x$ tends to 0 through rational values.) In fact, no matter what $c$ is, $f$ is constant in some neighbourhood of $c$ and so $f'(c) = 0$. Thus $f'$ is identically 0 on $Q$, but $f$ is not a constant function.

So Theorem (5.1) is certainly not trivial. Moreover, no matter how we go about proving it, we shall need completeness of the reals in one form or the other. We already proved a couple of results (Theorems (1.3) and (1.4)) about continuous functions using completeness of $I\!R$. It turns out that by doing just a little extra work we shall come within a striking distance of Theorem (5.1). The extra work is really an immediate consequence of the definition of a derivative. But since it will be needed later too, it is worth isolating as a separate theorem.

**5.2 Theorem :** Suppose a real-valued function $f$ attains its maximum on an interval $[a, b]$ at some interior point $c$ of $[a, b]$. Suppose $f'(c)$ exists. Then $f'(c) = 0$. A similar assertion holds if $f$ attains its minimum at $c$.

**Proof :** As $c$ is an interior point of $[a, b]$, there exists some $\delta_1 > 0$ such that $(c - \delta_1, c + \delta_1) \subset [a, b]$. Now suppose $f'(c) \neq 0$. Then $f'(c) > 0$ or $f'(c) < 0$. Assume that the first possibility holds. Choose $\epsilon = \dfrac{f'(c)}{2}$ and get a $\delta$ (which may be assumed to be between 0 and $\delta_1$) such that $\left| \dfrac{f(x) - f(c)}{x - c} - f'(c) \right| < \dfrac{f'(c)}{2}$ whenever $0 < |x - c| < \delta$. In particular, taking $x = c + \dfrac{\delta}{2}$, we get,

$$-\frac{\delta}{2}\frac{f'(c)}{2} < f(c + \frac{\delta}{2}) - f(c) - \frac{\delta}{2}f'(c) < \frac{\delta}{2}\frac{f'(c)}{2} \tag{1}$$

The first inequality in (1) implies $f(c + \dfrac{\delta}{2}) > f(c)$, contradicting that $f$ attains its maximum on $[a, b]$ at $c$ (note that $c + \dfrac{\delta}{2} \in [a, b]$).

If $f'(c) < 0$, we get a similar contradiction by taking $\epsilon = -\dfrac{f'(c)}{2}$ and considering $f(c - \dfrac{\delta}{2})$ for a suitable $\delta$. So we must have $f'(c) = 0$.

The last assertion follows by a similar argument. Or we may apply the argument to the function $-f$ which attains its maximum wherever $f$ attains its minimum. ∎

The proof may appear a bit technical. But the essential idea is very simple.

If $f'(c) > 0$ then to the immediate right of $c$, $f$ has a greater value than at $c$.If $f'(c) < 0$ then to the immediate left of $c$, $f$ has a greater value than at $c$. (cf. Exercise (3.4)) The assumption that $c$ is an interior point ensures that points near $c$, on either side of $c$ lie within $[a, b]$ and this gives the desired contradiction. It is vital that $c$ is an interior point. Take for example $f(x) = x$ and $[a, b] = [0, 1]$. Then $f$ attains its maximum at 1, minimum at 0 but $f'$ vanishes nowhere. It is also imporant that the existence of $f'(c)$ has to be given beforehand. The function $f(x) = |x|$ attains its minimum on $[-1, 1]$ at the interior point 0. But $f'(0)$ does not exist.

This theorem does not assert the existence of a maximum or a minimum of $f$ on $[a, b]$. All it says that in case there is a maximum (or minimum) at $c$ and $c$ happens to be an interior point then $f'(c)$ either fails to exist or vanishes. To apply this theorem, we need a source of functions which are guaranteed to have a maximum or a minimum on $[a, b]$. Theorem (1.4) gives one such source. Combining the previous theorem with it we get a theorem, called **Rolle's Theorem** which is the first among a series of related results.

**5.3 Theorem :** Suppose $f$ is a real-valued function which is continuous on a closed interval $[a, b]$ and differentiable on the open interval $(a, b)$. (In other words, $f'(c)$ exists at every interior point. No assumption is made regarding differentiability at an end point.) Assume further $f(a) = f(b) = 0$. Then there exists some $c \in (a, b)$ such that $f'(c) = 0$.

**Proof :** If $f$ is identically 0 on $[a, b]$ then $f'$ vanishes identically on $(a, b)$ and so we may take $c$ to be any point of $(a, b)$. If $f$ is not identically 0 on $[a, b]$ then there is some $x_0 \in (a, b)$ such that $f(x_0) \neq 0$. So either $f(x_0) > 0$ or $f(x_0) < 0$. Without loss of generality, assume that the first possibility holds. (Otherwise, as usual, replace $f$ by $-f$.) Now, by Theorem (1.4), there exists some point $c$ in $[a, b]$ such that $f(c) \geq f(x)$ for all $x \in [a, b]$. In particular $f(c) \geq f(x_0) > 0$. So $c$ must be an interior point of $[a, b]$, since $f(a) = f(b) = 0$. In other words $f$ attains its maximum on $[a, b]$ at the point $c \in (a, b)$. By assumption, $f'(c)$ exists. So by the theorem above $f'(c) = 0$. ∎

The theorem in the title, viz. **Lagrange's Mean Value Theorem** follows as a corollary.

**5.4 Theorem :** Suppose $f$ is continuous on $[a, b]$ and differentiable on $(a, b)$. Then there exists some $c \in (a, b)$ such that $f'(c) = \dfrac{f(b) - f(a)}{b - a}$.

**Proof :** The assumptions about continuity and differentiability are exactly same as Rolle's theorem. However, it is not given that $f(a) = f(b) = 0$. To apply Rolle's theorem we construct a new function $g$ obtained by modifying $f$ slightly. Specifically, let

$$g(x) = f(x) + Ax + B \tag{2}$$

where $A, B$ are some constants to be chosen suitably. Since $Ax + B$ is differentiable (and hence continuous) for all $x \in \mathbb{R}$, we see that $g$ is continuous / differentiable wherever $f$ is so. That is, $g$ is continuous on $[a, b]$ and differentiable on $(a, b)$. To apply Rolle's theorem we need $g(a) = 0$ and $g(b) = 0$. This can be arranged by putting $x = a$ and $x = b$ in (2) and solving the resulting system of equations, viz.,

$$f(a) + Aa + B = 0 \quad \text{and} \quad f(b) + Ab + B = 0 \tag{3}$$

yielding

$$A = \frac{f(b) - f(a)}{a - b} \quad \text{and} \quad B = \frac{f(b) - f(a)}{b - a} a - f(a). \tag{4}$$

(The actual value of $B$ is unimportant as long as $g(a) = 0$ and $g(b) = 0$.) Applying Rolle's theorem, we get some $c \in (a, b)$ such that $g'(c) = 0$. But from (2), $g'(x) = f'(x) + A$. So from (4), $f'(c) = g'(c) - A = -A = \dfrac{f(b) - f(a)}{b - a}$ as desired. ∎

Rolle's theorem clearly follows as a special case of Lagrange's MVT (Mean Value Theorem). However, since the former easily implies the latter as we just showed, they are of the same level of strength. In applications, it is the MVT that is generally preferred. Before discussing any such applications, a few comments are in order. The term 'mean value' may give the wrong impression that the number $c$ (whose existence is asserted by the theorem) is the arithmetic mean of $a$ and $b$ (i.e., $\dfrac{a + b}{2}$) or some other mean such as the geometric or the harmonic mean. That this is not the case can be seen from simple examples like $f(x) = x^3$ and $[a, b] = [0, 1]$. Here $f'(x) = 3x^2$ and $\dfrac{f(b) - f(a)}{b - a} = 1$. So the only point $c$ satisfying the conclusion of the MVT is $c = \dfrac{1}{\sqrt{3}}$. There is, in fact, no general formula to express $c$ in terms of $a, b$ and the function $f$. The term 'mean' denotes here the average rate of change. As we move from $a$ to $b$ the net change in the function is $f(b) - f(a)$ (which could be positive, negative or zero). So $\dfrac{f(b) - f(a)}{b - a}$ is the average or the mean rate of change over the interval $[a, b]$. Lagrange's theorem asserts that under the conditions specified, there exists at least one $c$ such that the instantaneous rate of change at $c$ equals the average rate of change over $[a, b]$. Naturally, this point $c$ depends as much on the function $f$ as the points $a$ and $b$. If we take a smaller interval, say, $[a, b]$, contained in $[a, b]$ then the MVT would give a point $c_1 \in (a_1, b_1)$ such that $f'(c_1) = \dfrac{f(b_1) - f(a_1)}{b_1 - a_1}$. But $c_1$ is in general different from $c$.

Geometrically, if we draw the graph of $y = f(x)$ for $a \leq x \leq b$, then $\dfrac{f(b) - f(a)}{b - a}$ is precisely the slope of the chord joining the end-points $A = (a, f(a))$ and $B = (b, f(b))$. Langrange's theorem asserts the existence of a point on the graph at which the tangent is parallel to the chord $AB$ (see Fig. 4.5.1).

As seen from the figure, there may be several such points. In other words, the mean value $c$ given by the MVT need not be unique. In fact examples can be given (see Exercise (5.5)) where the function $f$ is not a constant but has infinitely many mean values. The exercises will also give examples to show that the conditions in the hypothesis are vital.

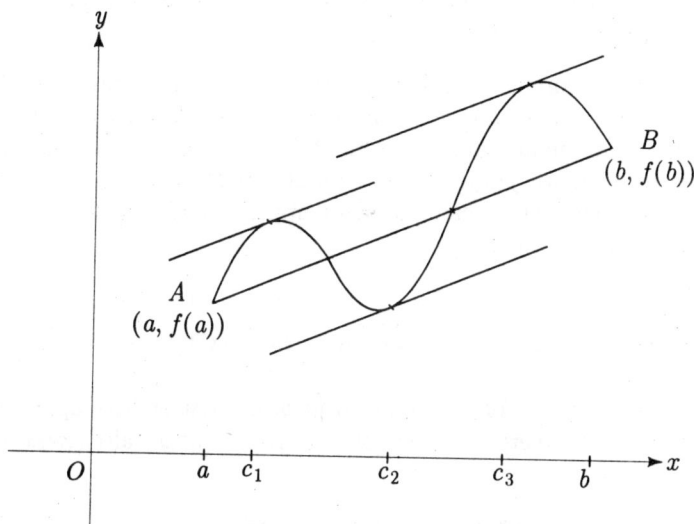

Figure 4.5.1 : Geometric Interpretation of Lagrange's MVT

The proof of Theorem (5.4) uses Theorem (1.4). The latter hinges crucially on the completeness of real numbers. So, as is the case with many other results based on completeness, the MVT merely gives the existence of the number $c$ without giving any algorithm to actually find $c$. This may seem like a demerit. But it turns out that mere existence of a mean value is sometimes enough in theoretical applications. As an illustration, we can now give a completely rigorous proof of Theorem (5.1). In fact, the argument needed also establishes the following more general result which completely subsumes Theoem (5.1).

**5.5 Theorem :** Suppose $f$ is continuous on $[a, b]$ and differentiable on $(a, b)$. Then

(i) If $f'(x) \geq 0$ for all $x \in (a, b)$, then $f$ is monotonically increasing on $[a, b]$ (i.e. $f(x_1) \leq f(x_2)$ whenever $a \leq x_1 < x_2 \leq b$).

(ii) If $f'(x) > 0$ for all $x \in (a, b)$ then $f$ is strictly monotonically increasing on $[a, b]$ (i.e. $f(x_1) < f(x_2)$ whenever $a \leq x_1 < x_2 \leq b$).

Similar assertions hold if $f' \leq 0$ throughout $(a, b)$ or if $f' < 0$ throughout $(a, b)$. Finally if $f' \equiv 0$ on $(a, b)$ then $f$ is constant.

**Proof :** Let $a \leq x_1 < x_2 \leq b$. Then Langrange's theorem is applicable on $[x_1, x_2]$. So there exists some $c \in (x_1, x_2)$ such that

$$f(x_2) - f(x_1) \; = \; f'(c)(x_2 - x_1) \tag{5}$$

Since $x_2 - x_1 > 0$, by (5), the sign of $f(x_2) - f(x_1)$ will be determined by that of $f'(c)$. If (i) holds, then $f(x_2) - f(x_1) \geq 0$, while in (ii), we have $f(x_2) - f(x_1) > 0$. Similarly the other three assertions follow. ∎

Although this theorem alone is evidence enough why Lagrange's MVT is such a basic result, it does not reveal its full strength. We used (5) above only qualitatively, viz., to deduce only the sign of $f(x_2) - f(x_1)$. A far more common use of (5) is to estimate $f(x_2) - f(x_1)$ quantitatively. Suppose, for example, that our data is such that we can find some upper bound, say $M$, on $|f'(x)|$ for all $x$ in $(a, b)$. Then (5) implies

$$|f(x_2) - f(x_1)| \; \leq \; M|x_2 - x_1| \tag{6}$$

for all $x_1, x_2$ in $[a, b]$. (If $x_1 > x_2$ we apply the MVT to the interval $[x_2, x_1]$ instead of $[x_1, x_2]$.)

This estimate is valid for all $x_1, x_2$ in $[a, b]$ no matter how apart they are from each other. The mere existence of a derivative at $x_2$ also gives a similar estimate, viz.,

$$|f(x_1) - f(x_2)| \; < \; N|x_1 - x_2| \tag{7}$$

where $N$ may be taken to be any number greater than $|f'(x_2)|$ (e.g. $|f'(x_2)| + 1$). The proof of this simple fact is left as an exercise. But (7) is valid only if $x_1$ is sufficiently close to $x_2$ (depending upon how close $N$ is to $|f'(x_2)|$). So we may say (7) is a pointwise or local estimate while (6) is a uniform, or global estimate of the difference $|f(x_1) - f(x_2)|$ in terms of the difference $|x_1 - x_2|$. Note, however, that the hypothesis behind (6) is much stronger than that behind (7).

A question naturally arises whether we can have a stronger version of the MVT if instead of the first order derivative, we are given some higher order derivative at all points of an interval. The answer is in the affirmative as the following theorem shows. A proof will be indicated in the exercises.

**5.6 Theorem :** Suppose $n$ is a positive integer and $f : [a, b] \longrightarrow \mathbb{R}$ is a function for which the $(n-1)^{\text{th}}$ derivative $f^{(n-1)}(x)$ exists and is continuous at every point of $[a, b]$, while the $n^{\text{th}}$ derivative $f^{(n)}(x)$ exists for all $x$ in $(a, b)$. Then there exists $c \in (a, b)$ (depending on $a, b, f$ and $n$) such that

$$\begin{aligned} f(b) \; = \; & f(a) + f'(a)(b - a) + \tfrac{1}{2}f''(a)(b - a)^2 + \cdots \\ & + \frac{f^{(n-1)}(a)}{(n-1)!}(b - a)^{n-1} + \frac{f^{(n)}(c)}{n!}(b - a)^n. \; \blacksquare \end{aligned}$$

This theorem is often called **Taylor's theorem.** Clearly the case $n = 1$ corresponds to the Lagrange's theorem. The last term, viz., $\dfrac{f^{(n)}(c)}{n!}(b - a)^n$ is often called the **remainder** (or more specifically, the remainder after the $n^{\text{th}}$ term). We know that $\dfrac{(b-a)^n}{n!} \to 0$ as $n \to \infty$ (see Exercise (3.9.4)). In fact, because of the term $n!$ in the denominator, $\dfrac{(b-a)^n}{n!}$ is small even for relatively small values of $n$. If the nature of the function $f$ is such that the $n^{\text{th}}$ derivative $f^{(n)}(x)$ is bounded on $(a, b)$, then regardless of what $c$ is, the remainder is very small and so the sum of the first $n$ terms, i.e. $f(a) + f'(a)(b-a) + \cdots + \dfrac{f^{(n-1)}(a)}{(n-1)!}(b-a)^{n-1}$ gives a very good approximation to $f(b)$, called **Taylor's approximation**. The remainder is nothing but the error in this approximation.

A few applications of the Lagrange's MVT will be given in the exercises and subsequent sections. We shall also need it later in proving what is called the fundamental theorem of calculus, so called because it establishes the connection between the two most fundamental processes of calculus, viz., differentiation and integration.

## EXERCISES

5.1 Define $f : Q \to Q$ by $f(x) = 1$ if $x \leq 0$ or if $x > 0$ and $x^2 < 2$ and $f(x) = 2$ otherwise. Show that for every $c \in Q$, $\lim\limits_{\substack{x \to c \\ x \in Q}} \dfrac{f(x) - f(c)}{x - c}$ exists and equals 0. But $f$ is not constant.

5.2 Prove that Theorem (5.1) is not true if the domain of the function is not an interval. For example, let $f(x) = 1$ or $-1$ according as $x > 0$ or $x < 0$. Here the domain of $f$ is the union of two intervals, $(-\infty, 0)$ and $(0, \infty)$. $f$ is constant on each one of them but not on the entire domain.

5.3 Let $f(x) = |x|$ and $[a, b] = [-1, 1]$. Prove that Lagrange's theorem does not hold, i.e., there is no $c \in (-1, 1)$ such that $f'(c) = \dfrac{f(1) - f(-1)}{2}$. What goes wrong?

5.4 Prove that Lagrange's theorem also fails for $[0, 1]$ if $f$ is defined by $f(x) = 0$ for $x \neq 1$ and $f(1) = 1$.

5.5 Let $f(x) = x \sin \dfrac{1}{x}$ for $x \neq 0$ and $f(0) = 0$. (cf. Exercise (3.14)). Prove that in the interval $[0, \dfrac{1}{\pi}]$, there are infinitely many values of $x$ where $f'$ vanishes.

*(5.6) Suppose $f$ is continuous on $[a, b]$ and differentiable on $(a, b)$. Assume further that $f(b) - f(a) = b - a$. Prove that for every positive integer $n$, there exist distinct points $c_1, c_2, \ldots, c_n$ in $(a, b)$ such that $f'(c_1) + f'(c_2) + \cdots + f'(c_n) = n$.

**5.7** Suppose $f$ is continuous on $[a, b]$ and differentiable on $(a, b)$. Assume that $\lim\limits_{x \to a^+} f'(x)$ exists and equals $L$. Prove that the right derivative of $f$ at $a$, i.e., $f'_+(a)$ also exists and equals $L$. [*Hint*: For every $x \in (a, b)$, there exists some $c_x \in (a, x)$ such that $\dfrac{f(x) - f(a)}{x - a} = f'(c_x)$. As $x \to a^+, c_x \to a^+$.]

**5.8** Prove that a derived function, i.e., a function which is the derivative of some other function cannot have a simple discontinuity (see Section 4.1 for a definition).

**5.9** Suppose $f(x)$ is a function expressed as some determinant, say,

$$f(x) = \begin{vmatrix} f_1(x) & f_2(x) & f_3(x) \\ g_1(x) & g_2(x) & g_3(x) \\ h_1(x) & h_2(x) & h_3(x) \end{vmatrix} \quad \text{for all} \quad x$$

where the entries $f_1(x), f_2(x)$ etc. are all differentiable functions of $x$. Prove that in general

$$f'(x) \neq \begin{vmatrix} f_1'(x) & f_2'(x) & f_3'(x) \\ g_1'(x) & g_2'(x) & g_3'(x) \\ h_1'(x) & h_2'(x) & h_3'(x) \end{vmatrix}.$$

Prove, however that if all functions in any two of the rows (or two columns) of the determinant are constants, than $f'(x)$ equals the determinant obtained by differentiating the remaining row (or column). For example, if the second and the third rows contain only constant entries $g_1, g_2, g_3$, and $h_1, h_2, h_3$, then

$$f'(x) = \begin{vmatrix} f_1'(x) & f_2'(x) & f_3'(x) \\ g_1 & g_2 & g_3 \\ h_1 & h_2 & h_3 \end{vmatrix}.$$

In general, show that $f'(x)$ is a sum of three determinants, each obtained by differentiating the entries in one row (or in one column) at a time.

**5.10** Suppose $f, g, h$ are continuous on $[a, b]$ and differentiable on $(a, b)$. Prove that there exists $c \in (a.b)$ such that

$$\begin{vmatrix} f'(c) & g'(c) & h'(c) \\ f(a) & g(a) & h(a) \\ f(b) & g(b) & h(b) \end{vmatrix} = 0.$$

**5.11** Using the last exercise prove **Cauchy's Mean Value Theorem** viz., if $f, g$ are continuous on $[a, b]$, differentiable on $(a, b)$ and $g(b) \neq g(a)$ then there exists $c \in (a, b)$ such that $\dfrac{f'(c)}{g'(c)} = \dfrac{f(b) - f(a)}{g(b) - g(a)}$. Is this result more general than Lagrange's MVT?

5.12 Can the result of the last exercise be proved by applying Lagrange's MVT to $f$ and $g$ and taking the ratio?

5.13 Prove that any two antiderivatives of a function on an interval differ by a constant. (Consequently, in order to identify the indefinite integral $\int f(x)dx$, it suffices to find any one antiderivative, say, $F$ or $f$. Then $\int f(x)dx$ is precisely the set of all functions of the form $F + c$ where $c$ is a constant which can take any (real) value. Therefore $c$ is often called the **constant of integration.** Note that this is just a name. We are yet to study integration!)

5.14 Suppose $F$ is an antiderivative of a function $f$ on an interval $[a, b]$. Let $x_1, \ldots, x_{n-1}$ be any points in $[a, b]$ with $x_1 < x_2 < \ldots < x_{n-1}$. Set $x_0 = a$ and $x_n = b$. (Any set of such points is called a **partition** of the interval $[a, b]$. Prove that there exist points $c_1, \ldots, c_n$ with $c_i \in (x_{i-1}, x_i)$ for $i = 1, \ldots, n$ such that $\sum_{i=1}^{n} f(c_i)(x_i - x_{i-1}) = F(b) - F(a)$. Deduce the result of Exercise (5.6) as a special case. (As we shall see later, this result is a crucial step in the proof of the fundamental theorem of calculus.)

5.15 Suppose $f$ is continuous on $[a, b]$ and differentiable on $(a, b)$. Suppose for some $c_1 \in (a, b), f'(c_1) < \dfrac{f(b) - f(a)}{b - a}$. Prove that there is some $c_2 \in (a, b)$ such that $f'(c_2) > \dfrac{f(b) - f(a)}{b - a}$. Why is this intuitively obvious? [*Hint* : First do this in the case where $f(a) = f(b) = 0$. Then reduce the general case to this one as in the proof of Theorem (5.4).]

5.16 Prove that between any two zeros of a differentiable function, there is at least one zero of its derivative. Deduce that a polynomial of degree $n$ can never have more than $n$ roots.

5.17 Suppose $f$ is continuous on $[a, b]$ and differentiable on $(a, b)$. Suppose $f'(x) \neq 0$ for all $x \in (a, b)$. Prove that either $f'(x) > 0$ for all $x \in (a, b)$ or $f'(x) < 0$ for all $x \in (a, b)$. [*Hint* : Apply a part of Exercise (1.9). Note that $f'$ is not given to be continuous on $(a, b)$. If it is, then an alternate proof can be given using the Intermediate Value Property. Actually, this exercise is equivalent to showing that every derived function, whether continuous or not, has IVP. In sharp contrast, it is not true that every derived function is necessarily bounded; see Exercise (5.25) below.]

5.18 Let $f(x) = x + \sin x$. Prove that $f$ is strictly increasing on $\mathbb{R}$ even though its derivative vanishes at some points. (More generally, show that this holds for any function $f$ whose derivative is non-negative everywhere and vanishes at only finitely many points in any closed interval.) (In other words the converse of part (ii) of Theorem (5.5) is not true.)

5.19 Let $f(x) = \dfrac{1}{2}x + x^2 \sin \dfrac{1}{x}$ for $x \neq 0$ and $f(0) = 0$. Prove that $f'(0) = \dfrac{1}{2} > 0$ but $f$ is not monotonically increasing in any interval containing 0. Does this contradict the result of Exercise (3.4)?

5.20 If $f$ is differentiable in a neighbouhood of a point $c$, $f'$ is continuous at $c$ and $f'(c) > 0$, prove that $f$ is strictly monotonically increasing in some interval containing $c$.

5.21 Show that a continuously differentiable funciton on a closed and bounded interval satisfies a Lipschitz condition (cf. Exercise (2.7)).

5.22 Suppose $n$ is a positive integer, $g$ is a funciton whose $(n-1)^{\text{th}}$ derivative, $g^{(n-1)}$, is continuous on $[a, b]$ and differentiable on $(a, b)$. Assume further that $g(b) = 0, g(a) = 0, g'(a) = 0, \ldots, g^{(n-1)}(a) = 0$. Prove that there is some $c \in (a, b)$ such that $g^{(n)}(c) = 0$. [*Hint* : Go on applying Rolle's theorem to $g, g', g'', \ldots$, till finally to $g^{(n-1)}$.]

5.23 Prove Taylor's theorem. [Hint : Let

$$g(x) = f(x) - f(a) - f'(x)(x - a) - \cdots - \frac{f^{(n-1)}(a)}{(n-1)!}(x - a)^n - K(x - a)^n$$

where $K$ is so chosen that $g(b) = 0$. Apply the result of the last exercise to $g$.]

5.24 Compare Theorem (3.3) with Taylor's theorem. Which has a stronger hypothesis? Which has a stronger conclusion? Are the two thereoems comparable (i.e. does either one of them imply the other)?

5.25 Define $f : [0, 1] \longrightarrow \mathbb{R}$ by $f(x) = x^2 \sin \left( \dfrac{1}{x^2} \right)$ if $x \neq 0$ and $f(0) = 0$.

Prove that $f$ is differentiable everywhere on $[0, 1]$ but $f'$ is not bounded on $[0, 1]$.

5.26 In Section 4.1, we defined piecewise continuity (see also Exercise (1.26)). Analogously, a real valued function $f$ is said to be **piecewise continuously differentiable** on an interval $[a, b]$ if there exist points $a = t_0 < t_1 < t_2 < \ldots < t_{n-1} < t_n = b$ such that

    (i) $f$ is continuously differentiable on $(t_{i-1}, t_i)$ for every $i = 1, 2, \ldots, n$,

    (ii) $\lim\limits_{t \to t_i^+} f'(t)$ exists for $i = 0, 1, \ldots, n - 1$, and

    (iii) $\lim\limits_{t \to t_i^-} f'(t)$ exists for $i = 1, 2, \ldots, n$.

(In view of Exercise (5.7), it follows from (ii) that $f$ is right differentiable at $t_0, t_1, \ldots, t_{n-1}$. Similarly (iii) implies that $f$ is left differentiable at $t_1, t_2, \ldots, t_n$. Hence the three conditions above can be replaced by a single one, viz. that $f$ is continuously differentiable on $[t_{i-1}, t_i]$ for $i = 1, 2, \ldots, n$. Note, however, that $f$ is not required to be differentiable at the $t_i$'s. So the definition is slightly weaker than saying that $f'$ is piecewise continuous on $[a, b]$. For example, the function $f(x) = |x|$ is picewise continuously differentiable on $[-1, 1]$ even though it is not differentiable at 0.)

Prove that the result of Exercise (5.21) continues to hold if $f$ is only piecewise continuously differentiable on $[a, b]$.

## 4.6     Term-by-term Differentiability of a Power Series

Let $f(x)$ represent the sum function of a power series $\sum\limits_{n=0}^{\infty} a_n x^n$. Let $R$ be the radius of convergence of $\sum\limits_{n=0}^{\infty} a_n x^n$. We assume $R > 0$. In Theorem (3.11.9) (see also Exercise (3.11.9)) we saw that the power series converges uniformly on $[-R', R']$ for every $0 < R' < R$ and remarked that this is the key to proving the local properties of the function $f(x)$ in $(-R, R)$. Accordingly, in Corollary (1.6), we proved that $f$ is continuous on $(-R, R)$. To recall the proof, for each $n \in \mathbb{N}$, we let $S_n(x)$ be the partial sum $\sum\limits_{k=0}^{n} a_k x^k$. Then $S_n \to f$ uniformly on $[-R, R']$, for every $0 < R' < R$. Also each $S_n$ is, being a polynomial, continuous everywhere. Since continuuity at a point $c$ passes over to the limit function under uniform convergence over a neighbourhood of $c$, we get that $f$ is continuous at $c$ for every $c \in (-R, R)$, the interval of convergence.

It is tempting to try a similar approach for differentiability of $f$ on $(-R, R)$. There is certainly no difficulty about differentiability of $S_n(x) = a_0 + a_1 x + a_2 x^2 + \ldots + a_n x^n$. Being a finite sum of differentiable functions, its derivative is obtained by differentiating each term and adding, that is,

$$S_n'(x) \;=\; a_1 + 2a_2 x + 3a_3 x^2 + \ldots + n a_n x^{n-1} \tag{1}$$

for all $x \in \mathbb{R}$, and hence in particular for all $x \in (-R, R)$. We already know that $S_n \to f$ uniformly on $[-R', R']$ for all $0 < R' < R$. If we could only show

$f'(x) = \lim_{n \to \infty} S'_n(x)$, then we would get $f'(c) = \sum_{n=1}^{\infty} n a_n c^{n-1}$, i.e., term-by-term differentiation would be valid.

The trouble is that, in general, even if a sequence of functions, say, $\{f_n(x)\}_{n=1}^{\infty}$ converges uniformly to a function $f(x)$ in some neighbourhood $U$ of a point $c$ and each $f_n$ is differentiable at $c$, it does not follow that $f$ is differentiable at $c$. And even if it is, it does not follow, that $f'(c) = \lim_{n \to \infty} f'_n(c)$. Actual examples of such pathological behaviour will be given in the exercises. But it is not hard to see exactly where the things could go wrong. For each $x \neq c$ in $U$, we define $g_n(x) = f_n(x) - f_n(c)$ and $h_n(x) = \dfrac{f_n(x) - f_n(c)}{x - c}$. Let $g(x) = f(x) - f(c)$ and $h(x) = \dfrac{f(x) - f(c)}{x - c}$. Since $f_n(x) \to f(x)$ for all $x \in U$, it is obvious that $g_n(x) \to g(x)$ and $h_n(x) \to h(x)$ for all $x \in U - \{c\}$. That is $g_n \to g$ and $h_n \to h$ pointwise on the deleted neighbourhood $U - \{c\}$ of $c$. Now the uniform convergence of $f_n$ to $f$ on $U - \{c\}$ easily implies that $g_n \to g$ uniformly on $U - \{c\}$. Still, it does not follow that $\dfrac{g_n(x)}{x - c} \to \dfrac{g(x)}{x - c}$ uniformly on $U - \{c\}$ (cf. Exercise (3.11.11)). In other words, we may not have $h_n \to h$ uniformly on $U - \{c\}$. If we did, things would be nice, because then by applying Theorem (3.11.3), we would get

$$\lim_{x \to c} (\lim_{n \to \infty} h_n(x)) = \lim_{n \to \infty} (\lim_{x \to c} h_n(x)) \tag{2}$$

which says precisely that

$$f'(c) = \lim_{n \to \infty} f'_n(c). \tag{3}$$

So, if we could have some hypothesis that would ensure that $h_n(x) \to h(x)$ uniformly on $U - \{c\}$, we would be through. The trouble is that this is a very strong condition. In fact, if $x$ is very close to to $c$ then $h_n(x)$ is very close to $f'_n(c)$. So assuming the uniform convergence of $h_n$ is almost like assuming the uniform convergence of $f'_n$. This puts us in a dilemma. We set out to prove that $f'_n \to f'$ pointwise. But in order to do it, we have to assume something stronger, viz., the uniform convergence of $f'_n$ to $f'$!

Fortunately, there is a way out. Instead of starting with a sequence $\{f_n\}$ of functions which converges uniformly to a function $f$ we shall start with a sequence $\{f_n\}$ of functions, for which the derived sequence $\{f'_n\}$ converges uniformly to some function, say $g$. We shall then show that under a certain mild additional condition, the original sequence $\{f_n\}$ must converge uniformly to some function $f$. Finally we shall show that this function $f$ is differentiable and its derivative is precisely $g$.

As the proof is a bit longish, we give it in two parts.

**6.1 Lemma :** Suppose $\{f_n\}$ is a sequence of real-valued functions defined on an interval $[a, b]$. Suppose each $f_n$ is differentiable on $[a, b]$ and that $\{f'_n\}$ converges to a function $g$ uniformly on $[a, b]$. Assume further that for some $x_0 \in [a, b]$

the sequence $\{f_n(x_0)\}$ is convergent. Then $\{f_n\}$ converges to some function $f$ uniformly on $[a, b]$.

**Proof :** We shall use the uniform version of the Cauchy criterion for convergence given in Exercise (3.11.10). Since $\{f_n'\}$ is uniformly convergent (to $g$) on $[a, b]$, it is a uniformly Cauchy sequence. From this, we want to show that $\{f_n\}$ is also a uniformly Cauchy sequence on $[a, b]$. That is, given $\epsilon > 0$, we shall show that there is some $p \in I\!N$ such that for all $m \geq n \geq p$, and for all $x \in [a, b]$,

$$|f_m(x) - f_n(x)| < \epsilon \tag{4}$$

The idea of the proof is as follows. By assumption, $\{f_n(x_0)\}$ is a convergent and hence a Cauchy sequence. So $|f_m(x_0) - f_n(x_0)|$ can be made small if $m, n$ are sufficiently large. Now,

$$\begin{aligned}|f_m(x) - f_n(x)| &\leq |(f_m(x) - f_n(x) - (f_m(x_0) - f_n(x_0))| \\ &+ |f_m(x_0) - f_n(x_0)|\end{aligned} \tag{5}$$

So if we can ensure that $|(f_m(x) - f_n(x)) - (f_m(x_0) - f_n(x_0))|$ is small then we would be through. For this we shall apply Lagrange's Mean Value Theorem to the difference function $f_m - f_n$. (It will not be of much avail to apply the MVT separately to $f_m$ and to $f_n$, because we do not know much about the size of $f_m'$ and $f_n'$ individually. But the difference $f_m' - f_n'$ is uniformly small if $m, n$ are large.)

Now to the details. Let $\epsilon > 0$ be given. Since $\{f_n'\}$ is a uniformly Cauchy sequence on $[a, b]$, there exists $p \in I\!N$ such that

$$|f_m'(x) - f_n'(x)| < \frac{\epsilon}{2(b - a)} \tag{6}$$

for all $m \geq n \geq p$ and for all $x \in [a, b]$.

Further, $\{f_n(x_0)\}$ is also a Cauchy sequence and so, replacing $p$ by a larger value if necessary, we may suppose

$$|f_m(x_0) - f_n(x_0)| < \frac{\epsilon}{2} \text{ for all } m \geq n \geq p. \tag{7}$$

Note that (6) gives an upper bound for $|f_m'(x) - f_n'(x)|$ on $[a, b]$. We now apply Lagrange's MVT (or rather, its consequence given in (6) of the last section) to the function $f_m - f_n$. Then for all $x_1, x_2 \in [a, b]$,

$$|(f_m(x_2) - f_n(x_2)) - (f_m(x_1) - f_n(x_1))| \leq \frac{\epsilon}{2(b - a)}|x_2 - x_1| \tag{8}$$

In particular taking $x_1 = x_0$ and $x_2 = x$, and noting that $|x - x_0| \leq b - a$ for all $x \in [a, b]$, we get

$$|(f_m(x) - f_n(x)) - (f_m(x_0) - f_n(x_0))| \leq \frac{\epsilon}{2(b - a)}(b - a) = \frac{\epsilon}{2} \tag{9}$$

for all $x \in [a, b]$ and $m \geq n \geq p$.

Using (5), (7) and (9), we see that (4) is true. Thus $\{f_n\}$ is a uniformly Cauchy sequence on $[a, b]$. By the result of Exercise (3.11.10), $f_n \to f$ uniformly on $[a, b]$ where the function $f : [a, b] \to \mathbb{R}$ is defined as $f(x) = \lim_{n \to \infty} f_n(x)$ for $x \in [a, b]$. ∎

We now come to the more delicate part of showing that the fucntion $f$ is differentiable on $[a, b]$ and that for every $c \in [a, b]$, $f'(c) = \lim_{n \to \infty} f_n'(c)$. For this, we construct the functions $h_n$ and $h$ as indicated before. Let $S = [a, b] - \{c\}$. For every $x \in S$ and $n \in \mathbb{N}$, let

$$h_n(x) = \frac{f_n(x) - f_n(c)}{x - c} \quad \text{and} \quad h(x) = \frac{f(x) - f(c)}{x - c} \tag{10}$$

As we remarked earlier, if we could show that $h_n \to h$ uniformly on $S$, then Theorem (3.11.3) would permit us to conclude (2) and hence (3). (In case $c = a$ or $b$, then instead of letting $x \to c$, we let $x \to c^+$ or $x \to c^-$ as the case may be.)

Since $f_n(x) \to f(x)$ for all $x \in S$, there is no difficulty in showing that $h_n(x) \to h(x)$ pointwise in $S$. The real difficulty is in showing that $h_n \to h$ *uniformly* on $S$. For this we shall again begin by showing that $\{h_n\}$ is a uniformly Cauchy sequence on $S$. Exercise (3.11.10) would then imply that $h_n$ converges uniformly to *some* function, say $\phi$, on $S$. In particular, for every $x \in S$, $h_n(x) \to \phi(x)$ as $n \to \infty$. But we already know that $h_n(x) \to h(x)$ as $n \to \infty$. So, by the uniqueness of limtis (cf. Theorem (2.6.1), modified for limits of sequences), we must have $h(x) = \phi(x)$ for all $x \in S$. In other words, $h_n \to h$ uniformly on $S$.

Thus we are ultimately reduced to having to prove that $\{h_n\}$ is a uniformly Cauchy sequence on $S$. A direct calculation shows that for all $x \in S$ and $m, n \in \mathbb{N}$,

$$|h_m(x) - h_n(x)| = \left| \frac{f_m(x) - f_n(x) - (f_m(c) - f_n(c))}{x - c} \right| \tag{11}$$

Once again, we apply Lagrange's MVT to the difference function $f_m - f_n$ over the interval $[x, c]$ or $[c, x]$ (depending upon whether $x < c$ or $x < c$). First, given $\epsilon > 0$, we choose $p$ so that for all $m \geq n \geq p$ and for all $x \in [a, b]$

$$|f_m'(x) - f_n'(x)| < \epsilon.$$

By the same argument used in establishing (8) we get that whenever $m \geq n \geq p$,

$$|(f_m(x_2) - f_n(x_2)) - (f_m(x_1) - f_n(x_1))| < \epsilon |x_2 - x_1| \tag{12}$$

for all $x_1 \neq x_2$ in $[a, b]$. Taking $x_1 = c, x_2 = x$ and using (11), we get that

$$|h_m(x) - h_n(x)| < \epsilon \text{ for all } x \in S \text{ and } m \geq n \geq p.$$

This is exactly what we wanted to show.

We have thus proved

**6.2 Theorem :** With the assumptions of Lemma (6.1), the function $f$ in its conclusion is differentiable and moreover $f' = g$ on $[a, b]$. ∎

Having developed the necessary tool, let us go back to the power series $\sum_{n=0}^{\infty} a_n x^n$ with radius of convergence $R$ $(> 0)$, and sum function $f(x)$ defined on $(-R, R)$. If we differentiate this power series term by term, we get a new power series $\sum_{n=1}^{\infty} n a_n x^{n-1}$, called the **derived power series**. Here the index variable starts with $n = 1$. To bring it to the standard form we can replace $n - 1$ with $n$ and write the derived power series as $\sum_{n=0}^{\infty} (n+1) a_{n+1} x^n$. Differentiating again we get the second derived power series as $\sum_{n=2}^{\infty} n(n-1) a_n x^{n-2}$ or as $\sum_{n=0}^{\infty} (n+1)(n+2) a_{n+2} x^n$. Repeating this process we can define the $k^{\text{th}}$ derived power series for every $k \in I\!N$.

Of course, we have not yet shown that $f'(x) = \sum_{n=1}^{\infty} n a_n x^{n-1}$. That, in fact, is the goal we are heading to. Still, we can consider the derived power series $\sum_{n=1}^{\infty} n a_n x^{n-1}$ (or $\sum_{n=0}^{\infty} (n+1) a_{n+1} x^n$) as a power series in its own rights and inquire about its radius of convergence. The answer turns not to be surprisingly simple.

**6.3 Theorem :** The radius of convergence of $\sum_{n=1}^{\infty} n a_n x^{n-1}$ is the same as that of the power series $\sum_{n=0}^{\infty} a_n x^n$.

**Proof :** It is clear that for $x \neq 0$, $\sum_{n=1}^{\infty} n a_n x^{n-1}$ is convergent if and only if $\sum_{n=1}^{\infty} n a_n x^n$ is convergent. For $x = 0$ also, both are convergent. In the second series, we may as well add the term $0 a_0 x^0$ and write it as $\sum_{n=0}^{\infty} n a_n x^n$ which is in the standard form for power series. So it suffices to show that the radius of convergence of the power series $\sum_{n=0}^{\infty} n a_n x^n$ equals that of $\sum_{n=0}^{\infty} a_n x^n$. From

Theorem (1.10.1) (or rather, its extension in Exercise (3.10.1)), the latter equals $1/L$ where $L = \varlimsup_{n\to\infty} |a_n|^{1/n}$. Similarly, the radius of convergence of $\sum_{n=0}^{\infty} na_n x^n$ is $1/L^{'}$ where $L^{'} = \varlimsup_{n\to\infty} |na_n|^{1/n}$. So we have to prove that $L = L^{'}$. Since $L$ is a limit point (in fact the largest limit point) of $\{|a_n|^{1/n}\}$, there exists, by Exercise (3.5.1), a subsequence, say $\{a_{n_k}\}$ of $\{a_n\}$ such that $\lim_{k\to\infty} |a_{n_k}|^{1/n_k} = L$. (Note that because of our assumpiton that $R > 0, L \neq \infty$.) Now, by Theorem (3.4.5) $n^{1/n} \to 1$ as $n \to \infty$ and hence $n_k^{1/n_k} \to 1$ as $k \to \infty$. So,

$$\lim_{k\to\infty} |n_k a_{n_k}|^{1/n_k} = \lim_{k\to\infty} |a_{n_k}|^{1/n_k} = 1 \cdot L = L \qquad (13)$$

Thus $L$ is also a limit polint of the sequence $\{|na_n|^{1/n}|\}$. But $L^{'}$ is the largest limit point of $\{|na_n|^{1/n}\}$. So $L \leq L^{'}$. The other way inequality, viz. $L^{'} \leq L$ is proved by essentially reversing the reasoning, i.e. we start with a subsequence of $\{|na_n|^{1/n}\}$ converging to $L^{'}$ and use (13). So $L = L^{'}$ and, as noted before, this completes the proof. ∎

We now have all the machinery to prove the validity of term-by-term differentiation of a power series.

**6.4 Theorem :** Let a power series $\sum_{n=0}^{\infty} a_n x^n$ have radius of convergence $R$ ($> 0$) and sum function $f(x)$ in the interval $(-R, R)$. Then for every $c$ in this interval, $f^{'}(c)$ exists and equals $\sum_{n=1}^{\infty} na_n c^{n-1}$.

**Proof :** For $n \in I\!N$, we let $S_n(x)$ be the partial sum $\sum_{k=0}^{n} a_k x^k$. Then $S_n^{'}(x)$ is given by (1). But $S_n^{'}(x)$ is precisely a partial sum of the derived power series $\sum_{n=1}^{\infty} na_n x^{n-1}$ or $\sum_{n=0}^{\infty} (n+1)a_{n+1} x^n$. Let $g(x)$ denote the sum function of this derived series. Let $c \in (-R, R)$. Find $R^{'}$ so that $c \in (R^{'}, R^{'})$ and $[-R^{'}, R^{'}] \subset (-R, R)$. Now, by Theorem (6.3), the sequence $\{S_n^{'}\}$ converges to $g(x)$ for all $x \in (-R, R)$ and by Theorem (3.11.5) (and its extension in Exercise (3.11.9)), this convergence is uniform on $[-R^{'}, R^{'}]$. We now apply Theorem (6.2) to the sequence $\{S_n\}$ and the interval $[-R^{'}, R^{'}]$. We already know that $\{S_n(x)\}$ converges to $f(x)$ for all $x \in [-R^{'}, R]$ (in fact for all $x \in (-R, R)$). So this part of the conclusion is not new to us. What really matters is that the function $f$ is differentiable on $[-R^{'}, R^{'}]$ and further $f^{'}(c) = g(c)$. But, by definition $g(c) = \sum_{n=1}^{\infty} na_n c^{n-1}$, proving the theorem. ∎

Thus we see that uniform convergence plays a very crucial role in proving that a power series can be differentiated term-by-term. Of course, we applied Theorem (6.2) only for a speical type of sequences, viz., the sequence of partial sums of a power series. That is why, the first part, i.e., the conclusion of Lemma (6.1), did not give us anything new. It is, in fact, possible to bypass uniform convergence completely and give a direct proof of Theorem (6.4). This will be indicated in the Exercises. We invite the percepient reader to pinpoint which part of the alternate argument corresponds to uniform convergence.

Since the derived series of a power series is also a power series with the same radius of convergence as the original power series, we can apply the preceding theorem to the derived series and get a formula for the second derivative of the sum function of the original power series. If we repeat this process, we get,

**6.5 Theorem :** Let $f(x)$ be the sum function of a power series $\displaystyle\sum_{n=0}^{\infty} a_n x^n$ with radius of convergence $R > 0$. Then $f$ is infinitely differentiable (i.e., has derivatives of all orders) in $(-R, R)$. Moreover, for every $k = 0, 1, 2, \ldots$,

$$
\begin{aligned}
f^{(k)}(x) &= \sum_{n=k}^{\infty} n(n-1)\ldots(n-k+1)a_n x^{n-k} \\
&= \sum_{n=0}^{\infty} (n+1)(n+2)\ldots(n+k)a_{n+k} x^n
\end{aligned}
$$

for all $x \in (-R, R)$. In particular, $a_k = \dfrac{f^{(k)}(0)}{k!}$ for $k = 0, 1, 2, \ldots$.

**Proof :** As remarked earlier, the first part follows by repeated applications of the last theorem. For the last part simply put $x = 0$, with the understanding that $0° = 1$. ∎

As an interesting consequence we get the **uniqueness theorem for power series** which was mentioned (and tacitly used) in Section 3.10.

**6.6 Theorem :** Suppose two power series $\displaystyle\sum_{n=0}^{\infty} a_n x^n$ and $\displaystyle\sum_{n=0}^{\infty} b_n x^n$ have the same sum function in some open interval containing 0. Then $a_k = b_k$ for all $k$.

**Proof :** Let $\displaystyle\sum_{n=0}^{\infty} a_n x^n = \sum_{n=0}^{\infty} b_n x^n = f(x)$ for all $x$ in some open interval containing 0. This interval may be taken to be $(-R, R)$ for some $R > 0$. Let $R_1, R_2$ be the radii of convergence and $f_1(x), f_2(x)$ be the sum functions of $\displaystyle\sum_{n=0}^{\infty} a_n x^n$ and $\displaystyle\sum_{n=0}^{\infty} b_n x^n$ respectively. Then $R_1 \geq R$ and $R_2 \geq R$. Also both $f_1$ and $f_2$ agree with $f$ on $(-R, R)$ which is a neighbourhood of 0. Hence $f_1^{(k)}(0) = f^{(k)}(0) = f_2^{(k)}(0)$

for all $k$. But then by the last theorem, $\dfrac{a_k}{k!} = \dfrac{b_k}{k!}$ for all $k$. ∎

Besides proving Theorem (6.6), there are many other applications of Theorem (6.4). For example, we see that if we define the exponential, the sine and the cosine functions by power series as in Exercise (3.3.14), then these functions are infinitely differentiable for all $x \in \mathbb{R}$ (since $R$, the radius of convergence is infinite in each case, see the comments after Theorem (3.10.1)). Algebraic manipluations with these series yield the various identities about these functions (cf. Exercise (3.10.3)). Now, because of Theorem (6.4), we can differentiate them term-by-term, and get, for all $x \in \mathbb{R}$

$$\frac{d}{dx}(e^x) = e^x, \quad \frac{d}{dx}(\sin x) = \cos x \text{ and } \frac{d}{dx}(\cos x) = -\sin x \qquad (14)$$

We also used term-by-term differentiation of power series in Exercise (3.10.7).

So far we considered power series of the form $\sum\limits_{n=0}^{\infty} a_n x^n$. More generally, we could consider power series with some other centre, say $c$, i.e., power series of the form $\sum\limits_{n=0}^{\infty} a_n (x-c)^n$. The last four theorems apply for such series also with hardly any change. In particular, such a power series can be differentiated term-by-term in the interval $(c - R, c + R)$ where $R$ is its radius of convergence.

So far we started with a power series and proved some properties of its sum function. Frequently we have to go the other way. That is, we have some function $f$ and some point $c$ in its domain. The problem is to find a power series centred at $c$ such that $f(x) = \sum\limits_{n=0}^{\infty} a_n (x-c)^n$, for all $x$ in the domain of $f$. This is known as **expanding** or **developing** $f$ by a power series in a neighbourhood of the point $c$. We already saw an instance of this in the answer to Q.3.10 where, in order to find a closed form expression for the Fibonacci numbers, we had to expand the function $\dfrac{x}{1 - x - x^2}$ as a power series in $x$.

In view of Theorem (6.6), a necessary condition for the existence of such a power series is that $f$ must be infinitely differentiable in some neighbourhood of 0. Moreover, we have no choice in constructing the power series $\sum\limits_{n=0}^{\infty} a_n (x-c)^n$.

We *have* to set $a_n = \dfrac{f^{(n)}(c)}{n!}$ for every $n$. The trouble is that even after this power series is constructed, its sum function, say $g(x)$, may not equal $f(x)$ for all values of $x$. This is, of course, to be expected because, since differentiation is a local operation, the successive derivatives of $f$ at $c$ depend only on the values of $f(x)$ for $x$ in some neighbourhood of $c$. Outside this neighbourhood, we can change $f(x)$ arbitrarily and then there is no reason to suppose that $g(x)$ will equal $f(x)$ outside this neighbourhood.

So the maximum we can hope for is that $g(x)$ should coincide with $f(x)$ in some neighbourhood of $c$. But even this is not always true. A classic counter-example is the function $f : \mathbb{R} \longrightarrow R$ defined by

$$f(x) = \begin{cases} e^{-1/x^2} & \text{if } x \neq 0 \\ 0 & \text{if } x = 0. \end{cases}$$

The graph of $f(x)$ is shown in Fig.4.6.1. It is often called an inverted **bell-shaped curve**.

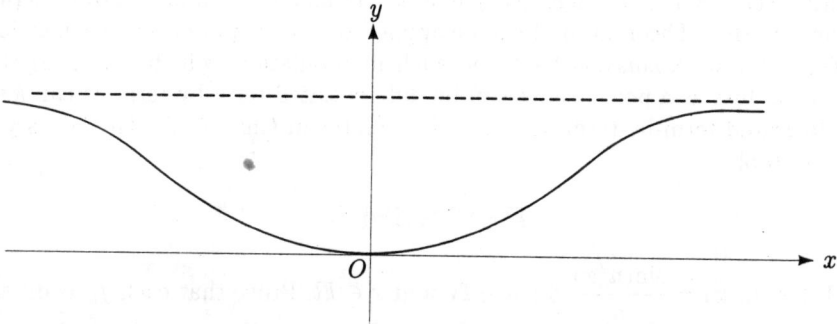

Figure 4.6.1 : An Infinitely Differentiable Non-analytic Function

As $x \to 0, \dfrac{1}{x^2} \to \infty$, whence $e^{-1/x^2} \longrightarrow 0$. So $f$ is continuous at 0. Continuity and differentiability of $f$ at $x \neq 0$ is immediate. Indeed, by (14) and the chain rule, for $x \neq 0, f'(x) = \dfrac{2}{x^3}e^{-1/x^2} = \dfrac{1}{x^3 e^{1/x^2}}$. It is easy to show (see Exercise (6.22) that as $x \to 0, e^{1/x^2}$ tends to $\infty$ much more rapidly than $x^3$ (or any polynomial in $x$) tends to 0. So the product $x^3 e^{1/x^2}$ tends to $\infty$. Hence $\lim\limits_{x \to 0} f'(x) = 0$. Therefore, by Exercise (5.8) $f'(0) = 0$. (This can also be verified directly.) An essentially similar reasoning shows that $f$ is infinitely differentiable over the entire real line and $f^{(n)}(0) = 0$ for all $n$. So if we let $a_n = \dfrac{f^{(n)}(0)}{n!}$, then the power series $\sum\limits_{n=0}^{\infty} a_n x^n$ converges to the identically 0 function. But $f$ never vanishes except at 0. So there is no neighbourhood of 0 in which $f$ can be expanded as a power series.

Thus we see that developability as a power series near a point is a considerably stronger requirement than infinite differentiability. Functions having the former property are often called **analytic** at that point. In the last section, we studied Taylor's aproximation of the $n^{\text{th}}$ order. If the nature of the function is such that the remainder after the $n^{\text{th}}$ term (i.e., the error of the approximation) tends to 0 as $n \to \infty$, then such a function can be shown to be analytic as will be indicated in the exercises, where a few basic properties of analytic functions will also be given.

Even when a function $f(x)$ is known to be analytic at a point $c$, it is generally not practicable to find its power series $\sum_{n=0}^{\infty} a_n(x - c)^n$ by evaluating $a_n$ as $\dfrac{f^{(n)}(c)}{n!}$. More efficient methods are often available. Usually one tries to express $f(x)$ in terms of some functions whose power series expansions are already known. This is what we did for $\dfrac{x}{1 - x - x^2}$ earlier in Section 3.10. If we can identify $f(x)$ as $F'(x)$ where $F(x)$ is analytic and has a known power series expansion, then Theorem (6.5) can be applied to get the power series expansion of $f(x)$. A few expansions based on such manipulations will be given in the exercises. Just as a power series can be differentiated term-by-term, it can also be integrated term-by-term, as we shall see later in Chapter 5. This gives yet another tool.

## EXERCISES

6.1  Let $f_n(x) = \dfrac{\sin(n^2 x)}{n}$ for $n \in \mathbb{N}$ and $x \in \mathbb{R}$. Prove that each $f_n$ is differentiable on $\mathbb{R}$ and $\{f_n\}$ converges uniformly on $\mathbb{R}$ to the identically zero function, but $\{f_n'(x)\}$ does not converge for some values of $x$ (e.g. $x = 0$).

6.2  For $n \in \mathbb{N}$, define $f_n : \mathbb{R} \longrightarrow \mathbb{R}$ by

$$f_n(x) = \begin{cases} |x| & \text{if } |x| > 1/n \\ \frac{1}{2}(nx^2 + \frac{1}{n}) & \text{if } |x| \le \frac{1}{n}. \end{cases}$$

  (i)  Prove that each $f_n$ is differentiable on $\mathbb{R}$. [*Hint* : Sketch the graph of $f_n$.]

  (ii)  Let $f(x) = |x|$ for all $x \in \mathbb{R}$. Prove that $f_n \to f$ uniformly on $\mathbb{R}$, but $f$ is not differentiable on $\mathbb{R}$. [*Hint* : For every $x \neq 0$, there is some $m$ such that $f_n(x) = |x|$ for all $n \ge m$.]

6.3  For $n \in \mathbb{N}$, let $f_n(x) = \dfrac{x^n}{n}$. Prove that $\{f_n\}$ converges uniformly to the function $f(x) = 0$ on $[0, 1]$. Prove, further that $\{f_n'\}$ also converges pointwise on $[0, 1]$ but that its limit function is not $f'$.

6.4  In Lemma (6.1), prove that the hypothesis that $\{f_n(x_0)\}$ is convergent for some $x_0$ is essential. [*Hint* : There is a very trivial counter-example.]

6.5  If $a, b$ are any two distinct real numbers, and $n \ge 2$ is an integer, show that $\dfrac{b^n - a^n}{b - a} - na^{n-1}$ equals $(b - a)(b^{n-2} + 2b^{n-3}a + 3b^{n-4}a^2 + \cdots + (n - 2)ba^{n-3} + (n - 1)a^{n-2})$. Deduce that if $|a| < M$ and $|b| < M$ then $\left| \dfrac{b^n - a^n}{b - a} - na^{n-1} \right| \le |b - a| n(n - 1) M^{n-2}$.

6.6 Using the last exercise give an alternate proof of Theorem (6.4). [*Hint* : If $c \in (-R, R)$ and $|\Delta x| < R - |c|$, then take $a = c, b = c + \Delta x, M = |c| + |\Delta x|$ in the last exercise. $\sum_{n=2}^{\infty} n(n-1)|a_n|M^{n-2}$ is convergent by repeated application of Theorem (6.3).]

6.7 From (14) and the properties of derivatives obtain the derivatives of other trigonometric functions such as $\tan x$, $\sec x$ etc. Also show that the derivatives of $\sinh(x)$ and $\cosh(x)$ (see Exercise (3.3.14)) are $\cosh(x)$ and $\sinh(x)$ respectively.

6.8 Prove that a function represented by a power series is analytic not only at its centre but at every point of its interval of convergence. [*Hint* : Let $\sum_{n=0}^{\infty} a_n x^n$ have sum function $f(x)$ and radius of convergence $R > 0$. If $|c| < R$ and $|x - c| < R - |c|$, express $\sum_{n=0}^{\infty} a_n x^n$ as $\sum_{n=0}^{\infty} a_n[(x - c) + c]^n$. Expand by binomial theorem and setting $\binom{n}{r} = 0$ for $n < r$ write this as a doubly infinite series $\sum_{n=0}^{\infty} \sum_{r=0}^{\infty} a_n \binom{n}{r} (x - c)^r c^{n-r}$. Apply Exercise (3.8.7).]

6.9 Prove that the power series $\sum_{n=0}^{\infty} x^n$ has $(-1, 1)$ as its interval of convergence but that its sum function $\dfrac{1}{1-x}$, is analytic at every $x \neq 1$. For example, show that near 3 it can be expanded as $\sum_{n=0}^{\infty} (-\frac{1}{2})^{n+1}(x - 3)^n$, which has radius of convergence 2.

6.10 Give an example to show that in Exercise (6.8), the power series of $f(x)$ at $c$ may have a higher radius of convergence than $R - |c|$.

6.11 Suppose $f, g$ are analytic at a point $c$. Prove that for any constants $\alpha, \beta$, $\alpha f + \beta g$ and $fg$ are analytic at $c$, (cf. Theorem (3.10.2)).

6.12 If $f$ is analytic at $c$ and $f(c) \neq 0$, prove that $\dfrac{1}{f}$ is analytic at $c$ and that its power series expansion at $c$, say, $\sum_{n=0}^{\infty} b_n (x - c)^n$ can be obtained from that of $f$ at $c$, say, $\sum_{n=0}^{\infty} a_n (x - c)^n$ by solving the following system of equations successively for $b_0, b_1, b_2, \ldots$.

$$a_0 b_0 = 1$$
$$a_0 b_1 + a_1 b_0 = 0$$
$$a_0 b_2 + a_1 b_1 + a_2 b_0 = 0$$
$$\cdots$$
$$a_0 b_n + a_1 b_{n-1} + \cdots + a_{n-1} b_1 + a_n b_0 = 0$$
$$\cdots\cdots$$

Obtain the expansion of $\dfrac{1}{1+x}$ at 0 by this method.

6.13 Let $h$ be a rational number. Prove that $(1+x)^h$ is analytic in $\mathbb{R}$ if $h$ is a positive integer and in $(-1,1)$ otherwise. Obtain its power series expansion near 0 (cf. Exercise (3.10.4) (iii)).

6.14 Let $\displaystyle\sum_{n=0}^{\infty} a_n x^n$ be the power series expansion of $f(x)$ at 0. For $n = 0, 1, \ldots$,

let $b_n = a_0 + a_1 + \ldots + a_n$. Prove that $\displaystyle\sum_{n=0}^{\infty} b_n x^n$ is the power series expansion of $\dfrac{f(x)}{1-x}$ at 0. Hence obtain the power series expansion of $\dfrac{1}{(1-x)^3}$ starting from that of $\dfrac{1}{(1-x)}$. Also expand $\dfrac{1}{(1-x)^3}$ by the last exercise (or using Theorem (6.4)). Hence show that for every positive integer $n, 1 + 2 + \ldots + n = \dfrac{n(n+1)}{2}$.

6.15 A function $f$ defined on a non-empty subset $S$ which is symmetric about 0 (i.e., $x \in S \Rightarrow -x \in S$ for all $x \in \mathbb{R}$) is said to be **even** if $f(x) = f(-x)$ for all $x \in S$ and **odd** if $f(x) = -f(-x)$ for all $x \in S$. Interpret these conditions in terms of the graph. Also prove that :

(i) every function can be expressed uniquely as a sum of an even and an odd function. (ii) the derivative of an even function (wherever it exists) is odd and vice versa (iii) if $f(x) = \displaystyle\sum_{n=0}^{\infty} a_n x^n$ then $f$ is even iff $a_n = 0$ for all odd $n$ and odd iff $a_n = 0$ for all even $n$.

6.16 Suppose $f$ is infinitely differentiable in some neighbourhood $U$ of a point $c$ and that there exists some $M$, such that for all $x \in U$ and all $n \in \mathbb{N}, |f^{(n)}(x)| < M$. Prove that $f$ is analytic at $c$, and has $\displaystyle\sum_{n=0}^{\infty} \dfrac{f^{(n)}(c)}{n!}(x-c)^n$ as its power series expansion at $c$. [*Hint* : Apply Taylor's theorem. This series is called, appropriately, the **Taylor series** of $f$ at $c$. If $c = 0$, it is also called the **Maclaurin** series of $f$.]

6.17 Prove the result of the last exercise under some weaker hypothesis about the bound on the $n^{\text{th}}$ derivative. Prove, however, that it is false if we merely assume $|f^{(n)}|$ is bounded in $U$ for all $n$.

6.18 Suppose $f$ is analytic at $c$ and not identically 0 in some neighbourhood of $c$. Assume $c$ is a zero of $f$, i.e., $f(c) = 0$. Prove that there exists a unique positive integer $m$ and an analytic function $g$ in a neighbourhood of $c$ such that $g(c) \neq 0$ and $f(x) = (x - c)^m g(x)$ holds in a neighbourhood of $c$. [*Hint* : Let $m$ be the least integer such that $f^{(m)}(c) \neq 0$. The integer $m$ here is said to be the **order** of the zero $c$ of $f$. A zero of order 1 is also called a **simple** zero.]

6.19 Suppose $f, g$ are analytic and have zeros of order $m, n$ respectively at $c$. Prove that $h = fg$ has a zero of order $m + n$ at $c$. [*Hint*: Use Exercise (4.9).]

6.20 By considering functions like $\sqrt{|x|}, x^2 \sin \dfrac{1}{x}$ and the function graphed in Fig. 4.6.1, show that analyticity is essential in the last exercise.

6.21 Verify that $\sin x$ has a simple zero at 0. What is the order of the zero at 0 of the functions $1 - \cos x$ and $\sin x - x$?

6.22 Suppose $p(x)$ is a non-zero polynomial in $x$. Prove that

(i) $\lim\limits_{x \to \infty} \dfrac{p(x)}{e^x} = 0$

(ii) $\lim\limits_{x \to 0} \dfrac{1}{p(x)e^{1/x^2}} = 0$.

[*Hint:* Note that for all $x > 0, e^x > \dfrac{x^n}{n!}$ for all $n \in I\!N$.]

6.23 Using the definition of $\ln x$ in Exercise (1.22), Exericse (9.8) and (14), show that $\dfrac{d}{dx}(\ln x) = \dfrac{1}{x}$ for all $x > 0$.

## 4.7    Derivatives of the Trigonometric Functions

We already found the derivatives of the sine and the cosine functions in (14) of the last section. But that derivation was based on the power series definitions of $\sin x$ and $\cos x$. These are hardly the usual definitions of $\sin x$ and $\cos x$. The most familiar definitions of these functions, taught even in high schools, are through geometry. In fact, these functions are among the oldest mathematical functions which had geometric definitions. The same is true about the number $\pi$. After the discovery of the wheel and the bow, it was but natural that a circle would be the most important curve (other than a straight line) for man. Its perfect symmetry and the observation that the orbits of most of the celestial bodies are (approximately) circular only served to add to the fascination about the circle. The trigonometric functions provided a convenient way to study the circle. The identity $\sin^2 \theta + \cos^2 \theta = 1$ (which is an immediate consequence of the Pythagorus theorem) is probably the oldest non-algebraic identity known to the mankind.

So it is natural to inquire if the relation $\dfrac{d}{dx}(\sin x) = \cos x$ has a purely geometric proof. Such a proof is indeed possible. But it relies heavily on geometric intuition. As remarked at the beginning of Section 3.5, such proofs cannot be a substitute for a rigorous analytical proof. But that was a situation where a purely analytical result was to be proved. When the result to be proved itself deals with concepts defined through geometric intuition, any proof of it must necessarily appeal somewhere to geometric intuition. So, with this understanding, we shall obtain the derivative of the sine function in a geometric way. In the next section, we shall see how the analytical definitions of the trigonometric functions (using power series) corelate with the geometric definitions.

It is hardly necessary to begin with the geometric definitions of the sine and the cosine functions. Familiarity with them as well as with their elementary properties (e.g. the addition formulas) will be taken for granted.

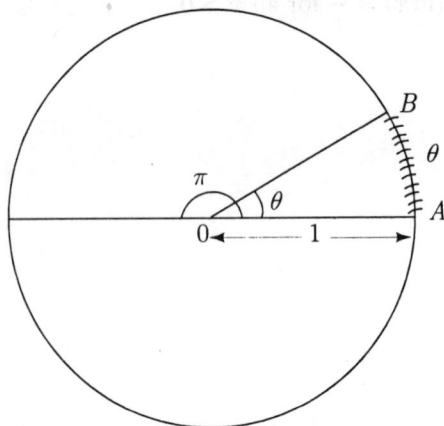

Figure 4.7.1 : Measurement of an Angle

We shall also denote the angles not by $x, y$ etc. but by Greek symbols like $\theta, \phi, \alpha, \beta$ which are more customary in trigonometry. However, some comment is in order regarding the measurement of the angles. It is a common practice while considering trigonometric functions to measure angles in radians. But how do we define a unit radian ? There are several ways to do so. The most standard method is to declare that an angle is equal to the length of the subtending arc of a unit circle, centred at the vertex of the angle (see Fig. 4.7.1). In particular an arc of length 1 of a unit circle subtends an angle of unit radian at the centre. The number $\pi$ is *defined* as half the length of the complete arc of a unit circle. In other words, $\pi$ is the angle subtended by the arc of a semi-circle. (At the end of this section, we shall briefly discuss another approach to measuring an angle, based on the area of the sector of a circle.)

Obviously, this method of measuring an angle makes sense only if we know what is meant by the length of an arc. In old days this was taken as an intuitively clear concept. More generally, the length of any curve $C$ was taken as a primitive term. A physical interpretation of it could be given as follows. Place a string along $C$. Then straighten it and measure its length. The length of $C$ is this length ! Or, place a string of uniform linear density $\rho$ (= mass per unit length) along $C$. Then its mass divided by $\rho$ will give the length of $C$.

While such methods are surely useful in *finding* the length of the curve $C$, they can hardly be taken as the definition of the length of $C$ much the same way as an experimental verification cannot be taken as a proof of a mathematical result. The first method above, for example, is based on the assumption that the length of a string remains unchanged when it is straightened. But that pre-supposes that we already have some such thing as the length of a curve.

Depending upon how we want to treat the concept of the length of the arc of a circle, the proof of $\dfrac{d}{d\theta}(\sin \theta) = \cos \theta$ takes various forms :

(I) If we take the arc length as a primitive concept then we shall have to assume as an axiom some basic property of it.

(II) At the other extreme, we can give a completely analytical definition of the length of an arc (or the length of any curve for that matter). We shall do so later when we study integration.

(III) As a via media, we can take the length of a line segment as a primitive term, as is done in the classical euclidean geometry. However, for any other curve, we can define its length as a certain supremum. This will enable us to give a more acceptable proof than in the first approach.

First, we give a proof in which the concept of an arc length is taken as a primitive term. Later we shall show how it can be modified to follow approach (III).

There is no difficulty in starting the proof. We calculate the incrementary ratio $\dfrac{\sin(\theta + \triangle\theta) - \sin\theta}{\triangle\theta}$. By a well-known trigonometric identity this comes as

$$\frac{\sin(\theta + \triangle\theta) - \sin\theta}{\triangle\theta} = \frac{2\cos(\theta + \frac{\triangle\theta}{2})\sin\frac{\triangle\theta}{2}}{\triangle\theta}$$

$$= \frac{\cos(\theta + \phi)\sin\phi}{\phi} \quad \text{(where } \phi = \frac{\triangle\theta}{2}\text{)} \qquad (1)$$

As $\triangle\theta$ tends to 0 so does $\phi$ and so $\lim\limits_{\triangle\theta \to 0} \dfrac{\sin(\theta + \triangle\theta) - \sin\theta)}{\triangle\theta}$ is the same as $\lim\limits_{\phi \to 0} \dfrac{\cos(\theta + \phi)\sin\theta}{\phi}$. Hence if we can show that

$$\cos(\theta + \phi) \to \cos\theta \text{ as } \phi \to 0 \qquad (2)$$

and

$$\frac{\sin\phi}{\phi} \to 1 \text{ as } \phi \to 0. \qquad (3)$$

then we would get

$$\frac{d}{d\theta}(\sin\theta) = \lim_{\triangle\theta \to 0} \frac{\sin(\theta + \triangle\theta) - \sin\theta}{\triangle\theta}$$

$$= \lim_{\phi \to 0} \cos(\theta + \phi) \lim_{\phi \to 0} \frac{\sin\phi}{\phi}$$

$$= \cos\theta \cdot 1$$

$$= \cos\theta \qquad (4)$$

which is the desired conclusion. So the problem is reduced to establishing (2) and (3).

Note that (2) says precisely that the cosine function is continuous everywhere, i.e., at all values of $\theta$. Geometrically, this is very obvious. Following the notation of Fig. 4.7.2, as $\phi \to 0, C$ approaches $B$ and so the projection $Q$ of $C$ on $OA$ approaches $P$, the projection of $B$ on $OA$. So $\cos(\theta + \phi) \to \cos\theta$ as $\phi \to 0$, i.e., (2) holds.

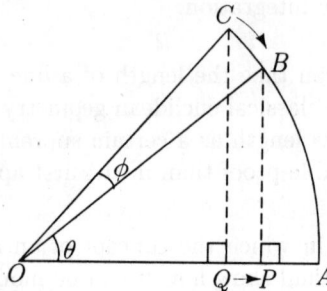

Figure 4.7.2 : Continuity of Cosine Function

The proof of (3) is not so straightforward. Since $\sin(-\phi) = -\sin(\phi)$, we have $\dfrac{\sin(-\phi)}{-\phi} = \dfrac{\sin\phi}{\phi}$ and so to establish (3), it suffices to let $\phi$ approach 0 through positive values. (Regarding the sign of an angle we follow the usual convention based upon clockwise and anticlockwise rotation.) Depending upon how we define the measure of an angle, the proof of (3) will differ. In the approach we have taken, an angle is measured by the length of the subtending arc. Then with the notations in Fig. 4.7.3, $2\phi$ equals the length of the arc $BC$ while $2\sin\phi$ equals the length of the straight line segement $BC$. Hence

$$\frac{\sin\phi}{\phi} = \frac{\text{length of segment } BC}{\text{length of arc } BC} \qquad (5)$$

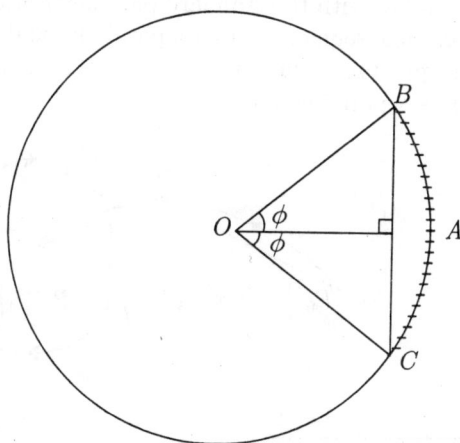

Figure 4.7.3 : Limit of $\dfrac{\sin\phi}{\phi}$ as $\phi$ tends to 0

Since the straight line distance between two points is the shortest distance between them, the ratio on the right in (5) is less than 1. However, if $\phi$ is very small, then the arc $BC$ is approximately equal to the straight line segment $BC$. And so the ratio is nearly 1. So in the limit as $B$ and $C$ both tend to $A$, it equals 1. This is a basic property of lengths applicable not just to a circle but to any curve. It is impossible to prove this statement, because we have not defined length. We are taking it as a primitive term. Later on (in Section 6.2) we shall give an analytical definition of the length of a curve and then we shall indeed prove that the ratio of the length of a portion of a curve to the distance between its end-points tends to 1 as the two points approach each other. But when length is taken as a primitive term, a statement like this has to be taken on faith. If we do so, (5) implies (3), which, as shown earlier, completes the proof that $\dfrac{d}{d\theta}(\sin\theta) = \cos\theta$.

The property of arc length just mentioned virtually amounts to taking (3) as an axiom. Let us now see what improvement would arise by following approach

(III) outlined above. That is, we take the length of a straight line segment as a primitive concept, and using it, define the length of an arc of a circle. We can in fact define the length of any curve $C$ with end points $A$ and $B$ as follows. We assume that the curve $C$ is **parametrised**, i.e., its points correspond to the values of some parameter, say, $t$. Let $A$ and $B$ correspond to $t = a$ and $t = b$. If $a < b$ we say the curve $C$ is traversed from $A$ to $B$. Every point, say $t$, of the interval $[a, b]$ determines a unique point, say, $P(t)$ of $C$. If the curve has a crossing, then two or more distinct values of $t$ give the same point of $C$. A curve with no crossings is called a **simple curve**.

(A formal definition of a parametrised curve will be given later. For a plane curve $C$, it simply means that both the $x$ and $y$ co-ordinates of a typical point on $C$ can be expressed as continuous functions of some variable, say $t$. We trust the reader is familiar with the standard parametrisations of some common curves such as straight line segments, circles, parabolas and ellipses. If a curve $C$ happens to be the graph of some function, say $y = f(x)$ then we can parametrise it by taking $x$ itself as a parameter.)

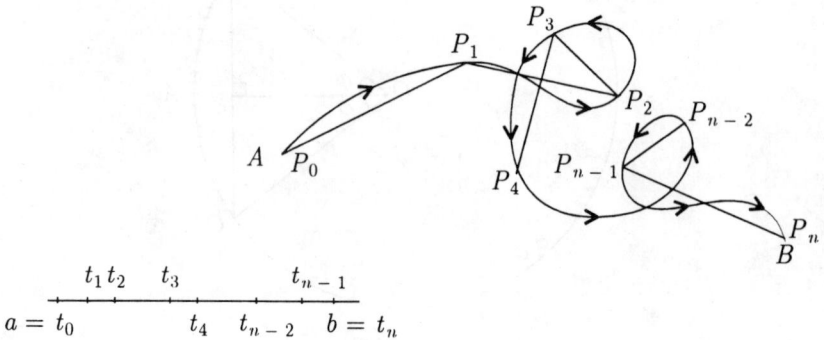

Figure 4.7.4 : Length of a Parametrised Curve

Now we can define the length of $C$ as follows. Take a finite number of points on $C$. Arrange them in the order in which they are encountered as the curve is traversed from $A$ to $B$. More precisely suppose these points are $P(t_1), P(t_2), \ldots, P(t_{n-1})$ where $a < t_1 < t_2 < \ldots < t_{n-1} < b$. For notational uniformity, set $t_0 = a, t_n = b$ and call $P(t_i)$ as $P_i$ for $i = 0, 1, \ldots, n$. The curve obtained by chaining together the straight line segments $P_0P_1, P_1P_2, \ldots, P_{n-1}P_n$ is called a **broken line path** $(P_0, P_1, \ldots, P_n)$ along $C$ from $A$ to $B$. Let $|P_{i-1}P_i|$ denote the length of the $i^{\text{th}}$ segment, $i = 1, \ldots, n$. We have not defined the length $l(C)$ of the curve $C$ yet. But it is clear that it should not be less than $\displaystyle\sum_{i=1}^{n} |P_{i-1}P_i|$ and moreover, if the adjacent points of a broken line path say $(Q_0, Q_1, \ldots, Q_m)$ along $C$ from $A$ to $B$ are mutually close, then $l(C)$ should be very close to $\displaystyle\sum_{i=1}^{m} |Q_{i-1}Q_i|$. This suggests the following definition.

**7.1 Definition :** The **length** $l(C)$ of a (parametrised) curve $C$ is the supremum of $\sum_{i=1}^{n} |P_{i-1}P_i|$, taken over the set of all broken line paths $(P_0, P_1, \ldots, P_n)$ along $C$ from $A$ to $B$. $C$ is called **rectifiable** if $l(C) < \infty$.

Before we use this definition to prove (4), a few comments are in order. First, this definition is consistent with the concept of the length of a line segment which we are taking as a primitive term. For suppose $C$ is a straight line segment from $A$ to $B$. Then for every broken line path $(P_0, P_1, \ldots, P_n)$ along $C$ from $A$ to $B$, the lengths of the segments $P_{i-1}P_i$ clearly add up to the length of the segment $AB$. So, $\sum_{i=1}^{n} |P_{i-1}P_i|$ is a constant, viz., $|AB|$ regardless of the broken line path. Hence $l(C) = |AB|$. Secondly, contrary to our intuition, not every curve is rectifiable. We have, of course not yet defined what is a curve. We shall do so later. But even with our intuitive understanding of a curve, an example can be given (Exercise (7.7)) of a non-rectifiable curve. The proof that its length is not finite rests crucially on the divergence of the harmonic series $\sum_{n=1}^{\infty} \frac{1}{n}$, another fact defying our intuition !

Let us now turn to proving (3). The result will fall out as an immediate consequence of the following inequality which is of some independent interest.

**7.2 Theorem :** If $\phi$ is the measure of an acute angle (i.e., if $0 < \phi < \frac{\pi}{2}$) then $\sin \phi \leq \phi \leq \tan \phi$.

**Proof :**

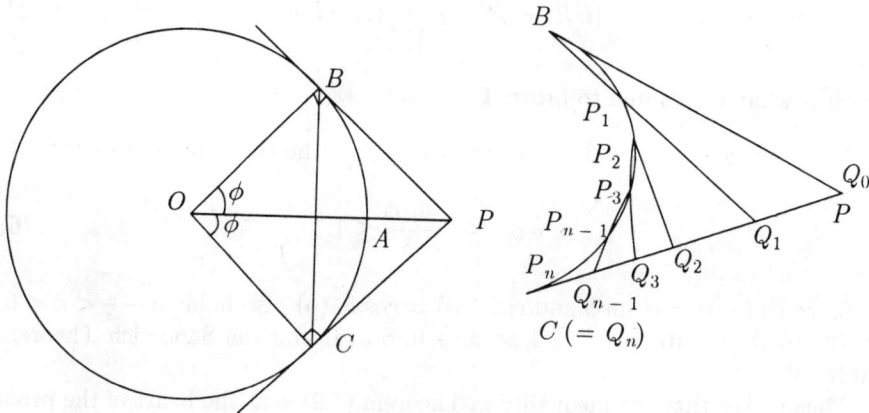

Figure 4.7.5 : A Trigonometric Inequality

Draw a circle of radius 1 centred at $O$ and let $\angle BOA$ and $\angle AOC$ be each equal to $\phi$. Let $BC$ meet $OA$ at $Q$. The tangents to the circle at $B$ and $C$ must meet at a point, say P, on $OA$ extended (see Fig. 7.4.5). Clearly $|BC| = 2|BQ| = 2\sin \phi$ while, by definition, $2\phi$ equals the length of the arc $\overset{\frown}{BC}$. From the very definition of the length, $l(\text{arc } BC) \geq |BC|$. So the first part of the inequality holds.

For the second part we note that $|BP| = |PC| = \tan\phi$, and so we would be done if we can show that the length of the arc $BC$ is at most $|BP| + |PC|$. From definition (7.1), this is equivalent to showing that for every broken line path, say, $(P_0, P_1, \ldots, P_n)$ from $B$ to $C$ along the arc $BC$, the sum $\sum_{i=1}^{n} |P_{i-1}P_i|$ is at most $|BP| + |PC|$.

For any such broken line path, let $P_{i-1}P_i$ meet $PC$ at $Q_i$ for $i = 1, \ldots, n$. Then the points $Q_1, Q_2, \ldots, Q_n (= B)$ lie on the segment $PC$ as we go from $P$ to $C$ (see the second part of Fig. 4.7.5 which is drawn a little distortedly for better illustration). So $\sum_{i=1}^{n} |Q_{i-1}Q_i| = |PC|$ where $Q_0$ is taken as $P$.

Now applying the triangle inequality successively to the triangles $BPQ_1 (= BQ_0Q_1), P_1Q_1Q_2, P_2Q_2Q_3, \ldots$ and finally to $P_{n-1}Q_{n-1}Q_n$, we have

$$
\begin{aligned}
|BP| + |PQ_1| &\geq |BP_1| + |P_1Q_1| = |P_0P_1| + |P_1Q_1| \\
|P_1Q_1| + |Q_1Q_2| &\geq |P_1P_2| + |P_2Q_2|
\end{aligned}
$$

$$\vdots$$

$$|P_{n-1}Q_{n-1}| + |Q_{n-1}Q_n| \geq |P_{n-1}P_n|$$

Let us add these inequalities and cancel the terms $|P_1Q_1|, \ldots, |P_{n-1}Q_{n-1}|$ from both the sides. Since $Q_1, Q_2, \ldots, Q_{n-1}$ lie in that order on the line segment $PC$, the lengths $|PQ_1|, |Q_1Q_2|, \ldots, |Q_{n-1}Q_n|$ add up to $|PC|$ and we get

$$|BP| + |PC| \geq \sum_{i=1}^{n} |P_{i-1}P_i|$$

which is what we wanted to prove. ∎

It is now easy to establish (3). For $0 < \phi < \frac{\pi}{2}$, the theorem above gives

$$\cos\phi \quad \leq \quad \frac{\sin\phi}{\phi} \leq 1. \tag{6}$$

Since $\sin(-\phi) = -\sin\phi$ and $\cos(-\phi) = \cos\phi$, (6) also holds if $-\frac{\pi}{2} < \phi < 0$. By (2) (with $\theta = 0$), $\cos\phi \to 1$ as $\phi \to 0$. So (6) and the Sandwich Theorem imply (3).

Thus we see that the inequality in Theorem (7.2) is at the heart of the proof of (4). It is instructive to mention an alternate derivation of this inequality. So far we have measured angles in terms of arc lengths. But they can also be measured in terms of areas of sectors. Specifically we say that the measure of an angle is $\theta$ (radians) if the area of the sector of a unit circle whose radii include that angle is $\frac{1}{2}\theta$. In this approach the area of a unit circle is $\pi$ by *definition* of $\pi$. In the earlier approach this would need a proof. In the new approach the price we have to pay is that we can no longer take the arc length of the unit circle as $2\pi$. This will have to be proved.

With this new approach, the inequality in Theorem (3) can be established simply by observing that in Fig. 4.7.5, the triangle $OBQ$ is a part of the sector $OAB$, which, in turn, is contained in the triangle $OBP$. So their areas satisfy :

$$\triangle OBQ \leq \text{ area (sector } OAB) \leq \triangle OBP. \tag{7}$$

The middle term is $\frac{1}{2}\theta$ by definition. The first and the last terms come out to be respectively, $\frac{1}{2}\sin\theta$ and $\frac{1}{2}\tan\theta$. So Theorem (7.2) holds.

This approach takes the area of a plane figure as an intuitively clear concept. To give it a rigorous foundation, we would have to start with the area of a rectanlge as the product of its sides and define the area of an arbitrary figure as the supremum of some sort just as we started with the length of a segment and defined the length of a curve as a certain supremum. This indeed can be done. But the verification that the area so defined coincides with the earlier definition for rectangles (and hence for triangles) is not as easy as the proof of the corresponding assertion about lengths of line segments. As with lengths, a rigorous definition of areas will be given later (see Chapter 6, Sections 2 and 3).

Having proved that $\frac{d}{d\theta}(\sin\theta) = \cos\theta$, it is now easy to find the derivatives of the remaining trigonometric functions. Since $\cos\theta = \sin(\frac{\pi}{2} - \theta)$, the chain rule immediately gives $\frac{d}{d\theta}(\cos\theta) = -\sin\theta$. The derivatives of other trinonometric functions can be obtained from those in the usual manner (cf. Exercise (6.7)).

## EXERCISES

7.1 Suppoe $P = (x_1, y_1)$ and $Q = (x_2, y_2)$ are two distinct points on a straight line $L$. Prove that $L$ can be parametrised by $x = x_1 + t(x_2 - x_1) = (1-t)x_1 + tx_2$ and $y = y_1 + t(y_2 - y_1) = (1-t)y_1 + ty_2$, where $t$ runs over all real numbers. Verify that if $t$ varies from 0 to 1, we get the segment from $P$ to $Q$.

7.2 Identify the curves parametrised by the following systems of equations by eliminating the parameter $t$ in each case.

    (i)   $x = a + r\cos t, \ y = b + r\sin t$   (ii)   $x = at^2, \ y = 2at$
    (iii)  $x = a\cos t, \ y = b\sin t$       (iv)   $x = a\cosh t, \ y = b\sinh t$.

7.3 Assuming that the length of a unit circle is $2\pi$, show that the length of a circle of radius $r$ is $2\pi r$. [*Hint :* Every broken line path along one circle corresponds to a broken line path along the other.]

7.4 Generalise the last exercise as follows. Let $C$ be a curve parametrised by $x = f(t), y = g(t), a \leq t \leq b$. Let $\alpha, \beta, r$ be some constants with $r > 0$. Let $C'$ be the curve parametrised by $x = rf(t) + \alpha, y = rg(t) + \beta, a \leq t \leq b$. Prove that $C'$ is rectifiable if and only if $C$ is rectifiable and further that $l(C') = rl(C)$.

**7.5** Prove that the length of the perimeter of a regular $n$-gon inscribed in a unit circle is $2n \sin \dfrac{\pi}{n}$ and that this tends to $2\pi$ as $n \to \infty$. Can this be taken as a proof that the length of the unit circle is $2\pi$?

**7.6** Evaluate :

(i) $\displaystyle\lim_{x \to 0} \dfrac{1 - \cos x}{x}$      (ii) $\displaystyle\lim_{x \to 0} \dfrac{1 - \cos(x/2)}{x^2}$      (iii) $\displaystyle\lim_{x \to 0} \dfrac{1 - \cos x}{x \sin x}$

(iv) $\displaystyle\lim_{x \to \frac{\pi}{4}} \dfrac{1 - \tan x}{x - \frac{\pi}{4}}$      (v) $\displaystyle\lim_{x \to 0} \dfrac{\cos(\sin x) - 1}{x^2}$      (vi) $\displaystyle\lim_{x \to 0+} \dfrac{\sqrt{x + \sin x}}{\sqrt{\sin x + \sin^2 x}}$ .

**7.7** Let $C$ be the graph of the function $y = f(x)$ where

$$f(x) = \begin{cases} x \sin(1/x) & \text{if } 0 < x \le \frac{1}{\pi} \\ 0 & \text{if } x = 0. \end{cases}$$

For every positive integer $n$, let $P_n$ and $Q_n$ be, respectively, the points $\left( \dfrac{1}{2n\pi}, 0 \right)$ and $\left( \dfrac{2}{(4n+1)\pi}, \dfrac{2}{(4n+1)\pi} \right)$ . Prove that:

(i)     $P_n$ and $Q_n$ lie on $C$ for all $n$.

(ii)    $|P_n Q_n| \ge \dfrac{2}{(4n+1)\pi}$

(iii)   The length of the broken line path $(P_1, Q_1, P_2, Q_2, \ldots, P_n, Q_n)$ is at least $\displaystyle\sum_{k=1}^{n} \dfrac{2}{(4n+1)\pi}$

(iv)   The curve $C$ is not rectifiable.

**7.8** In Theorem (7.2), show that both the inequalities are strict. [*Hint* : Apply the theorem to $\dfrac{\phi}{2}$ instead of $\phi$.]

**7.9** Using theorem (7.2) show that $3 < \pi < 4$. Obtain a better estimate using smaller values of $\phi$. Also show that $\dfrac{\sin \theta}{\theta}$ decreases strictly on $\left( 0, \dfrac{\pi}{2} \right)$. Hence prove the comment made in Exercise (2.1.3).

**7.10** Let $a_n = 2^n \sqrt{2 - \sqrt{2 + \sqrt{2 + \sqrt{2 + \sqrt{2} + \ldots}}}}$ there being in all $n$ radical signs. Prove that $a_n \to \pi$ as $n \to \infty$. This gives a method for approximating $\pi$ if we can compute square roots. [*Hint* : Show that for $n \ge 2$,

$$\cos \dfrac{\pi}{2^n} = \dfrac{1}{2} \sqrt{2 + \sqrt{2 + \sqrt{2 + \sqrt{2} + \ldots}}}$$ there being $n - 1$ radical signs.]

## 4.8 Advantages of the Power Series Definitions of $\sin x$ and $\cos x$

In the last two sections we saw that the derivatives of the sine and cosine functions can be obtained in two ways depending upon how they are defined, i.e., whether geometrically or using power series. In this section, we shall go further and show that the two definitions are completely equivalent. It would then be natural to ask what are the merits and demerits of each.

Since $\frac{d}{d\theta}(\sin \theta) = \cos \theta$ and $\frac{d}{d\theta}(\cos \theta) = -\sin \theta$, it follows that both $\sin \theta$ and $\cos \theta$ are infinitely differentiable over the entire real line. Moreover, the derivatives keep recurring in cycles of length 4. That is, $\frac{d^n}{d\theta^n}(\sin \theta) = \frac{d^m}{d\theta^m}(\sin \theta)$ whenever the difference $n - m$ is a multiple of 4 and similarly for the derivatives of $\cos \theta$. (More specifically, $\frac{d^n}{d\theta^n}(\sin \theta) = \sin \theta$ if $n$ is of the form $4k$ for a non-negative integer $k$, $\frac{d^n}{d\theta^n}(\sin \theta) = \cos \theta$ if $n = 4k + 1$, $\frac{d^n}{d\theta^n}(\sin \theta) = -\sin \theta$ if $n = 4k + 2$ and $\frac{d^n}{d\theta^n}(\sin \theta) = -\cos \theta$ if $n = 4k + 3$.) Moreover, $|\sin \theta|$ and $|\cos \theta|$ are bounded by 1 for all $\theta \in \mathbb{R}$. (In the geometric approach, this is immediate from the definition. In the definition using power series this follows as a consequence of Exercise (3.10.3).) This information coupled with several other results we have proved so far, enables us to show that the geometric and the analytic (i.e., the power series) definitions of the trigonometric functions are equivalent.

**8.1 Theorem :** For any angle $\theta$, we have

$$\sin \theta = \theta - \frac{\theta^2}{3!} + \frac{\theta^5}{5!} - \frac{\theta^7}{7!} + \ldots = \sum_{n=0}^{\infty} \frac{(-1)^n \theta^{2n+1}}{(2n + 1)!} \qquad (1)$$

and

$$\cos \theta = 1 - \frac{\theta^2}{2!} + \frac{\theta^4}{4!} - \frac{\theta^6}{6!} + \ldots = \sum_{n=0}^{\infty} \frac{(-1)^n \theta^{2n}}{(2n)!}. \qquad (2)$$

**Proof :** Assume $\sin \theta$ and $\cos \theta$ are defined geometrically. Let $f(\theta) = \sin \theta$. As noted above $f$ has derivatives of all order. Moreover, $|f^{(n)}(\theta)|$ always equals either $|\sin \theta|$ or $|\cos \theta|$. Hence $|f^{(n)}(\theta)| \leq 1$ for all $n \in \mathbb{N}$ and for all $\theta \in \mathbb{R}$.

Now fix $\theta > 0$. Let $n \in \mathbb{N}$. By Taylor's theorem (Theorem (5.6)), there exists some $c_n \in (0, \theta)$ such that

$$f(\theta) = \sin \theta = f(0) + f'(\theta)\theta + \frac{1}{2!}f''(0)\theta^2 +$$

$$+ \cdots + \frac{f^{(n-1)}(0)}{(n-1)!}\theta^{n-1} + \frac{f^{(n)}(c_n)}{n!}\theta^n \qquad (3)$$

Regardless of what $c_n$ is, $|f^{(n)}(c_n)| \le 1$. So, we have for every $n \in I\!N$,

$$\left| \sin\theta - \sum_{k=0}^{n-1} \frac{f^{(k)}(a)}{k!} \theta^k \right| \le \frac{\theta^n}{n!}. \tag{4}$$

But $\dfrac{\theta^n}{n!} \to 0$ as $n \to \infty$, for every fixed $\theta$ (Exercise (3.9.4).) So the infinite series $\displaystyle\sum_{k=0}^{\infty} \frac{f^{(k)}(0)}{k!}\theta^k$ converges to $\sin\theta$. Since $f^{(2k)}(0) = 0$ and $f^{(2k+1)}(0) = (-1)^k$ for $k = 0, 1, \ldots$, we see that (1) holds for all positive $\theta$. That it also holds for $\theta < 0$ follows from $\sin(-\theta) = -\sin\theta$. For $\theta = 0$, (1) is immediate.
An essentially similar argument establishes (2). ∎

This theorem is important both practically and theoretically. Because of the rapid growth of factorials, it provides very rapidly converging series for $\sin\theta$ and $\cos\theta$, thereby enabling us to calculate these functions with a high degree of accuracy within relatively few steps. Because of the identities $\sin\left(\dfrac{\pi}{2} + \theta\right) = \cos\theta$ and $\cos\left(\dfrac{\pi}{2} + \theta\right) = -\sin\theta$, it suffices to calculate $\sin\theta$ and $\cos\theta$ only for $0 \le \theta \le \dfrac{\pi}{2}$. Further, the identity $\sin\left(\dfrac{\pi}{2} - \theta\right) = \cos\theta$, allows us to suppose $0 \le \theta \le \dfrac{\pi}{4}$. Since $\dfrac{\pi}{4} < 1$, it follows that for $\theta \in \left[0, \dfrac{\pi}{4}\right], \theta^n \to 0$ as $n \to \infty$ (Theorem (3.3.4)). So the convergence of (1) and (2) is even more rapid. As a concrete example, suppose we want the value of $\sin 34°$. We could, of course, draw a right angled triangle with one angle equal to 34 degrees. But to measure its sides we would need highly sophisticated instruments to get even a modest accuracy (say upto three places of decimals). But with (1) the job is easy. Since $34° = \dfrac{34}{180} \times \pi$ radians $\simeq 0.5934$ radians, we apply (1) with $\theta = .5934$. Since $\theta^7 < (.6)^7 < 0.0285$ and $\dfrac{1}{7!} = \dfrac{1}{840} < .0012$, we see from (1) (and the estimate for the remainder in the Taylor's theorem) that $\theta - \dfrac{\theta^3}{6} + \dfrac{\theta^5}{120}$ will give an accuracy of upto four places of decimals.
So, Theorem (8.1) is important practically. But it also has a theoretical significance. It shows that the geometric and the analytic definitions of the trigonometric functions coincide with each other. The geometric definitions are obviously easier to understand than the power-series definitions, because they appeal to our geometric intuition, which is, after all an undeniably valuable asset in applications of mathematics. Ironically, it is this very feature of these definitions which makes them unacceptable in a perfectionistic approach, where everything has to be defined in a rigorous manner without appealing to intuition. As we saw in Section 2.3, it is this insistence which motivated the present definition of a limit instead of the one based on the intuitive concept of an infinitesimal. So, from this point of view, the power series definitions of sine and cosine are superior to the geometric ones, because the latter are based on

the concept of an angle which is intuitively clear but hard to define analytically.

In fact, the power series definitions of the trigonometric functions can be used to *define* the measure of an angle analytically. The rest of the section will be devoted to elaborate how this is done.

As Theorem (8.1) shows, we can start from the geometric definitions of $\sin\theta$ and $\cos\theta$. After we prove their continuity and obtain their derivatives by geometric arguments, we can use the powerful machinery of calculus, especially the Taylor's theorem (which, in turn, is based on Rolle's theorem. See Exercise (5.23)) and ultimately get their power series expansions. It is interesting that this procedure can be turned backwards. That is, we can start with the power series definitions of the sine and the cosine and end up showing that geometrically they represent the ratios of certian sides of suitable right angled triangles. In this approach, absolutely nothing is left to geometric intuition. Everything is defined rigorously in terms of real numbers. For example, the euclidean plane is *defined* as the set $\mathbb{R} \times \mathbb{R}$ (also denoted by $\mathbb{R}^2$) consisting of all ordered pairs of real numbers. If $P = (x_1, x_2)$ and $Q = (y_1, y_2)$ are any two elements (or 'points') of the euclidean plane then the **(euclidean) distance** between them (denoted by $d(P,Q)$ or $|PQ|$) is defined as $\sqrt{(x_1 - y_1)^2 + (x_2 - y_2)^2}$. If $P \neq Q$, then the **( straight) line** passing through $P$ and $Q$ is defined as the set $\{(z_1, z_2) :$ there is some $t \in \mathbb{R}$ such that $z_i = x_i + t(y_i - x_i), i = 1, 2, \}$. The **segment** $PQ$ is defined as the set $\{(z_1, z_2) :$ there is some $t \in [0,1]$ such that $z_i = x_i + t(y_i - x_i), i = 1, 2.\}$ (cf. Exercise (7.1)). Three distinct points $P, Q, R$ are said to be **collinear** if each is in (or 'lies on') the line passing through the remaining two.

A few elementary results based on these definitions will be given as exercises. The first serious difficulty in this analytical approach is encountered when it comes to defining an angle. Suppose $P = (x_1, x_2), Q = (y_1, y_2)$ and $R = (z_1, z_2)$ are three distinct points. Let $\theta$ be the angle between the segments $PQ$ and $PR$. From elementary geometry, we know

$$\cos\theta = \frac{(y_1 - x_1)(z_1 - x_1) + (y_2 - x_2)(z_2 - x_2)}{\sqrt{(y_1 - x_1)^2 + (y_2 - x_2)^2}\sqrt{(z_1 - x_1)^2 + (z_2 - x_2)^2}} \tag{5}$$

In the analytical approach, we would like to reverse this definition. That is, we would like to follow the power series definitions of sine and cosine and would like to define $\theta$ as the number whose cosine is the expression on the right of (5). By Exercise (3.10.3), we know that both the sine and cosine are bounded by 1. So, first of all, it must be shown that the expression on the R.H.S. of (5) is bounded by 1 in absolute value. This is very easy to do by sqaring the numerator and the denominator. But it also follows as a consequence of a more general result, called **Cauchy Schwarz inequality**. It is one of the most famous and useful inequalities in mathematics.

**8.2 Theorem :** If $u_1, u_2, \ldots, u_n$ and $v_1, \ldots, v_n$ are any real numbers then

$$|u_1 v_1 + u_2 v_2 + \ldots + u_n v_n| \leq \sqrt{u_1^2 + u_2^2 + \ldots + u_n^2}\sqrt{v_1^2 + v_2^2 + \ldots + v_n^2}$$

$$\tag{6}$$

Further, equality holds if and only if there exist some $\alpha, \beta \in \mathbb{R}$, not both 0 such that $\alpha u_i + \beta v_i = 0$ for all $i = 1, \ldots, n$.

**Proof :** Define $f : \mathbb{R} \longrightarrow \mathbb{R}$ by

$$f(t) \;=\; \sum_{i=1}^{n} (u_i + tv_i)^2 \tag{7}$$

When expanded,

$$f(t) \;=\; At^2 + 2Bt + C \tag{8}$$

where $A = \sum_{i=1}^{n} v_i^2, B = \sum_{i=1}^{n} u_i v_i$ and $C = \sum_{i=1}^{n} u_i^2$.

The leading coefficient, $A$ of this quadratic is positive except when all $v_i$'s are 0, in which case the assertion of the theorem holds trivially. So assume $A > 0$. From (7), $f(t) \geq 0$ for all $t$. Hence $Af(t) \geq 0$ for all $t$. From (8) and completing squares,

$$Af(t) = (At + B)^2 + (AC - B^2)$$

In particular, for $t = -\dfrac{B}{A}, Af(t) = AC - B^2$. So $AC - B^2 \geq 0$, which is exactly the inequality in the assertion. If equality holds, then $Af(t)$ and hence $f(t)$ vanishes for $t = -\dfrac{B}{A}$. But from (7), vanishing of $f(t)$ forces $u_i + tv_i = 0$ for every $i$. So $\alpha u_i + \beta v_i = 0$ for all $i$, where $\alpha = 1$ and $\beta = -\dfrac{B}{A}$. Conversely, if $\alpha u_i + \beta v_i = 0$ holds for all $i = 1, \ldots, n$, then it is trivial to check that equality holds in (6). For, if say, $\alpha \neq 0$, then $u_i = kv_i$ for $i = 1, 2, \ldots, n$ where $k = -\dfrac{\beta}{\alpha}$ and both the sides of (6) equal $|k| \left( v_1^2 + v_2^2 + \ldots + v_n^2 \right)$. ∎

**8.3 Corollary :** The expression, say $R$, on the right in (5) always lies in the interval $[-1, 1]$.

**Proof :** Simply apply the Cauchy-Schwarz inequailty with $n = 2$ and $u_i = y_i - x_i, v_i = z_i - x_i$ for $i = 1, 2$. Note that since $P$ is different from $Q$ and also from $R$, the denominator is not zero. ∎

The next task is to show that for every $R$ in $[-1, 1]$, there is some $\theta$ such that $\cos \theta = R$. And finally, we must either show that such $\theta$ is unique, and if not, we must indicate how we choose it.

So, if we have to define geometry analycially, we must closely study how the cosine function behaves. Now that we are again defining the sine and cosine analytically (by Exercise (3.3.14)), we shall use $x, y$ etc. instead of $\theta, \phi$ etc. The following theorem does the spadework for the study of the cosine function.

**8.4 Theorem :** Let $\sin x = \sum\limits_{n=0}^{\infty} \dfrac{(-1)^n x^{2n+1}}{(2n+1)!}$ and $\cos x = \sum\limits_{n=0}^{\infty} \dfrac{(-1)^n x^{2n}}{(2n)!}$ for $x \in$ $I\!\!R$. Then $-1 \leq \cos x \leq 1$ for all $x \in I\!\!R$. The set $\{x \in I\!\!R : x > 0, \cos x = 0\}$ is non-empty and has a smallest element. (Worded differently, the cosine function has a smallest positive zero.)

**Proof :** The first part follows from the identity $\sin^2 x + \cos^2 x = 1$ for all $x$ (Exercise (3.10.3)). For the second part, let $S$ be the set in question. Assume $S = \phi$, i.e., $\cos x \neq 0$ for all $x > 0$. Now, $\cos 0 = 1$. Moreover, the cosine function is continuous everywhere. For this, we have to apply Corollary (1.6) and not (2) in the last section. (Why ?) So by the Intermediate Value Property (Theorem (1.3)), if $\cos z < 0$ for some $z > 0$, then $S \neq \phi$ contrary to the assumption. So we have $\cos x > 0$ for all $x > 0$. By (14) in Section 4.6, $\dfrac{d}{dx}(\sin x) = \cos x$. So by Theorem (5.5), $\sin x$ is a strictly increasing function of $x$ on $(0, \infty)$. Since $\sin 0 = 0$, we conclude that $\sin x > 0$ for all $x > 0$. Hence, again by Theorem (5.5), $\cos x$ is strictly decreasing on $(0, \infty)$.

Now to reach a contradiction fix any $a > 0$. For every $b > a$, we have, by Lagrange's Mean Value Theorem,

$$\cos b - \cos a \;=\; -(b - a)\sin c \text{ for some } c \in (a, b) \tag{9}$$

Since $\sin x$ is strictly increasing (and positive) and $\cos x$ is strictly decreasing and positive on $(0, \infty)$, we have $\sin c > \sin a > 0$ and $0 < \cos b < \cos a < 1$. So (9) gives,

$$
\begin{aligned}
\sin a \;&<\; \frac{\cos a - \cos b}{b - a} \\[2mm]
&<\; \frac{1}{b - a} \text{ for all } b > a. 
\end{aligned}
\tag{10}
$$

Since $\sin a$ is a fixed positive real number, (10) gives a contradiction for sufficiently large $b$, (specifically if $b > a + \sin a$). Hence the assumption that $S \neq \phi$ was wrong.

It only remains to prove that $S$ has a least element. In any case, $S$ is bounded below (by 0 for example). So by Theorem (2.6.3) along with Axiom (2.6.2), $S$ has an infimum, say $b$. By the result of Exercise (2.6.13), there exists a sequence $\{x_n\}$ in $S$ which converges to $b$. By continuity of the cosine function and Theorem (1.2), $\cos(x_n) \to \cos b$ as $n \to \infty$. But $\cos(x_n) = 0$ for all $n$. So $\cos b = 0$. Clearly $b \neq 0$ (since $\cos 0 = 1$) and so $b > 0$. Hence $b \in S$. Thus $b$ is the smallest element of $S$. ∎

The preceding proof, although not very long, is remarkable in the multitude of non-trivial results it uses. In the geometric approach we know that $\dfrac{\pi}{2}$ is the smallest positive zero of the cosine function. Now that we have established the existence of such a zero in a purely analytical manner, we can give a purely analytical definition of $\pi$.

**8.5 Definition :** The number $\pi$ is double the smallest positive zero of the cosine function.

So $\cos\left(\dfrac{\pi}{2}\right) = 0$ by definition. Since $\cos 0 = 1 > 0$ and cosine is a continuous function, we get that $\cos x > 0$ for all $x \in \left[0, \dfrac{\pi}{2}\right)$, as otherwise the Intermediate Value Property would give a positive zero of the cosine function which is even smaller than $\dfrac{\pi}{2}$. So by Theorem (5.5), $\sin x$ is a strictly increasing function on $[0, \frac{\pi}{2}]$. In particular $\sin\dfrac{\pi}{2} > 0$. But since $\cos\dfrac{\pi}{2} = 0$, we already have $\sin\dfrac{\pi}{2} = 1$ or $-1$. So we conclude $\sin\dfrac{\pi}{2} = 1$. It is now easy to show that $\cos\left(\dfrac{\pi}{2} - x\right) = \sin x, \cos\left(\dfrac{\pi}{2} + x\right) = -\sin x$ and many other familiar results about trigonometrical properties of $\pi$. Here we prove only the one which is needed in order to define the angle between two lines.

**8.6 Theorem :** For every $R$ in $[-1, 1]$, there is exactly one value of $\theta \in [0, \pi]$ such that $\cos\theta = R$.

**Proof :** We have $\cos 0 = 1$ and $\cos\pi = \cos^2\dfrac{\pi}{2} - \sin^2\dfrac{\pi}{2} = 0 - 1 = -1$. So by the Intermediate Value Property, for every $R \in [-1, 1]$ there is at least one $\theta \in [0, \pi]$ such that $\cos\theta = R$. For uniqueness, it would suffice to show that the cosine function is strictly decreasing on $[0, \pi]$, which in turn would follow if we can show that $\sin x > 0$ for all $x \in (0, \pi)$. We already showed this for $x \in \left(0, \dfrac{\pi}{2}\right]$. Now suppose $x \in \left(\dfrac{\pi}{2}, \pi\right)$. Put $y = \pi - x$. Then $y \in \left(0, \dfrac{\pi}{2}\right)$ and so, as observed above, $\sin y > 0$. But then $\sin x = \sin(\pi - y) = \sin\pi\cos y - \cos\pi\sin y = \sin y > 0$. So $\sin x > 0$ for all $x \in (0, \pi)$, and as shown above, this completes the proof. ∎

Going back to (5), we have already shown that the expression on the right lies in $[-1, 1]$. The theorem above now enables us to define the angle between $PQ$ and $PR$. Once the basic concepts of distance and angle are defined analytically, all other concepts in geometry can be defined analytically. In particular, it can be shown that $\sin\theta$ and $\cos\theta$ have their geometric meanings (Exercise (8.18)).

Thus, geometry can be comletely subsumed by real numbers. A whole book can be written on euclidean geometry without drawing a single diagram ! While pedagogically this is undesirable, the logical perfection reached does appeal to some people. A diagram can be an aid to but not a substitute for a rigorous proof. An improperly drawn diagram can, in fact, lead to a wrong result. A good example of this will be given in Exercise (8.11).

Another advantage of the analytical approach is that it generalises easily from $\mathbb{R}^2$ (i.e., $\mathbb{R} \times \mathbb{R}$) to $\mathbb{R}^n$ (i.e., the product $\mathbb{R} \times \mathbb{R} \times \ldots \times \mathbb{R}, n$ times) for any positive integer $n$. For $n = 2$ and $3$ we get the geometries of the plane and the space respectively and we can visualise these. For $n > 3$, we cannot visualise $\mathbb{R}^n$. But that is no reason to suppose that it has no real life meaning. Think of a worm which lives and moves only in a small part of the surface of a large sphere. For the worm, the world will appear as a replica of $I\!\!R^2$. If it keeps on moving in a particular direction and comes back to where it started then, of course, it would realise that its world is not a 'flat' $I\!\!R^2$, but something curved. If the worm is intelligent, then this realisation can come to it even without undertaking such a long journey. Similarly, it is possible that the world we humans live in is not $I\!\!R^3$ but something curved sitting in $I\!\!R^4$. The great mathematicain and scientist Gauss, did, in fact, try to find out experimentally if this is indeed so. With the crude measurements possible in his time, he concluded that our world was flat. Subsequent experiments show that it is curved!

There are also other practical reasons to study the euclidean spaces $I\!\!R^n$ for $n > 3$. Instead of dealing with a set of $n$ mutually independent real numbers, it is often convenient to deal with a single point of $I\!\!R^n$. Examples of this will be seen later (see Section 6.5.).

While dealing with the geometry of $I\!\!R^n$, it is convenient to use the notation and terminology of vectors. We are *not* studying vector spaces here. For our purpose an **n-dimensional vector** will simply mean a point of $I\!\!R^n$, i.e., an ordered $n$-tuple of real numbers. Such vectors will be denoted by bold face letters $\mathbf{x}, \mathbf{y}, \mathbf{u}, \mathbf{v}, \mathbf{a}, \mathbf{b}$ etc. If $\mathbf{x} = (x_1, x_2, \ldots, x_n)$ then for $i = 1, \ldots, n, x_i$ is called the $i^{\text{th}}$ **component** (or coordinate) of $\mathbf{x}$. The **zero vector 0** is the vector each of whose components is 0. (Note incidentally, that for $m \neq n$, the zero vector in $I\!\!R^m$ is different from that in $I\!\!R^n$.) If we let $O$ be the 'origin' of $I\!\!R^n$, i.e., the point $(0, 0, \ldots, 0)$, then for any other point, say $P = (u_1, \ldots, u_n)$ in $I\!\!R^n$, the vector $\mathbf{u}$ may be identified informally with the 'position vector' of $P$, i.e., the 'directed line segment' from $O$ to $P$. We are not defining these terms formally. But they enable us to (privately) visualise, for our own convenience, what goes on.

We now define some basic operations on vectors in $I\!\!R^n$.

**8.7 Definition :** Let $\mathbf{u} = (u_1, u_2, \ldots, u_n)$ and $\mathbf{v} = (v_1, \ldots, v_n)$ be two vectors in $I\!\!R^n$ and let $\lambda$ be a real number (sometimes called a **scalar**, so as to distinguish it from a vector). Then,

(i)   the **sum** of $\mathbf{u}$ and $\mathbf{v}$, denoted by $\mathbf{u} + \mathbf{v}$ is defined as the vector $(u_1 + v_1, u_2 + v_2, \ldots, u_n + v_n)$

(ii)  the **scalar multiple** of $\mathbf{u}$ by $\lambda$ is defined as the vector $(\lambda u_1, \lambda u_2, \ldots, \lambda u_n)$, denoted by $\lambda \mathbf{u}$,

(iii) the **scalar product**, or the **dot product**, or the **inner product** of $\mathbf{u}$ and $\mathbf{v}$ is defined as the number $u_1 v_1 + u_2 v_2 + \ldots + u_n v_n$ and denoted by $\mathbf{u} \cdot \mathbf{v}$ (or sometimes by $\langle \mathbf{u}, \mathbf{v} \rangle$) and

(iv)  the **length** or **norm** of $\mathbf{u}$, denoted by $\|\mathbf{u}\|$ or by $|\mathbf{u}|$, is defined as the non-negative square root of $\displaystyle\sum_{i=1}^{n} u_i^2$ (which also equals $\mathbf{u} \cdot \mathbf{u}$).

A few basic properties of these concepts will be indicated in the exercises. Note that the Cauchy-Schwarz inequality assumes the compact form that for all $\mathbf{u}, \mathbf{v}$ in $I\!R^n$, $|\mathbf{u} \cdot \mathbf{v}| \le |\mathbf{u}||\mathbf{v}|$. If $P, Q$ are points of $I\!R^n$ with position vectors $\mathbf{u}, \mathbf{v}$ respectively, then $d(P, Q)$ or $|PQ|$ as defiend earlier is precisely the length of the vector $\mathbf{v} - \mathbf{u}$ (i.e., the vector $\mathbf{v} + (-1)\mathbf{u}$).

Summing up, we need the power series expansions of $\sin x$ and $\cos x$ to give their accurate values. But to *define* them as power series serves a perfectionistic purpose. It purges the plane geometry of the intuitive element in it and allows us to extend it to $I\!R^n$ for any $n$.

We conclude by remarking that the power series are not the only means of defining the sine and the cosine functions analytically. Just as the exponential function can be defined by first defining the natural logarithm function in terms of integrals and then taking its inverse, the sine function can also be defined by first defining the inverse sine function in terms of a certain integral. This will be briefly indicated in Section 5.10 (see Exercise (5.10.24)).

## EXERCISES

8.1 Complete the proof of Theorem (8.1) by establishing (2).

8.2 Using Theorem (8.1) and a calculator which does only arithmetical operations, find $\cos 72°$ correctly upto four places of decimals. Compare with the exact value of $\cos 72°$, viz., $\dfrac{\sqrt{5} - 1}{4}$.

8.3 Prove that if $\theta$ is sufficiently small and positive, then $\theta - \dfrac{\theta^3}{6} < \sin\theta < \theta$ and $1 - \dfrac{\theta^2}{2} < \cos\theta < 1 - \dfrac{\theta^2}{2} + \dfrac{\theta^4}{24}$. Generalise this result.

8.4 Let $P = (x_1, x_2)$ and $Q = (y_1, y_2)$ be two points in $I\!R^2$. Define $L(P, Q) = \{(z_1, z_2) \in I\!R^2 : \exists\, t \in I\!R \text{ s.t. } z_i = x_i + t(y_i - x_i), i = 1, 2\}$.

Prove that $P, Q \in L(P, Q)$ and $L(P, Q) = L(Q, P)$. Prove further that $L(P, Q) = \{P\}$ if $P = Q$ while if $P \ne Q$, then for every $R = (z_1, z_2) \in L(P, Q)$, there exists a unique $t$ such that $z_i = x_i + t(y_i - x_i)$. In other words, every line in $I\!R^2$ can be thought of as a replica of the real line.

8.5 Let $P, Q, R, S \in I\!R^2$ with $P \ne Q$ and $R \ne S$. Prove that :

(i) $L(P, Q) = L(R, S)$ if and only if $R, S \in L(P, Q)$

(ii) if $L(P, Q) \ne L(R, S)$ then $L(P, Q) \cap L(R, S)$ has at most one point. What does this mean geometrically ?

More precisely, suppose $P = (x_1, x_2), Q = (y_1, y_2), R = (z_1, z_2)$ and $S = (w_1, w_2)$. Then $L(P, Q) \cap L(R, S)$ is empty iff the determinant

$$\begin{vmatrix} y_1 - x_1 & w_1 - z_1 \\ y_2 - x_2 & w_2 - z_2 \end{vmatrix}$$

is zero. Otherwise find the unique point of intersection of $L(P, Q)$ and $L(R, S)$. (If the first possibility holds, the lines $L(P, Q)$ and $L(R, S)$ are called **parallel**.)

8.6 Let $P = (x_1, x_2), Q = (y_1, y_2)$ and $R = (x_1 + t(y_1 - x_1), x_2 + t(y_2 - x_2))$ where $t \in \mathbb{R}$. Prove that $|PR| = |t||PQ|$ and $|RQ| = |1 - t||PQ|$. What happens if $t = \dfrac{1}{2}$? Also, find $R$ given that $|PR| : |RQ| = \lambda : \mu$ where $\lambda, \mu$ are positive real numbers.

8.7 (a) Given three distinct points, say, $P = (x_1, x_2), Q = (y_1, y_2)$ and $R = (z_1, z_2)$ prove that $P, Q, R$ are collinear if and only if the determinant

$$\begin{vmatrix} y_1 - x_1 & z_1 - x_1 \\ y_2 - x_2 & z_2 - x_2 \end{vmatrix}$$

vanishes. Hence show that in the definition of collinearity, it does not matter which of the three points lies on the line passing through the other two points. That is, if $P$ is in $L(Q, R)$, then $Q$ is in $L(P, R)$ and $R$ is in $L(P, Q)$.

(b) Prove analytically that the diagonals of a parallelogram bisect each other. (This illustrates how a geometric result can be proved in the analytical approach. All other geometric results can be similarly proved with varying degrees of difficulty.)

8.8 A function $f : \mathbb{R} \longrightarrow \mathbb{R}$ is said to be **periodic** with a period $T$ ($> 0$) if $f(x + T) = f(x)$ for all $x \in \mathbb{R}$. Note that if $T$ is a period of $f$ so is every whole multiple of $T$. Prove that

(i) $\sin x$ and $\cos x$ are periodic with $2\pi$ as a period, while $|\sin x|$ and $|\cos x|$ are periodic with $\pi$ as a period. (This is obvious with the geometric definitions of $\sin x$ and $\cos x$. Give analytical proofs based on Theorem (8.4) and Definition (8.5).)

(ii) For $\alpha \neq 0, \sin(\alpha x)$ and $\cos(\alpha x)$ are periodic with a period $\frac{2\pi}{|\alpha|}$

(iii) If $\alpha$ is a rational number, then $\cos x + \cos \alpha x$ is periodic.

(iv) If $\alpha$ is irrational then $\cos x + \cos \alpha x$ is not periodic. [*Hint* : There is only one value of $x$ for which this function equals 2.] Thus, the sum of two periodic functions need not be periodic.

(Periodic functions are especially important in engineering because the motion of a wheel is often periodic.)

*8.9 Prove that every continuous, non-constant, periodic function $f$ has a smallest positive period $T^*$ and that every period of $f$ is a multiple of $T^*$.($T^*$ is called *the* period of $f$.) Prove that $2\pi$ is the period of $\sin x$ and also of $\cos x$.

8.10 For $0 < x < \dfrac{\pi}{2}$, prove that $\sin(\cos x) < \cos(\sin x)$.

8.11 What is wrong in the following argument which purports to show that every triangle $ABC$ is equilateral ?

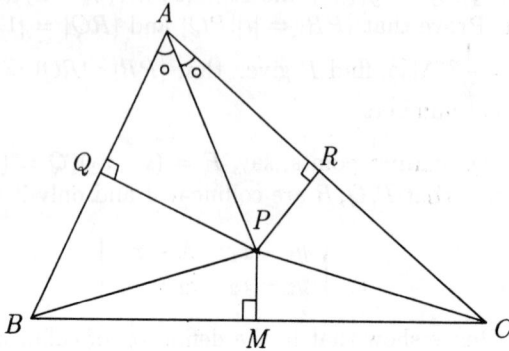

Figure 4.8.1 : Every Triangle is Equilateral!

Let the angle bisector of $\angle A$ meet the perpendicular bisector of the segment $BC$ at $P$. Draw perpendiculars $PQ$ and $PR$ to $AB, AC$ respectively (Fig. 4.8.1). Join $PB, PC$. The right angled trianlges $PAQ$ and $PAR$ are congrument as $\angle PAQ = \angle PAR$ and the side $PA$ in common. So $PQ = PR$ and $AQ = AR$. Also $PB = PC$ since $P$ lies on the perpendicular bisector of $BC$. Hence the right angled triangles $PQB$ and $PRC$ are also congruent. So $QB = RC$. Thus $AB = AQ + QB = AR + RC = AC$. Similarly $BC = AC$.

8.12 For every two vectors $\mathbf{u}$, $\mathbf{v}$ in $\mathbb{R}^n$ and any two scalars $\alpha, \beta$, prove the following properties and give them suitable names.

(i)   $\mathbf{u} + \mathbf{v} = \mathbf{v} + \mathbf{u}$
(ii)  $\alpha(\mathbf{u} + \mathbf{v}) = \alpha\mathbf{u} + \alpha\mathbf{v}$
(iii) $(\alpha + \beta)\mathbf{u} = \alpha\mathbf{u} + \beta\mathbf{u}$
(iv)  $0\mathbf{u} = \mathbf{0}$
(v)   $1\mathbf{u} = \mathbf{u}$
(vi)  $\mathbf{u} + (-\mathbf{u}) = \mathbf{0}$
(vii) $\alpha(\beta\mathbf{u}) = (\alpha\beta)\mathbf{u}.$

(All of these are, of course, trivial to establish and hence can be used without explicit mention. The only reason they are listed here is that by taking them as axioms, one can define the concept of an abstract vector space, just as by taking certain properties of the euclidean distance, one can define an abstract metric space, see Exercise (1.9.5).)

8.13 Prove that the dot product of vectors in $\mathbb{R}^n$ has the following properties for all $\mathbf{u}, \mathbf{v}, \mathbf{w} \in \mathbb{R}^n$ and $\alpha \in \mathbb{R}$ :

(i) $\mathbf{u} \cdot \mathbf{v} = \mathbf{v} \cdot \mathbf{u}$

(ii) $\mathbf{u} \cdot \mathbf{u} \geq 0$ with equality holding if and only if $\mathbf{u} = \mathbf{0}$.

(iii) $\mathbf{u} \cdot (\alpha\mathbf{v}) = (\alpha\mathbf{u}) \cdot \mathbf{v} = \alpha(\mathbf{u} \cdot \mathbf{v})$

(iv) $\mathbf{u} \cdot (\mathbf{v} + \mathbf{w}) = \mathbf{u} \cdot \mathbf{v} + \mathbf{u} \cdot \mathbf{w}$.

(Again, these properties are trivial but basic for defining an abstract inner product. Because of them the dot product is amenable to the familiar manipulations, e.g., $(\mathbf{u} + \mathbf{v}) \cdot (\mathbf{u} - \mathbf{v}) = |\mathbf{u}|^2 - |\mathbf{v}|^2$.)

8.14 Give both the statement and the proof of Theorem (8.2) in terms of the dot product on $\mathbb{R}^n$.

8.15 From the Cauchy-Schwarz inequality, deduce that for all $\mathbf{u}, \mathbf{v}$ in $\mathbb{R}^n$, $|\mathbf{u} + \mathbf{v}| \leq |\mathbf{u}| + |\mathbf{v}|$. When does equality hold ? Using it, define the angle between two non-zero vectors $\mathbf{u}, \mathbf{v}$ in $\mathbb{R}^n$.

8.16 Using the result of the last exercise prove that for any three points $P, Q, R$ in $\mathbb{R}^n$, $|PR| \leq |PQ| + |QR|$. (Because of this, the inequality in the last exercise is called the **triangle inequality**, cf. Exercise (1.9.9).)

8.17 Prove that the concepts of distance and angle are **translation invariant** i.e., suppose $P, Q, R, P', Q', R'$ are points with position vectors $\mathbf{u}, \mathbf{v}, \mathbf{w}$, $\mathbf{u}', \mathbf{v}', \mathbf{w}'$ respectivley. Suppose $\mathbf{u}' - \mathbf{u} = \mathbf{v}' - \mathbf{v} = \mathbf{w}' - \mathbf{w} = \mathbf{a}$ (say), i.e., $\mathbf{u}', \mathbf{v}', \mathbf{w}'$ are obtained by 'translating' $\mathbf{u}, \mathbf{v}, \mathbf{w}$ respectively each by the vector $\mathbf{a}$. Show that $|PQ|, |QR|, |PR|$ equal respectively $|P'Q'|, |Q'R'|$ and $|P'R'|$ and further that the angle between $PQ$ and $PR$ equals the angle between $P'Q'$ and $P'R'$. (This result enables us to choose the origin at will.)

8.18 Let $P, Q, R$ be three non-collinear points in $\mathbb{R}^2$ and suppose $PQ$ and $QR$ are at right angles. Prove that :

(i) $|PQ|^2 + |QR|^2 = |PR|^2$

(ii) $\dfrac{|PQ|}{|PR|} = \cos\theta$ where $\theta$ is the angle between $PQ$ and $PR$ as defined by (5)

(iii) $\sin\theta = \dfrac{|QR|}{|PR|}$.

[*Hint* : By the last exercise, $Q$ may be taken to be $(0, 0)$.]

((i) is, of course, the good old Pythagorus theorem while (ii) and (iii) are the starting points of the classical geometric approach to trigonometry. This exercise shows how all these things can be done analytically.)

## 4.9    L'Hôpital's Rule - a Useful but not a Golden Tool

There is something peculiar about the L'Hôpital's rule, apart from its name (which invariably invites a giggle from a beginner because of its resemblance with 'hospital', a most unlikely word to be heard in a mathematics course) which gives it a somewhat undeserved popularity. In terms of depth or ubiquitous applicability, L'Hôpital's rule ranks nowhere close to, say, the Lagrange's Mean Value Theorem. In fact many authors either do not mention it, or cover it only as a peripheral topic. But ask an average student who has recently left mathematics for good after a couple of courses in calculus. And chances are that he remembers the L'Hôpital's rule more than the Lagrange's MVT.

The reason perhaps lies in the purpose of applications. By its very nature, Lagrange's theorem is an existence theorem in that it merely gives the existence of something without giving an explicit construction for it. As we have already seen several times (e.g. Theorems (5.5), (6.2) and (8.4)), the very existence of such a mean value is theoretically important. But, for a beginner, it is not very exciting. He feels there is real stuff in what he is doing only when he calculates something or proves some formula rather than when he merely proves the existence of something. For example, a rigorous proof of the convergence of

$\sum_{n=1}^{\infty} \dfrac{1}{n^2}$ is not as appealing to him as is the fact that this sum equals $\dfrac{\pi^2}{6}$, even

though this may be done only by a plausible argument (see Exercise (1.4.6)). As elaborated in Section 2.7, one of the reasons the early calculus appears dull to a beginner is its lack of exciting problems resembling those in the pre-calculus mathematics.

On this background, the L'Hôpital's rule comes as a welcome change. As a typical problem based on it suppose we want to find $\lim_{x \to 0} \dfrac{\sin x - x}{x^3}$. Although evaluating the limit of a sequence or the sum of a series can be quite challenging, evaluation of limits of the form $\lim_{x \to c} f(x)$ is often an unexciting task. In fact when $f$ is continuous at $c$, this limit is simply $f(c)$. Inasmuch as many naturally occurring functions are continuous, this often serves to give the beginner the wrong impression that $\lim_{x \to c} f(x)$ is obtained simply by plugging the value $c$ into the formula for $f(x)$. There are of course, functions whose limits cannot be evaluated so simply. Such examples generally fall into two categories.

(I)   : some rather contrived examples such as $\lim_{x \to 0} x \sin \dfrac{1}{x}$ or those where $f(x)$ is defined by one formula on some subset $S$ of $\mathbb{R}$ and by another on the complement of $S$.

(II)  : examples wehre $f(x)$ is a ratio of two functions, say, $f(x) = \dfrac{g(x)}{h(x)}$ and $h(x) \to 0$ as $x \to c$, so that a direct application of the theorem for the limit of a ratio becomes impossible.

Examples of Type I are instructive but not very appealing. In Type II examples, if the numerator, i.e., $g(x)$, tends to $\infty$ or to some finite non-zero limit as $x \to c$, then $\lim\limits_{x \to c} \dfrac{g(x)}{h(x)}$ does not exist. But if $g(x)$ also tends to 0 as $x \to c$, the situation becomes interesting. Depending upon which of the two functions $g(x)$ and $h(x)$ tends to 0 more rapidly, the fate of $\lim\limits_{x \to c} f(x)$ changes. Especially interesting are the cases where the ratio $\dfrac{g(x)}{h(x)}$ tends to a finite, non-zero limit (which is also expressed by saying that $g(x)$ and $h(x)$ approach 0 equally rapidly) as $x \to c$. These cases can be further classified as follows :

(i)   the 'vanishing factor' $(x - c)$ (or some power of it) can be cancelled out both from the numerator and the denominator, either directly or with some algebraic manipulation, e.g., $\lim\limits_{x \to 2} \dfrac{x^3 - 8}{x^2 - 4}$ or $\lim\limits_{x \to 4} \dfrac{x - 4}{\sqrt{x} - 2}$

(ii)   limits of the type $\lim\limits_{x \to 0} \dfrac{\sin x}{x}$ or $\lim\limits_{x \to \frac{\pi}{4}} \dfrac{\tan^2 x - 1}{x^2 - \frac{\pi}{4} x}$, where the vanishing factor cannot be cancelled out, but the limit can be evaluated by recognising it or some part of it as a derivative of some function.

(iii)   limits of the form $\lim\limits_{x \to 0} \dfrac{x - \sin x}{x^3}$ where neither of the two methods above work. Naturally, such limits are more challenging.

It is limits of this last form where L'Hôpital's rule sometimes provides a very handy method. As we shall see, the answer comes with relatively easy work. Moreover, we get the actual answer and not just the existence of the limit. And herein probably lies the reason why L'Hôpital's rule appeals so much to a beginner.

There are various forms of the L'Hôpital's rule. Each of them deals with some limit which cannot be found or **determined** with the usual methods. For this reason, they are called **indeterminate forms**. The form which we just considered, viz. $\lim\limits_{x \to c} \dfrac{g(x)}{h(x)}$ where both $g(x)$ and $h(x)$ tend to 0 as $x \to 0$ is called an indeterminate form of type $0/0$. Similarly there are indeterminate forms of the types $\infty/\infty$ or $0.\infty$ or $1^\infty$. An example of the last indeterminate form is $\lim\limits_{n \to \infty} \left(1 + \dfrac{1}{n}\right)^n$ or $\lim\limits_{x \to 0} (1 + x)^{1/x}$. Similarly the limit of the sequence $n^{1/n}$ provides an example of an indeterminate form of type $\infty^0$.

We begin with the study of the $\dfrac{0}{0}$ form. So suppose $f(x) = \dfrac{g(x)}{h(x)}$ where $g, h$ are defined in a neighbourhood of a point $c$ and $\lim\limits_{x \to c} g(x) \doteq \lim\limits_{x \to c} h(x) = 0$. Clearly, this is equivalent to setting $g(c) = h(c) = 0$ and requiring that $g, h$ be continuous at $c$. Suppose now $g$ and $h$ are also differentiable at $c$. Then for $x \neq c$ we can write

$$f(x) = \frac{g(x)}{h(x)} = \frac{g(x)/(x - c)}{h(x)/(x - c)} = \frac{(g(x) - g(c))/(x - c)}{(h(x) - h(c))/(x - c)} \tag{1}$$

As $x \to c$, $\dfrac{g(x)}{x-c}$ and $\dfrac{h(x)}{x-c}$ tend to $g'(c)$ and $h'(c)$ by definition. Hence if $h'(c) \neq 0$, we see from (1) that

$$\lim_{x \to c} \frac{g(x)}{h(x)} = \frac{g'(c)}{h'(c)} \tag{2}$$

This is, of course, a very trivial result, requiring nothing besides the definition of a derivative. Note that we are requiring $g'$ and $h'$ to exist only at $c$ (and not necessarily in some neighbourhood of $c$).

If we assume further that $g'$ and $h'$ are continuous at $c$ (whcih, in particular also means they exist in some neighbourhood of $c$), then $g'(c) = \lim_{x \to c} g'(x)$ and $h'(c) = \lim_{x \to c} h'(x)$. If $h'(c) \neq 0$, then by Exercise (1.1), applied to $h'$, we get that $h'(x) \neq 0$ for all $x$ in some neighbourhood $N$ of $c$. But then the ratio $\dfrac{g'(x)}{h'(x)}$ is defined in $N$ and approaches the ratio $\dfrac{g'(c)}{h'(c)}$ as $x \to c$. So, we get,

**9.1 Theorem :** Suppose $g, h$ are real-valued functions defined in a neighbourhood of a point $c$ with $g(c) = 0 = h(c)$. If $g', h'$ are continuous at $c$ and $h'(c) \neq 0$ then

$$\lim_{x \to c} \frac{g(x)}{h(x)} = \lim_{x \to c} \frac{g'(x)}{h'(x)} = \frac{g'(c)}{h'(c)}. \tag{3}$$

This theorem is sometimes called the '**weak form**' of **L'Hôpital's rule.** The name is not inappropriate because its proof involves hardly anything besides the definitions of continuity and derivatives (and the general properties of limits). In fact, it may be argued that the additional hypothesis of Theorem (9.1) (viz., the continuity of $g'$ and $h'$ at $c$) is superfluous because, ultimately it gives the same conclusion as (2), which was derived merely from existence of $g'$ and $h'$ at $c$. This point is valid if we look at only the first and the last term of (3). But the real meat of the preceding theorem lies in the middle term of (3). It is the relationship between the first two terms of (3) which is the heart of L'Hôpital's rule. This relationship is of a recursive type. In essence it says that whenever you want to find the limit of a ratio of two functions and it is of the form $\dfrac{0}{0}$, try the ratio of their derived functions. But what if this new ratio is also of the form $\dfrac{0}{0}$? Then take the ratio of *their* derivatives, i.e., the second derivatives of the original functions. We keep repeating this process till we reach a higher order derivative of $h$ which does not vanish at $c$. (There is, however, no guarantee that this process will always terminate. Examples will be given in the exercises.) More precisely, we have the following theorem.

**9.2 Theorem :** Suppose $g, h$ are real-valued functions defined in a neighbourhood $N$ of a point $c$. Suppose for some positive integer $n$, both $g^{(n)}$ and $h^{(n)}$ are continuous at $c$. Assume further that

$$\text{(i) } g^{(k)}(c) = 0 = h^{(k)}(c) \text{ for all } k, 1 \leq k < n$$
$$\text{and} \quad \text{(ii) } h^{(n)}(c) \neq 0.$$

Then $\lim\limits_{x \to c} \dfrac{g(x)}{h(x)} = \dfrac{g^{(n)}(c)}{h^{(n)}(c)}$.

**Proof :** The case $n = 1$ is covered by Theorem (9.1). Assume now that $n = r > 1$ and that the assertion holds for all pairs of functions for $n = r - 1$. In particular we can apply it to the functions $g'$ and $h'$ to get

$$\lim_{x \to c} \frac{g'(x)}{h'(x)} = \frac{g'^{(n-1)}(c)}{h'^{(n-1)}(c)} \tag{4}$$

Now, $h'^{(n-1)}(c) = h^{(n)}(c) \neq 0$ by assumption. Similarly, $g'^{(n-1)}(c) = g^{(n)}(c)$. So the first equality of (3) along with (4) completes the inductive step and thereby proves the result. ∎

As an immediate application, we can now evaluate $\lim\limits_{x \to 0} \dfrac{x - \sin x}{x^3}$ mentioned above. Here $c = 0, g(x) = x - \sin x$ and $h(x) = x^3$. We begin by differentiating $g(x)$ and $h(x)$. $\lim\limits_{x \to 0} \dfrac{g'(x)}{h'(x)} = \lim\limits_{x \to 0} \dfrac{1 - \cos x}{3x^2}$ is of the $\dfrac{0}{0}$ form. Differentitating again, $\lim\limits_{x \to 0} \dfrac{g''(x)}{h''(x)} = \lim\limits_{x \to 0} \dfrac{-\sin x}{6x}$ which is still of the $\dfrac{0}{0}$ form. One more differentiation gives the limit as $-\dfrac{1}{6}$. (For the last part we can also use (4) in Section 4.7 and avoid differentiation.)

A few more examples of computations of such limits will be given as exercises. Let us now turn to the so-called strong form of L'Hôpital's rule. In Theorem (9.1), the second limit, viz., $\lim\limits_{x \to c} \dfrac{g'(x)}{h'(x)}$ exists because $\lim\limits_{x \to c} g'(x)$ and $\lim\limits_{x \to c} h'(x)$ both exist and the latter is non-zero. But there are situations where the ratio of two functions may have a limit even without either the numerator or the denominator having a limit. A trivial example is the ratio of $2\sin\dfrac{1}{x}$ and $\sin\dfrac{1}{x}$ as $x \to 0$. So it is natural to inquire if (3) still holds, it being only given that $\lim\limits_{x \to c} \dfrac{g'(x)}{h'(x)}$ exists. The answer is in the affirmative but far from trivial. The proof will be based on the Cauchy's Mean Value Theorem (Exercise (5.11)), which is a slight generalisation of the Lagrange's MVT.

**9.3 Theorem : (L'Hôpital's rule, strong form) :** Suppose $g, h$ are continuous in a neighbourhood $N$ of a point $c$ and are differentiable in the deleted neighbourhood $N - \{c\}$ of $c$. Assume further that $g(c) = 0 = h(c)$ but that $h$ does not vanish in $N - \{c\}$ and $\lim\limits_{x \to c} \dfrac{g'(x)}{h'(x)} = L$. Then $\lim\limits_{x \to c} \dfrac{g(x)}{h(x)}$ also equals $L$.

**Proof :** We shall first show that $\lim_{x \to c^+} \dfrac{f(x)}{g(x)} = L$. Let $x \in N$ and $x > c$. Since $h(x) \neq 0$ while $h(c) = 0$, we can apply Cauchy's MVT (Exercise (5.11)) to the pair $(g, h)$ (instead of $(f, g)$ as in Exercise (5.11)) to get some $c(x)$ (depending on $x$), in the interval $(c, x)$ such that

$$\frac{g(x)}{h(x)} = \frac{g(x) - g(c)}{h(x) - h(c)} = \frac{g'(c(x))}{h'(c(x))}. \tag{5}$$

As $x \to c^+, c(x) \to c^+$. So $\lim_{x \to c^+} \dfrac{g'(c(x))}{h'(c(x))}$ is the same as $\lim_{x \to c^+} \dfrac{g'(x)}{h'(x)}$. (To see this, merely observe that for $\delta > 0, c < x < c + \delta$ implies $c < c(x) < \delta$.) But the latter equals $L$. So by (5) $\lim_{x \to c^+} \dfrac{g(x)}{h(x)} = L$. A similar argument works for the left-handed limit, $\lim_{x \to c^-} \dfrac{g(x)}{h(x)}$, except that, we apply the Cauchy MVT to the interval $[x, c]$ rather than to $[c, x]$. Since both limits equals $L$ the proof is complete. ∎

It is instructive to compare Theorem (9.1) and Theorem (9.3). The latter uses a non-trivial theorem, which ultimately is based on the completeness of reals (see the comments in Section 4.5) while the former is very elementary. In actual problems involving limits, the weak form is often sufficient. But the strong form is needed in certain theoretical applications. Using it, we shall give a proof of Theorem (3.3). But first we prove an extension of Theorem (9.3).

**9.4 Theorem :** Suppose the real-valued functions $g, h$ have derivatives of order $n - 1$ in some neighbourhood $N$ of a point $c$ and derivatives of order $n$ in the deleted neighbourhood $N - \{c\}$. Assume further that for all $0 \le k < n, g^{(k)}(c) = 0 = h^{(k)}(c)$ and that $\lim_{x \to c} \dfrac{g^{(n)}(x)}{h^{(n)}(x)}$ exists. Then $\lim_{x \to c} \dfrac{g(x)}{h(x)}$ also exists and equals $\lim_{x \to c} \dfrac{g^{(n)}(x)}{h^{(n)}(x)}$.

**Proof :** This is derived from Theorem (9.3), exactly the same way as Theorem (9.2) was derived from Theorem (9.1). ∎

We are now ready to prove Theorem (3.3) along with a sort of converse to it. If $f'(c)$ exists, then we know that the function $f(c) + f'(c)(x - c)$ is a linear approximation to $f(x)$ in a neighbourhood of $c$. Note that this approximating function is a polynomial of degree 1 (unless $f'(c) = 0$, in which case it is a constant, i.e. a polynomial of degree 0). Theorem (3.3) says that if $f^{(n)}(c)$ exists then in a neighbourhood of $c$, $f(x)$ can be approximated by a polynomial in $x$ of degree $n$ (or less). We shall prove this and moreover also show that such a polynomial is unique.

**9.5 Theorem :** If $f^{(n)}(c)$ exists where $n \ge 1$, then we can write
$f(x) = q(x) + R_n(x)$, where

$$q(x) = f(c) + f'(c)(x - c) + \frac{1}{2}f''(c)(x - c)^2 + \ldots + \frac{1}{n!}f^{(n)}(c)(x - c)^n$$

and $\dfrac{R_n(x)}{(x - c)^n} \to 0$ as $x \to 0$. Moreover suppose $p(x)$ is a polynomial of degree $n$

(or less) such that $\dfrac{f(x) - p(x)}{(x - c)^n} \to 0$ as $x \to c$, then $p(x) = q(x)$.

**Proof[1]** : The existence of $f^{(n)}(c)$ implies that all derivatives of $f$ upto the $(n-1)^{th}$ order exist in some neighbourhood $N$ of $c$. Since $(x - c), (x - c)^2, \ldots,$ $(x - c)^n$ are infinitely differentiable everywhere, if we let

$$g(x) \;\; = \;\; f(x) - f(c) - f'(c)(x - c) - \cdots - \frac{f^{(n-1)}(c)}{(n - 1)!}(x - c)^{n-1} \qquad (6)$$

and

$$h(x) \;\; = \;\; (x - c)^n \qquad (7)$$

we see that both $g$ and $h$ have derivatives upto order $n - 1$ in the neighbourhood $N$ of $c$. A direct calculation shows that

$$g^{(n-1)}(x) \;\; = \;\; f^{(n-1)}(x) - f^{(n-1)}(c) \text{ for all } x \in N \qquad (8)$$

and

$$h^{(n-1)}(x) \;\; = \;\; n!(x - c) \text{ for all } x \in N \text{ ( in fact for all } x \in \mathbb{R}). \qquad (9)$$

Also at $c$, the derivatives of $g$ as well as of $h$ of all orders upto the $(n - 2)^{th}$ vanish. (In fact, the $(n - 1)^{th}$ derivatives also vanish. But we shall not need it.) Moreover, since $f^{(n)}(c)$ is given to exist, from (8) and (9) we get

$$\lim_{x \to c} \frac{g^{(n-1)}(x)}{h^{(n-1)}(x)} = \lim_{x \to c} \frac{f^{(n-1)}(x) - f^{(n-1)}(c)}{n!(x - c)} = \frac{1}{n!}f^{(n)}(c) \qquad (10)$$

We now apply Theorem (9.4) (with $n$ in its statement replaced by $n - 1$) and get from (10) that

$$\lim_{x \to c} \frac{g(x)}{h(x)} \;\; = \;\; \frac{1}{n!}f^{(n)}(c) \qquad (11)$$

or equivalently,

$$\lim_{x \to c} \frac{g(x) - \frac{1}{n!}f^{(n)}(c)h(x)}{h(x)} \;\; = \;\; 0 \qquad (12)$$

In view of (6) and (7), (12) says precisely that $\dfrac{R_n(x)}{(x - c)^n} \to 0$ as $x \to c$ where $R_n(x)$ is defined as $g(x) - \dfrac{1}{n!}f^{(n)}(c)(x - c)^n$.

---

[1]I am indebted to B.V. Limaye for this proof of the direct implication.

For the converse, we first note that if $p(x)$ is a polynomial of degree $n$ (or less) and $c \in \mathbb{R}$, then there exist unique $a_0, a_1, \ldots, a_n \in \mathbb{R}$ such that, for all $x \in \mathbb{R}$,

$$p(x) = a_0 + a_1(x - c) + a_2(x - c)^2 + \ldots + a_n(x - c)^n. \qquad (13)$$

(A proof of this fact will be indicated in Exercise (9.7).)

Now suppose

$$\frac{f(x) - p(x)}{(x - c)^n} \to 0 \text{ as } x \to 0. \qquad (14)$$

For $r = 0, 1 \ldots, n$ define

$$g_r(x) = f(x) - a_0 - a_1(x - c) - \cdots - a_r(x - c)^r \qquad (15)$$

and

$$h_r(x) = (x - c)^r \text{(with the understanding that } h_0 \equiv 1) \qquad (16)$$

Clearly, for every $r = 0, 1, \ldots, n$

$$\frac{g_r(x)}{h_r(x)} = \frac{f(x) - p(x)}{(x - c)^r} + a_{r+1}(x - c) + a_{r+2}(x - c)^2 + \cdots + a_n(x - c)^{n-r}$$

$$= \frac{f(x) - p(x)}{(x - c)^n}(x - c)^{n-r} + a_{r+1}(x - c) + \cdots + a_n(x - c)^{n-r}$$

So by (14) we see that

$$\frac{g_r(x)}{h_r(x)} \to 0 \text{ as } x \to c \text{ for every } r = 0, 1, \ldots, n. \qquad (17)$$

For $r = 0$, this simply means $f(x) \to a_0$ as $x \to c$. But by continuity of $f$ at $c, f(x) \to f(c)$ as $x \to c$. So,

$$a_0 = f(c). \qquad (18)$$

Hence $g_1(x) = f(x) - f(c) - a_1(x - c)$. Note that $g_1(c) = 0 = h_1(c)$. We apply Theorem (9.2) (with $n = 1$ in its statement), to get

$$\lim_{x \to c} \frac{g_1(x)}{h_1(x)} = \frac{g_1'(c)}{h_1'(c)} = \frac{f'(c) - a_1}{1}$$

But by (17), $\dfrac{g_1(x)}{h_1(x)} \to 0$ as $x \to c$. So

$$a_1 = f'(c) \qquad (19)$$

Hence $g_2(x) = f(x) - f(c) - f'(c)(x - c) - a_2(x - c)^2$. Applying Theorem (9.2) (with $n = 2$ this time) to $g_2$ and $h_2$ we get

$$\lim_{x \to c} \frac{g_2(x)}{h_2(x)} = \frac{g_2''(c)}{h_2''(c)} = \frac{f''(c) - 2a_2}{2}$$

which, because of (17) implies

$$a_2 = \frac{1}{2} f''(c) \tag{20}$$

We now repeat this procedure and show successively, by applying Theorem (9.2) everytime to a different pair of functions (and with a different value of $n$), that

$$a_r = \frac{1}{r!} f^{(r)}(c) \text{ for } r = 0, 1, \ldots, n - 1. \tag{21}$$

However, if $r = n$, we cannot apply Theorem (9.2), because $f^{(n)}(x)$ and hence $g_n^{(n)}(x)$ is given to exist only at $c$. Here we need the stronger Theorem (9.4) (with $n$ in its statement replaced by $n - 1$). Because of (21), we have that $g_n^{(r)}(c) = 0$ for $r = 0, 1, \ldots, n - 1$, and

$$g_n^{(n-1)}(x) = f^{(n-1)}(x) - f^{(n-1)}(c) - n! a_n(x - c) \tag{22}$$

for all $x$ in the neighbourhood $N$ of $c$.

Since $f^{(n)}(c)$ is given to exist, from (22) we see that

$$\lim_{x \to c} \frac{g_n^{(n-1)}(x)}{h_n^{(n-1)}(x)} = \lim_{x \to c} \frac{f^{(n-1)}(x) - f^{(n-1)}(c) - n! a_n(x - c)}{n!(x - c)}$$

$$= \frac{f^{(n)}(c)}{n!} - a_n \tag{23}$$

Hence by Theorem (9.4),

$$\lim_{x \to c} \frac{g_n(x)}{h_n(x)} = \frac{1}{n!} f^{(n)}(c) - a_n.$$

But by (17), $\lim\limits_{x \to c} \dfrac{g_n(x)}{h_n(x)} = 0$.

So $a_n = \dfrac{1}{n!} f^{(n)}(c)$. Thus (21) holds for $r = n$ too. In view of (13), this completes the proof. ∎

Because of this theorem, existence of $f^{(n)}(c)$ implies that $f$ can be approximated near $c$ by a polynomial of degree $n$. A strict converse of this statement would be that if $f$ can be so approximated then $f^{(n)}(c)$ exists. But this turns out not to be the case (see Exercise (9.8)).

Theorem (9.3) also holds if $c$ is $\infty$ or $-\infty$. Of course, in that case, continuity at $c$ makes little sense and the hypothesis that $g(c) = 0 = h(c)$ has to be replaced by $\lim_{x \to c} g(x) = 0 = \lim_{x \to c}$ (which was, in fact, our original hypothesis). This is a minor difficulty. The real difficulty is that Cauchy's MVT, which was crucially used in the proof of Theorem (9.3), has no analogue for intervals of the form $(x, \infty)$ or $(-\infty, x)$. Still, the proof can be salvaged by applying it to intervals of the form $(x, y)$ for $y$ sufficiently large (or to $(y, x)$ with $y$ sufficiently small) as we now show. We do only the case $c = \infty$ and leave the other as an exercise.

**9.6 Theorem :** Assume $g, h$ are differentiable in some deleted neighbourhood $N$ of $\infty$ (i.e., in an interval of the form $(R, \infty)$ for some $R$) and $\lim_{x \to \infty} g(x)$ and $\lim_{x \to \infty} h(x)$ are both 0. Assume further that $h'(x) \neq 0$ for all $x \in N$ and that $\lim_{x \to \infty} \dfrac{g'(x)}{h'(x)} = L$. Then $\lim_{x \to \infty} \dfrac{g(x)}{h(x)}$ also exists and equals $L$.

**Proof :** The hypothesis $h'(x) \neq 0$ in $(R, \infty)$, ensures by Lagrange's MVT that $h$ is one-to-one in $(R, \infty)$. So Cauchy's MVT is applicable for every interval $[x, y]$ with $R < x < y < \infty$.

Now, let $\epsilon > 0$ be given. There exists some $x_0$ such that for all $z > x_0$,

$$L - \frac{\epsilon}{2} < \frac{g'(z)}{h'(z)} < L + \frac{\epsilon}{2}. \tag{24}$$

We claim that for all $x > x_0$,

$$L - \frac{\epsilon}{2} \leq \frac{g(x)}{h(x)} \leq L + \frac{\epsilon}{2} \tag{25}$$

which would, of course prove the theorem since $L - \dfrac{\epsilon}{2} > L - \epsilon$ and $L + \dfrac{\epsilon}{2} < L + \epsilon$. To prove (25), fix any $x > x_0$. For every $y > x$, we get by Cauchy's MVT,

$$\frac{g(x) - g(y)}{h(x) - h(y)} = \frac{g'(z)}{h'(z)}$$

for some $z \in (x, y)$. Here $z$ could depend both on $x, y$. But no matter what $z$ is, $z > x > x_0$ and so (24) gives

$$L - \frac{\epsilon}{2} < \frac{g(x) - g(y)}{h(x) - h(y)} < L + \frac{\epsilon}{2} \tag{26}$$

By fixing $x$ and letting $y \to \infty$ in (26) we get (25). (Note that the strictness of the inequality may be lost in the limiting process.) As noted before, this finishes the proof. ∎

So far we dealt with the indeterminate form $\dfrac{0}{0}$. The indeterminate form $\dfrac{\infty}{\infty}$ can, in theory, be converted to the $\dfrac{0}{0}$ form merely by taking reciprocals. But this may not always be helpful in a particular problem. Suppose for example we want to find $\lim\limits_{x \to 0} \dfrac{\text{cosec } x}{\frac{1}{x}}$. We can write this as $\lim\limits_{x \to 0} \dfrac{x}{\sin x}$ and evaluate it by Theorem (9.1). But suppose we want $\lim\limits_{x \to \infty} \dfrac{\ln x}{x}$. We can write this as $\lim\limits_{x \to \infty} \dfrac{1/x}{1/\ln x}$ which is of the $\dfrac{0}{0}$ form. But if we go on applying Theorem (9.6), the ratios keep on getting more complicated. (See Exercises (1.21), (1.22) and (6.23) for definition and properties of natural logarithms.)

So it is better to have an independent proof for L'Hôpital's rule in the $\dfrac{\infty}{\infty}$ form. As is to be expected, the simple-minded approach of writing $\dfrac{g(x)}{h(x)}$ as $\dfrac{1/h(x)}{1/g(x)}$ is not going to work. (If it did, then we would have had no difficulty in the examples above.) Cauchy's MVT is once again the crucial tool.

**9.7 Theorem :** Suppose $g, h$ are differentiable in a deleted neighbourhood $N - \{c\}$ of a point $c \in \mathbb{R}$ and $h'$ never vanishes in $N - \{c\}$. If both $g(x) \to \infty$ and $h(x) \to \infty$ as $x \to c$ and $\lim\limits_{x \to c} \dfrac{g'(x)}{h'(x)} = L$, then $\lim\limits_{x \to c} \dfrac{g(x)}{h(x)}$ also equals $L$.

**Proof :** We shall first prove that the right handed limit $\lim\limits_{x \to c+} \dfrac{g(x)}{h(x)}$ equals $L$. As in Theorem (9.6) the hypothesis that $h' \neq 0$ ensures the applicability of Cauchy's MVT. But it cannot be applied to an interval of the form $[c, x]$ since $g(c)$ and $h(c)$ may not be defined. Once again, the difficulty will be overcome by applying it to intervals of the form $[x, y]$ where $c < x < y$ and $x$ is very close to $c$.

So, let $\epsilon > 0$ be given. First, choose $x_0 > c$ such that for all $z$ with $c < z < x_0$,

$$L - \frac{\epsilon}{2} < \frac{g'(z)}{h'(z)} < L + \frac{\epsilon}{2} \tag{27}$$

Further choose $x_1 < x_0$ so that for all $c < x < x_1, g(x), h(x)$ are positive and bigger than $g(x_0), h(x_0)$ respectively (This is possible since $g(x) \to \infty$ and $h(x) \to \infty$ as $x \to c+$.)

From (27) and Cauchy's MVT, we get, by an argument similar to that in the proof of Theorem (9.6) that for all $c < x < x_0$,

$$L - \frac{\epsilon}{2} < \frac{g(x) - g(x_0)}{h(x) - h(x_0)} < L + \frac{\epsilon}{2} \tag{28}$$

Now if $c < x < x_1$, then $g(x), h(x)$ are non-zero and so the middle term can be rewritten as $\dfrac{g(x)(1 - g(x_0)/g(x))}{h(x)(1 - h(x_0)/h(x))}$. Also since all the factors in this expression are positive, (28) becomes

$$\left(L - \frac{\epsilon}{2}\right) \frac{1 - h(x_0)/h(x)}{1 - g(x_0)/g(x)} < \frac{g(x)}{h(x)} < \left(L + \frac{\epsilon}{2}\right) \frac{1 - h(x_0)/h(x)}{1 - g(x_0)/g(x)} \tag{29}$$

Since $L, \epsilon, h(x_0), g(x_0)$ are fixed numbers, by taking limits as $x \to c^+$, the first and the last terms of (29) tend to $L - \dfrac{\epsilon}{2}$ and $L + \dfrac{\epsilon}{2}$ respectively. Consequently, there exists $x_2 < x_1$ such that for all $c < x < x_2$,

$$L - \epsilon < \left(L - \frac{\epsilon}{2}\right)\frac{1 - h(x_0)/h(x)}{1 - g(x_0)/h(x)} \text{ and } \left(L + \frac{\epsilon}{2}\right)\frac{1 - h(x_0)/h(x)}{1 - g(x_0)/y(x)} < L + \epsilon \quad (30)$$

(29) and (30) together imply that for all $c < x < x_2$.

$$L - \epsilon < \frac{g(x)}{h(x)} < L + \epsilon$$

which proves that $\displaystyle\lim_{x \to c^+} \frac{g(x)}{h(x)} = L$.

An entirely similar argument works for $\displaystyle\lim_{x \to c^-} \frac{g(x)}{h(x)}$. The proof is thus complete. ∎

This theorem, too, remains valid for $c = \infty$ or $-\infty$. We leave these extensions as exercises. We now see very effortlessly that $\displaystyle\lim_{x \to \infty} \frac{\ln x}{x} = 0$ because $\dfrac{d}{dx}(\ln x) = \dfrac{1}{x}$ and $\dfrac{d}{dx}(x) = 1$.

We omit the discussion of other indeterminate forms such as $0.\infty$ or $1^\infty$ or $\infty^0$. They do not arise as frequently as the $\dfrac{0}{0}$ or $\dfrac{\infty}{\infty}$ forms. When they do in any particular problem, they can usually be converted, by taking reciprocals or logarithms to the ones we have studied. For example, $\displaystyle\lim_{x \to \infty} x^{1/x}$ changes, after taking logarithms to $\displaystyle\lim_{x \to \infty} \frac{\ln x}{x}$ which we already found as 0. Because of continuity of the exponential function, this means $e^{\frac{\ln x}{x}} \to e^0$, i.e., $x^{1/x} \to 1$ as $x \to \infty$. Restricting $x$ only to integer values we get another proof of Theorem (3.4.5).

Despite the importance of L'Hôpital's rule in evaluating limits and also in proving theorems, it is not a such a golden tool as it appears. Its applicability is restricted, first of all, by the differentiability requirement. For example, it cannot be applied to $\displaystyle\lim_{x \to 0^+} \frac{\sin(\sqrt{x})}{\sqrt{x}}$, whereas a mere substitution $\sqrt{x} = y$ would give the answer by (3) in Section 4.7. It is also to be noted that for a limit like $\displaystyle\lim_{x \to 0} \frac{\sin x}{x}$, L'Hôpital's rule gets undeserved credit because to apply it you need the derivative of the sine function and to find the latter you need to evaluate $\displaystyle\lim_{x \to 0} \frac{\sin x}{x}$ anyway !

A more serious difficulty with the L'Hôpital's rule is the instability of its hypothesis. Suppose we want to evaluate $\displaystyle\lim_{x \to c} \frac{g(x)}{h(x)}$. This may be very easy and possible without L'Hôpital's rule. But let us multiply both $g(x)$ and $h(x)$ by some non-zero factor, say, $k(x)$. Letting $u(x) = g(x)k(x)$ and $v(x) = h(x)k(x)$, $\lim\limits_{x \to c} \dfrac{u(x)}{v(x)}$ is the same as $\lim\limits_{x \to c} \dfrac{g(x)}{h(x)}$. But if we fail to recognise $k(x)$ as a common factor of $u(x)$ and $v(x)$ then $\dfrac{u(x)}{v(x)}$ may be in an indeterminate form in which repeated applications of L'Hôpital's rule only serve to make the ratios more and more complicated.

A good example is given by $\lim\limits_{x \to \frac{\pi}{2}} \dfrac{\tan x}{\sec x}$ which is in the $\dfrac{\infty}{\infty}$ form. If we notice that $\sec x$ is a factor of $\tan x$ then cancelling it, this limit is simply $\lim\limits_{x \to \frac{\pi}{2}} \dfrac{\sin x}{1}$ which trivially equals 1, by continuity of the sine function. But if we apply L'Hôpital's rule to $\lim\limits_{x \to \frac{\pi}{2}} \dfrac{\tan x}{\sec x}$ the problem reduces to finding $\lim\limits_{x \to \frac{\pi}{2}} \dfrac{\sec^2 x}{\tan x \sec x}$. Now $\sec x$ is an obvious common factor. If we ignore it, the subsequent derivatives will get horrendous. Even if we cancel $\sec x$ now, it is not of much help. For, we would then have to find $\lim\limits_{x \to \frac{\pi}{2}} \dfrac{\sec x}{\tan x}$. Another application of L'Hôpital rule would give $\lim\limits_{x \to \frac{\pi}{2}} \dfrac{\sec x \tan x}{\sec^2 x}$. Cancelling $\sec x$ at this stage would get us back to the original problem! And this cycle would repeat endlessly.

L'Hôpital's rule is also not stable under a change of variable. For example $\lim\limits_{x \to \infty} \dfrac{x}{e^x}$ can be evaluated by L'Hôpital's rule quite easily, by an extension of Theorem (9.7) (to the case $c = \infty$). It is also tempting to try to evaluate this limit by putting $y = \dfrac{1}{x}$ and letting $y \to 0^+$. But if we do so, the limit in question becomes $\lim\limits_{y \to 0^+} \dfrac{e^{-1/y}}{y}$. L'Hôpital's rule will reduce this to $\lim\limits_{y \to 0^+} \dfrac{-e^{-1/y}}{y^2}$ which is more complicated than the original one and would become still more so if we apply it again.

Finally, L'Hôpital's rule may give wrong results if applied to limits which are not in the indeterminate forms. A simple counter-example is $\lim\limits_{x \to 0} \dfrac{x+1}{x+2}$ which is $\dfrac{1}{2}$. But L'Hôpital's rule would give it as $\lim\limits_{x \to 0} \dfrac{1}{1} = 1$.

Summing up, L'Hôpital's rule is a handy tool but requires some care in using.

## EXERCISES

9.1 Evaluate $\lim\limits_{x \to 4} \dfrac{x-4}{\sqrt{x}-2}$, $\lim\limits_{x \to \frac{\pi}{4}} \dfrac{\tan^2 x - 1}{x^2 - \frac{\pi}{4}x}$ and $\lim\limits_{x \to 0} \dfrac{1 - \cos x}{x^2}$ both without and with L'Hôpital's rule.

9.2 Evaluate the following limits using L'Hôpital's rule :

(i) $\lim\limits_{x \to 0} \dfrac{1}{x} - \dfrac{1}{\sin x}$ (ii) $\lim\limits_{x \to \frac{\pi}{4}} \dfrac{\sin x - \cos x}{x - \frac{\pi}{4}}$ (iii) $\lim\limits_{x \to 0} \dfrac{\cos x - 1 + \frac{x^2}{2}}{x^4}$

9.3 Let $g(x) = h(x) = e^{-1/x^2}$ for $x \neq 0$ and $g(0) = h(0) = 0$. Prove that $g, h$ have derivatives of all orders at 0, but Theorem (9.2) cannot be applied. (See Fig. (4.6.1).)

9.4 Suppose $g$ has derivatives of all orders at $c, g(c) = 0 = h(c)$ and $h$ is analytic at $c$ and not identically 0. Prove that Theorem (9.2) will work. (See Exercise (6.18).)

9.5 Prove that Theorems (9.3), (9.6) and (9.7) continue to hold even if $L = \infty$ or $-\infty$. (In fact, the proofs are a little simpler because instead of two-sided inequalities for $\dfrac{g(x)}{h(x)}$, only one-sided inequalities are involved.)

9.6 Prove Theorem (9.4).

9.7 Prove the assertion about polynomials of degree $n$ (or less) used in the proof of theorem (9.5). [*Hint* : For a purely algebraic proof, apply induction on $n$ after noting that $\dfrac{p(x) - p(c)}{x - c}$ is a polynomial of degree $n - 1$ (or less). For uniqueness observe that $a_r$ must equal $\dfrac{p^{(r)}(c)}{r!}$ or again apply induction.]

9.8 Let $f(x) = x^3 + x^4 \sin \dfrac{1}{x}$ with $f(0) = 0$. Prove that $f'(0)$ and $f''(0)$ exist, $f'''(0)$ does not exist, but the polynomial $p(x) = x^3$ satisfies $\dfrac{f(x) - p(x)}{x^3} \to 0$ as $x \to 0$. (Thus a strict converse to Theorem (3.3) is false.)

9.9 Complete the proof of Theorem (9.7) by showing that $\lim\limits_{x \to c^-} \dfrac{g(x)}{h(x)} = L$. [*Hint* : Instead of duplicating the argument replace $x$ by $-x$.]

9.10 Extend Theorem (9.7) to the cases $c = \infty$ and $c = -\infty$.

9.11 Suppose a real-valued function $f$ has derivatives of all orders at 0 and $f\left(\dfrac{1}{n}\right) = 0$ for every positive integer $n$. Prove that $f^{(n)}(0) = 0$ for all $n \geq 0$.

### 4.10. Use of Second Order Derivatives in Concavity Testing

In Theorem (5.5) we saw how the sign of the first derivative gives vital information as to whether the function is increasing or decreasing over a given interval. Of course, the *definition* of whether a function is increasing or decreasing over an interval has nothing to do with derivatives. It is a very natural concept, requiring no knowledge of derivatives or limits of any kind. A function can be increasing over an interval without being differentiable at every point of it. As a simple example define $f : [-1, 1] \to I\!\!R$ by

$$f(x) = \begin{cases} x & \text{if} \quad -1 \leq x < 0 \\ 2x & \text{if} \quad 0 \leq x \leq 1 \end{cases}$$

Then $f$ is strictly increasing on $[-1, 1]$, but is not differentiable at 0. In other words, the first derivative is only an instrument in *testing* if a function is increasing/decreasing, but not a vital prerequisite for *defining* what is an increasing or decreasing function.

The relationship between the second derivative of a function and concavity (upward or downward) of that function is quite similar. Unfortunately, several authors of elementary calculus books *define* concavity in terms of derivatives. This is quite unnecessary and in fact somewhat undesirable. The terms concave and convex are purely geometric and even a layman understands them because he is familiar with convex and concave mirrors and lenses. Moreover, they can be defined quite rigorously in elementary geometric terms. Although the second derivative is a convenient tool in testing the concavity of a function, there is no reason why it should enter the very definition of concavity. (To be sure, there are a few geometric concepts, e.g. tangency, whose rigorous definitions require derivatives. But concavity is not one of them.)

So we shall first define the concavity of a function $f$ purely in terms of a certain geometric property of the region bounded by the graph of $f$. This will require no knowledge of derivatives. Later we shall show how the second derivative of $f$, *in case it exists*, enables us to test if $f$ is concave upward or downward.

Intuitively, we call a body convex or concave depending upon whether its boundary is curved outside or inside. These terms are, of course, not mathematically precise. To give a precise meaning to them, let us first define the convexity of a subset $S$ of the plane $I\!\!R^2$. For any two points $P, Q$ in $I\!\!R^2$, let us denote by $\overline{PQ}$ the (straight) line segment joining them. Now supose $P, Q$ are in $S$. Then in general, the segment $PQ$ will not be completely in $S$, even though its end-points (viz. $P$ and $Q$) are in $S$. But, if $S$ is convex, with our intuitive understanding of the term, then the entire segtment $PQ$ must be contained in $S$. For otherwise it will cross the boundary of $S$ in at least two points, say, $A$ and $B$ as shown in Fig. 4.10.1 (a). But that means, somewhere in between these two points $A$ and $B$, the boundary of $S$ has bent inwards. So $S$ cannot be convex.

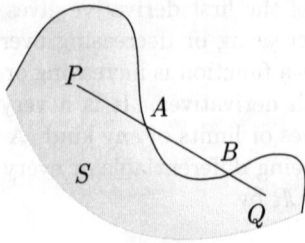

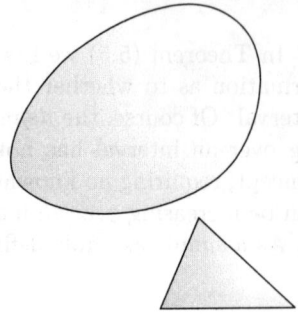

(a) Definition                                  (b) Convex

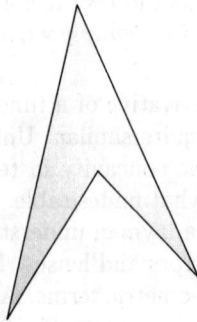

(c) Not convex

Figure 4.10.1 : Convex and Non-convex Figures

This intuitive reasoning leads to the following formal definition of convexity of a set (already given in Exercise (2.1.2) and illustrated in Fig. 2.1.2.)

**10.1 Definition :** A subset $S$ of $\mathbb{R}^2$ (or more generally of $\mathbb{R}^n$) is said to be **convex** if for every pair of points $P, Q$ in $S$, the segment $\overline{PQ}$ is contained in $S$.

As simple examples, the figures bounded by all triangles, circles, ellipses are convex. There are also unbounded convex sets. The entire plane itself, any half-plane, any quadrant, every straight line or line segment, a strip, i.e., a region bounded by a pair of parallel straight lines are also convex. On the other hand, an annulus (i.e., the region between two concentric circles), a horse-shoe type figure are not convex. Note that the circles and ellipses (without their interiors) are not convex.

The study of convex subsets is an important branch of gemoetry. A few simple results about them will be given as exercises.

We have not yet defined what is a concave subset. Informally, we may say it is the complement of a convex subset. But the term is rarely needed. Actually our interest is not so much in the convexity or concavity of sets as in that of functions, which we now proceed to define.

**10.2 Definition :** A real valued function $f$ is said to be **convex** or **concave upwards** on an interval $S$ (contained in the domain of $f$) if the region above the graph of $f$ over this interval, i.e., the set $S^f$ defined by $S^f = \{(x, y) \in \mathbb{R}^2 : x \in S, y \geq f(x)\}$ is convex. $f$ is said to be **concave downwards** if the region below the graph of $f$, i.e., the set $S_f$ defined by $S_f = \{(x, y) \in \mathbb{R}^2 : x \in S, y \leq f(x)\}$ is convex. (Occasionally we drop the 's' and write concave upward and downward.)

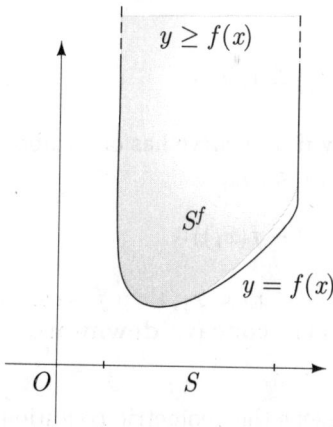

(a) Concave upward function

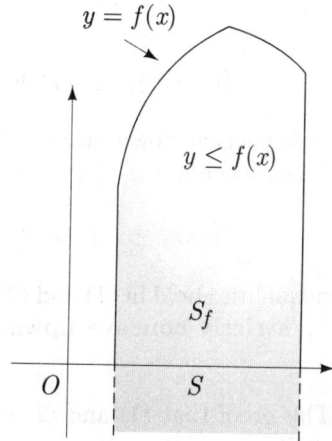

(b) Concave downward function

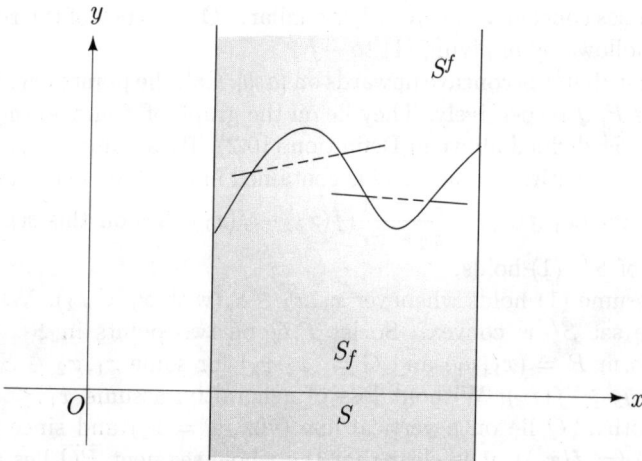

(c) Neither concave upwards nor concave downwards

Figure 4.10.2 : Concavity of a Function

Figure 4.10.2 shows the graphs of functions which are concave upward, concave downward and neither. The term 'convex' function is not used so frequently these days in this context as in the old literature. It is trivial to check that a function $f$ is concave upward on $S$ if and only if the function $-f$ is concave downward on $S$.

The following characterisation of concavity is very helpful. It is, in fact, taken as the definition if one wants to talk about concavity without mentioning convex sets.

**10.3 Theorem :** A function $f$ is concave upward on $S$ if and only if for all $x_1 < x_2$ in $S$, the portion of the curve $y = f(x)$ in between the points $(x_1, f(x_1))$ and $(x_2, f(x_2))$, lies on or below the chord joining these two points, i.e., for all $x_3 \in [x_1, x_2]$,

$$f(x_3) \; \leq \; f(x_1) + \frac{x_3 - x_1}{x_2 - x_1}(f(x_2) - f(x_1)) \tag{1}$$

Similarly, $f$ is concave downward on $S$ if and only if 'the curve lies on or above the chord', i.e., whenever $x_3 \in [x_1, x_2] \subset S$ (with $x_1 < x_2$),

$$f(x_3) \; \geq \; f(x_1) + \frac{x_3 - x_1}{x_2 - x_1}(f(x_2) - f(x_1)) \tag{2}$$

(If strict inequalities hold in (1) and (2) whenever $x_1 < x_3 < x_2$, then $f$ is called, respectively, **strictly concave upward** and **strictly concave downward** on $S$.)

**Proof :** The proof that (1) and (2) indeed represent the geometric conditions stated verbally above them is an easy exercise in co-ordinate geometry and is omitted. We shall prove that (1) characterises concavity upwards. The proof that (2) characterises concavity downward, is similar. (Or, in view of the remark made above, (2) follows by applying (1) to $-f$.)

So, suppose first that $f$ is concave upwards on $[a, b]$. Call the points $(x_1, f(x_1))$ and $(x_2, f(x_2))$ as $P, Q$ respectively. They lie on the graph of $f$ and so they belong to the region $S^f$ defined above in Definition (10.2). By assumption, $S^f$ is a convex subset. So the entire segment $\overline{PQ}$ is contained in $S^f$. A little calculation shows that the point $(x_3, f(x_1) + \frac{x_3 - x_1}{x_2 - x_1}(f(x_2) - f(x_1))$ lies on this segment. So, by definition of $S^f$, (1) holds.

Conversely, assume (1) holds whenever $x_1, x_2 \in S$ (with $x_1 < x_2$). We have to show that the set $S^f$ is convex. So let $P, Q$ be two points in $S^f$. Then $P, Q$ are of the form $P = (x_1, y_1)$ and $Q = (x_2, y_2)$ for some $x_1, x_2 \in S$ with $y_1 \geq f(x_1)$ and $y_2 \geq f(x_2)$. Without loss of generaltiy, assume $x_1 \leq x_2$. If $x_1 = x_2$, then both $P, Q$ lie on a vertical line (viz. $x = x_1$) and since $y_1, y_2$ are both $\geq f(x_1) (= f(x_2))$, it is clear that the whole segment $\overline{PQ}$ lies above $(x_1, f(x_1))$. Thus $\overline{PQ} \subset S^f$.

So assume $x_1 < x_2$. Any point, say $R$, on $\overline{PQ}$ is of the form $(x_3, y_3)$ where $x_1 \leq x_3 \leq x_2$ and

$$y_3 = y_1 + \frac{x_3 - x_1}{x_2 - x_1}(y_2 - y_1) \tag{3}$$

(To see this simply write down the equation of the straight line joining $P$ and $Q$.)

We want to prove that $R \in S^f$, i.e., $y_3 \geq f(x_3)$. We are given (1) which can be rewritten as

$$f(x_3) \quad \leq \quad \frac{x_2 - x_3}{x_2 - x_1} f(x_1) + \frac{x_3 - x_1}{x_2 - x_1} f(x_2) \qquad (4)$$

Since $\dfrac{x_2 - x_3}{x_2 - x_1}$ and $\dfrac{x_3 - x_1}{x_2 - x_1}$ are non-negative, and $f(x_1) \leq y_1, f(x_2) \leq y_2$, (4) gives

$$
\begin{aligned}
f(x_3) \quad &\leq \quad \frac{x_2 - x_3}{x_2 - x_1} y_1 + \frac{x_3 - x_1}{x_2 - x_1} y_2 \\
&= \quad y_1 + \frac{x_3 - x_1}{x_2 - x_1}(y_2 - y_1) \\
&= \quad y_3 \quad \text{(by (3).)}
\end{aligned}
$$

Thus we have proved that every point on $\overline{PQ}$ is in $S^f$, i.e., $S^f$ is a convex set. ∎

It is instructive to reformulate the conditions (1) and (2). Every $x_3 \in [x_1, x_2]$ can be expressed as $(1 - \lambda)x_1 + \lambda x_2$ for a unique $\lambda \in [0, 1]$. Indeed, $\lambda$ must be $\dfrac{x_3 - x_1}{x_2 - x_1}$. Note that $(1 - \lambda)$ and $\lambda$ are both non-negative and add up to 1. An expression of the form $\alpha x_1 + \beta x_2$ where $\alpha, \beta$ are non-negative and $\alpha + \beta = 1$ is called a **convex combination** of $x_1$ and $x_2$. (The justification for the name comes from the fact that the definition of convexity can be expressed in terms of convex combinations. See Exercise (10.5).) Now (1) can be paraphrased as

$$f(\alpha x_1 + \beta x_2) \quad \leq \quad \alpha f(x_1) + \beta f(x_2) \qquad (5)$$

whenever $\alpha \geq 0, \beta \geq 0$ and $\alpha + \beta = 1$. Verbally (5) says that a function $f$ is concave upwards if and only if the value of $f$ at a convex combination is at most the corresponding convex combination of the values of $f$. Similarly (2) can be paraphrased. The advantage of these definitions is that they are more general; that is, they make sense for real-valued functions defined over any convex set (and not just an interval). In fact in this more general context a function satisfying (5) is called a **convex function**. Although we shall have little need to study such general convex functions, we remark that they arise very naturally and are important in applicaions.

This theorem makes it a little easier to test if a function $f$ is concave upwards or downwards over an inteval. For example, the function $f(x) = |x|$ is concave upwards on $[-1, 1]$ (in fact on every interval) because whenever $\alpha, \beta \geq 0$ and $\alpha + \beta = 1$, we have $|\alpha x_1 + \beta x_2| \leq |\alpha x_1| + |\beta x_2| = \alpha|x_1| + \beta|x_2|$ and so (5) holds. Note that this function does not even have a first derivative at 0, let alone second derivative. This shows how we would lose generality if we defined concavity in terms of second derivatives instead of the geometric method above.

As probably the most 'practical' example of a concave upwards function, we invite the reader to show that the 'income tax function' pictured in Fig. 4.10.3

is concave upwards. As the name suggests, this function gives the tax you have to pay on a given income. The increasing slopes of the line segments in the graph reflect the fact that income in the higher slabs is taxed at a higher rate. An interesting application of the concave upwardness of this funciton will be given in Exercise (10.8).

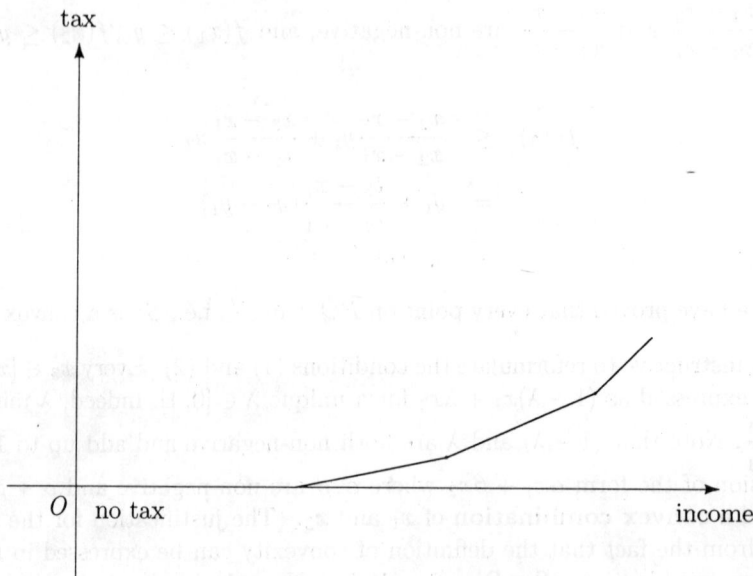

Figure 4.10.3 : Income Tax Function

It turns out, however, that functions which are concave upwards (or downwards) do behave rather nicely w.r.t. differentiation. Even though they may not be differentiable everywhere, they do have right and left handed derivatives at every interior point (as was the case in the examples above). Moreover these derivatives are increasing, as we see in the following basic result which relates concavity with derivatives.

**10.4 Theorem :** Let $f$ be concave upwards on an interval $S$. Then for every interior point $c$ of $S$ both $f'_+(c)$ and $f'_-(c)$ exist and moreover $f'_-(c) \leq f'_+(c)$. If $c, d$ are interior points of $S$ with $c < d$, then

$$f'_+(c) \quad \leq \quad \frac{f(d) - f(c)}{d - c} \quad \leq \quad f'_-(d) \tag{6}$$

Similar assertions hold if $f$ is concave downwards, except that the inequality (6) is reversed.

**Proof :** We first prove that whenever $x_1 < x_3 < x_2$ are in $S$,

$$\frac{f(x_3) - f(x_1)}{x_3 - x_1} \leq \frac{f(x_2) - f(x_1)}{x_2 - x_1} \leq \frac{f(x_2) - f(x_3)}{x_2 - x_3} \tag{7}$$

Geometrically this means that if $P_i = (x_i, f(x_i)), i = 1, 2, 3$ then the slope of $P_1 P_2$ lies in between that of $P_1 P_3$ and of $P_3 P_2$, see Fig. 4.10.4.

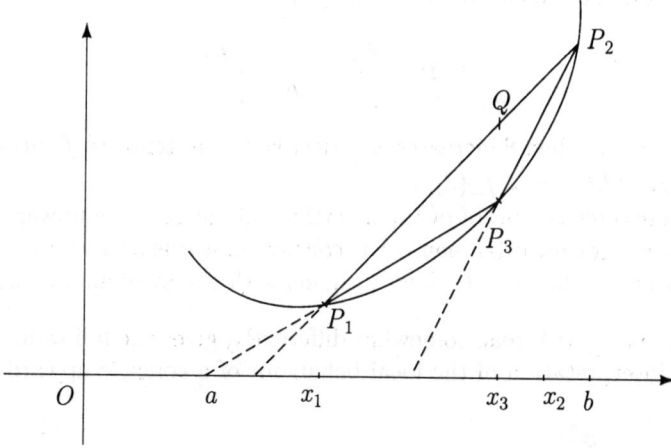

Figure 4.10.4 : Basic Property of Concave Upward Functions

The proof is also easier to see geometrically. Let the vertical line through $P_3$ cut the chord $P_1 P_2$ at $Q$. Since $f$ is concave upwards, $Q$ lies above $P_3$. So slope of $P_1 P_3 \leq$ slope of $P_1 Q$ and slope of $Q P_2 \leq$ slope of $P_3 P_2$. But $P_1 Q$ and $Q P_2$ have the same slope as $P_1 P_2$. This establishes (7). We leave it as an exercise to give this argument in a purely algebraic form.

Now suppose $c$ is an interior point of $S$. Then there exist $a, b$ such that $c \in (a, b)$ and $[a, b] \subset S$. For $0 < h < b - c$, let

$$\phi(h) = \frac{f(c + h) - f(c)}{h}$$

Clearly, $\phi(h)$ is the slope of the chord joining $(c, f(c))$ and $(c + h, f(c + h))$. Because of the first inequality in (7), it is clear that as $h$ decreases $\phi(h)$ also decreases. Moreover, because of the second inequality in (7), $\phi(h)$ is bounded below by $\frac{f(c) - f(a)}{c - a}$ (taking $x_1 = a, x_2 = c + h$ and $x_3 = c$). By completeness of the real line, $\lim_{h \to 0^+} \phi(h)$ exists. (See Theorem (2.6.5).) This means $f'_+(c)$, i.e., the right handed derivative of $f$ at $c$ exists and also that $f'_+(c) \leq \phi(h)$ for all $0 < h < b - c$. In particular, taking $h = d - c$, we get the first inequality in (6).

By a similar argument, if we let

$$\psi(h) = \frac{f(c) - f(c - h)}{h}$$

for $0 < h < c-a$, then $\psi(h)$ increases as $h$ decreases. Moreover, $\psi(h)$ is bounded above by $\dfrac{f(b) - f(c)}{b - c}$ and so $f'_-(c)$ exists. Further, if $0 < h < \min\{b - c, c - a\}$, then $\phi(h)$ and $\psi(h)$ are both defined and $\psi(h) \le \phi(h)$ as we see by taking $x_1 = c - h, x_2 = c + h$ and $x_3 = c$ in (7). So $\lim\limits_{h \to 0+} \psi(h) \le \lim\limits_{h \to 0+} \phi(h)$, i.e., $f'_-(c) \le f'_+(c)$.

Finally, for the second inequality in (6), let

$$\theta(h) = \frac{f(d) - f(d - h)}{h}$$

for $0 < h < c - a$. Then $\theta$ increases as $h$ decreases and tends to $f'_-(d)$ as $h \to 0^+$. In particular, $\theta(d - c) \le f'_-(d)$.

This completes the proof of the assertions about concave upwards function. If $f$ is concave downwards, then $-f$ is concave upwards and we can apply what we have proved so far to $-f$. Now (6) holds with the inequality signs reversed.∎

The inequality (6), read somewhat differently, gives the following interesting geometric interpretation of the local behaivour of a concave upwards function.

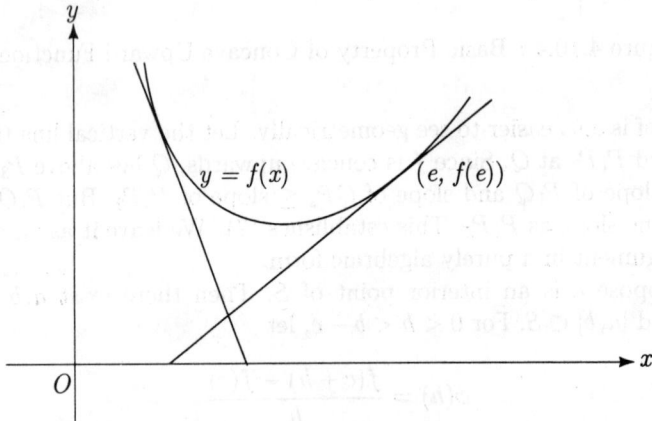

Figure 4.10.5 : Local Behaviour of Concave Upwards Function.

Suppose $f$ is concave upwards on $[a, b]$ and $e \in (a, b)$. We know $f'_-(e)$ and $f'_+(e)$ exist and $f'_-(e) \le f'_+(e)$. If $f$ is given to be differentiable, then of ocurse $f'_-(e) = f'_+(e) = f'(e)$ and the graph of $y = f(x)$ has a tangent at $(e, f(e))$ having slope $f'(e)$. We contend that in a neighbourhood of $e$, the graph of $f$ lies on or above the tangent (see Fig. 4.10.5). To see this, for $h > 0$, apply the first inequality in (6) with $c = e$ and $d = e + h$, while for $h < 0$, apply the second inequaltiy in (6) with $c = e + h$ and $d = e$. Note that it is sufficient if $f$ is concave upwards on $(c - \delta, c + \delta)$ for the same $\delta > 0$. Similarly it can be shown that if a funciton $f$ is differentiable at $e$ and concave downwards in a neighbourhood of $e$, then the graph of $f$ in this neighbourhood lies on or below the tangent at $(e, f(e))$. The converses of these observations are not true, see Exercise (10.9).

We are now in a position to tell when a differentiable funciton is concave upwards.

**10.5 Theorem :** Suppose $f$ is continuous on $[a, b]$ and differentiable on $(a, b)$. Then $f$ is concave upwrds on $[a, b]$ if and only if $f'$ is a monotonically increasing function on $(a, b)$.

**Proof :** The necessity of the condition follows immediately from (6) because under the hypothesis, $f'_+(c) = f'(c)$ and $f'_-(d) = f'(d)$.

For sufficiency, suppose $f'$ is monotonically increasing on $(a, b)$. Let $x_1, x_2$ and $x_3$ be any three points in $[a, b]$ with $x_1 < x_2$ and $x_1 \leq x_3 \leq x_2$. We write $x_3$ as $(1 - \lambda)x_1 + \lambda x_2$, where $\lambda = \dfrac{x_3 - x_1}{x_2 - x_1}$. Then $0 \leq \lambda \leq 1$. To show that $f$ is concave upwards on $[a, b]$ amounts to showing that

$$f(x_3) \leq (1 - \lambda)f(x_1) + \lambda f(x_2) \tag{8}$$

(See (5) and the accompanying comments.)

Applying Lagrange's MVT to the intervals $[x_1, x_3]$ and $[x_3, x_2]$ we get points $c_1 \in (x_1, x_3)$ and $c_2 \in (x_3, x_2)$ (see Fig. 4.10.4) such that

$$f(x_3) - f(x_1) \;=\; f'(c_1)(x_3 - x_1) \;=\; f'(c_1)\lambda(x_2 - x_1) \tag{9}$$

and

$$f(x_2) - f(x_3) \;=\; f'(c_2)(x_2 - x_3) \;=\; f'(c_2)(1 - \lambda)(x_2 - x_1) \tag{10}$$

Now $c_1 < c_2$ and so by hypothesis, $f'(c_1) \leq f'(c_2)$. So if we multiply (9) by $(1 - \lambda)$ and (10) by $\lambda$ (both of which are non-negative) we get

$$(1 - \lambda)(f(x_3) - f(x_1)) \leq \lambda(f(x_2) - f(x_3))$$

which gives (8) as desired. ∎

Even though we have stated the theorem for a closed and bounded interval, it remains true for any interval $S$, even an unbounded one. All we have to assume is that $f$ is continuous on $S$ and differentiable at every interior point of $S$. The proof goes through unchanged because Lagrange's MVT is applicable for every closed subinterval contained in $S$.

Like Theorem (10.4), Theorem (10.5) too has an analogue for concavity downwards, provable by replacing $f$ with $-f$.

We are now in a position to give the criterion for convavity, which, as noted earlier, is sometimes taken as the very definition.

**10.6 Theorem :** Suppose $f$ is continuous in $[a, b]$ and twice differentiable in $(a, b)$. If $f'' \geq 0$ throughout $(a, b)$ then $f$ is concave upwards on $[a, b]$. If $f''$ is continuous on $(a, b)$ then the converse is also true, i.e., $f$ is concave upwards on $[a, b]$ if and only $f''(x) \geq 0$ for all $x \in (a, b)$. (Similar assertions held for concavity downwards.)

**Proof :** If $f'' \geq 0$ on $(a, b)$, then by Theorem (5.5) (applied to $f'$), $f'$ is monotonically increasing on $(a, b)$ and so by the last theorem, $f$ is concave upwards on $[a, b]$. For the converse, assume $f$ is concave upwards on $[a, b]$ and that $f''$ is continuous on $(a, b)$. Let, if possible, $f''(c) < 0$ for some $c \in (a, b)$. Then by Exercise (1.1), there is some $h > 0$ such that $f'' < 0$ on $(c - h, c + h) \subset (a, b)$. But then $f'$ would be strictly decreasing on $(c - h, c + h)$ (again by Theorem (5.5)) contradicting the last theorem, which says that $f'$ is increasing *throughout* $(a, b)$. So we must have $f''(x) \geq 0$ for all $x \in (a, b)$. ∎

We can now see very effortlessly that the function $f(x) = x^2$ is concave upwards on every interval. More generally, this is the case for any even power of $x$. For odd powers, however, something interesting happens. Let $f(x) = x^3$, for example. Then $f''(x) = 6x$ which is positive for $x > 0$ and negative for $x < 0$. So by Theorem (10.6) $f$ is concave downwards on $(-\infty, 0)$ and concave upwards on $(0, \infty)$. By the observations we made after Theorem (10.4), the portion of the graph of $y = x^3$, on the right of 0 lies above the tangent at $(0, 0)$ while the portion on the left of 0 lies below the tangent (which is the $x$-axis in this case). In other words, at $(0, 0)$, the graph of $y = x^3$ crosses its own tangent! This strange behaviour is known as **inflection** and occurs whenever there is a change of concavity, either from upwards to downwards or vice versa. A formal definition is the following.

**10.7 Definition :** Suppose $f$ is defined in a neighbourhood of a point $c$. Then $c$ is called a **point of inflection** of $f$ if there is some $h > 0$ such that $f$ is concave upwards on $[c - h, c]$ and concave downwards on $[c, c + h]$ or vice versa.

We emphasize once again that this is a purely geometric concept, not vitally dependent on the derivatives of $f$. Of course if $f''$ exists at $c$ then it is easy to show that $f''(c) = 0$ if $c$ is a point of inflection (see Exercise (10.12)). That the converse is false is shown by $f(x) = x^4$ with $c = 0$.

Summing up, the signs of the first and the second derivatives of a function are instrumental in testing, respectively, on which intervals the function is increasing/decreasing and concave upwards/downwards. The information so gathered is valuable in sketching the graph of the function. Assuming $f''$ exists everywhere, one first identifies the intervals on which $f$ is concave upwards / downwards and hence also the points of inflection of $f$. If $c$ is such a point, one plots $(c, f(c))$ and, knowing $f'(c)$, draws a small line segment with this slope, passing thorugh $(c, f(c))$. Then the curve is drawn so as to cross these segments at the plotted points. An illustration of this procedure will be given in Exercise (10.14).

# EXERCISES

**10.1** Prove that the intersection of any number of convex sets is convex. Show by an example that the union of two convex sets need not be convex.

**10.2** Prove that a disc (i.e. the region bounded by a circle) and more generally every solid ball in $I\!R^n$ is convex. [*Hint* : Denote points of $I\!R^n$ by vectors. Apply the triangle inequality, Exercise (8.15).]

**10.3** Prove that the region bounded by an ellipse is convex.

**10.4** Denote points of $I\!R^n$ by vectors. Given $\mathbf{u}_1, \ldots, \mathbf{u}_m \in I\!R^n$, by a **convex combination** of them we mean a vector of the form $\alpha_1 \mathbf{u}_1 + \alpha_2 \mathbf{u}_2 + \cdots + \alpha_m \mathbf{u}_m$, where $\alpha_1, \ldots, \alpha_m$ are non-negative real numebrs with $\alpha_1 + \alpha_2 + \cdots + \alpha_m = 1$. Let $P, Q, R$ be three non-collinear points in $I\!R^n$ with position vectors $\mathbf{u}_1, \mathbf{u}_2, \mathbf{u}_3$ respectively. Prove that:

    (i) every point on the segment $\overline{PQ}$ is a unique convex combination of $\mathbf{u}_1$ and $\mathbf{u}_2$. (This is, in fact, the analytical definition of a line segment.)

    (ii) every point of the triangle $PQR$ is a unique convex combination of $\mathbf{u}_1, \mathbf{u}_2, \mathbf{u}_3$. (Again, in an analytical approach, this is taken as the definition of a triangle.)

    (iii) Using (ii) give an analytical proof that every triangle is a convex set.

**10.5** Prove that if $S$ is a convex subset of $I\!R^n$ then for every $\mathbf{u}_1, \ldots, \mathbf{u}_m \in S$, every convex combination of $\mathbf{u}_1, \ldots, \mathbf{u}_m$ is in $S$ (or in other words, $S$ is closed under convex combinations). [*Hint* : Apply induction on $m$. The case $m = 2$ is just the definition. For the inductive step note that if $\alpha_m < 1$ then $\alpha_1 \mathbf{u}_1 + \cdots + \alpha_m \mathbf{u}_m = (1 - \alpha_m) \left( \frac{\alpha_1}{1 - \alpha_m} \mathbf{u}_1 + \cdots + \frac{\alpha_{m-1}}{1 - \alpha_m} \mathbf{u}_{m-1} \right) + \alpha_m \mathbf{u}_m.$]

**10.6** Let $f$ be concave upwards on an interval $S$. Prove that for every convex combination $\alpha_1 x_1 + \cdots + \alpha_m x_m$ where $x_1, \ldots, x_m \in S, f(\alpha_1 x_1 + \cdots + \alpha_m x_m) \leq \alpha_1 f(x_1) + \cdots + \alpha_m f(x_m)$. In particular, show that

$$m f \left( \frac{x_1 + \cdots + x_m}{m} \right) \leq f(x_1) + \cdots + f(x_m).$$ When does equality hold for $f$ strictly concave upwards on $S$?

**10.7** Show that the tax benefit due to averaging of income is a consequence of the last exercise applied to the income tax function. (For more applications see Exercises (10.15) and (10.16) below.)

**10.8** Suppose $f''(c) > 0$. Prove that in some neighbourhood of $c$, the graph of $f$ lies above the tangent at $(c, f(c))$. [*Hint* : Apply Theorem (3.3).]

10.9 Let $f(x) = \frac{1}{4}x^2 + x^4 \sin\frac{1}{x}$ for $x \neq 0$ and $f(0) = 0$. Prove that in a neighbourhood of 0, the graph of $f$ lies above the tangent at $(0,0)$, but there is no neighbourhood of 0 in which $f$ is concave upwards. (This exercise is of the same spirit as Exercise (5.19). Note that $f''$ is not continuous at 0.)

10.10 In contrast to the last exercise, show that if $f : (a,b) \longrightarrow \mathbb{R}$ is differentiable and the graph of $f$ lies above the tangent at *every* point of $(a,b)$ then $f$ is concave upwards on $(a,b)$. [*Hint :* Show that (6) holds.]

10.11 Find all functions which are both concave upwards and concave downwards at the same time over an interval $S$. What are their points of inflection?

10.12 Suppose $c$ is a point of infelction of $f$. Prove that if $f''(c)$ exists then $f''(c) = 0$. [*Hint :* Apply Theorem (5.2) to $f'$.]

10.13 Let $f(x) = x^2$ for $x \geq 0$ and $f(x) = -x^2$ for $x < 0$. Prove that $f'$ exists everywhere and 0 is a point of inflection of $f$ but $f''(0)$ does not exist.

10.14 Let $f(x) = x + \sin x$. (See Exercise (5.18)). Identify the points of inflection of $f$. Hence sketch the graph of $f$ by the procedure indicated at the end of the text. [*Hint :* For large $x$, $\sin x$ is insiginficant as compared to $x$. So it will be a good idea to first draw the line $y = x$ as a pilot graph.]

10.15 Prove that the exponential function $e^x$ is strictly concave upwards on $\mathbb{R}$. Using this fact, logarithms and Exercise (10.6) prove the A.M.-G.M. inequality for any positive real numbers.

10.16 Prove that $y = \tan x$ is strictly concave upwards on $(0, \frac{\pi}{2})$. Hence show that in an acute angled triangle $ABC$,

$$\tan A + \tan B + \tan C \geq 3\sqrt{3}$$

with equality holding iff the triangle is equilateral. Similarly obtain characterisations of equilateral triangles in terms of sines and cosines of its angles.

## 4.11 Use of the Second Order Derivative Test in Problems of Finding Maxima/minima

Of all the problems to be solved in an elementary calculus course, those of the type where one wants to minimize the cost of, say, constructing tin cans of some size or to maximize the profit out of something are, by far, the most popular, at least with the average students. The reasons are multifold. First, like the problems of evaluating limits using L'Hôpital' a rule (see the comments at the beginning of Section 4.9), these problems give the student the satisfaction of actually doing something, rather than merely proving the existence of something. Secondly, by their very nature, optimisation problems (a term applied both to maximisation and minimisation) appeal to us as something very familiar. In our day-to-day shopping, for example, we either try to minimize the cost of the product we have in our mind or to maximise the quality we can get for the money we have in our pocket! And finally, the maxima/minima problems are also relatively easy to tackle by following a systematic procedure. They are, in fact, among the sure-shot questions in examinations. Typically, in such problems you first express the thing to be optimised as a function of some variable associated with the problem (sometimes a function of several such variables). Thanks to the inherent simplicity of Mother Nature, these functions are usually very nice, simple-minded functions, in sharp contrast to the crazy, pathological specimens cooked up by fussy mathematicians just to prove that sometimes things can go wrong!

So, typically, in order to solve a maxima/minima problem, a student differentiates this function (which is an easy, almost mechanical task) and sets the derivative equal to 0. The resulting equation usually has only one solution. Even if there are any others, they can usually be ruled out an practical grounds, e.g. if such solutions would imply some box to have an infinite volume or some object to have a negative mass! The lone solution that is left is then subjected to the so-called **second derivative test**, i.e., the second derivative of the function to be optimized is found (another mechanical, although sometimes a bit laborious task) and evaluated at this point where the first derivative vanished. Depending upon whether the second derivative is positive or negative the point is declared as the point where the function has a minimum or maximum.

While this proceudre is not totally baseless, the mechanical and sometimes hasty manner in which it is applied deserves some rebuttal. The last step especially, i.e. the second derivative test, is inadequate by itself. What is more, even in those problems where it works, it is dispensable and can be replaced by a careful study of the first derivative (which can usually be done with just a little more work than that involved in locating the zeros of the first derivative).

Before illustrating these comments with actual problems, let us first see what role derivatives have to play in finding the maxima and minima of functions. As with concavity of a function (discussed in Section 4.10), the *concept* of a maximum or a minimum of a real-valued function $f$ defined on a domain $S$, has nothing to do with derivatives. In fact, these concepts make sense even when the domain $S$ of $f$ is not a subset of $\mathbb{R}$. For example, $S$ could be the set of all students in a class and $f$ the height function, in which case, it is meaningless to talk of the continuity or the differentiability of $f$. In many problems, however, $S$ is indeed some subset of $\mathbb{R}$, in fact usually an interval and so in this context it does make sense to inquire whether the continuity or differentiability of $f$ implies anyting regarding the maximum or minimum value of $f$ on $S$. In fact we have already proved two such results. Let us recall them here.

In Theorem (1.4) we proved that if $S$ is a closed, bounded interval, then $f$ has both a minimum and a maximum on $S$. We gave examples to show that these hypotheses are vital. Further we remarked that mere continuity of $f$ does not give any method for *finding* the extremum of $f$. (An **extremum** is a term used for either a maximum or minimum. The term *optimum* means the same but has come in vogue more recently. A rather common abuse of language is to say that a point $x_0 \in S$ is a maximum (or a minimum) of $f$, when, in fact, a correct statement would be that a maximum of $f$ *occurs at* or is *attained at* $x_0$, because strictly speaking, the maximum is the maximum value $f(x_0)$ and not the point $x_0$ where it occurs.)

As is to be expected, if $f$ satisfies some condition stronger than continuity, then we may be able not only to prove the existence of an extremum but also to give a method for finding it. Differentiability is one such condition. In Theorem (5.2) we proved that if the maximum of $f$ over an interval occurs at an interior point, say $c$, of $S$ and $f'(c)$ exists then $f'(c) = 0$. (We stated Theoerem (5.2) for a closed and bounded interval. But the argument works for any interval as long as $c$ is an interior point of it. )

Let us paraphrase Theroem (5.2) with a new terminology. A point $c$ in the domain of a function $f$ is called a **critical point** of $f$ if either $f'(c)$ does not exist or $f'(c)$ exists and equals 0. For example, 0 and $\dfrac{1}{2}$ are the critical points of $f(x) = |x|(x-1)$ because $f'(0)$ does not exist while $f'(\frac{1}{2}) = 0$. (Note that in a small neighbourhood of $\dfrac{1}{2}$, $f(x) = x^2 - x$.)

With this terminology, Theorem (5.2) reads as:

**11.1 Theorem :** The maximum (as well as the minimum) of a function $f$ over an interval $S$, if it exists, must occur either at a critical point of $f$ (in $S$) or at an end-point of $S$ (if any). ∎

This theorem is hardly profound. But it narrows down the search for extreme values considerably. We first identify the critical points of $f$. By definition, these are points where $f'$ either fails to exist or vanishes. For functions occuring in real life problems, the first possibility usually does not arise. So, basically all we have to do is to find the roots of the equation $f'(x) = 0$ in $S$. Solving this equation may not always be an easy task. But assuming it can somehow be managed, we now know that the maximum (and also the minimum) of $f$, if it exists, must occur at one of these roots or at one of the end-points of $S$ (in case they belong to $S$). Usually the set of zeros of $f'$ in $S$ is a finite set and so the search is finally reduced to finding, out of a finite number candidates, where $f$

has the greatest (or the least) value. And this can be done simply by computing the values of the function at these points and comparing them.

Let us illustrate this technique with a simple problem.

**11.2 Problem :** A piece of wire of length $L$ is to be divided into two parts, one of which is to be bent into a circle and the other into a square. Find the maximum and the minimum total area enclosed by the two pieces.

**Solution :** Let $x$ and $L - x$ denote the lengths of the two segments. Here $0 \le x \le L$. The area enclosed by a circle of perimeter $x$ is $\pi \left( \dfrac{x}{2\pi} \right)^2 = \dfrac{x^2}{4\pi}$ while that enclosed by a square of perimeter $L - x$ is $\dfrac{(L-x)^2}{16}$. So, the problem amounts to find the maximum as well as the minimum of the function

$$f(x) \;=\; \frac{x^2}{4\pi} + \frac{(L-x)^2}{16} \tag{1}$$

over the interval $[0, L]$.

Since $f$ is continuous and the interval is closed and bounded, the maximum and the minimum exist. Moreover, $f$ is differentiable everywhere. Differentiating (1) and setting $f'(x) = 0$ gives $x = \dfrac{\pi L}{\pi + 4}$ as the only critical point of $f$. This point lies in the interval $[0, L]$ since $0 < \dfrac{\pi}{\pi + 4} < 1$. So, by the theorem above, the extreme values of $f$ must occur in the set consisting of $f(0), f(L)$ and $f \left( \dfrac{\pi L}{\pi + 4} \right)$. A direct calculation gives $f(0) = \dfrac{L^2}{16}, f(L) = \dfrac{L^2}{4\pi}$ and $f \left( \dfrac{\pi L}{\pi + 4} \right) = \dfrac{L^2}{4(\pi + 4)}$. A straight comparison (basd on approximate value of $\pi$) shows that $f$ attains its minimum at $\dfrac{\pi L}{\pi + 4}$ and its maximum at $L$. So $\dfrac{L^2}{4\pi}$ and $\dfrac{L^2}{4(\pi + 4)}$ are the maximum and the minimum sums of areas enclosed by the circle and the square. ∎

In this problem, the domain of the function to be optimised was a closed and bounded interval and so there were two end-points where the function had to be evaluated. This increased the work a little, but the advantage was that because of Theorem (1.4), the maximum and the minimum were sure to exist. If the interval $S$ on which $f$ is to be optimized does not contain two of its end-points then while finding the maximum or minimum of $f$, we also have to give it a thought to see if they exist. Sometimes this can be done simply by considering the behaviour of $f(x)$ as $x$ approaches the end points of $S$ (which may be $-\infty$ or $\infty$ as well as a finite real number). We illustrate this in the following problem.

**11.3 Problem :** Among all rectnagles with a given area find which has the smallest perimeter.

*Solution* : Let $x, y$ and $A$ be respectively the length, the breadth and the area of the rectangle. Its perimeter is $2(x + y)$ which is a function of two variables, $x$ and $y$. However, because of the conditions of the problem, these two variables are not independent of each other. They have to satisfy not only the inequalities $x > 0$ and $y > 0$ (dictated by common sense) but also the equation

$$xy = A \tag{2}$$

Such equations (or other conditions) are called **constraints.** So the present problem consists of minimising $2(x + y)$ **subject to** the constraint (2) (and also the inequality constraints $x > 0, y > 0$). The theory of functions of two (or more) variables gives certain standard methods for solving such 'constrained optimisation' problems. But it is beyond our scope. Nevertheless, in the present case we can easily solve (2) for either $x$ or $y$ and thereby make the perimeter $2(x+y)$ a function of just one variable. Solving (2) for $y$ we see that the problem reduces to minimising the function

$$f(x) = 2\left(x + \frac{A}{x}\right) \tag{3}$$

Here the doamin $S$ of $f$ is $(0, \infty)$, an unbounded open interval. Throughout $S, f$ is differentiable, and

$$f'(x) = 2 - \frac{2A}{x^2} \tag{4}$$

Solving $f'(x) = 0$, gives $x = \pm\sqrt{A}$. But $-\sqrt{A} \notin S$. So $\sqrt{A}$ is the only critical point of $f$ in $S$. The end-points $0$ and $\infty$ are not in $S$. But from (3) it is clear that $f(x) \to \infty$ as $x \to 0^+$ and also as $x \to \infty$. So clealry $f$ can have no maximum on $S$. This does not, by itself, imply that $f$ has a minimum at $\sqrt{A}$. Some argument is needed.

There are two ways to handle this. The first method is to show that $f$ attains its minimum on $S$. Since $S$ contains no boundary polints, the minimum must occur at an interior point, which by Theorem (5.2), must be a critical point. The fact that $\sqrt{A}$ is the only critical point of $f$ in $S$ would then show that $f(\sqrt{A})$ is the minimum value of $f$ on $S$.

In order to show that $f$ attains its minimum on $S$, we cannot apply Theorem (1.4) directly to $S$. But we can apply it to an interval of the form $[r, R]$ where $r > 0$ and $R > r$. Since $f(x) \to \infty$ as $x \to 0$, we may suppose $r$ is such that $f(x) > f(\sqrt{A})$ for all $0 < x \le r$. Similarly, since $f(x) \to \infty$ as $x \to \infty$, we may suppose $R$ is so large that $f(x) > f(\sqrt{A})$ for all $x \ge R$. Now, by Theorem (1.4), $f$ does attain its minimum on $[r, R]$. This minimum is at most $f(\sqrt{A})$. Since $f(x) > f(\sqrt{A})$ for all $x \in S - [r, R]$, it is clear that the minimum of $f$ on $[r, R]$ is also the minimum of $f$ on $S$. Hence $f$ attains its minimum on $S$ (at some point in $[r, R]$).

Another way to show that $f$ attains its minimum on $S$ at $\sqrt{A}$ is to look at the sign of $f'(x)$ for $x \in S$. From (4) it is clear that $f'(x) < 0$ if $0 < x < \sqrt{A}$ and

$f'(x) > 0$ if $\sqrt{A} < x$. So by Lagrange's MVT, $f$ is strictly decreasing on $(0, \sqrt{A})$ and strictly increasing on $(\sqrt{A}, \infty)$. (See Theorem (5.5), which was proved for a closed and bounded interval. But the same argument applies for any interval.) So $f(x) > f(\sqrt{A})$ for all $x \in S$ $x \neq \sqrt{A}$. Hence $f$ attains its minimum on $S$ at $\sqrt{A}$. When $x = \sqrt{A}$, $y$ is also $\sqrt{A}$. So the rectangle with the minimum perimeter is a square. ∎

The first argument given above to show that $f$ had a minimum in $(0, \infty)$ is needed rather frequently in many optimisation problems. So it is well worth isolating as a theorem. The proof, which is analogous to the argument given above is omitted.

**11.4 Theorem :** Suppose $f$ is continuous on an interval $S$ (which could be open, closed or semi-open). Suppose there exist $r, R, x_0 \in S$ with $r \leq x_0 \leq R$ such that

$$f(x) \leq f(x_0) \text{ for all } x \leq r \text{ and for all } x \geq R. \tag{5}$$

Then $f$ attains its maximum on $S$ at some point in $[r, R]$. Similarly $f$ attains its minimum on $S$ at some point in $[r, R]$ if instead of (5) we have

$$f(x) \geq f(x_0) \text{ for all } x \leq r \text{ and for all } x \geq R. \tag{6}$$

A couple of comments are in order. First, it is tempting to replace (5) by the slightly weaker hypothesis that $f$ is bounded above outside the interval $(r, R)$, that is, to assume that there is some $M$ such that $f(x) \leq M$ for all $x \in S - (r, R)$. But this will not suffice as can be seen by taking $S = (0, 1), r = 1/3, R = 2/3$ and $f(x) = x$ for all $x$. The crux of the matter is that the upper bound $M$ must be of the form $f(x_0)$ for some $x_0 \in [r, R]$. Secondly, even though (5) implies that $f$ attains its maximum on $S$ at some point in $[r, R]$, this point may be different from $x_0$. As a simple example, let $S = (-\infty, \infty), r = -1, R = 1$ and

$$f(x) = \begin{cases} 1 - |x| & \text{if } |x| \leq 1 \\ 0 & \text{if } |x| > 1 \end{cases}$$

Then (5) is satisfied with *any* $x_0 \in [r, R]$. But the maximum of $f$ occurs only at 0.

Similar comments apply for (6).

In essence this theorem extends Theorem (1.4) to the case where the interval $S$ is any interval. This extension is essentially a matter of common sense once Theorem (1.4) is proved, which is the real subtle theroem (the proof requiring the completeness of $\mathbb{R}$). In an elementary course, where the emphasis is more on finding the maxima / minima than on proving their exitence, such theorems are often taken for granted. And that is why, in a problem like the last one, an illogical claim is made that because $f$ has no maximum, $\sqrt{A}$ must give its minimum.

Another lapse of logic occurs when the second derivative test is resorted to without adequate justification. Note that in the solution to the last problem, the alternate argument we gave for showing that $f$ has its minimum on $\sqrt{A}$ is based soley on the change of sign of the first derivtive. The second derivative is nowhere needed. It is therefore truly surprising that many persons try to use it almost as an essential step in the maxima/minima problems. The reason for its popularity is probably simply that finding a derivative usually involves only mechanical work with which the average student feels more comfortable.

However, before condenmning the second derivative test too strongly, let us first see what connection the second derivative has with maxima and minima. We begin by defining a new term which, although conceptually related to a maximum, is not quite the same.

**11.5 Definition :** A function $f$ is said to have a **local maximum** (or a **relative maximum**) at a point $c$ if (i) $f$ is defined on some neighbourhood, say $N$, of $c$ and (ii) $f$ attains its maximum *on $N$*, at $c$, i.e., $f(c) \geq f(x)$ for all $x \in N$.

The rather peculiar name is justified by the fact that we are localising the concept of a maximum by requiring that the condition in it be satisfied only for points near $c$, rather like calling the best writer in a town as the local Shakespear (or local Kalidasa, if you prefer). A **relative** or **local minimum** is defined in an analogous manner. To emphasise their distinction from the ordinary maxima and minima, the latter are sometimes called **global** or **absolute** maxima and minima. In view of condition (ii) in the definition, it may appear that a global maximum is a stronger concept than a local maximum, much the same way as being a national chess champion means a lot mroe than being a state level champion. But this is not quite true, because condition (i) requires that a local maximum can occur only at an interior point of the domain of $f$ while a global maximum can occur at an end point. A global maximum, if it occurs at an interior point is, of course, also a local maximum. The converse is false. Consider for example, the function $f(x) = x^3 - 3x + 1$ whose graph is pictured in Fig. 4.11.1(a) If we let $S = (-\infty, \infty) = \mathbb{R}$, then $f$ has no global maximum or minimum over $S$, but there is a local minimum at 1 and a local maximum at $-1$. If we take $S = [0, 3]$ then the global maximum occurs at the end-point 3, while at 1 there is a local minimum which is also the global minimum.

Another point of difference between local and global maxima is that while a global maximum of $f$ over a set $S$ can occur at several points $S$ the values of $f$ at each of these points must be the same, i.e., the maximum value of $f$ on $S$ is unique. Take for example $f(x) = x^3 - 3x + 1$ again with $S = [-2, 2]$. Then $f$ assumes its maximum on $S$ at the interior point $-1$ and also again at the boundary point 2. But $f(-1) = f(2) = 3$. With local maxima and minima, this need not be the case. As a rather extreme example, the funciton $f(x) = x \sin \dfrac{1}{x}$ has infinitely many local maxima and also infinintely many local minima in the interval $(0, 1)$ (see Fig. 4.11.1(b)) and the values of $f$ at them are all different!

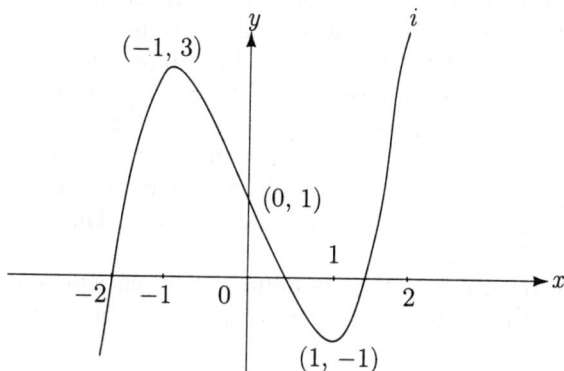

(a) $f(x) = x^3 - 3x + 1$

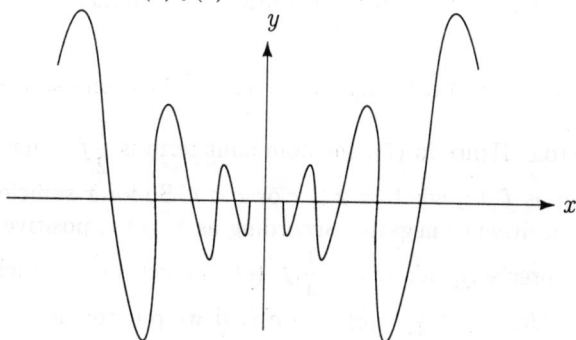

(b) $f(x) = x \sin \dfrac{1}{x}$

Figure 4.11.1 : Local Maximum and Minimum

Let us now study the relationship between derivatives of a function $f$ and the points of local maxima/minima of $f$ in an interval $S$. Let $c$ be a local maximum (or minimum) of $f$. Applying Theorem (11.1) to the neighbourhood $N$ in Definition (11.5), we see that $c$ must be a critical point of $f$. So, if $f$ is given to be differentiable on $S$, then $f'(c) = 0$ at every local maximum and local minimum. But the converse is not true. For example if $f(x) = x^3$, then $f'(0) = 0$. but there is neither a local maximum nor a local minimum at 0.

The **second derivative test** which we now prove formally, is a criterion to classify critical points.

**11.6 Theorem :** Suppose $c$ is an interior point of an interval $S$ on which a function $f$ is defined. Assume $f''(c)$ exists and $f'(c) = 0$ then

      (i)   $f$ has a local maximum at $c$ if $f''(c) < 0$

and  (ii)  $f$ has a local minimum at $c$ if $f''(c) > 0$.

**Proof :** The existence of $f''(c)$ implies, in particular, that $f'(x)$ exists for all $x$ in some neighbourhood, say, $N$ of $c$. Suppose $f''(c) > 0$. Then by Theorem (5.2) (applied to $f'$ rather than $f$) we see that there is a neighbourhood, say $(c - \delta, c + \delta)$ of $c$ such that $f'(x) > f'(c)$ for all $x \in (c, c + \delta)$ and $f'(x) < f'(c)$ for all $x \in (c - \delta, c)$. But we are given that $f'(c) = 0$. So we conclude that $f'$ is positive on $(c, c + \delta)$ and negative on $(c - \delta, c)$. Let $x \in (c, c + \delta)$. Applying Lagrange's MVT to the interval $(c, x)$ we see that $f(x) > f(c)$. Similarly for $x \in (c - \delta, c)$ we get $f(x) > f(c)$. (Evidently we are duplicating a part of the proof of Theorem (5.5).) So $f$ has a local minimum at $c$. Thus we have proved (ii). The proof of (i) is similar.

It is instructive to give an alternate argument which will combine (i) and (ii) together. Since $f'(c) = 0$, by Theorem (3.3) we can find a function $R_2(x)$ so that

$$f(x) - f(c) \;=\; \frac{1}{2} f''(c)(x - c)^2 + R_2(x) \qquad (7)$$

where $\dfrac{R_2(x)}{(x - c)^2} \to 0$ as $x \to c$. So $R_2(x)$ is very small as compared with $(x - c)^2$ if $x$ is very close to $c$. Hence in (7), the dominant term is $\dfrac{1}{2} f''(c)(x - c)^2$, which has the same sign as $f''(c)$ whether $x > c$ or $x < c$. So for $x$ sufficiently close to $c$, $f(x) - f(c)$ is positive or negative according as $f''(c)$ is positive or negative. To put this more precisely, taking $\epsilon = \dfrac{1}{4}|f''(c)|$, we get a $\delta > 0$ such that for all $x \in (c - \delta, c + \delta), |R_2(x)| < \frac{1}{4}|f''(c)|(x - c)^2$. If we put this in (7) we get that for all $x \in (c - \delta, c + \delta)$,

$$f(x) - f(c) \;\geq\; \tfrac{1}{4} f''(c)(x - c)^2 \qquad \text{if } f''(c) > 0$$
$$\text{and} \qquad f(x) - f(c) \;\leq\; -\tfrac{1}{4} f''(c)(x - c)^2 \qquad \text{if } f''(c) < 0.$$

These inequalities prove (ii) and (i) respectively. ∎

Like many tests for convergence of series, the second derivative test is not always conclusive. When $f''(c) = 0$, nothing can be concluded. Note also that the test is inapplicable if $f''(c)$ fails to exist. Examples of such situations will be given in the exercises.

The major inadequacy of the second derivative test is that even when it is applicable, it gives only *local* information. This is, of course, to be expected because its hypothesis itself is of a local type. So, when it comes to finding the global maximum or minimum of a function $f$ over an interval $S$, the second derivative test is not, by itself, an adequate tool. Let us go back to Problems (11.2) and (11.3), where the critical points of the function $f$ were , respectively, $\dfrac{L}{\pi + 4}$ and $\sqrt{A}$. An easy calclation shows that in both cases $f''$ is positive at these points. But that would only tell us that $f$ has a *local* minimum at each of them. The argument given in the solution to Problem (11.3), based on the sign of $f'$, on the other hand, showed that $f$ had a *global* minimum at $\sqrt{A}$. Even when it comes to finding a local maximum or a local minimum, the first proof of Theorem (11.6) shows that, wherever the second derivative test works, an argument based on the sign of the first derivative will also work.

So, the second derivative test is *always* dispensable. Still, the ease with which the second derivative can sometimes be found (as was the case in Problems (11.2) and (11.3)) makes it worthwhile to inquire if with some additional hypothesis, the second derivative can be made to give global extrema. We prove one such result, which is useful in applications.

**11.7 Theorem :** Suppose $S$ is any interval (closed, open or semi-open) and $f$ is differentiable on $S$. Suppose $c$ is the only interior point of $S$ at which $f'$ vanishes. Then if $f''(c) > 0$, $f$ attains its (global) minimum on $S$ at $c$ while if $f''(c) < 0$, then $f$ attains its (global) maximum on $S$ at $c$.

**Proof :** Assume $f''(c) > 0$. If $c$ is not a point of global minimum of $f$ then there exists some $b \in S$ such that $f(b) < f(c)$. Without loss of generality we may suppose $b > c$. Now by Theorem (11.6), $f$ has a local minimum at $c$. So if $a$ is sufficiently close to $c$ then $f(a) \geq f(c)$. We choose such $a$ so that $c < a < b$. Now we have $f(a) \geq f(c) > f(b)$. So by the Intermediate Value Property, there is some $c_1 \in (a, b)$ such that $f(c_1) = f(c)$. Note that $c_1 \neq c$ since $c \notin (a, b)$. We now apply Lagrange's MVT to the interval $[c, c_1]$ to get some $c_2 \in (c, c_1)$ such that $f'(c_2) = 0$. Clearly $c_2 \neq c$, contradicting that $c$ is the only interior point of $S$ where $f'$ vanishes. So $f$ attains its global minimum on $S$ at $c$. The proof of the other assertion is exactly similar. (Or, apply the familiar trick of replacing $f$ with $-f$.) ∎

This theorem works for finding the global minimum in Problems (11.2) and (11.3). Note, however, that it does not give the global maximum in Problem (11.2) since it occurs at an end-point.

The hypothesis that $f'$ should vanish only once in the interior of $S$ is a bit too restrictive and is generally not satisfied by familiar functions like polynomials (of degree 3 or more). It turns out, however that in many practical problems of finding maxima and minima, this hypothesis does hold. Typically, as $x$ approaches either of the end points of $S$, $f(x)$ behaves more and more undesirably and the optimum is reached somewhere in between. It is as if Mother Nature believes that the middle station is the best !

Summing up, the second derivative test comes handy only in a limited context, which nevertheless applies to many practical problems. It can always be dispensed with by considering the sign of the first derivative.

In fact, in some real life problems of finding maxima/minima, even the first derivative can be dispensed with. In Problem (11.2), the function $f(x)$ is a quadratic in $x$ and so we could have optimised it simply by completing the square. Similarly, Problem (11.3) could have been done purely algebraically had we used the A.M. - G.M. inequality. (As other examples of this method see Exercises (2.1.8) and (2.1.10).) Problems involving optimisation of trigonometric functions can often be done using merely the maximum and minimum values of $\sin\theta$ and $\cos\theta$. In geometric optimisation, an ingeneous application of certain simple facts from geometry can save laborious calculations of derivatives. Even in those problems where derivatives are necessary to optimise some quantity, say $y$, by expressing $y$ as a function of some independent variable, a clever choice of this independent variable can simplify the solution considerably. Examples to illustrate these comments will be given in the exercises.

Finally, even though we can do without the second derivatives in maxima/minima problems, they are important for other reasons, such as concavity testing, as we saw in the last section. Similarly even though, in general, local maxima and minima say nothing about global maxima and minima of a function, their knowledge provides vital information about the graph of the functions as they represent the 'crests' and 'valleys' on a graph.

With our knowledge of derivatives, the following general procedure can be laid down for skecthing the graph of a function $y = f(x)$. It is assumed, of course, that the function $f$ is well-behaved in that it has only finitely many points of discontinuity and only finitely many critical points.

(i)     Identify and plot a few 'obvious' points on the graph, e.g., points where it intersects co-ordinate axes. Also look for special features such as symmetry (which reduces half the work).

(ii)    Identify points of discontinuity and for each such point, say $x_0$, see how $f(x)$ behaves as $x \to x_0^+$ and $x \to x_0^-$. Similarly study what happens to $f(x)$ as $x \to \pm\infty$.

(iii)   Identify and plot critical points of $f$. Using second derivative test or otherwise classify them and thereby draw small portions of the graph near them.

(iv)    Identify and plot points of inflection of $f$ and sketch small portions of the graph near them as explained at the end of the last section.

(v)     Join the points and the portions of the graph drawn so far smoothly to get a qualitative sketch of the graph.

## EXERCISES

11.1 Prove Theorem (11.4).

11.2 Verify that the function $(x) = x\sin\dfrac{1}{x}$ has infinitely many local maxima and also infinitely many local minima in $(0, 1)$.

11.3 When is a local maximum also a local minimum for the same function ?

11.4 A **strict local maximum** of a function is a local maximum in which strict inequality holds in condition (ii) of Definition (11.5) for $x \neq c$. Prove that between any two strict local maxima of a continuous function, there must be at least one local minimum. As a result, the (strict) local maxima and minima must alternate with each other.

11.5 Let $f(x) = x^2$ for $x < 0$ and $f(x) = 2x^2$ for $x \geq 0$. Prove that $f$ has a local minimum at 0 which can be detected by the change of sign of the first derivative, but $f''(0)$ does not exist.

11.6 Let $f(x) = x^n$ for all $x$ where $n \ (> 3)$ is an even integer. Prove that $f$ has a local minimum at 0 which can be detected by the change of sign of the first derivativg but not by the second derivative test.

11.7 Let $f(x) = x^n$ where $n$ is an odd positive integer. Prove that $f$ has no local maxima or minima even though $f''$ vanishes at 0.

11.8 Generalise Theorem (11.6) by showing that if $n$ is the smallest positive integer such that $f^{(n)}(c) \neq 0$ then

    (i)  if $n$ is odd $f$ has neither a local minimum nor a local maximum at $c$

    (ii)  if $n$ is even then $f$ has a local minimum or a local maximum at $c$ according as $f^{(n)}(c) > 0$ or $f^{(n)}(c) < 0$.

State and prove a similar criterion for $c$ to be a point of inflection.

11.9  (a)  Give an example of a function which has a local minimum whcih cannot be detected even by the generalised result of the last exercise.

    (b)  Let $f(x) = x^2 \sin^2 \frac{1}{x}$ for $x \neq 0$ and $f(0) = 0$. Prove that $f$ is differentiable everywhere and has a local minimum at 0 which is not a strict local minimum. Prove also that there is no interval of the form $(0, \delta)$ in which $f$ is increasing nor any interval of the form $(-\delta, 0)$ in which $f$ is decreasing. (In other words, the local minimum at 0 cannot be detected by a change of sign of the first derivative.)

11.10 Find the maximum and minimum values of $3 \sin \theta + 4 \cos \theta$ as $\theta$ ranges over (i) $(-\infty, \infty)$ (ii) $[0, \frac{\pi}{2}]$. Do the problem both with and without derivatives.

11.11 Given a rectangular card-board of sides $a$ and $b$, a box (with no top) is to be formed by cutting off equal squares from its corners and folding the flaps along the sides of the inner rectangle. What is the maximum volume of such a box ?

11.12 Verify that for a point on a circle, the diametrically opposite point is also the farthest point. Does this hold true for an ellipse ? [Hint : In maximising a distance, it is easier to maximise its square.]

11.13 A swimmer is located at a point $P$ at a distance $a$ inside the sea from a point $O$ on a straight shore and wants to reach a point $R$ on the shore at a distance $b$ from $O$ (see Fig. 4.11.2).

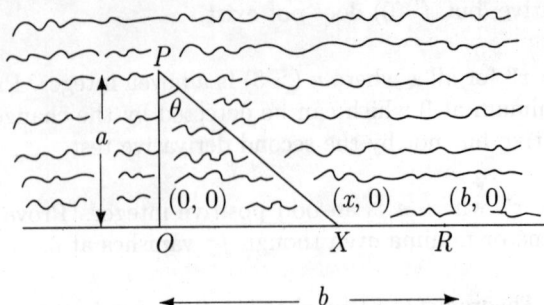

Figrue 4.11.2 : Quickest Path for a Swimmer

The swimmer can swim in water and run on shore with uniform speeds $u, v$ respectively. What is the shortest time the swimmer can reach $R$? What happens if $u > v$ (which is unlikely for a human swimmer but not for marine animals like walruses) ? [*Hint* : The swimmer must pass through some point $X = (x, 0)$ as shown. The working is a little easier if the time taken is expressed as a function of the angle $\theta$ instead of the distance $x$.]

11.14 What is wrong in the following reasoning for the solution to the last exercise? "The journey is partly in water and partly in land. The first part is done in shortest time if the swimmer swims straight from $P$ to $O$. And once at $O$, the second part is shortest if he runs from $O$ to $R$. So in all, the quickest path is to swim to $O$ and then to run to $R$."? (Quite appropriately, this approach to optimisation is called the **greedy approach.** More generally, the term is applied wherever a problem is divided into several steps and at each stage we make a choice which is best for that step. This approach is simple but does not usually give the optimum for the whole problem).

11.15 Among all triangles $ABC$ with a given base $BC$ and a given angle $A$, find the one with the maximum area. [*Hint* : A simple fact from geometry will indicate a convenient choice of the independent variable. In fact the problem can be done purely with geometry.]

11.16 Among all triangles $ABC$ with a given base $BC$ and a given area, find the one for which the angle $A$ is maximum. Comment on the relationship between this and the last exercise.

11.17 A pumping station is to be constructed at a point $P$ along the bank of a straight river and to be joined by straight pipelines to each of two townships situated on the same side of the river. Find where $P$ should be located so as to minimise the cost of the pipelines. Do the problem by calculus methods and also geometrically by considering the reflection of one of the townships in the river.

**\*11.18** Given an acute angled triangle $ABC$ find the point $P$ inside it for which the sum of the lengths $|PA| + |PB| + |PC|$ is minimum. What happens if the triangle is not acute angled ?

**11.19** Two straight roads one going east-west and the other north-south meet at a point $O$. A car starts at a distance $a$ due west of $O$ and moves eastward on the first road at a speed $u$. Simultaneously, another car starts at a distance $b$ due north of $O$ and moves southward on the second road at a speed $v$. Find the shortest distance between the cars.

**11.20** $PQ$ is a diameter of a unit circle $C$. With $Q$ as the centre a variable circle is drawn to cut $PQ$ in $R$ and to cut $C$ at $S$ and $T$. Find the maximum area of the traingle $QSR$.

**11.21** If 100 mango trees are gown per acre, the average yield per tree is 963 fruits. Every additional tree decreases the average yield by 5 per tree. How many trees should be planted per acre to maximise the total fruit yield? (This problem comes under what is called **discrete optimisation,** because the number of trees is a discrete variable. Still, the calculus methods can sometimes be applied with suitable modifications.)

**11.22** Let $f(x) = ax^3 + bx^2 + cx + d$ where $a, b, c, d$ are constants and $a \neq 0$. Prove that $f$ has no local maximum or minimum if $b^2 \leq 3ac$ and otherwise $f$ has exactly one local maximum and one local minimum. Prove further that $f$ always has one point of inflection. Sketch the graph of $f$.

**11.23** Sketch the graph of $y = f(x)$ given that $f(0) = 1$ and $f'(x) = x(x - 2)$ for all $x$.

**11.24** Sketch the graphs of (i) $y = \dfrac{x}{x - 1}$ and (ii) $y = \dfrac{1}{x(x - 1)}$ following the procedure given in the text.

**11.25** Find the best possible constants $\alpha, \beta$ such that $\alpha x \leq \sin x \leq \beta x$ for all $x \in [0, \dfrac{\pi}{2}]$.

## 4.12    Implicit Differentiation and Parametric curves

We have seen that the derivatives of a function $y = f(x)$ shed important light on certain features of the graph of $f$, such as its tangents, its points of inflection. But these concepts make sense for all plane curves, not necessarily only for those which occur as the graphs of some functions. Take for example the ellipse $C$ in the standard form, centred at the origin. As a set of points in the plane, $C$ consists of those points $(x, y)$ which satisfy the equation

$$\frac{x^2}{a^2} + \frac{y^2}{b^2} = 1 \tag{1}$$

At every point $P(x_0, y_0)$ of $C$, there is a tangent. But how do we find it? If we could write $C$ as the graph of some function then we could, of course, differentiate this function. But this is impossible because one of the cardinal requirements of a function is that it must be single-valued. Translated into geometric terms this means that no vertical line can cut the graph in moe than one point. But this does not hold for $C$. In fact, unless $P$ happens to be $(a, 0)$ or $(-a, 0)$, the vertical line through $P$ cuts $C$ again at a point $Q \neq P$ (see Fig. 4.12.1).

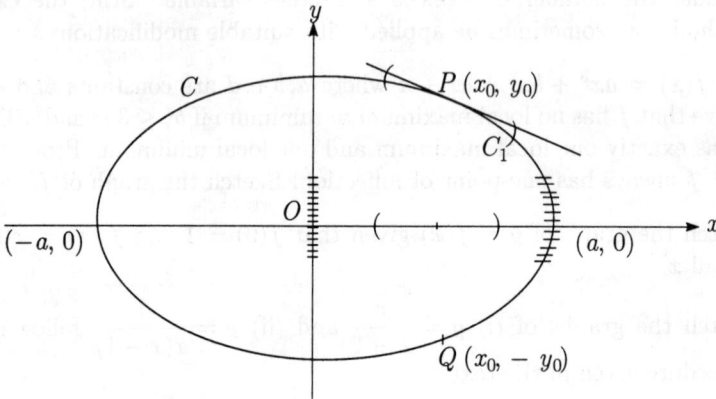

Figure 4.12.1 : Tangent to an Implicitly Defined Curve

Fortunately, the difficulty is not a serious one. Tangency is a local concept and so is a derivative. If we take a small portion $C_1$ of $C$ containing $P$, we see it can indeed be written in the form $y = f(x)$. In fact assuming $y_0 > 0$ (which implies $-a < x_0 < a$), we can solve (1) for $y$ and get

$$y = b\sqrt{1 - \frac{x^2}{a^2}} \quad (= f(x), \text{ say}) \tag{2}$$

which is valid for all $x$ in some neighbourhood of $x_0$. (In fact (2) is valid for all $x$ in $(-a, a)$, but that is not very vital here.)

Differentiating (2),

$$\left(\frac{dy}{dx}\right)_{x=x_0} = \frac{-bx_0}{a^2\sqrt{1-\frac{x_0^2}{a^2}}} = f'(x_0) \tag{3}$$

from which the equation of the tangent can be written down. If $P$ is $(a, 0)$ or $(-a, 0)$, no portion of $C$ containing $P$ in its interior can be expressed in the form $y = f(x)$. But we can express it as the graph of $x = g(y)$. Then we can find $\dfrac{dx}{dy}$ which comes out as 0. So the tangent is parallel to the $y$-axis at $(\pm a, 0)$.

The success of this approach was in solving (1) for $y$. This was a simple matter because (1) is a quadratic equation in $y$. But what if instead of (1) we had a complicated equation like

$$y^5 - 2y^4 + x^6 y + x - 3 = 0 \tag{4}$$

or worse still,

$$\cos(xy) + y^2 - \frac{2x}{\pi} = 0 \tag{5}$$

and we want the tangents at the points $(1, 2)$ and $\left(\dfrac{\pi}{2}, 1\right)$ respectively? (1) was a quadratic in $y$ (for each fixed $x$). Quadratic equations are easy to solve by a well-known formula. Somewhat more intricate formulas exist for solving equations if degree 3 and 4. But, in general, for an equation of higher degree, even though it has solutions, it is not always possible to express them by an explicit formula like (2) (see Section 1.13 for more elaboration). The equation (5) is not even an algebraic equation. Both (4) and (5) represent certain plane curves. For a fixed $x$, each becomes a numerical equation in $y$ and several methods are available to find their roots approximately. (One such method, called binary search was mentioned in Section 4.1. Another one, called the Newton-Raphson method will be studied in the next section.)

The point is that for (4) and (5) no expression like (2) is possible even in small neighbourhoods of the points $(1, 2)$ and $\left(\dfrac{\pi}{2}, 1\right)$ respectively. Equations (1), (4) and (5) are said to define $y$ as an **implicit function** of $x$. For (1), the equation (2) expresses $y$ as an **explicit** function (i.e., an ordinary function) of $x$. There is a theorem called the **Implicit Function Theorem** which says that under certain rather general conditions, every implicitly defined function (i.e., every equation of the form $\phi(x, y) = 0$ where $\phi$ is a function of two variables) indeed defines $y$ as a differentiable function of $x$ in some neighbourhood of a given point $(x_0, y_0)$ such that $\phi(x_0, y_0) = 0$. In other words, the implicit function theorem says that if the function $\phi$ and the point $(x_0, y_0)$ satisfy certain conditions then there exists a differentiable function $f$ defined on some neighbourhood $N$ of $x_0$ such that $f(x_0) = y_0$ and for all $x \in N, \phi(x, f(x)) = 0$. The derivative, $\dfrac{dy}{dx}$ at $x_0$ will then simply be $f'(x_0)$.

[As a concrete example, (1) can be written as $\phi(x,y) = 0$ where $\phi(x,y)$ $= \dfrac{x^2}{a^2} + \dfrac{y^2}{b^2} - 1$. This defines $y$ as an implicit function of $x$. Given any $(x_0, y_0)$ which satisfies $\phi(x_0, y_0) = 0$, (2) defines $y$ explicitly as a function of $x$ in a neighbourhood of $x_0$, provided $y_0 > 0$. If $y_0 < 0$, then $y = -b\sqrt{1 - \dfrac{x^2}{a^2}}$ defines $y$ as an explicit function of $x$ in a neighbourhood of $x_0$. If $y_0 = 0$ then $x_0 = \pm a$ and $y$ cannot be defined as a (single-valued) function of $x$ in any neighbourhood of $x_0$. However, $x$ can be defined as an explicit function of $y$ in a neighbourhood of 0, viz., $x = a\sqrt{1 - \dfrac{y^2}{b^2}}$ if $x_0 = a$ and $x = -\sqrt{1 - \dfrac{y^2}{b^2}}$ if $x_0 = -a$. ]

The trouble is that the implicit function theorem is an existence theorem. It only proves the existence of the function $f$. It gives no formula for it in general. And indeed, as we saw in the examples above, sometimes no such formula exists. In essence, all that the Implicit Function Theorem does is to show that $\dfrac{dy}{dx}$ at $x_0$ is a meaningful quantity, viz. the derivative of some well-defined function of $x$. It gives you no formula for this function, much the same way as merely proving that a series is convergent does not give a method for evaluating its sum.

But, as we have seen, the very fact that a series, say $\displaystyle\sum_{n=0}^{\infty} a_n$, is convergent enables us to draw certain inferences (e.g. that $a_n \to 0$ as $n \to \infty$, see Exercise (3.1.8)), regardless of what its sum is. Or, as a social analogy, the very fact that a person is married confers upon him/her certain status regardless of who the spouse is (even when sometimes, the spouse is dead or missing). Similarly the very fact that a relation of the form

$$\phi(x,y) \;\; = \;\; 0 \tag{6}$$

along with some point $(x_0, y_0)$ satisfying it defines a differentiable function $y = f(x)$ in a neighbourhood $N$ of $x_0$ such that

$$\phi(x, f(x)) \;\; = \;\; 0 \tag{7}$$

for all $x \in N$, enables us to proceed further. The moment $y$ becomes a differentiable function of $x$, any differentiable function of $y$ also becomes a differentiable function of $x$ and can be differentiated by the chain rule. So, even if we do not know $f(x)$ explicilty (except at $x_0$) in $N$, we may be able to find $f'(x_0)$.

We are not in a position even to state the Implicit Function Theorem at this stage, much less to prove it. But let us illustrate how it works. We first apply it to (1), where we already have the answer (3) by solving for $y$ explicitly. Substituting $y = f(x)$ (which we assume as not explicitly known) in (1) we get

$$\frac{x^2}{a^2} + \frac{[f(x)]^2}{b^2} \;\; = \;\; 1 \text{ for all } x \in N \tag{8}$$

We differentiate the left hand side using the chain rule. Since the right hand side is a constant function, we get

$$\frac{2x}{a^2} + \frac{2f(x)f'(x)}{b^2} = 0 \qquad (9)$$

This holds for all $x \in N$. In particular, it holds at $x = x_0$. But $f(x_0)$ is $y_0$. So (9) gives,

$$\frac{2x_0}{a^2} + \frac{2y_0 f'(x_0)}{b^2} = 0 \qquad (10)$$

which, after solving and noting $y_0 \neq 0$, gives

$$f'(x_0) = \frac{-x_0 b^2}{a^2 y_0} \qquad (11)$$

(which is the same as (3) since $y_0 = b\sqrt{1 - \frac{x_0^2}{a^2}}$ ).

Thus we see that even if we did not have (2), we could start right from (1) and get (10) and solve it finally to get $f'(x_0)$, which, geometrically, represents the slope of the tangent to the curve $C$ (defined by (1)) at the point $P = (x_0, y_0)$ on it.

Let us now apply this procedure to (4), where finding $f(x)$ explicitly is not possible, even though it exists by the Implicit Function Theorem. It is hardly necessary to replace $y$ by $f(x)$ everywhere before differentiating. We leave $y$ as it is and wherever it occurs, we differentiate that expression w.r.t. $y$ and multiply by $\frac{dy}{dx}$ (as per the chain rule). Thus (4) gives

$$5y^4 \frac{dy}{dx} - 8y^3 \frac{dy}{dx} + x^6 \frac{dy}{dx} + 6x^5 y + 1 = 0 \qquad (12)$$

Putting $x = 1$ and $y = 2$ gives $\frac{dy}{dx} = -\frac{13}{17}$. So this is the slope of the tangent at the point $(1,2)$ on the curve represented by (4). What if we want the second order derivative $\frac{d^2y}{dx^2}$ at $x = 1, y = 2$ ? We can solve (12) to get

$$\frac{dy}{dx} = -\frac{6x^5 y + 1}{5y^4 - 8y^3 + x^6} \qquad (13)$$

differentiating which,

$$\frac{d^2y}{dx^2} = \frac{(6x^5 y + 1)(20y^3 \frac{dy}{dx} - 24y^2 \frac{dy}{dx} + 6x^5) - (5y^4 - 8y^3 + x^6)(30x^4 y + 6x^5 \frac{dy}{dx})}{(5y^4 - 8y^3 + x^6)^2}$$

$$(14)$$

We can now put $x = 1, y = 2$ and $\frac{dy}{dx} = -\frac{13}{17}$ in (14) to get $\left(\frac{d^2y}{dx^2}\right)_{x=1}$

$= -\frac{25504}{4913} < 0$. Since $\frac{d^2y}{dx^2}$ is continous at 1, we see that the curve is concave downwards in a neighbourhood of the point $(1, 2)$. [Note that in this example, we could solve (12) for $\dfrac{dy}{dx}$ to get (13) and then (14). This is not possible in all examples. Nor is it necessary. We could have differentiated (12) directly and after regrouping the terms got

$$(5y^4 - 8y^3 + x^6)\frac{d^2y}{dx^2} + (20y^3 - 24y^2)\left(\frac{dy}{dx}\right)^2 + 12x^5\frac{dy}{dx} + 30x^4y = 0$$

and put $x = 1, y = 2, \dfrac{dy}{dx} = -\dfrac{13}{17}$ to get $\dfrac{d^2y}{dx^2}$ at $x = 2$. ]

Thus we see that even with implicit differentiation we get as much information as with explicit differentiation when the latter is possible. And when it is not, implicit differentiation is the only go.

The concept of a parametrised curve lies in between an explicitly and an implicitly defined functions, but leaning more towards the former. We already referred to it in Section 4.7 when we talked of the length of a curve. (See also Exercise (4.7.2).) There are two ways to look at a plane curve, say $C$. One is to treat it as a 'solution set' i.e., as the set of all points $(x, y)$ in $I\!\!R^2$ which satisfy some equation of the form

$$\phi(x, y) = 0 \tag{15}$$

where $\phi$ is a function of two variables, $x$ and $y$. Many of the curves we are familiar with are of this form. For example, the ellipse in (1) arises if we take $\phi(x, y) = \dfrac{x^2}{a^2} + \dfrac{y^2}{b^2} - 1$. As we just saw, implicit differentiation is a satisfactory tool for curves defined this way.

But there are situations where this form of specifying a curve $C$ is not very convenient. For example, how do we define its length ? This is where a parametrised curve comes very handy as we already saw. Similarly, when we want to study the motion of a particle on $C$, in effect, we are parametrising it with time as the parameter.

In essence, a parametrisation of a curve $C$ is a pair of functions $(f, g)$ where, for each $t$ in some interval, say, $[a, b]$, the point $(f(t), g(t))$ lies on $C$, and conversely, every point of $C$ is of this form for some $t \in [a, b]$. The very idea of motion forces us to require that each of the functions $f$ and $g$ be continuous on $[a, b]$. In a suggestive language, we may say that $C$ is the path traced out by a moving piont $(f(t), g(t))$ in the plane as the 'parameter' $t$ 'moves' from $a$ to $b$. Frequently, instead of $f$ and $g$ we use the symbols $x$ and $y$ themselves and write $x = x(t), y = y(t)$ instead of $x = f(t), y = g(t)$.

Given a curve in a parametric form

$$x = f(t) = x(t) , \quad y = g(t) = y(t) ; \quad a \le t \le b \tag{16}$$

it is possible, in theory, to eliminate $t$ and get an equation like (15). But this elimination is not always easy. The conversion from (15) to (16) is even more difficult.

The Implicit Function Theorem mentioned above gives a local solution to the latter problem, i.e. given a point $(x_0, y_0)$ satisfying (15), if the function $\phi$ satisfies certain conditions, then we can express a part of the solution set of (15) in the form $y = f(x)$, which is a parametrisation of a special type, the parameter being $x$ itself. Near some points, a parametrisation of the form $x = g(y)$ may work. But the Implicit Function Theorem does not give a parametrisation of the entire solution set, all at one time. For a particular function $\phi$, it may be possible to go from (15) to (16) using some *ad-hoc* methods and ingenuity. (For example, in the case of a hyperbola given implicitly by the equation $\dfrac{x^2}{a^2} - \dfrac{y^2}{b^2} = 1$, an ingeneous use of the identity $\tan^2 \theta + 1 = \sec^2 \theta$ leads to the parametrisation, $x = a \sec \theta$, $y = b \tan \theta$.)

Note also that the conversion from (15) to (16) is not unique. That is, the same curve can be parametrised in many different ways. For example, the semi-circle in the upper half of the $x$-$y$ plane

$$\{(x, y) : x^2 + y^2 = 1, y \geq 0\} \tag{17}$$

can be parametrised either by

$$x = \cos \theta, \quad y = \sin \theta; \quad 0 \leq \theta \leq \pi \tag{18}$$

or also by

$$x = x, \quad y = \sqrt{1 - x^2}; \quad -1 \leq x \leq 1. \tag{19}$$

The geometric properties of a curve such as its length, directions of the tangents at various points are independent of a particular parametrisation. But when it comes to the motion of a particle moving along the curve, attributes such as its speed and acceleration depend crucially on the parametrisation. It is therefore convenient to incorporate parametrisation as an integral part of the definition of a curve.

**12.1 Definition :** A **plane curve** (or **path**) (i.e., a curve in $I\!\!R^2$) is an ordered pair of continuous real-valued functions $(f, g)$, each defined on some closed and bounded interval, say, $[a, b]$.

A few comments are in order. First of all, instead of dealing with a pair of real-valued functions $(f, g)$, it is convenient to deal with a single function, say, $\alpha$ from $[a, b]$ to $I\!\!R^2$, defined by $\alpha(t) = (f(t), g(t))$, for all $t \in [a, b]$. $f, g$ are called the **component functions** of $\alpha$. We say $\alpha$ is continuous iff its component functions $f, g$ are both continuous. With this terminology, a plane curve is nothing but a continuous function, say, $\alpha$, from an interval of the form $[a, b]$ to $I\!\!R^2$. Some authors require this interval to be $[0, 1]$. But this is not very vital, because with a simple change of parameter this can always be arranged (see Exercise (12.3)). The restriction that the domain of $\alpha$ be a closed and bounded interval is also not very vital for our purpose. (It is important in certain branches of mathematics.) It could just as well be any interval, open or semi-open, bounded or unbounded. In fact it could be the whole real line, as we have in the case of the entire parabola, $y^2 = 4ax$ $(a > 0)$ parametrised by $x = at^2, y = 2at$. When the domain of $\alpha$ is of the form $[a, b]$ then the points $\alpha(a)$ and $\alpha(b)$ in the plane are called, respectively, the **initial** (or **starting**) **point** and the **terminal** (or **end**) point of the curve. We also say that the curve is **oriented** from $\alpha(a)$ to $\alpha(b)$. It may happen that $\alpha(a) = \alpha(b)$ in which case $\alpha$ is called a **closed curve** or sometimes a **cycle**. Notice that if we define $\beta : [a, b] \to I\!\!R^2$ by $\beta(t) = \alpha(a+b-t)$, then $\beta$ is also a curve in $I\!\!R^2$ whose initial and terminal points are precisely the terminal and the initial points of $\alpha$. We say $\alpha$ and $\beta$ are **oppositely oriented**. (In diagrams, orientations are generally shown by arrows.) In such a case $\beta$ is often denoted by $-\alpha$. Note that $-\beta = \alpha$. (Prove !)

The crucial point to note is that according to Definition (12.1), a plane curve is a certain function into a plane, and not a subset of the plane. In particular a curve $\alpha$ should never be confused with the range of $\alpha$, which is indeed a subset of $I\!\!R^2$, consisting of all points of the form $\alpha(t), a \leq t \leq b$. It is entirely possible to have two different curves with the same range set, e.g., the curves defined by (18) and (19) above. Note that $\alpha$ and $-\alpha$ always have the same range set. As a more poignant example, let $\alpha(\theta) = (\cos\theta, \sin\theta); 0 \leq \theta \leq 2\pi$ and $\beta(\theta) = (\cos\theta, \sin\theta); 0 \leq \theta \leq 3\pi$. Then both $\alpha$ and $\beta$ have the unit circle as their range. But they are different curves because even though the formulas for $\alpha$ and $\beta$ are the same, their domains are different. In intuitive language we may say $\alpha$ 'wraps' the circle once but $\beta$ wraps it $1\frac{1}{2}$ times! (See Fig. 4.12.2, where in the range of $\beta$ the upper semi-circle is shown thicker to indicate that it is covered twice.)

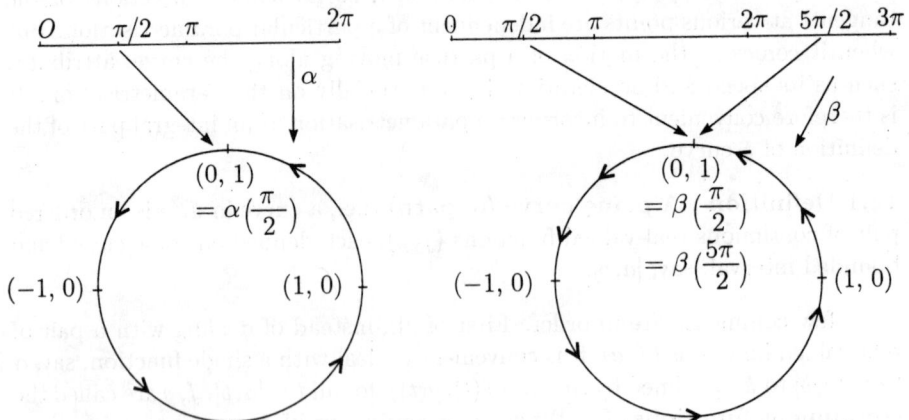

Figure 4.12.2 : Different Curves with same Ranges

Definition (12.1) also forces us to discard our intuitive understanding of a curve as a 'thin', i.e., 'one-dimensional' subset of $I\!\!R^2$. Already we saw that a curve is conceptually different from its range. So a proper question to ask is whether the range of a curve (and not the curve itself) is a one-dimensional subset of $\mathbb{R}^2$. And the answer is no. The only requirement which Definition (12.1) lays down on a plane curve $\alpha$ is that its component functions, say $f$ and $g$, are continuous. As discussed in Section 4.3, continuity is too general and hence too weak a concept. Exercise (3.10) gives an example of a continuous function which is not differentiable anywhere. By doing something similar, it is possible to construct a curve $\alpha : [0,1] \to \mathbb{R}^2$, whose range set is the entire closed unit disc ! Such **space-filling curves** (sometimes called **Peano curves**) are, like the divergent series $\sum\limits_{n=1}^{\infty} \dfrac{1}{n}$, among those mathematical objects where our intuition fails miserably.

The reason for this failure is that our intuition is based on curves which are not only continuous, but satisfy some stronger condition, viz., differentiability. A curve $\alpha : [a,b] \to \mathbb{R}^2$ is said to be **differentiable** if the component functions of $\alpha$, say $f$ and $g$, are both differentiable on $[a,b]$. If further, $f'$ and $g'$ are continuous on $[a,b]$ and for every $t \in [a,b]$, at least one of the numbers $f'(t), g'(t)$ is non-zero, (which is, of course, equivalent to saying that the vector $(f'(t), g'(t))$ is non-zero) then the curve $\alpha$ is said to be **smooth**. (The reader is cautioned that this word is also used with various other shades of restriction. Some authors, for example, require $f, g$ to be infinitely differentiable everywhere. In consulting other literature, it is best to check exactly in what sense it is used.)

Note that (18) is a smooth curve, but (19) is not because the function $g(x) = \sqrt{1 - x^2}$ is not differentiable[2] at the end-points $-1$ and $1$.

The geometric significance of smoothness is that the curve has a well-defined tangent at every point as we now show. The proof is based on the concept a unit vector, i.e., a vector of length 1. In essence, a unit vector is a way of specifying a direction. (See Exercise (12.5) for those not familiar with unit vectors.)

**12.2 Theorem :** Let $\alpha(t) = (f(t), g(t)), a \leq t \leq b$ a smooth curve. Then for every $t_0 \in [a,b], \alpha$ has a tangent at the point $\alpha(t_0)$ and this tangent is parallel to the vector $(f'(t_0), g'(t_0))$.

**Proof :** Let $P_0 = \alpha(t_0)$ and $P = \alpha(t_0 + \Delta t)$ where $|\Delta t|$ is so small that $t_0 + \Delta t \in [a,b]$. (As usual, if $t_0 = a$ or $b$ then we consider only positive (or negative) values of $\Delta t$.)

The directed line segment $\overrightarrow{P_0 P}$ has no well-defined direction if $P = P_0$. We claim that if $\Delta t$ is sufficiently small (and non-zero) then $P \neq P_0$. We are given that at least one of $f'(t_0)$ and $g'(t_0)$ is non-zero. Without loss of generality assume that $f'(t_0) \neq 0$ and further $f'(t_0) > 0$. (The argument is similar if $g'(t_0) \neq 0$.) Then by Theorem (5.2), there exists some $\delta > 0$ such that for $0 < |\Delta t| < \delta, f(t_0 + \Delta t) > f(t_0)$ if $\Delta t > 0$ and $f(t_0 + \Delta t) < f(t_0)$ if $\Delta t < 0$. In either case, $\Delta t \neq 0$ implies $\Delta x \neq 0$ where $\Delta x = f(t_0 + \Delta t) - f(t_0)$. So regardless ot what $\Delta y = 0$ is (where we define $\Delta y = g(t_0 + \Delta t) - g(t_0)$), $P \neq P_0$, and

---

[2] If we allow $\infty$ and $-\infty$ as possible values, then $g(x)$ is differentiable even at $-1$ and $1$ respectively. This is consistent with the fact that the tangents to the semi-circle at $(-1, 0)$ and $(1, 0)$ are vertical. So it is not a bad idea even to declare (19) as a smooth curve and some authors do adopt it.

so $\overrightarrow{P_0P}$ is the non-zero vector $(\triangle x, \triangle y)$. A unit vector along its direction is

$$\left( \frac{\triangle x}{\sqrt{(\triangle x)^2 + (\triangle y)^2}}, \frac{\triangle y}{\sqrt{(\triangle x)^2 + (\triangle y)^2}} \right).$$

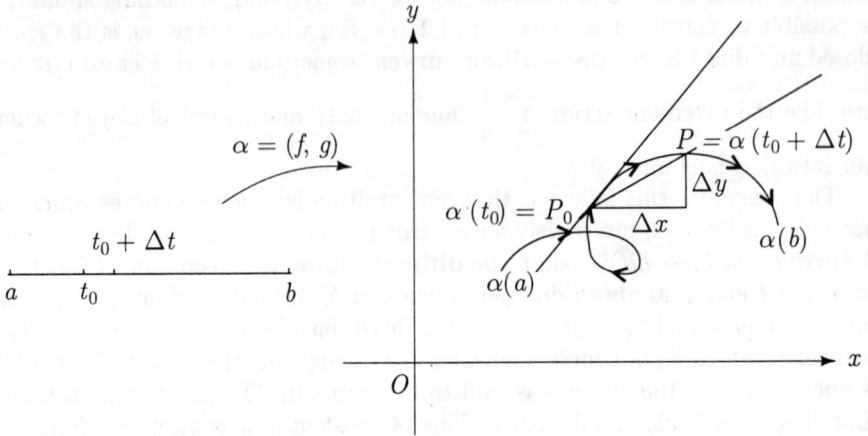

Figure 4.12.3 : Tangent to a Plane Curve

Let us rewrite this as

$$\left( \frac{\triangle x/\triangle t}{\sqrt{\left(\frac{\triangle x}{\triangle t}\right)^2 + \left(\frac{\triangle y}{\triangle t}\right)^2}}, \frac{\triangle y/\triangle t}{\sqrt{\left(\frac{\triangle x}{\triangle t}\right)^2 + \left(\frac{\triangle y}{\triangle t}\right)^2}} \right) \tag{20}$$

Now, by definition, $\frac{\triangle x}{\triangle t} \to f'(t_0)$ and $\frac{\triangle y}{\triangle t} \to g'(t_0)$ as $\triangle t \to 0$. So by elementary properties of limits, the vector in (20) tends to the vector

$$\left( \frac{f'(t_0)}{\sqrt{[f'(t_0)]^2 + [g'(t_0)]^2}}, \frac{g'(t_0)}{\sqrt{[f'(t_0)]^2 + [g'(t_0)]^2}} \right) \tag{21}$$

which is a unit vector. But on the other hand, as $\triangle t \to 0$, the secant $P_0P$ tends, by definition, to the tangent at $P_0$. In other words, the tangent at $P_0$ is the line through $P_0$ whose direction is the limit of the direction of the secant $P_0P$. So the line through $P_0$ in the direction of the unit vector in (21) is tangent to the curve at $P_0$. Obviously this is parallel to the vector $(f'(t_0), g'(t_0))$. ∎

Let us now go back to the problem of finding the tangent to a point $(x_0, y_0)$ on the ellipse (1). In (3), we found it by solving (1) explicitly for $y$. Later, in (11), we found it by implicit differentiation. As a third method, we can parametrise the entire ellipse $C$ (not just its upper half) by

$$x = a\cos\theta, y = b\sin\theta; \ \ 0 \le \theta \le 2\pi. \tag{22}$$

Let $(x_0, y_0)$ correspond to $\theta = \theta_0$. Then by the theorem just proved, the tangent at $(x_0, y_0)$ would be parallel to the vector

$$(-a\sin\theta_0, b\cos\theta_0) \tag{23}$$

If $y_0 = 0$, then $\sin\theta_0 = 0$ and the vector (23) is parallel to the $y$-axis. In other words, the tangents to the ellipse (1) at $(a, 0)$ and $(-a, 0)$ are 'vertical'. If on the other hand $y_0 \neq 0$, then from (23), the slope of the tangent is $\dfrac{b\cos\theta_0}{-a\sin\theta_0}$. Since $a\cos\theta_0 = x_0$ and $b\sin\theta_0 = y_0$, we can rewrite this as $-\dfrac{b^2 x_0}{a^2 y_0}$, which is the same as (11).

Thus we see that parametrisation of a curve provides an alternative (and a slightly more general) method than implicit differentiation for finding the tangent. Parametrisation, of course, has many other uses. We already saw in Section 4.7 how the arc length of a parametrised curve can be defined as the supremum of the lengths of broken line paths. We shall take this up further in Section 6.2.

Another advantage of parametric curves is that they generalise very easily to higher dimensions. Thus, a (parametric) curve in $I\!R^n$ is simply a continuous function from $[a, b]$ to $I\!R^n$. (Here again, by continuity of $\alpha$ we mean that of each of its $n$ component functions.) Theorem (12.2), as well as its proof, go through with hardly any change (other than replacing 2 with $n$).

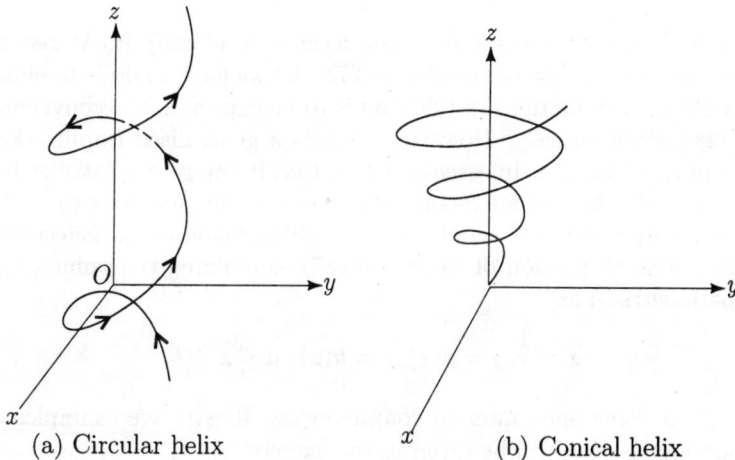

(a) Circular helix      (b) Conical helix

Figure 4.12.4 : Curves in $I\!R^3$

Figure 4.12.4 shows two space curves (i.e., curves in $I\!R^3$) a circular helix and a conical helix. The first is parametrised by

$$x = \cos\theta, \ y = \sin\theta, \ z = c\theta \ (c \text{ constant}), \ -\infty < \theta < \infty \tag{24}$$

while the second is parametrised by

$$x = \theta \cos \theta, \ y = \theta \sin \theta, \ z = \theta, \quad 0 \le \theta < \infty. \tag{25}$$

If we want to deal with curves in $I\!R^3$ by means of equations in $x, y, z$ of the form

$$\phi(x, y, z) \ = \ 0 \tag{26}$$

analogous to (15), then the first difficulty is that in general an equation like (26) represents a surface (e.g. a sphere, a plane, a cylinder, a cone etc. ) and not a 'curve'. A curve in $I\!R^3$ can be specified as the intersection of two surfaces. That is, it is the set of points $(x, y, z)$ satisfying a pair of simultaneous equations of the form

$$\phi_1(x, y, z) = 0 \quad \text{and} \quad \phi_2(x, y, z) = 0 \tag{27}$$

Note that these functions $\phi_1$ and $\phi_2$ are not unique. The same curve can be expressed as the intersection of two surfaces in many different ways. For example the same circle can be expressed as the intersection of a sphere and a plane, or of two spheres or of a plane and a cylinder and so on. So the functions $\phi_1$ and $\phi_2$ in (27) are not uniquely determined by the given curve. Secondly, it is not always easy to visualise a curve specified as the intersection of two surfaces. A parametrisation of the curve, on the other hand, given by, say,

$$x = f(t), \ y = g(t), \ z = h(t); \ a \le t \le b \tag{28}$$

'traces' the curve as $t$ moves and makes it easier to identify it. Moreover, if at all one wants, (28) can be converted to (27). All we have to do is to eliminate $t$ in any two pairs of equations in (28). As is to be expected, the conversion from (27) to (28) is not so easy. However, there is a generalised Implicit Function Theorem in this case too. In essence it says that if $(x_0, y_0, z_0)$ satisfies both the equations in (27), then under certain conditions on the functions $\phi_1$ and $\phi_2$, we can write $y$ and $z$ as functions of $x$ in a neighbourhood of $x_0$. Effectively, this means that a small portion of the curve (27), containing the point $(x_0, y_0, z_0)$ can be parametrised as

$$x = x, y = g(x), z = h(x); \ a \le x \le b.$$

where $(a, b)$ is some open interval containing $x_0$. Illustrative examples of such 'local' parametrisation will be given in the exercises.

Going back to the title, what implicit differentiation and parametrisation of curves have in common is that both are techniques designed to enable us to apply the facts about functions of one variable to situations where the data is in the form of functions of two (or more) variables.

# EXERCISES

12.1 For the function defined implicitly by (5), find the equation of the tangent at the point $\left(\dfrac{\pi}{2}, 1\right)$.

12.2 For the function defined by (4), find $\dfrac{dx}{dy}$ by implicit differentiation at the point $(1, 2)$. Then check that it is indeed the reciprocal of $\dfrac{dy}{dx}$ at that point.

12.3 Let $\alpha : [a, b] \to I\!\!R^2$ be a plane curve. Define $f : [0, 1] \to [a, b]$ by $f(t) = a + t(b - a)$. Let $\beta = \alpha \circ f$. Prove that the curves $\alpha$ and $\beta$ are in general different, but have substantially similar properties. For example, show that they always cover exactly the same sets of points of $I\!\!R^2$. Further, $\beta$ is rectifiable if and only if $\alpha$ is so and when this is the case, both have equal lengths. Similarly, show that $\beta$ is smooth if and only if $\alpha$ is so and that for every $t \in [0, 1]$, the direction of the tangent to $\beta$ at $\beta(t)$ is the same as that of the tangent to $\alpha$ at $\alpha(f(t))$. (The significance of this exercise is that in studying parametrised curves, without loss of generality, we may suppose that the parameter always ranges over the unit interval $[0, 1]$.)

12.4 Prove that for any curve $\alpha : [a, b] \to I\!\!R^2$, the oppositely oriented curve $-\alpha$ is smooth if and only if $\alpha$ is smooth. How are the directions of their tangents related?

12.5 Two non-zero vectors $\mathbf{u}$ and $\mathbf{v}$ in $I\!\!R^2$ (or more generally in $I\!\!R^n$) are said to have the same direciton, if there exists some $\lambda > 0$ such that $\mathbf{u} = \lambda \mathbf{v}$. Prove that this is an equivalence relation on the set of all non-zero vectors and that each equivalence class contains exactly one unit vector. (Adopting the definition trick, mentioned in Section 1.8, we may formally *define* the direction of a non-zero vector either as an equivalence class under the equivalence relation just defined or as the unique unit vector representing that class. Yet another instance where an intuitively clear concept requires a clumsy definition for the sake of precision!)

12.6 A **cycloid** is defined as the locus of a point $P$ on the circumference of a circle which moves without slipping in its plane always touching a fixed line $L$ (for example the wheel of a car moving on a straight road). Assuming that the speed is constant and radius is $a$ show that a cycloid may be parametrized by (see Fig. 4.12.5) $x = a(\theta - \sin\theta)$, $y = a(1 - \cos\theta)$.

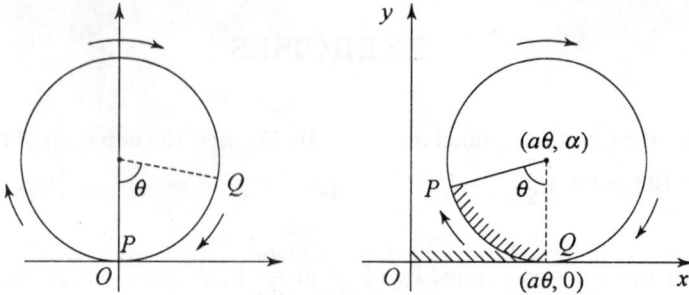

Figure 4.12.5 : A Cycloid

Sketch the portion of the cycloid for $0 \leq \theta \leq 2\pi$. (A cycloid is a very interesting curve and appears unexpectedly in certain optimisation problems. For example, suppose $A, B$ are two points in a vertical plane, but not in the same vertical line, with $A$ at a higher level than $B$. They are joined by a smooth curve $C$ and a particle moves under gravity from $A$ to $B$, along the curve $C$. It can be shown that the shortest time taken is not when $C$ is a straight line, as one would expect, but when $C$ is an inverted arc of a cyloid! This is known as the **brachistochrone problem**.)

12.7    Analogous to the last exercise, obtain the loci of $P$ if the circle moves along another fixed circle of radius $R$ touching it (i) externally (ii) internally (assuming $R > a$ here). Prove that each locus is a closed curve if and only if $R/a$ is a rational number. (These loci are called, respectively, an **epicycloid** and a **hypocycloid**. The special case of the epicycloid in which $R = a$ is called a **cardioid**. See Section 8.6 for an alternate definition and a sketch of it.)

12.8    Prove that in $\mathbb{R}^3$, the two planes having equations $a_1 x + b_1 y + c_1 z = d_1$ and $a_2 x + b_2 y + c_2 z = d_2$ intersect in a straight line if and only if the expressions $a_1 b_2 - a_2 b_1$, $b_1 c_2 - b_2 c_1$ and $c_1 a_2 - c_2 a_1$ are not all zero. Assume this condition is satisfied and $L$ is the straight line where they intersect. If $(x_0, y_0, z_0)$ is any point of $L$, prove that the entire line $L$ can be parametrized by

$$
\begin{aligned}
x &= x_0 + t(b_1 c_2 - b_2 c_1) \\
y &= y_0 + t(c_1 a_2 - c_2 a_1) \\
z &= z_0 + t(a_1 b_2 - a_2 b_1)
\end{aligned}
$$

When can $L$ be parametrised in the form $y = g(x), z = h(x)$?

12.9    Show that the curve $C$ in $\mathbb{R}^3$ parametrized by $x = \cos\theta$, $y = \sin\theta$, $z = a\cos\theta + b\sin\theta + c$, where $a, b, c$ are some constants, is the intersection of the right circular cylinder $S_1 = \{(x,y,z) : x^2 + y^2 = 1\}$ and the plane $S_2 = \{(x,y,z) : z = ax+by+c\}$. By setting a suitable co-ordinate system $(x', y')$ in the plane $S_2$, show that $C$ is a circle if $r = 0$ and an ellipse if $r > 0$ where $r = \sqrt{a^2 + b^2}$. [*Hint* : For $r > 0$, let $P = (\frac{a}{r}, \frac{b}{r}, r + c)$ and $Q = (-\frac{b}{r}, \frac{a}{r}, c)$. Take the axes along $O'P$ and $O'Q$. To find the $(x', y')$-coordinates of a point $(x, y, ax + by + c)$ in $S_2$, it is helpful to consider unit vectors along the axes.]

12.10 The intersection of a right circular cone and a plane is called a **conic**. Suppose the vertex (or the **apex**) of the cone is at the origin and that its axis is the $z$-axis. Further, for simplicity, take the cone as $z^2 = x^2 + y^2$. (Essentially the same argument applies to any other cone.) Then, without loss of generality the plane may be taken as $z = mx + c$ (except when it is parallel to the $z$-axis). Prove that if $c \neq 0$, then the conic is a circle for $m = 0$, an ellipse if $0 < |m| < 1$, a parabola if $m = \pm 1$ and a hyperbola if $|m| > 1$ (or if the plane is parallel to the $z$-axis). In each case parametrise it.

12.11 Let $C$ be the intersection of the surfaces $z = xy^2$ and $z \sin x + y = \frac{\pi}{12} + 1$. Prove that the point $P\left(\frac{\pi}{6}, 1, \frac{\pi}{6}\right)$ lies on $C$. Parametrise $C$ locally at $P$ in the form $y = g(x), z = h(x)$.

## 4.13 Binary Search and Newton's Method for Approximate Solutions of Equations

The problem of identifying the zeros of a function, i.e. the points where the function vanishes crops up many times. (In case the function happens to be a polynomial, the zeros are also called roots.) For example in maxima / minima problems, we need the zeros of the derived function. Similarly for identifying the points of inflection of a twice differentiable fraction $f$, we have to solve the equation $f''(x) = 0$. As we saw in the last section, when a function is given implicitly, to find its possible values for a given $x$, say $x_0$, we set $x = x_0$ in it and solve the resulting equation in $y$.

It is very rare that such equations can be solved exactly. Even for the relatively simple case of a polynomial equation, there is no general formula to express the roots in terms of radicals (see Section 1.13 for more elaboration). And even when such a formula exists, to evaluate radicals like $\sqrt{3}$, $\sqrt[3]{-5}$ etc. in numerical terms, we have to content ourselves with only approximate values. So we have to look for methods which will give approximate values of zeros. There are various such methods. In Section 4.1, we studied one such method, based on binary search. We shall study another one, in this section.

The merits and demerits of such methods are tested by applying various criteria listed below. (In each case we assume that $\alpha$ is a zero of a function $f$ while $\alpha'$ is an approximate zero given by the method.)

(i) **Applicability**: Under what conditions on the function $f$ is the method applicable?

(ii) **Efficiency**: What is the amount of work needed in finding $\alpha'$? How much time does it take?

(iii) **Efficacy**: How effective is the method? This itself can be measured in two ways, viz., how close is $f(\alpha')$ to 0 or how close is $\alpha'$ to $\alpha$?

(iv) **Stability**: If the function $f$ is altered slightly to get a new function $g$, and the method is applied to find an approximate zero, say $\alpha_1'$, of $g$, will $\alpha_1$ be close to $\alpha'$? Also is it possible to use some of the work already done in finding $\alpha'$ or does the method have to start all over again to find $\alpha_1'$?

These are, of course, the general criteria for measuring the merit of any numerical method. A quantitative study of this type comes under a vast topic called **analysis of algorithms.** We can hardly go into it. But a few qualitative remarks can be made. Considerations of stability are well beyond our scope. But regarding the three other yardsticks, it is to be expected, by sheer common sense, that a method which is more widely applicable will generally not score very high in terms of efficiency or efficacy. The latter two qualities are, in fact, related. We may compare two methods either by comparing the time they take to yield approximations with equal degrees of accuracy or by comparing the degrees of accuracy they achieve for the same amount of work. These two yardsticks are therefore said to be dual to each other.

What do we mean by the 'work' involved in a particular method? The answer depends upon what types of numerical calculations are involved and also on how many times they are carried out. The numerical calculations are of the simple type and hence all of them may be assumed to take more or less the same amount of time. So what matters most is the number of times they are carried out. The various methods for finding approximate zeros are usually **iterative** in nature. They don't give you the answer right away. Instead they ask you to start with some 'guess', say $x_0$ of your choice and set $\alpha' = x_0$. If $f(\alpha') = 0$, we are lucky. If not, the method tells you how to 'improve' your guess, i.e., to obtain some $x_1$ from $x_0$, by some calculaitons (depending on the mehtod) and set $\alpha' = x_1$. If $f(\alpha') = 0$ now, we are done. Otherwise we replace $x_1$ by $x_2$ which is obtained from $x_1$ exactly the way $x_1$ was obtained from $x_0$. And this procedure is repeated (hence the name 'iterative procedure') again and again till we reach either an exact solution or a solution which is within a desired degree of accuracy. Since the amount of work done in each iteration is the same, a good measure of the total work done is the number of iterations (i.e. the number of times the iterative procedure is executed.)

For example, in the case of the binary search applied to a continuous funciton $f$ on a closed and bounded interval $[a, b]$ (where we assume $f(a)$ and $f(b)$ are of opposite signs) we take $x_0 = \dfrac{a+b}{2}$. If we follow the proof of Theorem (1.3), we get a sequence $\{x_n\}$ which may be a finite sequence or an infinite sequence converging to a zero, say, $\alpha$, of $f$ in $[a, b]$. Moreover $|x_n - \alpha| \leq \dfrac{|b-a|}{2^{n+1}}$ for all $n$. So, given any $\delta > 0$, we know in advance how many iterates at most are needed to get an approximate zero $\alpha'$ so that $|\alpha' - \alpha| < \delta$. Note that from this we cannot say anything about how close $f(\alpha')$ will be to $f(\alpha)$ (i.e., to 0). If, however, $f$ satisfies some additional hypothesis (than mere continuity), such as a Lipschitz condition (see Exercise (2.7) for a definition and Exercise (5.21) for a condition implying it) then we can tell in advance in how many iterates $|f(\alpha')|$ will be smaller than a given $\epsilon > 0$. (See Exercise (13.1))

It was mentioned in Section 4.3 that differentiability is stronger than continuity. So it is to be expected that if a function $f$ is differentiable, then methods more efficient than the binary search (which requires mere continuity of $f$) should be possible. We shall study one such method, called **Newton's** or **Newton-Raphson's method**.

This method is based on the concept of the first order, or linear approximation to a differentiable function. Suppose we want to find a zero, say $\alpha$, of a function $y = f(x)$. Take a point, say, $x_0$. If $f(x_0) = 0$, we can take $\alpha = x_0$. If $f(x_0) \neq 0$ and $f'(x_0)$ exists, then in a small neighbourhood of $x_0$, the function $f(x)$ behaves very much like its first order linear approximation, viz.,

$$L(x) = f(x_0) + f'(x_0)(x - x_0) \qquad (1)$$

Geometrically, the graph of $L(x)$ is precisely the tangent to the graph of $y = f(x)$ at $(x_0, f(x_0))$. We assume $f'(x_0) \neq 0$, i.e. the tangent is not horizontal. Then by a simple calculation, it will intersect the $x$-axis at $(x_1, 0)$ where

$$x_1 = x_0 - \frac{f(x_0)}{f'(x_0)} \qquad (2)$$

(See Fig. 4.13.1.)

By very construction of $x_1$, $L(x_1) = 0$. If luckily $f(x_1)$ were also 0 then we could take $\alpha$ as $x_1$. The point to note is that if $x_1$ is close to $x_0$ then $L(x_1)$ will be very close to $f(x_1)$. And so, even if $f(x_1)$ is not 0, it will be fairly close. So a zero of $f(x)$ is likely to be located in a small neighbourhood of $x_1$. We now apply the construction above to $x_1$, i.e. we assume $f'(x_1)$ exists and is non-zero, and let $(x_2, 0)$ be the point of intersection of the $x$-axis and the tangent at $(x_1, f(x_1))$. Then, in exact analogy with (2), we get,

$$x_2 = x_1 - \frac{f(x_1)}{f'(x_1)} \tag{3}$$

Once again, $f(x_2)$ may or may not be 0. If it is, we take $\alpha = x_2$. Otherwise we continue the procedure above.

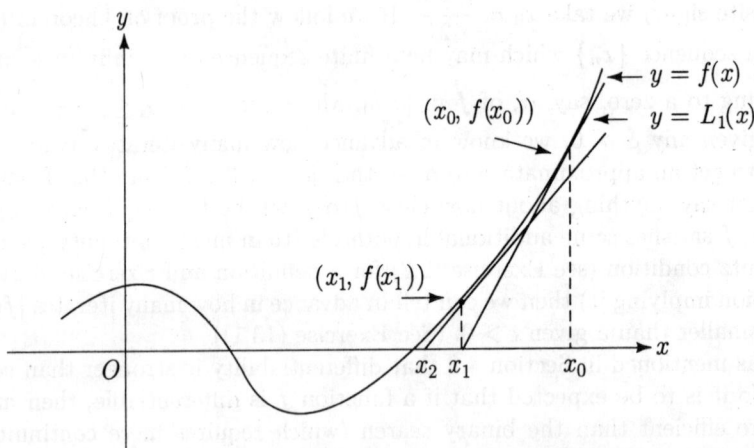

Figure 4.13.1 : Newton's Method

As a simple example, let us find the cube root of 2. This is equivalent to finding a zero of the function $f(x) = x^3 - 2$. Since $f(1) < 0$ and $f(2) > 0$, by Intermediate Value Property $f$ has at least one zero in the interval $[1, 2]$. So a good starting point is $x_0 = 1.5$. If we are cleverer we could notice that $f(1)$ is a lot closer to 0 than $f(2)$ is and so $x_0 = 1.25$ might be a better starting point. Such *ad-hoc* improvements are often possible in a particular problem, done by an (intelligent!) human being. But here we want to illustrate a certain procedure which is so mechanical that it can be done on a (dumb!) computer. Even with a hand-calculator it is easy to check that if we follow the general rule that

$$x_n = x_{n-1} - \frac{f(x_{n-1})}{f'(x_{n-1})} = x_{n-1} - \frac{x_{n-1}^3 - 2}{3x_{n-1}^2} = \frac{2x_{n-1}^3 + 2}{3x_{n-1}^2} \tag{4}$$

for $n = 1, 2, 3, \ldots$ then starting from $x_0 = 1.5$, we get successively,

$$x_1 = 1.2962963$$
$$x_2 = 1.2609322$$
$$x_3 = 1.2599219$$

$$x_4 \;\; = \;\; 1.2599210$$
$$x_5 \;\; = \;\; 1.2599210$$

There is no point in proceeding further since the same values will keep recurring. So we take 1.259921 as a zero of $x^3 - 2$. It is, of course, not an exact answer since $\sqrt[3]{2}$ is not a rational number but we get an accuracy of $10^{-7}$. (On an ordinary calculator $(1.259921)^3$ will come out to be 2. A more precise calculator will give 1.9999998. To get more accurate values of $\sqrt[3]{2}$, recourse must be had to a higher precision device or, to laborious hand calculations.)

Let us now compare this with binary search. We know $f(1) < 0$ and $f(2) > 0$. So we take $x_0 = 1.5$. A direct calculation gies $f(1.5) = 1.375 > 0$. So $f$ has a zero in $[1, 1.5]$. Thus $x_1$ is the mid-point of this interval, i.e., $x_1 = 1.25$. Calculating again, $f(x_1) = 0.046875$. This is fairly close to 0 and so the next value to try should perhaps be 1.26. But if we are implementing the binary search method mechanically, then we have to take $x_2$ as 1.375, the mid-point of $[1.25, 1.50]$. Continuing in this manner, we get

$$x_1 \;\; = \;\; 1.25$$
$$x_2 \;\; = \;\; 1.375$$
$$x_3 \;\; = \;\; 1.3125$$
$$x_4 \;\; = \;\; 1.28125$$
$$x_5 \;\; = \;\; 1.265625$$
$$x_6 \;\; = \;\; 1.2578125$$
$$x_7 \;\; = \;\; 1.26171875$$

So even after 7 iterations we are not even within $10^{-3}$ of the exact answer. Thus, at least in this example, we see that the Newton-Raphson mehtod, based on first order approximation, is far superior to the binary search, which is based on mere continuity. It should not, however, be supposed that the former always beats the latter. The basic difficulty with the Newton method is that a lot depends on the starting point $x_0$. If $x_0$ is such that at some stage we get $f'(x_n) = 0$, then we cannot define $x_{n+1}$. We could of course change $x_0$ slightly (i.e., 'perturb' it) to ensure that the new $x_n$ will not have a horizontal tangent. Still if $f'(x_n)$ is numerically small, then the point of intersection of the tangent at $(x_n, f(x_n))$ with the $x$-axis will be so far away from $x_n$ that the linear approximation may not be valid at $x_{n+1}$, i.e. $f(x_{n+1})$ may differ considerably from 0, thereby rocking the very foundation of Newton's method. In such cases, even though we get a well-defined sequence $\{x_n\}_{n=0}^{\infty}$ of iterates, it will not be convergent.

A bizarre example is the function defined by

$$f(x) \;\; = \;\; \begin{cases} \sqrt{x} & \text{if} \;\; x \geq 0 \\ -\sqrt{-x} & \text{if} \;\; x < 0 \end{cases} \tag{5}$$

Here $f$ has only one zero, viz. 0. $f$ is diferentiable everywhere except at 0. Suppose we start with some $x_0 > 0$. Then $x_1 = x_0 - \dfrac{\sqrt{x_0}}{\frac{1}{2\sqrt{x_0}}} = -x_0$. Since

$x_1 < 0, f'(x_1) = \dfrac{1}{2\sqrt{-x_1}}$ and so $x_2 = x_1 - \dfrac{f(x_1)}{2\sqrt{-x_1}}$ which comes out to be $x_0$.
So the sequence $\{x_n\}_{n=0}^{\infty}$ is simply the sequence whose terms are alternately $x_0$ and $-x_0$. It does not converge. Starting with a negative $x_0$ will be met with a similar fare. If we start with $x_0 = 0$ then there is no need to apply Newton's method (which will fail anyway since $f'(0)$ does not exist).

A few other examples of failure of Newton's method will be given in the exercises. Comparatively, the binary search is a slow but surer method. Newton's method is fast when it works, but there is no guarantee that it will work. Therefore it becomes desirable to find sufficient conditions on the function $f$ (besides its differentiability) that would ensure that Newton's method will work. We proceed to find one such condition.

Before doing so, it is convenient to look at the Newton's method from a new view-point. Suppose $f$ is differentiable and $f'$ never vanishes. Then we can define a new function $g$ by

$$g(x) \;=\; x - \frac{f(x)}{f'(x)}. \tag{6}$$

Then $\alpha$ is a zero of $f$ if and only if $g(\alpha) = \alpha$, or in the language of Exercise (1.14), $\alpha$ is a **fixed point** of the function $g$. So looking for the zeros of $f$ is as good as looking for fixed points of $g$. This is not a very profound observation. What lends weight to it is the fact that the sequence $\{x_n\}_{n=0}^{\infty}$ in the Newton's method is obtained by starting with any arbitrary $x_0$ and applying $g$ to it again and again, that is, $x_1 = g(x_0)$, $x_2 = g(x_1) = g(g(x_0)), \ldots$ and in general

$$x_n \;=\; g(x_{n-1}) \text{ for all } n \geq 1 \tag{7}$$

More generally, we can take any function $g$, any point $x_0$ in its domain and try to construct a sequence defined recursively by (7). The trouble, of course, is that for some $m$ it may happen that $g(x_m)$ does not lie in the domain of $g$. Suppose, however, that somehow this difficulty is overcome. Then we get an infinite sequence $\{x_n\}_{n=0}^{\infty}$, which depends, naturally, on the starting term $x_0$ and also on the funciton $g$. There is no guarantee that this sequence will converge. In fact, in the case of the function $g(x)$ associated with the function $f(x)$ in (5), we already saw that if $x_0 \neq 0$, then the sequence $\{x_n\}$ alternatively takes the values $x_0$ and $-x_0$. Suppose, however, that we put suitable restrictions on $g$ so as to ensure that this sequence is convergent. Let $L$ be its limit. That is, $L = \lim\limits_{n \to \infty} x_n$. There is no guarantee that $L$ will be in the domain of $g$. (An easy counter-example is to take $g(x) = \dfrac{1}{2}x$, and the domain of $g$ as the open interval $(0, 1)$.) But, once again, suppose that this is arranged, and further, also that $g$ is continuous at $L$. Then we can apply Theorem (1.2) and get $g(L) = \lim\limits_{n \to \infty} g(x_n)$. But by (7), $g(x_n) = x_{n+1}$. As $n \to \infty$, $x_{n+1}$ also tends to $L$. So, putting it all together, we see that if the various assumptions we have been making are true, then

$$L \;=\; g(L) \tag{8}$$

or in other words, $L$ is a fixed point of the function $g$. Moreover, the fact that $L = \lim\limits_{n\to\infty} x_n$ allows us to calculate $L$ with any pre-assigned degree of accuracy, by computing $x_n$ for a sufficiently large $n$. The latter is a mechanical task consisting of repeatedly evaluating the function $g$. Thus, we have a recipe (or an algorithm, to use a more formal phrase) for finding an approximate fixed point of $g$. As noted earlier, if $g$ is constructed from some other function $f$ as in (6), then this gives a recipe for finding an approximate zero of the function $f$.

For this method to work, the various assumptions we have made regarding the funciton $g$ and the sequence $\{x_n\}$ have to be satisfied. It turns out that this can be ensured if the function $g$ satisfies a rather strong condition, which we define below.

**13.1 Definition:** A real-valued funciton $g$ defined on a subset $S$ of $\mathbb{R}$ is called a **contraction** (on $S$) if there exists some $\alpha$ such that $0 \le \alpha < 1$ and for all $x, y \in S$

$$|g(x) - g(y)| \le \alpha|x - y|. \tag{9}$$

The name 'contraction' comes from the fact that in view of (9), the images of $x$ and $y$ under $g$ are brought closer or 'contracted' by application of $g$. Evidently (9) is a special case of a Lipschitz condition (see Exercise (2.7)), with the difference that in a Lipschitz condition, the constant $\alpha$ is not necessarily less than 1. This makes it easy to give examples of contractions and to derive their general properties, for example, that they are always continuous, in fact uniformly continuous. The property of contractions which is most crucial for us is proved in the following theroem, called the **contraction Fixed Point Theorem.**

**13.2 Thereom:** Suppose $g$ is a contraction and maps some closed and bounded interval $[a, b]$ into itself. Then $g$ has a unique fixed point $x^*$ in $[a, b]$. Moreover, for any $x_0 \in [a, b]$, if $x_n$ is defined recursively by (7), then the sequence $\{x_n\}_{n=1}^{\infty}$ converges to $x^*$.

**Proof:** We already noted that because of (9), $g$ is continuous. Since it is given that $g$ maps $[a, b]$ into itself, we can apply Exercise (1.14) to get that $g$ has at least one fixed point $x^*$ in $[a, b]$. If $x^{**}$ were another fixed point of $g$ in $[a, b]$ then by (9),

$$|x^* - x^{**}| = |g(x^*) - g(x^{**})| \le \alpha|x^* - x^{**}| < |x^* - x^{**}|$$

(where the last inequality follows because $|x^* - x^{**}| > 0$ and $\alpha < 1$). But this is a contradiction. So $g$ can have no fixed point other than $x^*$.

For the second part of the theroem, let $x_0 \in [a, b]$ be arbitrary. Since $g$ maps $[a, b]$ into itself, there is no difficulty in defining $x_n$ recursively by $x_n = g(x_{n-1})$ for $n = 1, 2, \ldots$. It only remains to prove that $x_n \to x^*$ as $n \to \infty$. For this we directly estimate $|x_n - x^*|$ for $n \ge 1$. From (9), we have

$$|x_n - x^*| = |g(x_{n-1}) - g(x^*)| \le \alpha|x_{n-1} - x^*|. \tag{10}$$

Applying (10) with $n = 1, 2, 3, \ldots$, successively (or equivalently, by induction on $n$), it follows that for all $n \geq 1$,

$$|x_n - x^*| \leq \alpha^n |x_0 - x^*| \tag{11}$$

Now, $|x_0 - x^*|$ is a fixed real number. By Theorem (3.3.4), $\alpha^n \to 0$ as $n \to \infty$. So $x_n \to x^*$ as $n \to \infty$. ∎

In the proof above we appealed to Exercise (1.14), which was based on the Inermediate Value Property. It is possible to bypass this and prove the existence of a fixed point of $g$ directly by showing that the sequence $\{x_n\}$ above is a Cauchy sequence. This will be indicated in Exericse (13.10).

It can be shown by simple examples, that the hypotheses that $g$ be a contraction and that the interval be closed are essential for the theroem to hold. (We already had one such example where $g(x) = \frac{1}{2}x$ with domain $(0, 1)$.)

The Contraction Fixed Point Theorem stated above readily extends to more general situations and is an invaluable tool in proving the existence of solutions to many problems, after the problems are so paraphrased that their solutions are precisely the fixed points of some suitably defined functions. We shall not go into it. We remark, however, that the Implicit Function Theorem mentioned in the last section follows as a consequence of (a generalised) Contraction Fixed Point Theorem. Similarly many differential equations can be shown to have solutions.

In order to prove that a funciton $g$ is a contraction, a sufficient condition is that its derivative $g'$ be defined and satisfy $|g'(x)| \leq \alpha$ for all $x$ in some interval (cf. Exercise (5.21)). But the other part of the hypothesis of Theorem (13.2) is somewhat nagging. That is, it has to be given beforehand that $g$ maps $[a, b]$ into itself, in order to ensure that this interval contains a fixed point $x^*$ of $g$. If instead of $[a, b]$, we take an inteval which is symmetric about $x^*$, say $[x^* - r, x^* + r]$, then for any $x$ in this interval we have $|x - x^*| \leq r$ and hence by (9), $|g(x) - g(x^*)| \leq \alpha r$. But $g(x^*) = x^*$ and $\alpha r \leq r$. So, we have $|g(x) - x^*| \leq r$ whenever $|x - x^*| \leq r$. Thus, $g$ maps $[x^* - r, \ x^* + r]$ into itself. The trouble, of course, is that we do not know $x^*$ to begin with. In fact, our goal is to find it! Still, this shows that if the starting point $x_0$ is chosen sufficiently close to $x^*$, then we need not worry about the terms of the sequence $\{x_n\}$ lying in the domain of $g$. (This point is further elaborated in the example below.)

In order to apply Theorem (13.2), to find a zero of a function $f$, we construct $g$ from $f$ in such a way that the zeros of $f$ will coincide with the fixed points of $g$. The simplest way to do this is to let

$$g(x) \quad = \quad x + f(x). \tag{12}$$

The Newton-Raphson method, on the other hand, is based on letting

$$g(x) \quad = \quad x - \frac{f(x)}{f'(x)}. \tag{13}$$

(which is same as (6))

Many other choices for $g(x)$ are possible. For example, we may take

$$g(x) \;=\; x - \frac{f(x)}{m} \tag{14}$$

where $m$ is some fixed (non-zero) cosntant. Depending upon how $g(x)$ is defined from $f(x)$, we get different methods for finding zeros, of $f$. The mehtod based on (12) is known as **Picard's method.**

Once the choice of $g(x)$ is made, suitable conditions on $f$ are imposed so that $g$ will satisfy the hypothesis of Theorem (13.2).

For example, for the Newton-Raphson method, we get the following theorem.

**13.3 Theorem:** Suppose $f : [a, b] \to \mathbb{R}$ is such that

(i) $f'$ is continuous on $[a, b]$, differentiable on $(a, b)$ and never vanishes in $[a, b]$

(ii) for every $x \in [a, b], x - \dfrac{f(x)}{f'(x)} \in [a, b]$, and

(iii) there exists some $\alpha, 0 \leq \alpha < 1$ such that for all $x \in (a, b)$,

$$\frac{|f(x) f''(x)|}{|f'(x)|^2} \;\leq\; \alpha. \tag{15}$$

Then for every $x_0 \in [a, b]$, the Newton method will converge to a zero of $f$ in $[a, b]$. Moreover, $f$ has only one zero in $[a, b]$.

**Proof:** Define $g$ by (13) above. Condition (ii) ensures that $g$ maps $[a, b]$ into itself. By (i), $g$ is continuous on $[a, b]$ and differentiable on $(a, b)$, with

$$g'(x) \;=\; 1 - \frac{[f'(x)]^2 - f(x) f''(x)}{[f'(x)]^2}$$

$$=\; \frac{f(x) f''(x)}{[f'(x)]^2} \quad \text{for all } x \in (a, b). \tag{16}$$

Lagrange Mean Value Theorem, combined with (15) and (16) shows that $g$ satisfies (9) for all $x, y \in [a, b]$. So Theorem (13.2) is applicable. The result follows since the fixed points of $g$ are precisely the zeros of $f$. ∎

Let us now go back to our earlier example where we found an approximate value of the cube root of 2 by expressing it as a zero of the function $f(x) = x^3 - 2$. Here $f'(x) = 3x^2$ and $f''(x) = 6x$, so that for $x \neq 0$,

$$\frac{f(x) f''(x)}{[f'(x)]^2} \;=\; \frac{2}{3} \left( 1 - \frac{2}{x^3} \right) \tag{17}$$

For all $x \geq 1, |1 - \dfrac{2}{x^3}| \leq 1$ and so (15) is satisfied with $\alpha = \dfrac{2}{3}$ over any interval of the form $[1, c]$ for $c > 1$. It only remains to find $a, b$ so that (ii) holds, i.e. so that $g$ maps $[a, b]$ into itself. Because of the remarks we made after Theorem (13.2), if $x^*$ is a fixed point of $g$ (which in this case, means $x^* = \sqrt[3]{2}$,) then any interval of the form $[x^* - r, x^* + r]$ on which $g$ is a contraction will have this property. In particular, we may take $r = x^* - 1$, since $g$ is a contraction on $[1, c]$ for every $c > 1$. This choice gives $a$ as 1 and $b$ as $2x^* - 1$. Thus $g$ is a contraction on $[1, 2x^* - 1]$ and maps this interval into itself. Note that we have not actually calculated $x^*$ and hence $b$ either. But that hardly matters. In order for Theorem (13.2) to apply, we merely have to show the *existence* of an interval $[a, b]$ satisfying the hypothesis. Theorem (13.2) now entitles us to start the Newton method with any $x_0 \in [1, 2x^* - 1]$. In particular we can take $x_0 = 1$. Earlier we did the calculations with $x_0 = 1.5$. To ensure that this starting point lies within $[1, 2x^* - 1]$ we need only check that $1.5 \leq 2x^* - 1$, or equivalently, $x^* > 1.25$. A simple calculation shows that this is indeed the case since $(1.25)^3 = 1.953125 < 2$.

Going back to the title question, the Newton's method works faster than binary search because it is based on the first order approximation which is stronger than the zeroth order approximation (which is all that continuity can guarantee). But its success is conditioned by the starting point and the nature of the funciton $f$. And, as shown above, sometimes it may fail. Binary search is slow but sure. In practice, usually a combination of the two leads to better results. That is, we first apply the Intermediate Value Property to identify an interval which contains at least one zero, say $x^*$, of $f$. Then we calculate $f', f''$ and find a sub-interval, say, $[c, d]$ containing $x^*$ in its inierior such that the function $g$ associated with $f$ (defined by (13)) is a contraction on $[c, d]$. Finally, binary search is applied starting with $c_0 = c, d_0 = d$ till we get a subinterval $[c_m, d_m]$ (for some positive integer $m$) such that $c_0 < c_m \leq x^* \leq d_m < d_0$. Note that $c_m$ and $d_m$ are the end-points of some sub-interval of $[c_0, d_0]$ when the latter is divided into $2^m$ equal parts. So $c_m - c_0$ and $d_0 - d_m$ are both positive multiples of $\dfrac{d_0 - c_0}{2^m}$. Hence if we let $r = \dfrac{d_0 - c_0}{2^m}$, then no matter where $x^*$ lies in $[c_m, d_m]$, the interval $[x^* - r, \ x^* + r]$ will be contained in $[c, d]$ and will contain both $c_m$ and $d_m$. We can now apply Theorem (13.2) and Newton's method can be started with $x_0 = c_m$ or with $x_0 = d_m$.

## EXERCISES

13.1   Suppose $|f(x) - f(y)| \leq M|x - y|$ for all $x, y \in [a, b]$ and $f(a)f(b) < 0$. Let $\epsilon > 0$ be given. Assume $M > 0$. Let $n$ be an integer such that $\dfrac{1}{2^n} < \dfrac{\epsilon}{M(b - a)}$. Prove that within at most $n$ iterates the binary search will give a point $\alpha' \in (a, b)$ such that $|f(\alpha')| < \epsilon$.

13.2   Find approximate values of $\sqrt{2}$ and $5^{1/3}$ using binary search as well as the Newton-Raphson method. Compare the performance.

13.3 Find an approximate value of $\cos\dfrac{\pi}{7}$ by first finding a cubic equation satisfied by it. Then find it using the power series for $\cos\dfrac{\pi}{7}$ (cf. Exercise (3.3.14)). Compare the performance. Why is the second method easier?

13.4 Estimate $\pi$ by solving $\tan x = 1$ and also by solving $\sin x = \dfrac{1}{2}$.

13.5 Let $h(x) = x^{1/3}$. Prove that if Newton's method is applied with any starting point other than 0, the resulting sequence of iterates is unbounded.

13.6 The funciton $f(x) = x^3 - 12x^2$ has two roots, 0 and 12. If Newton's method is applied with $x_0 = 7$ show that the resulting sequence converges to 0 and not to 12 even though the starting point 7 is closer to 12 than to 0. In other words, Newton's method may give a root but it may not be the 'desired' or the nearest root. Can you guess where the starting point should be to give 12 as a root? [*Hint*: A sketch of the graph will help.]

13.7 The notion of a fixed point can be defined, more generally, for any function $f : X \to X$ where $X$ is any set. For a positive integer $m$, let $f^m$ denote the composite of $f$ with itself $m$ times. (Thus $f^2 = f \circ f$, $f^3 = f \circ f \circ f$ etc. $f^1$ is just $f$ and by convention $f^0$ is taken as the identity function from $X$ to itself.)

    (a) Prove that a fixed point of $f$ is also a fixed point of $f^m$, for every $m \geq 0$.

    (b) If $m > 1$, show by an example that a fixed point of $f^m$ need not be a fixed point of $f$.

    (c) Prove, however, that if $f^m$ has a unique fixed point then $f$ has a unique fixed point. [*Hint* : If $x_0$ is a fixed point of $f^m$, show that so is $f(x_0)$. This simple but elegant argument is due to Kolmogorov.]

13.8 Define $g : [0,1] \to \mathbb{R}$ by $g(x) = \dfrac{1}{2}x + 5$. Prove that $g$ is a contraction but has no fixed polint in $[0,1]$. Does this contradict Theorem (13.2)?

13.9 Define $g : [1,\infty) \to [1,\infty)$ by $g(x) = x + \dfrac{1}{x}$. Prove that $|g(x) - g(y)| < |x - y|$ for all $x, y \in [1,\infty)$, with $x \neq y$ but still, $g$ is not a contraction. Starting with any $x_0 \in [1,\infty)$, if $x_n$ is defined recursively as in (7), i.e. as $x_n = g(x_{n-1})$, prove that $x_n \to \infty$ as $n \to \infty$. In particular, $g$ has no fixed point.

13.10 Give an alternate proof of Theorem (13.2) by showing that the sequence $\{x_n\}$ in it is a Cauchy sequence. [*Hint*: For every $n \geq 0$ show by induction that $|x_{n+1} - x_n| \leq \alpha^n |x_1 - x_0|$. Now for $n > m$, $|x_n - x_m| \leq \displaystyle\sum_{i=m}^{n-1} |x_{i+1} - x_i|$.

Finally use the convergence of the series $\displaystyle\sum_{n=0}^{\infty} \alpha^n$.]

13.11 Show that the Contraction Fixed Point Theorem holds for contractions from $I\!R$ to $I\!R$. (Thus the alternate proof in the last exercise has an edge over the proof given in the text, which is applicable only for contractions whose domains are closed and bounded intervals.)

# Chapter 5

# INTEGRATION

## 5.1. Integration - Conceptually Different from the Opposite of Differentiation

Normally, the study of any new concept should begin by explaining what it is rather than what it is not. But integration is an exception. Treating the integral (of a function of one variable) the same as its anti-derivative is one of the most common and deep-rooted misconceptions cherished by the average student. If he is asked what is, say, $\int_1^2 x^2 dx$, he will first find an antiderivative of the function $f(x) = x^2$. (By Exercise (4.4.2) (ii), $\frac{1}{3}x^3$ is one such antiderivative.) He will then evaluate it at the end-points 2 and 1 and subtract, so as to yield $\frac{8}{3} - \frac{1}{3}$, i.e., $\frac{7}{3}$ as the answer.

There is, indeed, nothing wrong in this procedure. In fact, as we shall see later, the most standard method of evaluating definite integrals is by finding anti-derivatives. So, if the problem was merely to *evaluate* the definite integral $\int_1^2 x^2 dx$ then there is nothing wrong in using anti-derivatives (or any other device for that matter). But it is wrong to think that *conceptually* $\int_1^2 x^2 dx$ is the same as the difference $\frac{1}{3}(2)^3 - \frac{1}{3}(1)^3$. The two are quite different indeed.

The situation is analogous to the difference between the limit of a function and the value of a function. If $f$ is a function defined in a neighbourhood of a point $c$, then $\lim_{x \to c} f(x)$ is conceptually quite different from $f(c)$. If $f$ is continuous at $c$ then the two are equal. (This, indeed, is the very definition of continuity of $f$ at $c$.) In applications, one often deals only with continuous functions and this serves to strengthen the misconception that $\lim_{x \to c} f(x)$ is the same as $f(c)$.

Similarly, suppose $f(x)$ is a function defined on some interval $[a, b]$ and has $F(x)$ as an anti-derivative on $[a, b]$ i.e., $F'(x) = f(x)$ for all $x \in [a, b]$. If $f$

is continuous on $[a, b]$ then the definite integral $\int_a^b f(x)dx$ happens to equal $F(b) - F(a)$. This is an important theorem (called the Fundamental Theorem of Calculus) and requires a proof. The expression $\int_a^b f(x)dx$ has a definition of its own which we shall give shortly and which is quite independent of the concept of a derivative.

The nomenclature is partly to be blamed for the confusion. As pointed out in Exercise (4.4.3), the term indefinite integral and the symbol $\int f(x)dx$ are often used to indicate the class of all anti-derivatives of a function $f(x)$. This name as well as the notation are very apt to be confused with the definite integral $\int_a^b f(x)dx$. Indeed, the impression a beginner gathers is that an indefinite integral and a definite integral are the same entities except for the 'superficial' difference that in the latter, the integral sign bears two extra symbols, one at the top and one at the bottom. Rather like the mustaches of a man. Whether he keeps them or shaves them off, the man underneath remains the same !

But the reality is different. The definite and the indefinite integrals are not like the same person with and without mustaches. If at all an analogy is needed, the two are like twin brothers, separated from each other in their early childhood (as happens typically in popular Hindi films), brought up in vastly different surroundings but looking absolutely identical nevertheless ! The Fundamental Theorem of Calculus is like an old personage who, based on some secret evidence, establishes that the two look-alikes are, in fact, twin brothers. But even then they are not the same person. Each has his own identity and separate personality.

Possibly with a view to avoid confusion with definite integrals, the indefinite integrals are nowadays more customarily called antiderivatives, which is a more appropriate name, truly reflecting what they really are.

Having said what on an integral *is not*, it is now time to tell what it *is*. As with most other basic concepts in calculus, integrals are certain limits. More precisely, they are limits of certain types of sums. The peculiar integral sign $\int$ originates in an elongated $S$ which (like $\Sigma$) was sometimes used to denote summation. So we may say that integration is a refined form of summation. In a (finite) summation, the individual summands retain their own identity. But in integration they are all merged together so as to form a whole, unified expression. This is also consistent with the meanings of such real-life expressions as 'integrated circuits' or 'national integration '. The word 'integer' itself emphasises wholeness, absence of fractions or parts. Incidentally, the word 'integral' has two meanings in mathematics. When used as an adjective, it indicates some (possibly remote) relationship with integers. For example, to say that $n$ assumes only integral values means the possible values of $n$ are integers. In algebra, one defines and studies such things as 'integral domains' called that way because the set of integers provides a foremost example of such a domain. When used as a noun, however, an 'integral' always means a result of the process of integration which we are about to study.

Although the ideas involved in defining integrals date as far back as to Archimedes, it was Riemann who gave a systematic formulation to the theory. So, many concepts are named in his honour.

For simplicity, we shall begin by defining integrals of the form $\int_a^b f(x)dx$ where $f$ is a real-valued function defined on an interval $[a, b]$. Later on we shall study other integrals. But they, too, will, without exception, be limits of certain sums. In fact, once the integrals of the type we shall consider here are clearly understood, the remaining integrals are straightforward extensions of the same basic idea and hence will be easier to understand.

There is, however, some difficulty a beginner encounters when he first sees a definite integral defined as the limit of a sum, even when he has thoroughly understood limits of other types such as $\lim_{x \to c} f(x)$ or $\lim_{n \to \infty} a_n$. Even though the formal definition of these limits limit is clumsy (for reasons explained in Section 2.3), the expressions whose limits are considered are themselves not clumsy in either of these limits. In the former we take the limit of $f(x)$, which is simply the function $f$ evaluated at $x$ while in the latter we take the limit of the $n^{th}$ term of a sequence. In the case of an infinite series $\sum_{n=1}^{\infty} a_n$, we take the limit of the $n^{th}$ partial sum, viz., $\sum_{k=1}^{n} a_k$ as $n$ tends to infinity.

In sharp contrast, in the case of the (definite) integral $\int_a^b f(x)dx$, the sum whose limit is taken is itself a highly clumsy expression. To be specific, $\int_a^b f(x)dx$ is a limit of sums of the following form. Subdivide the interval into $n$ subintervals (of not necessarily equal lengths) by inserting points, say, $x_1, \ldots, x_{n-1}$ into $(a, b)$ with $x_1 < x_2 < \ldots x_{n-1}$. Any such subdivision of $[a, b]$ is called a **partition** of $[a, b]$. The points $x_1, \ldots, x_{n-1}$ are called the **nodes** of this partition. For notational uniformity, we set $x_0 = a$ and $x_n = b$. For brevity such a partition will be denoted by $(x_0, x_1, \ldots, x_n)$. For $i = 1, \ldots, n$, the interval $[x_{i-1}, x_i]$ will be called its $i^{\text{th}}$ interval (or, more accurately, its $i^{\text{th}}$ sub-interval). Its length equals $x_i - x_{i-1}$ and will often be denoted by $\Delta x_i$. (This is a single symbol and its value depends not only on $x_i$ but also on $x_{i-1}$). Now choose any points $\xi_i$ in $[x_{i-1}, x_i]$ for $i = 1, \ldots, n$. We consider the sum $\sum_{i=1}^{n} f(\xi_i)\Delta x_i$, which depends not only on the integer $n$ but also on the points $x_1, \ldots, x_{n-1}$ and the points $\xi_i$. This sum is called the **Riemann sum** for the partition $(x_0, \ldots, x_n)$ and for the choice of points $(\xi_1, \ldots, \xi_n)$. The definite integral $\int_a^b f(x)dx$ is the limit of such Riemann sums.

Obviously, on the face of it, it is far from clear why on earth would anybody want to form such clumsy sums, much less to take their limits. Secondly, we have not yet said anything as to how these limits are taken, that is, limit of the Riemann sum as what tends to what. It is tempting to say that we should take the limit as $n$ tends to $\infty$. But this turns out to be not quite correct. For a fixed $n$, the Riemann sum $\sum_{i=1}^{n} f(\xi_i)\Delta x_i$ depends on the choice of the nodes $x_i, \ldots, x_{n-1}$ and also on the choice of the points $\xi_1, \ldots, \xi_n$. If we specify a particular choice of the $x$'s (for example that they be equally spaced, so that $x_i = a + \dfrac{i(b-a)}{n}$ for $i = 1, \ldots, n-1$) and a particular choice of the $\xi$'s (for example that $\xi_i$ be the mid-point of the $i^{th}$ subinterval, i.e., $\xi_i = \dfrac{x_{i-1} + x_i}{2}$ for $i = 1, \ldots, n$), then for each $n \in I\!N$, there will be a unique Riemann sum and we could define $\displaystyle\int_a^b f(x)dx$ as the limit of this sequence of Riemann sums. But the class of such Riemann sums turns out to be too specialised. So we have to allow all possible Riemann sums and in that case it is not enough to merely let $n$ tend to $\infty$. Because then there may not be a unique limit.

We postpone these difficulties to the next section where we shall not only define integrals as limits of Riemann sums but give ample evidence why they are worth a study despite their apparent clumsiness. For the time being, we instead define definite integrals in terms of the suprema and infima of certain sets of real numbers associated with the function $f$ and the interval $[a, b]$. Even this approach requires some motivation.

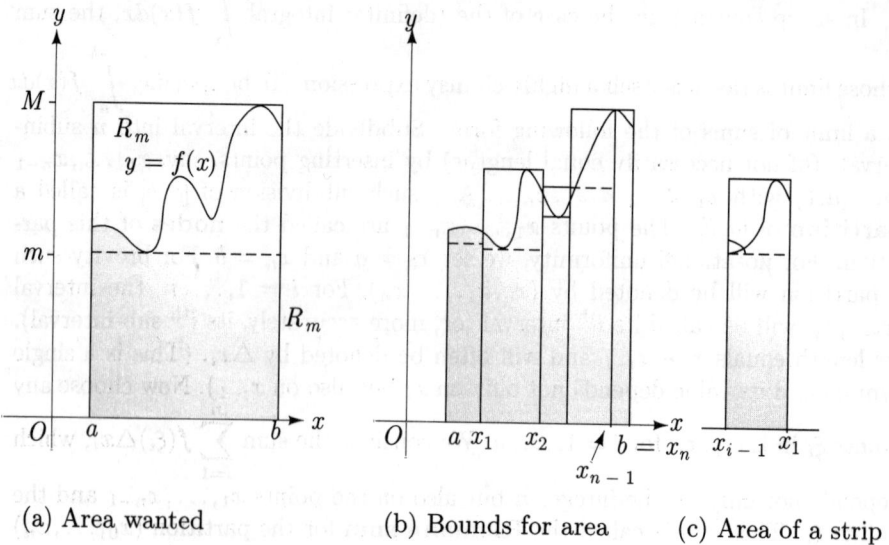

(a) Area wanted   (b) Bounds for area  (c) Area of a strip

Figure 5.1.1 Area below the Graph of a Function

The most standard motivation comes from the problem of finding the area below the graph of the function $y = f(x)$. We assume that the function $f$ is bounded on the integral $[a, b]$. For simplicity we assume also that $f$ takes only nonnegative values so that the graph of $f$ lies entirely on or above the $x$-axis. Our interest is in finding the area of the region $R$ bounded above by the graph of $f$, below by the x-axis, and on the left and the right by the lines $x = a$ and $x = b$ respectively. It is shown by shading in Fig.5.1.1(a). (We have not formally defined the area of a plane figure. But that does not matter here. As we are using this problem only to motivate the definition of an integral, any intuitive understanding of area will do.)

If the function $f$ were constant on $[a, b]$, say $f(x) = c$ for all $x \in [a, b]$, then our problem would be trivial because then the region $R$ would be a rectangle with height $c$ and base $b - a$ so that its area would be simply $c(b - a)$. More generally, if $f(x)$ is of the form $px + c$, for some constants $p$ and $c$, then the graph of $f$ would be a straight line, whence the region $R$ would be a trapezeum whose area can be evaluated easily. We leave it to the reader to check that it comes out to be $c(b - a) + \frac{1}{2}p(b^2 - a^2)$.

In general, the function $f$ may, of course be quite arbitrary and so the area of $R$ cannot be found so easily. So, as the next best thing, we try to estimate it by finding some upper and lower bounds on it. Our assumption that $f$ is bounded on $[a, b]$ implies that there exist real numbers $m$ and $M$ such that

$$m \leq f(x) \leq M, \text{ for every } x \in [a, b] \qquad (1)$$

Geometrically, this means that the region $R$ contains the rectangle $R_m$ with base $[a, b]$ and height $m$ and also that it is contained in the rectangle $R_M$ with the same base and height $M$ as shown in Fig.5.1.1 (a). So the area of $R$ lies in between the areas of these two rectangles, that is,

$$m(a - a) \leq \text{ area of } R \leq M(b - a) \qquad (2)$$

Thus $m(b - a)$ is a lower bound white $M(b - a)$ is an upper bound for the area of $R$. The lower bound will be improved if $m$ can be increased. The best lower bound will be obtained if $m$ is the largest number satisfying (1), that is, if $m$ equals the infimum of $f(x)$ on $[a, b]$. (Note that we are saying infimum and not minimum. As explained in Section 4.1, $f$ need not always have a minimum on $[a, b]$. But the infimum of $f$ on $[a, b]$ does exist by completeness of $\mathbb{R}$, since $f$ is given to be bounded on $[a, b]$). Similarly if $M$ is taken as the supremum of $f$ on $[a, b]$, i.e., $M = \sup_{a \leq x \leq b} f(x)$, then the upper bound in (2) is the best possible.

Of course, (2) is still a very crude inequality, especially if the difference between $M$ and $m$ is large. It is about as loose as trying to estimate the total rain water falling on a large country, knowing merely the lowest and the highest rainfall over the country. To get a more accurate estimate, we have to divide the country into several geographical regions over each of which the variation in the rainfall is not so large.

Similarly (2) can be improved considerably if we partition the interval $[a, b]$ and thereby divide the region $R$ into several vertical strips (Fig. 5.1.1 (b)) and find bounds (similar to (2)) for the area of each of these strips. In absence of any other information about the function $f$, it is impossible to foretell which partition will give sharper bounds. So we take any general partition, say, $(x_0, x_1, \ldots, x_n)$ of $[a, b]$. For $i = 1, \ldots, n$ we let $R_i$ be the portion of $R$ lying over the $i^{th}$ subinterval $[x_{i-1}, x_i]$. As mentioned earlier we denote $x_i - x_{i-1}$ by $\triangle x_i$. Now if we let $m_i$ and $M_i$ be respectively the infimum and the supremum of $f$ on $[x_{i-1}, x_i]$, then using exactly the same reasoning as for (2), we get

$$m_i \triangle x_i \leq \text{ area of } R_i \leq M_i \triangle x_i, \quad i = 1, \ldots, n. \tag{3}$$

(See Fig.5.1.1 (c))

It is clear that if we add the areas of the strips $R_i$, we get the area of the region $R$. So from (3),

$$\sum_{i=1}^{n} m_i \triangle x_i \leq \text{ area of } R \leq \sum_{i=1}^{n} M_i \triangle x_i \tag{4}$$

Let us now compare (2) and (4). Evidently, $m_i \geq m$ and $M_i \leq M$ for every $i = 1, \ldots, n$. Further, since $\sum_{i=1}^{n} \triangle x_i = \sum_{i=1}^{n}(x_i - x_{i-1}) = x_n - x_0 = b - a$, we have

$$\sum_{i=1}^{n} m_i \triangle x_i \geq \sum_{i=1}^{n} m \triangle x_i = m \sum_{i=1}^{n} \triangle x_i = m(b - a).$$

Similarly $\sum_{i=1}^{n} M_i \triangle x_i \leq M(b - a)$. Hence (4) is definitely not worse than (2). But in reality it may be much better, because for some $i$'s, the $m_i$ may be much larger than $m$ and/or the $M_i$ may be much smaller than $M$. If the difference $M_i - m_i$ is small for each $i$, then in (4) the first and the last terms differ very little from each other and so either can be taken as an approximate value of the middle term. Our intuition tells us that the chances of this happening are better, the smaller is each of the subintervals. Note that it is not enough merely that the number of subintervals be large. For example, if we divide $[a, b]$ into two equal halves and then further subdivide the right half into a large number of intervals, leaving the left half intact, then the number of intervals of the partition is large, but the two bounds in (4) will always differ by at least $\frac{1}{2}(M_1 - m_1)(b - a)$ where $M_1$ and $m_1$ are, respectively, the supremum and the infimum of $f$ on the left half of $[a, b]$.

Note that (4) resulted by subdividing the interval $[a, b]$ into $n$ intervals, $[x_0, x_1]$, $[x_1, x_2]$, $\ldots, [x_{n-1}, x_n]$. What if we subdivide some (possibly all) of these intervals further into still smaller subintervals ? We naturally expect that just as (4) is an improvement over (2), we would now get something which is even better than (4). This guess comes out to be quite correct. But before giving a formal proof, let us introduce some terminology and notation.

Let us denote the set of all possible partitions of the interval $[a, b]$ by $\mathcal{P}[a, b]$. Individual partitions will be denoted by symbols like $P, Q$ etc. They will be of the form $(x_0, x_1, \ldots, x_{n-1}, x_n)$ for some $n$ and for some $x_1, \ldots, x_{n-1} \in (a, b)$ with $a = x_0 < x_1 < x_2 < \ldots < x_{n-1} < x_n = b$. For any such partition $P = (x_0, \ldots x_n)$, the sums $\sum_{i=1}^{n} m_i \triangle x_i$ and $\sum_{i=1}^{n} M_i \triangle x_i$ defined above are called respectively, the **lower** and **upper Riemann sums** of the function $f$ for the partition $P$ and denoted, respectively, by $L(P; f)$ and $U(P; f)$. Obviously,

$$L(P; f) \leq U(P; f) \text{ for all } P \in \mathcal{P}[a, b]. \tag{5}$$

What is not so obvious is that even for two different partitions $P$ and $Q$, $L(P; f) \leq U(Q; f)$. In other words, not only is the lower Riemann sum for a partition, less than (or equal to) the upper Riemann sum for the same partition, but *every* lower Riemann sum is less than (or equal to) *every* upper Riemann sum, no matter which partitions they come from. To prove this, we first need to introduce an important concept.

**1.1 Definition:** Given two partitions, say, $P = (x_0, x_1, \ldots, x_n)$ and $Q = (y_0, y_1, \ldots, y_m)$ of $[a, b]$ (so that $x_0 = y_0 = a$ and $x_n = y_m = b$), we say that $Q$ is **finer** than (or a **refinement** of ) $P$ or that $P$ is **coarser** than $Q$ if every node of $P$ is also a node of $Q$, or in other words there exist integers $j_0, j_1, \ldots, j_n$ such that $0 = j_0 < j_1 < \ldots < j_n = m$ and $x_i = y_{j_i}$ for every $i = 0, 1, \ldots, n$.

Geometrically, the condition simply means that the partition $Q$ is obtained from $P$ by adding a few more nodes, or equivalently by further dividing the intervals of $P$. Figure 5.1.2 shows two partitions $P, Q$ of an interval with $Q$ a refinement of $P$. The intervals $[x_0, x_1], [x_2, x_3]$ and $[x_3, x_4]$ are subdivided into 3,4 and 2 subintervals. $[x_1, x_2]$ remains as it is as no node of $Q$ lies in between $x_1$ and $x_2$.

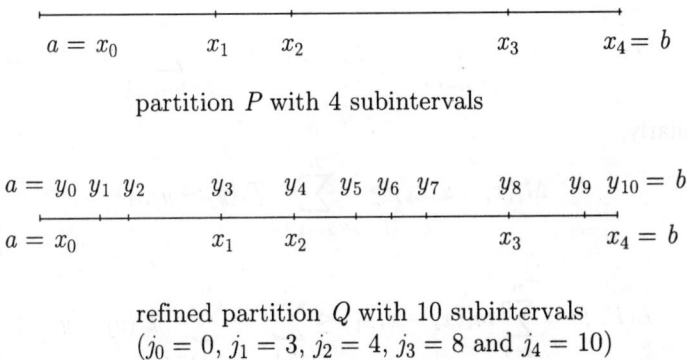

partition $P$ with 4 subintervals

refined partition $Q$ with 10 subintervals
$(j_0 = 0, j_1 = 3, j_2 = 4, j_3 = 8$ and $j_4 = 10)$

Figure 5.1.2 : Refinement of a Partition

The following simple observation is the key to the effect refinement has on upper and lower Riemann sums.

**1.2 Theorem:** If $P, Q \in \mathcal{P}[a, b]$ and $Q$ is a refinement of $P$ then for every bounded function $f$ on $[a, b]$,

$$L(P; f) \leq L(Q; f) \tag{6}$$
$$\text{and} \qquad U(P; f) \geq L(Q; f). \tag{7}$$

Verbally, refinement increases lower Riemann sum and decreases the upper Riemann sum and thereby brings them closer to each other.

**Proof:** Except for notational intricacy, the argument is a duplication of why (4) is a sharper statement than (2) above. Let $P = (x_0, \ldots, x_n), Q = (y_0, \ldots, y_m)$ with $x_i = y_{j_i}$ for $i = 0, 1, \ldots, n$. Denote by $m_i$ and $M_i$ the infimum and the supremum of $f$ on $[x_{i-1}, x_i]$, respectively, for $i = 1, \ldots, n$. Similarly, for $j = 1, \ldots, m$, let $t_j$ and $T_j$ denote, respectively, the infimum and the supremum of $f$ on $[y_{j-1}, y_i]$. Then, by definition

$$L(P; f) = \sum_{i=1}^{n} m_i(x_i - x_{i-1}), \quad U(P; f) = \sum_{i=1}^{n} M_i(x_i - x_{i-1})$$

$$L(Q; f) = \sum_{j=1}^{m} t_j(y_j - y_{j-1}), \quad U(Q; f) = \sum_{j=1}^{m} T_j(y_j - y_{j-1})$$

Consider the $i^{\text{th}}$ interval of $P$, viz., $[x_{i-1}, x_i]$, which is the same as $[y_{j_{i-1}}, y_{j_i}]$. In the partition $Q$, it gets further subdivided into the intervals $[y_{j_{i-1}}, y_{j_{i-1}+1}]$, $[y_{j_{i-1}+1}, y_{j_{i-1}+2}], \ldots, [y_{j_i-1}, y_{j_i}]$. Hence $m_i \leq t_j$ and $M_i \geq T_j$ for every $j = j_{i-1}+1, j_{i-1}+2, \ldots, j_i-1, j_i$. Note also that $x_i - x_{i-1}$ is simply $\displaystyle\sum_{j=j_{i-1}+1}^{j_i} y_j - y_{j-1}$.
So we have for every $i = 1, \ldots, n$

$$m_i(x_i - x_{i-1}) = \sum_{j=j_{i-1}+1}^{j_i} m_i(y_j - y_{j-1}) \leq \sum_{j=j_{i-1}+1}^{j_i} t_j(y_j - y_{j-1})$$

and similarly,

$$M_i(x_i - x_{i-1}) \geq \sum_{j=j_{i-1}+1}^{j_i} T_j(y_j - y_{j-1})$$

Hence

$$L(P; f) = \sum_{i=1}^{n} m_i(x_i - x_{i-1}) \leq \sum_{i=1}^{n} \sum_{j=j_{i-1}+1}^{j_1} t_j(y_j - y_{j-1})$$

which is precisely $\displaystyle\sum_{j=1}^{m} t_j(y_j - y_{j-1})$ since $j_0 = 0$ and $j_n = m$. But this sum is $L(Q; f)$ by definition. So (6) holds. Similarly (7) follows. ∎

As a corollary, we can now prove that no lower Riemann sum can exceed any upper Reimann sum.

**1.3 Corollary:** Let $P, Q$ be any two partitions of $[a, b]$. Then for every bounded function $f$ on $[a, b]$,

$$L(P; f) \leq U(Q; f)$$

**Proof:** The trick is to find a common refinement of $P$ and $Q$. The simplest way to do this is by **superimposing** the two partitions. Let the nodes of $P$ be $x_i, \ldots, x_{n-1}$ and those of $Q$ be $y_i, \ldots, y_{m-1}$. (Some of the nodes may be common.) Consider the partition whose nodes are $x_1, \ldots, x_{n-1}, y_1, \ldots, y_{m-1}$ arranged in an ascending order and omitting duplications if any. Denote this partition by $P \cup Q$. Clearly it is a refinement of $P$ as well as of $Q$. So by the theorem above

$$L(P; f) \leq L(P \cup Q; f) \text{ and } U(P \cup Q; f) \leq U(Q; f).$$

Since $L(P \cup Q; f) \leq U(P \cup Q; f)$ is true trivially, the Corollary follows. ∎

This Corollary enables us to define $\displaystyle\int_a^b f(x)dx$. Let $\mathcal{U}_a^b(f)$ and $\mathcal{L}_a^b(f)$ be, respectively, the sets of all possible upper and lower Riemann sums of $f$ for various partitions of $[a, b]$. In symbols,

$$\mathcal{U}_a^b(f) = \{U(P; f) : P \in \mathcal{P}[a, b]\}$$

and

$$\mathcal{L}_a^b(f) = \{L(P; f) : P \in \mathcal{P}[a, b]\}.$$

Because of the Corollary, every member of $\mathcal{L}_a^b(f)$ is a lower bound for $\mathcal{U}_a^b(f)$ and every member of $\mathcal{U}_a^b(f)$ is an upper bound for $\mathcal{L}_a^b(f)$. By completeness of the real number system (see Section 2.6), $\mathcal{U}_a^b(f)$ has a greatest lower bound and $\mathcal{L}_a^b(f)$ has a least upper bound. We call them the **upper** and the **lower**

**Riemann integrals** of $f$ over $[a, b]$, and denote them respectively by $\displaystyle\int_a^{\overline{b}} f(x)dx$

and $\displaystyle\int_{\underline{a}}^b f(x)dx$. The corollary above implies (cf. Exercise (1.2)), that

$$\int_{\underline{a}}^b f(x)dx \leq \int_a^{\overline{b}} f(x)dx \tag{8}$$

Verbally, the lower Riemann integral can never exceed the upper Riemann integral. Examples of functions can be given for which strict inequality holds in (8) (see Exercise (1.8)). Obviously such functions are very weird. For most well-behaved functions equality holds in (8). We call such a function **Riemann integrable** over $[a, b]$ and the common value of its lower and upper integrals its **Riemann** or **definite integral** (or simply its integral) over $[a, b]$. It is denoted by[1] $\int_a^b f(x)dx$ or occasionally by $\int_{[a,b]} f(x)dx$. As explained earlier, the integral

sign $\int$ comes from an elongated $S$ used for summation. The term $dx$ obviously connotes the limit of $\triangle x_i$'s. Note however, that we have not defined integrals in terms of limits but in terms of certain suprema and infima. As mentioned before, a definition of integrals as limits of sums will be given in the next section. Note also that the variable $x$, like the index variable of a summation, is a dummy variable. So, it can be replaced by any other variable. Thus $\int_a^b f(y)dy$ or

$\int_a^b f(z)dz$ is exactly the same as $\int_a^b f(x)dx$. It is even permissible to write

$\int_a^b f(a)da$ with the understanding that here $a$ has two roles. It is a constant when it appears below the integral sign. But it is a dummy variable inside the integral sign. A beginner is advised against such confusing usage.[2]

Note that in defining the Riemann integral, we only used that $f$ is bounded on $[a, b]$. The non-negativity assumption on $f$ was used only to give a motivation in terms of area. There, too, it is not vital. But if $f$ is negative on some portions of $[a, b]$, then the corresponding portions of the areas will have to be taken as negative.

The geometric significance of the integrals is noteworthy. Assume $f$ is non-negative and bounded. Let $R$ be the region below the graph of $f$ (Fig. 5.1.1(a) again). Since (4) holds for every portion $P$ of $[a, b]$, it follows that

$$\int_{\underline{a}}^{b} f(x)dx \leq \text{ area of } R \leq \int_a^{\overline{b}} f(x)dx \qquad (9)$$

So if $f$ is integrable on $[a, b]$ then $\int_a^b f(x)dx$ equals the area of $R$. If $f$ is not integrable on $[a, b]$, then $R$ cannot be assigned a unique area in terms of Riemann integrals. This sounds crazy and contrary to our intuition. Our intuition is based on continuous, or at least, piecewise continuous functions. In section 4, we shall prove that all such functions are integrable and so this kind of an anomaly does not arise for them.

---

[1]When the function $f(x)$ is expressed as a fraction, it is customary to write $f(x)dx$ by

putting $dx$ in the numerator. Thus, $\int_a^b \dfrac{p(x)}{q(x)}dx$ is often written as $\int_a^b \dfrac{p(x)dx}{q(x)}$. In the same

vein we write $\int_a^b \dfrac{dx}{g(x)}$ for $\int_a^b \dfrac{1}{g(x)}dx$.

[2]Some authors prefer to dispense with the dummy variable completely and write $\int_a^b f$

instead of $\int_a^b f(x)dx$ etc.

Throughout the definition of $\int_a^b f(x)dx$, we have assumed that $f$ is bounded and the interval $[a, b]$ is of finite length. When either of these conditions is violated, the integral can still be defined sometimes. It is called an improper integral. We shall study it later (Section 5.10).

Finally, we note that the definition of $\int_a^b f(x)dx$ given here is a theoretical one. Integrals are rarely evaluated by actually computing the upper and lower Riemann sums and finding their infima and suprema. This approach is feasible only for a very restricted class of functions (see Exercise (1.17)). For other functions, other methods are needed. And as mentioned at the beginning of this section, the fundamental theorem of calculus is, by far, the most common tool. We shall study it in Section 5.5. However, many of the theoretical properties of definite integrals can be established simply from the definition. A few such properties will be given as exercises, including very handy characterisations of integrability and the integral (Exercise (1.5) and (1.6)).

It will be noticed that in defining the definite integral $\int_a^b f(x)dx$, nowhere was the concept of derivatives needed. So it is conceptually wrong to say that integration is the opposite of differentiation.

## EXERCISES

1.1      Among all partitions of an interval $[a, b]$, which is the coarsest partition ? Which is the finest partition ? Given two partitions $P$ and $Q$ of $[a, b]$, is it always true that $P$ and $Q$ are comparable, i.e., either $P$ is a refinement of $Q$ or $Q$ is a refinement of $P$?

1.2      (i) Prove that every constant function is integrable over any interval. What is its integral ?

          (ii) If $f$ is integrable over $[a, b]$ and $c$ is any constant, prove that $cf$ is integrable over $[a, b]$ and that $\int_a^b cf(x)dx = c\int_a^b f(x)dx$. [*Caution* : The case $c < 0$ needs a little careful handling.]

1.3      Let $A$ and $B$ be non-empty subsets of $\mathbb{R}$ with the property that $x \le y$ for all $x \in A$ and $y \in B$. Let $a^* = \sup A$ and $b^* = \inf B$. Prove that $a^* \le b^*$. [*Hint* : If not, apply Exercise (2.6.12) and its analogue for infima with $\epsilon = \frac{1}{2}(a^* - b^*)$.]

1.4      In the last exercise, show that $a^* = b^*$ if and only if for every $\epsilon > 0$ there exist $x \in A$ and $y \in B$ with $y - x < \epsilon$.

1.5      Using the last exercise show that a function $f$ which is bounded on $[a, b]$ is integrable on $[a, b]$ if and only if for every $\epsilon > 0$ there exists a partition $P$ of $[a, b]$ for which $U(P; f) - L(P; f) < \epsilon$. [*Hint*: In the direct implication use the trick in the proof of Corollary (1.3).]

1.6    Suppose $f$ is integrable over $[a, b]$ and $I$ is a real number. Prove that
$\int_a^b f(x)dx = I$ if and only if $I$ has the property that for every $\epsilon > 0$,
there exists a partition $P$ of $[a, b]$ such that both the sums $U(P; f)$ and
$L(P; f)$ lie in the interval $(I - \epsilon, I + \epsilon)$. (This and the last exercise are
very convenient in establishing elementary properties of definite integrals.
See, for example, the next exercise.)

1.7    Let $f$ be bounded on $[a, b]$ and suppose $a < c < b$. Prove that $f$ is
integrable over $[a, b]$ if and only if it is integrable over $[a, c]$ and over $[c,b]$
and further,

$$\int_a^b f(x)dx = \int_a^c f(x)dx + \int_c^b f(x)dx.$$

Generalise. [*Hint* : A partition of $[a, c]$ and a partition of $[c, b]$ give, by
juxtaposition, a partition of $[a, b]$. Conversely every partition of $[a, b]$ has
a refinement which is of this form.]

1.8    Define $f : [0, 1] \longrightarrow I\!\!R$ by

$$f(x) = \begin{cases} 1 & \text{if} \quad x \text{ is rational} \\ 0 & \text{if} \quad x \text{ is irrational.} \end{cases}$$

Prove that for every $P \in \mathcal{P}[0, 1], U(P; f) = 1$ and $L(P; f) = 0$ and hence
that $f$ is not Riemann integrable over [0,1]. [*Hint*: The set of rationals
and that of irrationals are both dense.]

1.9    Let $f, g$ be bounded on $[a, b]$ and let $h = f + g$. Prove that $h$ is bounded on
$[a, b]$ and that for every partition $P$ of $[a, b], L(P; f) + L(P; g) \leq L(P; h)$
and $U(P; h) \leq U(P; f) + U(P; g)$. Deduce that if $f, g$ are integrable over

$[a, b]$, so is $h$ and $\int_a^b h(x)dx = \int_a^b f(x)dx + \int_a^b g(x)dx$. (It is also true
that the product of two integrable functions is integrable. But this is
not so easy to prove and is deferred to Theorem (4.5). Nor is there any
formula to express the integral of $fg$ in terms of those of $f$ and $g$.)

More generally, show that for any constants $\alpha$ and $\beta$, $\int_a^b \alpha f(x) + \beta g(x)dx$

$$= \alpha \int_a^b f(x)dx + \beta \int_a^b g(x)dx.$$

1.10   Show that if $f, g$ are integrable on $[a, b]$ and $f(x) \leq g(x)$ for all $x \in [a, b]$
then $\int_a^b f(x)dx \leq \int_a^b g(x)dx.$

1.11   Let $A$ be a non-empty, bounded subset of $I\!\!R$ with infimum $m$ and supremum $M$. Let $T$ and $t$ be respectively the supremum and the infimum of
the set $\{|x| : x \in A\}$. Show that $T - t \leq M - m$. Give an example where
the two are unequal.

1.12 Using the last exercise, show that if $f$ is integrable over $[a, b]$, so is $|f|$ and $\left| \int_a^b f(x)dx \right| \leq \int_a^b |f(x)|dx$. Deduce that if $|f(x)| \leq M$ for all $x \in [a, b]$ then $\left| \int_a^b f(x)dx \right| \leq M(b - a)$.

1.13 Give an example of a bounded function $f$ on $[0,1]$ such that $f$ is not integrable on $[0,1]$ but $|f|$ is.

1.14 If $f$ is monotonically increasing on $[a, b]$ and $P = (x_0, \ldots, x_n)$ is a partition of $[a, b]$, prove that

$$U(P; f) - L(P; f) = \sum_{i=1}^{n} (f(x_i) - f(x_{i-1}))(x_i - x_{i-1})$$

1.15 Prove that every monotonic function is integrable. [*Hint* : Given $\epsilon > 0$, let $P$ be a partition of $[a, b]$ into $n$ equal parts where $n$ is so large that $(b - a)|f(b) - f(a)| < n\epsilon$.]

1.16 Let $f(x) = x$ and $P$ be a partition of an interval $[a, b]$ into $n$ equal parts. Prove that

$$U(P; f) = \frac{b - a}{2n}(an + bn + b - a)$$

$$\text{and} \quad L(P; f) = \frac{b - a}{2n}(an + bn - b + a).$$

1.17 Using the last exercise and Exercise (1.6) show that $\int_a^b x dx = \frac{1}{2}(b^2 - a^2)$. Similarly, show that $\int_a^b x^2 dx$ equals $\frac{1}{3}(b^3 - a^3)$.

1.18 Find the area of a trapezeum by the geometric formula and also by expressing it as the area under the graph of a suitable non-negative function or as the difference of two such areas.

1.19 In the definition of $\int_a^b f(x)dx$, it is implicit that $a < b$. It is sometimes convenient to consider $\int_a^b f(x)dx$ for $a > b$. This is *defined* as $-\int_b^a f(x)dx$. (If $a = b$, we set $\int_a^b f(x)dx = 0$.) Show that the results of Exercise (1.7) holds for all $a, b, c$ lying in an interval over which $f$ is Riemann integrable. Do the results of Exercises (1.9), (1-10) and (1.12) holds if $b < a$?

## 5.2 Riemann Sums - a Better Approach to the Definition of a Definite Integral

In the last section we already defined the concept of a Riemann sum and remarked that Riemann integrals could be defined as limits of Riemann sums. But we abandoned the approach temporarily and instead defined the definite integral $\int_a^b f(x)dx$ as the simultaneous supremum of the set $\mathcal{L}_a^b(f)$ of all lower Riemann sums of $f$ on $[a, b]$ and the infimum of the set $\mathcal{U}_a^b(f)$ of all upper Riemann sums of $f$ on $[a, b]$. It is not always true that upper and lower Riemann sums are themselevs Riemann sums. This is linguistically hard to swallow but can happen because of the subtle distinction between a supremum and a maximum or between an infimum and a minimum. A typical Riemann sum of $f$ on $[a, b]$ is of the form $\sum_{i=1}^{n} f(\xi_i)(x_i - x_{i-1})$ where $(x_0, x_1, \ldots, x_n)$ is a partition of $[a, b]$ and $\xi_i$ is a point of $[x_{i-1}, x_i]$ for $i = 1, \ldots, n$. The upper Riemann sum of $f$ for this partition, viz. $U(P; f)$, on the other hand, is $\sum_{i=1}^{n} M_i(x_i - x_{i-1})$ where $M_i$ is the supremum of the set $\{f(x) : x_{i-1} \leq x \leq x_i\}$. If a point $\eta_i$ exists in $[x_{i-1}, x_i]$ such that $f(\eta_i) = M_i$, then $M_i$ would, in fact, be the maximum of $f$ on $[x_{i-1}, x_i]$ (see Section 4.1). If this happens for every $i = 1, \ldots, n$, then $\sum_{i=1}^{n} M_i(x_i - x_{i-1})$ would indeed be the Riemann sum $\sum_{i=1}^{n} f(\eta_i)(x_i - x_{i-1})$. But in general, it may happen that for some $i$, $f$ has a supremum but no maximum on $[x_{i-1}, x_i]$. And so, in general, an upper Riemann sum need not be a Riemann sum ! Similarly, a lower Riemann sum need not be a Riemann sum. · If $f$ is continuous on $[a, b]$, then in view of Theorem (4.1.4), applied to each $[x_{i-1}, x_i]$, both the upper and lower Riemann sums are, in fact, Riemann sums.

Even though the upper and lower Riemann sums may not themselves be Riemann sums, they nevertheless provide, respectively, upper and lower bounds on the set of all possible Riemann sums that one can get from a given partition. This is very trivial. Let $P = (x_0, x_1, \ldots, x_n)$ be a partition of an interval $[a, b]$ and suppose $f$ is a bounded function on $[a, b]$. Then for any choice of $\xi_1, \ldots, \xi_n$ in $[x_0, x_1], [x_1, x_2], \ldots, [x_{n-1}, x_n]$ respectively, we have, $m_i \leq f(\xi_i) \leq M_i$ for every $i = 1, \ldots, n$. Multiplying by $\Delta x_i$ and adding up, we get

$$L(P; f) = \sum_{i=1}^{n} m_i \Delta x_i \leq \sum_{i=1}^{n} f(\xi_i) \Delta x_i \leq \sum_{i=1}^{n} M_i \Delta x_i = U(P; f) \qquad (1)$$

As we proved in Theorem (1.2), as the partition becomes finer and finer, its upper and lower Riemann sums come closer and closer to each other. If $f$ is integrable over $[a, b]$, then by Exercise (1.5), given any $\epsilon > 0$, both $L(P; f)$ and $U(P; f)$ are within an $\epsilon$−neighbourhood of $\int_a^b f(x)dx$ for some partition $P$ of $[a, b]$ and hence (by Theorem (1.2) again), for every partition $Q$ which is finer than $P$. By (1) (applied to $Q$), the same is true for every Riemann sum obtainable from the partition $Q$. So we may say that if $f$ is integrable on $[a, b]$ then the Riemann sums tend to $\int_a^b f(x)dx$ as the partition on which they are taken becomes finer and finer, regardless of which intermediate points ($\xi$'s) are used in forming these sums.

This suggests an alternate definition of the Riemann integral. Given a partition $P = (x_0, x_1, \ldots, x_n)$ of $[a, b]$ and points $\xi_1, \ldots, \xi_n$ with $\xi_i \in [x_{i-1}, x_i]$ for $i = 1, \ldots, n$, we denote by $\boldsymbol{\xi}$, the ordered $n$-tuple $(\xi_1, \ldots, \xi_n)$. We call $P$ the underlying partition and say $\boldsymbol{\xi}$ is based on $P$. We denote the Riemann sum $\sum_{i=1}^n f(\xi_i)\Delta x_i$ by $R(P, \boldsymbol{\xi}; f)$ for short. We can then define $\int_a^b f(x)dx$ as the limit of $R(P, \boldsymbol{\xi}; f)$ as $P$ becomes finer and finer. Now, as $P$ becomes finer and finer, what does it tend to ? The most logical answer would be the finest partition of $[a, b]$. But (cf. Exercise (1.1)), there is no partition which is finest among all. This hardly need worry us. There is no largest positive integer but still we consider limits of the form $\lim a_n$ as $n$ gets larger and larger. Symbolically, we write $\lim_{n \to \infty} a_n$. But this is nothing more than a way of writing, because as explained in Section 2.4, infinity is a non-entity. Similarly, in the present case, either we can use some symbol (say $P_\infty$) for the (non-existent) finest partition and write 'as $P \to P'_\infty$' or we can retain the phrase as '$P$ gets finer and finer'. We prefer the latter. Our theorem can be stated as follows.

**2.1 Theorem:** Let $f$ be a bounded function on an interval $[a, b]$. Then $f$ is Riemann integrable on $[a, b]$ if and only if there exists a real number $I$ such that the Riemann sum $R(P, \boldsymbol{\xi}, f)$ tends to $I$ as $P$ becomes finer and finer, that is, for every $\epsilon > 0$, there exists $Q \in \mathcal{P}[a, b]$ such that whenever $P$ is a refinement of $Q$ and $\boldsymbol{\xi}$ is based on $P$, $|R(P, \boldsymbol{\xi}, f) - I| < \epsilon$. Moreover, when this happens, the number $I$ equals $\int_a^b f(x)dx$.

**Proof:** The direct implication in the first statement was already proved abvoe (except for an interchange of $P$ and $Q$ in the notation). For the converse, assume that the given condition holds. We have to prove that $f$ is integrable over $[a, b]$. We shall do so by appealing to Exercise (1.5). For this, let $\epsilon > 0$ be given. We have to find some partition $P$ of $[a, b]$ for which

$$U(P; f) - L(P; f) < \epsilon. \tag{2}$$

To find such a partition we proceed as follows. By applying the given condition with $\frac{\epsilon}{4}$ (instead of $\epsilon$), there exists a partition $Q$ of $[a, b]$ such that for every refinement $P$ of $Q$ and for every $\boldsymbol{\xi} = (\xi_1, \xi_2, \ldots, \xi_n)$ based on $P$,

$$I - \frac{\epsilon}{4} < R(P, \boldsymbol{\xi}; f) < I + \frac{\epsilon}{4}. \tag{3}$$

In particular this holds if $P$ is taken to be $Q$ itself. So we denote $Q$ by $P$ and claim that with this choice, (2) holds.

Actually, we shall prove two slightly stronger inequalities, which will be needed later on. That is, we shall prove,

$$I - \frac{\epsilon}{2} < U(P; f) < I + \frac{\epsilon}{2} \tag{4}$$

$$\text{and} \qquad I - \frac{\epsilon}{2} < L(P; f) < I + \frac{\epsilon}{2} \tag{5}$$

which together would imply (2).

Consider (4). In order to derive it from (3), we shall make a judicious choice of $\boldsymbol{\xi}$. Let $P = Q = (x_0, x_1, \ldots, x_n)$. As usual let $M_i = \sup\{f(x) : x_{i-1} \le x \le x_i\}$ for $i = 1, \ldots, n$. Now, $\dfrac{\epsilon}{4(b-a)}$ is a positive real number and so by Exercise (2.6.12), there exists $\xi_i \in [x_{i-1}, x_i]$ such that

$$M_i < f(\xi_i) + \frac{\epsilon}{4(b-a)} \tag{6}$$

We let $\boldsymbol{\xi} = \{\xi_1, \ldots, \xi_n\}$. Multiplying (6) by $\triangle x_i$ summing over for $i = 1$ to $n$ and noting that $\displaystyle\sum_{i=1}^{n} \triangle x_i = b - a$, we get

$$U(P; f) < R(P, \boldsymbol{\xi}; f) + \frac{\epsilon}{4} \tag{7}$$

which, in particular, means $|U(P; f) - R(P, \boldsymbol{\xi}; f)| < \dfrac{\epsilon}{4}$. But by (2) we also have $|R(P, \xi, f) - I| < \dfrac{\epsilon}{4}$. Using triangle inequality it follows that $|U(P; f) - I| < \dfrac{\epsilon}{4} + \dfrac{\epsilon}{4} = \dfrac{\epsilon}{2}$, which is exactly (4).

Similarly, (5) can be proved by applying (2) to a suitable $\boldsymbol{\xi} = (\xi, \ldots, \xi_n)$ (possibly different from the $\boldsymbol{\xi}$ in (7)) for which $L(P; f) > R(P; f) - \dfrac{\epsilon}{4}$. The argument is completely dual and left as an exercise. As noted before, we have now proved (2) and hence, by Exercise (1.5) $f$ is Riemann integrable on $[a, b]$. This completes the proof of the first statement in the Theorem.

For the second statement, we note that (4) and (5) imply that for every $\epsilon > 0$, there is some $P \in \mathcal{P}[a, b]$ such that both $L(P; f)$ and $U(P; f)$ are within an $\epsilon$-neighbourhood (in fact an $\dfrac{\epsilon}{2}$-neighbourhood) of the number $I$. So, by Exercise (1.6), $\displaystyle\int_a^b f(x)dx$ must equal $I$. $\blacksquare$

Interesting as this theorem is, it leaves someting to be desired. We remarked earlier that the expression 'as the partition $P$ gets finer and finer' is analogous to the expression 'as the integer $n$ gets larger and larger'. But there is an important

important difference. Given any two integers $m$ and $n$, we can always compare them, i.e., either $m \leq n$ or $n \leq m$. Let us see what effect this has on limits of sequences. Suppose $a_n \to L$ as $n \to \infty$ or as $n$ gets larger and larger. Let $\epsilon > 0$ be given. Then there is some $m$ such that for all $n \geq m, L - \epsilon < a_n < L + \epsilon$. Since every integer is comparable with $m$, the expression 'for all $n \geq m$' means a lot. It means 'for all except finitely many values of $n$' or 'for almost all $n$', (cf. Section 3.2). So if $n$ is 'sufficiently' large we are sure that $a_n$ will be very close to $L$. With partitions this is not quite the case. Given two partitions $P$ and $Q$ of an interval $[a, b]$, it may happen that neither $P$ is a refinement of $Q$ nor the other way (cf. Exercise (1.1)). This tarnishes the utility of the preceding theorem somewhat. Suppose $f$ is integrable on $[a, b]$ and $I = \int_a^b f(x)dx$. Let $\epsilon > 0$ be given. The theorem above says that there exists a partition $Q$ of $[a, b]$ such that for every partition $P$, *which is a refinement* of $Q$, every Riemann sum obtainable from $P$ lies within an $\epsilon-$neighbourhood of $I$. What if we have a partition $P$ which is 'fine' in some other sense but which is not a refinement of $Q$? As a concrete example of this, suppose $a, b$ are rational but $Q$ has an irrational node. Let $P_n$ be the partition of $[a, b]$ into $n$ equal subintevals. Then all the nodes of $P_n$ are rational. If $n$ is large, then undoubtedly, $P_n$ is a fine partion of $[a, b]$, but because of the irrational node of $Q, P_n$ can never be finer than $Q$. As a result, for large $n$, any Riemann sum based on $P_n$ may very well be close to $I$, but we would not be able to conclude this from the theorem above, because $P_n$ is not a refinement of $Q$.

This anomaly can be corrected by introducing a new measure of fineness of a partition which is not based on a comparision of two partitions but on the magnitude of a certain real number associated with that partition. We define it first.

**2.2 Definition:** Let $P = (x_0, x_1, \ldots, x_n)$ be a partition of an interval $[a, b]$. Then the **mesh** or the **norm** of $P$, denoted by $\mu(P)$ or by $\|P\|$, is the length of the largest subinterval of $[x_{i-1}, x_i]$ (there could, of course, be more than one such subintervals). Or in other words,

$$\mu(P) = \|P\| = \max\{x_i - x_{i-1} : 1 \leq i \leq n\}.$$

The term mesh is usually used in connection with nets, sieves or filters. It means the size of the largest object than can pass through. When used in connection with partitions, we expect that the smaller the mesh the finer would be a partition. But this is not quite the case. It is certainly true that if $P$ is a refinement of $Q$ then $\mu(P) \leq \mu(Q)$. This is trivial to prove since the subintervals of $P$ are contained in those of $Q$. But the converse is false. We already had a counter-example above where $a, b$ were rational and $Q$ had an irrational node. Obviously, $\mu(P_n) = \dfrac{b - a}{n}$ and this can be made less than $\mu(Q)$ by taking $n$ sufficiently large. But no matter how large $n$ is, $P_n$ is not a refinement of $Q$.

The concept of a mesh provides us with another definition of limits of Riemann sums. Instead of saying 'as the partition $P$ becomes finer and finer', we can now say 'as the mesh of $P$ tends to 0'. Even though the statement '$P$ is finer then $Q$' is much stronger than the statement '$\mu(P) \leq \mu(Q)$', it turns out, rather suprisingly, that in their limiting behaviour they are equivalent as we now show.

**2.3 Theorem:** Let $f$ be a bounded function on an interval $[a, b]$ and $I$ be a real number. Then the following two statements are equivalent to each other:

(i)  the Riemann sum $R(P, \boldsymbol{\xi} \, ; f)$ tends to $I$ as the partition gets finer and finer; that is, for every $\epsilon > 0$, there exists a partition $Q$ of $[a, b]$, such that for every partition $P$ of $[a, b]$ which is a refinement of $Q$, and for every $\boldsymbol{\xi}$ based on $P$, $|R(P, \boldsymbol{\xi} \, ; f) - I| < \epsilon$.

(ii) the Riemann sum $R(P, \boldsymbol{\xi} \, ; f)$ tends to $I$ as $\mu(P)$ tends to 0; that is, for every $\epsilon > 0$, there exists $\delta > 0$ such that for every partition $P$ of $[a, b]$ with $\mu(P) < \delta$ and for evey $\boldsymbol{\xi}$ based on $P$, $|R(P, \boldsymbol{\xi} \, ; f) - I| < \epsilon$.

**Proof:** Assume first that (i) holds. We have to prove (ii). Let $\epsilon > 0$ be given. Since by (i), every Riemann sum is trapped in between the upper and the lower Riemann sums for its underlying partition, it would suffice to find some $\delta > 0$ such that for all $P$ with $\mu(P) < \delta$,

$$U(P; f) < I + \epsilon \tag{8}$$

$$\text{and} \qquad I - \epsilon < L(P; f) \tag{9}$$

By (i), we know that there exists some partition $Q$ of $[a, b]$ such that whenever $P$ is a refinement of $Q$, and $\boldsymbol{\xi}$ is based on $P$,

$$I - \frac{\epsilon}{4} < R(P, \boldsymbol{\xi} \, ; f) < I + \frac{\epsilon}{4}.$$

which is same as (3) in the proof of Theorem (2.1). So by following exactly the same reasoning as used there, we get,

$$I - \frac{\epsilon}{2} < U(P; f) < I + \frac{\epsilon}{2} \tag{10}$$

$$\text{and} \qquad I - \frac{\epsilon}{2} < L(P; f) < I + \frac{\epsilon}{2} \tag{11}$$

which are same as (4) and (5) respectively. These inequalities hold whenever $P$ is a refinement of $Q$.

Now suppose $Q = (y_0, y_1, \ldots, y_k)$ with $a = y_0 < y_1 < y_2 < \ldots < y_{k-1} < y_k = b$. We assume $k > 1$, as otherwise every partition of $[a, b]$ is automatically a refinement of $Q$ and we would be done. We now let $M$ be a positive upper bound on $|f(x)|$ for $a \leq x \leq b$. $M$ exists since $f$ is bounded on $[a, b]$. We let $\delta = \dfrac{\epsilon}{4M(k - 1)}$. The reason for this rather clumsy choice will be clear in the course of the proof.

Let $P$ be a partition of $[a, b]$ with $\mu(P) < \delta$. If luckily $P$ is a refinement of $Q$, then (10) and (11) hold and hence (8) and (9) also hold as desired. If $P$ is not a refinement of $Q$, the argument is a little complicated. First consider $P \cup Q$, the partition obtained by superimposing $P$ and $Q$. Then $P \cup Q$ is a refinement of $Q$ (and also of $P$) and so (10) and (11) would hold if $P$ is replaced by $P \cup Q$. So, we have

$$I - \frac{\epsilon}{2} < U(P \cup Q; f) < I + \frac{\epsilon}{2} \tag{12}$$

$$I - \frac{\epsilon}{2} < L(P \cup Q; f) < I + \frac{\epsilon}{2} \tag{13}$$

In order to derive (8) and (9) from these, we need to estimate the difference between $U(P \cup Q; f)$ and $U(P; f)$ and also that between $L(P \cup Q; f)$ and $L(P; f)$. Since $P \cup Q$ is a refinement of $P$, we already know, by Theorem (1.2), that $U(P; f) \geq U(P \cup Q; f)$. But we claim that $U(P; f)$ is not substantially bigger than $U(P \cup Q; f)$.

Let $P = (x_0, x_1, \ldots, x_n)$. Take a typical subinterval $[x_{i-1}, x_i]$ of $P$. If none of the nodes $y_1, \ldots, y_{k-1}$ of $Q$ lies in its interior, then $[x_{i-1}, x_i]$ will also be a subinterval of $P \cup Q$. Otherwise it will get split into two or more subintervals. For notational simplicity, in the first case, let us say that the subinterval $[x_{i-1}, x_i]$ as well as the index $i$ are 'good' and in the second case that they are 'bad'.

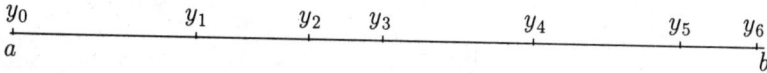

(a) partition $Q$

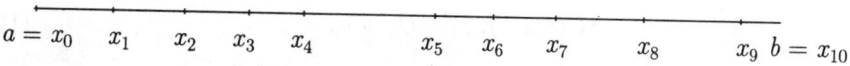

(b) partition $P$

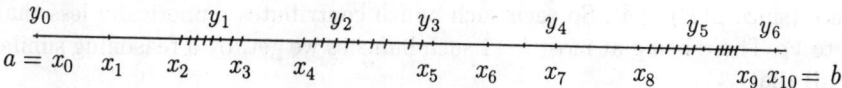

(c) partition $P \cup Q$ (split intervals are shaded)

Figure 5.2.1 : Splitting of Subintervals in Superimposition of Partitons

Figure 5.2.1 shows a situation in which $k = 6, n = 10$. The bad subintervals of $P$ are $[x_2, x_3], [x_4, x_5]$ and $[x_8, x_9]$ which get split into, respectively, 2,3 and 2 subintervals of $P \cup Q$. Note that $y_4 = x_7$ is a common node of $P$ as well as of $Q$. So the subintervals $[x_6, x_7]$ and $[x_7, x_8]$ are not bad. Although $y_4$ is an end-point of them, it is not an interior point.

Clearly, no two distinct bad subintervals of $P$ can have a common node of $Q$ inside them, as their interiors are disjoint. Since $Q$ has only $k-1$ nodes, it follows that in $P$ there are at most $k-1$ bad subintervals. This simple observation is the key to showing that $U(P; f)$ cannot differ much from $U(P \cup Q; f)$. Let us actually compare the two. By definition, $U(P; f) = \sum_{i=1}^{n} M_i \triangle x_i$ where, as usual, $M_i = \sup\{f(x) : x \in [x_{i-1}, x_i]\}$.

Let us split this sum into two parts as

$$U(P; f) = U_g + U_b \tag{14}$$

where $U_g = \sum_{i\ good} M_i \triangle x_i$, the sum being taken over good indices

and $\quad U_b = \sum_{i\ bad} M_i \triangle x_i$, the sum being taken over bad indices.

In $U_b$ there are at most $k-1$ terms and since $|M_i| \leq M$ for all $i$, and $\mu(P) < \delta$, each term is less than $M\delta$ in absolute value. Since $\delta \leq \dfrac{\epsilon}{4M(k-1)}$, we get

$$|U_b| < \frac{\epsilon}{4} \tag{15}$$

Let us now tackle the other sum, viz., $U_g = \sum_{i\ good} M_i \triangle x_i$. Evey term in this sum also occurs in the uper Rieman sum $U(P \cup Q; f)$ because every good sub-interval of $P$ is also a sub-interval of $P \cup Q$. So let us write

$$U(P \cup Q; f) = U_g + V_b \tag{16}$$

where $V_b$ is the sum of the remaining terms in $U(P \cup Q; f)$. These terms occur in bunches, each bunch corresponding to the parts into which a bad subinterval of $P$ is split by the nodes of $Q$. The sum of the lengths of these parts (in any one bunch) is the length of their parent bad subinterval (of $P$) and hence is less than $\delta$ (since $\mu(P) < \delta$). So each such bunch contributes numerically less than $M\delta$ to $V_b$. There being at most $k-1$ such bunches we get, by a reasoning similar to (15) that

$$|V_b| < \frac{\epsilon}{4} \tag{17}$$

From (14) to (17), we now have

$$
\begin{aligned}
U(P; f) - U(P \cup Q; f) &= U_b - V_b \\
&\leq |U_b| + |V_b| \\
&< \frac{\epsilon}{4} + \frac{\epsilon}{4} = \frac{\epsilon}{2}
\end{aligned}
\tag{18}
$$

(12) and (18) together imply (8). By an exactly similar reasoning, which we leave to the reader, we get (9), starting from (13) and showing subsequently that $L(P \cup Q; f) - L(P; f)$ is at most $\dfrac{\epsilon}{2}$.

Thus we have shown that (8) and (9) are true for every partition $P$ for which $\mu(P) < \delta$, and as noted before this completes the (rather lengthy) proof of the implication (i) $\Rightarrow$ (ii).

The implication (ii) $\Rightarrow$ (i) is trivial. Suppose (ii) holds. Let $\epsilon > 0$ be given. Then there exists a $\delta$ such that $|R(P, \boldsymbol{\xi}\,; f) - I| < \epsilon$ for every $\boldsymbol{\xi}$ based on $P$ whenever $\mu(P) < \delta$. Let $n$ be an integer greater than $\dfrac{b-a}{\delta}$ and let $Q$ be the partition of $[a, b]$ into $n$ equal parts. Then $\mu(Q) = \dfrac{b-a}{n} < \delta$. Now if $P$ is finer than $Q$, then $\mu(P) \le \mu(Q) < \delta$. And so $|R(P, \xi; f) - I| < \epsilon$. Thus (i) holds. ∎

Combining this theorem with Theorem (2.1), we immediately get

**2.4 Corollary:** Let $f$ be a bounded function on an interval $[a, b]$. Then $f$ is Riemann integrable over $[a, b]$ if and only if there exists a real number $I$ such that the Riemann sum $R(P, \boldsymbol{\xi}\,; f)$ tends to $I$ as $\mu(P)$ tends to 0, i.e., for every $\epsilon > 0$, there exists $\delta > 0$ such that $|R(P, \boldsymbol{\xi}\,; f) - I| < \epsilon$ for every $\boldsymbol{\xi}$ based on $P$ whenever $\mu(P) < \delta$. Moreover, when this happens, $I = \displaystyle\int_a^b f(x)dx$. ∎

Because of this corollary, we get an alternate definition of the Riemann integral $\displaystyle\int_a^b f(x)dx$, viz., as the limit of the Riemann sums as the mesh of their underlying partitions tends to 0. Although this definition is a little clumsy as compared with the definition in terms of the upper and lower Riemann sums, it has certain advantages. We list them below.

(i) Given a partition $P = (x_0, \ldots, x_n)$ it may not be easy to find $M_i$ and $m_i$. So calculating the upper and the lower Riemann sums itself can be a problem. With Riemann sums, however, all we need to do is to evaluate $f$ at some points $\xi_1, \ldots, \xi_n$ in the subintervals of $P$. It does not matter which points are chosen since all Riemann sums tend to the same limit. In fact, as we shall see later, the freedom we have in choosing $\xi_i$ is crucially needed in the proof of the Fundamental Theorem of Calculus.

(ii) In defining $\displaystyle\int_a^b f(x)dx$ through upper and lower Riemann sums, it was vital to assume that $f$ was bounded on $[a, b]$. In defining a Riemann sum this assumption is not necessary. This makes the definition a little more general. Suppose for example that $f$ is defined on [0,1] by $f(0) = 0$ and $f(x) = \dfrac{1}{\sqrt{x}}$ for $x > 0$. Then $f$ is unbounded on [0,1]. It can, however, be shown that the Riemann sums tend to the limit 2 and so $\displaystyle\int_0^1 f(x)dx$ comes out as 2. With the earlier definition in terms of upper and lower Riemann sums, the integral has to be studied as an improper integral as remarked at the end of the last section.

(iii) There is also some advantage in considering the limit in terms of the mesh of the partition rather than in terms of relative fineness. The mesh gives an absolute measure of how fine a partition is and thereby a quantitative rather than a qualitative idea of approaching the (hypothetical) finest partition. This helps considerably in evaluating $\int_a^b f(x)dx$ if it is already known to exist. We can take a sequence $P_n$ of partitions such that $\mu(P_n) \to 0$ as $n \to \infty$. (One of the simplest such sequences is one where $P_n$ is the partition of $[a, b]$ into $n$ equal parts.) We choose $\xi_i$'s by some prescribed rule (e.g. by taking the mid-point of each subinterval). We thus get a sequence, say $\{a_n\}_{n=1}^{\infty}$ of Riemann sums of $f$. Then $\lim\limits_{n\to\infty} a_n$ equals $\int_a^b f(x)dx$. In some cases we can explicitly evaluate this limit. Otherwise we can take $a_n$ (for a large $n$) as an approximation to $\int_a^b f(x)dx$. We shall study such 'approximate integration' in Section 5.9.

Sometimes the tables can be turned around. Instead of evaluating Riemann integrals as limits of sequences of Riemann sums, the limits of certain sequences can be evaluated by first expressing their terms as Riemann sums of suitable functions over suitable intervals and then by evaluating the integral of the function by some other method (e.g. using the fundamental theorem of calculus). Illustrations of this will be given in Exercise (2.6).

(iv) The strongest justification for the Riemann sums, however, is that despite their clumsy form, they are very natural sums to consider in many real-life applications. In the last section, we saw that if $f$ is non-negative on $[a, b]$ then the upper and the lower sums of $f$ on $[a, b]$ provide upper and lower bounds for the area below the graph of $f$. By a similar reasoning, each Riemann sum is an approximation to this area and the exact area is the limit of these Riemann sums.

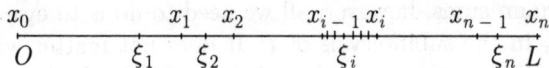

Figure 5.2.2 : Mass of a Wire as a Riemann Integral

As another example, suppose we have a piece of wire of length $L$. We are given its linear density $\rho$ (= mass per unit length) and want to find its total mass, say, $M$. This is trivial if $\rho$ is constant all over the wire. Then clearly the answer is simply $\rho L$. But what if $\rho$ is not a constant ? That is, suppose the linear density at a point which is $x$ units away from one end of the wire is $\rho(x)$. Suppose we have some method of finding $\rho(x)$ for a given $x$. In such a case we divide the wire into a large number of segments. If each segment is short enough, then its mass is approxmately the product $\rho(\xi)\triangle x$ where $\triangle x$ is the length of the segment and $\xi$ is any point in it. Adding these masses we get an approximation to $M$. A look at Figure 5.2.2 shows that this is nothing but the Riemann sum $R(P, \boldsymbol{\xi} ; \rho)$ for a suitable partition $P$ of the interval $[0, L]$ and a suitable $\boldsymbol{\xi}$ based on $P$. As $\mu(P) \to 0$, this approximation will be more and more accurate and so the exact mass $M$ will equal $\displaystyle\int_0^L \rho(x)dx$.

The procedure illustrated in this example is so typical and widely applicable that it deserves to be stated explicitly as 'recipe of integration'.

**Step 1:** Do the simple case. (In the example above this is the case where $\rho$ is constant.)

**Step 2:** Chop the general case into a large number of small pieces.

**Step 3:** Form sums, treating each piece approximately as a simple case.

**Step 4:** Express these sums as Riemann sums of a suitable function over a suitable interval.

**Step 5:** Take the limit of these sums as each of the pieces shrinks in size to 0. Express this limit as a definite integral.

**Step 6:** Evaluate the integral (by whatever method) to get the exact answer.

In the Exercises, a few problems will be given where this procedure is needed. More examples will come in the next chapter (which will be devoted to applications of integrals). After a little practice, it is, of course, unnecessary to work out these steps explicitly. The result of Step 5 in a particular example can often be written down by inspection once Step 1 is over.

The limits of Riemann sums $\displaystyle\sum_{i=1}^n f(\xi_i)\Delta x_i$ are rather curious types of limits. As $\mu(P) \to 0$, every $\Delta x_i \to 0$ and $n \to \infty$. So it is a limit in which each summand tends to 0 but the number of summands tends to $\infty$. To some extent this is analogous to the indeterminate form $0.\infty$ which we considered in Section 4.9.

Estimation of the rain-water falling over a country is an excellent example of a Riemann sum, except that the country being a subset of $\mathbb{R}^2$ rather than $\mathbb{R}$, the function involved (viz. the rainfall function) is a function of two variables. But the essential ideas are the same. It is impossible to put a rain gauge at every point of the country. So a few places are chosen as representative samples of various smaller regions. The rainfall at one such place, multiplied by the area of the region represented by it provides the approximate rain-water falling over that region. Adding such products, we get the estimated rain water falling over the whole country. The smaller these regions, the more accurate is this estimate. The exact answer is the limit of these sums. (cf. Exercise (2.2.7)). We shall briefly study such limits as **double integrals** in the next chapter.

The fact that integration is a limit of summation is reflected in the notation as well as in some terminology. We already mentioned that the integral sign $\displaystyle\int$

originates in an elongated $S$ formerly used to denote summation. Similarly, just as the terms summed in a summation are called summands, in an integral, the function to be integrated is called the **integrand**. As a further analogy, just as in a summation $\sum\limits_{k=1}^{n}$, the variable $k$ is merely an index (or a 'dummy') variable and could be replaced by any other variable, in the definite integral, $\int_a^b f(x)dx$, the variable $x$ has little role to play and, as explained in the last section, can be replaced by any other variable. So, $\int_a^b f(x)dx$ is the same as $\int_a^b f(y)dy$ or $\int_a^b f(t)dt$. Because of this it is becoming more common to denote the integral simply by $\int_a^b f$. But we shall stick to the more traditional notation, both because it truly reflects the genesis of the integral and also because occasionally we need to integrate a function of one variable variable w.r.t. some other (related) variable. For example we come across integrals of the form $\int_a^b f(x)ds$ where the variable $s$ is a (differentiable) function of $x$, say $s = g(x)$. By definition, this integral is the same as $\int_a^b f(x)\frac{ds}{dx}dx$ i.e., $\int_a^b f(x)g'(x)dx$ or $\int_a^b fg'$. But the way it arises in applications makes $\int_a^b f(x)ds$ a more natural notation.

## EXERCISES

2.1   Let $f$ be a bounded function on $[a, b]$ and $P$ a fixed partition of $[a, b]$. Let $\mathcal{R}(P)$ be the set of all Riemann sums of $f$ based on $P$. Prove that $U(P; f)$ and $L(P; f)$ are respectively the supremum and the infimum of $\mathcal{R}(P)$. (This was, in fact, proved, although not explicitly, in the proof of Theorem (2.1).) Show by an actual example that they need not be the maximum and the minimum elements of $\mathcal{R}(P)$.

2.2   Complete the proof of the implication (i) $\Rightarrow$ (ii) in theorem (2.3) by deriving (9) from (13).

2.3   Let $f$ be bounded on $[a, b]$. Prove that as $\mu(P) \to 0$, the upper Riemann sum $U(P; f)$ tends to the upper Riemann integral $\overline{\int_a^b} f(x)dx$ and $L(P; f)$ tends to $\underline{\int_a^b} f(x)dx$.

2.4   Let $f$ be bounded on $[a, b]$. Prove that $f$ is Riemann integrable over $[a, b]$ if and only if $U(P; f) - L(P; f) \to 0$ as $\mu(P) \to 0$.

2.5 Let $f$ be bounded on $[a, b]$. Prove that $f$ is Riemann integrable over $[a, b]$ if and only if for every $\epsilon > 0$, there exists $\delta > 0$, such that whenever $P$ is a partition of $[a, b]$ with $\mu(P) < \delta$, $|R(P, \boldsymbol{\xi}\ ; f) - R(P, \boldsymbol{\eta}; f)| < \epsilon$ for every $\boldsymbol{\xi}$ , $\boldsymbol{\eta}$ based on $P$. (Verbally, this condition says that as $\mu(P) \to 0$, the difference between any two Riemann sums based on $P$ tends to 0. This result is conceptually analogous to the Cauchy criterion for the convergence of a sequence, Theorem (3.5.4).) [*Hint* : Apply the last exercise and Exercise (2.1) for the converse implication.]

2.6 Express the limits of the following sequences as definite integrals of suitable functions over suitable intervals.

(i) $\dfrac{1}{n}(\dfrac{1}{n} + \dfrac{2}{n} + \ldots + \dfrac{n}{n})$ (ii) $\dfrac{1}{n} \sum\limits_{r=1}^{n} \sin(\dfrac{\pi r}{n})$

(iii) $\dfrac{1}{n} \sum\limits_{r=1}^{2n} \dfrac{r}{\sqrt{n^2 + r^2}}$ (iv) $\dfrac{1}{n} \ln \dfrac{(2n)!}{n^n n!}$

2.7 A straight 2 kilometer road runs from $A$ to $B$. The population density at a point $x$ k.m. away from $A$ is $\rho(x)$. Express the total population of the road as a definite integral.

2.8 In the last exercise suppose $\sigma(x)$ is the average per capita income of the people living at a distance $x$ k.m. from $A$. Find (i) the total income of the road population; (ii) the average per capita income for the road population.

2.9 The electrical resistance of a wire is inversely proportional to its area of cross-section and directly to its length. Express, as a definite integral, the resistance of a wire of length $L$ whose cross section at a point $x$ from one end is a disc of radius $\sigma(x)$.

2.10 Suppose $f(t)$ is the speed of a particle moving in a straight line at time $t$. $f(t)$ is positive or negative according as the direction of motion is forward or backward at time $t$, w.r.t. some fixed choice. What does $\int_a^b f(t)dt$ represent physically?

## 5.3    Integrability of a Function - a Global Concept

In Section 3.11, after Theorem (3.11.5) we remarked that many concepts in calculus are of a local nature in the sense that they depend only on the values of a function in a neighbourhood of a point. For example, continuity and differentiablity of a function are local concepts. We talk of continuity or differentiablity of a function *at a point* first. Then we define continuity or differentiablity *on an interval S* to mean continuity or differentiablity at each point of $S$. On the other hand, certain concepts such as uniform continuity are global in nature. We talk of uniform continuity *on a set S* (usually an interval). Similarly uniform convergence is a global property while pointwise convergence is a local property. Monotonicity and concavity (whether upwards or downwards) of a function over an interval are also global properties. (Monotonicity of a function *at* a point, as defined in Exercise (4.3.4) is a local property but it is different from monotonicity over an interval, as shown by Exercise (4.5.19).)

The concept of integrability of a function is also a global concept. Suppose $f$ is defined and bounded on an interval $[a, b]$. We defined the (Riemann) integrability of $f$ on $[a, b]$ in terms of certain sums obtained by first partitioning $[a, b]$. Instead of $[a, b]$, we could have taken any subinterval $S$ of $[a, b]$. By Exercise (1.7), $f$ would be integrable on $S$ too, if it is integrable on $[a, b]$. It is tempting to let the interval $S$ shrink to a point $c$ and thereby define the integral of $f$ at the point $c$ by taking a limit of the form $\lim\limits_{\delta \to 0} \int_{c-\delta}^{c+\delta} f(x)dx$. But it is not hard to show (cf. Exercise (3.1)) that this limit is always 0. So there is no such thing as integrability *at* a point $c$ or the integral of $f$ on a singleton set $\{c\}$, except in a trivial sense. These concepts are meaningful only over an interval. So, in this sense, Riemann integrability is a global concept like uniform continuity or monotonicity.

There is, however, a stronger sense in which integrability is a global concept. To see it, suppose $f$ is uniformly continuous on $[a, b]$ and $c$ is a point of $[a, b]$. If we change $f(c)$, without changing $f$ at any other point of $[a, b]$, then $f$ would no longer be uniformly continuous on $[a, b]$, because for $x$ near $c$, $|f(x) - f(c)|$ would no longer be small. Similarly, if $f$ is monotonically increasing on $[a, b]$ and we change $f(c)$, without changing $f$ at any other point of $[a, b]$, then the new function may not be monotonically increasing on $[a, b]$ (cf. Exercise (3.3)). In other words, even though uniform continuity and monotonicity are global properties, every single point counts in a negative sense, viz., changing the function at that point can destroy that property over the entire interval, rather like piercing an inflated balloon at one point.

With integrability the picture is completely different. Suppose $f$ is Riemann integrable on $[a, b]$ and we define a new function $g$ by altering $f$ at just one point say, $c$, of $[a, b]$. Then $g$ is integrable on $[a, b]$ and moreover $\int_a^b g(x)dx = \int_a^b f(x)dx$. Verbally, the value of a function at a single point has absolutely no role in determining the integrability or the value of the definite integral of that function. These things depend on the value of the function at the totality of all points of the interval $[a, b]$. A single point does not count at all.

Before giving a formal proof of the statement just made, let us see why it is intuitively clear. By Corollary (2.2), $\int_a^b f(x)dx$ is the limit of of a Riemann sum, say, $\sum_{i=1}^n f(\xi_i)\Delta x_i$ as the mesh of its underlying partition, say $P$, tends to 0. In a sum of this kind, the point $\xi_i$ may be thought of as a representative of the $i^{\text{th}}$ sub-interval $[x_{i-1}, x_i]$ of $P$. The length of this sub-interval, viz. $\Delta x_i$, is like the importance or 'weightage' given to $\xi_i$. As $\mu(P) \to 0$, each $\Delta x_i \to 0$. The total number of representatives increases while the individual importance of each representative decreases so that even if $\xi_i$ is one of the representatives and $f(\xi_i)$ is large, its contribution $f(\xi_i)\Delta x_i$ becomes neglibly small. Ultimately, in the limit, it vanishes. This, of course, does not mean that $\int_a^b f(x)dx$ vanishes, because as $\mu(P) \to 0$, even though each individual summand tends to 0, the number of summands tends to infinity. It is difficult to give a perfect real-life analogy, because in real life situations the variables encountered, such as population, are discrete. Still, as a somewhat jovial analogy, a definite integral is like a largely populated communist country where an individual has no strength but the country, as a whole, is a formidable power. As a less jovial analogy, consider a piece of wire of length $L$ and let $f(x)$ denote its linear density at a point at a distance $x$ from one of its ends. As mentioned at the end of the last section, $\int_a^b f(x)dx$ is the total mass of the wire, which will, in general, be some positive real number. If $P$ is a point on the wire at a distance $\xi$ from the end, then the mass of $P$ is 0, no matter how high $f(\xi)$ is. A single point can have no mass. This is because, the mass of $P$ is to be distinguished from the linear density at $P$. The former is always 0. But the latter is defined as the mass per unit length of the wire. For a layman, it is the ratio of the mass of a very small piece of wire containing $P$ to its length; for yesterday's calculus student it is the average mass density of an infinitesimally small piece of wire; for today's calculus student, it is the limit of the average mass density of a piece of wire (containing $P$) as its length tends to 0. (This is very analogous to the concept of the instantaneous speed and can similarly be expressed as a certain derivative.)

We now give a formal proof that in an integral a singleton point does not count. Actually, by induction we get a slightly stronger result.

**3.1 Theorem:** Let $f, g$ be functions defined on an interval $[a, b]$. Suppose there is a finite set of points, say, $c_1, \ldots, c_k$ of $[a, b]$ such that $f(x) = g(x)$ for all $x \in [a, b] - \{c_1, \ldots, c_k\}$. Then $f$ is Riemann integrable over $[a, b]$ if and only if $g$ is so and when this happens,

$$\int_a^b f(x)dx = \int_a^b g(x)dx.$$

**Proof:** We proceed by induction on $k$. Suppose $k = 1$. First assume $c_1$ equals one of the end-points of $[a, b]$, say $c_1 = a$. Suppose $f$ is integrable on $[a, b]$. Then $f$ is, first of all, bounded on $[a, b]$. Let $M$ be a (positive) upper bound on $|f(x)|$ in $[a, b]$. Then since $g(x) = f(x)$ for all $x \neq c_1$, it is clear that $M + |g(c_1)|$ is an upper bound on $|g(x)|$ for all $x \in [a, b]$, call this number as $M'$. So $g$ is certainly bounded on $[a, b]$. To show that it is integrable on $[a, b]$, we shall use Exercise

(1.5). Let $\epsilon > 0$ be given. Then by Exercise (1.5), there exists a partition $P$ of $[a, b]$ such that

$$U(P; f) - L(P; f) < \frac{\epsilon}{3} \tag{1}$$

where $U(P; f)$ and $L(P; f)$ are, respectively the upper and lower Riemann sums of $f$ for $P$. Let $P = (x_0, x_1, \ldots, x_n)$. By adding an extra node to $P$ (which does not affect the inequality in (1)), if necessary, we may suppose that the first sub-interval, $[x_0, x_1]$ is so small that $\Delta x_1 < \dfrac{\epsilon}{3(M + M')}$. Let us now compare the upper Riemann sums $U(P; f)$ and $U(P; g)$. As usual, let $M_i$ and $M_i'$ be, respectively, the suprema of $f$ and $g$ on $[x_{i-1}, x_i]$. Note that for $i > 1$, $M_i = M_i'$. So all except possibly one term in these sums match with each other, giving

$$U(g; f) - U(P; f) = (M_1' - M_1)\Delta x_1$$

Since $M_1' \leq M'$ and $M_1 \geq -M$, and $(M + M')\Delta x_1 < \frac{\epsilon}{3}$, we get

$$U(g; f) - U(P; f) < \frac{\epsilon}{3} \tag{2}$$

Similarly, comparing the lower Riemann sums of $f$ and $g$ we get

$$L(P; f) - L(P; g) < \frac{\epsilon}{3} \tag{3}$$

From (1), (2), (3), the triangle inequality and the fact that $L(P; f) \leq U(P; f)$, we get

$$U(P; f) - L(P; f) < \epsilon. \tag{4}$$

Thus we have shown that for every $\epsilon > 0$, there is some partition $P$ of $[a, b]$ for which (4) holds. By Exercise (1.5), this means $g$ is integrable over $[a, b]$. Interchanging the roles of $f$ and $g$, we get that integrability of $g$ implies that of $f$. To prove the equality of the integrals $\int_a^b f(x)dx$ and $\int_a^b g(x)dx$ we appeal to Exercise (1.6). Let $I = \int_a^b f(x)dx$. Given $\epsilon > 0$, we find a partition $P$ as above. Since $L(P; f) \leq I \leq U(P; f)$, (1) implies that $U(P; f)$ and $L(P; f)$ both lie in the neighbourhood $(I - \frac{\epsilon}{3}, I + \frac{\epsilon}{3})$ of $I$. But then by (2) and (3), $U(P; f)$ and $L(P; g)$ lie inside $(I - \epsilon, I + \epsilon)$. So by Exercise (1.6), $\displaystyle\int_a^b g(x)dx = I$.

Thus we have proved the theorem in the case $k = 1$ and $c_1 = a$. A similar argument holds if $c_1 = b$, the other end-point of $[a, b]$. Finally, if $a < c_1 < b$, we split $[a, b]$ into $[a, c_1]$ and $[c_1, b]$. For each of these, the assertion holds. Appealing to Exercise (1.7), it holds for $[a, b]$.

So far we have proved the case $k = 1$. Now to prove the inductive step, assume $k > 1$. Define $h : [a, b] \longrightarrow \mathbb{R}$ by

$$h(x) = \begin{cases} f(x) & \text{if } x \neq c_1 \\ g(x) & \text{if } x = c_1. \end{cases}$$

Then $f$ and $h$ differ only at $c_1$ while $g$ and $h$ differ only at the $k-1$ points $c_2, \ldots, c_k$. Applying what we have just proved to the pair $f$ and $h$ and the induction hypothesis to the pair $h$ and $g$ completes the inductive step and the proof. ∎

Because of this theorem, when we talk of the integrability of a bounded function $f$ on a (bounded) interval, it does not matter whether the interval is closed, open or semi-open. Suppose, for example, $f$ is defined and bounded on an open interval $(a, b)$. Then we can define $f(a)$ and $f(b)$ arbitrarily. If the function so defined is integrable on $[a, b]$ for some choice of $f(a), f(b)$ then it would be integrable for any other choice of $f(a)$ and $f(b)$. Morevoer, the integral $\int_a^b f(x)dx$ will also be the same.

This theorem has a negative significance. It says that if we know that a function $f$ is integrable on $[a, b]$, and also know $\int_a^b f(x)dx$, nothing can be concluded about the value of $f$ at a particular point, say $c$, of $[a, b]$. In particular, no conclusion can be drawn regarding the continuity, differentiablity or any other local behavior of $f$ at $c$. Actually continuity is a property which is much stronger than integrability. Every continuous function on a closed, bounded interval is integrable, as we shall show in the next section. The converse is false. As an easy example, a function which agrees with a constant function except at a finite number of points is integrable (by the theorem above or even directly) but not continuous.

It is worthwhile here to point out that there are situations where we would like a single point to count. Suppose for example, that a force (which may be mechanical or of some other kind) acts on an object for some time interval, say, $a \leq t \leq b$. We assume that the direction of this force remains the same throughout this interval but that its magnitude is a function of time, say, $f(t)$. Then its **impulse** is defined as the value of the integral $\int_a^b f(t)dt$. Physically, the impulse measures the total effect of the force on the object. Sometimes the force acts instantaneously, for example, when a cricket ball is hit with a bat. Here $f(t)$ is non-zero only for one value of $t$ and so the impulse ought to be zero no matter how high this value is. But obviously, this is inconsistent with reality. The situation can be explained by saying that an instantaneous force is only a fiction, and that in reality, every force acts over a time interval of positive length, however small it may be. The trouble is that it is next to impossible to determine $f(t)$ over this time interval. The impulse, however, can be measured easily, for example, by noting the change of velocity of the object. So a more practical approach is to allow $f(t)$ to be infinite for juat one value of $t$, say, $t = c$ and 0 everywhere else. If we do so, then $f$ is no longer a real valued function and so it makes no sense to talk of its integral. It is customary to call $f$ a **distribution** (or a **Dirac distribution**) in such a case. It is an extended real-valued function and the theory of integrals can be extended to distributions too. But we shall not go into it. So we shall stick to integrals of bounded real valued functions where the value at a single point does not count.

Let us now see how integrability behaves w.r.t. convergence of sequences of functions. Let $\{f_n\}_{n=1}^{\infty}$ be a sequence of functions all defined on an interval $[a, b]$. In Section 3.11, we defined two types of convergence for such a sequence, viz. the pointwise and the uniform convergence. Uniform convergence is a global concept. Pointwise convergence is a local concept; in fact, it is even more local in nature than concepts like continuity, because pointwise convergence of $\{f_n\}$ at a point $x_0$ depends not even on what happens in a neighbourhood of $x_0$, but only on what happens at $x_0$. Accordingly, we saw in Sections 4.1 and 4.6 that pointwise convergence does not, in general, preserve continuity or differentiablity. At the same time we also proved that continuity at $x_0$ is preserved if the convergence is uniform in a neighbourhood of $x_0$.

As integration is a global concept, we can hardly expect it to be preserved under pointwise convergence. Here we give an example of a sequence $\{f_n\}$ of integrable functions on $[0, 1]$, converging pointwise to a function $f$ which is not integrable on $[0, 1]$. To construct such a sequence, we first construct a sequence $\{q_n\}_{n=1}^{\infty}$ of rational numbers so that every rational number in $[0, 1]$ appears in it exactly once. We begin by setting $q_1 = 0$ and $q_2 = 1$. Every other rational number in $[0, 1]$ can be expressed uniquely in its reduced form as $\frac{m}{n}$ where $m, n$ are positive integers with no common factor other than $\pm 1$. The idea is to take rationals with increasing $n$. For each $n$ the rationals with $n$ as the denominator are to be ordered in ascending order. For $n = 2$, $\frac{1}{2}$ is the only rational in $[0, 1]$. So we set $q_3 = \frac{1}{2}$. Now for $n = 3$, we have $\frac{1}{3}$ and $\frac{2}{3}$ which we call as $q_4$ and $q_5$. Next we get $q_6 = \frac{1}{4}$ and $q_7 = \frac{3}{4}$ and continuing, the sequence is

$$0, 1, \frac{1}{2}, \frac{1}{3}, \frac{2}{3}, \frac{1}{4}, \frac{3}{4}, \frac{1}{5}, \frac{2}{5}, \frac{3}{5}, \frac{4}{5}, \frac{1}{6}, \frac{5}{6}, \frac{1}{7}, \frac{2}{7}, \ldots$$

A formula for the general $k^{th}$ term $q_k$ of this sequence can be given, but will be rather complicated. Nor is it necessary. What matters is that there exists a sequence of rationals in $[0, 1]$ in which every such rational appears exactly once. There are many such sequences and we have specified just one of them. (In effect, we have given a solution to the last part of Exercise (3.9.7)).

Having obtained (and fixed) such a sequence $\{q_n\}_{n=1}^{\infty}$, we define for $n = 1, 2, \ldots$, a function $f_n : [0, 1] \longrightarrow \mathbb{R}$ by

$$f_n(x) = \begin{cases} 1 & \text{if } x = q_k \text{ for some } k = 1, \ldots, n \\ 0 & \text{otherwise.} \end{cases}$$

Each $f_n$ differs from the constant function 1 only at a finite number of points, viz., $q_1, \ldots, q_n$. So by the theorem above $f_n$ is integrable over $[0, 1]$ and further, $\int_0^1 f_n(x)dx = 1$ for all $n$. Now consider the limit function, say $f$ of $\{f_n\}$. Let $x \in [0, 1]$. If $x$ is irrational then $f_n(x) = 0$ for every $n$ and so $f(x) = \lim_{n \to \infty} f_n(x) = 0$. Suppose $x$ is rational. Then $x = q_k$ for a unique integer $k$. It is clear that for $n \geq k$, $f_n(x) = 1$ and so $f(x) = 1$. Thus the limit function $f$ is 0 at all irrationals and 1 at all rationals in $[0, 1]$. By Exercise (1.8), $f$ is not Riemann integrable over $[0, 1]$.

Thus the pointwise limit of a sequence of integrable functions need not be integrable. Examples can also be given (cf. Exercise (3.4)) where the limit function, say $f$, is integrable but its integral $\int_0^1 f(x)dx$ is not the limit of $\int_0^1 f_n(x)dx$ as $n \to \infty$.

As is to be expected, with uniform convergence, the situation is better on both the fronts.

**3.2 Theorem.** Suppose a sequence $\{f_n\}$ of functions converges to a function $f$ uniformly on an interval $[a, b]$. If each $f_n$ is Riemann integrable over $[a, b]$, then so is $f$ and moreover,

$$\int_a^b f(x)dx = \lim_{n \to \infty} \int_a^b f_n(x)dx. \tag{5}$$

**Proof:** We first assert that $f$ is bounded on $[a, b]$. This follows from Exercise (3.11.2). To prove $f$ is Riemann integrable on $[a, b]$, we shall again use Exercise (1.5). Let $\epsilon > 0$ be given. We have to find a partition $P$ of $[a, b]$ for which

$$U(P; f) - L(P; f) < \epsilon. \tag{6}$$

The basic idea is to show that because of uniform convergence, for large $n$, $U(P; f)$ and $L(P; f)$ are very close to $U(P; f_n)$ and $L(P; f_n)$ respectively for any partition $P$ and then to use the Riemann integrability of $f_n$ to choose a suitable $P$.

First, by uniform convergence of $\{f_n\}$, there exists $N$ such that for all $n \geq N$, $|f_n(x) - f(x)| < \dfrac{\epsilon}{3(b-a)}$ all $x \in [a, b]$. In particular,

$$f_N(x) - \frac{\epsilon}{3(b-a)} < f(x) < f_N(x) + \frac{\epsilon}{3(b-a)} \tag{7}$$

for all $x \in [a, b]$. Now let $P = (x_0, x_1, \ldots, x_n)$ be any partition of $[a, b]$. For $i = 1, \ldots, n$, let $M_i(f)$ and $M_i(f_N)$ be the suprema of $f$ and $f_N$ respectively on $[x_{i-1}, x_i]$ and similarly let $m_i(f)$ and $m_i(f_N)$ be the infima of $f$ and $f_N$ respectively on $[x_{i-1}, x_i]$. From (7), it is clear that

$$m_i(f_N) - \frac{\epsilon}{3(b-a)} \leq m_i(f) \leq m_i(f_N) + \frac{\epsilon}{3(b-a)}$$

$$\text{and} \quad M_i(f_N) - \frac{\epsilon}{3(b-a)} \leq M_i(f) \leq M_i(f_N) + \frac{\epsilon}{3(b-a)}$$

(Note that the strictness of the inequality in (7) may be lost.) Multiplying these inequalies throughout by $\Delta x_i$ and summing over from $i = 1$ to $n$, gives

$$|L(P; f) - L(P; f_N)| \leq \frac{\epsilon}{3} \quad \text{and} \quad |U(P; f) - U(P; f_N)| \leq \frac{\epsilon}{3} \tag{8}$$

This holds for *every* partition $P$ of $[a, b]$. We now make a particular choice of $P$. As $f_N$ is Riemann integrable on $[a, b]$, by Exercise (1.5), there exists a partition $P$ of $[a, b]$, for which

$$U(P; f_N) - L(P; f_N) < \frac{\epsilon}{3} \tag{9}$$

(8) and (9), together with the triangle inequality imply (6), thereby showing the Riemann integrability of $f$ over $[a, b]$.

It remains to show that $\lim_{n \to \infty} \int_a^b f_n(x)dx = \int_a^b f(x)dx$. For this, given $\epsilon > 0$, by uniform convergence, there exists $m$ such that for all $n \geq m$, and for all $x \in [a, b]$,

$$|f_n(x) - f(x)| < \frac{\epsilon}{2(b-a)}.$$

So, by Exercises (1.9) and (1.12), we get for all $n \geq m$,

$$
\left| \int_a^b f_n(x)dx - \int_a^b f(x)dx \right| = \left| \int_a^b f_n(x) - f(x)dx \right|
$$
$$
\leq \int_a^b |f_n(x) - f(x)|dx
$$
$$
\leq \int_a^b \frac{\epsilon}{2(b-a)}dx
$$
$$
= \frac{\epsilon}{2(b-a)}(b-a) = \frac{\epsilon}{2} < \epsilon.
$$

It follows that as $n \to \infty$, $\int_a^b f_n(x)dx$ tends to $\int_a^b f(x)dx$. ∎

Like other theorems involving uniform convergence of functions, (e.g. Theorem (3.11.5)), this theorem can also be interpreted as a result dealing with commutativity of limits, that is, a change of order of two types of limiting processes. We saw in Section 5.2, that a Riemann integral can be thought of as the limit of a Riemann sum $R(P, \xi\,;f)$ as $\mu(P)$, the mesh of $P$, tends to 0. So (5) can be rewritten as,

$$
\lim_{\mu(P) \to 0} [R(P, \xi\,;\lim_{n \to \infty} f_n)] = \lim_{n \to \infty} [\lim_{\mu(P) \to 0} R(P, \xi\,;f_n)].
$$

This theorem is sometimes useful in a negative manner in testing if a sequence $\{f_n\}$ converges uniformly to $f$. If $\int_a^b f_n(x)dx$ does not tend to $\int_a^b f(x)dx$ then the answer is 'no', without having to show directly that there is some $\epsilon > 0$ for which there is no $N$ that will work for all $x \in [a, b]$. This test is especially convenient when the functions involved are such that their integrals can be calculated, or at least estimated, easily. Note, however, that it is not a positive test. That is, even if $\int_a^b f_n(x)dx$ tends to $\int_a^b f(x)dx$, it does not follow that $f_n \to f$ uniformly on $[a, b]$. In fact, by changing $f$ at a finite number of points, and applying Theorem (3.1), we see that we cannot even conclude pointwise convergence of $\{f_n\}$, being given that $\int_a^b f_n(x)dx \to \int_a^b f(x)dx$.

The preceding theorem, combined with Theorem (3.11.5) allows us to integrate a power series term-by-term on any closed interval contained in its interval of convergence. This will be given as an exercise (Exercise (3.6)).

We close this section with yet another ramification of the fact that integration is a global construction. Suppose $f$ is integrable on an interval $[a, b]$. Then it is integrable over $[a, c]$ for every $c \in [a, b]$, by Exercise (1.7). The value of the integral $\int_a^c f(x)dx$ will, naturally, change as $c$ changes. Denote it by $F(c)$. Then $F$ itself is another function from $[a, b]$ to $\mathbb{R}$, with $F(a) = 0$. As we normally use $x, y$ etc. to denote variables instead of $a, b, c$ etc. (which usually denote constants), we shall consider $F(x)$ for $x \in [a, b]$. It is awkward (although perfectly legitimate) to use the same variable $x$ as the dummy variable of integration. So we denote $F(x)$ by $\int_a^x f(t)dt$ or $\int_a^x f(u)du$ etc. instead of $\int_a^x f(x)dx$. Such a function is called a **function defined by integration**. As a concrete example, suppose $[a, b] = [1, 2]$ and $f(x) = x$. Then $F : [1, 2] \longrightarrow \mathbb{R}$ is defined by $F(x) = \int_1^x t\,dt$. By Exercise (1.17), this integral is $\dfrac{x^2}{2} - \dfrac{1}{2}$. So in this case $F(x) = \dfrac{x^2}{2} - \dfrac{1}{2}$.

In the definition of $F(x)$ as $\int_a^x f(t)dt$, the lower limit of integration (viz. $a$) was constant and the upper limit (viz. $x$) varied. We could, of course, consider a function of the form $\int_x^b f(t)dt$ where the upper limit is fixed. More generally, if $u(x)$ and $v(x)$ are any two functions of $x$ taking values in $[a, b]$ then we can treat $\int_{u(x)}^{v(x)} f(t)dt$ as a function of $x$. But this can always be expressed as $\int_a^{v(x)} f(t)dt - \int_a^{u(x)} f(t)dt$ by Exercise (1.19) and hence as $F(v(x)) - F(u(x))$ where $F$ is as defined above. So it suffices to study only $F$. It is easy to give geometric or physical interpretations to this new function $F$. Geometricolly $F(x)$ is the area below the graph of $f$ between the vertical lines through $a$ and $x$. If $f(t)$ is the linear density at a point $t$ units away from one end of a piece of wire then $F(x) = \int_0^x f(t)dt$ is the mass of the portion of the wire from that end upto the point $x$ units away from that end.

Naturally, we expect that the properties of the function $F$ should depend on those of the integrand function $f$ from which it is obtained, just as we expect the properties of a derived function $f'$ to depend on those of the original function $f$. In the case of the derived function, we know that when we go from $f$ to $f'$, a good deal of smoothness[3] may be lost. For example even if $f$ is differentiable (so that $f'$ makes sense) everywhere, the derived function $f'$ need not even be continuous (see, e.g. Exercise (4.3.14)). In the case of functions defined by integrals, precisely the opposite happens. That is, there is an increase of smoothness as we go from $f$ to $F$. Later (in Section 5.5) we shall show, for example, that $F$ is differentiable whenever $f$ is continuous. Here we prove that regardless of what $f$ is, as long as it is integrable, the function $F$ is always continuous.

**3.3 Theorem:** Suppose $f$ is integrable on $[a, b]$. Define $F : [a, b] \longrightarrow \mathbb{R}$ by

---

[3]In Section 4.12 we defined smoothness for plane curves. Here we use the term in a less precise sense, meaning only some sort of regularity, avoiding such jaggy behaviour as jumps and kinks.

$F(x) = \int_a^x f(t)dt$ for $x \in [a, b]$, Then $F$ is continuous on $[a, b]$.

**Proof:** We shall, in fact, show that $F$ is uniformly continuous on $[a, b]$. (This would, of course, also follow from theorem (4.2.2) once we show $F$ is continuous at every point of $[a, b]$. But in the present case it is very easy to establish directly that $F$ is uniformly continuous on $[a, b]$.) Let $\epsilon > 0$ be given. As $f$ is integrable on $[a, b]$, it is, in particular bounded on $[a, b]$. So let $M$ be any positive upper bound on $|f(x)|$ for $x \in [a, b]$. Now let $0 < \delta < \dfrac{\epsilon}{M}$. We assert that whenever $x, y \in [a, b]$ and $|x - y| < \delta, |F(x) - F(y)| < \epsilon$. First assume $y \leq x$. Then by Exercises (1.7) and (1.12)

$$
\begin{aligned}
|F(x) - F(y)| &= \left| \int_a^x f(t)dt - \int_a^y f(t)dt \right| \\
&= \left| \int_y^x f(t)dt \right| \\
&\leq \int_y^x |f(t)|dt \\
&\leq M(x - y) < M\delta < \epsilon.
\end{aligned}
$$

A similar reasoning works if $y \geq x$. ∎

The theorem above can be looked at somewhat philosophically. Consider a bounded function $f : [a, b] \longrightarrow \mathbb{R}$. The graph of $f$ could be very jaggy with sudden jumps and other discontinuities. If there are too many of these, then $f$ may not be Riemann integrable. Exercise (1.8) provides an example of such a function, which is discontinuous everywhere (see Exercise (4.1.25)(iv)). But if such irregularities are within certain limits (in a sense which will be made precise in the next section; see the comments after Theorem (4.2)) then $f$ is integrable on $[a, b]$. And once this happens, the new function $F$ that we get by integrating $f$ is continuous everywhere. The jaggedness of $f$ disappears in the process of integration. In other words, integration is a smoothening process. It serves to mask the identity of individual points and thereby lessen the effect of those that are not well-behaved. As a social analogy, a hopelessly divided mob in which everybody goes his own way has little chance of developing into an organised society. But if such individual aberrations are within certain limits then the mob can be integrated into a coherent culture in which the individual's fancies and whims are toned down. So it is also in this sense that integration is a global process.

## EXERCISES

3.1    Suppose $f$ is Riemann integrable over $[a, b]$. Let $M$ be any bound on $|f(x)|$ for $x \in [a, b]$. Prove that if $c \in (a, b)$ and $\delta > 0$ is so small that $[c - \delta, c + \delta] \subset [a, b]$ then $\left| \displaystyle\int_{c-\delta}^{c+\delta} f(x)dx \right| \leq 2M\delta$. Hence show that

$$\lim_{\delta \to 0^+} \int_{c-\delta}^{c+\delta} f(x)dx = 0.$$

3.2 Suppose $f$ is monotonically increasing on $[a, b]$. For every $c \in (a, b)$, prove that $\lim_{x \to c^-} f(x)$ and $\lim_{x \to c^+} f(x)$ both exist and $\lim_{x \to c^-} f(x) \leq \lim_{x \to c^+} f(x)$. [*Hint* : Use Exercise (1.3)].

3.3 In the last exercise suppose $L$ is some real number and a new function $g$ is defined by setting $g(c) = L$ and $g(x) = f(x)$ for all $x \neq c$. For which values of $L$ is $g$ monotonically increasing on $[a, b]$ ?

3.4 For $n \in \mathbb{N}$, define $f_n : [0, 1] \to \mathbb{R}$ by

$$f_n(x) = \begin{cases} 2^n & \text{if } \dfrac{1}{2^n} \leq x \leq \dfrac{1}{2^{n-1}} \\ 0 & \text{otherwise.} \end{cases}$$

Prove that $\int_0^1 f_n(x)dx = 1$ for all $n$. Show further that $f_n(x) \longrightarrow f(x)$ pointwise on $[0, 1]$ where $f$ is the identically zero function. Hence $\{f_n\}$ is an example of a sequence of integrable functions converging pointwise to an integrable function $f$ for which $\int_0^1 f_n(x)dx \not\to \int_0^1 f(x)dx$. (Another example is provided by the sequence $f_n(x) = nx^{n-1}$ but to calculate its integral, we have to wait till Section 5.5).

3.5 Prove that for every non-negative integer $n$, the function $f(x) = x^n$ is Riemann integrable over every interval $[a, b]$. [*Hint*: Apply Exercise (1.15). Split the interval at 0 if necessary.]

3.6 Suppose $\sum_{n=0}^{\infty} a_n x^n$ is a power series with sum function $f(x)$ and radius of convergence $R > 0$. Prove that for every interval $[a, b]$ contained in $(-R, R)$, $f$ is integrable over $[a, b]$ and further that $\int_a^b f(x)dx = \sum_{n=0}^{\infty} a_n \int_a^b x^n dx$. In other words, a power series can be integrated term-by-term within its interval of convergence. [*Hint* : Apply Theorem (3.2) along with the last exercise and Theorem (3.11.5). Note that it is not asserted, nor is it always true, that $f$ is integrable on $[-R, R]$.]

3.7 A function $f : [a, b] \longrightarrow \mathbb{R}$ is called a **step function** if there exists a partition $P = (x_0, x_1, \ldots, x_n)$ of $[a, b]$ such that $f$ is constant on each of the open intervals $(x_{i-1}, x_i)$ for $i = 1, \ldots, n$. Show that the function $f(x) = [x]$ (that is, the integral part of $x$) is a step function on every closed interval. Prove that the sums, differences, products and composites (when defined) of step functions are step functions.

3.8    Show that every step function is Riemann integrable. What is its integral ? [*Hint*: Apply Exercise (1.7) along with the comment made after Theorem (3.1).]

3.9    Let $f : [a, b] \longrightarrow \mathbb{R}$ be a bounded function and $P$ be any partition of $[a, b]$. Prove that the upper and the lower Riemann sums $U(P; f)$ and $L(P; f)$ are nothing but the integrals of certain step functions $h$ and $g$ on $[a, b]$ such that $g(x) \leq f(x) \leq h(x)$ for all $x \in [a, b]$. (Although hardly profound, this result has a certain significance. Because of the last exercise, the integrals of step functions can be defined without a limiting process. The present exercise shows that given an arbitrary function $f(x)$, we can get very close to its integral if we approximate $f(x)$ both from below and above, by step functions. So step functions are like the prototypes in the theory of integration.)

3.10    Let $f(x) = x^n$ and $F(x) = \displaystyle\int_0^x f(t)dt$. For $n = 0, 1, 2$ verify that $F$ is an antiderivative of $f$ (see Exercise (1.17)).

3.11    Let $f(x) = [x]$ and $F(x) = \displaystyle\int_0^x f(t)dt$ for $0 \leq x \leq 3$. Sketch the graph of $F$. Prove that $F$ is differentiable everywhere except at $x = 1$ and $x = 2$. What guess can you make from this and the last exercise regarding the derivative of a function defined by an integral?

## Q.5.4 Integrability of Continuous Functions

It was mentioned in the last section that a function which is continuous on an interval $[a, b]$, is Riemann integrable on $[a, b]$. In this section, we shall prove this result and a number of other results where continuity of certain functions implies the Riemann integrability of some related functions.

Before giving a formal proof that continuity implies Riemann integrability, it is instructive to look at the latter in a slightly different way. Let $f$ be a bounded function defined on an interval $[a, b]$. Riemann integrability of $f$ on $[a, b]$ means that the difference $U(P; f) - L(P; f)$ between the upper and the lower Riemann sums can be made sufficiently small if the partition $P$ of $[a, b]$ is sufficiently fine. As usual, if $P$ is $(x_0, x_1, \ldots, x_n)$ then this difference is the sum $\sum_{i=1}^{n}(M_i - m_i)\Delta x_i$, where $M_i$ and $m_i$ are, respectively, the supremum and the infimum of $f(x)$ on $[x_{i-1}, x_i]$. Denote $M_i - m_i$ by $F_i$. $F_i$ may be called the **fluctuation** of $f$ on $[x_{i-1}, x_i]$ for $i = 1, \ldots, n$. ('Variation' is perhaps a better term than 'fluctuation'. But the phrase 'variation of a function over an interval' is already used in mathematics in a somewhat different sense. So we prefer 'fluctuation'.)

Geometrically, $\sum_{i=1}^{n} F_i \Delta x_i$ is the difference of areas of the 'upper' and the 'lower' approximations by rectangles to the area below the graph of $y = f(x)$ (see Fig.5.4.1(b)). It is intuitively clear that if $f$ is continuous on $[a, b]$ and the mesh of $P$ is small, then each $F_i$ is also small and so in the sum $\sum_{i=1}^{n} F_i \Delta x_i$, even though the number of terms is large, the size of each sum is small of a higher order, being the product of two quantities, $F_i$ and $\Delta x_i$, each of which is itself small. As a result, we expect $\sum_{i=1}^{n} F_i \Delta x_i$ to tend to 0.

In the proof which we give below, instead of the difference between the upper and the lower Riemann sums, we consider the difference between any two Riemann sums with the same underlying partition. But the basic idea is the same. Note, moreover, that continuity is, in general, much stronger a condition than is needed for integrability. Indeed, we can take any continuous function $f$ on $[a, b]$, make it discontinuous by changing its values at a finite number of points in $[a, b]$. Because of Theorem (3.1), the new function will also be integrable on $[a, b]$.

**4.1 Theorem:** If $f$ is continuous on $[a, b]$ then it is Riemann integrable on $[a, b]$.

**Proof:** By Theorem (4.1.4), $f$ is bounded on $[a, b]$. Also by Theorem (4.2.2), $f$ is uniformly continuous on $[a, b]$. This fact will be crucially needed in proving that $f$ is Riemann integrable on $[a, b]$. To do this, we appeal to Exercise (2.5). So let $\epsilon > 0$ be given. By uniform continuity of $f$ on $[a, b]$ there exists $\delta > 0$ such that for all $x, y \in [a, b]$

$$|f(x) - f(y)| < \frac{\epsilon}{b - a} \tag{1}$$

whenever $|x - y| < \delta$. Let $P = (x_0, x_1, \ldots, x_n)$ be a partition of $[a, b]$ with $\mu(P) < \delta$. Consider any two Riemann sums, say $R(P, \boldsymbol{\xi} ; f)$ and $R(P, \boldsymbol{\eta}; f)$ based on $P$. Let $\boldsymbol{\xi} = (\xi_1, \ldots, \xi_n)$ and $\boldsymbol{\eta} = (\eta_1, \ldots, \eta_n)$ with $\xi_i, \eta_i, \in [x_{i-1}, x_i]$ for $i = 1, \ldots, n$. Note that $|\xi_i - \eta_i| \leq \Delta x_i \leq \mu(P) < \delta$ for all $i = 1, \ldots, n$. So we have

$$|R(P, \boldsymbol{\xi} ; f) - R(P, \boldsymbol{\eta}; f)| = |\sum_{i=1}^{n}(f(\xi_i) - f(\eta_i))\Delta x_i|$$

$$\leq \sum_{i=1}^{n}|f(\xi_i) - f(\eta_i)|\Delta x_i$$

which, in view of (1), implies

$$|R(P, \boldsymbol{\xi} ; f) - R(P, \boldsymbol{\eta}; f)| \quad < \quad \frac{\epsilon}{b-a} \sum_{i=1}^{n} \Delta x_i = \frac{\epsilon}{b-a}(b-a) = \epsilon.$$

Thus we have shown that whenever $\mu(P) < \delta$, any two Riemann sums of $f$ based on $P$ differ from each other by less than $\epsilon$. By Exercise (2.5), this means $f$ is Riemann integrable on $[a, b]$. ∎

In this proof continuity of $f$ on $[a, b]$ was used twice. First to show that $f$ is bounded on $[a, b]$ (so that it is at least meaningful to talk about integrability of $f$.) But more crucially, it was used in estimating the difference between two Riemann sums based on the same partition $P$. Here what was really needed was the uniform continuity of $f$, which is a global property. This is to be expected, because integrability is a global property and so to derive it we must have at hand some global property of the function.

As mentioned earlier, continuity is a far stronger condition than needed for integrability and so it is not surprising that the theorem above can be extended to a wider class of functions. We prove two such extensions here. In the first one we show that it is all right if the function $f$ has a finite number of discontinuities in $[a, b]$. It is tempting to try to derive this by constructing a function $g$ which is continuous on $[a, b]$ and which agrees with $f$ except at a finite number of points of $[a, b]$ and then applying the last theorem along with Theorem (3.1). But this approach will work only if the discontinuities of $f$ are removable (see Section 4.1). Otherwise it will fail. For example, suppose $f(x) = \sin\dfrac{1}{x}$ for $x \neq 0$. Define $f(0)$ arbitrarily. Then there is no continuous function on $[-1, 1]$ which will agree with $f$ at all except finitely many points because $\lim\limits_{x \to 0} f(x)$ does not exist.

So, a more careful argument is needed. The idea is to enclose the points of discontinuities within very small intervals and to apply the preceding theorem to the rest of the interval.

**4.2 Theorem:** Suppose $f : [a, b] \longrightarrow \mathbb{R}$ is bounded and discontinuous only at a finite number of points, say, $c_1, \ldots, c_k$. Then $f$ is Riemann integrable on $[a, b]$.

**Proof:** Clearly we may assume $c_1 < c_2 < \cdots < c_k$. Let $M$ be a positive number such that $|f(x)| < M$ for all $x \in [a, b]$. To prove the integrability of $f$ on $[a, b]$, we shall use Exercise (1.5). So let $\epsilon > 0$ be given. We need to find a partition $P$ of $[a, b]$ for which $U(P; f) - L(P; f) < \epsilon$. We proceed as follows. We assume $a < c_1$ and $c_k < b$. If either $c_1 = a$ or $c_k = b$, then the proof needs minor modifications (of a notational, and not conceptual type) which we leave to the reader.

Let $r$ be a positive real number and set $a_i = c_i - r$ and $b_i = c_i + r$ for $i = 1, \ldots, k$. If $r$ is sufficiently small then the $k$ intervals $[a_1, b_1], [a_2, b_2], \ldots, [a_k, b_k]$ are mutually disjoint and contained in $[a, b]$. (This is where we need $c_1 > a$ and $c_k < b$.) For notational uniformity call $a$ as $b_0$ and $b$ as $a_{k+1}$. Now the interval

$[a, b]$ gets chopped into $2k + 1$ subintervals in all, those of the form $[a_i, b_i]$ for $i = 1, \ldots, k$ and those of the form $[b_i, a_{i+1}]$ for $i = 0, 1, \ldots, k$. By assumption $f$ is continuous on each $[b_i, a_{i+1}]$ for $i = 0, 1, \ldots, k$ (see Fig. 5.4.1.).

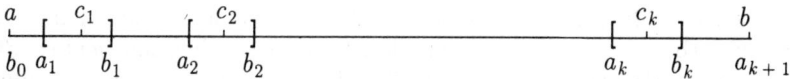

Figure 5.4.1 : Enclosing Discontinuities by Small Intervals

Now, given $\epsilon > 0$, taking $r$ still smaller if necessary, we may suppose

$$4Mkr < \frac{\epsilon}{2} \tag{2}$$

By the theorem above, $f$ is Riemann integrable on the interval $[b_i, a_{i+1}]$ for $i = 0, 1, \ldots, k$. So by Exercise (1.5), there exists a partition, say $P_i$, of $[b_i, a_{i+1}]$ such that

$$U(P_i; f) - L(P_i; f) < \frac{\epsilon}{2(k + 1)} \tag{3}$$

Now let $P$ be the partition of $[a, b]$ obtained by juxtaposing these partitions $P_0, P_1, \ldots, P_k$, along with the subintervals $[a_i, b_i], i = 1, 2, \ldots, k$. We claim that

$$U(P; f) - L(P; f) < \epsilon. \tag{4}$$

The subintervals of $P$ are of two types: (I) those of the form $[a_i, b_i], i = 1, \ldots, k$ and (II) those which occur as subintervals of some $P_i$ for $i = 0, 1, \ldots, k$. Let us see how much each type contributes to the difference $U(P; f) - L(P; f)$. Since $-M < f(x) < M$ for all $x \in [a_i, b_i]$, it is clear that an interval of the first type will contribute at most $(M - (-M))(2r)$, i.e., $4Mr$. So by (2), the total contribution from intervals of type I is less than $\frac{\epsilon}{2}$.

Subintervals of type II fall into $k + 1$ groups, depending upon which $P_i$ they come from. By (3), the contribution from each such group is less than $\frac{\epsilon}{2(k + 1)}$ and so the total contribution to $U(P; f) - L(P; f)$ from type II intervals is less than $\frac{\epsilon}{2}$. Adding the two, we get (4). Thus for every $\epsilon > 0$, we found a partition $P$ of $[a, b]$ for which (4) holds. By Exercise (1.5), $f$ is integrable on $[a, b]$. ∎

This theorem is important from a practical as well as a theoretical point of view. By Exercise (4.1.26) every piecewise continuous function on a closed, bounded interval is bounded. With this, the theorem above implies that every such function is integrable. This is important from a practical point of view, because as mentioned at the end of Section 4.1, in real life situations we have to deal with piecewise continuous functions.

Theoretically, the theorem above is important because it raises an interesting question, viz., to what extent can we weaken the hypothesis of the continuity of a function $f$ and still retain its integrability. Let us paraphrase this quantitatively. Suppose $f$ is a bounded function on $[a, b]$. Let $D_f$ be the set of discontinuities of $f$, i.e., the set of points in $[a, b]$ where $f$ is not continuous. Clearly $f$ is continuous on $[a, b]$ if and only if $D_f = \phi$, the empty set. So, the size of $D_f$ is a good measure of how discontinuous $f$ is. The theorem above says that if the set $D_f$ is finite then $f$ is Riemann integrable. But that is only a sufficient condition. The problem is to find a condition on $D_f$ which is both necessary and sufficient for integrability of $f$. This problem has been solved completely. The answer is that $f$ is Riemann integrable on $[a, b]$ if and only if for every $\epsilon > 0$, the set $D_f$ can be covered by a sequence, say, $I_1, I_2, I_3, \ldots$, of intervals such that the sum of their lengths is less than $\epsilon$. This is also expressed by saying that the Lebesgue measure of $D_f$ is 0. It is beyond our scope to prove this result. We only remark in passing that questions like this one led to a theory of integration which is far more general than Riemann integration.

Let us now turn to the other extension of Theorem (3.2). A perceptive reader must have noticed that so far we have said nothing regarding the integrability of the composite of two integrable functions. It turns out that unlike continuity and differentiablity, Riemann integrability is not always preserved under compositions. Suppose $f : [a, b] \longrightarrow \mathbb{R}$ is bounded, say, $m \leq f(x) \leq M$ for all $x \in [a, b]$. Then for any $g : [m, M] \longrightarrow \mathbb{R}$, the composite function $g \circ f$ makes sense. But even if $f, g$ are Riemann integrable, $g \circ f$ need not be so. A rather complicated counter example will be given in Exercise (4.8). It turns out, however, that if $g$ is continuous then the integrability of $f$ implies that of $g \circ f$. This generalises Theorem (4.2), because by taking $f(x) = x$, we get that every continuous function is Riemann integrable. So it is to be expected that the proof will resemble, but will be more intricate than that of Theorem (4.1).

**4.3 Theorem:** Suppose $f : [a, b] \longrightarrow \mathbb{R}$ is Riemann integrable and $g$ is defined and continuous on a closed interval containing the range of $f$. Let $h(x) = g(f(x))$ for $x \in [a, b]$. Then $h$ is Riemann integrable on $[a, b]$.

**Proof:** Let $m$ and $M$ be respectively the infimum and the supremum of $f$ on $[a, b]$. We are assuming that $g$ is continuous on $[m, M]$. Then by Theorem (4.1.4) $g$ is bounded on $[m, M]$ and hence $h$ is bounded on $[a, b]$. Let $K$ be a positive upper bound on $|g(x)|$ for $x \in [m, M]$. Then $K$ is also an upper bound for $|h(x)|$ on $[a, b]$.

To prove that $h$ is Riemann integrable on $[a, b]$, once again we appeal to Exercise (1.5). So let $\epsilon > 0$ be given. We have to find a partition $P$ of $[a, b]$ for which

$$U(P; h) - L(P; h) < \epsilon \qquad (5)$$

To construct $P$, we proceed as follows. First, as $g$ is continuous on $[m, M]$, it is uniformly continuous on $[m, M]$ by Theorem (4.2.2). So there exists $\delta > 0$ such that for all $x, y \in [m, M]$,

$$|g(x) - g(y)| < \frac{\epsilon}{2(b-a)} \tag{6}$$

whenever $|x - y| < \delta$. We further assume that $\delta < \frac{\epsilon}{4K}$. The reason for this restriction will be apparent later.

Having chosen such $\delta > 0$, the integrability of $f$ on $[a, b]$ gives, by Exercise (1.5), a partition, say $P = (x_0, x_1, \ldots, x_n)$, of $[a, b]$ for which

$$U(P; f) - L(P; f) < \delta^2. \tag{7}$$

We contend that for any such partition $P$, (5) also holds. Note that the term on the right in (7) is $\delta^2$ and not $\delta$ (as might appear more natural). The reason for this peculiar choice will also be clear later.

Now let us estimate $U(P; h) - L(P; h)$. As usual, for $i = 1, \ldots, n$, let $M_i$ and $m_i$ be the supremum and the infimum of $f$ on $[x_{i-1}, x_i]$ while let $K_i$ and $k_i$ be, respectively, the supremum and the infimum of $h$ on $[x_{i-1}, x_i]$. Then

$$U(P; h) - L(P; h) = \sum_{i=1}^{n} (K_i - k_i) \Delta x_i \tag{8}$$

As in the proof of Theorem (2.3), we shall split the terms in this sum into two parts, the 'good' and the 'bad' ones. The basis of classification will, however, be quite different. Consider a typical subinterval $[x_{i-1}, x_i]$ of the partition $P$. Let us call it (and the index $i$) good if $M_i - m_i < \delta$ and bad otherwise. In other words, we classify the subintervals of $P$ depending upon how much $f$ fluctuates on them. If the function $f$ were continuous then by taking $\mu(P)$ sufficiently small, we could have ensured that all its subintervals were good. But $f$ is not given to be continuous and so we have to deal with bad subintervals as well.

Let us see what is so good about a good subinterval, say $[x_{i-1}, x_i]$. Since $f([x_{i-1}, x_i]) \subset [m_i, M_i]$, the fluctuation of $g$ on $f([x_{i-1}, x_i])$ is at most the fluctuation of $g$ on $[m_i, M_i]$. The former is, by definition, $K_i - k_i$; while the latter is less than $\frac{\epsilon}{2(b-a)}$ by (6) because $M_i - m_i < \delta$. So,

$$\sum_{i \text{ good}} (K_i - k_i) \Delta x_i < \sum_{i \text{ good}} \frac{\epsilon}{2(b-a)} \Delta x_i \leq \frac{\epsilon}{2(b-a)} \sum_{i=1}^{n} \Delta x_i = \frac{\epsilon}{2} \tag{9}$$

Thus the contribution to (8) from good subintervals is less than $\frac{\epsilon}{2}$. For bad subintervals this reasoning breaks down. But we claim that the lengths of all the bad subintervals add up to at most $\delta$. (This is where we need $\delta^2$, rather than $\delta$ in (7)). Indeed, since $M_i - m_i \geq \delta$ if $i$ is bad, we have

$$\delta \sum_{i \text{ bad}} \Delta x_i \leq \sum_{i \text{ bad}} (M_i - m_i) \Delta x_i \leq \sum_{i=1}^{n} (M_i - m_i) \Delta x_i$$

As the last term equals $U(P; f) - L(P; f)$, by (7) we get,

$$\delta \sum_{i \text{ bad}} \Delta x_i < \delta^2$$

Cancelling $\delta$ (which is positive) we get

$$\sum_{i \text{ bad}} \Delta x_i < \delta \qquad (10)$$

Now, $K_i - k_i \leq 2K$ for all $i$, whether good or bad. This fact, along with (10) and our choice of $\delta$ as less than $\dfrac{\epsilon}{4K}$ implies,

$$\sum_{i \text{ bad}} (K_i - k_i)\Delta x_i < 2K\delta < \frac{\epsilon}{2}. \qquad (11)$$

From (8), (9) and (11), it now follows that (5) holds and, as noted before, this completes the proof. ∎

This theorem merely asserts the integrability of $h = g \circ f$ when $f$ is Riemann integrable and $g$ is continuous. It does not give any formula for $\displaystyle\int_a^b h(x)dx$. There is in fact no formula which expresses the integral of the composite $g \circ f$ in terms of those of $g$ and $f$ as there is for its derivative (viz., the chain rule). In particular, it is not true that $\displaystyle\int_a^b (g \circ f)(x)dx = \int_a^b f(x)dx \int_m^M g(x)dx$ as can be shown by simple examples (see Exercise (4.3)).

This theorem has a rather surprising application which is worth mentioning. In Exercise (1.9), it was shown that the sum $f + g$ of two integrable functions (over the same interval) is integrable and remarked without proof that the same holds for the product function $fg$. (Do not confuse the product $fg$ with the composite $f \circ g$.) The difficulty in proving this directly is that the supremum and the infimum of $fg$ cannot be easily related to those of $f$ and $g$. However, a proof can now be given using the theorem above. We first settle the case $f = g$ as a lemma. Here the product $fg$ is denoted by $f^2$ (which is confusing, because frequently $f^2$ is indeed used for the composite $f \circ f$, see e.g. Exercise (4.13.7)).

**4.4 Lemma:** If $f$ is Riemann integrable over $[a, b]$ then so is $f^2$, defined by $f^2(x) = [f(x)]^2$ for $x \in [a, b]$.

**Proof:** Define $g : \mathbb{R} \longrightarrow \mathbb{R}$ by $g(x) = x^2$ for $x \in \mathbb{R}$. Then $g$ is continuous everywhere and so by the last theorem, the composite function $g \circ f$ is Riemann integrable over $[a, b]$. But $g \circ f$ is precisely $f^2$. ∎

A direct proof of this lemma is also possible (cf. Exercise (4.5)). The result we are after now follows by an algebraic trick.

**4.5 Theorem:** If $f, g$ are Riemann integrable over $[a, b]$ so is their product $fg$ defined by $(fg)(x) = f(x)g(x)$ for $x \in [a, b]$.

**Proof:** By Exercise (1.9), $f + g$ is integrable. Hence by the lemma above $(f + g)^2, f^2$ and $g^2$ are all integrable over $[a, b]$. By writing

$$fg = \frac{1}{2}[(f + g)^2 - f^2 - g^2]$$

and using Exercise (1.9) again, it follows that $fg$ is Riemann integrable over $[a, b]$. ∎

Theorem (4.3) asserts that $g \circ f$ is Riemann integrable if $f$ is Riemann integrable and $g$ is continuous. Here continuity of $g$ is needed everywhere. Even if $g$ fails to be continuous at just one point, the theorem may not hold (see Exercise (4.8)). In other words, we cannot attempt an extension of Theorem (4.3) the way Theorem (4.2) is an extension of Theorem (4.1).

What happens if in the composite $g \circ f$, $g$ is integrable and $f$ continuous ? Here it turns out that $g \circ f$ need not be Riemann integrable. But a counter-example is beyond our scope. There is, in fact, a counter-example where $f$ is continuous, $g$ is discontinuous only at one point and $g \circ f$ is not integrable. Note that if $g$ were continuous everywhere then by Exercise (4.1.5), the composite $g \circ f$ would be continuous everywhere in its domain and hence integrable by Theorem (4.1). So even though a finite number of discontinuities of a function do not matter as far as integrability of that function is concerned, when it comes to the integrability of a composite function, even a single discontinuity can make things go wrong.

The integrability of continuous functions provides us with an important method for generating new functions from old ones. (The power series is another tool to define new functions. It is interesting that both the tools involve limits of sums, in one form or the other). In the last section we saw that if $f$ is Riemann integrable on $[a, b]$ and we define $F(x) = \int_a^x f(t)dt$ for $x \in [a, b]$, then this new function $F$ is continuous on $[a, b]$. But then by Theorem (4.1), $F$ itself is Riemann integrable on $[a, b]$. So we can define yet another function $G$ on $[a, b]$ by $G(x) = \int_a^x F(t)dt$ and this process can be continued *ad infinitum:* As was remarked at the end of the last section, integration is a smoothening process and so in this construction we keep on getting functions which are progressively more well-behaved. But to prove this we need the Fundamental Theorem of Calculus, which we shall study in the next section.

Among functions defined by integrals, there is one that stands out as one of the most frequently encountered functions of mathematics. It is the natural logarithm. We already defined $\ln x$ as the inverse function of the exponential function (cf. Exercise (4.1.22)). But the tables can be turned around. For $x > 0$, we can define $\ln x$ as $\int_1^x \frac{1}{t}dt$. As the function $f(t) = \frac{1}{t}$ is continuous for all $t > 0$, it is integrable over every closed, bounded subinterval of it, by Theorem (4.1). So the integral $\int_1^x \frac{1}{t}dt$ is a definite real number for each $x > 0$.

(Recall by Exercise (1.19) that for $0 < x < 1$, $\int_1^x \frac{1}{t} dt$ means $-\int_x^1 \frac{1}{t} dt$.) The properties of logarithms can also be derived from this definition and certain theorems about integration. The exponential function can then be defined as the inverse of the natural logarithm. We shall do so in Section 5.7, along with a brief discussion of the relative merits and demerits of the two approaches.

## EXERCISES

4.1    Give an alternate (but essentially similar) proof of Theorem (4.1) by showing that for every $\epsilon > 0$, there is some partition $P$ of $[a, b]$ for which $U(P; f) - L(P; f) < \epsilon$.

4.2    Indicate what modifications will be needed in the proof of Theorem (4.2) if $c_1 = a$ or $c_k = b$.

4.3    Let $f(x) = x^2$ and $g(x) = x$ for $x \in \mathbb{R}$. Then $f$ and $g$ map the interval $[0, 2]$ onto the intervals $[0, 4]$ and $[0, 2]$ respectively. Here $f, g$ and the composite function $g \circ f$ are continuous everywhere and hence integrable. But show that the numbers $\int_0^2 (g \circ f)(x) dx$, $\int_0^2 g(x) f(x) dx$ and $\int_0^4 g(x) dx \int_0^2 f(x) dx$ are all different. [*Hint*: Use Exercise (1.17).]

4.4    Let $M$ and $m$ be respectively the supremum and the infimum of $f(x)$ on $[a, b]$. Consider the function $f^2(x) = [f(x)]^2$. Prove that if $0 \notin [m, M]$ then the fluctuation of $f^2$ on $[a, b]$ is $|M^2 - m^2|$ while if $0 \in [m, M]$ then it is at most $\max\{M^2, m^2\}$.

4.5    Using the last exercise, give a direct proof of Lemma (4.4).

*4.6    Define $f : [0, 1] \longrightarrow \mathbb{R}$ by

$$f(x) = \begin{cases} 0 & \text{if } x = 0 \text{ or if } x \text{ is irrational} \\ \frac{1}{q} & \text{if } x = \frac{p}{q} \text{ in the reduced form, } p, q \in \mathbb{N} \end{cases}$$

Prove that for every $x_0 \in [0, 1]$, $\lim_{x \to x_0} f(x) = 0$. Deduce that the set $D_f$ of points of discontinuity of $f$ is precisely the set of non-zero rationals in $[0, 1]$. [*Hint*: For every $n \in \mathbb{N}$, there are only finitely many values of $x$ for which $f(x) \geq \frac{1}{n}$. Choose a deleted neighbourhood of $x_0$ which contains no such $x$.]

4.7    Prove that the function $f$ of the last exercise is integrable over $[0, 1]$. [*Hint*: Following the hint of the last exercise, given $\epsilon > 0$, find a suitable partition in which the finitely many 'bad' values of $x$ are enclosed in very small subintervals.]

4.8   Let $g : [0,1] \longrightarrow \mathbb{R}$ be $g(x) = 0$ if $x = 0$ and $g(x) = 1$ if $x > 0$. Prove that if $f$ is as in the last two exercises then the composite $g \circ f$ is not Riemann integrable over $[0,1]$ (cf. Exercise (1.8)).

4.9   Suppose $f : [a,b] \longrightarrow \mathbb{R}$ is continuous and monotonically increasing with $f(a) = c$ and $f(b) = d$. Prove that for every $y_0 \in [a,b]$, $f^{-1}(\{y_0\})$ (i.e., the set $\{x \in [a,b] : f(x) = y_0\}$) consists either of a single point or is a closed subinterval of $[a,b]$.

4.10  In the last exercise, suppose $g : [c,d] \longrightarrow \mathbb{R}$ is a bounded function with only finitely many discontinuities. Prove that $g \circ f$ is integrable over $[a,b]$. [*Hint*: Show that each point $y_0$ where $g$ is not continuous gives rise to at most two points of discontinuity of $g \circ f$.]

## 5.5   The Fundamental Theorem of Calculus

Before stating the theorem in the title, it is perhaps worthwhile to ponder a bit on what a fundamental theorem is. Mathematics abounds in theorems of various degrees of utility, depth and significance. A most frequently used result need not be a deep theorem, as we see from elementary facts from arithmetic (e.g. commutativity and associativity of addition) which even an illiterate person uses (without knowing the words 'commutativity' and 'associativity'). Even theorems which are both frequently used and non-trivial to prove do not necessarily qualify to be 'fundamental'. Lagrange's Mean Value Theorem is undoubtedly deep (with the proof requiring completeness of the real line) and important (as we have already applied it several times). Despite this, it cannot be called a fundamental theorem, because whether a theorem is fundamental or not is decided by its significance in terms of the philosophy of the subject. The fundamental Theorem of Calculus, which we already mentioned in Section 5.1, is undoubtedly deep (the proof making a crucial use of Lagrange's MVT) and highly useful (being the most standard way of evaluating a definite integral). But the real reason it is called fundamental is something else.

In Section 2.2, we elaborated that the limiting process is the heart of calculus. The limiting process occurs in many connections. But from the point of view of applications, the two limits which, by far, stand out are the derivatives (limits of the incrementary ratio) and the Riemann integrals (limits of certain sums). Traditionally, calculus is divided into two halves, the differential calculus and the integral calculus. Such a division has certain disadvantages from the pedagogical as well as other points of view. Nevertheless, the very fact that such an arrangement prevailed for some time indicates the pre-emptive position of derivatives and integrals in calculus. Conceptually, these two halves are independent of each other. Either can be studied without knowing anything about the other. There is, in fact, nothing in the definition of a derivative and that of a definite integral to suggest that the two are even remotely related.

Naturally, any theorem which establishes a vital link between these two fundamental concepts fully deserves to be called a theorem of fundamental significance. The Fundamental Theorem of Calculus does precisely that and hence its name.

Basically, the Fundamental Theorem of Calculus says that integration and differentiation are opposite processes, like ascending and descending or like oxidation and reduction. Put differently, it says that the definite integrals and anit-derivatives are closely related (so closely, in fact, as to confer a different name, viz. indefinite integrals, on the latter, very much like a girl Miss X marrying a man Mr. Y and thereafter being known as Mrs. Y.)

There are two versions, or 'forms' of the Fundamental Theorem. The first form, which is more useful in practice asserts that if a function $f$ is integrable over an integral $[a, b]$, then its (definite) integral can be evalauted very easily if we know an anti-derivative, say $F$, of $f$ on $[a, b]$. Specifically, it says that if $F' = f$ then

$$\int_a^b f(x)dx = F(b) - F(a). \tag{1}$$

Before we give a formal proof of (1), let us see why it is not such an unexpected result. Consider the motion of a particle $P$ on a straight line $L$. Fix a reference point $O$ on the line and one direction along $L$ as positive and the opposite direction as negative. Let $x$ denote time and $F(x)$ be the distance of $P$ from $O$ at time $x$. (Note that the distance could be positive or negative depending on which side of $O, P$ lies.) Then $F(b) - F(a)$ is the net displacement of $P$ over the time interval $[a, b]$. But we can also calculate this displacement using Exercise (2.10). Let $f(x)$ be the speed of $P$ at time $x$. Then the net displacement of $P$ as $x$ varies from time $a$ to time $b$ is $\int_a^b f(x)dx$. Since the net displacement is independent of the manner in which it is calculated, (1) must hold.

Since the instantaneous speed is the derivative of the distance function, it is tempting to think that we now have a proof of (1). But this argument is based on a physical interpretation of a derivative. And so, however convincing it is to a layman, it canot pass as a mathematical proof. (See the comments made after Theorem (4.5.1).) The difficulty is that both the terms 'displacement' and 'speed' have intuitive meanings and both the statements 'the speed is the derivative of displacement' and 'displacement is the integral of the speed' conform to our intuitive understanding of these concepts. In a mathematical approach we have a choice. Either we can take displacement as an undefined but intuitively clear term and define the speed as its derivative. But then we must give a rigourous proof that by integrating the speed we get back the displacement. Alternatively, we can take speed as an intuitively clear concept and define displacement as the definite integral of speed. But then we would have to give a rigourous proof that by differentiating the displacement, we get the speed. So either way, we have to do some hard work. Interestingly, these two approaches correspond precisely to the two forms of the Fundamental Theorem of Calculus.

We now formally state and prove the **first form**.

**5.1 Theorem:** Suppose $F$ is differentiable on $[a, b]$ and $F'$ is Riemann integrable on $[a, b]$. Then

$$\int_a^b F'(x)dx = F(b) - F(a).$$

**Proof:** By Corollary (2.4), $\int_a^b F'(x)dx$ is the limit of the Riemann sum of $F'$ based on a partition $P$ of $[a, b]$, as the mesh of $P$ tends to 0. Since this limit is already given to exist, to evaluate it we are free to take any sequence of partitions with their meshes tending to 0 and to take any Riemann sums of our choice. By Exercise (4.5.14), for each partition, say $P = (x_0, \ldots, x_n)$ of $[a, b]$, there is some $\xi = (\xi_1, \ldots, \xi_n)$ based on $P$ such that the Riemann sum $R(P, \xi ; f)$ equals $F(b) - F(a)$. By approaching $\int_a^b F'(x)dx$ through such constant Riemann sums we get the result. ∎

In the theorem above the integrability of $F'$ has to be given beforehand. That is, mere differentiability of $F$ on $[a, b]$ does not automatically imply that $F'$ is Riemann integrable on $[a, b]$. As Exercise (4.5.25) shows, the derived function $F'$ need not be bounded. It can also happen that $F'$ is bounded but not Riemann integrable on $[a, b]$. A counter-example to this effect is rather complicated and we omit it. However we mention this point because this is one of the major inadequacies of Riemann integrals which led to a more general theory of integration, called the Lebesgue integral. If a function $f$ is Riemann integrable then it is also Lebesgue integrable and its integrals in the two senses are equal. But the converse is not true. For example the function in Exercise (1.8) is not Riemann integrable but can be shown to be Lebesgue integrable. In fact, the class of Lebesgue integrable functions turns out to be closed not only under uniform convergence but even under pointwise convergence and so certain anomalies which occur for Riemann integrals (see, for example, the discussion before Theorem (3.2)), do not occur in the case of Lebesgue integrals.

Interesting as the Lebesgue theory is, we do not pursue it. For real life applications, the Riemann integrals are generally sufficient, because as mentioned in Section 4.1, functions arising in real life situations are usually either continuous or piecewise continuous. By Theorem (4.2) all such functions are Riemann integrable anyway. In fact, combining the preceding theorem with Theorem (4.1), we immediately get,

**5.2 Corollary:** If $F$ is continuously differentiable on $[a, b]$ then $\int_a^b F'(x)dx$ $= F(b) - F(a)$. Put differently, if $f$ is continuous and has $F$ as an anti-derivative on $[a, b]$ then

$$\int_a^b f(x)dx = F(b) - F(a). \quad \blacksquare$$

Note the crucial use of Lagrange's MVT (through Exercise (4.5.14)) in the proof of Theorem (5.1). As elaborated in Section 4.5, Lagrange's theorem is itself a deep result, requiring the completeness of the real line. So, apart from being fundamentally significant, the Fundamental Theorem is also a non-trivial result. In addition to these two virtues (which do not always co-exist), the Fundamental Theorem is also a highly useful resutl. It reduces the task of evaluating an integral $\int_a^b f(x)dx$ to that of finding an anti-derivative for $f$. Once such an anti-derivative, say $F$, is found, it is customary to write

$$\int_a^b f(x)dx = F(x) \Big|_a^b = F(b) - F(a).$$

As an example, in Exercise (1.17), the integrals $\int_a^b x dx$ and $\int_a^b x^2 dx$ were to be evaluated, from the definition. We can now evaluate them effortlessly. Since $\frac{x^2}{2}$ and $\frac{x^3}{3}$ are anti-derivatives of $x$ and $x^2$ respectively, we have

$$\int_a^b x dx = \frac{1}{2}x^2 \Big|_a^b = \frac{1}{2}(b^2 - a^2)$$

$$\text{and} \quad \int_a^b x^2 dx = \frac{1}{3}x^3 \Big|_a^b = \frac{1}{3}(b^3 - a^3).$$

More generally, for every integer $n \neq -1$, $\frac{x^{n+1}}{n+1}$ is an anti-derivative of $x^n$ and so we can evaluate $\int_a^b x^n dx$ (assuming, $0 \notin [a, b]$, if $n < 0$). Similarly we can easily evaluate $\int_a^b \sin x dx$ and $\int_a^b \cos x dx$. (See Exercise (5.1) for a direct evaluation of these integrals.)

We remarked in Section 5.2 that certain infinite series can be expressed as definite integrals. If we can evaluate the latter using the Fundamental Theorem, then we have a method for evaluation of certain series, which, in general, is quite a difficult task. Even for finite sums, if it is possible to recognise them in terms of Riemann sums of suitable functions, then the theorem above provides a method for evaluating them approximately. See Exercises (5.7) to (5.9) for illustrations.

For the Corollary above to work, two crucial questions arise. First, does every integrable function have an anti-derivative? And in case it does, how do we find it? It can be shown by simple examples that the answer to the first question is in the negative in general (cf. Exercise (5.2)). However, as the next theorem shows, the situation is better for continuous functions. This theorem is often called the **second form** of the fundamental theorem of calculus. As remarked earlier, it corresponds to the fact that if we integrate the speed function we get displacement which is an anti-derivative of speed. The proof, too, is based on this idea.

**5.3 Theorem:** Every continuous function $f$ on a closed interval $[a, b]$ has an anti-derivative. In fact one such anti-derivative is $F(x) = \int_a^x f(t)dt$, i.e., the function defined by integrating $f$.

**Proof:** By Theorem (4.1), $f$ is Riemann integrable on $[a, x]$ for every $x \in [a, b]$. So the function $F$ is certainly defined on $[a, b]$. To show it is an anti-derivative of $f$, let $c \in [a, b]$. For any $x \in [a, b]$, we have, by Exercise (1.19),

$$F(x) - F(c) = \int_a^x f(t)dt - \int_a^c f(t)dt = \int_c^x f(t)dt$$

and so, by Exercise (1.9) and (1.12)

$$
\begin{aligned}
|F(x) - F(c) - (x - c)f(c)| &= \left| \int_c^x f(t)dx - \int_c^x f(c)dx \right| \\
&= \left| \int_c^x f(t) - f(c)dt \right| \\
&\leq \left| \int_c^x |f(t) - f(c)|dt \right|.
\end{aligned}
\tag{2}
$$

By continuity of $f$ at $c$, given an $\epsilon > 0$, we can find some $\delta > 0$ such that $|f(t) - f(c)| < \dfrac{\epsilon}{2(b-a)}$ whenever $|t - c| < \delta$ and $t \in [a, b]$. So the integrand in the last term is at most $\dfrac{\epsilon}{2(b-a)}$ everywhere if $|x - c| < \delta$. Since $|x - c| \leq b - a$ whenever $x \in [a, b]$, we have, from (2), that whenever $x \in [a, b]$ and $|x - c| < \delta$,

$$\left| \frac{F(x) - F(c)}{x - c} - f(c) \right| \leq \frac{\epsilon(b-a)}{2(b-a)} < \epsilon \ (\text{for } x \neq c) .$$

In other words, $F'(c)$ exists and equals $f(c)$. ∎

Unlike the first form, the second form of the Fundamental Theorem is mostly of theoretical importance. When it comes to evaluating $\int_a^x f(t)dt$, unless we already know some anti-derivative of $f$, we are helpless. So, in such a case, merely knowing that $\int_a^x f(t)dt$ is an anti-derivative of $f$ is just about as helpful as knowing that $X$'s daughter is married to $X$'s son-in-law!

But even though the theorem above is not very helpful in evaluating an integral, it is important in studying the properties of functions defined by integrals. As a concrete example, consider the function $f(t) = \dfrac{1}{t}$ which is continuous on $(0, \infty)$. At the end of the last section we defined $\ln x$ as $\displaystyle\int_1^x \dfrac{dt}{t}$ for $x > 0$. By the theorem above we now know that $\ln x$ is an anti-derivative of $\ln x$ for $x > 1$. For $0 < x < 1$, $\displaystyle\int_1^x \dfrac{dt}{t}$ is, by definition, (cf. Exercise (1.19)), $-\displaystyle\int_x^1 \dfrac{dt}{t}$. Here the upper limit of integration is constant and the lower one is a variable. It is easy to show (cf. Exercise (5.5)) that in such a case $\dfrac{d}{dx}\left(\displaystyle\int_x^1 \dfrac{dt}{t}\right) = -\dfrac{1}{x}$ and so $\ln x$ is an anti-derivative of $\dfrac{1}{x}$ for all $x > 0$. This information is of no help in evaluating an integral like $\displaystyle\int_3^4 \dfrac{dx}{x}$. For although we could apply Corollary (5.2) and get $\displaystyle\int_3^4 \dfrac{dx}{x} = \ln x \Big|_3^4 = \ln 4 - \ln 3$, the very meaning of $\ln 4 - \ln 3$ is $\displaystyle\int_1^4 \dfrac{dt}{t} - \displaystyle\int_1^3 \dfrac{dt}{t}$, which by Exercise (1.17), equals $\displaystyle\int_3^4 \dfrac{dt}{t}$, exactly what we started with. So to evaluate $\displaystyle\int_3^4 \dfrac{dt}{t}$ we have to content ourselves with an approximate answer. A few methods for such approximate integration will be given in Section 5.9.

What exactly, then, do we gain by calling $\displaystyle\int_1^x \dfrac{dt}{t}$ as $\ln x$? The answer is that once we include this function in our bag of standard functions, we can prepare (by various methods) tables that give its approximate values, similar to the tables for the sine and cosine function. And these tables can be used in the integration of not only the function $\frac{1}{x}$, but many other functions. For example, the chain rule gives $\dfrac{d}{dx}(\ln \sec x) = \dfrac{1}{\sec x} \sec x \tan x = \tan x$. And so, $\ln \sec x$ is an anti-derivative of $\tan x$. So if we want to evaluate, say, $\displaystyle\int_{\pi/6}^{\pi/3} \tan x\, dx$, we apply Theorem (5.2) and get the answer as $\ln \sec x \Big|_{\pi/6}^{\pi/3}$, i.e., as $\ln 2 - \ln\left(\dfrac{2}{\sqrt{3}}\right)$ which can be found from the ready-made tables for $\ln$. It can, of course, be argued that we could just as well have prepared tables for $\displaystyle\int_0^x \tan t\, dt$. But obviously it is better to prepare the tables for just one function and use them for other related functions, just as it is more economical to make one long distance telephone call to a far-away friend and ask him to communicate your message to other friends in his locality than to make a separate call to each. (Antiderivatives of $\cot x$, $\sec x$ and $\operatorname{cosec} x$ can also be expressed using logarithms. They are, respectively, $-\ln \operatorname{cosec} x$, $\ln(\tan x + \sec x)$ and $-\ln(\cot x + \operatorname{cosec} x)$. These are very helpful because with suitable substitutions, many antiderivatives can be reduced to those of trigonometric functions. See the exercises for examples.)

Thus, although Theorem (5.3) guarantees an anti-derivative for every continuous function, it is of little help in evaluating definite integrals by finding anti-derivatives. There is, in fact, no golden method for doing the latter. The general strategy is to remember the anti-derivatives of a few standard functions and then to try to reduce the given function to one of the standard ones by various tricks such as algebraic manipulations, clever substitutions or by resorting to integration by parts. The latter two were already given as Exercises (4.4.5) and (4.4.6) respectively. The rule of integration by substitution is essentially the chain rule for derivatives. But when applied to evaluate definite integrals, it has a special theoretical significance which will be discussed in the next section. From a practical point of view, the success of the rule of substitution lies in coming up with the correct substitution. This is a skill one acquires with practice. But a few general hints can be given. A substitution like $x = a \sin t$ enables one to get rid of a term like $\sqrt{a^2 - x^2}$, with a radical sign in the integrand. In many situations, it is better to express the substitution in the form $t = h(x)$ rather than $x = g(t)$. For example, in the indefinite integral $\displaystyle\int \frac{x\,dx}{x^2 + 1}$, we put $t = x^2 + 1$, instead of saying $x = \sqrt{t - 1}$. The conversion from $dx$ to $dt$ is best done by following the formal rules for differentials, which, as mentioned in Section 4.4, are very convenient even though meaningless in a strict sense. Thus $t = x^2 + 1$ gives $dt = 2x\,dx$, or, $x\,dx = \dfrac{1}{2}dt$. The integral $\displaystyle\int \frac{x\,dx}{x^2 + 1}$ thus becomes $\displaystyle\int \frac{dt}{2t}$ which is $\dfrac{1}{2}\ln t$, i.e., $\dfrac{1}{2}\ln(x^2 + 1)$. It is customary to add a constant, say $c$, called the constant of integration and write the answer as $\dfrac{1}{2}\ln(x^2 + 1) + c$. See Exercise (4.5.13) for its role.

As the purpose of this book is to concentrate more on the why's than on the how's of calculus, we shall not go into an elaborate discussion of the so-called 'techniques of integration'. Nevertheless, in the exercises we shall give a few problems where anti-derivatives are to be found. For the ambitious student, such problems often contain a certain challenge and charm. For the average student, they are among the bread-and-butter problems from an examination point of view.

Despite all the techniques of integration at our disposal and all the skill one can muster, it is not always possible to express the anti-derivative of a continuous function in a closed form, i.e. in terms of some familiar functions. There is, of course, no unanimous answer to what one means by 'familiar' functions. But generally the term means all polynomials, rational functions, trigonometric, exponential and logarithm functions and their composites. As a simple example, let $f(x) = \sin(x^2)$. Then $f$ is continuous, and so by Theorem (5.3), has an anti-derivative, say $F(x)$. We can, in fact, write $F(x)$ as $\displaystyle\int_0^x \sin(t^2)dt$. But this is of little practical use. There is no way to express $F(x)$ in terms of familiar functions (a proof of this fact is far from trivial, just like the proof of the impossibility of solving a polynomial equation by radicals, see Section 1.13).

So to evaluate an integral like $\int_0^1 \sin(x^2)dx$ we have to resort to approximate integration. Occassionally, a definite integral can be evaluated exactly using special tricks even when the integrand is a function whose anti-derivative cannot be expressed in a closed form. This will be illustrated in the next section.

Summing up, even though the Fundamental Theorem of Calculus does not work everytime, when it works it is by far the best method to evaluate a definite integral. It is also a deep result. What gives it its name, however, is the link it establishes between the two major, and apparently unrelated, branches of calculus.

## EXERCISES

5.1  For every positive integer $n$ and $\theta \in (0, 2\pi)$, prove that $\sin\theta + \sin 2\theta + \ldots + \sin n\theta = \dfrac{\sin(\frac{(n+1)\theta}{2})\sin\frac{n\theta}{2}}{\sin\frac{\theta}{2}}$. Hence evaluate $\int_a^b \sin x dx$. Similarly prove an identity for the sum of cosines of angles which are in an arithmetic progression and using it evaluate $\int_a^b \cos x dx$.

5.2  Let $f(x) = \begin{cases} 0 & \text{if} \quad x \in [0,1], x \neq \frac{1}{2} \\ 1 & \text{if} \quad x = \frac{1}{2} \end{cases}$ . Prove that $f$ has no anti-derivative on $[0,1]$ but is integrable on $[0,1]$. [*Hint* : See Exercise (4.5.8).]

5.3  Obtain an explicit formula for an anti-derivative of the function $f(x) = |x| + |x - 1|, x \in \mathbb{R}$.

5.4  Suppose $f$ is integrable on $[a, b]$. Let $F(x) = \int_a^x f(t)dt$. For $c \in [a, b]$, prove that if $f$ is continuous at $c$ then $F$ is differentiable at $c$. Is the converse true? Visit Exercise (3.11) again to see if your guess was correct. (This is a slightly more general version of Theorem (5.3). But the same proof goes through.)

5.5  Prove that if $f$ is continuous on $[a, b]$ and $G(x) = \int_x^b f(t)dt$ for $a \leq x \leq b$ then $G'(x) = -f(x)$ for all $x \in [a, b]$. More generally suppose $u, v$ are differentiable functions of a real variable $x$ such that for every $x$, both $u(x)$ and $v(x)$ lie in $[a, b]$. Let $H(x) = \int_{u(x)}^{v(x)} f(t)dt$. Then prove that $H'(x) = v'(x)F(v(x)) - u'(x)F(u(x))$.

5.6  Find   (i) $\displaystyle\lim_{x \to x_0} \dfrac{1}{x^2 - x_0^2} \int_{x_0}^x \sin(t^2)dt$   (ii) $\displaystyle\lim_{x \to 0} \dfrac{1}{x^6} \int_0^{x^2} \dfrac{t^2 dt}{t^6 + 1}$.

5.7  Evaluate the limits of the first three sequences in Exercise (2.6).

5.8 Discuss the behaviour of $\sum_{k=1}^{n} \sin(\frac{k^2}{n^2}), \frac{1}{n} \sum_{k=1}^{n} \sin\left(\frac{k^2}{n^2}\right)$ and $\frac{1}{n^2} \sum_{k=1}^{n} \sin\left(\frac{k^2}{n^2}\right)$ as $n \to \infty$.

5.9 Express the sum $\sqrt{1} + \sqrt{2} + \ldots + \sqrt{10,000}$ in terms of a suitable Riemann sum. Hence find its approximate value and an upper bound on the error.

5.10 Prove that if $f_1(x)$ and $f_2(x)$ are two continous functions on $[a, b]$ then the area of the region between the graphs of these two functions and the lines $x = a, x = b$ is $\int_a^b |f_1(x) - f_2(x)| dx$. State and prove a similar formula for the area bounded by curves of the form $x = g_1(y), x = g_2(y), y = c$ and $y = d$ where $g_1, g_2$ are continuous functions on an interval $[c, d]$.

5.11 Find the area bounded by a circle of radius $a$. If two such circles have their centres separated by a distance $R$ ($< 2a$) find the area common to both.

5.12 Sketch and find the area bounded by the curves $x^2 + y^2 = 4, x^2 = -\sqrt{2}y$ and $x = y$.

5.13 Let $C_1$ and $C_2$ be the graphs of the functions $y = x^2$ and $y = 2x, 0 \le x \le 1$ respectively. Let $C_3$ be the graph of a continuous function $y = f(x), 0 \le x \le 1$, with $f(0) = 0$. Suppose for every point $(a, a^2)$ on $C_1$, the region bounded by portions of $C_2, C_3$ and the lines $x = a$ and $y = a^2$ is divided by $C_1$ into two regions of equal areas. Determine $f(x)$.

5.14 Prove the rule of integration by parts for definite integrals. That is, if $f'$ and $g$ are continuous on $[a, b]$ with $G$ being an anti-derivative of $g$, then

$$\int_a^b f(x)g(x)dx = f(b)G(b) - f(a)G(a) - \int_a^b f'(x)G(x)dx.$$

[*Hint*: Let $H(x) = \int_a^x f(t)g(t)dt - f(x)G(x) + \int_a^x f'(t)G(t)dt$. Show that $H'(x) = 0$ for all $x \in [a, b]$.]

5.15 Use integration by parts to find anti-derivatives of $\ln x, \tan^{-1} x$ and $x^2 \sin x$.

5.16 Try to find $\int \sec^3 \theta d\theta$ by parts. You get the same term on the right hand side, but with a minus sign. Bring it to the left and then evaluate $\int \sec^3 \theta d\theta$. Using it obtain $\int \sqrt{a^2 + x^2} dx$, $a$ being a constant.

5.17 (i) Find $\int \sin^2 x dx$ by a method similar to that in the last exercise and also by writing $\sin^2 x$ in terms of $\cos 2x$. Why can't $\int \sin(x^2) dx$ be evaluated using integration by parts?

(ii) Analogously, find $\int \sin x \cos x dx$ by parts in two different ways. Also find it by writing the integrand as $\frac{1}{2} \sin 2x$. You get three different answers. Is anything wrong?

**5.18** If $f$ is continuously differentiable, prove that $\int e^x(f(x) + f'(x))dx$

$= e^x f(x)$. Find $\int e^x(\sin(x^2) + 2x\cos(x^2))dx$. Note that neither

$\int e^x \sin(x^2)dx$ nor $\int xe^x \cos(x^2)dx$ has a closed form expression.

**5.19** Find $\int \dfrac{xdx}{ax^2 + bx + c}$ and $\int \dfrac{dx}{ax^2 + bx + c}$ where $a \neq 0$. Consider various

cases depending upon whether $b^2 - 4ac$ is positive, negative or 0. (These

integrals and integrals of the form $\int \dfrac{dx}{ax + b}$ are the prototypes of a method

based on what are called **partial fractions**, used for integrating rational

functions, i.e. functions of the form $\dfrac{p(x)}{q(x)}$ where $p(x)$ and $q(x)$ are polyno-

mials in $x$. Because of the division algorithm for polynomials it suffices to

assume that $p(x)$ is of a lower degree than $q(x)$. We may also assume $q(x)$

is monic, i.e., the coefficient of its highest degree term is 1. Because of a

famous theorem called the Fundamental Theorem of Algebra, $q(x)$ can be

uniquely factored into factors which are linear (i.e. of the form $x + a$) or

quadratic (i.e. of the form $x^2 + bx + c$ where $a, b, c$ are real and $b^2 < 4c$.)

If we assume that no factor is repeated then $\dfrac{p(x)}{q(x)}$ can be written uniquely

as a sum of **partial fractions**, each such fraction being a function of the

form $\dfrac{A}{x + a}$ or of the form $\dfrac{Bx + C}{x^2 + bx + c}$ where $A, B, C$ are constants. There

are various methods for finding these constants. One of them is to add

the partial fractions and write down a system of equations by comparing

coefficients of the like terms in the numerator. The case where $q(x)$ has

some repeated factors is a little more complicated and we omit it.)

**5.20** Evaluate $\int \dfrac{3x^2 - 1}{x^4 + 2x^3 + 2x^2 + x} dx$ by partial fractions.

**5.21** Find the following anti-derivatives using the various techniques of integration, including the rule of substitution.

(i) $\displaystyle\int \dfrac{\sin x}{\sqrt{1 + \cos x}} dx$      (ii) $\displaystyle\int \sqrt{1 + \sin x}\, dx$

(iii) $\displaystyle\int \dfrac{dx}{(x^2 + a^2)^{3/2}}$      (iv) $\displaystyle\int \sin\sqrt{1 - x}\, dx$

(v) $\displaystyle\int \dfrac{dx}{(\sin x - 5)\cos x}$      (vi) $\displaystyle\int \dfrac{\tan^{-1} x}{x^2} dx$

(vii) $\displaystyle\int e^{ax} \cos(bx + c)dx$      (viii) $\displaystyle\int e^{ax} \sin(bx + c)dx$

(ix) $\displaystyle\int \dfrac{dx}{\sqrt{e^{2x} + 1}}$      (x) $\displaystyle\int (x + 1)e^x \ln x\, dx$.

5.22 When the integrand is an expresion involving trigonometric functions of a variable, say $x$, the substitution $t = \tan \dfrac{x}{2}$ is often useful in converting the expression (by identities like $\sin x = \dfrac{2t}{t^2 + 1}$ and $\cos x = \dfrac{1 - t^2}{1 + t^2}$) to a rational function in $t$, which can then be tackled which partial fractions. Use this method to find $\displaystyle \int \frac{dx}{2 + \cos x}$. Can this method be used for $\displaystyle \int \frac{dx}{2 + \cos^2 x}$?

### 5.6. Why the Fundamental Theorem of Calculus Is Sometimes Inadequate

Having extolled the Fundamental Theorem of Calculus in the last section, it may come as a surprise that now we are talking of its inadequacies. But then, like every human being, every mathematical theorem has its plus and minus points. There is no single theorem which will answer all the needs in a particular area. If such a theorem exists, then mathematics would be a dull subject indeed!

We already mentioned at the end of the last section that the Fundamental Theorem is not of much help in evaluating a definite integral when an anti-derivative of the integrand cannot be found in a closed form. The other in-adequacies of the Fundamental Theorem are not so much inadequacies of the theorem *per se* as of the tendency to apply it heavily and indiscriminately to the extent of converting integral calculus into differential calculus. Although definite integrals are generally not easy to evaluate straight form definition, some of their properties can be established directly, i.e. without any reference to anti-derivatives. In fact such proofs often bring out the true spirit of the result. A proof based on the Fundamental Theorem may be slicker and shorter. But it may miss the true significance of the result. Also such a proof is often less general in terms of applicability, because the fundamental theorem requires the integrand to have an anti-derivative, while the result may be true even without this hypothesis.

As a simple example, we prove an elementary result about definite integrals both with and without the Fundamental Theorem of Calculus.

**6.1 Theorem:** Suppose $f$ is integrable over an interval $[a, b]$. Then so is the function $g : [a, b] \longrightarrow \mathbb{R}$ defined by $g(x) = f(a + b - x)$. Moreover,

$$\int_a^b f(x)dx = \int_a^b g(x)dx.$$

*Proof:* If we assume $f$ is continuous then a quick proof can be given using the Fundamental Theorem of Calculus and the rule of integration by substitution (which, as remarked earlier, is essentially the chain rule). Let $F$ be an anti-derivative of $f$ which exists by Theorem (5.3). Then by Corollary (5.2), $\int_a^b f(x)dx = F(b) - F(a)$. Now put $x = a + b - t = h(t)$ (say). Note that $h$ maps $[a, b]$ into itself and is continuous. So the composite $f \circ h$ is also continuous and hence integrable on $[a, b]$ by Theorem (4.1). Note also that $f \circ h$ is precisely $g$. Since $h$ is differentiable (with $h'(t) = -1$ for all $t$), by the Chain Rule, $F \circ h$ is also differentiable and $(F \circ h)'(t) = F'(h(t))h'(t) = -F'(x) = -f(x) = -f(h(t)) = -g(t)$. So again by Corollary (5.2),

$$\int_a^b g(t)dt = -(F \circ h)(t)\Big|_a^b = F(h(a)) - F(h(b)) = F(b) - F(a)$$

which gives the result.

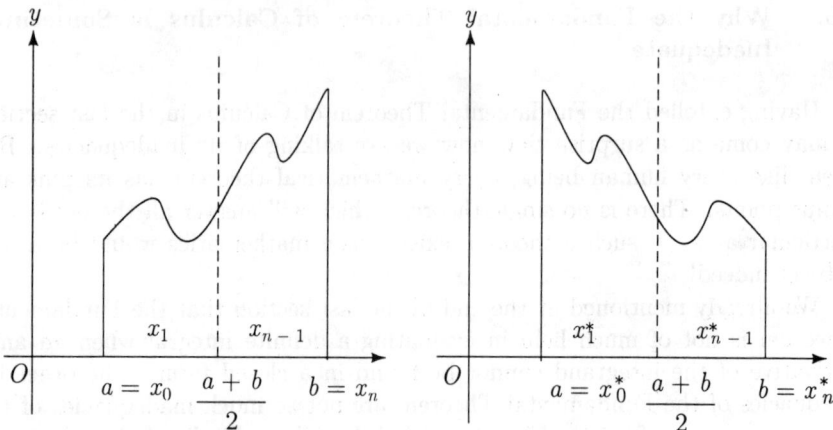

Figure 5.6.1 : A Property of Definite Integrals.

We now give a proof which does not use the Fundamental Theorem and requires merely that $f$ be integrable on $[a, b]$. The mapping $h$ defined above is nothing but the reflection of the interval $[a, b]$, w.r.t. its mid-point, $\dfrac{a+b}{2}$. So it is clear that the graphs of $f$ and $g \, (= f \circ h)$ will be mirror images of each other w.r.t. the vertical line through $\dfrac{a+b}{2}$. The same will also be true for the areas below their graphs (see Fig.5.6.1). Since areas are preserved under reflection, it is very clear that $\int_a^b f(x)dx = \int_a^b g(x)dx$. ∎

It may be objected that this argument is based on geometric intuition and hence not mathematically acceptable (for the same reason that the argument in the last section, based on speed and displacement cannot constitute a mathematical proof of the Fundamental Theorem). This objection can be answered by giving a mathematical definition of area and proving its invariance under reflections. But we are not yet in a position to do so. Nevertheless, there is another way out to give a mathematical form to the argument above so that it uses strictly the definition of a Riemann integral.

For every partition, say $P = (x_0, x_1, \ldots, x_n)$ of the interval $[a, b]$, we define a new partition $P^*$, called the reflection of $P$. Indeed, let $P^* = (x_0^*, x_1^*, \ldots, x_n^*)$ where for $i = 0, 1, \ldots, n, x_i^* = a + b - x_{n-i}$. For $i = 1, \ldots, n$, let $M_i = \sup\{f(x) : x_{i-1} \leq x \leq x_i\}$ and $M_i^* = \sup\{g(x) : x_{i-1}^* \leq x \leq x_i^*\}$. Then it is clear that $M_i^* = M_{n+1-i}$ and $x_i^* - x_{i-1}^* = x_{n+1-i} - x_{n-i}$. (See again Fig. 5.6.1. However, now we are referring to the figure only for clarifying the ideas and not as a source of a proof.) So the upper Riemann sum $U(P^*; g) = \sum_{i=1}^{n} M_i^* (x_i^* - x_{i-1}^*)$

equals $\sum_{i=1}^{n} M_{n+1-i}(x_{n+1-1} - x_{n-i})$. The terms of this sum are precisely those in the upper Riemann sum $U(P; f)$ except in the opposite order. Since the roles of $P$ and $P^*$ can be interchanged, it follows that every upper Riemann sum of $f$ is also an upper Riemann sum of $g$ and vice versa. Consequently, the upper Riemann integrable of $f$ and $g$ are equal, since both are infima of the same set of real numbers. By a similar argument, the lower Riemann integral of $f$ equals that of $g$. Hence if $f$ is Riemann integrable over $[a, b]$, so is $g$ and their Riemann integrals are equal. ∎

As a special case, $\int_0^a f(x)dx = \int_0^a f(a - x)dx$. If the nature of the function $f$ is such that $f(a - x)$ can be related to $f(x)$ in some way, then this simple little result enables us to evaluate a definite integral in a rather novel way. We illustrate it with a simple problem.

**6.2 Problem:** Evaluate $\int_0^{\pi/2} \sin^2 x dx$.

*Solution:* The identity $\sin^2 x = \frac{1}{2}(1 - \cos 2x)$ gives $\frac{x}{2} - \frac{1}{4}\sin 2x$ as an anti-derivative of the integrand and so by the Fundamental Theorem,

$$\int_0^{\pi/2} \sin^2 x dx = \frac{x}{2} - \frac{1}{4}\sin 2x \Big|_0^{\pi/2} = \frac{\pi}{4}.$$

Let us now calculate this integral using the theorem above. Let $I$ denote $\int_0^{\pi/2} \sin^2 x dx$. Then since $\sin\left(\frac{\pi}{2} - x\right) = \cos x$, the theorem above gives

$$I = \int_0^{\pi/2} \cos^2 x dx. \text{ Adding, } 2I = \int_0^{\pi/2} \sin^2 x + \cos^2 x dx = \int_0^{\pi/2} 1 dx = \frac{\pi}{2} \text{ (with-}$$

out the Fundamental Theorem, by Exercise (1.2)). So $I = \dfrac{\pi}{4}$. ∎

This problem provides a good illustration of the strength and the weakness of the Fundamental Theorem. The first method is certainly more powerful and is applicable to evaluate $\displaystyle\int_a^b \sin^2 x\,dx$ for any $a$ and $b$. The second method is special. But it works even when we cannot find an anti-derivative of the integrand. A few problems where the second method is mandatory, or at least highly preferable, will be given in the exercises.

Returning to Theorem (6.1), the first proof used a special case of the rule of substitution for integrals. Following essentially the same reasoning as there, we get the general **rule of substitution** (also known as the **change of varaible formula**) for definite integrals.

**6.3 Theorem:** Let $f : [a,b] \longrightarrow I\!R$ be continuous. Suppose $h : [c,d] \longrightarrow [a,b]$ is a continuously differentiable function with $h(c) = a$ and $h(d) = b$. Then

$$\int_a^b f(x)dx = \int_c^d f(h(t))h'(t)dt = \int_c^d f(x)\frac{dx}{dt}dt \qquad (1)$$

**Proof:** Once again, let $F$ be an anti-derivative of $f$. Apply the chain rule to the composite function $F \circ h$ and then the fundamental theorem of calculus to both $F'$ and $(F \circ h)'$. This gives the equality of the first two terms in the conclusion. The third term is just a different notation for the second term. ∎

This proof is a good example of how a result about integrals can be proved by borrowing a corresponding result about derivatives and then using the linkage provided by the Fundamental Theorem of Calculus. Although logically unassailable, such a proof often masks the true spirit of the result. For example, in the theorem just proved, the proof does not reveal what role is played by the factor $\dfrac{dx}{dt}$ in the last integral. It may be claimed that we need it because we want to convert integration w.r.t. $x$ to integration w.r.t. $t$ and so $dx$ has to be replaced by $\dfrac{dx}{dt}dt$. But this is a superficial explanation. The symbol $dx$ is just a part of the notation and has no meaning of its own (see the answer to Q.4.4).

A far more satisfactory explanation comes if we go to the very definition of definite integrals. By Corollary (2.4), $\displaystyle\int_a^b f(x)dx$ is the limit of the Riemann sums $\displaystyle\sum_{i=1}^n f(\xi_i)\Delta x_i$ where $\boldsymbol{\xi} = (\xi_1, \ldots, \xi_n)$ is based on a partition, say $P = (x_0, x_1, \ldots, x_n)$ of $[a,b]$. Here $\Delta x_i$ stands for the length of the $i^{\text{th}}$ subinterval $[x_{i-1}, x_i]$ of $P$. Unlike the differential $dx$, $\Delta x_i$ is a well-defined mathematical entity and amenable to all the rules of algebra. An equation like $dx = \dfrac{dx}{dt}dt$ is meaningless (although convenient). But an equation like $\Delta x_i = \dfrac{\Delta x_i}{\alpha}\alpha$ is perfectly meaningful if $\alpha$ is any non-zero real number. In particular, if we let $\alpha = \Delta t_i$ (where $\Delta t_i$ will be suitably defined) then $\Delta x_i = \dfrac{\Delta x_i}{\Delta t_i}\Delta t_i$.

Since under suitable conditons, $\dfrac{\Delta x_i}{\Delta t_i} \to \dfrac{dx}{dt}$, this simple equation suggests an alternate and more revealing proof of Theorem (6.3), analogous to the second proof of Theorem (6.1). Another advantage of such a proof is that later on we come across a change of variables formula for double integrals[4]. For double integrals, there is no direct analogue of the Fundamental Theorem of Calculus and so the proof given above will break down. But the alternate proof we are about to give pulls through after suitable modifications.

In Theorem (6.1), the function $h$ was defined by $h(t) = a + b - t$ and the key idea of the proof was to get a one-to-one correspondence between the Riemann sums of $f$ and those of $g$. This was simplified because of the geometric significance of the function $h$, viz. as a reflection. In the general situation all we know about $h$ is that it is a continuosly differentiable function which maps $[c, d]$ onto $[a, b]$ with $h(c) = a$ and $h(d) = b$. For every partition, say, $P = (x_0, x_1, \ldots, x_n)$ of $[a, b]$, we want to define a corresponding partition, say $Q$, of $[c, d]$. The easiest way to do so would be to let $Q = (t_0, t_1, \ldots, t_n)$ where $t_i$'s are so chosen that for each $i, h(t_i) = x_i$. Since $h$ is continuous, such $t_i$'s do exist by the Intermediate Value Property. But they need not be unique. Also if we are not careful in choosing them, we may have $t_{i-1} > t_i$ even though $x_{i-1} < x_i$. We also want that for every partition, say $(s_0, s_1, \ldots, s_m)$, of $[c, d]$, $(h(s_0), h(s_1), \ldots, h(s_m))$ should be a partition of $[a, b]$, which again may fail if $h$ is not monotonically increasing.

We eliminate all these difficulties by making a flat assumption that $h'(t) > 0$ for every $t \in (c, d)$. Theorem (4.5.5) then ensures that $h$ is strictly monotonically increasing on $[c, d]$. We remark that this is not such a stringent restriction. We could as well prove the result when $h$ is strictly monotonically decreasing. (As was, in fact, the case in Theorem (6.1)). In general we can chop the interval $[a, b]$ into a (possibly infinite) number of subintervals over each of which $h$ is monotonic.

So we assume $h : [c, d] \longrightarrow [a, b]$ is continuously differentiable and strictly monotonically increasing with $h(c) = a$ and $h(d) = b$. Then for every partition, say $P = (t_0, \ldots, t_n)$ of $[c, d]$, there is a partition, $(h(t_0), \ldots, h(t_n))$ of $[a, b]$. We denote it by $h(P)$ (see Figure 5.6.2.). If we call $h(t_i)$ as $x_i$, then for every $\xi_i \in [x_{i-1}, x_i]$, there is a unique $\alpha_i \in [t_{i-1}, t_i]$ such that $h(\alpha_i) = \xi_i$. For a function $f : [a, b] \longrightarrow \mathbb{R}$, let $g : [c, d] \longrightarrow \mathbb{R}$ be the composite function $f \circ h$.

Then $f(\xi_i) = g(\alpha_i)$ and so $\displaystyle\sum_{i=1}^{n} f(\xi_i)\Delta x_i = \sum_{i=1}^{n} g(\alpha_i)\Delta x_i$. The left hand side is a typical Riemann sum of $f$ based on the partition $h(P)$. But the right hand side is not a Riemann sum of $g$, based on $P$. This is because we are not given that

---

[4]The double integrals belong properly to the theory of functions of two variables, which is beyond our scope. Still, we shall study double integrals as a tool for evaluating volumes etc. in Section 6.4. A special, but important, case of the change of variables formula for double integrals will be given in Section 6.8.

$\Delta x_i = \Delta t_i$. It is this point which needs a careful handling.

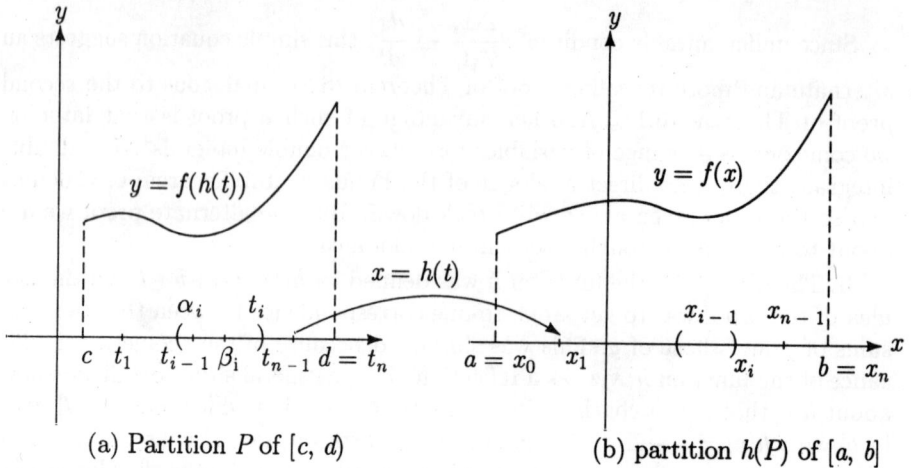

(a) Partition $P$ of $[c, d]$          (b) partition $h(P)$ of $[a, b]$

Figure 5.6.2 : Change of variables for integration

Let us rewrite $\displaystyle\sum_{i=1}^{n} g(\alpha_i)\Delta x_i$ as $\displaystyle\sum_{i=1}^{n} g(\alpha_i)\frac{\Delta x_i}{\Delta t_i}\Delta t_i$. Now $\dfrac{\Delta x_i}{\Delta t_i}$ is nothing but $\dfrac{x_i - x_{i-1}}{t_i - t_{i-1}}$, i.e. $\dfrac{h(t_i) - h(t_{i-1})}{t_i - t_{i-1}}$ which, by Lagrange's Mean Value Theorem, equals $h'(\beta_i)$ for some $\beta_i \in (t_{i-1}, t_i)$. So,

$$\sum_{i=1}^{n} f(\xi_i)\Delta x_i = \sum_{i=1}^{n} g(\alpha_i)\Delta x_i = \sum_{i=1}^{n} g(\alpha_i)h'(\beta_i)\Delta t_i \qquad (2)$$

If luckily, $\alpha_i = \beta_i$ for every $i = 1, \ldots, n$, then the last term in (2) would be a Riemann sum of the product function $gh'$ based on the partition $P$ of $[c, d]$ and by taking its limit as $\mu(P) \to 0$, we would get $\displaystyle\int_{c}^{d} g(t)h'(t)dt$. We leave it as an exercise (Exercise (6.3)) to show that as $\mu(P) \to 0$, $\mu(h(P))$ also tends to 0. So our conclusion will follow from (2) by taking limits as $\mu(P) \longrightarrow 0$.

The only catch is that $\alpha_i$ may be different from $\beta_i$. Still, both belong to the sub-interval $[t_{i-1}, t_i]$ of $P$. So, if $\mu(P)$ is small then $|\alpha_i - \beta_i|$ is small and so, by uniform continuity o f $g$, the difference $g(\alpha_i) - g(\beta_i)$ can be made sufficiently small. As a result, the sum $\displaystyle\sum_{i=1}^{n} g(\alpha_i)h'(\beta_i)\Delta t_i$ does not differ much from the sum $\displaystyle\sum_{i=1}^{n} g(\beta_i)h'(\beta_i)\Delta t_i$ which is a genuine Riemann sum of the function $gh'$. In the limit the two will be equal.

We make this argument precise and isolate the result since it is of an independent interest and applicable in other connections too. It is called **Bliss' theorem.**

**6.4 Theorem:** Suppose $u, v$ are both integrable on $[c, d]$ and at least one of them is continuous. For a partition $P = (t_0, t_1, \ldots, t_n)$ of $[c, d]$, let $\alpha_i, \beta_i \in [t_{i-1}, t_i]$ for $i = 1, \ldots, n$. Then as $\mu(P) \to 0$, the sum $\sum_{i=1}^{n} u(\alpha_i)v(\beta_i)\Delta t_i$ tends to $\int_c^d u(t)v(t)dt$.

**Proof:** By Theorem (4.5) we already know $uv$ is integrable on $[c, d]$. Without loss of generality assume $u$ is continuous. Let $M$ be any positive upper bound on $|v(t)|$ for $t \in [c, d]$. Let $\epsilon > 0$ be given. We have to find some $\delta > 0$ such that whenever $P = (t_0, t_1, \ldots, t_n)$ is a partition of $[c, d]$ with $\mu(P) < \delta$ and $\alpha_i, \beta_i \in [t_{i-1}, t_i]$ for $i = 1, \ldots, n$,

$$\left| \sum_{i=1}^{n} u(\alpha_i)v(\beta_i)\Delta t_i - \int_c^d u(t)v(t)dt \right| < \epsilon. \tag{3}$$

By Corollary (2.3), there exists $\delta_1 > 0$ such that for every partition $P = (t_0, \ldots, t_n)$ with $\mu(P) < \delta_1$ and for every choice of points $\beta_i \in [t_{i-1}, t_i], i = 1, \ldots, n$ we have,

$$\left| \sum_{i=1}^{n} u(\beta_i)v(\beta_i)\Delta t_i - \int_c^d u(t)v(t)dt \right| < \frac{\epsilon}{2}. \tag{4}$$

Since $u$ is uniformly continuous on $[c, d]$ (by Theorem (4.2.1)), there exists $\delta_2 > 0$ such that for all $\alpha, \beta \in [c, d]$,

$$|u(\alpha) - u(\beta)| < \frac{\epsilon}{2M(d - c)}. \tag{5}$$

Now let $\delta = \min\{\delta_1, \delta_2\}$. Let $P = (t_0, \ldots, t_n)$ be any partition of $[c, d]$ with $\mu(P) < \delta$ and suppose $\alpha_i, \beta_i \in [t_{i-1}, t_i]$ for $i = 1, 2, \ldots, n$. Then by applying (5) with $\alpha = \alpha_i, \beta = \beta_i$ and adding,

$$
\begin{aligned}
\left| \sum_{i=1}^{n} (u(\alpha_i) - u(\beta_i))v(\beta_i)\Delta t_i \right| &\leq \sum_{i=1}^{n} |u(\alpha_i) - u(\beta_i)||v(\beta_i)|\Delta t_i \\
&\leq \sum_{i=1}^{n} \frac{\epsilon}{2M(d - c)} M \Delta t_i \\
&= \frac{\epsilon}{2(d - c)} \sum_{i=1}^{n} \Delta t_i = \frac{\epsilon}{2}
\end{aligned} \tag{6}
$$

(4) and (6), along with the triangle inequality, imply (3). ∎

We are now in a position to give a proof of Theorem (6.3) which does not use the Fundamental Theorem of Calculus. Because of the extra restrictions we have imposed on the function $h$, we state it as a separate theorem.

**6.5 Theorem:** Suppose $h : [c,d] \longrightarrow [a,b]$ is continuously differentiable and $h'(t) > 0$ for all $t \in (c,d)$. Suppose also $h(c) = a$ and $h(d) = b$. Then for every continuous function $f : [a,b] \longrightarrow \mathbb{R}$, (1) holds.

**Proof:** In (2), let $\mu(P) \to 0$. Then $\mu(h(P))$ also tends to 0 by Exercise (6.3). So by Corollary (2.3), the first term tends to $\int_a^b f(x)dx$. By the last theorem, the last term tends to $\int_c^d f(h(t))h'(t)dt.$ ∎

It is now possible to explain the true role of the factor $h'(t) = \dfrac{dx}{dt}$ in the integrand of $\int_c^d f(h(t))h'(t)dt$. The function $h$ maps the interval $[c,d]$ continuously onto $[a,b]$. We may think of the ratio of the lengths of these onto intervals, viz., the ratio $\dfrac{b-a}{d-c}$ as the (average) index of magnification of $h$. This is in anology with the index of magnification of a microscope, defined as the ratio of the image size to the object size. Here, of course, $b-a$ may be smaller than $d-c$, in which case it is perhaps incorrect to use the word 'magnification'. But we use it with the understanding that sometimes the index of magnification may be less than 1. If $h$ is a linear function of the form $h(t) = \lambda t + \mu$ then all portions of the domain interval $[c,d]$ get uniformly magnified by the same factor, viz. $\lambda$. But in general this need not be so. Let $t^* \in [c,d]$. Then for $\Delta t$ small, $\dfrac{\Delta x}{\Delta t}$ (where $\Delta x = h(t^* + \Delta t) - h(t^*)$) may be taken as the (approximate) index of magnification of $h$ at $t^*$. Its limit, viz. $h'(t^*)$ will be the precise index of magnification at $t^*$ just as the instantaneous speed is the limit of the average speed. In the proof of above, the $i^{th}$ subintervals of $P$ and $h(P)$ are of lengths $\Delta t_i$ and $\Delta x_i$ respectively. So $\dfrac{\Delta x_i}{\Delta t_i}$ is the average index of magnification of $h$ over the $i^{th}$ subinterval. By Lagrange's MVT, this equals the index of magnification of $h$ at some $\beta_i \in (t_{i-1}, t_i)$. If $\Delta t_i$ is very small then this may also be taken as $h'(\alpha_i)$. In a language which is no longer used in mathematics but which is still popular among those who apply mathematics, the transformation $x = h(t)$ maps the 'length element' $dt$ onto the length element $dx$. The rule of change of variables for double integrals is a direct analogue of the theorem above, except that a length element (or 'elemental length') is replaced by an area element.

It is instructive to compare Theorems (6.3) and (6.5). Both deal with the rule of substitution, or change of variables. But Theorem (6.3) is essentially a theorem about derivatives translated into a theorem about integrals through the Fundamental Theorem of Calculus, while Theorem (6.5) is a genuine result about integrals. As remarked earlier, the hypothesis $h' > 0$ is not so crucial. On the other hand, in Theorem (6.3) continuity of $f$ (and hence of the composite function $g = f \circ h$) is vitally needed to apply the Fundamental Theorem. In Theorem (6.5), it is not so crucial. We need only assume that $f$ and $g$ are Riemann integrable and the entire proof goes through because we are assuming $h'$ to be continuous anyway. (Integrability of the product $gh'$ follows from that of $g$ by Theorem (4.5). And in order to apply Theorem (6.4), continuity of one of the functions $g$ and $h'$ is sufficient.)

Thus, Theorem (6.5) is a little more general than Theorem (6.3), even though its proof is considerably longer. Note, by the way, that both the proofs use Lagrange's Mean Value Theorem. Theorem (6.5) uses it directly while Theorem (6.3) uses it indirectly because the proof of the Fundamental Theorem requires Lagrange's MVT. The proof of Theorem (6.3) appears short because it uses a powerful theorem which subsumes a lot of work needed. The price one often has to pay for such readymade devices is a little loss of generality. In Exercise (6.6), we shall indicate how Theorem (4.6.2) which is a result from differential calculus, can be derived quite easily from Theorem (3.2), a result about integrals, by using the Fundamental Theorem of Calculus. Once again the price tag is that an additional hypothesis has to be introduced in order to make the Fundamental Theorem applicable.

Whether such loss of generality can be considered as an inadequacy of the Fundamental Theorem is debatable, especially so because the additional hypothesis of continuity which the Fundamental Theorem necessitates is usually satisfied in applications any way. Perhaps it is better to look at it as the toll one has to pay for driving on an express highway.

## EXERCISES

6.1 Suppose $f : \mathbb{R} \longrightarrow \mathbb{R}$ is periodic with period $T(> 0)$ (see Exercise (4.8.8)). Prove that if $f$ is integrable over $[0, T]$, then it is integrable over every interval of length $T$ and its integral over any such interval equals $\displaystyle\int_0^T f(x)dx$.

6.2 In the last exercise if $f$ is continuous, give an easier proof using the Fundamental Theorem of Calculus. [*Hint* : For $x \in \mathbb{R}$, let $G(x) = \displaystyle\int_x^{x+T} f(t)dt$.]

6.3 Let $h : [c, d] \longrightarrow [a, b]$ be continuously differentiable and strictly monotonically increasing with $h(c) = a$ and $h(d) = b$. For a partition $P = (t_0, t_1, \ldots, t_n)$ of $[c, d]$, let $h(P)$ be the partition $(h(t_0), h(t_1), \ldots, h(t_n))$ of $[a, b]$. Prove that as $\mu(P) \longrightarrow 0, \mu(h(P)) \longrightarrow 0$. [*Hint* : Use Exercise (4.5.21).]

6.4 Prove an analogue of Theorem (6.5) when $h'(t) < 0$ for all $t \in (c, d)$.

6.5 Prove analogues of Theorem (6.4) where, instead of sums of the form $\displaystyle\sum_{i=1}^n u(\alpha_i)v(\beta_i)\Delta t_i$, we consider

(i)    sums of the form $\displaystyle\sum_{i=1}^{n}(u(\alpha_i)+v(\beta_i))\Delta t_i$

(ii)   sums of the form $\displaystyle\sum_{i=1}^{n}(u^2(\alpha_i)+v^2(\beta_i))\Delta t_i$

*(iii) sums of the form $\displaystyle\sum_{i=1}^{n}\sqrt{u^2(\alpha_i)+v^2(\beta_i)}\Delta t_i$. [*Hint* : Use uniform continuity of the square-root function (Exercise (4.2.9)).]

(Collectively, all such results are called **Bliss' theorem**. (iii) above will be needed when we consider lengths of curves in the next chapter.)

6.6   Give an alternate proof of Lemma (4.6.1) (and also of Theorem (4.6.2)) applicable in the case where each $f'_n$ is continuous on $[a,b]$. [*Hint* : Apply the Fundamental Theorem of Calculus and Theorem (3.2) to the integrals $\displaystyle\int_{x_0}^{x} f'_n(t)dt$ for $x \in [a,b]$.]

6.7   If $f$ is continuous on $[0,1]$, prove that

$$\int_0^{\pi/2} f(\sin 2x)\sin x\,dx = \sqrt{2}\int_0^{\pi/4} f(\cos 2x)\cos x\,dx.$$

6.8   Evaluate the following definite integrals:

(i) $\displaystyle\int_0^{\pi/2} \frac{x\sin x\cos x}{\cos^4 x + \sin^4 x}dx$   (ii) $\displaystyle\int_0^{\pi} \frac{x\,dx}{1 + a\sin x}$   $(|a| < 1)$

(iii) $\displaystyle\int_0^{\pi/4} \ln(1+\tan x)dx.$   (iv) $\displaystyle\int_0^1 \tan^{-1}(\frac{1}{1-x+x^2})dx.$

6.9   Evaluate $\displaystyle\int_0^1 (tx+1-x)^n dx$ where $n$ is a positive integer and $t$ is a parameter independent of $x$. Hence show that

$$\int_0^1 x^k(1-x)^{n-k}dx = \frac{1}{\binom{n}{k}(n+1)}$$

for $k = 0, 1, \ldots, n$. (For a more straightforward proof, see Exercise (6.16).)

6.10  Suppose we have a family of related functions. It happens sometimes that the integral of a typical member of this family cannot be evaluated directly, but can be easily expressed in terms of the integral of some other member of the family. Often the family is indexed by an integer parameter, say $n \geq 0$. A formula which expresses (or 'reduces') the integral of the $n^{\text{th}}$ function in terms of that of the $(n-1)^{\text{th}}$ (or lower indexed) function is called a **reduction formula**. The situation is similar to that of a recurrence relation for a sequence (see Q.3.1). The integrals of the functions corresponding to the lower values of a parameter $n$ (say $n = 0$ or $1$) can often be evaluated directly. Repeated application of the reduction formula then gives the value of the integral of the $n^{th}$ function for a general $n$. For the family of functions $\sin^n x, n = 1, 2, \ldots$ prove the reduction formula

$$\int_0^{\pi/2} \sin^n x\, dx = \frac{n-1}{n} \int_0^{\pi/2} \sin^{n-2} x\, dx, \quad \text{for } n \geq 2.$$

Calculate $\displaystyle\int_0^{\pi/2} \sin^0 x\, dx$ and $\displaystyle\int_0^{\pi/2} \sin x\, dx$ directly. Hence show that :

$$\int_0^{\pi/2} \sin^{2m} x\, dx = \frac{(2m)!}{[2^m m!]^2} \frac{\pi}{2}$$

and

$$\int_0^{\pi/2} \sin^{2m+1} x\, dx = \frac{[2^m m!]^2}{(2m+1)!}.$$

6.11 Letting $I_n = \displaystyle\int_0^{\pi/2} \sin^n x\, dx$, prove that for all $m \geq 0$

$$1 \leq \frac{I_{2m}}{I_{2m+1}} \leq 1 + \frac{1}{2m}.$$

[*Hint*: Compare the integrands of $I_{2m+1}, I_{2m}$ and $I_{2m-1}$ and use the reduction formula.]

6.12 Using the last two exercises prove that

$$\lim_{n \to \infty} \frac{2.2}{1.3} \cdot \frac{4.4}{3.5} \cdot \frac{6.6}{5.7} \cdots \frac{(2n)(2n)}{(2n-1)(2n+1)} = \frac{\pi}{2}.$$

(This result is called **Wallis formula**. It is not a very efficient way of calculating $\pi$ approximately, as compared with other formulas like $\displaystyle\sum_{n=1}^{\infty} \frac{1}{n^2} = \frac{\pi^2}{6}$. But it is useful in deriving another important result called Sterling approximation, to be studied later.)

6.13 Derive a reduction formula for $\displaystyle\int_0^{\pi/2} \cos^n \theta\, d\theta$ and using it prove that $\displaystyle\int_0^1 (1 - x^2)^n dx = \frac{2^{2n}(n!)^2}{(2n+1)!}$. Deduce that

$$1 - \frac{1}{3}\binom{n}{1} + \frac{1}{5}\binom{n}{2} - \frac{1}{7}\binom{n}{3} + \cdots + \frac{(-1)^n}{2n+1}\binom{n}{n} = \frac{2^n(n!)^2}{(2n+1)!}.$$

6.14 Let $I_n = \int_0^{\pi/2} x \sin^n x\, dx$. Prove that $I_n = \dfrac{n-1}{n} I_{n-2} + \dfrac{1}{n^2}$ for $n \geq 2$.

6.15 Prove that $\int_0^{\pi} \dfrac{1 - \cos mx}{1 - \cos x}\, dx = m\pi$ for $m = 0, 1, 2 \ldots$.

6.16 Let $I_{m,n} = \int_0^1 x^m (1-x)^n\, dx$ where $m, n$ are non-negative integers. Prove that for all $m, n$,

(i)   $I_{m,0} = \dfrac{1}{m+1}$

(ii)  $I_{m,n} = \dfrac{n}{m+1} I_{m+1,n-1}$ if $n > 0$

(iii) $I_{m,n} = \dfrac{1}{m+n+1} \dfrac{1}{\binom{m+n}{m}}$.

6.17 Using the last exercise (or Exercise (6.9)), show that the sum of the coefficients in the expansion of $(1-x)^n$ is $0$ if $n$ is odd and $\dfrac{2n+2}{n+2}$ if $n$ is even. (This is an interesting application of integrals to binomial coefficients. A direct proof is easy for odd $n$, but not so easy for even $n$.)

## 5.7   Why is $e = e = e$?

Of all the questions to be answered in this book, the present one may easily seem the silliest. But then maybe not. The very fact that the number $e$ appears in it three times and not just twice, indicates that there is something more in the question than meets the eye. And if we see what it is, then the question is far from silly, or trivial.

Let us first see what the three $e$'s stand for. More generally we could take any real number $x$ and define $e^x$ in three different ways. The question, in essence, asks why these definitions are equivalent. In fact, it is only after showing this that we would be justified in using the same notation for all the three. To avoid confusion, we shall use three symbols $e^x, \exp(x)$ and $E(x)$ to denote three possibly different things. The first two of these we have already defined, at least partially. The third one will be defined and it will be shown that they are all equal.

The most natural meaning to be given to the symbol $e^x$ is the number $e$ raised to the number $x$, or the $x^{\text{th}}$ power of $e$. For this, we must first define the number $e$. There are two standard definitions in terms of sequences and series. One is the limit of the sequence $(1 + \frac{1}{n})^n$, which can easily be shown to be monotonically increasing and bounded above (see Exercise (7.1)). We take this as the *definition* of $e$. We can now define $e^2$ as $e.e$, $e^3$ as $e.e.e$, and more generally $e^n$ as the product $e.e.\ldots.e$ taken $n$ times, when $n$ is a positive integer. We let $e^0 = 1$ and $e^n = \frac{1}{e^{-n}}$ when $n$ is a negative integer. This way $e^x$ is defined for all integer exponents $x$ and the usual laws of indices (e.g. $e^{x+y} = e^x e^y$) hold for integer exponents. We can then extend the definition of $e^x$ to the case where $x$ is rational in the manner indicated in Exercise (3.4.2). The laws of indices continue to hold for rational exponents. We mention here an important consequence of these laws, which we shall need later.

**7.1 Proposition:** Let $x, y$ be rational numbers with $x < y$. Then $e^x < e^y$.

**Proof:** Note first that since $e > 1, e^h > 1$ for all rational $h > 0$. Letting $h = y - x$, we get $e^y = e^{x+h} = e^x e^h > e^x$ since $e^x > 0$ for all rational $x$. ∎

Having defined $e$, defining $e^x$ for a rational $x$ was a relatively easy matter because when $x = p/q$ where $p, q$ are positive integers, $e^x$ has a sort of physical interpretation. It is that real number which, when multiplied by itself $q - 1$ times, gives the same number which results when $e$ is multiplied by itself $p - 1$ times. (Note that $q - 1$ and $p - 1$ are the actual numbers of multiplications respectively.) If $p < 0$, we replace $p$ by $-p$ and take reciprocals.

When $x$ is irrational, $e^x$ has no such nice interpretation in terms of repeated multiplications. Nevertheless the proposition above suggests that $e^x$ should, on one hand, be bigger than $e^y$ whenever $y$ is a rational less than $x$ and, at the same time, that $e^x$ should be smaller than $e^z$ whenever $z$ is a rational bigger than $x$. That such a real number exists is not obvious and requires the completeness of the real number system. We give here a proof here which is based on complements in the form of the least upper bound axiom (Axiom (2.6.2) along with Theorem (2.6.3)). An alternate treatment based on Cauchy form of completeness (Theorem (3.5.4)) is also possible and will be indicated in the exercises.

**7.2 Theorem:** Let $x$ be a real number. Let $A = \{e^y : y \leq x, \ y \text{ rational }\}$ and $B = \{e^z : x \leq z, \ z \text{ rational}\}$. Then $\sup A = \inf B$.

**Proof:** Because of the proposition above, $a \leq b$ whenever $a \in A$ and $b \in B$. So in view of Exercises (1.3) and (1.4), the proof will be complete if for every $\epsilon > 0$, we can find rationals $y, z$ with $y \leq x \leq z$ such that $e^z - e^y < \epsilon$. For this, we shall use the fact that the set of rationals is dense (Corollary (3.3.2)). First fix some rational number $z_0 > x$. Let $\epsilon > 0$ be given. By Theorem (3.4.4), $e^{1/n} \to 1$ as $n \to \infty$. So we can fix some (sufficiently large) positive integer $n$ such that $e^{1/n} - 1 < \dfrac{\epsilon}{e^{z_0}}$. Now, by Corollary (3.3.2), there exist rationals $y \in (x - \dfrac{1}{2n}, x)$ and $z \in (x, x + \dfrac{1}{2n})$. Then $y < x < z$ and $z - y < \dfrac{1}{n}$. Note that $y < z_0$. So

$$
\begin{aligned}
e^z - e^y &= e^y(e^{z-y} - 1) \\
&< e^{z_0}(e^{1/n} - 1) \quad \text{(by the proposition above)} \\
&< \epsilon.
\end{aligned}
$$

As noted before this completes the proof. ∎

For any real number $x$, we now define $e^x$ as the common value of $\sup A$ (where $A = \{e^y : y \text{ rational}, y \leq x\}$) and $\inf B$ (where $B = \{e^z : z \text{ rational } z \geq x\}$). Note that if $x$ itself is rational that $e^x$ is the only common element of these two sets and since $a \leq b$ for all $a \in A, b \in B$, this common element equals $\sup A$ as well as $\inf B$. In other words, for a rational $x$, the new defintion of $e^x$ coincides with the old one. We now have a definition of $e^x$ (read 'e raised to $x$') for every real $x$.

We can now extend the laws of indices to all real exponents, e.g. we can prove $e^{x_1 + x_2} = e^{x_1} e^{x_2}$ for all real $x_1, x_2$. We can also prove that the function $f(x) = e^x$ is continuous, in fact differentiable everywhere and $f'(x) = e^x$ for all $x$. But direct proofs of these results turn out to be somewhat awkward, because everytime we have to deal with $e^x$ for $x$ irrational, we have to approximate it by $e^y$ for a rational $y$ and use properties of the latter. The other two approaches to the exponential function, viz. $\exp(x)$ and $E(x)$ are more convenient in that the arguments work equally well for rational as well as irrational $x$. The advantage of the definition of $e^x$ as given above, on the other hand, is that it conforms to the notation.

We first study $\exp(x)$. We define this to be the value of the power series $\sum\limits_{n=0}^{\infty} \dfrac{x^n}{n!}$. In Exercises (2.7.2) and (3.3.14) we already considered this series but denoted its sum by $e^x$. Since we are using $e^x$ for something else, we shall use $\exp(x)$ to denote the sum of this series until we prove that it also coincides with $e^x$ as defined above. Exercise (2.7.2) in fact attempts to give a proof of this. But that argument lacks rigour completely. We shall now give an argument which is based on various properties of the $\exp(x)$ function which were derived from the hard work done on power series. In particular we shall use:

| | | |
|---|---|---|
| (i) | $\exp(x + y) = \exp(x)\exp(y)$ for all real $x, y$ | (1) |
| (ii) | $\exp(x)$ is a continuous function of $x$ for all $x$ | (2) |
| (iii) | $\exp(x)$ is strictly increasing and postive for all $x$ | (3) |

These are consequences of Exercise (3.10.3), Corollary (4.1.6) and Exercise

(4.1.21) respectively. (An equation like (1), which relates the values assumed by a function at related points is called a **functional equation** .

The proof that $e^x = \exp(x)$ begins by considering the case $x = 0$, where both the sides are equal to 1. The next case, viz. $x = 1$ deserves to be singled out as it is also often taken as the definition of the number $e$.

**7.3 Theorem:** The number $e$, defined as $\lim\limits_{n\to\infty} (1 + \frac{1}{n})^n$ also equals $\sum_{n=0}^{\infty} \frac{1}{n!}$, i.e. $1 + 1 + \frac{1}{2!} + \frac{1}{3!} + \frac{1}{4!} + \ldots + \frac{1}{n!} + \ldots$. In other words $\exp(1) = e = e^1$.

**Proof:** Let $S_n = \sum\limits_{k=0}^{n} \frac{1}{k!}$. $S_n$ is nothing but the $n^{\text{th}}$ partial sum of the series for

$\exp(1)$. We have to show $S_n \to e$ as $n \to \infty$. Let $a_n = (1 + \frac{1}{n})^n$. Expanding by the binomial theorem as in Exercise (2.7.1), we have, for every $n \geq 1$,

$$
\begin{aligned}
a_n &= 1 + n\frac{1}{n} + \frac{(1 - \frac{1}{n})}{2!} + \frac{(1 - \frac{1}{n})(1 - \frac{2}{n})}{3!} + \ldots + \frac{(1 - \frac{1}{n})\ldots(1 - \frac{n-1}{n})}{n!} \\
&\leq 1 + 1 + \frac{1}{2!} + \ldots + \frac{1}{n!} \\
&= S_n.
\end{aligned}
$$

Taking limits as $n \to \infty$ we get $e \leq \exp(1)$.

For the other way inequality, fix a positive integer $m$. For $n \geq m$ we expand $\left(1 + \frac{1}{n}\right)^n$ again by the binomial theorem. But we retain only the terms upto $\binom{n}{m} \frac{1}{n^m}$. Since the terms dropped are all positive, we get

$$a_n \geq 1 + n\frac{1}{n} + \frac{(1 - \frac{1}{n})}{2!} + \frac{(1 - \frac{1}{n})(1 - \frac{2}{n})}{3!} + \ldots + \frac{(1 - \frac{1}{n})(1 - \frac{2}{n})\cdots(1 - \frac{m-1}{n})}{m!}.$$

We now fix $m$ and let $n \to \infty$. Since there are only finitely many terms on the right of the inequality, we can take term-by-term limit. (It is precisely at this point that the naive argument in Exercise (2.7.1) breaks down.) We thus get,

$$e \geq S_m \quad \text{for all} \quad m.$$

Now we let $m \to \infty$ and get $e \geq \exp(1)$. This completes the proof. ∎

We are now in a position to establish that $e^x = \exp(x)$ for all $x$. The plan is to start with the cases $x = 0$ and 1, which are already proved and then go on to extend the equality of the two functions to the cases where $x$ is a positive integer, an integer, a rational and finally any real number.

**7.4 Theorem:** For every real number $x, e^x = \exp(x)$.

**Proof:** First assume $x$ is a positive integer, say $n$. We apply induction on $n$. The last theorem is precisely the case $n = 1$. Suppose $e^k = \exp(k)$ for a positive integer $k$. Then by (1) above, $\exp(k+1) = \exp(k)\exp(1) = e^k e^1$ by induction hypothesis. So $\exp(k+1) = e^{k+1}$, completing the inductive step. If $x = 0$, $e^0$ and $\exp(0)$ are equal each being equal to 1. Now suppose $n$ is a negative integer. Then again by (1) above

$$\exp(n)(\exp(-n)) = \exp(0) = 1$$

Also $e^n e^{-n} = e^0 = 1$. Since $e^{-n} = \exp(-n)$, $-n$ being a positive integer, we must have $\exp(n) = \dfrac{1}{\exp(-n)} = \dfrac{1}{e^{-n}} = e^n$.

Next, assume $x = p/q$ where $p, q$ are positive integers. Then, by definition, $e^x$ is the $q^{th}$ root of $e^p$. But by (1), applied repeatedly,

$$
\begin{aligned}
(\exp(\frac{p}{q}))^q &= \exp(p/q)\exp(p/q)\ldots\exp(p/q)(q \text{ times}) \\
&= \exp(p/q + p/q + \ldots + p/q) = \exp(p) = e^p \text{ as proved earlier.}
\end{aligned}
$$

Thus $\exp(p/q)$ is also a positive $q^{th}$ root of $e^p$. By uniqueness of roots (cf. Theorem (3.4.1)), we get that $e^{p/q}$ and $\exp(p/q)$ are equal. Thus the assertion of the theorem holds if $x$ is a positive rational. If $x$ is a negative rational, we replace $x$ with $-x$ and take reciprocals, just as we did for the case where $n$ is negative integer.

Finally, suppose $x$ is an irrational number. Then, by definition, $e^x = \sup A = \inf B$ where $A = \{e^y : y \le x, \ y \text{ rational}\}$ and $B = \{e^z : z \ge x, \ z \text{ rational}\}$. (Since $x$ is irrational, we could as well write $y < x$ and $z > x$ instead of $y \le x$ and $z \ge x$ respectively.) By (3) above, $\exp(x)$ is an upper bound of the set $A$ and hence $e^x \le \exp(x)$. But by (3), $\exp(x)$ is also a lower bound of the set $B$ and so $\exp(x) \le e^x$. Thus $\exp(x) = e^x$. ∎

From now onwards, we are fully justified in using $e^x$ and $\exp(x)$ interchangeably. We now turn to the third interpretation of $e^x$, which will be denoted by $E(x)$ for the time being, until we have proved its equality with $e^x$. In Exercise (4.1.21) we defined the natural logarithm as the inverse of the exponential function. In other words, for every positive real number $x$, its natural logarithm, $\ln x$ is the unique real number $y$ for which $e^y = x$. In Exercise (4.1.22), some elementary properties of $\ln$ were derived from the corresponding properties of the exponential function. In Exercise (4.6.23), it was shown that $\dfrac{d}{dx}(\ln x) = \dfrac{1}{x}$ for all $x > 0$. So, by the first form of the Fundamental Theorem of Calculus (Theorem (5.5)) and the fact that $\ln 1 = 0$, we get, for all $x > 0$,

$$\int_1^x \frac{dt}{t} = \ln x \qquad (4)$$

If one wants, this entire procedure can be completely reversed. Instead of taking (4) as a theorem, it can be taken as a *definition* of $\ln x$ for $x > 0$. Properties of the natural logarithm can then be derived using results about definite integrals. The function $E(x)$ is then defined as the inverse function of the natural logarithm. That is, for a real number $x, E(x)$ is that unique real number whose natural logarithm is $x$. The properties of $E(x)$ can now be derived from those of the logarithm function. Finally, the derivatives of $E(x)$ are obtained and from these, one gets the power series expansion of $E(x)$ as $\sum_{n=0}^{\infty} \frac{x^n}{n!}$, valid for all real $x$, thereby establishing that $E(x) = \exp(x) = e^x$.

We have encountered something similar before. The trigonometric functions $\sin x$ and $\cos x$ can be defined geometrically and then their derivatives, and subsequently, their power series expansions can be obtained from geometric intuition. Or one can define $\sin x$ and $\cos x$ as certain power series and then define all geometric terms such as angles in a strictly analytical manner (see Sections 4.7 and 4.8). The first approach is easier to grasp but naive while the second one is mathematically perfect. This disparity does not arise in the present situation. Here both the approaches (viz., first defining the exponential function and then the logarithm as its inverse or first defining logarithm and then the exponential as its inverse) are equally rigorous and so from a mathematical point of view, it is purely a matter of taste as to which of them one prefers. Nevertheless, the two can be compared from a pedagogical point of view. The first approach, if attempted rigorously, requires considerable spadework on infinite series and uniform convergence. In the good old days when such degree of rigour was not insisted upon, this approach was popular. The approach in which logarithms are defined as certain definite integrals, also, of course, requires some spadework on definite integrals. But in a calculus course, integrals have to be studied anyway. On a comparative scale, the infinite series are not so indispensable, at least in an elementary calculus course. So the trend now-a-days is to define the logarithms first and exponentials later. We shall now present this approach somewhat tersely.

So we begin by taking (4) as the very definition of the natural logarithm, $\ln x$, of $x$. (Incidentally, the peculiar notation ln comes from the first letters of the words in the French expression '*logarithm naturelle*' used by their originator Napier. This, of course, does not explain what is 'natural' about $\ln x$. We shall a little later define logarithms with respect to various bases. A plausible explanation will then be given.)

The following theorem establishes the basic properties of logarithm. Note that we have to prove them from the properties of definite integrals.

**7.5 Theorem:** $\ln : (0, \infty) \to \mathbb{R}$ is a strictly increasing, differentiable function with $\frac{d}{dx}(\ln x) = x$ for all $x > 0$. For any two $x, y > 0$,

$$\ln(xy) = \ln x + \ln y \qquad (5)$$

There exists a function $E : \mathbb{R} \to (0, \infty)$ which is inverse to ln, i.e. $E(\ln x) = x$ for all $x \in (0, \infty)$ and $\ln E(x) = x$ for all $x \in \mathbb{R}$.

**Proof:** The differentiability of $\ln x$ and the assertion that $\frac{d}{dx}(\ln x) = x$ for all

$x \in (0, \infty)$ follow from the second form of the Fundamental Theorem of Calculus (Theorem (5.5)). Since $\frac{1}{x} > 0$ for all $x \in (0, \infty)$, by Theorem (4.5.5) it follows that $\ln x$ increases strictly on $(0, \infty)$.

To prove (5) let $x > 0, y > 0$. Then by Exercise (1.19),

$$\ln(xy) = \int_1^{xy} \frac{dt}{t} = \int_1^x \frac{dt}{t} + \int_x^{xy} \frac{dt}{t} \tag{6}$$

By definition, the first integral on the right is $\ln x$. The substitution $u = t/x$ reduces the second integral to $\int_1^y \frac{du}{u}$, which equals $\ln y$. So (5) holds.

For the last assertion, since $\ln$ is strictly increasing, it is a one-to-one function. So the proof will be complete if we can show that its range is $(-\infty, \infty)$. For this, we show that $\ln x \to \infty$ as $x \to \infty$ and $\ln x \to -\infty$ as $x \to 0^+$. For the first limit, take any number, say 2, bigger than 1. Then $\ln 2$ is a fixed positive number. By repeated applications of (5), $\ln(2^n) = \ln(2.2.....2)(n \text{ times}) = n \ln 2$. Given $R > 0$, there exists $n$ such that $n \ln 2 > R$. (It is here that we need $2 > 1$.) Then for all $x \geq 2^n$, by monotonicity of $\ln$, $\ln(x) \geq \ln(2^n) > R$. So $\ln x \to \infty$ as $x \to \infty$. Finally, as $x \to 0^+$, $\frac{1}{x} \to \infty$. By (5) $\ln x = -\ln\frac{1}{x}$ (since $\ln x + \ln\frac{1}{x} = \ln 1 = 0$). So, by what we have proved, $\ln x \to -\infty$ as $x \to 0^+$. As noted before, this completes the proof. ∎

We are now in a position to define the function $E(x)$ as the inverse function of $\ln$. We do so and derive its basic properties in the following theorem.

**7.6 Theorem:** The function $E : \mathbb{R} \to \mathbb{R}$, defined as the inverse function of $\ln$, is strictly increasing with range $(0, \infty)$. For all $a, b \in \mathbb{R}$, we have

$$E(a + b) = E(a)E(b) \tag{7}$$

The function $E$ is differentiable everywhere with

$$E'(x) = E(x) \text{ for all } x \in \mathbb{R}. \tag{8}$$

**Proof:** The first assertion follows directly from the last theorem. As for (7), let $x = E(a), y = E(b)$. Then $x, y$ are positive and, by definition, $\ln x = a, \ln y = b$. So by (5), $\ln(xy) = a + b$ which again translates as $E(a + b) = xy$. So (7) holds.

For the last assertion we apply Exercise (4.4.8). If $y = E(x)$ then $x = \ln y$. By the theorem above, $\frac{dx}{dy} = \frac{1}{y}$. So $\frac{dy}{dx} = \frac{1}{dx/dy} = y = E(x)$. Thus $E'(x) = E(x)$ for all $x$. ∎

Note the ease with which (8) is derived, starting from (4). For the same result about $\exp(x)$, or $e^x$, we first had to establish the validity of term-by-term differentiation of a power series (see (14) in Section 4.6), which entailed quite some work. It may be argued that the derivation above required the Fundamental Theorem of Calculus which also entails some work. This is true. But the Fundamental Theorem has to be studied anyway. The journey thereafter is not very long and since the equation (8) above is, by far, the most important property of the exponential function from the point of view of applications, a definition which leads to it with minimum efforts, without sacrificing rigour is preferred in modern courses in elementary calculus.

To establish the equality of $E(x)$ and $\exp(x)$ for all real $x$, we proceed as in Theorem (7.4), where the equality of $e^x$ with $\exp(x)$ was proved step-by-step starting with the equality for the special cases $x = 0$ and $x = 1$. Since $\ln 1 = 0, E(0) = 1 = \exp(0)$. For $x = 1$, we have the following result, analogous to Theorem (7.3).

**7.7 Theorem:** The number $e$ (defined as $\lim\limits_{n\to\infty} \left(1 + \dfrac{1}{n}\right)^n$) equals $E(1)$. Hence $\exp(1) = E(1)$.

**Proof:** The assertion amounts to showing that $\ln e = 1$. Since the ln function is continuous, we have by Theorem (4.1.2), $\ln e = \ln\left(\lim\limits_{n\to\infty}\left(1 + \dfrac{1}{n}\right)^n\right)$

$= \lim\limits_{n\to\infty} \ln\left[\left(1 + \dfrac{1}{n}\right)^n\right]$. By repeated application of (5), $\ln\left[\left(1 + \dfrac{1}{n}\right)^n\right]$ equals

$n \ln\left(1 + \dfrac{1}{n}\right)$, which we rewrite as $\dfrac{\ln\left(1 + \frac{1}{n}\right)}{1/n}$. Now, by differentiability of the ln

function, $\lim\limits_{h\to 0} \dfrac{\ln(1 + h) - \ln(1)}{h} = \dfrac{d}{dx}(\ln x)$ at $x = 1$, which equals $\dfrac{1}{1} = 1$. Since

$\ln(1) = 0$ and $\dfrac{1}{n} \to 0^+$ as $n \to \infty$, we get $\lim\limits_{n\to\infty} \dfrac{\ln(1 + \frac{1}{n})}{1/n} = 1$. Thus $\ln e = 1$, as

was to be shown. The second assertion follows from Theorem (7.3). ∎

This theorem and Theorem (7.3) together give a complete answer to the question in the title. The number $e$ can now be defined in three different ways,

viz., as $\lim\limits_{n\to\infty} \left(1 + \dfrac{1}{n}\right)^n$ or as $\sum\limits_{n=0}^{\infty} \dfrac{1}{n!}$ or as the unique real number whose natural

logarithm is 1. We can go further and show the equality of $e^x, \exp(x)$ and $E(x)$ for all $x$.

**7.8 Theorem:** For all real $x, e^x = \exp(x) = E(x)$.

**Proof:** The equality of the first two was proved in Theorem (7.4). Following essentially the same argument as in its proof, starting from Theorem (7.7) instead of Theorem (7.3) and using (1) and (7), it follows that $E(x) = \exp(x)$ for all rational values of $x$. To show that equality holds even when $x$ is irrational, we can argue as in the proof of Theorem (7.4) and show that $E(x)$ is simultaneously the supremum of the set $\{E(y) : y \text{ rational}, y \leq x\}$ as well as the infimum of the set $\{E(z) : z \text{ rational}, z \geq x\}$. All we need is that $E$ is a strictly increasing function. Alternatively we can use continuity of both the functions $E$ and $\exp$ and apply Exercise (4.1.4).

As a still more sophisticated proof, we see from (8) that the function $E$ is infinitely differentiable and $E^{(n)}(x) = E(x)$ for all $x \in \mathbb{R}$ and for all $n \in \mathbb{N}$. We can then use Exercise (4.6.18) to get the Maclaurin series of $E$. Since $E^{(n)}(0) = 1$ for all $n$, this series comes out to be $\sum_{n=0}^{\infty} \frac{1}{n!} x^n$, which is precisely $\exp(x)$. ∎

The fact that the exponential and the natural logarithm functions are inverse to each other can be expressed by the equations $e^{\ln a} = a$ for all $a > 0$ and $\ln(e^b) = b$ for all $b \in \mathbb{R}$. The first equation motivates the definition of $a^x$ for any positive real number $a$. Since $a = e^{\ln a}$, $a^x$ must equal $(e^{\ln a})^x$. If the laws of indices are to hold, then this expression must equal $e^{(\ln a)x}$ or $e^{x \ln a}$. Since $e^y$ has been defined for any real exponent $y$, we *define* $a^x$ as $e^{x \ln a}$. Note that if $a = e$, then $\ln a = 1$ and so $e^{x \ln a}$ coincides with $e^x$. We had already done this in Exercise (4.1.23). But that time the only available defintion of $e^x$, applicable for all real $x$, was as $\exp(x)$. Now that we can also define $e^x$ as $E(x)$, we can establish some of the properties of the function $f(x) = a^x$ a little more easily. A few such properties will be given as exercises.

Having defined arbitary powers, we can define logarithms w.r.t. any positive base. Let $a > 0$ be fixed. If $y = a^x$ we say $x$ is the **logarithm** of $y$ w.r.t. the **base** $a$ and write $x = \log_a y$. The familiar algebraic properties of logarithms (including the rule for change of base) can be established from the properties of $\ln x$ and will be given as exercises. The natural logarithm $\ln x$ is the same as $\log_e x$, i.e., the logarithm with base $e$. Instead of the number $e$ (which is not even rational, by Exercise (2.7.5)), it would appear more natural to use the term 'natural logarithm' for logarithms w.r.t. some familiar base such as 10 or 2. These bases are indeed more convenient when we are dealing with numbers written in the decimal and binary systems respectively. Logarithms w.r.t. the base 10 are sometimes called **common logarithms** and denoted by simply $\log x$ instead of $\log_{10} x$. The reader is cautioned, however, that many old books on calculus use $\log x$ to denote $\log_e x$ which is nowadays denoted generally by $\ln x$.

If $a = 1$, then $\log_a x$ makes no sense. If $0 < a < 1$, then the function $f(x) = \log_a x$ decreases as $x$ increases. This is sometimes inconvenient. So, generally while studying logarithms, the base is tacitly assumed to be bigger than 1. For every $a > 1$, we can define $E_a(x)$ as the inverse function of the function $\log_a x$, $E_e$ is precisely $E$. It is clear that $E_a(x) = a^x$. We leave it as an exercise to show that $E_a'(x) = (\ln a)E_a(x)$. So $E_a$ has the magic property that it equals its own derivative if and only if $a = e$. This is probably the reason why $e$ is considered the natural base for taking logarithms.

## EXERCISES

**7.1** Prove that the sequence $a_n = \left(1 + \dfrac{1}{n}\right)^n$ is monotonically increasing and bounded.

**7.2** Assume $e^x$ has been defined for all rational values of $x$. Suppose $\{x_n\}$ is a Cauchy sequence of rational numbers. Prove that $\{e^{x_n}\}$ is a Cauchy sequence (of real numbers) and hence a convergent sequence (in $I\!\!R$). Prove further that if $\{x_n\}$ and $\{y_n\}$ are sequences of rationals converging to the same limit (which may be irrational) then the sequences $\{e^{x_n}\}$ and $\{e^{y_n}\}$ converge to the same limit (in $I\!\!R$).

7.3 Using the last exercise, give a definition of $e^x$ for $x$ real, which extends the definition of $e^x$ for $x$ rational (see the comments at the end of Section 3.5.).

7.4 Suppose $f : I\!\!R \to I\!\!R$ is continuous and satisfies the condition that for all $x, y \in I\!\!R$, $f(x + y) = f(x) + f(y)$. (A relationship of this type among the values assumed by a function at mutually related points is often called a **functional equation**.) Prove that there is some $c \in I\!\!R$ such that $f(x) = cx$ for all $x \in I\!\!R$. [*Hint* : Begin by showing that $f(0) = 0$. Guess $c$ if at all it exists. Then imitate the proof of Theorem (7.8).]

7.5 In the last exercise show, in fact, that it suffices to assume continuity of $f$ at any one point.

7.6 Suppose $f : (0, \infty) \to I\!\!R$ is continuous at 1 and satisfies the functional equation $f(xy) = f(x) + f(y)$ for all $x, y \in (0, \infty)$ Prove that there is some $c$ such that $f(x) = c \ln x$ for all $x \in (0, \infty)$.

7.7 Let $a > 0$. Prove the laws of indices for $a^x$. Also prove that $\dfrac{d}{dx}(a^x)$
$= (\ln a)a^x$ for all $x \in I\!\!R$.

7.8 Let $r$ be any real number. Prove that for $x > 0, \dfrac{d}{dx}(x^r) = rx^{r-1}$.

7.9 Show that the series $\displaystyle\sum_{n=1}^{\infty} \dfrac{1}{n^r}$ is convergent if $r > 1$ and divergent if $r \le 1$.

(For rational $r$, this was already proved in Theorem (3.6.3) and Exercise (3.6.2). This and the last exercise are illustrations of how most of the results proved earlier for rational exponents continue to hold for real exponents too.)

7.10 Suppose $f : I\!\!R \to (0, \infty)$ is differentiable everywhere and there is some constant $k$ such that $f'(x) = kf(x)$ for all $x \in I\!\!R$. Prove that there is some $a > 0$ and some $b > 0$ such that $f(x) = ba^x$ for all $x$.

7.11 Suppose $f : I\!\!R \to I\!\!R$ is non-negative, differentiable at 0 and satisfies the functional equation $f(x + y) = f(x)f(y)$ for all $x, y \in I\!\!R$. Prove that either $f$ is identically zero or else there is some $a$ such that $f(x) = a^x$ for all $x$.

7.12 Prove that for all $|x| < 1$,

$$\ln(1 + x) = x - \frac{x^2}{2} + \frac{x^3}{3} - \frac{x^4}{4} + \ldots + \frac{(-1)^{n+1}x^n}{n} + \ldots$$

(By Leibnitz test the series converges even for $x = 1$ and it is tempting to conclude that $\sum_{n=1}^{\infty} \frac{(-1)^{n+1}}{n} = \ln 2$. This is indeed true, but requires some argument. See Exercise (7.20) below.)

7.13 Using the last exercise obtain the value of $\ln(1.04)$ correct to five places of decimals.

7.14 Let $a > 0, a \neq 1$. Prove that

(i)   $\log_a(xy) = \log_a(x) + \log_a(y)$ for all $x > 0, y > 0$

(ii)  $\log_a(x^y) = y \log_a(x)$ for all $x > 0, y \in \mathbb{R}$.

(iii) $\log_x(y) = \dfrac{\log_a(y)}{\log_a(x)}$ for all $x > 0, y > 0$. (This is the change of base rule.)

7.15 Prove that for all real $x$, $e^x = \lim_{n \to \infty} (1 + \frac{x}{n})^n$.

7.16 Let $f(x) = \ln|x|$ for $x \neq 0$. Prove that $f$ is differentiable at every $x \neq 0$ and $f'(x) = \frac{1}{x}$. (So $\log|x|$ is an antiderivative of $\frac{1}{x}$. In evaluating definite integrals, this is useful only if the function whose logarithm is involved never vanishes in the interval of integration. It is wrong, for example, to say that $\int_{-2}^{2} \frac{dx}{x} = \ln|x|\Big|_{-2}^{2} = 0$.)

7.17 Show that the function $x^{1/x}, (x > 0)$ attains its maximum at $e$. Determine which of the numbers $3^\pi$ and $\pi^3$ is bigger.

7.18 Evaluate the limit in Exercise (2.6) (iv).

7.19 Evaluate (i) $\lim_{n \to \infty} \left( \frac{1}{n} + \frac{1}{n+1} + \ldots + \frac{1}{2n} \right)$ (ii) $\lim_{n \to \infty} \int_0^1 \frac{n x^{n-1}}{1+x} dx$.

7.20 Using (i) of the last exercise, show that $\sum_{n=1}^{\infty} \frac{(-1)^{n+1}}{n} = \ln 2$. [*Hint* : By induction on $n$, or otherwise, prove that $\frac{1}{n+1} + \frac{1}{n+2} + \ldots + \frac{1}{2n}$ equals the partial sum $\sum_{k=1}^{2n} \frac{(-1)^{k+1}}{k}$.]

7.21 Let $m$ and $x_1, x_2, \ldots, x_n$ be positive real numbers. Prove that if $0 < m < 1$, then $n^{1-m}(x_1 + x_2 + \ldots + x_n)^m \geq x_1^m + x_2^m + \ldots + x_n^m$ while if $m > 1$ then $(x_1 + x_2 + \ldots + x_n)^m \leq n^{m-1}(x_1^m + x_2^m + \ldots + x_n^m)$. What happens if $m = 1$?

7.22 Let $\alpha_1, \alpha_2, \ldots, \alpha_n$ be non-negative real numbers such that $\alpha_1 + \alpha_2 + \ldots + \alpha_n = 1$ and $x_1, x_2, \ldots, x_n$ be any positive real numbers. Prove that

$$x_1^{\alpha_1} x_2^{\alpha_2} \ldots x_n^{\alpha_n} \leq \alpha_1 x_1 + \alpha_2 x_2 + \ldots + \alpha_n x_n$$

with equality holding if and only if all $x_i$'s are equal. Deduce the A.M.-G.M. inequality from this. (For this reason, this result is called the **generalised A.M.-G.M. inequality**. It also implies some other useful inequalities. For example, by taking $n = 2$ and $\alpha_1 = \frac{1}{p}, \alpha_2 = \frac{1}{q}$ where $p, q$ are positive real numbers with the property that their reciprocals add up to 1, we get **Young's inequality** which states that for such $p, q$ and for any positive $x, y$, $xy \leq \dfrac{x^p}{p} + \dfrac{y^q}{q}$.)

7.23 If $f$ is integrable over an interval $[0, 2a]$ and symmetric about its midpoint $a$ (i.e. $f(a + u) = f(a - u)$ for all $u \in [0, a]$) then show that $\int_0^{2a} f(x)dx = 2 \int_0^a f(x)dx$. From this fact and Exercise (4.11.25) deduce that $\int_0^\pi e^{-R \sin \theta} d\theta \leq \dfrac{\pi}{R}$ for all $R > 0$. (This is called the **Jordan inequality** and is useful in estimation of certain integrals in complex analysis.)

## 5.8    The Logarithmic and Exponential Growths

Besides the rational or 'algebraic' functions, the most commonly occurring functions in mathematics are the trigonometric, the inverse trigonometric, the exponential and the logarithm functions. Together they (and other functions arising from them by taking various composites) are called **elementary functions**. The trigonometric and inverse trigonometric functions have obvious physical applications. It is not immediately clear from the definitions what practical applications the logarithm and the exponential functions have and hence the definitions are likely to appear somewhat arbitrary. Why for example, define $\ln x$ as the integral of the function $\frac{1}{x}$? Why not some other function? Similarly, what is so great about the power series $\sum_{n=0}^\infty \frac{x^n}{n!}$, whose sum defines $e^x$? Why not some power series like $\sum_{n=0}^\infty \frac{\sin n}{n!} x^n$?

Probably the most ubiquitous elementary application of logarithms is in computations. It is much easier to add than to multiply. Suppose we want to multiply two real numbers $x$ and $y$. Without loss of generality we assume them to be positive. Fix any base $a > 1$. (Usually $a$ is taken as 10). Then

$$\log_a(xy) = \log_a x + \log_a y$$

From readymade tables for (approximate) values of $\log_a$, we find $\log_a x$, $\log_a y$ and add to get, say $z$. Now we look at the antilogarithm tables. (Or we can scan through the $\log_a$ table itself since $\log_a$ is a strictly increasing function.) We then find $a^z$, which is (approximately) $xy$. Instead of tables, one can take a scale which is marked logarithmically. That is, the point marked $x$ is actually at a distance $\log_a x$ from some reference point. One can then slide pointers along this scale to perform ordinary addition. Such 'slide rules' used to be an indispensable tool for an engineering student till a generation ago. The cheap, efficient and handy electronic calculators have made such devices obsolete today. But in the past, logarithms have saved thousands of man-hours of labourious computation.

There are, however, other applications of logarithms which are not out-dated by calculators. Probably the most important among these is that logarithms provide a new scale of measuring the growth of a function. It was shown in theorem (7.5) that $\ln x \to \infty$ as $x \to \infty$. However, by L'Hôpital's rule (in the form of theorem (4.9.7), or rather its extension in Exercise (4.9.10)) it follows that $\dfrac{\ln x}{x} \to 0$ as $x \to \infty$. Essentially the same argument shows that for every positive real number $r$, no matter how small, the ratio $\dfrac{\ln x}{x^r} \to 0$ as $x \to \infty$. In other words $\ln x$ grows negligibly slowly as compared with $x^r$. This fact is expressed by saying that the **logarithmic growth** is **slower than algebraic growth**. More generally we make the following definition.

**8.1 Definition:** Suppose $f, g$ are functions defined in a deleted neighbourhood of a point $c$ (which could be $\infty$ or $-\infty$ as well). Assume $f(x) \to \infty$ and $g(x) \to \infty$ as $x \to c$. Then we say $f(x)$ **tends to $\infty$ (or grows) slower** than $g(x)$ and write $f(x) = o(g(x))$ (read as "$f(x)$ equals **small oh** of $g(x)$") if $\lim\limits_{x \to c} \dfrac{f(x)}{g(x)} = 0$. We say $f(x)$ tends to $\infty$ (or **grows**) **no faster** than $g(x)$ if $\dfrac{f(x)}{g(x)}$ is bounded in some (deleted) neighbourhood of $c$, and write $f(x) = O(g(x))$ (read "$f(x)$ equals **big oh** of $g(x)$").

In this terminology, we can write $\ln x = o(x^r)$ for every $r > 0$. Note that if $0 < r < s$, then $x^r$ grows slower than $x^s$ as $x \to \infty$. To say $f(x) = o(g(x))$ is in general much stronger than to say $f(x) = O(g(x))$. If $f(x) = (2 + \sin x)x$ and $g(x) = x$ then $f(x) = O(g(x))$, but $f(x) \neq o(g(x))$. If $f(x) = O(g(x))$ and $g(x) = O(f(x))$ are both true then we say $f(x)$ and $g(x)$ grow at **comparable rates**. This, for example, is the case when $\dfrac{f(x)}{g(x)}$ tends to a finite non-zero limit as $x \to c$.

It should be noted that the $o$ or $O$ notations are merely convenient short forms for certain statements. By themselves they do not mean anything definite. When we write $(a+b)(a-b) = a^2 - b^2$, both sides have independent meanings and we are stating that they are equal. But when we say $f(x) = O(g(x))$, we are merely making a statement about the behaviour of $f(x)/g(x)$. We are not saying that $f(x)$ (which has a definite meaning) equals $O(g(x))$ (which has no definite meaning) just as when we say 'mother-in-law' we do not mean a mother in law! The reason to emphasize this point is that if we indiscriminately treat the symbols $o(g(x))$ and $O(g(x))$ as algebraic expressions, we easily run into absurdities. For example, $x^2 = O(x^2)$ and $2x^2 = O(x^2)$ and so we would get $x^2 = 2x^2$.

The fact that the logarithmic growth is very slow is also borne out by the graph of the function $y = \ln x$ (see Figure 5.8.3). Its slope at $(x, \ln x)$ is $\frac{1}{x}$. So, for large $x$ the graph is almost flat. There are, of course, functions which grow even more slowly than $\ln x$. We can, for example, take $\ln(\ln x)$ which is defined for all $x > 1$. Continuing in this manner, we get a whole spectrum of functions which tend to $\infty$ more and more slowly. The slower a function grows, the faster grows its inverse function. This is obvious from the fact that the derivative of the inverse function is the reciprocal of the derivative of the original function. So if $\ln x$ grows very slowly then we expect $e^x$ to grow very rapidly. This indeed turns out to be the case. From Exercise (4.6.22), we know that $\frac{p(x)}{e^x} \to 0$ as $x \to \infty$ where $p(x)$ is any polynomial in $x$. This fact is expressed by saying that the **exponential growth is stronger than algebraic growth.** There are, of course, functions which grow even more rapidly than $e^x$. For example $e^{x^2}, e^{x^3}$ etc. (Note that an expression like $a^{b^c}$ is always to be interpreted as $a^{(b^c)}$ and never as $(a^b)^c$. By laws of indices, the latter equals $a^{bc}$ which is, in general, different from $a^{(b^c)}$.) If we let the base of the power also increase, we get still higher growth. For example, $e^{x^2} = o(x^{x^2})$ as $x \to \infty$. While comparing two powers it is best first to express them with a common base, usually $e$. For example, to compare $x^{x^2}$ and $(x^2)^x$ we write them, respectively, as $e^{x^2 \ln x}$ and $e^{2x \ln x}$ and see that the latter grows slower than the former. A few comparisons of various pairs of functions will be given as exercises.

We now have an infinite variety of functions growing at different rates. Our measurement of rate of growth is, of course, comparative and not absolute. The whole idea is to find, given some function $f(x)$ which tends to $\infty$, some familiar function $g(x)$ whose rate of growth is comparable to that of $f(x)$. (It is somewhat like having an impressive collection of musical instruments, tuned to different frequencies, and then, for a given singer, to try to find the one which matches his pitch.) The branch of mathematics which deals with such problems is known as **asymptotic estimation.** The peculiar name comes from the asymptotes of a hyperbola. Suppose the equation of a hyperbola in the standard form is $\frac{x^2}{a^2} - \frac{y^2}{b^2} = 1$. Then its asymptotes[5] are the lines $y = \pm\frac{b}{a}x$ (see Fig.5.8.1). Each

---

[5]The popular description of an asymptote, viz. that it is a tangent at infinity is very

of them comes very close to the curve for large values of $|x|$. This is because,
as $x \to \pm\infty$, the difference between $\dfrac{b}{a}x$ and $b\sqrt{\dfrac{x^2}{a^2} - 1}$ tends to $0$ (Exercise
(8.1)). Consequently, the two functions $f(x) = b\sqrt{\dfrac{x^2}{a^2} - 1}$ and $g(x) = \dfrac{b}{a}x$ grow
at comparable rates. So $g(x)$, which is obviously simpler to study than $f(x)$, is
an asymptotic approximation to the latter.

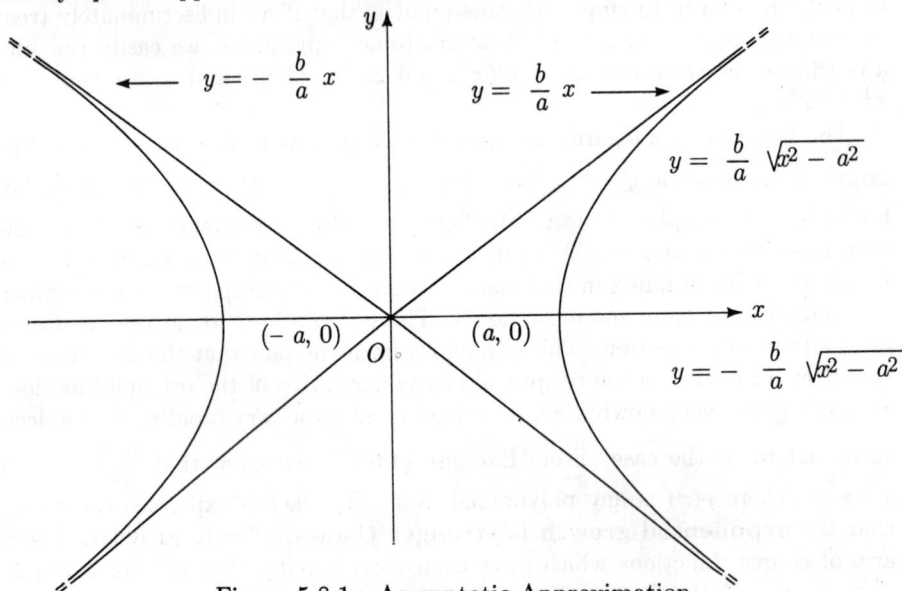

Figure 5.8.1 : Asymptotic Approximation

Obviously, there is some point in having a measuring scale of growth only if
there are some naturally occurring situations where the growth is comparable to
that scale. We shall show that this is indeed the case for the logarithmic growth
(represented by the function $\ln x$) and for exponential growth (represented by
the function $e^x$, or more generally, by $a^x$ where $a > 1$). We illustrate these with
one example each.

The example we choose for logarithmic growth may come as a surprise,
because the function involved in it is not of a real variable $x$, but of an integer
variable $n$. So let $n$ be a positive integer and let $H_n = 1 + \dfrac{1}{2} + \ldots + \dfrac{1}{n}$ be the
$n^{\text{th}}$ harmonic number. Then $\{H_n\}$ is a monotonically increasing sequence which
tends to $\infty$ as $n \to \infty$, because by Theorem (3.6.1), the series $\displaystyle\sum_{n=1}^{\infty} \dfrac{1}{n}$ is divergent.
We claim that $H_n$ grows at a rate comparable to that of $\ln n$. Actually we
shall show that for large $n$, $H_n$ is very close to $\ln n$. Even though both $\ln n$
and $H_n$ tend to $\infty$ as $n \to \infty$, the difference between them tends to a constant.
Moreover, this constant is surprisingly small. (Specifically, $H_n - \ln n \to \gamma$ where

---

suggestive here.

$\gamma \simeq 0.57721566$.) So, for large $n$, we may take $\ln n$ as a very good approximation to $H_n$. If we want, say, $H_{1000000}$ then we find, either from the tables or from a calculator, that $\ln(1000000)$ is (approximately) 13.815511. So we get

$$
\begin{aligned}
H_{1000000} &= \frac{1}{1} + \frac{1}{2} + \frac{1}{3} + \dots + \frac{1}{999999} + \frac{1}{1000000} \\
&\simeq 13.815511 + \gamma \\
&\simeq 14.392726
\end{aligned}
$$

Anybody would agree that this is far better than finding and adding one million reciprocals!

Thus logarithms give us a very vivid idea of how slowly the harmonic series $\sum_{n=1}^{\infty} \frac{1}{n}$ diverges. We must, of course, cite some real-life situations where this series, or rather the harmonic numbers $H_n$ occur, otherwise we would be guilty of trying to establish the usefulness of something by showing how it leads to something useless !

Interestingly, the harmonic numbers do not figure very frequently in classical mathematics. But recently, because of the phenomenal growth of computer science, a new branch of mathematics, called **analysis of algorithms**, has evolved. In a nutshell, it deals with the problem of determining how efficient an algorithm is. In this branch, the harmonic numbers arise naturally. For example, in the simple-minded algorithm (cf. Exercise (2.1.11)) for finding the maximum of a finite set by comparing its elements one-by-one, the tentative maximum has to be revised whenever the new element creates a 'record', i.e., when it is bigger than all the preceding elements. Estimating how many times this is likely to happen is an important consideration in determining the efficiency of that algorithm. As a similar, but less technical situation, suppose the annual rainfall at a place varies randomly within certain bounds. Let us assume that starting from some year, we keep a record of the annual rainfall. It can be shown from probabilistic considerations (see Exercise (6.5.4)) that for every positive integer $n, H_n$ is the expected number of record-breaking years during the first $n$ years. So, during the first century there would be about 5 years with record-breaking rainfalls, since $H_{100} \simeq \ln 100 + \gamma \simeq 5.18238$. But during the next century this figure dwindles down to $H_{200} - H_{100}$ which is less than 1.

The relationship between $H_n$ and $\ln n$ may come as a surprise if $\ln$ is defined as the inverse of the exponential function, the latter being defined by power series. But if $\ln n$ is defined as the integral $\int_1^n \frac{dx}{x}$ then the relationship is very natural, because $H_n$ is the sum of the values of the integrand at the points $1, 2, \dots, n$ in the interval of integration, $[1, n]$. In fact the proof that $H_n - \ln n$ tends to a finite limit is fairly short and geometric as we now show.

**8.2 Theorem:** As $n \to \infty, H_n - \ln n \to \gamma$, where $\gamma$ is a fixed real number.

**Proof:** Consider the graph of the function $y = \frac{1}{x}$ for $x \geq 1$ as shown in Figure

5.8.2. It is clear that $H_{n-1}$ represents the sum of the areas of the rectangles with unit bases and heights $1, \frac{1}{2}, \ldots, \frac{1}{n-1}$. Obviously, this is greater than the area below the curve $y = \frac{1}{x}$, the $x$-axis and the lines $x = 1, x = n$, which equals $\ln n$. So $H_{n-1} - \ln n$ equals the shaded area in the figure and increases with $n$. On the other hand, if we consider the lower rectangles, we get, $\frac{1}{2} + \frac{1}{3} + \ldots + \frac{1}{n} < \ln n$, i.e. $H_n - \ln n < 1$ and hence $H_{n-1} - \ln n < 1$ (since $H_{n-1} < H_n$). So the sequence $\{H_{n-1} - \ln n\}_{n=1}^{\infty}$ is monotonically increasing and bounded above by 1. Such a sequence is always convergent. Let $\gamma = \lim_{n \to \infty} (H_{n-1} - \ln n)$. Since $H_n = H_{n-1} + \frac{1}{n}$ and $\frac{1}{n} \to 0$ as $n \to \infty$, it follows that $H_n - \ln n$ also converges to $\gamma$. ∎

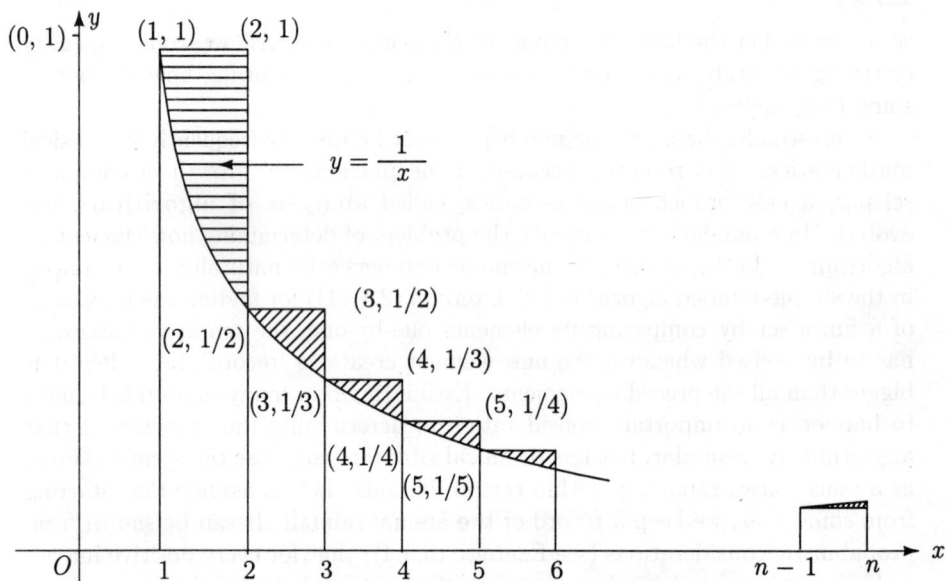

Fig. 5.8.2 : Harmonic Numbers and Natural Logarithm

The number $\gamma$ is called **Euler's constant**. Clearly $0 < \gamma \le 1$, although with a little work (see Exercise (8.6)), more accurate estimates can be obtained. For example, it is easy to show that $\gamma > \frac{1}{2}$. The approximate value of $\gamma$ is 0.57721556. It is not known if $\gamma$ is rational.

This example illustrates how the ln function enters into asymptotic estimation. As an even more important example, we shall obtain an asymptotic estimate for $n!$, popularly known as **Stirling's formula**. The factorials are, by far, the most frequently occurring numbers in combinatories (the branch of mathematics which includes things like permutations and combinations). They grow very fast as $n$ grows. In fact their growth is faster than even the exponential growth. Given any (fixed) real number $R$, no matter how large, we have $\frac{R^n}{n!} \to 0$

as $n \to \infty$ (see Exercise (3.9.5)). However, if we compare $n!$ with $n^n$, we see that $n! = o(n^n)$, because $\dfrac{n!}{n^n} = \dfrac{1}{n}\dfrac{2}{n} \ldots \dfrac{n}{n} \leq \dfrac{1}{n}$ for all $n \in I\!N$. Stirling's formula tells us where $n!$ fits in as far as its rate of growth is concerned. Specifically, it says $\displaystyle\lim_{n\to\infty} \dfrac{n!e^n}{\sqrt{2\pi n}\,n^n} = 1$. In particular, it follows that $n! = O\left(\sqrt{2\pi n}\dfrac{n^n}{e^n}\right)$. Note, however, that Stirling's formula does *not* say that $\displaystyle\lim_{n\to\infty}\left(n! - \sqrt{2\pi n}\dfrac{n^n}{e^n}\right) = 0$.

This is false. In fact, the difference $n! - \sqrt{2\pi n}\dfrac{n^n}{e^n}$ tends to $\infty$ as $n \to \infty$. In this respect, Stirling's formula is not as sharp as the asymptotic estimation of $H_n$ by $\ln n$, where the difference between the two tends to a fixed (and a very small) number $\gamma$. What the formula does say is that $\displaystyle\lim_{n\to\infty} \dfrac{n! - \sqrt{2\pi n}\frac{n^n}{e^n}}{n!} = 0$. In other words, even though the error in estimating $n!$ as $\sqrt{2\pi n}\dfrac{n^n}{e^n}$ is large, it is negligible as compared with $n!$. For example, for $n = 10, n! = 3,628,800$ and $\sqrt{2\pi n}\dfrac{n^n}{e^n}$ is approximately 3,598,696. Here the error is 30104 but the percentage (or relative) error is only $\dfrac{5}{6}\%$. For large quantities it is this relative error that matters more. And it is in this sense that the Stirling formula provides a good approximation to $n!$ for large $n$.

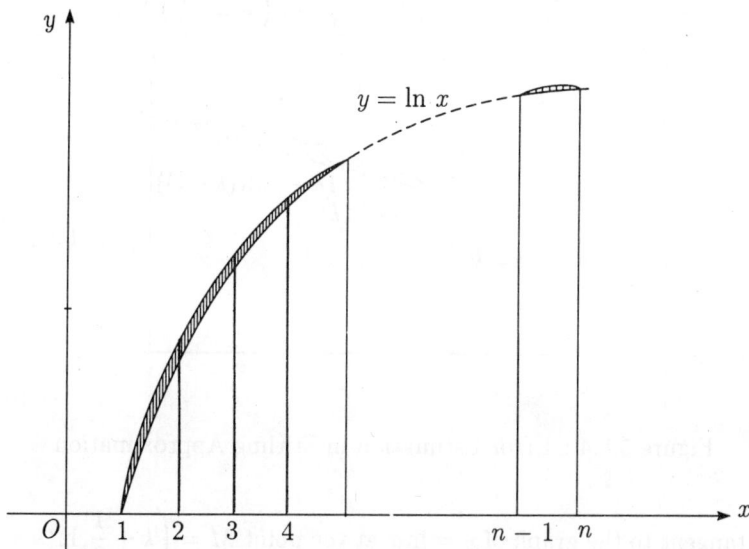

Figure 5.8.3 : Logarithmic Form of Stirling Approximation

Now, coming to the proof of the Stirling formula, we note that $n!$ is the product and not the sum of the first $n$ positive integers. So a method, analogous to that used above in estimating $H_n$ cannot be applied directly. But this difficulty can be easily overcome by taking natural logarithms, because $\ln(n!) = \ln 1 +$

$\ln 2 + \ldots + \ln n$. If we denote this sum by $A_n$, then Stirling formula is equivalent to proving that $\lim\limits_{n \to \infty} \left( A_n - (n + \frac{1}{2}) \ln n + n \right) = \frac{1}{2} \ln(2\pi)$. We therefore consider the function $f(x) = \ln x$. Then $\int_1^n \ln x \, dx = (x \ln x - x) \Big|_1^n = n \ln n - n + 1$ represents the area, say $B_n$, bounded by the curve $y = \ln x$, the $x$ - axis and the lines $x = 1$ and $x = n$. Let $T_n$ be the sum of the areas of the $n - 1$ trapezia shown in Fig.5.8.3 (the first trapezium really reduces to a triangle). It is readily seen that $T_n = \sum\limits_{k=1}^{n-1} \frac{1}{2}[\ln k + \ln(k + 1)] = A_n - \frac{1}{2} \ln n$. The difference $B_n - T_n$ represents the shaded area in Fig.5.8.3.

It is easily checked (from the second derivative) that the function $\ln x$ is concave downwards on $(0, \infty)$ and so it follows that all these trapezia lie below the graph. Consequently, $B_n - T_n$ is positive for all $n > 1$ and increases with $n$. We claim, however, that the sequence $B_n - T_n$ is bounded above. To see this, consider a typical trapezium with sides along $x = k$ and $x = k + 1$, shown in Fig.5.8.4. The shaded area is now $(B_{k+1} - B_k) - (T_{k+1} - T_k)$. (For large $k$ this area is very small and so in order to show it perceptibly, the scales on the two axes are taken differently.)

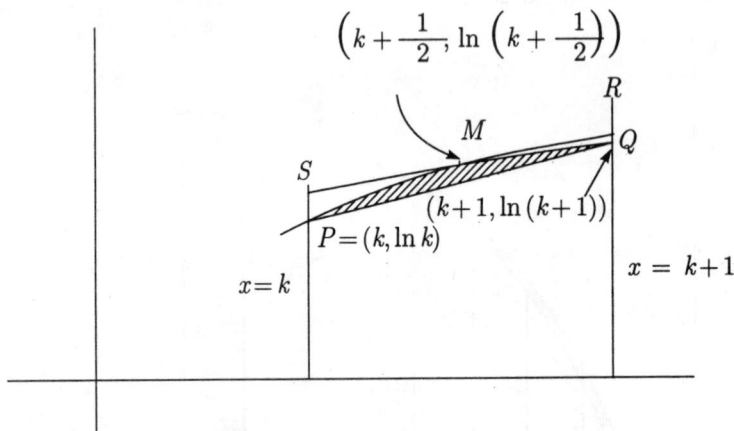

Figure 5.8.4 : Error Estimation in Stirling Approximation

Draw a tangent to the graph of $y = \ln x$ at the point $M = \left( k + \frac{1}{2}, \ln(k + \frac{1}{2}) \right)$. Since the function $\ln x$ is concave downwards on $(0, \infty)$, the graph lies entirely below the tangent (see Fig. 4.10.5 and the accompanying comments). Consequently the shaded area is less than the area of the trapezium $PQRS$. By a direct computation (see Exercise (8.7)), the latter comes out as $\frac{1}{2} \ln(1 + \frac{1}{2k})$ $-\frac{1}{2}\ln(1 + \frac{1}{2k + 1})$ which itself is less than $\frac{1}{2} \ln(1 + \frac{1}{2k}) - \frac{1}{2} \ln(1 + \frac{1}{2k + 2})$.

Putting $k = 1, 2, \ldots, n - 1$ in the inequality

$$(B_{k+1} - B_k) - (T_{k+1} - T_k) \ < \ \frac{1}{2}\ln(1 + \frac{1}{2k}) - \frac{1}{2}\ln(1 + \frac{1}{2k+2})$$

and summing over, we get for all $n > 1$,

$$B_n - T_n \ < \ \frac{1}{2}\ln\frac{3}{2} - \frac{1}{2}\ln(1 + \frac{1}{2n}) \ < \ \frac{1}{2}\ln\frac{3}{2}.$$

Hence the sequence $B_n - T_n$ is monotonic and bounded above. Denote its limit by $\sigma$. Substituting for $B_n$ and $T_n$ we thus get that $\sum\limits_{k=1}^{n}\ln k - (n + \frac{1}{2})\ln n + n$ tends to the finite limit $1 - \sigma$, or equivalently $\dfrac{n!e^n}{n^n\sqrt{n}}$ tends to the finite non-zero limit $\lambda$ where $\lambda = e^{1-\sigma}$. We have thus proved the Stirling formula except for showing that $\lambda = \sqrt{2\pi}$. In most applications, we are interested only in the order of magnitude of $n!$ and so the exact value of $\lambda$ is not important. Nevertheless, for the sake of completeness, we indicate, in the exercises, a method to compute it. It will be based on Wallis formula for $\pi$, given in Exercise (6.12).

We record the work so far as a theorem.

**8.3 Theorem:** As $n \to \infty$, $\dfrac{n!}{\sqrt{n}\left(\frac{n}{e}\right)^n} \to \sqrt{2\pi}$. ∎

The theoretical as well as practical utility of Stirling's formula is easy to appreciate. In combinatorial problems (involving, for example, analysis of algorithms), we often have to get an idea how big a binomial coefficient, say $\binom{n}{k}$ is. We assume $0 \leq k \leq n$. If $k$ (or $n - k$) is small then we can tell directly about how big $\binom{n}{k}$ is. When $n$ is large and $k$ is not close either to $0$ or to $n$, a direct calculation of $\binom{n}{k}$ is prohibitive. But writing it as $\dfrac{n!}{k!(n-k)!}$ and using the Stirling formula (or rather its logarithmic form) for each of the three factorials, gives a good idea of the order of magnitude of $\binom{n}{k}$. A few problems based on this method will be given in the exercises.

Having illustrated how logarithmic growth arises naturally, let us now turn to its inverse, viz., the exponential growth, i.e. growth of the form $a^x$ where $a > 1$ is fixed. (Note that for $a > b > 1, b^x = o(a^x)$ and so not all exponential growths are comparable. Still, it is customary to say, $a^x$ grows exponentially for every $a > 1$). At the end of Section 3.3 (see also Exercise (3.3.13)), we already remarked about the growth of $a^n$ as $n$ approahes $\infty$ through integer values. Situations of exponential growth where the exponent tends to $\infty$ through real values are not far to seek. In fact, exponential growth is the most natural growth to be encountered in real life.

Let us see why. Suppose $t$ is time and let $x$ be some quantity which grows with $t$. We assume that the growth is reproductive and accumulative in nature. This means that the quantity existing at any time reproduces some quantity of the same type. This new quantity is added to the old one and also starts reproducing along with the old one. Most biological growths (such as that of a colony of bacteria, or even human population) are of this type. (For simplicity, we are ignoring here the negative growth caused by death. Examples of negative growths, also called decay, will be given later. In a realistic example, involving life forms, we must, of course, consider both births and deaths. In such cases, we may interpret growth to mean the net effect of the two.) As a man-made example, we may take the money invested in accumulative saving schemes. The interest accrued is added to the principal and starts earning interest along with the original principal. As an example of a growth which is not of this type, suppose $x$ is the distance travelled by a moving car. Here the new distance covered by the car has nothing to do with the distance already covered. The speed, that is, the rate at which the distance increases w.r.t. time, depends on other things such as the fuel supply to the engine and the condition of the road at that time but not on how much distance the car has travelled till that time.

Biological growth is, strictly speaking, a discrete and not a continuous process. It takes some time interval, say $T$, for new offsprings to be born and to gain maturity to produce newer offsprings. So in a biological growth the quantity $x$ remains constant for a period of length $T$ and then suddenly increases in a moment. This period $T$ can vary from a few minutes, (for certain fungi) to several decades (for certain mammals). For an artificial growth like that of money in a bank account, this period $T$ is precisely the time period between two successive compoundings (see Section 2.1).

It would thus appear that the methods of calculus are not applicable to study natural growth, because in calculus the variables involved are continuous and not discrete (see Section 2.2). But when the period $T$ is very small, we may regard the discrete growth approximately as continuous growth. This is analogous to treating a very frequently compounded interest as continuously compounded (see again Section 2.1). Even when the period $T$ is not small, we may often think of the biological growth as continuous if the quantity $x$ is large. This is because the quantity $x$ usually consists of a very large number of individuals. Even though the period of growth for each individual is (approximately) $T$, the individuals are at various stages of development, so that at any one time a certain proportion of the population is producing new offsprings. In a large population it is statistically safe to assume that this variation is uniformly distributed and so the growth can be considered as continuous.

Thus we assume that the quantity $x$ which grows with time $t$ is a continuous, in fact a differentiable, function of $t$. Then $\dfrac{dx}{dt}$ is the rate of growth. It is obvious that $\dfrac{dx}{dt}$ should be proportional to $x$. Other conditions being equal, it stands to reason that a colony of bacteria which is twice as big as another should grow twice as fast. This observation also tallies with experiments (which is, after all, the acid test of acceptance in physical sciences) and leads to what is known as the **law of exponential growth**. The reason behind this name is as follows. Suppose $\dfrac{dx}{dt} = kx$ where $k$ is a constant. Let $x_0$ be the value of $x$ at some time say $t_0$. (A data like this is called an **initial condition**.) Let $x_1$ be the value of $x$ at some other time $t_1$. Then from the rule of substitution (which is valid since $x$ is a strictly increasing and continuously differentiable function of $t$), we get

$$\int_{x_0}^{x_1} \frac{1}{x} dx = \int_{t_0}^{t_1} \frac{1}{x} \frac{dx}{dt} dt = \int_{t_0}^{t_1} k\, dt$$

which immediately gives

$$\ln\left(\frac{x_1}{x_0}\right) = k(t_1 - t_0)$$

or $x_1 = \lambda e^{kt_1}$ where $\lambda = \dfrac{x_0}{e^{kt_0}}$.

Replacing $x_1, t_1$ by $x, t$ respectively, we thus see that $x = \lambda a^t$ where $a = e^k$. Thus $x$ grow exponentially with time $t$. Because of the enormity of the exponential growth, it would appear that a colony of bacteria will eventually grow bigger than the whole earth! In practice, this does not happen, of course, because the other resources needed for the growth are limited. So the growth equation $\dfrac{dx}{dt} = kx$ does not hold with the same value of $k$ after some time. A steady state may be reached (corresponding to $k = 0$) or there may even be a decline (corresponding to $k < 0$). A more realistic model of continuous biological growth, called the **logistic equation**, given by $\dfrac{dx}{dt} = kx(1 - \dfrac{x}{C})$ has been proposed. Here the constant $C$ represents the **carrying capacity** of the surroundings. For small $x$ this is close to the model we studied. This model was found to be adequate for bacterial growth but not for more complex life forms.

The case where $k < 0$, is often called a **decay**. Certain radioactive substances obey the exponential law of decay. A most exciting application of this fact occurs in 'dating' fossils, i.e. determining how old they are. This will be illustrated in the exercises (Exercise (8.15)).

## EXERCISES

8.1 Let $a > 0, b > 0$. Prove that as $x \to \infty$, $\left| \dfrac{b}{a}x - b\sqrt{\dfrac{x^2}{a^2} - 1}\right| \to 0$. Hence show that the functions $\dfrac{b}{a}x$ and $b\sqrt{\dfrac{x^2}{a^2} - 1}$ grow at comparable rates as $x \to \infty$.

8.2 Let $1 < a < b$. Prove that the functions $\ln_a x$ and $\ln_b x$ grow at comparable rates as $x \to \infty$, but their inverse functions, viz., $a^x$ and $b^x$ don't.

8.3 Compare the growth rates of the following functions and arrange them in ascending order of growth :
$$2^x, e^x, x^x, e^{x/2}, \sqrt{x}, \log_3 x, \ln x, \ln(\ln x), (\ln x)^x, x^{\ln x}, x^e, (ex)^x, x^{2x}.$$

8.4 Give an example of a pair of functions $f(x), g(x)$, each tending to $\infty$ as $x \to \infty$ such that neither $f(x) = O(g(x))$ nor $g(x) = O(f(x))$. (In other words, we may not always be able to compare two functions in terms of growth rates.)

8.5   Prove that the series $\displaystyle\sum_{n=2}^{\infty} \frac{1}{n \ln n}$ is divergent. Hence give an alternate proof that $\ln n$ does not grow comparably to $n^\alpha$ for any $\alpha > 0$, as $n \to \infty$ (through integral values).

8.6   Prove that the sequence $\{H_{n-1} - \ln n\}_{n=2}^{\infty}$ is monotonically increasing while the sequence $\{H_n - \ln n\}_{n=1}^{\infty}$ is monotonically decreasing, but that both converge to $\gamma$. (This gives a simple method for calculating $\gamma$ to any degree of accuracy. For example, $n = 10$ gives $0.5263 < \gamma < 0.6264$.)

8.7   Verify that the area of the trapezium $PQRS$ in Fig.5.8.4 is indeed
$$\frac{1}{2} \ln \left(1 + \frac{1}{2k}\right) - \frac{1}{2} \ln \left(1 + \frac{1}{2k+1}\right).$$

8.8   Let $\lambda_n = \dfrac{n! e^n}{\sqrt{n}\, n^n}$ for $n \geq 1$ and let $\lambda = \lim\limits_{n \to \infty} \lambda_n$ (which exists as shown before the statement of Theorem (8.3)). Prove that $\dfrac{\sqrt{n}\lambda_n^2}{\sqrt{2}\lambda_{2n}} = \dfrac{4^n (n!)^2}{(2n)!}$.

8.9   Using the last exercise and Wallis formula (Exercise (6.12)), show that
$$\frac{\pi}{2} = \lim_{n \to \infty} \frac{1}{2n+1} \frac{n\lambda_n^4}{2\lambda_{2n}^2} \quad \text{and hence that } \lambda = \sqrt{2\pi}.$$

8.10  Estimate $\displaystyle\binom{100}{30}$ using Stirling's approximation. Compare this with the value $2.937234 \times 10^{25}$ (obtainable from a programmable calculator).

8.11  The numbers $\dfrac{1}{4^n}\displaystyle\binom{2n}{n}$ figure rather frequently in mathematics. Prove that
$$\frac{1}{4^n}\binom{2n}{n} = \frac{1.3.5.\ldots(2n-1)}{2.4.6.\ldots 2n}.$$ Determine for which values of $p$ the series
$$\sum_{n=1}^{\infty} \left(\frac{1.3.5.\ldots(2n-1)}{2.4.6.\ldots 2n}\right)^p \text{ converges.}$$

8.12  Do Exercise (1.3.1).

8.13  Suppose $x_i = \lambda a^{t_i}, i = 1, \ldots, n$, where $a, \lambda$ are constants with $\lambda > 0, a > 1$. Prove that if $t_1, \ldots, t_n$ are in A.P. then $x_1, \ldots, x_n$ are in G.P. (This is very trivial, of course. But the significance is that in an exponential growth if readings are taken at regular time intervals, they will form a geometric progression, very much like the annual totals in compound interest.)

8.14  Suppose a radio-active substance obeys the law of decay $\dfrac{dx}{dt} = -\lambda x$ where $\lambda > 0$ is a constant. Prove that the time taken for any quantity of the substance to reduce to half its size is a constant, independent of the quantity. (This constant is called the **half-life** of that substance and can vary from a fraction of a second to a few centuries.)

8.15 Suppose the ratio of the radio-active carbon isotope ($C_{14}$) to the non-radio-active carbon ($C_{12}$) is the same in all living organisms. Let this ratio be $\alpha$. After death, $C_{14}$ decays exponentially, while $C_{12}$ remains intact. Suppose $\beta$ is the ratio of $C_{14}$ to $C_{12}$ in a fossil, today. If the half-life of $C_{14}$ is $T$, show that the 'age' of the fossil is $\dfrac{(\ln \alpha - \ln \beta)\, T}{\ln 2}$. (This method is popularly known as **carbon dating**.)

8.16 The food deficit of a country is 10 per cent. Its population grows contin-uously at the rate of 3% p.a. while every year its annual food production is 4% more than that of the last year. Assuming that the per capita food requirement remains the same, find after how many years the country will be self-sufficient in food.

8.17 One of the rules of thumb adopted in banking circles is the **seventy per-cent rule** which says that if interest at the rate of $\alpha$ p.c.p.a. is compounded annually, then the principal will be doubled in about $\dfrac{70}{\alpha}$ years. What is the mathematical basis for this rule? Show that the rule is valid for small values of $\alpha$, e.g. $\alpha = 8$ or 10 but that it fails if $\alpha$ is large, e.g. $\alpha = 70$. (Of course, no bank gives interest at such a fabulous rate!).

8.18 Generalise Exercise (7.1) by showing that the function $\left(1 + \dfrac{1}{x}\right)^{x}$ is increas-ing on $(0, \infty)$. Prove, however, that the function $\left(1 + \dfrac{1}{x}\right)^{x+1}$ is decreasing on $(0, \infty)$ and hence that the sequence $\left(1 + \dfrac{1}{n}\right)^{n+1}$ is monotonically de-creasing. Thus for all $n$, $\left(1 + \dfrac{1}{n}\right)^{n} < e < \left(1 + \dfrac{1}{n}\right)^{n+1}$. (This result is of the same spirit as that of Exercise (8.6) and can be used to obtain approxi-mate values of $e$ with a desired degree of accuracy. But the partial sums of the rapidly convergent series $\displaystyle\sum_{n=0}^{\infty} \dfrac{1}{n!}$ provide a far more efficient method.)

8.19 If $|x| \le \frac{1}{2}$, prove that $|\ln(1 + x)| \le 2|x|$. Deduce that if a series $\displaystyle\sum_{n=1}^{\infty} b_n$ is absolutely convergent, then the infinite product $\displaystyle\prod_{n=1}^{\infty}(1 + b_n)$ is convergent. (See the comments about infinite products in Exercise (3.9.15). There the logarithms had base 10, but that does not matter.)

**5.9.   The Trapezoidal Rule and the Simpson's Rule for Approximate Integration**

Like other basic concepts in calculus, definite integrals are certain limits, specifically limits of Riemann sums. By its very nature, a limit is an infinitistic concept, that is, it is generally not possible to evaluate a limit exactly in a finite number of steps, even when we know it exists. In real life applications, mere existence of the limit is not much help. Integrals have many real life applications. Even though the Fundamental Theorem of Calculus makes it possible to evaluate the definite integrals, it is inadequate when we cannot find an anti-derivative of the integrand. Occassionally, some special methods work as illustrated in Section 5.6. But they are usually of limited use, being too special. So, to get the numerical value of a definite integral, we often have to resort to approximation. For this reason approximate integration is also often called **numerical integration**. The problem of evaluating an integral is often called **quadrature**, apparently because historically areas were measured by specifying squares of equal areas and evaluation of an integral amounts to evaluating a certain area.

Throughout, we assume $I = \int_a^b f(x)dx$ where $f$ is an integrable function on an interval $[a, b]$. To find the approximate value of $I$, we chop the interval $[a, b]$ into, say, $n$ subintervals by a partition, say , $P = (a = x_0, x_1, x_2 \ldots, x_n = b)$. For $i = 1, \ldots, n$, let $I_i = \int_{x_{i-1}}^{x_i} f(x)dx$. Then $I = I_1 + \ldots + I_n$. The approximation methods we shall study consist of approximating each $I_i$ by some number, say $J_i$ and adding. That is, $J_1 + J_2 + \ldots + J_n$ will be an approximate value of $I$. Denote this sum by $J$. The error of this approximation is obviously at most $\sum_{i=1}^{n} |I_i - J_i|$. Therefore to get an upper bound on the magnitude of the error $I - J$, we seek an upper bound on each $|I_i - J_i|$. Depending upon how $J_i$ is obtained, we get different estimates on the error.

In absence of any other information about the function, we may as well assume that the nodes of the partition $P$ are equally spaced; i.e. all subintervals of $P$ are equal in length. We denote this length by $h$. So, $h = \dfrac{b - a}{n}$ and $x_i = a + hi$, for $i = 1, 2 \ldots, n$. We shall denote by $y_i$ the value of $f$ at $x_i$. This notation will be used throughout.

In the methods we shall study each $J_i$ will be of the form $\int_{x_{i-1}}^{x_i} g_i(x)dx$ where $g_i(x)$ is some approximation to $f(x)$ for $x_{i-1} \leq x \leq x_i$. Clearly,

$$|I_i - J_i| = |\int_{x_{i-1}}^{x_i} f(x) - g_i(x)dx| \leq \int_{x_{i-1}}^{x_i} |f(x) - g_i(x)|dx \qquad (1)$$

If the function $f$ satisfies some smoothness conditions (such as differentiability of a certain order) and the functions $g_i$ are chosen carefully, then we can get certain upper bounds on the integrand and hence on the last integral in (1). Adding these we get an upper bound on the magnitude of the error $I - J$, for

$$|I - J| = |\sum_{i=1}^{n} I_i - \sum_{i=1}^{n} J_i| = |\sum_{i=1}^{n}(I_i - J_i)| \leq \sum_{i=1}^{n} |I_i - J_i|. \tag{2}$$

What about the approximating functions $g_i(x)$? In the methods we shall present, they will all be polynomials and so their integrals can be evaluated very easily. As is to be expected, the higher the degrees of these polynomials, the more work they will entail. But with a clever choice of these polynomials, it turns out that the calculations of their integrals (viz. $J_i$'s) can be done very efficiently.

We shall discuss three methods of quadrature of this type. They will progressively involve polynomials of higher degrees and will yield progressively more accurate estimations. To illustrate this, we shall compute $\int_0^1 \dfrac{dx}{1 + x^2}$ using each of these methods and compare the answers with the exact answer, viz. $\tan^{-1}(1)$, i.e. $\dfrac{\pi}{4}$, which is approximately 0.7853981.

The simplest polynomial $g_i(x)$ which approximates $f(x)$ on $[x_{i-1}, x_i]$ is of degree 0, i.e., a constant polynomial. Which constant should this be ? If we want $g_i$ to be a good approximation to $f$, then it is natural to expect that $g_i$ should agree with $f$ at at least one point, say, $\xi_i$ in $[x_{i-1}, x_i]$. So we take $g_i(x) = f(\xi_i)$ for all $x \in [x_{i-1}, x_i]$. In this case, $J_i$, that is, $\int_{x_{i-1}}^{x_i} g_i(x)dx$ is simply $f(\xi_i)(x_i - x_{i-1})$ (which equals $hf(\xi)$) and $J$, the approximate integral, is $\sum_{i=1}^{n} J_i = \sum_{i=1}^{n} f(\xi_i)(x_i - x_{i-1}) = h\sum_{i=1}^{n} f(\xi_i)$. This is, of course, exactly a Riemann sum of $f$ based on the partition $P = (x_0, x_1 \ldots, x_n)$. This approxiamtion is called a rectangular approximation because each $J_i$ is the area of a rectangle of height $f(\xi_i)$ and base $[x_{i-1}, x_i]$ (see Fig.5.9.1(a)).

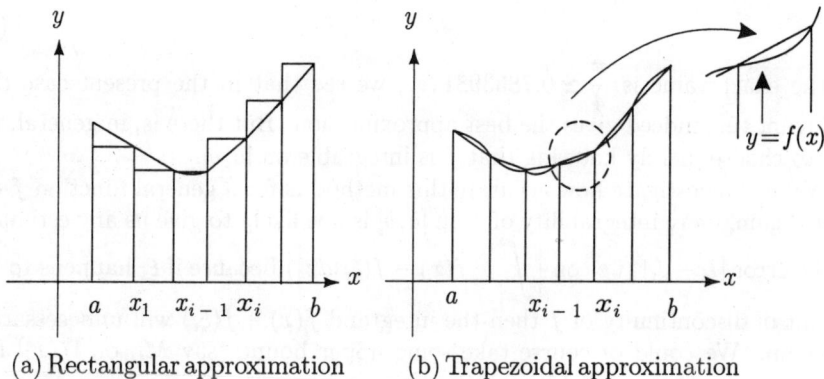

(a) Rectangular approximation    (b) Trapezoidal approximation

Figure 5.9.1 : Rectangular and Trapezoidal Approximations

How accurate is the rectangular approximation ? The answer depends largely on the choice of the point $\xi_i$'s. If $f$ is continuous on $[a, b]$, then it can be shown (see Theorem (6.1.2)) that for each $i$, there exists some $\xi_i \in [x_{i-1}, x_i]$ such that $\int_{x_{i-1}}^{x_i} f(x)dx = f(\xi_i)(x_i - x_{i-1})$. With this choice, $J = \sum_{i=1}^{n} f(\xi_i)(x_i - x_{i-1})$ will equal $I$ (by Exercise (1.7)) i.e., the approximation will be exact! The trouble, of course, is that in general we have no way of finding this 'magic' point $\xi_i$. There is, in fact, no general method to guide the choice of $\xi_i$. We have to declare it by some rule, such as $\xi_i = x_i$ (right end-point)or $\xi_i = x_{i-1}$ (left end-point) or $\xi_i = \dfrac{x_{i-1} + x_i}{2}$ (mid-point) and depending upon which choice is made we get various rules for quadrature called the **right end-point rule** or the **mid-point rule** etc. Given a particular $f$, a particular rule may work out better than another. For example, in our sample problem, viz, $\int_0^1 \dfrac{dx}{1+x^2}$, the integrand is decreasing throughout. So, the right end-point rule will give us a value which is too small, the left end-point rule will give a value which is too large and the mid-point rule is more likely to give a better estimate. If we take $n = 10$ then $x_i = \dfrac{i}{10}$ and $y_i = \dfrac{1}{1+x_i^2} = \dfrac{1}{1+i^2/100}$, for $i = 0, 1, \ldots n$. Also $\dfrac{x_{i-1} + x_i}{2} = \dfrac{2i-1}{20}$. Using a hand-calculator, it is easy to check that we get the following approximate values for $\int_0^1 \dfrac{dx}{1+x^2}$.

$$\text{Right end-point rule} : h \sum_{i=1}^{10} y_i = \frac{1}{10} \sum_{i=1}^{10} \frac{100}{i^2 + 100} \simeq 0.7599815 \qquad (3)$$

$$\text{Left end-point rule} : h \sum_{i=1}^{10} y_{i-1} = \frac{1}{10} \sum_{i=1}^{10} \frac{100}{(i-1)^2 + 100} \simeq 0.8099815 \,(4)$$

$$\text{Mid-point rule} : h \sum_{i=1}^{10} \frac{1}{1 + (\frac{2i-1}{20})^2} = \frac{1}{10} \sum_{i=1}^{10} \frac{400}{400 + (2i-1)^2} \simeq 0.7842604.$$

$$(5)$$

As the exact value is $\dfrac{\pi}{4} \simeq 0.7853981\ldots$, we see that in the present case the mid-point rule indeed gives the best approximation. But there is, in general, no way to choose merely knowing that $f$ is integrable on $[a, b]$.

We now investigate how accurate this method is for a general function $f$ on $[a, b]$. Again, mere integrability of $f$ on $[a, b]$ is not likely to give us any estimate on the error $|I_i - J_i|$ (i.e. on $\left| \int_{x_{i-1}}^{x_i} f(x) - f(\xi_i)dx \right|$) because if $\xi_i$ happens to be a point of discontinuity of $f$ then the integrand $f(x) - f(\xi_i)$ will unnecessarily shoot up. We could of course take some upper bound, say $M_i$, on $|f(x)|$ for $x \in [x_{i-1}, x_i]$ and then $\left| \int_{x_{i-1}}^{x} f(x) - f(\xi_i)dx \right|$ will be at most $2M_i(x_i - x_{i-1})$, i.e. $2M_i h$. But this is a very crude estimate. To get a sharper estimate, we need to assume some regularity condition on $f$. Mere continuity of $f$ is not of much help. But if we assume $f$ to be differentiable then things improve drastically because instead of an estimate like $2M_i h$ (with a factor of $h$ in it), we can get an estimate with a factor $h^2$ in it. When $h$ is small, this is a significant improvement. For simplicity, we give the argument in the case of the left end-point rule. With minor modifications it will work for the right end-point or the mid-point rule too.

**9.1 Theorem:** Suppose $f$ is differentiable on $[x_{i-1}, x_i]$ and $|f'(x)| \le M_i$ for all $x \in [x_{i-1}, x_i]$. Then

$$\left| \int_{x_{i-1}}^{x_i} f(x)dx - y_{i-1}h \right| \le \frac{1}{2}M_i h^2 \tag{6}$$

**Proof.** For each $x \in [x_{i-1}, x_i]$, we apply the Lagrange Mean Value Theorem to the interval $[x_{i-1}, x]$ to get some $c$ (depending on $x$) such that

$$f(x) - f(x_{i-1}) = f'(c)(x - x_{i-1}) \tag{7}$$

which immediately gives,

$$|f(x) - f(x_{i-1})| \le M_i(x - x_{i-1}) \text{ for all } x \in [x_{i-1}, x_i]. \tag{8}$$

We now have,

$$
\begin{aligned}
\left| \int_{x_{i-1}}^{x_i} f(x)dx - y_{i-1}h \right| &= \left| \int_{x_{i-1}}^{x_i} f(x) - f(x_{i-1})dx \right| \\
&\le \int_{x_{i-1}}^{x_i} |f(x) - f(x_{i-1})|dx \\
&\le \int_{x_{i-1}}^{x_i} M_i(x - x_{i-1})dx \\
&= M_i \left( \frac{x^2}{2} - x_{i-1}x \right) \Big|_{x_{i-1}}^{x_i} \\
&= M_i \frac{h^2}{2}
\end{aligned}
$$

as can be seen by an elementary computation (keeping in mind that $h = x_i - x_{i-1}$). ∎

As an immediate corollary, we get,

**9.2 Corollary:** If $f$ is differentiable on $[a, b]$ with $|f'(x)| \le M$ for all $x \in [a, b]$ and $J$ is the approximate integral of $f$ given by the left end-point rule, then

$$|I - J| \le \frac{1}{2}(b - a)Mh \tag{9}$$

**Proof:** With the notations introduced earlier,

$$|I - J| = \left| \sum_{i=1}^{n} I_i - \sum_{i=1}^{n} J_i \right| \leq \sum_{i=1}^{n} |I_i - J_i|$$

$$\leq \sum_{i=1}^{n} \frac{1}{2} M_i h^2 \text{ (by the theorem above)}$$

$$\leq \frac{n}{2} M h^2 = \frac{1}{2}(b-a)Mh \text{ (since } nh = b - a)$$

as asserted. ∎

In our sample problem, $f(x) = \dfrac{1}{1+x^2}$ and $|f'(x)| = \dfrac{2x}{(1+x^2)^2}$. An easy upper bound for $|f'|$ is provided by the fact that the numerator $2x$ is at most 2 while the denominator is at least 1 for $x \in [0,1]$. So we take $M = 2$. With more work, we could get a smaller $M$, but that is not so vital. What matters more is the ease with which $M$ is found. With this choice of $M$, and $a = 0, b = 1$, we see from the corollary that $n = 10$ (i.e. $h = \dfrac{1}{10}$) would yield an error which is at most $\dfrac{1}{10}$. The calculations we have made (specifically, (4) above) show that the actual error is even smaller (viz. 0.0245834). This is to be expected because there is some extravagance in selecting $M$. We could, in fact, get a much better estimate for the error if instead of Corollary (9.2), we applied Theorem (9.1). In that case, instead of a single upper bound $M$ on $|f'(x)|$ for $x \in [a, b]$, we would have to compute 10 separate upper bounds $M_1, \ldots, M_{10}$ on the 10 subintervals. This would entail more calculations. Moreover, there is a vital difference between the applicability of Theorem (9.1) and Corollary (9.2). The latter tells us *a priori* how small $h$ should be (and hence how large $n$ should be) to ensure a certain degree of accurary, while Theorem (9.1) comes into the picture only after $n$ is determined. So even though it gives a closer estimate of the error $|I - J|$, it is really the Corollay which is more useful.

Although neither Theorem (10.1) nor Corollary (10.2) is very profound, we have presented them in some detail because they illustrate the technique. Note that in the left end-point rule, for each $i = 1, \ldots, n$, the function $g_i(x)$ is simply the constant function $f(x_{i-1})$, for $x \in [x_{i-1}, x_i]$. So we can rewrite (7) as

$$f(x) - g_i(x) = f'(c)(x - x_{i-1}). \tag{10}$$

In the other quadrature methods we shall study, the approximating function $g_i(x)$ on $[x_{i-}, x_i]$ will be a different one and so instead of (10) we shall get some other equation. But thereafter, the derivation will be essentially similar.

We first consider a method, popularly known as the **trapezoidal rule**. As the name suggests, in this method, we approximate each strip of the area by a trapezeum, (see Fig.5.9.1(b), with the tip of one such trapezeum shown enlarged in the circle). It is clear that the area of the $i^{\text{th}}$ trapezeum is $\dfrac{h}{2}(y_{i-1} + y_i)$ and so the total area which is customarily denoted by $T_n$ is

$$T_n = \sum_{i=1}^{n} \frac{h}{2}(y_{i-1} + y_i) = h\left(\frac{1}{2}y_0 + \frac{1}{2}y_n + \sum_{k=1}^{n-1} y_k\right). \tag{11}$$

It is clear that $T_n$ is simply the arithmetic mean of the approximations given by the left and right end-point rules. So, in our sample problem, viz. $\int_0^1 \frac{dx}{1+x^2}$, we see from (3) and (4) that the trapezoidal rule with $n = 10$ gives

$$T_{10} = \frac{1}{2}(0.7599815 + 0.8099815) = 0.7849815 \tag{12}$$

which is much better than either of (3) or (4) (or even (5) for that matter).

That this is not an accident is also borne out by the error estimation. Here, the approximating function $g_i(x)$ on $[x_{i-1}, x_i]$ has a straight line graph and so is of the form $Ax + B$ for some contants $A, B$. The conditions $g_i(x_{i-1}) = y_{i-1}$ and $g_i(x_i) = y_i$ determine $A, B$ uniquely. Or, we can write down the equation of a staright line to get

$$g_i(x) = y_{i-1} + \frac{y_i - y_{i-1}}{h}(x - x_{i-1}), \text{ for } x_{i-1} \le x \le x_i. \tag{13}$$

In order to get the analogue of (10), we proceed as follows. Although the derivation is a little more intricate, the basic idea is the same, viz. Lagrange's MVT. Actually, we shall use only the special case, viz. Rolle's theorem.

**9.3 Lemma:** Assume $f$ is twice differentiable on $[x_{i-1}, x_i]$. Then for every $x \in [x_{i-1}, x_i]$, there exists $c \in (x_{i-1}, x_i)$ such that

$$f(x) - g_i(x) = \frac{1}{2}f''(c)(x - x_{i-1})(x - x_i). \tag{14}$$

**Proof:** If $x = x_i$ or $x_{i-1}$ then both the sides of (14) vanish and we may take $c$ to be any point of $(x_{i-1}, x_i)$. So assume $x_{i-1} < x < x_i$. Now define a new function $K : [x_{i-1}, x_i] \to \mathbb{R}$ by

$$K(y) = f(y) - g_i(y) - (f(x) - g_i(x))\frac{(y - x_{i-1})(y - x_i)}{(x - x_{i-1})(x - x_i)}. \tag{15}$$

(Note that here $x$ is a constant and $y$ is the variable.)

Since $f$ is twice differentiable, so is $K$. It is clear that $K$ vanishes at $y = x_{i-1}, y = x_i$ and also at $y = x$. Hence by Rolle's theorem, $K'$ vanishes at at least two distinct points (one in the interval $(x_{i-1}, x)$ and the other in $(x, x_i)$). Hence by Rolle's theorem again, applied to $K'$, we see that $K''(c) = 0$ for some $c \in (x_{i-1}, x_i)$. To calculate $K''$ from (15), we note that $g_i''$ vanishes identically since $g_i$ is a polynomial of degree 1. Also the last term in (15) is a quadratic in $y$ and so only the coefficient of $y^2$ matters in the second derivative.

With these simplifying observations, it is easy to see that $K''(c) = 0$ means precisely (14). ∎

As remarked earlier, the rest of the journey follows the same pattern as from (7) to (9). This gives the following error estimation for the trapezoidal rule.

**9.4 Theorem:** Suppose $f$ is twice differentiable on $[a, b]$ with $|f''(x)| \leq M$ for all $x \in [a, b]$. Then for $T_n$, the trapezoidal approxiamtion to $I = \int_a^b f(x)dx$ with $n$ subintervals, we have,

$$|I - T_n| \leq \frac{M(b-a)}{12} h^2. \tag{16}$$

**Proof:** With our earlier notations, $I = \sum_{i=1}^{n} I_i$ and $T_n = J = \sum_{i=1}^{n} J_i$. In the present case, for $i = 1, \ldots, n$

$$|I_i - J_i| = \left| \int_{x_{i-1}}^{x_i} f(x) - g_i(x)dx \right| \leq \int_{x_{i-1}}^{x_i} |f(x) - g_i(x)|dx \tag{17}$$

where $g_i$ is given by (13).

From (14), using our hypothesis about $f''$ and keeping in mind that for $x \in [x_{i-1}, x_i], x - x_{i-1} \geq 0$ and $x_i - x \geq 0$, we get

$$|f(x) - g_i(x)| \leq \frac{M}{2}(x - x_{i-1})(x_i - x) \tag{18}$$

An elementary calcuation which is left as an exercise (Exercise (9.1)), shows that

$$\int_{x_{i-1}}^{x_i} (x - x_{i-1})(x_i - x)dx = \frac{1}{6}(x_i - x_{i-1})^3 = \frac{1}{6}h^3. \tag{19}$$

Since $|I - T_n| \leq \sum_{i=1}^{n} |I_i - J_i|$, and $nh = b - a$, we get the result from (17), (18) and (19). ∎

It is instructive to compare Corollary (9.2) and Theorem (9.4). Note that $M$ denotes two different things in them. In Corollary (9.2) it is a numerical upper bound for the first derivative while in Theorem (9.4), it is an upper bound for the second derivative. In general these two upper bounds are unrelated to each other. However, assuming that they are of comparable order of magnitude, we see that (16) is decidedly superior to (9) because of the factor $h^2$ in it.

It means that to get comparable accuracy, where the left end-point rule (or any other method of rectangular approximation) will need 100 subintervals, the trapezoidal rule can do the job with just 10 subintervals. These are, of course, general observations. As remarked earlier, if we are lucky in choosing the $\xi_i$'s, the rectangular approximation may be exact while the trapezoidal one will have some error. What we are concerned with is not such fortuitous accuracy but a guaranteed accuracy.

To apply the preceding theorem to our sample problem $\int_0^1 \dfrac{dx}{1+x^2}$, we see

that $f''(x) = \dfrac{6x^2 - 2}{(1+x^2)^3}$ and so we may take $M = 4$. So the error is at most $\dfrac{1}{300}$.

In reality it is much less (even less than $\dfrac{1}{2000}$) as we already saw.

The higher degree of accuracy in the trapezoidal rule (even though the calcuations involved are essentially the same as in the rectangular approximation) is due to the fact that the approximating function $g_i(x)$ on $[x_{i-1}, x_i]$ was taken to be a polynomial of degree 1 (instead of a polynomial of degree 0 as in the case of a rectangular approximation). Logically, we expect that if we take as $g_i(x)$ a polynomial of degree 2, then we should get even greater accuracy, with a factor of $h^3$ instead of $h^2$. This guess turns out to be quite correct and is the basis of what is called the Simpson's rule for approximate integration. In fact, as we shall see later, there is a bonus. That is, Simpson's rule gives an error estimate not only with a factor $h^3$, but, in fact, with a factor of $h^4$.

Before stating the **Simpson's rule** we record a lemma which gives a handy formula for the integral of a quadratic polynomial over an interval.

**9.5 Lemma:** Suppose $g(x)$ is a quadratic polynomial (i.e. a polynomial of degree 2). Then for every $\alpha \in \mathbb{R}$ and $h > 0$,

$$\int_{\alpha-h}^{\alpha+h} g(x)dx = \frac{h}{3}[g(\alpha - h) + 4g(\alpha) + g(\alpha + h)] \tag{20}$$

**Proof:** Let $g(x) = Ax^2 + Bx + C$, where $A, B, C$ are some constants. Then $\int_{\alpha-h}^{\alpha+h} g(x)dx = [A\dfrac{x^3}{3} + B\dfrac{x^2}{2} + C]\Big|_{\alpha-h}^{\alpha+h}$. The result follows by a straight-forward calculation, which is left as an exercise. ∎

The significance of this lemma is that in order to find the integral of a quadratic polynomial over an interval we need only know its length and the values of the polynomial at the two end-points and the mid-point of that interval. In particular, if the polynomial $g(x)$ is so constructed as to have prescribed values at these three points, then we can instantaneously write down its integral without finding $g(x)$ explicitly. This is the basic idea in Simpson's rule. As before, given a function $f : [a, b] \to \mathbb{R}$ we divide $[a, b]$ into $n$ equal parts at points $a = x_0 < x_1 < \ldots < x_{i-1} < x_n = b$ and set $h = \dfrac{b-a}{n}$. We are now going

to group two adjacent sub-intervals at a time and apply the preceding lemma to
each pair. For this we need to assume $n$ is even. For every odd $i, 0 < 1 < n$, we
let $g_i : [x_{i-1}, x_{i+1}] \to I\!R$ be the unique quadratic polynomial which assumes the
values $y_{i-1}, y_i$ and $y_{i+1}$ at the points $x_{i-1}, x_i, x_{i+1}$, respectively. (The existence
and uniqueness of such a polynomial is easy to establish, see Exercise (9.3)). In
other words $g_i$ agrees with $f$ at $x_{i-1}, x_i$ and $x_{i+1}$. We approximate $\int_{x_{i-1}}^{x_{i+1}} f(x)dx$
by $\int_{x_{i-1}}^{x_{i+1}} g_i(x)dx$. Since $g_i(x)$ is a quadratic in $x$, its graph is a parabola. So
this approximation is also called **parabolic approximation**. Fig. 5.9.2 (along
with an enlargement of a part of it) shows the graph of the original $f$ and that
of $g_i$.

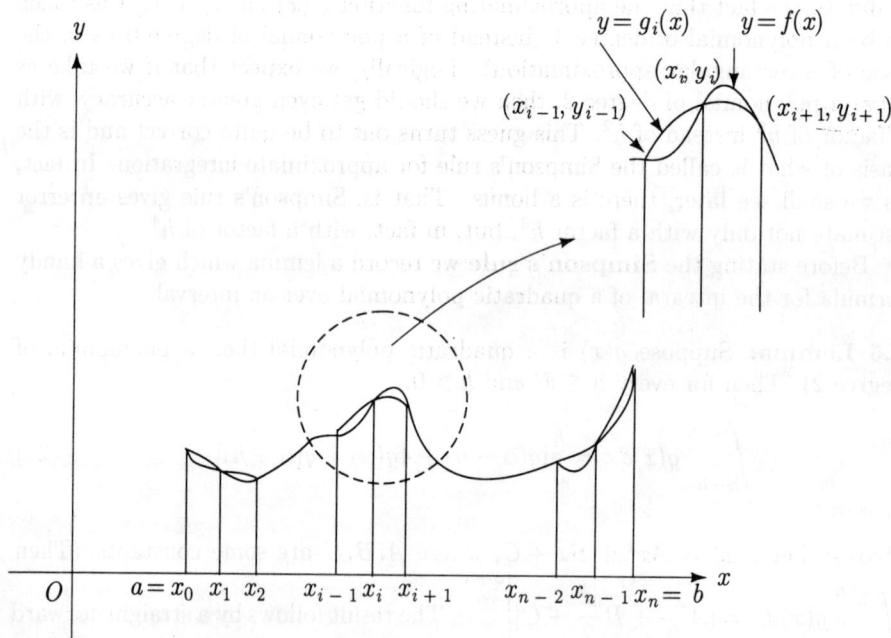

Figure 5.9.2 : Simpson's Rule

Each $\int_{x_{i-1}}^{x_{i+1}} g_i(x)dx$ can be evaluated using Lemma (9.5) and comes out
as $\dfrac{h}{3}(y_{i-1} + 4y_i + y_{i+1})$. Summing this over for all odd values of $i$ between 1
and $n - 1$, we get the Simpson's approximation to $\int_a^b f(x)dx$, with $n$ equal

subintervals, commonly denoted by $S_n$. Thus

$$S_n = \sum_{\substack{i=1 \\ i \text{ odd}}}^{n-1} \frac{h}{3}(y_{i-1} + 4y_i + y_{i+1})$$

$$= \frac{h}{3}[y_0 + y_n + 2\sum_{\substack{i=2 \\ i \text{ even}}}^{n-2} y_i + 4\sum_{\substack{i=1 \\ i \text{ odd}}}^{n-1} y_i] \tag{21}$$

Let us calculate this for our sample problem $\int_0^1 \frac{dx}{1+x^2}$ with $n = 10$. The values $y_0, y_1, \ldots, y_{10}$ were already calculated. So the work involved is little more than that with the rectangular or the trapezoidal rule. With a hand calculator, $S_n$ in this case comes out to be 0.7853981. The exact answer is $\frac{\pi}{4}$, which also equals 0.7853981 approximately. Of course $S_n$ cannot actually equal $\frac{\pi}{4}$ as the latter is irrational. But the error is less than $10^{-7}$. In other words Simpson's rule is so good that it can be used to evaluate $\frac{\pi}{4}$ (and hence $\pi$) approximately.

Let us now analyse the secret of such phenomenal accuracy of the Simpson's rule. As remarked earlier, this time the approximating polynomial $g_i(x)$ is of degree 2 instead of degree 1 as in the trapezoidal rule. If $f$ is thrice differentiable on $[x_{i-1}, x_{i+1}]$, then it is not hard to prove the analogue of Lemma (9.3) with three rather than two, applications of Rolle's theorem. Specifically, one can show that in such a case, for every $x \in [x_{i-1}, x_{i+1}]$, there exists some $c \in (x_{i-1}, x_{i+1})$ such that

$$f(x) - g_i(x) = \frac{1}{6}f'''(c)(x - x_{i-1}(x - x_i)(x - x_{i+1}). \tag{22}$$

We can then proceed as in Theorem (9.4) and obtain an upper bound on $|I - S_n|$ which will have a factor $h^3$. We leave this as an exercise (cf. Exercise (9.6)), partly because it will enable the reader to master the technique but more because, a better error estimate is possible for the Simpson's rule because of a clever trick, viz., to replace the quadratic polynomial $g_i(x)$ by a cubic polynomial $G_i(x)$ which has all the desirable properties of $g_i(x)$ plus something extra. It is this extra feature which enables us to get an upper bound on $|I - S_n|$ which has a factor of $h^4$.

The following lemma gives the explicit formula and the vital properties of this magic polynomial $G_i(x)$. The formula may appear contrived. There is, in fact, a way to show how the formula arises naturally. But it would take us too far afield to discuss it.

**9.6 Lemma:** Suppose $f$ is differentiable at $x_i$. Denote $f'(x_i)$ by $y_i'$. Define $G_i : [x_{i-1}, x_{i+1}] \to \mathbb{R}$ by

$$G_i(x) = \frac{y_{i+1} - y_{i-1} - 2y_i'h}{2h^3}(x-x_i)^3 + \frac{y_{i-1} + y_{i+1} - 2y_i}{2h^2}(x-x_i)^2 + y_i'(x-x_i) + y_i.$$

Then the cubic polynomial $G_i(x)$ has the following properties:

(i) $G_i(x_i) = y_i, G_i(x_{i-1}) = y_{i-1}, G_i(x_{i+1}) = y_{i+1}, G_i'(x_i) = y_i'$

(ii) $\displaystyle\int_{x_{i-1}}^{x_{i+1}} G_i(x)dx = \frac{h}{3}(y_{i-1} + 4y_i + y_{i+1}).$

**Proof:** All four parts of (i) are proved by straightforward computation. For (ii), make the substitution $x = u + x_i$. Then,

$$\int_{x_{i-1}}^{x_{i+1}} G_i(x)dx = \int_{-h}^{h} Au^3 + Bu^2 + Cu + Ddu$$

where $A, B, C, D$ are the coefficients of the descending powers of $(x - x_i)$ in $G_i(x)$. As the interval of integration is symmetric about 0, only the terms $Bu^2$ and $D$ matter and the integral becomes $\dfrac{2Bh^3}{3} + 2Dh$, which, after substituting for $B$ and $D$ reduces to the desired value. ∎

Part (ii) of this lemma shows that in defining the Simpson's rule we could as well use $G_i(x)$ instead of $g_i(x)$. This may sound foolish because $G_i(x)$ is more complex than $g_i(x)$. But actually it pays to do so, because using $G_i$ we get a sharper error estimate than with $g_i$. As regards its greater complexity, it is hardly a drawback, because neither $g_i(x)$ nor $G_i(x)$ figures in the actual computation of $S_n$ (which, as (21) shows, depends only on the values $y_0, \ldots, y_n$ of $f$ at $x_0, \ldots, x_n$ respectively). Both are needed only as tools in the error estimation, and not in the formula for the upper bounds on errors.

We are now ready to prove the analgoues of Lemma (9.3) and Theorem (9.4) for Simpson's rule. As is to be expected, we now require the fourth derivative of $f$.

**9.7 Lemma:** Suppose the fourth derivative $f^{(4)}$ of $f$ exists on $[x_{i-1}, x_{i+1}]$. Then for every $x \in [x_{i-1}, x_{i+1}]$, there exists $c \in (x_{i-1}, x_{i+1})$ such that

$$f(x) - G_i(x) = \frac{1}{24}f^{(4)}(c)(x - x_{i-1})(x - x_i)^2(x - x_{i+1}) \qquad (23)$$

**Proof:** The proof runs parallel to that of Lemma (9.3). For $x = x_{i-1}, x_i$ or $x_{i+1}$, (23) holds with any value of $c$. For any other $x \in [x_{i-1}, x_{i+1}]$, define a function $K : [x_{i-1}, x_{i+1}] \to \mathbb{R}$ by

$$K(y) = f(y) - G_i(y) - (f(x) - G_i(x))\frac{(y - x_{i-1})(y - x_i)^2(y - x_{i+1})}{(x - x_{i-1})(x - x_i)^2(x - x_{i+1})} \qquad (24)$$

[Note again that $x$ is a constant and $y$ is the variable].

By part (i) of the preceding lemma, $K$ vanishes at three points, viz., $x_{i-1}, x_i$ and $x_{i+1}$. Also $K(x) = 0$. So $K$ has at least four distinct zeros in $[x_{i-1}, x_{i+1}]$. By Rolle's theorem, $K'$ has at least three distinct zeros, one in each of the open intervals marked by these four points. In addition, $K'(x_i) = 0$, because $f'(x_i) = G'_i(x_i)$ by the last lemma and because of the multiple factor $(y - x_i)^2$ in the last term of (24). So $K'$ has at least four zeros in $(x_{i-1}, x_{i+1})$. By repeated applications of Rolle's theorem, $K^{(4)}$ vnishes at some $c \in (x_{i-1}, x_{i+1})$. And this leads to (23) sine the fourth derivative of $(y - x_{i-1})(y - x_i)^2(y - x_{i+1})$ is simply 24 everywhere. ∎

The stage is now set for an upper bound on the error in Simpson's rule.

**9.8 Theorem:** Suppose $f^{(4)}(x)$ exists and $|f^{(4)}(x)| \leq M$ for all $x \in [a, b]$. Then for $S_n$, the Simpson's approximation to $I \ (= \int_a^b f(x)dx)$, with even $n$, we have

$$|I - S_n| \leq \frac{M(b - a)}{180} h^4. \tag{25}$$

**Proof:** Once again, the proof runs parallel to that of Theorem (9.4). For each odd $i$ between 1 and $n - 1$, we apply the last lemma to get that for every $x \in [x_{i-1}, x_{i+1}]$

$$|f(x) - G_i(x)| \leq \frac{M}{24}(x - x_{i-1})(x - x_i)^2(x_{i+1} - x) \tag{26}$$

which is the analgoue of (18). If we integrate the right hand side of (26) on $[x_{i-1}, x_{i+1}]$, we get $\dfrac{Mh^5}{90}$ (cf. Exercise (9.1)). Now $I$ and $S_n$ are each a sum of $\dfrac{n}{2}$ integrals. Specifically, $I - S_n = \displaystyle\sum_{\substack{i=1 \\ i \text{ odd}}}^{n-1} \int_{x_{i-1}}^{x_{i+1}} f(x) - G_i(x)dx$. So,

$$|I - S_n| \leq \frac{n}{2}\frac{Mh^5}{90} = \frac{M(b - a)}{180}h^4. \ ∎$$

In our sample problem, $f(x) = \dfrac{1}{1 + x^2}$ and by direct calculation, $f^{(4)}(x) = \dfrac{24(1 - 10x^2 + 5x^4)}{(1 + x^2)^5}$. We may take $M = 24$ as an upper bound on $|f^{(4)}(x)|$ for $0 \leq x \leq 1$ (see Exercise (9.9)). The theorem above then gives $\dfrac{24}{180}10^{-4}$, or 0.00001333 as an upper bound on the accuracy of the Simpson's rule with $n = 10$. As we already saw, the actual accuracy is much higher, in fact almost total. But even the guaranteed accuracy is quite impressive because of the $h^4$ factor. To get a comparable accuracy with the trapezoidal rule, we would need 100 subintervals !

Just as the trapezoidal rule can be interpreted as the arithmetic mean of the left end-point and the right end-point rules, it is possible to interpret (cf.

Exercise (9.7)) the Simpson's rule as a certain weighted average of the left, the right and the mid-point rules. And that is why, the arithmetical work in the Simpson's rule is little more than in any rectangular quadrature. It is remarkable indeed that a suitable linear combination of the three rectangular quadratures gives such a high degree of accuracy at little extra cost.

The similarity in the statements and proofs of Corollary (9.2), Theorem (9.4) and Theorem (9.8) suggests that a unified treatment must be possible. This is indeed the case. All the three can be derived as special cases of a general construction. In fact, this general construction makes it possible to give other methods of quadrature which can yield higher accuracy than the Simpson's rule. But they lack the simplicity of the Simpson's rule. The Simpson's rule stands out as probably the best method of quadrature, at least at an elementary level. So we omit the general construction.

# EXERCISES

9.1   Prove (19). [*Hint*: Put $u = x - x_{i-1}$].

Similarly show that $\displaystyle\int_{x_{i-1}}^{x_{i+1}} (x - x_{i-1})(x - x_i)^2(x_i - x)dx = \frac{4h^5}{15}$ where

$h = \dfrac{x_{i+1} - x_{i-1}}{2}$ and $x_i = \dfrac{x_{i-1} + x_{i+1}}{2}$.

9.2   What can be said about the sign of $I - T_n$ when $f$ is (i) concave upwards (ii) concave downwards on $[a, b]$?

9.3   Given three distinct real numebrs $x_{i-1}, x_i$ and $x_{i+1}$ and three (not necessarily distinct) real numbers $y_{i-1}, y_i, y_{i+1}$ prove that there exists a unique polynomial $g_i(x)$ of degree 2 (or less) such that $g_i(x_{i-1}) = y_{i-1}, g_i(x_i) = y_i$ and $g_i(x_{i+1}) = y_{i+1}$. [*Hint* : Let $g_i(x) = Ax^2 + Bx + C$. Write three equations for $A, B, C$ and solve simultaneously.]

9.4   As a generalisation of the last exercise, show that given any dinstinct real numbers $x_0, \ldots, x_k$ and any real numbers $y_0, \ldots, y_k$, there exists a unique polynomial $h(x)$ of degree $\leq k$ such that $h(x_i) = y_i$ for $i = 0, 1, \ldots, k$. Show, in fact, that $h(x)$ is given explicitly by $\displaystyle\sum_{i=0}^{k} y_i h_i(x)$ where for $i = 0, \ldots, k$,

$h_i(x) = \dfrac{(x - x_0) \ldots (x - x_{i-1})(x - x_{i+1}) \ldots (x - x_k)}{(x_i - x_0) \ldots (x_i - x_{i-1})(x_i - x_{i+1}) \ldots (x_i - x_k)}$. (This famous result is called **Lagrange's Interpolation Formula**.)

9.5   If $f$ is a polynomial of degree $\leq 3$, prove that $S_n$ coincides with $\displaystyle\int_a^b f(x)dx$ for any $a, b$.

9.6   Prove (22) and then using it, show that if $|f^{(3)}(x)| \leq M$ for all $x \in [a, b]$, then for the Simpson rule, $|I - S_n| \leq \dfrac{M}{24}(b - a)h^3$.

9.7    Suppose $n$ is even and $[a, b]$ is divided into $\dfrac{n}{2}$ equal parts. Let $L, R$ and $M$ be the approximations to $\displaystyle\int_a^b f(x)dx$ given, respectively, by the left end-point, the right end-point and the mid-point rules. Prove that $S_n = \dfrac{L}{6} + \dfrac{R}{6} + \dfrac{2M}{3}$. Hene show that $S_n \to \displaystyle\int_a^b f(x)dx$ as $n$ tends to infinity through even values.

9.8    Find $\ln 2$ approximately by evaluating $\displaystyle\int_1^2 \dfrac{dx}{x}$ using the rectangular, the trapezoidal and the Simpson's rules with $n = 10$. Obtain upper bounds on the errors. (The actual value of $\ln 2$ is 0.693147 )

9.9    For $f(x) = \dfrac{1}{1+x^2}$, verify that 24 is an upper bound on $|f^{(4)}(x)|$ for $x \in [0, 1]$. Prove, in fact, that it is the maximum.

## 5.10    Improper Integrals

In Section 5.2, it was remarked that certain notations and terminology about definite integrals owe their origins to the fact that integration is a limit of summation. When an integral cannot be evaluated exactly in a closed form we have to settle for an approximate answer in the form of a summation (a few methods for this were given in the last section). On the other hand, Exercise (5.9) shows how certain finite sums can be approximated by definite inetgrals (see also Exercises (2.6), (5.7) and (7.18)).

There is, however, an essential difference between a (finite) sum, say, $\displaystyle\sum_{k=1}^n a_k$ and a definite integral, say, $\displaystyle\int_a^b f(x)dx$. In the former, the index variable $k$ is a discrete variable taking only the integer values from 1 to $n$, while in the latter, the variable of integration, viz. $x$, is a continuous variable taking all real values from $a$ to $b$. Certain concepts such as continuity and derivatives make no sense for discrete variables. Consequently, it would appear that results about integrals, such as the Fundamental Theorem of Calculus where derivatives are crucially involved, can have no analogues for finite sums. Surprisingly, such analogues sometimes do exist. Quite understandably, such analogues are generally not as profound as the corresponding results about integrals. But their very existence is interesting and sometimes leads one to explore more similarities.

The key idea is to replace the concept of a derivative with something which comes closest to it in some sense. Suppose $f$ is a real-valued function defined in a neighbourhood of a point $c$. Then $f'_+(c) = \lim\limits_{h \to 0+} \dfrac{f(c+h) - f(c)}{h}$ and $f'_-(c) = \lim\limits_{h \to 0+} \dfrac{f(c) - f(c-h)}{h}$. Here $h$ tends to 0 through positive real values. For small $h$, the incrementary ratios $\dfrac{f(c+h) - f(c)}{h}$ and $\dfrac{f(c) - f(c-h)}{h}$ are close to $f'_+(c)$ and $f'_-(c)$ respectively. There is, of course, no such thing as the smallest positive real number and that is exactly why we have to resort to limits. Suppose, however, just for a moment, that there were a smallest positive real number, say $h_0$. In that case $f'_+(c)$ and $f'_-(c)$ would equal, respectively, $\dfrac{f(c + h_0) - f(c)}{h_0}$ and $\dfrac{f(c) - f(c - h_0)}{h_0}$.

A sequence is, by definition, a function defined on $I\!N$, the set of positive integers. Suppose $\{a_n\}_{n=1}^{\infty}$ is a sequence of real numbers. Let us write $f(n)$ for $a_n$. Here $n$ is a discrete variable ranging over $I\!N$. Let $m \in I\!N$. Since $n$ is a discrete variable, expressions like $\lim\limits_{n \to m} f(n)$ or $\lim\limits_{n \to m} \dfrac{f(n) - f(m)}{n - m}$ have no meanings. However, the closest $n$ can came to $m$ (without actually equalling $m$) is when $n = m \pm 1$. So, in view of what is said in the last paragraph, $\dfrac{f(m+1) - f(m)}{1}$, i.e., $a_{m+1} - a_m$ is the closest we can come to the concept of the right handed derivative of $f$ at $m$. Conceptually, $a_{m+1} - a_m$ requires only simple arithmetic for its definition, unlike the far more subtle concept of a derivative which requires the concept of a limit. So on the face of it, it may seem a shallow and an unimportant concept. But the concept appears frequently enough in certain branches of mathematics to be given a name. We give it along with that of the dual concept corresponding to the left handed derivative.

**10.1 Definition:** Let $\{a_n\}_{n=1}$ be a sequence. Then for $m \in I\!N$, $a_{m+1} - a_m$ and $a_m - a_{m-1}$ are called, respectively, the **forward** and the **backward divided differences** at $m$.

The backward divided difference is undefined for $m = 1$. (By express convention we may set is as $a_1 - 0$, i.e., as $a_1$ itself). For $m > 1$, it is clear that the backward divided difference at $m$ is precisely the forward divided difference at $m - 1$. The two concepts being so intimately related, it is customary to call the forward divided difference, simply as divided difference or even more simply, just 'difference'. (The term divided difference is used in a more general situation where the points in the domain of a function $f$ are separated by whole multiples of some unit, say $h > 0$. In that case the forward divided difference of $f(x)$ at $c$ is $\dfrac{f(c+h) - f(c)}{h}$. Here $f(c+h) - f(c)$ is the difference and $\dfrac{f(c+h) - f(c)}{h}$ is the divided difference. In the case of a sequence, $h = 1$ and so the divided difference coincides with the difference.) The new sequence $\{b_n\}_{n=1}^{\infty}$ where $b_n = a_{n+1} - a_n$ for $n \in I\!N$ is called the **derived sequence** of the original sequence $\{a_n\}_{n=1}^{\infty}$ and often denoted by $\{\triangle a_n\}_{n=1}^{\infty}$, the name clearly coming from that of the analogous concept of a derived function. Unlike derivatives, there is absolutely no condition for the existence of the derived sequence. *Every* (infinite) sequence has a derived sequence. So, in particular, one can define the second and the third derived sequences, and in general the derived sequence of any order of a given sequence. In certain respects they do behave like derivatives. A few such properties will be given in the exercises. There is a branch of mathematics called **calculus of differences**, which studies properties of derived sequences.

For a finite sequence, say, $\{a_k\}_{k=1}^{h}$, the forward difference is defined for $k = 1, \ldots, n-1$ while the backward difference is defined for $k = 2, \ldots, n$. Evidently, $\sum_{k=1}^{n-1} (a_{k+1} - a_k) = a_n - a_1$. This fact is too trivial to deserve a mention. But it is the exact analogue of the first form of the Fundamental Theorem of Calculus, viz., $\displaystyle\int_a^b f'(x)dx = f(b) - f(a)$, because the derivative corresponds to the difference and integration to summation. The discrete analogue of the second form of the Fundamental Theorem is equally trivial. The concept corresponding to the function defined by an integral is clearly, what we have been calling as the partial sum of a sequence (or series). If $\{a_n\}_{n=1}^{\infty}$ is a sequence and $A_n = \displaystyle\sum_{k=1}^{n} a_k$ is its partial sum, then $A_n - A_{n-1} = a_n$ and this corresponds to $F'(x) = f(x)$ where $F(x) = \displaystyle\int_a^x f(t)dt$. (Note that here we are taking the backward difference at $n$.)

The utter triviality of the discrete analogue of the Fundamental Theorem of Calculus may suggest that it is probably unrewarding to pursue such analogies. But there are exceptions. The following result on finite sums, called **summation by parts**, is the discrete analogue of integration by parts. In fact, although this analogy is not needed in the proof (which is elementary, albeit a little tricky), without such analogy, the statement of the result may appear too weird. The result has some interesting applications as we shall see.

**10.2 Theorem:** Let $\{a_n\}_{n=1}^{\infty}$ and $\{b_n\}_{n=1}^{\infty}$ be any two sequences. For $n \in I\!N$ put $A_n = \displaystyle\sum_{k=1}^{n} a_k$ (with $A_0 = 0$). Then for any integers $p$ and $q$ with $1 \le p \le q$, we have,

$$\sum_{n=p}^{q} a_n b_n = A_q b_q - A_{p-1} b_p - \sum_{n=p}^{q-1} A_n (b_{n+1} - b_n) \tag{1}$$

**Proof:** Since $a_n = A_n - A_{n-1}$, we have

$$\sum_{n=p}^{q} a_n b_n = \sum_{n=p}^{q} (A_n - A_{n-1}) b_n = \sum_{n=p}^{q} A_n b_n - \sum_{n=p}^{q} A_{n-1} b_n \qquad (2)$$

By a change of index, the last sum in (2) equals $\sum_{n=p-1}^{q-1} A_n b_{n+1}$. So,

$$\sum_{n=p}^{q} a_n b_n = A_q b_q + \sum_{n=p}^{q-1} A_n b_n - \sum_{n=p}^{q-1} A_n b_{n+1} - A_{p-1} b_p$$

which gives the result. ∎

As a consequnce, we get the following useful result regarding convergene of certain series. It is called **Dirichlet's test.**

**10.3 Theorem:** Suppose the partial sums $A_n$ of a series $\sum_{n=1}^{\infty} a_n$ are all bounded (by a bound independent of $n$), and $\{b_n\}_{n=1}^{\infty}$ is a monotonically decreasing sequence converging to 0. Then $\sum_{n=1}^{\infty} a_n b_n$ is convergent.

**Proof:** We shall apply the Cauchy criterion (Theorem (3.5.5)) for convergence of series. Let $\epsilon > 0$ be given. By the first part of the hypothesis, there exists $M > 0$ such that $|A_n| \leq M$ for all $n \in \mathbb{N}$. Fix such $M$. From the second part of the hypothesis, there exists $m \in \mathbb{N}$ such that $0 \leq b_n < \dfrac{\epsilon}{2M}$ for all $n \geq m$. Now from the last theorem and the fact that $|b_{n+1} - b_n| = b_n - b_{n+1}$ for all $n \in \mathbb{N}$, we get that whenever $q \geq p \geq m$,

$$\left| \sum_{n=p}^{q} a_n b_n \right| \leq |A_q b_q| + |A_{p-1} b_p| + \sum_{n=p}^{q-1} |A_n| (b_n - b_{n+1})$$

$$\leq M \left( b_q + b_p + \sum_{n=p}^{q-1} (b_n - b_{n+1}) \right)$$

$$= M(2b_p) < M \frac{2\epsilon}{2M} = \epsilon.$$

So by Theorem (3.5.5) (beware of the changes in notation) the series $\sum_{n=1}^{\infty} a_n b_n$ is convergent. ∎

If we take $a_n = (-1)^{n+1}$, then $A_n = 1$ or $0$ for each $n$ and so the theorem just proved implies that whenever $\{b_n\}$ is a monotonically decreasing sequence converging to 0, the series $\sum_{n=1}^{\infty}(-1)^{n+1}b_n$ is convergent. This is the well-known alternating series test (also called Leibnitz test) in Exercise (3.8.1). Yet other interesting application of Theorem (10.2) will be given in Exercises (10.4) and (10.5).

Probably the most striking instances of the analogy between summation and integration occur when we compare infinite series and what are called improper integrals. So far, in our definition of a definite integral $\int_a^b f(x)dx$, we assumed that $f$ was bounded on $[a, b]$. Also the interval $[a, b]$ is of finite length, viz. $b - a$. This fact was crucially used in many of the results we proved about definite integrals. Integrals where the integrand and/or the interval of integration is unbounded are called **improper integrals**. Like finite series, the 'proper', i.e. the ordinary integrals are generally sufficient in day-to-day real life applications. But like infinite series, the improper integrals are occasionally necessary, especially for theoretical purposes[6] (see, for example, Exercises (10.23) and (10.24)). Moreover the approach to them is very similar. Just as an infinite series is handled as the limit of its partial sums (which are finite sums), an improper integral is handled as a limit of ordinary definite integrals. (Note, however, that unlike a finite sum, a definite integral is itself a limit of a finite sum. So in an improper integral, the limiting process enters twice).

For convenience, let us first tackle the impropriety arising out of the domain being an unbounded interval.

**10.4 Definition:** Suppose $f$ is a real-valued function defined on an interval of the form $[a, \infty)$, for some $a \in \mathbb{R}$. Assume $f$ is integrable on $[a, b]$ for every $b \geq a$. Denote $\int_a^b f(x)dx$ by $F(b)$. Then $F$ is a well-defined real-valued function on $[a, \infty)$. If $\lim_{b \to \infty} F(b)$ exists we say that the **improper integral** $\int_a^\infty f(x)dx$ **converges** (or **exists**) and define its value to be this limit.

This definition is in direct analogy with the definition of convergence of $\sum_{n=1}^{\infty} a_n$ as $\lim_{n \to \infty} S_n$ where $S_n = \sum_{k=1}^{n} a_k$, for $n \in \mathbb{N}$. Because of this resemblance, improper integrals of this type behave like infinite series in many respects. There is, however, an important difference. In an infinite series the variable is discrete while in an improper integral it is continuous. As a result, the analogy is not perfect as we shall see later.

Before giving examples of similarities and dissimilarities, let us first take some illustrative examples. Consider $\int_1^\infty \frac{dx}{1 + x^2}$. For $b \geq 1$, we have $\int_1^b \frac{dx}{1 + x^2}$

---

[6]Sometimes improper integrals arise in connection with proper integrals too, as for example, when we convert an integral using trigonometric substitutions, see e.g. Exercise (6.8).

$= \tan^{-1}(b) - \dfrac{\pi}{4}$. As $b \to \infty$, $\tan^{-1}(b) \to \dfrac{\pi}{2}$. So $\displaystyle\int_1^\infty \dfrac{dx}{1 + x^2}$ is convergent with

value $\dfrac{\pi}{4} \left(= \dfrac{\pi}{2} - \dfrac{\pi}{4}\right)$. On the other hand, $\displaystyle\int_0^\infty e^x\,dx$ is divergent because $\displaystyle\int_0^b e^x\,dx$

$= e^b - 1 \to \infty$ as $b \to \infty$. In both these examples, we could settle the convergence of $\displaystyle\int_a^\infty f(x)\,dx$ by directly calculating $\displaystyle\int_a^b f(x)\,dx$ in a closed form and then studying its behaviour as $b \to \infty$. In this respect, the improper integrals behave slightly differently (and more pleasantly) than infinite sums. As was pointed out in Section 3.9, it is generally impossible to decide the convergence or otherwise of a series $\displaystyle\sum_{n=1}^\infty a_n$ directly from the definition because it is very rare that the partial sum $\displaystyle\sum_{h=1}^n a_k$ can be evaluated in a closed form. And that is why we need so many tests for convergence of series.

Even with improper integrals things are not, of course, always so rosy. Take, for example, the integrals $\displaystyle\int_0^\infty e^{x^2}\,dx$ and $\displaystyle\int_0^\infty e^{-x^2}\,dx$. Neither $e^{x^2}$ nor $e^{-x^2}$ has an antiderivative in a closed form and so we cannot settle the convergence of either of these two improper integrals, straight from the definition. As with series, we need various tests for this purpose. Using some of these tests it is not hard to show that $\displaystyle\int_0^\infty e^{x^2}\,dx$ is divergent while $\displaystyle\int_0^\infty e^{-x^2}\,dx$ is convergent. But to actually evaluate the latter requires fairly advanced methods; see Exercise (10.23)(iii). This behaviour, too, is analogous to that encountered with infinite series. For example, in Theorem (6.3) it was shown that $\displaystyle\sum_{n=1}^\infty \dfrac{1}{n^2}$ is convergent. But rigourously showing that its value is $\dfrac{\pi^2}{6}$ is fairly non-trivial. (For a non-rigourous argument, see Exercise (1.4.9).)

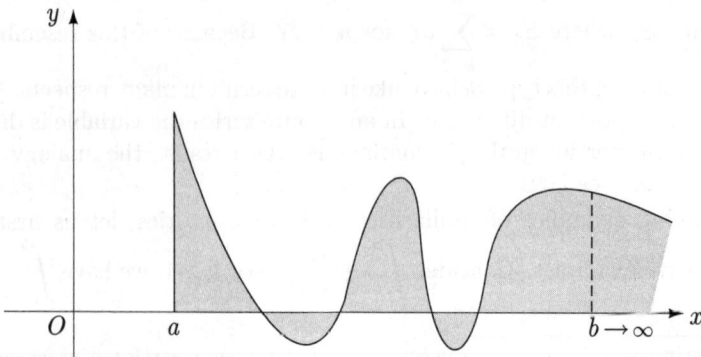

Figure 5.10.1 : Geometric Interpretation of Improper Integral

The geometric interpretation of an improper integral $\int_a^\infty f(x)dx$ is straight-forward but sometimes paradoxical to a beginner. Just as $\int_a^b f(x)dx$ represents the area bounded by the $x$-axis, the graph $y = f(x)$ and the vertical lines $x = a$ and $x = b$, the integral $\int_a^\infty f(x)dx$ represents the area of the region bounded by the $x$ - axies, the graph $y = f(x)$ and the line $x = a$ on the left, (with the understanding that areas below the $x$ - axis are to be treated negative), see Fig. 5.10.1. Note that this region is not bounded on the right.

A beginner often feels that since this region is unbounded, it can never have a finite area. To dispel such a thought, note that even if one side of a rectangle is very large, its area can be quite small if the other side is sufficiently small. In the case of an improper integral, we have a region with an infinitely long base, but varying height and if this height decreases very rapidly, then it is quite possible that the region in question has a finite area. That is exactly what happens in the case of the integral $\int_1^\infty \dfrac{dx}{1+x^2}$ which converges to $\dfrac{\pi}{4}$ as we already saw. Even when there are 'pockets', i.e. subintervals of $[a, \infty)$ over which $f(x)$ is not small, the area may still be finite if the widths of these pockets are very small, see Exercise (10.6)(a). The geometric interpretation often makes it easier to understand and to construct counter-examples about improper integrals.

Let us now come down to developing certain tests for the convergence of $\int_a^\infty f(x)dx$, analogous to the tests for the convergence of an infinite series $\sum\limits_{n=1}^{\infty} a_n$. Exercise (3.1.8) shows that a necessary condition for the latter is that $a_n \to 0$ as $n \to \infty$. It is then tempting to think that a necessary condition for convergence of $\int_a^\infty f(x)dx$ is that $f(x) \to 0$ as $x \to \infty$. But this is not so. The reason is that, as discussed in Section 5.3, integration is a global construction and so the existence of an integral implies nothing about the values of the integrand at particular points. A trivial counter-example is to take $f(x) = 0$ whenever $x \notin I\!\!N$. Then regardless of how $f$ is defined on $I\!\!N$, $\int_a^b f(x)dx = 0$ for all $a, b \in I\!\!R$ and so $\int_a^\infty f(x)dx$ converges to 0. Less trivial examples (e.g. where $f$ is continuous) can also be given (see Exercise (10.6)(b)).

Despite this, the analogue of Cauchy criterion (Theorem (3.5.5)) holds. The proof of the direct implication remains basically the same. But the converse implication requires a little more careful handling.

**10.5 Theorem:** The improper integral $\int_a^\infty f(x)dx$ is convergent if and only if for every $\epsilon > 0$, there exists $R$ $(\geq a)$ such that whenever $R \leq b_1 \leq b_2$, we have

$$\left| \int_{b_1}^{b_2} f(x)dx \right| < \epsilon.$$

**Proof.** Assume $\int_a^\infty f(x)dx$ is convergent with value $I$. Given $\epsilon > 0$, there exists $R \geq a$ such that $\left| \int_a^b f(x)dx - I \right| < \frac{\epsilon}{2}$ for all $b \geq R$. In particular, whenever $R \leq b_1 < b_2$ we have $\left| \int_a^{b_1} f(x)dx - I \right| < \frac{\epsilon}{2}$ and $\left| \int_a^{b_2} f(x)dx - I \right| < \frac{\epsilon}{2}$, which by triangle inequality (and symmetry), gives $\left| \int_a^{b_2} f(x)dx - \int_a^{b_1} f(x)dx \right| < \frac{\epsilon}{2} + \frac{\epsilon}{2} = \epsilon.$ Exercise (1.7) shows that $\left| \int_{b_1}^{b_2} f(x)dx \right| < \epsilon.$

For the converse, consider the sequence $\{I_n\}_{n=1}^\infty$ where $I_n = \int_a^{a+n} f(x)dx$ for $n \in I\!N$. The hypothesis easily implies that $\{I_n\}$ is a Cauchy sequence (see Exercise (10.7)) and so by Theorem (3.5.4), there exists some $I$ such that $I_n \to I$ as $n \to \infty$. We claim that $\int_a^\infty f(x)dx$ converges to $I$, that is $\lim_{b \to \infty} \int_a^b f(x)dx = I$. Let $\epsilon > 0$ be given. By hypothesis, there is some $R \geq a$ such that whenever $R \leq b_1 \leq b_2$,

$$\left| \int_{b_1}^{b_2} f(x)dx \right| < \frac{\epsilon}{2} \tag{3}$$

Also since $I_n \to I$ as $n \to \infty$, there exists $m \in I\!N$ such that for all $n \geq m$, $|I_n - I| < \frac{\epsilon}{2}$. In particular $|I_m - I| < \frac{\epsilon}{2}$, i.e.,

$$\left| \int_a^{a+m} f(x)dx - I \right| < \frac{\epsilon}{2} \tag{4}$$

We may suppose $m$ is so large that $a + m \geq R$. Then for all $b \geq a + m$, we have (by putting $b_1 = a + m$ and $b_2 = b$ in (3))

$$\left| \int_{a+m}^b f(x)dx \right| < \frac{\epsilon}{2} \tag{5}$$

Now 
$$\left| \int_a^b f(x)dx - I \right| = \left| \int_a^{a+m} f(x)dx - I + \int_{a+m}^b f(x)dx \right|$$
$$\leq \left| \int_a^{a+m} f(x)dx - I \right| + \left| \int_{a+m}^b f(x)dx \right|$$
$$< \frac{\epsilon}{2} + \frac{\epsilon}{2} = \epsilon.$$

So $\lim_{b \to \infty} \int_a^b f(x)dx = I$ as was to be proved. ∎

The Cauchy crietrion makes it easy to prove the following properties of improper integrals. (The first property is also easy to establish directly from the definition.) The proofs are left as exercises because they resemble the proofs of the corresponding properties of infinite series.

**10.6 Theorem:** Suppose $a \in \mathbb{R}$ and $f, g$ are real-valued functions which are integrable over $[a, b]$ for every $b \geq a$. Then,

(i)  (*linearity*) if $\int_a^\infty f(x)dx$ and $\int_a^\infty g(x)dx$ both exist then for any constants $\alpha, \beta, \int_a^\infty \alpha f(x) + \beta g(x)dx$ is convergent and convergences to $\alpha \int_a^\infty f(x)dx + \beta \int_a^\infty g(x)dx,$

(ii)  (*absolute convergence implies convergence*) if $\int_a^\infty |f(x)|dx$ is convergent then so is $\int_a^\infty f(x)dx,$

(iii)  (*comparison test*) if $0 \leq f(x) \leq g(x)$ for all $x \geq a$ then convergence of $\int_a^\infty g(x)dx$ implies that of $\int_a^\infty f(x)dx,$ and

(iv)  if $f(x) \geq 0$ and $g(x) \geq 0$ for all $x \geq a$ and $\dfrac{f(x)}{g(x)}$ tends to a finite non-zero limit as $x \to \infty$ then either $\int_a^\infty f(x)dx$ and $\int_a^\infty g(x)dx$ both converge or both diverge. $\blacksquare$

As in the case of infinite series, the converse of (ii) is false. In fact, the standard counter-example, viz. the series $\sum\limits_{n=1}^\infty (-1)^{n-1}\dfrac{1}{n}$ can be modified for this purpose. All we need to do is to define $f : [1, \infty) \to \mathbb{R}$ by $f(x) = (-1)^{[x]+1}\frac{1}{[x]}$, where, as usual, $[x]$ denotes the integral part of $x$. Here $f$ is a constant equal to $(-1)^{n+1}\dfrac{1}{n}$ on $[n, n+1)$ for $n \in \mathbb{N}$. So, for $b \in [m, m+1), \int_1^b f(x)dx$ lies between $\sum\limits_{n=1}^m (-1)^{n+1}\dfrac{1}{n}$ and $\sum\limits_{n=1}^{m+1}(-1)^{n+1}\dfrac{1}{n}$ (which of these two numbers is bigger depends on whether $m$ is even or odd) and differs from each by at most $\dfrac{1}{m+1}$. (This is much easier to see from the graph of $f(x)$ sketched in Figure 5.10.2.) This observation, along with the convergence of $\sum\limits_{n=1}^\infty (-1)^{n+1}\dfrac{1}{n}$ implies the convergence of $\int_1^\infty f(x)dx$. The proof that $\int_1^\infty |f(x)|dx$ is divergent is similar. This example may appear a little artificial. (A more 'natural' example will be given in Exercise (10.12).) But it illustrates a handy construction which is sometimes useful in converting counter-examples about infinite series to those about improper integrals (for details see Exercise (10.9)).

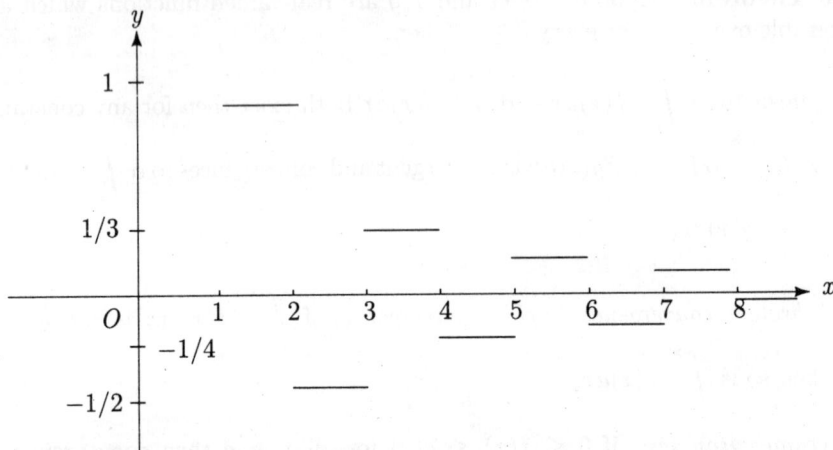

Figure 5.10.2 :   Convergent but not Absolutely Convergent
                  Improper Integral

As in the case of convergence of sequences and series (see Section 3.2), eventuality plays an important role in convergence of improper integrals. In other words, following the notation of Definition (10.4), the improper integral $\int_a^\infty f(x)dx$ is convergent if we can find some $c > a$ such that $\int_c^\infty f(x)dx$ is convergent. This increases the applicability of parts (iii) and (iv) of the last theorem because it suffices if the conditions in them are satisfied eventually, i.e., for all $x$ after some stages.

Actually, the comparison test (and its consequences) are, in a way, more applicable for integrals than for series. The comparison test requires that we should know the convergence of something else beforehand. In the case of infinite series, the geometric series are just about the only naturally occurring series whose convergence can be settled directly. As noted earlier, for improper integrals, the situation is more pleasant because of the Fundamental Theorem of Calculus. In fact, the following simple result gives a good source of some standard integrals with which the given integrand can be compared.

**10.7 Theorem:** Let $p$ be any real number. Then $\int_1^\infty \dfrac{dx}{x^p}$ is convergent if $p > 1$ and divergent otherwise.

**Proof:** For $b \geq 1$, $\int_1^b \dfrac{dx}{x^p} = \int_1^b x^{-p}dx = \dfrac{1}{1-p}(b^{1-p} - 1)$ if $p \neq 1$. For $p > 1$,

$b^{1-p} = \dfrac{1}{b^{p-1}} \to 0$ as $p \to \infty$, while for $p < 1$, $b^{p-1} \to \infty$ as $b \to \infty$. This proves the theorem for $p \neq 1$. The case $p = 1$ follows from the fact that $\ln b \to \infty$ as $b \to \infty$ (see the proof of Theorem (7.5)). ∎

For example, consider $\displaystyle\int_1^\infty \dfrac{x^2 - 7x}{10x^5 + \sin x - 5} dx$. For large $x$, $\sin x$ and $5$ are insignificant as compared to $10x^5$ and $x^2$ dominates $7x$. So if we take $f(x)$ as $\dfrac{x^2 - 7x}{10x^5 + \sin x - 5}$ and $g(x)$ as $\dfrac{1}{x^3}$, we see that $\dfrac{f(x)}{g(x)} \to \dfrac{1}{10}$ as $x \to \infty$. From the last theorem, $\displaystyle\int_1^\infty g(x)dx$ converges. So from (iv) in Theorem (10.6), the given integral converges.

The comparison test also enables us to prove that $\displaystyle\int_0^\infty e^{x^2} dx$ is divergent from the divergence of $\displaystyle\int_0^\infty e^x dx$ (which follows directly from the definition, as noted above). All we need to do is to note that for $x \geq 1$, $e^{x^2} \geq e^x \geq 0$. Similarly from the convergence of $\displaystyle\int_0^\infty e^{-x} dx$ and the fact that $0 \leq e^{-x^2} \leq e^{-x}$ for $x \geq 1$, we see that $\displaystyle\int_0^\infty e^{-x^2} dx$ is convergent.

The exercises will give further drill in the use of the various tests for convergence of improper integrals. By now it should be clear enough why the improper integrals behave like infinite series in some (but not all) respects. We now turn to the question of relating the convergence of an integral of the form $\displaystyle\int_a^\infty f(x)dx$ with that of a series of the form $\displaystyle\sum_{n=1}^\infty f(x_n)$, where $\{x_n\}$ is a monotonically increasing sequence in $[a, \infty)$, such that $x_n \to \infty$ as $n \to \infty$ (e.g. $x_n = a + n$ or $x_n = a + n^2$ etc.) The fact that integration is a global process (Section 5.3) cautions us against expecting much. And simple counter-examples (to be given in the Exercises), in fact, show that all sorts of situations are possible, e.g. where $\displaystyle\int_a^\infty f(x)dx$ is convergent but $\displaystyle\sum_{n=1}^\infty f(x_n)$ is convergent for some choices of $\{x_n\}$ but divergent for some other choices. What matters is what the function $f$ is and how the $x_n$'s are chosen. The following theorem gives one important case where with suitable assumptions of this type, we get a positive result.

**10.8 Theorem:** Suppose $f$ is monotonically decreasing and non-negative on $[1, \infty)$. Then $\displaystyle\int_1^\infty f(x)dx$ is convergent if and only if the series $\displaystyle\sum_{n=1}^\infty f(n)$ is convergent.

**Proof:** By Exercise (1.15), the integral $\int_1^b f(x)dx$ is defined for every $b \geq 1$ and it increases as $b$ increases by non-negativity of the integrand. Also $\sum_{n=1}^{\infty} f(n)$ is a series of non-negative terms and so the partial sum $\sum_{k=1}^{n} f(k)$ increases as $n$ increases. So it suffices to show that $\int_1^b f(x)dx (1 \leq b < \infty)$ and $\sum_{k=1}^{n} a_k (n \in \mathbb{N})$ are either both bounded or both unbounded. For this, suppose $1 \leq b < \infty$. Let $m = [b]$. Then $m \leq b < m + 1$ and by non-negativity of $f$,

$$\int_1^m f(x)dx \leq \int_1^b f(x)dx \leq \int_1^{m+1} f(x)dx. \tag{6}$$

Now, $\int_1^m f(x)dx = \sum_{k=2}^{m} \int_{k-1}^{k} f(x)dx$. By monotonicity of $f$, for every $k = 2, \ldots,$ $m$, $f(k-1) \geq f(x) \geq f(k)$ for all $x \in [k-1, k]$ and so

$$f(k) = \int_{k-1}^{k} f(k)dx \leq \int_{k-1}^{k} f(x)dx \leq \int_{k-1}^{k} f(k-1)dx = f(k-1). \tag{7}$$

From (6) and (7) it follows that

$$\sum_{k=2}^{m} f(k) \leq \int_1^b f(x)dx \leq \sum_{k=1}^{m} f(k). \tag{8}$$

This inequality (whose derivation is basically the same as showing that $H_n - 1 \leq \ln n \leq H_n$, where $H_n$ is the $n^{th}$ harmonic number) shows that $\int_1^b f(x)dx$ and $\sum_{k=1}^{m} f(k)$ are both bounded or both unbounded and thereby completes the proof.∎

This theorem is known as **Maclaurin Integral Test** for series, because although theoretically it could be used to test the convergence of an improper integral when we already know the convergence of the corresponding series, in practice, it is generally used the other way. Once again, the reason is that it is comparatively more common to evaluate a definite integral in a closed form than a partial sum. As a good illustration, it was proved through Theorem (3.6.3), Exercise (3.6.2) and Exercise (7.9) that the series $\sum_{n=1}^{\infty} \frac{1}{n^r}$ is convergent if $r > 1$ and divergent otherwise. But the proof required Cauchy's condensation test, (whose underlying reasoning was analogues to (7) above). Using the theorem above and Theorem (10.7), we get the result effortlessly for $r \geq 0$, because for these values of $r$, the function $f(x) = \dfrac{1}{x^r}$ is monotonically decreasing and positive on $[1, \infty)$. The case $r < 0$ is trivial anyway since $n^r \not\to 0$ as $n \to \infty$ if $r < 0$.

So far we discussed the basic properties of integrals of the form $\displaystyle\int_a^\infty f(x)dx$. An entirely analogous treatment holds for improper integrals of the form $\displaystyle\int_{-\infty}^a f(x)dx$. If one wants, one can convert such integrals to those of the type we have discussed using the subsituation $u = -x$. So we omit their discussion. Improper integrals of the form $\displaystyle\int_{-\infty}^\infty f(x)dx$ where the interval of integration is unbounded at both the ends are also handled similarly. But a word of caution is in order. In such integrals it is assumed that $f$ is integrable over $[a, b]$ for all $a, b \in \mathbb{R}$. By definition, $\displaystyle\int_{-\infty}^\infty f(x)dx$ is the limit of $\displaystyle\int_a^b f(x)dx$ as $a \to -\infty$ and $b \to \infty$ independently of each other. Alternatively, we can take any 'break point', say $c$, and define $\displaystyle\int_{-\infty}^\infty f(x)dx = \int_{-\infty}^c f(x)dx + \int_c^\infty f(x)dx$, to mean that the left hand side exists only when each of the two integrals on the right exists. It is not hard to show that if this condition holds for one value of $c$ then it holds for any other value of $c$. So it does not matter which break point we take. In particular, taking $c = 0$, we can write $\displaystyle\int_{-\infty}^\infty f(x)dx = \int_{-\infty}^0 f(x)dx + \int_0^\infty f(x)dx$. Thus, for example, $\displaystyle\int_{-\infty}^\infty \frac{dx}{x^2 + 2x + 2}$ equals $\displaystyle\int_{-\infty}^0 \frac{dx}{x^2 + 2x + 2} + \int_0^\infty \frac{dx}{x^2 + 2x + 2}$. Theorem (10.7) along with the comparison test shows that each of these integrals exist. We caution the reader against defining $\displaystyle\int_{-\infty}^\infty f(x)dx$ as $\displaystyle\lim_{a \to \infty} \int_{-a}^a f(x)dx$.

Certainly, if $\displaystyle\int_{-\infty}^\infty f(x)dx$ exists then this limit also exists and can be used to evaluate $\displaystyle\int_{-\infty}^\infty f(x)dx$. But mere existence of $\displaystyle\lim_{a \to \infty} \int_{-a}^a f(x)dx$ does not imply that $\displaystyle\int_{-\infty}^\infty f(x)dx$ exists. For example, for every $a > 0$, $\displaystyle\int_{-a}^a \sin x\,dx = 0$ and so $\displaystyle\lim_{a \to \infty} \int_{-a}^a \sin x\,dx = 0$. But neither $\displaystyle\int_0^\infty \sin x\,dx$ nor $\displaystyle\int_{-\infty}^0 \sin x\,dx$ exists (since $\displaystyle\lim_{b \to \infty} \cos b$ and $\displaystyle\lim_{b \to -\infty} \cos b$ do not exist) and so $\displaystyle\int_{-\infty}^\infty \sin x\,dx$ does not exist.

Having discussed improper integrals where the interval of integration is unbounded, let us now turn to the other kind of impropriety, viz., that the integrand is unbounded. So suppose $f : [a, b] \to \mathbb{R}$ is any function, not necessarily bounded. How do we define $\displaystyle\int_a^b f(x)dx$? There are several ways to do this. In the answer to Q 5.2 we remarked that if we define definite integrals as limits of

Riemann sums then the definition makes sense even without the hypothesis of boundedness of the integrand. Another approach is to express the integrand, viz. the function $f$, as a pointwise limit of a sequence of bounded functions. For example, for each $n \in \mathbb{N}$ define $f_n : [a, b] \to \mathbb{R}$ by

$$f_n(x) = \begin{cases} f(x) & \text{if} \quad -n \leq f(x) \leq n \\ n & \text{if} \quad f(x) > n \\ -n & \text{if} \quad f(x) < -n. \end{cases}$$

Geometrically, the graph of $f_n$ is obtained by truncating the graph of $f$ by the line $y = n$ at the top and the line $y = -n$ at the bottom. Even if $f$ is unbounded, $f_n$ is bounded for every $n$. And $f$ is bounded on $[a, b]$ if and only if $f = f_n$ for all sufficiently large $n$. In general we always have $f(x) = \lim\limits_{n \to \infty} f_n(x)$ for every $x \in [a, b]$. So we may define $\int_a^b f(x)dx$ as $\lim\limits_{n \to \infty} \int_a^b f_n(x)dx$, provided this limit exists.

Although both these approaches are perfectly general, neither is very satisfactory in practice unless the function $f$ is particularly well-behaved, say, piecewise continuous on $[a, b]$. The functions occurring in practical problems are generally of this type and for such functions, instead of either of the two approaches given above, a simpler approach, analogous to that of defining $\int_a^\infty f(x)dx$ as $\lim\limits_{b \to \infty} \int_a^b f(x)dx$, is possible. That is, instead of changing the function $f$, we integrate it over a smaller interval and then let this interval grow larger and larger. We first consider the case where the impropriety lies only at the left end-point, $a$, that is; $f$ is unbounded on $[a, b]$ but bounded on $[a + \delta, b]$ for every $\delta > 0$. (Obviously we are interested only in $\delta < b - a$).

**10.9 Definition:** Assume $f$ is defined on $(a, b]$ and integrable on $[a + \delta, b]$ for every $\delta$ $(0 < \delta < b - a)$. Then $\int_a^b f(x)dx$ is said to converge (or exist) if $\lim\limits_{\delta \to 0^+} \int_{a+\delta}^b f(x)dx$ exists.

Note that if $f$ is already defined and integrable on $[a, b]$ then by continuity of functions defined by integrals (cf. Theorem (3.3), or rather, its analogue where the upper limit of integration is fixed and the lower one varies), $\lim\limits_{\delta \to 0^+} \int_{a+\delta}^b f(x)dx$ exists and equals $\int_a^b f(x)dx$ (which already exists as a 'proper' integral). So the definition above does not lead to any ambiguities. It merely extends the definition of $\int_a^b f(x)dx$ when it has to be treated as an improper integral. A similar approach is taken when the impropriety lies at the other end, viz. $b$. That is, $\int_a^b f(x)dx$ is defined as $\lim_{\delta \to 0^+} \int_a^{b-\delta} f(x)dx$. In case there are improprieties at both the ends (e.g. the integral $\int_0^1 \frac{dx}{x(1-x)}$), we take any intermediate point, say $c$, in $(a,b)$ and define $\int_a^b f(x)dx$ as $\int_a^c f(x)dx + \int_c^b f(x)dx$.

As examples, we see that for $p > 0$, the function $\frac{1}{x^p}$ is unbounded at 0. Here $\int_\delta^1 \frac{dx}{x^p} = \frac{1}{1-p}(1 - \delta^{1-p})$ for $p \neq 1$. If $0 < p < 1, \delta^{1-p} \to 0$ as $\delta \to 0^+$ and so $\int_0^1 \frac{dx}{x^p}$ converges with value $\frac{1}{1-p}$. But, for $p > 1, \int_0^1 \frac{dx}{x^p}$ is divergent because $\lim_{\delta \to 0^+} \frac{1}{1-p}\left(1 - \frac{1}{\delta^{p-1}}\right)$ does not exist. For $p = 1$ also $\int_0^1 \frac{dx}{x}$ diverges since $\ln \delta \to -\infty$ as $\delta \to 0^+$.

If $f$ has a finite number of improprieties, say, $c_1 < c_2 \ldots < c_k$ in $(a,b)$, then $\int_a^b f(x)dx$ is said to exist if each $\int_{c_{i-1}}^{c_i} f(x)dx$ (where we set $c_0 = a$ and $c_{k+1} = b$) exists for $i = 1, \ldots, k+1$ and their sum is taken as the value of $\int_a^b f(x)dx$. As an example, take $[a,b] = [0, 2\pi]$ and $f(x) = \tan x$. Here the improprieties occur at $\frac{\pi}{2}$ and $\frac{3\pi}{2}$. Since $\int_0^{\pi/2} \tan x dx = \lim_{\delta \to 0^+} \int_0^{\pi/2-\delta} \tan x dx = \lim_{\delta \to 0^+} \ln \sec(\frac{\pi}{2} - \delta)$ does not exist, we see that $\int_0^{2\pi} \tan x dx$ does not exist either. A blind application of the Fundamental Theorem of Calculus would, however, give $\int_0^{2\pi} \tan x dx = \ln \sec(2\pi) - \ln \sec(0) = 0 - 0 = 0$.

If the integrand has an infinite number of improprieties in $[a,b]$, the definition above makes no sense. There are, in fact, functions which are unbounded on every interval, no matter how small (see Exercise (10.16)). Fortunately such bizarre situations rarely arise in real life applications and so the two kinds of improprieties we have considered are adequate. We have convered the improper integrals of the type $\int_a^\infty f(x)dx$ in detail. The properties and tests of convergence of improper integrals of the other type are similar and will be given in the exercises.

To conclude the section we consider one more aspect of the analogy between infinite series and improper integrals. In Section 3.11 we defined uniform convergence of a series, say, $\sum_{n=1}^\infty f_n(x)$ of real valued functions all defined on a common domain, say $S$. We saw that uniform convergence is a powerful condition (see, for example, Corollary (4.1.6)). There is an analogue of uniform convergence for improper integrals too. Let us write $f(x,n)$ instead of $f_n(x)$.

Here in essence, there are two variables, one variable $x$ ranging over the set $S$ and the other variable $n$ ranging over $I\!N$, the set of positive integers. To define uniform convergence for improper integrals, we replace the discrete variable $n$ by a continuous variable, say $y$, varying over some interval, say $[a, \infty)$ and consider a real-valued function, say $f(x, y)$ of two variables, the variable $x$ ranging over some set $S$. For each fixed value of $x$, $f(x, y)$ is a function of $y$ only and we can talk of the convergence or otherwise of the improper integral $\displaystyle\int_a^\infty f(x, y)dy$. (Here the integration is to be carried w.r.t. $y$, treating $x$ as a constant. In Section 6.4 we shall see more instances where we integrate a function of two variables w.r.t. one of them, holding the other variable as a constant.) Suppose for every $x \in S$, this integral converges. Then its value will, in general, depend on $x$. Denote it by $g(x)$. Here $g$ is a real valued function on the set $S$ and we say that the improper integral $\displaystyle\int_a^\infty f(x, y)dy$ **converges pointwise** on $S$ to $g(x)$. Here, for every $\epsilon > 0$, and for every $x \in S$, we can find some $R$ (which could depend both on $\epsilon$ and $x$) such that for all $b \geq R,$ $\left| \displaystyle\int_a^b f(x, y)dy - g(x) \right| < \epsilon$. As is to be expected, in the case of **uniform convergence**, we require that this $R$ should depend only on $\epsilon$.

Evidently, uniform convergence implies pointwise convergence but not conversely. The theory of unifrom convergence of improper integrals can be developed analogously to that of uniform convergence of series. But in order to get any interesting results, the set $S$ has to be a subset of $I\!R$ (in fact, usually an interval). In that case $f$ would be a function of two real variables, viz. $x$ and $y$. As the therory of functions of several real variables is beyond our scope, we do not pursue this line further.

## EXERCISES

10.1   Suppose $\{a_n\}_{n=0}^\infty$ and $\{b_n\}_{n=0}^\infty$ are sequences of real numbers.

(a) Prove that $\Delta(a_n \pm b_n) = \Delta(a_n) \pm \Delta(b_n)$ where $\Delta$ is the (forward) difference.

(b) Is it true that $\Delta(a_n b_n) = (\Delta a_n)b_n + a_n \Delta b_n$? Similarly see if $\Delta\left(\dfrac{a_n}{b_n}\right)$ equals $\dfrac{b_n \Delta a_n - a_n \Delta b_n}{b_n^2}$.

10.2   For every integer $k \geq 0$, let $\Delta^k$ denote the $k^{\text{th}}$ difference. Prove that

(i) $\Delta^k a_n = \displaystyle\sum_{i=0}^k (-1)^{k-i} \binom{k}{i} a_{n+i}$ (ii) $a_n = \displaystyle\sum_{k=0}^n \binom{n}{k} \Delta^k a_0$

[*Hint* : Apply induction on $k$ for (i) and on $n$ for (ii) and a well-known identity about the binomial coefficients. Here $\Delta^k a_i$ denotes the $i^{\text{th}}$ term of the $k^{\text{th}}$ derived sequene of $\{a_n\}_{n=0}^\infty$].

10.3 Prove that $\Delta a_n \equiv 0$ if and only if $\{a_n\}$ is a constant sequence. More generally, show that $\Delta^k a_n \equiv 0$ if and only if $a_n$ is a polynomial (in $n$) of degree at most $k-1$.

10.4 Suppose $\displaystyle\sum_{n=1}^{\infty} a_n$ has bounded partial sums, $b_n \to 0$ as $n \to \infty$ and further $\displaystyle\sum_{n=1}^{\infty}(b_n - b_{n+1})$ converges absolutely. Prove that $\displaystyle\sum_{n=1}^{\infty} a_n b_n$ is convergent.

10.5 Suppose $\displaystyle\sum_{n=1}^{\infty} a_n$ is convergent and $\{b_n\}$ is a monotonic, convergent sequence (not necessarily converging to 0). Show that $\displaystyle\sum_{n=1}^{\infty} a_n b_n$ is convergent. (This is known as **Abel's test**).

10.6 (a) Define $f : [0,\infty) \to I\!R$ by

$$f(x) = \begin{cases} n & \text{if } x \in [n - \frac{1}{2n^3}, n + \frac{1}{2n^3}] \text{ for some } n \in I\!N \\ 0 & \text{otherwise.} \end{cases}$$

Sketch the graph of $f$. Show that for every $n \in I\!N$, $\displaystyle\int_0^n f(x)dx$ equals

$$\sum_{k=1}^{n-1} \frac{1}{k^2} + \frac{1}{2n^2}$$ and hence that $\displaystyle\int_0^\infty f(x)dx$ is convergent with value

$\displaystyle\sum_{n=1}^{\infty} \frac{1}{n^2}$. Note, however, that $f(x) \not\to 0$ as $a \to \infty$. In fact $f$ is not bounded on $[0,\infty)$.

(b) Modify the example in (a) to get an example of a continuous, unbounded, nonnegative function $f$ on $[0,\infty)$ for which $\displaystyle\int_0^\infty f(x)dx$ is convergent.

10.7 Prove that the sequence $\{I_n\}_{n=1}^{\infty}$ constructed in the proof of the converse implication in Theorem (10.5) is a Cauchy sequence. [*Hint* :For every $R > a$, there is some $m \in I\!N$ such that $a + m > R$. ]

10.8 Prove Theorem (10.6).

10.9 Let $\displaystyle\sum_{n=1}^{\infty} a_n$ be a series. Define $f : [1,\infty) \to I\!R$ by $f(x) = a_n$ if $n \le x < n+1, n \in I\!N$. Prove that $\displaystyle\int_1^\infty f(x)dx$ is convergent if and only if $\displaystyle\sum_{n=1}^{\infty} a_n$ is convergent.

*10.10 Prove that in Theorem (10.6), part (iv), the hypothesis about non-negativity of $f$ and $g$ cannot be dropped. [*Hint* : First construct a counter-example for the corresponding result for infinite series. Then convert it using the last exercise.]

10.11 Let $I_n = \displaystyle\int_{n\pi}^{(n+1)\pi} \frac{\sin x}{x} dx$ for $n \in \mathbb{N}$. Prove that $I_n > 0$ if $n$ is even and $I_n < 0$ if $n$ is odd and that for all $n, |I_{n+1}| \leq |I_n|$. Prove also that $I_n \to 0$ as $n \to \infty$. Hence show that the improper integral $\displaystyle\int_{\pi}^{\infty} \frac{\sin x}{x} dx$ is convergent. (For a short but tricky proof, integrate $\displaystyle\int_{\pi}^{b} \frac{\sin x}{x} dx$ by parts and show the convergence of $\displaystyle\int_{\pi}^{\infty} \frac{\cos x}{x^2} dx$ using comparison test.)

10.12 Show that $\displaystyle\int_{\pi}^{\infty} \frac{\sin x}{x} dx$ is not absolutely convergent. [*Hint* : with the notation of the last exercise, show that $|I_n| \geq \dfrac{2}{n\pi}$ for all $n \in \mathbb{N}$.]

10.13 Prove that $\displaystyle\int_{0}^{\infty} \frac{\sin x}{x} dx$ is convergent but not absolutely convergent. (Note that there is no impropriety at 0. This well-known integral, often called **Dirichlet's integral**, has value $\dfrac{\pi}{2}$. But there is no easy proof.)

10.14 In Exercise (10.6)(a), show that even though $\displaystyle\int_{1}^{\infty} f(x)dx$ is convergent, $\displaystyle\sum_{n=1}^{\infty} f(n)$ is not convergent, but $\displaystyle\sum_{n=1}^{\infty} f\left(n + \frac{1}{2}\right)$ is convergent.

10.15 For the function $f(x)$ whose graph is sketched in Fig.5.10.2, show that even if $\displaystyle\int_{1}^{\infty} |f(x)|dx$ is divergent, $\displaystyle\sum_{n=1}^{\infty} |f(n^2)|$ is convergent while $\displaystyle\sum_{n=1}^{\infty} |f(n)|$ is divergent.

10.16 Define $f : [1, 2] \to \mathbb{R}$ by

$$f(x) = \begin{cases} 0 & \text{if } x \text{ is irrational} \\ q & \text{if } x \text{ is a rational equal to } \frac{p}{q} \text{ in its reduced form.} \end{cases}$$

Prove that $f$ is unbounded on every (non-degenerate) subinterval of $[1,2]$.

10.17 Prove the analogues of Theorems (10.5) and (10.6) for improper integrals of the type $\displaystyle\int_{a}^{b} f(x)dx$, where $f$ is unbounded at $a$, but integrable on $[a + \delta, b]$ for every $\delta > 0$.

**10.18** Suppose $p(x)$ and $q(x)$ are polynomials of degrees $m$ and $n$ respectively. Assume $q(x)$ has no real roots in $[a, \infty)$. Prove that $\displaystyle\int_a^\infty \frac{p(x)}{q(x)}dx$ is convergent if and only if $n \geq m + 2$.

**10.19** Prove that the series $\displaystyle\sum_{n=2}^\infty \frac{1}{n(\ln n)^p}$ is convergent if and only if $p > 1$.

**10.20** Test the convergence of the following improper integrals :

(i) $\displaystyle\int_0^\infty e^{-x}(x^7 - 5x^6 + 9)dx$ (ii) $\displaystyle\int_{-\infty}^\infty e^{-x}(x^7 - 5x^6 + 9)dx$

(iii) $\displaystyle\int_0^\infty \frac{\cos x}{x}dx$ (iv) $\displaystyle\int_1^\infty \frac{\cos x}{\sqrt{x}}dx$

(v) $\displaystyle\int_\pi^\infty \cos(x^2)dx$ (vi) $\displaystyle\int_0^1 \ln x\, dx$

(vii) $\displaystyle\int_\pi^\infty \sin(\frac{1}{x})dx$ (viii) $\displaystyle\int_0^1 \sin(\frac{1}{x})dx$

(ix) $\displaystyle\int_0^\infty \frac{1}{(1+x)\sqrt{x}})dx$.

**\*10.21** Prove that $\displaystyle\int_0^\infty \frac{dx}{1 + x^4 \sin^2 x}$ is convergent but $\displaystyle\int_0^\infty \frac{dx}{1 + x^2 \sin^2 x}$ is divergent. [*Hint* : Break into a series of integrals over intervals of the form $[n\pi, (n+1)\pi]$, $n \in I\!N$. Use the fact that $\displaystyle\int_0^\pi \frac{dx}{1 + A^2 \sin^2 x} = \frac{\pi}{\sqrt{1 + A^2}}$ where $A$ is any constant. ]

**10.22** Prove that the integral $\displaystyle\int_0^\infty e^{-x}x^{\alpha-1}dx$ converges for $\alpha > 0$. (The value of this integral depends upon $\alpha$ and is commonly denoted by $\Gamma(\alpha)$. The function $\Gamma$ so defined is called the **gamma function**. It is one of the most frequently occurring functions in applications.)

**10.23** Prove that the gamma function defined in the last exercise satisfies the following properties:

(i) $\Gamma(\alpha + 1) = \alpha\Gamma(\alpha)$ for every $\alpha > 0$

(ii) $\Gamma(n) = (n-1)!$ for every $n \in I\!N$ (For this reason, the gamma function is often regarded as an extension of factorials).

(iii) $\displaystyle\int_0^\infty e^{-x^2}dx = \frac{1}{2}\Gamma(\frac{1}{2})$. (It can be shown that $\Gamma(\frac{1}{2}) = \sqrt{\pi}$. But the proof requires advanced techniques. See Exercise (6.8.22)(v).)

**10.24** It is possible to define the trigonometric functions by first defining the inverse trigonometric functions through integrals. This can be done in several ways. We indicate in this exercise the approach in terms of the sine functions.

(i)   Show that the improper integral $\int_0^1 \dfrac{dy}{\sqrt{1-y^2}}$ is convergent (without using trignometric functions !) and equals $\int_{-1}^0 \dfrac{dy}{\sqrt{1-y^2}}$. [*Hint* : Put $u = 1 - y$ and compare with $\int_0^1 \dfrac{dv}{\sqrt{u}}$.]

(ii)  Define $F(y) = \int_0^y \dfrac{dt}{\sqrt{1-t^2}}$ for $y \in [-1,1]$. Call $\int_{-1}^1 \dfrac{dt}{\sqrt{1-t^2}}$ as $\pi$.
      Prove that $F$ is a strictly monotonically increasing function which maps $[-1,1]$ onto $[-\dfrac{\pi}{2}, \dfrac{\pi}{2}]$.

(iii) Define the sine function on $\left[-\dfrac{\pi}{2}, \dfrac{\pi}{2}\right]$ as the inverse function of $F$. That is, for $x \in \left[-\dfrac{\pi}{2}, \dfrac{\pi}{2}\right]$, $\sin x$ is that unique real number $y$ in $[-1,1]$ for which $\int_0^y \dfrac{dt}{\sqrt{1-t^2}} = x$. Prove that $\sin(-x) = \sin x$ and $\sin x$ is continuous for all $x \in \left[-\dfrac{\pi}{2}, \dfrac{\pi}{2}\right]$.

(iv)  For $x \in \left[\dfrac{\pi}{2}, \dfrac{3\pi}{2}\right]$, define $\sin x = -\sin(x - \pi)$ and then extend the sine function periodically for all $x \in \mathbb{R}$. (That is, given $x \in \mathbb{R}$, find a unique integer $k$ such that $x - 2\pi k \in \left[-\dfrac{\pi}{2}, \dfrac{3\pi}{2}\right)$ and then set $\sin x = \sin(x - 2\pi k)$.) Prove that $\sin x$ is differentiable for all $x$ and further that $\dfrac{d}{dx}(\sin x) = \pm\sqrt{1 - \sin^2 x}$. Decide for which $x$ the $+$ sign holds. [*Hint* : For differentiability when $x$ is an odd multiple of $\dfrac{\pi}{2}$ use Exercise (4.5.7) (and its analogue for left handed derivatives).]

(v)   Define $\cos x = \dfrac{d}{dx}(\sin x)$. Prove that $\sin^2 x + \cos^2 x = 1$ for all $x$. Prove that $\cos x$ is continuous for all $x$, is periodic with period $2\pi$ and vanishes precisely when $x$ is an odd multiple of $\dfrac{\pi}{2}$.

(vi)  Prove that $\cos x$ is differentiable for all $x$ and further that $\dfrac{d}{dx}(\cos x) = -\sin x$ for all $x \in \mathbb{R}$. [*Hint* : Once again, differentiability when $x = $ odd multiple of $\dfrac{\pi}{2}$, needs a little different treatment.]

*(vii) For all $x, y$ prove that $\cos(x + y) = \cos x \cos y - \sin x \sin y$. [*Hint* : Fix $y$ and let $f(x) = \cos(x + y) - \cos x \cos y + \sin x \sin y$. Prove that $f''(x) + f(x) = 0$ for all $x$ and hence that $[f'(x)]^2 + [f(x)]^2$ is a constant. Now note that $f(0) = 0$ and $f'(0) = 0$.]

(viii) Prove that $\cos\left(\dfrac{\pi}{2} - x\right) = \sin x$ and $\cos\left(\dfrac{\pi}{2} + x\right) = -\sin x$ for all $x$. Prove other familiar identities about the sine and cosine functions.

# Chapter 6

# APPLICATIONS OF INTEGRALS

## 6.1    Average Values of Functions

Of all the elementary constructions of arithmetic, the concept of an average (or mean, or more precisely, the arithmetic mean as it is sometimes called) is probably the most popular one. Even young children keep track of the batting averages of their favourite cricketers, car owners always talk of the running averages of their cars. (On an average, the owner of a new car talks about it 6.3 times a day!) And the unreliability of averages is the gist of a few jokes, for example where a cook claims his preparation as 'well cooked on the average' when in fact, half of it is uncooked and the other half is almost charred!

The strange thing about the average is that it need not be an actual value. Thus, if the average height of students in a class is, say, 1.65 meters, there need not be any student in that class whose height is exactly 1.65 meters. In fact there need not even be a student whose height is close to 1.65 meters. This is more likely to happen in a small or a highly polarised class (with, say, half the students of height 1.50 and the other half of 1.80 meters). In a large class, on the other hand, with the height of the student well spread out over some interval, we indeed expect that there will be some student whose height will lie within, say, 1 cm. of the average height. The larger the class, the more this margin can be narrowed down. And in the limiting case, we expect that there would always be a student whose height is exactly the average height.

The trouble is how to define the limiting case here. Let us formulate the problem mathematically. Suppose $f$ is a real-valued function on a non-emtpy set $S$. Here $S$ could be any set (not necessarily a subset of $I\!R$). For example, $S$ could be the set of students in a class and $f$ could be the height function. Then the average of $f$, on the set $S$, is simply the ratio of $\sum_{s \in S} f(s)$ to the number of elements in the set $S$. When the set $S$ is infinite, this definition breaks down for two reasons. First of all, how does one define the sum $\sum_{s \in S} f(s)$? One possible answer in terms of unordered summation was given in Definition (3.8.5). But the function $f : S \to \mathbb{R}$ is usually not summable in that sense. And, secondly, when you divide $\sum_{s \in S} f(s)$ by the number of elements in $S$, the answer will be either 0 (when $f$ is summable over $S$, i.e. when $\sum_{s \in S} f(s)$ is finite) or it will be a meaningless expression. Neither is satisfacotry.

So, to give a satisfactory definition of the average value of a real-valued function $f$, defined on an infinite set $S$, we must have some way of measuring how big $S$ is, other than the number of elements in it. There is a branch of mathematics, known as measure theory which does precisely this. It is well beyond our scope to study measures in general. We shall therefore define the average of a function $f : S \to \mathbb{R}$, only in the case where the set $S$ is an interval, say $[a, b]$ and the size or the 'measure' of an interval is simply its length. Later, we shall see how the concept of an average can be extended to slightly more general situations.

**1.1 Definition:** Suppose $f : [a, b] \to \mathbb{R}$ is integrable. Then the **average** (or **mean**) value of $f$ (on $[a, b]$) is defined as

$$\frac{1}{(b-a)} \int_a^b f(x)dx.$$

It is easy to motivate this definition using any interpretation of the definite integral. If, for example, we take $x$ to be time and $f(x)$ to be the speed at time $x$, then, as we have seen before (see Exercise (5.2.10)), the integral $\int_a^b f(x)dx$ is the net distance travelled in the time interval $[a, b]$, and so $\frac{1}{b-a} \int_a^b f(x)dx$ is indeed the average speed over this time interval. Geometrically, $\int_a^b f(x)dx$ is the area of the region below the graph of $y = f(x)$. (In case $f$ is negative at some points, then the corresponding portion of the graph is actually below the $x$-axis, so its contribution to the area is negative.) Here $\frac{1}{b-a} \int_a^b f(x)dx$ is the average height of the region. Equivalently, it is the height of a rectangle with base $[a, b]$ having the same area as the region below the graph of $y = f(x), a \leq x \leq b$.

The concept of average of a function $f : S \to \mathbb{R}$ where $S$ is a finite set (not necessarily a subset of $\mathbb{R}$) can also be made to correspond, somewhat artificially, to the average we have defined. Indeed, let the elements of $S$ be $s_1, \ldots, s_n$. Consider the interval $[0, n]$. We let $s_i$ correspond to the point $i$ of $[0, n]$ for $i = 1, \ldots, n$. Define $f : [0, n] \to \mathbb{R}$ by

$$f(x) = \begin{cases} 0 & \text{if } x = 0 \\ f(s_i) & \text{if } i - 1 < x \leq i, \ i = 1, \ldots, n. \end{cases}$$

It is easily seen (preferably with the aid of a graph) that $f$ is a step function and its integral over $[0, n]$, i.e. $\int_0^n f(x)dx$ is precisely $\sum_{i=1}^n f(s_i)$. So the average of $f$ on $S$ is the same as that of $f$ on $[0, n]$. Note that the function $f$ is in general not continuous on $[0, n]$. And since the only values it attains are $0, f(s_1), \ldots, f(s_n)$, it may very well happen that the average is not attained. This anomaly does not arise if the function $f$ is continuous. A quick proof can be given using the Fundamental Theorem of Calculus (cf. Exercise (5.9.5)). But a direct proof is almost as easy since an average is, after all, an intermediate value.

**1.2 Theorem:** If $f : [a, b] \to \mathbb{R}$ is continuous, then there exists $c \in [a, b]$ such that $f(c) = \dfrac{1}{b-a} \int_a^b f(x)dx$. Verbally, $f$ attains its average value on $[a, b]$, at some point of $[a, b]$.

**Proof:** By Theorem (5.4.1), $f$ is integrable over $[a, b]$ and so there is certainly no difficulty in defining the average of $f$ on $[a, b]$. By Theorem (4.1.4), $f$ attains its extrema on $[a, b]$, i.e. there exist point $\alpha, \beta \in [a, b]$ such that

$$f(\alpha) \le f(x) \le f(\beta) \quad \text{for all} \quad x \in [a, b]. \tag{1}$$

Applying Exercise (1.10) to (1), we get

$$\int_a^b f(\alpha)dx \le \int_a^b f(x)dx \le \int_a^b f(\beta)dx \tag{2}$$

Since $f(\alpha), f(\beta)$ are constants, we get $\int_a^b f(\alpha)dx = f(x)(b-a)$ and $\int_a^b f(\beta)dx = f(\beta)(b-a)$ (cf. Exercise (5.1.2)). Putting this in (2) and dividing throughout by $b - a$ (which is positive), we get

$$f(\alpha) \le \frac{1}{b-a} \int_a^b f(x)dx \le f(\beta) \tag{3}$$

In other words, the average of $f$ on $[a, b]$, lies between $f(\alpha)$ and $f(\beta)$. Since $f$ is continuous on the interval $[\alpha, \beta]$ (or the interval $[\beta, \alpha]$ in case $\beta < \alpha$), the Intermediate Value Property (Theorem (4.1.3)), implies that there is some $c \in [\alpha, \beta]$ such that $f(c) = \dfrac{1}{b-a} \int_a^b f(x)dx$. Since $[\alpha, \beta] \subset [a, b], c \in [a, b]$ and the proof is complete. ∎

This theorem is often called the **Mean Value Theorem for definite integrals**. It resembles, and can in fact be derived from, the Lagrange's Mean Value Theorem (see Exercise (1.5)). Note that the point $c$ at which $f$ attains its mean, is in general not unique (see Exercise (1.2)). Also little can be said about its location in the interval $[a, b]$.

The definition of average given above is often inadequate and has to be replaced by what is called weighted average. Suppose, for example, we want to get an idea of the average rainfall, say $\overline{R}$, for a country. We place $n$ rain-gauges at selected, representative places in the country and measure the rainfalls, say $R_1, \ldots, R_n$ recorded by them. Let $R$ be the average of $R_1, \ldots, R_n$, i.e. $R = (R_1 + R_2 + \ldots R_n)/n$. Can we put $\overline{R} = R$? The answer is 'no' for two reasons. First, the rainfall varies so much from place to place. Let us, however, ignore this point and assume that the rainfall is constant in the region represented by each rain-gauge. Even then, it would be misleading to take $R$ as the average rain-fall for the country because in doing so, we are ignoring the differences in the sizes of the region. Larger regions should get proportionally higher representation or 'weightage' in computing the average. The correct thing to do is to let $A_1, \ldots, A_n$ be the areas of the regions represented by the rain-gauges. Let $A = A_1 + \ldots A_n$. Then $\overline{R}$ can be taken to be $\frac{R_1 A_1 + R_2 A_2 + \ldots + R_n A_n}{A}$. Here for each $i = 1, \ldots, n, A_i$ is the weight and $A_i/A$ is the **relative weight** given to the $i^{\text{th}}$ rain-gauge and $\overline{R}$ is the **weighted average** of the rainfalls recorded by them. When all the $A_i$'s are equal, $\overline{R}$ coincides with $R$. But in general, it is different and gives a more satisfactory definition of average than $R$. (Note that $\overline{R}$ depends only on the relative weights.)

It is easy to define the concept of a weighted average for a function $f : [a, b] \rightarrow \mathbb{R}$. We must be given some other function, say, $g : [a, b] \rightarrow \mathbb{R}$. For $x \in [a, b], g(x)$ is called the weight of $x$, and so $g$ is called the **weight function**. We assume $g(x) \geq 0$ for all $x \in [a, b]$, because in applications, we rarely have to consider weight functions with negative values. We further assume that the total weight, i.e. the integral $\int_a^b g(x)dx$ is positive, as otherwise, the weight function would have no effect.

A good example of a weight function is the population density $\rho(x)$ at a point on a road as in Exercise (5.2.7). The total weight, in that case, is the population of the road. Suppose, now, (cf. Exercise (5.2.8)) that $\sigma(x)$ is the average per capita income at a point $x$. Then the total income of the population on the road is not $\int_0^2 \sigma(x)dx$ but $\int_0^2 \sigma(x)\rho(x)dx$. And to get the average per capita income for the road, we must divide this by the total population, viz., $\int_0^2 \rho(x)dx$.

More generally, we make the following definition.

**1.3 Definition:** Let $f, g$ be integrable functions on $[a, b]$, with $g \geq 0$ everywhere and $\int_a^b g(x)dx > 0$. Then the **weighted average** of $f$ (w.r.t. the **weight function** $g$) is the number $\dfrac{\int_a^b f(x)g(x)dx}{\int_a^b g(x)dx}$.

Note that the average of $f$ depends not only on $f$ but also on the weight function. If $g$ is a (positive) constant, then the average of $f$ w.r.t. $g$, is the same as its (unweighted) average. So Definition (1.3) is a generalisation of Definition (1.1). It is but natural to inquire if theorem (1.2) has a generalisation for weighted averages. This indeed is the case and, quite appropriately, the result is called the **Generalised Mean Value Theorem** for definite integrals.

**1.4 Theorem:** Suppose $f, g : [a, b] \to \mathbb{R}$, $g \geq 0$, $\int_a^b g(x)dx > 0$ and $f$ is continuous. Then there exists some $c \in [a, b]$ such that $f(c)$ is the weighted average of $f$ on $[a, b]$ (w.r.t. the weight function $g$) i.e.,

$$f(c) = \frac{\int_a^b f(x)g(x)dx}{\int_a^b g(x)dx}.$$

**Proof:** The proof is essentially a duplication of that of Theorem (1.2). Instead of (2), we have (since $g(x) \geq 0$ for all $x$).

$$\int_a^b f(\alpha)g(x)dx \leq \int_a^b f(x)g(x)dx \leq \int_a^b f(\beta)g(x)dx$$

and instead of dividing throughout by $(b - a)$ we divide by $\int_a^b g(x)dx$ (which is assumed to be positive). ∎

Taking the example of the income of a road population, the theorem above says that if richness of a locality varies continuously w.r.t. distance (which often is the case) then there is some locality on the road, occupied by people of average income.

We now consider the problem of extending the definition of the average value of a function $f : S \to \mathbb{R}$, when the set $S$ is not an interval and, in fact, not even a subset of $\mathbb{R}$. As observed earlier, there is no problem when the set $S$ is finite. But when $S$ is infinite, we need the concept of a measure. This difficulty can sometimes be overcome by taking weighted averages, where the relative weights attached are certain probabilities. (Strictly speaking, this does not solve the problem, because a rigorous approach to probability requires measure. But here we assume an intuitive understanding of this concept which we use even in day-to-day conversation.)

Let us first see how the average over a finite set $S$ can be paraphrased in terms of probabilities.

**1.5 Theorem:** Let $f : S \to \mathbb{R}$ be a function with $S$ a non-empty finite set. Let $S = S_1 \cup \ldots \cup S_k$ be a decomposition of $S$ into a finite number of mutually disjoint non-empty subsets. For $i = 1, \ldots, k$ let $p_i$ be the probability that an element of $S$ (taken at random) belongs to $S_i$, and let $m_i$ = average value of $f$ on $S_i$. Then $p_1 + \ldots + p_k = 1$ and the average value of $f$ on $S$ is $\sum_{i=1}^k p_i m_i$.

**Proof:** Let the number of elements in $S_i$ be $n_i$, for $i = 1, \ldots, k$. Then $n$, the number of elements in $S$ is $\sum_{i=1}^k n_i$. Clearly for each $i$, $p_i$ equals $\dfrac{n_i}{n}$. Also

$$m_i = \frac{\sum_{s \in S_i} f(s)}{n_i} \text{ and so } \sum_{s \in S_i} f(s) = n_i m_i. \text{ Clearly } \sum_{i=1}^k p_i = \frac{1}{n} \sum_{i=1}^k n_i = \frac{n}{n} = 1.$$

Further, the average of $f$ on $S$

$$= \frac{1}{n} \sum_{s \in S} f(s)$$

$$= \frac{1}{n} \sum_{i=1}^{k} \sum_{s \in S_i} f(s) = \sum_{i=1}^{k} \frac{n_i m_i}{n} = \sum_{i=1}^{k} p_i m_i.$$

as was to be proved.∎

Although this theorem is hardly profound, it has an interesting significance. We are free to choose the partition anyway we like. Suppose, for example, $t_1, \ldots, t_k$ are the distinct values assumed by $f$ and we let $S_i = f^{-1}(\{t_i\}) = \{s \in S : f(s) = t_i\}$. Then $f$ is constant on each $S_i$ with value $t_i$ and so $m_i = t_i$. Also $p_i$ is now the probability that the value of $f$ equals $t_i$. In other words, the function $f$ assumes the values $t_1, t_2, \ldots, t_k$ with probabilities $p_1, \ldots, p_k$ respectively. The theorem asserts that the average value of $f$ is $\sum_{i=1}^{k} p_i t_i$, which is the weighted average of $t_1, \ldots, t_k$ with weight $p_1, \ldots, p_k$. (Note that $p_1 + p_2 + \ldots + p_k = 1$). For this reason, the average value is also called the **expected value**. Note that in this formulation, the size of $S$ is immaterial. So even if $S$ is an infinite set but the function $f : S \to \mathbb{R}$ assumes only finitely many values, say $t_1, \ldots, t_k$ with probabilities $p_1, \ldots, p_k$, then the average or expected value of $f$ is the number $\sum_{i=1}^{k} p_i t_i$. In fact, even when the values assumed by $f$ are the terms of some infinite sequence, say $\{t_n\}_{n=1}^{\infty}$, this definition will make sense provided we have some way of summing up $\sum_{n=1}^{\infty} p_n t_n$, where for each $n$, $p_n$ is the probability that $f$ assumes the value $n$. When the $t_n$'s are all non-negative, we know (cf. Corollary (3.7.2)) that the sum of the infinite series $\sum_{n=1}^{\infty} p_n t_n$ can be defined unambigously (i.e. is independent of the order in which the terms are added.) So, we take $\sum_{n=1}^{\infty} p_n t_n$ as the definition of the average in such a case. To illustrate it, we calculate it explicitly in one problem.

**1.6 Problem:** Suppose a coin which has probability $p$ $(> 0)$ of showing a head is tossed till a head shows. Find the average (or expected) number of tosses.

**Solution:** We give two solutions. In the first solution, we let $S$ be the set of all sequences in which a (possibly empty) string of $T$'s (for tails) is followed by a lone terminal $H$ (for head). Thus, $H, TH, TTH, TTTH$, are some of the members of the set $S$. We define $f : S \to \mathbb{R}$ to be the length function, i.e. $f(H) = 1, f(TTH) = 3$ etc. Clearly the problem amounts to finding the expected value of $f$ on $S$. Note that the function $f$ assumes only the positive integers as its values and that for every positive integer $n$, the only element of $S$ at which $f$ assumes the value $n$ is the sequence $TT\ldots TH$ in which $(n-1)$ $T$'s are followed by $H$. Since the probability of $T$ is $1-p$ and the tosses are independent, the probability of an occurrence of this string is $(1-p)^{n-1}p$. So $f$ assumes the value $n$ with probability $(1-p)^{n-1}p$. Therefore, the expected number of tosses is $\displaystyle\sum_{n=1}^{\infty} np(1-p)^{n-1}$.

To evaluate this sum, we note that the power series $\displaystyle\sum_{n=1}^{\infty} nx^{n-1}$ is obtained by term-by-term differentiation of the power series $\displaystyle\sum_{n=0}^{\infty} x^n$ which has radius of convergence 1 and sum function $\dfrac{1}{1-x}$ (see Theorem (3.3.5)). By Theorem (4.6.4), for $|x| < 1$, the series $\displaystyle\sum_{n=1}^{\infty} nx^{n-1}$ converges to $\dfrac{d}{dx}\left(\dfrac{1}{1-x}\right)$, i.e. to $\dfrac{1}{(1-x)^2}$.

Since $|1-p| < 1$, $\displaystyle\sum_{n=1}^{\infty} np(1-p)^{n-1} = p\sum_{n=1}^{\infty} n(1-p)^{n-1} = p\dfrac{1}{(1-(1-p))^2} = \dfrac{1}{p}$.

So the expected number of tosses till a head shows is $\dfrac{1}{p}$.

The second solution we give is somewhat slicker in that it avoids having to sum an infinite series. But it is not completely rigorous, because it is based on certain statements which are plausible but not provable with the machinery we have. Let us look at the problem somewhat playfully. Assume each toss of coin costs 1 rupee and a player is said to win when he gets a head. The problem is to find the average cost of winning the game. Denote this cost by $E$. We shall obtain a simple equation for $E$ and solve it.

Let us suppose that the player keeps on tossing the coin even after getting a head. These subsequent tosses are dummy tosses which cost him nothing. Nor do they earn him any more wins. The reason we consider such dummy tosses is that instead of finite sequences of $T$'s and $H$'s of varying lengths, we consider only infinite sequences of $T$'s and $H$'s. Let $S$ be the set of all such sequences. $S$ is an infinite set. A typical member of $S$ can be written as $(X_1, X_2, \ldots, X_n, \ldots)$ where each $X_n$ is either $T$ of $H$. (Such sequences are often called **binary sequences** because each term can assume only two possible values.) Now define $f : S \to \mathbb{R}$ by $f(X_1, X_2, \ldots, X_n, \ldots) = k$, if $k$ is the least integer such that $X_k = H$. If each $X_n = T$, then $f$ is undefined. But the probability of this happening is 0. (This is plausible because $(1-p)^n \to 0$ as $n \to \infty$. But a rigorous proof would require the concept of a certain measure on $S$ and we omit it.) The average cost of winning the game is clearly the average value of $f$ on the set $S$.

Now divide the set $S$ into two mutually disjoint subsets, $A$ and $B$, where $A$ consists of all infinite sequences beginning with $H$ and $B$ consists of all infinite sequences beginning with $T$. Let $E_A$ and $E_B$ be the average values of $f$ on $A$

and $B$ respectively. Then by Theorem (1.5),

$$E \;=\; p_A E_A + p_B E_B \tag{4}$$

where $p_A$ (respectively, $p_B$) is the probability that a sequence in $S$ belongs to $A$ (respectively, to $B$.) Since this depends only on the outcome of the first toss, we get,

$$p_A = p \quad \text{and} \quad p_B = 1 - p. \tag{5}$$

The value of $f$ at every element in $A$ is 1 and so $E_A = 1$. The calculation of $E_B$ is a little tricky. Each sequence in $E_B$ begins with a $T$, which costs the player one rupee. The player gains nothing by this toss, because after the first tail, it is as if the whole game begins fresh again. The average cost of winning it is, by assumption, $E$. The first wasted toss increases it by 1 and so we get

$$E_B \;=\; E + 1 \tag{6}$$

Putting (5) and (6) (and $E_A = 1$) into (4) we get

$$E \;=\; p + (1 - p)(E + 1) \tag{7}$$

solving which we get $pE = 1$, or $E = \frac{1}{p}$, exactly the same answer as before. ∎

Analogous problems where the first solution will be cumbersome but the second one will work will be given in the exercises. The problem itself has an interesting 'practical' significance which will be brought out in Exercise (1.13).

We close with yet another method of defining the average of a function $f : S \to I\!R$ where $S$ is an infinite set. If $S$ is some interval, we already know how to do this. Otherwise we try to construct some other function, say $h$, from $S$ onto some interval $[a, b]$. Let us write $z = h(s)$ for $s \in S$. In general, the function $h$ need not be one-to-one and so the inverse function $h^{-1} : [a, b] \to S$ is not well-defined. Suppose, however, that we are able to construct $h$ in such a way that whenever $h(s_1) = h(s_2)$, we also have $f(s_1) = f(s_2)$. In other words, the value of the given function $f$ at a point $s \in S$ depends only on the value of the function $h$ at $s$. In that case we can unambiguously define a function $g : [a, b] \to I\!R$ by $g(z) = f(s)$ where $s \in S$ is such that $h(s) = z$. (In case $h$ is one-to-one, $g$ is merely the composite of the two functions $h^{-1} : [a, b] \to S$ and the function $f : S \to I\!R$).

We can now define the average value of $f$ to be the average value of the function $g$, i.e., the number $\dfrac{1}{b-a} \displaystyle\int_a^b g(x)\,dx$. The basic idea in this method is to let the variable $s$, which varies over the set $S$, correspond to a real variable $z$, which varies over some interval $[a, b]$. The underlying presumption is that as $s$ varies 'uniformly' over $S$, $z$ varies 'uniformly' over $[a, b]$. There is no way, of course, to define what is 'uniform variation', without having some measure on the set $S$. So the reliability of the average defined by this method is largely an article of faith. By choosing a different function than $h$, the variable $s$ can be made to correspond to an entirely different variable, say $w$, taking values in some interval, say $[c, d]$. We would then get some other average value of the same function $f : S \to \mathbb{R}$.

We illustrate this procedure with a simple problem. Suppose we want to find the average length of a chord of a unit circle $C$, i.e., a circle of radius 1. Here $S$ is the set of all chords of $C$ and $f : S \to \mathbb{R}$ is the length function. The length of the chord is uniquely determined by its (perpendicular) distance from $O$, the center of $C$. Indeed, if this distance is $z$, then the length of the chord is $2\sqrt{1 - z^2}$ (see Fig. 6.1.1 (a)). Clearly $z$ varies over the interval $[0, 1]$. So, if we let $h : S \to [0, 1]$ be the function which associates to each cord, its perpendicular distance from $O$, then the function $g : [0, 1] \to \mathbb{R}$, defined above, becomes $g(z) = 2\sqrt{1 - z^2}$. The substitution $z = \sin\theta$ gives,

$$\int_0^1 2\sqrt{1 - z^2}dz = \int_0^{\pi/2} 2\cos^2\theta d\theta = \int_0^{\pi/2} (1 + \cos 2\theta)d\theta = \left[\theta + \frac{\sin 2\theta}{2}\right]\Big|_0^{\pi/2}$$
$$= \frac{\pi}{2}.$$ So, the average length of a chord of a unit circle is $\frac{\pi}{2}$.

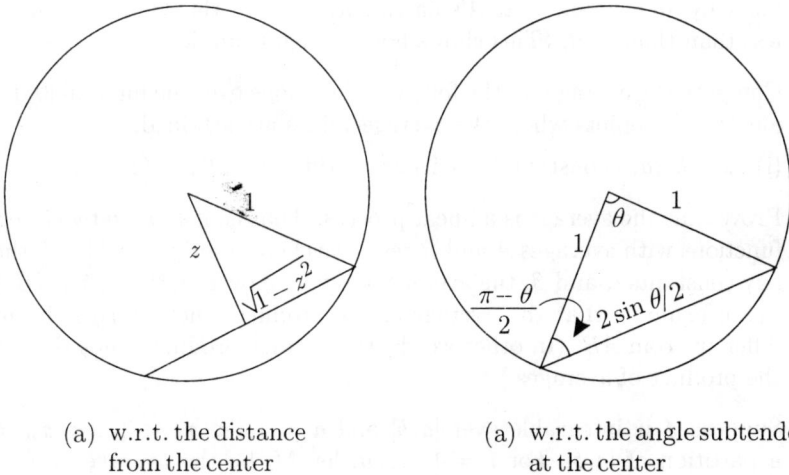

(a) w.r.t. the distance        (a) w.r.t. the angle subtended
    from the center                at the center

Figure 6.1.1 : Average Length of a Chord of Unit Circle.

Suppose, however, that instead of the distance from $O$, we consider the angle, say $\theta$, subtended by the chord at $O$. As the chord varies over $S$, $\theta$ varies over $[0, \pi]$. The length of the chord can also be expressed in terms of $\theta$. Indeed, from Fig.6.1.1(b), it is clear that the length is $2\sin\dfrac{\theta}{2}$. So the average length of a chord would come out to be $\dfrac{1}{\pi}\int_0^{\pi} 2\sin\dfrac{\theta}{2}d\theta = \dfrac{1}{\pi}(-4\cos\dfrac{\theta}{2})\Big|_0^{\pi} = \dfrac{4}{\pi}$.

As yet another approach, we can take $h$ to be $f$ itself. That is, we determine the length of a chord, by its length itself! In that case since the length varies between 0 to 2, the average length is simply $\dfrac{1}{2}\int_0^2 xdx = 1$.

In fact, there are many other ways to choose the function $h : S \to \mathbb{R}$. Each choice will, in general, give a different value of the average length of a chord. So unless there is some particular choice which stands out as the most natural, this method of defining an average of a real-valued function defined on an arbitrary set $S$ is open to criticism. A similar objection can be levelled against other definitions of an average. A definition of an average based on a measure on $S$ is also ambiguous, because on the same set, many different measures can be put. A definition based on probability fares no better. There is, in fact, no absolute definition of probability. It is subject to the choice of a measure. Even the classical definition of probability as the ratio of the number of favourable cases to the total number of cases is based on the tacit assumption that all cases are equally likely. And such an assumption amounts to declaring a certain measure. No wonder then, that averages can lie both in real life and in mathematics!

## EXERCISES

1.1   Suppose $A$ and $B$ are two cricketers and that both during the first and the second half of some cricket season, $A$'s batting average is more than $B$'s. Show by an example that $A$'s batting average for the whole season may be less than that of $B$. (This shows how averages can lie).

1.2   Compute the averages of the following functions over the interval $[0, 1]$. Also identify the points where the average values are attained.

   (i) $ax + b$ $(a, b$ constants)   (ii) $x^2$   (iii) $x^3$   (iv) $x(1 - x)$.

1.3   Prove that the average is a linear process. That is, if $f, g$ are two integrable functions with averages $A$ and $B$ respectively over an interval $[a, b]$, then for any constants $\alpha$ and $\beta$, the average of $\alpha f + \beta g$ over $[a, b]$ is $\alpha A + \beta B$. Show by an example that the average of the product function $fg$ is in general different from $AB$. (In other words, the average product is not the same as the product of averages.)

1.4   Suppose $f$ is integrable over $[a, b]$ and $a = x_0 < x_1 < \ldots < x_n = b$ is a partition of $[a, b]$. For $i = 1, \ldots, n$, let $M_i$ be the average of $f$ on the subinterval $[x_{i-1}, x_i]$. Prove that the average of $f$ on $[a, b]$ is the weighted average of $M_1, M_2, \ldots, M_n$ with weights $\Delta x_1, \ldots, \Delta x_n$ respectively.

1.5   Give an alternate proof of Theorem (1.2) using Fundamental Theorem of Calculus.

1.6   Show by an example that the hypothesis of continuity of $f$ in Theorem (1.2) cannot be dropped.

1.7   Give an alternate proof of Theorem (1.4) in case $g$ is also continuous, using the Fundamental Theorem of Calculus and Cauchy's Mean Value Theorem (Exercise (4.5.11)).

1.8   Let $f(x) = -x$ for all $x$ and

$$g(x) = \begin{cases} 1 & \text{if } 0 \le x \le 2 \\ -1 & \text{if } 2 < x \le 3. \end{cases}$$

Prove that $f$ is continuous on $[0,3]$ but does not attain its average w.r.t. the weight function $g$ on $[0,3]$. (In other words, the non-negativity of the weight function is vital for Theorem (1.4) to hold.)

1.9 Let $f(x) = x$ for all $x$. Find the mean value of $f$ on $[0,1]$ w.r.t. each of the following weight functions. Identify the point(s) where the average is attained.

(i) $x$      (ii) $1 - x$      (iii) $x(1 - x)$.

1.10 State and prove the analogue of Exercise (1.4) for weighted averages.

1.11 Suppose $f, g : [a, b] \to \mathbb{R}$ are functions with $f$ monotonically increasing, $g$ integrable and non-negative on $[a, b]$. Prove that there exists $c \in [a, b]$ such that

$$\int_a^b f(x)g(x)dx = f(b) \int_a^c g(x)dx + f(a) \int_c^b g(x)dx.$$

Show that if $W_1$ and $W_2$ are the weights of the subintervals $[a, c]$ and $[c, b]$ respectively, then the mean value of $f$ on $[a, b]$ (w.r.t. the weight function $g$) divides the interval $[f(a), f(b)]$ in the ratio $W_1 : W_2$. (Put differently, the point $c$ divides the interval $[a, b]$ 'weightwise' in the same ratio as the average of $f$ divides the interval $[f(a), f(b)]$. So in this sense $c$ is a mean value even though $f(c)$ may not be the average value. For this reason this result is called **Second Mean Value Theorem for integrals.**)

[*Hint* : Consider the function $H : [a, b] \to \mathbb{R}$ defined by

$$H(x) = f(b) \int_a^x g(t)dt + f(a) \int_x^b g(t)dt.$$

Apply the Intermediate Value Property to $H$. ]

1.12 Suppose the linear density (= mass per unit length) and the specific heat (= number of calories per unit mass to raise its temperature by $1°C$) of a wire of length $L$ at a point $x$ units away from one of its ends are $\rho(x)$ and $\sigma(x)$ respectively. What is the average specific heat of the whole wire ?

1.13 Suppose the probability of a baby being a boy is $p$. If a couple keeps producing babies till getting a son, what is the expected number of children it has? (If $p = \frac{1}{2}$, the answer is 2, although in reality it will be less because no couple will produce more than, say, 20 children. The population growth in certain countries is attributed, partly, to a desire to have a son. This exercise shows that this is incorrect if $p = \frac{1}{2}$ or more).

1.14 What will be the answer to Problem (1.6) if the coin is unbiased (i.e. $p = \frac{1}{2}$) and tossed till a head shows twice in succession? [*Hint* : For the first solution, you will need, for each positive integer $n$, the number of winning sequences of length $n$, i.e., sequences of length $n$ in which the only successive $H$'s occur in the last two places. This can be done by establishing a Fibonacci relation (see Exercise (3.1.4) and also Section 3.10). The second solution goes through without much difficulty.]

1.15 In a casino, gambling machines are kept, where at each round a player pays one rupee and tosses a fair coin, and is given a reward when three heads show in succession. The machines are meant only to attract customers and the management wants to wind up even on them. What amount of reward will ensure this in the longer run?

*1.16 Suppose $f : \mathbb{R} \to \mathbb{R}$ is differentiable and has the property that for every interval $[a, b]$, the average value of $f$ on $[a, b]$ is $f(\frac{a+b}{2})$. Prove that $f$ is linear, i.e. there exist some constants $\lambda$ and $\mu$ such that for all $x$, $f(x) = \lambda x + \mu$. [*Hint* : Fix $a \in \mathbb{R}$. For $h > 0$, apply the hypothesis to the invervals $[a, a+h]$, $[a-h, h]$ and $[a-h, a+h]$ to get that $f(a+\frac{h}{2}) + f(a-\frac{h}{2}) = 2f(a)$. Differentiate w.r.t. $h$ and show $f'$ is constant.]

1.17 What is the average area of a right angled triangle inscribed in a unit circle, assuming (i) that the length of its shortest altitude varies uniformly (ii) that the length of its shortest side varies uniformly?

1.18 Suppose $f : [a, b] \to \mathbb{R}$ is continuous and $h : [c, d] \to [a, b]$ is continuous, strictly monotonically increasing with $f(c) = a$, $f(d) = b$. Show by an example that the average of $f$ on $[a, b]$ may differ from that of $f \circ h$ on $[c, d]$. (In other words, averages are not invariant under a change of variables. This is the major difficulty in defining the average of a function in terms of some real variable associated with the domain.)

1.19 The concept of a weighted average of a finite set has a geometric interpretation. Let $\mathbf{u}_1, \ldots, \mathbf{u}_m$ be vectors in a plane (or in some other euclidean space $\mathbb{R}^n$, see Section 4.8), and $\lambda_1, \ldots, \lambda_m$ be some non-negative numbers not all zero (so that $\sum_{i=1}^{m} \lambda_i > 0$). Then the weighted average, say $\mathbf{u}$, of $\mathbf{u}_1, \ldots, \mathbf{u}_m$ with weights $\lambda_1, \ldots, \lambda_m$ respectively, is the vector $\frac{1}{\sum_{i=1}^{m} \lambda_i} \sum_{i=1}^{m} \lambda_i \mathbf{u}_i$. Note that this vector depends only the relative proportions of the $\lambda$'s. It is therefore customary to assume $\sum_{i=1}^{m} \lambda_i = 1$. We do so. Prove that :

(i) if $\mathbf{u}_1, \mathbf{u}_2$ are position vectors of two points, say $A$ and $B$ respectively, then every weighted average of $\mathbf{u}_1$ and $\mathbf{u}_2$ is the position vector of a unique point on the straight line segment joining $A$ and $B$.

(ii) a subset $S$ of the plane is convex (cf. Definition (4.10.1)) if and only if it has the property that for every $\mathbf{u}_1, \ldots, \mathbf{u}_m$ in $S$, every weighted average of them is in $S$. (For this reason, a weighted average of points in $\mathbb{R}^m$ is often called a **convex combination** of them.)

1.20 Let $\mathbf{a}, \mathbf{b}, \mathbf{c}$ be the position vectors of the vertices $A, B, C$ of a triangle $ABC$ in a plane. Prove that the position vector of every point of the triangle is a unique convex combination of $\mathbf{a}, \mathbf{b}, \mathbf{c}$. That is, if $\mathbf{u}$ is the position vector of a point $P$ of the triangle $ABC$, then there exist unique non-negative real numbers $\alpha, \beta, \gamma$ with $\alpha + \beta + \gamma = 1$ and $\mathbf{u} = \alpha\mathbf{a} + \beta\mathbf{b} + \gamma\mathbf{c}$. These numbers are called the **barycentric coordinates** of $P$. (The name will be justified later when we study centres of mass (see Exercise (4.13)). Identify the points of the triangle whose barycentric coordinates are

(i) $(1, 0, 0)$    (ii) $(\frac{1}{3}, \frac{1}{3}, \frac{1}{3})$    (iii) of the form $(\alpha, 0, 1 - \alpha)$

(iv) $\left( \dfrac{a}{a+b+c}, \dfrac{b}{a+b+c}, \dfrac{c}{a+b+c} \right)$ where $a, b, c$ are the lengths of the sides opposite to $A, B, C$ respectively).

(v) $\left( \dfrac{\tan A}{k}, \dfrac{\tan B}{k}, \dfrac{\tan C}{k} \right)$ where $k = \tan A + \tan B + \tan C$ and $ABC$ is acute-angled.

1.21 If a coin with probability $p$ of showing a head is tossed $n$ times, show that the expected number of heads is $pn$.

(This exercise suggests another attempt to define probability as the average proportion of occurrence of an event during a large number of trials; yet another proof of how the concepts of an average and probability are inexorably related to each other.)

1.22 Horizontal strips of width $r$ are fitted on a large window so that the width of the gap between two consecutive strips is $R$. If a ball of radius $a$ is thrown horizontally at random to the window $n$ times, what is the expected number of times it will pass through the window?

1.23 Suppose $\phi$ is continuous on an interval $[a, b]$ and takes values in an interval $[c, d]$. Assume further that $f$ is a function which is concave upwards on $[c, d]$. Prove that

$$ f\left( \frac{1}{b-a} \int_a^b \phi(x)dx \right) \leq \frac{1}{b-a} \int_a^b (f(\phi(x)))dx. $$

Verbally stated, the value of $f$ at the average value of $\phi$ cannot exceed the average value of the composite function $f \circ \phi$. (This result, called **Jensen's inequality**, is clearly the continuous version of Exercise (4.10.6). This observation also suggests a proof.)

## 6.2    Arc Length as a Definite Integral

We have emphasized several times that integrals are limits of sums. In fact, in Section 5.2, we gave a 'recipe of integration' using which we can express many things such as area, mass, electrical resistance of a wire, the population of a road etc. in terms of suitable integrals. In this and the next two sections we study in greater detail the applications of definite integrals to evaluate arc lengths, areas, volumes and mass. But before doing so, it is worthwhile pondering a little on exactly what makes integrals applicable here.

A quantitative attribute of sets is called **additive** if for any set $S$, you can obtain it by obtaining it for any subset, say $A$, of $S$ and also for its complement $S - A$ and adding the two. Instead of just two mutually complementary subsets, we could, of course, have any finite number of mutually disjoint subsets whose union is $S$ (or, in other words, a partition of $S$ into a finite number of subsets). Many quantities we come across are additive in nature. These include length, area, volume, mass, specific heat etc. If a stone is broken into two parts, the sum of their masses is obviously the mass of the original stone. The same is true of the length or the electric resistance of a piece of wire. So the length of an arc is an additive attribute. By contrast, the length of the boundary (i.e., the perimeter as it is often called) of a plane region is not additive. When a region $S$ is divided into two, say $S_1$ and $S_2$, the common boundary of $S_1$ and $S_2$ adds to the perimeter of each of them and so their sum exceeds the perimeter of $S$, see Fig. 6.2.1(a).

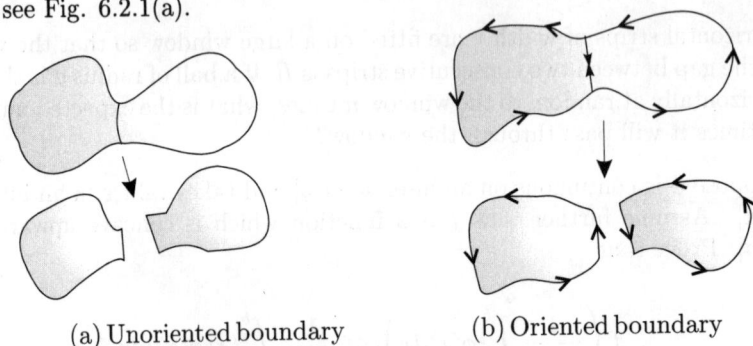

(a) Unoriented boundary          (b) Oriented boundary

Fig. 6.2.1 : Non-additivity of the Perimeter

(For certain purposes it is convenient to orient the boundary of the region according to some convention, say, so that the region always lies on the left as the boundary is traversed. In such a case when a region $S$ is divided into $S_1$ and $S_2$, the common boundary of $S_1$ and $S_2$ gets counted twice but with opposite orientations, as shown in Fig.5.2.1(b). So in some sense it gets cancelled and therefore the boundary is additive. Still its length is not.)

Quantities which depend on the inter-relationship between the various parts of an object are generally not additive. Thus, for example, the strength of an object is generally more than the total of the strength of its parts. If a diamond is broken into two parts, it loses its value considerably. So the market value is not always an additive attribute. On the other hand, when the value depends only on weight (as in the case of pure gold), it is an additive attribute, because weight (or more precisely the mass) is an additive attribute.

The rigorous mathematical definitions of arc lengths, areas and volumes have a common feature. In each case we have some prototypes. In the case of arc length these prototypes are straight line segments, while in the case of areas and volumes they are rectangles and rectangular boxes respectively. The length (or area or volume) of each prototype is defined by a simple arithmetical formula. The set, say $S$, whose length (or area etc.) is to be defined is split into a large number of pieces. Each piece is treated approximately as a prototype and their lengths (or areas etc.) are added. These sums usually fall too short of the exact value of the attribute under question which is defined as the supremum of all approximating sums. As the individual approximating prototypes shrink in size, their sum is a closer approximation to the exact answer. This is also the underlying idea behind a Riemann integral, which is defined simultaneously as the supremum of all lower Riemann sums and the infimum of all upper Riemann sums. So it is hardly surprising that even though integrals are not needed to *define* arc lengths, areas or volumes, they are an invaluable tool to evaluate them.

We proceed to elaborate these statements. In this section we deal with the arc length of a curve $C$. Aras and volumes will be taken up later. For simplicity, we assume $C$ to be a plane curve, although essentially the same arguments work for a space curve, or more generally, for a curve in the euclidean space $\mathbb{R}^n$, for any $n$. The only difference, perhaps, is that the length of a prototype, viz. a segment, say $PQ$, is taken as an intuitively clear concept for a plane or space segment, while in $\mathbb{R}^n$ for $n > 3$ (and even for $n = 1, 2$, or $3$ if our approach is to be analytical) $|PQ|$ has to be *defined* as $\sqrt{\sum_{i=1}^{n}(x_i - y_i)^2}$, where $P$ and $Q$ represent, respectively, the ordered $n$-tuples $(x_1, \ldots, x_n)$ and $(y_1, \ldots, y_n)$ in $\mathbb{R}^n$. (See Section 4.8 for more details of the analytical approach.)

In terms of the length of the prototype, we defined the arc length of the curve $C, l(C)$ (cf. Definition (4.7.1) and Figure (4.7.9)) as the supremum of all sums of the form $\sum_{i=1}^{n} |P_{i-1} P_i|$ where $(P_0, P_1, \ldots, P_n)$ is a broken line path along $C$ from $A$ to $B$ ($A, B$ being the end-points of $C$). Suppose now that $C$ is a parametrised curve. As explained in Section 4.12, parametrisation is an integral part of the very concept of a curve (see the comments preceding Definition (4.12.1)). Assume $C$ is parametrised by $\alpha(t) = (f(t), g(t)), a \leq t \leq b$ where $f$ and $g$ are some continuous real-valued functions defined on the interval $[a, b]$. In terms of this parametrisation, $A = \alpha(a), B = \alpha(b)$ and a broken line path, say, $(P_0, P_1, \ldots, P_n)$ along $C$ corresponds to a partition, say, $(t_0, t_1, \ldots, t_n)$ of $[a, b]$ with $\alpha(t_i) = P_i$ for $i = 0, 1, \ldots, n$. So, in this notation,

$$l(C) = \sup\left\{\sum_{i=1}^{n}\sqrt{[f(t_i) - f(t_{i-1})]^2 + (g(t_i) - g(t_{i-1}))^2}\right\} \text{ as } (t_0, t_1, \ldots, t_n)$$

ranges over all partitions of $[a, b]$. For brevity, denote $f(t_i) - f(t_{i-1})$ by $\Delta x_i$ and $g(t_i) - g(t_{i-1})$ by $\Delta y_i$ for $i = 1, \ldots, n$. As usual let $\Delta t_i = t_i - t_{i-1}$. Then, recalling that $\mathcal{P}[a, b]$ stands for the set of all partitions of the interval $[a, b]$, we can write the length of $C$ more compactly as

$$l(C) = \sup\left\{\sum_{i=1}^{n}\sqrt{\left(\frac{\Delta x_i}{\Delta t_i}\right)^2 + \left(\frac{\Delta y_i}{\Delta t_i}\right)^2}\,\Delta t_i : (t_0, \ldots, t_n) \in \mathcal{P}[a, b]\right\} \qquad (1)$$

So far, we assumed $f$ and $g$ to be continuous functions of $t \in [a, b]$. Suppose now they are differentiable on $[a, b]$ too. Then by Lagrange's Mean Value Theorem, for every $i = 1, \ldots, n$, we get points $\xi_i$ and $\eta_i$ in $(t_{i-1}, t_i)$ such that $\dfrac{\Delta x_i}{\Delta t_i} = f'(\xi_i)$ and $\dfrac{\Delta y_i}{\Delta t_i} = g'(\eta_i)$. So,

$$\sum_{i=1}^{n}\sqrt{\left(\frac{\Delta x_i}{\Delta t_i}\right)^2 + \left(\frac{\Delta y_i}{\Delta t_i}\right)^2}\,\Delta t_i = \sum_{i=1}^{n}\sqrt{[f'(\xi_i)]^2 + [g'(\eta_i)]^2}\,\Delta t_i \qquad (2)$$

Our goal is to show that the Riemann integral $\displaystyle\int_a^b \sqrt{[f'(t)]^2 + [g'(t)]^2}\,dt$, which we denote by $I$, equals $l(C)$. It is tempting to try to do so by substituting (2) into (1) and then taking limits as the mesh of the partition $P$ of $[a, b]$ tends to 0. Intuitively the justification is that as the partition $P = (t_0, t_1, \ldots, t_n)$ gets finer and finer the broken line path $(P_0, P_1, \ldots, P_n)$ gives a better and better approximation to the curve $C$, and at the same time the sum on the right hand side of (2) gets closer and closer to the integral $I$. But in a rigourous approach we must overcome the following difficulties :

(1) Do the quantities $l(C)$ and $I$ really exist? In other words is the curve $C$ rectifiable and is the function $\sqrt{[f'(t)]^2 + [g'(t)]^2}$ integrable on $[a, b]$?

(2) Even if the integral $I$ exists, it is the limit of Riemann sums of the form

$$\sum_{i=1}^{n}\sqrt{[f'(\xi_i)]^2 + [g'(\xi_i)]^2}\,\Delta t_i.$$

But in the sum of the right hand side of (2), the points $\xi_i$ and $\eta_i$ need not be the same and so this sum is not quite a Riemann sum of the function $\sqrt{[f'(t)]^2 + [g'(t)]^2}$ based on the partition $(t_0, t_1, \ldots, t_n)$.

(3) By definition, $l(C)$ is the supremum and not the limit of the lengths of the broken line paths. So some work is needed to show that $l(C)$ indeed equals the limit of $\displaystyle\sum_{i=1}^{n}|P_{i-1}P_i|$ as the partition $(t_0, t_1, \ldots, t_n)$ gets finer and finer.

We tackle these difficulties one-by-one. As we shall see below if we assume both $f'$ and $g'$ to be continuous on $[a, b]$ then (1) is taken care of. Actually, it will suffice if $f'$ and $g'$ are only piecewise continuous on $[a, b]$ and most of the commonly encountered curves satisfy this condition. So (1) is not a serious difficulty. This assumption of continuity of $f'$ and $g'$ also enables us to overcome the second difficulty, in view of Exercise (5.6.5). (Actually, for this purpose the continuity of either one of $f'$ and $g'$ and mere integrability of the other would suffice). So it is really the third difficulty which needs some work. We begin with a lemma which is of some independent interest, as it is the precise expression of the intuitively obvious fact that a finer partition of $[a, b]$ yields a broken line path which is a better approximation to the curve $C$.

**2.1 Lemma :** Let $P = (t_0, t_1, \ldots, t_n)$ and $Q = (s_0, s_1, \ldots, s_m)$ be partitions of $[a, b]$ with $Q$ being a refinement of $P$. Let $P_i = \alpha(t_i) = (f(t_i), g(t_i))$ for $0 \le i \le n$ and $Q_j = \alpha(s_j) = (f(s_j), g(s_j))$ for $0 \le j \le m$. Then

$$\sum_{i=1}^{n} |P_{i-1}P_i| \le \sum_{j=1}^{m} |Q_{j-1}Q_j|.$$

**Proof :** The proof is extremely easy and involves little more than an application of the triangle inequality.

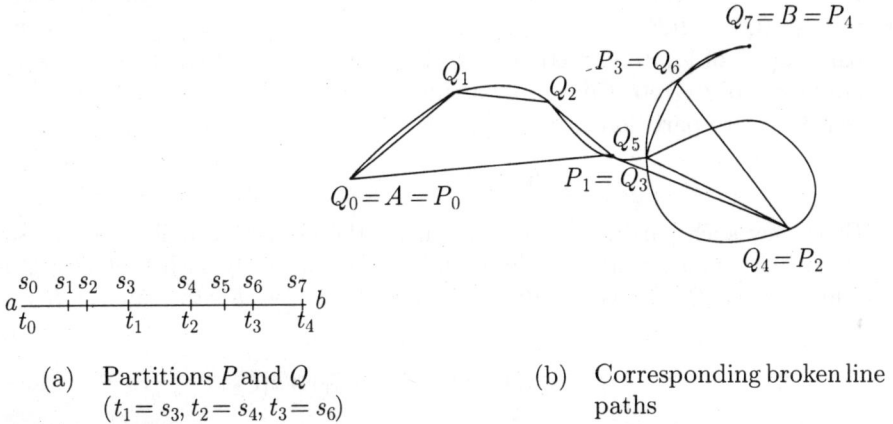

(a) Partitions $P$ and $Q$      (b) Corresponding broken line
($t_1 = s_3,\ t_2 = s_4,\ t_3 = s_6$)      paths

Figure 6.2.2 : Partitions and Broken Line Paths

Since $Q$ is a refinement of $P$, there exist integers $0 = j_0 < j_1 < \ldots < j_n = m$ such that for every $r = 0, \ldots, n, t_r = s_{j_r}$. Now, for each $i = 1, \ldots, n$, $|P_{i-1}P_i|$ is simply $|Q_{j_{i-1}}Q_{j_i}|$. By triangle inequality, the latter is at most equal to the sum

$$\left|Q_{j_{i-1}}Q_{j_{i-1}+1}\right| + \left|Q_{j_{i-1}+1}Q_{j_{i-1}+2}\right| + \cdots + \left|Q_{j_i-1}Q_{j_i}\right|$$

which appears as a block of consecutive terms in the sum $\displaystyle\sum_{j=1}^{m} |Q_{j-1}Q_j|$. These

blocks are mutually disjoint and exhaust all the terms in $\sum\limits_{j=1}^{m}|Q_{j-1}Q_j|$ and thus
we get the desired inequality. (For clarification of the underlying idea, see Fig. 6.2.2). ∎

This lemma is of the same spirit as the part of Theorem (5.1.2) dealing with the lower Riemann sums of a function. (Note, however, that there is no analogue of the part dealing with upper Riemann sums.) So the following corollary, which is of the same spirit as Exercise (5.1.7) should come as no surprise.

**2.2 Corollary :** Arc length is additive. In other words suppose $C$ is a curve, parametrised by $\alpha(t) = (f(t), g(t)), a \leq t \leq b$ and let $a < c < b$. Let $C_1, C_2$ be the portions of $C$ from $\alpha(a)$ to $\alpha(c)$ and from $\alpha(c)$ to $\alpha(b)$ respectively. Then $C$ is rectifiable if and only if $C_1, C_2$ are both rectifiable and when this happens,

$$l(C) = l(C_1) + l(C_2).$$

**Proof :** Let us introduce the notation that if $P$ is a partition (of $[a, b]$ or of $[a, c]$ or $[c, b]$) then $l(P)$ denotes the length of the corresponding broken line path. If $P, Q$ are partitions of $[a, c]$ and of $[c, b]$ respectively, then denote by $P + Q$ the partition of $[a, b]$ obtained simply by juxtaposing $P$ and $Q$. Clearly, $l(P) + l(Q) = l(P + Q) \leq l(C)$ and so in particular, $l(P) \leq l(C)$ for every partition $P$ of $[a, c]$. So rectifiability of $C$ implies that of $C_1$, and by a similar argument, of $C_2$ too. Conversely assume $C_1, C_2$ are both rectifiable. We claim that for every partition $R$ of $[a, b]$

$$l(R) \leq l(C_1) + l(C_2). \tag{3}$$

Given any such partition $R$ of $[a, b]$, insert the point $c$ in it if necessary so as to get partitions $P$ and $Q$ of $[a, b]$ and $[c, b]$ respectively such that $P + Q$ is a refinement of $R$. (If $c$ is already a node of $R$ then $P + Q$ is the same as $R$.) The lemma above gives,

$$l(R) \leq l(P) + l(Q) \leq l(C_1) + l(C_2)$$

and thus (3) holds. It follows that if $C_1, C_2$ are rectifiable then so is $C$. Moreover, since $l(C)$ is the least upper bound of the set $\{l(R) : R$ a partition of $[a, b]\}$, we also get $l(C) \leq l(C_1) + l(C_2)$. To show that equality holds, let, if possible, $l(C) < l(C_1) + l(C_2)$. Let $\epsilon = \frac{1}{2}(l(C_1) + l(C_2) - l(C))$. Then $\epsilon > 0$. By Exercise (2.6.12), there exist partition $P$ and $Q$ of $[a, c]$ and $[c, b]$ respectively such that $l(P) > l(C_1) - \epsilon$ and $l(Q) > l(C_2) - \epsilon$. But then $P + Q$ is a partition of $[a, b]$ for which $l(P + Q) = l(P) + l(Q) > l(C_1) + l(C_2) - 2\epsilon$. Thus, $l(P + Q) > l(C)$, a contradiction. Hence we must have $l(C) = l(C_1) + l(C_2)$. ∎

The lemma also enables us to express $l(C)$ as a limit rather than as a supremum. We follow the notations in the proof just given.

**2.3 Theorem :** Let $C$ be a rectifiable curve. Then $l(C)$ is the limit of $l(P)$ as the partition $P$ of $[a, b]$ gets finer and finer. That is, for every $\epsilon > 0$, there exists a partition $P$ of $[a, b]$ such that whenever $Q$ is a refinement of $P$, $|l(Q) - l(C)| < \epsilon$. Conversely, if this limit exists, then $C$ is rectifiable and its length, $l(C)$, equals this limit.

**Proof :** By definition, $l(C)$ is the least upper bound of the set $\{l(P) : P$ a partition of $[a, b]\}$. So by Exercise (2.6.12), given $\epsilon > 0$, there exists a partition $P$ of $[a, b]$ for which $l(P) > l(C) - \epsilon$. Now let $Q$ be any refinement of $P$. Then by Lemma (2.1), $l(Q) \geq l(P)$. Since we always have $l(Q) \leq l(C)$, we get $l(C) - \epsilon < l(Q) \leq l(C)$ and hence $|l(Q) - l(C)| < \epsilon$.

For the converse, let $L$ be the limit of $l(P)$ as $P$ gets finer and finer. Let $S$ be the set $\{l(P) : P$ a partition of $[a, b]\}$. We claim first that for every $\epsilon > 0$, $L + \epsilon$ is an upper bound of $S$. Indeed, given any such $\epsilon > 0$, by assumption there exists $P$ such that whenever $Q$ is a refinement of $P$, $|l(Q) - L| < \epsilon$. Now, given any arbitrary partition, say $R$, of $[a, b]$ we let $P \cup R$ be the partition obtained by superimposing $P$ and $R$ (see the proof of Corollary (5.1.3)). Then $P \cup R$ is a refinement of $R$ and so $|l(P \cup R) - L| < \epsilon$, which, in particular, means $l(P \cup R) < L + \epsilon$. But since $P \cup R$ is also a refinement of $P$, the lemma above gives $l(R) \leq l(P \cup R)$. So we have $l(R) < L + \epsilon$ for every partition $P$ of $[a, b]$. So the set $S$ is bounded above and hence the curve $C$ is rectifiable. In fact since $\epsilon > 0$ is arbitrary in this reasoning, we get $l(C) \leq L$. To claim $l(C)$ actually equals $L$, suppose once again, that $l(C) < L$. Taking $\epsilon = L - l(C)$, and applying the hypothesis we get a partition $P$ of $[a, b]$ for which $|l(P) - L| < \epsilon$ and hence $l(P) > L - \epsilon = l(C)$, a contradiction. ∎

An alert reader will notice that in the theorem just proved the limit is taken as the partition $P$ becomes finer and finer and not as its mesh tends to 0. In other words the theorem is of the spirit of Theorem (5.2.1) and not of Corollary (5.2.4). As explained after Theorem (5.2.1), there is a subtle difference between the two and their equivalence was not trivial (Theorem (5.2.3)). Logically we should now aim at a theorem which equates $l(C)$ with $\lim_{\mu(P) \to 0} l(P)$. Fortunately, for the type of curves we are interested in, it is not necessary to do so as we are already in a position to equate $l(C)$ with a Riemann integral.

**2.4 Theorem :** Suppose $C$ is a (parametric) plane curve $\alpha(t) = (f(t), g(t))$, $a \leq t \leq b$ where the functions $f$ and $g$ are continuously differentiable on $[a, b]$. Then $C$ is rectifiable and further, its arc length is given by

$$l(C) = \int_a^b \sqrt{[f'(t)]^2 + [g'(t)]^2}\, dt. \tag{4}$$

**Proof :** Since both $f'$ and $g'$ are continuous on $[a, b]$, they are bounded. Let $M$ be a common upper bound for $|f'(t)|$ and $|g'(t)|$ for $t \in [a, b]$. Let $P = (t_0, t_1, \ldots, t_n)$ be a partition of $[a, b]$ and $(P_0, P_1, \ldots, P_n)$ be the corresponding

broken line path along $C$. Then $l(P)$ is given by (2) and since $|f'(\xi_i)| \leq M$
and $|g'(\xi_i)| \leq M$ regardless of where $\xi_i$ and $\eta_i$ lie, we have $l(P) \leq \sqrt{2} M \sum_{i=1}^{n} \Delta t_i$
$= \sqrt{2} M (b-a)$. So the set of lengths of all broken line paths along $C$ is bounded
and hence $l(C)$ is rectifiable.

Continuity of $f'$ and $g'$ ensures that of $\sqrt{(f')^2 + (g')^2}$ (since the square-root
function is also continuous) and so by Theorem (5.4.1), the integral
$\int_a^b \sqrt{[f'(t)]^2 + [g'(t)]^2} dt$ exists. Let us call it $I$. By Corollary (5.2.4), $I$ equals
the limit of the Riemann sum of the form $\sum_{i=1}^{n} \sqrt{[f'(\xi_i)]^2 + [g'(\xi_i)]^2} \Delta t_i$ based
on a partition, say $P = (t_0, t_1, \ldots, t_n)$ of $[a,b]$ as $\mu(P) \to 0$. But since both $f'$
and $g'$ are continuous, by Exercise (5.6.5), part (iii), $I$ also equals the limit of
the sums of the form $\sum_{i=1}^{n} \sqrt{[f'(\xi_i)]^2 + [g'(\eta_i)]^2} \Delta t_i$ where $\xi_i$ and $\eta_i$ are any two
(independently chosen) points of the $i^{\text{th}}$ subinterval $[t_{i-1}, t_i]$ of the partition
$P = (t_0, t_1, \ldots, t_n)$ as $\mu(P) \to 0$. From this it follows very easily that $I$ is also
the limit of sums of the form $\sum_{i=1}^{n} \sqrt{[f'(\xi_i)]^2 + [g'(\eta_i)]^2} \Delta t_i$ as the underlying
partition $P$ gets finer and finer. (The proof of this fact is similar to that of the
implication (ii) $\Rightarrow$ (i) in Theorem (5.2.3) and is left as an exercise.) By the last
theorem, $l(C)$ is the limit of $l(P)$ as $P$ gets finer and finer. By (2), every sum of
the form $l(P)$ can be written as a sum of the form $\sum_{i=1}^{n} \sqrt{[f'(\xi_i)]^2 + [g'(\eta_i)]^2} \Delta t_i$,
but not necessarily conversely. So the situation is somewhat like that of a sub-
sequence of a sequence. As shown in Section 3.5, a subsequence of a convergent
sequence must have the same limit as the parent sequence. So if the subsequence is already known to be convergent then the two limits must coincide by
uniqueness of limits (by the analogue of Theorem (2.6.1) for sequences).

The proof that $l(C) = I$ is similar. For, otherwise, let $\epsilon = \frac{1}{2}|l(C) - I|$. Then
there exist partitionss $P, P'$ of $[a,b]$ such that whenever $Q$ is a refinement of $P$,
we have $|l(Q) - l(C)| < \epsilon$ and whenever $Q$ is a refinement of $P'$, we have
$|\sum_{i=1}^{n} \sqrt{[f'(\xi_i)]^2 + [g'(\eta_i)]^2} \Delta t_i - I| < \epsilon$ for every $\boldsymbol{\xi}$ and $\boldsymbol{\eta}$ based on $Q$. We let
$Q$ be any common refinement of $P$ and $P'$ (for example, $Q = P \cup P'$) and
choose $\xi_i, \eta_i$ as in (2). Then $l(Q)$ equals $\sum_{i=1}^{n} \sqrt{[f'(\xi_i)]^2 + [g'(\eta_i)]^2} \Delta t_i$. So we get
$|l(Q) - l(C)| < \epsilon$ and also $|l(Q) - I| < \epsilon$, which by triangle inequality, gives
$|l(C) - I| < 2\epsilon = |l(C) - I|$, a contradiction. This completes the proof. ∎

This theorem enables us to calculate the arc length very efficiently, provided we can evaluate the Riemann integral appearing in it. As a simple example, let us verify that the length of a line segment $C$, joining points $P = (x_1, y_1)$ and $Q = (x_2, y_2)$ is indeed $\sqrt{(x_2 - x_1)^2 + (y_2 - y_1)^2}$. A parametrisation for $C$ is $\alpha(t) = ((1 - t)x_1 + tx_2, (1 - t)y_1 + ty_2), 0 \le t \le 1$ (cf. Exercise (4.7.1)). So $f'(t) = x_2 - x_1$ and $g'(t) = y_2 - y_1$ for all $t \in [0, 1]$. Thus

$$l(C) = \int_0^1 \sqrt{(x_2 - x_1)^2 + (y_2 - y_1)^2} dt = \sqrt{(x_2 - x_1)^2 + (y_2 - y_1)^2}.$$ Of course,

we know this already. But now we know that for the prototypes with which we started, the formula agrees with the old definition. (In a fussy approach, the old length would be called something like a 'tentative length'. The 'length' of a curve $C$ would then be defined as the supremum of the tentative lengths of broken line paths. What we have just shown is that the length of a line segment coincides with its tentative length.) We can also now 'officially' calculate the length of a circle of radius $a$. Taking the standard parametrisation $x = a \cos t, y = a \sin t, 0 \le t \le 2\pi$, we immediately get the length as $\int_0^{2\pi} \sqrt{a^2 \sin^2 t + a^2 \cos^2 t} \, dt = 2\pi a$. In the geometric approach to trigonometry, this information is not new, because an angle is measured in terms of arc length. But if we take the analytical approach (given in Section 4.8 or Exercise (5.10.24)) then we now have a formula for the arc length of a circle, and more generally, for the arc length of any arc of a circle.

A few more examples of calculations of arc lengths will be given in the exercises. It is perhaps fair to point out that the theorem above is not such a golden tool as it appears. The reason is not that its hypothesis is too stringent. On the contrary, most of the familiar curves satisfy it. The trouble is that the integrand in (4) involves a radical sign and rarely allows an antiderivative in a closed form. Even for a familiar curve like an ellipse, parametrised by $x = a \cos t$, $y = b \sin t$ $(0 \le t \le 2\pi)$, the arc length is $\int_0^{2\pi} \sqrt{a^2 \sin^2 t + b^2 \cos^2 t} \, dt$. Unless $a = b$ (in which case the ellipse reduces, to a circle) there is no way to evaluate the integral in a closed form. We have to resort to some such methods as the Simpson's rule. Integrals of this form are called **elliptic integrals** and arise in many applications, often completely unrelated to ellipses (e.g. see Exercise (8.20)).

Nevertheless, the theorem above has a theoretical significance. Suppose the curve $C$ is parametrised by $\alpha(t) = (x(t), y(t)), a \le t \le b$. Let $s : [a, b] \to \mathbb{R}$ be the function defined by the integral, that is, $s(t) = \int_a^t \sqrt{[f'(u)]^2 + [g'(u)]^2} du$ for $t \in [a, b]$. Obviously $s(0) = 0$ and $s(b) = l(C) =$ the arc length of $C$. Since the integrand is non-negative, it is clear that $s$ is a monotonically increasing function. (This is, of course, evident even otherwise since $s(t)$ is nothing but the length of the portion of $C$ from $\alpha(a)$ to $\alpha(t)$ and, because of additivity, will increase as $t$ increases.) Using the second form of the Fundamental Theorem of Calculus (Theorem (5.3)), it follows that $\dfrac{ds}{dt} = \sqrt{[f'(t)]^2 + [g'(t)]^2}$ for all

$t \in [a, b]$. Recalling that $x = f(t)$ and $y = g(t)$, this equation becomes

$$\frac{ds}{dt} = \sqrt{\left(\frac{dx}{dt}\right)^2 + \left(\frac{dy}{dt}\right)^2} \tag{5}$$

This formula, which gives the derivative of the arc length function in terms of those of the co-ordinate functions, is a very useful formula. As already remarked in Section 4.4, it is often written by dropping the differential $dt$, as

$$ds = \sqrt{(dx)^2 + (dy)^2} \quad \text{or} \quad (dx)^2 = (dx)^2 + (dy)^2. \tag{6}$$

Although strictly speaking (6) has no meaning of its own, its very formulation is suggestive. As shown in Fig. 6.2.3, if the points $t$ and $t+\Delta t$ in $[a, b]$ correspond to the points $P$ and $Q$ on the curve $C$ and $\Delta x, \Delta y$ are the increments in $x$ and $y$, then the length of the chord is $\sqrt{(\Delta x)^2 + (\Delta y)^2}$. Let $\Delta s$ denote the length of the arc $PQ$. In general, $\Delta s$ exceeds $\sqrt{(\Delta x)^2 + (\Delta y)^2}$. But (6) shows that $\lim_{\Delta s \to 0} \dfrac{\sqrt{(\Delta x)^2 + (\Delta y)^2}}{\Delta s} = 1$. This fulfils a claim made in Section 4.7 (and used in the proof of the fact that $\lim_{\phi \to 0} \dfrac{\sin \phi}{\phi} = 1$) to the effect that the length of a small portion of a curve is approximately the length of the chord joining its end-points. (We are tacitly assuming here that $\Delta s > 0$ if $|\Delta t|$ is sufficiently small and positive. This would be so if at least one of $\Delta x$ and $\Delta y$ is non-zero. The proof of Theorem (4.12.2) shows that this is indeed the case for a smooth curve.)

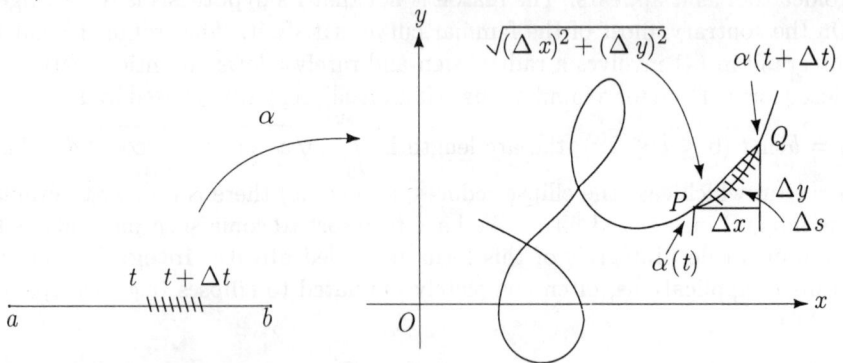

Figure 6.2.3 : Line Element

In a less precise but suggestive formulation, (6) says that if $\Delta x$ and $\Delta y$ are 'infinitesimally small' then the chord $PQ$ may be taken to be equal to the arc $PQ$. Although a rigorous mathematical definition of an infinitesimally small quantity is not easy (and therefore the term is generally avoided in modern textbooks), it is a very useful concept from a practical point of view. So it is customary to call an infinitesimally small portion of a curve as a **line element** and treat it as a straight line segment. This approach will, in fact, be followed in the next section for finding the area of a surface of revolution.

Formula (5) also has a physical significance when the variable $t$ denotes time and $\alpha(t)$ denotes the position of a moving particle at time $t$. In this situation it is customary to denote derivatives by putting a dot on the top of the variable which is differentiated. Thus in this new notation, (5) becomes

$$\dot{s} = \sqrt{(\dot{x})^2 + (\dot{y})^2} \tag{7}$$

The left hand side is obviously the speed since $s$ is the distance travelled by the particle. The right hand side is the length or the absolute value of the vector $(\dot{x}, \dot{y})$. It is customary to introduce unit vectors $\mathbf{i}, \mathbf{j}$ along the positive $x$ and $y$ axes respectively. Then the position vector of the particle at time $t$ is $x\mathbf{i} + y\mathbf{j}$ where $x = f(t)$ and $y = g(t)$. It is customary to denote this by $\mathbf{r}$. Thus $\mathbf{r} = \alpha(t)$. This gives a somewhat more convenient parametrisation of the curve not only because it saves space but more importantly because it can also be used for curves in the three dimensional space, or more generally for curves in $I\!R^n$ for any $n \geq 1$. (For $n = 3$ it is customary to write $\mathbf{r} = x\mathbf{i} + y\mathbf{j} + z\mathbf{k}$ where $\mathbf{k}$ is a unit vector along the positive $z$-axis.) Let $\dot{\mathbf{r}}$ denote the vector $\dot{x}\mathbf{i} + \dot{y}\mathbf{j}$ (or the vector $\dot{x}\mathbf{i} + \dot{y}\mathbf{j} + \dot{z}\mathbf{k}$ in case of a motion in the three dimensional space). This vector is called **velocity** of the particle at time $t$ because as we shall see later, it can also be obtained by differentiating the function $\mathbf{r} = \alpha(t)$ w.r.t. $t$. (In essence, this was already done in Theorem (4.12.2). But we shall do it more systematically in Section 6.6.) We thus see that the speed is nothing but the magnitude of the velocity. A layman often uses the two terms interchangeably and for a straight line motion in one direction the two may be taken as equal since each determines the other. However, for a motion in a plane or in a higher dimensional space, velocity and speed should not be confused with each other. The former is a vector and so has a magnitude (which is the speed) and also a direction. The speed is a scalar. It is possible that the speed is constant but the velocity is not because its direction is changing. For example, this is what happens for a particle moving in a circle with uniform speed. We shall study such motions in Section 6.6.

For applications it is important to extend Theorem (2.4) slightly. Suppose for example, we want to find the perimeter of a semi-circular region. The boundary, say $C$, consists of two parts, the arc of a semi-circle ($C_1$) and a diameter of the circle ($C_2$) (see Fig.6.2.4). There is no single parametrisation of the entire $C$ which is smooth everywhere. (Intuitively this is so because the curve has two sharp corners, one at the point $(a, 0)$ and the other at $(-a, 0)$, and so there is no unique tangent at either of them.) Nevertheless each of the portions $C_1$ and $C_2$ can be parametrised separately so that Theorem (2.4) can be applied to each. In fact we have already given such parametrisations above. So we can get $l(C_1)$ and $l(C_2)$. But by Corollary (2.2), $l(C) = l(C_1) + l(C_2)$. In order to really apply Corollary (2.2), technically we must ensure that the domain intervals over which $C_1$ and $C_2$ are parametrised are adjacent to each other. But this is easy to arrange (cf. Exercise (4.12.3)). For example, we may consider the parametrisation $\alpha(t) = (f(t), g(t))$, for $t \in [0, 2]$ given by

$$f(t) = \begin{cases} a \cos \pi t, & 0 \le t \le 1 \\ -a + 2a(t-1) & 1 \le t \le 2 \end{cases}$$

and

$$g(t) = \begin{cases} a \sin \pi t & 0 \le t \le 1 \\ 0 & 1 \le t \le 2 \end{cases}$$

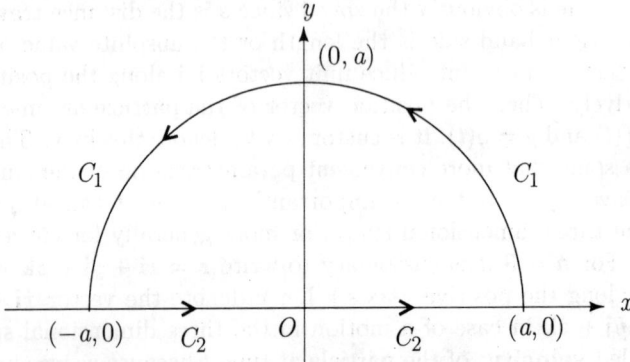

Figure 6.2.4 : A Piecewise Continuously Differentiable Curve.

Here, both $f$ and $g$ are continuous at every point of $[0, 2]$. But while $g$ is differentiable everywhere on $[0, 2]$, $f'(1)$ does not exist. Still $f$ has both a right and a left derivative at 0. So if we split $[0, 2]$ into $[0, 1]$ and $[1, 2]$ then Theorem (2.4) becomes applicable to each part.

Generalising this reasoning, we get an extension of Theorem (2.4) for curves which are piecewise continuously differentiable. It will be given as an exercise (Exercise (2.10)).

## EXERCISES

2.1    Which of the following quantities are additive in nature?

   (i)    Specific heat (i.e., the heat needed to raise the temperature by a unit degree celcius)

   (ii)   period of oscillation of a pendulum (as a function of its length)

   (iii)  the charge for a long distance telephone call

   (iv)   the population of a region

   (v)    the average distance of a point on a straight road from its nearer end.

2.2    Prove that if $f : [a, b] \to \mathbb{R}$ is continuously differentiable, then the length of its graph is $\displaystyle\int_a^b \sqrt{1 + [f'(x)]^2} \, dx$.

**2.3** A **catenary** is defined as a curve of the form $y = a \cosh \dfrac{x}{a}$ where $a > 0$ is some constant. The name derives from the Latin word *catena* meaning a chain, because it is not hard to show that if a uniform heavy chain is suspended between two points (not necessarily in the same horizontal plane), then with a suitable choice of coordinate system, as shown in Fig.6.2.5, the chain lies along a catenary. Interestingly, the curve also figures in some other contexts. For example, the shape of a soap film bounded by two circles of different radii lying in parallel planes with their centres lying on an axis perpendicular to these planes, is a surface obtained by revolving a catenary around this axis.

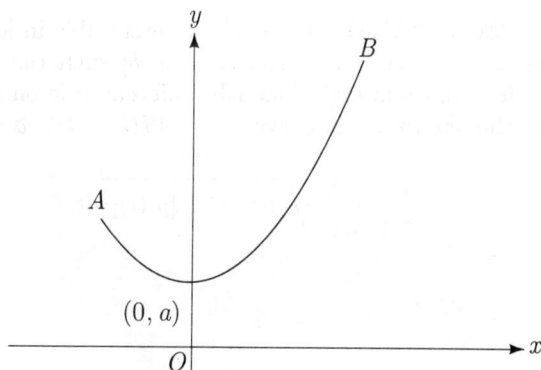

Figure 6.2.5 : A Catenary

Find the length of the catenary if the coordinates of $A, B$ are given.

**2.4** Find the arc length of one arch of the cycloid (cf. Exercise (4.12.6)) $x = a(\theta - \sin\theta), y = a(1 - \cos\theta), 0 \le \theta \le 2\pi$.

**2.5** For the epicycloid considered in Exercise (4.12.7), find the arc length of the portion of it lying between two consecutive positions of the moving point when it lies on the fixed circle. What happens as $R \to \infty$ ? (The catenary, cycloid and epicycloid are among those naturally occurring curves whose lengths can be evaluated in a closed form.)

**2.6** Find the arc lengths of the following curves

(i) $\quad x = \dfrac{t}{2}\sqrt{1 + t^2} + \dfrac{1}{2}\ln(t + \sqrt{1 + t^2}), y = \dfrac{2\sqrt{2}}{3}t^{3/2}; 0 \le t \le 1$

(ii) $\quad x = t, y = \displaystyle\int_0^t \sqrt{\cos 2u}\, du\ ;\ 0 \le t \le \dfrac{\pi}{4}$

(These are artificial examples where the arc length can be evaluated in a closed form.)

**2.7** Sketch the curve $y^2 = x^3, 0 \le x \le 1$ and find its length.

2.8    Generalise Theorem (2.4) to curves in $I\!R^3$. (It will first be necessary to prove an appropriate analogue of Exercise (5.6.5) part (iii) for sums of the form

$$\sum_{i=1}^{n} \sqrt{[u(\xi_i)]^2 + [v(\eta_i)]^2 + [w(\zeta_i)]^2}\,\Delta t_i$$

where $u, v, w$ are continuous and $\xi_i, \eta_i, \zeta_i$ vary independently over $[t_{i-1}, t_i]$.)

2.9    Find the arc lengths of the portions of the circular helix $x = \cos\theta, y = \sin\theta, z = \theta; 0 \le \theta \le 2\pi$ and the conical helix $x = \theta\cos\theta, y = \theta\sin\theta, z = \theta; 0 \le \theta \le 2\pi$. (See Fig.4.12.3 for a sketch.)

2.10   Suppose $f, g$ are piecewise continuously differentiable in $[a, b]$. Prove that there exists a partition $(t_0, t_1, \ldots, t_n)$ of $[a, b]$ such that for every $i = 1, \ldots, n$, both $f$ and $g$ are continuously differentiable on $[t_{i-1}, t_i]$. Prove further that the length of the curve $\alpha(t) = (f(t), g(t)); \ a \le t \le b$ equals

$$\sum_{i=1}^{n} \int_{t_{i-1}}^{t_i} \sqrt{[f'(t)]^2 + [g'(t)]^2}\,dt.$$

## 6.3    Areas and Volumes by the Method of Slicing

In the last section it was remarked that like arc lengths, areas and volumes are additive in nature and that this is the basic reason why definite integrals figure so frequently in their evaluation. In the case of the arc length of a (parametrised) curve, we defined it first as a certain supremum, starting from some prototypes (viz. line segments), then in Corollary (2.2) proved its additivity and finally showed in Theorem (2.4) how it can be expressed as a definite integral, at least under certain regularity conditions.

A similar approach is possible for areas and volumes. We define them first starting from certain prototypes (which may be called 'boxes'). Then we can prove that they are additive in nature and finally, with suitable additional hypothesis they can be expressed as certain integrals. Although in spirit this approach runs parallel to that taken for arc lengths, it is fraught with certain difficulties which are not easy to resolve in an elementary course. We mention these difficulties and then, instead of surmounting them, take the naive approach of taking the basic properties of areas and volumes for granted. For simplicity we stick to areas. The discussion of volumes is similar.

There is no difficulty in defining the area of a subset, say $S$, of the plane. Let $a, b, c, d$ be any real numbers. By the rectangle $R(a, b; c, d)$ we mean the set $\{(x, y) : a \leq x \leq b; c \leq y \leq d\}$. (We assume tacitly that $a \leq b$ and $c \leq d$ for otherwise the rectangle degenerates into an empty set.) This rectangle is also often denoted by $[a, b] \times [c, d]$. Note that it includes all its boundary points and hence it also often called a closed rectangle (cf. Exercise (3.5.6)).

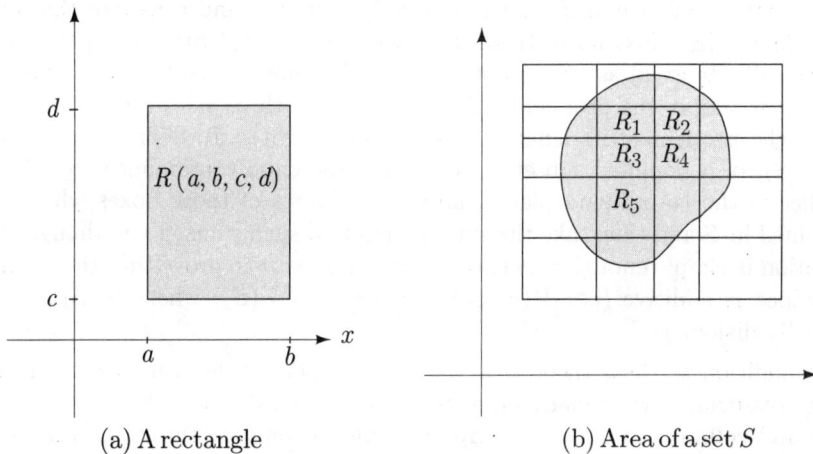

(a) A rectangle  (b) Area of a set $S$

Figure 6.3.1 : Area of a Plane Figure

Fig.6.3.1(a) shows a typical rectangle. The intervals $[a, b]$ and $[c, d]$ on the $x$ and the $y$-axis respectively are called, the sides of the rectangle $R(a, b; c, d)$ (which may be denoted simply by $R$ when its sides are understood). Such rectangles are the prototypes in the definition of an area. We define the area (or, if we prefer, the tentative area) of $R$ to be simply the product of the lengths of its sides, i.e., as $(b - a)(d - c)$ and denote it by $A(R)$. Now suppose $S$ is a bounded subset of the plane. Then $S$ is contained in some big rectangle, say $R = R(A, B; C, D)$. We chop $R$ by vertical and horizontal lines into smaller rectangles as shown in Fig.6.3.1(b). Some of these rectangles may be disjoint from $S$. Let $R_1, \ldots, R_n$ be the rectangles which are completely contained in $S$. Then $S$ contains their union and so it is clear that the area of $S$ should not be less than $\sum_{i=1}^{n} A(R_i)$. Further, if the chopping is very fine then this sum will differ from the area of $S$ very little. It is therefore logical to define the area of $S$, denoted by $A(S)$ to be the supremum of all such sums. (If $S$ is unbounded, then the plane is decomposed into an infinite number of rectangles, say by lines of the form $x = m$ and $y = n$ where $m, n$ are integers. The set $S$ intersects infinitely many of these rectangles and the area of $S$ is defined by adding the areas of these intersections.)

This approach considerably resembles the definition of an arc length. Unfortunately, the similarity does not extend to the proofs of the basic properties. For example, it was easy to prove that for a line segment, the length coincides with its tentative length (see the comments after Definition (4.7.1)) or that the length is additive (Corollary (2.3)). But in the case of areas, the corresponding additivity property is surprisingly non-trivial to establish. That is, if $R_1, R_2$ are mutually disjoint plane figures, then it is not easy to prove that $A(R_1 \cup R_2) = A(R_1) + A(R_2)$.

The definition of the volume of a three dimensional figure (and more generally, that of a figure in $\mathbb{R}^n$ for any $n > 2$) is similar and runs into the same basic difficulties. Instead of rectangles, we use 'boxes' (with sides parallel to the co-ordinate axes) as our prototypes. The volume (or the tentative volume) of a box is simply the product of the lengths of its three sides (or, in layman's terms, the product of its length, breadth and height). To define the volume, say $V(S)$, of a bounded subset $S$, we chop some cube containing it by planes parallel to the co-ordinate planes, add the volumes of those boxes which are contained in $S$ and then take the supremum of all such sums. Even though this definition is simple enough conceptually, it is not easy to prove that the volume so defined is additive (i.e. $V(S_1 \cup S_2) = V(S_1) + V(S_2)$ whenever $S_1, S_2$ are mutually disjoint).

In addition to these theoretical difficulties, there is the usual practical difficulty (associated with many other basic concepts as well) of how to *compute* areas and volumes. It is usually impracticable to compute things directly from the definition, when the definition involves suprema or infima over infinite sets of real numbers. In the case of the arc length, which was defined as a certain supremum, Theorem (2.4) gave us a handy formula to express it as a definite integral and thereby permit its calculation (at least approximately) under a rather mild hypothesis. It is natural to inquire if there is a similar formula for computing areas and volumes.

The method of slicing is an attempt to kill two birds in one stone, in that it answers both the question of defining area and that of finding a formula for it. The technique is not new to us. In fact we used it to motivate the concept of Riemann integral. To recall it, suppose $R$ is the set $\{(x, y) : a \leq x \leq b, 0 \leq y \leq f(x)\}$ where $f$ is a non-negative real-valued function defined on interval $[a, b]$ (see Fig.6.3.2(a)). In Sections 5.1 and 5.2 we saw how a partition of $[a, b]$ gives rise to a slicing of $R$ by vertical lines (i.e. lines parallel to the $y$-axis). By treating each piece approximately as a rectangle we get an approximate value for $A(R)$. The limit of these approximations, viz. $\int_a^b f(x)dx$ is the exact value of $A(R)$. That time, we assumed that we had an intuitive understanding of area and the integral $\int_a^b f(x)dx$ merely gave a formula for it. Now that we are

are still fumbling with a rigorous definition of area, here is a bright idea. Why not *define* $A(R)$ as $\int_a^b f(x)dx$? In doing so, we are appealing to the 'definition trick' mentioned in Section 1.8. If we observe that for any $x_1 \in [a,b], f(x_1)$ is simply the length of the cross-section of $R$ with the vertical the $y = x_1$ (see Fig. 6.3.2(a)), then this trick can be extended to other subsets of $R$ as well. For simplicity assume $R$ is bounded. Then there exist real numbers $a, b$ such that $a \leq x \leq b$ for all $(x,y) \in R$. Geometrically, $R$ lies completely in the strip bounded by the parallel lines $x = a$ and $x = b$. Now consider the cross-section or the slice of $R$ by any vertical line lying in between these two lines. As shown in Fig.6.3.2(b), this cross-section may be empty (e.g. the line $x = x_1$), or a single point (e.g. $x = x_4$), or a single segment (e.g. the line $x = x_2$), or a union of more than one segments (e.g. $x = x_3$). We assume that the set $R$ is such that every vertical line, say $x = c$, cuts it into a finite number of mutually disjoint segments. Denote by $l(c)$ the sum of the lengths of these segments. Then $l$ is a bounded function defined on $[a, b]$. We define the area, $A(R)$, of $R$ to be the value of the integral $\int_a^b l(x)dx$.

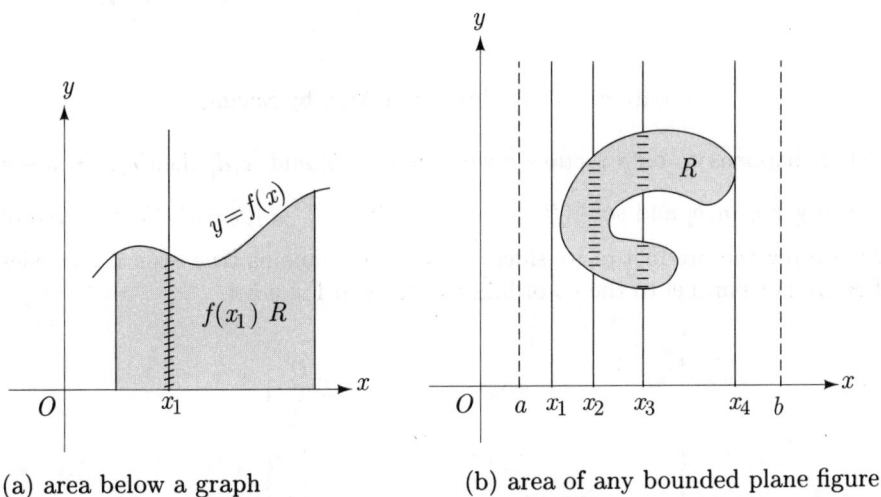

(a) area below a graph                          (b) area of any bounded plane figure

Figure 6.3.2 : Area by the Method of Slicing

It can easily be seen that Exercise (5.10) is a special case of this construction. With this definition of area, it is easy to prove that area is additive. Suppose for example that $R$ is divided into $R_1$ and $R_2$. For $c \in [a, b]$, let $l_1(c)$ and $l_2(c)$ be the sums of the lengths of the line segments which form the cross section of $R_1$ and $R_2$ respectively with the line $y = c$. It is clear (see Fig.6.3.3) that these segments together add up to form the segments which form the cross section of $R$ with $x = c$. Hence $l(c) = l_1(c) + l_2(c)$ for all $c \in [a, b]$.

Since $A(R) = \int_a^b l(x)dx, A(R_1) = \int_a^b l_1(x)dx$ and $A(R_2) = \int_a^b l_2(x)dx$, it is clear that $A(R) = A(R_1) + A(R_2)$.

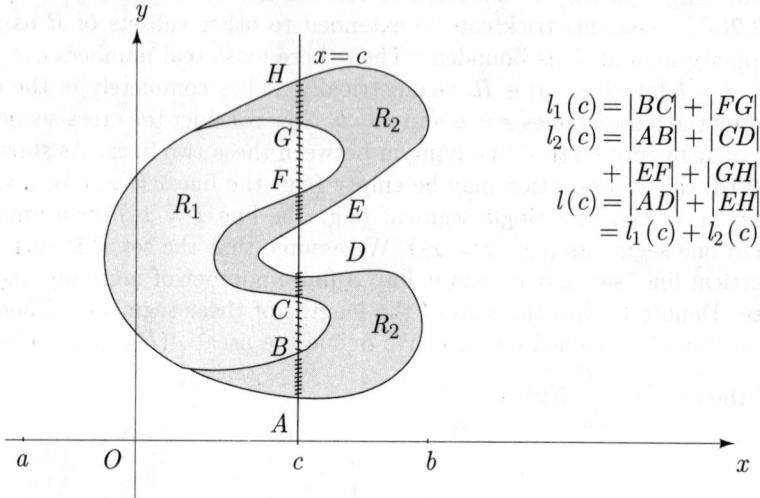

$$l_1(c) = |BC| + |FG|$$
$$l_2(c) = |AB| + |CD|$$
$$+ |EF| + |GH|$$
$$l(c) = |AD| + |EH|$$
$$= l_1(c) + l_2(c)$$

Figure 6.3.3 : Additivity of Area by Slicing

If $R$ happens to be a rectangle with sides $[a, b]$ and $[c, d]$ then $l(x) = d - c$ for every $x \in [a, b]$ and so $A(R) = \int_a^b d - c \, dx = (d-c)(b-a)$. So the area of $R$ is simply the product of its sides. In fact this remains true even if the sides of $R$ are not parallel to the co-ordinate axes as in Fig.6.3.4.

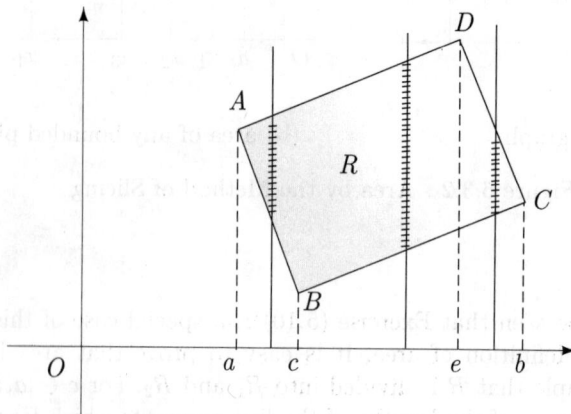

Figure 6.3.4 : Area of a Rectangle

We leave a proof of this as an exercise (Exercise (3.5)). Here to carry out the integration $\int_a^b l(x)dx$ it is necessary to split the interval of integration by points, say $c$ and $e$ which are the projections of the vertices of $R$, because even though each vertical cross section is just one line segment, the formula for its length varies depending upon where the line cuts the $y$-axis. By a similar reasoning we can prove that the area of any triangle equals half the product of its base and height. Since area is already proved to be additive and any polygon can be decomposed into a finite number of triangles, it follows that for any polygon, the area defined by slicing coincides with our intuitive understanding of the area of the polygon.

Even for some other familiar figures such as a circle, we can verify that the areas of regions bounded by them agree with those obtained by slicing. In fact, with essentially the same amount of work we can compute the area of the region $R$ bounded by an ellipse in the standard form $\dfrac{x^2}{a^2} + \dfrac{y^2}{b^2} = 1$, shown in Fig.6.3.5. (The case $a = b$ gives the area of a circle).

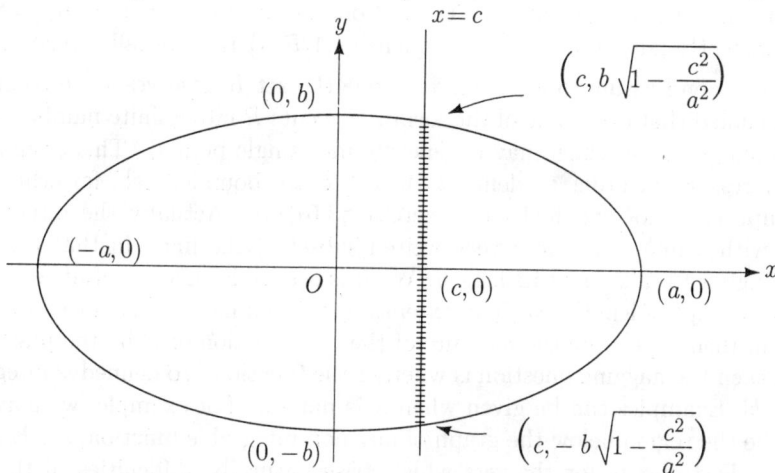

Figure 6.3.5 : Area Bounded by an Ellipse

For $-a \le c \le a$, the section of $R$ with $x = c$ is a segment of length $2b\sqrt{1 - \dfrac{c^2}{a^2}}$. So $l(c) = 2b\sqrt{1 - \dfrac{c^2}{a^2}}$ and hence $l(x) = 2b\sqrt{1 - \dfrac{x^2}{a^2}}$ So

$$
\begin{aligned}
A(R) &= 2b \int_{-a}^{a} \sqrt{1 - \frac{x^2}{a^2}}\, dx \\
&= 2b \int_{-\pi/2}^{\pi/2} \cos\theta\, a \cos\theta\, d\theta \text{ where } x = a\sin\theta, \frac{-\pi}{2} \le \theta \le \frac{\pi}{2} \\
&= ab \int_{-\pi/2}^{\pi/2} 1 + \cos 2\theta\, d\theta = ab \left[\theta + \frac{1}{2}\sin 2\theta\right]\Big|_{-\pi/2}^{\pi/2} = \pi ab. \quad (1)
\end{aligned}
$$

A few more problems of calculating areas will be given as exercises. It is interesting to note that the method of slicing is essentially the same as that used by Archimedes for calculating the area under a parabola, more than 2000 years ago, even though a rigorous theory of integration, due to Riemann (and a few others) is less than two centuries old.

Despite the theoretical and practical success of the method of slicing, it raises a few disturbing questions. First, the various restrictions put on the subset $R$ of the plane. We assumed above that $R$ is bounded. But this is not a very serious restriction. As noted earlier, when $R$ is unbounded, we can chop the whole plane into unit squares (e.g. by lines of the form $x = m$ and $y = n$ where $m, n$ are integers) find the area of the portion of $R$ which lies in each square and add (as an infinite sum). Alternatively, let $R_n$ be the portion of $R$ lying in the square with vertices $(\pm n, \pm n)$, where $n \in \mathbb{N}$. Clearly each $R_n$ is bounded and $R_n \subset R_{n+1}$ for every $n$. So $\{A(R_n)\}_{n=1}^{\infty}$ is a monotonically increasing sequence (cf. Exercise (3.9)). We can now define $A(R)$ as $\lim_{n \to \infty} A(R_n)$ with the understanding that this limit could be finite or $\infty$. (In case $R$ is itself bounded, this definition is consistent with the old one because in that case $R_n = R$ for all sufficiently large $n$ and so the sequence $\{A(R_n)\}$ is eventually constant.)

The second assumption we made about the set $R$ deserves some comment. We assumed that every line of the form $x = c$ cuts $R$ into a finite number of line segments (some of which may degenerate into single points). This assumption was necessary in order to define $l(c)$. There are bounded sets for which this assumption is not satisfied (see Exercise (3.6)(a)). Actually the intersection of $R$ with a line $c$ can be a very weired subset of the line. In that case it is meaningless to talk about its length. What is needed is an appropriate extension of the concept of length. Such an extension, called a measure is indeed possible. We can then let $l(c)$ be the measure of the cross section of $R$ by the line $x = c$. Even then the nagging question is whether the function $l$ so defined is integrable on $[a, b]$. Examples can be given when it is not so. (For example, we may take $R$ to be the region below the graph of any non-integrable function, see Exercise (1.8)). Fortunately for the sets which arise naturally, difficulties of this sort rarely arise and so we need not worry about them.

There is, however, one difficulty which is serious indeed from a theoretical point of view. In defining the area $A(R)$ of $R$, the set $R$ was sliced by straight lines parallel to the $y$-axis. What if we had, instead sliced by horizontal lines, i.e. by lines of the form $y = k$, for some constant $k$? The answer is some integral of the form $\int_c^a m(y)dy$ where $c, d$ are such that $R$ lies between the lines $y = c$ and $y = d$ and for every $k \in [c, d]$, $m(k)$ is the sum of the lengths of the segments into which $R$ is out by the line $y = k$ (see Fig.6.3.6). Let us call this integral $B(R)$. Is $B(R)$ always equal to $A(R)$? The question is of the same spirit as the equality of double summation (see Corollary (3.7.5)). For particular figures such as a rectangle, a triangle or an ellipse, equality of $A(R)$ and $B(R)$ can be verified by direct computation. But to prove it for an arbitrary subset of the plane is far from easy. This is an important problem because for some figures it may be more convenient to slice horizontally than vertically. There is, in fact, nothing special about these two types of slicing. We could slice $R$ by slant lines, i.e. lines parallel to some other fixed line. In fact, for a figure such as the rectangle in Fig.6.3.4, slicing by lines parallel to one of the sides will be decidedly advantageous. But in order to do it legitimately, we must know beforehand that we get the same answer.

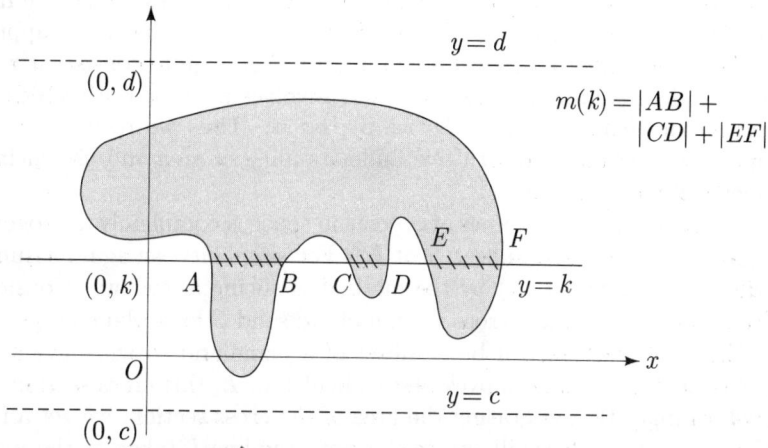

Figure 6.3.6 : Area by Horizontal Slicing

The reason we would like $A(R)$ to be independent of the direction of slicing lines is that we expect the area to satisfy a very natural requirement. In high school geometry we study congruent figures (without defining congruence rigorously) and convince ourselves that congruent figures have equal areas. Now suppose $R$ is a plane region and let $R'$ be the region obtained from $R$ by a counterclockwise rotation through some an angle say $\alpha$ (see Fig.6.3.7). Then the figures $R$ and $R'$ are mutually congruent and we would like to have $A(R) = A(R')$. If we have the freedom to choose the direction of slicing, then this is very easy to do. All we need to do is to slice $R$ by vertical lines like $L$ and $R'$ by lines like $L'$, obtained by rotating $L$ through the angle $\alpha$. The corresponding cross sections will be congruent and hence the integrals for $A(R)$ and $A(R')$ will be exactly the same. If, however, we are required to slice both $R$ and $R'$ vertically, then it is not easy to prove that $A(R) = A(R')$.

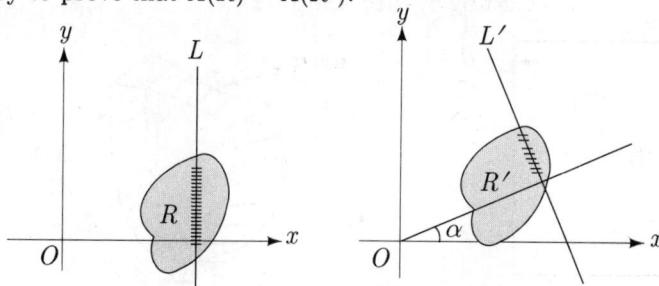

Figure 6.3.7 : Invariance of Area under a Rotation

This difficulty takes us back to square one. If we already have a definition of the area of a plane figure and a particular slicing method is only a means of evaluating it, then this difficulty would not arise. Thus we have a difficult choice. Either we must first define area without slicing, establish its basic properties and show that it can be evaluated by slicing parallel to any line or else we can define it in terms of a particular slicing and do a lot of work to establish the freedom of choosing a more convenient slicing. It turns out that both the approaches are, in fact, equivalent and yield a definition which is quite consistent with our intuitive understanding of area. But that involves a lot of work which is well beyond an elementary course. So we bypass it. Thus we revert back to the scenario where we have an intuitive understanding of area and use slicing only as a method of evaluating it.

The situation about volumes of objects in space is completely analogous. Let $V(S)$ be the volume of a subset $S$ of $\mathbb{R}^3$. For simplicity we again assume $S$ is bounded. We can find $V(S)$ by the method of slicing as follows. Consider any one fixed line $L$ and take a cross section of the solid $S$ by a plane perpendicular to $L$. This cross-section will be a subset of a plane and so will have a certain area. As the slicing plane moves perpendicular to $L$, this cross section and its area will change. By integrating the area of the cross section, we get a formula for $V(S)$. In Fig. 6.3.8 we illustrate this with the line $L$ taken as the $y$-axis.

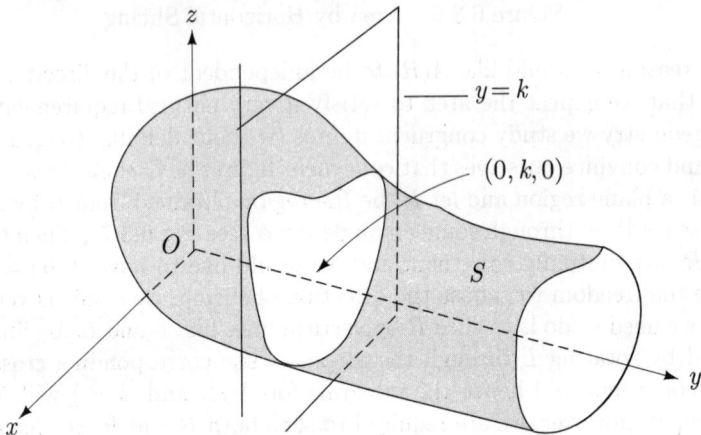

(a) Cutting by the plane $y = k$

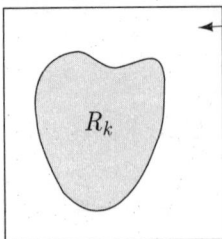

(b) Cross-section $R_k$

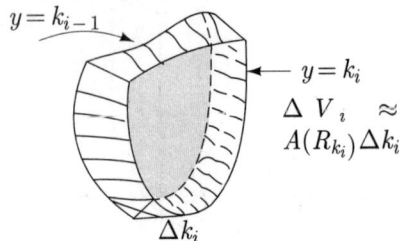

(c) Portion between adjacent planes

Figure 6.3.8 : Volume by Slicing by Parallel Planes

(a) shows the cutting plane $y = k$ and the cross-section, say $R_k$, of $S$ by this plane is shown in (b). Let $A(R_k)$ be the area of $R_k$ (which we assume to have been already defined by some approach). For a fixed $S, A(R_k)$ will vary with $k$ and $V(S) = \displaystyle\int_c^d A(R_y)dy$ where $c$ and $d$ are so chosen that $S$ lies entirely between the planes $y = c$ and $y = d$. As in the case of the area, we can look at this equation either as a definition of $V(S)$ or as a formula for $V(S)$ assuming we already have an intuitive understanding of volume. We adopt the latter approach. Some plausible justification, of course, needs to be given why $\displaystyle\int_c^d A(R_y)dy$ indeed equals $V(S)$. But this is easy following the recipe of integration given in Section 5.2. We chop the solid $S$ by planes of the form $y = k_i$, where $(k_0, k_1, \ldots, k_n)$ is a partition of the interval $[c, d]$. This is very much like cutting a loaf of bread into slices. Fig.6.3.8(c) shows the portion between two adjacent planes. Its thickness is $k_i - k_{i-1}$ for some $i$. Call this $\triangle k_i$. Note that the two 'faces' $R_{k_i}$ and $R_{k_{i-1}}$ need not be congruent to each other and may have different areas. But if $\triangle k_i$ is small then we may suppose $A(R_{k_i})$ and $A(R_{k_{i-1}})$ do not differ much and therefore the volume enclosed between these two parallel planes is approximately $A(R_{k_i})\triangle k_i$. (This is the intuitive part. We cannot prove it, because we have no rigorous definition of a volume. In essence, what we are assuming here is that if a solid has a uniform cross-sectional area then its volume is the product of this area and the thickness of the solid. A rectangular box and a right circular cylinder are examples of such solids. More generally, the term 'right cylinder' is used for any set obtained by moving any plane figure perpendicular to the plane. A theory of volume can be developed using all such right cylinders, and not just the rectangular boxes, as prototypes. But the volume so defined turns out to be the same as the volume defined using only rectangular boxes, with edges parallel to the axes, as prototypes.) So $V(S)$ is approximately $\displaystyle\sum_{i=1}^n A(R_{k_i})\triangle k_i$ which tends to $\displaystyle\int_c^d A(R_y)dy$ as the mesh of $(k_0, k_1, \ldots, k_n)$ tends to 0. Therefore this integral must give the exact value of $V(S)$.

As simple illustrations, we calculate the volume of a solid sphere of radius $r$ and a right pyramid of height $h$ with a square base of side $r$, shown respectively in Fig.6.3.9 (a) and (b). In the case of the sphere, a cross section by the plane $y = k$ $(-r \leq k \leq r)$ is a circular disc centered at $(0, k, 0)$ and radius $\sqrt{r^2 - k^2}$. The area of this disc is $\pi(r^2 - k^2)$. Replacing $k$ by $y$, $A(R_y) = \pi(r^2 - y^2)$. Hence,

$$\text{volume of solid sphere} = \int_{-r}^r \pi(r^2 - y^2)dy$$

$$= \pi\left[r^2 y - \frac{y^3}{3}\right]_{-r}^r = \pi(2r^3 - \frac{2r^3}{3}) = \frac{4\pi r^3}{3}. \quad (2)$$

For the pyramid shown in Fig.6.3.9(b), it is advantageous to use slicing by planes perpendicular to the $z$-axis. A simple calculation based on similar triangles shows that for every $k \in [0, h]$, the cross-section of the pyramid by the plane $z = k$ is a square of side $\left(\dfrac{h-k}{h}\right) r$. So $A(R_z) = \left(\dfrac{h-z}{h}\right)^2 r^2$. Hence,

$$\text{Volume of pyramid} = \int_0^h \left(\frac{h-z}{h}\right)^2 r^2 dz$$

$$= \frac{r^2}{h^2}\left[h^2 z - hz^2 + \frac{z^3}{3}\right]\Big|_0^h = \frac{hr^2}{3}. \tag{3}$$

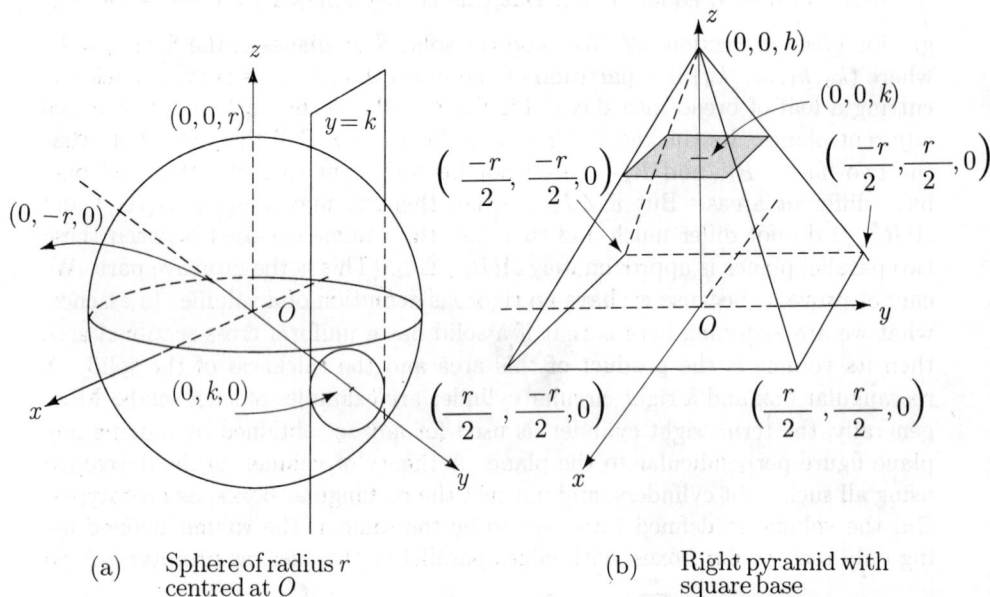

(a)   Sphere of radius $r$ centred at $O$    (b)   Right pyramid with square base

Figure 6.3.9 : Volume of a Sphere and a Pyramid

The success of the method of slicing depends, partly, on how easy it is to find the area of a typical cross section. In the two examples just considered, the cross sections were very familiar figures. In more complicated problems, it sometimes helps, to choose a particular slicing over the others as we did above for the pyramid. A few other examples of computation of volume by the method of slicing will be given in the exercises.

There is an important class of solids where the method of slicing works very nicely. They are the solids of revolution obtained by revolving a plane figure, say $R$ in space around an axis which is a line, say $L$, lying in the plane as shown in Fig.6.3.10, where the plane is taken as the $y$-$z$ plane and the line $L$ is the $z$-axis. Every point such as $P$ in $R$ traces a circle in the space whose radius is the (perpendicular) distance of $P$ from $L$ and whose center is the foot of the perpendicular drawn from $P$ to $L$.

Note that the cross section of $S$, the solid of revolution, by the $y$-$z$ plane contains two replicas of $R$, one $R$ itself and the other its reflection in the line

*L*. (In case the axis *L* passes through *R*, these two replicas overlap partly or sometimes even wholly. In such cases instead of 'revolution' it would be better to call 'rotation'. But the term 'revolution' is generally applied to cover all cases. In the discussion below, it is tacitly assumed that *R* lies entirely on one side of the axis *L*. The case where it does not will be given in the exercises (Exercise (3.19).). With our assumption the cross section of *S* with every plane passing through the axis will be a pair of replicas of *R*.

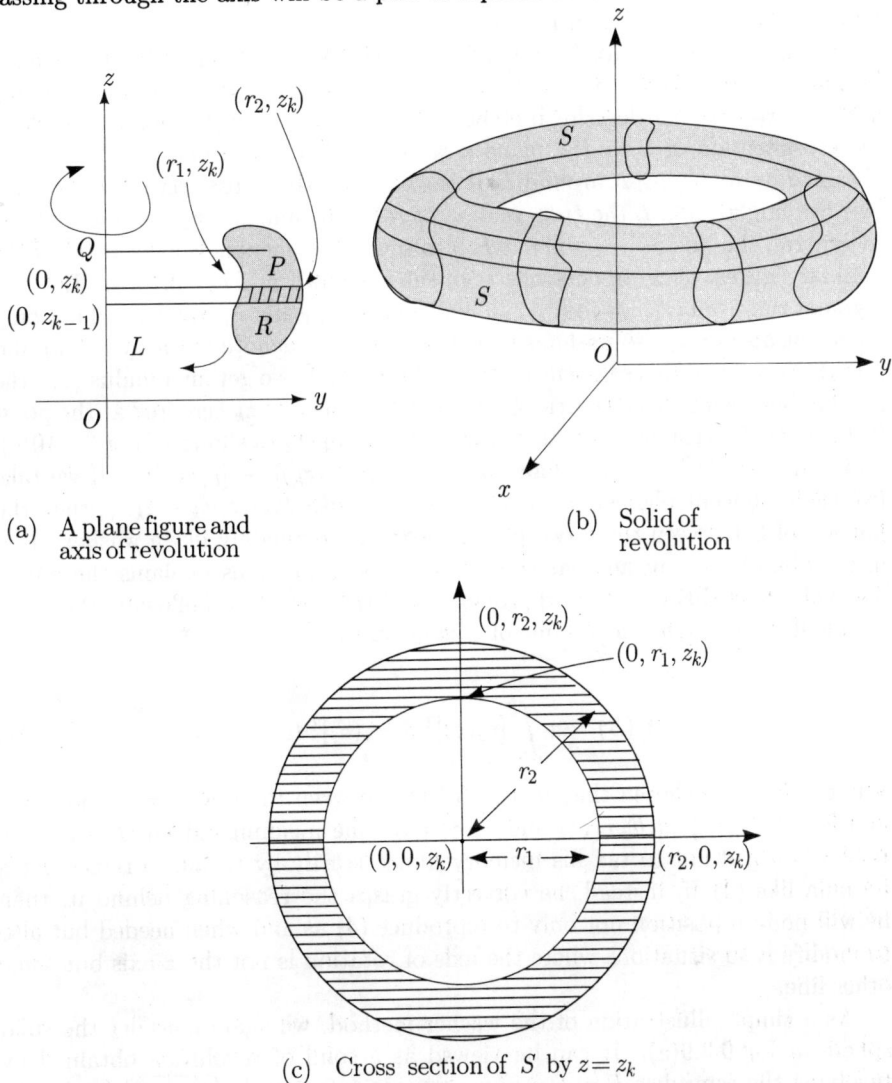

(a) A plane figure and axis of revolution

(b) Solid of revolution

(c) Cross section of *S* by $z = z_k$

Figure 6.3.10 : Volume of a Solid of Revolution by Washer Method

Even though rather special, the solids of revolution are very common in our daily life. Familiar solids like spheres, cylinders, tori (the tennicoid rings used in a popular sport of yesterday or doughnuts), right circular cones are all solids of revolution. Possibly because originally they were made on a potter's wheel, many of the household items such as cups, bowls, jugs, utensils, bottles etc. are often surfaces of revolution (i.e. surfaces obtained by revolving a plane curve around an axis in the plane) and so the solids enclosed by them are solids of revolution. Hence the problem of finding the volume of a solid of revolution (as well as the problem of finding the area of a surface of revolution, which will be taken up later in this section) is a very practical one. We give below two methods for solving it, with rather peculiar names.

The first method, called the **washer method** consists of slicing the solid $S$ by planes perpendicular to the axis $L$ of revolution. As the method is already familiar, we need not describe it elaborately except to justify its name. Consider the cross-section of $S$ by the plane $z = z_k$ (see Fig. 6.3.10 again). Since $S$ is obtained by revolving $R$ around $L$, it is clear that this cross section is obtained by revolving around $L$ the cross section of $R$ by the line $z = z_k$ in the $y$-$z$ plane. In general, the line $z = z_k$ will cut $R$ into several segments. But for most of the familiar figures, there is only one segment as shown in Fig.6.3.10(a). Let this segment run from $(r_1, z_k)$ to $(r_2, z_k)$. The numbers $r_1$ and $r_2$ will depend not only on $R$ but also on $z_k$, so perhaps it will be better to denote them by $r_1(z_k)$ and $r_2(z_k)$. Now when this segment is revolved around $L$, we get an annulus (i.e. the portion between two concentric circles) in the plane $z = z_k$ centered at the point $(0, 0, z_k)$ on $L$, with inner radius $r_1$ and outer radius $r_2$ as shown in Fig.6.3.10(c). So its area is $\pi(r_2^2 - r_1^2)$, or more precisely, $\pi([r_2(z_k)]^2 - [r_1(z_k)]^2)$. If we take two such adjacent planes $z = z_k$ and $z = z_{k-1}$ with $\Delta z_k = z_k - z_{k-1}$, then the portion of $S$ between these two planes looks approximately like a washer (used in plumbing), i.e. an annular ring of thickness $\Delta z_k$. This explains the name. The volume of this washer is approximately $\pi(r_2^2 - r_1^2)\Delta z_k$. Following the usual recipe of integration, the volume of $S$ is given by

$$V(S) = \pi \int_a^b [r_2(z)]^2 - [r_1(z)]^2 dz \tag{4}$$

where $a, b$ are so chosen that $R$ lies entirely between the lines $z = a$ and $z = b$ and for every $z \in [a, b], r_1(z)$ and $r_2(z)$ have the meanings given earlier. The reader is urged not to tax his memory unnecessarily by trying to remember a formula like (4) If, instead, he correctly grasps the reasoning behind it, then he will be in a position not only to reproduce (4) as and when needed but also to modify it to situations where the axis of rotation is not the $z$-axis but some other line.

As a simple illustration of the washer method, we again consider the solid sphere in Fig.6.3.9(a). It can be viewed as a solid of revolution obtained by revolving the semi-disc $R = \{(y, z) : -r \leq z \leq r, 0 \leq y \leq \sqrt{r^2 - z^2}\}$ in the $y$-$z$ plane around the $z$-axis. Here for every $z$ in $[-r, r]$, the cross sectional annulus reduces to a disc because the inner radius $r_1(z)$ is 0. The outer radius $r_2(z)$ is $\sqrt{r^2 - z^2}$ for every $z \in [-r, r]$ (see Fig.6.3.11(a)). We take $a = -r, b = r$. Substituting these in (4), we get the volume of the sphere as $\pi \int_{-r}^r (r^2 - z^2) dz$

which comes out to be $\dfrac{4\pi}{3}r^3$, tallying with (2). (Actually, the method used for deriving (2) is essentially the washer method if we treat the sphere as a solid of revolution around the $y$-axis instead of the $z$-axis).

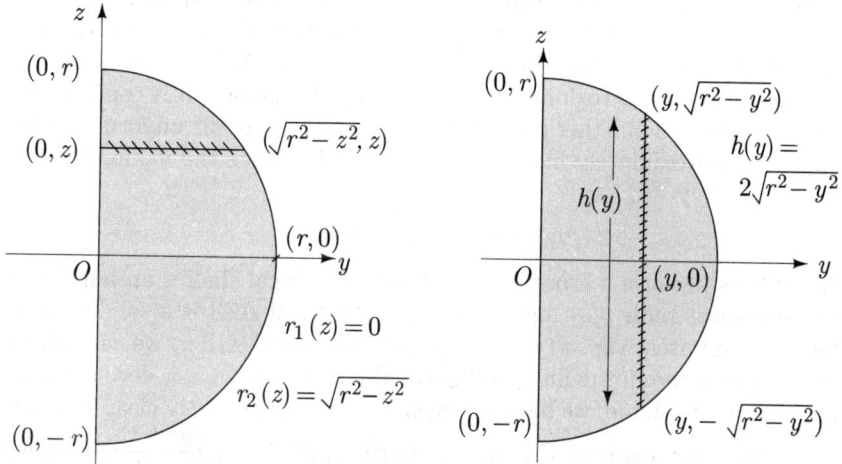

(a) Washer method          (b) Shell method

Figure 6.3.11 : Volume of a Sphere by Washer and Shell Methods

The washer method is just a special case of the general method of finding the volume of any solid by slicing it by a family of parallel planes. There is another method called the **shell method**, which works only for solids of revolution. Here the slicing is not done by a family of planes. Instead, it is done by a family of coaxial right circular cylinders, each having its axis along the axis $L$ of revolution. Each such cylinder is a surface of revolution obtained by revolving around $L$, some line parallel to $L$.

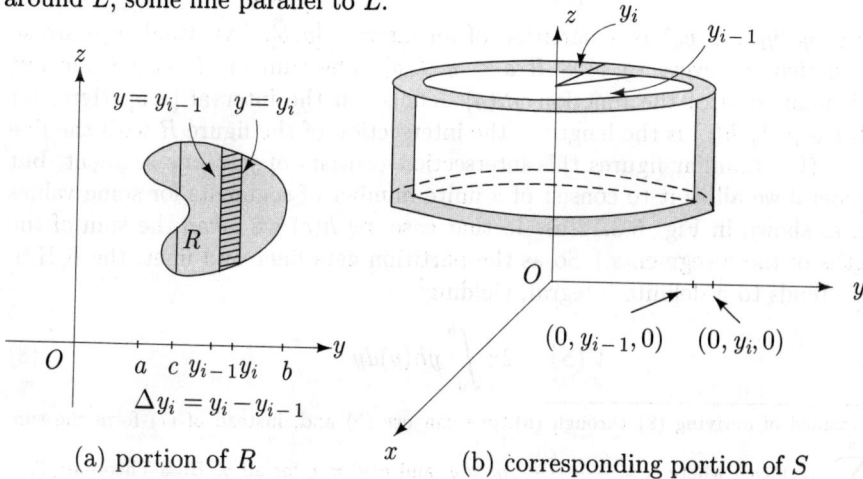

(a) portion of $R$          (b) corresponding portion of $S$

Figure 6.3.12 : The Shell Method for Volume of Solid of Revolution

Going back to the situation where $S$ is obtained by revolving a figure $R$ in the $y$-$z$ plane around the $z$-axis, Fig.6.3.12(a) shows the portion of $R$ between two parallel lines $y = y_{i-1}$ and $y = y_i$ while (b) shows the corresponding portion of $S$. This is the portion between the two co-axial cylinders, the inner one having radius $y_{i-1}$ and the outer one $y_i$. This portion is not of constant height. Let its outer height be $h(y_i)$ while the inner height be $h(y_{i-1})$. Here $h$ is a function of $y$ and depends on the region $R$. For most familiar regions, $h$ is continuous and $y_i - y_{i-1}$ being small, this portion may be taken to be of uniform height. It then looks approximately like a cylindrical shell. Hence the name. Its volume is approximately

$$\triangle V_i \simeq \pi(y_i^2 - y_{i-1}^2)h(y_i) \tag{5}$$

(This follows because a cross section of the cylindrical shell is an annulus with inner and outer radii $y_{i-1}$ and $y_i$. So the cylindrical shell is a solid of uniform cross sectional area, viz., $\pi(y_i^2 - y_{i-1}^2)$. As mentioned earlier, we can take such solids as prototypes for defining volumes. But since we are not developing a rigorous theory of volume, we have to accept (5) as an intuitively clear statement.)

The right hand side of (5) can be rewritten as $2\pi h(y_i)\dfrac{y_i + y_{i-1}}{2}\triangle y_i$ where $\triangle y_i = y_i - y_{i-1}$. Now $\dfrac{y_i + y_{i-1}}{2}$ is the mid-point of $[y_{i-1}, y_i]$ and so by continuity of $h$ and smallness of $\triangle y_i$, we may take $h(y_i)$ and $h(\dfrac{y_{i-1} + y_i}{2})$ as approximately equal. Then instead of (5), we have

$$\triangle V_i \simeq 2\pi \frac{y_{i-1} + y_i}{2} h\left(\frac{y_{i-1} + y_i}{2}\right)\triangle y_i \tag{6}$$

and so for the total volume $V(S)$, we have

$$V(S) \simeq 2\pi \sum_{i=1}^{n} \frac{y_{i-1} + y_i}{2} h(\frac{y_{i-1} + y_i}{2})\triangle y_i \tag{7}$$

where $(y_0, y_1, \ldots, y_n)$ is a partition of an interval $[a, b]$. (As usual, $a, b$ are so chosen that for every $(y, z) \in R, a \leq y \leq b$). The sum in (7) is nothing but a Riemann sum of the function $yh(y)$ defined on the interval $[a, b]$. Here, for each $c \in [a, b], h(c)$ is the length of the intersection of the figure $R$ with the line $y = c$. (For familiar figures this intersection consists of just one segment, but in general we allow it to consist of a finite number of segments for some values of $c$, as shown in Fig. 6.3.12(a). In that case, by $h(c)$ we mean the sum of the lengths of these segments.) So as the partition gets finer and finer, the R.H.S. of (7) tends to a definite integral, yielding[1]

$$V(S) = 2\pi \int_a^b yh(y)dy \tag{8}$$

---

[1]Instaed of deriving (8) through (6), one can use (5) and, instead of (7) form the sum $2\pi \sum_{i=1}^{n} u(\xi_i)h(\eta_i)$ where $\xi_i = \dfrac{y_{i-1} + y_i}{2}, \eta_i = y_i$ and $u(y) = y$ for all $y$. Bliss Theorem (Theorem (5.6.4)) can now be applied.

Once again, the reader is advised to remember the derivation rather than the formula. It is a little easier to do so if we recognise $2\pi y h(y)$ as the area of a right circular cylinder of radius $y$ and height $h(y)$. (The general theory of areas of surface of revolution will be taken up later in this section.)

As an illustration, let us once more calculate the volume of a solid sphere $S$ of radius $r$, but this time by the shell method. As in the washer method, $S$ is obtained by revolving the semi-disc $R = \{(y, z) : -r \leq z \leq r, 0 \leq y \leq \sqrt{r^2 - z^2}\}$. From Fig. 6.3.11(b), it is clear that $h(y) = 2\sqrt{r^2 - y^2}$ for every $y \in [0, r]$. So taking $a = 0$ and $b = r$, (8) gives

$$
\begin{aligned}
V(S) &= 2\pi \int_0^r 2y\sqrt{r^2 - y^2}\,dy \\
&= 2\pi \left[ -\frac{2}{3}(r^2 - y^2)^{3/2} \right]_0^r \\
&= \frac{4\pi}{3}r^3
\end{aligned}
$$

which again tallies with (2).

In essence the only difference between the washer and the shell method is how the plane figure $R$ is sliced. In the washer method it is sliced by straight lines perpendicular to the axis of revolution while in the shell method it is sliced by straight lines parallel to the axis. Although both methods give the same volume, in a particular problem, one method may be better than the other. The reasoning used in the shell method will also be needed in finding what is called the moment around a line, as we shall see in the next section.

Let us now turn to the problem of finding the area of a surface of revolution. Such a surface is obtained by revolving a plane curve, say $C$, around an axis, say $L$, lying in the plane of the curve. For simplicity we again assume that the curve $C$ lies entirely on one side of the line $L$. (Note that here we are using the word 'curve' in a layman's sense and not in the technical sense as a parametric curve. The latter will, of course, be needed in computations.) Spheres, right circular cylinders, right circular cones and tori are common examples of surfaces of revolution. As noted earlier, we encounter many such surfaces in day-to-day life. These surfaces are not planar and so the methods given earlier for finding areas do not apply to them. The general theory of areas of non-planar surfaces (or '**curved surfaces**' as they are often called) is beyond our scope. But the special case of surfaces of revolution can be tackled by the method of slicing. The approach, once again, will be plausible, because we do not have a rigorous definition of even planar areas, let alone curved areas.

The prototype for a surface of revolution will be the curved surface of a right circular cone shown in Fig.6.3.13(a). (Unlike in the case of a solid of revolution, for surfaces of revolution we take the plane to be the $x$-$y$ plane and the axis of revolution as the $x$-axis. This is purely for the sake of the ease in drawing diagrams. The reasoning remains exactly same no matter which plane and axis are chosen.) It is obtained by revolving a line segment $OP$ of length $l$ so that the point $P$ traces a circle of radius $r$ (in the plane $x = r$, perpendicular to the $x$-axis) with centre $M$. The lengths $l$ and $r$ are called respectively the slant height and the base radius of the cone. (Usually a cone is drawn so that its axis is vertical and the vertex (or 'apex' as it is often called) $O$ lies above the base and this justifies the name.)

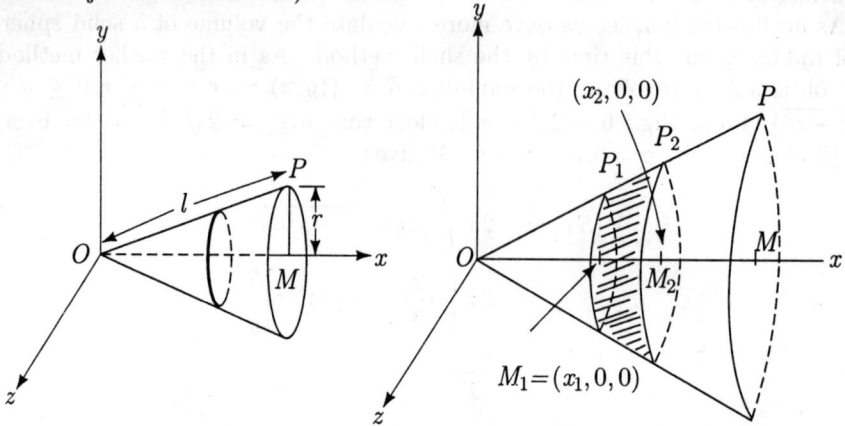

(a) Right circular cone　　　　　　　　(b) Band of a cone

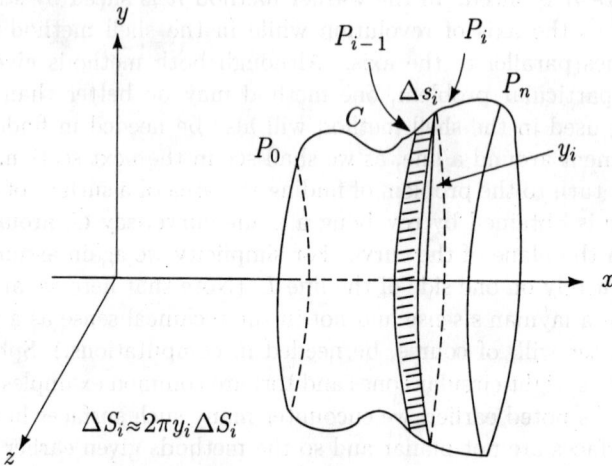

(c) Slicing of a surface of revolution

Figure 6.3.13 : Area of a Surface of Revolution

We shall assume without proof that

$$\text{area of curved surface of a cone } = \pi r l = \pi |OP||PM| \qquad (9)$$

As we have no formal definition of the area of a surface, we cannot prove (9) rigorously. But a plausible argument can be made as follows. Suppose the surface is cut along the line $OP$ and spread on a plane. Then it will occupy a sector of a disc of radius $l$ and central angle $\frac{2\pi r}{l}$ (since $2\pi r$ is the length of the perimeter of the base of the cone). It is easily seen that the area of this sector is $\pi r l$ (cf. Exercise (3.10)).

Our interest is more in the area of a band of this conical surface enclosed between two planes perpendicular to the axis of the cone as shown in Fig.6.3.13(b). Let these planes cut $OM$ at $M_1$ and $M_2$ and $OP$ at $P_1$ and $P_2$. The area of the band is the difference of the areas of the 'outer' cone and the 'inner' cone, each of which can be found using (9). Using similarity of the triangles $OM_1P_1$ and $OM_2P_2$ it is easy to show from (9) (cf. Exercise (3.20)) that

$$\text{area of conical band} = 2\pi |P_1P_2| \left( \frac{|P_2M_2| + |P_1M_1|}{2} \right) \tag{10}$$

We are now ready to derive the formula for the area of a surface, say $S$, obtained by revolving a curve, say $C$, in the $x$-$y$ plane around the $x$-axis. We assume $C$ is the form $y = f(x)$, for $a \leq x \leq b$. Let $(x_0, x_1, \ldots, x_n)$ be a partition of $[a, b]$ and let $(P_0, P_1, \ldots, P_n)$ be the corresponding broken line path along $C$, where $P_i = (x_i, f(x_i))$ for $i = 0, 1, \ldots, n$. We chop $S$ into $S_1, S_2, \ldots, S_n$ where for $i = 1, \ldots, n$, $S_i$ is the surface obtained by revolving around $x$-axis the portion of $C$ between $P_{i-1}$ and $P_i$. Let $\triangle S_i$ be the area of $S_i$. We approximate the portion of $C$ between $P_{i-1}$ and $P_i$ by the line segment $P_{i-1}P_i$. Then $\triangle S_i$ can be approximated by the area of the band of the right circular cone obtained by revolving the segment $P_{i-1}P_i$ around the $x$-axis, which can be evaluated using (10). (The line through $P_i$ and $P_{i-1}$ need not meet the $x$-axis at $O$, but (10) remains valid regardless of it.) Noting that $|P_{i-1}M_{i-1}| = f(x_{i-1})$ and $|P_iM_i| = f(x_i)$ (assuming $C$ lies on the positive side of the $x$-axis)

$$\text{area of } S = \sum_{i=1}^{n} \triangle S_i \simeq \sum_{i=1}^{n} 2\pi \left( \frac{f(x_{i-1}) + f(x_i)}{2} \right) \triangle s_i \tag{11}$$

where $\triangle s_i = |P_{i-1}P_i|$.

If we assume $f$ is continuous (which is an inherent requirement for a curve) then $f(x_{i-1}) \simeq f(x_i)$ if the mesh of the partition is small and so (11) can be replaced by

$$\text{area of } S \simeq \sum_{i=1}^{n} 2\pi f(x_i) \triangle s_i \tag{12}$$

To reduce this further we assume $f$ is continuously differentiable on $[a, b]$. Then by arguing exactly as in the derivation of (2) in the last section, we get $\triangle s_i = \sqrt{1 + [f'(\xi_i)]^2} \triangle x_i$ for some $\xi_i \in (x_{i-1}, x_i)$. Hence (12) becomes

$$\text{area of } S \simeq \sum_{i=1}^{n} 2\pi f(x_i) \sqrt{1 + [f'(\xi_i)]^2} \triangle x_i \tag{13}$$

Since $\xi_i$ need not equal $x_i$, the sum in (13) is not exactly a Riemann sum of the function $\pi f(x) \sqrt{1 + [f'(x)]^2}$. But this can be taken care of[2] by Bliss

---

[2]As in the alternate derivation of (8), here too, one can bypass (12) and instead of (13)

theorem (Exercise (5.6.5)). So, by taking the limit as the partition gets finer and finer, we have

$$\text{area of } S = \int_a^b 2\pi f(x)\sqrt{1 + [f'(x)]^2}\,dx \qquad (14)$$

as the formula to evaluate the area of the surface of revolution. The reader is advised to remember (12) rather than (14). The factors in the terms in (12) have a vivid geometric interpretation. $\Delta s_i$ is the length of the $i^{\text{th}}$ 'line element' on the curve while $f(x_i)$ is the perpendicular distance of this line element (actually a point on it) from the axis of revolution. Keeping this in mind, the reasoning above yields a formula applicable for any continuously differentiable curve (not necessarily a plane curve either). This will be given as an exercise, (Exercise (3.21)).

Before applying (14), let us verify that it gives the right answer for our prototype, viz., the right circular cone in Fig.6.3.13(a). The line $OP$ has equation $y = \dfrac{r}{\sqrt{l^2 - r^2}}x;\ 0 \le x \le \sqrt{l^2 - r^2}$ and so $f'(x) = \dfrac{r}{\sqrt{l^2 - r^2}}$ for all $x$. Hence from (14), the area of the curved surface of the cone is

$$2\int_0^{\sqrt{l^2-r^2}} \pi \frac{r}{\sqrt{l^2-r^2}}x\frac{l}{\sqrt{l^2-r^2}}\,dx = \frac{2\pi rl}{l^2 - r^2}\int_0^{\sqrt{l^2-r^2}} x\,dx$$

which is indeed $\pi rl$.

If $f(x)$ is a constant, say $k$, then (14) gives the area of a right circular cylinder as $2\pi(b-a)k$. As a less trivial example, let us calculate the surface area of a sphere of radius $r$. This can be obtained by revolving the upper semi-circle $y = \sqrt{r^2 - x^2}; -r \le x \le r$ around the $x$-axis. Taking $f(x) = \sqrt{r^2 - x^2}, a = -r$ and $b = r$ in (14), we get the area of a sphere of radius $r$ as,

$$\int_{-r}^{r} 2\pi\sqrt{r^2 - x^2}\sqrt{1 + \frac{x^2}{r^2 - x^2}}\,dx = \int_{-r}^{r} 2\pi r\,dx = 4\pi r^2. \qquad (15)$$

As in the case of the arc length, the utility of (14) is somewhat marred by the presence of the radical in the integrand. Still, in a few cases it gives the exact answer in a closed form as will be indicated in the exercises.

---

consider the sum $\displaystyle\sum_{i=1}^{n} 2\pi f(\eta_i)\sqrt{1 + [f'(\xi_i)]^2}\Delta x_i]$ where $\eta_i = \dfrac{x_{i-1} + x_i}{2}$ and apply Bliss' Theorem.

# EXERCISES

3.1 Suppose we define the area of a bounded plane figure $R$ as the supremum of sums of the form $\sum_{i=1}^{n} A(R_i)$ as indicated at the beginning of the section. Prove that the area of a rectangle $R(a, b; c, d)$ equals $(b - a)(d - c)$. Prove a similar result about the volume of a rectangular box with sides parallel to the co-ordinate axes.

3.2 Exactly what difficulty would arise in proving the additivity of areas and volumes if we follow the definition in the last exercise ?

3.3 Prove that the area defined by the method of vertical slicing is invariant under translation. That is, suppose a plane figure $R'$ is obtained by translating (i.e. shifting) every point of another plane figure $R$ by the same vector, say $(\alpha, \beta)$. In other words, $R' = \{(x, y) : (x - \alpha, y - \beta) \in R\}$. Prove that $A(R) = A(R')$. [*Hint* : First do this when $R$ is bounded. Suppose $a \leq x \leq b$ for all $(x, y) \in R$. Show that for every $c \in [a, b]$ the intersection of $R$ with $y = c$ and that of $R'$ with $y = c + \alpha$ are congruent to each other and so have the same lengths.]

3.4 If $R$ is a triangle with vertices at $(x_1, y_1), (x_2, y_2)$ and $(x_3, y_3)$, prove that $A(R)$ equals half the absolute value of the determinant

$$\begin{vmatrix} x_1 & y_1 & 1 \\ x_2 & y_2 & 1 \\ x_3 & y_3 & 1 \end{vmatrix}$$

[*Hint* : Without loss of generality assume $x_1 \leq x_2 \leq x_3$. The last exercise allows $(x_2, y_2)$ to be taken as $(0, 0)$. Carry out the integration over $[x_1, 0]$ and $[0, x_3]$ and add.]

3.5 Prove that the area of any rectangle equals the product of its length and breadth. [*Hint* : See Fig.5.13.4 and the comments accompanying it or use the last exercise.]

3.6 (a) Let $R = \{(x, y) : 0 < y \leq 1, \sin \dfrac{1}{y} \leq x \leq 2\} \cup \{(x, 0) : -1 \leq x \leq 2\}$.

   Sketch $R$. Show that if $-1 \leq c \leq 1$ then the line $x = c$ cuts $R$ into infinitely many line segments. Prove, however, that every horizontal line $y = c$ for $0 < c \leq 1$ cuts $R$ into a single line segment of length $2 - \sin \dfrac{1}{c}$.

   (b) Let $R = \{(x, y) : 0 < y \leq 1, y \sin \dfrac{1}{y} \leq x \leq 2\} \cup \{(x, 0) : 0 \leq x \leq 2\}$.

   Prove that $R$ has finite area but the boundary of $R$ has infinite length. [*Hint* : See Exercise (4.7.7).]

3.7    In defining the area of a bounded plane figure $R$ by vertical slicing as $\int_a^b l(x)dx$, all that was required of the numbers $a, b$ was that for all $(x, y) \in R, a \leq x \leq b$. So we could replace $a$ by any smaller number $a'$ and $b$ by any larger number $b'$. Why does this not change the integral? (A similar comment holds for many other definitions and formulas in the text.)

3.8    For the area enclosed by the ellipse in Fig.6.3.5, verify that the horizontal slicing gives the same area as the vertical one.

3.9    Prove that areas and volumes are monotonic in the sense that whenever $R_1 \subset R_2, A(R_1) \leq A(R_2)$ and similarly whenever $S_1 \subset S_2, V(S_1) \leq V(S_2)$.

3.10   Let $R$ be a sector (i.e. a portion of a disc between two radii) of radius $r$ and central angle $\theta$ (where $0 \leq \theta \leq 2\pi$). Prove that the area of $R$ is $\frac{1}{2}\theta r^2$. [*Hint* : First assume $\theta \leq \frac{\pi}{2}$. Divide the sector into two parts by a perpendicular from the end of one radius to the other.]

3.11   Find the area of the region bounded by the curve $y = \tan x$, the tangent to it at $(\frac{\pi}{4}, 1)$ and the $x$-axis.

3.12   Let $R$ be the set of all points of a unit square which are closer to the centre than to any of its edges. Sketch $R$ and find its area.

3.13   Prove that the volume of the pyramid with a square base in Fig.6.3.9(b) remains the same even if it is not a 'right' pyramid, i.e. its vertex is not at $(0, 0, h)$ but at some other point $(a, b, h)$ of the same height. Does its surface area also remain the same ?

3.14   Prove that the volume of a right circular cone of height $h$ and base radius $r$ is $\frac{1}{3}\pi r^2 h$. Is this true even if the cone is not 'right' ?

3.15   Suppose $R$ is a bounded plane region and a solid $S$ is obtained by joining every point of $R$ to a fixed point, say $V$, outside the plane. (Such a solid is also called the cone on $R$ with vertex $V$. Right circular cones and pyramids are special cases of this construction.) Prove that the volume of such a soild equals $\frac{1}{3}A(R)h^2$ where $h$ is the 'height' of the cone, i.e. the perpendicular distance of $V$ from the plane containing $R$.

3.16   Find the volume enclosed by the ellipsoid $\dfrac{x^2}{a^2} + \dfrac{y^2}{b^2} + \dfrac{z^2}{c^2} = 1$. [*Hint* : A cross section by the plane $z = k$ is an ellipse. Use (1) to find the area bounded by it.]

3.17   A torus is obtained by revolving a circle of radius $b$ around a line in its plane at a distance $a$ $(> b)$ from the centre. Find the volume enclosed by the torus both by the washer and the shell method. Also find the area of the torus surface.

3.18  A fancy cake has a circular base of radius $r$ and is so made that whenever it is cut by a plane perpendicular to a fixed diameter of the base, the cross section is an isosceles triangle with base along a chord of the base circle. Find the volume of the cake.

3.19  Suppose a solid $S$ is obtained by revolving a plane figure $R$ around a line $L$ passing through $R$ (see Fig.6.3.14). Let $R_1$ and $R_2$ be the portions of $R$ on the two sides of $L$ and let $R_1'$ and $R_2'$ be their reflections in $L$ (in the opposite sides). Let $S_1, S_2, S_3$ be the solids generated by revolving $R_1, R_2$ and $R_1 \cap R_2'$ respectively around $L$. Prove that $V(S) = V(S_1) + V(S_2) - V(S_3)$.

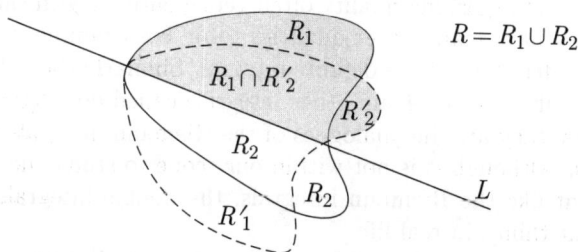

Figure 6.3.14 : Axis of Revolution Passing through a Plane Figure

3.20  Derive (10) from (9).

3.21  Let $C : \alpha(t) = (f(t), g(t), h(t)); a \leq t \leq b$ be a continuously differentiable curve in space and let $L$ be a fixed straight line in the space. For each $t \in [a, b]$, let $\sigma(t)$ be the perpendicular distance of $\alpha(t)$ from $L$. Assume that when $C$ is revolved around $S$, no two distinct points of $C$ trace the same circle. Prove that the area of the surface of revolution is given by

$$\int_a^b 2\pi\sigma(t)\sqrt{[f'(t)]^2 + [g'(t)]^2 + [h'(t)]^2}\,dt.$$

Derive (14) as a special case.

3.22  Find the surface area of a 'cycloidal bowl' obtained by revolving one arch of a cycloid (see Exercise (4.12.6), except that the curve there should be turned upside down) around its line of symmetry, viz. $x = a\pi$. Also find the volume it encloses.

3.23  Find the volume as well as the surface area of a 'spherical cap' of height $h$ and base radius $r$. [*Hint* : First show that the radius of the sphere is $\dfrac{r^2 + h^2}{2h}$.]

## 6.4.    Double Integrals as Certain Volumes

It is a peculiarity of mathematics (and a major source of its strength) that the same mathematical concept often corresponds to many diverse things in real life. For example, the product $xy$ of two numbers $x$ and $y$ represents (a) the area of a rectangle with sides $x$ and $y$ units, (b) the distance run by a car at speed $x$ for a time interval $y$, (c) the cost (in rupees) of $x$ grams of gold sold at the price of $y$ rupees per gram and many other things. Similarly a Riemann integral represents not only the area below a graph, but many other things such as the mass of a wire, the population of a road and so on depending upon what the integrand stands for. Sometimes it so happens that a particular interpretation of a mathematical concept dominates others and gets identified with it (just as a person with a distinguishing quality often gets identified with that quality, even though he may have many other qualities too). For example, the area below a graph is often identified with a definite integral. Similarly the volumes of certain solids get identified with certain other integrals called double integrals. As the name suggests, they are the analogues of the Riemann integrals for functions of two variables. Although it is not within our scope to study the latter *per se*, it is evident that like the Riemann integrals, the double integrals also stand for many different things in real life.

In the last section, we showed how the method of slicing can be used to find the volumes of solids in space, with a special reference to the solids of revolution. We now turn to certain other types of solids which are bounded below by the $x$-$y$ plane and above by some surface. Pyramids, cones, hemispheres are some examples of such solids. But there are many others. Unlike the solids of revolutions, these solids are not generally manufactured for daily use. In fact they are rarely seen as geometric objects as such. Still they are studied and their volumes calculated because their volumes often stand for other important things such as mass, heat etc., which, on the face of it, have nothing to do with volumes.

Let us begin with an illustration. Suppose we want to find the average rainfall for a country in the shape of a plane region, say $R$. Taking a few sample readings and their average may not give a correct answer. As explained in Section 6.1, their weighted average will be a better indicator. That is, if we chop $R$ into subregions $R_1, \ldots, R_n$, put one rain-gauge in each and let $f_1, f_2, \ldots, f_n$ be the rainfalls recorded by them, then

$$\text{average rainfall for } R \simeq \frac{f_1 A(R_1) + f_2 A(R_2) + \ldots + f_n A(R_n)}{A(R)} \tag{1}$$

where, as usual, $A$ stands for area. In fact, exact equality would hold in (1) if the rainfall is constant all over each subregion $R_i$ (although not necessarily so over the entire country $R$). The chances of this are more if each $R_i$ is small. So the approximation becomes closer and closer if all $R_i$'s are small and exact if we take the limit of the right-hand side of (1) as each $R_i$ approaches 0 in size. It is important to note here that by smallness of $R_i$ we mean that $R_i$ is contained in a rectangle both sides of which are small. It is not enough merely that $A(R_i)$ is small. If, for example, $R_i$ is a long, thin strip then $A(R_i)$ may be small. But there is no reason to suppose that the rainfall will be more or less uniform on $R_i$ and in that case the two sides of (1) may differ considerably.

Where do volumes come in? Before answering this question, let us first write (1) a little differently. For each point $(x, y)$ in the region $R$ let $f(x, y)$ be the rainfall at $(x, y)$. In general, $f(x, y)$ will vary as $(x, y)$ varies over the region $R$. So $f(x, y)$ is a real-valued function with domain $R$. It is a function of two variables, viz. $x$ and $y$. A systematic study of such functions is beyond our scope. But here we need only one basic idea, which is easy to grasp because of its analogy with the corresponding idea for functions of one variable.

Returning to (1), for each $i = 1, \ldots, n, f_i$ will be of the form $f(x_i, y_i)$ for some point $(x_i, y_i) \in R_i$ (To be precise, $(x_i, y_i)$ is the point where the rain-gauge in the subregion $R_i$ is located). So the sum in (1) can also be written as $\sum_{i=1}^{n} f(x_i, y_i) A(R_i)$. This is highly analogous to the Riemann sum of a function, except that now the function is of two variables and the domain $R$ is partitioned into subregions instead of subintervals. A sum like this is in fact called a Riemann sum of $f$. And, with analogy for functions of one variable, its limit as the size of each subregion tends to 0 is called the **double integral** of $f$ over $R$, denoted by $\iint_R f(x, y) dA$, or more appropriately, by $\iint_R f(x, y) d(x, y)$. We can define triple integrals for a function of three variables in an analogous manner. The theory of double and triple integrals is analogous to (although a little more complex) then that of the integral of a function of one variable and forms an integral part of the calculus of several variables. We do not pursue it. Nevertheless, using double integrals we can give an exact formula for the average rainfall (and more generally, the average value of any function defined over a bounded plane region). The limit of the sum in (1) is $\iint_R f(x, y) dA$ as we already saw. As for the denominator, we observe that if we integrate a constant function, say $f(x, y) = k$, over $R$ then every Riemann sum of $f$ will be $kA(R)$ and so $\iint_R k dA$ is simply $kA(R)$. Keeping this in mind we can replace (1) by the more exact,

$$\text{average rainfall for } R = \frac{\iint_R f(x, y) dA}{\iint_R 1 dA} \qquad (2)$$

where $f(x, y)$ is the rainfall at $(x, y)$.

Formula (2) would be an excellent example of an exact but useless formula (a criticism sometimes hurled at mathematics itself !) unless there is some easy method to evaluate the double integrals in it. The denominator, as we just saw, is simply the area of $R$ which can be evaluated by the method of slicing as we saw in the last Section. But what about the numerator $\iint_R f(x, y) dA$? To answer this question, we need to look at the geometric interpretation of double integrals, which, again is highly analogous to the geometric interpretation of a definite integral $\int_a^b f(x) dx$ as the area below the curve $y = f(x)$.

Just as we draw the graph of a function of one variable as a curve of the form $y = f(x)$ in the $x$-$y$ plane, we can draw the graph of a function of two variables as a surface of the form $z = f(x, y)$ in the three dimensional $(x, y, z)$-space. Obviously such graphs cannot be drawn exactly on a piece of paper or on a blackboard or on any other two dimensional instrument. To draw them convincingly needs both mathematical and artistic ability. However a few such graphs are shown in Fig.6.4.1, where (a) is a plane, (b) a hemi-sphere and (c) a surface obtained by revolving the parabola $z = y^2$ around the $z$-axis. The domain $R$ is the entire $x$-$y$ plane in (a) and (c). In (b) it is the unit disc in the $x$-$y$ plane, centered at $O$.

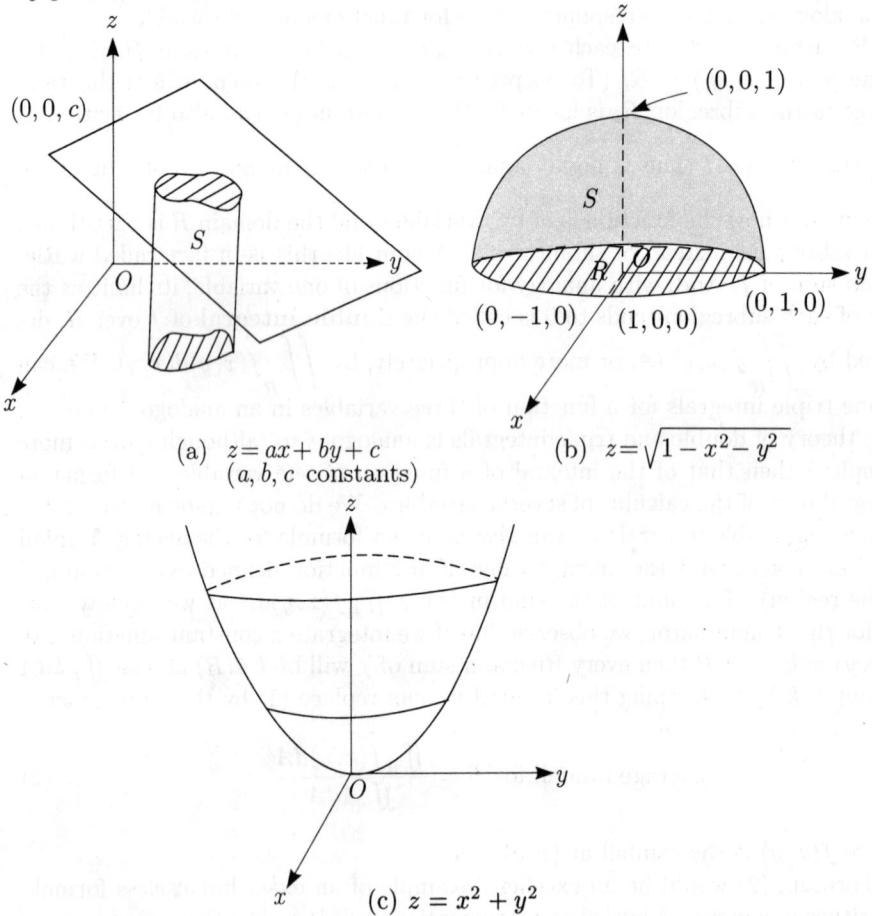

(a) $z = ax + by + c$
($a, b, c$ constants)

(b) $z = \sqrt{1 - x^2 - y^2}$

(c) $z = x^2 + y^2$

Figure 6.4.1 : Graphs of Functions of Two Variables

We assert that if $f(x, y) \geq 0$ for all $(x, y) \in R$ then $\iint_R f(x, y) dA$ equals $V(S)$, the volume of the solid $S$, bounded below by the domain $R$ in the $x$-$y$ plane and above by the surface $z = f(x, y)$. In precise notation $S$ is the set $\{(x, y, z) : (x, y) \in R, 0 \leq z \leq f(x, y)(\}$. $S$ can also be thought of as the portion of the vertical solid cylinder over $R$ truncated by the surface $z = f(x, y)$. For example, in Fig.6.4.1 (b), if we take $R$ to be the unit disc in the $x$-$y$ plane centered at $O$ then $S$ is precisely the solid upper hemisphere of radius 1 and center $O$. In (a), if we let $R$ be a bounded subset of the $x$-$y$ plane such that $ax + by + c \geq$ for all $(x, y) \in R$ then $S$ is the solid shown. (Such a solid is sometimes called a **frustrum** of a cylinder.)

Inasmuch as we do not have a rigorous definition of either a volume or of a double integral (which involves area), we can hardly give a rigorous proof that $\iint_R f(x, y) dA$ equals $V(S)$. All we can do is to give a plausible argument to show why this equality is reasonable to expect. And the argument can be given a little more vividly in the case of the rainfall function, although the essential ideas remain the same for any non-negative, real-valued function. (Actually non-negativity of the function $f$ is not so vital if we agree to treat volumes below the $x$-$y$ plane as negative. But the functions arising in real life applications are generally non-negative anyway.)

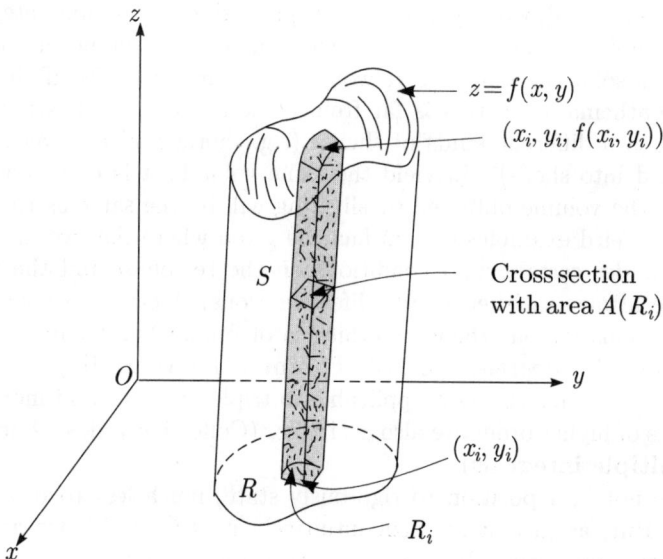

Figure 6.4.2 : Double Integral as Volume below Graph

So, suppose $f(x, y)$ is the rainfall at a point $(x, y) \in R$. Suppose by some magic, the rainwater, instead of flowing along the ground, stayed and accumulated at every point it fell upon. (This possibility is not so far-fetched if the precipitation were in the form of snow rather than rain.) Then at every point $(x, y) \in R$, a vertical column of water will form with its bottom at $(x, y, 0)$ and top at $(x, y, f(x, y))$. These columns together constitute the solid $S$. So $V(S)$ is the total rain water (by measure) that falls on $R$. To see that it equals the double integral of $f$ over $R$, we note that the latter is, by definition, the limit of Riemann sums of the form $\sum_{i=1}^{n} f(x_i, y_i) A(R_i)$, where $R$ is partitioned into subregions $R_1, \ldots, R_n$. The portions of $S$ above these subregions constitute a partition of the solid $S$. If $R_i$ is small, then $f$ may be taken to be approximately a constant over it. So the portion of the solid $S$ which lies over $R_i$ is approximately a vertical column of height $f(x_i, y_i)$ and cross sectional area $A(R_i)$ (see Fig.6.4.2). Hence its volume is approximately $f(x_i, y_i)A(R_i)$. Adding these, we see that $V(S)$ is approximately the Riemann sum $\sum_{i=1}^{n} f(x_i, y_i)A(R_i)$ and, once again, taking limit as each $R_i$ shrinks in size to 0, we get that $V(S) = \iint_R f(x, y)dA$.

The reasoning given above resembles closely that used for showing that an integral of the form $\int_a^b f(x)dx$ equals the area below the graph of $y = f(x)$. Note that here the volume of the solid $S$ was obtained by slivering the solid into vertical columns of small cross-sectional areas (and taking limit as these areas tend to 0). This is conceptually different from the method of slicing (used in the last section) to chop a solid into thin slices (and taking limit as the thickness of each slice tends to 0). That is why in the last section we could express a volume as a definite integral, while just now we expressed it as a double integral. From a layman's point of view, the volume will remain the same no matter by what method, the solid is chopped[3]. It would be a miracle indeed if this were not so. But mathematically this is far from obvious. Just as it is non-trivial to prove that two different kinds of slicing (e.g., horizontal and vertical or into washers and into shells) will yield the same volume, it is quite non-trivial to prove that the volume obtained by slivering will be the same as that obtained by slicing. Weird examples can, in fact, be given where the two are not equal. However, under certain mild conditions on the region $R$ and the function $f$ (which are always satisfied in real life situations) it can be shown that they are equal. This famous theorem, which is of fundamental importance in the theory of double integrals, is called **Fubini's theorem.** Higher dimensional analogues of Fubini's theorem applicable to triple integrals and more generally to integrals of higher order are also available. (Collectively all such integrals are called **multiple integrals**).

We are not in a position to rigorously state, much less to prove, Fubini's theorem. But, as just mentioned, intuitively it is very obvious and we shall assume it without proof. It provides us with the missing link in the following sequence of steps to calculate some quantity (in our example the total rainwater) associated with a plane region $R$.

Step 1:     Express the desired quantity as the double integral $\iint_R f(x, y)dA$ of a suitable real-valued function $f$ defined on $R$.

Step 2:     Equate $\iint_R f(x, y)dA$ with the volume of the solid $S$ lying between the region $R$ in $x$-$y$ plane and the surface $z = f(x, y)$.

Step 3:     Evaluate this volume by the method of slicing, applied in any form.

As a concrete example, we do the following problem.

---

[3]Whether you eat a potato in the form of French fries or in the form of wafers, the potato gives you the same number of calories!

**4.1 Problem:** Suppose the annual rainfall at a point on a circular island of radius 100 km. is proportional to the distance of that point from the center of the island. If the rainfall at the coast is 200cm., find the average rainfall for the island.

**Solution:** We take the island to be the unit disc $R = \{(x,y) : x^2 + y^2 \leq 1\}$ in the $x$-$y$ plane. In other words we are taking 100 km. as a unit of length. Now let $f(x,y)$ be the rainfall at a point $(x,y)$ in $R$. Since $\sqrt{x^2 + y^2}$ is the distance of this point from the center of the island, the data of the problem implies

$$f(x,y) = a\sqrt{x^2 + y^2} \text{ for all } (x,y) \in R$$

where the constant $a$ equals $\dfrac{200 \text{ cm}}{100 \text{ km}} = 2 \times 10^{-5}$. The total rainfall is the double integral $\displaystyle\iint_R a\sqrt{x^2 + y^2}\, dA$, which equals $a \displaystyle\iint_R \sqrt{x^2 + y^2}\, dA$.

(Such elementary properties of double integrals are analogous to the corresponding properties of ordinary integrals and are proved by similar methods. We shall use them without explicit mention.)

To evaluate $\displaystyle\iint_R \sqrt{x^2 + y^2}\, dA$ we sketch the solid $S$ above $R$ in Fig.6.4.3(a). It is not hard to see that the surface $z = \sqrt{x^2 + y^2}$ is an inverted right circular cone with vertex at $O$, obtained by revolving the line $y = z$ in the $y$-$z$ plane around the $z$-axis.

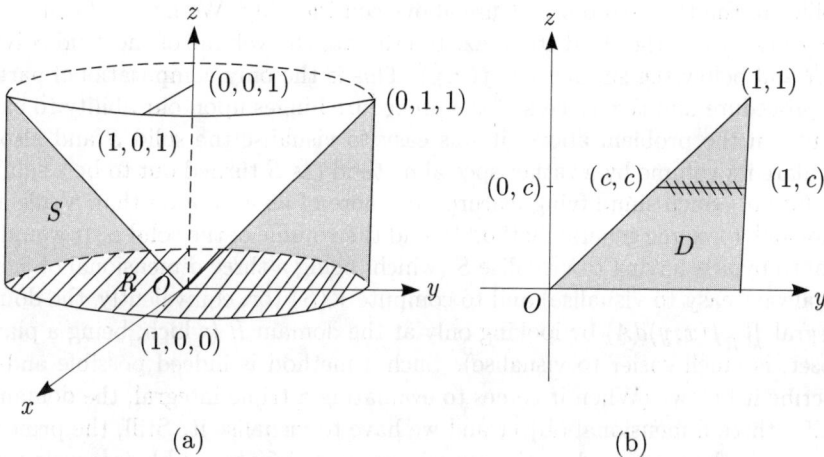

(a)                    (b)

Figure 6.4.3 : Rainfall on a Circular Island

So the solid $S$ is precisely the solid of revolution obtained by revolving the triangular region $D$ shown in Fig.6.4.3(b). The volume of $S$ can be found by the washer method. The cross section of $S$ by the plane $z = c$ will be an annulus with inner radius $c$ and outer radius 1. The area of this cross section is $\pi(1 - c^2)$.

So,

$$\iint_R \sqrt{x^2 + y^2} dA = V(S) = \pi \int_0^1 (1 - z^2)dz = \pi - \frac{\pi}{3} = \frac{2\pi}{3}10^6 \text{ km}^3.$$

(We could also get this by subtracting the volume of the cone from the volume of the cylinder. But that amounts to the same work. The figure $10^6$ comes since 100 km is the unit of length and so one unit volume is $10^6$ km$^3$).

As the area of $R$, $A(R)$ is $\pi 10^4$ km$^2$ and $a = 2 \times 10^{-5}$, we get

$$\text{average rainfall} = \frac{V(S)}{A(R)} = \frac{2\pi}{3\pi} \frac{10^6}{10^4} \frac{2}{10^5} \text{ km} = \frac{400}{3} \text{ cm.} \blacksquare$$

In this problem $f$ was the rainfall function and $\iint_R f(x,y)dA$ is the total rainwater (by measure) falling on $R$. In other situations $f$ may have other interpretation and accordingly the interpretation of $\iint_R f(x,y)dA$ will change. Suppose, for example, that $R$ is a sheet of metal and for $(x,y) \in R$, $f(x,y)$ is the laminar density (that is, mass per unit area) at $(x,y)$. Then $\iint_R f(x,y)dA$ will represent the total mass of $R$. The reasoning is essentially a duplication of that given for obtaining the mass of a piece of wire (see Fig.5.2.2 and the accompanying comments) with the difference that there it was the linear density while in the case of a plane region (or a 'lamina' as it is often called) it is the laminar density (also sometimes called the areal density). Similarly if $f(x,y)$ is the population density at a point $(x,y)$ in a country $R$ then $\iint_R f(x,y)dA$ is the total population of the country.

Thus we see that the volumes of certain solids can represent many different things, often things having little to do with geometry. This enhances the applicability of the three-step procedure above considerably. We must, of course, be able to carry out the third step, viz. to evaluate the volume of the solid $S$ lying on $R$ and below the surface $z = f(x,y)$. This is the only computational part in the procedure and the success of the procedure hinges upon our ability to carry it out. In the problem above, it was easy to visualise the solid $S$ and also to calculate its volume by a rather special method (as $S$ turned out to be a solid of revolution.) Such simplifying features are more of an exception than a rule and so we look for some general method to find the volume of the solid $S$. It would be ideal to bypass having to visualise $S$ (which, being a three dimensional object is not always easy to visualise) and to compute $V(S)$ (or, equivalently, the double integral $\iint_R f(x,y)dA$) by looking only at the domain $R$ (which, being a planar subset, is much easier to visualise). Such a method is indeed possible and we describe it below. (When it comes to evaluating a triple integral, the domain is itself a three dimensional object and we have to visualise it. Still, the principle involved is the same and so the experience gained from double integrals turns out to be useful for triple and other multiple integrals as well.)

The basic idea is to integrate w.r.t. one variable at a time, holding the other one as a constant. In order to focus on this idea, we first do the case where $R = R(a,b;c,d)$, that is $R$ is a rectangle with sides $[a,b]$ and $[c,d]$. As before, let $f$ be a non-negative real valued function on $R$ and let $S$ be the solid $\{(x,y,z) : a \leq x \leq b, c \leq y \leq d, 0 \leq z \leq f(x,y)\}$. Our interest is in finding

$V(S)$. We do this by the method of slicing. The solid $S$ looks rather like a loaf of a bread with a rectangular base and we shall cut it by planes parallel to one of the sides of the base as shown in Fig.6.4.4(a). A typical slice obtained by a plane of the form $y = h$ (where $c \leq h \leq d$) is shown in (b).

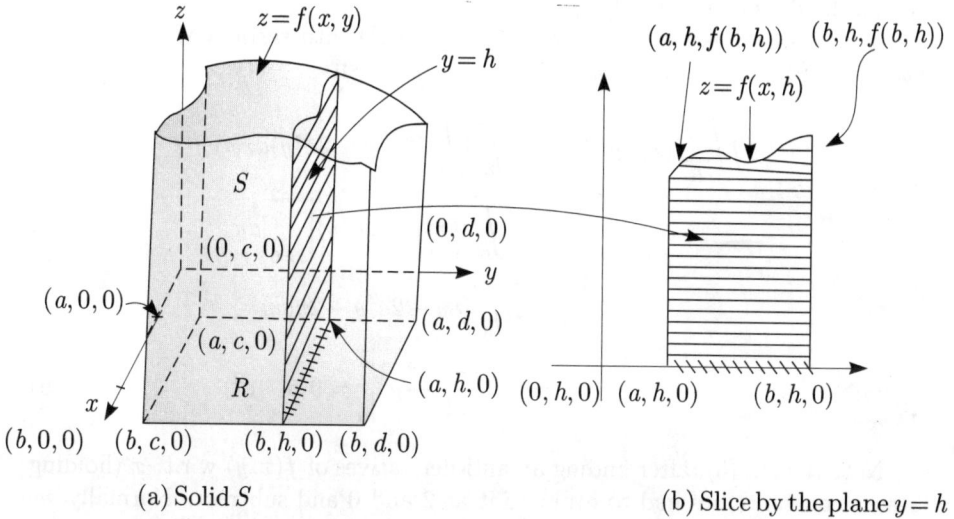

(a) Solid $S$      (b) Slice by the plane $y = h$

Figure 6.4.4 : Double Integral over a Rectangular Domain

It is clear that the top of this slice is the intersection of the surface $z = f(x,y)$ with the plane $y = h$ and so it is the curve $z = f(x,h), a \leq x \leq b$. Here, even though $f$ is a function of two variables, the variable $y$ is held to be a constant $h$ and so effectively $f$ is a function of just one variable $x$. (A function so obtained is sometimes denoted by $f(-,h)$, the bar indicating that the entry to be put there is a variable. With this practice, the original function will be $f(-,-)$ where the two bars represent two mutually independent variables.) Moreover it will be a different function of $x$ for different values of $h$. (This is where the analogy with a loaf of bread is a little misleading. In the case of a loaf, the cross sections are mutually congruent.)

Let $I(h)$ be the area of this cross-section. $I$ is a function of $h \in [a,b]$. From the theory of integrals of functions of one variable, we know that

$$I(h) = \int_a^b f(x,h)dx \qquad (3)$$

where the integral is the ordinary definite integral of a function of one variable. Changing $h$ to $y, I(y) = \int_a^b f(x,y)dx$ for every $y \in [c,d]$. Since the solid $S$ lies entirely between the planes $y = c$ and $y = d$, we get (by the method of slicing studied in the last section),

$$V(S) = \iint_R f(x,y)dA = \int_c^d I(y)dy = \int_c^d [\int_a^b f(x,y)dx]dy \qquad (4)$$

The last term in (4) deserves a closer look. Riemann integration appears twice in it. The 'inner' integration is w.r.t. $x$, and the variable $y$ is treated as a constant. So the outcome of this inner integration is a function of $y$, which we denoted by $I(y)$. Here $x$ does not appear at all (not even as a constant). And the outer integration is w.r.t. $y$, the outcome of which is a real number which equals the double integral $\iint_R f(x,y)dA$.

As a simple illustration, let $R$ be the rectangle with vertices at $(0,0)$, $(2,0)$, $(2,3)$ and $(0,3)$ and let $f(x,y) = 3x^2 y - 2e^x y$. Then

$$
\begin{aligned}
\iint_R f(x,y)dA &= \int_0^3 [\int_0^2 (3x^2 y - 2e^x y)dx]dy \\
&= \int_0^3 [(x^3 y - 2e^x y)\Big|_{x=0}^{x=2}]dy \qquad (5) \\
&= \int_0^3 [8y - 2e^2 y + 2y]dy \\
&= \frac{10 - 2e^2}{2} y^2 \Big|_0^3 = 9(5 - e^2). \qquad (6)
\end{aligned}
$$

Note that in (5), after finding an anti-derivatives of $f(x,y)$ w.r.t. $x$ (holding $y$ as a constant), we had to evaluate it at 2 and 0 and subtract. Normally we indicate this by a vertical bar like $\Big|_0^2$. In the case of double integrals it is a good idea to specify which variable takes the given values and write $\Big|_{x=0}^{x=2}$ instead of $\Big|_0^2$ because more than one variable occur in the antiderivative. In (6), it is not necessary to do so, because $y$ is the only variable there.

We could also have evaluated the volume of the solid $S$ in Fig.6.4.4(a) by slicing by planes perpendicular to the $x$-axis. In that case the roles of $x$ and $y$ would be interchanged. The inner integration would be w.r.t. $y$ and instead of (4) we would get

$$
V(S) = \iint_R f(x,y)dA = \int_a^b \left[ \int_c^d f(x,y)dy \right] dx. \qquad (7)
$$

In the last terms of (4) and (7), the process of integration (w.r.t. one variable at a time) occurs repeatedly. So they are often called **iterated integrals**. (Iteration simply means repetition.) We leave it to the reader to verify by direct calculation that in the example just worked out, the iterated integral in (7) comes out to tally with (4). In general, this is far from obvious and we need Fubini's theorem to show the equality of the two iterated integrals, each being equal to the double integral $\iint_R f(x,y)dA$. Examples can be given where one of the interared integrals is easier to evaluate than the other (see Exercise (4.2)).

We now turn to the evaluation of $\iint_R f(x,y)dA$ where $R$ is not necessarily a rectangle. (Problem (4.1) was actually one such case. But there the evaluation of $V(S)$ i.e. of $\iint_R f(x,y)dA$ was done by a special method as $S$ was a solid of revolution.) The method essentially remains the same as for a rectangular domain viz. converting the double integral to one of the two iterated integrals. But the details get more complicated. In the case of a rectangle $R = R(a, b; c, d)$ every plane of the form $y = h$ (where $c \leq h \leq d$) cuts $R$ in a segment, specifically, the segment from $(a, h)$ to $(b, h)$, as shown in Fig.6.4.5(a). (Note that since $R$ lies in the $x$-$y$ plane, instead of taking the plane $y = h$ in space, we could as well take the line $y = h$ in the $x$-$y$ plane.)

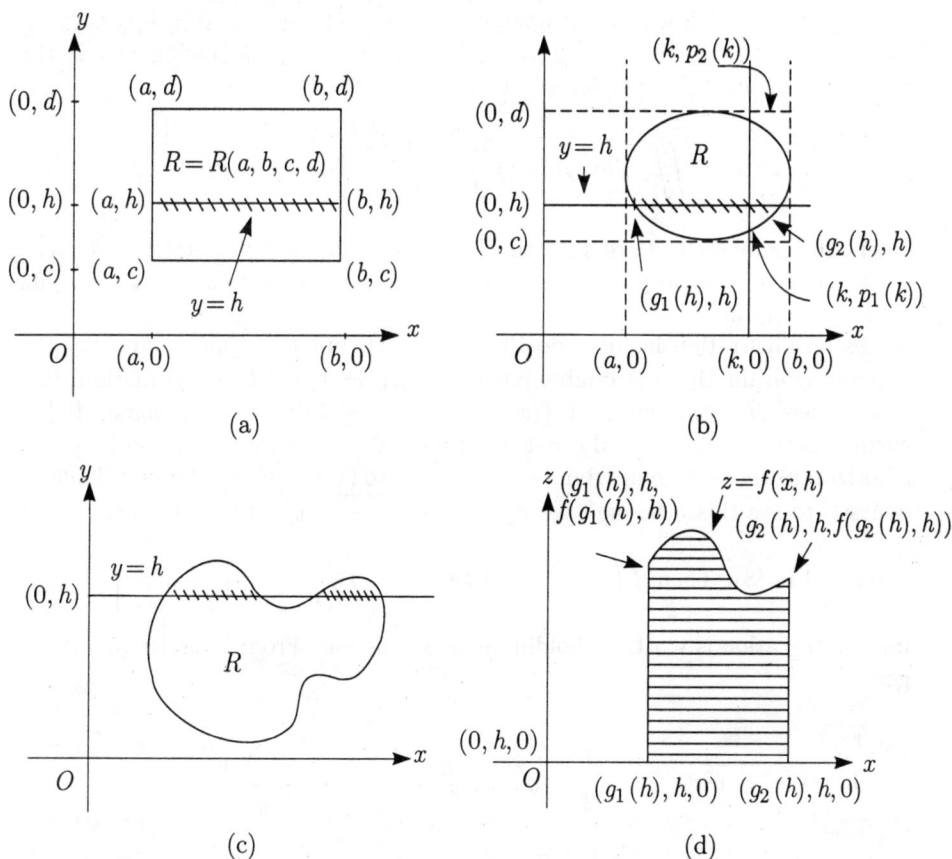

Figure 6.4.5 : Evaluation of a Double Integral as an Iterated Integral

These segments are all congruent to each other and hence have equal lengths. In fact, they are all stacked one above the other. The $x$ - coordinates of their endpoints are the same (viz. $a$ and $b$) for all $h$ in $[c, d]$. For a more general domain $R$ in the plane this need not be so. In fact, as shown in Fig.6.4.5(c), for some values of $h$, the line $y = h$ may cut $R$ in more than one segment. We ignore such domains and confine to the relatively simple case where for each $h$ in some interval $[c, d]$, the line $y = h$ cuts $R$ in exactly one segment, as in (b). (There is really not much loss of generality in assuming this. Domains such as those in (c), can generally be decomposed into a finite number of subdomains for each of which the simplyfing assumption holds. Since integrals are additive, one can evaluate $\iint f(x,y)dA$ over each of these subdomains and add. So we shall not consider this complication.)

The essential difference between (a) and (b) is, as just said, that unlike in (a), the $x$-coordinates of the segment cut off by the line $y = h$ in (b) vary with $h$. Let us call them $g_1(h)$ and $g_2(h)$ respectively, with $g_1(h) \le g_2(h)$. Then the segment runs from $(g_1(h), h)$ to $(g_2(h), h)$. The corresponding slice of the solid $S$ by the plane $y = h$ will be as shown in (d). It is similar to that in Fig.6.4.4(b), except that $a$ and $b$ are replaced by $g_1(h)$ and $g_2(h)$. So by following exactly the same steps as in the derivation of (4), we get,

$$V(S) = \iint_R f(x,y)dA = \int_c^d \left[ \int_{g_1(y)}^{g_2(y)} f(x,y)dx \right] dy \tag{8}$$

where the numbers $c, d$ are such that the domain $R$ of $f$ lies entirely between the the lines $y = c$ and $y = d$. Thus we have expressed the double integral as an iterated integral.

As an illustration let us redo Problem (4.1), but this time computing the volume by evaluating the double integral $\iint_R \sqrt{x^2 + y^2}\, dA$ by the method just given. Here $R$ is the unit disc $\{(x,y) : x^2 + y^2 \le 1\}$ in the $x$-$y$ plane. It lies entirely between $y = -1$ and $y = 1$. For each $h \in [-1, 1]$, the line $y = h$ cuts it in a horizontal segment going from $(-\sqrt{1 - h^2}, h)$ to $(\sqrt{1 - h^2}, h)$. (Draw a diagram yourself to see this.) So we take $g_1(y) = -\sqrt{1 - y^2}, g_2(y) = \sqrt{1 - y^2}, c = -1$ and $d = 1$ in (8). Then $\iint_R \sqrt{x^2 + y^2}dA = \int_{-1}^{1} \left[ \int_{-\sqrt{1-y^2}}^{\sqrt{1-y^2}} \sqrt{x^2 + y^2}dx \right] dy$. The

inner integration is w.r.t. $x$, holding $y$ as a constant. From Exercise (5.16), we get

$$\int_{-\sqrt{1-y^2}}^{\sqrt{1-y^2}} \sqrt{x^2 + y^2}dx = \frac{1}{2}\left[ x\sqrt{x^2+y^2} + y^2\ln(\frac{x + \sqrt{x^2+y^2}}{y}) \right]\Bigg|_{x=-\sqrt{1-y^2}}^{x=\sqrt{1-y^2}} \tag{9}$$

$$= \sqrt{1-y^2} + \frac{y^2}{2}\ln(\frac{1 + \sqrt{1-y^2}}{1 - \sqrt{1-y^2}}) \tag{10}$$

(Strictly speaking, (9) and (10) are valid only for $y \ne 0$. The case $y = 0$ requires a special handling. Here the integral on the left is 1 by direct calculation. It is not hard to show (see Exercise (4.3)) that $y^2\ln\left(\frac{1+\sqrt{1-y^2}}{1-\sqrt{1-y^2}}\right) \to 0$ as $y \to 0$. So we may take (10) to be valid for all $y \in [-1, 1]$, including $y = 0$.)

To complete the evaluation of $\iint_R \sqrt{x^2 + y^2}dA$ we now need to evaluate $\int_{-1}^{1} \sqrt{1-y^2} + \frac{y^2}{2}\ln(\frac{1+\sqrt{1-y^2}}{1-\sqrt{1-y^2}})dy$. The integral $\int_{-1}^{1}\sqrt{1-y^2}dy$ is easy to evaluate with the substitution $y = \sin\theta$ and comes out to be $\frac{\pi}{2}$ (which is also obvious geometrically since the integral represents the area of the right half of the unit disc). As for the second term, we apply integration by parts and get,

$$\int_{-1}^{1} \frac{y^2}{2} \ln\left(\frac{1+\sqrt{1-y^2}}{1-\sqrt{1-y^2}}\right) dy = I_1 + I_2$$

where $\qquad I_1 = \frac{y^3}{6} \ln\left(\frac{1+\sqrt{1-y^2}}{1-\sqrt{1-y^2}}\right)\Bigg|_{-1}^{1} = 0$

and $\quad I_2 = \int_{-1}^{1} \frac{y^3}{6}\left[\frac{y}{(1+\sqrt{1-y^2})\sqrt{1-y^2}} + \frac{y}{(1-\sqrt{1-y^2})\sqrt{1-y^2}}\right] dy$

$$= \frac{1}{3}\int_{-1}^{1}\frac{y^2}{\sqrt{1-y^2}}dy$$

$$= \frac{1}{3}\int_{-\pi/2}^{\pi/2} \sin^2\theta\, d\theta \quad \text{where } y = \sin\theta$$

$$= \frac{1}{3}\int_{-\pi/2}^{\pi/2}\frac{1-\cos 2\theta}{2}d\theta = \frac{\theta}{6} - \frac{\sin 2\theta}{12}\Bigg|_{-\pi/2}^{\pi/2} = \frac{\pi}{6}. \tag{11}$$

Hence $\iint_R \sqrt{x^2+y^2}dA = \frac{\pi}{2} + \frac{\pi}{6} = \frac{2\pi}{3}$, exactly as before.

Instead of (8), we could equate $\iint_R f(x,y)dA$ with the other iterated integral. This time we assume that $R$ lies between two vertical lines, say $x = a$ and $x = b$ and that for every $k \in [a,b]$, the line $x = k$ cuts $R$ in a segment from $(k, p_1(k))$ to $(k, p_2(k))$ where $p_1, p_2$ are some functions of $k$ (depending on what the region $R$ is). Then carrying out the inner integration w.r.t. $y$, we get

$$V(S) = \iint_R f(x,y)dA = \int_a^b\left[\int_{p_1(x)}^{p_2(x)} f(x,y)dy\right]dx. \tag{12}$$

As with rectangles, it may happen that for some function $f(x,y)$ one of the iterated integrals is easier to evaluate than the other. In fact, this is more likely to happen for non-rectangular domains. There are examples where one of the iterated integrals cannot be evaluated because the antiderivative of the function $f(x,y)$ w.r.t. one of the two variables $x$ and $y$ may not be expressible in a closed form but the anti-derivative of $f(x,y)$ w.r.t. the other variable is quite easy to obtain. A simple example is, $f(x,y) = e^{x^2}$. In such cases, we may be able to convert the given iterated integral into the other iterated integral by going through a double integral over a suitable domain. This is known as **change of order of integration**. The theoretical justification for it lies, once again, in Fubini's theorem and so we omit it. But we illustrate the technique with a problem.

**4.2 Problem.** Evaluate $\displaystyle\int_0^1 \int_y^1 e^{x^2}\,dx\,dy$ by changing the order of integration.

**Proof:** We first have to identify a domain $R$ such that the given integral equals the double integral $\iint_R e^{x^2}\,dA$. Since the outer integration is w.r.t. $y$ and has 0 and 1 as the lower and upper limits, $R$ lies between the lines $y = 0$ and $y = 1$. Now, for any $h \in [0,1]$, the interval of the inner integration ranges from $h$ to 1. This means that the line $y = h$ cuts $R$ in the segment going from $(h, h)$ to $(1, h)$. These points lie on the straight lines $y = x$ and $x = 1$ respectively. So it is clear that $R$ is the portion of the strip $0 \le y \le 1,$, lying between the lines $y = x$ and $x = 1$. $R$ is sketched in Fig.6.4.6(a). It is the triangle with vertices at $(0,0), (1,0)$ and $(1,1)$.

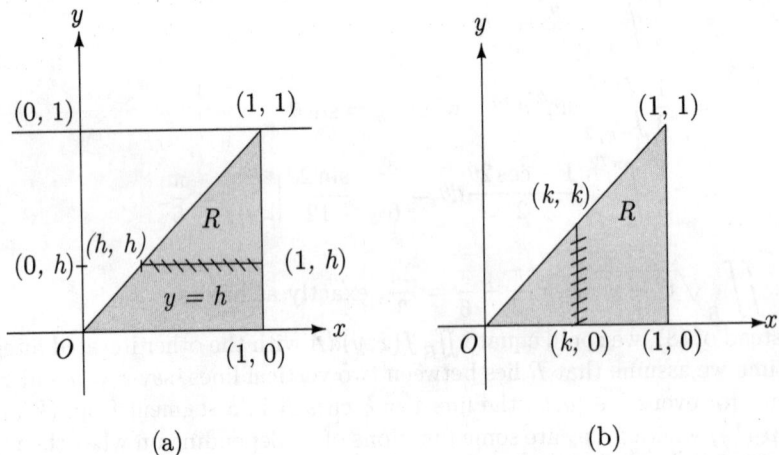

Figure 6.4.6 : Change of Order of Integration

Having identified $R$, writing $\iint_R e^{x^2}\,dA$ as the other iterated integral is easy. $R$ lies between the lines $x = 0$ and $x = 1$. For each $k \in [0,1]$, the line $x = k$ cuts $R$ in the segment from $(k, 0)$ to $(k, k)$ as shown in Fig.6.4.6(b). So, the given iterated integral, after changing the order of integration, becomes $\displaystyle\int_0^1 \int_0^x e^{x^2}\,dy\,dx$. Note that now the inner integration being w.r.t. $y, e^{x^2}$ is a constant and so $\displaystyle\int_0^x e^{x^2}\,dy$ is simply $xe^{x^2}$. This gives $\displaystyle\int_0^1 \int_0^x e^{x^2}\,dy\,dx$ as $\displaystyle\int_0^1 xe^{x^2}\,dx$. The advantage is that now while integrating w.r.t. $x$, the integrand is $xe^{x^2}$ and not $e^{x^2}$ as in the given iterated integral. The function $xe^{x^2}$ has an antiderivative in a closed form, viz. $\dfrac{1}{2}e^{x^2}$. So, the original iterated integral comes out as $\dfrac{1}{2}e^{x^2}\Big|_0^1 = \dfrac{1}{2}(e-1).$ ∎

A few more problems of evaluating double integrals as iterated integrals will be given as exercises. The double integrals have theoretical as well as practical applications. An ordinary Riemann integral is an instance of the limiting process. In an iterated integral, the limiting process enters twice. The equality of the two iterated integrals, (which is analogous in spirit to the change of order of summation for doubly infinite series) is, therefore, yet another instance of the communitativity of two limiting process (see the comments made after Corollary (3.7.5), and also those before Theorem (3.11.3).) Not surprisingly, results of this type have powerful theoretical applications. However, we do not pursue them here because our interest is not in the double integrals *per se* (which properly belong to the study of functions of two variables, which we are not undertaking.) Nevertheless, in Exercise (8.22), we shall indicate how double integrals lead to an evaluation of the improper integral $\int_0^\infty e^{-x^2} dx$.

Instead, we turn to the practical applications of double integrals. As remarked earlier, the double integral $\iint_R f(x,y)dA$ stands for many different things depending upon what physical interpretation the function $f(x,y)$ has. Usually, these things represent some intrinsic properties of $R$ such as its mass, rainfall, heat etc. That is, these properties depend only on $R$ and not on anything else. But there are certain attributes which depend not only on $R$ but also on the position $R$ has relative to something else. An important class of such attributes comprises various of types of moments. The concept of the moment of a force is important in physics for studying rotational motion. We shall not go into it. But it can be illustrated with a very simple example.

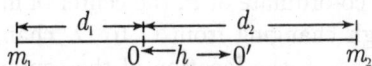

(a) Moment of point masses about a point

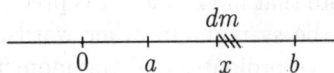

(b) Moment of linear mass distribution

Figure 6.4.7 : Moments of Masses about Points

Suppose a rod of negligible mass is hinged at a point $O$ on it and masses $m_1$ and $m_2$ are placed on the rod at distances $d_1$ and $d_2$ from $O$, on opposite sides of $O$ (see Fig.6.4.7(a)). Then the rod will be in equilibrium not necessarily when $m_1 = m_2$ but when $m_1 d_1 = m_2 d_2$. (In a fair balance $d_1 = d_2$ and so $m_1 = m_2$ is a necessary condition for equilibrium. But in an illegal balance or in a see-saw, objects of different masses can balance each other because $d_1 \neq d_2$.) The expressions $m_1 d_1$ and $m_2 d_2$ are called the **moments of the masses** $m_1$ and $m_2$ respectively about the point $O$. For equilibrium, these moments must be equal in magnitude but opposite in the sense in which they tend to rotate the rod (viz., one clockwise and the other counter-clockwise). This can be expressed a little more conveniently by co-ordinating the rod with $O$ as the origin. Let the masses $m_1, m_2$ be at $x_1, x_2$ respectively. (In Fig.6.4.7(a), $x_2 = d_2$ but $x_1 = -d_1$ since $d_1 > 0$ but $x_1 < 0$.) Then $m_1 x_1 + m_2 x_2$ is the total moment of the system. If it is positive, it will cause the rod to swing one way (specifically, so that $m_2$ moves downdards), the other way if $m_1 x_2 + m_2 x_2 < 0$ and be in equilibrium if $m_1 x_1 + m_2 x_2 = 0$. The same holds if, instead of two, we have a system of $k$ masses, say $m_1, \ldots, m_k$ located at points with co-ordinates $x_1, \ldots, x_k$ respectively. The total moment of this system about $O$ (the origin) is $m_1 x_1 + \ldots + m_k x_k$.

Now let $M = m_1 + \ldots + m_k$ be the total mass of the system and let $\bar{x} = \frac{1}{M}(m_1 x_1 + \ldots + m_k x_k)$. Then $M\bar{x} = \sum_{i=1}^{k} m_i x_i$. In other words, the total moment of the system of the $k$ masses located at $x_1 \ldots, x_k$ about $O$ is the same as the moment about $O$ if the entire mass $M$ of the system were located at a point $P$ whose co-ordinate w.r.t. $O$ is $\bar{x}$. For this reason $\bar{x}$ is called the **center of mass** of the system. In fact this holds even if instead of $O$, we take moments about some other point, say $O'$, on the line at a distance $h$ from $O$ (Fig.6.4.7(a)). For suppose $x_i'$ is the co-ordinate of $m_i$ w.r.t. $O'$. Then $x_i' = x_i - h$ for $i = 1, \ldots, k$. So the moment of the system about $O'$ comes out as

$$\sum_{k=1}^{k} m_i x_i' = \sum_{i=1}^{k} m_i(x_i - h) = \sum_{i=1}^{k} m_i x_i - h\sum_{i=1}^{k} m_i$$

$$= M\bar{x} - Mh = M(\bar{x} - h) \tag{13}$$

$$= \text{moment of } M, \text{ located at } \bar{x}, \text{ about } O'. \tag{14}$$

Note that in (13), $\bar{x} - h$ is precisely the new co-ordinate of $P$, the center of mass of the system. In other words, even though changing from $O$ to $O'$ changes the co-ordinates and the moments of the masses, the location of the center of mass of the system remains the same. Put differently, the center of mass of a system depends only on the relative distances between the masses and stays invariant under a change of co-ordinates. The center of mass of a system is therefore an important tool in analysing what is called a **rigid body motion**, i.e. a motion in which the relative distances between every two points of the body remains unchanged (or, in other words, there is no deformation within the body). Note that, by definition, the $x$-coordinate of the center of mass is the weighted average of the $x$-coordinates of the masses, the 'weights' being equal to the respective masses. (This, indeed, is the reason why it is called a 'weighted' average).

Suppose now that instead of a 'discrete' mass distribution, we have a continuous mass distribution along a line segment, say an interval $[a, b]$ on the $x$-axis. Suppose $\rho(x)$ is the linear mass density at $x, a \leq x \leq b$. We already saw in Section 5.2 that the total mass, say $M$, equals $\int_a^b \rho(x)dx$. Following the recipe of integration, the moment about $O$ is $\int_a^b x\rho(x)dx$. (It is a little easier to see formulas like this if one thinks of a small 'mass element' $dm$ located at $x$, as shown in Fig.6.4.7(b). Then its moment about $O$ is $xdm$ which equals $x\rho(x)dx$. (It is, in fact, customary to express the moment as $\int_a^b xdm$. But this is a bad practice because the variable $m$ does not vary between $a$ to $b$. It is better to write $\int_a^b x\dfrac{dm}{dx}dx$ or $\int_{x=a}^{x=b} xdm$.) So the **centre of mass** of the segment $[a, b]$ is $\bar{x} = \dfrac{\int_a^b x\rho(x)dx}{\int_a^b \rho(x)dx}$. If the segment has uniform mass density $\rho$, then

$$\bar{x} = \frac{\rho\int_a^b xdx}{\rho(b-a)} = \frac{1}{2}(a+b).$$ In other words it is the mid-point of the interval $[a, b]$.

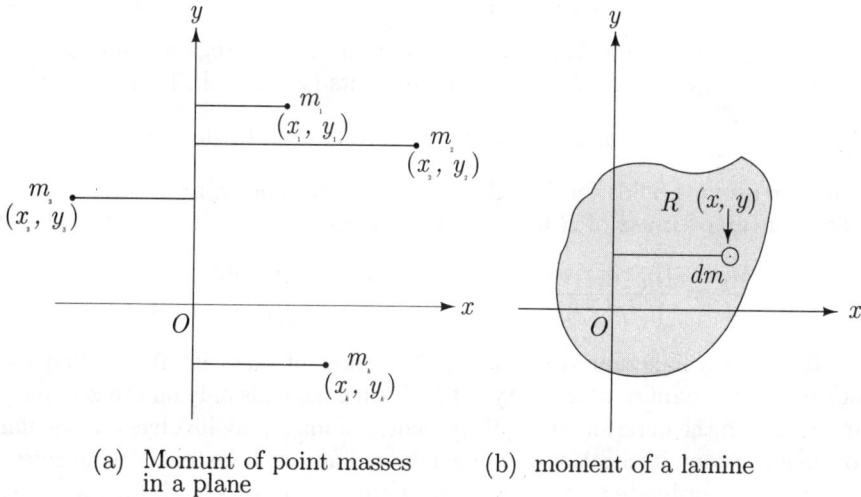

(a) Momunt of point masses in a plane

(b) moment of a lamine

Figure 6.4.8 : Moments of Masses about Straight Lines

For a mass distribution in a plane, the situation is entirely analogous, except that instead of taking moments about points, we now take moments about straight lines in the plane. It suffices to consider the moments about the two co-ordinate axes, because it is not hard to show (cf. Exercise (4.10)) that these two moments determine the moment about any other line. Fig.6.4.8(a) shows the moment of a system of finitely many masses $m_1, \ldots, m_k$ located at, say, $(x_1, y_1), \ldots, (x_k, y_k)$. The (total) moment $M_y$ about the $y$-axis is $\sum_{i=1}^k m_i x_i$.

Similarly the moment $M_x$ about the $x$-axis is $\sum_{i=1}^n m_i y_i$. The **center of mass** of the system is defined as the point $(\bar{x}, \bar{y})$ where

$$\bar{x} = \frac{M_y}{M} = \frac{\sum_{i=1}^{h} m_i x_i}{\sum_{i=1}^{k} m_i}, \quad \bar{y} = \frac{M_x}{M} = \frac{\sum_{i=1}^{k} m_i y_i}{\sum_{i=1}^{k} m_i} \tag{15}$$

(Note that in defining $\bar{x}$ we are taking the ratio $\dfrac{M_y}{M}$ and not the ratio $\dfrac{M_x}{M}$ as might appear more natural at first sight. The thing to keep in mind is that $\bar{x}$ is the average value of $x$ (w.r.t. the mass) and $x$ is the distance of a point $(x, y)$ from the $y$-axis and not from the $x$-axis.)

As in the case of a linear mass distribution, the center of mass of a planar lamina is obtained by replacing the sums in (15) by integrals, with the difference that now we need double integrals. Thus suppose $R$ is a bounded plane region whose mass density at a point $(x, y)$ is $\rho(x, y)$. Then the moment of a typical 'mass element' $dm$ around $y$-axis, shown in Fig.6.4.8(b), is $xdm$, or $x\rho(x, y)dA$, where $dA$ is the 'area element' and $(x, y)$ is any point of it. And so $M_y$, the moment of $R$ about the $y$-axis is $\displaystyle\iint_R x\rho(x, y)dA$. (Those who consider working with such mass elements as too sloppy, are free to chop the region $R$ into a finite number of subregions, say $R_1, \ldots, R_n$, take points $(x_i, y_i) \in R_i$ for $i = 1, \ldots, n$, form the sum $\displaystyle\sum_{i=1}^{n} x_i\rho(x_i, y_i)\Delta(R_i)$ and take limit as each $R_i$ shrinks to 0 in size.) A similar reasoning holds for $M_x$, the moment of $R$ about $x$-axis.

The **center of mass** of $R$ is the point $\bar{x}, \bar{y}$ where

$$\bar{x} = \frac{M_y}{M} = \frac{\iint_R x\rho(x, y)dA}{\iint_R \rho(x, y)dA} \quad \text{and} \quad \bar{y} = \frac{M_x}{M} = \frac{\iint_R y\rho(x, y)dA}{\iint_R \rho(x, y)dA} \tag{16}$$

when the density function is constant, the center of mass is often called the **centroid** (or the **centre of gravity**) of $R$. It then depends only on the geometry of $R$. Although the determination of the center of mass may involve a particular co-ordinate system, its location is independent of any such system. A convenient choice of the co-ordinate system and considerations of symmetry often simplify the determination of the center of mass. Suppose for example, the region $R$ and the density function $\rho(x, y)$ are symmetric about the $y$-axis. This means that for every $(x, y) \in R, (-x, y)$ is also in $R$ and further that $\rho(-x, y) = \rho(x, y)$. Then the moment $M_x$ of $R$ about $y$-axis is 0 and so the center of mass will lie on the $y$-axis. Intuitively this is obvious since for every area element $dm$ there will be a symmetrically situated area element of equal mass but with opposite moment. In a formal proof, we partition $R$ by a partition which is symmetric w.r.t. the $y$-axis. As a simple illustration, we do a problem.

**4.3 Problem:** Find the center of mass of a half-disc assuming

(i) that the laminar mass density is uniform

(ii) that the laminar density is proportional to the distance from the diameter of the half-disc.

**Solution:** We represent the half-disc by the region $R = \{(x, y) : x^2 + y^2 \le a^2, y \ge 0\}$, where $a$ is its radius.

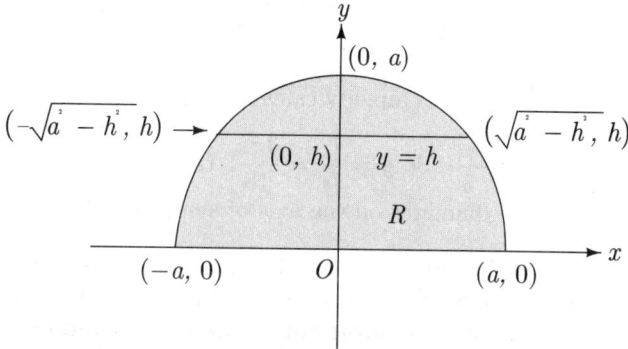

Figure 6.4.9 : Determination of Centre of Mass

In both (i) and (ii), $R$ as well as the density function are symmetric about the $y$-axis and so $\bar{y} = 0$. For (i) we take $\rho = 1$ while for (ii) we let $\rho(x, y) = ky$ where $k$ is some constant. Hence in both cases, it is convenient to carry out the inner integration w.r.t. $x$ rather than w.r.t. $y$. A look at Fig. 6.4.9 shows that for (i),

$$
\begin{aligned}
\bar{y} &= \frac{\iint_R y\, dA}{\iint_R dA} \\[2mm]
&= \frac{1}{\text{area of } R} \times \int_0^a \int_{-\sqrt{a^2-y^2}}^{\sqrt{a^2-y^2}} y\, dx\, dy \\[2mm]
&= \frac{2}{\pi a^2} \int_0^a 2y\sqrt{a^2 - y^2}\, dy = \frac{2}{\pi a^2}\left[-\frac{2}{3}(a^2 - y^2)^{3/2}\right]\Big|_0^a \\[2mm]
&= \frac{2}{\pi a^2}\frac{2a^3}{3} = \frac{4}{3\pi}a.
\end{aligned}
$$

Thus the center of mass lies at a distance $\frac{4a}{3\pi}$ from the diameter and on the axis of symmetry.

Similarly, for (ii),

$$
\bar{y} = \frac{\iint_R ky^2\, dA}{\iint_R ky\, dA} = \frac{\iint_R y^2\, dA}{\iint_R y\, dA} = \frac{I_1}{I_2} \quad \text{(say).} \quad \text{Then,}
$$

$$
\begin{aligned}
I_1 &= \int_0^a \int_{-\sqrt{a^2-y^2}}^{\sqrt{a^2-y^2}} y^2\, dx\, dy \\[2mm]
&= \int_0^a 2y^2\sqrt{a^2 - y^2}\, dy \\[2mm]
&= \int_0^{\pi/2} (2a^2\sin^2\theta)(a^2\cos^2\theta)\, d\theta \quad \text{where } y = a\sin\theta
\end{aligned}
$$

$$= \frac{a^2}{2} \int_0^{\pi/2} \sin^2 2\theta d\theta = \frac{a^4}{4} \int_0^{\pi/2} 1 - \cos 4\theta d\theta$$

$$= \frac{a^4}{4} [\theta - \frac{1}{2} \sin 4\theta] \Big|_0^{\pi/2} = \frac{\pi a^4}{8}.$$

The integral $I_2 = \iint_R y dA$ was already calculated above (in the answer to part

(i)) and came out to be $\frac{2a^3}{3}$. So $\bar{y} = \frac{I_1}{I_2} = \frac{3\pi a}{16}$. Hence the center of mass is at

a distance $\frac{3\pi a}{16}$ from the diameter on the axis of symmetry. ∎

In some physical applications, it is not the distance but the square of the distance of a mass that matters. The moments so obtained are called the **moments of the moment** or the **second moments of mass** or **moments of inertia** and are denoted by $I$ rather than by $M$. Thus for example, the second moment of a plane lamina $R$ about the $x$-axis will be $I_x$ and equals $\iint_R y^2 \rho(x,y) dA$. Their calculation is similar and a few problems based on them will be given as exercises.

The main theme of this section was to show how the volumes of certain solids represent double integrals of functions of two variables and hence, indirectly, many different things such as mass, moments, heat etc. Suppose now that instead of a planar region $R$ we have a bounded solid $S$ in the three dimensional space $\mathbb{R}^3$. Suppose $f(x,y,z)$ is a real-valued function defined at all points $(x,y,z) \in S$. Then the triple integral of $f$ on $S$, denoted by $\iint_S f(x,y,z) dV$ is defined in an entirely analogous manner as a double integral. Geometrically, it represents the 'hyper-volume' of a certain four dimensional object, viz. the set $\{(x,y,z,w) : 0 \le w \le f(x,y,z), (x,y,z) \in S\}$, where we have, for simplicity, assumed that $f \ge 0$, although this is not a very vital assumption. It is non-trivial to show that under certain general conditions on $S$ and $f$, the triple integral $\iint_S f(x,y,z) dV$ equals an iterated integral in which there are three Riemann integrations, one w.r.t. each of $x, y$ and $z$. Since these three variables can be permuted in six different ways, there are, in all, 6 possible iterated integrals and in some problems some of them may be advantageous over the rest.

As in the case of double integrals, depending upon the physical interpretation of the integrand, the triple integrals represent various quantities such as masses, moments or moments of inertia. A few problems based on triple integrals, including those of calculating the centers of mass of certain solid objects, will be given as exercises. Note that we have said nothing so far about the centers of mass of pieces of wire (except when they are straight line segments) or of surfaces. These will be taken up in Section 6.6.

## EXERCISES

4.1   Obtain (6) by evaluating $\iint_R f(x,y) dA$ as the other iterated integral.

4.2   Let $R$ be the rectangle $R(a,b;c,d)$ where $a > 0, c > 0$. Evaluate $\iint_R y \cos(xy) dA$ using both iterated integrals. Which is easier?

4.3 Prove that $y^2 \ln(\dfrac{1 + \sqrt{1-y^2}}{1 - \sqrt{1-y^2}}) = 2y^2 \ln(1 + \sqrt{1-y^2}) - 2y^2 \ln y$ and hence

that $y^2 \ln \left( \dfrac{1 + \sqrt{1-y^2}}{1 - \sqrt{1-y^2}} \right) \to 0$ as $y \to 0$.

4.4 The volume of the right pyramid in Fig.6.3.9(a) came out to be $\dfrac{hr^2}{3}$. Obtain the same answer by double integrals. (Note that the function $f(x,y)$ has to be defined by four different formulas, depending upon where $(x,y)$ lies in the base square.)

4.5 Prove, using double integrals, that the volume of a right tetrahedron (i.e. a tetrahedron three of whose edges meet each other at right angles) equals $\frac{1}{3}Ah$ where $A$ is the area of one of its faces and $h$ is the altitude on that face. [*Hint* : Choose a suitable co-ordinate system. The result is actually true for *any* tetrahedron but the computations are somewhat messy.]

4.6 Let $R = R(a, b; c, d)$ and assume $f(x,y)$ is of the form $g(x)h(y)$ where $g, h$ are real-valued functions of one variable. Prove that

$$\iint_R f(x,y)dA = \int_a^b g(x)dx \int_c^d h(y)dy.$$

4.7 Evaluate $\iint_R \sin(x+y)dA$ where $R = R(a,b;c,d)$ using the last exercise and also directly.

4.8 Evaluate $\iint_R f(x,y)dA$ where

(i) $f(x,y) = x^2 y$ and $R$ is the triangle with vertices at $(0,0), (1,0)$ and $(-1,2)$

(ii) $f(x,y) = x - y$ and $R$ is the region enclosed by the curves $y = x^2$ and $y = x + 2$.

4.9 Evaluate the following iterated integrals by changing the order of integration.

(i) $\displaystyle\int_0^{16} \int_{\sqrt{y}}^4 \sin(x^3)dxdy$     (ii) $\displaystyle\int_0^{\pi/2} \int_x^{\pi/2} \frac{\sin y}{y} dy dx$

4.10 Let $L$ be a straight line in the $x$-$y$ plane with equation $ax + by + c = 0$. Prove that the moment of a mass distribution (with total mass $M$) about $L$ equals $\dfrac{aM_y + bM_x + cM}{\sqrt{a^2 + b^2}}$. [*Hint* : Note that the distance of a point $(x_0, y_0)$ from $L$ is $\dfrac{ax_0 + by_0 + c}{\sqrt{a^2 + b^2}}$, which is positive on one side of $L$ and negative on the other.]

4.11 Prove that the moment of a planar mass distribution about a line in its plane is zero if and only if the line passes through its center of mass.

4.12 A half-disc of uniform mass density hangs freely on a string attached to one of the ends of its diameter. Find the angle between the string and the diameter.

4.13 Let $ABC$ be a triangle and $P$ be a point in it whose barycentric co-ordinates are $\alpha, \beta, \gamma$ (see Exercise (1.20)). Prove that $P$ is the center of mass of a system of three masses $\alpha, \beta, \gamma$ placed at $A, B, C$ respectively. (This justifies the name as *barys* in Greek means 'heavy'.)

4.14 Suppose a planar mass distribution of total mass $M$ is split into $k$ distributions of masses $M_1, \ldots, M_k$ with centres of mass located at $(\bar{x}_1, \bar{y}_1), \ldots, (\bar{x}_k, \bar{y}_k)$ respectively. Prove that the center of mass $(\bar{x}, \bar{y})$ of the entire distribution is given by

$$\bar{x} = \frac{1}{M}\left(\sum_{i=1}^{k} M_i\bar{x}_i\right), \bar{y} = \frac{1}{M}\left(\sum_{i=1}^{k} M_i\bar{y}_i\right).$$

4.15 Prove that the centroid of a triangular lamina in fact lies at its centroid, i.e. the intersection of its medians. (This fact coupled with the last exercise gives a method for finding the centroid of a quadrilateral, or more generally of any polygon.)

4.16 The base of an isosceles triangle lies along the diameter of a half disc of radius $r$, both made of the same uniform material. If the half-disc and the triangle lie in the same plane but on opposite sides of the diameter and the center of mass of the system lies on the diameter, determine the height of the triangle.

4.17 In Problem (4.1), find all points of average rainfall. Why are they not located halfway between the coast and the center of the island?

4.18 In Problem (4.3), why is the center of mass in case (ii) closer to the diameter than that in case (i)?

4.19 In Problem (4.3), find the moments of inertia of the half disc about each of the two co-ordinate axes.

4.20 Prove that if $S$ is the solid of revolution obtained by revolving a plane region $R$ around a line $L$, not passing through its interior, then $V = 2\pi r A$ where $V = $ volume of $S$, $A = $ area of $R$ and $r$ is the perpendicular distance of the center of mass of $R$ from $L$. In other words, the volume is the same as if the whole area $A$ were concentrated at the center of mass. [*Hint* : Use Exercise (4.10). For a direct proof, obtain $V$ by the shell method. This result is called the **first theorem of Pappus**.].

4.21 Use the last result to give an alternate solution to part (i) of Problem (4.3).

4.22 Let $S = \{(x, y, z) : x^2 + y^2 \leq 1, 0 \leq z \leq 1 - x^2 - y^2\}$. Sketch $S$. Express the triple integral $\iiint_S xy + z\, dV$ in the form of an iterated integral in 6 different ways. Evaluate.

4.23 For a solid $S$ in $\mathbb{R}^3$, its **moment** about the $y$-$z$ plane, denoted by $M_{yz}$, is defined as the triple integral $\iint_S x\rho(x, y, z)dV$ where $\rho(x, y, z)$ is the density at $(x, y, z)$. The moments $M_{yz}$ and $M_{zx}$ are defined similarly. The **center of mass** is the point $(\bar{x}, \bar{y}, \bar{z})$ where $\bar{x} = \dfrac{M_{yz}}{M}$ etc. $M$ being the mass of $S$. Find the centers of mass of a hemisphere, a right circular cone and a right circular cylinder assuming (i) that the mass density is uniform (ii) that the mass density is proportional to the distance from the base.

4.24 For a bounded solid $S$ and a line $L$ in space, the **moment of inertia** (or the **second moment of mass**) of $S$ about $L$, denoted by $I_L$, is defined as $\iiint_S [r(x, y, z)]^2 \rho(x, y, z)dV$ where $r(x, y, z)$ is the perpendicular distance of a point $(x, y, z)$ from $L$ and $\rho(x, y, z))$ is, as usual, the density at $(x, y, z)$. The **radius of gyration** of $S$ about $L$ is defined as $\sqrt{\dfrac{I_L}{M}}$, where $M$ is the mass of $S$. In other words, it is the root of the average (w.r.t. mass) of the distance square from $L$. Find the radius of gyration of a uniform rod of length $h$ and cross-sectional radius $r$ about (i) its own axis (ii) a line perpendicular to the rod, at the center of the rod (iii) a line perpendicular to the rod at one end of it.

4.25 Let $S, L$ be as in the last exercise and let $L'$ be the line parallel to $L$ and passing through the center of mass of $S$. Prove that $I_L = I_{L'} + Mh^2$ where $h$ is the distance between $L$ an $L'$. [*Hint* : Without loss of generality take the center of mass to be the origin and $L'$ as the $z$-axis. This result, called the **theorem of parallel axes** reduces the determination of radius of gyration about an arbitrary axis to the case where the axis passes through the center of mass.]

## 6.5.    Integrals In Determination of Probabilities

The uncertainties of life (the only certain thing about which is death!) gen-rerate, even in a layman, a lot of interest in probability (and, in many cases, in astrology too!). Thus a farmer wants to know how likely it is to rain, a gambler (and a warrior) wants to assess his chances of winning and insurance companies (or rather, the brains behind them, called actuaries) try to estimate the proba-bility that an insured will die in a given span of time. In Section 6.1 we showed how the concept of an average is intimately related to probability. As integrals are sometimes needed for the former, it should come as no surprise that they arise in determining the latter.

The probability which is studied in schools (and junior colleges) comes mostly under what is called discrete probability theory. As a typical problem, suppose we have an urn (a word rarely encountered outside probability problems !) containing, say, 5 white and 3 black halls. We draw two balls at random without replacement. What is the probability that both the balls are white? To answer this, we first observe that there are $\binom{8}{2} = 28$ ways of drawing 2 balls from a collection of 8 balls. Out of these, there are $\binom{5}{2} = 10$ ways in which both the balls are white. So the desired probability is $10/28$, i.e. $5/14$.

The formula which underlies such calculations is often expressed as

$$\text{desired probability} \; = \; \frac{\text{number of favourable cases}}{\text{total number of cases}}. \tag{1}$$

This simple-minded formula is often given a somewhat formal expression as follows. We let $S$ be the set of all possible cases. In the problem just considered $S$ is the set of all selections of 2 balls from the given 8 balls, or equivalently, $S$ is the set of all 2-element subsets of the given set of balls. The set $S$ is sometimes given a rather fancy name, the **sample space**. We then let $A$ be the subset of $S$ consisting of the favourable cases (in the present problem, the set of those selections where both the balls selected are white). The subset $A$ is called an **event**[4] in the sample space $S$, and the probability of its occurrence is denoted by $P(A)$. For any (finite) set $X$, we denote the number of elements in it (also called the **cardinality** of $X$), by $|X|$. (The symbols $n(X)$ or $\#(X)$ are also used

---

[4]In day-to-day life, an event is an occurrence, a happening of some kind. To think of it as a set can be confusing to a beginner. Here, once again, we appeal to the definition trick in Section 1.8. In probability problems, an 'event' (in layman's sense) is expressed by a statement about a variable taking values in the sample space. In the illustration given here, the variable is a pair of balls and the 'event' is that they are of the same colour. This is equivalent to considering the set, say $A$, of all pairs in which both the balls are of the same colour and saying that the given pair belongs to this set. So we might as well identify the event with the set corresponding to it and it is in this sense that we call a subset of the sample space as an event. When this is clearly understood, no harm arises in using the word interchangeably in both the senses, the layman's as well as the technical. For example, when the sample space is a part of the real line, then for $a, b, c \in \mathbb{R}$, we can write $P(a \le x \le b)$ for $P([a,b])$ and $P(x = c)$ for $P(\{c\})$.

sometimes). With these new notations and terminology, (1) becomes

$$P(A) = \frac{\text{cardinality of } A}{\text{cardinality of } S} = \frac{|A|}{|S|} \tag{2}$$

and can be illustrated with a Venn diagram (Fig.6.5.1(a)).

Although such reformulations and illustrations are hardly profound, they serve an important pedagogical purpose of visualising probability in terms of the size of a set. Certain elementary laws of probability become crystal clear when expressed in terms of sets and Venn diagrams. For example, Fig.6.5.1(b) and (c) illustrate, respectively, the law for probability that at least one of the two events occurs and the law for conditional probability.

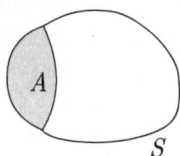

$$P(A) = \frac{|A|}{|S|}$$
(a) probability of
an event

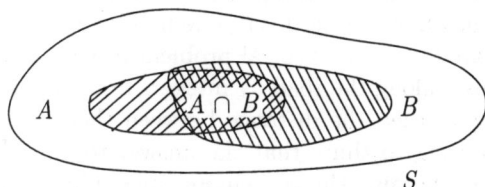

$$P(A \cup B) = P(A) + P(B) - P(A \cap B)$$
(b) probability of disjunction
of two events

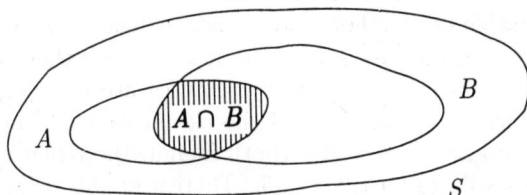

$$P(A|B) = \frac{|A \cap B|}{|B|} = \frac{P(A \cap B)}{P(B)}$$
(c) conditional probability

Figure 6.5.1 : Venn Diagrams for Probability

When the sample space $S$ is infinite, the formulas (1) and (2) become meaningless. As remarked in Section 6.1 (see the comments preceding Theorem (1.4)), we need the concept of a measure to define probability. For, in that case we merely need to replace the word 'cardinality' in (2) by 'measure' and let

$$P(A) = \text{probability of event } A = \frac{\text{measure of } A}{\text{measure of } S}. \tag{3}$$

The crucial question, of course, is how to define a measure. A systematic study of measure theory is well beyond our scope. We remark, however, that in probability, the most crucially needed property of a measure is its additivity.

As we have seen in the past three sections, arc lengths, areas and volumes are additive in nature. So sometimes, they can be used as measures. By way of illustration, we do here one problem where the desired probability is found by identifying the sample space $S$ as a subset of the plane $\mathbb{R}^2$ and then using the area of a plane figure as the measure. To highlight the contrast as well as the interplay between the discrete and the continuous probabilities, we actually do two problems of a similar spirit, one the **discrete house problem** and the other the **continuous house problem.**

Let us suppose there is a straight road, one kilometer long, on which there is a house at every one-tenth of a kilometer. There would thus be eleven houses on the road (including the houses at the two ends of it). Suppose we pick two (distinct) houses at random. What is the probability that the distance between them will not exceed, say, one half kilometer? Because the houses are discretely located, this is a typical problem in discrete mathematics. Let us number the houses along the road by integers from 0 to 10. If $x$ and $y$ are two houses (with $y \geq x$) then the distance between them is simply $\frac{1}{10}(y - x)$ kilometers. It is tempting to think that the answer to our problem is $1/2$. But this is clearly wrong because there are more pairs of houses that are closely located than those that are remotely located. It is easy to see that there are in all $\frac{1}{2}(11.10) = 55$ pairs of distinct houses. This is, therefore, the total number of cases. As with all probability problems, we have to proceed on the assumption that each of these 55 cases is equally likely. (This is, in fact, the interpretation given to the phrase 'at random'.) The next step in our solution is to count the so-called 'number of favourable cases'. For convenience, whenever we consider a pair of houses, say, $(x, y)$ let us always suppose that $x < y$. We then want the number of such pairs with $x < y$ (since the two houses are distinct) and for which $\frac{1}{10}(y - x) \leq \frac{1}{2}$. Here $x$ can take any of the eleven values from 0 to 10. When $x$ is 0, $y$ has to be between 1 and 5 (both included). When $x$ is 1, $y$ can vary from 2 to 6. This will go on till $x$ is 5. Thereafter the variation of $y$ will be restricted in its upper bound. For example when $x$ is 7, $y$ has to be either 8,9 or 10 (see Figure 6.5.2). Keeping track of the number of possible values $y$ can take for various values of $x$ and adding, we get the number of favourable cases as $5 + 5 + 5 + 5 + 5 + 5 + 4 + 3 + 2 + 1$, that is, 40. Thus the answer to our problem, namely the probability that the distance between two randomly selected distinct houses be less than or equal to $1/2$ kilometer, is $40/55$ or $8/11$.

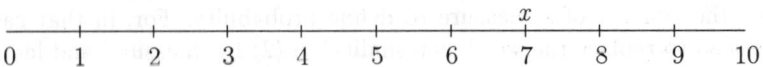

$$x$$

| | | | | | | | | | | |
|0|1|2|3|4|5|6|7|8|9|10|

Figure 6.5.2 : The Discrete House Problem

Let us now try the continuous version of this problem. Let us suppose that our one kilometer long road is literally packed with houses in the extreme; that is, there is a house at every point of it. (This may sound like an unrealistic problem, but if the road happens to be in a city like Mumbai, it is awfully close to reality !) We now ask the same question, namely, if two distinct houses are picked at random, what is the probability that the distance between them does not exceed 1/2 kilometer ? The method of solution to the earlier problem breaks down completely, because the total number of cases and the number of favourable cases are both infinite now and so the answer would be $\dfrac{\infty}{\infty}$ which is meaningless. How do we tackle the problem then ?

There are two methods for doing this. One is to regard the continuous version as a limiting case of the discrete version. Let us, therefore, revisit the discrete house problem, assuming this time that there is a house at every $\dfrac{1}{n}^{\text{th}}$ kilometer where $n$ is a positive integer. (In the problem that we just solved $n$ was 10.) Let $p_n$ be the probability that the distance between two distinct, randomly selected houses is at most 1/2. Now as $n$ tends to infinity the distance between consecutive houses tends to zero, the number of houses tends to infinity and the problem becomes more and more like the problem we are trying to solve. It is therefore reasonable to expect that if we solve the problem for each $n$, that is, if we compute $p_n$ for every $n$ and then take the limit of the sequence $\{p_n\}$ as $n$ tends to infinity, then this limit would be the answer to our problem. Even if we are not able to actually evaluate this limit (a fairly common situation with sequences), still, for sufficiently large $n, p_n$ would at least give us an approximate answer.

If we take this approach, then in order to compute $p_n$ we have to make two cases depending upon whether $n$ is even or odd. (It will be a good exercise for the reader to pin-point the reason for this.) If $n$ is even, say $n = 2k$ where $k$ is a positive integer, then it is easy to show, by reasoning similar to above, that $p_n = \dfrac{3k + 1}{4k + 2}$; while for $n$ odd, say, $n = 2k + 1, p_n$ comes out to be $\dfrac{3k}{4k + 2}$. As $k \to \infty$ both these expressions tend to a common limit, namely $3/4$. Thus $p_n$ converges to $3/4$ and this is the required probability.

There is also another method to solve the continuous house problem which does not involve its approximation by the discrete version. We begin by taking a new look at the solution given above for our original problem, with eleven houses on the road. In Figure 6.5.2, we pictured them with dots on a straight line. Let us now picture the ordered pairs of houses by dots in a square as in Figure 6.5.3. We need not explain in detail how this is done, because the idea is precisely that of cartesian coordinates. For example, the point $(\frac{3}{10}, \frac{8}{10})$ (the circled point in the figure) represents the pair consisting of the third and the eigth house. Because of our convention that the house with a smaller number will be listed first, we have drawn only the upper triangular half of the full square. There are in all 66 dots in this triangle. Since we want pairs of distinct houses, we ignore the dots on the diagonal. This leaves 55 dots, the same as the total number of cases. The favourable cases correspond to the dots on and below the line $y = x + \frac{1}{2}$. They are shown by enclosing them with a curve $C$.

Their count comes out to be 40, exactly as before.

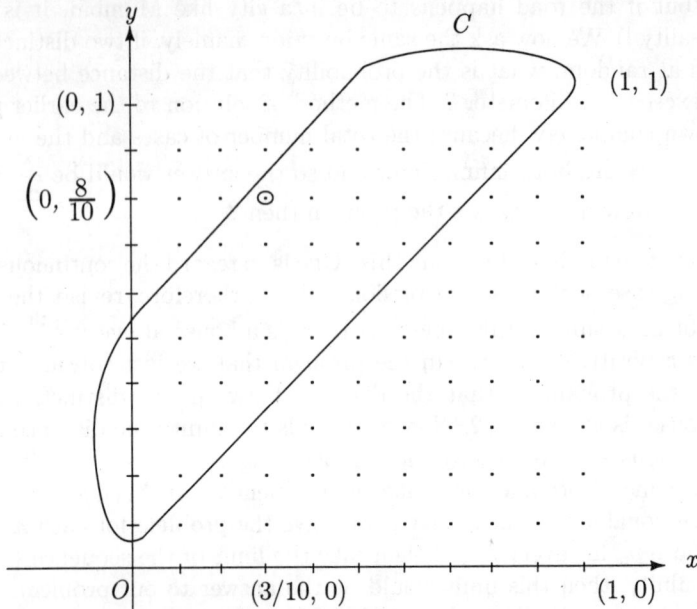

Figure 6.5.3 : Re-interpretation of the Solution to the Discrete House Problem

So far we have not done anything new. We have merely given a geometric interpretation to our earlier solution. But in this new formulation, the solution can be easily adapted for the continuous house problem. Once again, we picture pairs of houses by points in a square and consider only the upper triangular half of it as in Figure 6.5.4. This corresponds to the set of all possible cases. The set of favourable cases corresponds to the trapezoidal region between the parallel lines $y = x$ and $y = x + \frac{1}{2}$. Both the sets contain infinitely many points. So we cannot compare them on that ground. But there is another way to compare them. We simply take their areas ! The area of the upper triangle is 1/2 square kilometers while that of the trapezeum is 3/8 square kilometers. By taking the ratio, we get 3/4 as the required probability. This is of course the same as the answer obtained by the first method. Note incidentally, that in the continuous version it does not matter whether the two houses picked are distinct or not. Points corresponding to pairs of identical houses lie on the line $y = x$. The line itself has no area and so the area of the trapezeum is unaffected whether points of the boundary are included in it or not.

In this problem, we were dealing with unordered pairs of houses. That is, we treated $(x, y)$ the same as $(y, x)$ (and hence could assume without loss of generality that $x \le y$). If we deal with ordered pairs, then the sample space $S$ would be the entire unit square with area 1 (instead of just the triangle above the diagonal) and the 'favourable region' would be the portion of $S$ lying between the two parallel lines $y = x + \frac{1}{2}$ and $y = x - \frac{1}{2}$. Its area would be twice that of the favourable region shown in Fig.6.5.4. So the probability, being the ratio of the areas, would remain the same, viz. $\frac{3}{4}$.

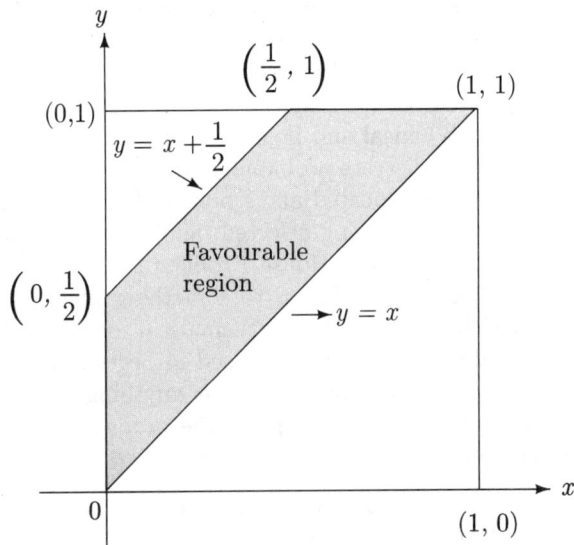

Figure 6.5.4 : Solution to the Continuous House Problem

The crucial point in the second solution to the Continuous House Problem was that an (ordered) pair of real numbers corresponds to a unique point of the cartesian plane $\mathbb{R}^2$. Hence any event concerning such pairs corresponds to a subset of $\mathbb{R}^2$. (In the present problem the event was that $|x - y| \leq \frac{1}{2}$ and the corresponding subset of $\mathbb{R}^2$ was the portion of $S$ lying between the lines $y = x + \frac{1}{2}$ and $y = x - \frac{1}{2}$.) If instead of two, the problem involves three points, say $x, y, z$ in $\mathbb{R}$ then the event would be a subset of $\mathbb{R}^3$ and to find its probability we would have to consider its volume (which, like area, is a particular measure). In the Continuous House Problem, the area of the favourable region, being a trapezeum, could be calculated easily without integrals. But in more complicated problems, the areas and volumes have to be calculated by first realising them as certain multiple integrals and then converting them into iterated integrals as elaborated in the last section. Incidentally, these applications also show why multiple integrals in euclidean spaces of dimensions higher than 3 also have real-life applications. The physical applications such as mass, moments and heat may give the impression that, since the world we live in is three dimensional (an assumption not quite tenable while studying things like relativity), higher order multiple integrals have little practical relevance. But this is not so.

Actually, integrals are needed in probability theory not only while dealing with more than one continuous variables (such as $x$ and $y$ in the Continuous House Problem) but even while dealing with a single continuous variable. In the Continuous House Problem, we assumed that there is a house at every point of the road. It was implicit in the problem that the houses were uniformly distributed over the entire road. In reality, this, of course, need not be so. Even on a continuously populated road, some parts may be more densely populated than others. Still, if we take any particular house (i.e. a point on the road), then the probability of its being selected is the same viz. 0, regardless of whether it lies in a densely populated part or a sparsely populated part of the road.

This may appear paradoxical and is often confusing to a beginner, because it is in sharp contrast with discrete probability. In the Discrete House Problem, for example, the houses were located at 11 points of the road, and each had a certain positive probability of being selected, and these probabilities together added up to 1. In the Continuous House Problem each house has probability 0 of selection and still they all add up to something non-zero, viz. 1. The integrals provide us with just the right mathematical tool needed to handle this apparently paradoxical situation. As discussed in Section 5.3, integration is a global concept. A single point does not count anything. But an interval (of positive length) does. So, if we can express the probability that a variable $x$ takes values between $a$ and $b$, denoted by $P(a \leq x \leq b)$, as the integral of some function, say, $\int_a^b p(x)dx$, then we indeed have $P(x = c) = 0$ for all $c \in [a, b]$ and still $P(a \leq x \leq b) > 0$.

A function $p(x)$ which does this trick is called a probability density function of the variable $x$. The variable $x$ is called a **random variable** or a **stochastic variable**. Note that $p(c)$ does *not* give the probability that $x$ equals $c$. In fact, as just mentioned, for a continuous variable, this probability is always 0. The situation is analogous to a (linear) mass density function, say, $\rho(x)$ of a piece of wire. The mass of each individual point is 0. $\rho(c)$ does not give the mass of $c$, but the mass per unit length at $c$. To be more precise, $\rho(c)$ is the limit of the average mass density over an interval containing $c$, just as the instantaneous speed is the limit of the average speed. In symbols

$$\rho(c) = \lim_{h \to 0} \frac{\text{mass of the piece of wire from } c \text{ to } c + h}{h} \tag{4}$$

Similarly, the **probability density** of a (real) variable $x$ at a point $c$, denoted by $p(c)$, is defined as

$$p(c) = \lim_{h \to 0} \frac{P(c \leq x \leq c + h)}{h} \tag{5}$$

where we allow $h$ to be negative as well (in which case it would be better to write $P(c + h \leq x \leq c)$ instead of $P(c \leq x \leq c + h)$.)

Let us define a new function $H : \mathbb{R} \longrightarrow \mathbb{R}$ by $H(k) = P(x \leq k) = P((-\infty, k])$ for $k \in \mathbb{R}$. Since the interval $(-\infty, c + h]$ is the disjoint union of the intervals $(-\infty, c])$ and $(c, c + h]$, and $P$ is an additive function (being a probobility function), we have $P((c, c + h]) = P((-\infty, c + h]) - P((-\infty, c]) = H(c + h) - H(c)$. But by additivity of $P$ and the fact that $P(\{c\}) = 0$ we also have $P([c, c + h]) = P((c, c + h]) + P(\{c\}) = P((c, c + h])$. Putting these equations together, we can write the numerator of (5) as $H(c + h) - H(c)$. Then it is clear that

$$p(c) = H^{'}(c) \tag{6}$$

where $H$ is the function defined by $H(k) = P(x \leq k)$ for $k \in \mathbb{R}$. Note that $H$ is a monotonically increasing function of $k$ and so $p(c) \geq 0$ for all $c$. We assume $p$ is piecewise continuous. Then, from (6) and the Fundamental Theorem of Calculus, we get that for any $c_1, c_2$ with $c_1 < c_2$,

$$\int_{c_1}^{c_2} p(x)dx = H(c_2) - H(c_1) = P(c_1 \leq x \leq c_2). \tag{7}$$

The probability density function $p(x)$ of a variable $x$ is in general defined over the entire interval $(-\infty, \infty)$. Since $P(-\infty < x < \infty) = 1$, we must have

$$\int_{-\infty}^{\infty} p(x)dx = 1 \tag{8}$$

In many situations, the random variable $x$ is constrained to lie in some bounded interval, say $[a, b]$. In the Continuous House Problem, for example, $[a, b]$ was $[0, 1]$. In such cases, $H(k) = 0$ for all $k \leq a$ and $H(k) = 1$ for all $k \geq b$. So from (6) $p(x)$ vanishes on $(-\infty, a)$ and also on $(b, \infty)$. Also (8) can be replaced by

$$\int_{a}^{b} p(x)dx = 1 \tag{9}$$

In such a case we say that the functions $p$ and $H$ are **concentrated** on $[a, b]$.

Before proceeding further, we give a few examples of probability density functions.

**Example 1:** In the Continuous House Problem, we tacitly assumed that the houses were uniformly spread over a one kilometer long road. This amounts to saying that the probability density function is a constant and is concentrated on the unit interval. The requirement that $\int_{0}^{1} p(x)dx = 1$ sets this constant as 1. More generally, for any interval $[a, b]$, the function defined by

$$p(x) = \begin{cases} \dfrac{1}{b-a} & \text{if } a \leq x \leq b \\ 0 & \text{otherwise} \end{cases} \tag{10}$$

is the probability density function of a random variable which is concentrated and uniformly spread over the interval $[a, b]$. Thus, for example, if in the Continuous House Problem, the road were 3 km. long and housing density uniform then for any portion, say, from $c_1$ to $c_2$ kms. from one end of it, the probability that a house chosen at random from the road comes from that portion is $\int_{c_1}^{c_2} \dfrac{1}{3}dx = \dfrac{1}{3}(c_2 - c_1)$. Note that this depends only on the length $c_2 - c_1$ of that portion and not on where it is located in the road.

**Example 2:** Define $p : I\!R \longrightarrow I\!R$ by

$$p(x) = \begin{cases} 3x^2 & \text{if } 0 \le x \le 1 \\ 0 & \text{otherwise.} \end{cases} \tag{11}$$

Since $p(x) \ge 0$ for all $x$ and $\displaystyle\int_{-\infty}^{\infty} p(x)dx = \int_0^1 3x^2 dx = 1$, we see that $p(x)$ is a probability density function. It is concentrated on $[0, 1]$. But it is not uniform. For $0 \le c_1 < c_2 \le 1$, $\displaystyle\int_{c_1}^{c_2} p(x)dx = c_2^3 - c_1^3$ which depends not only on $c_2 - c_1$ but on $c_1$ and $c_2$. For example, the intervals $[0, \frac{1}{2}]$ and $[\frac{1}{2}, 1]$ are of equal lengths. But if a point is picked at random from $[0, 1]$ then it is 7 times more likely to belong to $[\frac{1}{2}, 1]$ than to $[0, \frac{1}{2}]$ because $\displaystyle\int_{1/2}^1 p(x)dx = \frac{7}{8}$ while $\displaystyle\int_0^{\frac{1}{2}} p(x)dx = \frac{1}{8}$. This is to be expected because $p(x)$ has very low values for $x$ near 0. The 'dense' part of the interval is near 1.

**Example 3:** If we define $p : I\!R \to I\!R$ by

$$p(x) = \begin{cases} \frac{3}{2}x^2 & \text{if } 0 \le x \le 1 \\ \frac{3}{2}(2 - x)^2 & \text{if } 1 \le x \le 2 \\ 0 & \text{otherwise.} \end{cases} \tag{12}$$

We get another probability density function. It is similar to the last example except that it is concentrated on $[0, 2]$ and the densest point is 1. It is symmetric about the point 1 and becomes rarer and rarer as we move away from it on either side.

**Example 4:** Define $p : I\!R \to I\!R$ by $p(x) = \dfrac{1}{\pi(1 + x^2)}$. It was shown in Section 5.10 that the improper integral $\displaystyle\int_{-\infty}^{\infty} \dfrac{dx}{1 + x^2}$ converges to $\pi$. So $p(x)$ is a probability density function. Note that it is not concentrated on any (finite) interval, as was the case in the last three examples.

**Example 5:** Fix any $a > 0$ and define

$$p(x) = \begin{cases} ae^{-ax} & \text{if } x \ge 0 \\ 0 & \text{if } x < 0. \end{cases} \tag{13}$$

Since $\displaystyle\int_a^{\infty} ae^{-ax} = \lim_{R \to \infty}(1 - e^{-aR}) = 1$, $p(x)$ is the probability density function of a random variable $x$. It is called an **exponential probability density function.**

**Example 6:** At the end of Section 6.1, the problem of finding the average length of a chord of a unit circle was considered and it was remarked that the answer would change depending upon which parameter associated with the

chord (e.g. its distance from the centre, the angle it subtends at the centre etc.)
is taken to vary uniformly. We can now express this in terms of probability
density functions. Let $x$ be the length of a chord of a unit circle. Then $x$
is a real variable which can take any value in the interval $[0, 2]$. We want to
find the probability density function, say $p(x)$, of $x$. If $x$ itself is taken to vary
uniformly on $[0, 2]$ then we are in the situation of Example 1 above and $p(x)$
is the constant function $\frac{1}{2}$. Suppose, however, that we assume that the angle,
say $\theta$, subtended by the chord at the centre of the circle varies uniformly over
$[0, \pi]$. Under this assumption, to find $p(x)$, we first find the associated function
$H$. That is, for each $c \in \mathbb{R}$, we find $P(x \le c)$ i.e. the probability that $x$
is at most $c$. Evidently, $H(c) = 0$ if $c < 0$ and $H(c) = 1$ if $c > 2$. To find
$H(c)$ for $c \in [0, 2]$, we note that $x = 2 \sin \frac{\theta}{2}$ (see Fig.6.1.1(b)). Here $\theta$ varies
uniformly over $[0, \pi]$. A simple calculation based on properties of the inverse
sine function gives, $0 \le x \le c$ if and only if $0 \le \theta \le 2 \sin^{-1}(\frac{c}{2})$. As $\theta$ is
assumed to vary uniformly over $[0, \pi]$, $P(\theta \le 2 \sin^{-1}(\frac{c}{2}))$ is simply $\frac{2}{\pi} \sin^{-1}(\frac{c}{2})$.
So $H(c) = \frac{2}{\pi} \sin^{-1}(\frac{c}{2})$ for all $c \in [0, 2]$. Using (6), we get

$$p(x) = \begin{cases} \dfrac{d}{dx}(\dfrac{2}{\pi} \sin^{-1}(\dfrac{x}{2})) = \dfrac{1}{\pi\sqrt{1 - \frac{x^2}{4}}}, & 0 \le x \le 2 \\ 0 & \text{otherwise} \end{cases} \tag{14}$$

(Strictly speaking, (14) is undefined for $x = 2$. This is because the $\sin^{-1}$ function
is not differentiable at 1. However, $p(2)$ may be defined arbitrarily, because, once
again, integration is a global process in which the value of the integrand at any
one point makes no difference.)

We leave it to the reader (Exercise (5.5)) to formulate the probability density
function of $x$, the length of a chord of a unit circle, under the assumption that
the perpendicular distance of the chord from the centre of the circle varies
uniformly.

**Example 7:** Lest it appear that probability density functions are very special,
we remark that any non-negative, piecewise continuous function, say $f(x)$, over
any interval, say $[a, b]$ can be easily converted into a probability density function
$p(x)$ concentrated over $[a, b]$. Let $I = \int_a^b f(x)dx$. Then $I \ge 0$. If $I = 0$ then it is
not hard to show (Exercise (5.9)(a)) that $f$ vanishes at all except finitely many
points of $[a, b]$. Such a function is, of course, uninteresting. If $I > 0$, we let
$p(x) = \dfrac{f(x)}{I}$ for $a \le x \le b$ and get a probability density function concentrated
on $[a, b]$.

The analogy between probability density and linear mass density extends
to other concepts as well. If $\rho(x)$ is the linear mass density at a point $x$ of
a piece of wire stretching over an interval, say $[a, b]$ then $\int_a^b \rho(x)dx$ is its total mass. If $\sigma(x)$ is some other real valued function on $[a, b]$ (e.g. specific
heat, cf. Exercise (1.12)) then the mass-average of $\sigma$ over $[a, b]$ is the ratio $\int_a^b \sigma(x)p(x)dx / \int_a^b p(x)dx$. Now suppose we have a random variable $x$ with

probability density function $p(x)$, concentrated on $[a, b]$, and let $f(x)$ be some other function on $[a, b]$. Then the **average** or the **mean** or **expected value** of $f$ on $[a, b]$ is defined as $\int_a^b f(x)p(x)dx$. (Note that it is not necessary to divide by $\int_a^b p(x)dx$ since this integral is 1 anyway.) The justification for the term 'expected' value is that if $P = (x_0, x_1, \ldots, x_n)$ is a partition of $[a, b]$ and $\xi_i \in [x_{i-1}, x_i]$ for $I = 1, \ldots, n$ then $p(\xi_i)\triangle x_i$ is (approximately) the probability that $x$ lies in $[x_{i-1}, x_i]$ and so $\sum_{i=1}^n f(\xi_i)p(\xi_i)\triangle x_i$ is approximately the expected value of the (discrete) set $\{f(\xi_i), \ldots, f(\xi_n)\}$. As usual, the exact value is the limit of this sum as the partition becomes finer and finer. The expected value of $f(x)$ is often denoted by $E(f(x))$ or simply by $E(f)$.

The average value of the variable $x$ itself is simply $\int_a^b xp(x)dx$. In analogy with a linear mass distribution, it is often called the **(first) moment** of $x$ (about 0). It is denoted by $E(x)$. Similarly the **second moment** of $x$ (about 0) is defined as $\int_a^b x^2 p(x)dx$ and denoted by $E(x^2)$. It is useful in defining certain important concepts such as the standard deviation which is a measure of fluctuation (see Exercise (5.10)). But we shall not go into it. In all these definitions, we can drop the assumption that $p(x)$ is concentrated on some interval $[a, b]$. But then the integrals would be of the form $\int_{-\infty}^{\infty}$ and one has to worry about their convergence.

As simple examples, we leave it to the reader to check that for the probability density functions given by (10), (11), (12) and (13), the average values of $x$ are, respectively, $\frac{a+b}{2}, \frac{3}{4}, 1$ and $\frac{1}{a}$. As regards Example (7), we already found in Section 6.1 that the average length of a chord of a unit circle is $\frac{4}{\pi}$ if we regard the angle it subtends at the centre as uniformly varying. The same result can also be obtained from (14) above. Indeed, $E(x) = \int_0^2 \dfrac{xdx}{\pi\sqrt{1 - \frac{x^2}{4}}}$. This is an improper integral. But the integrand has $-\dfrac{4}{\pi}(1 - \dfrac{x^2}{4})^{1/2}$ as an antiderivative on $[0, 2 - \delta]$ for every $\delta > 0$. So

$$\int_0^2 \frac{xdx}{\pi\sqrt{1 - \frac{x^2}{4}}} = \lim_{\delta \to 0+} \int_0^{2-\delta} \frac{xdx}{\pi\sqrt{1 - \frac{x^2}{4}}} = \lim_{\delta \to 0+} [\frac{4}{\pi} - \frac{4}{\pi}(1 - (\frac{2 - \delta)^2}{4})] = \frac{4}{\pi},$$

the same answer as before.

Just as the concept of mass density can be defined for planar regions, the concept of a probability density function also has an analogue for random variables ranging over the plane $\mathbb{R}^2$. Any such variable can be denoted by a joint variable $(x, y)$ where $x, y$ take real values. Let $(h, k)$ be any point in $\mathbb{R}^2$. Let

$\triangle R$ be a small bounded region containing $(h, k)$. Let $P((x,y) \in \triangle R)$ be the probability that $(x, y)$ lies in $\triangle R$. Then the **(laminar) probability density** of $(x, y)$ at $(h, k)$ denoted by $p(h, k)$ is defined as

$$p(h, k) = \lim \frac{P((x, y) \in \triangle R)}{\text{area of } \triangle R} \qquad (15)$$

the limit being taken as the size of the region $\triangle R$ tends to 0. (Recall that this means that $\triangle R$ is contained in a rectangle both of whose sides tend to 0.)

Just as the mass of any plane region $R$ is obtained by taking the double integral over $R$ of the (laminar) density function, the probability that $(x, y) \in R$ is obtained by integrating the probability density function.

$$P((x, y) \in R) = \iint_R p(x, y) dA \qquad (16)$$

Again, concepts like the average value of a function $f(x, y)$ can be defined by taking suitable double integrals. That is,

$$E(f(x, y)) = \iint_{I\!R^2} f(x, y) p(x, y) dA \qquad (17)$$

(Strictly speaking, this is an improper double integral. We have considered double integrals only over bounded regions. So to make sense out of (17), we may suppose that $p(x, y)$ is concentrated on some bounded subset of $I\!R^2$, i.e. it vanishes outside it.)

We could at this stage give examples of laminar probability density functions similar to those of linear probability density functions given above. But we omit them. Instead, we turn to an important special class of laminar probability density functions. Suppose $x$ and $y$ are two random variables taking real values. To say that they are mutually independent means, intuitively, that the values which either of them assumes do not put any restriction on the values the other can assume. In other words suppose $x$ is known to take some value, say $h$. Then for any $k_1$ and $k_2$ with $k_1 < k_2$, the probability that $y$ takes the values between $k_1$ and $k_2$, i.e. $P(k_1 \leq y \leq k_2)$ is unaffected by what $h$ is. Similarly, for any fixed $k$, and any $h_1, h_2$ with $h_1 < h_2$, the probability that $h_1 \leq x \leq h_2$, given that $y = k$ is independent of $k$. This leads to the following formal definition.

**5.1 Definition:** Two random real variables $x$ and $y$ are said to be **independent** of each other if for all $h_1, h_2, k_1, k_2$ with $h_1 < h_2$ and $k_1 < k_2$, the events $(h_1 \leq x \leq h_2)$ and $(k_1 \leq x \leq k_2)$ are mutually independent, that is,

$$P(h_1 \leq x \leq h_2 \text{ and } k_1 \leq y \leq k_2) = P(h_1 \leq x \leq h_2) P(k_1 \leq y \leq k_2). \qquad (18)$$

For example, the lengths of the two arms of a right angled triangle are mutually independent. But the length of the hypotense and that of one of the arms are not independent of each other as the former is always bigger than the other. Similarly the perimeter and the largest angle of a triangle are mutually independent. But the perimeter and the area are not, because the perimeter puts a restriction on the area (although it does not determine it uniquely). It is difficult to give non-mathematical examples, because whether two things are really independent or not can never be decided conclusively. Still, we may say that the birthday of a person (which is to be taken as a continuous variable depending upon the actual moment of birth) is independent of the height of the person, while the height and the shoe size of a person are related at least to some extent and hence are not independent of each other. And, finally, whether the type of spouse you will get is dependent or independent of the position of Mars at the time of your birth is dependent upon your belief in astrology !

Let us now see what the condition (18) of mutually independent random real variables $x$ and $y$ implies regarding the probability density function of the joint variable $(x, y)$ ranging over the plane $\mathbb{R}^2$. Let $p_1(x), p_2(y)$ and $p(x, y)$ be the probability density functions of the variables $x, y$ and $(x, y)$ respectively. We want to express $p$ in terms of $p_1$ and $p_2$. We assume $p_1$ and $p_2$ are continuous.

Let $(h, k)$ be any point in $\mathbb{R}^2$ and let $\triangle x, \triangle y$ be positive real numbers. Let $\triangle R$ be the rectangle $R(h, h + \triangle x; y, y + \triangle y)$. That is, $\triangle R = \{(x, y) : h \leq x \leq h + \triangle x, k \leq y \leq \triangle y\}$. The area of $\triangle R$ is $\triangle x \triangle y$. Now since $x, y$ are mutually independent, by (18) we have

$$
\begin{aligned}
P((x, y) \in \triangle R) &= P(h \leq x \leq h + \triangle x)P(k \leq y \leq k + \triangle y) \\
&= \int_h^{h+\triangle x} p_1(x)dx \int_k^{k+\triangle y} p_2(y)dy.
\end{aligned}
$$

and hence

$$
\frac{P((x, y) \in \triangle R)}{\text{area of } \triangle R} = (\frac{1}{\triangle x}\int_h^{h+\triangle x} p_1(x)dx)(\frac{1}{\triangle y}\int_k^{k+\triangle y} p_2(y)dy) \qquad (19)
$$

In (19) we take the limit as $\triangle x, \triangle y$ both tend to 0. Then by (15), the left hand side tends to $p(h, k)$. The first factor on the right hand side tends to $p_1(h)$, by continuity of $p_1$ and the Fundamental Theorem of Calculus. Similarly the second factor tends to $p_2(k)$. So we get,

$$
p(h, k) = p_1(h)p_2(k) \qquad (20)
$$

and hence (replacing $h$ by $x$ and $k$ by $y$), $p(x, y) == p_1(x)p_2(y)$. We record this as a theorem.

**5.2 Theorem:** Suppose $x, y$ are mutually independent random variables with probability density functions $p_1(x)$ and $p_2(y)$, concentrated over intervals $[a, b]$ and $[c, d]$ respectively. Assume $p_1, p_2$ are continuous over $[a, b]$ and $[c, d]$ respectively. Then the probability density function of the (joint) variable $(x, y)$ is given by

$$
p(x, y) = p_1(x)p_2(y) \qquad (21)
$$

**Proof:** First suppose $(x, y) \in R(a, b; c, d)$. If $x < b$ and $y < d$ then there exist positive $\triangle x, \triangle y$ such that the rectangle $R(x, x + \triangle x; y, y + \triangle y)$ is contained in the rectangle $R(a, b; c, d)$ (see Fig.6.5.5). Then (20) implies (21). If $x = b$ and/or $y = d$, then $(x, y)$ lies on the boundary of $R(a, b; c, d)$. In that case we have to let $\triangle x$ and/or $\triangle y$ be negative. But the argument for (20) goes through with minor modifications. (For example, if $\triangle x < 0$, then instead of the interval $[x, x + \triangle x]$ we take the interval $[x + \triangle x, x]$.) So (21) holds for all $(x, y) \in R(a, b; c, d)$.

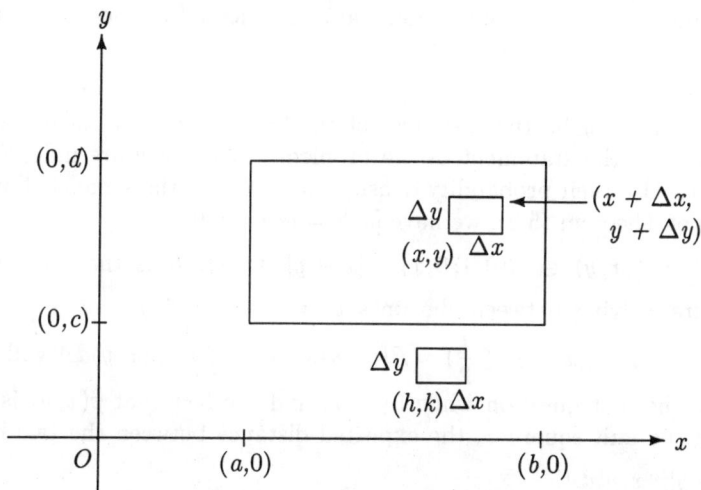

Figure 6.5.5 : Probability Density Function of Joint Variable

Now suppose $x = h, y = k$ with $(h, k) \notin R(a, b; c, d)$. Then we can find $\triangle x > 0, \triangle y > 0$ such that $R(h, h + \triangle x, k, k + \triangle y)$ is disjoint from $R(a, b; c, d)$. Now (19) holds with both the sides 0. So taking limits as $\triangle x, \triangle y$ both tend to 0, we get (20) again with both sides vanishing. Hence (21) holds in all cases. ∎

The hypothesis that $p_1, p_2$ be concentrated over $[a, b], [c, d]$ respectively is really not vital. We have put it in only because we have not considered double integrals over unbounded sets. As regards the continuity condition on $p_1(x)$ and $p_2(x)$, we can allow each of these functions to have a finite number of discontinuities. Suppose $a_1, \ldots, a_m$ and $c_1, \ldots, c_n$ are points of discontinuity of $p_1, p_2$ respectively. Then the argument above will still hold for $(x, y)$ when $x \notin \{a_1, \ldots, a_m\}$ and $y \notin \{c_1, \ldots, c_n\}$. The points where $x = a_i$ (say) all lie on a vertical line. A line has area 0. Just as a Riemann integral is unaffected by what happens at a point, a double integral is unaffected by the values of the integrand over any subset whose area is 0. Since the 'bad' points lie on a union of finitely many horizontal and vertical lines, it does not matter what value we assign to $p(x, y)$ when $x = a_i$ for some $i$ or $y = c_j$ for some $j$.

With this theorem at our disposal, we are now ready to do the Continuous House Problem under a more general (and realistic) assumption about the density of the houses along the road. We also find the mean distance between the houses picked.

**5.3 Problem:** Suppose two houses are selected at random from a 1 km. long road. What is the probability that the distance between them is at most 1/2? What is the average or the expected distance between them? Do the problem under two different assumptions, viz.

(i)  the houses are uniformly distributed over the road

(ii)  the probability density of the house at a distance $x$ from one end of the road is $2x$.

**Solution:** Let $x, y$ be the distances of the houses picked from one end (the same end as in the statement of the problem). Then $x, y$ are independent of each other. Also their probability density functions are the same. Following the notations of Theorem (5.2), we have $[a, b] = [c, d] = [0, 1]$.

Let $R = \{(x, y) \in R(0, 1; 0, 1) : |x - y| \leq \frac{1}{2}\}$. $R$ is the portion of the unit square $S$ lying between the lines $y = x + \frac{1}{2}$ and $y = x - \frac{1}{2}$ shown in Fig.6.5.6. Then $P(|x - y| \leq \frac{1}{2}) = P((x, y) \in R) = \iint_R p(x, y)dA$ will give the answer to the first question while by (17) and the fact that $p(x, y)$ is concentrated on the unit square $S$, the expected distance between the two houses is $\iint_S |x - y|p(x, y)dA$.

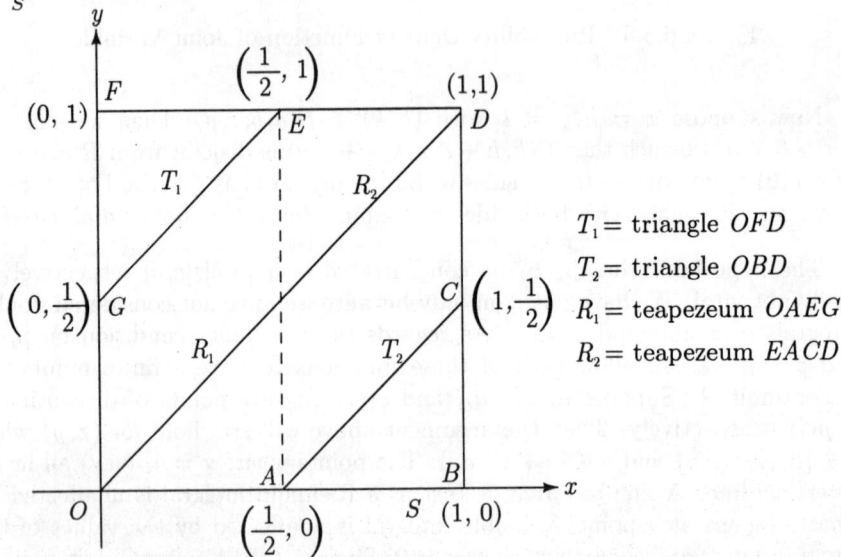

$T_1 =$ triangle $OFD$
$T_2 =$ triangle $OBD$
$R_1 =$ teapezeum $OAEG$
$R_2 =$ teapezeum $EACD$

Figure 6.5.6 : Generalised Continuous House Problem

Now, under the assumption (i), $p_1(x) \equiv p_2(y) \equiv 1$ and so by Theorem (5.2), $p(x, y) = 1$ for all $(x, y) \in S$. So $\iint_R p(x, y)dA$ is simply the area of $R$, which is $\frac{3}{4}$ as calculated earlier. This is the same answer that we had in the

Continuous House Problem and the method is basically the same. As for the average distance, we divide $S$ into two regions $T_1$ and $T_2$, $T_1$ being the upper triangle, i.e. the triangle above the diagonal $OD$, and $T_2$ the lower triangle. Note that $|x - y| = y - x$ in $T_1$ and $|x - y| = x - y$ in $T_2$. By additivity of integrals,

$$\iint_S |x - y| dA = \iint_{T_1} y - x \, dA + \iint_{T_2} x - y \, dA \tag{22}$$

To calculate $\iint_{T_1} y - x \, dA$ we write it as an iterated integral. For $0 \le h \le 1$, the line $x = h$ cuts $T_1$ in the segment from $(h, h)$ to $(h, 1)$. So,

$$
\begin{aligned}
\iint_{T_1} y - x \, dA &= \int_0^1 \int_x^1 y - x \, dy dx \\
&= \int_0^1 \frac{1}{2} - \frac{x^2}{2} - x + x^2 \, dx = \frac{x}{2} - \frac{x^2}{2} + \frac{x^3}{6} \Big|_0^1 = \frac{1}{6}.
\end{aligned}
$$

By a similar calculation, which is left as an exercise, $\iint_{T_2} x - y \, dA$ also comes out as $\frac{1}{6}$. So from (22), the average distance between the houses is $\frac{1}{3}$.

Before tackling the problem under assumption (ii), we verify that since $\int_0^1 2x dx = 1$, $p_1(x) = 2x$ is indeed a probability density function. We also have $p_2(y) = 2y$ for $0 \le y \le 1$. So by Theorem (5.2) again, $p(x, y) = 4xy$ for all $(x, y) \in S$. The probability that $|x - y| \le \frac{1}{2}$ now is $\iint_R 4xy dA$. To evaluate this double integral, it is convenient to divide $R$ into two parts $R_1$ and $R_2$ by the line $x = \frac{1}{2}$, as shown in Fig. 6.5.6. Each $R_1$ and $R_2$ is a trapezeum. By additivity of double integrals,

$$\iint_R 4xy dA = \iint_{R_1} 4xy dA + \iint_{R_2} 4xy dA \tag{23}$$

By conversion to an iterated integral,

$$
\begin{aligned}
\iint_{R_1} 4xy dA &= \int_0^{1/2} \int_0^{x+\frac{1}{2}} 4xy dy dx \\
&= \int_0^{1/2} 2x(x + \frac{1}{2})^2 dx \\
&= \int_0^{1/2} 2x^3 + 2x^2 + \frac{x}{2} dx = \frac{1}{32} + \frac{1}{12} + \frac{1}{16} = \frac{17}{96}.
\end{aligned}
$$

Similarly, the double integral $\iint_{R_2} 4xy dA$ can be converted to the iterated integral $\int_{1/2}^{1}\int_{x-\frac{1}{2}}^{1} 4xy\,dy\,dx$ which can then be evaluated to give,

$$\iint_{R_2} 4xy\,dA = \int_{1/2}^{1} 2x(1-(x-\frac{1}{2})^2)dx$$

$$= \int_{0}^{1/2} (2u+1)(1-u^2)du \quad \text{where } u = x - 1/2$$

$$= \int_{0}^{1/2} 2u+1-2u^3-u^2\,du = \frac{1}{4}+\frac{1}{2}-\frac{1}{32}-\frac{1}{24} = \frac{65}{96}.$$

Putting these values in (22), the probability that the two houses randomly selected from the road are at most 1/2 km. apart is $\frac{17}{96}+\frac{65}{96} = \frac{41}{48}$. Note tnat this is slightly higher than the corresponding probability under assumption (i). It is an instructive exercise to pin-point the reason for this. (Exercise (5.12)).

Finally, to find the average distance between the houses we have to evaluate $\iint_{S} |x-y|4xy\,dA$. As before, we divide $S$ into two triangles $T_1$ and $T_2$ and write

$$\iint_{S} |x-y|4xy\,dA = \iint_{T_1} 4xy(y-x)dA + \iint_{T_2} 4xy(x-y)dA \qquad (24)$$

This time we calculate the second integral on the R.H.S. and leave the first one as an exercise. For $0 \le h \le 1$, the line $x = h$ cuts $T_2$ in the segment from $(h, 0)$ to $(h, h)$. So

$$\iint_{T_2} 4xy(x-y)dA = \int_{0}^{1}\int_{0}^{x} 4x^2y - 4xy^2\,dy\,dx$$

$$= \int_{0}^{1} 2x^4 - \frac{4x^4}{3}\,dx = \int_{0}^{1} \frac{2}{3}x^4\,dx = \frac{2}{15}.$$

By a similar calculation, $\iint_{T_1} 4xy(y-x)dA$ is also $\frac{2}{15}$. (The same thing was true for the two integrals in the R.H.S. of (22). Actually this is not a coincidence. It is a consequence of the fact that $T_1$ and $T_2$ are reflections of each other in the line $y = x$ and the integrands also behave accordingly. Just as a Riemann integral can be transformed into another, there is a similar rule for transformation of double integrals. So the equality of the two double integrals in the right hand sides of (22) or of (24) can be proved by showing that they can be transformed into each other. But we shall not go into it, because our interest is not in double integrals *per se*.)

So from (24), we see that the expected distance between the two houses picked at random under assumption (ii) is $\frac{4}{15}$, which is less than the corresponding answer (viz. $\frac{1}{3}$) under the assumption (i). Again, we leave it as an exercise to fathom the reason for this. ∎

The reader will notice that the Continuous Home Problem was done in two ways : directly and also by approximating it with a discrete problem. It is natural to inquire whether the problem just solved could have been tackled by the second approach. This is indeed possible. The procedure is as follows. Suppose $p(x)$ is the probability density function of a random variable $x$, concentrated over an interval $[a, b]$. Divide the interval $[a, b]$ into $n$ equal parts by points $x_i, i = 0, \ldots, n$, where $x_i = a + \frac{i}{n}(b - a)$. We assume $p$ is continuous. Then for every $i$, there exists, by Theorem (1.2), some $\xi_i \in [x_{i-1}, x_i]$ such that

$$P(x_{i-1} \leq x \leq x_i) = \int_{x_{i-1}}^{x_i} p(x)dx = p(\xi_i)(x_i - x_{i-1}) = \frac{b - a}{n}p(\xi_i) \qquad (25)$$

We can now replace $x$ with a discrete variable $\xi$ which assumes only finitely many values $\xi_1, \ldots, \xi_n$ with probabilities $\frac{b - a}{n}p(\xi_1), \ldots, \frac{b - a}{n}p(\xi_n)$ respectively. (That these probabilities indeed add up to 1 is a consequence of (25).) We then solve this discrete problem. The answer will, in general, depend on $n$. By taking its limit as $n \to \infty$, we get the answer to the continuous problem.

The trouble with this procedure is that although Theorem (1.2) guarantees the existence of a $\xi_i$ which satisfies (25), actually finding it is not always an easy job (see the comments after Theorem (1.2)). A considerably simpler procedure is to take $\xi_i$ to be just any point of $[x_{i-1}, x_i]$. For example, let us take $\xi_i = x_i$ for $i = 1, \ldots, n$. We consider a discrete variable $x$ which assumes only the values $x_1, \ldots, x_n$. Note that we cannot now let $\frac{b - a}{n}p(x_i)$ be the probability that $x$ assumes the value $x_i$, because $\sum_{i=1}^{n} \frac{b - a}{n}p(x_i)$ need not be 1. We call this number $W_n$. For $n$ sufficiently large, $W_n$ will be positive (see Exercise (5.9(b))). So we may assign the probability $\frac{b - a}{nW_n}p(x_i)$ to $x_i$. Now we indeed have a discrete random variable. The answer to the discretised version of the problem will give the answer to the continuous problem by taking limit as $n \to \infty$.

To illustrate, consider the second part of Problem (5.3) under assumption (ii). Here $[a, b] = [0, 1]$ and $p(x) = 2x$. Dividing $[0, 1]$ into $n$ equal parts gives $x_i = \frac{i}{n}$ and so $p(x_i) = \frac{2i}{n}$. By an easy calculation, $W_n = \sum_{i=1}^{n} \frac{b - a}{n}p(x_i)$ comes out as $\frac{1}{n^2}\sum_{i=1}^{n} 2i = \frac{n(n + 1)}{n^2} = \frac{n + 1}{n}$. So we consider a discrete variable which assumes the value $\frac{i}{n}$ with probability $\frac{p(x_i)}{nW_n}$, i.e. $\frac{2i}{n(n + 1)}$. If $y$ is another such variable and $x, y$ are mutually independent, then

$$E(|x - y|) = \frac{4}{n^3(n + 1)}\sum_{j=1}^{n}\sum_{i=1}^{n} ij|i - j| \qquad (26)$$

This double sum can be evaluated by splitting it into two parts, i.e. by taking terms for which $i > j$ and those for which $i \leq j$. The evaluation of the sum requires formulas for summations like $\displaystyle\sum_{j=1}^{n} j^4$ and hence is relegated to the exercises. Note that the steps involved are strikingly similar to those in the continuous version. In fact, examples like this suggest a symbolic equation,

discrete mathematics + limiting process = continuous mathematics.

The exercises will provide more drill in the evaluation of probabilities using integrals. We also mention that there is a theory of integrals called **Riemann-Stielje integrals** which are more general than the Riemann integrals and can handle both continuous and discrete random variables. But they are beyond our scope.

## EXERCISES

5.1 Three random real variables $x, y, z$ are called **collectively independent** if for all $a_1, a_2, b_1, b_2, c_1, c_2$ (some of which could possibly be $\pm\infty$) we have

$$P(a_1 \leq x \leq a_2, b_1 \leq y \leq b_2, c_1 \leq z \leq c_2)$$
$$= P(a_1 \leq x \leq a_2)P(b_1 \leq y \leq b_2)P(c_1 \leq z \leq c_2). \quad (27)$$

The variables $x, y, z$ are called **pairwise independent** if every two of them are mutually independent. Prove that:

(i) collective independence implies pairwise independence.

(ii) the lengths of the three sides of a rectangular box are collectively independent.

(iii) the lengths of the three sides of a triangle are pairwise independent but not collectively independent. (Thus, the two concepts are not quite the same. When used without qualification, the term 'independent' means 'collectively independent'. The independence of more than three variables is defined by an obvious extension of (27).)

5.2 If three points are picked at random from $[0,1]$, find the probability that no two of them are at a distance $1/2$ or more apart. Do the problem under the assumption that the probability density function is (i) a constant (ii) given by $p(x) = 2x$ for $x \in [0, 1]$. [*Hint* : Begin by dividing the set $\{(x, y, z) : 0 \leq x \leq 1, 0 \leq y \leq 1, 0 \leq z \leq 1, |x - y| \leq \frac{1}{2}, |y - z| \leq \frac{1}{2}, |x - z| \leq \frac{1}{2}\}$ into 6 parts depending upon whether $x \leq y \leq z$ or $x \leq z \leq y$ etc.]

5.3 For problems involving more than three (collectively) independent variables, we need the concept of the '$n$-dimensional volume' for $n > 3$, sometimes called **hypervolume**. This can be defined as well as evaluated by the method of slicing using induction. Let $V_n(S)$ denote the $n$-dimensional volume of a (bounded) subset $S$ of $\mathbb{R}^n$. As $S$ is bounded, there exist $a, b$ such that for all $(x_1, \ldots, x_n) \in S, a \leq x_n \leq b$. For each $c \in [a, b]$, the hyperplane $x_n = c$ cuts $S$ into a subset $S_c$, say. This $S_c$ is a bounded subset of the $(n-1)$-dimensional hyperplane $x_n = c$ and so its $(n-1)$ dimensional volume, $V_{n-1}(S_c)$ is defined. The $n$-dimensional volume of $S$ is defined by integrating this. That is,

$$V_n(S) = \int_a^b V_{n-1}(S_{x_n}) dx_n$$

Find the $n$-dimensional volume of the solid $n$-dimensional ball of radius 1. [*Hint* : Use similarity of figures and the reduction formula in Exercise (5.6.13).]

5.4 Suppose we go on picking real numbers $x_1, x_2, x_3, \ldots, x_n, \ldots$ randomly from the interval $[0, 1]$, with uniform probability density. A **record** is said to occur at the $n^{\text{th}}$ stage if $x_n > x_i$ for all $i = 1, \ldots, n-1$. Prove that the probability that a record will occur at the $n^{\text{th}}$ stage is $\frac{1}{n}$ for every $n = 1, 2, \ldots$. Deduce that during the first $k$ selections, the expected number of records is the $n^{\text{th}}$ harmonic number $H_n$. (See Section 5.8.)

5.5 Formulate the probability density function of the length of a chord of a unit circle under the assumption that its perpendicular distance from the centre of the circle varies uniformly. Then verify that the average chord length under this assumption is $\frac{\pi}{2}$.

5.6 Calculate the first and the second moments for the probability density functions given by (10), (11), (12) and (13).

5.7 Give examples of continuous probability density functions for which

(i) $E(x)$ is undefined (i.e. not finite)

(ii) $E(x)$ is defined but $E(x^2)$ is not. (Obviously such functions cannot be concentrated over bounded intervals.)

5.8 Prove that if the probability density function $p(x)$ is symmetric about a point $a$, i.e. $p(a+h) = p(a-h)$ for all $h > 0$, then $E(x)$, if it exists, is $a$. Does this apply to the function $p(x) = \dfrac{1}{\pi(1+x^2)}$, $-\infty < x < \infty$?

5.9 (a) Suppose $f$ is continuous and non-negative on $[a, b]$ and $\displaystyle\int_a^b f(x)dx = 0$. Prove that $f \equiv 0$ on $[a, b]$. [*Hint* : Use Exercise (4.1.1).] What can be said if $f$ is given to be only piecewise continuous?

(b) Suppose $f$ is continuous and non-negative on $[a, b]$ and $\displaystyle\int_a^b f(x)dx = 1$. Prove that there exists a positive integer $m$ such that for all $n \geq m$,

$$\sum_{i=1}^n f(x_i) > 0 \text{ where } x_i = a + \frac{i(b-a)}{n}$$

(c) Suppose $g, h$ are continuous, real-valued functions on an interval $[a, b]$ and $g(x) \leq h(x)$ for all $x \in [a, b]$ with strict inequality holding at at least one point of $[a, b]$. Prove that $\int_a^b g(x)dx < \int_a^b h(x)dx$.

5.10 Let $\mu = E(x)$ be finite where $x$ is a continuous random variable. Then $E((x-\mu)^2)$ is called the **variance** of $x$ denoted by $\mathrm{Var}(x)$ and its square root is called the **standard deviation** of $x$. Prove that $\mathrm{Var}(x) = E(x^2) - \mu^2$. Prove that if $p(x)$ is continuous then $\mathrm{Var}(x) = 0$ if and only if $p(x)$ is constant.

5.11 Complete the calculations of integrals in the solution to Problem (5.3).

5.12 Prove that in Problem (5.3), under assumption (ii), a point picked at random from $[0,1]$ is three times more likely to lie in $[\frac{1}{2}, 1]$ than in $[0, \frac{1}{2}]$. How does this explain the difference in the answers to the problem under the two assumptions?

5.13 Find approximate answers to Problem (5.3) by dividing $[0,1]$ into 4 equal parts and replacing the continuous variables with discrete ones taking the values $\frac{1}{4}, \frac{1}{2}, \frac{3}{4}$ and 1.

5.14 For every positive integer $n$, prove that

$$\sum_{j=1}^{n} j^4 = \frac{1}{5}\{(n+1)^5 - \frac{5}{2}(n+1)^4 + \frac{5}{3}(n+1)^3 - \frac{1}{6}(n+1)\}.$$

[*Hint* : Induction on $n$ is, of course, possible. An alternate approach is to write $j^4$ as $24\binom{j}{4} + 6j^3 - 11j^2 + 6j$, use the well-known formulas for $\sum_{j=1}^{n} j^r$ for $r = 1, 2$ and 3 and note that $\sum_{j=1}^{n} \binom{j}{4} = \binom{n+1}{5}$ which can be proved combinatorially.]

5.15 Using the last exercise, find the limit of the right hand side of (26) as $n \to \infty$. [*Hint* : Only the coefficient of $n^5$ matters for finding the limit.]

5.16 A commuter leaves home at a time which varies uniformly from 8:00 to 8:10 a.m. The time it takes him to reach the station varies uniformly (and independently of his starting time from home) between 15 to 20 minutes. From the station there are two trains, one departing at 8:20 a.m. and the other at 8:27 a.m. Find the probability that the commuter misses the first train but makes the second. When this happens, how long, on the average, does he wait (for the second train)?

## 6.6    Arc Length - a Natural but an Inconvenient Parameter

In Section 4.12 it was emphasised that there is a difference between a lay-man's understanding of a curve and its mathematical conception. Intuitively, we think of a curve as a thin, one-dimensional object. But in mathematics, what matters more is how this object is traced by a moving particle. Therefore, mathematically, a curve is a certain function and should not be confused with the range of that function (which is a set of points in the plane, or in space etc.) Therefore it is mathematically meaningless to talk of 'a parametrisation of a curve', because parametrisation is an integral part of the very identity of a curve. A curve cannot even exist without a parametrisation. Similarly it is silly to say 'different parametrisations of the same curve', because the moment you change the parametrisation, you change the curve.

Still, even in mathematics we sometimes talk of a curve as if it were just a point-set. This ambivalence can be confusing to a beginner. For example we consider the area of the surface of revolution obtained by revolving a curve around a straight line. To evaluate this area we may use a particular parametri-sation of that curve. But we obviously expect the answer to be independent of which parametrisation is used. An even more basic example is the concept of the length of a curve. We would naturally expect it to be an invariant as-sociated intrinsically with the geometry of the curve and not with a particular parametrisation, which is somewhat extrinsic to the curve. The same thing holds for other geometric attributes of a curve, such as its curvature (which is a quantitative measure of the degree of the bend in it) and torsion (which measures how twisted the curve is).

This really poses a dilemma. The situation is somewhat (but not quite) like this. The name of a person is, strictly speaking, an extraneous thing attached to that person (often without taking his/her will into account!). It is not an intrinsic part of the personality of that person. Still, in any civilized society, we need a name to identify a person and often the same person is known by different names in different circles resulting in a sometimes comic confusion. It would be ideal if each person is given some 'official' name (hopefully reflecting the vital features of his/her personality) which is to be used in legal dealings. Moreover, a dictionary is to be maintained which will give all the popular names of a person in addition to his official name. This way we can have the convenience of the names without their domination over the intrinsic identity of that person.

We shall adopt a similar strategy for curves. We shall continue to regard a curve as a certain function rather than as a mere geometric object. But we shall define a certain equivalence relation for curves so that curves which are equivalent to each other have essentially the same underlying geometric object. Moreover, among all curves belonging to the same equivalence class, we shall select one which gives the geometric properties of this object in a most natural way. (This corresponds to assigning an official name to a person.)

We recall that by a curve in $I\!R^n$ we mean a continuous function $\alpha : [a, b] \to I\!R^n$, or equivalently an ordered $n$-tuple of real-valued continuous functions, say,

$\alpha_1, \ldots, \alpha_n$ each defined on $[a, b]$. For each $t \in [a, b], \alpha(t) = (\alpha_1(t), \ldots, \alpha_n(t))$ is a point on the curve. $\alpha(a)$ and $\alpha(b)$ are called the initial and the terminal points of $\alpha$. The curve $\alpha$ is called closed if $\alpha(a) = \alpha(b)$ i.e. if the initial and the terminal points (together called the end-points) of $\alpha$ coincide. The range of the function $\alpha$, i.e. the set $\{(x_1, \ldots x_n) \in {I\!\!R}^n : $ there is some $t \in [a, b]$ such that $x_i = \alpha_i(t)$ for $i = 1, \ldots n\}$ is called the range (or sometimes the **trace**) of $\alpha$. For $n = 2$ and $3$ it is customary to denote points of ${I\!\!R}^n$ by $(x, y)$ and $(x, y, z)$ instead of $(x_1, x_2)$ and $(x_1, x_2, x_3)$ respectively. Still more preferable is to denote them by their position vectors. The postion vector is almost always denoted by **r**. For ${I\!\!R}^2$, we fix unit vector **i**, **j** along the $x$- and $y$-axis respectively and write $(x, y)$ as $\mathbf{r} = x\mathbf{i} + y\mathbf{j}$. In ${I\!\!R}^3$, we take a third unit vector **k** along $z$-axis so that $(\mathbf{i}, \mathbf{j}, \mathbf{k})$ forms a right handed orthonormal system (i.e. if **i** points along the index finger and **j** along the middle finger of the right hand, then **k** points along the thumb).

With this notation, a curve is simply a continuous vector-valued function $\mathbf{x} = \alpha(t)$ defined on some closed and bounded interval. We say $\alpha$ is differentiable if each of the co-ordinate functions $x, y, z$ (or $x_1, \ldots, x_n$) are differentiable as functions of $t$ and denote the derivative $\alpha'(t)$ of **r** w.r.t. $t$ by the vector $\dfrac{d\mathbf{r}}{dt} = \dfrac{dx}{dt}\mathbf{i} + \dfrac{dy}{dt}\mathbf{k}$. $\alpha$ is said to be **smooth** at $t_0 \in [a, b]$ if $\alpha'$ is continuous and non-zero in some neighbourhood of $t_0$. It was proved in Theorem (4.12.2) that if $\alpha$ is smooth at $t_0$, then the vector $\alpha'(t_0)$ is parallel to the tangent to the curve at the point $\alpha(t_0)$. Therefore if we denote by $\mathbf{u}(t_0)$ a unit vector along the tangent to the curve, then we have

$$\mathbf{u}(t_0) = \frac{\alpha'(t_0)}{|\alpha'(t_0)|} = \frac{d\mathbf{r}/dt}{|d\mathbf{r}/dt|} = \frac{\dot{\mathbf{r}}}{|\dot{\mathbf{r}}|} \tag{1}$$

where $|\;|$ denotes the length of a vector, and (following the notation introduced in Section 6.2), the dot $(\cdot)$ above a variable indicates differentiation w.r.t. the parameter $t$. Somootness at the end-points $a$ and $b$ is, as usual, only in terms of the right or left handed derivative. A curve $\alpha$ is said to be smooth if it is smooth at every point of $[a, b]$. Finally, $\alpha$ is called **piecewise smooth** if there exists a partition, $(t_0, t_1, \ldots, t_k)$ of $[a, b]$, such that $\alpha$ is smooth[5] on $(t_{i-1} t_i)$ for every $i = 1, \ldots, k$. In such a case, at $\alpha(t_i)$ for $i = 1, \ldots, k - 1$, the portions of the curve upto $\alpha(t_i)$ and after $\alpha(t_i)$ may sometimes have tangents, but the two tangents may be different. When these two tangents have different directions, the picture of the curve looks as if there is a sharp corner or a kink at $\alpha(t_i)$ as in Fig.6.6.1(a). (Another example is the curve sketched in Fig.6.2.4). When these two tangents have opposite directions, as in Fig.6.6.1(b) or (c) we say there is a **cusp** at $\alpha(t_i)$. At a cusp, the curve reverses its direction. (For actual examples

---

[5]Note that even for piecewise smooth curves, we are not relaxing the continuity requirement of $\alpha$ at $t_i$. The relaxation is only about differentiability. In other words, even for a piecewise smooth curve $\alpha$, we require that $\alpha$ is continuous (and not just piecewise continuous) on $[a, b]$

of cusps see Exercises (6.3) and (6.4).)

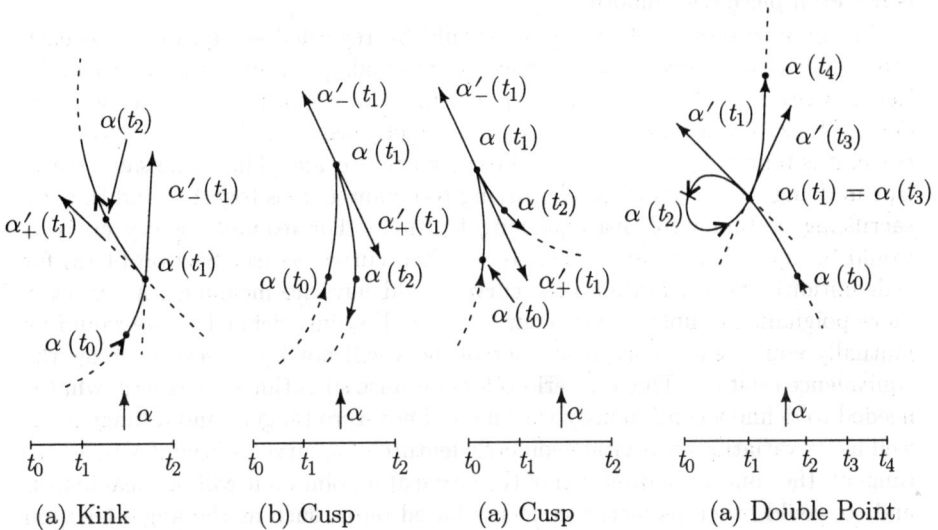

Figure 6.6.1 : Piecewise Smooth and Smooth Curves

Thus the curves in Fig. 6.6.1(a), (b) and (c) are piecewise smooth but not smooth. The curve in (d), however is smooth everywhere. Here $\alpha(t_1) = \alpha(t_3)$ but $\alpha'(t_1) \neq \alpha'(t_3)$. So in the picture of the curve it appears as if the curve has no unique tangent at the point $P$. This is indeed true if a curve is viewed purely as a geometric object. But in our conception of a curve, it also matters how this geometric object is traversed. As $t$ varies from $t_0$ to $t_3$, the point $\alpha(t)$ traverses through $P$ twice, once when $t = t_1$ and then again when $t = t_2$. So $P$ is called a **node** or a **multiple point** (or, more specifically, a **double point**, because $\alpha(t)$ equals $P$ for only two values of $t$). Every time the curve passes $P$, it has a non-zero tangent. So $\alpha$ is smooth at $t_1$ as well as at $t_2$. The only thing is that $\alpha'(t_i) \neq \alpha'(t_2)$. An actual example of a smooth double point will be given in the exercises. (Exercise (6.5)).

An example of a curve which is not even piecewise smooth can be constructed using any continuous, nowhere differentiable function (see, for example, Exercise (4.3.10)). But considering that smoothness can also be violated by the vanishing of $\alpha'$ (and not just by its non-existence), a much simpler example can be given. Consider two curves $\alpha_1, \alpha_2$ both defined from $[0, 1]$ to $I\!\!R^2$ by $\alpha_1(t) = t\mathbf{i}$ for all $t \in [0, 1]$ and

$$\alpha_2(t) = \begin{cases} \mathbf{0} & \text{if } 0 \leq t \leq \frac{1}{2} \\ (2t - 1)\mathbf{i} & \text{if } \frac{1}{2} < t \leq 1 \end{cases} \tag{2}$$

It is easy to describe $\alpha_1$ and $\alpha_2$ verbally, in terms of the motions of a particle. Both cover the same range, viz., the portion of the $x$-axis from $(0,0)$ to $(1,0)$. But $\alpha_1$ is a uniform motion and is smooth everywhere since $\alpha_1'(t) = \mathbf{i}$ for all $t \in [0, 1]$, while in $\alpha_2$, the particle stands still for the first half and then runs at double-speed for the second half. Here $\alpha_2'(t) = \mathbf{0}$ for $0 \le t < \frac{1}{2}$ and so the curve is not even piecewise smooth.

Let us now decide which curves should be regarded as equivalent to each other. Let $\alpha, \beta$ be two curves defined on intervals $[a, b]$ and $[c, d]$ respectively. Let us write $\alpha \sim \beta$ to mean $\alpha$ is equivalent to $\beta$, a relation which we want to define. Since the geometric object most directly associated with a curve is its trace, it is tempting to define $\alpha \sim \beta$ to mean that $\alpha$ and $\beta$ have the same trace. But in doing so, we would be identifying too many curves together and thereby sacrificing certain vital distinguishing features. For example, every curve $\alpha$ would be equivalent to its oppositely oriented curve $-\alpha$ (see Section (4.12) for a definition). So orientation of a curve would have no meaning. As an even more poignant example, the two curves $\alpha_1$ and $\alpha_2$ just defined above would be mutually equivalent. This means smoothness will not be preserved under the equivalence relation. This is a serious loss because smoothness is exactly what is needed to define a continuously varying and non-zero tangent and a tangent (as well as its variation) is a vital geometric feature of a curve. Without a non-zero tangent, the concept of direction of the curve at a point on it will be meaningless and so would other geometric concepts based on it, such as the angle between two curves (see Exercise (6.9)) or curvature, which, as we shall see in the next section, is the rate at which the curve changes its direction.

So we should be more restrictive in defining the equivalence of curves so that under this equivalence the vital features such as smoothless and the direction of the tangent are preserved. It is helpful at this stage to recall Exercise (4.12.3) where it was shown that without loss of generality we may suppose that every curve is parametrised over the unit interval $[0, 1]$. This was done by defining a suitable function $f : [0, 1] \to [a, b]$, viz. $f(t) = a + t(b - a)$. This function is linear with derivative $b - a$. In the solution of that exercise what is crucially needed is not the linearity of the function $f$, but the fact that its derivative is positive everywhere. This motivates the following definition.

**6.1 Definition.** Two curves $\alpha$ and $\beta$ defined respectively on the intervals $[a, b]$ and $[c, d]$ are said to be **equivalent** (written $\alpha \sim \beta$) if there exists a continuous function $f : [a, b] \to I\!\!R$ such that

(i)  $f(a) = c, f(b) = d$ and $f'$ is continuous and positive for all except finitely many values of $t$ (or, in other words, there exists a partition, say $(t_0, t_1, \ldots, t_k)$ of $[a, b]$ such that for $i = 1, \ldots, k, f$ is continuously differentiable on $(t_{i-1}, t_i)$ and $f'(t) > 0$ for all $t \in (t_{i-1}, t_i)$.

(ii)  $\beta \circ f = \alpha$, or in other words, $\beta(f(t)) = \alpha(t)$ for all $t \in [a, b]$.

The function $f : [a, b] \to I\!\!R$ is called an **equivalence** between $\alpha$ and $\beta$.

In view of Lagrange's Mean Value Theorem (see Exercise (4.5.18)), (i) implies that $f$ is strictly monotonically increasing and maps $[a, b]$ onto $[c, d]$. So if we call $f(t)$ as $u$ then as $t$ moves from $a$ to $b$ tracing $\alpha$, $u$ moves from $c$ to $d$ tracing $\beta$

by (ii). This is illustrated in Fig.6.6.2 where the two curves $\alpha, \beta$ are equivalent.

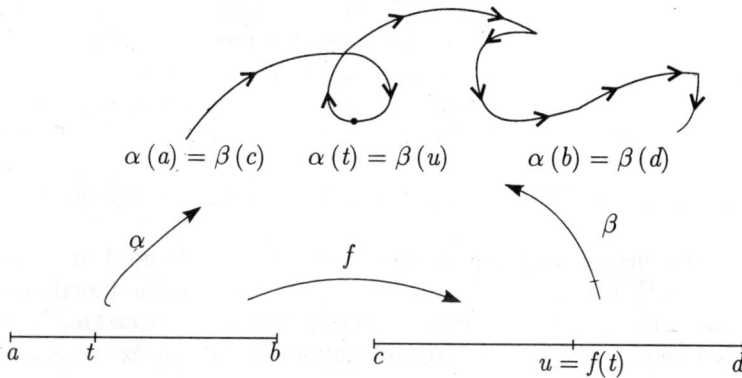

$$\alpha\,(a) = \beta\,(c) \qquad \alpha\,(t) = \beta\,(u) \qquad \alpha\,(b) = \beta\,(d)$$

Figure 6.6.2 : Equivalent Curves

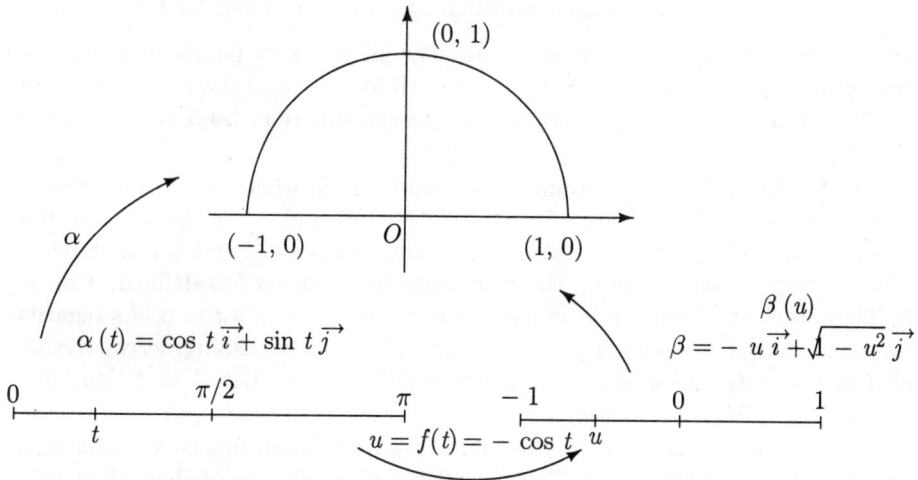

Figure 6.6.3 : Equivalent Parametrisations

Exercise (4.12.3) already gives examples of equivalent curves, because the function $f$ in it satisfies the conditions in the definition above. But now that we are not requiring $f$ to be linear, many other examples are possible. Note that in condition (i) of the Definition (6.1), we are not requiring $f'$ to exist and be continuous and positive everywhere. If we do so, we would get an equivalence relation which is unnecessarily restrictive, because certain pairs of curves which we would like to be equivalent will not be so under the more stringent condition. A simple example is provided by the upper arc of a unit semi-circle, $\{(x, y) : x^2 + y^2 = 1, y \geq 0\}$. Two standard parametrisations $\alpha$ and

$\beta$ of it are shown in Fig.6.6.3. Their natural equivalence comes through the function $f(t) = -\cos t$, whose derivative is positive everywhere on $[0, \pi]$ except at the two points 0 and $\pi$.

The equivalence relation $\sim$ given by Definition (6.1) turns out to be just the right one because, as we shall soon show, it preserves the vital features such as the arc length and directions of tangents. But before we do so, it would be in the order of things to verify that $\sim$ is indeed an equivalence relation, i.e. that it is reflexive, symmetric and transitive.

**6.2 Theorem.** The relation $\sim$ is an equivalence relation on the set of all curves.

**Proof.** For reflexivity, i.e. for showing $\alpha \sim \alpha$, we merely need to take $f(t) = t$ for all $t \in [a, b]$. For symmetry, suppose $\alpha \sim \beta$ and $f$ is the function given by Definition (6.1). As already observed $f$ is strictly monotonic and maps $[a, b]$ onto $[c, d]$. By Exercise (4.1.9), the inverse function, say $g$, of $f$ exists, is continuous and maps $[c, d]$ onto $[a, b]$. Now suppose $f'$ exists, is continuous and non-zero for all $t$ except possibly for $t = t_0, \ldots, t_k$ (say). Let $u_i = f(t_i)$ for $i = 0, \ldots, k$. Then for every $u \in [c, d], u \neq u_0, \ldots, u_k$, by Exercise (4.4.8), $g'(u)$ exists and equals $\dfrac{1}{f'(g(u))}$. Since $g$ is continuous everywhere, $f'(t)$ is continuous for $t = t_0, \ldots, t_k$ and the reciprocal function is continuous at all non-zero numbers, it follows that $g'$ is continuous at all $u \neq u_0, \ldots, u_k$. Moreover, since $\beta \circ f = \alpha$, we have $\beta \circ f \circ g = \alpha \circ g$. But $f \circ g$ is simply the identity function. So $\beta = \alpha \circ g$. Thus $\beta \sim \alpha$.

Finally, for transitivity, assume $\alpha \sim \beta$ and $\beta \sim \gamma$, where $\alpha, \beta, \gamma$ are defined on $[a, b], [c, d]$ and $[p, q]$ (say). Let $f : [a, b] \to I\!R$ and $g : [c, d] \to I\!R$ be the functions which give the equivalence of $\alpha$ and $\beta$ and of $\beta$ and $\gamma$ respectively. Since $f$ maps $[a, b]$ onto $[c, d]$, the composite function $g \circ f$ is defined. Call it $h$. Then $h$ is continuous on $[a, b]$ and $\gamma \circ h = \gamma \circ g \circ f = \beta \circ f = \alpha$. As regards the behaviour of $h'$, assume $f, g$ have continuous, non-vanishing derivatives at all $t \neq t_1, \ldots, t_k$ and $u \neq u_0, \ldots, u_r$ respectively (say). Let $t_i^* = f^{-1}(u_i)$ for $i = 0, \ldots, r$. Then for all $t \neq t_0, \ldots, t_k, t_0^*, \ldots, t_r^*$, both $f'(t)$ and $g'(f(t))$ are continuous and non-zero. So by the chain rule, $h'(t)$ is continuous and non-zero for all $t \neq t_0, \ldots, t_k, t_0^*, \ldots, t_r^*$. Thus $h$ satisfies all conditions of Definition (6.1) w.r.t. $\alpha$ and $\gamma$. So $\alpha \sim \gamma$ as desired. ∎

We now turn to checking which features of a curve are preserved under the equivalence relation $\sim$. It is very easy to show that if $\alpha \sim \beta$ then the rectifiability of either one of them implies that of the other. Indeed, for this purpose all we need to know about the function $f$ in Definition (6.1) is that it maps $[a, b]$ onto $[c, d]$ in a strictly monotonically increasing manner and that $\beta \circ f = \alpha$. For, these properties imply that every broken line path, say $P_0 - P_1 - \ldots - P_n$ along $\alpha$ is also a broken line path along $\beta$ and vice versa. This follows because under $f$ any partition of $[a, b]$ corresponds to a unique partition of $[c, d]$. So the set of lengths of broken line paths along $\alpha$ and $\beta$ is the same. Hence $\alpha$ is rectifiable if and only if $\beta$ is and moreover they have the same lengths when this happens.

Next, we consider the preservation of smoothness of a curve. It may be noted that since we are allowing the function $f$ in Definition (6.1) to behave badly at a few points, smoothness *per se* is not preserved under $\sim$. For example, the two cures $\alpha$ and $\beta$ in Fig.6.6.3 are equivalent to each other. But while $\alpha$ is smooth everywhere, $\beta$ is not so. Specifically $\beta$ is not smooth at $-1$ and $1$. These are also the points where the inverse function $f^{-1}(u) = \cos^{-1}(-u)$ fails to be differentiable. So the leeway allowed for $f$ in Definition (6.1) may change a smooth curve to one which is not smooth. But this leeway is available only for finitely many points. And so, things turn out to be better for piecewise smooth curves as we now show.

**6.3 Theorem:** If two curves are mutually equivalent and one of them is piecewise smooth, so is the other.

**Proof.** We follow the notation of Definition (6.1) and suppose $\alpha \sim \beta$. Since the relation $\sim$ is symmetric, we may suppose $\beta$ is piecewise smooth and shall prove that $\alpha$ is piecewise smooth. We know $\alpha = \beta \circ f$ where $f : [a,b] \to \mathbb{R}$ satisfies condition (i) in Definition (6.1). So, there exists a partition, say $(t_0, t_1, \ldots, t_k)$, of $[a,b]$ such that $f'$ is continuous and positive on $(t_{i-1}, t_i)$ for $i = 1, \ldots, k$. Also since $\beta$ is piecewise smooth, there exists a partition, say $(u_0, u_1, \ldots, u_r)$, of $[c,d]$ such that $\beta$ is smooth on $(u_{j-1}, u_j)$ for every $j = 1, \ldots, r$, i.e. $\beta'$ is continuous and non-zero at every $u \in (u_{j-1}, u_j)$. Let $t_j^* = f^{-1}(u_j)$ for $j = 0, \ldots, r$. Then since $f$ is strictly monotonically increasing, $(t_0^*, t_1^*, \ldots, t_r^*)$ is a partition of $[a,b]$. Now let $(s_0, s_1, \ldots, s_m)$ be the partition of $[a,b]$ obtained by superimposing the two partitions $(t_0, t_1, \ldots, t_k)$ and $(t_0^*, t_1^*, \ldots, t_r^*)$. Then for each $p = 1, \ldots, m, (s_{p-1}, s_p) \subset (t_{i-1}, t_i)$ for some (unique) $i$ and $(f(s_{p-1}), f(s_p)) \subset (u_{j-1}, u_j)$ for some (unique) $j$. So for all $t \in (s_{p-1}, s_p), f'$ is continuous at $t$ and $\beta'$ is non-zero at $f(t)$. So the vector $\beta'(f(t))f'(t)$ is continuous and non-zero. But by the Chain Rule this is precisely the vector $\alpha'(t)$. (The chain rule was proved for the composite of two real-valued functions. Here we need it for the composite of one real-valued and one vector valued function. The chain rule in this form, will be given in Exercise (6.6)). Thus on $(s_{p-1}, s_p), \alpha'$ is continuous and non-zero, for every $\beta = 1, \ldots, m$. Hence $\alpha$ is piecewise smooth. ∎

This theorem is one of the reasons why the class of piecewise smooth curves stands out among all curves. As explained in Section 6.2, the smoothness requirement is a bit too restrictive in that many naturally occurring curves such as the boundaries of various regions have no smooth parametrisations. They often have a piecewise smooth parametrisation and the theorem above says that any equivalent parametrisation will be piecewise smooth. As far as local attributes such as curvature, and torsion are concerned, piecewise smoothness is as good as smoothness, except at a finite number of points. And as far as global attributes (such as arc length) are concerned, they are additive and so by chopping the domain interval, say $[a,b]$, of a piecewise smooth curve $\alpha$ into a finite number of subintervals, we may suppose $\alpha$ is smooth except possibly at the two end points.

We postpone the discussion of local properties to the next section. In the present section we consider additive attributes of curves. The foremost among them is arc length. We shall show that the formula for arc length given in Theorem (2.4) is also valid for a piecewise smooth curve. We begin with the case where the curve $\alpha$ is smooth everywhere except possibly at the end points $a$ and $b$.

**6.4 Theorem.** Suppose a plane curve $\alpha = (f, g)$ defined on $[a, b]$ is smooth on $(a, b)$. Then the integral $\int_a^b |\alpha'(t)| dt$, i.e. the integral $\int_a^b \sqrt{[f'(t)]^2 + [g'(t)]^2} dt$, which may be improper at one or both the ends, is convergent if and only if $\alpha$ is rectifiable and when this is the case, the integral converges to the length of $\alpha$.

**Proof.** Theorem (2.4) deals with the case when there is no impropriety in the integral. Suppose now that $\alpha$ is smooth at $a$ but not $b$. Then for every $\delta > 0$ (and $\delta < b - a$), Theorem (2.4) applies to $\alpha$ restricted to the interval $[a, b - \delta]$. So

$$\int_a^{b-\delta} \sqrt{[f'(t)]^2 + [g'(g)]^2} dt = l(b - \delta) \qquad (3)$$

where $l(b - \delta)$ means the arc length of the restriction of $\alpha$ to $[a, b - \delta]$. By additivity of arc length, the R.H.S. of (3) increases as $\delta$ decreases. By additivity of definite integral (and the positivity of the integrand), the L.H.S. of (3) also increases as $\delta$ decreases. So as $\delta \to 0^+$, the limit of one side of (3) exists if and only if the limit of the other side exists and when this happens the two limits are equal. By definition, the limit of the L.H.S. as $\delta \to 0^+$ is $\int_a^b \sqrt{[f'(t)]^2 + [g'(t)]^2} dt$. So we would be through if we can show that $\lim_{\delta \to 0^+} l(b - \delta)$ exists if and only if $\alpha$ is rectifiable and that when this happens, the arc length of $\alpha$ equals this limit.

First assume $\alpha$ is rectifiable with length $L$. Then by Corollary (2.2), $l(b-\delta) \leq L$ for all $\delta > 0$ and so, as $l(b - \delta)$ increases with decreasing $\delta$, $\lim_{\delta \to 0} l(b - \delta)$ exists by completeness of the real number system. (See Theorem (2.6.5). This use of completeness of $I\!\!R$ is analogous to that in the proof of Theorem (4.10.4).) Let this limit be $M$. Then $M \leq L$. To prove that equality holds, let, if possible, $L > M$. Put $\epsilon = \frac{L-M}{2}$. Then $\epsilon > 0$ and so, by continuity of $\alpha$ at $b$, there exists $h > 0$ such that whenever $t \in [b - h, b]$, both $|f(t) - f(b)|$ and $|g(t) - g(b)|$ are less than $\frac{1}{2}\epsilon$ each. Since

$$|\alpha(t) - \alpha(b)| = \sqrt{(f(t) - f(b))^2 + (g(t) - g(b))^2},$$

we get

$$|\alpha(t) - \alpha(b)| < \sqrt{\frac{\epsilon^2}{4} + \frac{\epsilon^2}{4}} < \epsilon \text{ whenever } t \in [b - h, b]. \qquad (4)$$

By definition of the arc length, $L$ is the supremum of the lengths of all broken line paths along $\alpha$. So there exists a partition, say $P = (t_0, t_1, \ldots, t_n)$ of $[a, b]$ such that

$$l(P) = \sum_{i=1}^{n} |P_{i-1}P_i| > L - \epsilon \qquad (5)$$

where, as usual, $P_i = \alpha(t_i)$ for $i = 0, 1, \ldots, n$. We have $t_0 = a$ and $t_n = b$. We may suppose $t_{n-1} \geq b - h$, for if this is not so, we can insert $b - h$ as another node to get a new partition $P^*$ which is finer than $P$. So by Lemma (2.1), $l(P^*) \geq l(P)$ and hence (5) will hold with $P$ being replaced by $P^*$. Now the last term in the sum in (5) is $|\alpha(t_{n-1}) - \alpha(t_n)|$ which, by (4) is at most $\epsilon$. So,

$$\sum_{i=1}^{n-1} |P_{i-1}P_i| > L - \epsilon - \epsilon = M \qquad (6)$$

If we let $\delta = b - t_{n-1}$, then the sum in (6) is the length of a broken line path along the restriction of $\alpha$ to $[a, b - \delta]$. So $L(b - \delta) \geq \sum_{i=1}^{n} |P_{i-1}P_i|$. But then we get $l(b - \delta) > M$ which contradicts that $l(b - \delta)$ increases monotonically to $M$ as $\delta \to 0^+$.

Thus, we have shown that if $\alpha$ is rectifiable with length $L$ then $\lim_{\delta \to 0^+} l(b - \delta)$ exists and equals $L$. Conversely, suppose $\lim_{\delta \to 0^+} l(b - \delta)$ exists and denote it again by $M$. Then (taking $\epsilon = 1$ in the definition of a limit), there exists $h_1 > 0$ such that

$$M - 1 < l(b - \delta) < M + 1 \quad \text{whenever} \quad 0 < \delta < h_1 \qquad (7)$$

Also, by a reasoning similar to that for obtaining (4), there exists $h_2 > 0$ such that

$$|\alpha(t) - \alpha(b)| < 1 \text{ whenever } t \in [b - h_2, b]. \qquad (8)$$

Let $h = \min\{h_1, h_2\}$. Let $P = (t_0, t_1, \ldots, t_n)$ be any partition of $[a, b]$. We may again suppose that $t_{n-1} > b - h$. Then letting $\delta = b - t_{n-1}$, we have

$$\begin{aligned} l(P) &= \sum_{i=1}^{n-1} |P_{i-1}P_i| + |P_{n-1}P_n| \\ &< l(b - \delta) + 1 \text{ by (8)} \\ &< M + 2 \text{ by (7)} \end{aligned}$$

Hence $l(P)$ is bounded above by $M + 2$ for all partitions $P$ of $[a, b]$. So $\alpha$ is rectifiable. As already shown, $M$ equals length of $\alpha$.

Thus we have proved the theorem when the integral in its statement is improper at $b$, because of lack of smoothness at $b$. An entirely similar argument applies if $\alpha$ is smooth at $b$ but not smooth at $a$. Finally, if $\alpha$ is smooth neither at $a$ nor at $b$, take any $c$ in $(a, b)$. Apply the argument to $[a, c]$ and $[c, b]$ separately and use the additivity of the arc length and of (possibly improper) integrals. ∎

Additivity techniques also establish the following generalisation[6].

**6.5 Theorem** : The preceding theorem also holds if $\alpha$ is piecewise smooth on $[a, b]$.

**Proof.** By definition, there is a partition, say $P = (t_0, t_1, \ldots, t_n)$ of $[a, b]$ such that $\alpha$ is smooth on $(t_{i-1}, t_i)$ for all $i = 1, \ldots, n$. Apply the theorem above to the restriction of $\alpha$ to each $[t_{i-1}, t_i]$. Note that $\alpha$ is rectifiable if and only if each of these restrictions is rectifiable and when so, their lengths add up to the length of $\alpha$. But this is also the case for the integral $\int_a^b \sqrt{[f'(t)]^2 + [g'(t)]^2} \, dt$ which exists if and only if $\int_{t_{i-1}}^{t_i} \sqrt{[f'(t)]^2 + [g'(t)]^2} \, dt$ exists for $i = 1, \ldots, n$ and when so, they add up to $\int_a^b \sqrt{[f'(t)]^2 + [g'(t)]^2} \, dt$. ∎

It should be noted that a piecewise smooth curve is not always rectifiable. For example, define $\alpha : (0, \frac{1}{\pi}] \to I\!R^2$ by

$$\alpha(t) = \begin{cases} t\mathbf{i} + t\sin(\frac{1}{t})\mathbf{j} & \text{if } 0 < t \leq \frac{1}{\pi} \\ 0 & \text{if } t = 0. \end{cases} \tag{9}$$

Then $\alpha$ is smooth everywhere except at the end-point $0$. But it is not rectifiable (cf. Exercise (4.7.7)). This is consistent with Theorem (6.4) because in this case the improper integral $\int_0^{1/\pi} |\alpha'(t)| \, dt$ i.e. the integral,

$$\int_0^{1/\pi} \sqrt{1 + \sin^2 \frac{1}{t} - \frac{2}{t} \sin \frac{1}{t} \cos \frac{1}{t} + \frac{1}{t^2} \cos^2(\frac{1}{t})} \, dt$$

can directly be shown to be divergent (see Exercise (6.10)).

The leverage which Theorem (6.4) has over Theorem (2.4) can be illustrated by considering the two parametrisations $\alpha$ and $\beta$ of the upper semi-circle shown in Fig.6.6.3. Here $\alpha$ is smooth and we get its length by evaluating the integral $\int_0^\pi |\alpha'(t)| \, dt = \int_0^\pi \sqrt{\sin^2 t + \cos^2 t} \, dt = \int_0^\pi 1 \, dt = \pi$. But, $\beta$ is not smooth at $-1$ and $1$. The integral $\int_{-1}^1 |\beta'(u)| \, du$ comes out to be $\int_{-1}^1 \sqrt{1 + \frac{u^2}{1-u^2}} \, du$, i.e. $\int_{-1}^1 \frac{du}{\sqrt{1-u^2}}$ which is improper at both the ends and can be evaluated. Here Theorem (6.4) is applicable but Theorem (2.4) fails.

---

[6]The essential difference between this theorem and the result of Exercise (2.10) is that there the functions $f$ and $g$ were required to have right as well as left handed derivatives at the nodes of the partition. In the present theorem, no such assumption is made. We, of course, assume that $f, g$ are continuous everywhere and hence, in particular, at these nodes.

Although the theorems above are stated for plane curves, they hold for curves in $\mathbb{R}^n$ for any $n$ with only marginal changes in the proof. Instead of the two function $f$ and $g$, there will be $n$ real-valued functions on $[a, b]$. We shall use these extensions especially for $n = 3$, i.e. for space curves. Note that in the proofs above, only the continuity of $\alpha'$ (except possibly at a finite number of points) was used. The fact that $|\alpha'| > 0$ was never used. Its significance is apparent from the theorem below.

**6.6 Theorem:** Suppose $\alpha$ is a piecewise smooth curve defined on an interval $[a, b]$. Assume $\alpha$ is rectifiable and has length $L$. Define $l : [a, b] \to \mathbb{R}$ by $l(c) = \displaystyle\int_a^c |\alpha'(t)| dt$. Then $l$ is a strictly increasing continuous function which maps $[a, b]$ onto $[0, L]$. Also $l'$ is continuous and positive except possibly at a finite number of points in $[a, b]$. (Specifically, $l'(t) = |\alpha(t)|$ except for finitely many values of $t$.)

**Proof:** By the last theorem, for every $c \in [a, b], l(c)$ is the length of the restriction of $\alpha$ to $[a, c]$. So $l$ is certainly a monotonically increasing function which maps $[a, b]$ onto $[0, L]$. The remaining properties are essentially a consequence of the Fundamental Theorem of Calculus. But we have to be a little careful in applying it since the integrand $|\alpha'(t)|$ is not given to be continuous on the entire interval $[a, b]$. We are given that there is a partition, say $(t_0, t_1, \ldots, t_n)$ of $[a, b]$ such that for every $c \neq t_0, t_1, \ldots, t_n, |\alpha'(c)|$ is positive and continuous. We prove that $l'$ exists and is continuous and positive at every $c \neq t_0, t_1, \ldots, t_n$. Fix any such $c$. Then there is a unique $k$ such that $c \in (t_{k-1}, t_k)$. Fix any $c_1 \in (t_{k-1}, c)$ and $c_2 \in (c, t_k)$. Then $t_{k-1} < c_1 < c < c_2 < t_k$. Now the possibly improper integral $\displaystyle\int_a^c |\alpha'(t)| dt$ equals, by definition, $\displaystyle\sum_{i=1}^{k-1} \int_{t_{i-1}}^{t_i} |\alpha'(t)| dt + \int_{t_{k-1}}^c |\alpha'(t)| dt$. This last integral can be further split as $\displaystyle\int_{t_{k-1}}^{c_1} |\alpha'(t)| dt + \int_{c_1}^c |\alpha'(t)| dt$. So for $c_1 < x < c_2$ we can write

$$l(x) \;=\; I + F(x) \tag{10}$$

where

$$I = \sum_{i=1}^{k-1} \int_{t_{i-1}}^{t_i} |\alpha'(t)| dt + \int_{t_{k-1}}^{c_1} |\alpha'(t)| dt \text{ and } F(x) = \int_{c_1}^x |\alpha'(t)| dt.$$

Here $I$ is a constant and $F(x)$ is a function defined by an integral for $x \in [c_1, c_2]$. Note that in $F(x)$ the integrand $|\alpha|$ is continuous everywhere and there is no impropriety. So by the second form of the Fundamental Theorem of Calculus (Theorem (5.5.3)), $F'(x)$ exists and equals $|\alpha'(x)|$ for all $x \in [c_1, c_2]$. But since $I$ is a constant, by (10) we have $l'(x) = F'(x)$. So $l'(x) = |\alpha'(x)|$ for all $x \in [c_1, c_2]$. Since $[c_1, c_2] \subset (t_{k-1}, t_k), \alpha'$ and hence $|\alpha'|$ is continuous and positive by smoothness of $\alpha$ on $(t_{i-1}, t_i)$. In particular $l'$ is continuous and positive at *c*. (The reason we had to insert $c_1$ and $c_2$, was merely that we could not apply Theorem (5.5.3), in the form it is stated, directly to $[t_{k-1}, t_k]$. But we can apply it to $[c_1, c_2]$, which is sufficient for our purpose because differentiability and continuity are local properties. The trick adopted here is similar in spirit to that adopted in the proofs of Theorem (4.6.4) and Corollary (4.1.6), where, to prove the continuity of the sum function of a power series at a point $c$ in its interval of convergence $(-R, R)$, we had to first find $R' < R$ such that $c \in (-R', R')$, because the convergence of the power series is uniform on $[-R', R']$ but not necessarily on $(-R, R)$.)

We still have to prove the continuity of $l$ at the 'bad' points $t_0, t_1, \ldots, t_n$. Given any $k$, fix $c_1, c_2$ so that $t_{k-1} < c_1 < t_k < c_2 < t_{k+1}$. (If $k = 0$, then $c_1$ is undefined since we need consider only the right handed continuity at $t_0$. Similarly if $k = n$, $c_2$ is undefined.) Then, by additivity of arc length,

$$l(t_k) \;\; = \;\; \int_a^{t_k} |\alpha'(t)| dt = l(c_1) + \int_{c_1}^{t_k} |\alpha'(t)| dt. \tag{11}$$

The last integral is, by definition, $\displaystyle \lim_{\delta \to 0+} \int_{c_1}^{t_k - \delta} |\alpha'(t)| dt$. So it is continuous from left at $t_k$. But then, since $l(c_1)$ is a constant, (11) shows that $l$ is continuous from left at $t_k$. Continuity from right at $t_k$ is proved similarly, by writing

$$l(t_k) = l(c_2) - \int_{t_k}^{c_2} |\alpha'(t)| dt. \tag{12}$$

Thus $l$ is continuous everywhere on $[a, b]$. It only remains to be proved that $l$ is strictly increasing on $[a, b]$. This follows from Exercise (4.5.18). ∎

We are now ready to prove that every piecewise smooth curve is equivalent to a curve which is parametrised by arc length.

**6.7 Theorem:** Suppose $\alpha$ is a piecewise smooth, rectifiable curve of length $L$, defined on $[a, b]$. Then for every $s \in [0, L]$, there is a unique $c \in [\alpha, \beta]$ such that $s = \int_a^c |\alpha'(t)| dt$. If $\beta(s)$ is defined as $\alpha(c)$ then $\beta$ is a piecewise smooth curve defined on $[0, L]$ and $\alpha \sim \beta$.

**Proof:** This follows immediately from the various properties of the function $l : [a, b] \to \mathbb{R}$ defined in the last theorem. The first assertion follows from the fact that $l$ is strictly monotonically increasing and maps $[a, b]$ onto $[0, L]$. Also $l$ is continuous everywhere and $l'$ is continuous and positive everywhere except possibly at a finite number of points. Finally $\beta(l(c)) = \beta(s) = \alpha(c)$ for all $c \in [a, b]$. So $\beta \circ l = \alpha$. Hence $l$ has all the properties of the function $f$ in Definition (6.1). So $\beta$ is equivalent to $\alpha$. ∎

**6.8 Definition:** A rectifiable curve $\beta$ of length $L$ is said to be **parametrised w.r.t. arc length** if $\beta$ is defined on $[0, L]$ and for every $s \in [0, L]$, $s$ also equals the length of the restriction of $\beta$ to $[0, s]$.

For example, the curve $\alpha_1 : [0,1] \to I\!\!R^2$ defined by $\alpha_1(s) = s\mathbf{i}$ is parametrised w.r.t. arc length. Instead of $\mathbf{i}$, we could take any unit vector $\mathbf{u}$ in $I\!\!R^2$ (or in $I\!\!R^n$ for that matter). More generally, given any two distinct vectors $\mathbf{v}$ and $\mathbf{w}$ in $I\!\!R^n$, if we let $L = |\mathbf{v} - \mathbf{w}|$ and define $\beta : [0, L] \to I\!\!R^n$ by $\beta(s) = \mathbf{v} + s(\mathbf{w} - \mathbf{v})$ then $\beta$ is parametrised w.r.t. arc length as is very easy to see. Note that the trace of $\beta$ is the line segment joining the points with position vectors $\mathbf{v}$ and $\mathbf{w}$ (see Fig.6.6.4(a)). Similarly suppose $h, k, r$ are constants with $r > 0$. Define $\beta : [0, 2\pi r] \to I\!\!R^2$ by $\beta(s) = (h + r\cos(\frac{s}{r}))\mathbf{i} + (k + r\sin(\frac{s}{r}))\mathbf{j}$. Then $\beta$ is parametrised w.r.t. the arc length. The trace of $\beta$ is the circle of radius $r$ with centre $(h, k)$ (see Fig.(6.6.4)(b)).

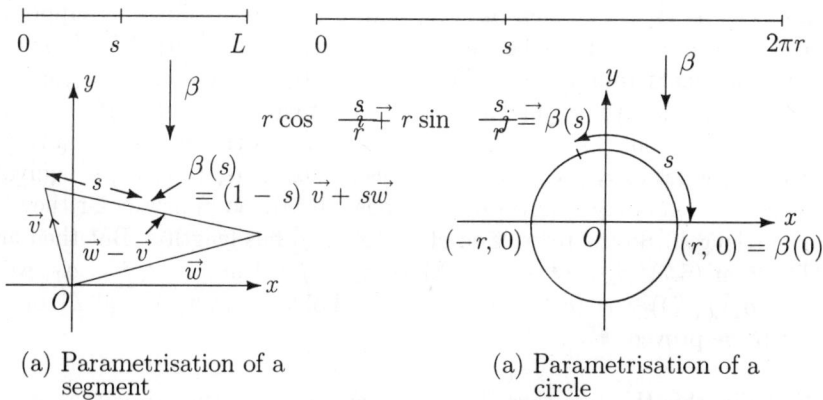

(a) Parametrisation of a
segment

(a) Parametrisation of a
circle

Figure 6.6.4 : Arc Length Parametrisation

If we identify together curves which are mutually equivalent (under the relation $\sim$) then the theorem above says that every piecewise smooth rectifiable curve can be reparametrised w.r.t. arc length. In the analogy made at the beginning of this section, this arc length parametrisation of a curve corresponds to the official name of a person, while the other parametrisations correspond to the other names by which the person is known.

Note that the curve $\alpha_2$ defined by (2) above is not piecewise smooth. The function $l$ associated with it is given by

$$l(c) = \begin{cases} 0 & \text{if} \quad 0 \le c \le \frac{1}{2} \\ 2c - 1 & \text{if} \quad \frac{1}{2} < c \le 1 \end{cases}$$

The curve has arc length 1. The function $l$ is not one-to-one. In fact, it is constant on $[0, \frac{1}{2}]$. So, for $s = 0$, there is no unique value of $c$ such that $l(c) = 0$. Hence $\beta(0)$ cannot be fefined. Thus the curve $\alpha_2$ has no equivalent arc-length parametrisation. Of course, if we identify a curve with its trace then $\alpha_1$ is an arc length parametrisation where $\alpha_1(t) = t\mathbf{i}$ for $0 \le t \le 1$. But as we have already seen, it is not a good idea to identify two curves merely because they have the same trace.

Theorem (6.7) says that every piecewise smooth, rectifiable curve has an arc length parametrisation. We now show that if two such curves are mutually equivalent then the associated arc length parametrisation is the same for both. To appreciate what the theorem says, suppose $\beta_1, \beta_2$ are the arc length parametrisations of $\alpha_1, \alpha_2$ respectively. Then we already know $\alpha_1 \sim \beta_1$ and $\alpha_2 \sim \beta_2$. So if we are further given that $\alpha_1 \sim \alpha_2$, then from the fact that $\sim$ is an equivalence relation, we get $\beta_1 \sim \beta_2$. But we can assert more. That is, not only are $\beta_1$ and $\beta_2$ mutually equivalent, but they are equal on the nose.

**6.9 Theorem:** Suppose $\alpha_1, \alpha_2$ are two piecewise smooth, rectifiable curves with $\beta_1, \beta_2$ their respective arc length parametrisations. Then $\beta_1 = \beta_2$.

**Proof:** Let $\alpha_1, \alpha_2$ be defined on $[a, b]$ and $[c, d]$ respectively, and suppose $f : [a, b] \to \mathbb{R}$ is an equivalence between them. Then $\alpha_2 \circ f = \alpha_1$. We already saw (see the comments after Theorem (6.2)) that $\alpha_1, \alpha_2$ have the same length. Let $L$ be this common length. From Theorem (6.7), for each $s \in [0, L], \beta_1(s) = \alpha_1(c_1)$, where $c_1$ is the unique point in $[a, b]$ such that the restriction of $\alpha_1$ to $[a, c_1]$ has length $s$. Now let $c_2 = f(c_1)$. Then the restriction of $\alpha_1$ to $[a, c_1]$ is equivalent to the restriction of $\alpha_2$ to $[c, c_2]$ (see Exercise (6.8)). So in particular they have the same length. So the restriction of $\alpha_2$ to $[c, c_2]$ has length $s$. But then again by Theorem (6.7), $\beta_2(s) = \alpha_2(c_2)$. Since $c_2 = f(c_1)$ and $\alpha_2 \circ f = \alpha_1$, we get $\beta_2(s) = \alpha_2(f(c_1)) = \alpha_1(c_1) = \beta_1(s)$. As this holds for every $s \in [0, L], \beta_1 = \beta_2$ as was to be proved. ∎

Verbally, this theorem says that as far as piecewise smooth curves are concerned, each equivalence class contains a unique curve which is parametrised by arc length. In terms of the analogy with names, this means every person has a unique official name! Let us stretch this simile a little. We remarked that it would be ideal if the official name of a person reflects his/her vital features. So, in analogy, we should require that the arc length parametrisation should reflect the geometric properties of the curve, or in other words that the arc length is a geometric property of the curve. This is intuitively obvious. But, for a formal proof, we need to first define what is a geometric property. It is no use to say that it is a property studied in geometry, because this immediately raises the question "What is geometry?". A better approach is to define a geometric property first and then define geometry as the study of geometric properties. A **geometric property** is defined as a property which is invariant (i.e. unchanged) under a congruence, where a **congruence** is defined as a transformation which preserves all distances. (Examples of congruences are translations, rotations and reflections. See Exercises from (6.19) to (6.21) for more on this.) So, to show that the arc length of a curve is a geometric property of the curve we must show that congruent curves have equal lengths. This is easy to establish because the length of a line segment is invariant under a congruence since it is the distance between its end-points. So every broken line path along one curve corresponds to a broken line path of equal length along the other curve and vice versa. Since arc length is defined in terms of the lengths of broken line paths along the curve, it is clear that it is invariant under congruence and hence a geometric property.

In applications, a curve $C$ is often specified as a geometric object, usually defined by an equation like $\phi(x, y) = 0$ for plane curves (or by a system of equations for curves in higher dimensional euclidean spaces). And we have to answer some questions regarding its physical properties. For example, we may be asked to find the mass or the moment of a piece of wire in the form of a semi-circular arc, say the upper arc of a unit circle. Or we may be asked to find the point where the curve bends most abruptly. It is very difficult to answer these questions when the curve is given by an equation like $\phi(x, y) = 0$. So we need to parametrise the curve. As mentioned in Section 4.12, such a parametrisation is generally not unique. Among all parametrisations of $C$, the arc length parametrisation stands out as the most natural one to answer questions depending on the geometric properties of $C$, because as noted above, the arc length is a geometric invariant. So it is customary to use the arc length parametrisation (with the parameter almost invariably denoted by $s$ rather than by $t$ or by some other variable) and define the desired attribute in terms of integration or differentiation of certain functions w.r.t. $s$. Problems of the latter type will be taken up in the next section. Here we consider problems where integration w.r.t. $s$, the arc length parameter, is involved.

Suppose, for example, that we are given a piece of wire laid along some plane curve $C$ and we want to find its mass and the centre of mass, being given the density function, say $\rho$. Let $L$ be arc length of $C$ and suppose $\beta : [0, L] \to I\!\!R^2$ is an arc length parametrisation of $C$. Then for each $s \in [0, L], \beta(s)$ is a point on $C$. The density at this point is usually not given directly in terms of $s$. Instead it is far more common to give it in the form $\rho(x, y)$ where $x, y$ are the $x$ - and $y$ - coordinates of the point $\beta(s)$ and $\rho(x, y)$ is a function of two variables. In this case the mass density at $\beta(s)$ is $\rho(x(s), y(s))$. So the total mass of the wire will be $\int_0^L \rho(x(s), y(s)) ds$. (In a formal approach, this is the very definition of the mass of the wire. The argument that it is indeed a reasonable definition comes from the usual recipe of integration, since a 'line element' $ds$ at $\beta(s)$ contributes $\rho(x(s), y(s)) ds$ to the mass). For brevity, this integral is also often written as $\int_0^L \rho(x, y) ds$ where it is to be clearly understood that $x$ and $y$ and hence $\rho(x, y)$ are functions of $s$. Similarly, $M_x$, the moment of the piece of wire about the $x$-axis is $\int_0^L y\rho(x, y) ds$, or more precisely, $\int_0^L y(s)\rho(x(s), y(s)) ds$.

As an illustration, we find the centre of mass of a semi-circle.

**6.10 Problem:** Find the centre of mass of a piece of wire laid along the semi-circle $C = \{(x, y) : x^2 + y^2 = r^2, y \geq 0\}$ assuming that the linear mass density $\rho(x, y)$ at $(x, y)$ is (i) uniform (ii) $y$.

**Solution:** An arc length parametrisation of $C$ was already given in Fig.6.6.4(b), except that $L$ now is $\pi r$ since we are dealing with a semi-circle. Thus $\beta :$ $[0, \pi r] \to I\!\!R^2$ is given by $\beta(s) = (x(s), y(s))$ where

$$x(s) = r\cos(\frac{s}{r}) \quad \text{and} \quad y(s) = r\sin(\frac{s}{r}) \qquad (13)$$

When the density is uniform, we may take $\rho(x,y) = 1$ everywhere. In that case the mass $M$ and the moments $M_y, M_x$ are, respectively, the integrals $\int_0^{\pi r} 1\,ds$, $\int_0^{\pi r} r\cos(\frac{s}{r})ds$ and $\int_0^{\pi r} r\sin(\frac{s}{r})ds$. Easy calculations, using the substitution $s = \theta r$, give these integrals as $\pi r, 0$ and $2r^2$ respectively. So the centre of mass $(\overline{x},\overline{y})$ is given by $\overline{x} = M_y/M = 0$ and $\overline{y} = M_x/M = \dfrac{2r^2}{\pi r} = \dfrac{2r}{\pi}$. Thus the centre of mass lies at $(0, 2r/\pi)$.

If $\rho(x,y) = y$, the calculations are similar. Leaving the details as exercises,

$$
\begin{aligned}
M &= \int_0^{\pi r} \rho(x,y)ds = \int_0^{\pi r} y\,ds = \int_0^{\pi r} r\sin(\frac{s}{r})ds = 2r^2. \\
M_y &= \int_0^{\pi r} x\rho(x,y)ds = \int_0^{\pi r} xy\,ds = \int_0^{\pi r} r^2\cos(\frac{s}{r})\sin(\frac{s}{r})ds = 0. \\
M_x &= \int_0^{\pi r} y\rho(x,y)ds = \int_0^{\pi r} y^2\,ds = \int_0^{\pi r} r^2\sin^2(\frac{s}{r})ds = \frac{\pi}{2}r^3.
\end{aligned}
$$

So the centre of mass in this case is at $(0, \frac{\pi}{4}r)$. ∎

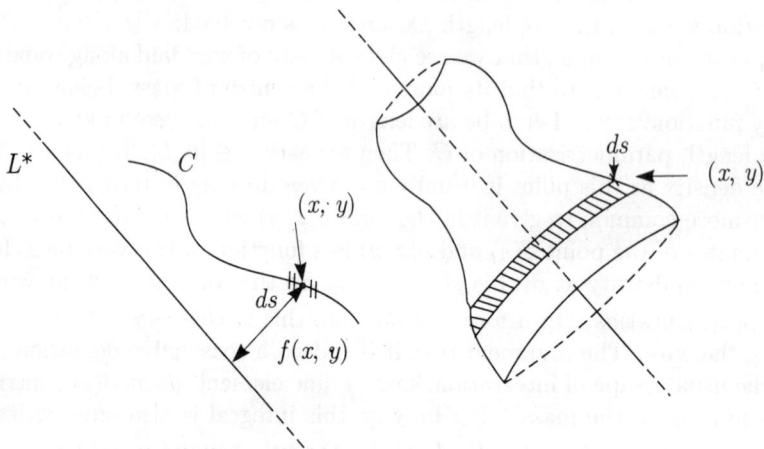

Figure 6.6.5 : Area of a Surface of Revolution.

An essentially similar approach was taken in Section (6.3) for finding the area of the surface, say $S$, of revolution obtained by revolving a curve $C$ around a straight line which we denote by $L^*$ since $L$ is needed for the arc length of $C$. Let $f(x,y)$ be the perpendicular distance of a point $(x,y)$ from $L^*$. (An elementary formula for $f(x,y)$, in terms of the equation of $L^*$ is given in elementary coordinate geometry.) Then the contribution to the area made by a line element $ds$ is $2\pi f(x,y)ds$ and so the surface area of $S$ is $2\pi \int_0^L f(x(s),y(s))ds$ (see Fig. 6.6.5).

The trouble with using the arc length as a parameter is that although every piecewise smooth curve can, in theory, be parametrised (by Theorem (6.7)) by arc length, in practice, this parametrisation is often not easily expressible in terms of closed form expressions of the parameter $s$. In Fig.6.6.4 we gave the arc length parametrisations of a line segment and a circle. But these are just about the only common curves which admit such handy parametrisations. As was remarked in Section 6.2, the integrand in the formula for the arc length of a curve is under a radical sign and often does not admit an evaluation in a closed form. Since the arc length parametrisation is obtained by inverting the function defined by this integral, it, too, rarely admits a closed form expression.

Take the case of the upper half of an ellipse $C = \{(x,y) : \dfrac{x^2}{a^2} + \dfrac{y^2}{b^2} = 1, y \geq 0\}$. The standard parametrisation of $C$ is $x = a\cos\theta, y = b\sin\theta, 0 \leq \theta \leq \pi$. But this is not in terms of the arc length $s$. To get the arc length parametrisation, we need to express $\theta$ in terms of $s$. The formula which links $s$ and $\theta$ is given by $\dfrac{ds}{d\theta} = \sqrt{(\dfrac{dx}{d\theta})^2 + (\dfrac{dy}{d\theta})^2}$, i.e. $\dfrac{ds}{d\theta} = \sqrt{a^2\sin^2\theta + b^2\cos^2\theta}$. So, integrating (and noting that $s = 0$ when $\theta = 0$), we get

$$s = \int_0^\theta \sqrt{a^2\cos^2 t + b^2\sin^2 t}\, dt \tag{14}$$

To express $\theta$ in terms of $s$, we first need to evaluate this integral as a closed form expression in $\theta, l(\theta)$, and then to solve the equation $s = l(\theta)$ for $\theta$. Here the first task is impossible and even in those cases where it is possible, the second one is not always easy (see Exercise (6.1)).

Thus the arc length parameter, although a very natural one from a theoretical point of view, is not a very convenient one from a practical point of view. It is therefore necessary to be able to tackle problems about curves which are parametrised in some other way. For piecewise smooth curves this is very easy to do. Suppose $\alpha : [a, b] \to \mathbb{R}^2$ is such a curve, say $C$. Let $\beta : [0, L] \to \mathbb{R}^2$ be the arc length parametrisation of $C$ (which exists and is unique by Theorems (6.7) and (6.9)). Then for each $t \in [a, b]$, there is a unique $s \in [0, L]$ such that $\alpha(t) = \beta(s)$. And, following the proof of theorem (6.7), the relationship between $s$ and $t$ is such that $\dfrac{ds}{dt} = l'(t) = |\alpha'(t)| = \sqrt{(\dfrac{dx}{dt})^2 + (\dfrac{dy}{dt})^2}$. Now, typically in a problem dealing with $C$, the answer is of the form $\int_0^L f(x,y)ds$ where $f$ is some function of two variables $x$ and $y$. This integral really means $\int_0^L f(\beta(s))ds$. Now $\beta(s) = \alpha(t)$. As $t$ varies from $a$ to $b$, $s$ varies from $0$ to $L$. So, by the rule of substitution for integrals (Theorem (5.6.3)), $\int_0^L f(\beta(s))ds$ equals $\int_a^b f(\alpha(t))\dfrac{ds}{dt}dt$, i.e. $\int_a^b f(\alpha(t))|\alpha'(t)|dt$. (Strictly speaking, Theorem (5.6.5) is applicable only when the integrals are proper and $|\alpha'(t)|$ is continuous everywhere. What we need here is an extension of the rule of substitution. It will be given as an exercise, Exercise (6.17) and (6.18)).

The major advantage of using parametrisations other than those in terms of the arc length is that they are often more natural and easier to work with. Another (although a relatively minor) advantage is that certain simplifying features of the problem may be more apparent with some other parameter than with the arc length. Consider, for example, Problem (6.10) above. The semi-circle as well as the density function are symmetric about the $y$-axis. So it is obvious that the centre of mass should lie on the $y$-axis, as it indeed comes out to be. But from the arc length parametrisation, this is not quite so obvious. If, instead, we had parametrised the upper semicircle as $x = r\sin\theta, y = r\cos\theta, \frac{\pi}{2} \le \theta \le \frac{\pi}{2}$, then the integral for $M_y$ (the moment about the $y$-axis) would have been $\int_{-\pi/2}^{\pi/2} \rho(r\sin\theta,$ $r\cos\theta)r^2\sin\theta d\theta$. The fact that $\rho$ is symmetric (in both (i) and (ii)) w.r.t. the $y$-axis implies that the integrand is an odd function of $\theta$ and since the interval of integration, viz. $[-\pi/2, \pi/2]$ is symmetric about 0, it is clear that the integral vanishes, even without calculation.

We already encountered a formula (see (14) in Section 6.3) for the areas of a surface of revolution in terms of parameters other than arc length (see also Exercise (6.3.21)). As another illustration of the use of such parameters, we do a problem which is a variation of Problem (6.10).

**6.11 Problem:** Do Problem (6.10) for the upper half of an ellipse $C = \{(x,y) :$ $\frac{x^2}{a^2} + \frac{y^2}{b^2} = 1, y \ge 0\}$.

**Solution :** Parametrise $C$ by $x = a\cos\theta, y = b\sin\theta, 0 \le \theta \le \pi$. The reasoning and calculations are similar to those in the solution of Problem (6.10). In (i), $\rho(x,y) = 1$ and so $M = \int_0^\pi \frac{ds}{d\theta}d\theta = \int_0^\pi \sqrt{a^2\sin^2\theta + b^2\cos^2\theta}d\theta$. This integral cannot be evaluated exactly. As for $M_y$ we have, $M_y = \int_0^\pi x\frac{ds}{d\theta}d\theta = a\int_0^\pi \cos\theta\sqrt{a^2\sin^2\theta + b^2\cos^2\theta}d\theta$. By Theorem (5.6.1), this integral also equals

$$a\int_0^\pi \cos(\pi - \theta)\sqrt{a^2\sin^2(\pi - \theta) + b^2\cos^2(\pi - \theta)}d\theta.$$

But this means $M_y = -M_y$. So $M_y = 0$. For $M_x$, we have

$$M_x = b\int_0^\pi \sin\theta\sqrt{a^2\sin^2\theta + b^2\cos^2\theta}d\theta$$

which transforms into $ab\int_{-1}^1 \sqrt{1 - e^2u^2}du$ where $u = \cos\theta$ and $e^2 = (a^2 - b^2)/a^2$. ($e$ is the eccentricity of the ellipse but that is incidental here.) This integral can be evaluated by parts (or otherwise) and comes out as $ab(\sqrt{1 - e^2} + \frac{\sin^{-1}e}{e})$. So, in the case of a uniform mass density, the centre of mass of $C$ lies at $(\overline{x}, \overline{y})$ where $\overline{x} = 0$ and $\overline{y} = \dfrac{ab(\sqrt{1 - e^2} + \frac{\sin^{-1}e}{e})}{\int_0^\pi \sqrt{a^2\sin^2\theta + b^2\cos^2\theta}d\theta}$.

When $\rho(x,y) = y$, the mass $M$ is $\int_0^\pi b\sin\theta\sqrt{a^2\sin^2\theta + b^2\cos^2\theta}d\theta$ which was just calculated as $ab(\sqrt{1-e^2} + \dfrac{\sin^{-1}e}{e})$. We leave it to check that $M_y = 0$ in this case too (which is also obvious from symmetry considerations). Finally

$$
\begin{aligned}
M_x &= \int_0^\pi y\rho(x,y)\frac{ds}{d\theta}d\theta = \int_0^\pi y^2\frac{ds}{d\theta}d\theta \\
&= \int_0^\pi b^2\sin^2\theta\sqrt{a^2\sin^2\theta + b^2\cos^2\theta}\,d\theta.
\end{aligned}
$$

This integral, too, cannot be evaluated exactly. So leaving it as it is, the centre of mass is at $(0,\bar{y})$ where $\bar{y} = \dfrac{b}{a}\dfrac{\int_0^\pi \sin^2\theta\sqrt{a^2\sin^2\theta + b^2\cos^2\theta}\,d\theta}{e\sqrt{1-e^2} + \sin^{-1}e}$. ∎

It may be argued that in the problem above, even with a parameter (viz. $\theta$) of our choice, we ended up with integrals which could not be evaluated in closed forms. But this is not as bad as what would happen with an arc length parameter. With an arc length parametrisation we could not even get started, because we could not express $x$ and $y$ (nor their derivatives $\dfrac{dx}{ds}$ and $\dfrac{dy}{ds}$) in terms of $s$. With $\theta$ as a parameter, we could not only start but also finish the solution. The fact that certain definite integrals had to be left unevaluated need not really worry us. Efficient methods (such as the Simpson's rule, see Section 5.9) are available to calculate such integrals with any desired degree of accuracy. And in real life problems, this is all one can really hope for. Even in the case of a circle of radius $r$, to say that its perimeter admits an exact expression, viz. $2\pi r$, is really an illusion, because the number $\pi$ has to be defined by some infinitistic process, such as the sum of some series or as some integral such as $\int_0^1 \dfrac{4dx}{1+x^2}$ or $\int_{-1}^1 \dfrac{dx}{\sqrt{1-x^2}}$. In real life problems $\pi$ is approximated by some rational number, such as $\dfrac{22}{7}$ or 3.1416. (There have been instances where certain legislative bodies have 'legally' equated $\pi$ with $\dfrac{22}{7}$!).

In terms of the analogy at the beginning of this section, the arc length parametrisation of a curve is like the official name of a person while other parametrisations are like the other names given to him. It so happens sometimes that hardly anybody knows a person by his official name. If you inquire by that name, even the next door neighbour may draw a blank. But if you inquire by his popular name, you may easily find out his whereabouts. You may not actually meet that person there, but you will get the information you need!.

A few other problems involving calculations of certain attributes of curves will be given in the exercises. The next higher dimensional analogue of a curve is a (two-dimensional) surface. We saw that a curve in $I\!\!R^2$ can be specified either as the set of all points of the form $\phi(x,y) = 0$ where $\phi$ is a function of two variables or parametrically as a function of the form $\alpha : [a,b] \to I\!\!R^2$. Similarly, a surface in $I\!\!R^3$ can be specified either in the form $\{(x,y,z) : \phi(x,y,z) = 0\}$

where $\phi$ is a function of three variables, or parametrically, as a function of the form $\alpha : D \to \mathbb{R}^3$ where $D$ is some subset of the plane. Let, for example, $S$ be the surface of the sphere, of radius $r$ centred at $(0,0,0)$. (This language is a legacy of the past. Nowadays, mathematicians always reserve the term 'sphere' for a 'hollow' or 'thin' sphere, i.e. for the boundary of the sphere. However, mathematicians of yesterday and non-mathematicians of today often use the term 'sphere' to mean the solid bounded by the sphere. The context, of course, usually makes it clear what is meant.) Then $S = \{(x, y, z) : \phi(x, y, z) = 0\}$ where $\phi(x, y, z) = x^2 + y^2 + z^2 - r^2$, for $(x, y, z) \in \mathbb{R}^3$.

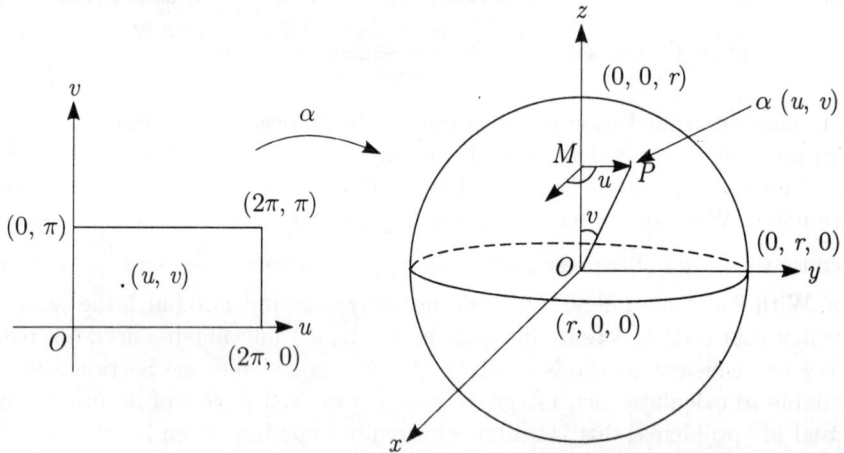

Figure 6.6.6 : Parametrisation of a Sphere

A parametrization for $S^2$ can be given as follows. Denote the co-ordinates of a point in $\mathbb{R}^2$ by $u$ and $v$ and those of a point in $\mathbb{R}^3$ by $x, y$ and $z$. Now let $D$ be the rectangle $\{(u, v) : 0 \leq u \leq 2\pi, 0 \leq v \leq \pi\}$ in the $(u, v)$-plane. Define $\alpha : D \to \mathbb{R}^3$ by $\alpha(u, v) = (x, y, z)$ where $x = r \sin v \cos u, y = r \sin v \sin u, z = r \cos u$. By eliminating $u$ and $v$ in these three equations, it is easy to show that $\alpha(u, v)$ lies on $S$ for all $(u, v) \in D$. Conversely, suppose $P$ is a point on $S$. Let the perpendicular from $P$ to the $z$ - axis fall at $M$. Let $v$ be the angle between the ray $\overrightarrow{OP}$ and the positive $z$-axis. Then $0 \leq v \leq \pi$ and the $z$-coordinate. of $P$ is $r \cos v$ (see Fig. 6.6.6.). Also $P$ lies on a horizontal circle centred at $M = (0, 0, r \cos v)$ and radius $r \sin v$. So if $u$ denotes the angle between the ray $\overrightarrow{MP}$ and the positive $x$-axis then $0 \leq u \leq 2\pi$ and the $x$ and $y$ co-ordinates of $P$ are $r \sin v \cos u$ and $r \sin v \sin u$ respectively. So $P$ is the image of $(u, v)$ under $\alpha$. It we want only the upper hemi-sphere, we only need to restrict $\alpha$ to the rectangle $\{(u, v) : 0 \leq u \leq 2\pi, 0 \leq v \leq \frac{\pi}{2}\}$. Of course, in this case we also have an easier alternate parametrisation $\beta : D \to \mathbb{R}^3$ where $D = \{(u, v) : u^2 + v^2 \leq r^2\}$ and $\beta(u, v) = (u, v, \sqrt{r^2 - u^2 - v^2})$ for $(u, v) \in D$.

The theory of parametrised surfaces can be developed analogously to that of parametrised curves. Instead of a 'line element', one considers what is called an 'area element', which, in essence is the image (under $\alpha$) of a rectangle in

$D$ whose sides are infinitesimally small, say $du$ and $dv$. The desired attributes such as surface areas, mass, moments etc. of $S$ can then be expressed as double integrals (over $D$) of suitable real-valued functions and can be evaluated by the method of iterated integrals given in Section 6.4. But the determination of the area element requires the differentiation of functions of two variables and so it is beyond our scope. So we omit problems such as finding the centres of mass of surfaces, even though the role of integration in them is exactly that played by it in similar problems about laminas, solids and curves.

There are, however, two classes of surfaces for which our methods sometimes suffice. One is planar surfaces. These are, of course, subsets of a plane and can be handled by the methods in Sections 6.3 and 6.4. The other class of surfaces is the surface of revolution. Let $S$ be a surface of revolution obtained by revolving a plane curve $C$ around a line $L^*$. We already obtained a formula for the surface area of $S$ (see Fig.6.6.5). Suppose now $\rho(x, y, z)$ is the areal (also called laminar) mass density (i.e. mass per unit area) at a point $(x, y, z)$ on $S$. Then our methods will not in general suffice to obtain an expression for the mass of $S$. But suppose the nature of the function $\rho(x, y, z)$ is such that it is constant on each of the circles which result by revolving the various points of $C$ around $L^*$. Then the contribution to the mass made by the band obtained by revolving a line element $ds$ around $L^*$ (see Fig.6.6.5 again) can be easily determined. And the mass of $S$ can then be found by integration w.r.t. $s$, or w.r.t. some other parameter for $C$.

We illustrate this technique in the following problem.

**6.12 Problem:** Find the centre of mass of the upper hemisphere $S = \{(x, y, z) : x^2 + y^2 + z^2 = r^2, z \geq 0\}$ assuming that the areal mass density at $(x, y, z)$ is (i) a constant (ii) proportional to $z$.

**Solution:**

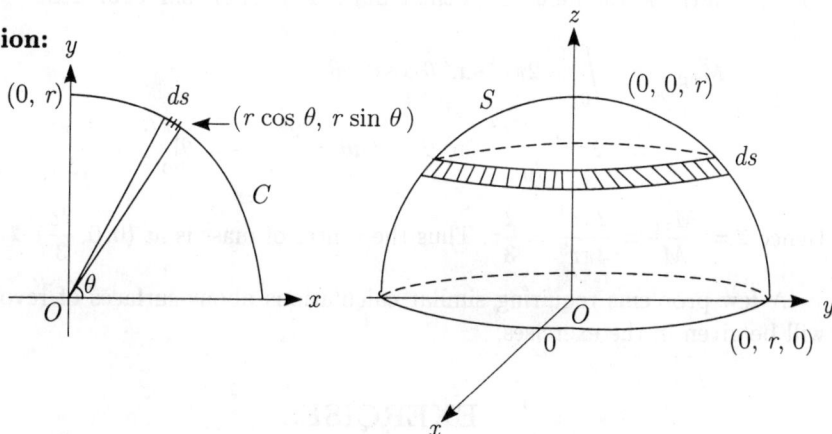

Figure 6.6.7 : Centre of Mass of a Hemispherical Surface

Let $C$ be the arc of the circle $\{(y, z) : y^2 + z^2 = r^2, y \geq 0, z \geq 0\}$ lying in the $y$-$z$ plane. Then $S$ is obtained by revolving $C$ around the $z$-axis. We parametrise $C$ by $y = r\cos\theta, z = r\sin\theta, 0 \leq \theta \leq \frac{\pi}{2}$ (see Fig.6.6.7). Let $\rho(x, y, z)$ be the mass density at $(x, y, z)$. In (i) we may assume $\rho \equiv 1$ everywhere and in (ii) $\rho(x, y, z) = z$. Then both in (i) and (ii), the centre of mass will lie, by symmetry consideration, on the $z$-axis. So we need to calculate only $\bar{z}$ which equals $M_{xy}/M$, where $M$ is the mass of $S$ and $M_{xy}$ is its moment about the $x$-$y$ plane.

Now in (i) $M$ is simply the area of $S$, which was already found to be $2\pi r^2$ (see the end of Section 6.3). To calculate $M_{xy}$, we must consider an area element, take the product of its mass with $z$ (i.e. the height of that element above the $x$-$y$ plane), add and take limit. Let us add together the contribution to $M_{xy}$ made by all area elements lying in the band obtained by revolving a line element $ds$ of the curve $C$. Taking the element $ds$ to be approximately a segment, we assume this band as conical. All points in this band lie (approximately) at the same height $z = r\sin\theta$ above the $x$-$y$ plane. From (10) in Section 6.3, the area of the band is (again approximately) $2\pi r\cos\theta ds$. So, following the usual recipe of integration,

$$M_{xy} = \int_0^{\pi/2} 2\pi r \sin\theta\, r\cos\theta \frac{ds}{d\theta} d\theta$$

$$= \pi r^3 \int_0^{\pi/2} \sin 2\theta d\theta = \frac{\pi r^3}{2}(\cos 0 - \cos \pi) = \pi r^3.$$

Hence $\bar{z} = \dfrac{M_{xy}}{M} = \dfrac{\pi r^3}{2\pi r^2} = \dfrac{r}{2}$. So the centre of mass is at $(0, 0, \frac{r}{2})$.

In (ii) the calculations are similar. In fact since $\rho(x, y, z) = z$, the mass $M$ now equals $M_{xy}$ for (i), which is $\pi r^3$ as just calculated. To find $M_{xy}$ for (ii), the calculations are similar except that to find the mass of the conical band we have to multiply its area by the density $z$ (which is approximately a constant on the band). So the moment of the band is $2\pi r\cos\theta r^2 \sin^2\theta ds$. Hence

$$M_{xy} = \int_0^{\pi/2} 2\pi r^3 \sin^2\theta \cos\theta r d\theta$$

$$= 2\pi r^4 \int_0^{\pi/2} \sin^2\theta \cos\theta d\theta = \frac{2\pi r^4}{3} \sin^3\theta \Big|_0^{\pi/2} = \frac{2\pi r^4}{3}.$$

Hence $\bar{z} = \dfrac{M_{xy}}{M} = \dfrac{2\pi r^4}{3\pi r^3} = \dfrac{2}{3}r$. Thus the centre of mass is at $(0, 0, \frac{2r}{3})$. ∎

A few problems requiring similar calculations about surfaces of revolution will be given in the exercises.

## EXERCISES

6.1  Prove that the arc length parametrisation of one arch of the cycloid $x = a(\theta - \sin\theta), y = a(1 - \cos\theta) : 0 \leq \theta \leq 2\pi$ (see Exercise 4.12.6) is given by

$$x = 2a\cos^{-1}(\frac{4a - s}{4a}) + \frac{(s - 4a)\sqrt{s(8a - s)}}{8a}, y = \frac{s(8a - s)}{8a}; 0 \leq s \leq 8a.$$

(This example illustrates how an arc length parametrisation of a curve, even when explicitly possible, can be a lot clumsier than some other parametrisation).

6.2 Instead of one arch of the cycloid in the last exercise, suppose we consider two arches, that is, the curve

$$x = a(\theta - \sin\theta), y = a(1 - \cos\theta); 0 \le \theta \le 4\pi.$$

What will be the arc length parametrisation now?

6.3 Define $\alpha : [0, 2\pi] \longrightarrow \mathbb{R}^3$ by

$$\alpha(\theta) = \begin{cases} \cos\theta\mathbf{i} + \sin\theta\mathbf{j} & 0 \le \theta \le \pi \\ (-2 + \cos(\theta - \pi))\mathbf{i} + \sin(\theta - \pi)\mathbf{j} & \pi < \theta \le 2\pi. \end{cases}$$

Sketch $\alpha$. Verify that it is piecewise smooth and has a cusp at $\theta = \pi$, i.e. at $(-1, 0)$.

6.4 Define $\alpha : [0, 2\pi] \to \mathbb{R}^2$ by

$$\alpha(\theta) = \begin{cases} \cos\theta\mathbf{i} + \sin\theta\mathbf{j} & 0 \le \theta \le \pi \\ (1 - 2\cos(\theta - \pi))\mathbf{i} + 2\sin(\theta - \pi)\mathbf{j} & \pi < \theta \le 2\pi. \end{cases}$$

Sketch $\alpha$. Verify that it is piecewise smooth and has a cusp at $\theta = \pi$, i.e. at $(-1, 0)$. How does this cusp differ from that in the last example?

6.5 (a) Prove that the function $f : [-1, \infty) \to \mathbb{R}$ defined by $f(t) = -\sqrt{t^2 + t^3}$ if $-1 \le t \le 0$ and $f(t) = \sqrt{t^2 + t^3}$ if $t > 0$ is differentiable at $0$, but not at $-1$. If we let $g(t) = \sqrt{t^2 + t^3}$ for all $t \ge -1$, is $f$ differentiable at $0$?

(b) Sketch the curve $C$ given (implicitly) by $y^2 = x^2 + x^3$. Show that $C$ can be parametrised by

$$\alpha(t) = \begin{cases} (-t - 2)\mathbf{i} - \sqrt{(t+2)^2 - (t+2)^3}\mathbf{j} & -\infty < t \le -2 \\ (-t - 2)\mathbf{i} + \sqrt{(t+2)^2 - (t+2)^3}\mathbf{j} & -2 \le t < 1 \\ t\mathbf{i} - \sqrt{t^2 + t^3}\mathbf{j} & -1 \le t < 0 \\ t\mathbf{i} + \sqrt{t^2 + t^3}\mathbf{j} & 0 \le t < \infty. \end{cases}$$

(c) Using (a), show that the curve $\alpha$ in (b) is not smooth at $t = -1$, but that it has a smooth double point at $O$. Find the two values of $t$ for which $\alpha(t) = \mathbf{0}$, the zero vector.

6.6 Let $\alpha, \beta$ be vector valued functions from $\mathbb{R}$ to $\mathbb{R}^n$. By differentiability of $\alpha$ we mean that if $\alpha = (\alpha_1, \dots, \alpha_n)$ then each coordinate function $\alpha_i : \mathbb{R} \to \mathbb{R}$ is differentiable. Similarly for $\beta$. Let $f : \mathbb{R} \to \mathbb{R}$ any function. Prove that

(i) If $\alpha, \beta$ are differentiable at $t_0 \in \mathbb{R}$ then so are $\alpha + \beta$ and $\alpha \cdot \beta$. Further, $(\alpha + \beta)' = \alpha' + \beta'$ and $(\alpha.\beta)' = \alpha'.\beta + \alpha \cdot \beta'$. (Here $(\alpha.\beta)(t)$ denotes the dot product of the vectors $\alpha(t)$ and $\beta(t)$. That is, if $\alpha = (\alpha_1, \ldots, \alpha_n)$ and $\beta = (\beta_1, \ldots, \beta_n)$ then $(\alpha \cdot \beta)(t) = \sum_{i=1}^{n} \alpha_i(t)\beta_i(t)$.)

(ii) if $f, \alpha$ are differentiable at $t_0$, so is $f\alpha$ and further $(f\alpha)' = f'\alpha + f\alpha'$. (Here, too, $(f\alpha)(t)$ means the vector $f(t)\alpha(t)$, i.e. the multiple of $\alpha(t)$ by the scalar $f(t)$.)

(iii) if $f$ is differentiable at $t_0$ and $\alpha$ is differentiable at $f(t_0)$, then the composite $(\alpha \circ f)(t) = \alpha(f(t))$ is differentiable at $t_0$ and $(\alpha \circ f)'(t_0) = f'(t_0)\alpha'(f(t_0))$. (Chain rule). [*Hint*: All three results follow from the corresponding results applied to the coordinate functions.]

6.7   Define $\alpha : \mathbb{R} \to \mathbb{R}^2$ by $\alpha(t) = \cos t \mathbf{i} + \sin t \mathbf{j}$ for $t \in \mathbb{R}$. Prove that $\alpha(0) = \alpha(2\pi)$ but that there is no $c \in (0, 2\pi)$ for which $\alpha'(c) = \mathbf{0}$. (In other words, the analogue of Rolle's and Lagrange's Mean Value Theorem fail for vector valued functions. Can you explain exactly what goes wrong, when everything is defined in terms of the derivatives of co-ordinate functions?)

6.8   Following the notation of Definition (6.1), show that for every $c_1 \in (a, b)$, the restriction of $\alpha$ to $[a, c_1]$ is equivalent to the restriction of $\beta$ to $[c, f(c_1)]$ and similarly, the restriction of $\alpha$ to $[c_1, b]$ is equivalent to the restriction of $\beta$ to $[f(c_1), d]$. (Verbally, corresponding portions of equivalent curves are equivalent.) Conversely show that if $\alpha, \beta$ are curves with domains $[a, b]$ and $[c, d]$ resperctively, and $c_1 \in (a, b), c_2 \in (c, d)$ are such that $\alpha/[a, c_1] \sim \beta/[c, c_2]$ and $\alpha/c_1, b] \sim \beta/[c_2, d]$ then $\alpha \sim \beta$. (In other words, when equivalent pairs of curves are concatenated together, we get equivalent curves.)

6.9   The angle between two curves intersecting at a point is defined as the angle between their tangents at that point (asuming that they exist and are nonzero). Give a precise statement of the fact that this concept is preserved under equivalence of curves. Prove it.

*6.10  Show directly that the improper integral

$$\int_0^\pi \sqrt{1 + \sin^2 \frac{1}{t} - \frac{2}{t} \sin \frac{1}{t} \cos \frac{1}{t} + \frac{1}{t^2} \cos^2 \frac{1}{t}} \, dt$$

is divergent. [*Hint*: Consider intervals of the form $\left[ \dfrac{1}{(2n+1)\pi}, \dfrac{1}{(2n+\frac{2}{3})\pi} \right]$, $n \in \mathbb{N}$. Obtain lower bounds on the integrand over these intervals. ]

6.11  Find the centre of mass of one arc of a cycloid (Exercise (6.1)) assuming that the density is (i) uniform (ii) proportional to $y$.

6.12 Suppose $S$ is a surface obtained by revolving a rectifiable, piecewise smooth plane curve $C$ around a line $L^*$ in its plane. Assume $C$ lies entirely on one side of $L^*$. Prove that the surface area of $S$ equals $2\pi L^* a$ where $L^*$ is the length of $C$ and $a$ is the perpendicular distance of the centre of gravity of $C$ from $L^*$. (This result, which is analogous to that in Exercise (4.20), is also due to Pappus.)

6.13 Use the last exercise to find the surface area of a torus (see Exercise (6.3.17)).

6.14 Find the centre of mass of the curved surface of a right circular cone assuming that the density at a point is (i) uniform (ii) proportional to its distance from the apex.

6.15 Find the centre of gravity of one half of a torus surface using Pappus theorem (Exercise (6.12)). (Note that there are two ways to divide a torus into two equal halves).

6.16 Give parametrisations of (i) a torus surface  (ii) a right circular cylinder of radius $a$.

6.17 Suppose $h : [c,d] \to [a,b]$ is continuous with $h'$ continuous and positive on $(c,d)$. Assume $h(c) = a$ and $h(d) = b$. Suppose $f : (a,b) \to \mathbb{R}$ is continuous. Prove that the integrals $\int_a^b f(x)dx$ and $\int_c^d f(h(t))h'(t)dt$ either both converge or both diverge and moreover that when they converge their values are the same. (This is a slight extension of Theorem (5.6.5). Note that one or both the intervals could be infinite. When this is so, minor modifications are needed in the notations and in the proof. For example, if $c = -\infty$, then by $h(c)$ we mean $\lim\limits_{x \to -\infty} h(x)$.)

6.18 Extend the last exercise to the case when $h'$ is positive and continuous except at a finite number of points of $(c,d)$.

6.19 Let $0 \le \alpha < 2\pi$. Let $T_1$ be the counterclockwise rotation of the $x$-$y$ plane (around the origin) through the angle $\alpha$ and let $T_2$ be the reflection of the plane in the line which makes an angle $\alpha/2$ with the (positive) $x$-axis. Prove that the images of a point $(x,y)$ in the plane under $T_1$ and $T_2$ are the points $(x\cos\alpha - y\sin\alpha, x\sin\alpha + y\cos\alpha)$ and $(x\cos\alpha + y\sin\alpha, x\sin\alpha - y\cos\alpha)$ respectively. (It is much easier to see this using polar coordinates to be studied in Section 6.8.)

6.20 Prove that any point in a plane is uniquely determined by its distances from any three non-collinear points in that plane. That is, if $A, B, C$ are non-collinear and $P, Q$ are such that $|PA| = |QA|, |PB| = |QB|$ and $|PC| = |QC|$ then $P$ and $Q$ must coincide with each other.

*6.21 Suppose $T$ is a congruence of the plane which fixes the origin. Prove that $T$ is either a rotation or a reflection followed by a rotation. [*Hint* : Apply the last exercise taking $A, B, C$ to be the images under $T$ of the points $(0,0), (1,0)$ and $(0,1)$.] Deduce that the only congruences of the plane are composites of rotations, reflections and translations. (As a result, any property of a planar set which is invariant under these three basic transformations is a geometric property. Congruences of the three dimensional space can also be characterised. But that is not so easy.)

6.22 Assume that the Earth is spherical and that the population density at a point on it decreases linearly with the latitude from 1 (at points on the equator) to 0 (at the poles). What is the probability that a person on earth happens to be from the tropics (i.e. latitudes from $23^{1/2°}$ South to $23^{1/2°}$ North)?

## 6.7    Curvature and Torsion of a Curve

Our very idea of a curve is that it is not straight. Even though we include straight lines among curves, we tend to think of them as something degenerate. (It would be a pity indeed if the 'curves' of a beauty queen turn out to be straight lines!) A 'true' curve has to have a bend in it. No bend, no curve. Just as 'temperature' measures how hot an object is, 'curvature' measures the degree of bend in a curve. Or, by a verbal quibble, the curvature of a curve measures how curved a curve is ! So, as a concept which deals with the most vital feature of a curve (viz. its bending), it stands to reason that the curvature is a very important, probably the most important, feature of a curve if only from a linguistic point of view!

There are, of course, less trivial reasons. The most important among them, probably, is that, as we shall see later, the curvature plays a crucial role in studying the motion along a curve. This is also evident in day-to-day life. Driving along a curved road entails more care than along a straight road. Especially, where the road bends abruptly, one has to reduce the speed lest the vehicle goes off the road. These things can be stated very precisely in terms of the curvature of the road. Like slope, curvature is an important safety consideration in the design of mountain roads.

Let us now define the curvature mathematically. Throughout we shall assume we have a space curve, i.e. a curve in $\mathbb{R}^3$, $\alpha : [a, b] \to \mathbb{R}^3$. We shall denote the parameter by $t$. For $t \in [a, b]$, $\alpha(t)$ is a vector in $\mathbb{R}^3$. We shall denote it by $\mathbf{r}$. We let $\mathbf{i}$, $\mathbf{j}$, $\mathbf{k}$ denote unit vectors along the positive $x$-, $y$- and $z$-axes respectively. Then we can write $\mathbf{r} = x\mathbf{i} + y\mathbf{j} + z\mathbf{k}$. Differentiation w.r.t. the parameter $t$, will be denoted by a dot. Thus we have

$$\frac{d\mathbf{r}}{dt} = \dot{\mathbf{r}} = \dot{x}\mathbf{i} + \dot{y}\mathbf{j} + \dot{z}\mathbf{k} = \frac{dx}{dt}\mathbf{i} + \frac{dy}{dt}\mathbf{j} + \frac{dz}{dt}\mathbf{k} \qquad (1)$$

provided $x, y, z$ are all differentiable. We assume $\alpha$ is piecewise smooth. Then by Theorem (6.7), there is an equivalent arc length parametrisation $\beta : [0, L] \to \mathbb{R}^3$ where $L$ is the arc length of $\alpha$. So we shall think of $x, y, z$ and hence $\mathbf{r}$ as functions of $s \in [0, L]$. Since $\beta$ is also piecewise smooth, these functions are continuously differentiable except at a finite number of points of $[0, L]$. Differentiation w.r.t. the arc length parameter $s$ will be denoted by a dash ($'$). Thus, analogous to (1) we have

$$\frac{d\mathbf{r}}{ds} = \mathbf{r}' = x'\mathbf{i} + y'\mathbf{j} + z'\mathbf{k} = \frac{dx}{ds}\mathbf{i} + \frac{dy}{ds}\mathbf{j} + \frac{dz}{ds}\mathbf{k} \qquad (2)$$

wherever $\dfrac{d\mathbf{r}}{ds}$ exists. As was mentioned in the last section, the curvature and torsion of a curve are local attributes, that is, we talk of a curvature of a curve *at a point* on it. There is no such thing as the curvature or torsion of the entire curve as a whole. This is in sharp contrast with the length of a curve which is a global concept. There is no such thing as the length at a point. Being a local concept, the study of curvature mostly entails differentiation rather than integration. So it is a little ironic to study it in the chapter on applications of integration. But, as emphasized in Section 5.5, the differential and the integral calculus need not be thought of as water-tight compartments. Secondly, we did need integrals to express the arc length parametrisation in terms of more convenient parametrisations, as was done in the last section. In fact, in the last section we saw that the parameter $s$, as a function of the parameter $t$, is given by

$$s = s(t) = \int_a^t |\alpha'(u)| du = \int_a^t \sqrt{(\frac{dx}{du})^2 + (\frac{dy}{du})^2 + (\frac{dz}{du})^2} du, \text{ for } t \in [a, b] \quad (3)$$

and hence, by the second from of the Fundamental Theorem of Calculus,

$$\frac{ds}{dt} = \dot{s} = \sqrt{(\frac{dx}{dt})^2 + (\frac{dy}{dt})^2 + (\frac{dz}{dt})^2} = \sqrt{\dot{\mathbf{r}} \cdot \dot{\mathbf{r}}} = |\dot{\mathbf{r}}| \qquad (4)$$

for all values of $t$ where $\dfrac{dx}{dt}, \dfrac{dy}{dt}, \dfrac{dz}{dt}$ are all continuous. As $\alpha$ is piecewise smooth, this condition is satisfied for all $t$, except possibly for a finite number of values of $t$. Throughout our discussion, we shall assume that these values are excluded. Since the concepts to be defined and studied in this section are local in nature, we may therefore suppose that the curve $\alpha$ is smooth everywhere, because this would be the case in some neighbourhood of the point of our consideration. Then the curve $\beta$ is also smooth as a function of $s$, the arc length parameter. We shall not need (3) which expresses $s$ as a function of $t$. However, we shall frequently need its consequence (4). In fact as an immediate application of (4) we get the following simple result.

**7.1 Theorem.** If $\alpha$ is a smooth curve and $\beta$ is its arc length parametrisation, then $\beta(s) = \mathbf{r}'$ (i.e. $\dfrac{d\mathbf{r}}{ds}$) is the unit tangent vector to it at the point $\beta(s)$.

**Proof.** This was essentially proved already in Theorem (4.12.2). (There we considered a plane curve, but the argument goes through for a space curve too.) Thus $\dot{\mathbf{r}}$ is a vector parallel to the tangent vector at $\alpha(t) = \beta(s)$. By the chain rule, (see Exercise (6.6)) $\mathbf{r}' = \dfrac{d\mathbf{r}}{ds} = \dfrac{d\mathbf{r}}{dt} / \dfrac{ds}{dt}$. (This is meaningful, because the curve being smooth, $\dfrac{ds}{dt}$ is positive throughout.) But by (4), $\dfrac{ds}{dt}$ is precisely $\left| \dfrac{d\mathbf{r}}{dt} \right|$. Hence $\mathbf{r}'$ is a unit vector. As $\mathbf{r}'$ and $\dot{\mathbf{r}}$ are positive scalar multiples of each other, they have the same direction. So $\mathbf{r}'$ is the unit tangent vector at $\beta(s)$. ∎

It is customary to denote $\mathbf{r}'$ by $\mathbf{u}$, the 'u' coming from a 'unit vector'. As $s$ changes, so does $\mathbf{u}$. So we can think of $\mathbf{u}$ as a function of $s$ (and hence of any other parameter w.r.t. which the curve can be parametrised). Instead of introducing some other symbol to denote this function, it is customary to write $\mathbf{u} = \mathbf{u}(s)$ or $\mathbf{u} = \mathbf{u}(t)$ etc. (In fact this practice is common for other functions of the parameter too. Thus we write $x = x(t), y = y(t)$ etc. instead of $x = f(t), y = g(t)$ etc. This is a little confusing initially, because when you write $x = x(t)$ and $x = x(s)$, strictly speaking, they are different functions, of different variables. Perhaps the best way to avoid this confusion is to read an expression like $\mathbf{u}(s)$ as "$\mathbf{u}$, treated as a function of $s$" rather than the usual "$\mathbf{u}$ of $s$." But with a little practice, the confusion can be avoided even without such elaborate reading.)

We are now ready to define the curvature of a curve at a point. Intuitively, it is a measure of how rapidly the curve bends. In particular we expect that a straight line should have no (i.e. zero) curvature at every point, that the curvature of a circle should be the same at all its points and further that the larger the radius of the circle, the smaller be its curvature. This expectation is based on the everyday experience that when we jog on a circular running track, we are conscious of the change of direction, but if we jog along a ring road around a big city, we feel as if we are jogging along a straight road. For a curve like an ellipse, we expect that its maximum and minimum curvature should occur at the ends of its major and minor axes respectively. Howsoever we define the curvature, it must meet these minimum requirements.

Mathematically, the curvature is the rate of change of the direction of the curve. The direction of a curve at any point is, of course, the direction of the tangent to the curve at that point. Since we are interested only in the direction and not the magnitude of the tangent, we take a unit vector along it. We have already denoted it by **u**. The rate of change is mathematically represented by a derivative. So the curvature is the derivative of **u**. This raises two difficulties. First, since **u** is a vector valued function, its derivative will also be a vector, while we want the curvature to be a scalar. But this difficulty can be resolved simply by letting the curvature be the magnitude of the derivative of **u**. The second difficulty is to decide the variable w.r.t. which **u** should be differentiated. Suppose a smooth curve is parametrised as $\mathbf{r} = \mathbf{r}(t)$. Here **u** can be expressed as a function of $t$, specifically, $\mathbf{u} = \dot{\mathbf{r}}/|\dot{\mathbf{r}}|$. So it is natural to define the curvature as $|d\mathbf{u}/dt|$, assuming, of course, that the derivative exists. But if we do so, then the curvature will not be an intrinsic geometric property of the curve, but will depend on a particular parametrisation of it. In more precise terms, two curves which are mutually equivalent in the sense of Definition (6.1), may have different curvatures. As explained in the last section, for intrinsic geometric properties of a curve, the arc length parametrisation is the ideal (although not necessarily a practicable) choice. So while defining curvature, we differentiate **u** w.r.t. $s$, the arc length, that is, we consider $\dfrac{d\mathbf{u}}{ds}$ instead of $\dfrac{d\mathbf{u}}{dt}$.

(There are situations where $\dfrac{d\mathbf{u}}{dt}$ is more relevant than $\dfrac{d\mathbf{u}}{ds}$. Suppose for example $t$ denotes the time. Then $\dfrac{d\mathbf{u}}{dt}$ is the time rate of change of direction. For a person driving along a curved road, it is like the apparent curvature for him. It will depend not only on the road but also upon the speed. We feel this in everyday life. A person walking along a curved road hardly notices its curvature. A cyclist feels it but needs no special caution (unless the bend is very sharp). But a car driver has to exercise care to avoid an accident. Despite this difference, we still talk of the curvature as an intrinsic geometric attribute of the road, just like its length. So it still makes sense to define curvature in terms of $d\mathbf{u}/ds$. At any rate, the chain rule allows us to obtain the apparent curvature, viz. $d\mathbf{r}/dt$ easily from $d\mathbf{r}/ds$ by merely multiplying by the speed $ds/dt$.)

**7.2 Definition.** Suppose **u**, the unit tangent vector, is a differentiable function of $s$, the arc length (which means, each of the components $\dfrac{dx}{ds}, \dfrac{dy}{ds}$ and $\dfrac{dz}{ds}$ is differentiable w.r.t. s). Then its **curvature** denoted by $\kappa(s)$ (read "kappa of $s$") is the magnitude of the vector $|d\mathbf{u}/ds|$. That is,

$$\kappa(s) = |d\mathbf{u}/ds| = |\mathbf{u}'| = \sqrt{\left(\frac{d^2x}{ds^2}\right)^2 + \left(\frac{d^2y}{ds^2}\right)^2 + \left(\frac{d^2z}{ds^2}\right)^2}. \tag{5}$$

Before studying the curvature further, let us verify that this definition satisfies our expectations about the curvature of a straight line and of a circle. For a straight line, the vector **u** is a constant vector (along the direction of that line) and so $d\mathbf{u}/ds$ vanishes identically. Hence $\kappa = 0$ at all points. For a circle of radius $a$, centreed at the origin, in the $x$-$y$ plane, an arc length parametrisation is given by (cf. Fig.6.6.4 (b))

$$\mathbf{r} = a \cos \frac{s}{a}\mathbf{i} + a \sin \frac{s}{a}\mathbf{j} + 0\mathbf{k} \tag{6}$$

Differentiating, $\mathbf{u} = d\mathbf{r}/ds = -\sin\frac{s}{a}\mathbf{i} + \cos\frac{s}{a}\mathbf{j}$ (which is a unit vector as it ought to be). Differentiating again,

$$\kappa = |d\mathbf{u}/ds| = \left| -\frac{1}{a}\cos\frac{s}{a}\mathbf{i} - \frac{1}{a}\sin\frac{s}{a}\mathbf{j} \right| = \frac{1}{a} \tag{7}$$

Thus the curvature of a circle is uniform all over and inversely proportional to its radius. For this reason, for any point on any curve (not necessarily a circle), where the curvature $\kappa$ is non-zero, $\frac{1}{\kappa}$ is known as the **radius of curvature** at that point. The peculiar name suggests that as far as the curvature is concerned, the curve behaves locally like a circle. In fact, more is true as we shall see shortly.

The calculation of the curvature at a point on an ellipse will be given in the exercises (Exercise (7.3)). It is more complicated than the calculation of curvature for a line or a circle $T$. The reason, of course, is that an ellipse does not admit an arc length parametrisation in a closed form, as explained in the last section. For such curves, it is necessary, first of all, to derive a formula for the curvature in terms of parameters other than the arc length. This, too, will be given in the exercises (Exercise (7.2)).

Since the curvature $\kappa$ is, by definition, the length of the vector $\mathbf{u}'$, wherever $\kappa \neq 0$, we can write

$$\mathbf{u}' = \kappa\mathbf{p} \tag{8}$$

where $\mathbf{p}$ is a unit vector in the direction of $\mathbf{u}'$. (If $\kappa = 0$, then $\mathbf{u}'$ is the zero vector and $\mathbf{p}$ is undefined). This vector $\mathbf{p}$ is called the **principal normal** or simply the **normal** to the curve at the point $\beta(s)$. The name is justified because, as we now show, it is perpendicular to $\mathbf{u}$, which is a unit vector along the tangent to the curve.

**7.3 Theorem:** The principal normal vector, wherever defined, is perpendicular to the tangent to the curve at that point, that is

$$\mathbf{u} \cdot \mathbf{p} = 0 \tag{9}$$

**Proof.** Since $\mathbf{u}$ is a unit vector, we have

$$\mathbf{u} \cdot \mathbf{u} = 1. \tag{10}$$

We now apply the result of Exercise (6.6(i)), to the derivative of (10) w.r.t. $s$. Since $\mathbf{u}' = d\mathbf{u}/ds = \kappa\mathbf{p}$, we get

$$0 = \frac{d}{ds}(\mathbf{u} \cdot \mathbf{u}) = \frac{d\mathbf{u}}{ds} \cdot \mathbf{u} + \mathbf{u} \cdot \frac{d\mathbf{u}}{ds} = 2\kappa\mathbf{u} \cdot \mathbf{p}.$$

But $\kappa \neq 0$ (or else $\mathbf{p}$ is undefined). So (9) holds. ∎

We emphasise again that like **u**, the normal vector **p** changes from point to point on the curve. So **p** is a vector valued function of $s$. When we merely write **p**, we mean **p** at a particular point on the curve. If the curve is a circle, then from (7), we see that $\mathbf{p}(s) = -\cos\frac{s}{a}\mathbf{i} - \sin\frac{s}{a}\mathbf{j}$. So, the normal to the circle lies along its radius and towards the centre, as expected.

A particular consequence of (9) is that the two vectors **u** and **p** at any point $P$ on the curve are not collinear, i.e. are not multiples of each other. So they span a plane. This plane is called the **plane of the curve** or **osculating plane** at that point. (The latter name is peculiar and will be explained shortly.) If the entire curve lies in a plane (e.g. a circle or an ellipse) then at every point, the plane of the curve is the same, viz. the plane containing the entire curve. But for a non-planar curve, such as a helix (see Fig. 4.12.3), the plane of the curve changes from point to point. For such a curve, we may say that an infinitesimally small portion of the curve containing the point $P$ lies in the plane of the curve at $P$. The plane through $P$, perpendicular to the vector **u** is called the **normal plane** of the curve at $P$, because every vector in it is perpendicular to **u**. Clearly **p** also lies in the normal plane. In fact, it is easy to see that the normal plane and the osculating planes are always at right angles and **p** lies along the line of their intersection. (If the original curve is a straight line, then there is no unique plane containing it. But the normal plane is still defined.) Figure 6.7.1 shows these two planes at a point $P$ on a non-planar curve $C$.

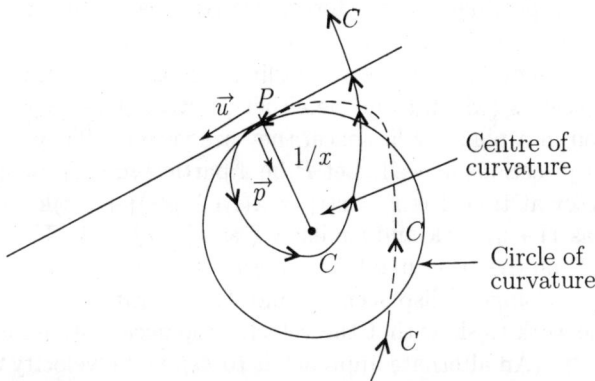

Fig.6.7.1 : Circle of Curvature and Plane of the Curve.

Now suppose we walk along the principal normal **p** starting from the point $P$, for a distance $\frac{1}{\kappa}$ (i.e. the radius of curvature). The point so reached has position vector $\mathbf{r} + \mathbf{u}'$. It is called the **centre of curvature** of the curve at the point $P$, and the circle of radius $\frac{1}{\kappa}$, with this as the centre, drawn in the plane of the curve (i.e. the plane spanned by **u** and **p**) is called the **circle of curvature** of the curve $C$ at $P$. (See Fig.6.7.1). Denote it by $S$. Then $P$ lies on $S$ and the tangent to $S$ at $P$ would be perpendicular to the radius through $P$. But that means that the tangent to $S$ is along $\mathbf{u}$, which is also the direction of the tangent to the original curve $C$ at $P$. In other words, the circle of curvature touches the curve at $P$. There are, of course, many other curves which do that. But the circle of curvature stands out in that, by its very construction, $S$ has the same radius of curvature (viz. $\frac{1}{\kappa}$) at $P$ as $C$ does. In other words, not only $S$ touches $C$ at $P$ but it also bends at the same rate as $C$. For this reason, the circle of curvature is given a rather romantic name, the **osculating circle**. (Osculate means to kiss !). And the plane of the curve (in which the osculating circle lies) is also called the **osculating plane.** The tangent line provides a first order approximation to a curve. But the osculating circle provides a second order approximation to it, because it not only shares with $C$ its direction but also the rate of change of the direction.

It is, therefore, to be expected that in situations where only derivatives upto the second order are involved, the tangent line and the circle of curvature should capture the essential local features of what happens on a curve. As mentioned at the end of Section 4.3, derivatives of order higher than the second are rarely encountered in physical applications. It is for this reason that the curvature is such an important attribute of a curve.

As an immediate illustration of these comments we study the motion of a particle on a curve, because in the study of motion the first and the second derivatives correspond, respectively, to the velocity and acceleration. The real life importance of velocity is all too apparent. Acceleration is important because, according to Newton's second law of motion, a force acting on an object causes an acceleration proportional to the force. Derivatives of higher order almost never figure in the study of motion.

In the case of a motion along a straight line, we may regard the velocity and acceleration as scalars (which could be either positive or negative). But when we study motion in a plane or in space, they are vectors. They have not only magnitudes but directions as well. Let $P$ be a particle moving in space so that its position vector at time $t$ is $\mathbf{r} = \mathbf{r}(t) = x(t)\mathbf{i} + y(t)\mathbf{j} + z(t)\mathbf{k}$. Note that we have *defined* $\dot{\mathbf{r}}$ as $\dot{x}\mathbf{i} + \dot{y}\mathbf{j} + \dot{z}\mathbf{k}$ and similarly $\ddot{\mathbf{r}}$ as $\ddot{x}\mathbf{i} + \ddot{y}\mathbf{j} + \ddot{z}\mathbf{k}$. The velocity and acceleration of $P$, on the other hand, are physical concepts defined, respectively, as the time rate of change of displacement and the time rate of change of velocity. So it takes some work to show that the velocity and acceleration indeed equal $\dot{\mathbf{r}}$ and $\ddot{\mathbf{r}}$ respectively. (An alternate approach is to *define* the velocity as $\dot{\mathbf{r}}$ and the acceleration as $\ddot{\mathbf{r}}$. But then some work is needed to show that $\dot{\mathbf{r}} = \dot{x}\mathbf{i} + \dot{y}\mathbf{j} + \dot{z}k$ and $\ddot{\mathbf{r}} = \ddot{x}\mathbf{i} + \ddot{y}\mathbf{j} + \ddot{z}\mathbf{k}$. It is really a matter of taste as to which approach one follows, since they ultimately entail the same work. The reason we prefer to define $\dot{\mathbf{r}}$ as $\dot{x}\mathbf{i} + \dot{y}\mathbf{j} + \dot{z}\mathbf{k}$ is that elementary results about the derivatives of vector valued functions, such as those mentioned in Exercise (6.6), can be obtained quickly in terms of the derivatives of the component function. We already used one such result in the proof of Theorem (7.3).)

The proof of the equivalence of the two approaches is simplified by the following lemma which is also of some independent interest.

**7.4 Lemma:** Let $\mathbf{w} = \mathbf{w}(h) = f_1(h)\mathbf{i} + f_2(h)\mathbf{j} + f_3(h)\mathbf{k}$ be a vector valued function of a real variable $h$, defined in a deleted neighbourhood of a point $h_0$. Then $\lim_{h \to h_0} \mathbf{w}(h)$ exists if and only if $\lim_{h \to h_0} f_i(h)$ exists for $i = 1, 2, 3$ and further, when this happens,

$$\lim_{h \to h_0} \mathbf{w}(h) = \left[\lim_{h \to h_0} f_1(h)\right] \mathbf{i} + \left[\lim_{h \to h_0} f_2(h)\right] \mathbf{j} + \left[\lim_{h \to h_0} f_3(h)\right] \mathbf{k}. \tag{11}$$

**Proof.** First assume $\lim_{h \to h_0} f_i(h)$ exists for $i = 1, 2, 3$. Call it $L_i$. Then we have to show that $\lim_{h \to h_0} \mathbf{w}(h) = L_1 \mathbf{i} + L_2 \mathbf{j} + L_3 \mathbf{k}$. The definition of a limit of a vector-valued function is analogous to that of the limit of a scalar-valued function, except that instead of the difference between two real-numbers, we take the magnitude of the difference between two vectors. Thus let $\epsilon > 0$ be given. Then there exist $\delta_1, \delta_2, \delta_3$, all positive such that for $i = 1, 2, 3$, $|f_i(h) - L_i| < \dfrac{\epsilon}{\sqrt{3}}$ whenever $0 < |h - h_0| < \delta_i$ Let $\delta = \min\{\delta_1, \delta_2, \delta_3\}$. Then

$$|f_i(h) - L_i| < \frac{\epsilon}{\sqrt{3}} \text{ for } i = 1, 2, 3, \text{ whenever } 0 < |h - h_0| < \delta \tag{12}$$

Now

$$
\begin{aligned}
|\mathbf{w}(h) &- (L_1 \mathbf{i} + L_2 \mathbf{j} + L_3 \mathbf{k})| \\
&= |(f_1(h) - L_1)\mathbf{i} + (f_2(h) - L_2)\mathbf{j} + (f_3(h) - L_3)\mathbf{k}| \\
&= \sqrt{(f_1(h) - L_1)^2 + (f_2(h) - L_2)^2 + (f_3(h) - L_3)^2} \tag{13} \\
&< \sqrt{\frac{\epsilon^2}{3} + \frac{\epsilon^2}{3} + \frac{\epsilon^2}{3}} = \epsilon \text{ whenever } 0 < |h - h_0| < \delta \text{ by (12)}
\end{aligned}
$$

So $\lim_{h \to h_0} \mathbf{w}(h) = L_1 \mathbf{i} + L_2 \mathbf{j} + L_3 \mathbf{k}$.

Conversely, suppose $\lim_{h \to h_0} \mathbf{w}(h)$ exists and equals $L_1 \mathbf{i} + L_2 \mathbf{j} + L_3 \mathbf{k}$ (say) where $L_1, L_2, L_3$ are some real numbers. We contend that $\lim_{h \to h_0} f_i(h) = L_i$ for $i = 1, 2, 3$. Again let $\epsilon > 0$ be given. Then there exists $\delta > 0$ such that whenever $0 < \delta < |h - h_0|$, we have $|\mathbf{w}(h) - (L_1 \mathbf{i} + L_2 \mathbf{j} + L_3 \mathbf{k})| < \epsilon$. In view of (13) this means

$$\sqrt{(f_1(h) - L_1)^2 + (f_2(h) - L_2)^2 + (f_3(h) - L_3)^2} < \epsilon,$$
$$\text{whenever } 0 < |h - h_0| < \delta. \tag{14}$$

Since the terms in the sum under the radical sign are all non-negative, we have

$$
\begin{aligned}
|f_1(h) - L_1| &= \sqrt{(f_1(h) - L_1)^2} \\
&\leq \sqrt{(f_1(h) - L_1)^2 + (f_2(h) - L_2)^2 + (f_3(h) - L_3)^2} \\
&< \epsilon \text{ whenever } 0 < |h - h_0| < \delta \text{ (by (14))}.
\end{aligned}
$$

So $\lim_{h \to h_0} f_1(h) = L_1$. Similarly, $\lim_{h \to h_0} f_2(h) = L_2$ and $\lim_{h \to h_0} f_3(h) = L_3$. This completes the proof of the first assertion of the theorem and also establishes (11) in the course of the proof. ∎

The reader is urged not to get carried away by the technicalities of the proof above. The essential idea of the proof is merely that the length of a vector is small if and only if the magnitude of each of its components is small. This fact, applied to the difference vector $\mathbf{w}(h) - (L_1\mathbf{i} + L_2\mathbf{j} + L_3\mathbf{k})$ gives the proof.

This lemma establishes the expected relationship between the limit of a vector valued function (of a real variable) and the limits of its component functions. Verbally it says that the components of the limit of a vector valued function are precisely the limits of the component functions. Obviously, the same thing will hold for anything which is based on the concept of a limit of a function, such as continuity, derivatives etc. The next theorem, although proved only for velocity and acceleration, really holds for the derivative of any vector-valued function.

**7.5 Theorem:** Let $\mathbf{r} = \mathbf{r}(t) = x\mathbf{i} + y\mathbf{j} + z\mathbf{k}$ be the position of a particle $P$ at time $t$. Assume $x, y, z$ have derivatives of order two w.r.t. $t$. Then the velocity $\mathbf{v}$, speed $v$ and acceleration $\mathbf{a}$ of $P$ at time $t$ are given by

$$\mathbf{v} \;=\; \mathbf{v}(t) = \dot{\mathbf{r}} = \dot{x}\mathbf{i} + \dot{y}\mathbf{j} + \dot{z}\mathbf{k} \tag{15}$$

$$v \;=\; |\mathbf{v}(t)| = |\dot{\mathbf{r}}| = \sqrt{\dot{x}^2 + \dot{y}^2 + \dot{z}^2} = \dot{s} \tag{16}$$

$$\text{and } \mathbf{a} \;=\; \mathbf{a}(t) = \ddot{\mathbf{r}} = \ddot{x}\mathbf{i} + \ddot{y}\mathbf{j} + \ddot{z}\mathbf{k}. \tag{17}$$

**Proof.** In (15), $\mathbf{v}$ means the derivative of the vector-valued function $\mathbf{r}$ w.r.t. $t$. In other words,

$$\mathbf{v}(t) = \lim_{\Delta t \to 0} \frac{\mathbf{r}(t + \Delta t) - \mathbf{r}(t)}{\Delta t}$$

If we write the numerator in terms of $\mathbf{i}$, $\mathbf{j}$ and $\mathbf{k}$, we get

$$\mathbf{v}(t) = \lim_{\Delta t \to 0} \left( \frac{x(t + \Delta t) - x(t)}{\Delta t}\mathbf{i} + \frac{y(t + \Delta t) - y(t)}{\Delta t}\mathbf{j} + \frac{z(t + \Delta t) - z(t)}{\Delta t}\mathbf{k} \right)$$

Since the coefficients of $\mathbf{i}$, $\mathbf{j}$, $\mathbf{k}$ tend to $\dot{x}(t), \dot{y}(t)$ and $\dot{z}(t)$ respectively, as $\Delta t \to 0$, the lemma above (with $h = \Delta t$ etc.) gives (15). From (15) we get (17) in a similar manner. Finally (16) except for the last part, also follows from (15) if we define the speed as the magnitude of the velocity vector. But the speed can also be defined independently as the time rate of the distance traveled by the particle $P$. In that case, the distance traveled being $s$, the speed is simply $ds/dt$, or $\dot{s}$ and (16) is a consequence of (4). ∎

Since the vector $\dot{\mathbf{r}}$ lies along the tangent to the path traced by $P$, and $\mathbf{u}$ equals $\mathbf{r}'$, we can write

$$\mathbf{v} = \dot{\mathbf{r}} = \frac{d\mathbf{r}}{dt} = \frac{d\mathbf{r}}{ds}\frac{ds}{dt} = v\mathbf{u}. \tag{18}$$

We can also get the acceleration $\mathbf{a}$ by differentiating (18) w.r.t. $t$. Using Exercise (6.6(ii)), we get,

$$\mathbf{a} = \frac{d\mathbf{v}}{dt} = \ddot{\mathbf{r}} = \dot{v}\mathbf{u} + v\dot{\mathbf{u}} \tag{19}$$

But $\dot{\mathbf{u}} = \dfrac{d\mathbf{u}}{dt}$, which by the chain rule, equals $\dfrac{d\mathbf{u}}{ds}\dfrac{ds}{dt}$, i.e. $v\mathbf{u}'$. By (8), $\mathbf{u}'$ equals $\kappa\mathbf{p}$ provided $\kappa \neq 0$. Substituting this into (19) we get the following simple but important result.

**7.6 Theorem:** For a particle satisfying the assumptions of the last theorem, the velocity $\mathbf{v}$ lies along the tangent vector to its path. The acceleration $\mathbf{a}$ lies in the osculating plane and is given by

$$\mathbf{a} = \dot{v}\mathbf{u} + \kappa v^2\mathbf{p} = \ddot{s}\mathbf{u} + \kappa v^2\mathbf{p}. \tag{20}$$

**Proof:** we already derived (20) under the assumption that $\kappa \neq 0$. If $\kappa = 0$, then $\mathbf{u}' = \mathbf{0}$ (the zero vector). In this case $\dot{\mathbf{u}} = \mathbf{u}'v$ (by the chain rule) is also $\mathbf{0}$. So from (19) we get $\mathbf{a} = \dot{v}\mathbf{u}$. So even though $\mathbf{p}$ is undefined at such a point, we may treat (20) as valid with the understanding that the second term is 0. Finally, since the osculating plane is the plane spanned by $\mathbf{u}$ and $\mathbf{p}$, (20) also shows that $\mathbf{a}$ lies in it. ∎

This result is often called the resolution of the acceleration along the tangent and the normal. It is quite different from resolving along the fixed vectors $\mathbf{i}, \mathbf{j}$ and $\mathbf{k}$, because the vectors $\mathbf{u}$ and $\mathbf{p}$ are changing from point to point. So a resolution like this is an example of what is called a resolution along a moving frame of reference. The components $\ddot{s}$ and $\kappa v^2$ are called respectively the **tangential** and the **normal components** of the acceleration. Each has an interesting significance. Suppose the motion is entirely along a straight line. If we let $\mathbf{u}$ be a unit vector along that line, then obviously the velocity and the acceleration are $\dot{s}\mathbf{u}$ and $\ddot{s}\mathbf{u}$ respectively. In other words, the tangential component of the acceleration is the same as the acceleration would have been if the motion were momentarily on a straight line.

The normal component $\kappa v^2$ is not so easy to recognize. To interpret it, we consider motion of a particle $P$ along a circle of radius $a$ (say), with uniform angular velocity say $\omega$. This means that the radius through $P$ turns by an angle $\omega t$ in any time interval of length $t$. It follows that the speed $v$ of $P$ is constant and equals $a\omega$. (Note, however, that the velocity $\mathbf{v}$ of $P$ is not constant as its direction is changing). Taking the circle to lie in the $x$-$y$ plane with centre $O$ and supposing that $P$ was at $(a, 0)$ when $t = 0$, (see Fig.6.7.2(a)), we get $\mathbf{r} = \mathbf{r}(t) = a\cos\omega t\mathbf{i} + a\sin\omega t\mathbf{j}$, whence

$$\mathbf{a} = \ddot{\mathbf{r}} = -a\omega^2\cos\omega t\mathbf{i} - a\omega^2\sin\omega t\mathbf{j}. \tag{21}$$

Keeping in mind that $\omega = \frac{v}{a}$, that $-\cos\omega t\mathbf{i} - \sin\omega t\mathbf{j}$ is precisely the unit normal vector $\mathbf{p}$ at $P$, and also that for a circle of radius $a$, $\kappa = \frac{1}{a}$ at all points, we can rewrite (21) as

$$\mathbf{a} = \kappa v^2\mathbf{p} \tag{22}$$

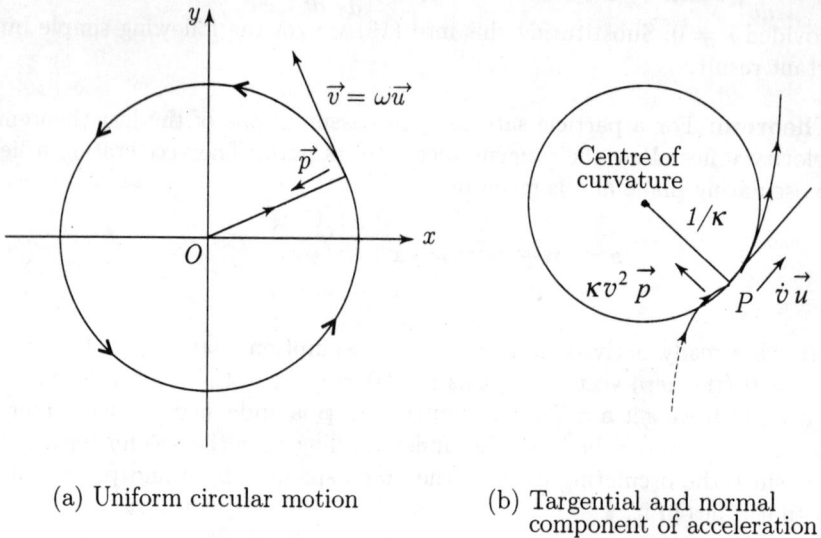

(a) Uniform circular motion

(b) Targential and normal component of acceleration

Figure.6.7.2 : Resolution of Acceleration.

We conclude that in a uniform circular motion there is no tangential acceleration. The acceleration is directed towards the centre. For this reason it is often called the **centripetal acceleration**. In order to keep the particle $P$ in the uniform circular motion, this acceleration has to be provided by some centripetal force such as gravity (as happens in the case of planetary motion) friction or the tension in some string. If this force suddenly ceases to exist, the particle will start moving along the tangent to the circle at the point where it was at that time. For an observer sitting at $P$, however, it will appear that he is thrown away from the centre. This virtual force experienced by him is called the **centrifugal force**. It is experienced by passengers sitting in a bus which turns abruptly. It is used in the process of churning to drive the fat particles towards the periphery of a cylindrical vessel.

Returning to (20), we see that locally, the motion of the particle $P$ is like a superimposition of two motions, one a straight line motion along the tangent at $P$ and the other a uniform circular motion along a circle. This circle is precisely the circle of curvature at $P$. Thus curvature plays an important role in the study of the motion of a particle along a curve. If a person is driving along a curved road, then at a sharp bend, $\kappa$ is very high and the speed $v$ must be reduced to make the product $\kappa v^2$ small enough so that the necessary centripetal acceleration can be provided by the friction or sometimes by a tilt designed in the road. While designing the road a civil engineer is faced with the dual problem. That is, in order to allow vehicles upto a desired speed limit $v$, he has to ensure that $\kappa$ is kept low enough.

Finding the velocity and acceleration of a particle, given its position at time $t$ is a simple matter involving differentiation. Far more intricate (and important) is to go the other way. In many real life problems, the acceleration **a** of a particle is known from the knowledge of the forces acting on it. The equation giving **a** is called the **equation of motion.** (When **a** is resolved into **i**, **j** and **k**, we get a system of three equations. Often these equations are called the equations of motion.) For example, for a particle moving under gravity, the equation of motion is $\mathbf{a} = -g\mathbf{k}$ where **k** is a vertically upward unit vector and $g$ is the constant of gravity. (In metric units $g = 9.8$ meters/sec$^2$.) Besides, some conditions, called **initial conditions**, which specify the position and the velocity of the particle at some particular moment are given. The problem is to solve the given equation of motion subject to the given initial condition, that is, to determine **r** as a function of $t$ so that its acceleration will satisfy the equation of motion and the initial conditions will also hold. The solution requires solving differential equations and is beyond our scope. Moreover, such problems have little to do with curvature, which is our present topic of study. Nevertheless, a couple of simple problems of this type (where the differential equations involved can be solved just by inspection) will be given in the exercises (see Exercises (7.10) to (7.13)). More interesting (and challenging) problems of solving equations of motion are studied in a branch of mathematics called **mechanics** (which actually borders with physics). In the next section, we shall study a classic application to planetary motion.

We now turn to another important geometric attribute of a curve, called its torsion. It is a measure of the degree of twist in a curve. Curvature can be looked at as a measure of the non-linearity of a curve. If a curve is a straight line then its curvature is identically zero and it is not hard to show (see Exercise (7.14)) that the converse is also true. Similarly the torsion of a curve is a measure of its non-planarity. A planar curve has zero torsion. For a curve to have (non-zero) torsion, its plane must turn in space just as for a curve to have a non-zero curvature, its tangent line must change direction. So informally, the torsion of a curve $C$ at a point $P$ on it is the rate of change of its osculating plane at $P$. As with curvature, this rate is obtained by taking the derivative w.r.t. the arc-length parameter $s$. The difficulty is to decide what to differentiate. What do we mean by the derivative of the osculating plane ? This is taken care of by finding something associated with the osculating plane and differentiating it instead. Note that a plane in space passing through a given point $P$ is uniquely determined by its normal, which is a straight line through $P$. And this straight line, in turn, is uniquely determined by a unit vector parallel to it. (For the same line, there are in fact two such unit vectors, they are oppositely directed.) So, to define the torsion of a curve at a point $P$, we take a unit vector **b** perpendicular to the osculating plane through $P$, and differentiate **b** as a (vector valued) function of the arc length $s$, exactly the way we defined curvature at $P$ by differentiating **u** as a function $s$.

It is not hard to define **b**. We already know that the osculating plane is spanned by the unit vector **u** and the principal normal vector **p**. These vectors are of unit length each and perpendicular to each other by (9). So if we take their cross product, we shall get a unit vector perpendicular to the osculating plane. We trust that the reader is familiar with the definition and the standard properties of the cross product of vectors in $I\!R^3$. (Exercise (7.15) will help those who are not.) Note that, in taking the cross-product, the order of the vectors matters. We let **b** be **u** × **p** (and not **p** × **u**). Then **b** is perpendicular to **u** (and also to **p**) and hence lies in the normal plane to the curve at $P$. The vector **b** is called the **binormal** to the curve at $P$. The three unit vectors **u**, **p** and **b** form a right handed system. The plane spanned by **u** and **b** is called the **rectifying plane** of the curve at $P$. Fig. 6.7.3 shows these vectors at a point on a curve. (It is tacitly assumed in the definition of the binormal that **p** is defined at $P$, that is, **u**$'$ is non-zero. Where **u**$' = 0$, there is no binormal.)

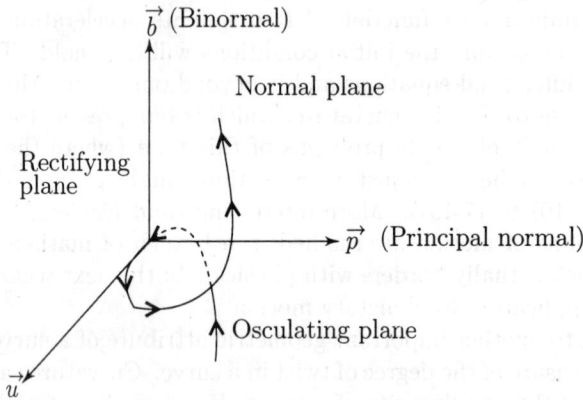

Figure. 6.7.3 : Binormal to a Curve.

Now to differentiate **b** w.r.t. $s$, we need a formula for differentiating a cross product of two vector valued functions. The formula is the expected one and will be given in the exercises (Exercise (7.17)). So, since

$$\mathbf{b} = \mathbf{u} \times \mathbf{p} \qquad (23)$$

we get $\dfrac{d\mathbf{b}}{ds} = \dfrac{d\mathbf{u}}{ds} \times \mathbf{p} + \mathbf{u} \times \dfrac{d\mathbf{p}}{ds}$. Here we need to assume that **p** is a differentiable function of $s$. Since **p** itself was obtained by differentiating **u**, this means we must assume that **u** is a twice differentiable vector valued function of $s$. Now $\dfrac{d\mathbf{u}}{ds}$ is parallel to **p** (by definition of **p**) and so $\dfrac{d\mathbf{u}}{ds} \times \mathbf{p} = \mathbf{0}$. Hence we have

$$\frac{d\mathbf{b}}{ds} = \mathbf{u} \times \frac{d\mathbf{p}}{ds} \quad \text{or} \quad \mathbf{b}' = \mathbf{u} \times \mathbf{p}' \qquad (24)$$

In particular, this means **b**$'$ is perpendicular to the vector **u**. But since **b** · **b** $= 1$(**b** being a unit vector), we get by differentiating w.r.t. $s$ that $2\mathbf{b} \cdot \mathbf{b}' = 0$. So **b**$'$ is perpendicular to **b** also. Now **u**, **p**, **b** are three mutually orthogonal vectors. Since **b**$'$ is perpendicular to both **u** and **b**, it is parallel to **p**. Hence it is a scalar multiple of **p**. In the case of curvature also **u**$'$ was a scalar multiple of **p** and we defined this scalar as the curvature. For the torsion also we follow a similar approach except that for some technical reasons to be explained later we define the **torsion** $\tau$ (read 'tau') to be the negative of this scalar. That is, $\tau$ is defined so that

$$\mathbf{b}' = -\tau \mathbf{p}. \tag{25}$$

Like $\kappa$, the torsion $\tau$ is a function of $s$ and can change from point to point. But while the curvature $\kappa$, wherever defined, is always non-negative, the torsion $\tau$ can be either positive or negative. If the curve is planar then at every point $\mathbf{u}$ and $\mathbf{p}$ always lie in the plane of the curve and so $\mathbf{b}$ is a constant vector. (We are not saying that either $\mathbf{u}$ or $\mathbf{p}$ is a constant. But their cross product $\mathbf{u} \times \mathbf{p}$ is always a unit vector perpendicular to the plane of the curve and hence is a constant vector.) So $\mathbf{b}' = \mathbf{0}$ and thus we see that every planar curve has zero torsion at every point. The converse is also true but not so easy to prove (see Exercise (7.14)).

So, to get a non-trivial example of torsion, we must take a non-planar curve. Probably the simplest and the most commonly occurring curve of this type is a circular helix shown in Fig. 4.12.3(a). Certain winding staircases and springs are in the shape of helices. (See Exercise (7.22) for a popular definition of a helix.) The DNA molecule, important in genetics, has the shape of a double helix. The simplest parametrisation of a helix is by

$$\mathbf{r} = \mathbf{r}(t) = a \cos t\mathbf{i} + a \sin t\mathbf{j} + ct$$

where $a, c$ are positive constants. (In Fig.4.12.3(a), $a = 1$). To get the arc length parametriszation, we have $\dot{s} = |\dot{\mathbf{r}}| = |-a \sin t\mathbf{i} + a \cos t\mathbf{j} + c| = \sqrt{a^2 + c^2}$. So taking $s = 0$ when $t = 0$, (i.e. measuring the arc length from the point $(a, 0, 0)$) we have $s = \sqrt{a^2 + c^2}t$. (For $t < 0$, we take the arc length to be negative, indicating it is measured backwards from $(a, 0, 0)$.) So the arc length parametrisation of the helix is

$$\mathbf{r} = \mathbf{r}(s) = a \cos\left(\frac{s}{\sqrt{a^2 + c^2}}\right)\mathbf{i} + a \sin\left(\frac{s}{\sqrt{a^2 + c^2}}\right)\mathbf{j} + \frac{cs}{\sqrt{a^2 + c^2}}\mathbf{k} \tag{26}$$

(We remark again that such an explicit arc-length parametrisation is generally not possible for many curves. The calculation of their torsion requires us to express it in terms of derivatives w.r.t. other parameters. A formula of this type will be given in the exercises (see Exercise (7.20)).)

Differentiating (26) twice w.r.t. s, we get

$$\mathbf{u} = \mathbf{r}' = \frac{1}{\sqrt{a^2 + c^2}}\left(-a \sin\left(\frac{s}{\sqrt{a^2 + c^2}}\right)\mathbf{i} + a \cos\left(\frac{s}{\sqrt{a^2 + c^2}}\right)\mathbf{j} + c\mathbf{k}\right) \tag{27}$$

$$\text{and} \quad \frac{d\mathbf{u}}{ds} = \mathbf{u}' = \frac{a}{a^2 + c^2}\left(-\cos\left(\frac{s}{\sqrt{a^2 + c^2}}\right)\mathbf{i} - \sin\left(\frac{s}{\sqrt{a^2 + c^2}}\right)\mathbf{j}\right)$$

Since the vector in the parentheses is a unit vector and $a > 0$ we get

$$\kappa = \frac{a}{a^2 + c^2} \quad \text{and} \quad \mathbf{p} = -\cos(\frac{1}{\sqrt{a^2 + c^2}})\mathbf{i} - \sin(\frac{s}{\sqrt{a^2 + c^2}})\mathbf{j} \tag{28}$$

Hence

$$\mathbf{b} = \mathbf{u}\times\mathbf{p} = \frac{1}{\sqrt{a^2+c^2}} \begin{vmatrix} \mathbf{i} & \mathbf{j} & \mathbf{k} \\ -a\sin\left(\frac{s}{\sqrt{a^2+c^2}}\right) & a\cos\left(\frac{s}{\sqrt{a^2+c^2}}\right) & c \\ -\cos\left(\frac{s}{\sqrt{a^2+c^2}}\right) & -\sin\left(\frac{s}{\sqrt{a^2+c^2}}\right) & 0 \end{vmatrix}$$

$$= \frac{1}{\sqrt{a^2+c^2}}\left(c\sin\left(\frac{s}{\sqrt{a^2+c^2}}\right)\mathbf{i} - c\cos\left(\frac{s}{\sqrt{a^2+c^2}}\right)\mathbf{j} + a\mathbf{k}\right)$$

Differentiating w.r.t. $s$,

$$\mathbf{b}' = \frac{c}{a^2+c^2}\left(\cos\left(\frac{s}{\sqrt{a^2+c^2}}\right)\mathbf{i} + \sin\left(\frac{s}{\sqrt{a^2+c^2}}\right)\mathbf{j}\right)$$

$$= -\frac{c}{a^2+c^2}\mathbf{p} \quad \text{(by 28)} \tag{29}$$

Hence from (25), we get $\tau = \dfrac{c}{a^2+c^2}$ at all points of the helix. Since $c > 0$, this helix is going upwards as $t$ increases and winds counterclockwise around the $z$-axis (as viewed from above). Since the anticlockwise sense and upward direction are customarily associated with positivity, we would like the torsion of such a curve to be positive and this explains the negative sign in (25). If the helix winds clockwise as it climbs up then instead of (26), its arc length parametrisation would be

$$\mathbf{r} = \mathbf{r}(s) = a\cos\left(\frac{s}{\sqrt{a^2+b^2}}\right)\mathbf{i} - a\sin\left(\frac{s}{\sqrt{a^2+c^2}}\right)\mathbf{j} + \frac{cs}{\sqrt{a^2+c^2}}\mathbf{k} \tag{30}$$

We leave it to the reader to show, by taking steps similar to (27)-(29) that the torsion now is $\dfrac{-c}{a^2+c^2}$. (The curvature $\kappa$ remains the same, viz., $\dfrac{a}{a^2+c^2}$.)

We could have, of course, defined the torsion $\tau$ merely as the magnitude of the vector $\mathbf{b}'$, just as we defined the curvature $\kappa$ as the magnitude of the vector $\mathbf{u}'$. In that case $\tau$ would always be non-negative. But then we would lose the extra information that we get because of its sign. (For example, we would not be able to distinguish between the two helices above on the basis of torsion). It is tempting to try to assign a sign to the curvature too, so as to make it reveal more information about the curve. Unfortunately, for curves in space there is no way to do this, as there is for the torsion. It is instructive to see the reason for this difference. In the case of a torsion, the vector $\mathbf{p}$ is already defined and by differentiating $\mathbf{b}(= \mathbf{u}\times\mathbf{p})$ we get a vector which is a scalar multiple of $\mathbf{p}$. This multiple could either be positive or negative. In the case of curvature, we *define* $\mathbf{p}$ as the unit vector in the direction of $\mathbf{u}'$. So $\mathbf{u}'$ is necessarily a positive multiple of $\mathbf{p}$. It is tempting to try to circumvent this difficulty by first defining $\mathbf{p}$ as a vector normal to $\mathbf{u}$. The trouble is that in $I\!\!R^3$, there are infinitely many unit vectors perpendicular to $\mathbf{u}$. Out of these, we must chose $\mathbf{p}$ to be the one which lies in the plane of the curve. There are actually two such vectors, opposite to each other. But there is no canonical way in space to decide which of them should be taken as **p** and which as −**p**. (This difficulty does not arise in defining the binormal **b**. There are two unit vectors perpendicular to the plane spanned by **u** and **p**. The cross product enables us to distinguish among them. One of them is **u**×**p** and the other is **p**×**u**. We define **b** as **u**×**p**. Then **p**×**u** is −**b**.)

For planar curves the situation is better. For such curves there is no torsion. But the curvature can be given a sign as follows. For simplicity, we assume the curve lies in the $x$-$y$ plane. Now let **u** be the unit tangent vector at a point $P$. Then we can define **p**, the unit normal vector at $P$, even without differentiating **u**. There are two vectors in the $x$-$y$ plane which are perpendicular to **u**. Each is obtained by rotating **u** through 90°. We let **p** be the one which is obtained by a counterclockwise rotation as shown in Fig. 6.7.4. In symbols, let $\mathbf{u} = \cos\theta\mathbf{i} + \sin\theta\mathbf{j}$, where $\theta$ is the angle **u** makes with the positive $x$-axis $(0 \leq \theta \leq 2\pi)$. Then we take **p** to be $\cos\left(\theta + \frac{\pi}{2}\right)\mathbf{i} + \sin\left(\theta + \frac{\pi}{2}\right)\mathbf{j}$, or equivalently, $-\sin\theta\mathbf{i} + \cos\theta\mathbf{j}$.

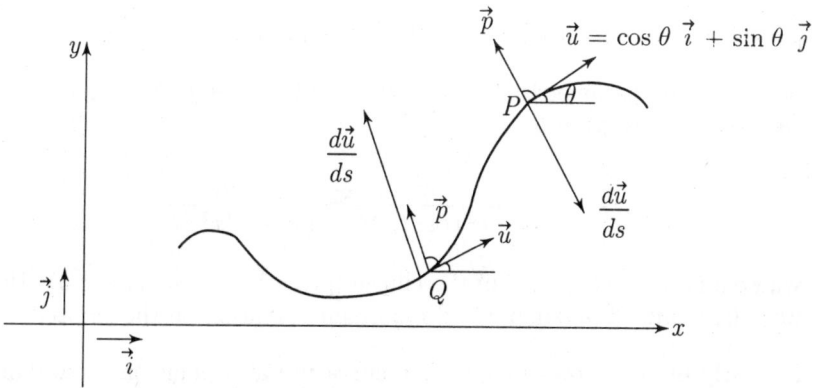

Figure 6.7.4 : Signed Curvature for a Plane Curve.

It now follows, exactly as before that $\mathbf{u}'$ is perpendicular to **u**. Hence $\mathbf{u}' = \kappa\mathbf{p}$ for a unique scalar $\kappa$. But now $\kappa$ will be either positive, negative or even 0. It can be shown that in Fig. 6.7.4, $\kappa < 0$ at a point like $P$ but $\kappa > 0$ at a point like $Q$. The reader must have guessed that this is related to the concavity of the curve. This indeed turns out to be so, as will be shown in Exercises (7.6) and (7.7) .

The curvature and the torsion are two important geometric attributes of a curve. It can in fact be shown (although the proof is non-trivial) that these two invariants completely determine the curve upto a congruence. In other words two curves of equal arc length, say $L$, such that for every $s \in [0, L], \kappa(s)$ for the two curves is the same and $\tau(s)$ is also the same must be congruent to each other. As remarked in the last section, such a pair of curves has identical geometric properties. Thus the curvature and the torsion are like the prime attributes of a curve. Every other geometric property of a curve can in theory, be derived from them. For planar curves the torsion vanishes identically. And so the curvature (with a sign) determines the curve uniquely. No wonder it is such a fundamentally important attribute of the curve.

# EXERCISES

**7.1** Suppose $\mathbf{w} = \mathbf{w}(s)$ is a non-vanishing, differentiable vector valued function of a real parameter (not necessarily the arc length). Prove that :

(i) $\dfrac{d}{ds}(|\mathbf{w}|) = |\mathbf{w}|' = \dfrac{\mathbf{w} \cdot \mathbf{w}'}{|\mathbf{w}|}$

(ii) $\dfrac{d}{ds}\left(\dfrac{\mathbf{w}}{|\mathbf{w}|}\right) = \left(\dfrac{\mathbf{w}}{|\mathbf{w}|}\right)' = \dfrac{(\mathbf{w} \cdot \mathbf{w})\mathbf{w}' - (\mathbf{w} \cdot \mathbf{w}')\mathbf{w}}{(\mathbf{w} \cdot \mathbf{w})^{3/2}}$

**7.2** For a twice differentiable smooth curve $\mathbf{r} = \mathbf{r}(t)$, prove that

$$\kappa = \frac{\sqrt{(\dot{\mathbf{r}} \cdot \dot{\mathbf{r}})(\ddot{\mathbf{r}} \cdot \ddot{\mathbf{r}}) - (\dot{\mathbf{r}} \cdot \ddot{\mathbf{r}})^2}}{(\dot{\mathbf{r}} \cdot \dot{\mathbf{r}})^{3/2}} .$$

[*Hint* : Apply the last exercise with $\mathbf{w} = \dot{\mathbf{r}}$ and the chain rule.]

**7.3** For an ellipse parametrised by $\mathbf{r}(t) = a\cos t\mathbf{i} + b\cos t\mathbf{j}$, show that the curvature at $\mathbf{r}(t)$ is given by

$$\kappa = \frac{\sqrt{1 - e^2}}{a(1 - e^2\cos^2 t)^{3/2}} = \frac{ab}{(a^2 - e^2 x^2)^{3/2}}$$

where $e$ is the eccentricity of the ellipse (given by $b^2 = a^2(1 - e^2)$). Hence find the points of maximum and minimum curvature on the ellipse.

**7.4** Similarly obtain expressions for the curvature at a point on a parabola, a hyperbola and a conical helix (see (25) in Section 4.12).

**7.5** Suppose $\mathbf{r} = \mathbf{r}(t) = x(t)\mathbf{j} + y(t)\mathbf{j} + z(t)\mathbf{j}$ is a curve in space. If $\alpha, \beta, \gamma$ are the angles made by the tangent vector with the positive $x$-,$y$- and $z$-axes respectively, prove that $\cos\alpha = \dfrac{\dot{x}}{|\dot{\mathbf{r}}|}$, $\cos\beta = \dfrac{\dot{y}}{|\dot{\mathbf{r}}|}$ and $\cos\gamma = \dfrac{\dot{z}}{|\dot{\mathbf{r}}|}$. (Here $\alpha, \beta, \gamma$ all lie in $[0, \pi]$. The numbers $\cos\alpha, \cos\beta$ and $\cos\gamma$ are called the **direction cosines** of the tangent line.) What happens if $t$ is the arc length parameter?

**7.6** Let $\mathbf{r} = \mathbf{r}(t) = x(t)\mathbf{i} + y(t)\mathbf{j}$ be a twice differentiable smooth plane curve. Let $\phi$ be the angle which the unit tangent vector $\mathbf{u}$ makes with the positive $x$-axis. Here $\phi$ can take any value in $[0, 2\pi)$. Prove that :

(i) $\cos\phi = \dfrac{\dot{x}}{|\dot{\mathbf{r}}|}, \sin\phi = \dfrac{\dot{y}}{|\dot{\mathbf{r}}|}, \mathbf{u} = \cos\phi\mathbf{i} + \sin\phi\mathbf{j}$

(ii) the vector $\dfrac{-\dot{y}\mathbf{i} + \dot{x}\mathbf{j}}{|\dot{\mathbf{r}}|}$ is a unit vector obtained by rotating $\mathbf{u}$ counterclockwise through $90°$ and equals $\dfrac{d\mathbf{u}}{d\phi}$.

(iii) the vector $\mathbf{u}' = \dfrac{d\mathbf{u}}{ds}$ equals $\dfrac{d\mathbf{u}}{d\phi}\dfrac{d\phi}{ds}$; if $\dfrac{d\phi}{ds} \geq 0$, it equals the curvature $\kappa$, otherwise it equals $-\kappa$.

(iv) the **signed curvature** of the curve, defined as $\dfrac{d\phi}{ds}$, equals $\dfrac{\dot{x}\ddot{y} - \dot{y}\ddot{x}}{(\dot{x}^2 + \dot{y}^2)^{3/2}}$.

7.7 For a plane curve of the form $y = f(x)$, where $f$ is twice differentiable, show that the signed curvature equals $\dfrac{f''(x)}{(1 + (f'(x))^2)^{3/2}}$. How is the signed curvature related to concaity?

7.8 Find the point on the curve $y = e^x$ for which the radius of curvature is minimum.

7.9 A road on a mountain slope is parametrised by $x = 2t, y = \frac{1}{2}t^2, z = \frac{4}{3}t^{3/2}$ where $z$ denotes the height about a horizontal plane and $\frac{1}{2} \leq t \leq 3$. Find the points where (i) the road is steepest (ii) it bends most abruptly.

7.10 For a particle moving freely under gravity show that its height above ground at time $t$ is $-\frac{1}{2}gt^2 + ut + h$ where $h$ = its height at time $t = 0$, $u$ = its speed (upwards) at time $t = 0$ and $g$ is the acceleration due to gravity.

7.11 Suppose a particle $P$ is projected from a point $O$ on the ground with an initial speed $v_0$, in a direction making an angle $\alpha$ with the ground plane $(0 < \alpha < \frac{\pi}{2})$. Prove that the path traced by $P$ is a parabola lying in a vertical plane. Find when, where and with what speed the particle hits the ground again.

7.12 According to Newton's law of gravitation two bodies of masses $m_1$ and $m_2$, separated by a distance $r$ attract each other with a force equal to $\dfrac{Gm_1m_2}{r^2}$ in magnitude, where $G$ is a constant called the gravitational constant. Prove that a rocket fired upwards from the surface of the earth will never return to earth if its initial speed exceeds $\sqrt{\dfrac{2GM}{R}}$ where $M$ and $R$ are, respectively, the mass and the radius of the earth. Neglect any other forces acting on the rocket. Also assume that the mass of the rocket is negligible as compared with $M$, so that the earth remains stationary. [*Hint*: Show that the magnitude of the acceleration can be written as $\dfrac{dv}{dz}\dfrac{dz}{dt}$ where $v$ is the speed and $z$ is the distance from the centre of the earth.]

7.13 A string of length $\ell$ can withstand a tension upto $T$. A particle of mass $m$ is attached to one end of the srting and is rotated about the other in a circle of radius $\ell$. If there are no other forces acting on the particle show that the maximum angular speed it can reach is $\sqrt{\dfrac{T}{m\ell}}$.

7.14 Assuming sufficient differentiability, prove that

(a)   a curve is a stright line if and only if its curvature vanishes identically.

*(b)   a curve is planar if and only if its torsion vanishes identically.

[*Hint* : Apply Lagrange's Mean Value Theorem.]

7.15   Let $\mathbf{u} = u_1\mathbf{i} + u_2\mathbf{j} + u_3\mathbf{k}$ and $\mathbf{v} = v_1\mathbf{i} + v_2\mathbf{j} + v_3\mathbf{k}$ be two vectors. Their **cross product** denoted by $\mathbf{u} \times \mathbf{v}$ is defined as the vector $(u_2v_3 - u_3v_2)\mathbf{i} + (u_3v_1 - u_1v_3)\mathbf{j} + (u_1v_2 - u_2v_1)\mathbf{k}$, which is easier to remember as a formal determinant

$$\begin{vmatrix} \mathbf{i} & \mathbf{j} & \mathbf{k} \\ u_1 & u_2 & u_3 \\ v_1 & v_2 & v_3 \end{vmatrix}$$

Prove that

(i)    the cross product is bilinear and anticommutative, that is, for any three vectors $\mathbf{u}$, $\mathbf{v}$ and $\mathbf{w}$ and scalars $\alpha, \beta$, $\mathbf{u} \times (\alpha\mathbf{v} + \beta\mathbf{w}) = \alpha(\mathbf{u} \times \mathbf{v}) + \beta(\mathbf{u} \times \mathbf{w})$, $(\alpha\mathbf{u} + \beta\mathbf{v}) \times \mathbf{w} = \alpha(\mathbf{u} \times \mathbf{w}) + \beta(\mathbf{v} \times \mathbf{w})$ and $\mathbf{v} \times \mathbf{u} = -(\mathbf{u} \times \mathbf{v})$.

(ii)   $\mathbf{u} \times \mathbf{v}$ is at right angles to both $\mathbf{u}$ and $\mathbf{v}$.

(ii)   $|\mathbf{w}| = |\mathbf{u}|\,|\mathbf{v}|\sin\theta$, where $\theta$ is the angle $(0 \le \theta \le \pi)$ between the directions of $\mathbf{u}$ and $\mathbf{v}$. ($\theta$ is undefined if $\mathbf{u}$ or $\mathbf{v}$ is $\mathbf{0}$ in which case $\mathbf{w}$ is $\mathbf{0}$.)

(iv)   if $\mathbf{u}$, $\mathbf{v}$ are non-zero and not multiples of each other then $\mathbf{u}$, $\mathbf{v}$ and $\mathbf{u} \times \mathbf{v}$ form a right handed system, that is, if the index finger and the middle finger of the right hand point in the directions of $\mathbf{u}$ and $\mathbf{v}$ respectivley then the thumb will point in the direction of $\mathbf{u} \times \mathbf{v}$, or equivalently, if a person walks in the plane spanned by $\mathbf{u}$ and $\mathbf{v}$ with his head towards $\mathbf{u} \times \mathbf{v}$ then to go from $\mathbf{u}$ to $\mathbf{v}$ along the shortest arc, he will have to move counterclockwise. (We assume that $(\mathbf{i}, \mathbf{j}, \mathbf{k})$ is a right handed system.)

(v)    If $\mathbf{p}, \mathbf{q}, \mathbf{r}$ are any three mutually perpendicular unit vectors with $\mathbf{r} = \mathbf{p} \times \mathbf{q}$ and $\mathbf{u}$, $\mathbf{v}$ are resolved along them as, say, $\mathbf{u} = a_1\mathbf{p} + a_2\mathbf{q} + a_3\mathbf{r}$ and $\mathbf{v} = b_1\mathbf{p} + b_2\mathbf{q} + b_3\mathbf{r}$ then

$$\mathbf{u} \times \mathbf{v} = \begin{vmatrix} \mathbf{p} & \mathbf{q} & \mathbf{r} \\ a_1 & a_2 & a_3 \\ b_1 & b_2 & b_3 \end{vmatrix}.$$

7.16   If $\mathbf{u}$, $\mathbf{v}$, $\mathbf{w}$ are any three vectors in space then their **scalar triple product** is defined as $\mathbf{u} \cdot (\mathbf{v} \times \mathbf{w})$ and denoted by $(\mathbf{u}\ \mathbf{v}\ \mathbf{w})$. Prove that :

(i)    if $\mathbf{u} = u_1\mathbf{p} + u_2\mathbf{q} + v_3\mathbf{r}, \mathbf{v} = v_1\mathbf{p} + v_2\mathbf{q} + v_3\mathbf{r}$ and $\mathbf{w} = w_1\mathbf{p} + w_2\mathbf{q} + w_3\mathbf{r}$, where $\mathbf{p}, \mathbf{q}, \mathbf{r}$ are as in para (v) of the last exercise then

$$(\mathbf{u}\ \mathbf{v}\ \mathbf{w}) = \begin{vmatrix} u_1 & u_2 & u_3 \\ v_1 & v_2 & v_3 \\ w_1 & w_2 & w_3 \end{vmatrix}$$

(ii) $|(\mathbf{u}\ \mathbf{v}\ \mathbf{w})|$ equals the volume of the parallelopiped with sides $\mathbf{u}$, $\mathbf{v}$ and $\mathbf{w}$ (for this reason, $(\mathbf{u}\ \mathbf{v}\ \mathbf{w})$ is sometimes also called the **box product**).

7.17 If $\mathbf{u}$ and $\mathbf{v}$ are differentiable vector valued functions of a real variable $t$, prove that so is $\mathbf{u}\times\mathbf{v}$ and further that $\dfrac{d}{dt}(\mathbf{u}\times\mathbf{v}) = \dfrac{d\mathbf{u}}{dt}\times\mathbf{v} + \mathbf{u}\times\dfrac{d\mathbf{v}}{dt}$.

7.18 Prove that if $\mathbf{r} = \mathbf{r}(s)$ is thrice differentiable and $\mathbf{r}'' \neq \mathbf{0}$ (so that $\mathbf{p}$ and $\mathbf{b}$ are defined) then

$$\frac{d\mathbf{p}}{ds} = \mathbf{p}' = -\kappa\mathbf{u} + \tau\mathbf{b}. \tag{31}$$

[*Hint* : Note that $\mathbf{p} = \mathbf{b}\times\mathbf{u}$. Apply the last exercise, along with (8) and (25). The formulas (8), (25) and (31) are called **Frenet formulae** or **Serret-Frenet formulae**.]

7.19 With the conditions of the last exercise, prove that $\tau = (\mathbf{r}'\ \mathbf{r}''\ \mathbf{r}''')/\kappa^2$.

7.20 Assuming sufficient differentiability and non-vanishing of derivatives, prove that the curvature and the torsion of a curve parametrised by $\mathbf{r} = \mathbf{r}(t)$ are given by

$$\kappa = \frac{|\dot{\mathbf{r}}\times\ddot{\mathbf{r}}|}{|\dot{\mathbf{r}}|^3} \quad \text{and} \quad \tau = \frac{(\dot{\mathbf{r}}\ \ddot{\mathbf{r}}\ \dddot{\mathbf{r}})}{|\dot{\mathbf{r}}\times\ddot{\mathbf{r}}|^2}$$

7.21 Obtain the torsion of a conical helix (see Exercise (7.4)).

7.22 A helix is popularly defined as a curve on a cylindrical surface which becomes a straight line if the cylinder is unfolded into a plane. Equivalently, it is the image of a straight line in a plane when the plane is 'wrapped' around a cylinder without stretching or compressing. Prove that this description is correct.

7.23 Suppose the tangent at any point of a curve makes a constant angle with a fixed direction. Prove that $\kappa/\tau$ is a constant. Prove also the converse. Show that the circular helix has this property.

7.24 Prove that the normal to the parabola $y^2 = 4ax$ at any point $P = (at^2, 2at)$ makes the same angle with the axis of the parabola as with the line $PF$, where $F = (a, 0)$ is the focus of the parabola. (Consequently, all light rays coming parallel to the axis pass through the focus after being reflected by the curve and vice versa. Because of this 'focusing property' of a parabola, paraboloidal mirrors are used in search lights.) [*Hint* : Put $t = \tan\theta$.] State, prove and interpret a similar result about ellipses.

7.25 Let $P_0 = (x_0, y_0)$ be a point on the curve $y = f(x)$. Let $y_0' = f'(x_0)$, $y_0'' = f''(x_0)$ and assume $y_0'' \neq 0$. Let $\rho$ be the radius of curvature at $P_0$, and $C_0$ the centre of curvature at $P_0$. Let $\sigma = \sigma(\Delta x)$ be the distance between $C_0$ and a nearby point $P = (x_0 + \Delta x, f(x_0 + \Delta x))$ on the curve. Prove that :

(i) $\sigma^2 \simeq \rho^2$ upto the 0-th order of approximation

(ii) $\sigma^2 \simeq \rho^2 + (1 + y_0'^2)(\triangle x)^2$ upto the 1st order of approximation (why is this reasonable to expect?)

(iii) $\sigma^2 \simeq \rho^2 + y_0' y_0''(\triangle x)^3 + \frac{1}{4}(y_0'')^2(\triangle x)^4$ upto the 2nd order of approximation.

Hence show that unless $y_0' = 0$, the circle of curvature at $P_0$ crosses the curve at $P_0$, that is, $P$ lies inside or outside of $C_0$, ·depending upon the sign of $\triangle x$, if $|\triangle x|$ is sufficiently small. This is in sharp contrast with the behaviour of the tangent line at $P_0$, which does not cross the curve except at a point of inflection.

## 6.8     Polar Co-ordinates

The history of mathematics is replete with new ideas of varying depth and importance. But, as is to be expected, very few of these can be said to have truly revolutionised mathematics. Normally one would think that a concept which achieves such a distinction must be highly profound and subtle. For example, the concept of a limit, which is the very heart of calculus as was elaborately explained in Section 2.2, is a very subtle concept, a precise definition of which eluded mathematicians for more than a century. Another example is the concept of zero, attributed to India (thereby leading to the double entendre "India's contribution to the world of mathematics is zero !"). Although we may think there is nothing in it, it is precisely for this reason that philosophically it is a very profound concept. In mathematics, the introduction of 0 made it possible to represent numbers w.r.t. a given base such as 10, 2, 3 etc. More importantly, it introduced the concept of the identity of addition and opened the door for negative numbers.

Ironically, sometimes some very simple ideas have had an astonishing effect on mathematics. A good example is provided by the concept of plane co-ordinates. It is far from subtle. Even a layman uses it. For example, when he gives his address as, say, three blocks to the east and two to the south from some central place in the town, in essence he is telling you the cartesian co-ordinates of his house, with the central place as the origin. Similarly identifying an object in a rectangular array by telling which numbered row and column it lies in amounts to specifying its co-ordinates. The latitude and the longitude of a place on earth are also its co-ordinates. (But they are not cartesian co-ordinates. They come under what are called spherical co-ordinates as we shall see later).

The impact on mathematics of co-ordinates, especially the cartesian co-ordinates, is truly profound. Because a point, say $P$, in a plane corresponds uniquely to an ordered pair, say $(x, y)$, of real numbers, anything about $P$ can be paraphrased in terms of $x$ and $y$. Many problems in geometry can be tackled by first converting them to algebraic problems about the co-ordinates and solving them by the methods of algebra. For example, the concurrence of the three altitudes of a triangle can be proved by showing that a certain system of three linear equations has a solution (see Exercise (1.12.4)). In the usual geometric proofs, one has to draw diagrams to make the proofs intelligible. Often, separate diagrams are needed to dispose of the various cases such as an acute angled triangle or an obtuse angled one. And an improperly drawn diagram can lead to a wrong result (see Exercise (4.8.11) for an example). In a proof using co-ordinates, diagrams are hardly needed. This advantage becomes all the more apparent in the three dimensional space $I\!R^3$ where visualising and drawing accurate diagrams is far more challenging than in $I\!R^2$. In the analytical approach, all we need is to attach one more co-ordinate, say $z$. This does not substantially increase the complications. In fact, with co-ordiantes, it is almost as easy to study the $n$-dimensional space $I\!R^n$ for any positive integer $n$, as outlined in Section 4.8. Even though we may not be able to visualise them, as shown in Section 6.5, they do have practical applications.

In fact, the advantage gained by co-ordinates over the naive, intuitive approach of classical geometry has prompted mathematicians to define a point in a plane as an ordered pair of real numbers. One can then define the basic concepts of geometry such as line segments, their lengths and the angles between them in a strictly analytical manner as shown in Section 4.8. This way geometry is completely subsumed by co-ordinates, even though originally they were meant only as an aid to geometry. Rather like a servant subsequently taking over as the master ! While this is a great achievement from the view point of logical perfection, it has also made geometry abstract and thereby alienated a beginner from its visual charm and the elegance of some of its proofs (see Exercises (1.1.4) and (1.1.5) for examples of problems whose solutions require considerable ingenuity without co-ordinates but which become rather routine with co-ordinates).

We leave aside this debate whether the co-ordinates have done more harm than good. Let us go back to the classical approach where a plane (or a space) is an intuitively clear object and the co-ordinates are only a means of labeling its points. We have already heavily used the cartesian co-ordinates. All the formulas that we derived, for example those for the arc length, the curvature, areas and volumes etc. were in terms of cartesian co-ordinates. In this section we shall see that the cartesian co-ordinates are not the only means of specifying points in a plane. There are other co-ordinate systems and one of them, called the polar co-ordinate system is sometimes more convenient than the cartesian system. By a different co-ordinate system, we mean one which is qualitatively different. If for example, we shift the origin and/or rotate the axes, we do get a different system which might simplify some problems. But it is still a system of cartesian co-ordinates. If, however, we take the axes as inclined at an angle $\theta$ (not necessarily a right angle) then we do get a system which is more general than the cartesian co-ordinates. Such co-ordinates, called **oblique co-ordinate**, were popular at one time. The cartesian co-ordinates in which $\theta = \dfrac{\pi}{2}$, were called **rectangular cartesian co-ordinates**. Old books on co-ordinates geometry contain formulas for equations of a straight line, a circle etc. in oblique co-ordinates. As is to be expected, these formulas are in general more complicated than the corresponding formulas for rectangular cartesian co-ordinates (see, for example, Exercise (8.1)). It turns out that the advantage gained by allowing the axes to be inclined at an arbitrary angle is not worth such complications. So today nobody talks about oblique co-ordinates and 'cartesian co-ordinates' nowadays always means what formerly was called 'rectangular cartesian co-ordinates'. (The term 'cartesian' comes from the name of their inventor, Descartes.)

As an important example of a co-ordinate system for the plane which is genuinely different from the cartesian system we consider the polar co-ordinates. In this system some point, say $O$, (called the 'pole', whence the name) and some 'initial ray' (i.e. a half line) originating at $O$ are fixed, as shown in Fig.6.8.1 (a). Given any point $P$ in the plane other than $O$, we let $r$ be the distance between $O$ and $P$ and let $\theta$ be the angle which the ray $\overrightarrow{OP}$ makes with the fixed direction we have chosen.

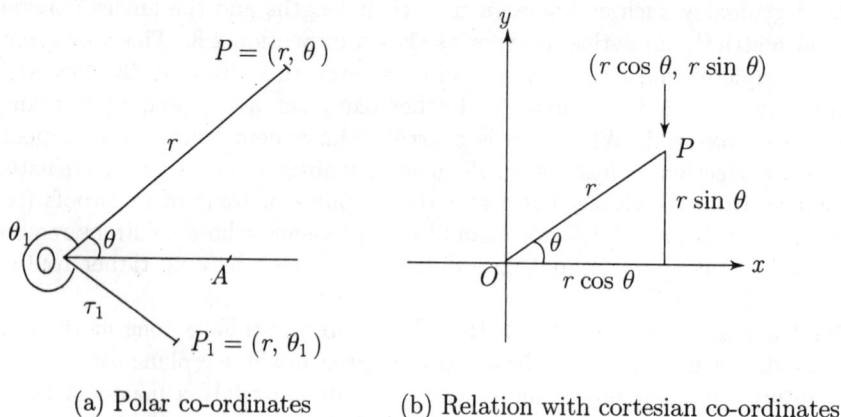

(a) Polar co-ordinates            (b) Relation with cortesian co-ordinates

Figure 6.8.1 : Polar and Cartesian Co-ordinates

The numbers $r$ and $\theta$ are called the polar co-ordinates of $P$. We write $P = (r, \theta)$. Here $\theta$ can lie in any semi-open interval of length $2\pi$. Usually we take this interval to be $[0, 2\pi)$. Values of $\theta$ outside this interval are permissible. But they will not give any new points. Thus $(2, \frac{19\pi}{3})$, and $(2, \frac{-11\pi}{3})$ represent the same point as $(2, \frac{\pi}{3})$. In general, for any $r > 0$, the pairs $(r, \theta_1)$ and $(r, \theta_2)$ represent the same point if and only if $\theta_1 - \theta_2$ is an integral multiple of $2\pi$. So unlike cartesian co-ordinates, the polar co-ordinates of a point are not unique. Sometimes this is a nuisance, but sometimes it is a boon as we shall see later. When $P = O, r = 0$. Now $\theta$ has no unique value. Thus $(0, \theta)$ represents the same point, viz. $O$, for any value of $\theta$. (Note that we always have $r \geq 0$. Some authors allow $r$ to be negative as well. In that case $(r, \theta)$ and $(-r, \theta + \pi)$ will represent the same point.) The co-ordinates $r$ and $\theta$ are called, respectively, the **length** (or sometimes the **magnitude**) and the **polar angle** or the **argument** of $P = (r, \theta)$. (The cartesian co-ordiantes $x$ and $y$ also have names. They are called, respectively, the **abscissa** and the **ordinate** of $P = (x, y)$. But these terms are rarely used these days.) As mentioned above, the argument is not unique and for $r > 0$, any two arguments of the same point differ by an integral multiple of $2\pi$. The unique value, which lies in a chosen semi-open interval, say $[0, 2\pi)$, of length $2\pi$ is sometimes called the **principal argument** of $P$. These terms are often used when points of the plane are identified with complex numbers. We shall have little occasion to use them.

It is easy to establish the relationship between the cartesian and polar co-ordinates of the same point. Choose the pole as the origin, the initial ray as the positive $x$-axis and the ray at $\theta = \frac{\pi}{2}$ as the positive $y$-axis as shown in Fig.6.8.1 (b). An elementary calculation (which, in effect, is the very definition of $\sin \theta$ and $\cos \theta$) gives

$$x = r \cos \theta; \; y = r \sin \theta \tag{1}$$

These equations express the cartesian co-ordinates in terms of the polar ones. The other way conversion requires a little care. There is no difficulty about $r$ which equals $\sqrt{x^2 + y^2}$. It is customary to get $\theta$ by eliminating $r$ from the two equations in (1), thereby giving $\tan \theta = \frac{y}{x}$. But to conclude $\theta = \tan^{-1} \frac{y}{x}$ from this is not quite correct. First of all, $\frac{y}{x}$ is meaningless if $x = 0$. Secondly, even when $x \neq 0, \tan^{-1}(\frac{y}{x})$ will, by definition, always lie between $\frac{-\pi}{2}$ and $\frac{\pi}{2}$. Even if we add multiples of $2\pi$, we still do not get all possible values of $\theta$. For example the points $(-1, -1)$ and $(1, 1)$, in the cartesian co-ordinate system will both correspond to $\left(\sqrt{2}, \frac{\pi}{4}\right)$ which is clearly wrong, since $(-1, -1)$ obviously corresponds to $\left(\sqrt{2}, \frac{\pi}{4} + \pi\right)$. A better way out is to let $\theta$ be $\cos^{-1}(\frac{x}{\sqrt{x^2 + y^2}})$ if $y \geq 0$. This gives a unique value of $\theta$ in $[0, \pi]$ and for this value of $\theta$, we indeed have $x = r \cos \theta$ and $y = r \sin \theta$. For $y < 0$, we let $\theta = \pi + \cos^{-1}(\frac{-x}{\sqrt{x^2 + y^2}})$. We are, of course, assuming here that $\sqrt{x^2 + y^2} \neq 0$. If $\sqrt{x^2 + y^2} = 0$, then $P = O$

and $\theta$ can be anything. Summing up,

$$r = \sqrt{x^2 + y^2}; \quad \theta = \begin{cases} \cos^{-1}\left(\dfrac{x}{\sqrt{x^2+y^2}}\right) & \text{if } y \geq 0, x^2 + y^2 > 0 \\[3mm] \pi + \cos^{-1}\left(\dfrac{-x}{\sqrt{x^2+y^2}}\right) & \text{if } y < 0, x^2 + y^2 > 0 \\[3mm] \text{any value} & \text{if } x^2 + y^2 = 0. \end{cases} \qquad (2)$$

Because of (1) and (2) anything that can be done using the cartesian co-ordinates can also be done using polar co-ordinates and vice-versa. For example the equation of a straight line in polar co-ordinates will be of the form $ar\cos\theta + br\sin\theta + c = 0$ where $a, b, c$ are constants, not all 0. This can be rewritten as $r\cos(\theta - \alpha) + c' = 0$ where $\sin\alpha = \dfrac{b}{\sqrt{a^2+b^2}}, \cos\alpha = \dfrac{a}{\sqrt{a^2+b^2}}$ and $c' = \dfrac{c}{\sqrt{a^2+b^2}}$. Thus the general equation of a straight line in polar co-ordinates is of the form $r = A\sec(\theta - \alpha)$ for some constants $A$ and $\alpha$. This can also be derived directly (see Exercise (8.2)). In fact a direct derivation is often preferable. Getting an equation in the cartesian co-ordinates and then mechanically converting it into an equation in polar co-ordinates using (1) is rather like thinking in one language and then translating your thoughts into another, word by word, using a dictionary. Although this way we may be able to make ourselves understood, in doing so we may miss certain peculiar features of that second language and the convenience they provide. (In real life, such translations are often laughable especially when idioms of one language are translated literally.) For example, in English there is no single word for 'mother's brother'. But Hindi and many other Indian languages do have such words. In that case, translating 'mother's brother' as such into Hindi, although technically correct, ignores the availability of a single word to express the same concept and will easily betray the non-nativity of the speaker.

The same is true for polar co-ordinates. A simple feature which stands out immediately from the definition is that no work is needed to find the distance of a point $P$ from the pole. Simply read the first polar co-ordinate of $P$, viz. $r$. The second co-ordinate, $\theta$, in essence tells you the direction. So the polar co-ordinates are nothing but a refinement of a layman's statement like, "My hometown is 300 km. to the northeast from here".

Exercise (6.19) shows how sometimes polar coordinates are more convenient than the cartesian coordinates. As another illustration, certain curves can be expressed more simply in the polar co-ordinates than in cartesian co-ordinates. For example, the polar equation of a circle of radius $a$, centered at the pole, is simply $r = a$. (If the centre is somewhere else, then it is fairly complicated). Curves of the form $r = f(\theta)$ where $f$ is some (non-negative) function of $\theta$, are especially easier to sketch directly in polar co-ordinates than by converting them to cartesian co-ordinates. We take a ray rotating around the pole $O$ and as it rotates, go on marking points on it at distances $f(\theta)$ from $O$. Fig. 6.8.2 shows three such curves, the **spiral**, the **exponential spiral** and the **cardioid** the last name coming because of the heartlike shape of the curve. (The cardioid can also be obtained as a special case of the epicycloid, see Exercise (8.7)(a).)

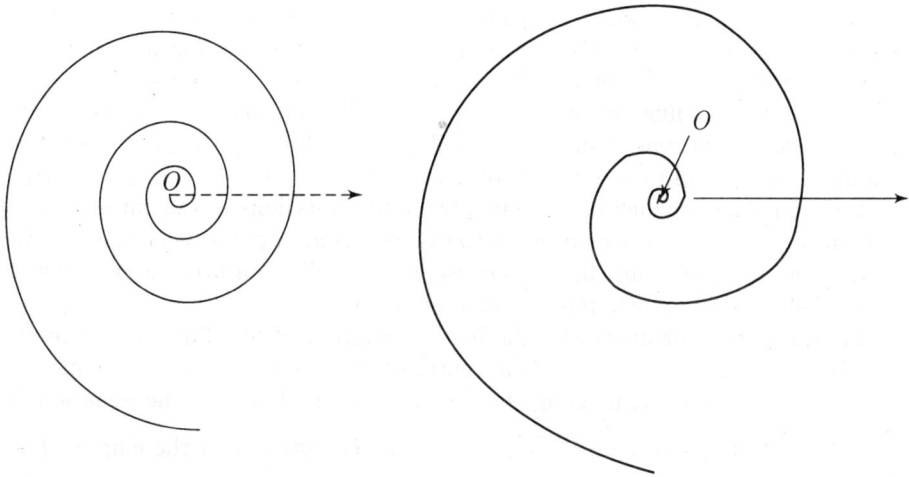

(a) Spiral $r = \theta$, $(\theta \geq 0)$         (b) Exponential spiral $r = e^{\theta}$

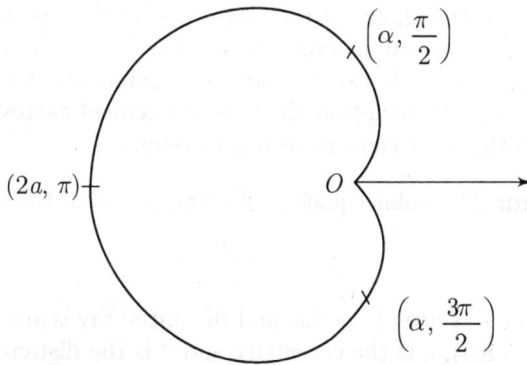

(c) Cardioid $r = a\,(1 - \cos\theta)$, $(a > 0)$

Figure 6.8.2 : Curves of the Form $r = f(\theta)$

As is to be expected, just as curves of the form $y = f(x)$ are a little easier to analyse in cartesian co-ordinates than curves given implicitly as, say, $\phi(x, y) = 0$, in polar co-ordinates, too, curves of the form $r = f(\theta)$ are easier to deal with. When converted to cartesian co-ordinates such curves do not generally assume the form $y = f(x)$ or $x = g(y)$ (although, of course, this can be arranged locally by the Implicit Function Theorem, mentioned in Section 4.12). The two spirals above are good examples of this. In fact for such curves, the multivaluedness of the argument $\theta$ comes in handy. For a curve of the form $y = f(x)$, a line $x = c$ can cut it in at most one point. But the ray $\theta = \alpha$ cuts the spirals in infinitely many points (which are equally spaced for the spiral in (a) but farther and farther apart as we move away from $O$ for the exponential spiral in (b)). For every positive integer $k$, the ray $\theta = \alpha + 2\pi k$ is the same as the ray $\theta = \alpha$. But the values of $r$ are different and so we get two different points on the same ray. The cartesian equations of such curves will be very clumsy to say the least. For example, the spiral in (a) will be represented by $x = \sqrt{x^2 + y^2} \cos(\sqrt{x^2 + y^2})$.

Even for curves which have relatively simple equations in the cartesian co-ordinates, it is sometimes convenient to express them in polar co-ordinates. We do this for the case of an ellipse, because as we shall see a little later, using it we shall show why planets move in elliptic orbits.

In cartesian co-ordinates to obtain the equation of an ellipse in a simplified form, we take the origin as the centre of the ellipse, the $x$-axis along the major axis and the $y$-axis along the minor axis. In that case the equation is $\dfrac{x^2}{a^2} + \dfrac{y^2}{b^2} = 1$, where $b = a\sqrt{1 - e^2}$, $e$ being the eccentricity of the ellipse. The foci are at $(\pm ae, 0)$ and the corresponding directrices are the lines $x = \pm\frac{a}{e}$. We could convert this to the polar form using (1). The centre would then be the pole. But it turns out that it is more convenient to take one of the foci as the pole. This point should always be kept in mind while switching to a new co-ordinate system. There are infinitely many cartesian co-ordinate systems (depending upon the choice of the origin and of the axes) and similarly there are infinitely many polar co-ordinate systems. The relations (1) and (2) are the conversion formulas between a cartesian system and its corresponding polar system. But it can happen that the most convenient cartesian system need not correspond to the most convenient polar system.

**8.1 Theorem:** The polar equation of an ellipse, is of the form

$$r = \frac{ed}{1 + e \cos \theta} \tag{3}$$

where the pole is at one of the foci and the initial ray is along the major axis (as shown in Fig.6.8.3), $e$ is the eccentricity and $d$ is the distance between the focus and the corresponding directrix.

**Proof :** If we take the centre $C$ of ellipse as the origin and the axes as co-ordinate axes, then its equation is

$$\frac{x^2}{a^2} + \frac{y^2}{b^2} = 1 \text{ or equivalently, } (1 - e^2)x^2 + y^2 = a^2(1 - e^2) \tag{4}$$

where $a, b$ are its semi-major and semi-minor axes. The foci are at $(ae, 0)$ and $(-ae, 0)$. We let the pole $O$ be the focus $(ae, 0)$. Then shifting the origin to $O$, the equation (4) becomes

$$(1 - e^2)(x + ae)^2 + y^2 = a^2(1 - e^2) \tag{5}$$

We can now put $x = r\cos\theta, y = r\sin\theta$ so as to get a quadratic in $r$, viz.,

$$(1 - e^2\cos^2\theta)r^2 + 2(1 - e^2)(ae\cos\theta)r - a^2(1 - e^2)^2 = 0 \tag{6}$$

If we solve this quadratic for $r$ and keep in mind that $r \geq 0$, we get (3), because

$$d = \frac{a}{e} - ae = \frac{a(1 - e^2)}{e}.$$

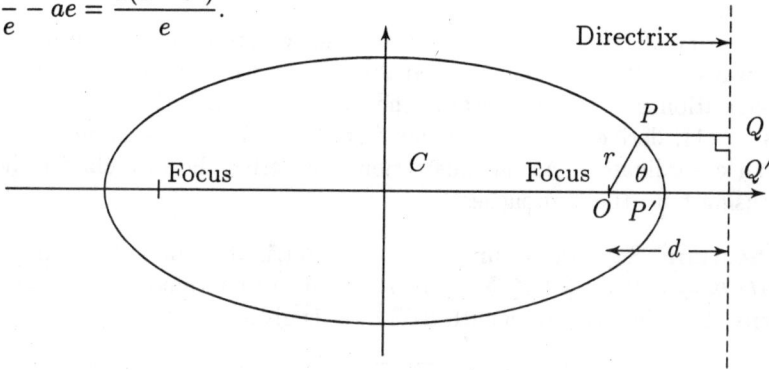

Figure 6.8.3 : Equation of an Ellipse in Polar Co-ordinates

There is an alternate and a more instructive way of deriving (3) because it is also applicable to a parabola and a hyperbola as well as to an ellipse. These three curves have a property in common, viz., for every point $P$ on any of them the ratio of its distance from a focus to its distance from the corresponding directrix is a constant, equal to $e$. (The curve is an ellipse, parabola or a hyperbola according as $e < 1, e = 1$ or $e > 1$. A parabola has only one focus). Let $P$ be a point on the ellipse (or the parabola or the hyperbola). Let $Q$ be the foot of the perpendicular form $P$ to the directrix. Let $P', Q'$ be the feet of the perpendiculars from $P, Q$ respectively to the initial ray. (The initial ray is the perpendicular from $O$ to the directrix for all three curves. The case of the ellipse is shown in Fig.6.8.3. We leave it as an exercise to draw similar figures for a hyperbola and a parabola.) Then, by the property mentioned above,

$$r = OP = ePQ = eP'Q' = e(OQ' - OP') = e(d - r\cos\theta)$$

solving which we get (3). (Notice that had we taken the pole at the other focus, then instead of (3) the equation would have been $r = \dfrac{ed}{1 - e\cos\theta}$ .) ∎

An additional advantage of the alternate derivation of (3) is that it shows that for a fixed focus and directrix (and consequently for a fixed $d$), as the eccentricity $e$ tends to 1 (from below), the equation of the ellipse tends to that of the parabola. From Exercise (8.3), it follows that as $e \to 1^-$, the other focus of the ellipse tends to the point at infinity. So we may treat a parabola as a limiting case of an ellipse as one of its foci (and the corresponding directrix) remains fixed but the other focus moves to infinity. This is not so easily apparent in the cartesian co-ordinates. (A parabola can also be thought of as the limiting case of a hyperbola.)

A few more examples of curves in polar co-ordinates will be given in the exercises. So far we considered non-parametric curves. For a parametric curve in polar co-ordinates, $r$ and $\theta$ are given as some (continuous) functions of some parameter, say $t$. Again, for brevity we write $r = r(t)$ and $\theta = \theta(t)$ instead of $r = f(t), \theta = g(t)$ etc. For curves of the form $r = f(\theta)$, we can take $\theta$ itself as a parameter.

The theory of parametrised curves in polar co-ordinates can be developed in two ways. Either we can proceed directly or we can assume a cartesian parametrisation $x = x(t), y = y(t)$ of the same curve with the same parameter and using (1), derive a result in polar form from the corresponding result in cartesian co-ordinates. As an illustration, we derive the formula for the arc length using both the approaches.

**8.2 Theorem:** Suppose a curve $C$ is parametrised in polar coordinates by $r = r(t), \theta = \theta(t), a \le t \le b$ where $r$ and $\theta$ are continuously differentiable functions of $t$. Then the arc length $l(C)$ of $C$ is given by

$$l(C) = \int_a^b \sqrt{\left(\frac{dr}{dt}\right)^2 + r^2 \left(\frac{d\theta}{dt}\right)^2}\, dt = \int_a^b \sqrt{(\dot{r})^2 + (r\dot{\theta})^2}\, dt \tag{7}$$

**Proof:** First we give a proof using cartesian co-ordiantes. A cartesian parametrisation of $C$ is given (using (1)) by

$$x = x(t) = r(t) \cos(\theta(t)), y = y(t) = r(t) \sin(\theta(t)), a \le t \le b \tag{8}$$

As the trigonometric functions are continuously differentiable everywhere and the products and composites of continuously differentiable functions are also continuously differentiable, it follows from (8) that $x, y$ are continuously differentiable functions of $t$ and further,

$$\dot{x} = \frac{dx}{dt} = \dot{r} \cos\theta - r(\sin\theta)\dot{\theta} \text{ and } \dot{y} = \dot{r} \sin\theta + r(\cos\theta)\dot{\theta} \tag{9}$$

If we substitute these in the formula given by Theorem (2.4), we get,

$$l(C) = \int_a^b \sqrt{(\dot{x})^2 + (\dot{y})^2}\, dt = \sqrt{(\dot{r})^2 + r^2(\dot{\theta})^2}\, dt$$

as desired.

For a direct proof, consider a typical point $P = (r(t), \theta(t))$ on the curve $C$ and let $Q = (r(t+\Delta t), \theta(t+\Delta t))$ be a nearby point on $C$. Denote $r(t+\Delta t) - r(t)$ and $\theta(t+\Delta t) - \theta(t)$ by $\Delta r$ and $\Delta\theta$ respectively. Denote the length of the segment $PQ$ by $\Delta s$. (See Fig.6.8.4(a), where $\Delta\theta$ is shown as positive. The argument goes through even if $\Delta\theta$ is negative or zero.)

Now applying the cosine rule to the triangle $OPQ$ we have

$$\Delta s = \sqrt{(r + \Delta r)^2 + r^2 - 2r(r + \Delta r) \cos\Delta\theta} \tag{10}$$

which, after dividing by $\Delta t$ and a little simplifaction, becomes

$$\frac{\Delta s}{\Delta t} = \sqrt{\frac{4r^2 \sin^2(\frac{\Delta\theta}{2})}{(\Delta t)^2} + \frac{4r\Delta r \sin^2(\frac{\Delta\theta}{2})}{(\Delta t)^2} + (\frac{\Delta r}{\Delta t})^2} \tag{11}$$

Now as $\Delta t \to 0$, $\dfrac{\Delta r}{\Delta t} \to \dfrac{dr}{dt}(= \dot{r})$. Also

$$\lim_{\Delta t \to 0} \frac{\sin(\Delta\theta/2)}{\Delta t} = \lim_{\Delta t \to 0} \frac{\sin(\Delta\theta/2)}{\Delta\theta/2} \frac{\Delta\theta/2}{\Delta t} = 1 \cdot \frac{1}{2}\frac{d\theta}{dt}.$$

The limit of the middle term under the radical sign in (11), is similar to that of the first term. But there is the factor $\Delta r$ which also tends to 0 as $\Delta t \to 0$. So,

$$\lim_{\Delta t \to 0} \frac{\Delta s}{\Delta t} = \sqrt{(\dot{r})^2 + (r\dot{\theta})^2} \tag{12}$$

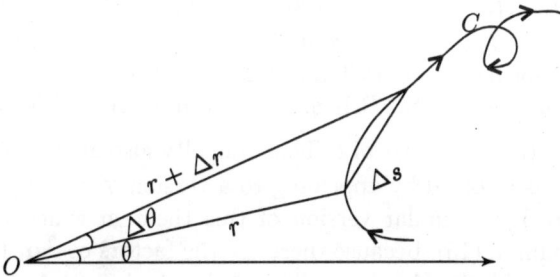

Figure 6.8.4 : Arc length in Polar Co-ordinates.

To derive (7) from (12), there are two ways. The naive approach, which is similar to that for cartesian the co-ordinates, (see Fig.6.2.3 and the accompanying comments) is to recognise that if $\Delta t$ is infinitesimally small then so is $\Delta s$ and in that case, $\Delta s$ is also the length of the portion of the curve from $P$ to $Q$ (the 'line element' as it is called). So $\lim\limits_{\Delta t \to} \dfrac{\Delta s}{\Delta t}$ would indeed be the derivative of the arc length function $s$, w.r.t. the parameter $t$. So,

$$\frac{ds}{dt} = \sqrt{(\dot{r})^2 + (r\dot{\theta})^2} \tag{13}$$

The assertion of the theorem now follows from the fundamental theorem of calculus (the first form). Those who do not like this naive approach will have to do essentially the same work as that needed for establishing Theorem (2.4). Fortunately, Lemma (2.1), Corollary (2.2) and Theorem (2.3) remain intact since they are purely geometric and hence independent of a particular co-ordinate system. The only change is that given a partition, say $R = (t_0, t_1, \ldots, t_n)$ of $[a, b]$, the length of the corresponding broken line path along $C$ will have to be obtained using (10). Writing $r_i$ for $r(t_i)$, $\theta_i$ for $\theta(t_i)$, $\Delta r_i$ for $r_i - r_{i-1}$ and $\Delta\theta_i$ for $\theta_i - \theta_{i-1}$, we get

$$l(R) = \sum_{i=1}^{n} \sqrt{(r_i + \Delta r_i)^2 + r_i^2 - 2r_i(r_i + \Delta r_i)\cos\Delta\theta_i} \tag{14}$$

Dividing and multiplying by $\Delta t_i (= t_i - t_{i-1})$ and working as in (11), the $i^{\text{th}}$ term of this summation can be written as

$$\sqrt{r_i^2 \left(\frac{\sin(\Delta\theta_i/2)}{\Delta\theta_i/2}\right)^2 \left(\frac{\Delta\theta_i}{\Delta t_i}\right)^2 + \left(\frac{\Delta r_i}{\Delta t_i}\right)^2 + r_i \frac{\Delta r_i}{\Delta t_i} \left(\frac{\sin(\Delta\theta_i/2)}{\Delta\theta_i/2}\right)^2 \frac{\Delta\theta_i}{\Delta t_i} \Delta t_i} \, \Delta t_i$$

If we now apply the Lagrange Mean Value Theorem to the interval $[t_{i-1}, t_i]$ for the functions $r$ and $\theta$, (and to the interval $[0, \frac{\Delta\theta_i}{2}]$ for the function $\sin\theta$) we get

$$l(R) = \sum_{i=1}^{n} \left[r_i^2(\cos^2\alpha_i)(\dot\theta(\xi_i))^2 + (\dot r(\eta_i))^2 + r_i \dot r(\eta_i)(\cos^2\alpha_i)(\dot\theta(\xi_i))^2 \Delta t_i\right]^{1/2} \Delta t_i$$

(15)

for some $\xi_i \in (t_{i-1}, t_i)$, $\eta_i \in (t_{i-1}, t_i)$ and $\alpha_i \in (0, \frac{\Delta\theta_i}{2})$. (It is tacitly assumed here that $\Delta\theta_i > 0$. If $\Delta\theta_i < 0$, we apply the MVT to the interval $[\frac{\Delta\theta_i}{2}, 0]$. If $\Delta\theta_i = 0$, we do not need the MVT and take $\alpha_i = 0$.)

The trouble now is that the R.H.S. of (15) is not quite a Riemann sum of the function $\sqrt{(r(t)\dot\theta(t))^2 + (r(t))^2}$ of $t$. This difficulty also arose with cartesian coordinates and was resolved by appealing to a version of a theorem called Bliss' theorem. There is no similar version of this theorem which can be directly applied to the sum in (15), because there are the factors $\cos^2\alpha_i$ (which are very close to 1 if $\Delta t_i$ and hence $\Delta\theta_i$ is small) and also the last term under the radical sign has a factor $\Delta t_i$. Still, the argument used in the proof of Bliss' theorem can be modified. Thus using uniform continuity of $\dot r, \dot\theta$ and the fact that $\alpha_i$ is close to 0, it can be shown (cf. Exercise (8.8)) that

$$l(R) \approx \sum_{i=1}^{n} \sqrt{(r(t_i)\dot\theta(t_i))^2 + (\dot r(t_i))^2} \, \Delta t_i \qquad (16)$$

in the sense that the difference between the two sides of (16) can be made as small as we like if the partition $R$ is sufficiently fine. We now apply Theorem (2.3). As $R$ gets finer and finer, $l(R)$ approaches $l(C)$. But, on the other hand, the sum on the right of (16) approaches $\int_a^b \sqrt{(\dot r(t))^2 + (r(t)\dot\theta(t))^2} \, dt$ as $R$ gets finer and finer, by Theorem (5.2.1). So (7) holds. ∎

The second proof, especially in its rigorous form, is considerably more complicated than the first proof. This is to be expected because the first proof merely consisted of translating the formula for the arc length in cartesian co-ordinates into a formula in the polar co-ordinates. And this saved the hard work already done to establish the former. This approach is therefore preferable wherever we are dealing with attributes which are independent of a particular co-ordinate system. The arc length is one such attribute. The *formula* for the arc length is different in different systems. But the concept of arc length is the same.

Taking analogy with languages, the word for 'sugar' may be different in some other language. But the substance 'sugar' is the same and independent of any language. So if a scientist has studied properties of sugar and expressed them in English, then to express them in French, all we need is a good translator. It is hardly necessary that a French scientist should study those properties again and record them in French ! (This, of course does not apply to literary works where the particular word 'sugar' matters, for example, in a pun or to make a rhyme.)

We follow this translatory method to obtain formulas for the velocity and acceleration of a particle, say $P$, moving in a plane. Let $\mathbf{r} = \mathbf{r}(t)$ be the position vector $\overrightarrow{OP}$ of $P$ (where $O$ is the pole), at time $t$. Then the velocity $\mathbf{v}$ and acceleration $\mathbf{a}$ of $P$ are, respectively $\dot{\mathbf{r}}$ and $\ddot{\mathbf{r}}$. We defined the derivative of a vector valued function in terms of the derivatives of the component functions w.r.t. the cartesian system. But in Theorem (7.5) (which was based on Lemma (7.4)) we showed that the derivative of a vector valued function has a meaning independent of the co-ordinate system. So we are entitled to use any results about such derivatives, even though they were proved using cartesian co-ordinates (e.g. Exercise (6.6)).

In the cartesian co-ordinates, $\mathbf{v} = \dot{\mathbf{r}} = \dot{x}\,\mathbf{i} + \dot{y}\,\mathbf{j}$ and $\mathbf{a} = \ddot{\mathbf{r}} = \ddot{x}\,\mathbf{i} + \ddot{y}\,\mathbf{j}$. (Note that we are dealing with a planar motion and so there is no $z$-component.) To get formulas for $\mathbf{v}$ and $\mathbf{a}$ in polar co-ordinates, we first have to decide which unit vectors should play the role of the vectors $\mathbf{i}$ and $\mathbf{j}$ in the cartesian co-ordinates. For this we first have to look at these vectors a little differently than just a fixed pair of mutually orthogonal unit vectors. Let $P_0 = (x_0, y_0)$ be a point in the plane. Of all the curves passing through $P_0$, there are two curves which stand out as having the simplest possible equations. They are (i) $y = y_0$ and (ii) $x = x_0$. Both are straight lines. On $y = y_0$, only the $x-$ co-ordinate changes while on $x = x_0$, only the $y$- co-ordinate changes. For this reason, these two lines are called the **co-ordinate curves**. The vectors $\mathbf{i}$ and $\mathbf{j}$ are nothing but unit tangent vectors to the co-ordinate curves at $P_0$, in the directions in which the co-ordinates $x$ and $y$ increase respectively (the other one remaining constant). See Fig.6.8.5 (a). To stress this fact let us denote $\mathbf{i}$ and $\mathbf{j}$ by $\mathbf{e}_x$ and $\mathbf{e}_y$ respectively.

We can similarly consider co-ordinate curves in polar co-ordinates. Let $P_0 = (r_0, \theta_0)$. (We assume $P$ is different from the pole $O$, as otherwise $\theta_0$ is not uniquely defined.) Then the two co-ordinate curves through $P_0$ are (i) $r = r_0$ which is a circle, say $C$ and (ii) $\theta = \theta_0$, which is a ray, say $L$. On $C$, only $\theta$ changes and we let $\mathbf{e}_\theta$ be a unit tangent to $C$ in the direction in which $\theta$ increases. Similarly, on $L$, only $r$ changes and we let $\mathbf{e}_r$ be the unit tangent vector in the direction in which $r$ increases. Note that $\mathbf{e}_r$ and $\mathbf{e}_\theta$ are at right angles to each other, as in the case of the co-ordinate curves for cartesian co-ordinates. But there is an important difference. In cartesian co-ordinates, the unit tangent vectors $\mathbf{e}_x$ and $\mathbf{e}_y$ to the co-ordinate curves are the same at every point (viz., $\mathbf{i}$ and $\mathbf{j}$). But in the polar co-ordinates their directions vary from point to point (see Fig.6.8.5(b)).

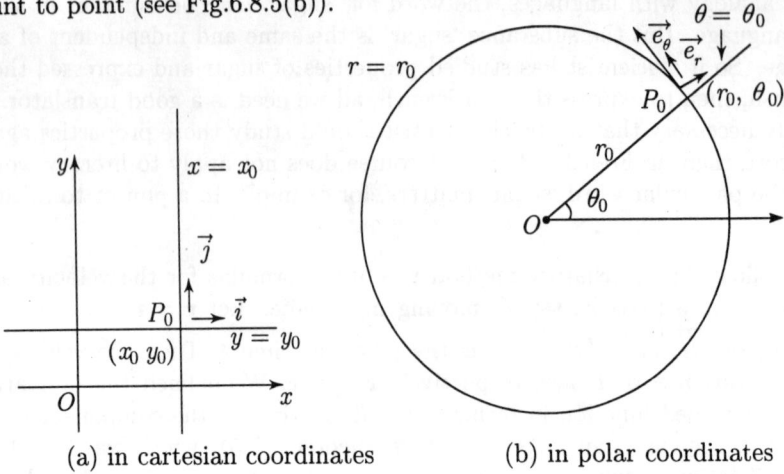

(a) in cartesian coordinates        (b) in polar coordinates

Figure 6.8.5 : Co-ordinate Curves and their Tangents

Despite this difference, it turns out that the vectors $e_r$ and $e_\theta$ provide a convenient basis for studying what happens near $P_0$. (The idea of a variable basis of vectors was already encountered in the last section where we saw that the triple $(\mathbf{u}, \mathbf{p}, \mathbf{b})$ changed from point to point of a space curve.) As we shall see shortly, it is especially convenient to resolve the velocity and acceleration along these vectors $e_r$ and $e_\theta$. When so done, the components along these vectors are called **radial** and **transverse** components respectively. (The vectors $e_r$ and $e_\theta$ are called, the radial and transverse vectors at $P_0$.)

Some care has to be taken in doing so. Suppose $\mathbf{u} = \mathbf{u}(t)$ is a vector valued function of a real parameter $t$. In the cartesian co-ordinates, if we resolve it as $\mathbf{u} = u_1(t)\mathbf{i} + u_2(t)\mathbf{j}$, then its derivative $\dot{\mathbf{u}} = \dfrac{d\mathbf{u}}{dt}$ will simply be $\dot{u}_1(t)\mathbf{i} + \dot{u}_2(t)\mathbf{j}$. This is in fact how we have *defined* the derivative of a vector valued function. But as Theorem (7.5) shows, $\dot{\mathbf{u}}$ has its own meaning and then using the rules given in Exercise (6.6), we get $\dot{\mathbf{u}} = \dfrac{d}{dt}(u_1\mathbf{i}) + \dfrac{d}{dt}(u_2\mathbf{j}) = \dot{u}_1\mathbf{i} + u_1\dfrac{d\mathbf{i}}{dt} + \dot{u}_2\mathbf{j} + u_2\dfrac{d\mathbf{j}}{dt} = \dot{u}_1\mathbf{i} + \dot{u}_2\mathbf{j}$, since $\mathbf{i}$ and $\mathbf{j}$ are constant vectors. If we resolve $\mathbf{u} = \mathbf{u}(t)$ in polar co-ordinate as $u_3(t)e_r + u_2(t)e_\theta$ (say) and try to adopt a similar procedure, then since $e_r$ and $e_\theta$ are no longer constant vectors, we have to take into account their derivatives w.r.t. $t$. A careful observation will show that both $e_r$ and $e_\theta$ depend only on $\theta$ (and not on $r$). Hence to differentiate them w.r.t. $t$, we can differentiate them w.r.t. $\theta$ and then multiply by $\dfrac{d\theta}{dt}$, i.e. by $\dot{\theta}$.

It is possible to differentiate $e_r$ and $e_\theta$ w.r.t. $\theta$ by a direct geometric argument (cf. Exercise (8.13)). But it is easier to do so in cartesian co-ordinates and then to translate the results into polar co-ordinates. We now let $e_r$ and $e_\theta$ be the radial and transverse unit vectors at a general point $P = (r, \theta)$. Then clearly $e_r$ makes an angle $\theta$ with the positive $x$-axis while $e_\theta$, is obtained by rotating $e_r$ counter-clockwise through a right angle (see Fig.6.8.5 (b)). So it is clear that

$$\mathbf{e}_r = \cos\theta\mathbf{i} + \sin\theta\mathbf{j} \quad \text{and} \quad \mathbf{e}_\theta = -\sin\theta\mathbf{i} + \cos\theta\mathbf{j} \tag{17}$$

which, upon differentiating w.r.t. $\theta$ immediately gives

$$\frac{d}{d\theta}(\mathbf{e}_r) = \mathbf{e}_\theta \quad \text{and} \quad \frac{d}{d\theta}(\mathbf{e}_\theta) = -\mathbf{e}_r \tag{18}$$

If the differentiation is w.r.t. some parameter $t$, then we have,

$$\dot{\mathbf{e}}_r = \frac{d}{dt}(\mathbf{e}_r) = \dot{\theta}\mathbf{e}_\theta \quad \text{and} \quad \dot{\mathbf{e}}_\theta = -\dot{\theta}\mathbf{e}_r \tag{19}$$

We are now in a position to derive formulas for velocity and acceleration in polar co-ordinates.

**8.3 Theorem:** Let $(r,\theta)$ be the position of a point $P$ at time $t$ where $r = r(t), \theta = \theta(t)$ are assumed to be twice differentiable. Then except when $P$ is at the pole $O$, its velocity and acceleration are given by

$$\mathbf{v} = \dot{r}\mathbf{e}_r + r\dot{\theta}\mathbf{e}_\theta \tag{20}$$

and

$$\mathbf{a} = (\ddot{r} - r\dot{\theta}^2)\mathbf{e}_r + (2\dot{r}\dot{\theta} + r\ddot{\theta})\mathbf{e}_\theta \tag{21}$$

**Proof:** Let $\mathbf{r} = \mathbf{r}(t)$ be the position vector of $P$ at time $t$. Since $r = |\mathbf{r}|$ and $\mathbf{e}_r$ is a unit vector in the direction $\overrightarrow{OP}$, it is clear that

$$\mathbf{r} = r\mathbf{e}_r. \tag{22}$$

Since $\mathbf{v} = \dot{\mathbf{r}}$ (by Theorem (7.5)), (20) follows immediately from (22), using the first equation in (19). To get the acceleration $\mathbf{a}$, we differentiate (20) and use (19) again to get (21) after combining the like terms. ∎

As a classic application of the simple formulas in (21), it can be shown that Newton's laws of motion along with the law of gravitation imply that planets move around the Sun in elliptic orbits with the Sun at one of the foci. This fact, along with two others, was observed by the astronomer Kepler in the early seventeenth century. The three together are known as **Kepler's laws** of planetary motion. Later in the same century, Newton propounded his famous theory of gravitation. The nature of Kepler's laws is such that in order for them to hold, the force acting on the planet has to be directed towards the Sun and inversely proportional, in magnitude, to its distance from the Sun. So here is an excellent example of how a theory is developed in order to explain observed facts of nature. We shall prove here that each planet moves in an elliptic orbit and give the other two laws as exercises.

Let $O$ denote the position of the Sun. We assume $O$ is fixed (which is not actually true, but is a tenable assumption since we are studying the motion relative to the Sun). We take $O$ as the pole. Let $M$ be the mass of the Sun. Consider a planet of mass $m$. Then when the planet is at a point $P$ in its orbit, the gravitational force $\mathbf{F}$ on it is in the direction $\overrightarrow{PO}$. Such a force is called a **central** force. By Newton's law of gravitation its magnitude is $\dfrac{GMm}{r^2}$ where $G$ is a constant and $r = |OP|$. So if we let $\mathbf{e}_r$ be a unit vector along $\overrightarrow{OP}$ then by Newton's second law of motion, the acceleration $\mathbf{a}$ at $P$ is given by

$$\mathbf{a} = -\frac{k}{r^2}\mathbf{e}_r \tag{23}$$

where $k$ is a constant (specifically, $k = GM$) and $\dfrac{k}{r^2} > 0$.

From this simple equation we shall show that the planet moves in an ellipse, using (21). In order to apply (21), it is necessary to know beforehand that the motion lies entirely in a plane. We assume this without proof here because the kind of reasoning needed for this has little to do with what we are now doing. Note that although we have fixed the pole $O$ at the Sun, we have not yet chosen an initial ray. We fix it as follows. Let $A$ be a point in the orbit which is closest to the Sun. (That such a point exists is non-trivial and requires completeness of the real number system. We omit the proof.) Clearly $A \neq O$, since the planet can never hit the Sun! So $\overrightarrow{OA}$ is a well-defined ray and we take it as the initial ray (see Fig.6.8.6). Let $a_0 = |OA|$. Then the polar co-ordinates of $A$ are $(a_0, 0)$. It can be shown that in a small neighbourhood of $A$, $r$ can be expressed as a differentiable function of $\theta$ (the proof is again omitted).

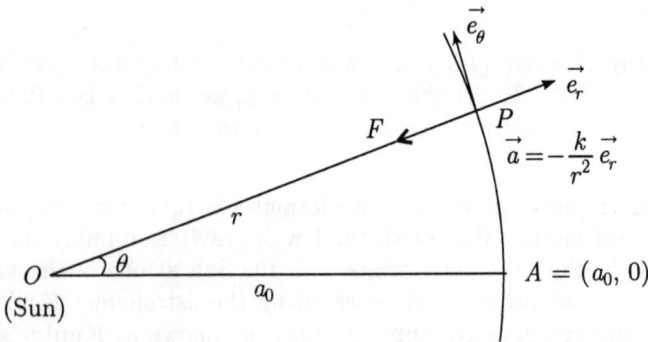

Figure 6.8.6 : Orbit of a Planet around the Sun.

Since $r$ attains its minimum at $\theta = 0$, we get $\dfrac{dr}{d\theta} = 0$ at $\theta = 0$ (by Theorem (4.11.1)). So

$$r = a_0 \text{ and } \frac{dr}{d\theta} = 0 \text{ at } \theta = 0. \tag{24}$$

Conditions such as these are popularly called **initial conditions.** We shall convert (23) into a differential equation about the derivatives of $r$ w.r.t. $\theta$ and then solve it so that the initial conditions in (24) are satisfied. This will give us $r$ as a function of $\theta$ and will thus determine the orbit of the planet. The general theory of differential equations is beyond our scope. But the differential equation we shall run into can be solved rather easily using certain tricky substitutions. So the derivation of Kepler's law from Newton's laws not only illustrates the use of polar co-ordiantes but also gives a glimpse at some of the techniques used for solving differential equations.

Because of (21), we get two equations from (23) viz.,

$$\ddot{r} - r\dot{\theta}^2 = -\frac{k}{r^2} \tag{25}$$

and

$$2\dot{r}\dot{\theta} + r\ddot{\theta} = 0 \tag{26}$$

Multiplying (26) by $r$, we can rewrite it as $\dfrac{d}{dt}(r^2\dot{\theta}) = 0$. Hence $r^2\dot{\theta}$ is a constant, say $c$. So

$$\dot{\theta} = \frac{c}{r^2} \tag{27}$$

We want to eliminate $t$ between (25) and (27). For this, we need to express $\ddot{r}$ in terms of derivatives of $r$ w.r.t. $\theta$. By the chain rule,

$$\dot{r} = \frac{dr}{dt} = \frac{dr}{d\theta}\dot{\theta} \quad \text{and} \quad \ddot{r} = \frac{d}{dt}(\frac{dr}{d\theta})\dot{\theta} + \frac{dr}{d\theta}\ddot{\theta} \tag{28}$$

which, using (26), becomes

$$\ddot{r} = \frac{d}{d\theta}(\frac{dr}{d\theta})\dot{\theta}^2 - 2\frac{dr}{d\theta}\frac{\dot{r}\dot{\theta}}{r} = \frac{d^2r}{d\theta^2}\dot{\theta}^2 - 2(\frac{dr}{d\theta})^2\frac{\dot{\theta}^2}{r} \tag{29}$$

Substituting (29) into (25) and using (27), we get

$$r\frac{d^2r}{d\theta^2} - 2(\frac{dr}{d\theta})^2 - r^2 = -\frac{k}{c^2}r^3 \tag{30}$$

This is a differential equation involving derivatives of $r$ w.r.t. $\theta$. If we could solve it, we would get some curve which will represent the orbit of the planet in non-parametric form. A direct solution of (3) is not so easy. So we change the variable $r$ to a new variable $v$, defined by

$$v = \frac{1}{r} \tag{31}$$

Easy calculations give $\dfrac{dr}{d\theta} = -\dfrac{1}{v^2}\dfrac{dv}{d\theta}$ and $\dfrac{d^2r}{d\theta^2} = -\dfrac{1}{v^2}\dfrac{d^2v}{d\theta^2} + \dfrac{2}{v^3}\left(\dfrac{dv}{d\theta}\right)^2$. Putting these into (30), we get

$$\frac{d^2v}{d\theta^2} + v = \frac{k}{c^2} \tag{32}$$

which can be simplified still further if we put $u = v - \dfrac{k}{c^2}$, giving,

$$\frac{d^2 u}{d\theta^2} + u = 0 \tag{33}$$

Essentially the same differential equation was encountered in Exercise (5.10.24) part (viii), and we adopt the same trick used there. We multiply (33) by $2\dfrac{du}{d\theta}$ and write it as

$$\frac{d}{d\theta}\left( \left(\frac{du}{d\theta}\right)^2 + u^2 \right) = 0 \tag{34}$$

which, forces $(\dfrac{du}{d\theta})^2 + u^2$ to be a constant. This constant is determined by (24). At $\theta = 0, u = \dfrac{1}{a_0} - \dfrac{k}{c^2}$ while $\dfrac{du}{d\theta} = \dfrac{dv}{d\theta} = -v^2\dfrac{dr}{d\theta} = 0$ since $r = \dfrac{1}{v}$. So

$$\left(\frac{du}{d\theta}\right)^2 + u^2 = \left(\frac{1}{a_0} - \frac{k}{c^2}\right)^2 \tag{35}$$

whence $\dfrac{du}{d\theta} = \pm\sqrt{(\dfrac{1}{a_0} - \dfrac{k}{c^2})^2 - u^2}$. The $+$ or $-$ holds depending upon whether $u$ increases or decreases with $\theta$, which depends upon whether $r(= \dfrac{1}{u + \frac{k}{c^2}})$ decreases or increases with $\theta$. Since $\theta = 0$ is a point of minimum of $r$ (and hence maximum of $u$), $\dfrac{du}{d\theta}$ changes sign at 0, with $\dfrac{du}{d\theta} > 0$ for $\theta < 0$ and $\dfrac{du}{d\theta} < 0$ for $\theta > 0$, in some neighbourhood of $\theta = 0$. Writing $\alpha$ for $\dfrac{1}{a_0} - \dfrac{k}{c^2}$, we have

$$\frac{du}{d\theta} = -\sqrt{\alpha^2 - u^2} \text{ for } \theta > 0 \quad \text{and} \quad \frac{du}{d\theta} = \sqrt{\alpha^2 - u^2} \text{ for } \theta < 0 \tag{36}$$

To solve the first of these equations, we rewrite it as $\dfrac{-du}{\sqrt{\alpha^2 - u^2}} = d\theta$ (or, more cleanly, as $\dfrac{1}{\sqrt{\alpha^2 - u^2}}\dfrac{du}{d\theta} = -1$ and integrate w.r.t. $\theta$) to get

$$\sin^{-1}\left(\frac{u}{\alpha}\right) = -\theta + \lambda \tag{37}$$

where $\lambda$ is some constant. Although (37) holds for all sufficiently small positive values of $\theta$, by continuity considerations it also holds for $\theta = 0$. In fact $u = \dfrac{1}{r} - \dfrac{k}{c^2}$ equals $\alpha$ when $\theta = 0$. So the constant $\lambda$ in (37) can be taken as $\dfrac{\pi}{2}$. Thus from (37) we get,

$$u = \alpha \sin\left(\frac{\pi}{2} - \theta\right) = \alpha \cos\theta = \left(\frac{1}{a_0} - \frac{k}{c^2}\right)\cos\theta \tag{38}$$

This is valid for all $\theta$ in some semi-open interval of the form $[0, h)$. We leave it to the reader to check that if we start with the second equation in (36), then we get that (38) holds for all $\theta$ in an interval of the form $(-h_1, 0]$ for some $h_1 > 0$. Thus (38) is a solution of the differential equation (33) in some neighbourhood of $\theta = 0$. To show that (38) holds for *all* values of $\theta$, a different technique is needed. A direct calculation shows that (38) satisfies (33) for all $\theta$. Moreover, the initial conditions (24) (converted to $u = \dfrac{1}{a_0} - \dfrac{k}{c^2} (= \alpha)$ and $\dfrac{du}{d\theta} = 0$ at $\theta = 0$) are also satisfied by (38). It can then be shown (cf. Exercise (8.12)) that (38) is the only solution of (33) and hence of (30). Converting back in terms of $r$, (38) gives $\dfrac{1}{r} - \dfrac{k}{c^2} = (\dfrac{1}{a_0} - \dfrac{k}{c^2}) \cos \theta$, or,

$$r = \frac{c^2/k}{1 + (\dfrac{c^2}{ka_0} - 1) \cos \theta} \tag{39}$$

which is of the same form as (3) if we let $e = \dfrac{c^2}{ka_0} - 1$ and $d = \dfrac{c^2}{ke}$. This does *not* conclusively prove that the orbit of the planet is an ellipse because as mentioned in the proof of Theorem (8.1), (3) is also the equation of a parabola or a hyperbola. These possibilities are, in fact, not ruled out by Newton's law. Thus, theoretically, a planet could move along a parabola or a hyperbola, with the Sun as a focus. But in that case the planet would be receding away from the Sun and would never return to the same position again. Since this is inconsistent with observed facts, the orbit must be an ellipse. For the common planets in the solar system, these ellipses have eccentricities very close to 0 and so the orbits are nearly circular. However, comets have high eccentricities and so their distances from the Sun (and hence also from the Earth) vary considerably. Still they "visit" us (i.e. come within visible range from the Earth) with absolute and predictable regularity. The most well-known comet, Halley's comet has an eccentricity nearly 0.97 and comes close to Earth (with eccentricity 0.02) approximately every 76 years. The eccentricity of the orbit of comet Kohoutek is 0.9999.

So far we have discussed arc lengths and velocity and acceleration in polar co-ordinates. Related topics such as the areas of surfaces of revolution can be similarly handled (see Exercise (8.10)). In the exercises (Exercise (8.11)) we shall give an important result which expresses, in terms of an integral, the area swept out by the position vector of point $P$ moving along a given curve, of the form $r = f(\theta)$. In polar co-ordinates there is no exact analogue of the concept of 'area below the graph' of a function, say $y = f(x)$ in cartesian co-ordinates. The area subtended at the pole (which is another way of looking at the area swept by the vector $\overrightarrow{OP}$, see Fig.6.8.7) comes closest to it. As we have already seen in Theorem (8.1), an ellipse can be described in the polar coordinates as a curve of this form. So this gives yet another method to find the area enclosed by an ellipse (exercise (8.12)(iii)).

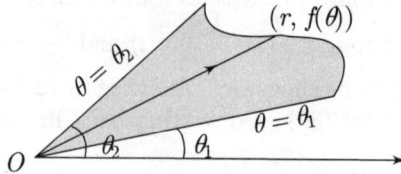

Figure 6.8.7 : Area subtended by an Arc at the Pole

We now turn to the double integral of a real-valued function which is expressed in polar co-ordinates, say $z = g(r, \theta)$. The study of double integrals comes properly under the theory of functions of two variables which is beyond our scope. But as discussed in Section 6.4, the double integrals represent the volumes of certain solids. And if we evaluate them by the method of slicing then it amounts to converting the double integral to an iterated integral.

In the case of the polar co-ordinates, too, the double integrals represent certain volumes and can be evaluated by the method of slicing. But the slicing is of a different type than in the case of cartesian co-ordinates. To understand how it is done, it is first of all necessary to introduce a co-ordinate system for the space, called **cylindrical polar co-ordinates**, (or simply cylindrical co-ordinates for short) which is an extension of the system of polar co-ordinates for a plane. This extension is done in a manner which is similar to the cartesian co-ordinatess where we add a third axis (viz. the $z$-axis) perpendicular to the $x$-$y$ plane and passing through the origin. In the case of polar co-ordinates we add an axis (also called the $z$-axis) perpendicular to the $r$-$\theta$ plane and passing through the pole $O$. Thus a point $P$ in space is specified by a triplet $(r, \theta, z)$ where $z$ is the 'height' above the $r$-$\theta$ plane and $(r, \theta)$ are the polar co-ordinates of the vertical projection of $P$ on this plane (see Fig.6.8.8(a)).

Since the $z$-co-ordinate in the cylindrical system is the same as for the cartesian system, the rules for conversion of the cylindrical to the cartesian system and vice versa are the same as (1) and (2), with a third equation $z = z$ thrown in. The name 'cylindrical' is explained if we consider what are called the co-ordinate surfaces, i.e. surfaces on which one of the co-ordinates is constant. In the cartesian co-ordinates $(x, y, z)$ all the three co-ordinate surfaces are planes. But in the cylindrical system, $r = c$ is a right circular cylinder of radius $c$ ($> 0$) and axis along the $z$-axis. The surface $\theta = \alpha$ is a vertical half plane bounded by the $z$-axis. A co-ordinate surface of the third type, viz., $z = h$ is a horizontal plane as in the cartesian co-ordinates.

Certain surfaces can be expressed more easily in polar co-ordinates than in the cartesian co-ordinates. For example the surface $z = \sqrt{x^2 + y^2}$ in the cartesian system represents an inverted right circular cone with vertex at the origin $O$ (see Fig.6.4.3 (a)). In the cylindrical co-ordinates, the equation is simply $z = r$. In this formulation it is much easier to see the shape of the surface. Since the equation is independent of $\theta$, whenever a point, say $P_0 = (r_0, 0, z_0)$ is on the surface, so is the entire circle $(r_0, \theta, z_0)$ for all $\theta \in [0, 2\pi]$. (This circle is one of the three co-ordinate curves passing through $P_0$). So it follows that $z = r$ is a surface of revolution obtained by revolving the curve $z = r$ in the half plane $\theta = 0$. But this curve is a half ray. So the surface $z = r$ is a right circular cone, with axis along the $z$-axis. Thus the cylindrical system is especially convenient for surfaces of revolution around the $z$-axis.

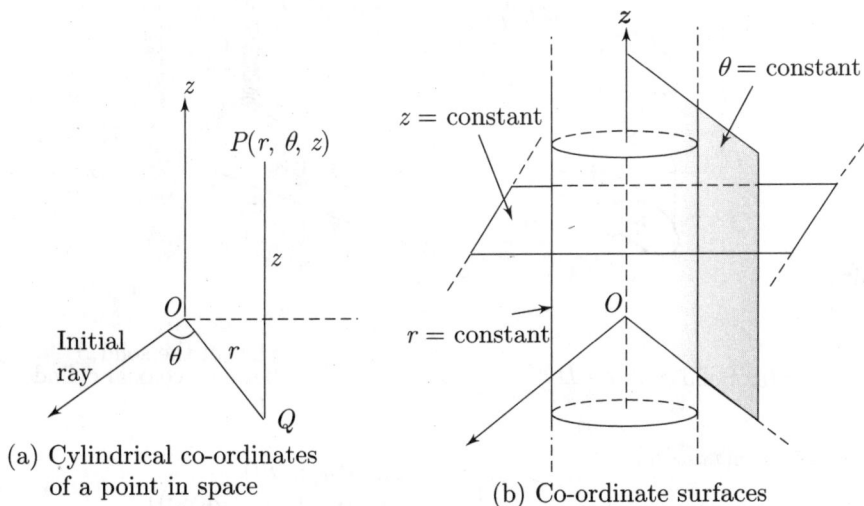

(a) Cylindrical co-ordinates of a point in space

(b) Co-ordinate surfaces

Figure 6.8.8 : Cylindrical Polar Co-ordinates

Returning to double integrals, suppose $R$ is a region in the $r$-$\theta$ plane and $g$ is a non-negative, real-valued function defined on $R$. Then the equation $z = g(r, \theta)$ defines a surface in space. As in the cartesian case, we let $S$ be the set of all points in the space of the form $(r, \theta, z)$, where $(r, \theta) \in R$ and $0 \le z \le g(r, \theta)$. Then $S$ is a solid which is bounded below by $R$ and above by the surface $z = g(r, \theta)$, see Fig.6.8.9(a). As in the cartesian case done in Section 6.4, the double integral of $g$ over $R$, written as $\iint_R g(r, \theta) dA$, equals the volume, say $V$ of the solid $S$. (If we allow $g$ to take negative values then the corresponding volume is to be counted as negative.) In the cartesian case we sliced $S$ by planes of the form $x = $ constant (or $g = $ constant). We now slice $S$ by surfaces of the form $r = $ a constant. These are co-axial cylinders. Figure 6.8.9(b) shows the slice of $S$ between two adjacent cylinders $r = r_0$ and $r = r_0 + \Delta r$. Denote the volume of this slice by $\Delta V$. This slice has a uniform thickness $\Delta r$. If this is small, we may treat the areas of the two inner and the outer cylindrical boundary surfaces as

approximately equal. So we get,

$$\Delta V \simeq A(r_0)\Delta r$$

where $A(r_0)$ is the area of the cross section, say $S_{r_0}$, of the solid $S$ by the cylinder $r = r_0$. So if we could express $A(r_0)$ in terms of $r_0$, then, following the usual recipe of integration (which we also used for finding volumes in Section 6.3), we can get $V$ by integrating $A(r)$ w.r.t. $r$, between appropriate limits (which would depend on the region $R$).

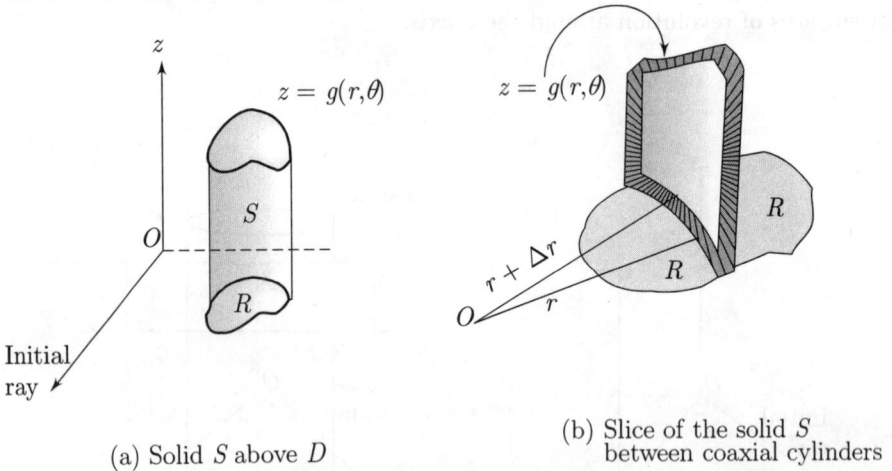

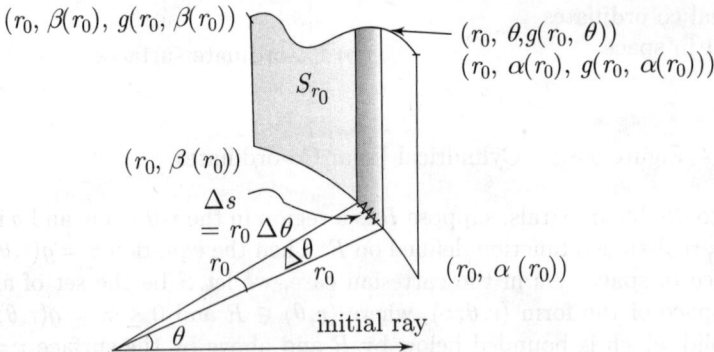

(a) Solid $S$ above $D$

(b) Slice of the solid $S$ between coaxial cylinders

(c) Area of a cross section of $S$ by a vertical cylinder

Figure 6.8.9 : Double Integrals in Polar Co-ordinates

To find $A(r_0)$, we note first that $S_{r_0}$ is not a planar surface. It is a part of the cylinder $r = r_0$, pictured in Fig.6.8.9 (c). Its lower boundary consists of all points of the form $(r_0, \theta, 0)$ where $\theta$ varies between two limits, say, $\alpha(r_0)$ and $\beta(r_0)$ which depend on $r_0$ and also on the region $R$ we started with. This is an arc of a circle of radius $r_0$. Similarly, its upper boundary consists of all points of the form $(r_0, \theta, g(r_0, \theta)), \alpha(r_0) \leq \theta \leq \beta(r_0)$. This is in complete analogy with what happens in cartesian co-ordinates (see Fig.6.4.4 and Fig. 6.4.5). But the difference is that $S_{r_0}$ is not a planar surface. It is a part of a cylinder. We have not studied methods for finding the areas of such 'curved' surfaces. So here we have to rely on intuition, just as we did for the formula for the surface area of a cone (see (9) in Section 6.3 and the subsequent comments). Thus suppose the cylinder is 'unfolded' into a planar surface without distorting its area. Then a vertical strip of $S_{r_0}$ lying above a portion of the base arc between, say, $\theta$ and $\theta + \Delta\theta$ (see Fig. 6.8.9(c)) will approximately be a rectangle with base $\Delta s$ where $\Delta s = r_0 \Delta\theta$. The area of this rectangular strip is approximately $g(r_0, \theta) r_0 \Delta\theta$. So, again by the recipe of integration,

$$A(r_0) = \int_{\alpha(r_0)}^{\beta(r_0)} g(r_0, \theta) r_0 d\theta \tag{40}$$

Changing $r_0$ to $r$ and using (40), we now get

$$V = \text{ volume of } S = \iint_R g(r, \theta) dA = \int_a^b \int_{\alpha(r)}^{\beta(r)} g(r, \theta) r d\theta dr \tag{41}$$

where $a$ and $b$ are such that the region $R$ lies entirely in the annulas $a \leq r \leq b$.

Let us verify (41) in an example where we already know the answer by other methods. In Problem (4.1), we had to find the volume of the solid $S$ lying below the surface $z = \sqrt{x^2 + y^2}$ and over the disc $R = \{(x, y) : x^2 + y^2 \leq 1\}$. Ignoring units, the answer came out to be $\frac{2\pi}{3}$. As explained above, in polar co-ordinates the equation of the cone becomes $z = r$. So $g(r, \theta) = r$. For the disc $R, r$ varies between 0 to 1 and for each such $r, \theta$ varies from 0 to $2\pi$. (For $r = 0, \theta$ is not uniquely defined. But what happens at a single point does not matter in integration, which is a global process as explained in Section 5.3). So, from (41)

$$V = \iint_R r dA = \int_0^1 [\int_0^{2\pi} r^2 d\theta] dr = \int_0^1 2\pi r^2 dr = \frac{2\pi}{3} \tag{42}$$

which is the same answer as before.

In fact this example also illustrates how double integrals in the cartesian co-ordinates can sometimes be simplified by conversion to polar co-ordinates. After finding the volume in Problem (4.1), later in the same section, the same answer was obtained by evaluating $\iint_R \sqrt{x^2 + y^2} dA$ by conversion to an iterated integral (see formulas (9), (10) and (11) in Section 6.4). But that turned out to be fairly complicated. However as (42) shows, in polar co-ordiantes the result comes out quite effortlessly. The reason behind this is two-fold. First, the integrand, viz. $\sqrt{x^2 + y^2}$, takes a simple form, viz. $r$, when expressed in polar co-ordinates, with the nagging radical sign gone. Secondly, in polar co-ordinates the unit disc $R$ is the set $\{(r, \theta) : 0 \leq r \leq 1, 0 \leq \theta \leq 2\pi\}$. If we treat $r$ and $\theta$ as two new variables but picture them as if they were the cartesian co-ordinates of a point, then the set $\{(r,\theta) : 0 \leq r \leq 1, 0 \leq \theta \leq 2\pi\}$ is the rect-angle $R(0,1;0,2\pi)$. And double integrals over rectangular domains are generally easier to evaluate than those over non-rectangular domains. The transforma-tion $x = r\cos\theta, y = r\sin\theta$ converts the rectangle, say, $D(= R(0,1;0,2\pi))$ in the $(r,\theta)$ plane to the unit disc $R = \{(x,y) : x^2 + y^2 \leq 1\}$ in the $x$-$y$ plane. The function $f(x,y) = \sqrt{x^2 + y^2}$ becomes $g(r,\theta) = r$ with this substitution. Note that the double integral $\iint_R \sqrt{x^2 + y^2} dA$ changes to $\iint_D r^2 dA$ which is trivial to evaluate. A similar calculation, with $z = \sqrt{a^2 - x^2 - y^2}$ and $D$ the disc of radius $a$ centred at $O$, expresses the volume of a hemisphere of radius $a$ as $\int_0^{2\pi}\int_0^a r\sqrt{a^2 - r^2}drd\theta$. This comes out to be $\frac{2\pi}{3}a^3$.

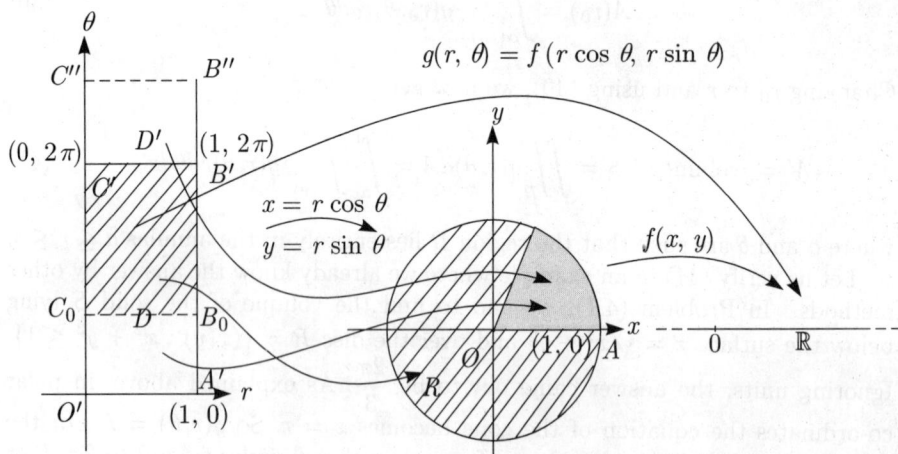

Figure 6.8.10 : Conversion of Double Integrals

More generally, this substitution can be used to convert any double inte-gral $\iint_R f(x,y)dA$ to a double integral $\iint_D rg(r,\theta)dA$ where $D$ is a region in the $(r,\theta)$-plane and $g(r,\theta) = f(r\cos\theta, r\sin\theta)$. All that is needed is that the transformation $x = r\cos\theta, y = r\sin\theta$ should map the region $D$ in a one-to-one manner onto the region $R$ in the $x$-$y$ plane. Again, it is all right if points on some 'curve' in $R$ have more than one pre-images, because curves have no areas and hence do not matter in double integrals. For instance, in the example just worked out, $D$ was the rectangle $\{(r,\theta) : 0 \leq r \leq 1, 0 \leq \theta \leq 2\pi\}$ and $R$ was the disc $\{(x,y) : x^2 + y^2 = 1\}$. The transformation $x = r\cos\theta, y = r\sin\theta$ is onto, but not one-to-one since the entire segment $O'C'$ (see Fig.6.8.10) is taken to the single point $O$. Similarly points on the segment $OA$ have two preimages each, one on the segment $O'A'$ and one on $C'B'$. But this causes no harm. If, however, we enlarge $D$ to a bigger rectangle $D' = O'A'B''C''$, then the transformation maps two rectangles (viz. $O'A'B_0C_0$ and $C'B'B''C''$) of positive areas onto the same image (viz. the sector OAB) and then the double integral $\iint_R f(x,y)dA$

will, in general, not equal $\iint_D rg(r,\theta)dA$.

It is interesting to note that even when the transformation to polar co-ordinates is one-to-one, the double integral $\iint_R f(x,y)dA$ does *not* equal the double integral $\iint_D g(r,\theta)dA$, (as might appear at first sight) but, instead, equals $\iint_D rg(r,\theta)dA$. It is instructive to see the role of the extra factor $r$ in the integrand. Here we can take a cue from the similar rule of change of variable for integrals of functions of one variables. Suppose $f : [a,b] \to \mathbb{R}$ is continuous and we put $x = h(t)$ where $h$ is a continuously differentiable function which maps an interval $[c,d]$ bijectively onto $[a,b]$. Then, by Theorem (5.6.3),

$$\int_a^b f(x)dx = \int_c^d h'(t)g(t)dt \tag{43}$$

where $g$ is the composite $f \circ h$, i.e. $g(t) = f(h(t))$. It was explained after the proof of Theorem (5.6.3), that $h'(t)$ is the magnification index of the substitution $x = h(t)$. That is, for infinitesimally small $\Delta t$, a subinterval of length $\Delta t$, say $[t, t+\Delta t]$, gets mapped onto a subinterval of length $h'(t)\Delta t$. For the substitution $x = r\cos\theta, y = r\sin\theta$, the role of $r$ is also the same, that is, it is ratio of the area of the image of an infinitesimally small portion of $D$ to the area of that portion.

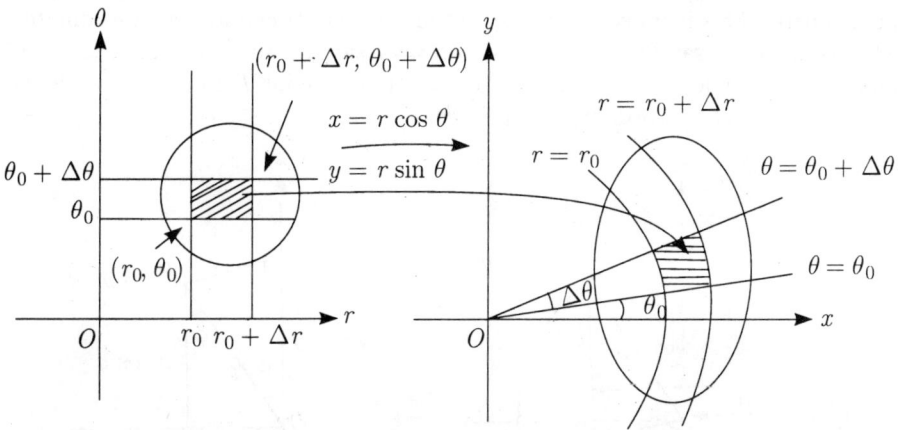

Figure 6.8.11 : Magnification Index of Polar Substitution

To see this consider a small rectangle, say $R(r_0, r + \Delta r; \theta_0 + \Delta\theta)$ at a point $(r_0, \theta_0)$ in the region $D$ in the $(r, \theta)$ plane (see Fig.6.8.11). Then its area is $\Delta r \Delta \theta$. Its image in the $x$-$y$ plane is the shaded portion between the two concentric circles $x^2 + y^2 = r_0^2$ and $x^2 + y^2 = (r_0 + \Delta r)^2$ and the two rays making angles $\theta_0$ and $\theta_0 + \Delta\theta$ with the $x$-axis. The image is often called a cell. The area of this cell is the difference between the area of a sector of the outer circle and that of the inner circle. So from Exercise (6.3.10),

$$\text{area of cell} = \frac{1}{2}\Delta\theta((r_0 + \Delta r)^2 - r_0^2) = r_0\Delta r\Delta\theta + \frac{1}{2}\Delta\theta(\Delta r)^2$$

and hence

$$\frac{\text{area of cell}}{\text{area of rectangle}} = r_0 + \frac{1}{2}\Delta r$$

As $\Delta r$ and $\Delta\theta$ each tends to 0, this ratio approaches $r_0$. So the magnification index for the area at $(r_0, \theta_0)$ is $r_0$. At $(r, \theta)$, it is $r$.

Instead of $r$ and $\theta$, one can use any pair of variables, say $u$ and $v$ such that $x, y$ are expressible as functions of $u$ and $v$. Then there is an analogous formula for converting a double integral over a region in the $x$-$y$ plane to that over a region in the $u$-$v$ plane. But we shall not go into it, as our interest is not in double integrals *per se* but only in showing how polar co-ordinates can sometimes be used to evaluate them. The exercises will provide more examples of this. Although we have not formally studied improper double integrals, in many respects they behave like improper integrals of functions of one variable. Using a few such properties and conversion to polar co-ordinates, it can be shown that the improper integral $\int_0^\infty e^{-x^2}\, dx$ has value $\frac{\sqrt{\pi}}{2}$ (see Exercise (8.22)).

We conclude the section with a brief discussion of yet another system of co-ordinates for space, called the **spherical polar co-ordinates** (or spherical co-ordinates for short). As the name suggests, one of the co-ordinate surfaces must be a sphere. This is indeed the case. As in the case of cylindrical co-ordinates, we fix one point, say $O$, in space and two rays, say, $L_1$ and $L_2$ originating at $O$ and at right angles to each other. Now given any point $P$ in space, we define its spherical polar co-ordiantes as follows.

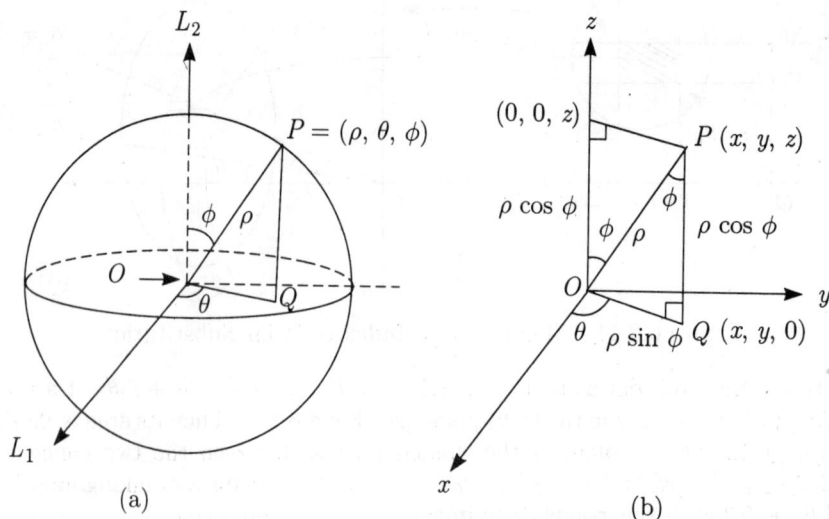

(a)                                      (b)

Figure 6.8.12 : Spherical Polar Co-ordinates

Let $\rho$ be the distance of $P$ from $O$. Let $\phi$ be the angle which $\overrightarrow{OP}$ makes with $L_2$. ($\phi$ varies between 0 to $\pi$ and is undefined if $P = O$.) To define $\theta$, we assume that $P$ does not lie on $L_2$ or on its opposite ray. Consider the plane through $O$ perpendicular to $L_2$. Since $L_1$ is assumed to be perpendicular to $L_2$, this plane contains the half ray $L_1$. Now let $Q$ be the projection of $P$ on this plane. Then $\theta$ is defined as the angle through which $L_1$ must be rotated counter-clockwise (as seen from $L_2$) to make it lie along $\overrightarrow{OQ}$ (see Fig.6.8.12(a)). Equivalently, $\theta$ is precisely the second polar co-ordinate of $Q$ if $O$ is taken as the pole and $L_1$ as the initial ray. $\theta$ can take values from 0 to $2\pi$.

It is easy to convert spherical polar co-ordinates to the cartesian ones. Take $O$ as the origin, $L_1$ as the positive $x$-axis, and $L_2$ as the positive $z$-axis. This determines the positive $y$-axis as the unique ray perpendicular to $L_2$, for which $\theta = \frac{\pi}{2}$. Then from Fig.6.8.12(b) it is clear that $OQ = \rho \sin \phi$ and $PQ = \rho \cos \phi$. Having known $OQ$, we get $x, y$ from $\theta$. Thus,

$$x = \rho \sin \phi \cos \theta, y = \rho \sin \phi \sin \theta, z = \rho \cos \phi \qquad (44)$$

Expressing $\rho, \theta, \phi$ in terms of $x, y, z$ is left as an exercise. If we call $\rho \sin \phi$ as $r$, then $r, \theta$ are precisely the polar co-ordinates of $Q$, the projection of $P$. Moreover, $r$ and $z$ are related to $P$ and $\phi$ by

$$r = \rho \sin \phi, z = \rho \cos \phi \qquad (45)$$

which is strikingly similar to (1). As a result, the cylindrical polar co-ordinates are midway between the cartesian and the spherical polar co-ordinates. First you get $r$ and $z$ from $\rho$ and $\phi$ and then you get $x$ and $y$ from $r$ and $\theta$. The conversion formulas are structurally similar in each of these two transitions. It is clear that the co-ordinate surface $\rho = c$ (where $c > 0$) is a sphere of radius $c$, centered at $O$. This explains the name and also suggests that spherical co-ordinates will be particularly convenient while dealing with problems involving spheres. This indeed turns out to be the case.

The determination of the other co-ordinate surfaces and co-ordinate curves will be given in the exercises (Exercise (8.24)). The rather clumsy formulas in (44) may give the impression that the spherical co-ordinates are somewhat artificial, especially to someone who is heavily used to cartesian co-ordinates. It is true that the cartesian system does have certain simplifying features, for example, the fact that the tangent vectors to the co-ordinate curves at every point are the same. But then, as we saw in the derivation of Kepler's laws (and even otherwise), sometimes the polar co-ordinates are more convenient. As regards which system is more natural, it is mostly a question of what one is used to, very much like languages. Ironically, we come across the spherical polar co-ordiantes in school long before we study cartesian co-ordinates ! In geography we learn to locate a point, say $P$, on the earth given its latitude and longitude. The latitude can range from $90°$ South to $90°$ North or in radians from $\frac{-\pi}{2}$ to $\frac{\pi}{2}$. If we add $\pi$ to it, it would range from 0 to $\pi$ and would give the angle which the vector $\overrightarrow{OP}$ ($O$ being the centre of the Earth) with $\overrightarrow{OS}$ ($S$

being the South pole). This is exactly $\pi - \phi$, where $\phi$ is the third spherical co-ordinate of $P$. Thus specifying the latitude of a point $P$ is just another way of specifying $\phi$. By a similar reasoning, the longitude (which varies from 180° East to 180° West) of $P$ is essentially the co-ordinate $\theta$. What about the third polar co-ordinate $\rho$? By definition, it is the distance $OP$. For a point on the surface of the Earth, $\rho = R$, where $R$ is the radius of the Earth. Hence $\rho$ is constant. But strictly speaking the earth is not a perfect sphere. There are mountains and plateaus. The points at a distance $R$ correspond to the mean sea level. For a point at an altitude $h$ (which is measured as the height above the mean sea level), $\rho$ equals $R + h$. Since $R$ is a fixed number, knowing $h$ is as good as knowing $\rho$. In summary, the spherical polar co-ordinates $\rho, \theta$ and $\phi$ describe, respectively, the altitude, the longitude and the latitude.

# EXERCISES

**8.1** Prove that in oblique co-ordinates with the axes inclined to each other at an angle $\alpha$ (where $0 < \alpha < \pi$), the distance between $(x_1, y_1)$ and $(x_2, y_2)$ is $\sqrt{(x_2 - x_1)^2 + (y_2 - y_1)^2 + 2(x_2 - x_1)(y_2 - y_1)\cos\alpha}$. Hence obtain the equation of a circle of radius $r$ and centre $(a, b)$. Prove that an equation of the form $Ax^2 + By^2 + 2Hxy + 2Fx + 2Gy + C = 0$ represents a circle if and only if $A = B(\neq 0)$ and $AB - H^2 = A^2 \sin^2 \alpha$. By contrast, prove that the condition for collinearity of three points in oblique co-ordinates is the same as in the rectangular cartesian co-ordinates and that every straight line is expressed by a linear equation.

**8.2** Prove directly that in polar co-ordinates, the general equation of a straight line not passing through the pole is of the form $r = A\sec(\theta - \alpha)$ for some constants $A$ and $\alpha$ with $A > 0$. What is the geometric significance of $A$ and $\alpha$? What is the equation of a line passing through the pole ?

**8.3** For the ellipse represented by (3), find (i) the length of its semi-major axis and (ii) the polar co-ordinates of its other focus. (The answers should be in terms of the distance $d$ and the eccentricity $e$.)

**8.4** Draw figures like Fig.6.8.3 for the case of a hyperbola and a parabola.

**8.5** Sketch the following curves of the form $r = f(\theta)$. (Wherever necessary, the domain of $f$ is restricted to those values for which $f$ is non-negative. As an aid to sketch these curves in some cases it is helpful to first draw the graph of $r = f(\theta)$, as if $\theta$ and $r$ were rectangular cartesian co-ordinates and then convert the portion of the graph for which $r \geq 0$, to polar co-ordinates.)

   (a) $r = 1 + \cos\theta$   (b) $r = 1 - \sin\theta$   (c) $r = \sin 2\theta$
   (d) $r = \cos 2\theta$      (e) $r = \sin 3\theta$      (f) $r = \cos 3\theta$

   Do you notice any pattern?

**8.6** As mentioned in the text, we are not allowing $r$ to take negative values. Sometimes it is convenient to allow negative values of $r$. How would the graphs in (c) to (f) of the last exercise change if we do so?

8.7 (a) Show that the cardioid in Fig. 6.8.2(c) is a special case of the epicy-cloid in Exercise (4.12.7) if both the fixed and the moving circle have diameter $a$ each.

(b) A **lemniscate** is defined as the locus of a point $P$ which moves in a plane so that the product of its distances from two fixed points $F_1$ and $F_2$ in the plane (called the foci) is a constant, say, $a^2 (a > 0)$. (This is in direct analogy with an ellipse which results if instead of the product we require that the sum of the distances be constant.) Let $O$ be the mid-point of the segment $F_1 F_2$. Taking $O$ as the pole and one of the rays $OF_1$ and $OF_2$ as the initial ray, show, both directly and using cartesian co-ordinates, that the polar equation of the lemniscate is $r^2 = 2a^2 \cos 2\theta$. Sketch the curve.

8.8 In the second proof of Theorem (8.2), prove (16). That is, given $\epsilon > 0$, prove that there exists $\delta > 0$, such that wherever $R = (t_0, t_1, \ldots, t_n)$ is a partition of $[a, b]$ with mesh $\mu(R) < \delta$, the difference between the right hand sides of (15) and (16) is less than $\epsilon$. [*Hint* : In view of Exercise (4.2.9), it suffices for each $i = 1, \ldots, n$, to make the difference between the quantities under the $i^{\text{th}}$ radical signs less than $\dfrac{\epsilon}{b-a}$. Let $M > 0$ be a constant such that for all $t \in [a, b], |r(t)|, |\dot{r}(t)|$ and $|\dot{\theta}(t)|$ are all less than $M$. Use uniform continuity of $\dot{r}^2, \dot{\theta}^2$ and the fact that $|AB - A'B'| \leq |A - A'||B| + |A'||B - B'|.$]

8.9 Show that the arc length of a curve of the form $r = f(\theta), \alpha \leq \theta \leq \beta$ is $\displaystyle\int_\alpha^\beta \sqrt{[f(\theta)]^2 + [f'(\theta)]^2}\, d\theta$. Hence find the arc lengths of (i) the cardioid $r = a(1 - \cos\theta)$ (ii) the lemniscate of Exercise (8.7) (iii) the spiral $r = \theta, 0 \leq \theta \leq 6\pi$.

8.10 Obtain an expression in polar co-ordinates for the areas of the surface ob-tained by revolving a curve $\mathbf{r} = \mathbf{r}(t), a \leq t \leq b$ around (i) the line along the initial ray (ii) the line perpendicular to the initial axis. [*Hint* : Consider a line element $ds$.] Determine these areas in the case of the lemniscate of Exercise (8.7).

8.11 Prove that the area shown in Fig.6.8.7 equals $\displaystyle\int_{\theta_1}^{\theta_2} \frac{1}{2}[f(\theta)]^2 d\theta$. What would be the answer if the curve is given parametrically instead of in the form $r = f(\theta)$? [*Hint* : Let $\Delta A$ be the area of a triangle $OPQ$ where $P = (r, \theta)$ and $Q(r + \Delta r, \theta + \Delta\theta)$. Consider $\displaystyle\lim_{\Delta\theta \to 0} \frac{\Delta A}{\Delta\theta}.$]

8.12 Using the last exercise determine the areas enclosed by

(i) the cardioid $r = a(1 - \cos\theta)$ (ii) the lemniscate $r^2 = 2a^2 \cos 2\theta$

(iii) the ellipse $\dfrac{x^2}{a^2} + \dfrac{y^2}{b^2} = 1$

8.13  (a) Let $P$ and $Q$ be points on a unit circle centred at $O$, so that $\angle QOP = \Delta\theta$ (see Fig.6.8.13.). Prove that as $\Delta\theta \to 0$, the vector $\dfrac{\overrightarrow{PQ}}{\Delta\theta}$ tends to the unit tangent vector to the the circle at $P$.

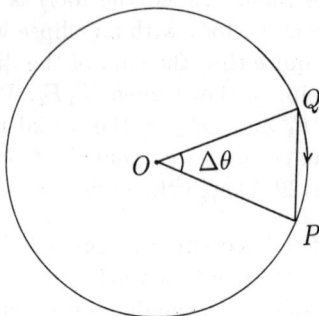

Figure 6.8.13 :    Derivatives of Radial and
                   Transverse unit vectors.

(b) Using (a) give a direct, geometric proof that $\dfrac{d}{d\theta}(\mathbf{e}_r) = \mathbf{e}_\theta$ where $\mathbf{e}_r, \mathbf{e}_\theta$ are the radial and transverse unit vectors respectively.

(c) Using (a) and (b), prove that $\dfrac{d}{d\theta}(\mathbf{e}_\theta) = -\mathbf{e}_r$. [*Hint* : Note that $\mathbf{e}_\theta$ at a point $(r, \theta)$ is the same as $\mathbf{e}_r$ at $(r, \theta + \frac{\pi}{2})$].

8.14  Consider the differential equation (33), viz. $\dfrac{d^2u}{d\theta^2} + u = 0$.

(a) Prove that if $u = f_1(\theta), u = f_2(\theta)$ are solutions of this differential equation, then so is $u = f_1(\theta) - f_2(\theta)$.

(b) Prove that if $u = f(\theta)$ is a solution with $f(0) = 0$ and $f'(0) = 0$ then $f(0) = 0$ for all $\theta$. (This is essentially what was used in Exercise (5.10.24) part (vii).)

(c) Prove that if $u = f_1(\theta), u = f_2(\theta)$ are solutions of $\dfrac{d^2u}{d\theta^2} + u = 0$ such that $f_1(0) = f_2(0)$ and $f_1'(0) = f_2'(0)$ then $f_1(\theta) = f_2(\theta)$ for all $\theta$.

8.15  In the text, when we compared (39) with (3), we tacitly assumed that $\dfrac{c^2}{ka_0} - 1 > 0$, i.e. $\dfrac{c^2}{k} > a_0$. Why is this assumption justified ?

8.16  Prove that for each planet, the line segment from the Sun to the planet sweeps out equal areas in equal time. (This is known as **Kepler's first law**.) Prove, in fact, that this law is equivalent to saying that the force on the planet is central (i.e. directed towards the pole which is also the Sun.)

**8.17** Prove that Kepler's first law, along with the fact that the orbit of the planet is an ellipse (known as **Kepler's second law**) implies that the force on the planet is towards the Sun and proportional to $\dfrac{1}{r^2}$. (This is essentially the converse of the derivation of (39) from Newton's law of gravitation.)

**8.18** Prove that the constants $a_0$ in (27) and $c$ in (24) equal, respectively $a(1-e)$ and $ka(1-e^2)$ where $a$ and $e$ are, respectively, the length of the semi-major axis and the eccentricity of the orbit of the planet. Hence show that the period of the planet (i.e the time taken for one complete revolution) equals $\pi a \sqrt{\frac{a}{k}}$.

**8.19** Prove that the square of the period is proportional to the cube of the major axis of the orbit, the constant of proportionality being the same for every planet (in the same solar system). (This is known as **Kepler's third law**.)

**8.20** Show that the equation of motion of a simple pendulum of length $L$ is given by $\ddot{\theta} + \dfrac{g}{L}\sin\theta = 0$ where $\theta$ is the angle the pendulum makes with the vertical and $g$ is the acceleration due to gravity. Can you solve this equation ?

**8.21** Evaluate the following double integrals over the unit disc by conversion to polar co-ordinates.

(i) $\displaystyle\iint x^2 dA$ (ii) $\displaystyle\iint x^2 y^2 dA$ (iii) $\displaystyle\iint \sin(x^2 + y^2) dA$

**8.22** For $R > 0$, let $D_R$ be the portion in the first quadrant of the disc of radius $R$ centred at the origin and let $S_R$ be the square with vertices $(0,0), (R,0), (R,R)$ and $(0,R)$. For every $R > 0$, prove that :

(i) $D_R \subset S_R \subset D_{\sqrt{2}R}$

(ii) $\displaystyle\iint_{D_R} e^{-(x^2+y^2)} dA \le \iint_{S_R} e^{-(x^2+y^2)} dA \le \iint_{D_{\sqrt{2}R}} e^{-(x^2+y^2)} dA$

(iii) $\displaystyle\iint_{S_R} e^{-(x^2+y^2)} dA = \left( \int_0^R e^{-x^2} dx \right)^2$ and

(iv) $\displaystyle\iint_{D_R} e^{-(x^2+y^2)} dA = \frac{\pi}{4}(1 - e^{-R^2})$

Show that $\displaystyle\lim_{R\to\infty} \iint_{D_R} e^{-(x^2+y^2)} dA$ and $\displaystyle\lim_{R\to\infty} \iint_{S_R} e^{-(x^2+y^2)} dA$ both exist and are equal. Finally, conclude that $\displaystyle\int_0^\infty e^{-x^2} dx = \frac{\sqrt{\pi}}{2}$.

**8.23** In spherical co-ordinates, identify the surfaces $\theta = \theta_0$ and $\phi = \phi_0$. Also identify the three co-ordinate curves passing through a given point $P_0 = (\rho_0, \theta_0, \phi_0)$, that is, curves along which only one of the co-ordinate changes and the other two remain constant. Prove that the tangents to these curves are mutually orthogonal.

*Appendix A*

# CONSTRUCTION OF REAL NUMBERS

In Section 2.6 we assumed the completeness of the real number system as an axiom (Axiom (2.6.2)). We showed how it implies, and indeed is equivalent to, many other properties of $I\!R$ (see Exercises (3.5.2) and (3.5.10)). All these properties are, therefore, different versions of the same property of $I\!R$ viz., its completeness.

As elaborated in Section 2.6, completeness is, by far, the most basic property of the real number system and gives it an edge over the rational number system. The rational numbers are familiar even to a layman. It is possible to construct real numbers from rational numbers in such a way that their completeness can be proved and hence does not have to be taken as an axiom. We took it as an axiom because we did not construct real numbers. There are two standard ways to construct reals from rationals, one due to Dedekind and the other to Cantor. Both give the same result. But the latter is a little more intuitive and we present it here briefly. It is based on the fact that every real number can be expressed as the limit of a sequence of rational numbers. Many different sequences of rationals can, of course, converge to the same real number. It should also be noted that not every sequence of rationals converges to some real number, e.g. the sequence $\{(-1)^n\}_{n=1}^{\infty}$.

We want to turn the tables around. That is, we assume we know absolutely nothing about real numbers and that we are living solely in the world of rational numbers. We can still define things like convergence of a sequence exactly the way it is defined for a sequence of real numbers, except that everywhere we have to put only rational numbers. Thus, we say that a sequence, say $\{x_n\}$ of rationals converges to a (rational) number $L$ if for every (rational) $\epsilon > 0$, there exists a positive integer $m$ such that for all $n \geq m, |x_n - L| < \epsilon$. The elementary results about limits of sequences (such as the uniqueness of limits, the fact that every convergent sequence is bounded, or that the sum/product of two convergent sequences converges to the sum/product of their limits) go through without any change. However, results which require completeness of the real number system break down. This is to be expected because the real number system is constructed precisely to cure the evils wrought out by incompleteness of $Q$, the rational number system.

As discussed in Section 2.6, the rational number system contains many sequences which *should* converge but don't, a good example being the sequence

$\left\{\left(1+\dfrac{1}{n}\right)^{n}\right\}$. As a sequence of real numbers it converges to $e$. But $e$ is irrational (see Exercise (2.7.5) and also Section 5.8 for a rigorous treatment.) And so it is not convergent in $Q$ (or else, as a sequence in $I\!R$ it would have two limits). (See Exercise (2.6.7) for another example of a sequence of rationals which should converge but does not.) Intuitively, every such sequence corresponds to a 'hole' and the real number system is obtained by filling up these holes.

What should we use to fill up these holes ? To appreciate the subtility of this question, suppose for example that we have the sequence $\{x_n\}$ where $x_n = \left(1+\dfrac{1}{n}\right)^{n}$ for all $n \in I\!N$. As we just saw, this sequence represents a hole in $Q$. The real number $e$ fills up this hole. But how do we *define* $e$ in the first place ? It is no use to define it as the limit of the sequence $\{x_n\}$ because in $Q$ this sequence has no limit. In $I\!R$ it does have a limit, but we are assuming we know nothing about $I\!R$. Indeed, we are yet to construct it. So we would be committing a cicious cycle by defining $e$ as the real number to which $\{x_n\}$ is convergent.

Thus we are in a dilemma. We *know* that the sequence $x_n = \left(1+\dfrac{1}{n}\right)^{n}$ has no limit in $Q$ and still we have to describe how to fill up the hole created by it, with the restriction that our description must not use anything beyond rational numbers. It is almost like asking a mute preson to name his infirmity orally! It may be argued that our knowledge that this sequence has no limit in $Q$ itself involved real numbers. But this can be easily countered by another example. The sequence $\{a_n\} = (\frac{1}{1},\frac{3}{2},\frac{7}{5},\frac{17}{12},\ldots)$ constructed in Exercise (2.6.7) is such that $a_n^2 \to 2$ as $n \to \infty$, see Exercise (3.1.3). So, if at all $\{a_n\}$ converges to some number $L$ in $Q$, then $L^2 = 2$. But there is no rational number with this property as shown in Exercise (2.6.4). Here the construction as well the proof of non-convergence of $\{a_n\}$ (within $Q$) requires no knowledge of any irrational number. (Note that there is a subtle difference between saying that the number 2 has no square root in $Q$ and saying that $\sqrt{2}$ is not a rational number. The latter presupposes that we already have in mind an entity $\sqrt{2}$. The former statement makes no such claim. In the argument above, only the first statement is used and that statement is perfectly within the world of rational numbers.)

There is an ingeneous way out of this dilemma. It is provided by the 'definition trick' introduced in Section 1.8. We are given the sequence $\{x_n\} = \left(1+\dfrac{1}{n}\right)^{n}$ and we want to fill the hole arising out of it, or mathematically, we want to define the real number $e$ to which this sequence should converge (in the real number systerm). Following the definition trick we take something else closely associated with $e$ and define $e$ as this something. Now, what could be more natural than to take this associated thing as the sequence $\{x_n\}$ itself ? This is a case of converting a problem itself into a solution, rather like silencing a staunch critique of a government department by appointing him as the minister in charge of that department!

This approach may appear evasive. But then the same is probably true of all other instances of the definition trick. At any rate, it serves the purpose, viz. it gives a way to construct the real numbers in terms of sequences of rational numbers. But before giving it in detail, a couple of points need to be noted. First, it is not true that *every* sequence of rational numbers will represent a real number. Secondly, two different sequences may correspond to the same real number. So a rigorous definition of a real number will be that it is an equivalence class under a certain equivalence relation on the set of certain sequences. This is of the same spirit as the rigorous definition of a vector as an equivalence class of directed line segments, where two such segments are considered to be equivalent if and only if they have equal lengths and are parallel.

We first introduce some notations. For brevity, sequences, say, $\{x_n\}$, $\{y_n\}$ etc. of rationals will be denoted simply by $\bar{x}, \bar{y}$ etc. The equivalence relation we shall define will be denoted by the symbol '$\sim$'. That is, instead of saying that the sequences $\bar{x}$ and $\bar{y}$ are equivalent to each other (under a relation which we have not yet defined), we sahll write $\bar{x} \sim \bar{y}$. Finally the equivalence class containing a given sequence $\bar{x}$ will be denoted by $[\bar{x}]$. By definition, this consists of all sequences which are equivalent to the sequence $\bar{x}$.

To construct real numbers, we do not start with all sequences of rational numbers, but only those which should converge. Some of these sequences may converge within $Q$ itself, for example the sequence $\{1 + \frac{1}{n}\}$ converges to 0, while some others, e.g. the sequence $\{\left(1 + \frac{1}{n}\right)^n\}$, do not converge in $Q$. A sequence like $\{(-1)^n\}$ also fails to converge in $Q$. But we do not consider it while constructing $\mathbb{R}$ because it is not a sequence which *should* converge.

How do we tell which sequences of rationals should converge and which should not ? Intuitively, the former are the sequences whose terms are getting mutually close as $n \to \infty$. This is precisely the idea behind Cauchy sequences as defined in Section 3.5. (See Definition (3.5.2).) But in that definition, the number $\epsilon$ could be real. As we are living in the world of rational numbers, we can let $\epsilon$ take only rational values. Because of the fact that $Q$ is dense in $\mathbb{R}$ (see Section 3.3), it is easy to show that if a sequence $\{x_n\}$ satisfies the condition in Definition (3.5.2) for every rational $\epsilon > 0$, then it does so for every (real) $\epsilon > 0$. But right now, we know nothing about real numbers. And so we change that definition and accordingly also change the name of the concept defined by it, even though later on the two definitions will come out to be equivalent to each other.

**A.1 Definition :** A sequence $\{x_n\}$ of rational numbers is called a **fundamental sequence** if for every (rational) $\epsilon > 0$, there exists some $p \in \mathbb{N}$ suach that $|x_n - x_m| < \epsilon$ whenever $m \geq p$ and $n \geq p$.

It is easy to see that every convergent sequence of rationals is fundamental. That the converse is false is shown by the sequence $\{\left(1 + \frac{1}{n}\right)^n\}$. Indeed, this is precisely why the rational number system is incomplete.

Now let $S$ be the set of all fundamental sequences of rational numbers. We define an equivalence relation on $S$ as follows. Let $\bar{x} = \{x_n\}, \bar{y} = \{y_n\}$ be two elements of $S$. We say $\bar{x}$ is **equivalent** to $\bar{y}$ (and write $\bar{x} \sim \bar{y}$) if for every rational $\epsilon > 0$, there exists some $m \in I\!N$ such that $|x_n - y_n| < \epsilon$ for all $n \geq m$. This definition is obviously motivated by the requirement that two sequences which are equivalent to each other should converge to the same real number. But since we are not allowed to use real numbers, we have paraphrased this requirement by saying that the $n$-th terms of the two sequences come very close to each other as $n \to \infty$. It is easy to see that if $\{x_n\}$ converges to a rational number $L$, then $\{y_n\}$ is equivalent to $\{x_n\}$ if and only if $\{y_n\}$ also converges to $L$.

It must, of course, be verified that the relation $\sim$ we have defined is indeed an equivalence relation, i.e. it is reflexive, symmetric and transitive. Of these, the first two are clear right from the definition. For transitivity, we need the traingle inequality, as an alert reader must have guessed. Suppose $\bar{x} \sim \bar{y}$ and $\bar{y} \sim \bar{z}$. Then, given a (rational) $\epsilon > 0$, there exist $m_1, m_2 \in I\!N$ such that $|x_n - y_n| < \frac{\epsilon}{2}$ for all $n \geq m_1$ and $|y_n - z_n| < \frac{\epsilon}{2}$ for all $n \geq m_2$. We let $m = \max\{m_1, m_2\}$. Then by the triangle inequality[1], for all $n \geq m$, $|x_n - z_n| \leq |x_n - y_n| + |y_n - z_n| < \frac{\epsilon}{2} + \frac{\epsilon}{2} = \epsilon$. That is, $\bar{x} \sim \bar{z}$. So, $\sim$ is indeed an equivalence relation on $S$, the set of all fundamental sequences of rational numbers. As mentioned earlier, we shall denote by $[\bar{x}]$ the equivalence class containing a given element $\bar{x} \in S$.

We are now ready for a formal definition of real numbers and their basic operations of addition and multiplication, which we continue to denote by the same symbols as for $Q$, viz. $+$ and $\cdot$, the latter often being suppressed from notation.

**A.2 Definition** : A **real number** is an equivalence class of fundamental sequences of rational numbers under the equivalence relation $\sim$. The set $\omega$ all real numbers will be denoted by $I\!\!R$. (Thus $I\!\!R$ is precisely the set of all equivalence classes in which the set $S$ is divided by the relation $\sim$. In set theory, such sets are suggestively denoted by symbols like '$S/\sim$' and are called **quotient sets**.) If $\alpha, \beta$ are two real numbers represented by sequences $\bar{x} = \{x_n\}$ and $\bar{y} = \{y_n\}$ respectively (i.e., $\alpha = [\bar{x}]$ and $\beta = [\bar{y}]$), then we define $\alpha + \beta$ to be the real number represented by the sequence $\bar{x} + \bar{y} = \{x_n + y_n\}$. Similarly, we define $\alpha \cdot \beta$ to be $[\bar{x}\bar{y}] = [\{x_n y_n\}]$. Finally, we say $\alpha \leq \beta$ (read '$\alpha$ is less than or equal to $\beta$) if for every (rational) $\epsilon > 0$, there exists $m \in I\!N$, such that, $x_n \leq y_n + \epsilon$ for all $n \geq m$.

Whenever a definition like this is made, a scruple invariably required to be followed in mathematics is to ensure that it is well defined, i.e. independent of the choice of the representatives. In the case of the definition above, suppose, for example, that the real numbers $\alpha, \beta$ are represented by the sequences $\bar{z} = \{z_n\}$

---

[1] Here by the triangle inequality, we mean the triangle inequality for rational numbers only. A little later we shall define absolute value for real numbers and then we can talk of the triangle inequality for real numbers.

and $\bar{w} = \{w_n\}$ respectively, instead of $\bar{x}$ and $\bar{y}$. Then $\alpha + \beta$ would be the real number represented by the sequence $\bar{z} + \bar{w}$ i.e. by the sequence $\{z_n + w_n\}$. This sequence will, in general, be different from the sequence $\bar{x} + \bar{y}$. If we can show that they are equivalent (under the relation $\sim$), then they would represent the same real number and the addition would be well-defined. In short, we must ensure that $\bar{x} + \bar{y} \sim \bar{z} + \bar{w}$ whenever $\bar{x} \sim \bar{z}$ and $\bar{y} \sim \bar{w}$. Once it is understood exactly what must be verified, doing it actually is often a routine task. Indeed, let $\epsilon > 0$ be given. Then there exist $m_1, m_2 \in I\!N$ such that $|x_n - z_n| < \frac{\epsilon}{2}$ for all $n \geq m_1$ and $|y_n - w_n| < \frac{\epsilon}{2}$ for all $n \geq m_2$. As usual, we let $m = \max\{m_1, m_2\}$. Then by the triangle inequality, we get that for all $n \geq m, |(x_n + y_n) - (z_n + w_n)| < \epsilon$ thereby proving that $[\bar{x} + \bar{y}] = [\bar{z} + \bar{w}]$ as desired.

That the multiplication of real numbers is well-defined can also be proved along similar lines. However, first we have to prove that every fundamental sequence is bounded. We leave this as an easy exercise. Given fundamental sequences $\bar{x}, \bar{y}, \bar{z}, \bar{w}$ as above, we let $M$ be a positive rational number such that for all $n \in I\!N, |y_n| < M$ and $|z_n| < M$. Rewriting $x_n y_n - z_n w_n$ as $(x_n - z_n)y_n + z_n(y_n - w_n)$ and using the triangle inequality we get $|x_n y_n - z_n w_n| \leq M(|x_n - z_n| + |y_n - w_n|)$ which can be made arbitrarily small by taking $n$ sufficiently large because of the hypothesis that $\bar{x} \sim \bar{z}$ and $\bar{y} \sim \bar{w}$. This shows that $\bar{x}\bar{y} \sim \bar{z}\bar{w}$ thereby proving that the multiplication of real numbers is well-defined.

The proof that the relation '$\leq$' for real numbers is also well-defined is slightly different. Note that when we say that $\bar{x} \leq \bar{y}$ it does *not* mean that $x_n \leq y_n$ for all $n$, not even for 'almost all' $n$ as might appear natural at first sight. For example, if $x_n = \frac{1}{n}$ and $y_n = 0$ for all $n \in I\!N$, then there is no value of $n$ for which $x_n \leq y_n$ and still we have $\bar{x} \leq \bar{y}$ because it is true that given any (rational) $\epsilon > 0$, there exists $m \in I\!N$ such that $x_n \leq y_n + \epsilon$ for all $n \geq m$. All we have to do is to write $\epsilon$ as a ratio (not necessarily in the reduced form) of positive integers, say, $\epsilon = \frac{p}{q}$ and set $m = q$. (Basically what we have proved amounts to saying that $\frac{1}{n} \to 0$ as $n \to \infty$ in the rational number system. Here we have crucially used the fact that $\epsilon$ is rational. As observed at the beginning of Section 3.3, in the *real* number system it is not so easy to prove that $\frac{1}{n} \to 0$ as $n \to \infty$. Even though it looks obvious, the proof requires the completeness of $I\!\!R$ which we are yet to establish.)

The leeway given by $\epsilon$ in the definition of the relation $\leq$ is also needed in proving that it is well-defined. Suppose as above that $\bar{x}, \bar{y}, \bar{z}, \bar{w}$ are fundamental sequences with $\bar{x} \sim \bar{z}, \bar{y} \sim \bar{w}$. We are given that $\bar{x} \leq \bar{y}$ and we have to prove that $\bar{z} \leq \bar{w}$. This time given an $\epsilon > 0$, we divide it into three, rather than two parts. Thus, there exist $m_1, m_2, m_3 \in I\!N$ such that :

$$
\begin{aligned}
|x_n - z_n| &< \frac{\epsilon}{3} \quad \text{for all } n \geq m_1 \\
|y_n - w_n| &< \frac{\epsilon}{3} \quad \text{for all } n \geq m_2 \\
x_n - y_n &< \frac{\epsilon}{3} \quad \text{for all } n \geq m_3.
\end{aligned}
$$

Let $m = \max\{m_1, m_2, m_3\}$. Then for all $n \geq m$, the first two inequalities imply in particular that $z_n < x_n + \frac{\epsilon}{3}$ and $y_n < w_n + \frac{\epsilon}{3}$. Combining these with the third inequality above, we get that $z_n < w_n + \epsilon$ for all $n \geq m$ and hence that $\bar{z} \leq \bar{w}$ as was to be proved.

So far we have constructed the real number system $I\!R$, defined two algebraic operations $+$ and $\cdot$ and an order relation $\leq$ on it. Now come four important tasks :

(i) To show that the operations $+$ and $\cdot$ have all the familiar properties such as commutativity, associativity, distributivity, the existence of additive and multiplicative inverses (which permit us to perform, respectively, the subtraction and division of real numbers). In technical terms, this is expressed by saying that $I\!R$ is a **field**.

(ii) To show that the relation $\leq$ satisfies the elementary properties of an order relation, viz. that it is reflexive, transitive, anti-symmetric and satisfies the law of trichotomy (the meanings of these terms will become clear later) and further that it is compatible with the algebraic operations on $I\!R$ in the sense that for any three real numbers $\alpha, \beta, \gamma$, we have $\alpha + \gamma \leq \beta + \gamma$ whenever $\alpha \leq \beta$ and $\alpha\gamma \leq \beta\gamma$ whenever $\alpha \leq \beta$ (and $0 \leq \gamma$). In technical terms, this amounts to saying that $I\!R$ is not only a field, but what is called an **ordered field**.

(iii) To show that the system $I\!R$ we have constructed is an extension of $Q$, we must show that the system from which we started i.e., the system $Q$ is a subsystem of $I\!R$. On the nose, of course, $Q$ is not even a subset of $I\!R$. So what we mean by a subsystem in this case is that there exists a one-to-one function $f : Q \longrightarrow I\!R$ which is compatible with the basic algebraic operations $+$ and $\cdot$ and also with the order relation $\leq$ in the sense that for all $x, y \in Q$, we have $f(x + y) = f(x) + f(y)$, $f(xy) = f(x)f(y)$ and $f(x) \leq f(y)$ if and only if $x \leq y$. Here, we have used the same symbols for the operarions in $Q$ and in $I\!R$. The first equality verbally means that if we start with any two rational numbers then whether we add them as rational numbers and then take the image of their sum (under the function $f$) or whether we first take their images in $I\!R$ and add them as real numbers, we get the same result. The other two statements can be paraphrased similarly. A function $f$ with these properties is called an **embedding**. Because of its properties, we may very well identify each rational number $x$ with its corresponding real number $f(x)$. If we do so then $Q$ gets identified with its image $f(Q)$ in $I\!R$. And it is in this sense that $I\!R$ can be thought of as an extension of the rational number system $Q$.

(iv) Finally, we must show that the real number system $I\!R$ we have constructed is complete, as otherwise, the very purpose of the construction would be lost ! So this is the most interesting part of the srory.

We proceed to these tasks one-by-one. (i) is a straight-forward verification. Most of the time, the desired property for $I\!R$ is established from the corresponding property for $Q$. Suppose for example, that we want to establish the commutativity of addition for $\mathbb{R}$. That is, for any two real numbers $\alpha, \beta$ we want to show that $\alpha + \beta = \beta + \alpha$. Let $\bar{x}, \bar{y}$ be sequences such that $\alpha = [\bar{x}]$ and $\beta = [\bar{y}]$. Then $\bar{x} + \bar{y}$ represents $\alpha + \beta$ while $\bar{y} + \bar{x}$ represents $\beta + \alpha$. But the sequences $\bar{x} + \bar{y}$ and $\bar{y} + \bar{x}$ are equal, by commutativity of addition for $Q$. So certainly they are equivalent under the relation $\sim$. But this means that they represent the same real number, i.e. $\alpha + \beta = \beta + \alpha$. Other verifications are equally routine. The only exception perhaps is in showing that every non-zero real number has a multiplicative inverse (i.e. a reciprocal). It is easy to see that the real number represented by $\bar{0}$, the identically zero sequence, does the job of the zero element (technically called the **identity of addition**). That is to say, for every real number $\alpha$, $\alpha + [\bar{0}] = \alpha$. Similarly the constant sequence $\bar{1}$ represents the unity element (technically called the **identity of multiplication**) in that $\alpha \cdot [\bar{1}] = \alpha$ for every real number $\alpha$. Now, suppose $\alpha$ is non-zero. We want to find a real number $\beta$ such that $\alpha\beta = [\bar{1}]$. It is natural to try to do so by first taking any sequence $\bar{x} = \{x_n\}$ representing $\alpha$ and then letting $\beta = [\bar{y}]$ where the sequence $\bar{y} = \{y_n\}$ is defined by $y_n = \frac{1}{x_n}$ for $n \in \mathbb{N}$. But care has to be taken to ensure that $x_n \neq 0$. Here, even though $[\bar{x}] \neq [\bar{0}]$, it does not follow that $x_n \neq 0$ for *every* $n$ and so we have to be careful in defining $y_n$. Note that $\alpha \neq [\bar{0}]$ means that the fundamental sequence $\{x_n\}$ is not equivalent (under the relation $\sim$) to the constant sequence $\bar{0}$. By definition of the relation $\sim$, this means that there is some (rational) $\epsilon > 0$, such that for every $m \in \mathbb{N}$, there exists $n \geq m$ such that $|x_n - 0| \geq \epsilon$. Here we are free to choose $m$ and we shall exercise this freedom shortly. Since $\{x_n\}$ is a fundamental sequence, there exists $p \in \mathbb{N}$ such that $|x_q - x_r| < \frac{\epsilon}{2}$ whenever $q \geq p$ and $r \geq p$. We choose $m$ to be this integer $p$. Then for some $n \geq p, |x_n| \geq \epsilon$. Fix any such $n$. Then for every $k \geq n$ the inequalities $|x_n - x_k| < \frac{\epsilon}{2}$ and $|x_n| \geq \epsilon$ together imply that $|x_k| \geq \frac{\epsilon}{2}$ (as otherwise the triangle inequality would imply $|x_n| = |(x_n - x_k) + x_k| \leq |x_n - x_k| + |x_k| < \frac{\epsilon}{2} + \frac{\epsilon}{2} = \epsilon$, a contradiction.) In particular it follow that $x_k \neq 0$ for all $k \geq n$. We now define $y_k = \frac{1}{x_k}$ for $k \geq n$. For $k < n$, we define $y_k$ arbitrarily. Then for all $k \geq n, x_k y_k = 1$ and so $[\bar{x}][\bar{y}] = [\bar{1}]$, i.e. $\alpha\beta = [\bar{1}]$ as was our goal.

Coming to our second task, viz., showing that the relation $\leq$ we have defined for real numbers satisfies certain properties, here, too, some of the verifications are routine. For example, the reflexivity and the transitivity of $\leq$ for $\mathbb{R}$ follow from the corresponding properties of the relation $\leq$ for $Q$. Another important property of an order relation is its **anti-symmetry**. This means that if two numbers are such that each is less than or equal to the other then they must be equal. We expect that antisymmetry of $\leq$ for $\mathbb{R}$ should follow directly from that for $Q$. But it is not quite so. Suppose $\alpha = [\bar{x}]$ and $\beta = [\bar{y}]$ are real numbers such that $\alpha \leq \beta$ and $\beta \leq \alpha$. As noted earlier, this does *not* mean that for every $n, x_n \leq y_n$ and $y_n \leq x_n$ and so we cannot apply the antisymmetry of $\leq$ for $Q$ directly. Nevertheless, we can proceed as follows. Note that we are not required to prove that $\bar{x} = \bar{y}$ but merely to prove that $[\bar{x}] = [\bar{y}]$, i.e. that $\bar{x} \sim \bar{y}$. To this end, let $\epsilon > 0$ be given. From the hypothesis, there exist integers $m_1, m_2$

such that $x_n \leq y_n + \epsilon$ for all $n \geq m_1$ and $y_n \leq x_n + \epsilon$ for all $n \geq m_2$. Once again, let $m = \max\{m_1, m_2\}$. Then for all $n \geq m$, we have both $x_n - y_n < \epsilon$ and $y_n - x_n < \epsilon$ which together imply that $|x_n - y_n| < \epsilon$ which shows that $\bar{x} \sim \bar{y}$ as desired.

Another important property of the relation $\leq$ which we must establish is what is called the **law of trichotomy**. The word lirerally means 'three branches' and its use is justified because it says that given two numbers exactly one of the three possibilities must hold, viz. either the first number is less than the second, or the second number is less than the first or the two numbers are equal. It is this property which is used instinctively when we say that the positive negation (see Section 1.11) of $x < y$ is $x \geq y$. If, instead of strict inequalities we allow possible equalities, then there are only two possibilities : either the first number is less than or equal to the second or vice versa. This is called the **law of dichotomy**. Both the laws, of course, convey the same idea. In essence they say that every pair of numbers is comparable. To appreciate the point involved, consider some relationships which do not obey these laws. For example, the subset relationship (often denoted by the symbol $\subset$) is certainly reflexive, transitive and antisymmetric. But the law of dichotomy fails. Given two sets , say $A$ and $B$, it is not always true that either $A \subset B$ or $B \subset A$. Similarly, given two positive integers $m$ and $n$ we cannot say that either $m$ divides $n$ or $n$ divides $m$.

To establish the law of dichotomy for real numbers, suppose $\alpha = [\bar{x}]$ and $\beta = [\bar{y}]$ are two real numbers. If they are equal, we are done. Otherwise, the sequences $\{x_n\}$ and $\{y_n\}$ are not equivalent to each other. This means there is some (rational) $\epsilon > 0$ such that for every $m \in I\!N$, there exists some $n \geq m$ for which $|x_n - y_n| \geq \epsilon$. (Here we have used the law of trichotomy for rational numbers.) Again, the choice of $m$ is ours and we shall exercise it shortly. Also, as both $\{x_n\}$ and $\{y_n\}$ are fundamental sequences, there exist $m_1, m_2$ such that $|x_p - x_q| < \frac{\epsilon}{4}$ whenever $p \geq m_1, q \geq m_1$ and $|y_p - y_q| < \frac{\epsilon}{4}$ whenever $p \geq m_2, q \geq m_2$. We let $m = \max\{m_1, m_2\}$. Then, as said above, there exists some $n \geq m$ such that $|x_n - y_n| \geq \epsilon$. Fix one such $n$. Then for all $k \geq n$, we have the inequalities $|x_n - x_k| < \frac{\epsilon}{4}$ and $|y_n - y_k| < \frac{\epsilon}{4}$. Also we have either $x_n - y_n \geq \epsilon$ or $y_n - x_n \geq \epsilon$. Depending upon which possibility holds, we have either $x_k - y_k \geq \frac{\epsilon}{2}$ for all $k \geq n$ or $y_k - x_k \geq \frac{\epsilon}{2}$ for all $k \geq n$. (This is seen most easily with a diagram.) In the first event, $\bar{y} \leq \bar{x}$, i.e. $\beta \leq \alpha$ while in the second event, $\alpha \leq \beta$. Thus the law of dichotomy holds.

Finally, the compatibility of the relation $\leq$ with the algebraic operations $+$ and $\cdot$ for $I\!R$ follows from the compatibility of $\leq$ for $Q$ with the algebraic operations. This is a routine verification and we omit it. We have thus proved that $I\!R$ is not only a field but an ordered field.

Now that we have the concept of an order relation on $I\!R$ we can classify real numbers as positive, negative or zero. We call a real number $\alpha$ as positive if $[\bar{0}] < \alpha$, (i.e. $[\bar{0}] \leq \alpha$ and $[\bar{0}] \neq \alpha$). We say $\alpha$ is negative if $\alpha < [\bar{0}]$. It is easy to show that $\alpha$ is positive if and only if $-\alpha$ is negative. Because of the compatibility of the order relation with addition, it follows that for any two real numbers $\alpha$ and $\beta$, $\alpha < \beta$ if and only if $\beta - \alpha$ is positive. The law of trichotomy is equivalent to saying that for every real number, there are precisely three mutually exclusive possibilities, viz., that it is positive, negative or 0. (All these statements may appear to be too trivial to deserve a mention. But we must keep in mind that since we have just constructed real numbers, any properties thereof must be deduced strictly from the construction, regardless of how familiar or predictable they appear to us.)

We can define the concept of an absolute value in the predicted way. That is, for a real number $\alpha$, $|\alpha|$ equals $\alpha$ if $\alpha$ is positive or zero and it equals $-\alpha$ if $\alpha$ is negative. The basic properties of absolute value, notably the triangle inequality can now be derived for real numbers also. Once again, the proofs usually follow right from the definitions or from the corresponding properties for rational numbers.

Having defined the concept of the absolute value of a real number, the field is now open for defining the concept of convergence of a sequence of real numbers. In fact, the definitions given earlier in this book apply unchanged since these definitions are independent of the construction of real numbers. The same is true for the elementary results and their proofs, such as the uniqueness of limits or the Sandwich Theorem or the fact that the limit of a sum of two convergent sequences of real numbers equals the sum of their limits. Note however, that a proof which uses directly or indirectly, the completeness of the real number system is not valid at this stage. Note in particular, that we cannot use the result that the sequence $\{\frac{1}{n}\}$ converges to 0 as a sequence of real numbers, because as shown at the beginning of Section 3.3, although this fact looks obvious, its proof requires completeness of $I\!R$ which we have not yet established.

The third task, viz., showing that $I\!R$ can be thought of as an extension of $Q$ is easy. As discussed above, we shall construct a one-to-one function $f : Q \to I\!R$ which is compatible with $+$, $\cdot$ and $\leq$. For any rational number $x$, let $\bar{x}$ be the constant sequence whose every term is $x$. Trivially, $\bar{x}$ is a fundamental sequence and hence a member of the set $S$. Further, if $x, y$ are two rational numbers, then $\bar{x} \sim \bar{y}$ if and only if $x = y$. So if we define $f(x) = [\bar{x}]$ for $x \in Q$, then we get a one-to-one function from $Q$ to $I\!R$. The verifications that for any two rationals $x$ and $y$, $f(x+y) = f(x) + f(y)$ etc. are extremely simple, since all the sequences involved are constant sequences. From now onwards, we identify a rational number $x$ with its corresponding real number, viz., $f(x)$ and thereby treat $Q$ as a subset of $I\!R$. No confusion is likely by doing so, because the properties of $f$ ensure that in whatever operations we do on rational numbers, it does not matter whether we treat them as rational numbers or as real numbers.

It is equally clear that if $x$ is a rational number, then its absolute value $|x|$ is also a rational number and further that $f(|x|) = |f(x)|$. Verbally, the absolute value of a rational number $x$ is the same whether $x$ is treated as a rational number or as a real number. Since convergence of sequences is defined in terms of absolute values, it is tempting to conclude that if a sequence, say $\{x_n\}$ of rationals converges to a number $x$ in the rational number system then it will also converge to $x$ in the real number system. This is indeed true, but the proof is not so immediate. The catch is that when we want to prove that $\{x_n\}$ converges to $x$ in $I\!R$, the condition in the definition of convergence has to be satisfied for every real $\epsilon > 0$ and the hypothesis that $\{x_n\}$ converges to $x$ in $Q$ only guarantees that it holds for every *rational* $\epsilon > 0$. To fill the gap we need that $Q$ as a subset of $R$ is dense. As this property will also be pivotal in proving the completeness of $R$, we give the proof in detail.

**A.3 Theorem:** The set of rationals, regarded as a subset of $R$ is dense in it. That is, given two real numbers $\alpha, \beta$ with $\alpha < \beta$ there exists a rational number $z$ such that $\alpha < z < \beta$ (or, a little pedantically, $\alpha < [\bar{z}] < \beta$ where $\bar{z}$ is the constant sequence with each term equal to $z$).

**Proof:** Suppose $\alpha = [\bar{x}]$ and $\beta = [\bar{y}]$ where $\bar{x} = \{x_n\}$ and $\bar{y} = \{y_n\}$ are fundamental sequences of rational numbers. We are given that $\alpha \neq \beta$ and, in fact, that $\alpha < \beta$. Duplicating the initial part of the argument in the proof of the law of trichotomy for $R$, we get that there is some rational $\epsilon > 0$ and some $n \in N$ such that : (i) $y_n \geq x_n + \epsilon$ and (ii) for all $k \geq n, |x_k - x_n| < \frac{\epsilon}{4}$ and $|y_k - y_n| < \frac{\epsilon}{4}$. Now let $z$ be the rational number $x_n + \frac{\epsilon}{2}$. Then $z + \frac{\epsilon}{2} \leq y_n$ by (i). Hence by (ii) and the triangle inequality, we see that for all $n \geq k, x_k < z - \frac{\epsilon}{4}$ and $z + \frac{\epsilon}{4} < y_k$. This shows $\alpha \leq [\bar{z}] \leq \beta$. That equality cannot hold at either end is proved by noting that for all $k \geq n$, both $|x_k - z|$ and $|z - y_k|$ exceed $\frac{\epsilon}{4}$. Thus we have $\alpha < [\bar{z}] < \beta$. ∎

As an immediate corollary we get that a sequence which is convergent in $Q$ is also convergent (and has the same limit) when thought of as a sequence in $R$. Indeed, given a real $\epsilon > 0$ in the definition of convergence, all we have to do is to apply the theorem above to get a rational $\delta$ with $0 < \delta < \epsilon$ and invoke the condition in the definition of convergence (in $Q$ ) for $\delta$. Recall that while proving that the relation $\leq$ for real numbers is well-defined, we proved that in $Q, \frac{1}{n} \to 0$ as $n \to \infty$. We can now say that the same thing holds in $R$ also. Earlier (see Theorem (3.3.1) we proved this using completeness of $R$. Now we have proved it as a consequence of the theorem above. This is not very surprising, because the theorem above is a vital step in the proof that the real number system we have constructed is complete. This, in fact, is our fourth and final task and we are almost at it. We need just one intermediate step, which is actually more like a paraphrase of the last theorem and is often taken as an alternate definition of denseness.

**A.4 Theorem:** Every real number can be approximated arbitrarily closely by a rational number. In symbols given any $\alpha \in R$ and any (real) $\epsilon > 0$, there exists some $z \in Q$ such that $|\alpha - z| < \epsilon$.

**Proof:** All we have to do is to put $\beta = \alpha + \epsilon$ and apply the theorem above. (Actually, we get a sharper conclusion than is asserted. That is, we get $\alpha < z < \alpha + \epsilon$ and not merely $\alpha - \epsilon < z < \alpha + \epsilon$.) ∎

If we apply this theorem repeatedly with $\epsilon = 1, \frac{1}{2}, \frac{1}{2}, \ldots, \frac{1}{n}, \ldots$, we get a sequence $\{z_n\}$ of rationl numbers such that $|\alpha - z_n| < \frac{1}{n}$ for every $n = 1, 2, 3, \ldots$. We already proved that in $R$, $\frac{1}{n} \to 0$ as $n \to \infty$. It then follows that $z_n \to \alpha$

as $n \to \infty$. Thus we have shown that every real number can be expressed as a limit of a sequence of rational numbers. This is taken as yet another expression of denseness of $Q$ in $I\!R$. As mentioned at the beginning, it is this fact which is the motivation for our construction of real numbers.

Finally, we prove that the real number system is complete. We shall first prove this in a form for which Theorem (A.4) comes in most handy.

**A.5 Theorem:** Every Cauchy sequence of real numbers is convergent (in $I\!R$).

**Proof:** Suppose $\{\alpha_n\}$ is a Cauchy sequence of real numbers. By the theorem above, for each $n \in I\!N$, there exists a rational number $z_n$ such that $|\alpha_n - z_n| < \frac{1}{n}$. We claim that the sequence $\bar{z} = \{z_n\}$ is a fundamental sequence. Let a rational $\epsilon > 0$ be given. Since the sequence $\{\alpha_n\}$ is a Cauchy sequence, there exists $p \in I\!N$ such that $|\alpha_m - \alpha_n| < \frac{\epsilon}{3}$ whenever $m \geq p$ and $n \geq p$. We may further suppose $p$ is so chosen that for all $q \geq p, \frac{1}{q} < \frac{\epsilon}{3}$. This is possible because we have already shown (without using completeness !) that $\frac{1}{p} \to 0$ in $I\!R$ as $p \to \infty$. Thus, for all $m, n \geq p$ all three inequalities viz., $|\alpha_m - z_m| < \dfrac{\epsilon}{3}, |\alpha_m - \alpha_n| < \dfrac{\epsilon}{3}$ and $|\alpha_n - z_n| < \dfrac{\epsilon}{3}$ hold true. Together they imply that $|z_m - z_n| < \epsilon$ whenever $m \geq p$ and $n \geq p$. Thus we have shown that $\bar{z}$, i.e. the sequence $\{z_n\}$ is a fundamental sequence of rational numbers. Let $\alpha = [\bar{z}]$ be the real number represented by it.

We contend that $\{\alpha_n\}$ converges to $\alpha$ (in $I\!R$). Instead of doing this directly, we shall show that the sequence $\{z_n\}$ (now regarded as a sequence of real numbers) converges to $\alpha$. This would complete the proof because we have chosen the $z_n$'s so that $|\alpha_n - z_n| < \frac{1}{n}$ for all $n \in I\!N$ and we have already shown that $\frac{1}{n} \to 0$ in $I\!R$ as $n \to \infty$. For we can write, $|\alpha_n - \alpha| \leq |\alpha_n - z_n| + |z_n - \alpha| \leq \frac{1}{n} + |z_n - \alpha|$ and note that each term in the last expression can be made arbitrarily small.

Thus we are reduced to proving that the sequence $\{z_n\}$ converges to $\alpha$ in $I\!R$. The proof is not difficult but somewhat subtle. Note that the real number $\alpha$ is, by definition, the equivalence class $[\bar{z}]$ where $\bar{z}$ is the sequence $\{z_n\}$ of rational numbers. In a somewhat loose language, we are proving that the sequence $\{z_n\}$ converges to itself ! This is the curious part of the construction as mentioned earlier, viz., we are filling the hole created by a (non-convergent but Cauchy) sequence by that sequence itself. The sequence $\{z_n\}$ thus plays a double role here. On one hand, it is a sequence in $I\!R$. But on the other hand, its terms are rationals and as a sequence of rationals it is a fundamental sequence and hence a representative of a real number $\alpha$. This double role can sometimes lead to a confusion. To avoid it, note that when the rational number $z_n$ is thought of as a real number it is really the equivalence class of the constant sequence $\bar{z}_n$. The correct notation for it is $[\bar{z}_n]$. So we have to prove that the sequence $\{[\bar{z}_n]\}$ of real numbers converges to the real number $[\bar{z}]$ in $I\!R$. Let a real $\epsilon > 0$ be given. In view of Theorem (A.3) we may suppose that $\epsilon$ is rational (as otherwise we replace it with a rational $\delta$ such that $0 < \delta < \epsilon$). We have to find some $m \in I\!N$ such that for all $n \geq m, |[\bar{z}_n] - [\bar{z}]| < \epsilon$. It suffices to find an $m$ such that for all $n \geq m, [\bar{z}] - \frac{\epsilon}{2} \leq [\bar{z}_n] \leq [\bar{z}] + \frac{\epsilon}{2}$. Here $\frac{\epsilon}{2}$ is represented by the constant sequence

$\{\frac{\epsilon}{2}\}$. And so $[\bar{z}] - \frac{\epsilon}{2}$ and $[\bar{z}] + \frac{\epsilon}{2}$ are real numbers represented respectively, by the sequences $\{u_n\}$ and $\{v_n\}$ where for each $n \in \mathbb{N}, u_n = z_n - \frac{\epsilon}{2}$ and $v_n = z_n + \frac{\epsilon}{2}$. (Here we are crucially using that $\epsilon$ and hence $\frac{\epsilon}{2}$ are rational.) Denoting these sequences by $\bar{u}$ and $\bar{v}$ as usual, our task is reduced to finding an $m \in \mathbb{N}$ such that for all $n \geq m, [\bar{u}] \leq [\bar{z}_n] < [\bar{v}]$. Recalling how the relation $\leq$ was defined for real numbers, and keeping in mind that $\bar{z}_n$ is a constant sequence with each term equal to $z_n$, the inequality $[\bar{u}] \leq [\bar{z}_n]$ means that for every rational $\delta > 0$, $u_k \leq z_n + \delta$ holds for almost all values of $k$ (i.e. for all $k$ bigger than or equal to some $r \in \mathbb{N}$, which could depend on $\delta$). We shall actually do a little better. That is, we shall show that regardless of what $\delta$ is, the inequality $u_k \leq z_n$ holds for almost all values of $k$. (In other words, the integer $r$ will be independent of $\delta$.) Similarly, we shall show that $z_n \leq v_k$ for almost all values of $k$.

We have now reduced the problem to a level where all we need is that the sequence $\{z_n\}$ is a fundamental sequence. This being the case, there is some $m \in \mathbb{N}$, such that for all $n \geq m, k \geq m$, we have $|z_n - z_k| < \frac{\epsilon}{2}$. Fix any $n \geq m$. Then for every $k \geq r, |z_k - z_n| < \frac{\epsilon}{2}$, or equivalently, $z_k - \frac{\epsilon}{2} < z_n < z_k + \frac{\epsilon}{2}$. But this says exactly that for all $k \geq m, u_k < z_n < v_k$. As shown earlier, this completes the proof. ∎

The theorem just proved gives one of the versions of completeness of the real line, often called **Cauchy completeness**. The version of completeness which we assumed as an axiom (Axiom (2.6.2)) is called the **order completeness**. We already derived the former from the latter (see Theorem (3.5.4)). To derive the order completeness of $\mathbb{R}$ from its Cauchy completeness, we appeal to Exercise (3.5.10). We of course need to show that the set $\mathbb{N}$ has no upper bound in $\mathbb{R}$. This is an easy consequence of the fact that $\frac{1}{n} \to \infty$ as $n \to \infty$ (in $\mathbb{R}$) which we proved above as a consequence of the denseness of $Q$ in $\mathbb{R}$. (In Section 3.3, the sequence of these implications was reversed. There we assumed the order completeness of $\mathbb{R}$, then showed that the set $\mathbb{N}$ has no upper bound in $\mathbb{R}$, then proved that $\frac{1}{n} \to 0$ as $n \to \infty$ and finally deduced the denseness of $Q$ in $\mathbb{R}$. (See the proofs of Theorem (3.3.1) and Corollary (3.3.2).)

To show that $\mathbb{N}$ is unbounded in $\mathbb{R}$, suppose to the contrary that $L$ is an upper bound (not necessarily the least upper bound) of $\mathbb{N}$ in $\mathbb{R}$. Clearly $L > 0$ and so $\frac{1}{L} > 0$. Since $\frac{1}{n} \to \infty$ as $n \to \infty$, there exists some $m \in \mathbb{N}$ such that $\frac{1}{n} < \frac{1}{L}$ for all $n \geq m$. But then we would have $m > L$, a contradiction. Thus we have shown that $\mathbb{N}$ is not bounded above in $\mathbb{R}$. This fact, along with the Cauchy completeness of $\mathbb{R}$ implies the order completeness of $\mathbb{R}$, by Exercise (3.5.10).

Thus, at long last, we have finished the construction and the proof of completeness of the real number system. As emphasized earlier, there is hardly a non-trivial result of calculus which does not require completeness directly or indirectly. Without completeness there will be no guarantee that a limit which we would like to exist will, in fact, exist. Figuratively, if the limiting process is the heart of calculus, then it is completeness which makes the heart tick !

So, completeness is certainly a big achievement and makes the construction of real numbers worthwhile. Note that no proper subset of $\mathbb{R}$ which contains $Q$

can be complete. For, if $A$ is any such set, then every Cauchy sequence in $A$ and hence, in particular, every Cauchy sequence of rationals will have a limit in $A$. But every real number is expressible as the limit of some sequence of rationals. So $A$ would have to contain every real number.

Put differently, if we want to cure the incompleteness of the rational number system by extending it, then we cannot stop short of $I\!R$. It is the smallest extension of $Q$ in which completeness holds. Unfortunately, this smallest extension is itself rather too big from certain points of view. For example, even though both $Q$ and $I\!R$ are infinite as sets, the former is denumerable while the latter is not (see Exercise (3.9.7)). In other words, we can list the rationals as a sequence and exhaust them. But this is impossible for reals. As a result, things like inductive proofs or recursive definitions make no sense for $I\!R$. Another, and a more serious difference is that the arithmatic of the rational numbers can be understood and handled even by a layman since it involves no limiting process. By contrast, even the definition of a real number involves the limiting process crucially. This makes them subtle and often useless in practice. If you go to a shopkeeper and ask for $e$ kilograms of rice, he will be dumbfounded. You will have to give him a rational approximation to it. The same thing also holds for so many other things such as integrals. They exist as certain real numbers. But to evaluate them we need approximation techniques such as the Simpson's rule.

Summing up, although the real number system is a completion (both mathematically and figuratively) of the rational number system, the latter is still the core of the former.

*Appendix B*

# CONSTRUCTION OF INFINITESIMALS

It is ironic that the most basic concept of calculus, viz. that of a limit should also be the hardest to define in rigorous mathematical terms. As explained in Section 2.3, the real difficulty is that the intuitively clear and highly convenient concept of an infinitesimally small change turns out to be extremely elusive. The present definition of a limit does not really overcome this difficulty but bypasses it by giving a logically perfect but highly clumsy formulation of the concept of a limit which totally avoids the word 'infinitesimal'. This word is in fact almost a taboo in today's chaste mathematics books, even though it is used informally by mathematicians and less inhibitively by others.

A rigorous definition of an 'infinitesimal' is relatively very recent and has laid the foundation of a new branch of mathematics called the **non-standard analysis** the name clearly indicating a departure from the 'standard' analysis which is based on the standard definition of a limit. (By the way, the word 'analysis' is not formally defined. When used in the present context, it means the theoretical aspect of calculus rather than its applications to finding volumes and centres of mass). Although it is unlikely that the non-standard analysis will replace the standard one in near future, it is worth taking a look at, if for no other reason then at least to be assured that the concept of an infinitesimal *can* be defined rigorously and thereby to shed the apologetic attitude in its use.

Although we shall not study non-standard analysis[1] in detail, we shall show how the real number system $I\!R$ can be extended to a system $I\!R^*$ called the **hyperreal number system** which contains not only a copy of the 'standard' real number system, but also certain positive hyperreal[2] numbers called infinitesimals which are smaller than every positive (standard) real number (or rather its copy in $I\!R^*$.) For $x \in I\!R$, we shall denote by $x^*$ its copy in $I\!R^*$. If $f : I\!R \longrightarrow I\!R$ is a (standard) function, then we shall construct a function $f^* : I\!R^* \longrightarrow I\!R^*$ which is an extension of $f$ in the sense that for every $x \in I\!R$, $(f^*)(x^*)$ will equal $(f(x))^*$. Finally we shall show that if $c$ and $L$ are any two (standard) real

---

[1] A good reference is the book 'Standard and Non-standard Analysis' by R. F. Hoskins.

[2] The prefix 'hyper' means 'above, beyond, in excess'. It is used here to indicate that we are going beyond the system of real numbers. We have already encountered it in words like a hypervolume or, more commonly, hyperbola. In the former, it connotes that we are going beyond the three dimensional world. In the case of a hyperbola, it indicates that the eccentricity exceeds 1, or if we think of a hyperbola as a conic section (see Exercise (4.12.10)), then that the cutting plane makes an angle with the base which is higher than that which a generator of the cone makes with the base.

numbers then $\lim_{x \to c} = L$ (according to our definition of a limit) if and only if for every infinitesimal $h$ in $I\!\!R^*$, the hyperreal number $f^*(c^* + h) - L^*$ is also an infinitesimal. This will prove how the basic concept of a limit can be defined in terms of infinitesimals. The theroty can then be developed to give similar alternate definitions of derivatives, integrals etc. and also to give a rigorous treatment of differentials (which, as explained in Section 4.4, is another thorny spot in the traditional approach to calculus). But we shall not go into it.

Even this limited goal of ours is fraught with a major difficulty as we shall see later. But first let us get down to the construction of hyperreal numbers from the real numbers. The motivation, as well as the methodology, resemble considerably the construction of the real numbers from the rational numbers given in Appendix A. Recall that we need to construct the real numbers because in $Q$, the rational number system, there are certain 'holes', each hole corresponding to a Cauchy sequence whose terms are in $Q$ but which does not converge in $Q$. And the trick was to patch up each such hole by that sequence itself. That is, we let $S$ be the set of all Cauchy sequences (called fundamental sequences for technical reasons), say, $\bar{x} = \{x_n\}_{n=1}^{\infty}$ of rationals, defined a certain equivalence relation '$\sim$' on the set $S$ and the real numbers were defined precisely as the equivalence classes of this relation. This was done in such a way that if $\{x_n\}_{n=1}^{\infty}$ is a Cauchy sequence of rationals, then, viewed as a sequence of real numbers, $\{x_n\}_{n=1}^{\infty}$ converges to the real number represented by its own equivalence class. (See the proof of Theorem (A.5).)

We shall follow a similar strategy for the construction of hyperreal numbers, which, as noted above, are an extension of $I\!\!R$, the real number system. The need for them arises because in the real number system, there is no way of rigorously stating what is infinitesimally small. Consider, for example, the sequence, $\{\left(\frac{1}{n}\right)\}_{n=1}^{\infty}$. We all agree that its terms are getting infinitesimally small as $n$ tends to infinity. But we cannot express this fact precisely since we do not have infinitesimals in $I\!\!R$. So here is a bright idea. *Why not call this sequence itself as an infinitesimal* exactly the way we called a Cauchy sequence of rationals (or rather, its equivalence class) as a real number ? This is precisely what we shall do. But we must answer two questions :

(i) Along with this particular sequence, we may have to let in many other sequences of real numbers, or else we shall not be able to perform additions and multiplications in the hyperreal number syatem. Just exactly which sequences of real numbers do we allow ? (Recall that in the construction of the reals from the rationals, we took only the Cauchy sequences of rationals to start with.)

(ii) What should be the equivalence relation on the set of these sequences of real numbers so that each equivalence class will correspond to a hyperreal number ?

Once these two basic questions are answered, the construction of the hyper-reals from reals will be carried out following the same rituals as those in the construction of the real numbers from the rational numbers. That is, we shall prove that there are well defined operations of addition and multiplication on $R^*$, the set of hyperreal numbers, so that they form a field. Further, an order relation for hyperreal numbers will be defined which is compatible with their addition and multiplication so that they form not only a field, but an ordered field. Moreover, the real number system $R$ will be embedded in $R^*$ exactly the way $Q$ is embedded in $R$, that is, to each real number $x$, we shall associate the hyperreal number (to be denoted by $x^*$) which is nothing but the equivalence class of the constant sequence $(x, x, x, \ldots)$.

The answers to the two basic questions raised above turn out to represent two extremes in terms of their degrees of difficulty. The first question has the simplest possible answer. We allow *all* possible sequences of real numbers in the construction of hyperreal numbers. In other words, we let $S$ be the set of all sequences of real numbers (without any restriction such as that it be a Cauchy sequence). We shall define a suitable equivalence relation on $S$ and the equivalence classes will be called hyperreal numbers.

The second question, viz. when should two sequences, say $\{x_n\}$ and $\{y_n\}$ be regarded as equivalent turns out to be surprisingly difficult. There is, of course, no dearth of equivalence relations on the set $S$ and we could pick any of them. But the resulting system $R^*$ may not have the properties we want it to have. For example, if we let $\{x_n\} \sim \{y_n\}$ to mean that $x_n = y_n$ for all $n \in N$, then $\sim$ is certainly an equivalence relation, but the resulting system $R^*$ will not be a field because the sequence $(1, 0, 0, 0, \ldots)$ will represent a non-zero hyperreal number but it would have no reciprocal in $R^*$. It is tempting to try to salvage this by modifying the definition of '$\sim$' slightly, for example, by requiring that $\{x_n\} \sim \{y_n\}$ to mean that they are eventually equal, that is, there exists some $m \in N$ such that $x_n = y_n$ for all $n \geq m$. It is not difficult to verify that this is indeed an equivalence relation. The sequence $(1, 0, 0, 0, \ldots)$ now represents the hyperreal number 0 and so no harm arises by the non-existence of its reciprocal. Unfortunately, the sequence $(1, 0, 1, 0, 1, 0, \ldots)$ represents a non-zero hyperreal number with no reciprocal and so the difficulty still remains.

Other attempts to define a suitable equivalence relation on $S$, the set of all sequences of real numbers, are also met with some failure or the other. There is, in fact, no easy known way of specifying such an equivalence relation. The only known method requires the use of what are called **ultrafilters**. To discuss them would take us too far afield. What is more, to prove the existence of desired ultrafilters, we need an important basic property of sets called the **axiom of choice**. Simply stated, it says that if we have a collection of mutually disjoint, non-empty sets then it is possible to form a set by picking one element from each of them. This is certainly consistent with our day-to-day experience. For example, we form a legislative council by selecting (or electing !) one representative from each constituency. Or we form a sample bouquet by picking one flower of each type. In fact, examples like these give the impression that the axiom of choice is obvious to the point of being trivial. But it is not so. Like many other examples from real life, the examples given above involves only finite sets. But when we have an infinite collection of sets, this process of just 'picking one element from each' will never end. If there is some rule by which we can simultaneously identify one element of each set, then we can certainly form a set consisting of the elements so identified. For example, if the collection consists of mutually disjoint, closed and bounded intervals, then we can select the right end-point of each such interval. As a real life example, if we have a hypothetical, infinite collection of pairs of shoes, then we can form the set consisting of all left shoes. Here we do not need the axiom of choice.

The trouble is that when all we are given is a collection of sets, such a rule may not always exist. To stress this point, suppose instead of shoes, we have an infinite number of pairs of (unworn) socks. For each such pair, we can (arbitrarily) call one of them as the left and the other as the right sock. But how to do so for *all* pairs is not obvious and requires the axiom of choice. Another reason that it has to be assumed as an axiom is that many of its consequences are far from obvious. Indeed, the existence of ultrafilters is one such consequence.

The axiom of choice has generated a lot of stir in mathematics and philosophy. Its use is shunned by some mathematicians as it merely asserts the existence of a certain set without giving any method for its construction. A proof based on it is therefore called a **non-constructive proof** and attempts were often made to replace such a proof by a 'constructive' one. In some cases, they were successful. In some others, there was a glorious retreat by proving the impossibility of a constructive proof. In many cases the search is inconclusive so far.

Interesting as all this is, we abandon the discussion of the axiom of choice as we shall need it just once. So instead of invoking the axiom of choice to get an ultrafilter, we shall directly assume as an axiom the particular consequence of the ultrafilter in which we are interested. The matter deals with the classification of the subsets of $I\!N$ into two mutually disjoint categories which we call as 'large' and 'small'. In other words we would like each subset of $I\!N$ to be either large or small but not both. Our intuition demands that a subset of a small set also be small or equivalently, every superset of a large set be large. We also expect that the union of two small sets be small. (Of course, it may not be as small as either of the two. But we have no degrees of smallness. If some subset is not small, then it has to be large and we would not like that the union of two or finitely many small sets be large.) Finally, we would like the empty set to be small as otherwise every subset would be large which is clearly unacceptable. At the other extreme we do not want every subset to be small either and this is ensured by requiring that the entire set $I\!N$ be large.

A little less intuitively, we want the intersection of two large sets to be large. This means that we must not confuse 'small' with 'finite' and 'large' with 'infinite'. Although there is some resemblance (for example, the union of two finite sets is finite), the intersection of two infinite sets need not be infinite. In fact it could be empty ! We do not want this to happen with large sets. A particular consequence of this requirement deserves mention. For a subset $A$ of $I\!N$ we denote by $A'$ its complement (in $I\!N$), that is, $A' = \{n \in I\!N : n \notin A\}$. Then out of this complementary pair of subsets $A$ and $A'$, exactly one is small and the other large. For, if both are small then their union $I\!N$ is small, a contradiction.

On the other hand, if both are large, then their intersection, which is the empty set, is large, also a contradiction.

A classification of subsets of $I\!N$ like this is by no means unique or rare. Several examples of such classifications exist. For example, fix any positive integer, say, $k$. Declare a subset $A$ of $I\!N$ as large if $k \in A$ and small otherwise. This may seem like a trivial classification. But it meets all the requirements laid down above. This classification is called an **atomic classification**, the integer $k$ being called the atom. Note that every positive integer gives rise to a different classification. A singleton set $\{r\}$ is large in the classification with atom $r$. In every other atomic classification, it is small.

What has classification of subsets of $I\!N$ got to do with the construction of hyperreal numbers ? The answer is that every such classification defines a binary relation on the set $S$ of all sequences of real numbers and as already discussed we are after a suitable equivalence relation on $S$. Suppose we have fixed some classification of subsets of $I\!N$ into large and small subsets satisfying the conditions above. Given two sequences, say $\{x_n\}$ and $\{y_n\}$ of real numbers, we declare them to be equivalent to each other if and only if their indices match on a large set, or in symbols, the set $\{n : x_n = y_n\}$ is large. It is obvious that this relation is symmetric. Its reflexivity follows from the assumption that the set $I\!N$ is large. For transitivity, consider three sequences $\{x_n\}, \{y_n\}$ and $\{z_n\}$. Let $A, B, C$ be, respectively, the sets $A = \{n : x_n = y_n\}, B = \{n : y_n = z_n\}$ and $C = \{n : x_n = z_n\}$. The equivalence of the first two sequences means $A$ is large while that of the second and the third means that $B$ is large. But then their intersection $A \cap B$ is large by assumption. Since $C$ is a superset of $A \cap B$, it follows that it is large, which means that $\{x_n\}$ is equivalent to $\{z_n\}$.

Note that this eqivalence relation depends very much on the way we are classifying the subsets of $I\!N$ as large and small. Suppose, for example, that we have an atomic classification with atom $k$. Then $\{x_n\}$ and $\{y_n\}$ are equivalent to each other if and only if their $k$-th terms are equal, i.e., $x_k = y_k$. Here a single term decides the entire equivalence class. Our intention is to call each equivalence class a hyperreal number. But with the atomic classification just mentioned, each equivalence class corresponds to a single real number (viz., the $k$-th term) and so the resulting hyperreal number system will be the same as the ordinary real number system.

To get a hyperreal number system which is a genuine exrension of $I\!R$, we need a different way of classifying the subsets of $I\!N$. A cue is provided by Section 3.2 where it was discussed why eventuality is so important for sequences. Recall that a subset $A$ of $I\!N$ is called eventual if it contains 'almost all' elements of $I\!N$, i.e., if there exists some $m \in I\!N$ such that $n \in A$ for all $n \geq m$. What if we define two sequences $\{x_n\}$ and $\{y_n\}$ to be equivalent whenever they are eventually equal, i.e., whenever the set $\{n : x_n = y_n\}$ is eventual? It is not difficult to verify that this is indeed an equivalance relation. In fact the proof is essentially the same as above if we treat evenual subsets as large. All that is needed is that the intersection of two eventual subsets is eventual and that every superset of an eventual subset is eventual.

The trouble is that even though we have declared all eventual subsets as large, this does not exhaust all large subsets. Let $A$ and $B$ be, for example, the sets of all even and all odd positive integers. Then exactly one of them is large and the other small (they being complements of each other). But neither of them is eventual.

Let us illustrate what difficulties can arise because of this. One such difficulty was already mentioned above, viz., that the hyperreal number represented by the sequence $(1, 0, 1, 0, 1, 0, \ldots)$ would fail to have a reciprocal. Another difficulty will lie in trying to define an order relation '$\leq$' for hyperreal numbers. With the equivalence relation defined in terms of eventual subsets, let $\alpha$ and $\beta$ be two equivalence classes. These will also be elements of the hyperreal number system we are after. We would like to define an order relation for hyperreal real numbers satisfying the same basic properties as the order relation for real numbers. For simplicity, we denote both these relations by the same symbol, viz. '$\leq$'. How do we define '$\alpha \leq \beta$' ? To do this we first need two sequences, say $\{x_n\}$ and $\{y_n\}$, which represent $\alpha$ and $\beta$ respevtively. Since equality of $\alpha$ and $\beta$ was defined in terms of eventual subsets (i.e., $\alpha = \beta$ if and only if the set $\{n : x_n = y_n\}$ is eventual), it is tempting to define $\alpha \leq \beta$ if and only if the set $\{n : x_n \leq y_n\}$ is eventual. In all such definitions made in terms of representatives of equivalence classes, care has to be taken to show that they are well-defined, i.e., independent of a particular choice of representatives. In the present case, this is very easy. Suppose, for example, that $\{z_n\}$ and $\{w_n\}$ are two sequences representing $\alpha$ and $\beta$ respectively. This means that $\{z_n\}$ is equivalent to $\{x_n\}$ and $\{w_n\}$ is equivalent to $\{y_n\}$, or in other words the sets $A = \{n : x_n = z_n\}$ and $B = \{n : y_n = w_n\}$ are both eventual. Now let $C$ and $D$ be respectively the sets $\{n : x_n \leq y_n\}$ and $\{n : z_n \leq w_n\}$. If we define '$\alpha \leq \beta$ in terms of the representatives $\{x_n\}$ and $\{y_n\}$ it means the set $C$ is eventual. But then $A \cap C \cap B$ is eventual. It is easy to see moreover, that $A \cap C \cap B$ is a subset of $D$, whence $D$ is also eventual. But that means $\alpha \leq \beta$ had we chosen $\{z_n\}$ and $\{w_n\}$ to represent $\alpha$ and $\beta$ respectively. Reversing the roles of $C$ and $D$ completes the proof that the order relation '$\leq$' is well-defined. The elementary properties such as reflexivity ($\alpha \leq \alpha$), transitivity ($\alpha \leq \gamma$ whenever $\alpha \leq \beta$ and $\beta \leq \gamma$) and antisymmetry ($\alpha \leq \beta$ and $\beta \leq \alpha$ together imply $\alpha = \beta$) are easily established. But there is one important property of the order relation on $I\!\!R$, called the **law of dichotomy** (which says that for any two real numbers $x$ and $y$ we must have either $x \leq y$ or $y \leq x$)[3]. With our definition of hyperreal numbers, this law fails. Suppose, for example, that $\alpha$, $\beta$ are represented, respectively, by the sequences $\{x_n\}$ and $\{y_n\}$ where $x_n = (-1)^n$ and $y_n = (-1)^{n+1}$ for all $n \in I\!\!N$. The terms of both the sequences are alternately 1 and $-1$. But they are not equivalent since the set $\{n : x_n = y_n\}$ is the empty set which is not eventual. Also the sets $A = \{n : x_n \leq y_n\}$ and $B = \{n : y_n \leq x_n\}$ are respectively the sets of all odd and of all even integers. Since neither of these sets is eventual, we see that neither $\alpha \leq \beta$ nor $\beta \leq \alpha$.

The law of dichotomy is needed so frequently that it is often used without

---

[3]A somewhat more popular version of this property is the law of trichotomy, which says that for any two real numbers $x$ and $y$, one of the three possibilities must hold : $x < y$, $x = y$ or $x > y$.

an explicit mention. If we want it to hold in the hyperreal number system, we must change our definition by changing the equivalence relation for sequences. As we just saw, the villain behind the failure of the law of dichotomy was that neither the set $A$ of odd positive integers nor its complement $B$ is an eventual subset. To overcome this anomaly, in our classification of subsets of $I\!N$, among 'large' sets we must not only include all eventual subsets of $I\!N$, but also one (but not both) of the sets $A$ and $B$. In fact, we would have to do this for every pair of complementary subsets of $I\!N$. The trouble is how to do it. So far nobody has been able to give an explicit example of such a classification. This is where we need the axiom of choice. As noted earlier, instead of deriving a desired classification from the axiom of choice, we prefer to take the existence of such a classification itself as an axiom.

**B.1 Axiom:** The totality of all subsets of $I\!N$, the set of positive integers, can be classified into two categories, 'large' and 'small' in such a way that :

(i) Every finite subset (and in particular the empty set $\emptyset$) is small and the entire set $I\!N$ is large.

(ii) The union of two (and hence any finite number of) small subsets is small.

(iii) Every subset of a small subset is small.

(iv) For every subset $A$ of $I\!N$, exactly one of $A$ and its complement $A'$ is small (and the other large).

(v) The intersection of two and hence any finite number of large sets is large. Every eventual set is large.

(vi) Every superset of a large set is large. ∎

(Strictly speaking the last two properties follow from the first four and hence need not be assumed as a part of the axiom. But they are listed for ready reference. We have already used both of them several times.)

Note that an atomic classification satisfies all the conditions above except the first. A classification in which all finite subsets of $I\!N$ are small and all other subsets large satisfies the first three conditions but not the fourth. The same is true of the classification in which all eventual subsets are large and the rest small. In fact, as pointed out earlier, so far nobody has been able to explicitly construct a classification satisfying all the conditions above. Its existence has to be assumed as an axiom (or derived from another axiom, viz., the axiom of choice).

In any classification given by the axiom, since all finite sets are small, all eventual subsets are large. There are, of course, many other large sets. As mentioned before, if $A$ is the set of all even positive integers, then either $A$ or its complement $A'$ is large. The axiom itself does not say which is the case since the classification whose existence is assumed in it is not unique. Using the axiom of choice we can, in fact, arrange things so that either of these two sets can be made large (and the other small). But even then there will be many sets whose fate we do not know. It is a little disturbing that the theory of hyperreal numbers we are going to develop should depend crucially on something we do not know completely. But that is a necessary evil one has to accept, whenever the axiom of choice is used.

We now fix any one classification given by the axiom and formally define a **hyperreal number** as an equivalence class of sequences of real numbers, under the equivalence relation arising out of this classification. We denote the set of all such hyperreal numbers by $I\!\!R^*$. On this set we define two basic binary operations, called the addition and multiplication. For convenience we denote them by the same usual symbols, viz. $+$ and $\cdot$, the latter often being suppressed from notation. Suppose $\alpha$ and $\beta$ are hyperreal numbers represented by the sequences $\{x_n\}$ and $\{y_n\}$ respectively. Then we define $\alpha + \beta$ and $\alpha \cdot \beta$ (or simply $\alpha\beta$ ) to be the hyperreal numbers represented by the sequences $\{x_n + y_n\}$ and $\{x_n y_n\}$ respectively. The proof that these operations are well-defined is standard and left as an exercise. Equally routine (to the point of being tedious) is the verification that these operations obey all the basic laws such as commutativity, associativity and distributivity. The only point where some care is perhaps necessary is in showing that every non-zero hyperreal number, say $\alpha$ has a reciprocal. To see this, suppose $\alpha$ is represented by a sequence $\{x_n\}$. The hyperreal number 0 is represented by the sequence $\{y_n\}$ where $y_n = 0$ for all $n \in I\!\!N$. (It would suffice, of course, if $y_n$ vanishes for all $n$ in some large subset of $I\!\!N$.) Similarly the hyperreal number 1 is represented by any sequence $\{w_n\}$ for which the set $\{n : w_n = 1\}$ is large. To say $\alpha \neq 0$ means that the set, say, $A = \{n : x_n = 0\}$ is small. We now define a sequence $\{z_n\}$ by $z_n = 0$ if $n \in A$ and $z_n = \dfrac{1}{x_n}$ if $n \in A'$. Let $\beta$ be the hyperreal number represented by $\{z_n\}$. Then $\alpha\beta$ is represented by the sequence $\{x_n z_n\}$ Clearly the set $\{n : x_n z_n = 1\}$ contains $A'$ and hence is large since $A$ is small. But then the sequences $\{x_n z_n\}$ and $\{w_n\}$ are equivalent and this means $\alpha\beta = 1$.

We already showed how the order relation could be defined for hyperreal numbers. The difficuly we had regarding the law of dichotomy is now gone since every subset of $I\!\!N$ is either large or small. Indeed, let $\alpha, \beta$ be hyperreal numbers represented by sequences $\{x_n\}, \{y_n\}$ respectively. Let $A = \{n : x_n \leq y_n\}$ and $B = \{n : y_n \leq x_n\}$. Because of the law of dichotomy for real numbers, $A \cup B = I\!\!N$. If neither $A$ nor $B$ is large, then both would be small. But then their union, $I\!\!N$ would also be small, a contradiction. If $A$ is large we have $\alpha \leq \beta$, while if $B$ is large then $\beta \leq \alpha$. (If both $A$ and $B$ are large then so is $A \cap B$ which would mean $\alpha = \beta$ since $A \cap B = \{n \in I\!\!N : x_n = y_n\}$ because of the antisymmetry property for real numbers.) This shows that the antisymmetry property for reals implies that for hyperreals, which is also true for most other properties.

Having defined the order relation for hyperreal numbers, we define the absolute value of a hyperreal number in exactly the same way as for real numbers. That is, given a hyperreal number $\alpha$ represented by a sequence $\{x_n\}$, we let $|\alpha|$

be $\alpha$ if $0 \leq \alpha$ and $-\alpha$ otherwise. Alternately, we can define $|\alpha|$ as the hyperreal number represented by the sequence $\{|x_n|\}$. It is easy to show that this is well-defined (i.e. independent of a particular representive of $\alpha$) and that the two ways of defining the absolute value are equivalent.

Let us now see how the real number system can be embedded into the hyperreal number system. Given any real number $x$, we consider the constant sequence $\{x_n\}$ whose every term is $x$ and denote by $x^*$ the equivalence class of this sequence. We call $x^*$ the copy of $x$ in $\mathbb{R}^*$. We already did this for the real numbers 0 and 1 and the numbers we got were precisely $0^*$ and $1^*$ respectively. They were also the zero and the unit element of $\mathbb{R}^*$. It is easily seen that the correspondence between $x$ and $x^*$ is one-to-one and preserves all the basic structure of $\mathbb{R}$. That is to say, for any two real numbers $x$ and $y$, we have $(x + y)^* = x^* + y^*, (xy)^* = (x^*)(y^*)$ and $x \leq y$ if and only if $x \leq y$. It is also trivial to check that the embedding of $\mathbb{R}$ into $\mathbb{R}^*$ is compatible with absolute values, that is, for every real number $x$, $|x^*| = (|x|)^*$. Verbally, it makes no difference whether you first embed in $\mathbb{R}$ and then take the absolute value or whether you first take the absolute value and then embed it in $\mathbb{R}$. Because of all these results, whatever manipulations we do in $\mathbb{R}$ will continue to hold if we do the same manipulations to the copy of $\mathbb{R}$ in $\mathbb{R}^*$. As a result, we often identify a real number $x$ with its copy $x^*$ in $\mathbb{R}^*$. This way we regard $\mathbb{R}$ as a subset of $\mathbb{R}^*$. A hyperreal number is called **standard** if it is a copy of some real number, i.e. it can be represented by a constant sequence of real numbers. (Note, however, that a standard real number can also be represented by a non-constant sequence. Indeed, we are free to arbitrarily change the terms whose indices come from any small subset of $\mathbb{N}$.)

Thus the hyperreal number system consists of the standard real numbers and some other 'non-standard real numbers.' We remark again that instead of the classification given by Axiom ( B.1), had we taken an atomic classification, then the hyperreal number system we would have got would have coincided with $\mathbb{R}$. That is, there would have been no non-standard real numbers. In the hyperreal number system we have constructed there are plenty of them and they include the elusive infinitesimals plus many other things. We begin by defining infinitesimals.

**B.2 Definition:** A hyperreal number $h$ is called **infinitesimal** if (i) it is positive (i.e. if $0 \leq h$ but $0 \neq h$ ) and (ii) it is less than every positive standard real number.

Note that if $h$ is a standard real number, then (i) and (ii) are mutually contradictory. So an infinitesimal cannot be a standard real number. This is just another way of saying that in the real number system there is no such thing as an infinitesimally small positive real number. Indeed it is precisely this inadequacy of $\mathbb{R}$ which led to the notoriously clumsy definition of a limit as explained in Section 2.3.

Let us see if things have improved with hyperreal numbers at hand. There is little difficulty in giving examples of infinitesimals. Indeed, let $h_1$ and $h_2$ be the hyperreal numbers represented by the sequences $\{\frac{1}{n}\}$ and $\{\frac{1}{n+1}\}$ respectively. Then $h_1$ is an infinitesimal. For clearly, $h_1 > 0$. Further, suppose $x$ is a standard positive real number, represented by a constant sequence $\{x\}$ where $x$ is a positive real number. Then there exists some $m \in \mathbb{N}$, such that $\frac{1}{n} < x$ for all $n \geq m$ by Theorem (3.3.1). But then the set $\{n : \frac{1}{n} < x\}$ is eventual and hence large. So $h_1 < x$. By a similar reasoning, $h_2$ (and more generally, any hyperreal number represented by a sequence of positive terms tending to 0) is an infinitesimal. Note however that $h_1 \neq h_2$ even though the sequence representing $h_1$ (viz. the sequence $\{\frac{1}{n+1}\}$ is obtained from the sequence $\{\frac{1}{n}\}$ (which represents $h_1$) by deleting only one term. In fact we have $h_2 < h_1$. By deleting more terms we get an infinite descending hierarchy of infinitesimals. There are infinitesimals which are still smaller, e.g. the one represented by the sequence $\{\frac{1}{n^2}\}$.

There are also hyperreal numbers which are infinitely large, that is, bigger than every standard real number. In fact the sequence $\{n\}$ represents one such number, while the sequence $\{n + 1\}$ represents one which is still bigger. It is easy to see that the latter equals the former plus the number 1. This is in sharp contrast with the imprecise but intuitively suggestive statemnet $\infty + 1 = \infty$. Just like infinitesimals, we have a hierarchy of infinitely large hyperreal numbers too. Actually, the two concepts are intimately related as the following theorem, whose proof is as predictable as its statement, shows.

**B.3 Theorem:** A (positive) hyperreal number $\alpha$ is an infinitesimal if and only if its reciprocal $\frac{1}{\alpha}$ is infinitely large.

**Proof:** Represent $\alpha$ by a sequence $\{x_n\}$ of real numbers. Let $A = \{n : x_n > 0\}$. Then $A$ is large since $\alpha > 0$. As shown earlier we can represent $\frac{1}{\alpha}$ by a sequence $\{y_n\}$ where $y_n = \frac{1}{x_n}$ for $n \in A$ and $y_n$ is arbitray for $n \notin A$. Now suppose $\alpha$ is an infinitesimal. Let $R$ be any positive standard real number. Then $\frac{1}{R}$ is a positive standard real number and so $\alpha < \frac{1}{R}$. But this means that the set $B = \{n : x_n < \frac{1}{R}\}$ is large. For all $n \in A \cap B, 0 < x_n < \frac{1}{R}$ and hence $y_n > R$ for all $n \in A \cap B$. As $A$, $B$ are both large, so is $A \cap B$, whence $\frac{1}{\alpha} > R$ proving that $\frac{1}{\alpha}$ is infinitely large. The converse follows essentially by reversing the argument.∎

This theorem gives a precise expression for the intuitively clear but meaningless statement that an infinitesimal is the reprocal of infinity. It also justifies the name 'infinitesimal'. Just as a decimal is the reprocal of ten (or 'deci' in Latin) an infinitesimal is the reciprocal of infinity.

Two hyperreal numbers $\alpha$ and $\beta$ are said to be **infinitely close** (or sometimes **infinitesimally close**) to each other if their difference $|\alpha - \beta|$ is either 0 (i.e. $\alpha = \beta$ ) or an infinitesimal. This is equivalent to saying that $|\alpha - \beta| < \epsilon$ for every positive standard hyperreal number $\epsilon$. A non-zero hyperreal number $\alpha$ is said to be infinitesimally small if it is infinitely close to 0, i.e. if $|\alpha|$ is an infinitesimal. Thus we now finally have a presise definition of this elusive term. Note that no two distinct standard real numbers can be infinitely close to each other. Indeed that is why this concept cannot be used to define a limit while staying within the real number system.

Let us now see if the infinitesimals do their promised job, viz., that of providing an alternate definition of a limit. Suppose $f : \mathbb{R} \longrightarrow \mathbb{R}$ is a function. We construct a new function $f^* : \mathbb{R}^* \longrightarrow \mathbb{R}^*$ which is an extension of $f$ in the sense that it coincides with $f$ for standard real numbers, that is, for every real number $x$, we have $f^*(x^*) = (f(x))^*$ where, $x^*$ and $(f(x))^*$ are copies of $x$ and of $f(x)$ in $\mathbb{R}^*$. Suppose $\alpha$ is a hyperreal nymber represented by a sequence $\{x_n\}$ (say) of real numbers. Let $\beta$ be the hyperreal number represented by the sequence $\{f(x_n)\}$. We define $f^*(\alpha)$ to be $\beta$. Again one has to show that this is independednt of the choice of the representative, i.e., if $\{y_n\}$ is some other sequence representing the hyperreal number $\alpha$, then $\{f(y_n)\}$ represents $\beta$. This is immediate from the definition of the equivalence relation we have defined for sequences. The function $f^*$ may be called the **non-standard extension** of $f$.

Thus every function $f : \mathbb{R} \longrightarrow \mathbb{R}$ gives rise to a function $f^* : \mathbb{R}^* \longrightarrow \mathbb{R}^*$. We are not claiming, nor is it true, that *every* function $g : \mathbb{R}^* \longrightarrow \mathbb{R}^*$ arises from some function from $\mathbb{R}$ to $\mathbb{R}$. A simple counterexample is a constant function $g : \mathbb{R}^* \longrightarrow \mathbb{R}^*$ which takes everything to some non-standard real number (e.g., the infinitesimal represented by the sequence $\{\frac{1}{n}\}$). This need not worry us. Our concern is not to study all possible functions from $\mathbb{R}^*$ to $\mathbb{R}^*$ but rather to show how certain concepts and properties of functions from $\mathbb{R}$ to $\mathbb{R}$ can be translated very succinctly in terms of certain concepts and properties of their extensions.

The following theorem shows how the most fundamental concept of a limit of a function translates in terms of infinitesimals.

**B.4 Theorem:** Suppose $f : \mathbb{R} \longrightarrow \mathbb{R}$ is a function and $c, L$ are some real numbers. Then $\lim_{x \to c} f(x) = L$ if and only if for every (non-zero,) infinitesimally small hyperreal number $\alpha$, the hyperreal number $f^*(c + \alpha) - f^*(c)$ is either 0 or infinitesimally small.

**Proof:** Assume first that $\lim_{x \to c} = L$. Let $\alpha$ be an infinitesimally small number. Represent $\alpha$ by some sequence $\{x_n\}$. The standard real number $c$ can be represented by the constant sequence $\{c\}$. Then $c + \alpha$ is represented by the sequence $\{c + x_n\}$. The hyperreal numbers $f^*(c + \alpha)$ and $L$ are represented, respectively, by the sequence $\{f(c + x_n)\}$ and the constant sequence $\{L\}$ and hence finally the sequence $\{|f(c + x_n) - L|\}$ represents the hyperreal number $|f^*(c + \alpha) - L|$. We have to show that this number is less than every positive standard hyperreal number. Let $\beta$ be a given positive standard hyperreal number. Then we may represent $\beta$ by a constant sequence of the form $\{\epsilon\}$

where $\epsilon$ is a positive real number. We are given that $\lim_{x \to c} f(x) = L$. Hence there exists $\delta > 0$ such that $|f(x) - L| < \epsilon$ whenver $0 < |x - c| < \delta$. We treat $\delta$ as a standard positive hyperreal number represented by the constant sequence $\{\delta\}$. Since $|\alpha|$ is an infinitesimal, the set, say, $A = \{n : 0 < |x_n| < \delta\}$ is large. For every $n \in A, |f(c + x_n) - L| < \epsilon$. Clearly $A = B \cup C$ where $B = \{n \in A : |f(c + x_n) - L| = 0\}$ and $C = \{n \in A : 0 < |f(c + x_n) - L| < \epsilon\}$. Since $A$ is a large set, at least one (and in fact, exactly one) of the sets $B$ and $C$ is large. If $B$ is large then $f^*(c + \alpha) - L$ is 0 while if $C$ is large then it is infinitesimally small as was to be proved.

Conversely, suppose that the given condition holds. We have to show that $\lim_{x \to c} f(x) = L$. We argue by contradiction. The argument will resemble the proof of the converse implication in Theorem (3.1.1). A proof using that theorem is in fact possible, but we prefer to give a direct argument. Let a real $\epsilon > 0$ be given. Suppose we are unable to find a $\delta$ answering the definition of $\lim_{x \to c} f(x) = L$. This means for every $\delta > 0$, there exists some $x$ (depending on $\delta$) such that $0 < |x - c| < \delta$ but $|f(x) - L| \geq \epsilon$. In particular, for every $n \in I\!N$, (taking $\delta$ as $\frac{1}{n}$) there exists some real number $x_n$ such that $0 < |x_n - c| < \frac{1}{n}$ and $|f(x_n) - L| \geq \epsilon$. Now let $y_n = x_n - c$ and $\alpha$ be the hyperreal number represented by the sequence $\{y_n\}$. From its very construction, it is clear that $\alpha$ is infinitesimally small. So by our hypothesis, the hyperreal number $f^*(c + \alpha) - L$ (which we call $\beta$) is either 0 or infinitesimally small. This hyperreal number is represented by the sequence $\{f(c + y_n) - L\}$ i.e., by the sequence $\{f(x_n) - L\}$. We are given that $|\beta|$ is either 0 or an infinitesimal. In either case, the set $\{n : |f(x_n) - L| < \epsilon\}$ is a large set. But by the very choice of $x_n$, this set is the empty set, which cannot be large. This contradiction proves that $\lim_{x \to c} f(x) = L$. ∎

Thus the clumsy $\epsilon$-$\delta$ definition of a limit can be replaced by one based on infinitesimals. (It may be argued that $\epsilon$ does enter into the definition of an infinitesimal, for when we say that a sequence $\{x_n\}$ of positive real numbers represents an infinitesimal, it means precisely that for every $\epsilon > 0$, the set $\{n : x_n < \epsilon\}$ is large. But this is not as bad as the expression 'for every $\epsilon > 0$, there exists a $\delta > 0$, such that for every ....' which is not only linguistically clumsy but also introduces too many quantifiers whose order is vital and necessitates certain care which often alienates the average student. ) Other basic concepts such as continuity and derivatives can also be stated very succinctly in terms of infinitesimals. Indeed the theorem above immediately implies that a function $f : I\!R \longrightarrow I\!R$ is continuous at $c$ if and only if it has the property that whenever $x$ is infinitely close to $c$, $f(x)$ is infinitely close to $f(c)$. In this formulation, the proof of the continuity of the composite of two continuous functions takes a very natural form, being a clear case of syllogism. (See Exercises (2.3.6) and (4.1.5). To appreciate the point involved here, recall that as emphasized in Section 2.3, the sentence '$x$ tends to $c$' has no meaning of its own. But the statement '$x$ is infinitesimally close to $c$' is perfectly meaningful.)

For derivatives we have to do a little more work. It can then be shown that if $f$ is differentiable at $c$ then, for every non-zero, infinitesimally small hyperreal number $\alpha$, the ratio $\dfrac{f^*(c+\alpha) - f(c)}{\alpha}$ equals $f'(c)$. In other words, the derivative is the incrementary ratio when the increment in the independent variable is infinitesimally small. We can then say, quite rigorously that the instantaneous speed is the average speed over an infinitesimally long time interval. Concepts such as 'line elements', or differentials can also be given presice meanings in terms of the concept of an infinitesimal. But we shall not go into it as we have limited ourselves only to the definition of a limit.

It is unlikely that this new approach will soon enter the calculus textbooks. First the maturity needed to handle it (especially if we want to derive Axiom (B.1) from the axiom of choice) is considerable. Another disturbing feature, as already pointed out is the uncertainty inherent in the axiom of choice. Note for example, that all we know from Axiom (B.1) is that either the set $A$ of all even positive integers or its complement $A'$, viz. the set of all odd positive integers is large. Depending upon which one of them is large, the hyperreal number represented by the sequence $\{(-1)^n\}$ will equal the standard real number 1 or the standard real number $-1$. But we cannot say with certainty which is the case. This type of uncertainty is different from, say, our inability to tell at present if the Euler's constant $\gamma$ (see Section 5.8) is rational or irrational. The distinction is that the rationality or otherwise of $\gamma$ is *unknown* at present but not permanantly *unknowable* as is the case with the answers to many questions arising out of the axiom of choice.

Even then, the mere knowledge that a rigorous treatment of infinitesimals is possible has a certain pedagogical value. The discipline of mathematics requires that every term that you use (other than the primitive terms) be first defined. In an elementary course, this discipline can hardly be followed to the hilt. So it is common to use a few terms without clear-cut definitions (and a few deep theorems without rigorous proofs). As long as we are assured that such terms (or theorems) *have* definitions (or proofs), omitting the same is an intentional compromise for practical reasons and not a lacuna which cuts at the very root.

# ANSWERS TO EXERCISES

These answers are not to be looked at as complete solutions. They are merely meant to help the reader *after* he has honestly tried a problem. If he feels he has a correct solution, he may compare it with the answer here. If the two match, most likely he is right. If they don't, he is advised to check his solution again and/or to discuss it with others. If he still believes the answer given here is incorrect, he is urged to write to the author. In the case of thought-oriented problems, the answers given are often in the form of extended hints. Again, if the reader believes there is a flaw in the answer given here or that he has a more elegant solution, the author would appreciate hearing from him.

In case the reader is unable to do a problem, he is strongly urged to resist the temptation to give up too soon and look up the answer. Keep trying. Do a few special cases of the problem to see if they give any clues. Discuss the problem with others. If nothing comes up, simply leave the problem temporarily. A flash may come after a few days. (In the personal experience of the author, some of the problems took several months.) Look up the solution only when you are convinced it is not worth your time to try further.

No answers are provided in some of the following cases:

(i) Where the problem asks for a straightforward verification of something or for a routine calculation

(ii) where the problem is intentionally vague as it asks the reader to give examples of a general nature or make comments.

(iii) where the hint given is either sufficiently detailed or sufficiently incisive to reduce the rest of the work to a routine.

# CHAPTER 1

## Section 1.1

1.  First take the goat, then the wolf. Bring back the goat. Then take the hay and finally the goat.

2.  First note that if the fake coin is known to be among three coins and also whether it is lighter or heavier than the rest, then it can be detected in just one weighing. Call this the finishing step. Now number the coins as $C_1, C_2, \ldots, C_{12}$. For $i = 1, \ldots, 12$, let $H_i$ be the statement '$C_i$ is heavier than the rest' and let $L_i$ be the statement '$C_i$ is lighter than the rest.' Exactly one of these 24 statements is true and we have to find which one it is. Begin by weighing $C_1, C_2, C_3, C_4$ in one pan against $C_5, C_6, C_7, C_8$ in another.

    > *Case 1:* The first pan is heavier. Then the true statement is in the set $\{H_1, H_2, H_3, H_4, L_5, L_6, L_7, L_8\}$. Take $C_1, C_2, C_3$ to the second pan, remove $C_5, C_6, C_7$ from the second pan, put $C_9, C_{10}, C_{11}$ in the first pan and weigh again. If the second pan is heavier then one of $H_1, H_2, H_3$ is true and apply the finishing step. If the two pans balance then one of $L_5, L_6, L_7$ is true and apply the finishing step. If the first pan is heavier then either $H_4$ or $L_8$ is true. Weigh $C_4$ (or $C_8$) against any other $C_i$.

    > *Case 2:* The second pan is heavier. Similar to case 1.

    > *Case 3:* The two pans are equal. Then the true statement is in $\{H_9, H_{10}, H_{11}, H_{12}, L_9, L_{10}, L_{11}, L_{12}\}$. Do the same shifting as in Case 1. Unequal pans lead to the finishing step. Otherwise the fake coin is $C_{12}$. Weigh against any $C_i$ to see if $H_{12}$ or $L_{12}$ holds.

3.  Take different numbers of tablets from different machines and weigh together. If the total weight falls short by $m$ grams, then the machine from which $m$ tablets were taken is faulty.

4.  For a purely geometric solution, note that $\triangle AED$ and $\triangle ADB$ are similar with $F$ and $M$ being mid-points of corresponding sides. So $\angle MAD = \angle FAE$ and hence $\angle FAM = \angle EAD$. Also $\triangle AEF$ and $\triangle ADM$ are similar, giving $\dfrac{AF}{AM} = \dfrac{AE}{AD}$. So $\triangle FAM$ and $\triangle EAD$ are similar. But then $\angle AFM =$

$\angle AED = 90°$. The solution is completed by noting that $FM$ is parallel to $BE$.

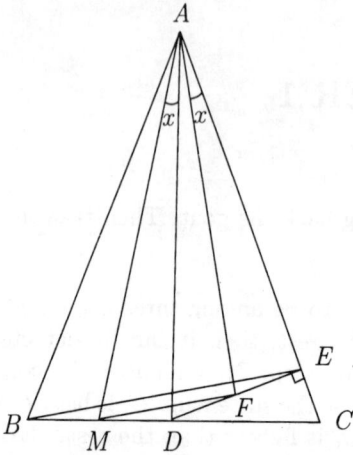

Exercise (1.4)

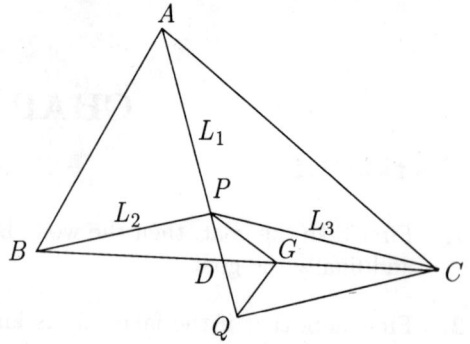

Exercise (1.5)

For a solution using coordinates, choose axes so that $D = (0,0), C = (a,0), B = (-a,0), A = (0,b)$. Let $F = (h,k)$. Then $E = (2h,2k)$. Perpendicularity of $DE$ and $AC$ gives $ah = bk$. Also since $E$ lies on $AC$, we get $a(2k - b) + 2bh = 0$.. A little calculation now gives $\frac{2k-b}{2h} \frac{2k}{2h+a} = -1$ which means $AF \perp BE$.

5. Let the perpendiculars from $A, B$ and $C$ to $BC, CA$ and $AB$ lie along the lines $L_1, L_2$ and $L_3$ respectively, meeting at $P$. Let the perpendiculars from $A', B', C'$ to $B'C', C'A', A'B'$ lie along $L_1', L_2', L_3'$ respectively. Let the line through $C$ parallel to $L_2$ meet $L_1$ at $Q$ and let the line through $Q$ parallel to $AB$ meet $BC$ at $G$. Let $L_1$ meet $BC$ at $D$.

Similarity of $\triangle$'s $PDB$ and $QDC$ gives $PD.DC = DQ.BD$. But similarity of $\triangle$'s $BDA$ and $GDQ$ gives $BD.DQ = DA.DG$. Hence $PD.DC = DA.DG$ or $\dfrac{DP}{DG} = \dfrac{DA}{DC}$, whence $PG//AC$.

To finish the proof show that $\triangle A'B'C'$ is similar to $\triangle CPQ$ and further that its vertical angles are divided by $L_1', L_2', L_3'$ exactly the same way as the vertical angles of $\triangle CPQ$ are divided by $CG, PG, QG$. Since $CG, PG, QG$ are concurrent so must be $L_1', L_2', L_3'$.

For a proof using co-ordinates, let the coordinates of $A, B, C, A', B', C'$ be, respectively, $(x_1, y_1), (x_2, y_2), (x_3, y_3), (x_1', y_1'), (x_2', y_2'), (x_3', y_3')$. Then the equation of $L_1$ is $(x - x_1)(x_2 - x_3) + (y - y_1)(y_2 - y_3) = 0$. Similarly write the equations of $L_2, L_3, L_1', L_2', L_3'$. Concurrency of $L_1, L_2, L_3$ implies that the determinant

$$\begin{vmatrix} x_2' - x_3' & y_2' - y_3' & x_1(x_2' - x_3') + y_1(y_2' - y_3') \\ x_3' - x_1' & y_3' - y_1' & x_2(x_3' - x_1') + y_2(y_3' - y_1') \\ x_1' - x_2' & y_1' - y_2' & x_3(x_1' - x_2') + y_3(y_1' - y_2') \end{vmatrix}$$

vanishes. Adding rows, this is equivalent to the vanishing of the sum $\sum x_1(x_2' - x_3') + y_1(y_2' - y_3')$. After re-arranging, this sum equals the sum $-(\sum x_1'(x_2 - x_3) + y_1'(y_2 - y_3))$, whose vanishing implies the vanishing of a similar determinant, proving concurrency of $L_1', L_2', L_3'$. (In problems like this $\sum$ is used to denote a sum of three terms, of which one is given and the other two are obtained from it by cyclic permutation of the indices. For example, $\sum x_1 y_2'$ means $x_1 y_2' + x_2 y_3' + x_3 y_1'$.)

A proof using vectors is the shortest. Let $\mathbf{a}, \mathbf{b}, \mathbf{c}, \mathbf{a}', \mathbf{b}', \mathbf{c}'$ be the position vectors of $A, B, C, A', B', C'$ from $P$. Since $L_1$ is perpendicular $B'C'$, we get $\mathbf{a}.(\mathbf{b}' - \mathbf{c}') = 0$, or $\mathbf{a}.\mathbf{b}' = \mathbf{a}.\mathbf{c}'$. Similarly $\mathbf{b}.\mathbf{a}' = \mathbf{b}.\mathbf{c}'$ and $\mathbf{c}.\mathbf{a}' = \mathbf{c}.\mathbf{b}'$. Now assume $L_1', L_2'$ meet at $R$. Let $\mathbf{r}$ be the position vector of $R$ (from $P$). Since $A', R$ lie on $L_1'$ which is $\perp$ to $BC$ we get $(\mathbf{a}' - \mathbf{r}).(\mathbf{b} - \mathbf{c}) = 0$, or $\mathbf{r}.(\mathbf{b} - \mathbf{c}) = \mathbf{a}'.\mathbf{b} - \mathbf{a}'.\mathbf{c}$. Similarly $\mathbf{r}.(\mathbf{c} - \mathbf{a}) = \mathbf{b}'.\mathbf{c} - \mathbf{b}'.\mathbf{a}$. Adding and using the earlier equations, $\mathbf{r}.(\mathbf{a} - \mathbf{b}) = \mathbf{c}'.(\mathbf{a} - \mathbf{b})$, whence $RC' \perp AB$.

## Section 1.2

1.  The assertion is trivial for $n = 2$. So assume $n > 2$ and consider all possible (non-degenerate) triangles with vertices among $P_1, P_2, \ldots, P_n$. Among these let $P_1 P_2 P_3$ (say) be one with the smallest shortest altitude. Let this altitude pass through $P_1$. Then the line through $P_2$ and $P_3$ can contain no other $P_i$; as otherwise two points on this line will be on the same side of the altitude through $P_1$ and, will form, along with $P_1$, a triangle with a shorter altitude.

2.  Without loss of generality, let the given segment, say $OA$, be of unit length.

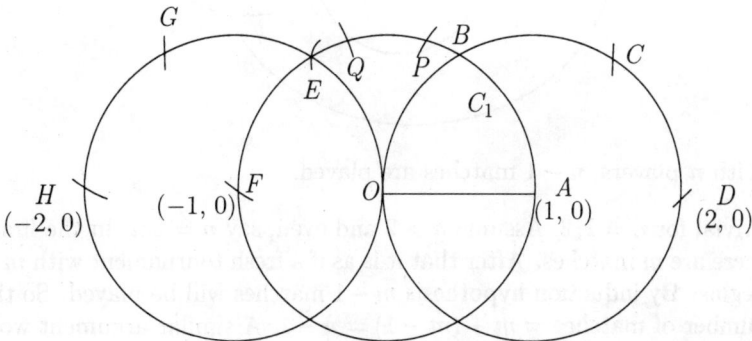

Draw a circle of radius 1 with $A$ as the centre and starting from $O$, mark off point $B, C, D$ with $OB = BC = CD = 1$. Similarly draw the unit circle, $C_1$, with centre at $O$ and get $E$ and $F$ on it with $AB = BE = EF$.

Finally draw a unit circle with $F$ as the centre and get $G, H$ on it with $OE = EG = GH = 1$. It is easily seen that taking $O$ as the origin and $A$ as $(1, 0)$, the points $D, H$ are $(2, 0)$ and $(-2, 0)$ respectively.

Now let the circles of radii 2 each with $D, H$ as centres meet $C_1$ at $P, Q$ respectively (all points of intersections are to be taken on the positive side of the $x$-axis). A little calculation gives $P = \left( \dfrac{1}{4}, \dfrac{\sqrt{15}}{4} \right)$ and $Q = \left( -\dfrac{1}{4}, \dfrac{\sqrt{15}}{4} \right)$.

So $PQ$ is a segment of length $\dfrac{1}{2}$. The circle with $O$ as the centre and radius equal to $PQ$ will cut the segment $OA$ at its midpoint.

## Section 1.3

**1.**   See Exercise (5.8.12).

## Section 1.4

**2.**   The prime 2 and all primes of the form $4n + 1$ where $n$ is a positive integer are expressible as sums of two perfect squares.

**3.**   Four colours will always suffice. A map for which three do not suffice is shown below.

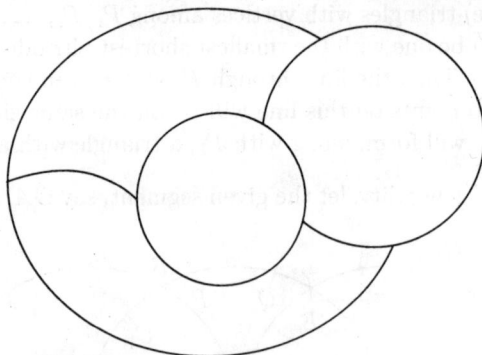

**4.**   With $n$ players, $n - 1$ matches are played.

**5.**   Trivial for $n = 1, 2$. Assume $n > 2$ and even, say $n = 2m$. In the first round there are $m$ matches. After that it is as if a fresh tournament with $m$ players begins. By induction hypothesis $m - 1$ matches will be played. So the total number of matches $= m + (m - 1) = n - 1$. A similar argument works if $n$ is odd.

**6.**   Every match has exactly one loser and every player, except the champion, loses exactly one match.

7. As the roots are all non-zero, $a_0 \neq 0$. Now, $a_0 x^n + a_1 x^{n-1} + \cdots + a_n = x^n f\left(\dfrac{1}{x}\right) = a_n(1 - \alpha_1 x) \ldots (1 - \alpha_n x)$. The roots of this polynomial are $\dfrac{1}{\alpha_1}, \ldots, \dfrac{1}{\alpha_n}$ and their sum equals $-\dfrac{a_1}{a_0}$ as seen by equating the coefficient of $x^{n-1}$ in the equation

$$a_0 x^n + a_1 x^{n-1} + \ldots + a_n = a_0 \left(x - \frac{1}{\alpha_1}\right)\left(x - \frac{1}{\alpha_2}\right) \ldots \left(x - \frac{1}{\alpha_n}\right).$$

8. The proof follows from the power series expansion of $\sin x$ as $\displaystyle\sum_{n=0}^{\infty} \frac{(-1)^n x^{2n+1}}{(2n+1)!}$ and other properties of the sine function (see Sections 4.7 and 4.8), specifically that $\sin y = 0$ iff $y = n\pi, n$ an integer.

9. Here $a_0 = 1, a_1 = -\dfrac{1}{6}$ and the roots are $(n\pi)^2, n = 1, 2, 3, \ldots$.

   So $\displaystyle\sum_{n=1}^{\infty} \frac{1}{n^2 \pi^2} = \frac{1}{6}$, i.e. $\displaystyle\sum_{n=1}^{\infty} \frac{1}{n^2} = \frac{\pi^2}{6}$.

## Section 1.5

1. Suppose $\triangle ABC \cong \triangle PQR$. To prove that their areas are equal, the intuitive proof in schools is to cut $\triangle PQR$ and place it on $\triangle ABC$ so that their points match exactly. A rigorous proof requires, first of all, a rigorous definition of area, see Section 6.3. For falsity of the converse consider two triangles with the same base and equal heights but with different verical angles.

2. Draw diagonal $BD$ of the quadrilateral $ABCD$. Congruency of the triangles ABD and CDB is the essential middle step in both the direct and the converse implications. In the direct implication it is proved by $A - S - A$ criterion and then $S - S - S$ is derived as a consequence. In the converse implication it is the other way.

3. Let the angle bisector of $\angle A$ meet $BC$ at $D$. In the converse implication the perpendicularity of $AD$ with $BC$ has to be established to prove congruency of $\triangle$'s $ADB$ and $ADC$. This is not necessary in the direct implication since the $S - A - S$ criterion is applicable.

4. For the direct implication, $\cos A, \cos B$ and $\cos C$ equal $\dfrac{1}{2}$ each. For the converse, the trigonometric identities $1 - \cos A = 2\sin^2 \frac{A}{2}$ and $\cos B + \cos C = 2\cos \frac{B+C}{2} \cos \frac{B-C}{2} = 2\sin \frac{A}{2} \cos \frac{B-C}{2}$ reduce the given condition to a quadratic in $\sin \frac{A}{2}$, viz., $\sin^2 \frac{A}{2} - \cos \frac{B-C}{2} \sin \frac{A}{2} + \frac{1}{4} = 0$. Existence of a real root implies $\cos^2 \frac{B-C}{2} \geq 1$ which forces $B = C$. Similarly $A = B$.

## Section 1.6

1. (i) The second statement is wrong. The circle on which $A, B, C, D$ lie may not be the one with $AC$ as a diameter.

   (ii) $\infty - \infty$ has no meaning. The assertion is correct and follows from $\sec x - \tan x = \dfrac{1}{\sec x + \tan x}$. As $x \to \dfrac{\pi}{2}, \sec x + \tan x \to \infty + \infty = \infty$.

   (iii) The implied justification that the limit of a sum of terms equals the sum of their limits is valid only when the sum involves a fixed number of terms. Here the number of terms summed increases with $n$. In fact, $\displaystyle\sum_{k=1}^{n} \frac{1}{n+k} \geq \sum_{k=1}^{n} \frac{1}{2n} = \frac{n}{2n} = \frac{1}{2}$ and so the limit, if it exists, has to be at least $\dfrac{1}{2}$. (The correct limit is evaluated in Exercise (5.7.19) (i).)

   (iv) See the beginning of Section 4.5.

   (v) Just because $AB$ and $AC$ have the same number of points, they need not have the same lengths. The length of a segment (or of any curve for that matter) is not the sum total of the lengths of its points. In fact the 'length' of every point is 0. If we want to show that $AB$ and $AC$ have the same lengths by establishing a one-to-one correspondence between them, then such a correspondence must not only be one-to-one, but also length preserving, which is not the case here.

## Section 1.8

1. If $x \neq y$, then $\{x, y\} = \{y, x\}$ but we want to distinguish $(x, y)$ from $(y, x)$. So we must add something to indicate which is the first element. (The second element would also do, of course. But once we make the choice, we have to stick to it.) If $x = y$, then $(x, x) = \{\{x\}, x\}$ which differs from both $x$ and $\{x\}$.

2. It is a function whose domain is the set $\{1, 2, \ldots, n\}$.

3. A finite sequence is defined as a certain function. But to define a function one needs the concept of an ordered pair. So although an ordered pair can be identified with a 2-tuple, it cannot be *defined* that way.

4. The terms 'array', 'row' and 'column' are intuitively clear but not mathematically defined. A correct definition of an $m \times n$ matrix is that it is a function whose domain is the set $\{1, 2, \ldots, m\} \times \{1, 2, \ldots, n\}$.

5. (i) and (iii) are mathematically imprecise. (ii) is wrong for two reasons. First it would make any two intersecting lines tangents to each other. Secondly, a 'true' tangent at a point may intersect the curve at some other point away from it.

## Section 1.9

1.  Let $g_1$ be a girl in the hint, $b_1$ a boy $g_1$ does not dance with and $g_2$ a girl $b_1$ dances with. Then there has to be a boy $b_2$ who dances with $g_1$ but not with $g_2$, as otherwise $B_{g_1}$ will be a proper subset of $B_{g_2}$. If $B$ is infinite, then some girls could be dancing with infinitely many boys and the hint is meaningless.

2.  For the first part, if the assertion is false then starting from any $B_{i_1}$, there would be a strictly ascending infinite chain of subsets
    $B_{i_1} \subsetneqq B_{i_2} \subsetneqq B_{i_3} \subsetneqq \ldots \subsetneqq B_{i_m} \subsetneqq \ldots$ which is impossible as there are only finitely many $B_i$'s. Now let $g_1, \ldots, g_k$ be the girls in the dance problem. Let $B_i = B_{g_i}, i = 1, \ldots, k$. If the assertion of the problem fails then there is some $r$ such that $B_i \subset B_r$ for all $i = 1, \ldots, k$. The solution to be last exercise now works, starting from $g_r$.

3.  Let $B = \{b_1, b_2, \ldots, b_n, \ldots, \}$ and $G = \{g_1, g_2, \ldots, g_n\}$. Let $g_i$ dance with $b_j$ for all $j \leq i$.

4.  Call the persons $P_1, \ldots, P_5$. Let $L$ be the set of all locks and $L_i$ be the subset of those which $P_i$ can open. Let $L_i'$ be the complement of $L_i$. The data implies, that for every $i \neq j, L_i' \cap L_j' \neq \emptyset$, but for every three distinct $i, j, k, L_i' \cap L_j' \cap L_k' = \emptyset$. For every $i \neq j$, let $x_{ij} (= x_{ji})$ be a lock which neither $P_i$ nor $P_j$ can open but which everybody else can open. This gives a system of 10 locks, with 3 keys to each lock, and each $P_i$ having 6 keys.

5.  Only (iii) needs some proof. Let $A = \{i : x_i \neq y_i\}, B\{i : y_i \neq z_i\}$ and $C = \{i : x_i \neq z_i\}$. Then $C \subset A \cup B$ and so $|C| \leq |A| + |B|$. But $|A|, |B|$ and $|C|$ are precisely, $d(\vec{x}, \vec{y}), d(\vec{y}, \vec{z})$ and $d(\vec{x}, \vec{z})$ respectively.

6.  $d(\vec{x}, \vec{y}) = r$ means $x_i \neq y_i$ for $r$ values of the index $i$. These $r$ values can be chosen in $\binom{10}{r}$ ways. Each choice gives a unique $\vec{y}$ because each entry has only two possible values. The answer follows by adding $\binom{10}{0}, \binom{10}{1}$ and $\binom{10}{2}$.

7.  Each student's answerbook is a binary sequence of length 10. For two such answerbooks $\vec{x}$ and $\vec{y}$, if less than 6 answers match then $d(\vec{x}, \vec{y}) \geq 5$. Hence by the triangle inequality the two sets $\{\vec{z} : d(\vec{x}, \vec{z}) \leq 2\}$ and $\{\vec{z} : d(\vec{y}, \vec{z}) \leq 2\}$ would be disjoint. These sets have 56 elements each. If this happens for every pair of students then there would be 20 mutually disjoint sets with 56 elements each. But the total number of binary sequences of length 10 equals $2^{10}$ which is less than $56 \times 20$.

8.  (i) and (ii) are trivial. For (iii), depending on the signs of $a$ and $b$, $|a + b|$ equals $a+b$, or $a-b$, or $-a+b$ or $-a-b$. The inequalities $a \leq |a|, -a \leq |a|, b \leq |b|$ and $-b \leq |b|$ imply the result in all four cases. For the second version of the triangle inequality, note that $a = (a - b) + b$ gives $|a| \leq |a - b| + |b|$ and hence, $|a| - |b| \leq |a - b|$. Similarly, $|b| - |a| \leq |b - a| = |a - b|$. The result follows since $| |a| - |b| |$ equals either $|a| - |b|$ or $|b| - |a|$.

**9.** The three conditions follow, respectively, from (i), (ii), (iii) of the last exercise. (For the triangle inequality, apply (iii) after taking $a = x - y$ and $b = y - z$.) Geometrically, to say that $|x - a| < |x - b|$ means $x$ is closer to $a$ than to $b$. So $x$ has to lie on the same side of their mid-point $\frac{a+b}{2}$ as $a$. Similarly, the second statement means that the distance between $x$ and $c$ cannot exceed $\delta$ and so $x$ must lie between $c - \delta$ and $c + \delta$. (Formal proofs of these statements would run into a number of cases depending upon the signs of $x - a, x - b, x - c$.)

**10.** Positivity and symmetry are immediate. The triangle inequality follows from that for each co-ordinate separately (i.e. $|x_1 - x_3| \leq |x_1 - x_2| + |x_2 - x_3|$ and similarly fot the other coordinate). $D$ consists of all points $(x, y)$ for which $|x| + |y| \leq 5$. This is a square bounded by the straight line $x + y = \pm 5$ and $x - y = \pm 5$. The set $S$ consists of points $(x, y)$ for which $|x| + |y| = |x - 3| + |y - 2|$. Its determination runs into 9 cases (as there are three possibilities for $x$, viz., $x < 0, 0 \leq x \leq 3$ and $3 < x$ and similarly there are three possibilities for $y$). Six of these cases give no points of $S$ (e. g. when $x < 0$ and $y < 0$ we have $-x - y = 5 - x - y$ which is impossible). Two of the remaining cases give infinite vertical rays while the case $0 \leq x \leq 3$; $0 \leq y \leq 2$ gives the line segment from $(\frac{1}{2}, 2)$ to $(\frac{5}{2}, 0)$.

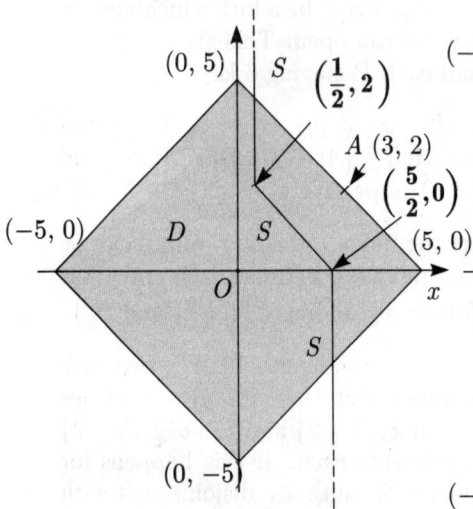

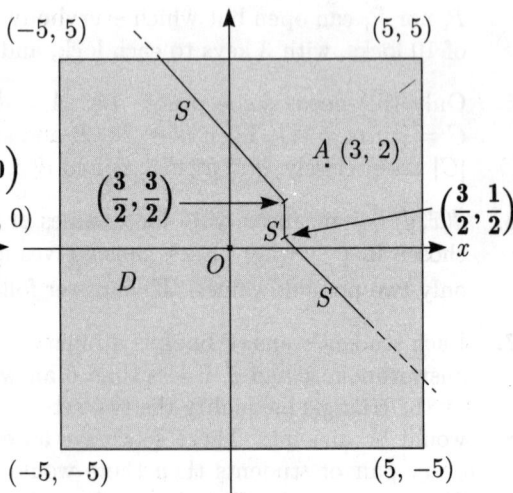

Exercise (9.10)                                    Exercise (9.11)

**11.** For the triangle inequality, note that $d((x_1, y_1), (x_3, y_3))$ equals either $|x_1 - x_3|$ or $|y_1 - y_3|$. Apply triangle inequalities (for absolute values). The set $D$ consists of points $(x, y)$ for which $|x| \leq 5$ and $|y| \leq 5$. This is a square with vertices at $(\pm 5, \pm 5)$. The set $S$ consists of points $(x, y)$ for which $\max\{|x|, |y|\} = \max\{|x - 3|, |y - 2|\}$. This runs into four cases depending upon which of the members in each set is greater. The case $|y| \geq |x|$ ; $|y - 2| \geq$

$|x - 3|$ is vacuous for in this case we must also have $|y| = |y - 2|$ which is possible only for $y = 1$. But then we get $|x| \leq 1$ and also $|x - 3| \leq 1$ contradicting the triangle inequality. Of the remaining three cases, the case $|y| \leq |x|$ ; $|y - 2| \leq |x - 3|$ leads to $|x| = |x - 3|$ i.e., $x = \frac{3}{2}$. Further, by exercise (9.9), $|y| \leq \frac{3}{2}$ and $|y - 2| \leq \frac{3}{2}$ together give $\frac{1}{2} \leq y \leq \frac{3}{2}$. The other two cases are a little more complicated. If $|x| \leq |y|$ and $|y - 2| \leq |x - 3|$ then $|y| = |x - 3|$ implies $x = 3 \pm y$. Further, $|x| \leq |x - 3|$ and $|y - 2| \leq |y|$ imply, respectively, $x \leq \frac{3}{2}$ and $y \geq 1$, again by Exercise (9.9). So we must have $x = 3 - y$ and not $x = 3 + y$. This gives a ray of slope $-1$ starting at $(\frac{3}{2}, \frac{3}{2})$. The fourth case, $|y| \leq |x|$ ; $|x - 3| \leq |y - 2|$ is handled similarly and gives the ray of slope $-1$ starting at $(\frac{3}{2}, \frac{1}{2})$.

## Section 1.10

1.  Let $F, A, C$ denote the sets of friends, actors and cricketers. The statements '$\exists\, x \in F$, s.t. $x \in A$' and '$\exists\, x \in F$, s.t. $x \in C$' do not together imply '$\exists\, x \in F$, s.t. $x \in A \cap C$'. Note that here $x$ is like a dummy variable and could be replaced by any other variable without changing the statement. If we rewrite the second statement as '$\exists\, y \in F$, s.t. $y \in C$' then there would be no confusion.

2.  Let $S$ be the set of all positive real numbers. The first statement reads " $\forall\, x \in S, \exists\, y \in S$ s.t. $y < x$" while the second says " $\exists\, y \in S$ s.t. $\forall\, x \in S, y < x$". Because of the change of order of quantifiers, the two statements are not equivalent. The second is stronger than the first.

## Section 1.11

1.  (i) There is a man who is rich and intelligent. (ii) There is a man who is rich but not intelligent. (iii) John is either poor or dumb. (iv) John is poor and dumb. (v) John is either (poor and dumb) or (rich and intelligent). (vi) For every woman there is a man she does not love. (vii) There exists a man whom every woman hates. (Hate $\equiv$ not love!) (viii) There exists a man such that for every woman there is a time when he asks her to dance and she refuses. (Less mechanically, there is a man with whom every woman refuses to dance sometimes.) (ix) There exists $\epsilon > 0$, such that for every positive integer $m$, there exists $n \geq m$ and $x \in S$ such that $|f_n(x) - f(x)| \geq \epsilon$. (x) For every $L$ there exists $\epsilon > 0$ such that for every $m$, there exists $n \geq m$ such that $|a_n - L| \geq \epsilon$.

2.  No, no, yes, no.

3.  'Half the men' refers to different men in the two statements. Indeed, the sets to which it refers are mutually disjoint and hence there is no contradiction.

4.  If in $\triangle ABC, \angle B > \angle C$ then $AB = AC$ would give $\angle B = \angle C$ while $AC < AB$ would give $\angle C > \angle B$. So $AC > AB$.

## Section 1.12

1. Let $(h, k)$ be the centre of the circle, say $C$. If $(x_1, y_1)$, $(x_2, y_2)$ and $(x_3, y_3)$ are three distinct rational points on $C$, express $(h, k)$ as the point of intersection of two lines of the form $a_1 x + b_1 x + c_1 = 0$ and $a_2 x + b_2 y + c_2 = 0$ where $a_1, b_1, c_1, a_2, b_2, c_2$ are all rational. Then $(h, k)$ is rational. But then for all positive integers $m$ and $n$, $(h + \frac{m^2 - n^2}{m^2 + n^2}, k + \frac{2mn}{m^2 + n^2})$ is a rational point on $C$.

2. Otherwise $4k - 1$ will factor as $(4n_1 + 1)(4n_2 + 1) \ldots (4n_r + 1)$ where $n_1, \ldots, n_r$ are (not necessarily distinct) positive integers. This product is of the form $4m + 1$. But $4k - 1 = 4m + 1$ implies $2(k - m) = 1$, a contradiction.

3. If $p_1, p_2, \ldots, p_r$ are primes of the form $4n - 1$, the prime factors of $4p_1 \ldots p_r - 1$ are all different from $p_1, \ldots, p_r$. At least one of them is of the form $4n - 1$.

4. The point $\left( \dfrac{\sum (x_1 x_2 + y_1 y_2)(y_2 - y_1)}{\sum x_1 y_2 - x_2 y_1}, \dfrac{\sum (x_1 x_2 + y_1 y_2)(x_1 - x_2)}{\sum x_1 y_2 - x_2 y_1} \right)$ is common to all the altitudes. (For the meaning of $\sum$, see the answer to Exercise (1.1.5).)

5. Consider the determinant of the coefficients in the equations of the three altitudes (cf. the answer to Exercise (1.1.5)) and add rows. This proof is slicker but does not give the orthocentre.

6. $\frac{5}{21}$ cc. of milk will come back along with $\frac{100}{21}$ cc. of water to the first glass. So $\alpha = \frac{100}{21}$. The second glass will retain $\frac{100}{21}$ cc. of milk. Hence $\beta = \frac{100}{21}$. So $\alpha = \beta$.

7. Since the volumes remain the same, as much milk goes into the second glass as water to the first. So $\alpha = \beta$.

8. See the beginning of Section 3.1. Using (3) there, let $n$ be the least integer such that $\left( \dfrac{19}{21} \right)^n < \dfrac{3}{5}$. So $n = 6$ is the answer. The reasoning in the last exercise will not work as it only tells that the percentages are equal but not how much they are.

9. A submatrix of a Hilbert matrix is in general not a Hilbert matrix, even when it consists of blocks of consecutive rows and columns.

11. Take $x_i = i$ and $y_j = j - 1$.

12.

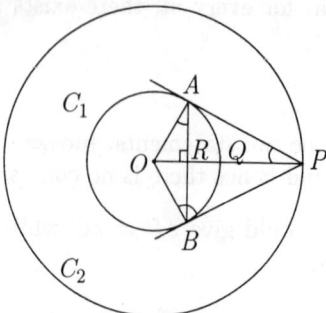

In the figure, $OP \perp AB$. Let $OP$ cut $C_1$ and $AB$ at $Q$ and $R$ respectively. In the right angled triangle $OAP$, $OA = \frac{1}{2}OP$. So $\angle OPA = 30°$. Hence in the right angled triangle $ORA$, $\angle OAR = 30°$. So $OR = \frac{1}{2}OA$. Hence $RP = \frac{3}{2}OA = \frac{3}{2}QP$, showing $Q$ is the centroid of $\triangle ABP$.

For a solution using co-ordinates, let $P = (x_2, y_2)$ and let $R = (x_1, y_1)$ be the mid-point of $AB$. Then the equation of $AB$ as the chord of contact of $P$, is $xx_2 + yy_2 = 1$. But in terms of the mid-point $(x_1, y_1)$, the equation of $AB$ is $xx_1 + yy_1 = x_1^2 + y_1^2$. Comparing the two equations, $\dfrac{x_1}{x_2} = \dfrac{y_1}{y_2} = \dfrac{x_1^2 + y_1^2}{1} = k$ (say). Thus $O, R, P$ are collinear and $x_2^2 + y_2^2 = 4$ determines $k = \dfrac{1}{4}$. So $R = \left(\dfrac{x_2}{4}, \dfrac{y_2}{4}\right)$. Since $Q$, the point of intersection of $OP$ with $C_1$ is $\left(\dfrac{x_2}{2}, \dfrac{y_2}{2}\right)$, it follows that $QP = 2QR$. So $Q$ is the centroid of $\triangle ABP$.

# CHAPTER 2

## Section 2.1

1.    For $r = 2, \ldots, n$, write $\dfrac{n(n-1)\ldots(n-r+1)}{n^r}$ as
$$\left(1 - \frac{1}{n}\right)\left(1 - \frac{2}{n}\right)\ldots\left(1 - \frac{r-1}{n}\right) \text{ and } \frac{(n+1)n(n-1)\ldots(n-r+2)}{(n+1)^r} \text{ as }$$
$$\left(1 - \frac{1}{n+1}\right)\left(1 - \frac{2}{n+1}\right)\ldots\left(1 - \frac{r-1}{n+1}\right). \text{ Now } 1 - \frac{k}{n} < 1 - \frac{k}{n+1} \text{ for } k = 1, 2, \ldots, r-1.$$

2.    Simply add the areas of the triangles $PA_iA_{i+1}, i = 1, 2, \ldots, n$ (with $A_{n+1} = A_1$). For the counter-example, let $A_1 = (2, 0), A_2 = (0, 2), A_3 = (-2, 0)$, $A_4 = (0, 1)$ and $P = (\frac{1}{2}, 1)$. The area of $A_1A_2A_3A_4$ is 2 but the sum on the right is $4\frac{1}{2}$. (The triangles $PA_2A_3, PA_3A_4$ and $PA_4A_1$ are not entirely within the quadrilateral $A_1A_2A_3A_4$.)

3.    Divide $P_n$ into $n$ mutually congruent triangles by joining its vertices to the centre. For the second part draw $P_{2n}$ from $P_n$ by inserting additional vertices at the mid-points of the arcs joining pairs of consecutive vertices of $P_n$. Then $P_n$ is a proper subset of $P_{2n}$.

4.    Approximately, radius $\times$ semi-perimeter of circle $= r\frac{1}{2}2\pi r = \pi r^2$.

5.    (i) For $a > 0, f(x) = a(x + \frac{b}{2a})^2 + c - \frac{b^2}{4a}$. So minimum occurs at $x = -\frac{b}{2a}$ and there is no maximum. For $a < 0$, there is no minimum but the maximum occurs at $x = -\frac{b}{2a}$. For $a = 0$, there is neither a maximum nor a minimum except in the trivial case when $b$ also vanishes and therefore $f$ is a constant function with every point both a maximum and a minimum. If instead of $\mathbb{R}$ we take a closed interval, say, $[p, q]$, the answer depends on whether $-\frac{b}{2a}$ is

in this interval. If it does, then for $a > 0$, the minimum (and for $a < 0$, the maximum) occurs there while the maximum (resp. the minimum) occurs at that end of $[p, q]$ which is farther from $-\frac{b}{2a}$. If $-\frac{b}{2a} \notin [p, q]$, both the maximum and the minimum occur at the end point.

(ii) The problem is trivial of $a = b = 0$. Otherwise let $\alpha$ be an angle such that $\sin \alpha = \dfrac{a}{\sqrt{a^2 + b^2}}$, $\cos \alpha = \dfrac{b}{\sqrt{a^2 + b^2}}$. Then $g(x) = \sqrt{a^2 + b^2} \sin(\alpha + x)$ has maximum at $x = \dfrac{\pi}{2} - \alpha + 2n\pi$ and minimum at $x = -\dfrac{\pi}{2} - \alpha - 2n\pi$ where $n$ is any integer. For an interval $[p, q]$, the answer depends upon whether it contains any points of this form, as otherwise the maximum/minimum occurs at one of the end-points.

6. Modifying the solution to Exercise (1.5.4) and noting $\sin \frac{A}{2} \geq 0$, $\cos A + \cos B + \cos C = 1 - 2\sin^2 \frac{A}{2} + 2\sin \frac{A}{2} \cos \frac{B-C}{2} \leq 1 - 2\sin^2 \frac{A}{2} + 2\sin \frac{A}{2} = \frac{3}{2} - 2(\sin \frac{A}{2} - \frac{1}{2})^2 \leq \frac{3}{2}$.

7. $(\tan A \tan B \tan C)^{1/3} \leq \dfrac{\tan A + \tan B + \tan C}{3} = \dfrac{\tan A \tan B \tan C}{3}$ gives $(\tan A \tan B \tan C)^2 \geq 27$. So the minimum value of $\tan A \tan B \tan C$ (and hence of $\tan A + \tan B + \tan C$) is $3\sqrt{3}$ or more. But if $A = B = C = 60°$ then it equals $3\sqrt{3}$. So $3\sqrt{3}$ is the minimum. Since in the A.M. - G.M. inequality, equality holds only when all numbers are equal, the minimum is attained only for an equilateral triangle.

8. Let $x, y, z$ denote the sides of the box. Then its surface area $= 2(xy + yz + zx) = A$ (say) and volume $xyz = V$. G.M. $\geq$ H.M. gives $V^{1/3} \geq \dfrac{3xyz}{xy + yz + zx}$ $= \dfrac{6V}{A}$. So $A \geq 6V^{2/3}$ with equality holding only for a cube.

9. Same reasoning gives $V \leq \dfrac{A^{3/2}}{6\sqrt{6}}$ with equality holding for a cube.

10. $V = 4uvz = $ constant. $A = 4(uv + vz + zu)$. By Exercise (1.8), $A$ is minimum when $u = v = z = \left(\dfrac{V}{4}\right)^{1/3}$, i.e. when $x = y = (2V)^{1/3}$ and $z = \left(\dfrac{V}{4}\right)^{1/3}$.

11. Step 1 :　Set $j = 1$, $M = x_1$.

　　Step 2 :　If $j = n + 1$, report $M$ as the maximum and stop.

　　Step 3 :　If $x_j > M$, call $x_j$ as $M$.

　　Step 4 :　Replace $j$ by $j + 1$ (i.e. increase the value of $j$ by 1) and go to Step 2.

(Whenever a condition under 'if' is not satisfied, by default go to the next step.)

## Section 2.2

1.  Let $a_n$ be the percentage of water in the first glass (and also the percentage of milk in the second) after $n$ exchanges. We expect $a_n \to 50$ as $n \to \infty$.

2.  The whole piece of paper will never be consumed. But as $n \to \infty$, the portion left after the $n^{\text{th}}$ day (which equals $(\frac{1}{2})^n$) should tend to 0.

3.  $\dfrac{dl}{dT}$ where $T =$ temperature and $l =$ length of a piece of wire (as a function of $T$) gives the rate of expansion. But the coefficient of expansion is the rate per unit length, i.e. $\dfrac{1}{l}\dfrac{dl}{dT}$.

4.  Let $M$ be the mid-point of $PQ$ and $O$ the centre of the circle. Then $OM \perp PQ$. So $\angle OPM + \angle POM = 90° =$ angle between $OP$ and the tangent at $P$. So the angle between $PQ$ and the tangent equals $\angle POM = \frac{1}{2}\angle POQ = \angle PAQ$. For the second proof, the angle between $PQ$ and the tangent at $P$ is the limit of $\angle QPR$ as $R$ approaches $P$ along the circle. But $\angle QPR = \angle QAR$ (angles in the same arc) and the latter approaches $\angle QAP$ as $R \to P$.

5.  Let $O = (0,0)$ and $P = (x_1, y_1)$. For the first proof, the slope of $OP$ being $\dfrac{y_1}{x_1}$, that of the tangent at $P$ is $-\dfrac{x_1}{y_1}$. This gives its equation since $x_1^2 + y_1^2 = a^2$. For the second derivation, let $Q = (x_2, y_2)$ be another point on the circle. Since $x_1^2 + y_1^2 = a^2 = x_2^2 + y_2^2$, the slope of the chord $PQ$ can be written as $-\dfrac{x_1 + x_2}{y_1 + y_2}$ which tends to $-\dfrac{x_1}{y_1}$ as $Q \to P$. A similar argument works for an ellipse or a hyperbola.

6.  An equation of the form $Ax + By + C = 0$ changes to $Aau + Bbv + C = 0$. So straight lines go to straight lines. By the first method in the last exercise, the tangent at $Q$ has equation $uu_1 + vv_1 = 1$ where $u_1 = \dfrac{x_1}{a}, v_1 = \dfrac{y_1}{b}$. Putting $u = \dfrac{x}{a}, v = \dfrac{y}{b}$, this becomes $\dfrac{xx_1}{a^2} + \dfrac{yy_1}{b^2} = 1$, which is the equation of the tangent to the ellipse at $P$.

7.  If the rainfall is a constant, say $k$, the volume of rain water is $k$ times the area of $C$. In general divide $C$ into subregions $C_1, \dots, C_n$ of areas $A_1, \dots, A_n$ respectively. Let $k_i$ be the rainfall at some point of $C_i$. Then the approximate rain water is $\sum_{i=1}^{n} k_i A_i$ and the exact answer is the limit of this as each $C_i$ shrinks to a point in size. But when this happens, $n \to \infty$ and each $A_i \to 0$.

8.  The mass is the limit of $\displaystyle\sum_{i=1}^{n} \rho_i V_i$ where the solid is cut into $n$ pieces of volumes $V_1, \dots, V_n$ and $\rho_i$ is the density at some point in the $i^{\text{th}}$ piece.

## Section 2.3

1. It is preferable to write $|f(x) - L| < \epsilon$ as $L - \epsilon < f(x) < L + \epsilon$. The idea then is to reduce it to an inequality of the form $c - \delta_1 < x < c + \delta_2 (x \neq c)$. This latter inequality should preferably be equivalent to the former, in which case $\min\{\delta_1, \delta_2\}$ will give the best (i.e. the largest) $\delta$ that will work for the given $\epsilon$. This is not always easy to achieve. It is all right if the inequality $c - \delta_1 < x < c + \delta_2$ is stronger than $L - \epsilon < f(x) < L + \epsilon$; but it must not be weaker. This time $\min\{\delta_1, \delta_2\}$ will give a $\delta$ which works, although a larger $\delta$ may also work for the same $\epsilon$.

   In (i) $L - \epsilon < f(x) < L + \epsilon$ is the same as $c - \epsilon < x < c + \epsilon$. Here $\delta_1 = \delta_2 = \epsilon$. So $\delta = \epsilon$ is the best $\delta$. In (ii) $11 - \epsilon < 3x + 5 < 11 + \epsilon$ is equivalent to $2 - \dfrac{\epsilon}{3} < x < 2 + \dfrac{\epsilon}{3}$. So again $\delta = \dfrac{\epsilon}{3}$ is the best $\delta$. In (iii) $8 - \epsilon < x^3 < 8 + \epsilon$ is equivalent to $(8 - \epsilon)^{1/3} < x < (8 + \epsilon)^{1/3}$, or to $2 - \delta_1 < x < 2 + \delta_2$, where $\delta_1 = 2 - (8 - \epsilon)^{1/3}$ and $\delta_2 = (8 + \epsilon)^{1/3} - 2$. So the minimum of these is the best $\delta$. From $(2 - \delta_1)^3 = 8 - \epsilon$ and $(2 + \delta_2)^3 = 8 + \epsilon$, it can be shown that $\delta_2 < \delta_1$ and so $\delta = \delta_2$. But if we merely want *some* $\delta$ which will work for the given $\epsilon$, a much easier way is to write $|x^3 - 8| = |x - 2||x^4 + 2x + 4|$ and to obtain an upper bound on $|x^4 + 2x + 4|$ valid in some neighbourhood of 2. This upper bound need not be very sharp. Any crude upper bound, found by inspection would do. For example take the neighbourhood $(2 - 1, 2 + 1)$, i.e. $(1, 3)$ of 2. For all $x \in (1, 3)$, $|x^2 + 2x + 4| \le |x|^2 + 2|x| + 4 \le 19$. So $x \in (2 - 1, 2 + 1)$ and $x \in (2 - \frac{\epsilon}{19}, 2 + \frac{\epsilon}{19})$ together would imply (the implication not being reversible) that $|x^3 - 8| = |x - 2||x^2 + 2x + 4| < \frac{\epsilon}{19} \cdot 19 = \epsilon$. Hence we can take $\delta = \min\{1, \frac{\epsilon}{19}\}$. A similar technique for (iv) gives $\delta = \min\{1, \frac{\epsilon}{34}\}$. (Many other choices are, in fact, possible, e.g. $\delta = \min\{2, \frac{\epsilon}{45}\}$.)

2. In each case take $\epsilon$ to be half of the difference between the old and the new values of $L$, say $L_1$ and $L_2$, respectively. Let $\delta_0 > 0$ be the delta which works for this $\epsilon$ w.r.t. $f$ and $L_1$. ($\delta_0$ may be different in each case.) Now given any $\delta > 0$, take any $x$ s.t. $0 < |x - c| < \min\{\delta_0, \delta\}$. Then $L_1 - \epsilon < f(x) < L_1 + \epsilon$ and so $f(x) \notin (L_2 - \epsilon, L_2 + \epsilon)$ since the intervals $(L_1 - \epsilon, L_1 + \epsilon)$ and $(L_2 - \epsilon, L_2 + \epsilon)$ are disjoint by the very choice of $\epsilon$. So for $f$ and $L_2$, no $\delta$ can be found for this particular $\epsilon$. Note that this argument also shows that the limit, whenever it exists, is unique. (See Theorem (6.1) and the diagram there which is drawn with a slightly different notation.)

3. Any (positive) $\epsilon < 1$ would do. For every $\delta > 0$, find $n \in I\!N$ such that $(4n + 1)\dfrac{\pi}{2} > \delta$. Then $f\left(\dfrac{1}{(4n + 1)\frac{\pi}{2}}\right) = 1$ while $f\left(\dfrac{1}{(4n + 3)\frac{\pi}{2}}\right) = -1$. The interval $(L - \epsilon, L + \epsilon)$, has length $2\epsilon$ which is less than 2 and hence cannot contain both 1 and $-1$ no matter what $L$ is. But $\dfrac{1}{(4n + 1)\frac{\pi}{2}}$ and $\dfrac{1}{(4n + 3)\frac{\pi}{2}}$ are both in $(0 - \delta, 0 + \delta)$, a contradiction.

4.  We are given some $\delta_0 > 0$ such that for all $x \in (c - \delta_0, c + \delta_0)$, $x \neq c$, $f(x) \leq g(x) \leq h(x)$. Let $\lim_{x \to c} f(x) = \lim_{x \to c} h(x) = L$. Let $\epsilon > 0$ be given. If we can trap both $f(x)$ and $h(x)$ in $(L - \epsilon, L + \epsilon)$ we would be through. So let $\delta = \min\{\delta_0, \delta_1, \delta_2\}$ where $\delta_1, \delta_2$ are the deltas, for the given $\epsilon$, for $f$ and $h$ respectively. In polished language, we write "Since $\lim_{x \to c} f(x) = L$, there exists $\delta_1 > 0$ such that $L - \epsilon < f(x) < L + \epsilon$ whenever $0 < |x - c| < \delta_1$" and similarly for $\delta_2$. Since $|\sin \frac{1}{x}| \leq 1$ for all $x \neq 0$, we have $-|x| \leq x \sin \frac{1}{x} \leq |x|$ for all $x \neq 0$. So the Sandwich Theorem applies. For the last part take
    $$x_1 = \frac{1}{(4n + 1)\frac{\pi}{2}} \quad \text{and} \quad x_2 = \frac{1}{2n\pi} \quad \text{where } n \text{ is a (sufficiently large) integer.}$$

5.  Let $\lim_{x \to c} f(x) = L$. Take a $\delta$ corresponding to $\epsilon = 1$ (an arbitrary choice). Then in the deleted $\delta$-neighbourhood of $c$, $L - 1 < f(x) < L + 1$. So $|f(x)|$ is bounded by $\max\{|L + 1|, |L - 1|\}$ and hence by $|L| + 1$.

6.  No. Neither (i) nor (ii) is an implication statement. The parts "As $x$ tends to $c$" etc. have no meanings of their own.

7.  Following the hint, we have (i) if $0 < |x - c| < \delta_1$ then $|f(x) - L| < \delta$ and (ii) if $|y - L| < \delta$ then $|g(y) - M| < \epsilon$. Note that in (ii) we have not only used $\lim_{y \to L} g(y) = M$ but also $g(L) = M$. Now if we put $y = f(x)$, we have a genuine case of syllogism and get $\lim_{x \to c} h(x) = M$.

8.  Using the same reasoning as in the proof of (v) of Theorem (3.1), there exists $\delta > 0$ such that $0 < |x - c| < \delta_0$ implies $f(x) > \frac{L}{2}$ and so $\sqrt{f(x)}$ is defined for these values of $x$. Writing $|\sqrt{f(x)} - \sqrt{L}|$ as $\dfrac{|f(x) - L|}{\sqrt{f(x)} + \sqrt{L}} \leq \dfrac{|f(x) - L|}{\sqrt{L}(1 + \frac{1}{\sqrt{2}})}$, the desired $\delta$ can be obtained the same way as in the proof of (v). A clever alternative is to note that $|\sqrt{f(x)} - \sqrt{L}| \leq \sqrt{|f(x) - L|}$, which can be proved using the A.M. - G.M. inequality (cf. Exercise (4.2.9)).

9.  Apply the last exercise to $f^2$. For a direct proof the triangle inequality gives (see Exercise (1.9.8)) $\left| |f(x)| - |L| \right| \leq |f(x) - L|$, . If $L = 0$, $\left| |f(x)| - |L| \right|$ is the same as $|f(x) - L|$ and the converse holds. To prove it is false in general, take $f(x) = 1$ if $x > 0$, $f(x) = -1$ if $x < 0$. (See Section 2.5.)

## Section 2.4

1.  The direct definition is "for every $R$, there exists $M$ such that for all $x > M$, $f(x) < R$." For equivalence of the two definitions, merely observe that '$f(x) < R$' is equivalent to '$-f(x) < -R$'. For the equivalence in the second part, use the equivalence of '$x > M$' with '$-x < -M$'.

2.  Put $h = x - \frac{\pi}{2}$. (A substitution of this type is often useful in studying limits as it focuses directly on the difference between the value of the independent variable and its limit. There are many results, especially inequalities, which hold for small values of this difference and they can be applied

sometimes. Even otherwise, it is convenient as in the present problem.) Then as $x \to \frac{\pi}{2}, h \to 0$. Since $\cos x = \cos(\frac{\pi}{2} + h) = -\sin h$ and $\sin h \to 0$ we have $\cos x \to 0$ as $x \to \frac{\pi}{2}$. The identity $\cos h = 1 - 2\sin^2 \frac{h}{2}$ shows that $\lim_{h \to 0} \cos h = 1$ (since $\frac{h}{2} \to 0$ as $h \to 0$). Hence $\lim_{x \to \frac{\pi}{2}} \sin x = \lim_{h \to 0} \sin\left(h + \frac{\pi}{2}\right) = \lim_{h \to 0} \cos h = 1$. Now for (i), given $R > 0$, suppose $\delta_1 > 0$ is such that $0 < |x - \frac{\pi}{2}| < \delta_1$ implies $|\cos x| < \frac{1}{R}$. Then $|\sec x| > R$ whenever $0 < |x - \frac{\pi}{2}| < \delta_1$. So $|\sec x| \to \infty$ as $x \to \frac{\pi}{2}$. For $|\tan x|$, given $R \geq 0$, take $\delta = \min\{\delta_1, \delta_2\}$ where $\delta_1 > 0$ is such that $0 < |x - \frac{\pi}{2}| < \delta_1$ implies $|\sin x - 1| < \frac{1}{2}$ and hence $|\sin x| > \frac{1}{2}$ and $\delta_2 > 0$ is such that $0 < |x - \frac{\pi}{2}| < \delta_2$ implies $|\cos x| < \frac{1}{2R}$. (ii) and (iii) follow from (i) and repeated application of the result that if $f(x) \to \infty$ as $x \to c$, then $(f(x))^2 \to \infty$ as $x \to c$. In (iv), we already have, $|x| \to \frac{\pi}{2}$ as $x \to \frac{\pi}{2}$ and so $\tan |x| \to \infty$ as $x \to \frac{\pi}{2}$. Now, $\left|\sin\left(\dfrac{1}{x - \frac{\pi}{2}}\right)\right|$ is bounded (by 1) in a deleted neighbourhood of $\frac{\pi}{2}$ and so $f(x) \geq \tan |x| - 1$ which tends to $\infty$, as $x \to \frac{\pi}{2}$.

**3.**  (ii) is immediate. For (i) note that since $\sin x > 0$ in a neighbourhood of $\frac{\pi}{2}$,
$$\left| \, |\sec x| - |\tan x| \, \right| = |\sec x - \tan x| = \frac{1}{|\sec x + \tan x|} = \frac{1}{|\sec x| + |\tan x|}.$$
(iii) follows from $\sec^4 x - \tan^4 x = \sec^2 x + \tan^2 x$. (iv) reduces to proving that $\lim_{h \to 0} \sin \frac{1}{h}$ does not exist. This was done in Exercise (3.3).

**4.**  The proof of the first assertion is routine in all cases except in (iv), where the Sandwich Theorem is needed for $x^2 \sin \frac{1}{x}$ (cf. Exercise (3.4)). The assertions about $\lim_{x \to 0} f(x)g(x)$ are also easy except that in (iv) we again need Exercise (3.3).

**5.**  Assume $f(x) \leq g(x) \leq h(x)$ for all $x > R$ and $\lim_{x \to \infty} f(x) = \lim_{x \to \infty} h(x) = L$. If $L \neq \pm\infty$, then given $\epsilon > 0$, let $M = \max\{M_1, M_2\}$ where $M_1, M_2$ are such that $f(x) \in (L - \epsilon, L + \epsilon)$ for all $x > M_1$ and $h(x) \in (L - \epsilon, L + \epsilon)$ for all $x > M_2$. Then $g(x) \in (L - \epsilon, L + \epsilon)$ for all $x > M_1$. So the change is that instead of $\delta_1, \delta_2$ we have $M_1, M_2$ and instead of setting $\delta = \min\{\delta_1, \delta_2\}$, we set $M = \max\{M_1, M_2\}$. If $L = \infty, h(x)$ is not needed. Given $R$, any $M$ which works for $f$ also works for $g$. Similarly if $L = -\infty, f$ is not needed. Situation for $\lim_{x \to -\infty}$ is similar.

**6.**  $(c - \delta_1, c + \delta_1) \cap (c - \delta, c + \delta_2) = (c - \delta, c + \delta)$ where $\delta = \min\{\delta_1, \delta_2\}$. $(M_1, \infty) \cap (M_2, \delta) = (M, \infty)$ where $M = \max\{M_1, M_2\}$ and similarly $(-\infty, M_1) \cap (-\infty, M_2) = (-\infty, M)$ where $M = \min\{M_1, M_2\}$. Now suppose $f(x) \leq g(x) \leq h(x)$ for all $x$ in some deleted neighbourhood, say $U_0$ of $c$ (where $c$ is possibly $\infty$ or $-\infty$). Assume $\lim_{x \to c} f(x) = \lim_{x \to c} h(x) = L$ (which too could equal $\pm\infty$). Then given any neighbourhood $V$ of $L$, let $U = U_1 \cap U_2$ where $U_1, U_2$ are deleted neighbourhoods of $c$ such that $f(x) \in V$ for all $x \in U_1$

and $h(x) \in V$ for all $x \in U_2$. Then for all $x \in U$, both $f(x), h(x)$ are in $V$ and hence, $V$ being an interval, $g(x)$ is also in $V$. This proves $\lim_{x \to c} g(x) = L$, i.e. the Sandwich Theorem holds. In essence, taking the intersection of (deleted) neighbourhoods of $c$ corresponds to a step like setting $\delta = \min\{\delta_1, \delta_2\}$ if $c$ is finite, to setting $M = \max\{M_1, M_2\}$ if $c = \infty$ and to setting $M = \min\{M_1, M_2\}$ if $c = -\infty$. So unified proofs can be given.

## Section 2.5

1.  The two proofs are essentially the same. But the present one looks slightly neater because it concentrates on the essential aspect a little more vividly.

2.  Let $c \in \mathbb{R}$. There exists $\delta_0 > 0$ such that $(c - \delta_0, c + \delta_0) - \{c\} \subset S_1 \cup S_2$. Given $\epsilon > 0$, find $\delta_i > 0$ such that for $x \in S_i, 0 < |x - c| < \delta_i$ implies $|f(x) - L| < \epsilon; i = 1, 2$. Take $\delta = \min\{\delta_0, \delta_1, \delta_2\}$. The case $c = -\infty$ is similar to $c = \infty$ (proved in the text) except that the inequalities involving $R, R_1$ and $R_2$ are reversed and $R' = \min\{R, R_1, R_2\}$.

3.  Set $\delta = \min\{\delta_0, \delta_1, \ldots, \delta_k\}$ or $R' = \max\{R, R_1, R_2, \ldots, R_k\}$ etc. For an inductive proof, call $S_2 \cup S_3 \cup \ldots \cup S_k$ as $S_2'$. Then by induction hypothesis, $\lim_{\substack{x \to c \\ x \in S_2'}} f(x) = L$. Now apply Theorem (5.1) to the two sets $S_1$ and $S_2'$.

4.  Let $S_1 = \{x : x < 1, x \text{ is rational}\}$ etc. Since $\lim_{x \to 1} x = 1, \lim_{\substack{x \to 1 \\ x \in S_1}} x = 1$, i.e. $\lim_{\substack{x \to 1 \\ x \in S_1}} f(x) = 1$. Similarly define $S_2, S_3, S_4$ and show $\lim_{\substack{x \to c \\ x \in S_i}} f(x) = 1$ for $i = 2, 3, 4$. So $\lim_{x \to c} f(x) = 1$.

5.  Let $x \in \mathbb{R}$. If $x = k$ for some positive integer $k$, then $x \in S_k$ and hence $x \in \bigcup_{n=1}^{\infty} S_n$. All other real numbers are in every $S_n$ and hence in $\bigcup_{n=1}^{\infty} S_n$. However, a union of the form $S_{n_1} \cup S_{n_2} \cup \ldots \cup S_{n_k}$ contains no positive integers except $n_1, \ldots, n_k$.

6.  If $x \in S_k$ then $f(x) = 0$, except when $x = k$. So given $\epsilon > 0$, take $R = k$. Then for all $x \in S_k, |f(x) - 0| = 0 < \epsilon$ whenever $x > R$. So $\lim_{\substack{x \to \infty \\ x \in S_k}} f(x) = 0$.

    For the second part, $\lim_{\substack{x \to \infty \\ x \in S_0}} f(x) = 0$ while $\lim_{\substack{x \to \infty \\ x \in \mathbb{N}}} f(x) = 1$.

7.  Let $V$ be the neighbourhood $(-1, 1)$ of $0$. In order for $f(x)$ to lie in $V$ for all $x \in S_k \cap U_k$, where $U_k$ is some deleted neighbourhood of $\infty$, it is necessary that $U_k$ is of the form $(R_k, \infty)$ for some $R_k \geq k$. But then $\cap_{k=1}^{\infty} U_k = \emptyset$.

8.  Given $\epsilon > 0$, let $\delta = \epsilon^2$.

**9.** Given $R(> 0)$, let $\delta = \frac{1}{R}$ for the first part. For the second part, given $R$ (which may be assumed to be negative), let $\delta = -\frac{1}{R}$.

## Section 2.6

**1.** The crucial point is to observe that whenever $L \neq L'$, there exist mutually disjoint neighbourhoods $V, V'$ of $L, L'$ respectively. Proof of Theorem (6.1) does this if at least one of $L, L'$ is finite. If $L = -\infty$ and $L' = \infty$ (say) take $V = (-\infty, 0)$ and $V' = (0, \infty)$. Now in all cases, find deleted neighbourhoods $U, U'$ of $c$ (which could equal $\pm\infty$ also) such that $x \in U$ implies $f(x) \in V$ and $x \in U'$ implies $f(x) \in V'$. Then for any $x \in U \cap U'$, $f(x) \in V \cap V'$, a contradiction.

**2.** Write $\binom{n}{r} \dfrac{1}{10^r n^r}$ as $\dfrac{1}{r!10^r} \cdot \dfrac{n}{n} \cdot \dfrac{n-1}{n} \cdots \dfrac{n-r-1}{n}$ for the first part. For the second, prove (by induction or otherwise) that $2^{r-1} \leq r!$, with strict inequality for $r > 2$.

**3.** $1 + \dfrac{1}{10} + \dfrac{1}{2.10^2} + \dots + \dfrac{1}{2^{n-1}10^n} = 1 + \dfrac{1}{10}\displaystyle\sum_{r=0}^{n-1}\dfrac{1}{(20)^r} = 1 + \dfrac{1}{10}\dfrac{1 - (1/20)^n}{1 - \frac{1}{20}}$

$< 1 + \dfrac{1}{10}\dfrac{1}{19/20} = 1 + \dfrac{2}{19} = \dfrac{21}{19}$.

**4.** (i) Cancelling common factors we may suppose $p, q$ are relatively prime i.e. have no common factor other than $\pm 1$. Now if $p^2 = 2q^2$ then $p^2$ and hence $p$ is even, say $p = 2k$. But then $q^2 = 2k^2$, forcing $q$ to be even too, a contradiction. For (ii), $x^2 > 2 \Rightarrow p^2 > 2q^2 \Rightarrow 2p^2 + 4pq + 2q^2 > p^2 + 4pq + 4q^2 \Rightarrow (p + 2q)^2 < 2(p + q)^2 \Rightarrow y^2 < 2$. (iii) is similar.

**5.** (i) $x^2 < 2 \Rightarrow y^2 > 2$ (by (iii) of the last exercise) $\Rightarrow z^2 < 2$ (by (ii) of it applied to $y$ and $z$ instead of $x$ and $y$). Similarly (ii). The rest follows by straight computation.

**6.** Note first that if $a, b$ are positive then $a^2 > b^2$ iff $a > b$. (Factorise $a^2 - b^2$.) So 2 is an upper bound for $A$. Suppose $\frac{p}{q}$ is the supremum of $A$, with $p, q$ positive integers. Then $\frac{p}{q} \notin A$, as otherwise by (i) of the last exercise, $\frac{3p+4q}{2p+3q}$ is also in $A$ and bigger than $\frac{p}{q}$. So $(\frac{p}{q})^2 > 2$. But then by (ii) of the last exercise, $\frac{3p+4q}{2p+3q}$ is a smaller upper bound for $A$.

**7.** No. But it is alternately increasing and decreasing. (If the $n^{\text{th}}$ term is $\frac{p_n}{q_n}$ then $p_n^2 - 2q_n^2 = (-1)^n$ for $n \geq 2$, as can be proved by induction on $n$. See Exercise (3.1.3).)

**8.** If $a, b$ are both maximum elements of $A$ then $a \leq b$ and $b \leq a$ both hold, forcing $a = b$. Clearly $a$ is an upper bound of $A$. If there were a smaller upper bound say $c$, then $a \leq c$ and also $c < a$, a contradiction. So $a = \sup A$. The interval $[0, 1)$ or the set $\{-1, \frac{1}{2}, -\frac{1}{3}, \dots, \frac{1}{n}, \dots\}$ has a supremum but no maximum. If $A$ is finite (and non-empty) then $A$ has a maximum element which can be found by an algorithm (cf. Exercise (1.11)). For the last assertion take reflections w.r.t. 0.

9.  $B \neq \emptyset$ as $A$ is bounded above. Any element of $A$ is a lower bound for $B$. Let $\alpha$ be the supremum of $A$. Then $\alpha \in B$. If $b \in B$ and $b < \alpha$, then $\alpha$ is not the *least* upper bound of $A$. So $\alpha \leq b$. (It pays to draw figures to understand and conceive such proofs. Just draw a line and mark appropriate points on it.) Thus, $\alpha$ is the least element, and hence also the infimum of $B$. Conversely suppose $\beta$ is the infimum of $B$. Then for every $a \in A, a \leq \beta$ as otherwise there would be some $b \in B$ with $\beta < b < a$, contradicting that every element of $B$ is an upper bound of $A$. Thus $\beta \in B$ and hence it is the least element of $B$. So $\beta$ is the l.u.b. of $A$. $A \cap B$ is $\emptyset$ if $A$ has no maximum. Otherwise $A \cap B$ consists of the (unique) maximum element of $A$.

10. The implication (ii) $\Rightarrow$ (i) follows from the last exercise.

11. Show that supremum and the infimum of the set $\{a_n : n \in I\!N\}$ are the limits for the first and the second assertion respectively.

12. If $y = \sup A$ and $\epsilon > 0$, then $y - \epsilon < y$ and so $y - \epsilon$ is not an upper bound of $A$. So there exists $x \in A$ such that $x > y - \epsilon$. Conversely if $y$ is an upper bound but not the l.u.b. of $A$, then there is a smaller upper bound, say $y_1$ of $A$. Apply (ii) with $\epsilon = y - y_1$ to get $x \in A$ such that $x > y - \epsilon = y_1$, a contradiction.

13. If $y$ is the maximum element of $A$, let $x_n = y$ for every $n$. Assume $A$ has no maximum element. Let $x_1 \in A$ be arbitrary. Let $x_2 \in A$ be such that $x_2 > x_1$ and $x_2 > y - 1$. Having chosen $x_2$, let $x_3 \in A$ be bigger than both $x_2$ and $y - \frac{1}{2}$. In general, having chosen $x_n$, let $x_{n+1} \in A$ be such that $x_{n+1} > x_n$ and $x_{n+1} > y - \frac{1}{n}$, (using (ii) of the last Exercise with $\epsilon = \min\{y - x_n, \frac{1}{n}\}$.)

## Section 2.7

1.  The catch is that the number of terms added is not independent of $n$. (cf. Exercise (1.1.6) (iii) for a similar fallacy.)

2.  $e^x = (\lim_{n \to \infty} (1 + \frac{1}{n})^n)^x = \lim_{n \to \infty} (1 + \frac{1}{n})^{nx} = \lim_{n \to \infty} \sum_{r=0}^{\infty} \frac{(nx) \dots (nx - r + 1)}{r! n^r}$

    $= \sum_{r=0}^{\infty} \lim_{n \to \infty} \frac{x(x - \frac{1}{n})(x - \frac{2}{n}) \dots (x - \frac{r-1}{n})}{r!} = \sum_{r=0}^{\infty} \frac{x^r}{r!}$.

3.  For $n = q + k$ (where $k \geq 1$), $n! = q!(q + 1) \dots (q + k) \geq q!(q + 1)^k$. So

    $$\sum_{n=q+1}^{\infty} \frac{1}{n!} \leq \frac{1}{q!} \sum_{k=1}^{\infty} \frac{1}{(q+1)^k}. \text{ But } \sum_{k=1}^{r} \frac{1}{(q+1)^k} = \frac{1}{q+1} \frac{1 - (\frac{1}{q+1})^r}{1 - \frac{1}{q+1}}$$

    $$= \frac{1}{q}(1 - (\frac{1}{q+1})^r) \leq \frac{1}{q} \text{ for every } r.$$

**4.** The error, $\sum_{n=11}^{\infty} \frac{1}{n!}$, is at most $\frac{1}{10!10} = \frac{1}{36,288,000} < \frac{1}{10^7}$.

**5.** Follow the hint and note that $q! \sum_{n=0}^{q} \frac{1}{n!}$ is an integer. So $q!e - \sum_{n=0}^{q} \frac{q!}{n!} = q! \sum_{n=q+1}^{\infty} \frac{1}{n!}$ would be an integer. But by Exercise (7.3), $0 < q! \sum_{n=q+1}^{\infty} \frac{1}{n!} < q! \frac{1}{q!q} = \frac{1}{q} \leq 1$, a contradiction.

**6.** Let $\alpha = \lim_{n\to\infty}(1+\frac{1}{10n})^n$. Then $\alpha^{10} = \lim_{n\to\infty}(1+\frac{1}{10n})^{10n} = \lim_{m\to\infty}(1+\frac{1}{m})^m = e$ (putting $m = 10n$). So $\alpha = e^{1/10}$.

# CHAPTER 3

## Section 3.1

**1.** $a_n = \dfrac{y_0(x_0 - 5)}{x_0(y_0 + 5)}a_{n-1} + \dfrac{500y_0}{x_0(y_0 + 5)}$ where $x_0, y_0$ are the initial volumes of milk and water respectively (assuming $x_0 \geq 5$). Solving, $a_n = \dfrac{100y_0}{x_0 + y_0}\left(1 - \left(\dfrac{y_0(x_0 - 5)}{x_0(y_0 + 5)}\right)^n\right)$. Intuitively, $\lim_{n\to\infty} a_n = \dfrac{100y_0}{x_0 + y_0}$.

**2.** Begin the induction with $n = 0$. Putting $a_{n-1} = 50(1 - (\frac{19}{21})^{n-1})$ in (1) will give (3), completing the inductive step. From (1), $a_n - a_{n-1} = \frac{2}{21}(50 - a_{n-1})$ so $a_n$ is monotonically increasing iff every term is $\leq 50$. For $n = 0, a_0 = 0 < 50$. The inductive step follows from (1).

**3.** The first part is by direct computation. For the second, $|a_n - \sqrt{2}|$
$$= \frac{|a_n^2 - 2|}{|a_n + \sqrt{2}|} = \frac{|p_n^2 - 2q_n^2|}{(a_n + \sqrt{2})q_n^2} \leq \frac{1}{(a_n + 1\sqrt{2})a_n^2}$$ since $a_n > 0$ for all $n$. As $n \to \infty, q_n$ and hence $q_n^2 \to \infty$. So $|a_n - \sqrt{2}| \to 0$, i.e. $a_n \to \sqrt{2}$.

**4.** Let $\alpha = \frac{1+\sqrt{5}}{2}$ and $\beta = \frac{1-\sqrt{5}}{2}$. Observe that $\alpha, \beta$ are roots of the quadratic $x^2 - x - 1 = 0$. So $\alpha^2 = \alpha + 1$ and hence $\alpha^n = \alpha^{n-1} + \alpha^{n-2}$ for all $n \geq 2$. Similarly for $\beta$. The inductive step is now easy.

**5.** Say $c = \infty$ and $U = (R, \infty)$. Suppose $\lim_{y\to\infty} h(y) = L$ and $b_n \to \infty$ as $n \to \infty$ (with $b_n \in U$ for all $n$). To prove $h(b_n) \to L$ as $n \to \infty$, let $\epsilon > 0$ be given. Find $M(\geq R)$ such that $h(y) \in (L - \infty, L + \epsilon)$ for all $y > M$. Now find $m \in \mathbb{N}$ such that $b_n > M$ for all $n \geq m$. Then $h(b_n) \in (L - \epsilon, L + \epsilon)$ for all $n \geq m$. For the converse, assume $h(y) \not\to L$ as $y \to \infty$. Then there exists $\epsilon > 0$ such that for every $M$, there is some $y > M$ such that $h(y) \notin (L - \epsilon, L + \epsilon)$. Put $M = R + n, n = 1, 2, 3, \ldots$, to get a sequence $\{b_n\}$ such that $b_n > R + n$ and $|h(b_n) - L| \geq \epsilon$ for all $n$. But then $b_n \to \infty$ as $n \to \infty$, a contradiction. For $c = -\infty$, take reflection.

**6.** Note that $\frac{1}{x-h} + \frac{1}{x+h} = \frac{2x}{x^2-h^2} > \frac{2x}{x^2} = \frac{2}{x}$. Using this fact and pairing the terms symmetrically located around the middle term gives $\frac{1}{10n} + \frac{1}{10n+2} + \frac{1}{10n+4} + \frac{1}{10n+6} + \frac{1}{10n+8} > \frac{5}{10n+4} = \frac{1}{2n+\frac{4}{5}} > \frac{1}{2n+1}$.

**7.** For $\sum_{n=1}^{\infty} n$, $S_n = \frac{n(n+1)}{2} \to \infty$ as $n \to \infty$. For $\sum_{n=1}^{\infty} (-1)^n$, $S_n = -1$ or $0$ according as $n$ is odd or even. So $\lim_{n\to\infty} S_n$ does not exist, because the restricted limits $\lim_{\substack{n\to\infty \\ n \text{ even}}} S_n$ and $\lim_{\substack{n\to\infty \\ n \text{ odd}}} S_n$ are different. For the third series write $\frac{1}{k(k+1)}$ as $\frac{1}{k} - \frac{1}{k+1}$. Then $S_n = \sum_{k=1}^{n} \frac{1}{k(k+1)} = (1 + \frac{1}{2} + \ldots + \frac{1}{n}) - (\frac{1}{2} + \frac{1}{3} + \ldots + \frac{1}{n} + \frac{1}{n+1}) = 1 - \frac{1}{n+1}$. So $S_n \to 1$ as $n \to \infty$. (A rigorous proof that $\frac{1}{n} \to 0$ as $n \to \infty$ will come in Theorem (3.1).) Hence $\sum_{n=1}^{\infty} \frac{1}{n(n+1)}$ converges with sum 1.

**8.** Write $a_n$ as $S_n - S_{n-1}$. As $n \to \infty$, both $S_n$ and $S_{n-1}$ tend to the finite limits $L$. So $a_n \to L - L = 0$.

**9.** Note that with the $\epsilon$ given in the hint, $x > y$ whenever $x \in (L - \epsilon, L + \epsilon)$ and $y \in (M - \epsilon, M + \epsilon)$. (Draw a figure to see such things easily. It is a good habit.) Now, get $m_1, m_2$ such that for all $n \geq m_1, a_n \in (L - \epsilon, L + \epsilon)$ and for all $n \geq m_2, b_n \in (M - \epsilon, M + \epsilon)$. Then for $n \geq \max\{m_1, m_2\}, a_n > b_n$, a contradiction. For the example, take $a_n = 0, b_n = \frac{1}{n}$.

**10.** No. In the Sandwich Theorem, it is given that $a_n \leq b_n \leq c_n$ for all $n$ and also that $\lim_{n\to\infty} a_n = \lim_{n\to\infty} c_n (= L$ say $)$. It is not given beforehand that $\lim_{n\to\infty} b_n$ exists, as would be needed if the last exercise were to apply. For a correct proof, imitate the solution to Exercise (2.3.4). Given $\epsilon > 0$, choose $m_1, m_2 \in \mathbb{N}$ such that for all $n \geq m_1, a_n \in (L - \epsilon, L + \epsilon)$ and for all $n \geq m_2, c_n \in (L - \epsilon, L + \epsilon)$. Now take $m = \max\{m_1, m_2\}$.

## Section 3.2

**1.** The only change needed is that instead of setting $m = \max\{m_1, m_2\}$ we now have to set $m = \max\{m_0, m_1, m_2\}$ where $m_0$ is an integer such that the inequalities $a_n \leq b_n$ etc. hold for all $n \geq m_0$.

**2.** A neighbourhood of $\infty$ is an interval of the form $(R, \infty)$ for some real number (not necessarily an integer) $R$. Let $m = \max\{1, [R] + 1\}$ where $[R]$ is the greatest integer not exceeding $R$. Then for $n \in \mathbb{N}, n \in (R, \infty)$ iff $n \neq 1, 2, \ldots, m$. The analogue of Theorem (2.1) would say if $f(x) = g(x)$ for all sufficiently large $x$ and $\lim_{x\to\infty} f(x) = L$, then $\lim_{x\to\infty} g(x) = L$ and vice versa.

**3.** Let $a_n = (-1)^n$. Then $\frac{a_1 + a_2 + \ldots + a_n}{n} = \frac{-1}{n}$ or $0$ according as $n$ is odd or even (cf. Exercise (1.7)). So given $\epsilon > 0$, if we choose $m$ to be any integer $> \frac{1}{\epsilon}$, then $|\frac{a_1 + a_2 + \ldots + a_n}{n} - 0| \leq \frac{1}{m} < \epsilon$ for all $n \geq m$, regardless of whether $n$ is even or odd. Thus $\dfrac{a_1 + a_2 + \ldots + a_n}{n} \to 0$ as $n \to \infty$. But $\lim_{n \to \infty} a_n$ does not exist since the restricted limit $\lim_{\substack{n \to \infty \\ n \text{ even}}} a_n$ equals 1 while $\lim_{\substack{n \to \infty \\ n \text{ odd}}} a_n$ equals $-1$.

## Section 3.3

1. Following the hint, in (i), $\dfrac{b_n}{n^r}$ is a sum of $k+1$ terms each tending to 0 while in (ii) it is a sum of $k$ such terms and the constant $p_k$. In (iii), $\dfrac{b_n}{n^r}$ can be written as $n^{k-r}(p_k + \frac{p_{k-1}}{n} + \ldots + \frac{p_0}{n^k})$. Since each term except the first one in parentheses tends to 0, $\frac{b_n}{n^r} \to \infty$ or $-\infty$ according as $p_k > 0$ or $p_k < 0$. In all cases, $\frac{c_n}{n^r} \to q_r$ as $n \to \infty$.

2. Let $P(n)$ be the statement in the hint. $P(1)$ is true trivially. Let $n > 1$. If $k \in S$ for some $k < n$, then $S$ will have a least element by the truth of $P(k)$. Otherwise $n$ itself is the least element of $S$.

3. For the case $x < 0$ in the hint, since $\mathbb{N}$ is not bounded above, there is some $n \in \mathbb{N}$ such that $n > -x$. Now, the interval $(x, 0)$ can contain at most the integers, $-1, -2, \ldots, -(n-1)$. To finish the proof, note that every non-empty finite subset of $\mathbb{R}$ has a smallest element.

4. Let $\alpha = \frac{1+\sqrt{5}}{2}$ and $\beta = \frac{1-\sqrt{5}}{2}$. Then $|\frac{\beta}{\alpha}| = |\frac{1-\sqrt{5}}{1+\sqrt{5}}| < 1$. So $\frac{\beta^n}{\alpha^n} \to 0$ as $n \to \infty$.
Now, $\dfrac{F_n}{F_{n-1}} = \dfrac{\alpha^n - \beta^n}{\alpha^{n-1} - \beta^{n-1}} = \alpha \dfrac{1 - (\frac{\beta}{\alpha})^n}{1 - (\frac{\beta}{\alpha})^{n-1}} \to \alpha \dfrac{1-0}{1-0} = \alpha$.

5. Let $L = \lim_{n \to \infty} \frac{F_n}{F_{n-1}}$. The recurrence relation gives $\frac{F_n}{F_{n-1}} = 1 + \frac{1}{F_{n-1}/F_{n-2}}$ and letting $n \to \infty$, $L = 1 + \frac{1}{L}$. The quadratic $L^2 - L - 1 = 0$ has $\frac{1+\sqrt{5}}{2}$ as the only positive root (the other root being $\frac{1-\sqrt{5}}{2}$). By Exercise (1.9), $L \geq 0$ and so $L = \frac{1+\sqrt{5}}{2}$.

6. The equality follows by a direct calculation using $F_{n+1} = F_n + F_{n-1}$ and $L^2 = L + 1$. Since $L, a_{n-1}$ are positive, $a_n - L$ and $L - a_{n-1}$ are of the same sign and so $L - a_n$ and $L - a_{n-1}$ are of opposite signs, which implies the second assertion. Also $a_{n-1} > 1$ for all $n \geq 2$, which, by induction on $n$, implies $|a_n - L| \leq \frac{1}{L^{n-1}} \to 0$, whence $|a_n - L| \to 0$ as $n \to \infty$. For the last assertion, note that $a_n - L$ changes sign but decreases in absolute value for every $n$.

7. Let the sides be $a, b$ with $a > b$. The condition implies that $\frac{a}{b} = \frac{b}{a-b}$, which gives $a^2 - ab - b^2 = 0$, or $(\frac{a}{b})^2 - (\frac{a}{b}) - 1 = 0$.

8. The angle between the two hands is $120°$ at 4.00 p.m. and decreases at the rate of $(6 - \frac{1}{2})°$ per minute. So the two hands will coincide at $\frac{120}{11/2}$, i.e. $\frac{240}{11}$

or 21.818 ... minutes past 4.00 p.m. To get the answer using infinite series, it takes 20 minuts for the minute hand to come to where the hour hand was initially (i.e. at 4.00 p.m.). By that time the hour hand moves some distance to cover which the minute hand needs $20 \times \frac{1}{12}$ minutes, since the minute hand is 12 times as fast as the hour hand. This process continues and gives an infinite geometric series $20 + 20.\frac{1}{12} + 20.\frac{1}{144} + \ldots$, the sum of which is $\frac{20}{1-\frac{1}{12}}$, i.e. $\frac{240}{11}$.

9. When the fly is moving towards $B$, the distance between them decreases at the rate of 24 km/hr, and the fly covers $\frac{20}{24}$th of their initial distance, and during this time $A$ moves by $\frac{5}{24}$th of that distance. A similar calculation holds when the fly starts moving towards $A$. So in the first round trip, the fly covers a distance of $\left(\frac{20}{24} + \frac{15}{24}.\frac{20}{25}\right)$ i.e. $\frac{4}{3}$ km. and at the end of it $A$ and $B$ are at a distance $\frac{15}{24} - \left(\frac{15}{24}.\frac{5}{25} + \frac{15}{24}.\frac{4}{25}\right)$, i.e. $\frac{2}{5}$ km. Applying the same reasoning to the subsequent round trips of the fly the total distance travelled by it is an infinite geometric series with the first term $\frac{4}{3}$ and common ratio $\frac{2}{5}$. So its sum is $\frac{20}{9}$.

10. The mathematicians will meet after $\frac{1}{9}$ hr. The fly, which never rests, travels $\frac{20}{9}$ km. in this time.

11. Let $\alpha, \beta, \gamma$ be respectively, the chances of winning for $A, B, C$. Now, $A$ can win on the 1st, or the 4th, or the 7th or the 10th or etc. round. The probabilities of these wins are, respectively $p, (1-p)^3 p, (1-p)^6 p, (1-p)^9 p, \ldots$ etc. Adding, $\alpha = \frac{p}{1-(1-p)^3} = \frac{1}{p^2-3p+3}$. By similar reasonings, $\beta = \frac{(1-p)p}{p^2-3p+3}$ and $\gamma = \frac{(1-p)^2 p}{p^2-3p+3}$. For a proof without infinite series, the probability of $A$'s losing on the 1st round is $1-p$. After this, it is as if a fresh game starts, except that the order of players is $B, C, A$ rather than $A, B, C$. So the probability of $B$'s winning this new game is $\alpha$. Hence the probability of $B$'s winning the original game is $(1 - p)\alpha$ i.e. $\beta = (1 - p)\alpha$. Similarly $\gamma = (1 - p)^2\alpha$. Determine $\alpha, \beta, \gamma$ from these two equations and $\alpha + \beta + \gamma = 1$.

12. Let $x = 1 - y$, where $0 < y < 1$. Then $n^k x^n > (n + 1)^k x^{n+1}$ will hold whenever $\left(\frac{n+1}{n}\right)^k < \frac{1}{x}$, i.e. $\left(1 + \frac{1}{n}\right)^k - 1 < \frac{1}{x} - 1 = \frac{y}{1-y}$. Using the binomial theorem, $\left(1 + \frac{1}{n}\right)^k - 1$ is a sum of $k$ terms each of which tends to 0 as $n \to \infty$. Make each less than $\frac{y}{k(1-y)}$.

13. By the $n^{\text{th}}$ day, $B$ gets a total of $500n(n + 1)$ rupees but gives matchsticks costing $\frac{2^{n+1}-1}{1000}$ rupees. As $2^{10} > 1000$, it is easily seen that the inequality $500,000n(n + 1) + 1 < 2^{n+1}$ holds for $n = 30$. (In fact, it holds even for $n = 28$, so even February is no good for $B$.)

14. (i) $\sinh x = \sum_{n=0}^{\infty} \frac{x^{2n+1}}{(2n + 1)!}$, $\cosh x = \sum_{n=0}^{\infty} \frac{x^{2n}}{(2n)!}$ (ii) $e^{i\theta} = \sum_{n=0}^{\infty} \frac{i^n \theta^n}{n!}$

$$= \sum_{n=0}^{\infty} \frac{i^{2n}\theta^{2n}}{(2n)!} + \sum_{n=0}^{\infty} \frac{i^{2n+1}\theta^{2n+1}}{(2n + 1)!} = \sum_{n=0}^{\infty} \frac{(-1)^n\theta^{2n}}{(2n)!} + i\sum_{n=0}^{\infty} \frac{(-1)^n\theta^{2n+1}}{(2n + 1)!}$$

$= \cos\theta + i\sin\theta$. From this (or from (i) and using a similar reasoning),

$$\cos\theta = \frac{e^{i\theta} + e^{-i\theta}}{2} = \cosh i\theta \text{ and } \sin\theta = \frac{e^{i\theta} - e^{-i\theta}}{2i} = -i\sinh i\theta.$$

(iii) Use $\frac{d}{dx}(x^n) = nx^{n-1}$.

15. The hint follows by writing $\alpha$ as $\frac{\alpha q}{q}$. Fix any positive irrational number, say, $\alpha = \sqrt{2}$ (Exercise (2.6.4) (i)). By Corollary (3.2), there exists $q \in \mathbb{Q}$ such that $\frac{a}{\alpha} < q < \frac{b}{\alpha}$. Also we may suppose $q \neq 0$, (as otherwise, we find another rational between 0 and $\frac{b}{\alpha}$ ). Take $x$ as $q\alpha$.

16. Apply Corollary (3.2) repeatedly, i.e., after finding some rational $x_1 \in (a, b)$, get another rational $x_2 \in (x_1, b)$, then $x_3 \in (x_2, b)$ and so on. Similarly the last exercise actually implies that there are infinitely many irrationals in $(a, b)$.

# Section 3.4

1. If $n$ is even, then $(-y)^n = y^n$ for all $y$. Hence $x < 0$ can have no $n^{\text{th}}$ root. Also if $x > 0$ and $y$ is the (unique) positive $n^{\text{th}}$ root of $x$, then $(-y)^n = x$ but if $z < 0$ and $(z)^n = x$ then $-z$ would be a positive $n^{\text{th}}$ root of $x$, forcing $z = -y$. If $n$ is odd and $x > 0$, then no negative number can be an $n^{\text{th}}$ root of $x$. There is only one positive $n^{\text{th}}$ root by Theorem (4.1). If $x < 0$, apply this reasoning to $-x$.

2. Let $u = (x^m)^{1/n}$ and $v = x^{1/n}$. Then $(v^m)^n = v^{mn} = v^{nm} = (v^n)^m = x^m = u^n$. So $v^m = u$ by uniqueness of the $n^{\text{th}}$ roots. So $u = (x^{1/n})^m$. (Here the laws of indices have been used only for integral exponents.) Now let $w = (x^p)^{1/q} = (x^{1/q})^p$. Then $w^q = x^p$ and so $w^{pn} = w^{qm} = x^{pm} = (x^m)^p = (u^n)^p = u^{np} = u^{pn}$. So $w = u$ by uniqueness of $(pn)^{\text{th}}$ roots. This shows that $x^y$ is independent of the manner in which $y$ is expressed as a ratio of two integers. From this the laws of indices for rational exponents follow easily from those for integral exponents. For example, to prove $x^{y_1+y_2} = x^{y_1}x^{y_2}$ bring $y_1$ and $y_2$ to a common denominator, say $y_1 = m/p$ and $y_2 = n/p$. Then $x^{y_1+y_2} = x^{\left(\frac{m+n}{p}\right)} = (x^{1/p})^{m+n} = (x^{1/p})^m(x^{1/p})^n = x^{y_1}x^{y_2}$. The other law is even easier. For the extension of Lemma 4.2, suppose $y = \frac{m}{n}$ where $m, n$ are positive integer and let $y_1, y_2$ be non-negative with $z_1 = y_1^{m/n}$ and $z_2 = y_2^{m/n}$. Then, $y_i^m = z_i^n, i = 1, 2$. By the lemma in its present form, $y_1 < y_2 \Leftrightarrow y_1^m < y_2^m \Leftrightarrow z_1^n < z_2^n \Leftrightarrow z_1 < z_2$.

3. Let $y = \frac{p}{q}$, where $p, q \in \mathbb{N}$. Given $R(> 0)$ let $m$ be any integer $> R^{q/p}$. Then for all $n \in \mathbb{N}, n \geq m \Rightarrow n^y > m^y > (R^{q/p})^y = R$ (by the extension of Lemma 4.2, given in the last exercise). If $y < 0$, then $-y > 0$ and $n^y = \frac{1}{n^{-y}} \to 0$ as $n \to \infty$.

4. $0 < \sqrt{n+1} - \sqrt{n} = \frac{1}{\sqrt{n+1}+\sqrt{n}} < \frac{1}{\sqrt{n}}$. By the last exercise and the Sandwich Theorem, $\sqrt{n+1} - \sqrt{n} \to 0$ as $n \to \infty$.

**5.** For the first part, write $p(n) = n^k(a_k + \frac{a_{k-1}}{n} + \ldots + \frac{a_0}{n^k})$ and observe that the bracketed expression tends to $a_k$ as $n \to \infty$. For the second part $\frac{p(n)}{n^{k+1}}$ is a sum of $k+1$ terms each tending to 0.

**6.** Since $p(n) \to \infty$ as $n \to \infty, p(n) > 0$ eventually, i.e. for all except finitely many values of $n$. So $[p(n)]^{1/n}$ makes sense. Let $m$ be such that for all $n \geq m, |\frac{a_{k-1}}{n}|, |\frac{a_{k-2}}{n^2}|, \ldots, |\frac{a_1}{n^{k-1}}|, |\frac{a_0}{n^k}|$ are each less than $\frac{a_k}{2k}$. Then $p(n) = n^k(a_k + \frac{a_{k-1}}{n} + \ldots + \frac{a_0}{n^k})$ lies between $\frac{1}{2}a_k n^k$ and $\frac{3}{2}a_k n^k$. So, $(\frac{1}{2}a_k)^{1/n}(n^{1/n})^k < [p(n)]^{1/n} < (\frac{3}{2}a_k)^{1/n}(n^{1/n})^k$ for all $n \geq m$. Apply Theorems (4.4), (4.5) and the Sandwich Theorem.

**7.** Take $\epsilon = \frac{x}{2}$, in the definition of $\lim_{k\to\infty} x_k = x$ to get $m$ such that $k \geq m \Rightarrow \frac{x}{2} < x_k < \frac{3x}{2}$, which in particular means $x_k > 0$ for $k \geq m$. For the second part, given $\epsilon > 0$, let $\delta_1 = (x^{1/n}+\epsilon)^n - x$ and $\delta_2 = x - (x^{1/n}-\epsilon)^n$. By Lemma (4.2), $\delta_1, \delta_2$ are positive. (We assume $\epsilon < x^{1/n}$, otherwise take a smaller $\epsilon$.) Let $\delta = \min\{\delta_1, \delta_2\}$. Find $r$ such that for all $k \geq r, |x_k - x| < \delta$. Then for all $k \geq r, x_k \in (x - \delta, x + \delta) \subset (x - \delta_2, x + \delta_1) = ((x^{1/n} - \epsilon)^n, (x^{1/n} + \epsilon)^n)$. So again by Lemma (4.2), $x^{1/n} - \epsilon < x_k^{1/n} < x^{1/n} + \epsilon$ for all $k \geq r$. Finally if $y = \frac{m}{n}$ where $m \in \mathbb{Z}, n \in \mathbb{N}$, then $x_k \to x$ implies $x_k^{1/n} \to x^{1/n}$ and hence, by the usual basic properties of limits of sequences, $(x_k^{1/n})^m \to (x^{1/n})^m$, i.e. $x_k^y \to x^y$.

## Section 3.5

**1.** For (i) $\Rightarrow$ (iii), start with $\epsilon = 1$ and $m = 1$ to get $n_1 \geq 1$ such that $|a_{n_1} - L| < 1$. Now take $\epsilon = \frac{1}{2}, m = n_1 + 1$, to get $n_2 > n_1$ such that $|a_{n_2} - L| < \frac{1}{2}$. Next, take $\epsilon = \frac{1}{3}, m = n_2 + 1$ to get suitable $n_3$. Continue. Then $a_{n_k} \to L$ as $k \to \infty$. The implications (iii) $\Rightarrow$ (ii) and (ii) $\Rightarrow$ (i) are trivial.

**2.** In the proof of Theorem (5.2), every neighbourhood of the point $L$ contains infinitely many terms of the sequence $\{x_n\}$. Hence it is a limit point. (Actually, from this point on, the proof of Theorem (5.1) duplicates the argument for the implication (i) $\Rightarrow$ (iii) in the last exercise.) To prove that Theorem (5.1) implies Theorem (2.6.4), suppose $\{x_n\}$ is monotonically increasing and bounded above. As $\{x_n\}$ is trivially bounded below (by $x_1$), by Theorem (5.2), it has a subsequence, say $\{x_{n_k}\}$, converging to, say $L$. The proof is completed by showing that $x_n \to L$ as $n \to \infty$. Given $\epsilon > 0$, find $r$ such that $k \geq r$ implies $|x_{n_k} - L| < \epsilon$, which by monotonicity of $\{x_n\}$ means $x_{n_k} \in (L - \epsilon, L]$. Put $m = n_r$. Then for all $n \geq m, L \geq x_n \geq x_m = x_{n_r} > L - \epsilon$, whene $|x_n - L| < \epsilon$. Finally, to derive Axiom (2.6.2) from Theorem (2.6.4), suppose $A$ is a non-empty subset of $\mathbb{R}$, with an upper bound, say $b_0$. Fix any $a_0 \in A$ and divide the interval $[a_0, b_0]$ at the mid-point $c_0$. If $c_0$ is an upper bound of $A$ call $a_0$ as $a_1$ and $c_0$ as $b_1$.

Otherwise there is some $a_1 \in A$ such that $c_0 < a_1 \leq b_0$. In that case call $b_0$ as $b_1$. In either case $a_1 \in A, b_1$ is an upper bound for $A$ and $b_1 - a_1 \leq \frac{1}{2}(b_0 - a_0)$. Repeating this argument get a sequence of nested intervals $\{[a_n, b_n]\}$ with $a_n \in A, b_n$ an upper bound of $A$, and $b_n - a_n \leq \frac{1}{2}(b_{n-1} - a_{n-1})$. Now show $\lim_{n \to \infty} a_n$ (which exists by Theorem (2.6.4)) equals sup $A$.

3.  All positive integers.

4.  Note that every rational in $(0,1]$ appears infinitely often in the sequence. Corollary (3.2) and Exercise (5.1) above imply the result.

5.  (i) If $x \in (a, b)$ then $(x - r, x + r) \subset (a, b)$ where $r = \min\{x - a, b - x\}$. So $(a, b)$ is an open set. For intervals bounded at one end, let $r$ be the distance of $x$ from that end. For $(-\infty, \infty)$, any $r > 0$ will work. If $x$ is in the union of open intervals then $x$ is in one of them, which must, therefore, also contain $(x - r, x + r)$ for some $r > 0$. But then so does the union. The empty set is vacuously open. (See the comments on vacuous truth in Section 1.5.)

    (ii) Let $x \in U \cap V$ where $U, V$ are open. Then there exist $r_1, r_2 > 0$ such that $(x - r_1, x + r_1) \subset U$ and $(x - r_2, x + r_2) \subset V$. Then $(x - r, x + r) \subset U \cap V$ where $r = \min\{r_1, r_2\}$. (Essentially, all that is used here is that the intersection of two neighbourhoods is a neighbourhood. The definition of an open set can be paraphrased as a set which contains a neighbourhood of each of its points.)

6.  (i) Write the complements of the given sets as unions of one or more open intervals. (ii) The interval $[a, b)$ contains no neighbourhood of $a$ and hence is not open. Its complement is $(-\infty, a) \cup [b, \infty)$ which contains no neighbourhood of the point $b$ in it and hence is not open either. A similar argument holds for $(a, b]$. (iii) If $B$ is closed, $a_n \to L$ with $a_n \in B$ for all $n$ and $L \in \mathbb{R} - B$, then there is some $\epsilon > 0$ such that $(L - \epsilon, L + \epsilon) \subset \mathbb{R} - B$, i.e. $(L - \epsilon, L + \epsilon) \cap B = \emptyset$. This $\epsilon$ violates the definition of $\lim_{n \to \infty} a_n = L$. Conversely if $B$ is not closed, then $\mathbb{R} - B$ is not open and so there exists some $x \in \mathbb{R} - B$ such that for every $r > 0$, $(x - r, x + r) \not\subset \mathbb{R} - B$, i.e. $(x - r, x + r) \cap B \neq \emptyset$. Taking $r = 1, \frac{1}{2}, \frac{1}{3}, \frac{1}{4}, \ldots, \frac{1}{n}$, get a sequence $\{a_n\}$ in $B$ such that $|a_n - x| < \frac{1}{n}$ for all $n$. But then $a_n \to x$, a contradiction. (iv) Let $B$ be the set of all limit points of $\{x_n\}$. If $x \in \mathbb{R} - B$, then there is some $\epsilon > 0$ such that $(x - \epsilon, x + \epsilon)$ contains $x_n$ for at most finitely many values of $n$. Given $y \in (x - \epsilon, x + \epsilon)$, by (i) of Exercise (5.5), there exists $r > 0$ such that $(y - r, y + r) \subset (x - \epsilon, x + \epsilon)$. But then $(y - r, y + r)$ too contains $x_n$ for at most finitely many values of $n$. Hence $y$ cannot be a limit point of $\{x_n\}$, i.e. $y \in \mathbb{R} - B$. So $(x - \epsilon, x + \epsilon) \subset \mathbb{R} - B$ showing that $\mathbb{R} - B$ is open.

7.  Following the hint, the supremum and the infimum belong to that set.

8.  If $a_n = L$ for infinitely many values of $n$, we are done. Otherwise without loss of generality, assume $a_n < L$ for infinitely values of $n$, say, $n = m_1, m_2, \ldots, m_r, \ldots$ with $m_1 < m_2 < m_3 < \ldots$. (We are not claiming $a_{m_1} < a_{m_2} < a_{m_3}$ etc.) Call $a_{m_k}$ as $y_k$. Then $\{y_k\}$, being a subsequence of

$\{a_n\}$, converges to $L$, and $y_k < L$ for all $k$. It suffices to find a monotonically increasing subsequence of $\{y_k\}$. Take $k_1 = 1$ and $\epsilon_1 = L - y_{k_1}$. Then there exists $k_2 > k_1$ such that $y_{k_2} \in (L - \epsilon_1, L + \epsilon_1)$, which can only mean $L - \epsilon_1 < y_{k_2} < L$. Hence $y_{k_1} < y_{k_2} < L$. Now take $\epsilon_2 = L - y_{k_2}$ to get $k_3 > k_2$ such that $y_{k_2} < y_{k_3} < L$. Continue.

9. If $\{a_n\}$ is not bounded above, take $n_1 = 1$. Then $\exists \, n_2 > n_1$ such that $a_{n_2} > a_{n_1}$, as otherwise $a_{n_1}$ would be an upper bound on $\{a_n\}$. Having chosen $n_2$, $\exists \, n_3 > n_2$ such that $a_{n_3} > a_{n_2}$ as otherwise $\max\{a_1, a_2, \ldots, a_{n_2}\}$ would be an upper bound on $\{a_n\}$. Continue. If $\{a_n\}$ is unbounded below, consider $\{-a_n\}$. Other cases are covered by the hint.

10. Suppose $\{a_n\}$ is monotonically increasing with an upper bound $L$. If $\{a_n\}$ is not Cauchy, then there exists $\epsilon > 0$ such that for all $p \in \mathbb{N}, \exists m, n \in \mathbb{N}$ with $p \le m < n$ and $|a_m - a_n| \ge \epsilon$. Giving $p$ larger and larger values, we get integers $m_1 < n_1 < m_2 < n_2 < \ldots < m_k < n_k < m_{k+1} < \ldots$, such that the intervals $(a_{m_k}, a_{n_k}), k = 1, 2, \ldots$ are mutually disjoint and of length $\ge \epsilon$ each. But they are all contained in $(a_1, L)$ which is of finite length. This would give an upper bound on $\mathbb{N}$, vis. $\frac{L - a_1}{\epsilon}$, a contradiction. The derivation of Axiom (2.6.2) from Theorem (2.6.4) was already given in Exercise (5.2).

11. Letting $S_n = \sum_{k=1}^{n} a_k$, simply observe that $\sum_{k=m}^{n} a_k = S_n - S_{m-1}$. The theorem merely says that $\{S_n\}$ is convergent iff it is a Cauchy sequence.

12. Direct part is trivial. For the converse, see the solution to Exercise (5.2).

## Section 3.6

1. Following the hint, since $2^m \le n < 2^{m+1}, 2^{m-1} H_n$ equals $\frac{1}{2} +$ a sum of rationals with odd denominators. If $2^{m-1} H_n$ were an integer then these rationals would add up to an odd multiple of $\frac{1}{2}$, a contradiction.

2. Argue as in the proof of Theorem (6.3), noting that for $y \le 1, \frac{1}{2^{y-1}} \ge 1$.

3. Fix $m \in \mathbb{N}$ such that $d \le ma$. Then $S_n = \sum_{k=1}^{n} \frac{1}{ka+d} \ge \sum_{k=1}^{n} \frac{1}{a(k+m)} = \frac{1}{a}(H_{m+n} - H_m) \to \infty$ as $n \to \infty$.

4. By Theorem (5.5), $\exists \, p$ such that $-\frac{\epsilon}{2} < \sum_{k=n+1}^{r} a_k < \frac{\epsilon}{2}$ whenever $p \le n < r$. Put $m = p$. Then for any $n \ge m$, by Exercise (1.9), $-\frac{\epsilon}{2} \le \lim_{r \to \infty} \sum_{k=n+1}^{r} a_k \le \frac{\epsilon}{2}$, whence $-\epsilon < \sum_{k=n+1}^{\infty} a_k < \epsilon$.

5. $(n+1)! + k$ is divisible by $k$ for $k = 2, 3, \ldots, n+1$.

6. Let the primes in $A_1$ be $p_1, p_2, p_3, \ldots$ in ascending order and let $A_2$ consist of $n_1, n_2, n_3, \ldots$ in ascending order. Then $\sum_{n \in A_2} \frac{1}{n} = \lim_{r \to \infty} \sum_{k=1}^{r} \frac{1}{n_k}$ and so it suffices to show that for every $r \in \mathbb{N}, \sum_{k=1}^{r} \frac{1}{n_k} \le L^2$. For any such fixed $r$, there exists $q$ such that the prime factors of $n_1, \ldots, n_r$ are

all contained in $\{p_1, p_2, \ldots, p_q\}$. Since each $n_k$ is of the form $p_i p_j$ for some $i, j \in \{1, 2, \ldots, q\}$, $\sum_{k=1}^{r} \frac{1}{n_k} \leq (\frac{1}{p_1} + \ldots + \frac{1}{p_q})(\frac{1}{p_1} + \ldots + \frac{1}{p_q})$ (Usually strict inequality will hold since $\frac{1}{p_i p_j}$ and $\frac{1}{p_j p_i}$ both appear in the R.H.S.) But $\sum_{i=1}^{q} \frac{1}{p_i} \leq L$. (Basically we are taking the product of $\sum_{n=1}^{\infty} \frac{1}{p_n}$ with itself. This will be formally defined in the next section.)

7. Similar to the last exercise. ($L < 1$ is not needed in these two exercises.)

8. Every element of $A_*$ is in $A_m$ for a unique positive integer $m$. So $\sum_{n \in A_*} \frac{1}{n} = \sum_{m=1}^{\infty} \sum_{n \in A_m} \frac{1}{n} \leq \sum_{m=1}^{\infty} L^m = \frac{L}{1-L}$ by the last exercise and Theorem (3.5).

9. No $p_i (1 \leq i \leq m)$ can divide $ka + 1$ (since $p_i$ divides $a$). So every element of $S$ is the product of primes coming from $A$ only.

10. Follow the hint. The inequality $\sum_{n \in S} \frac{1}{n} \leq \sum_{n \in A_*} \frac{1}{n}$ follows from the fact that $S \subset A_*$ and that $\frac{1}{n} > 0$ for all $n \in A_*$.

## Section 3.7

1. Follow the hint. Fix $n$. For $i = 1, \ldots, n, \exists k_i$ such that $a_i = b_{k_i}$. Let $m = \max\{k_1, \ldots, k_n\}$. Then every term in $S_n$ (i.e. in $a_1 + a_2 + \ldots + a_n$) appears in $T_m$ (i.e. in $b_1 + b_2 + \ldots + b_m$). So $\sum_{n=1}^{\infty} a_n = \lim_{n \to \infty} S_n \leq \sum_{j=1}^{\infty} b_j$ by Exercise (1.9). Interchange the roles of $a$'s and $b$'s to get the other way inequality.

2. Let $S = \{\sum_{x \in A} f(x) : A \subset X, A \text{ finite}\}$ and for $i = 1, 2, \ldots, k$ let $S_i = \{\sum_{x \in A} f(x) : A \subset X_i, A \text{ finite}\}$. For every finite subset $A$ of $X, A$ is the union of mutually disjoint subsets $A \cap X_1, \ldots, A \cap X_k$ and so $\sum_{x \in A} f(x) = \sum_{i=1}^{k} \sum_{x \in A \cap X_i} f(x) \leq \sum_{i=1}^{k} \sup S_i$. Hence $\sup S \leq \sum_{i=1}^{k} \sup S_i$.

   If strict inequality holds, let $\epsilon > 0$ be such that $\sup S + \epsilon < \sum_{i=1}^{k} \sup S_i$. For $i = 1, \ldots, k, \exists$ a finite subset $A_i$ of $X_i$ such that $\sum_{x \in A_i} f(x) + (\epsilon/k) > \sup S_i$. Let $A = A_1 \cup A_2 \cup \ldots \cup A_k$. Then $\sum_{x \in A} f(x) = \sum_{i=1}^{k} \sum_{x \in A_i} f(x) > \sum_{i=1}^{k} \sup S_i - k(\epsilon/k) > \sup S$, a contradiction.

3. $Y_k$ is the set of $k-1$ points $(1, k-1), (2, k-2), \ldots, (i, k-1-i), \ldots, (k-1, 1)$ all of which lie on the line $y = -x + k$.

4. Let $X = \mathbb{N} \times \mathbb{N} = Y_1 \cup Y_2 \cup Y_3 \cup \ldots$ be the decomposition in the last exercise. For positive integers $i, j$, let $f(i, j) = a_i b_j$. Then $c_k$ is simply

$$\sum_{(i,j) \in Y_k} f(i,j) \text{ and } \sum_{k=1}^{\infty} c_k \text{ is } \sum_{k=1}^{\infty} \sum_{(i,j) \in Y_k} f(i,j) \text{ which by Theorem (7.4) equals}$$

$$\sum_{(i,j) \in X} f(i,j). \text{ But by Theorem (7.6) the latter equals } (\sum_{m=1}^{\infty} a_m)(\sum_{n=1}^{\infty} b_n).$$

So if $\sum_{m=1}^{\infty} a_m$ and $\sum_{n=1}^{\infty} b_n$ are both convergent then so is $\sum_{n=1}^{\infty} c_n$ and equals $(\sum_{m=1}^{\infty} a_m)(\sum_{n=1}^{\infty} b_n)$.

5.   The essential idea is the same, but the notations are complex. Let $k$ $(\geq 2)$ be a positive integer and suppose $\sum_{n=1}^{\infty} a_n^{(1)}, \sum_{n=1}^{\infty} a_n^{(2)}, \ldots, \sum_{n=1}^{\infty} a_n^{(k)}$ are series of non-negative terms. Let $\theta$ be a permutation of the indices $1, 2, \ldots, k$. Formally, $\theta$ is a bijection of the set $\{1, 2, \ldots, k\}$ onto itself. Then the two multiply infinte sums $\sum_{n_1=1}^{\infty} \sum_{n_2=1}^{\infty} \cdots \sum_{n_k=1}^{\infty} a_{n_1}^{(1)} a_{n_2}^{(2)} \cdots a_{n_k}^{(k)}$ and

$$\sum_{n_{\theta(1)}=1}^{\infty} \sum_{n_{\theta(2)}=1}^{\infty} \cdots \sum_{n_{\theta(k)}=1}^{\infty} a_{n_1}^{(1)} a_{n_2}^{(2)} \cdots a_{n_k}^{(k)}$$ are equal. For a proof, first do the

case where $\theta$ merely interchanges one pair of indices, leaving all other indices unchanged. (For example, with $k = 4, \theta(1) = 1, \theta(2) = 4, \theta(3) = 3$ and $\theta(4) = 2$.) Such a permutation is known as a **transposition.** In this case the argument is essentially a duplication of the proof of Theorem (7.6). The general case requires a fact about permutations (provable by induction on $k$) that every permutation can be expressed as a composite of transpositions.

6.   Take $b_n = a_n$ for all $n$. Then $\sum_{m=1}^{\infty} \sum_{n=1}^{\infty} a_m a_n$ is finite, and so $\sum_{n=1}^{\infty} a_n a_n$, which is a sum over the subset $\{(n, n) : n \in \mathbb{N}\}$ of $\mathbb{N} \times \mathbb{N}$ is also convergent. For a direct proof, note that by Exercise (1.8), $\exists\, p$ such that $0 \leq a_n \leq 1$ for all $n \geq p$. But then for all $k \geq p, a_k^2 \leq a_k$ and hence for all $n \geq m \geq p, \sum_{k=m}^{n} a_k^2 \leq \sum_{k=m}^{n} a_k$. Apply Theorem (5.5).

7.   Let $Y = X_1 \cup X_2, Y_0 = X_1 \cap X_2, Y_1 = X_1 - Y_0$ and $Y_2 = X_2 - Y_0$.

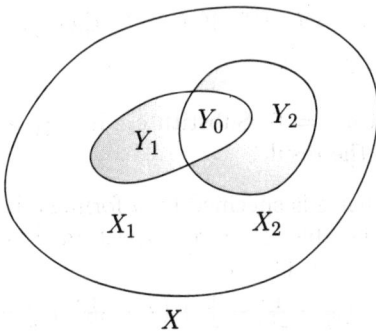

Then from the Venn diagram it is clear that $Y_0 \cup Y_1 \cup Y_2$ is a decomposition of $Y$ and $Y_0 \cup Y_i$ is a decomposition of $X_i$ for $i = 1, 2$ into mutually disjoint subsets. So $\sum_{x \in X_i} f(x) = \sum_{x \in Y_i} f(x) + \sum_{x \in Y_0} f(x)$ for $i = 1, 2$ and $\sum_{x \in Y} f(x) = \sum_{x \in Y_1} f(x) + \sum_{x \in Y_2} f(x) + \sum_{x \in Y_0} f(x)$. This implies the result easily. For three subsets $X_1, X_2, X_3$ of $X$, a similar Venn diagram would give

$$\sum_{x \in X_1 \cup X_2 \cup X_3} f(x) = \sum_{x \in X_1} f(x) + \sum_{x \in X_2} f(x) + \sum_{x \in X_3} f(x)$$
$$- \sum_{x \in X_1 \cap X_2} f(x) - \sum_{x \in X_2 \cap X_3} f(x) - \sum_{x \in X_3 \cap X_1} f(x)$$
$$+ \sum_{x \in X_1 \cap X_2 \cap X_3} f(x).$$

The result can be generalised for any (finite) number of subsets. It is called the **principle of inclusion and exclusion,** because for every $x \in X$, the term $f(x)$ appears several times (depending upon which subsets contain $x$), sometimes with a $+$ sign (inclusion) and sometimes with a $-$ sign (exclusion), the net count being once.

8.　$a_{m,n}$ equals 0 if $n > m$, $-1$ if $n = m$ and $2^{(n-m)}$ if $n < m$. For every fixed $n$, $\sum_{m=1}^{\infty} a_{m,n}$ is 0 and so $\sum_{n=1}^{\infty} \sum_{m=1}^{\infty} a_{m,n} = 0$. But, for a fixed $m$, $\sum_{n=1}^{\infty} a_{m,n}$ equals $-\frac{1}{2^{m-1}}$ and so $\sum_{m=1}^{\infty} \sum_{n=1}^{\infty} a_{m,n} = \sum_{m=1}^{\infty} -\frac{1}{2^{m-1}} = -2$.

9.　See Exercise (1.1.6) (iii).

## Section 3.8

1.　$S_{2n+2} = S_{2n} + a_{2n+1} - a_{2n+2} \geq S_{2n}$ and $S_{2n+3} = S_{2n+1} - (a_{2n+2} - a_{2n+3}) \leq S_{2n+1}$. Also $S_{2n} = a_1 - (a_2 - a_3) - (a_4 - a_5) - (a_4 - a_5) \ldots - (a_{2n-2} - a_{2n-1}) - a_{2n} \leq a_1$. So $\{S_{2n}\}$ is bounded above. Similarly $\{S_{2n+1}\}$ is bounded below (by 0). So both are convergent. Also $|S_{2n+1} - S_{2n}| = |a_{2n+1}| \to 0$ as $n \to \infty$. So their limits are equal. The proof is vividly pictured by

$$0 \leq S_2 \leq S_4 \leq \ldots \leq S_{2n} \ldots \to S \leftarrow \ldots S_{2n+1} \leq \ldots \leq S_3 \leq S_1 = a_1$$

which also shows that for all $n$, $S$ lies between $S_{n-1}$ and $S_n$ (or between $S_n$ and $S_{n-1}$) and hence $|S_n - S| \leq |S_n - S_{n-1}| \leq a_n$ for all $n$.

2.　If $a_n \to L$ then $\big||a_n| - |L|\big| \leq |a_n - L| \to 0$ as $n \to \infty$. So $|a_n| \to |L|$. Falsity of the converse is proved by taking $a_n = (-1)^n$. If $L = 0$, then $|a_n - L| = \big||a_n| - |L|\big|$ and so the converse holds.

3.　If $a \geq 0$, then $a^+ = a$, $|a| = a$ and $a^- = 0$. Substitute. If $a \leq 0$ then $a^+ = 0$, $|a| = -a$ and $a^- = -a$ imply the result.

4.　In problems of this type where a sequence is specified by a formula, it pays to write down the first few terms, to be able to see some pattern. Here the series $\sum_{n=0}^{\infty} a_n$ looks (keeping in mind that $0! = 1$),

$$\tfrac{1}{0!} + \left(\tfrac{1}{0!} - \tfrac{1}{2^{-1}}\right) + \tfrac{1}{1!} + \left(\tfrac{1}{1!} - \tfrac{1}{2^0}\right) + \tfrac{1}{2!} + \left(\tfrac{1}{2!} - \tfrac{1}{2^1}\right) + \tfrac{1}{3!} + \left(\tfrac{1}{3!} - \tfrac{1}{2^2}\right) + \tfrac{1}{4!} + \ldots.$$

So $a_n > 0$ if $n$ is even and $a_n \leq 0$ if $n$ is odd. The partial sums $\sum_{k=0}^{2n+1} a_k$ and $\sum_{k=0}^{2n+1} |a_k|$ equal, respectively, $2\sum_{k=0}^{n} \tfrac{1}{k!} - \sum_{k=0}^{n} \tfrac{1}{2^{k-1}}$ and $\sum_{k=0}^{n} \tfrac{1}{2^{k-1}}$ which tend to $\tfrac{2}{e} - 4$ and to 4 respectively. Since $e$ is irrational so is $\tfrac{2}{e}$ and hence $\tfrac{2}{e} - 4$ (or else $\tfrac{2}{e} - 4 + 4$ would be rational.)

5.　The result in the hint is proved by noting that since for every finite subset $A$ of $X$, $\sum_{x \in A} g(x) \leq \sum_{x \in A} h(x)$, any upper bound on the set of sums $\{\sum_{x \in A} h(x) : A \subset X, A \text{ finite}\}$ is also an upper bound for the set of sums $\{\sum_{x \in A} g(x) : A \subset X, A \text{ finite}\}$. Now, from Exercise (8.3), $f = f^+ - f^-$ and $|f| = f^+ + f^-$. So summability of $f$ implies that of $f^+$ and $f^-$ (by definition) and hence of $|f|$ by Theorem (7.3). Conversely $0 \leq f^+ \leq |f|$ and $0 \leq f^- \leq |f|$ and the result in the hint show that whenever $|f|$ is summable, so are $f^+$ and $f^-$. But then $f$ is summable.

**6.** Merely write $f$ as $f^+ - f^-$ and apply Theorem (7.4) to $f^+$ and $f^-$ separately.

**7.** This follows from the last exercise the same way as Corollary (7.5) follows from Theorem (7.4).

**8.** Convergence of $\sum_{n=1}^{\infty} \frac{(-1)^{n+1}}{\sqrt{n}}$ follows from Exercise (8.1). For divergence of

$\sum_{n=1}^{\infty} \frac{1}{\sqrt{n}}$, see Exercise (6.2). Now, $|c_n| = |\sum_{k=1}^{n-1} a_k a_{n-k}| = |\sum_{k=1}^{n-1} \frac{(-1)^{n+2}}{\sqrt{k}\sqrt{n-k}}|$

$= |\sum_{k=1}^{n-1} \frac{1}{\sqrt{k}\sqrt{n-k}}|$. From the fact that for positive $x, y$, $\frac{1}{\sqrt{xy}} \geq \frac{2}{x+y}$ it

follows that $|c_n| \geq \frac{2(n-1)}{n}$. So $c_n \not\to 0$ as $n \to \infty$ and hence by Exercise

(1.8), $\sum_{n=1}^{\infty} c_n$ is divergent.

**9.** Let $L = \sum_{m=1}^{\infty} |a_m|$ and $M = \sum_{n=1}^{\infty} |b_n|$. Then $L, M$ are finite. For the function $f$ in the hint, if $A$ is any finite subset of $\mathbb{N} \times \mathbb{N}$, then there exist integers $m, n$ such that $A \subset \{1, 2, \ldots, m\} \times \{1, 2, \ldots, n\}$. So $\sum_{x \in A} |f(x)| = \sum_{(i,j) \in A} |a_i b_j| \leq \sum_{i=1}^{m} \sum_{j=1}^{n} |a_i b_j| = (\sum_{i=1}^{m} |a_i|)(\sum_{j=1}^{n} |b_j|) \leq LM$. So $|f|$, and hence $f$ is summable over $\mathbb{N} \times \mathbb{N}$ by Exercise (8.5). By Exercise (8.6), the decomposition theorem holds for $f$ on $\mathbb{N} \times \mathbb{N}$. The Cauchy product corresponds to one such decomposition by Exercises (7.3) and (7.4).

## Section 3.9

**1.** $\dfrac{an+b}{n(n+1)(n+2)} = \dfrac{a}{(n+1)(n+2)} + \dfrac{b}{n(n+1)(n+2)}$. So it suffices to find

$\displaystyle\sum_{n=1}^{\infty} \frac{1}{(n+1)(n+2)}$ and $\displaystyle\sum_{n=1}^{\infty} \frac{1}{n(n+1)(n+2)}$. The former can be written as

a telescopic series with sum $\dfrac{1}{2}$. Writing $\dfrac{1}{n(n+1)(n+2)}$ as $\dfrac{1}{2}(\dfrac{1}{n} - \dfrac{1}{n+1}) -$

$\dfrac{1}{2}(\dfrac{1}{n+1} - \dfrac{1}{n+2})$ gives $\displaystyle\sum_{n=1}^{\infty} \frac{1}{n(n+1)(n+2)}$ as a sum of two telescoping

series with total value $\dfrac{1}{4}$.

**2.** By induction on $n$, $a_n \leq k b_n$ where $k = \frac{a_1}{b_1}$. Apply Comparison Test to $\sum a_n$ and $\sum k b_n$.

**3.** Suppose $y_n = \dfrac{x_1 + x_2 + \ldots + x_n}{n}$ where $x_n \to \infty$ as $n \to \infty$. Given $R \ (> 0)$,

first find $r$ such that for all $n \geq r$, $x_n > R+1$. Writing $y_n$ as $\dfrac{x_1 + \ldots + x_r}{n} +$

$\dfrac{x_{r+1} + \ldots + x_n}{n}$, it follows that for $n \geq r$, $y_n \geq \dfrac{\lambda}{n} + R + 1$ where $\lambda = x_1 + \ldots +$

$x_r - r(R + 1)$. Now find $m \geq r$ such that $|\dfrac{\lambda}{n}| < 1$ for

all $n \geq m$. Then $y_n > R$ for all $n \geq m$. Now let $a_n = (n!)^{1/n}$. Then

$\log a_n = \dfrac{1}{n}(\log 1 + \log 2 + \ldots + \log n)$. Since $\log n \to \infty$ as $n \to \infty$, it follows that $\log a_n$ and hence $a_n \to \infty$ as $n \to \infty$.

4.  For a direct proof, fix $r \in I\!N$ such that $|x| < r$. Let $k = \dfrac{|x|^r}{r!}$. Then for $n > r, |\dfrac{x^n}{n!}| = k |\dfrac{x}{r+1}||\dfrac{x}{r+2}|\ldots|\dfrac{x}{n-1}||\dfrac{x}{n}| \le \dfrac{|kx|}{n} \to 0$ as $n \to \infty$. By Ratio Test, $\displaystyle\sum_{n=0}^{\infty} \dfrac{x^n}{n!}$ converges and hence by Exercise (1.8), $\dfrac{x^n}{n!} \to 0$ as $n \to \infty$. The second proof is slicker because it uses a powerful theorem.

5.  Convergence of $\displaystyle\sum_{n=1}^{\infty} \dfrac{a_n}{r^n}$ follows by comparison with the geometric series $\displaystyle\sum_{n=1}^{\infty} \dfrac{1}{r^{n-1}}$, since $0 \le \dfrac{a_n}{r^n} < \dfrac{r}{r^n}$. Now suppose $\displaystyle\sum_{n=1}^{\infty} \dfrac{a_n}{r^n} = \sum_{n=1}^{\infty} \dfrac{b_n}{r^n}$ where the sequences $\{a_n\}_{n=1}^{\infty}$ and $\{b_n\}_{n=1}^{\infty}$ are not the same. Let $k$ be the smallest integer such that $a_k \ne b_k$ and assume $a_k > b_k$. Then $\displaystyle\sum_{n=k}^{\infty} \dfrac{a_n}{r^n} = \sum_{n=k}^{\infty} \dfrac{b_n}{r^n}$ and so $\dfrac{1}{r^k} \le \dfrac{(a_k - b_k)}{r^k} = \displaystyle\sum_{n=k+1}^{\infty} \dfrac{b_n - a_n}{r^n} \le \sum_{n=k+1}^{\infty} \dfrac{r-1}{r^n} = \dfrac{r-1}{r^{k+1}} \dfrac{1}{1-\frac{1}{r}} = \dfrac{1}{r^k}$ (using Theorem (3.5)). This forces equality throughout, i.e. $a_k = b_k + 1$ and $b_n - a_n = r - 1$ for every $n > k$, which can happen only when $b_n = r - 1$ and $a_n = 0$. So (i) holds. Similarly if $a_k < b_k$ then (ii) holds. Conversely if (i) or (ii) holds then by summing suitable geometric series it follows that $\displaystyle\sum_{n=1}^{\infty} \dfrac{a_n}{r^n} = \sum_{n=1}^{\infty} \dfrac{b_n}{r^n}$. Finally suppose $x \in [0, 1]$. If $x = 1$, take $a_n = r - 1$ for all $n \ge 1$. Otherwise there is a unique $k$ ($0 \le k \le r-1$) such that $x \in [\dfrac{k}{r}, \dfrac{k+1}{r})$. Put $a_1 = k$. Then $0 \le x - \dfrac{a_1}{r} < \dfrac{1}{r}$ and so there is a unique $k$ (possibly different from the earlier $k$), $0 \le k \le r-1$ such that $x - \dfrac{a_1}{r} \in [\dfrac{k}{r^2}, \dfrac{k+1}{r^2})$. Put $a_2 = k$. Now $x - \dfrac{a_1}{r} - \dfrac{a_2}{r^2}$ lies in $[0, \dfrac{1}{r^2})$ and hence in $[\dfrac{k}{r^3}, \dfrac{k+1}{r^3})$ for a unique $k, 0 \le k \le r-1$. Put $a_3 = k$. Continue *ad infinitum* to get the desired $\{a_n\}$.

6.  A real number $x$ is rational iff its fractional part ($= x - [x]$) is so. So it suffices to prove the assertion for $x \in [0, 1)$. Let $0.a_1 a_2 a_3 \ldots a_n \ldots$ be a decimal expansion of $x$. If it is terminating, clearly $x$ is rational (in fact, of the form $\dfrac{k}{10^q}$ for some integers $k, q$). Suppose the expansion is recurring, say, $x = 0.a_1 a_2 \ldots a_m \overline{a_{m+1} \ldots a_n}$. Then $10^m x = a_1 a_2 \ldots a_m . \overline{a_{m+1} \ldots a_n}$
$= y + \lambda(1 + \dfrac{1}{10^{n-m}} + \dfrac{1}{10^{2(n-m)}} + \ldots)$ where $y = a_m + 10a_{m-1} + \ldots + 10^{m-1}a_1$ which is an integer and $\lambda = \dfrac{a_{m+1}}{10} + \dfrac{a_{m+2}}{10^2} + \ldots + \dfrac{a_n}{10^{n-m}}$ which

is a rational. This gives $x = \dfrac{y}{10^m} + \lambda \dfrac{10^{n-m}}{10^n - 10^m}$, a rational. For the converse let $x = \frac{p}{q}$, in the reduced form. The process in the solution to the last exercise for obtaining the decimal expansion of $x$ reduces to the long division of $p$ by $q$. It will terminate iff the only primes dividing $q$ are 2 and 5. Otherwise, since the remainder every time is between 1 and $q - 1$, after some stage the same remainder will occur and the entire calculation will go on repeating in cycles, leading to a recurring decimal expansion. For radix $r$ (instead of 10), the expansion is recurring iff all prime divisors of $q$ also divide $r$.

7.  By Exercise (9.5), each real has at most two decimal expansions. Choose $a_n$ so that it differs from the digits in the $n^{\text{th}}$ place of decimal in both the expansions of $x_n$ by at least 2. Then the number $0.a_1 a_2 \ldots a_n \ldots$ differs from $x_n$ for every $n$, a contradiction. An enumeration of all rationals in $[0,1]$ is given by $x_1 = 0, x_2 = 1, x_3 = \frac{1}{2}, x_4 = \frac{1}{3}, x_5 = \frac{2}{3}, x_6 = \frac{1}{4}, x_7 = \frac{3}{4}, x_8 = \frac{1}{5}, x_9 = \frac{2}{5}, \ldots$ where we go on increasing the denominator and for a fixed denominator, increasing the numerator, ensuring that they are always relatively prime.

8.  (i) if the assertion is false then for some $\epsilon > 0$, the interval $[L + \epsilon, M]$ (where $M$ is any upper bound on $\{a_n\}$) will contain a subsequence, say, $\{a_{n_k}\}$. By Exercise (5.2), $\{a_{n_k}\}$ has a limit point, say $L'$, which is also a limit point of $\{a_n\}$. But $L' \geq L + \epsilon > L$, a contradiction. The analogous result for $\underline{\lim} a_n$ is that if $\underline{\lim}_{n \to \infty} a_n = L$ then $\forall \epsilon > 0$, $\exists m$ s.t. $\forall n \geq m, a_n > L - \epsilon$. This is proved analogously or by applying the earlier result to $\{-a_n\}$ and noting that $\underline{\lim}_{n \to \infty} a_n = -\overline{\lim}_{n \to \infty} - a_n$. (ii) Let $L = \underline{\lim}_{n \to \infty} a_n$ and $L' = \overline{\lim}_{n \to \infty} a_n$. Clearly $L \leq L'$. If $L < L'$, let $\epsilon = \frac{L' - L}{4}$. Then for every real number $M$, at least one of the intervals $(L - \epsilon, L + \epsilon)$ and $(L' - \epsilon, L' + \epsilon)$ is disjoint from $(M - \epsilon, M + \epsilon)$. But each contains infinitely many terms of $\{a_n\}$. So $\{a_n\}$ cannot converge to $M$. Conversely if $L = L'$, then for every $\epsilon > 0$, by (i) (and its analogue for $\underline{\lim}$) there exist $m_1, m_2$ such that $a_n < L + \epsilon$ for all $n \geq m_1$ and $L' - \epsilon < a_n$ for all $n \geq m_2$. Let $m = \max\{m_1, m_2\}$. Then $a_n \in (L - \epsilon, L + \epsilon)$ for all $n \geq m$. So $a_n \to L(= L')$ as $n \to \infty$.

9.  Let $L = \overline{\lim}_{n \to \infty} a_n^{1/n}$ and $L' = \overline{\lim}_{n \to \infty} \frac{a_{n+1}}{a_n}$. If $L > L'$, let $\epsilon = \frac{L - L'}{2} > 0$. By (i) of the last exercise, $\exists m$ such that $\frac{a_{n+1}}{a_n} < L'$ for all $n \geq m$. Writing $a_n$ as $\frac{a_n}{a_{n-1}} \cdot \frac{a_{n-1}}{a_{n-2}} \cdot \ldots \cdot \frac{a_{m+1}}{a_m} \cdot a_m$ and letting $K = \frac{a_m}{(L' + \epsilon)^m}$, we get $a_n \leq (L' + \epsilon)^n K$ for all $n \geq m$. Let $\{a_{n_k}^{1/n_k}\}$ be a subsequence of $\{a_n^{1/n}\}$ converging to $L$. Then $L = \lim_{k \to \infty} a_{n_k}^{1/n_k} \leq \lim_{k \to \infty} (L' + \epsilon) K^{1/n_k} = L' + \epsilon$. So $L \leq L' + \epsilon$, a contradiction. This proves the last inequality of the assertion. The proof of the first one is similar (or apply the last inequality to $\frac{1}{a_n}$, after noting that taking reciprocals reverses $\overline{\lim}$ and $\underline{\lim}$). The middle inequality is trivially true. For the sequence $1, 2^2, 1, 2^4, 1, 2^6, 1, 2^8, \ldots$ all three inequalities are strict as the respective quantities are $0, 1, 2$ and $\infty$.

**10.** The proof of Theorem (9.3) goes through using the properties of lim sup and lim inf in Exercise (9.8) (i).

**11.** For the first part, the proof of Theorem (9.5) goes through, again using Exercise (9.8) (i). For the second part, the hypothesis implies that there exists $\epsilon > 0$ and a subsequence $\{a_{n_k}\}$ such that $a_{n_k}^{1/n_k} > 1 + \epsilon$ for all $k$. But then $a_{n_k} > (1 + \epsilon)^{n_k} > 1$. Hence $a_n \not\to 0$ as $n \to \infty$. Apply Exercise (1.8).

**12.** The crucial task is to prove that $\lim_{n\to\infty} n(1-(1-\frac{1}{n})^y) = y$. If $y$ is an integer, this can be done by the binomial theorem. Otherwise more sophisticated techniques such as expanding $\left(1 - \frac{1}{n}\right)^y$ in powers of $\frac{1}{n}$ (see Exercise (4.6.13)) or L'Hôpital's rule (Section 4.9) are needed.

**13.** Let $l = \lim_{n\to\infty} n(1 - \frac{a_n}{a_{n-1}})$. If $l > 1$, find $y$ as in the hint and let $b_n = \frac{1}{n^y}$. Then $\exists\, m_1$ such that for all $n \geq m_1, n(1 - \frac{a_n}{a_{n-1}}) > y$ which gives $\frac{a_n}{a_{n-1}} < 1 - \frac{y}{n}$. Now $\frac{b_n}{b_{n-1}} = (1 - \frac{1}{n})^y$ can be shown to be approximately $1 - \frac{y}{n} + \frac{y(y-1)}{2n^2}$ and hence bigger than $1 - \frac{y}{n}$ for all sufficiently large $n$, using Taylor's theorem (Theorem (4.5.6)). In an informal approach this is done by expanding $(1 - \frac{1}{n})^y$ into powers of $\frac{1}{n}$, by the binomial theorem for a rational exponent (see Exercise (4.6.13)). So $\exists\, m(\geq m_1)$ such that for all $n \geq m, \frac{a_n}{a_{n-1}} \leq \frac{b_n}{b_{n-1}}$, i.e. the condition in Exercise (9.2) holds eventually. If $l < 1$, the argument is similar except that this time $\frac{a_n}{a_{n-1}} > 1 - \frac{y}{n}$ while $(1 - \frac{1}{n})^y \simeq 1 - \frac{y}{n} + \frac{y(y-1)}{2n^2} < 1 - \frac{y}{n}$ since $y(y-1) < 0$.

**14.** (i) divergent by condensation test (ii) convergent by alternating test (iii) convergent by ratio test (note that $n(\frac{1}{2})^n \to 0$ as $n \to \infty$ by Theorem (3.6)) (iv) absolutely convergent (by comparison with $\sum_{n=1}^{\infty} \frac{1}{n^2}$) and hence convergent (v) convergent by ratio test (vi) divergent since $\frac{n^n}{n!} \geq n \not\to 0$ as $n \to \infty$ (vii) divergent by comparison with $\sum_{n=1}^{\infty} \frac{1}{n}$ (viii) and (ix) convergent by root test.

**15.** Writing $1 - \frac{1}{k^2}$ as $\frac{(k-1).(k+1)}{k.k}$ for $k = 2, 3, \ldots, n$, $P_n$ is simply $\frac{n+1}{2n}$.

## Section 3.10

**1.** Use Exercise (9.11) instead of Theorem (9.5).

**2.** The first part follows from Theorems (4.5) and (10.1). The divergence of $\sum_{n=0}^{\infty} x^n$ for $x = \pm 1$ follows directly by calculating partial sums $S_n$ (which equal $n + 1$ if $x = 1$ and $\frac{1}{2}[(-1)^n + 1]$ if $x = -1$), $\sum_{n=1}^{\infty} \frac{x^n}{n}$ converges for

$x = -1$ by the alternating test (Exercise (8.1)) and diverges for $x = 1$ by Theorem (6.1). The last series $\sum_{n=1}^{\infty} \dfrac{x^n}{n^2}$ converges both for $x = 1$ and $x = -1$ by Theorems (6.3) and (8.1).

**3.** The Cauchy product of the two series $\displaystyle\sum_{n=0}^{\infty} \dfrac{x^n}{n!}$ and $\displaystyle\sum_{m=0}^{\infty} \dfrac{y^m}{m!}$ is $\displaystyle\sum_{n=0}^{\infty} c_n$ where

$$c_n = \sum_{i=0}^{n} \frac{x^i}{i!} \frac{y^{n-i}}{(n-i)!} = \frac{1}{n!} \sum_{i=0}^{n} \binom{n}{i} x^i y^{n-i} = \frac{(x+y)^n}{n!}.$$ Similarly, adding the Cauchy product of $\sin x \,\Big(= \displaystyle\sum_{n=0}^{\infty} \dfrac{(-1)^n x^{2n+1}}{(2n+1)!}\Big)$ and $\cos y \,\Big(= \displaystyle\sum_{n=0}^{\infty} \dfrac{(-1)^n y^{2n}}{(2n)!}\Big)$ to that of $\cos x$ and $\sin y$, we get $\sin x \cos y + \cos x \sin y = \sum_{n=0}^{\infty} c_n$, where

$$c_n = \sum_{i=0}^{n} \frac{(-1)^i x^{2i+1}}{(2i+1)!} \frac{(-1)^{n-i} y^{2n-2i}}{(2n-2i)!} + \frac{(-1)^i x^{2i}}{(2i)!} \frac{(-1)^{n-i} y^{2n-2i+1}}{(2n-2i+1)!}$$

$$= \frac{(-1)^n}{(2n+1)!} \left[ \sum_{i=0}^{n} \binom{2n+1}{2i+1} x^{2i+1} y^{2n-2i} + \binom{2n+1}{2i} x^{2i} y^{2n+1-2i} \right]$$

$$= \frac{(-1)^n (x+y)^{2n+1}}{(2n+1)!}$$

by the binomial theorem. This proves $\sin(x+y) = \sin x \cos y + \cos x \sin y$. The proof of the other identity is similar.

**4.** (i) $\frac{1}{|r|}$ (ii) 1 (see Exercise (4.6)) (iii) $\infty$ if $h$ is a non-negative integer (in which case $a_n = 0$ eventually) and 1 otherwise (since in this case $\frac{a_{n+1}}{a_n} \to -1$ as $n \to \infty$).

**5.** $F_0 = 0 \leq 1 = 2^0; F_1 = 1 \leq 2 \leq 2^1$. For $n \geq 2, F_{n-1} \leq 2^{n-1}$ and $F_{n-2} \leq 2^{n-2}$ by induction hypothesis. So $F_n = F_{n-1} + F_{n-2} \leq 2^{n-1} + 2^{n-2} \leq 2^{n-1} + 2^{n-1} = 2^n$. So $F_n^{1/n} \leq 2$ for all $n$, which means $\overline{\lim} F_n^{1/n} \leq 2$ and hence the radius of convergence of $\sum_{n=0}^{\infty} F_n x^n$ is at least $\frac{1}{2}$.

**6.** Calculation of $p_0, p_1, p_2$ is trivial. If $n \geq 3$, then for a win to occur on the $n^{\text{th}}$ toss (but not earlier), the first toss must be $T$ or else the first two tosses must be HT. In the first case it is as if a new game starts and is won on the $(n-1)^{\text{th}}$ round, while in the second, the new game starts after the second round and is won on (its) $(n-2)^{\text{th}}$ round. This gives (9). Since the partial sum $\sum_{k=0}^{n} p_k$ is the probability of winning on the $n^{\text{th}}$ toss or earlier, $\sum_{k=0}^{n} p_k \leq 1$ for all $n$ whence the series $\sum_{n=0}^{\infty} p_n$ of non-negative terms is convergent. Let $L$ be its sum. Summing (9) from 3 to $\infty$ gives $L - (p_0 + p_1 + p_2) = \frac{1}{2} = \frac{1}{2}(L - p_0 - p_1) + \frac{1}{4}(L - p_0)$ i.e. $L - \frac{1}{4} = \frac{1}{2}L + \frac{1}{4}L$, or $L = 1$. Intuitively this is obvious because if a coin is tossed infinitely often it is extremely unlikely that two consecutive heads will never occur. To prove

$P(x) = \dfrac{x^2}{4 - 2x - x^2}$, multiply (9) throughout by $x^n$ and sum from 3 to $\infty$,

to get $P(x) - p_0 - p_1 x - p_2 x^2 = \frac{x}{2}(P(x) - p_0 - p_1 x) + \frac{x^2}{4}(P(x) - p_0)$, i.e.

$P(x) - \frac{1}{4}x^2 = \frac{x}{2}P(x) + \frac{x^2}{4}P(x)$. Factoring $4 - 2x - x^2$ as $-(x - \alpha)(x - \beta)$

where $\alpha = -1 + \sqrt{5}$ and $\beta = -1 - \sqrt{5}$, $\dfrac{x^2}{4 - 2x - x^2}$ can be resolved as

$-1 + \dfrac{1}{\sqrt{5}}\left(\dfrac{3 - \sqrt{5}}{\alpha - x} - \dfrac{3 + \sqrt{5}}{\beta - x}\right)$. Expanding as geometric series gives $p_0 = 0$

and $p_n = \dfrac{1}{\sqrt{5}}\left[\dfrac{3 - \sqrt{5}}{\alpha^{n+1}} - \dfrac{3 + \sqrt{5}}{\beta^{n+1}}\right]$ for $n \geq 1$.

7. Differentiate $\sum_{n=0}^{\infty} x^n = \frac{1}{1-x}$ and multiply by $x$.

8. With $\alpha, \beta$ as in the answer to Exercise (10.6), $|\frac{1}{\alpha}| < 1$ and $|\frac{1}{\beta}| < 1$. So by the

last exercise $\displaystyle\sum_{n=1}^{\infty} \dfrac{n}{\alpha^n} = \dfrac{1/\alpha}{(1 - \frac{1}{\alpha})^2}$ and $\displaystyle\sum_{n=1}^{\infty} \dfrac{n}{\beta^n} = \dfrac{1/\beta}{(1 - \frac{1}{\beta})^2}$. Then $\sum_{n=1}^{\infty} n p_n$

comes out as $\dfrac{1}{\sqrt{5}}\left[\dfrac{3 - \sqrt{5}}{(\alpha - 1)^2} - \dfrac{3 + \sqrt{5}}{(\beta - 1)^2}\right]$, which, after substituting the values

of $\alpha$ and $\beta$, equals 6. To get this directly from (9) multiply it by $n$ and sum from $n = 3$ to $\infty$. Letting $A = \sum_{n=1}^{\infty} n p_n$, we get $A - 2p_2 = \frac{1}{2}A + \frac{1}{2}\sum_{n=3}^{\infty} p_{n-1} + \frac{1}{4}A + \frac{1}{2}\sum_{n=3}^{\infty} p_{n-2}$. Since both the infinite sums are 1, this gives $\frac{1}{4}A = \frac{3}{2}$, i.e. $A = 6$.

9. The analogue of (9) this time is $p_n = \frac{1}{2}p_{n-1} + \frac{1}{4}p_{n-2} + \frac{1}{8}p_{n-3}$ for $n \geq 4$, with $p_0 = p_1 = p_2 = 0$ and $p_3 = \frac{1}{8}$. This is proved by noting that the game goes on upto the $n^{\text{th}}$ round only if the first toss is $T$, or the first $H$ and the second $T$ or the first two $H$'s and the third $T$. The generating function $P(x)$ comes out as $\dfrac{x^3}{8 - 4x - 2x^2 - x^3}$.

10. The name is justified by the fact that the exponential function $e^x = \displaystyle\sum_{n=0}^{\infty} \dfrac{x^n}{n!}$

is the exponential generating function of the constant sequence $1, 1, 1, \ldots$.

Since $(n!)^{1/n} \to \infty$ (by Exercise (9.3)), $\overline{\lim}\left(\dfrac{|a_n|}{n!}\right)^{1/n}$ is in general much

smaller than $\overline{\lim}|a_n|^{1/n}$ and hence $\displaystyle\sum_{n=0}^{\infty} \dfrac{a_n}{n!}x^n$ has a larger radius of conver-

gence then $\displaystyle\sum_{n=0}^{\infty} a_n x^n$.

11. The first part is a repetition of Exercise (10.4) (i). If $0 < |\beta| < |\alpha|$, then for $x \in (-\frac{1}{|\alpha|}, \frac{1}{|\alpha|})$, both $\sum_{n=0}^{\infty} \alpha^n x^n$ and $\sum_{n=0}^{\infty} \beta^n x^n$ converge and hence so does $\sum_{n=0}^{\infty}(\alpha^n + \beta^n)x^n$. So the radius of convergence, say $R$, of $\sum_{n=0}^{\infty}(\alpha^n + \beta^n)x^n$ is at least $\frac{1}{|\alpha|}$. But it cannot be bigger, as otherwise for any

$x \in (\frac{1}{|\alpha|}, \min\{R, \frac{1}{|\beta|}\})$, $\sum_{n=0}^{\infty}(\alpha^n + \beta^n)x^n$ and $\sum_{n=0}^{\infty} \beta^n x^n$ would both converge, which would imply the convergence of their difference, $\sum_{n=0}^{\infty} \alpha^n x^n$, a contradiction.

**12.** $\dfrac{1}{(1-x)^k} = \left(\dfrac{1}{1-x}\right)^k$ is the Cauchy product of the series $1 + x + x^2 + \ldots + x^n + \ldots$ with itself taken $k$ times. A typical term of this product is $x^{n_1} x^{n_2} \ldots x^{n_k}$ which will equal $x^n$ if and only if $n = n_1 + n_2 + \ldots + n_k$. This corresponds to the (unique) selection of $n$ objects with $n_i$ objects of type $i$ for $i = 1, 2, \ldots, k$. Hence the coefficient of $x^n$ in the product is the number of all possible selections of $n$ objects from $k$ types of objects.

## Section 3.11

**1.** (ii) The direct implication is trivial. For the converse, given $\epsilon > 0$, for each $i = 1, 2, \ldots, k$, find $N_i$ which works for $T_i$ (i.e., $|f_n(x) - f(x)| < \epsilon$ for all $x \in T_i$ and for all $n \geq N_i$). Set $N = \max\{N_1, N_2, \ldots, N_k\}$. Then $N$ works for the entire set $S$. The converse is false for an infinite collection of subsets. Take $f_n(x) = x^n$, $S = [0, 1)$ and $T_1 = [0, 1/2), T_2 = [1/2, 2/3), T_3 = [2/3, 3/4) \ldots$.

**2.** Since $f_n \overrightarrow{\rightarrow} f$ on $S, \exists \, r$ such that $\sup_{x \in S} |f_n(x) - f(x)|$ is finite for all $n \geq r$. In particular, $\sup_{x \in S} |f_r(x) - f(x)|$ is finite, i.e. $\exists \, K$ such that for all $x \in S, |f_r(x) - f(x)| \leq K$. Since $|f_r(x)| \leq M_r$ for all $x \in S$, it follows, by triangle inequality, that $|f(x)| \leq |f(x) - f_r(x)| + |f_r(x)| \leq K + M_r$. So $f$ is bounded on $S$. Also since $\sup_{x \in S} |f_n(x) - f(x)| \to 0$ as $n \to \infty, \exists \, k \, (\geq r)$ such that $|f_n(x) - f(x)| \leq 1$ for all $x \in S$ and for all $n \geq k$. By triangle inequality again, this means $|f_n(x)| \leq 1 + |f(x)| \leq 1 + K + M_r$ for all $x \in S$ and $n \geq k$. So the functions $\{f_n\}$ for $n \geq k$ are uniformly bounded (by $1 + K + M_r$). To take care of the remaining functions, set $M = \max\{M_1, M_2, \ldots, M_{k-1}, 1 + K + M_r\}$. Then $|f_n(x)| \leq M$ for all $n$ and for all $x \in S$.

**3.** Here $f(x) = \lim_{n \to \infty} \frac{1}{x + \frac{1}{n}} = \frac{1}{x}$ is unbounded on $(0, 1)$. But $f_n(x) = \frac{1}{x + \frac{1}{n}}$ is bounded (by $n$) on $S$ for every $n$. The $f_n$'s are not uniformly bounded, since given $M \, (> 0)$ if $m$ is an integer $> M$, then $f_m(x) > M$ if $x$ is close to $0$ (specifically if $0 < x < \frac{1}{M} - \frac{1}{m}$).

**4.** (i) The limit function is $f(x) = \lim_{n \to \infty} \dfrac{n}{nx + 1} = \dfrac{1}{x}$. For each fixed $n$, the difference $|\dfrac{n}{nx + 1} - \dfrac{1}{x}| = \dfrac{1}{nx^2 + x}$ can be made very large by taking $x$ sufficiently close to $0$. But on intervals of the form $[\delta, 1)$ where $0 < \delta < 1$, this difference is at most $\dfrac{1}{n\delta^2 + \delta}$ which tends to $0$ as $n \to \infty$. So the convergence is uniform on such intervals but not on the entire interval $(0, 1)$. In (ii) and (iii), the limit function is identically $0$. In (ii), $|\dfrac{x}{nx + 1} - 0| \leq \dfrac{x}{nx} = \dfrac{1}{n} \to 0$ as $n \to \infty$ and so the convergence is uniform on $(0, 1)$. Arguing as in

(i) the convergence in (iii) is non-uniform on $(0,1)$ but uniform on $[\delta, 1)$ if $0 < \delta < 1$. In (iv) and (iv) the sum function $S(x)$, is $x$ and $\frac{x^2 - x}{1 + x}$ respectively. In (iv), $|S_n(x) - S(x)| = x^{n+1}$ and so, the convergence is uniform on $[0, 1 - \delta]$ for $\delta > 0$, but not on $(0,1)$. In (v), $|S_n(x) - S(x)|$ comes out as $\frac{x^{n+1}}{1+x}(1 - x)$ for $0 < x < 1$. Given $\epsilon > 0$, split $(0,1)$ into $(0, 1 - \epsilon]$ and $(1 - \epsilon, 1)$. (We assume $\epsilon < 1$, as otherwise take a smaller $\epsilon$.) Since $(1 - \epsilon)^n \to 0, \exists\ m$ such that $(1 - \epsilon)^n < \epsilon$ for all $n \geq m$. Now, for $x \in (0, 1 - \epsilon], |S_n(x) - S(x)| \leq x^{n+1} \leq (1 - \epsilon)^{n+1} < \epsilon$ if $n \geq m$, while, for $x \in (1 - \epsilon, 1),\ |S_n(x) - S(x)| \leq 1 - x < \epsilon$ since $x^{n+1} \leq 1$ and $1 + x \geq 1$. So the convergence is uniform on the entire interval $(0, 1)$.

5. From Exercise (8.1), $|S_n(x) - S(x)| \leq |S_n(x) - S_{n-1}(x)| = f_n(x)$ for all $x \in T$. So by (ii), $S_n \to S$ uniformly on $T$.

6. $|S_n(x)| = |1 + x + \ldots + x^n|$ is bounded (by $n + 1$) on $(-1, 1)$. But the sum function $S(x) = \frac{1}{1-x}$ is not bounded on $(-1, 1)$. For a direct argument, $|S_n(x) - S(x)| = \frac{|x|^{n+1}}{1 - x}$. For a fixed $n, x^{n+1} \geq (\frac{1}{2})^{n+1}$ for all $x \geq \frac{1}{2}$. So for $x \in [1 - (\frac{1}{2})^{n+1}, 1), |S_n(x) - S(x)| \geq 1$. Hence $\sup_{x \in (-1,1)} |S_n(x) - S(x)| \not\to 0$ as $n \to \infty$.

7. By Exercise (10.2) the interval of convergence is $(-1, 1)$. For every $x \in (-1, 1), |\frac{x^n}{n^2}| \leq \frac{1}{n^2}$. Since $\sum_{n=1}^{\infty} \frac{1}{n^2}$ is convergent by Theorem (6.3), the convergence is uniform on $(-1, 1)$ (in fact on $[-1, 1]$) by Theorem (11.4).

8. For the first part, again apply Exercise (10.2). For the second, from the Leibnitz test (Exercise (8.1)), it follows that for every $n, |S_n(x) - S(x)| \leq |S_n(x) - S_{n-1}(x)| = |f_n(x)| = \frac{|x|^n}{n} \leq \frac{1}{n}$ for all $x \in (-1, 1)$. So the convergence is uniform on $(-1, 1)$ (in fact also on $(-1, 1]$, even though at $x = 1$, the convergence is not absolute.)

9. Use Exercise (10.1) instead of Theorem (10.1)

10. Adapt the proof of Theorem (5.4). If $f_n \to f$ uniformly and $\epsilon > 0$, by Theorem (11.2) find $p$ such that for all $n \geq p, |f_n(x) - f(x)| < \frac{\epsilon}{2}$ for all $x \in S$. The triangle inequality gives that whenever $m \geq n \geq p, |f_m(x) - f_n(x)| < |f_m(x) - f(x)| + |f(x) - f_n(x)| < \epsilon$ for all $x \in S$. For the converse, the hypothesis implies in particular that for every $x \in S, \{f_n(x)\}$ is a Cauchy sequence and hence convergent by Theorem (5.4). Define $f : S \to \mathbb{R}$ by $f(x) = \lim_{n \to \infty} f_n(x)$, for $x \in S$. To prove $f_n \to f$ uniformly on $S$, let $\epsilon > 0$ be given. From the hypothesis, get $p$ such that for all $m \geq n \geq p$ and for all $x \in S, |f_m(x) - f_n(x)| < \frac{\epsilon}{2}$. Letting $m \to \infty$ (keeping $n$ fixed) and using Exercises (8.2) and (1.9), we get $|f(x) - f_n(x)| \leq \frac{\epsilon}{2} < \epsilon$ for all $x \in S$ and all $n \geq p$.

11. For the first part merely note $|f_n(x) \pm g_n(x) - (f(x) \pm g(x))| \leq |f_n(x) - f(x)| + |g_n(x) - g(x)|$. Make each term less than $\frac{\epsilon}{2}$. For the counter-example

for $f_n g_n$, let $S = (0,1), f_n(x) = \frac{x}{nx+1}$ and $g_n(x) = g(x) = \frac{1}{x}$ for all $n$. Then $g_n \rightrightarrows g$ trivially while $f_n \rightrightarrows 0$ on $(0,1)$ by (ii) of Exercise (11.4). But $|f_n(x)g_n(x) - f(x)g(x)| = \frac{1}{nx+1}$ which does not tend to 0 uniformly on $S$, as seen in the solution to (iii) of Exercise (11.4). For $f_n/g_n$, keep $f_n$ as it is but let $g_n(x) = g(x) = x$ for $x \in (0,1)$.

12. From the hypothesis, $\exists \ M(> 0)$ such that for all $x \in S, |f(x)| \leq M$ and $|g(x)| \leq M$. From the reasoning in the solution to Exercise (11.2), $\exists \ k$ such that for all $n \geq k$ and all $x \in S, |f_n(x)| \leq M+1$ and $|g_n(x)| \leq M+1$. Then, for all $x \in S$ and for all $n \geq k, |f_n(x)g_n(x) - f(x)g(x)| \leq |f_n(x)(g_n(x) - g(x))| + |(f_n(x) - f(x))g(x)| \leq (M+1)|g_n(x) - g(x)| + M|f_n(x) - f(x)|$ which can be made $< \epsilon$ by ensuring $|g_n(x) - g(x)| < \frac{\epsilon}{2(M+1)}$ and $|f_n(x) - f(x)| < \frac{\epsilon}{2M}$. So $f_n g_n \rightrightarrows fg$.

# CHAPTER 4

## Section 4.1

1. Taking $\epsilon = \frac{1}{2}f(c)$ in the definition of $\lim_{x \to c} f(x) = f(c)$, there exists $\delta > 0$ such that $0 < |x - c| < \delta$ implies $f(c) - \frac{f(c)}{2} < f(x) < f(c) + \frac{f(c)}{2}$, which in particular means $f(x) > \frac{1}{2}f(c) > 0$ for all $x \in U - \{c\}$ where $U = (c-\delta, c+\delta)$. Trivially this also holds for $x = c$. If $f(c) < 0$, taking $\epsilon = -\frac{1}{2}f(c)$, we get $f(x) < 0$ for all $x \in U$.

2. Let $Z$ be the set of zeros of a continuous function $f : \mathbb{R} \to \mathbb{R}$. Then $\mathbb{R} - Z$ is the union of the two sets $\{x \in \mathbb{R} : f(x) > 0\}$ and $\{x \in \mathbb{R} : f(x) < 0\}$. By the last exercise each of these two sets is open. By Exercise (3.5.5), $\mathbb{R} - Z$ is open, i.e., $Z$ is closed.

3. Merely apply the corresponding results about limits. For example, $\lim_{x \to c} f(x)g(x) = \lim_{x \to c} f(x) \lim_{x \to c} g(x)$. Since $\lim_{x \to c} f(x) = f(c)$ and $\lim_{x \to c} g(x) = g(c)$, we get $\lim_{x \to c} (f(x)g(x)) = f(c)g(c)$, i.e., $fg$ is continuous at $c$.

4. Suppose $f$ is continuous and vanishes at every rational. Given any $c \in \mathbb{R}$, let $\{x_n\}$ be a sequence of rationals converging to $c$. Then $f(c) = \lim_{n \to \infty} f(x_n) = \lim_{n \to \infty} 0 = 0$. For the second assertion, apply the first part to the difference of the two continuous functions.

5. Given $\epsilon > 0$, find $\delta_1 > 0$ such that $|y - f(c)| < \delta_1$ implies $|g(y) - g(f(c))| < \epsilon$. For this $\delta_1$, there exists $\delta > 0$ such that $|f(x) - f(c)| < \delta_1$ whenever $|x - c| < \delta$. Putting $y = f(x), |g(f(x)) - g(f(c))| < \epsilon$ whenever $|x - c| < \delta$.

**6.** For a given $\epsilon > 0, \delta = \epsilon$ will work for any $c \in \mathbb{R}$ for the identity function. For constant functions, *any* $\delta > 0$ will work for any $c$. For the second part, apply induction on the degree of the polynomial. Continuity of the absolute value function at a point $c \in \mathbb{R}$ follows from the inequality $\Big| |x| - |c| \Big| \leq |x - c|$ for all $x \in \mathbb{R}$ (see Exercise (1.9.8)).

**7.** The function $f(x) = x^n$ is continuous on $[0, 1]$ and $f(0) = 0, f(1) = 1$. So for every $m \in [0, 1]$, there exists $c \in [0, 1]$, such that $c^n = m$. If $m > 1$, apply this argument to $\frac{1}{m}$ and then take reciprocals.

**8.** Apply Exercise (3.4.7) along with Theorem (1.2). (An alternate argument is possible using the next exercise.)

**9.** Let $I = f([a, b])$. To show that $I$ is an interval, it suffices to show that whenever $y_1 < y_3 < y_2$, and $y_1, y_2$ are in $I$, $y_3$ is also in $I$. Let $y_i = f(x_i)$ for $i = 1, 2$, for some $x_1, x_2 \in [a, b]$. Then by Theorem (1.3), there exists $x_3 \in [x_1, x_2]$ (or $x_3 \in [x_2, x_1]$ as the case may be) such that $y_3 = f(x_3)$. Since $x_3 \in [a, b], y_3 \in I$. Thus $I$ is an interval. By Theorem (1.4), $I$ has a least and a greatest element. So it is of the form $[c, d]$ for some $c, d \in \mathbb{R}, c \leq d$. The function $f(x) = \sin x$ on $[0, 2\pi]$ shows that $c \, (= -1)$ and $d \, (= 1)$ may both differ from $f(a), f(b) \, (= 0$ here). If $f$ is one-to-one, strict monotonicity is the same as monotonicity. If $f$ is neither monotonically increasing nor monotonically decreasing on $[a, b]$, then there exist $x_1 < x_2 < x_3$ in $[a, b]$ such that either $f(x_1) < f(x_2) > f(x_3)$ or $f(x_1) > f(x_2) < f(x_3)$. In the first case depending upon which of $f(x_1)$ or $f(x_3)$ is greater, an application of Theorem (1.3) to the interval $[x_2, x_3]$ or $[x_1, x_2]$ would give a point which will violate that $f$ is one-to-one. The second case is handled similarly. For continuity of $f^{-1}$, let $y_0 \in (c, d)$. Then $y_0 = f(x_0)$ for a unique $x_0 \in (a, b)$. Let $\epsilon > 0$ be given. Choose $0 < \epsilon_1 \leq \epsilon$ such that $(x_0 - \epsilon_1, x_0 + \epsilon_1) \subset (a, b)$. Then $f(x_0 - \epsilon_1) < f(x_0) < f(x_0 + \epsilon_1)$ if $f$ is increasing and $f(x_0 - \epsilon_1) > f(x_0) > f(x_0 + \epsilon_1)$ if $f$ is decreasing. In either case set $\delta = \min\{|f(x_0 - \epsilon_1) - f(x_0)|, |f(x_0 + \epsilon_1) - f(x_0)|\}$. Then $f^{-1}(y_0 - \delta, y_0 + \delta) \subset (f^{-1}(y_0) - \epsilon_1, f^{-1}(y_0) + \epsilon_1) \subset (f^{-1}(y_0) - \epsilon, f^{-1}(y_0) + \epsilon.)$ Proof of continuity of $f^{-1}$ at the end-points $c, d$ is similar.

**10.** Let $f(x) = x^3 - 6x^2 + 3x + 1$. Then $f$ is continuous everywhere. In view of Theorem (1.3), it suffices to find $x_1 < x_2 < x_3 < x_4$ such that $f$ changes sign over each of the three subintervals. By trial, take $x_2 = 0, x_3 = 1$. Any sufficiently small $x_1$ would ensure $f(x_1) < 0$, since $\lim\limits_{x \to -\infty} f(x) = -\infty$, e.g., take $x_1 = -1$. Similarly $x_4 = 6$ gives $f(x_4) > 0$. If $p(x)$ is a polynomial of odd degree with a positive leading coefficient, then $p(x) \to \pm\infty$ as $x \to \pm\infty$. So there exist $x_1, x_2$ such that $p(x_1) < 0$ and $p(x_2) > 0$. Apply Theorem (1.3). If the leading coefficient of $p(x)$ is negative, replace $p(x)$ by $-p(x)$.

**11.** Assume $S$ lies entirely between the planes $z = a$ and $z = b$ (say). Define $f : [a, b] \to \mathbb{R}$ by $f(x_0)$ (where $a \leq x_0 \leq b$) to be the volume of the portion of $S$ between the planes $z = a$ and $z = x_0$. Then $f(a) = 0$ and $f(b) =$

volume of $S$. So Theorem (1.3) would imply the result if $f$ is shown to be continuous. This is intuitively clear but a rigorous proof requires a rigorous definition of volume (cf. Section 6.3).

**12.** Let $L'$ be the line through $O$, parallel to $L$. For $r \geq 0$, let $f(r)$ be the length of the chord of $C$ cut off by a line parallel to $L'$ at a distance $r$ from it. Then $f(0) \geq 2b$ while $f(r) = 0$ if $r \geq a$. So Theorem (1.3) would give the result if $f$ is shown to be continuous, which can be done by obtaining an explicit formula for $f(r)$.

**13.**

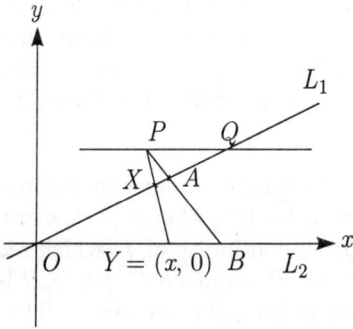

Let $L_1$ and $L_2$ meet at $O$. Take $O$ as the origin and $L_2$ as the $x$ - axis. Let a variable line through $P$ cut $L_1$ and $L_2$ at $X, Y$ respectively. Set $Y = (x, 0)$ and express the ratio $PX : XY$ as a continuous function of $x$, say $f(x)$. Then $f(x) \to \infty$ as $x \to 0^+$ while $f(x) \to 0$ as $x \to \infty$. So by Theorem (1.3), $f(c) = \frac{1}{2}$ for some $c$. To find the line geometrically, draw a line through $P$ parallel to $L_2$ and let it meet $L_1$ at $Q$. Take $A$ on $L_1$ so that $QA : AO = 1 : 2$. Let $AP$ cut $L_2$ at $B$.

If we let $g(x)$ be the length of the line segment $YX$, then $g$ is also a continuous function of $x$. Also, $g(x) \to 0$ as $x \to 0^+$ and $g(x) \to \infty$ as $x \to \infty$. So the existence of the points $C$ and $D$ again follows from the IVP. But there is no geometric construction for them using a ruler and compass only. The proof of this fact is not easy and requires the same techniques as those needed in the Angle Trisection Problem discussed in Section 1.13. So once again, proving the existence of something is not the same as constructing it.

**14.** $g$ is continuous, $g(a) = f(a) - a \geq 0$ (since $f(a) \in [a, b]$) and $g(b) \leq 0$. So by Theorem (1.3), $g(c) = 0$ for some $c \in [a, b]$. Geometrically, the result means that the graphs of the functions $y = f(x)$ and $y = x$ must intersect. The result fails for open intervals, e.g. take $f(x) = x^2$ for $x \in (0, 1)$.

**15.** For the first assertion, note that $x_0$ and $x_n$ are of different colours and so in the sequence $x_0, x_1, \ldots, x_n$ , a change of colour must occur at least once (more accurately, an odd number of times). For the second part, by Theorem (3.5.1), there exists a subsequence $\{y_{n_k}\}$ of $\{y_n\}$ and some $y^* \in [a, b]$ such that $y_{n_k} \to y^*$ as $k \to \infty$. Since $|z_{n_k} - y_{n_k}| < \frac{1}{n_k} \to 0$ as $k \to \infty$, it follows that $\{z_{n_k}\}$ also converges to $y^*$. (To see this, write $|z_{n_k} - y^*|$ as $|(z_{n_k} - y_{n_k}) + (y_{n_k} - y^*)|$ and apply the triangle inequality.) By Theorem (1.2), both $f(y_{n_k})$ and $f(z_{n_k})$ tend to $f(y^*)$ as $k \to \infty$. But since $f(y_{n_k}) > y_{n_k}$, for all $k$, by Exercise (3.1.9), $\lim_{k \to \infty} (y_{n_k}) \geq \lim_{k \to \infty} y_{n_k}$, i.e. $f(y^*) \geq y^*$. Similarly $f(z_{n_k}) < z_{n_k}$ gives $f(y^*) \leq y^*$. So $f(y^*) = y^*$, a contradiction.

16. Define $g : [0, \frac{1}{2}] \to \mathbb{R}$ by $g(x) = f(x + \frac{1}{2}) - f(x)$. Then $g(\frac{1}{2}) = -g(0)$ and so $g(0), g(\frac{1}{2})$ are of opposite signs. By Theorem (1.3), $g(c) = 0$ for some $c$.

17. Get a nested sequence $\{[a_n, b_n]\}_{n=0}^{\infty}$ of intervals such that $f$ is not bounded above on $[a_n, b_n]$ for all $n$ and $b_n - a_n = \frac{1}{2^n}(b_0 - a_0) \to 0$ as $n \to \infty$. Let $\{a_n\}, \{b_n\}$ converge to a common point $c$. Continuity of $f$ at $c$ implies that there is some neighbourhood $U$ of $c$ such that $f$ is bounded on $U$. But $U$ contains $[a_n, b_n]$ for all sufficiently large $n$, a contradiction.

18. Taking $L$ as the $x$ - axis and $P$ as $(0, 1)$, without loss of generality, a point $(x, 0)$ on $L$ is at a distance $\sqrt{x^2 + 1}$ from $P$. The function $f(x) = \sqrt{x^2 + 1}$ is continuous for all $x$ (by Exercises (1.6), (1.8) and (1.5)). It is obvious that $f$ has no maximum and attains its minimum at $x = 0$. To get this from Theorem (1.4), note that $f(x)$ is large for $|x|$ large. So if $R$ is sufficiently large then the minimum of $f$ on $[-R, R]$, which exists by Theorem (1.4), will also be its minimum on entire $\mathbb{R}$.

19. The problem is equivalent to proving that the maximum and the minimum of the function $f(\theta) = \sqrt{(a \cos \theta - h)^2 + (b \sin \theta - k)^2}$, where $a, b, h, k$ are some constants, exist on $[0, 2\pi]$. This follows from continuity of $f$ (which requires the continuity of trigonometric functions).If $P$ is not on $C$, let $Q, R$ be the points on $C$ which are closest to and farthest from $P$ respectively. The circle with $P$ as its centre and $PQ$ as a radius must touch $C$ as otherwise some points on $C$ will lie inside it and hence be closer to $P$ than $Q$ is. Similarly the circle centred at $P$ and $PR$ as a radius touches $C$.

20. Parametrise $C$ as $(a \cos \theta, b \sin \theta), 0 \leq \theta \leq 2\pi$. Let the given triangle be $ABD$ with $AB \leq BD \leq DA$. For a point $X = (a \cos \theta, b \sin \theta)$ on $C$, other than $P$, let $Y_1 = (x_1, y_1)$ and $Y_2 = (x_2, y_2)$ be the points in the plane of $C$ such that the triangles $PXY_1$ and $PXY_2$ are similar to $ABD$. Let $f(\theta) = (\frac{x_1^2}{a^2} + \frac{y_1^2}{b^2} - 1)(\frac{x_2^2}{a^2} + \frac{y_2^2}{b^2} - 1)$. Then $f$ is a continuous function of $\theta$. If $X$ is close to $P$ then one of $Y_1$ and $Y_2$ is inside $C$ and the other is outside $C$ and so $f(\theta) < 0$. But if $X$ is the point on $C$ which is farthest to $P$ then both $Y_1, Y_2$ are outside $C$ and so $f(\theta) > 0$. By Theorem (1.3), $f(\theta) = 0$ for some $\theta$, which means at least one of $Y_1, Y_2$ is on $C$.

21. (ii) $e^y = e^{x+h} = e^x e^h > e^x.1$ (since $e^h > 1$ and $e^x > 0$). (iii) The first assertion follows from $e^x > x$ for all $x > 0$. The second follows from the first assertion and $e^x = \frac{1}{e^{-x}}$. For (iv), given $x > 0$, by (iii) there exist $a, b$ such that $e^a < x < e^b$. Use continuity of the exponential function (from Corollary (1.6)) and the IVP.

22. The exponential funciton is strictly increasing by (ii) of the last exercise and hence its inverse function ln is also striclty increasing with domain equal to the range of the exponential function and range equal to its domain. By (iii) of the last exercise, these are, respectively, $(0, \infty)$ and $(-\infty, \infty)$. For continuity of the logarithm function, given $y_0 > 0$, with $e^{x_0} = y_0$, apply the last part of Exercise (1.9) to any interval containing $x_0$ in its interior, say

$[x_0 - 1, x_0 + 1]$. If $x, y$ are positive and $a = \ln x, b = \ln y$ then $e^a = x, e^b = y$ and so $xy = e^a e^b = e^{a+b}$, whence $a + b = \ln(xy)$.

23. Following the hint, if $x^y$ is defined as in Exercise (3.4.2), we have to prove that $\ln(x^y) = y \ln x$. This is clear for $y = 0$ and 1. For $y \in \mathbb{N}$, apply induction on $y$. Since $x^y = x^{y-1} x^1$, the last exercise gives $\ln(x^y) = \ln(x^{y-1}) + \ln x^1 = (y - 1) \ln x + \ln x = y \ln x$. If $y$ is a negative integer, then $0 = \ln(x^\circ) = \ln(x^{y+(-y)}) = \ln(x^y) + \ln(x^{-y})$ implies $\ln(x^{-y}) = -\ln(x^y) = (-y) \ln x$. Finally, suppose $y = m/n$, with $m \in \mathbb{Z}, n \in \mathbb{N}$. Then $(x^y)^n = (x)^m$ and so $n \ln(x^y) = m \ln x$, whence $\ln(x^y) = \frac{m}{n} \ln x = y \ln x$.

24. Even though each $x_n$ is in $(a, b)$, there is no guarantee that $\lim\limits_{k \to \infty} x_{n_k}$ is in $(a, b)$, since this limit could equal $a$ or $b$. Since $f$ is not given to be continuous at these two points, the proof breaks down. Similarly it fails for a semi-open interval. (For an actual example, take $f(x) = \frac{1}{x}$ on $(0, 1]$.)

25. (i) jump discontinuities at all integers from $-9$ to 9 ; simple discontinuity at 10

(ii) jump discontinuities at $\pm\sqrt{n}$ for $n = 1, 2, \ldots, 99$. Simple discontinuities at $\pm 10$

(iii) simple discontinuity at 1, discontinuity at 3

(iv) discontinuous at all points. (For any $c$, $f(x) \to 1$ as $x \to c^+$ through rationals and $f(x) \to 0$ as $x \to c^+$ through irrationals. So $\lim\limits_{x \to c^+} f(x)$ does not exist. Similarly, $\lim\limits_{x \to c^-} f(x)$ does not exist.

26. Suppose $f : [a, b] \longrightarrow \mathbb{R}$ is piecewise continuous. Following the notations in the text, for $i = 1, 2, \ldots, n + 1$, let $f_i : [t_{i-1}, t_i] \longrightarrow \mathbb{R}$ be the continuous function which coincides with $f$ on $(t_{i-1}, t_i)$. Then by Theorem (1.4), $\exists M_i$ such that $|f_i(x)| \leq M_i$ for all $x \in (t_{i-1}, t_i)$. Now let $M = \max\{M_1, M_2, \ldots, M_{n+1}, |f(t_0)|, |f(t_1)|, \ldots, |f(t_{n+1})|\}$. Then $|f(x)| \leq M$ for all $x \in [a, b]$. For the example, take $f(x) = x$ for $x \in (0, 1)$ and $f(0) = f(1) = \frac{1}{2}$.

### Section 4.2

1. Trivial (of the same spirit as Exercise (3.11.1)).

2. To get the desired covering, divide $S$ into $n$ subintervals of equal lengths where $n > \frac{l}{\delta}$, $l$ being the length of $S$. To show $f$ is bounded on each of these subintervals, fix any $x_i$ in the $i^{\text{th}}$ subinterval. Then for every $x$ in the $i^{\text{th}}$ subinterval, $f(x)$ lies in $(f(x_i) - 1, f(x_i) + 1)$.

3. Certainly $f$ is bounded, since $|\sin(\frac{1}{x})| \leq 1$ for all $x \in (0, 1)$. To show it is not uniformly continuous on $(0, 1)$, let $\epsilon = 1$. Then for every $\delta > 0$, choose $n \in \mathbb{N}$ so large that $\frac{2}{(4n+1)\pi} < \delta$. Then $\left| \frac{2}{(4n+1)\pi} - \frac{2}{(4n+3)\pi} \right| < \delta$

but $\left|f(\frac{2}{(4n+1)\pi}) - f(\frac{2}{(4n+3)\pi})\right| = |1 - (-1)| = 2 > \epsilon$. So no $\delta > 0$ satisfies Definition (2.1), for $\epsilon = 1$.

4.  Given $\epsilon > 0$, find $\delta$ satisfying Definition (2.1). With this $\delta$, find $p$ satisfying Definition (3.5.2), i.e., $|x_m - x_n| < \delta$ whenever $m \geq p$ and $n \geq p$. But by uniform continuity, $|x_m - x_n| < \delta$ implies $|f(x_m) - f(x_n)| < \epsilon$. So $|f(x_m) - f(x_n)| < \epsilon$ whenever $m \geq p$ and $n \geq p$. For the function $f(x) = \sin(\frac{1}{x})$, the sequence $\{\frac{2}{(2n+1)\pi}\}_{n=1}^{\infty}$ is a Cauchy sequence (since it converges to 0) but $f(\frac{2}{(2n+1)\pi}) = (-1)^n$ and so $\{f(\frac{2}{(2n+1)\pi}\}$ is not a Cauchy sequence. Hence $f$ is not uniformly continuous on $(0, 1)$.

5.  The first assertion follows from the Cauchy criterion (Theorem (3.5.4)) for sequences and the last exercise. For the second assertion, let $L = \lim\limits_{n\to\infty} f(x_n)$ and $L' = \lim\limits_{n\to\infty} f(y_n)$. If $L \neq L'$, let $\epsilon = \frac{1}{4}|L - L'|$. Then $|x - y| \geq \epsilon$ whenever $x \in (L - \epsilon, L + \epsilon)$ and $y \in (L' - \epsilon, L' + \epsilon)$. For this $\epsilon$, let $\delta > 0$ be as given by uniform continuity of $f$ on $(a, b)$. Then $\exists\ m_1$ such that for all $n \geq m_1, x_n \in (a, a + \delta)$ and $y_n \in (a, a + \delta)$, which implies $|x_n - y_n| < \delta$ for all $n \geq m_1$. Also since $f(x_n) \to L$ and $f(y_n) \to L', \exists\ m_2$ such that for all $n \geq m_2, f(x_n) \in (L - \epsilon, L + \epsilon)$ and $f(y_n) \in (L' - \epsilon, L' + \epsilon)$. But then for $n \geq \max\{m_1, m_2\}$ we have $|x_n - y_n| < \delta$ and yet $|f(x_n) - f(y_n)| \geq \epsilon$ a contradiction. So $L = L'$. That $\lim\limits_{x\to a^+} f(x)$ equals $L$ follows from Theorem (3.1.1) (or rather, its analogue for right handed limits.)

6.  For the direct implication, define $g : [a, b] \longrightarrow \mathbb{R}$ by $g(x) = f(x)$ for all $x \in (a, b)$, $g(a) = \lim_{n\to\infty} f(a + \frac{1}{n})$ and $g(b) = \lim_{n\to\infty} f(b - \frac{1}{n})$. By the last exercise $g$ is continuous at $a$ and at $b$. On $(a, b)$, $g$ is continuous since it coincides with $f$ which is continuous. So $g$ is continuous on $[a, b]$. For the converse, $g$ is uniformly continuous on $[a, b]$ by Theorem (2.2) and hence on its subset $(a, b)$ by Exercise (1.1). So $f$, which coincides with $g$ on $(a, b)$ is also uniformly continuous on $(a, b)$.

7.  Given $\epsilon > 0$, merely set $\delta = \frac{\epsilon}{M+1}$.

8.  If $x = h^2$ and $y = 4h^2$ where $0 < h < \frac{1}{2}$, then $\frac{|f(x)-f(y)|}{|x-y|} = \frac{2h-h}{3h^2} = \frac{1}{3h}$ which can be made larger than any given $M$ by taking $h$ sufficiently small. So no constant $M$ can exist so as to satisfy the Lipschitz condition, even on $(0, 1)$ and hence certainly not on $[0, 1]$. But by Exercise (1.8), the function $f(x) = \sqrt{x}$ is continuous at all $x > 0$. Continuity at $x = 0$ follows by taking $\delta = \epsilon^2$ in the definitiion of $\lim\limits_{x\to 0^+} \sqrt{x} = 0$. So by Theorem (2.2), $f$ is uniformly continuous on $[0, 1]$ (in fact, on any bounded interval contained in $[0, \infty)$). On $[1, \infty)$, $f$ satisfies Lipschitz condition with $M = \frac{1}{2}$ since for all $x \geq 1, y \geq 1, |\sqrt{x} - \sqrt{y}| = \frac{|x-y|}{\sqrt{x}+\sqrt{y}} \leq \frac{|x-y|}{1+1} = \frac{1}{2}|x - y|$. So $f$ is uniformly continuous on $[0, 1]$ and also on $[1, \infty)$. To prove uniform continuity on $[0, \infty)$, given $\epsilon > 0$, let $\delta_1, \delta_2$ satisfy the Definition (2.1) for $\frac{\epsilon}{2}$ on $[0, 1]$ and on $[1, \infty)$ respectively.

Let $\delta = \min\{\delta_1, \delta_2\}$. Let $x, y \in [0, \infty)$ satisfy $|x - y| < \delta$. If $x$ and $y$ both belong either to $[0, 1]$ or to $[1, \infty)$, then certainly $|f(x) - f(y)| < \frac{\epsilon}{2} < \epsilon$. Otherwise, without loss of generality, let $x \in [0, 1]$ and $y \in [1, \infty)$. Then $x \leq 1 \leq y$ and since $|x - y| < \delta$, we have $|x - 1| < \delta_1$ and $|y - 1| < \delta_2$, whence $|f(x) - f(1)| < \frac{\epsilon}{2}$ and $|f(y) - f(1)| < \frac{\epsilon}{2}$, which by the triangle inequality implies $|f(x) - f(y)| < \epsilon$.

9. Following the hint, the first assertion reduces to proving $\sqrt{y + h} \leq \sqrt{y} + \sqrt{h}$ whenever $y \geq 0, h \geq 0$. This follows directly by squaring both the sides. For the second assertion, merely take $\delta = \epsilon$ in the definition of uniform continuity.

## Section 4.3

1. Apply the basic properties of limits to $\displaystyle\lim_{\Delta x \to 0} \frac{h(c + \Delta x) - h(c)}{\Delta x}$ which

   equals $a\left(\dfrac{f(c + \Delta x) - f(c)}{\Delta x}\right) + b\left(\dfrac{g(c + \Delta x) - g(c)}{\Delta x}\right)$. The cases $a = 1$, $b = \pm 1$ show that the derivative of a sum/difference of two functions is the sum/difference of their derivatives.

2. The trouble is that $\dfrac{h(c + \Delta x) - h(c)}{\Delta x}$ is, in general, quite different from the

   product $\left(\dfrac{f(c + \Delta x) - f(c)}{\Delta x}\right)\left(\dfrac{g(c + \Delta x) - g(c)}{\Delta x}\right)$ and so the result about

   the product of the limits is inapplicable. To get the correct answer, note that $h(c + \Delta x) - h(c)$ equals $f(c + \Delta x)g(c + \Delta x) - f(c)g(c)$ which can be rewritten as $f(c + \Delta x)(g(c + \Delta x) - g(c)) + (f(c + \Delta x) - f(c))g(c)$. Dividing by $\Delta x$ and noting that $f(c + \Delta x) \to f(c)$ as $\Delta x \to 0$ by Corollary (3.2), it follows from the basic properties of limits that $h'(c)$ exists and equals $f(c)g'(c) + f'(c)g(c)$. To prove the quotient rule we can apply the product rule to the pair $f$ and $1/g$. So it suffices to show that the derivative of $\dfrac{1}{g}$ at $c$

   is $-\dfrac{g'(c)}{g(c)^2}$. For this, note first that by Corollary (3.2), $g$ is continuous at $c$ and hence by Exercise (1.1), $g(c + \Delta x) \neq 0$ if $|\Delta x|$ is sufficiently small. Rewrit-

   ing $\dfrac{1}{\Delta x}\left(\dfrac{1}{g(c + \Delta x)} - \dfrac{1}{g(c)}\right)$ as $-\left(\dfrac{g(c + \Delta x) - g(c)}{\Delta x}\right)\dfrac{1}{g(c)}\dfrac{1}{g(c + \Delta x)}$ and

   observing that $g(c + \Delta x) \to g(c)$ as $\Delta x \to 0$ gives the result.

3. By the binomial theorem, $\dfrac{1}{\Delta x}[(x + \Delta x)^n - x^n] = \displaystyle\sum_{r=1}^{n} \binom{n}{r} x^{n-r}(\Delta x)^{r-1}$. As

   $\Delta x \to 0$, each of the terms except the first tends to 0. For an inductive proof, the case $n = 1$ is trivial. Apply the product rule in the last exercise to $x^{n-1}x$ to get the inductive step. For $n = 0$, $x^n$ is a constant function ($0^0$ is taken as 1 here), and so its derivative 0 everywhere. If $n$ is a negative integer, then $x^n = \dfrac{1}{x^{-n}}$ and so by the earlier part and the quotient rule of the last exercise, $\dfrac{d}{dx}(x^n) = -\dfrac{1}{x^{-2n}}\dfrac{d}{dn}(x^{-n}) = -x^{2n}(-n)x^{-n-1} = nx^{n-1}$

4.  We have $\displaystyle\lim_{\Delta x \to 0} \dfrac{f(c+\Delta x) - f(c)}{\Delta x} = f'(c) > 0$. Taking $\epsilon = \dfrac{1}{2}f'(c), \exists\, \delta > 0$

such that, $f'(c) - \dfrac{1}{2}f'(c) < \dfrac{f(c+\Delta x) - f(c)}{\Delta x} < f'(c) + \dfrac{1}{2}f'(c)$ which, in

particular means $\dfrac{f(c+\Delta x) - f(c)}{\Delta x} > 0$ whenever $0 < |\Delta x| < \delta$. Thus

$f(c+\Delta x) - f(c)$ and $\Delta x$ have the same signs, which implies the result. If $f'(c) < 0$, then $f$ is strictly decreasing at $c$, i.e., $f(x) > f(c)$ if $x < c$ and $f(x) < f(c)$ if $x > c$ provided $|x - c|$ is sufficiently small.

5.  The first assertion follows simply from the defintions. That $f'(0) = 0$ is a special case of Exercise (3.3). For the last assertion, by essentially the same argument as in the example after Corollary (3.2), show that $g'_+(0) = 2$ and $g'_-(0) = 1$. So $g'(0)$ does not exist.

6.  Let $f_i(x) = |x - c_i|, i = 1,\ldots,n$. By the same argument as in the example after Corollary (3.2), $f_i$ is continuous everywhere and differentiable everywhere except at $c_i$. So $f = \sum\limits_{i=1}^{n} f_i$ is continuous everywhere and differentiable at every $x \neq c_1,\ldots,c_n$. For $x = c_j, f$ cannot be differentiable at $c_j$ for if it is then since $f_1,\ldots,f_{j-1},f_{j+1},\ldots,f_n$ are all differentiable at $c_j, f - (f_1 + \ldots + f_{j-1} + f_{j+1} + \ldots + f_n)$, which equals $f_j$, would be differentiable at $c_j$.

7.

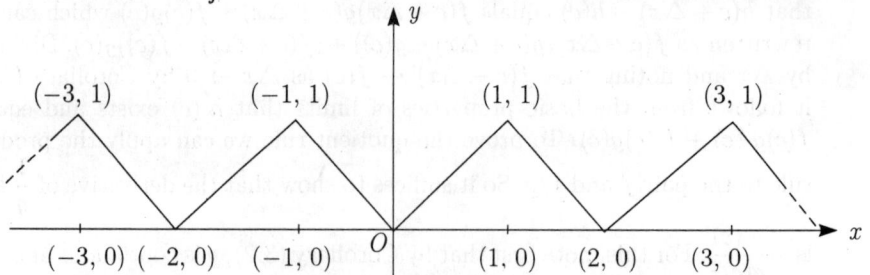

Since $\{x + 2\} = x + 2 - [x + 2] = x + 2 - [x] - 2 = \{x\}$ and $[x]$ and $[x] + 2$ are either both even or both odd, we have $f(x + 2) = f(x)$ for all $x$. The other properties of $f$ are clear from the graph. For an analytic proof, note that on an interval of the form $(n, n+1)$, where $n \in \mathbb{Z}, f(x)$ coincides with $x - n$ if $n$ is even and with $n + 1 - x$ if $n$ is odd. So $f$ is differentiable at every $x \in \mathbb{R} - \mathbb{Z}$. Continuity and non-differenitability at $n \in \mathbb{Z}$ have to be established by considering the right and left sided limits and derivatives. Note that for all $n \in \mathbb{Z}, f'_+(n) = (-1)^n$ and $f'_-(n) = (-1)^{n+1}$.

8.  Note that $[4^n x] = r$ if and only if $\dfrac{r}{4^n} \leq x < \dfrac{r}{4^n} + \dfrac{1}{4^n}$. From this, an explicit

formula for $f_n$ is given by

$$f_n(x) = \begin{cases} 4^n x - 2k & \text{if } \dfrac{2k}{4^n} \le x < \dfrac{2k}{4^n} + \dfrac{1}{4^n} \\ 2k + 2 - 4^n x & \text{if } \dfrac{2k+1}{4^n} \le x < \dfrac{2k+1}{4^n} + \dfrac{1}{4^n}. \end{cases}$$

By arguments similar to those in the last exercise it now follows that $f_n$ is continuous everywhere and differentiable everywhere except at points of the form $\dfrac{r}{4^n}$ for $r \in \mathbb{Z}$. An alternate way is to recognise that $f_n$ is the composite of the function $f$ of the last exercise with the function $g_n(x) = 4^n x$. As both these functions are continuous everywhere, so is $f_n$ by Exercise (1.5). There is also a rule, called the chain rule (see Exercise (3.12)) for derivatives of composite functions which would give a little easier solution to the parts involving $f_n'$. For the last assertion it suffices to show that $|f(x) - f(y)| \le |x - y|$ for all $x, y \in \mathbb{R}$. This is clear if $|x - y| \ge 1$, since $f(x), f(y)$ both lie in $[0, 1]$ for all $x, y$. Suppose $|x-y| < 1$ and without loss of generality that $x < y$. If the interval $[x, y]$ does not contain any integer, then $(x, f(x))$ and $(y, f(y))$ lie on a line of slope $\pm 1$ and so $|f(x) - f(y)| = |x-y|$. If $[x, y]$ contains an integer $n$ (which is unique since $|x - y| < 1$) then $(x, f(x)), (n, f(n))$ and $(y, f(y))$ are the vertices of a right-angled triangle, the sides meeting at $(n, f(n))$ having slopes 1 and $-1$. So $|f(x) - f(y)| \le \max\{|f(x) - f(n)|, |f(n) - f(y)|\} = \max\{n - x, y - n\} \le y - x$ (since $x \le n \le y$).

9.  $|f_n(x)| \le 1$ for all $n$ and for all $x$. So $|(\frac{3}{4})^n f_n(x)| \le (\frac{3}{4})^n$ and the series $\sum\limits_{n=0}^{\infty} (\frac{3}{4})^n f_n(x)$ converges uniformly to $g(x)$ on $\mathbb{R}$; i.e., the sequence of partial sums $\sum\limits_{k=0}^{n} (\frac{3}{4})^k f_k(x)$ converges uniformly to $g(x)$ on $\mathbb{R}$. As each partial sum is continuous (being a finite sum of continuous functions), by Theorem (1.5), $g$ is continuous everywhere on $\mathbb{R}$.

10. Since $f$ is periodic with period 2, $f(x) = f(4^n x)$ is periodic with period $\dfrac{2}{4^n}$ and the first assertion in the hint follows from the fact that for $n > m, \delta_m = \pm\dfrac{1}{2}\dfrac{1}{4^m} = \pm\dfrac{2.4^{n-m-1}}{4^n}$ and hence $\delta_m$ is a whole multiple of $\dfrac{2}{4^n}$. By the choice of the sign of $\delta_m$, $x + \delta_m$ and $x$ always lie in an interval on which the graph of $f_k$ is a straight line of slope $\pm 4^k$ for every $0 \le k \le m$. Hence $|f_k(x + \delta_m) - f_k(x)| = |4^k \delta_m|$ for $0 \le k \le m$. Now $|\sum\limits_{k=0}^{m-1} (\frac{3}{4})^k (f_k(x + \delta_m) - f_k(x))| = \sum\limits_{k=0}^{m-1} (\frac{3}{4})^k 4^k |\delta_m| = \dfrac{3^m - 1}{2}|\delta_m| < \dfrac{1}{2}3^m |\delta_m|$, while $(\frac{3}{4})^m (f_m(x + \delta_m) - f_m(x)) = \pm(\frac{3}{4})^m 4^m \delta_m = \pm 3^m \delta_m$. Hence $|g(x+\delta_m) - g(x)| \ge \dfrac{1}{2}3^m |\delta_m|$. This shows $g'(x)$ cannot exist. For if it does, then $\exists \delta > 0$ such that whenever

$0 < |h| < \delta, \left|\dfrac{g(x+h) - g(x)}{h} - g'(x)\right| < 1$, which implies $|g(x+h) - g(x)| <$

$|h|(|g'(x)| + 1)$. Fix such $\delta$ and find $m$ so large that $\dfrac{1}{2}\dfrac{i}{4^m} < \delta$ and $\dfrac{1}{2}3^m >$

$|g'(x) + 1|$. Setting $h = \delta_m$ gives a contradiction.

**11.** Merely set $\epsilon(x) = \begin{cases} \dfrac{f(x) - f(c)}{x - c} - f'(c) & \text{if } x \neq c \\ 0 & \text{if } x = c. \end{cases}$

Then as $x \to c$, $\dfrac{f(x) - f(c)}{x - c} - f'(c) \to f'(c) - f'(c) = 0$.

**12.** Let $\epsilon_1(x)$ and $\epsilon_2(x)$ be the functions given by the last exercise defined because of differentiability of $f$ at $c$ and of $g$ at $f(c)$ respectively. A straightforward calculation gives

$$\begin{aligned} h(x) = g(f(x)) &= g(f(c)) + g'(f(c))(f(x) - f(c)) + (f(x) - f(c))\epsilon_2(f(x)) \\ &= h(c) + g'(f(c))[f'(c)(x - c) + (x - c)\epsilon_1(x)] \\ &\quad + [f'(c)(x - c) + (x - c)\epsilon_1(x)]\epsilon_2(f(x)) \\ &= h(c) + g'(f(c))f'(c)(x - c) + (x - c)\epsilon_3(x) \\ \text{where } \epsilon_3(x) &= g'(f(c))\epsilon_1(x) + f'(c)\epsilon_2(f(x)) + \epsilon_1(x)\epsilon_2(f(x)). \end{aligned}$$

As $x \to c, \epsilon_1(x) \to 0$. Also as $x \to c, f(x) \to f(c)$ (by continuity of $f$ at $c$, which is a consequence of its differentiability at $c$) and hence $\epsilon_2(f(x)) \to 0$ as $x \to c$. So $\epsilon_3(x) \to 0$ as $x \to c$, which gives $\dfrac{h(x) - h(c)}{x - c} \to g'(f(c))f'(c)$ as $x \to c$. The Chain Rule is obvious if derivatives are interpreted as rates. If $y$ grows $\alpha$ times as fast as $x$ and $z$ grows $\beta$ times as fast as $y$, then $z$ grows $\alpha\beta$ times as fast as $x$. Take $\alpha = f'(c)$ and $\beta = g'(f(c))$.

**13.** If $c > 0$, then in some neighbourhood of $c, f(x) = x^2$ and so $f'(c) = 2c$ by Exercise (3.3). Similarly if $c < 0$ then $f'(c) = -2c$. $f'_+(0) = \lim\limits_{x \to 0^+} x = 0$ and

$f'_-(0) = \lim\limits_{x \to 0^-} \dfrac{-x^2}{x} = 0$. So $f'(0) = 0$. Thus $f$ is differentiable everywhere. But $f'(x) = 2|x|$ for all $x$, which shows $f$ is not twice differentiable at $0$. For a general $n$, let $f(x) = x^{n+1}$ if $x \geq 0$ and $f(x) = -x^{n+1}$ if $x < 0$.

**14.** For $x \neq 0, \dfrac{f_1(x) - f_1(0)}{x - 0} = \sin\dfrac{1}{x}$ which has no limit as $x \to 0$ by Exercise

(2.3.2). However, $\dfrac{f_2(x) - f_2(0)}{x - 0} = x \sin\dfrac{1}{x} \to 0$ as $x \to 0$ by Exercise (2.3.4).

So $f'_2(0)$ exists. At $x \neq 0, f'_2$ can be found using Exercise (3.2), (3.3), (3.12) and the derivative of the sine function. Thus, $f'_2(x) = 2x \sin\dfrac{1}{x} - \cos\dfrac{1}{x}$. As $x \to 0, 2x \sin\dfrac{1}{x} \to 0$ but $\cos\dfrac{1}{x}$ tends to no limit. So $f'_2$ is not continuous at $0$.

**15.** If $c > 0$, then in some deleted neighbourhood of $c$, $\dfrac{\sqrt{x} - \sqrt{c}}{x - c} = \dfrac{1}{\sqrt{x} + \sqrt{c}} \rightarrow$

$\dfrac{1}{2\sqrt{c}}$ as $x \rightarrow c$. But $\lim\limits_{x \to 0+} \dfrac{\sqrt{x} - \sqrt{0}}{x - 0} = \lim\limits_{x \to 0+} \dfrac{1}{\sqrt{x}}$ does not exist (as a finite real number).

**16.** $y = \sqrt{1 - x^2}$ gives, by the last exercise and chain rule, $\dfrac{dy}{dx} = \dfrac{-x}{\sqrt{1 - x^2}}$ if $|x| < 1$. So the slope of the tangent at a point $(x_0, y_0)$ is $\dfrac{-x_0}{\sqrt{1 - x_0^2}} = -\dfrac{x_0}{y_0}$.

The radius through $(x_0, y_0)$ has slope $\dfrac{y_0}{x_0}$. So they are perpendicular to each other. For the lower semi-circle, work with $y = -\sqrt{1 - x^2}$. At $(\pm 1, 0)$, the tangents are vertical and have no finite slope. $\dfrac{dy}{dx}$ also fails to exist at $x = \pm 1$.

**17.** For an ellipse $\dfrac{x^2}{a^2} + \dfrac{y^2}{b^2} = 1$, differentiate $y = b\sqrt{1 - \dfrac{x^2}{a^2}}$ in the upper half and $y = -b\sqrt{1 - \dfrac{x^2}{a^2}}$ in the lower half. If $P = (x_0, y_0)$ (with $y_0 > 0$) is on the ellipse then the slope of the tangent at $P$ is $\dfrac{-bx_0}{a\sqrt{a^2 - x_0^2}}$ which equals $-\dfrac{b^2 x_0}{a^2 y_0}$. The equation of the tangent comes out as $y - y_0 = -\dfrac{b^2 x_0}{a^2 y_0}(x - x_0)$ or as $\dfrac{x x_0}{a^2} + \dfrac{y y_0}{b^2} = 1$ (since $\dfrac{x_0^2}{a^2} + \dfrac{y_0^2}{b^2} = 1$) which tallies with that obtained in co-ordinate geometry. The cases of the other conics are similar.

## Section 4.4

**1.** The hint already shows that the restricted limit $\lim\limits_{\substack{\Delta x \to 0 \\ \Delta x \in S}} \dfrac{\Delta z}{\Delta x}$ equals $\dfrac{dz}{dy}\dfrac{dy}{dx}$. If $0$ is an isolated point of the complement of $S$ (i.e. if $\exists \, \delta > 0$ such that $0 < |\Delta x| < \delta \Rightarrow \Delta y \neq 0$, then $\lim\limits_{\substack{\Delta x \to 0 \\ \Delta x \notin S}} \dfrac{\Delta z}{\Delta x}$ is vacuously equal to anything and hence, in particular, to $\dfrac{dz}{dy}\dfrac{dy}{dx}$. Otherwise rewriting $\dfrac{\Delta z}{\Delta x}$ as in the hint, we get $\lim\limits_{\substack{\Delta x \to 0 \\ \Delta x \notin S}} \dfrac{\Delta z}{\Delta x} = 0 = \left(\dfrac{dz}{dy}\right)_{f(x_0)} \lim\limits_{\substack{\Delta x \to 0 \\ \Delta x \notin S}} \dfrac{\Delta y}{\Delta x} = \left(\dfrac{dz}{dy}\right)_{f(x_0)} \left(\dfrac{dy}{dx}\right)_{x_0}$. Since both the restricted limits are equal, by Theorem (5.1), $\lim\limits_{\Delta x \to 0} \dfrac{\Delta z}{\Delta x} = \dfrac{dz}{dy}\dfrac{dy}{dx}$.

**2.** Simply apply corresponding results about derivatives. (cf. Exercises (3.1), (3.3)).

**3.** (i) reads as $\int 0 dx$ includes the class of constant functions. (iii) would read $a\int f(x)dx + b\int g(x)dx = \int(af(x)+bg(x))dx$ and would mean that if $F \in \int f(x)dx$ and $G \in \int g(x)dx$ then $aF + bG \in \int af(x)+bg(x)dx$.

**4.** The last exercise already establishes the first part. For the second, fix any $F \in \int f(x)dx$ (which is given to be non-empty). Let $H \in \int h(x)dx$. Let $G = H - F$. Then $G'(x) = H'(x) - F'(x) = h(x) - f(x) = g(x)$. So $G \in \int g(x)dx$. Thus $H = F + G$ as desired.

**5.** Let $F \in \int f(x)dx$; i.e., $F'(x) = f(x)$ for all $x \in S$. Let $H = F \circ g$ be the composite function $H(t) = F(g(t))$. By the chain rule, $H'(t) = F'(g(t))g'(t) = f(g(t))g'(t)$. So $F(g(t)) \in \int f(g(t))g'(t)dt$.

**6.** If $U \in \int u(x)dx$, and $v$ is differentiable over $S$, and $G(x) \in \int U(x)v'(x)dx$ then $Uv-G \in \int u(x)v(x)dx$. For a proof, $\frac{d}{dx}(U(x)v(x)-G(x)) = U'(x)v(x)+ U(x)v'(x) - U(x)v'(x) = u(x)v(x)$.

**7.** Let $f(y) = \frac{1}{y^2}$. Then Exercise (4.5), with a slight change of notation (i.e. $t$ replaced by $x$ and $x$ by $y$) gives $\int \frac{1}{y^2}dy = \int f(g(x))\frac{dy}{dx}dx$. Let $F(y) = -\frac{1}{y}$. By Exercise (4.3) (ii), $F \in \int \frac{1}{y^2}dy$ and so $F(g(x))$, i.e., $-\frac{1}{g(x)}$ is an antiderivative of $\frac{g'(x)}{(g(x))^2}$. Since $\frac{g'(x)}{(g(x))^2} = 1, x$ is also an antiderivative of $\frac{g'(x)}{(g(x))^2}$ and so equating $F(y)$ with $x$ will give a solution of the differential equation. Thus $-\frac{1}{y} = x$, i.e, $xy = -1$ is a solution. (More generally, $(x+c)y = -1$ will also be a solution. The non-trivial part is to show that these are the only solutions, i.e., *every* solution is of the form $y = -\frac{1}{x+c}$ for some constant $c$. We shall not go into this.)

**8.** Note that as $f$ is a one-to-one function, $\Delta x \neq 0 \Rightarrow \Delta y \neq 0$. So we can consider $\frac{\Delta x}{\Delta y}$ and write it as $\frac{1}{\Delta y/\Delta x}$. Since the inverse function $g$ is continuous by Exercise (1.9), it follows that $\Delta x \to 0$ as $\Delta y \to 0$. So $\lim_{\Delta y \to 0} \frac{\Delta x}{\Delta y} = \lim_{\Delta x \to 0} \frac{1}{\Delta y/\Delta x} = \frac{1}{f'(x_0)}$. The function $f(x) = x^{1/n} (x > 0)$ where $n \in \mathbb{N}$ is the inverse function of $g(y) = y^n$. By Exercise (3.3) $g'(y) = ny^{n-1}$. So $\frac{d}{dx}(x^{1/n}) = \frac{1}{ny^{n-1}} = \frac{1}{nx^{(n-1)/n}} = \frac{1}{n}x^{\frac{1}{n}-1}$. Finally, let $n = \frac{p}{q}$ with $p \in \mathbb{Z}, q \in \mathbb{N}$ and $f(x) = x^n$ for $x > 0$. Then $f(x) = (x^{1/q})^p$. So by Exercise (3.3), the Chain Rule and the statement just proved, $f'(x) =$

$p(x^{1/q})^{p-1}\dfrac{1}{q}x^{\frac{1}{q}-1}$ which simplifies to $\dfrac{p}{q}x^{\frac{p}{q}-1}$, i.e., to $nx^{n-1}$.

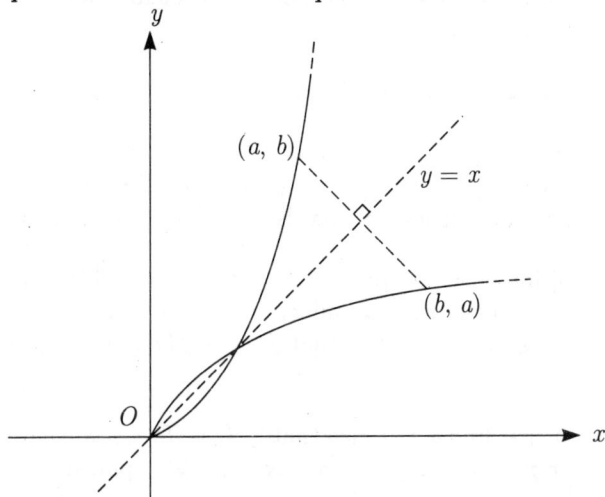

If $g$ is the inverse function of $f$ then the graphs of $y = f(x)$ and $y = g(x)$ are reflections of each other w.r.t. the line $y = x$. That is, a point $(a, b)$ lies on the first graph iff the point $(b, a)$ lies on the second. If $f$ is differentiable at $a$ and $f'(a) \neq 0$, then the tangent at $(a, b)$ has slope $f'(a)$, while that at $(b, a)$ has slope $\dfrac{1}{f'(a)}$. The two tangents are equally inclined to the line of symmetry $y = x$.

9.   $D^n(h) = \sum\limits_{r=0}^{n} \binom{n}{r} D^{(r)}(f) D^{(n-r)}(g)$. The case $n = 1$ is precisely the product rule for derivatives. For the inductive step, note that by the linearity property of derivatives, $D^{n+1}(h) = D(D^n(h)) = D(\sum\limits_{r=0}^{n} \binom{n}{r} D^r(f) D^{(n-r)}(g)) = \sum\limits_{r=0}^{n} \binom{n}{r}(D^{r+1}(f) D^{(n-r)}(g) + D^r(f) D^{(n-r+1)}(g))$ by the product rule. Regrouping the terms and using the binomial identity $\binom{n}{r} + \binom{n+1}{r} = \binom{n+1}{r+1}$ gives the inductive step. (Note the resemblance with the inductive proof of the binomial theorem.)

10. $2(f'(x))^2 + 2f(x)f''(x)$, $(f'(x))^2$ and $(f'(x))^2 + f(x)f''(x)$.

## Section 4.5

1.   For every $c \in Q, \exists\, \delta > 0$ such that the interval $(c - \delta, c + \delta)$ is either completely to the left or completely to the right of $\sqrt{2}$. (This requires irrationality of $\sqrt{2}$, see Exercise (2.6.4).) $f$ is constant on $(c - \delta, c + \delta) \cap Q$.

3. Here $f$ is not differentiable on $(-1, 1)$. (Actually $f$ is differentiable every-where except at 0. So even one point can make a whole lot of difference. That's mathematics!)

4. Here, discontinuity at one of the end-ponts, viz. 1, vitiates the matters.

5. Let $x_n = \frac{1}{n\pi}, n = 1, 2, 3, \ldots$. Then for every $n, f(x_{n+1}) = f(x_n) = 0$ and so by Rolle's theorem $\exists\, c_n \in (x_{n+1}, x_n)$ such that $f'(c_n) = 0$. These $c_n$'s belong to mutually disjoint intervals and hence are distinct.

6. Divide $[a, b]$ into $n$ equal parts by $x_0, x_1, \ldots, x_n$ where $x_i = a + i\frac{b-a}{n}$ for $i = 0, 1, \ldots, n$. (Thus $x_0 = a$ and $x_n = b$.) For each $i = 1, \ldots, n$, the MVT gives some $c_i \in (x_{i-1}, x_i)$ such that $f(x_i) - f(x_{i-1}) = f'(c_i)\frac{b-a}{n}$. Add these equations.

7. Given $\epsilon > 0$, first find $\delta > 0$ such that $|f'(x) - L| < \epsilon$ whenever $a < x < a + \delta$. For any such $x$, choose any $c_x$ given by the MVT applied to the interval $[a, x]$. Then $a < c_x < a + \delta$ and so $|f'(c_x) - L| < \epsilon$. But then $\left| \frac{f(x) - f(a)}{x - a} - L \right| < \epsilon$.

8. By the last exercise (and its analogue for left derivatives), if $\lim\limits_{x \to a^+} f'(x) = L = \lim\limits_{x \to a^-} f'(x)$, then $f'_+(a) = L = f'_-(a)$. So $f'(a)$ exists and $f'(a) = \lim\limits_{x \to a} f'(x)$. Thus $f'$ is continuous at $a$.

9. As a trivial counter-example, let $f(x) = \begin{vmatrix} 1 & 0 & 0 \\ 0 & x & 0 \\ 0 & 0 & 1 \end{vmatrix} = x$. Then $f'(x) = 1$ for all $x$. But the determinant of the derivatives has all except one entries 0. For the second part, expand $f(x)$ w.r.t. the first row and write it as a linear combination of $f_1(x), f_2(x), f_3(x)$ with constant coefficients. Then by Exercise (3.1), $f'(x)$ is the same linear combination of $f'_1(x), f'_2(x)$ and $f'_3(x)$. For the last part, write $f(x)$ as sum of 6 terms, apply the product rule to each term and regroup the resulting 18 terms into 3 groups of 6 each.

10. Let $\phi(x) = \begin{vmatrix} f(x) & g(x) & h(x) \\ f(a) & g(a) & h(a) \\ f(b) & g(b) & h(b) \end{vmatrix}$. Then $\phi(a) = \phi(b) = 0$ and so by Rolle's theorem, $\phi'(c) = 0$ for some $c \in (a, b)$. Apply the last exercise to write $\phi'(c)$ as a determinant.

11. Take $h(x) = 1$ for all $x$ and expand $\phi'(c)$ to get the result. Taking $g(x) = x$ gives Lagrange's MVT as a special case. So on the face of it, this result is more general than the MVT. But it is obtained from Rolle's theorem which itself is a special case of Lagrange's theorem. So all of them are essentially of the same strength.

**12.** No. In fact to think so is one of the most common mistakes. Application of Lagrange's MVT will give points $c_1, c_2 \in (a, b)$ for which $\dfrac{f'(c_1)}{g'(c_2)} = \dfrac{f(b) - f(a)}{g(b) - g(a)}$. But there is no guarantee that $c_1 = c_2$. (As an actual example, try $f(x) = x^2, g(x) = x^3$ on $[0, 1]$.)

**13.** If $F_1'(x) = F_2'(x) = f(x)$ for all $x \in [a, b]$ then $H'(x)$ vanishes identically on $[a, b]$ where $H(x) = F_1(x) - F_2(x)$. But then by Theorem (5.1), $H$ is a constant.

**14.** Apply the MVT to $[x_{i-1}, x_i]$ for $i = 1, \ldots, n$ and add, keeping in mind that $F' = f$. Exercise (5.6) results if $x_i - x_{i-1} = \frac{b-a}{n}$ for each $i$.

**15.** Suppose first $f(a) = f(b) = 0$. Let $f'(c_1) < 0$. If $f(c_1) > 0$ apply MVT to $[a, c_1]$ to get a desired $c_2$. If $f(c_1) < 0$, apply MVT to $[c_1, b]$. If $f(c_1) = 0$, then $f'(c_1) < 0$ implies by Exercise (3.4), that for some $h > 0, f(c_1 + h) < f(c_1) = 0$. Now apply MVT to $[c_1 + h, b]$. In the general case, consider $g(x) = f(x) - f(a) - \frac{f(b)-f(a)}{b-a}(x - a)$ (which is what the function in the proof of Theorem (5.4) reduces to). Then $g(a) = g(b) = 0$ and $g'(c) = f'(c) - \frac{f(b)-f(a)}{b-a}$ for all $c$. So apply the first part to $g$. Intuitively the result is obvious because treating a derivative as speed, if the instantaneous speed at some moment is less than the average speed over some interval, then to make up for it, at some other moment, it must be greater.

**16.** The first part follows directly from Rolle's theorem. For the second, if $p(x)$ is a polynomial of degree $n$ then $p'(x)$ has degree $n - 1$. So a proof by induction on $n$ is possible.

**17.** $f$ is one-to-one on $[a, b]$ as otherwise $f(x_1) = f(x_2)$ would give (by MVT) some $c$ for which $f'(c) = 0$. By Exercise (1.9) $f$ is either strictly increasing or strictly decreasing. So either $f'(x) \geq 0$ throughout or $f'(x) \leq 0$ throughout. From the hypothesis, strict inequality holds. To get the last statement, if $f'(a_1) > \alpha$ and $f'(b_1) < \alpha$, consider $g(x) = f(x) - \alpha x$.

**18.** $f'(x) = 1 + \cos x \geq 0$ for all $x$. So certainly $f(x_2) \geq f(x_1)$ whenever $x_1 < x_2$. To show strict inequality, note that the zeros of $f'$ (viz. all odd multiples of $\pi$) are isolated and so within $(x_1, x_2)$ we can find a subinterval $(y_1, y_2)$ which contains none of these zeros. Then $x_1 < y_1 < y_2 < x_2$ and $f(x_1) \leq f(y_1) < f(y_2) \leq f(x_2)$.

**19.** For the first part apply Exercise (3.14). Note that for $x \neq 0, f'(x) = \frac{1}{2} + 2x \sin \frac{1}{x} - \cos \frac{1}{x}$. As $x \to 0, \frac{1}{2} + 2x \sin \frac{1}{x} \to \frac{1}{2}$ and so $\exists \, \delta > 0$ such that $|\frac{1}{2} + 2x \sin \frac{1}{x}| < \frac{3}{4}$ whenever $0 < x < \delta$. The interval $(0, \delta)$ contains infinitely many points $x_n$ for which $\cos \frac{1}{x_n} = 1$ and also infinitely many points $y_n$ for which $\cos \frac{1}{y_n} = -1$. Around each $x_n, \exists$ a subinterval in which $f'(x) > \frac{1}{4} > 0$ and hence in which $f$ is strictly increasing. Similarly, $f$ is strictly increasing in some neighbourhood of each $y_n$. So there is no interval around 0 in which $f$ is either increasing throughout or decreasing throughout. The result of Exercise (3.4) is not contradicted because it deals with monotonicity at a point and not over an interval. Here $f$ is indeed monotonically increasing *at* 0, but not *over* $(-\delta, \delta)$ for any $\delta > 0$.

**20.** By Exericse (1.1), $\exists\, \delta > 0$ such that $f'(x) > 0$ for all $x \in (c - \delta, c + \delta)$. So $f$ is strictly increasing on $(c - \delta, c + \delta)$.

**21.** Let $f'$ be continuous on $[a, b]$. By Theorem (1.4), $\exists\, M \geq 0$ such that $|f'(c)| \leq M$ for all $c \in [a, b]$. Let $x_1, x_2 \in [a, b]$ with $x_1 \neq x_2$. Then by Lagrange's MVT applied to $[x_1, x_2]$ or to $[x_2, x_1]$ as the case may be, $\exists\, c \in (x_1, x_2)$ such that $f(x_2) - f(x_1) = f'(c)(x_2 - x_1)$. But then $|f(x_2) - f(x_1)| = |f'(c)||x_2 - x_1| \leq M|x_2 - x_1|$. So $f$ satisfies a Lipschitz condition.

**22.** $\exists\, c_1 \in (a, b)$ s.t. $g'(c_1) = 0$. Since $g'(a) = 0$, Rolle's theorem gives some $c_2 \in (a, c_1)$ such that $g''(c_2) = 0$. Since $g''(a) = 0, \exists\, c_3 \in (a, c_2)$ such that $g'''(c_3) = 0$. Continue. The desired $c$ is $c_n$.

**23.** Follow the hint and observe that in the $r^{\text{th}}$ derivative of $g(x)$ (where $1 \leq r \leq n - 1$, only the powers of $(x - a)$ from $(x - a)^r$ onwards matter. Of these, only $(x - a)^r$ gives a term (viz. $r!$) which is non-zero at $a$. So $g(a) = g'(a) = \ldots = g^{(n-1)}(a) = 0$. By the last exercise, $g^{(n)}(c) = 0$ for some $c \in (a, b)$. But $g^{(n)}(c) = f^{(n)}(c) - Kn!$. The condition $g(b) = 0$ determines $K$ uniquely as

$$\frac{1}{(b - a)^n}\left(f(b) - f(a) - f'(a)(b - a) - \ldots - \frac{f^{(n-1)}(a)}{(n - 1)!}(b - a)^{n-1}\right). \text{ Substitute.}$$

**24.** To compare the two theorems change the notation in Taylor's theorem slightly. Replace $[a, b]$ by $[c, c+h]$. We assume $h > 0$ for simplicity. Theorem (3.3) requires $f^{(n)}$ to exist at $c$ while Taylor's theorem requires it to exist on $(c, c + h)$. So the hypotheses of the two theorems are not comparable. To compare their strengths, bring them to a common platform by assuming $f^{(n)}$ to exist at $c$ and also on $(c, c + h)$. Both the theorems approximate $f(c + h)$ by the $f(c) + hf'(c) + \dfrac{h^2}{2!}f'(c) + \ldots + \dfrac{h^{n-1}}{(n - 1)!}f^{(n-1)}(c)$. But Theorem (3.3) says that the remainder is of the form $\dfrac{f^{(n)}(c)}{n!}h^n + R_n$ where $R_n$ is so small that $\dfrac{R_n}{h^n} \to 0$ as $h \to 0$, while Theorem (5.6) gives a formula for the remainder as $\dfrac{f^{(n)}(c + \theta h)}{n!}h^n$ for some $\theta \in (0, 1)$. If $f^{(n)}$ is continuous at $c$ then $\left(\dfrac{f^{(n)}(c + \theta h)}{n!}h^n - \dfrac{f^{(n)}(c)}{n!}h^n\right)/h^n \to 0$ as $h \to 0$ and we get the same information as Theorem (3.3). So under this additional hypothesis (which is often satisfied anyway), Theorem (5.6) is stronger.

**25.** Differentialility of $f$ at 0 follows by essentially the same argument as that for $f_2$ in Exercise (3.14). For $x \neq 0$, $f'(x) = 2x \sin\left(\dfrac{1}{x^2}\right) - \dfrac{2}{x}\cos\left(\dfrac{1}{x^2}\right)$.

So, for a positive integer $n$, $\left| f'\left(\dfrac{1}{\sqrt{n\pi}}\right) \right| = 2\sqrt{n\pi}$ which can be made as large as we like.

**26.** Following the notation in the question, by Exercise (5.21), there exist $M_1, M_2,$ $\ldots, M_n$ such that for any $x, y \in [t_{i-1}, t_i]$, $|f(x) - f(y)| \le M_i|x - y|$. Let $M = \max\{M_1, M_2, \ldots, M_n\}$. Then for all $x, y \in [a, b], |f(x) - f(y)| \le M|x - y|$. This is clear if $x, y$ are both in $[t_{i-1}, t_i]$ for some $i$. Otherwise, say, $t_{i-1} \le x \le t_i < t_{i+1} < \ldots < t_j \le y \le t_{j+1}$ for some $i, j$. Apply the Lipschitz condition to the pairs $(x, t_i), (t_i, t_{i+1}), \ldots, (t_j, y)$ and add.

## Section 4.6

**1.** $|f_n(x) - 0| \le \frac{1}{n}$ for all $x$ and for all $n$. So $\sup_{x \in \mathbb{R}} |f_n(x) - 0| \le \frac{1}{n} \to 0$ as $n \to \infty$. But $f_n'(x) = n \cos(n^2 x)$ which is not convergent for $x = k\pi$, where $k$ is an integer.

**2.** (i)

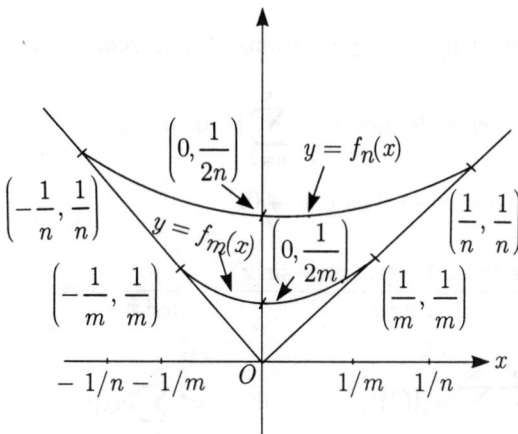

The graph of $f_n$ consists of two rays of slopes $\pm 1$, joined by an arc of the parabola $y = \frac{1}{2}(nx^2 + \frac{1}{n})$ from $(-\frac{1}{n}, \frac{1}{n})$ to $(\frac{1}{n}, \frac{1}{n})$. Show that the tangents to the arc at these points have the same slopes as the respective rays passing through them. (In other words, $(f_n')_-(\frac{1}{n}) = 1 = (f_n')_+(\frac{1}{n})$ and $(f_n')_+(-\frac{1}{n}) = -1 = (f_n')_-(-\frac{1}{n})$.) For any other $c, \exists$ a neighbourhood in which $f_n$ coincides with one of the functions $x, -x$ and $\frac{1}{2}(nx^2 + \frac{1}{n})$ and hence is differentiable at $c$.

(ii) Follow the hint. If suffices to take $m$ to be any integer $> \frac{1}{|x|}$. Then $\{f_n(x)\}$ is eventually constant, and converges to $|x|$. For $x = 0$, $f_n(0) = \frac{1}{2n}$ for all $n$ and so $f_n(0) \to 0 = |0|$. It is clear from the graph that $\sup_{x \in \mathbb{R}} |f_n(x) - |x|| = |f_n(0) - |0|| = \frac{1}{2n}$. So the convergence is uniform.

**3.** That $f_n \xrightarrow{} 0$ is clear since $|f_n(x) - 0| \le \frac{1}{n}$ for all $x \in [0, 1]$. $f_n'(x) = x^{n-1} \to 0$ if $x \in [0, 1)$ and $\to 1$ if $x = 1$. But $f'$ vanishes everywhere.

**4.** Take $f_n(x) = n$ for all $x \in [a, b]$.

**5.** For the first part, the sum on the right is $(b-a)\sum_{k=0}^{n-2}(k+1)a^k b^{n-k-2}$ which can be rewritten as $\sum_{k=0}^{n-2}(k+1)a^k b^{n-1-k} - \sum_{k=0}^{n-2}(k+1)a^{k+1}b^{n-2-k}$. With a change of index (from $k$ to $k+1$) in the second summation, this is $\displaystyle\sum_{k=0}^{n-2}(k+1)a^k b^{n-1-k}$

$-\displaystyle\sum_{k=1}^{n-1}ka^k b^{n-1-k} = b^{n-1} + \sum_{k=1}^{n-2}a^k b^{n-1-k} - (n-1)a^{n-1}$. To finish the proof use

the fact that $\dfrac{b^n - a^n}{b - a} = \displaystyle\sum_{k=0}^{n-1}a^k b^{n-1-k}$. For the second part, apply the triangle

inequality and note that $1 + 2 + \ldots + (n-1) = \dfrac{n(n-1)}{2} \leq n(n-1)$.

6.  Follow the hint. By Theorem (6.3), $\displaystyle\sum_{n=1}^{\infty}na_n c^{n-1}$ converges. Let $L$ denote its

sum. Then $\left|\dfrac{f(c + \Delta x) - f(x)}{\Delta x} - L\right| = \left|\displaystyle\sum_{n=1}^{\infty}a_n\left(\dfrac{(c + \Delta x)^n - c^n}{\Delta x} - nc^{n-1}\right)\right|$

$\leq \displaystyle\sum_{n=2}^{\infty}|a_n||\Delta x|n(n-1)M^{n-2}$ by the last exercise. The series

$\displaystyle\sum_{n=2}^{\infty}|a_n|n(n-1)M^{n-2}$ is convergent by two applications of Theorem (6.3)

and the fact that $|M| = |c| + |\Delta x| < R$. Let $K = \displaystyle\sum_{n=2}^{\infty}|a_n|n(n-1)M^{n-2}$.

Then $\left|\dfrac{f(c + \Delta x) - f(x)}{\Delta x} - L\right| \leq K|\Delta x| \to 0$ as $\Delta x \to 0$.

7.  $\dfrac{d}{dx}\tan x = \dfrac{d}{dx}\left(\dfrac{\sin x}{\cos x}\right) = \dfrac{(\cos x)(\cos x) - (\sin x)(-\sin x)}{\cos^2 x} = \dfrac{1}{\cos^2 x} = \sec^2 x$
    etc.

8.  Follow the hint. Note that $\displaystyle\sum_{n=0}^{\infty}\sum_{r=0}^{\infty}|a_n|\binom{n}{r}|x - c|^r|c|^{n-r} = \sum_{n=0}^{\infty}|a_n|(|x - $

    $c| + |c|)^n$ is convergent since $|x - c| + |c| < R$. So the doubly infinite sequence $\{a_{n,r}\}$ where $a_{n,r} = a_n\binom{n}{r}(x - c)^r c^{n-r}$ is summable over the set

    $\{0, 1, 2, \ldots\} \times \{0, 1, 2, \ldots\}$. By Exercise (3.8.7), $\displaystyle\sum_{n=0}^{\infty}\sum_{r=0}^{\infty}a_n\binom{n}{r}(x - c)^r c^{n-r} = $

    $\displaystyle\sum_{r=0}^{\infty}(x - c)^r\sum_{n=0}^{\infty}\binom{n}{r}a_n c^{n-r}$. Since $\binom{n}{r} = \dfrac{n(n - 1)\ldots(n - r + 1)}{r!}$ for $n \geq r$

    while $\binom{n}{r} = 0$ for $n < r$, $\displaystyle\sum_{n=0}^{\infty}\binom{n}{r}a_n c^{n-r}$ is the same as $\dfrac{1}{r!}\displaystyle\sum_{n=r}^{\infty}n(n - 1)\ldots(n - $

    $r + 1)a_n c^{n-r}$, which, by Theorem (6.5), equals $\dfrac{f^{(r)}(c)}{r!}$. Hence $\displaystyle\sum_{n=0}^{\infty}a_n x^n$ equals

    $\displaystyle\sum_{r=0}^{\infty}\dfrac{f^{(r)}(c)}{r!}(x - c)^r$, for $x \in (c - (R - |c|), c + R - |c|)$. That is, $f(x)$ can be expanded in powers of $(x - c)$ in a neighbourhood of $c$.

**9.** More generally, for any $c \neq 1$, $\dfrac{1}{1-x} = \dfrac{1}{(1-c)-(x-c)} = \dfrac{1}{1-c} \dfrac{1}{(1 - \frac{x-c}{1-c})}$

$= \displaystyle\sum_{n=0}^{\infty} \dfrac{(x-c)^n}{(1-c)^{n+1}}$ (by Theorem (3.3.5)). This is valid when $\left| \dfrac{x-c}{1-c} \right| < 1$.

Thus the radius of convergence $= |1 - c| > 0$ since $c \neq 1$.

**10.** Take $a_n = 1$ for all $n \geq 0$. Then $f(x) = \dfrac{1}{1-x}$. If $c = -\dfrac{1}{2}$, then $R - |c| = \dfrac{1}{2}$.

But by the last exercise, the radius of convergence of the power series expansion of $f(x)$ around $c$ is $1 - (-\dfrac{1}{2}) = \dfrac{3}{2}$.

**12.** Assume $c = 0$ and $a_0 = 1$. (Otherwise divide by $a_0$ which is non-zero since $f(0) = a_0$.) The system of equations can then be solved inductively as $b_0 = 1, b_1 = -a_1, b_2 = a_1^2 - a_2, b_3 = -a_1^3 + 2a_1 a_2 - a_3, \ldots$. Let $g(x)$ be the sum function of the power series $\displaystyle\sum_{n=0}^{\infty} b_n x^n$. Then $f(x)g(x)$ is the sum function of the Cauchy product of $\displaystyle\sum_{n=0}^{\infty} a_n x^n$ and $\displaystyle\sum_{n=0}^{\infty} b_n x^n$. But the system of equations ensures that this Cauchy product is simply 1. So $g(x) = \dfrac{1}{f(x)}$ is analytic at 0. The catch in this argument is that $\displaystyle\sum_{n=0}^{\infty} b_n x^n$ is not shown to have a positive radius of convergence. To do so directly by estimating $|b_n|^{1/n}$ is messy. A different approach is to note that by continuity, $f$ is non-zero in a neighbourhood $U$ of 0. For $x \in U$, write $\dfrac{1}{f(x)} = \dfrac{1}{1 + \displaystyle\sum_{n=1}^{\infty} a_n x^n}$. Since $\displaystyle\sum_{n=1}^{\infty} |a_n x^n| \to 0$ as $x \to 0$, we may assume $U$ is so small that $\displaystyle\sum_{n=1}^{\infty} |a_n||x|^n < 1$ for all $x \in U$. Now expand $\dfrac{1}{1 + \displaystyle\sum_{n=1}^{\infty} a_n x^n}$ as a

geometric series as $1 - \displaystyle\sum_{n=1}^{\infty} a_n x^n + \left( \displaystyle\sum_{n=1}^{\infty} a_n x^n \right)^2 - \left( \displaystyle\sum_{n=1}^{\infty} a_n x^n \right)^3 + \ldots$. Expand the powers as Cauchy products and treat the series as doubly infinite. Apply Theorem (3.7.4) (or rather, its extension for summable functions) after grouping together the like powers. The coefficient of $x^n$ comes out to be $b_n$. This is also somewhat messy to prove. However, the argument here shows that $\dfrac{1}{f(x)}$ can be expanded near 0 by *some* power series. Call it $\displaystyle\sum_{n=0}^{\infty} b_n x^n$. Then $\left( \displaystyle\sum_{n=0}^{\infty} a_n x^n \right) \left( \displaystyle\sum_{n=0}^{\infty} b_n x^n \right) = 1 = 1 + 0x + 0x^2 + 0x^3 + \ldots$

gives, by the uniqueness theorem for power series, the system of equations for $b_n$'s.

**13.** If $h \in I\!N$, then $(1+x)^h$ is a polynomial (of degree $h$). Treated as a power series, its radius of convergence is $\infty$ and so by Exercise (6.8) $(1+x)^h$ is analytic everywhere. If $h$ is only a rational then $(1+x)^h$ makes sense and is infinitely differentiable (by Exercise (4.8) applied repeatedly) for $1+x > 0$, i.e. for $x > -1$. But $f^{(r)}(0)$ comes out as $h(h-1)\ldots(h-r+1)$ and so by Exercise (3.10.4) (iii), $\displaystyle\sum_{n=0}^{\infty} \frac{f^{(n)}(0)}{n!} x^n$ converges for $x \in (-1,1)$ but not for $|x| > 1$.

**14.** Note that $\displaystyle\sum_{n=0}^{\infty} b_n x^n$ is the Cauchy product of $\displaystyle\sum_{n=0}^{\infty} a_n x^n$ and $\displaystyle\sum_{n=0}^{\infty} x^n$. The latter has sum $\dfrac{1}{1-x}$. Taking $f(x) = \dfrac{1}{1-x}$ we get $\displaystyle\sum_{n=0}^{\infty}(n+1)x^n$ as the power series expansion of $\dfrac{1}{(1-x)^2}$. From this, the power series expansion of $\dfrac{1}{(1-x)^3}$ is $\displaystyle\sum_{n=0}^{\infty} b_n x^n$ where $b_n = \displaystyle\sum_{k=0}^{n}(k+1)$. By either of the two other methods the power series expansion of $\dfrac{1}{(1-x)^3}(= (1-x)^{-3})$ is $\displaystyle\sum_{n=0}^{\infty} \frac{(n+2)(n+1)x^n}{2}$.

Equating $b_{n-1}$ with $\dfrac{(n+1)n}{2}$ gives the identity.

**15.** (i) Let $u(x) = \dfrac{f(x)+f(-x)}{2}$ and $v(x) = \dfrac{f(x)-f(-x)}{2}$. Then $u(x)$ is even, $v(x)$ is odd and $f(x) = u(x) + v(x)$. For uniqueness suppose $f = u_1 + v_1$ with $u_1$ even and $v_1$ odd. Then $u - u_1 = v_1 - v$ is both even and odd. The only such function is $0$. (ii) can be proved directly from the definition of a derivative. But a slicker way is to consider the function $g(x) = -x$. If $f$ is even then $f \circ g = f$ and so by the chain rule, $(f' \circ g)g' = f'$. But $g' = -1$ everywhere. So $f'(-x)(-1) = f'(x)$, i.e., $f'$ is odd. Similarly if $f$ is odd then $f \circ g = -g$ implies, by the chain rule, that $f'$ is even. (iii) $f(x) = \displaystyle\sum_{n \text{ even}} a_n x^n + \sum_{n \text{ odd}} a_n x^n$ while $f(-x) = \displaystyle\sum_{n \text{ even}} a_n x^n - \sum_{n \text{ odd}} a_n x^n$.

If $f$ is even then $\displaystyle\sum_{n \text{ odd}} a_n x^n = 0$. By the uniqueness theorem of power series $a_n = 0$ for all odd $n$. Similarly if $f$ is odd then $\displaystyle\sum_{n \text{ even}} a_n x^n = 0$ and hence $a_n = 0$ for all even $n$. Converses are trivial. Geometrically, the graph of an even function is symmetric about the $y$-axis while that of an odd function is symmetric w.r.t. the origin. This also gives a geometric proof of (ii) because lines which are images of each other w.r.t the $y$-axis have slopes which are negatives of each other and those which are images of each other w.r.t. the origin are parallel to each other.

**16.** The crucial point is that $M$ is independent of $n$. Apply Theorem (5.6) (with a slight change of notation). The remainder $R_n = \dfrac{f^{(n)}(c + \theta_n(x - c))}{n!}(x - c)^n$ for some $\theta_n \in (0, 1)$. Regardless of what $\theta_n$ is, $|R_n| \leq \dfrac{M|x - c|^n}{n!} \to 0$ as $n \to \infty$ by Exercise (3.9.4). So, as $n \to \infty$, $\displaystyle\sum_{k=0}^{n-1} \dfrac{f^{(k)}(c)}{k!}(x - c)^k \to f(x)$ for all $x \in U$.

**17.** If the bound, say $M_n$, on the $n^{th}$ derivative $f^{(n)}(x)$ (or rather, on its absolute value) for $x \in U$ is such that $\dfrac{M_n|x - c|^n}{n!} \to 0$ for every $x \in U$, the argument above goes through. This covers cases like $M_n = n$ or $M_n = n^2$ etc. But mere boundedness of $|f^{(n)}|$ is not enough as shown by the function in the text viz. $f(x) = e^{-1/x^2}$ if $x \neq 0$ and $f(0) = 0$. Here each $f^{(n)}$ is continuous and hence bounded on every bounded neighbourhood of $0$.

**18.** Let $f(x) = \displaystyle\sum_{n=0}^{\infty} a_n(x - c)^n$. If $f^{(n)}(c) = 0$ for all $n$, it would mean $a_n = 0$ for all $n$ and hence $f$ is identically $0$. So the integer $m$ in the hint exists. Clearly $a_k = 0$ for $k = 0, 1, \ldots, m - 1$. So $f(x) = \displaystyle\sum_{n=m}^{\infty} a_n(x - c)^n = (x - c)^m g(x)$ where $g(x) = \displaystyle\sum_{k=0}^{\infty} a_{m+k}(x - c)^k$. Then $g$ is analytic at $c$ and $g(c) = a_m = \dfrac{f^{(m)}(c)}{m!} \neq 0$. For uniqueness of $m$ (and also of $g(x)$) let, if possible, $f(x)$ also equal $(x - c)^r h(x)$ with $h(c) \neq 0$. Then $r > m$ gives $g(x) = (x - c)^{r-m} h(x)$ and hence $g(c) = 0$, a contradiction. Similarly $r < m$ gives a contradiction. So $r = m$ which further implies $g(x) = h(x)$.

**19.** Let $f(x) = (x - c)^m f_1(x), g(x) = (x - c)^n g_1(x)$ with $f_1, g_1$ analytic and non-vanishing at $c$. Then $h(x) = (x - c)^{m+n} h_1(x)$ where $h_1 = f_1 g_1$ is analytic and non-vanishing at $c$.

**20.** If $f(x) = (x - c)^m g(x)$ with $g$ analytic and $g(c) \neq 0$, then $f$ is analytic and $m$ is the least integer such that $f^{(m)}(c) \neq 0$. The first two of the given functions are not infinitely differentiable at $0$ (cf. Exercises (3.14) and (3.15) while the last function, derivatives of all orders vanish at $0$. So none of these functions is analytic. (The point is that no integral power of $x$ can be factored out from these functions so as to have the other factor non-vanishing at $0$.)

**21.** Let $f(x) = \sin x$. Then $f(0) = 0$, but $f'(0) = \cos 0 = 1 \neq 0$. So $0$ is a zero of ordre $1$. By similar reasonings $1 - \cos x$ and $\sin x - x$ have zeros of orders $2$ and $3$ respectively at $0$.

**22.** For (i) it suffices to show that for every (fixed) $n \in \mathbb{N}$, $\dfrac{x^n}{e^x} \to 0$ as $x \to \infty$.

For $x > 0$, $e^x > \dfrac{x^{n+1}}{(n+1)!}$ and so $0 < \dfrac{x^n}{e^x} < \dfrac{(n+1)!}{x} \to 0$ as $x \to \infty$. For

(ii) let $p(x) = a_r x^r + a_{r+1} x^{r+1} + \ldots + a_n x^n$ with $a_r \neq 0$. Then $p(x) = a_r x^r (1 + q(x))$ where $q(x) \to 0$ as $x \to 0$. Hence $\exists\, \delta > 0$ such that for

$0 < |x| < \delta, |p(x)| \geq \dfrac{1}{2} |a_r||x|^r$. Now $e^{1/x^2} = \displaystyle\sum_{k=0}^{\infty} \dfrac{1}{k! x^{2k}} > \dfrac{1}{(r+1)! x^{2r+2}}$. So

$\left| \dfrac{1}{p(x) e^{1/x^2}} \right| \leq \dfrac{2(r+1)! |x|^{2r+2}}{|a_r||x|^r}$ which tends to 0 as $x \to 0$.

**23.** Let $y = \ln x$ $(x > 0)$. Then $x = e^y$. So $\dfrac{dy}{dx} = \dfrac{1}{dx/dy} = \dfrac{1}{e^y} = \dfrac{1}{x}$.

### Section 4.7

**1.** $(x,y)$ lies on $L$ iff $\dfrac{y - y_1}{y_2 - y_1} = \dfrac{x - x_1}{x_2 - x_1} = t$ (say). This gives the parametric equations. (It is assumed here that $L$ is not parallel to either of the axes. If it is, simply delete one of the two ratios and the parametrisation still hods.) For $(x,y)$ to lie on the ray originating at $P$ and passing through $Q, t \geq 0$. As $t$ increases, $(x,y)$ moves away from $P$. It is at $Q$ when $t = 1$. So points on the segment correspond to $t \in [0,1]$.

**2.** (i) $(x - a)^2 + (y - b)^2 - r^2$; the circle of radius $r$ centred at $(a,b)$.

(ii) $y^2 = 4ax$; the parabola with focus at $(a,0)$ and directrix $x = -a$

(iii) $\dfrac{x^2}{a^2} + \dfrac{y^2}{b^2} = 1$; the ellipse with major axis $2a$ and foci at $(\pm\sqrt{a^2 - b^2}, 0)$

(iv) $\dfrac{x^2}{a^2} - \dfrac{y^2}{b^2} = 1$; the hyperbola with axis $2a$ and foci at $(\pm\sqrt{a^2 + b^2}, 0)$.

**3.** Without loss of generality assume the unit circle $C$ and the other circle $C'$ both have centres at $O$. For any pont $P$ on $C$, let $P'$ be the point where the ray $OP$ meets $C'$. Suppose $C_1 = (P_0, P_1, \ldots, P_n)$ is a broken-line path on $C$ (with $P_0 = P_n$). Then $C_1' = (P_0', P_1', \ldots, P_n')$ is a broken line path along $C'$. The triangles $OP_{i-1}P_i$ and $OP_{i-1}'P_i'$ are similar for $i = 1, 2, \ldots, n$ with $|P_{i-1}'P_i'|/|P_{i-1}P_i| = |OP_i'|/|OP_i| = r$. So $l(C_1') = rl(C_1)$. Conversely given $C_1'$, we can construct $C_1$ so that $l(C_1') = rl(C_1)$. Hence $\sup l(C_1')$ as $C_1'$ ranges over all possible broken line paths along $C'$ is $r$ times the corresponding supremum for $C$.

**4.** Here for $P = (x,y) = (f(t), g(t))$ on $C_1$, let $P' = (x', y') = (rf(t) + \alpha, rg(t) + \beta)$. Then $P'$ lies on $C'$. If $P_1 = (x_1, y_1)$ and $P_2 = (x_2, y_2)$ are on $C$ then $|P_1'P_2'| = \sqrt{(rx_1 - rx_2)^2 + (ry_1 - ry_2)^2} = r|P_1P_2|$. The rest of the reasoning is similar to that in the last exercise.

**5.** The angle subtended by each side at the centre of the circle is $\frac{2\pi}{n}$. So the length of each side is $2\sin\frac{\pi}{n}$. This proves the first part. As $n \to \infty$, $\sin\frac{\pi}{n} \to 0$ and so $\dfrac{\sin(\pi/n)}{\pi/n} \to 1$ by (3). But then $2n\sin(\pi/n) = 2\pi\dfrac{\sin(\pi/n)}{\pi/n} \to 2\pi$ as $n \to \infty$. This cannot be taken as a proof that the length of the unit circle is $2\pi$ because this fact is crucially needed in proving (3). So unless we have some other proof of (3), it would be a vicious cycle.

**6.** (i) $\dfrac{1 - \cos x}{x} = \dfrac{2\sin^2(x/2)}{x} = \sin\dfrac{x}{2}\dfrac{\sin(x/2)}{x/2} \to 0.1 = 0$ as $x \to 0$.

(ii) $\dfrac{1 - \cos(x/2)}{x^2} = \dfrac{2\sin^2\frac{x}{4}}{x^2} = \dfrac{1}{2}\left(\dfrac{\sin(x/2)}{x/2}\right)^2 \to \dfrac{1}{2}$ as $x \to 0$.

(iii) $\dfrac{1 - \cos x}{x\sin x} = \dfrac{\sin^2 x}{x\sin x(1 + \cos x)} = \dfrac{\sin x}{x}\dfrac{1}{1 + \cos x} \to \dfrac{1}{2}$ as $x \to 0$.

(iv) $-2$. ( $\lim\limits_{x\to\frac{\pi}{4}}\dfrac{\tan x - 1}{x - \frac{\pi}{4}}$ is the derivative of $\tan x$ at $\dfrac{\pi}{4}$. Or simplify after putting $h = x - \frac{\pi}{4}$, a useful trick in finding limits.)

(v) $\dfrac{\cos(\sin x) - 1}{x^2} = \dfrac{\cos(\sin x) - 1}{\sin^2 x}\dfrac{\sin^2 x}{x^2}$. As $x \to 0$, $\sin x \to 0$ and so the first factor has the same limit as $\dfrac{\cos x - 1}{x^2}$ which is $-\frac{1}{2}$. So the given limit is $-\frac{1}{2}$.

(vi) $\dfrac{\sqrt{x + \sin x}}{\sqrt{\sin x + \sin^2 x}} = \dfrac{\sqrt{\frac{x}{\sin x} + 1}}{\sqrt{1 + \sin x}} \to \sqrt{2}$ as $x \to 0^+$.

**7.** (i) a trivial verification.

(ii) $|P_nQ_n| = \sqrt{\left(\dfrac{1}{2n\pi} - \dfrac{2}{(4n+1)\pi}\right)^2 + \left(\dfrac{2}{(4n+1)\pi}\right)^2} \geq \dfrac{2}{(4n+1)\pi}$.

(iii) is immediate from (ii), but it must first be ensured that the values of the parameter $x$ corresponding to $P_1, Q_1, P_2, Q_2, \ldots, P_n, Q_n$ are monotonically decreasing (or increasing). These values are

$$\dfrac{1}{2\pi}, \dfrac{1}{2\pi + \frac{\pi}{2}}, \dfrac{1}{4\pi}, \dfrac{1}{4\pi + \frac{\pi}{2}}, \ldots, \dfrac{1}{2n\pi}, \dfrac{1}{2n\pi + \frac{\pi}{2}}.$$

So they are indeed decreasing and hence $(P_1, Q_1, \ldots, P_n, Q_n)$ is a broken line path-along $C$. (iv) By comparison with the harmonic series $\sum\limits_{n=1}^{\infty}\dfrac{1}{n}$, the series $\sum\limits_{n=1}^{\infty}\dfrac{2}{(4n+1)\pi}$ is divergent. So by (iii) the lengths of the broken line paths along $C$ are not bounded above.

**8.** $0 < \sin \frac{\phi}{2} \le \frac{\phi}{2}$ and $\cos \frac{\phi}{2} < 1$ together give $\sin \frac{\phi}{2} \cos \frac{\phi}{2} < \sin \frac{\phi}{2} \cdot 1 \le \frac{\phi}{2}$ and hence $\sin \phi < 2\frac{\phi}{2} = \phi$. Similarly $\frac{\phi}{2} \le \tan \frac{\phi}{2}$ and $\frac{1}{1 - \tan^2 \frac{\phi}{2}} > 1$ give $\phi = 2\frac{\phi}{2}$

$\le 2 \tan \frac{\phi}{2} < \frac{2 \tan \frac{\phi}{2}}{1 - \tan^2 \frac{\phi}{2}} = \tan \phi$.

**9.** $\sin \frac{\pi}{6} < \frac{\pi}{6}$ gives $3 < \pi$ while $\frac{\pi}{4} < \tan \frac{\pi}{4}$ gives $\pi < 4$. A better upper bound for $\pi$ (viz. $2\sqrt{3}$) is given by $\frac{\pi}{6} < \tan \frac{\pi}{6} (= \frac{1}{\sqrt{3}})$. For a better lower bound on $\pi$, use $\sin(\frac{\pi}{12}) < \frac{\pi}{12}$. Since $\sin \dfrac{\pi}{12} = \sin 15° = \dfrac{\sqrt{3} - 1}{2\sqrt{2}} \simeq 0.2588$ we get $\pi > 3\sqrt{2}(\sqrt{3} - 1) \simeq 3.1056$. For the last two assertions, note that for $0 < \theta < \frac{\pi}{2}$, $\dfrac{d}{d\theta}\left(\dfrac{\sin \theta}{\theta}\right) = \dfrac{\theta \cos \theta - \sin \theta}{\theta^2} = \dfrac{\cos \theta}{\theta^2}(\theta - \tan \theta) < 0$.

**10.** The hint follows by induction on $n$ and the identity $1 + \cos 2\theta = 2\cos^2 \theta$. From the hint, $a_n$ is $\pi \dfrac{\sin \theta_n}{\theta_n}$ where $\theta_n = \dfrac{\pi}{2^{n+1}}$.

## Section 4.8

**1.** The value of the $n^{\text{th}}$ derivations of $\cos \theta$ at 0 is 0 if $n$ is odd and $(-1)^k$ if $n = 2k$. Taking $f(\theta) = \cos \theta$, write an equation analogous to (3) and note that for all $c$, $\left|\dfrac{f^{(n)}(c)}{n!}\theta^n\right| \le \dfrac{\theta^n}{n!}$.

**2.** $\cos 72° = \sin 18° = \sin \dfrac{\pi}{10}$. Since $\dfrac{\pi}{10} < \dfrac{4}{10}$ and $(\dfrac{4}{10})^5 \dfrac{1}{5!} = \dfrac{1024}{120} \dfrac{1}{10^5} < 10^{-4}$, $\dfrac{\pi}{10} - \dfrac{\pi^3}{6.10^3}$ approximates $\sin \dfrac{\pi}{10}$ upto four places of decimal. With a hand claculator, this value is 0.3090 upto 4 places of decimals. $\dfrac{\sqrt{5} - 1}{4}$ also has the same (approximate) value. (We could have started with $\cos 72° = \cos \dfrac{2\pi}{5}$ and used (2) instead of (1). But since $\dfrac{2\pi}{5} > 1$, its powers get bigger and so to get the error less than $10^{-4}$, we would need many more terms.)

**3.** By Taylor's theorem, $\sin \theta = \theta - \dfrac{\theta^2}{2!} \sin \phi_1 = \theta - \dfrac{\theta^3}{6} + \dfrac{\theta^4}{4!} \sin \phi_2$ for some $\phi_1, \phi_2 \in (0, \theta)$. If $\theta \le \dfrac{\pi}{2}$, then $\sin \phi_1, \sin \phi_2$ are both positive and this gives the first inequality. More generally, we have, for every integer $n \ge 0$,

$$\theta - \frac{\theta^3}{3!} + \frac{\theta^5}{5!} - \cdots - \frac{\theta^{4n+3}}{(4n+3)!} < \sin \theta < \theta - \frac{\theta^3}{3!} + \frac{\theta^5}{5!} - \cdots + \frac{\theta^{4n+1}}{(4n+1)!}$$

and by a similar reasoning,

$$1 - \frac{\theta^2}{2!} + \frac{\theta^4}{4!} - \cdots - \frac{\theta^{4n+2}}{(4n+2)!} < \cos \theta < 1 - \frac{\theta^2}{2!} + \frac{\theta^4}{4!} - \cdots + \frac{\theta^{4n}}{(4n)!}.$$

**4.** Define a function $f_{P,Q} : \mathbb{R} \to \mathbb{R}^2$ by $f_{P,Q}(t) = (x_1 + t(y_1 - x_1), x_2 + t(y_2 - x_2))$. Then $L(P,Q)$ is simply the range of the function $f_{P,Q}$. If $P = Q$ then $f_{P,Q}$ is a constant function. Otherwise it is a one-to-one function since at least one of $(y_1 - x_1)$ and $(y_2 - x_2)$ is non-zero. This proves the last part. To show that $L(P,Q) = L(Q,P)$ notice that $f_{P,Q}(t) = f_{Q,P}(1 - t)$ for all $t \in \mathbb{R}$ and so both the functions have the same range.

**5.** Let $P = (x_1, x_2), Q = (y_1, y_2), R = (z_1, z_2)$ and $S = (w_1, w_2)$. For (i) assume first that $R, S \in L(P,Q)$. Then with the notation in the solution above, $R = f_{P,Q}(t_1)$ and $S = f_{P,Q}(t_2)$ for some $t_1, t_2 \in \mathbb{R}$. Now for every $t \in \mathbb{R}$, an easy calculation gives $f_{R,S}(t) = f_{P,Q}(t_1 + tt_2 - tt_1)$ which shows $L(R,S) \subset L(P,Q)$. For the other way inclusion note first that $t_1 \neq t_2$ since $R \neq S$. Easy calculations give $P = f_{R,S}\left(\dfrac{t_1}{t_1 - t_2}\right)$ and $Q = f_{R,S}\left(\dfrac{1 - t_1}{t_2 - t_1}\right)$. So $P, Q$ lie on $L(R,S)$ and hence by the earlier part, $L(P,Q) \subset L(R,S)$. Conversely, suppose $L(P,Q) = L(R,S)$. Then $R, S$ are in $L(R,S)$ and hence in $L(P,Q)$. This proves (i). (ii) follows from (i) since if $A, B$ are both in $L(P,Q) \cap L(R,S)$ with $A \neq B$, then by (i) $L(P,Q)$ and $L(R,S)$ both equal $L(A,B)$. Geometrically, two distinct lines have at most one point in common. For a non-empty intersection there must exist $t, u \in \mathbb{R}$ such that $f_{P,Q}(t) = f_{R,S}(u)$. Written in full form, this is equivalent to saying that the system of equations

$$x_1 + t(y_1 - x_1) = z_1 + u(w_1 - z_1)$$
$$x_2 + t(y_2 - x_2) = z_2 + u(w_2 - z_2)$$

has a solution (in $t$ and $u$). The non-vanishing of the determinant is precisely the necessary and sufficient condition for this.

**6.** The first part is a trivial verification. If $t = \dfrac{1}{2}$, then $R$ is $\left(\dfrac{x_1 + y_1}{2}, \dfrac{x_2 + y_2}{2}\right)$ and is the mid-point of the segment $PQ$. More generally if $t = \dfrac{\lambda}{\lambda + \mu}$ then $R$ divides $PQ$ in the ratio $\lambda : \mu$.

**7.** (a) Suppose $Q$ is in $L(P,R)$. Then $\exists\, t \in \mathbb{R}$ such that $y_i = x_i + t(z_i - x_i)$ for $i = 1, 2$. But then the first column of the determinant is a multiple of the second and so the determinant vanishes. Conversely if the determinant vanishes then, since $z_1 - x_1$ and $z_2 - x_2$ cannot both vanish (or else $P = R$), the first column is a multiple of the second, which is equivalent to $Q$ lying on $L(P,R)$. If we subtract the second column from the first, the determinant equals

$$\begin{vmatrix} y_1 - z_1 & z_1 - x_1 \\ y_2 - x_2 & z_2 - x_2 \end{vmatrix}$$

which would vanish iff $R \in L(P,Q)$. Thus $Q \in L(P,R)$ is equivalent to $R \in L(P,Q)$ and similarly to $P \in L(Q,R)$.

(b) Let $P, Q, R, S$ be the vertices of the parallelogram, with $S = (w_1, w_2)$ and $P, Q, R$ as above. It is given that $L(P, Q) \cap (L(R, S) = \emptyset$ and $L(R, S) \cap L(Q, R) = \emptyset$ and hence by the last exercise

$$\begin{vmatrix} y_1 - x_1 & w_1 - z_1 \\ y_2 - x_2 & w_2 - z_2 \end{vmatrix} = 0 = \begin{vmatrix} w_1 - x_1 & y_1 - z_1 \\ w_2 - x_2 & y_2 - z_2 \end{vmatrix}$$

Adding the second column to the first in each, we get

$$\begin{vmatrix} R_1 & w_1 - z_1 \\ R_2 & w_2 - z_2 \end{vmatrix} = 0 = \begin{vmatrix} R_1 & y_1 - z_1 \\ R_2 & y_2 - z_2 \end{vmatrix}$$

where $R_i = y_i + w_i - x_i - z_i$ for $i = 1, 2$. Expanding, we get a system of equations satisfied by $R_1$ and $R_2$, viz.,

$$(w_2 - z_2)R_1 - (w_1 - z_1)R_2 = 0$$
$$(y_2 - z_2)R_1 - (y_1 - z_1)R_2 = 0.$$

Now $Q, R, S$ are not collinear and so by (a),

$$\begin{vmatrix} w_2 - z_2 & w_1 - z_1 \\ y_2 - z_2 & y_1 - z_1 \end{vmatrix} \neq 0.$$

But then the system above has only the trivial solution, viz. $R_1 = 0 = R_2$. This means $x_i + z_i = y_i + w_i$, for $i = 1, 2$. Thus the mid-points of $PR$ and $QS$ are the same.

8.  (i) $\cos \frac{\pi}{2} = 0$ and $\sin \frac{\pi}{2} = 1$ gives $\sin \pi = 2 \sin \frac{\pi}{2} \cos \frac{\pi}{2} = 0$ and $\cos \pi = \cos^2 \frac{\pi}{2} - \sin^2 \frac{\pi}{2} = -1$. From this, $\sin(x + \pi) = \sin x \cos \pi - \cos x \sin \pi = -\sin x$ and similarly $\cos(x + \pi) = -\cos x$. So $\sin(x + 2\pi) = -\sin(x+\pi) = \sin x$ etc. (ii) $\sin(\alpha(x + \frac{2\pi}{|\alpha|})) = \sin(\alpha x \pm 2\pi) = \sin(\alpha x)$ etc. (iii) Let $\alpha = p/q$ with $p \in \mathbb{Z}, q \in \mathbb{N}$. Then $2\pi q$ is easily seen to be a period. (iv) Let, if possible, $T$ be a period. Then $\cos T + \cos(\alpha T) = \cos 0 + \cos(\alpha 0) = 2$ which forces $\cos T = 1$ and $\cos(\alpha T) = 1$. From this it is not hard show that $T = 2m\pi$ and $\alpha T = 2n\pi$ for some $m, n \in \mathbb{Z}$. (The next exercise makes this easier.) But then $\alpha = \frac{\alpha T}{T} = \frac{n}{m}$.

9.  First note that the difference of any two periods is also a period. Now if $f$ has no smallest positive period, then $\exists$ a sequence $T_1 > T_2 > \ldots > T_n > T_{n+1} > \ldots$ of positive periods. This sequence is convergent (not necessarily to 0). Let $U_n = T_n - T_{n+1}$. Then each $U_n$ is a period and $U_n \to 0$ as $n \to \infty$. So $f$ has arbitrarily small positive periods. Now fix any $x \in \mathbb{R}$. For every $n \in \mathbb{N}, \exists$ some positive period $U$ of $f$ such that $U < \frac{1}{n}$. The interval $(x, x + \frac{1}{n})$ contains at least one multiple of $U$, say $kU$. Then $f(kU) = f(0)$. Call $kU$ as $x_n$. Then $x_n \to x$ as $n \to \infty$ and so by continuity $f(x_n) \to f(x)$. But $f(x_n) = f(0)$ for all $n$. So $f(x) = f(0)$ for all $x$; i.e. $f$ is constant. To show $2\pi$ is the smallest period of $\sin x$, let, if possible, $0 < T < 2\pi$ be

also a period. Since $\cos x = -\sin\left(\frac{\pi}{2}+x\right)$, $T$ is also a period of $\cos x$. So $\cos T = \cos 0 = 1$. But then, $1 - 2\sin^2\dfrac{T}{2} = \cos T = 1$, whence $\sin\frac{T}{2} = 0$, i.e. $2\sin\frac{T}{4}\cos\frac{T}{4} = 0$. Since $\frac{T}{4} \in (0, \frac{\pi}{2})$, $\sin\frac{T}{4} > 0$. But then $\cos\frac{T}{4} = 0$ contradicting that $\frac{\pi}{2}$ is the smallest positive zero of $\cos x$.

10. $\sin x < x < \frac{\pi}{2}$ gives $\cos(\sin x) > \cos x$. Also since $\cos x > 0, \cos x > \sin(\cos x)$.

11. The point $P$ will lie outside the triangle $ABC$ and not inside as in Fig. 4.8.1. In a correctly drawn figure only one of $Q$ and $R$ will lie on the segment $AB, AC$ and one will lie outside. So one of $AB, AC$ will be the sum and the other the difference of $AQ \,(= AR)$ and $QB \,(= RC)$.

14. The statement would say $|\mathbf{u}\cdot\mathbf{v}| \le |\mathbf{u}||\mathbf{v}|$ for all $\mathbf{u}, \mathbf{v}$ in $I\!\!R^n$. The proof would begin by noting that $f(t) = |\mathbf{u} + t\mathbf{v}|^2 \ge 0$ for all $t$ and then expanding $f(t)$ as a quadratic in $t$, viz. $|\mathbf{v}|^2 t^2 + 2(\mathbf{u}\cdot\mathbf{v})t + |\mathbf{u}|^2$.

15. $|\mathbf{u}+\mathbf{v}|^2 = |\mathbf{u}|^2 + 2\mathbf{u}\cdot\mathbf{v} + |\mathbf{v}|^2 \le |\mathbf{u}|^2 + 2|\mathbf{u}||\mathbf{v}| + |\mathbf{v}|^2 = (|\mathbf{u}| + |\mathbf{v}|)^2$. Equality holds iff $\mathbf{u}\cdot\mathbf{v} = |\mathbf{u}||\mathbf{v}|$ which is the case iff one of $\mathbf{u}, \mathbf{v}$ is $0$ or they are positive scalar multiples of each other. The angle between non-zero vectors $\mathbf{u}, \mathbf{v}$ is $\cos^{-1}\left(\dfrac{\mathbf{u}\cdot\mathbf{v}}{|\mathbf{u}||\mathbf{v}|}\right)$.

16. Let $\mathbf{a}, \mathbf{b}, \mathbf{c}$ be the position vectors of $P, Q, R$ respectively. Then $|PR| = |(\mathbf{c}-\mathbf{a})| = |(\mathbf{c}-\mathbf{b}) + (\mathbf{b}-\mathbf{a})| \le |\mathbf{c}-\mathbf{b}| + |\mathbf{b}-\mathbf{a}| = |PQ| + |QR|$.

17. $|P'Q'| = |\mathbf{v}' - \mathbf{u}'| = |\mathbf{u} + \mathbf{a} - \mathbf{v} - \mathbf{a}| = |\mathbf{u} - \mathbf{v}| = |PQ|$ etc.

18. Following the hint, let $\mathbf{u}, \mathbf{w}$ be the position vectors of $P$ and $R$ respectively. Then $\mathbf{u}\cdot\mathbf{w} = \cos\dfrac{\pi}{2} = 0$. So $|PR|^2 = |\mathbf{w} - \mathbf{u}|^2 = |\mathbf{w}|^2 + |\mathbf{u}|^2 - 2(\mathbf{u}\cdot\mathbf{w}) = |PQ|^2 + |QR|^2$, proving (i). For (ii), $\cos\theta = \dfrac{(\mathbf{w} - \mathbf{u})\cdot(-\mathbf{u})}{|\mathbf{w}-\mathbf{u}||\mathbf{u}|} = \dfrac{|\mathbf{u}|^2 - \mathbf{u}\cdot\mathbf{w}}{|\mathbf{w}-\mathbf{u}||\mathbf{u}|} = \dfrac{|\mathbf{u}|}{|\mathbf{w}-\mathbf{u}|} = \dfrac{|PQ|}{|PR|}$. (iii) follows from (i), (ii) and the identity $\sin\theta = \sqrt{1 - \cos^2\theta}$ for $\theta \in [0, \pi]$.

## Section 4.9

1. $4, \frac{16}{\pi}$ and $\frac{1}{2}$.

2. (i) $0$ (ii) $\sqrt{2}$ (iii) $\frac{1}{24}$.

3. $h^{(n)}(0) = 0$ for all $n$. (See Section 4.6 for details.)

4. By Exercise (6.18), $\exists\, m$ such that $h^{(m)}(c) \ne 0$.

**5.** In Theorem (9.3), assume, instead, that $\dfrac{g'(x)}{h'(x)} \to \infty$ as $x \to c$, where $c \in \mathbb{R}$.

Given $R \in \mathbb{R}$, $\exists\, \delta > 0$ such that $\dfrac{g'(y)}{h'(y)} > R$ if $0 < |y - c| < \delta$. For any $x \in (c, c + \delta)$ (or $x \in (c - \delta, c)$) apply Cauchy's MVT to get $c_x \in (c, x)$ (or $c_x \in (x, c)$) such that $\dfrac{g(x) - 0}{h(x) - 0} = \dfrac{g'(c_x)}{h'(c_x)}$. Then $0 < |c_x - c| < \delta$ and so $\dfrac{g'(c_x)}{h'(c_x)} > R$, giving $\dfrac{g(x)}{h(x)} > R$. In the case of Theorem (9.6), assume $\dfrac{g'(x)}{h'(x)} \to \infty$. Given $R > 0$, first find $x_0$ such that for all $z > x_0$, $\dfrac{g'(z)}{h'(z)} > R + 1$.

Fix any $x > x_0$. Arguing as in the proof of Theorem (9.6), prove that for every $y > x$, $\dfrac{g(x) - g(y)}{h(x) - h(y)} > R + 1$ and hence, letting $y \to \infty$, $\dfrac{g(x)}{h(x)} \geq R + 1 > R$. In the case of Theorem (9.7), assume $\dfrac{g'(x)}{h'(x)} \to \infty$ as $x \to c$. Given $R$, first choose $x_0 > c$ such that $\dfrac{g'(z)}{h'(z)} > R + 1$ whenever $c < z < x_0$. Then choose $x_1 \in (c, x_0)$ such that $g(x) > \max\{0, g(x_0)\}$ and $h(x) > \max\{0, h(x_0)\}$ whenever $c < x < x_1$. Imitating the proof of Theorem (9.7), show that if $c < x < x_1$, then

$$\frac{g(x)}{h(x)} > (R + 1)\frac{1 - h(x_0)/h(x)}{1 - g(x_0)/g(x)}.$$

Finally, $\exists\, x_2 < x_1$ such that for all $c < x < x_2$, $\dfrac{g(x)}{h(x)} > (R + 1) - \dfrac{1}{2} = R + \dfrac{1}{2} > R$. If $L = -\infty$, apply the argument above to $\dfrac{-g}{h}$.

**7.** Following the hint, by induction hypothesis, $\dfrac{p(x) - p(c)}{x - c}$ can be written as $b_0 + b_1(x - c) + \ldots + b_{n-1}(x - c)^{n-1}$. So $p(x) = p(c) + b_0(x - c) + \ldots + b_{n-1}(x - c)^n$.

**8.** The troublesome term is $x^4 \sin \frac{1}{x} = g(x)$ (say). By the same reasoning as in Exercise (3.14), $g'(0) = g''(0) = 0$. However, for $x \neq 0$, $g''(x)$ comes out as (using Exercise (4.9) or otherwise) $12x^2 \sin \frac{1}{x} - 6x \cos \frac{1}{x} - \sin \frac{1}{x}$ which tends to no limit as $x \to 0$. So $g''$ is not continuous and hence not differentiable at 0. The same holds for $f''$. Still, $\dfrac{f(x) - x^3}{x^3} = x \sin \dfrac{1}{x} \to 0$ as $x \to 0$.

**10.** Assume $\lim\limits_{x \to \infty} \dfrac{g'(x)}{h'(x)} = L$ where $g'(x), h'(x)$ exist for all $x > M$ (for some fixed $M$), both $g(x)$ and $h(x)$ tend to $\infty$ as $x \to \infty$ and $h'(x) \neq 0$ for all $x > M$. Given $\epsilon > 0$, first choose $M_0(> M)$ such that for all $z > M_0$, $\dfrac{g'(z)}{h'(z)}$ lies between $L - \dfrac{\epsilon}{2}$ and $L + \dfrac{\epsilon}{2}$. Then choose $M_1 > M_0$ such that for all

$x > M_1, g(x) > \max\{g(M_0), 0\}$ and $h(x) > \max\{h(M_0), 0\}$. By Cauchy's MVT, for every $x > M_1$, we get (analogously to (29))

$$(L - \frac{\epsilon}{2}) \frac{1 - h(M_0)/h(x)}{1 - g(M_0)/g(x)} < \frac{g(x)}{h(x)} < (L + \frac{\epsilon}{2}) \frac{1 - h(M_0)/h(x)}{1 - g(M_0)/g(x)}.$$

As $x \to \infty$, the first and the last terms tend to $L - \dfrac{\epsilon}{2}$ and $L + \dfrac{\epsilon}{2}$ respectively.

As $L - \dfrac{\epsilon}{2} > L - \epsilon$ and $L + \dfrac{\epsilon}{2} < L + \epsilon$, $\exists\ M_2\ (> M_1)$ such that for all $x > M_2$, (taking steps similar to (29) and (30)), $L - \epsilon < \dfrac{g(x)}{h(x)} < L + \epsilon$. Thus

$$\lim_{x \to \infty} \frac{g(x)}{h(x)} = L. \text{ If } c = -\infty, \text{ replace } x \text{ by } -x.$$

**11.** Let $m$ be the least integer, if any, such that $f^{(m)}(0) \neq 0$. By Theorem (9.5), write $f(x) = \dfrac{f^{(m)}(0)}{m!} x^m + R_m(x)$ where $R_m(x)/x^m \to 0$ as $x \to 0$.

Taking $\epsilon = \dfrac{1}{2}|\dfrac{f^{(m)}(0)}{m!}|$, $\exists\ \delta > 0$, such that $|R_m(x)/x^m| < \dfrac{1}{2}|\dfrac{f^{(m)}(0)}{m!}|$ if $0 < |x| < \delta$. Consequently, $|\dfrac{f^{(m)}(0)}{m!} + \dfrac{R_m(x)}{x^m}| > \dfrac{1}{2}|\dfrac{f^{(m)}(0)}{m!}|$ and so $|f(x)| > \dfrac{1}{2}|\dfrac{f^{(m)}(0)}{m!}|\,||x^m| > 0$ for all $x \neq 0$ in $(-\delta, \delta)$. But $\dfrac{1}{n} \in (-\delta, \delta)$ if $n \in I\!N$ is large (specifically, if $n > \dfrac{1}{\delta}$), contradicting that $f(\dfrac{1}{n}) = 0$. For an alternate proof using simply the Mean Value Theorem, the hypothesis implies that for every $k \geq 0$, $\exists\ \delta_k > 0$ such that $f^{(k)}(x)$ is differentiable in $[0, \delta_k]$. Let $x_n^{(0)} = \dfrac{1}{n}, n \in I\!N$. Then almost all terms of $\{x_n^{(0)}\}_{n=1}^{\infty}$ lie in $(0, \delta_0)$. By Rolle's theorem, between $x_{n+1}^{(0)}$ and $x_n^{(0)}$, $\exists$ some $x_n^{(1)}$ such that $f'(x_n^{(1)}) = 0$. Again, almost all $x_n^{(1)}$'s lie in $(0, \delta_1)$. By Rolle's theorem, $\exists\ x_n^{(2)} \in (x_{n+1}^{(1)}, x_n^{(1)})$ such that $f''(x_n^{(2)}) = 0$. Continuing inductively, for every $k$, $\exists$ a strictly monotonically decreasing sequence $\{x_n^{(k)}\}$ (with $n \geq N_k$, say) converging to $0$ such that $f^{(k)}(x_n^{(k)}) = 0$ for all $n \geq N_k$. Continuity of $f^{(k)}$ at $0$ implies $f^{(k)}(0) = 0$.

## Section 4.10

**1.** Let $S = S_1 \cap S_2 \cap \ldots \cap S_n$ where each $S_i$ is convex. Let $P, Q \in S$. Then $P, Q \in S_i$ for every $i$ and so the segment $\overline{PQ}$ is contained in $S_i$. Hence $\overline{PQ} \subset \cap_{i=1}^{n} S_i = S$. Thus $S$ is convex. (The argument holds even for the intersection of infinitely many convex sets). If $S_1, S_2$ are two disjoint discs then they are both convex (see the next exercise). But $S_1 \cup S_2$ is not. (Take a segment with one end in $S_1$ and the other in $S_2$.)

**2.** Let $S = \{\mathbf{x} \in I\!R^n : |\mathbf{x}-\mathbf{a}| \le r\}$ be the ball with centre $\mathbf{a}$ and radius $r$. Suppose $\mathbf{u}, \mathbf{v} \in S$ and $0 \le \lambda \le 1$. Then $|(1-\lambda)\mathbf{u}+\lambda\mathbf{v}-\mathbf{a}| = |(1-\lambda)(\mathbf{u}-\mathbf{a})+\lambda(\mathbf{v}-\mathbf{a})| \le (1-\lambda)|\mathbf{u}-\mathbf{a}| + \lambda|\mathbf{v}-\mathbf{a}| \le (1-\lambda)r + \lambda r = r$. So $(1-\lambda)\mathbf{u} + \lambda\mathbf{v} \in S$.

**3.** Let $S$ be the region bounded by the ellipse be $\dfrac{x^2}{a^2} + \dfrac{y^2}{b^2} = 1$. Let $P = (x_1, y_1), Q = (x_2, y_2) \in S$. Then $\dfrac{x_i^2}{a^2} + \dfrac{y_i^2}{b^2} \le 1$ for $i = 1, 2$. Also by Cauchy-Schwartz inequality, $\dfrac{x_1 x_2}{a^2} + \dfrac{y_1 y_2}{b^2} \le \left(\dfrac{x_1^2}{a^2} + \dfrac{y_1^2}{b^2}\right)^{1/2} \left(\dfrac{x_2^2}{a^2} + \dfrac{y_2^2}{b^2}\right)^{1/2} \le 1$. Using these facts, it is easy to show that for every $\lambda \in [0, 1]$, $\dfrac{((1-\lambda)x_1 + \lambda x_2)^2}{a^2} + \dfrac{((1-\lambda)y_1 + \lambda y_2)^2}{b^2} \le 1$, whence $((1-\lambda)x_1 + \lambda x_2, (1-\lambda)y_1 + \lambda y_2)) \in S$.

**4.** (i) If a piont $A$ on $\overline{PQ}$ divides it in the ratio $\lambda : \mu$ (with $\lambda > 0, \mu > 0$) then the position vector of $A$ is $\dfrac{\mu\mathbf{u}_1 + \lambda\mathbf{u}_2}{\lambda + \mu}$ which is a convex combination of $\mathbf{u}_1$ and $\mathbf{u}_2$ since $\lambda > 0, \mu > 0$ and $\frac{\mu}{\lambda+\mu} + \frac{\lambda}{\lambda+\mu} = 1$. The only cases left are $A = P$ and $A = Q$. They correspond to $1\mathbf{u}_1 + 0\mathbf{u}_2$ and $0\mathbf{u}_1 + 1\mathbf{u}_2$ respectively. (ii) Let $A$ be a point in the triangle with position vector $\mathbf{a}$. If $A$ lies on any of the sides then by (i) $\mathbf{a}$ is a convex combination of two of $\mathbf{u}_1, \mathbf{u}_2, \mathbf{u}_3$. Add the third one too with coefficient $0$. If $A$ is in the interior of the triangle, let $PA$ meet $QR$ at $B$. By (i), $\mathbf{b} = (1-\lambda)\mathbf{u}_2 + \lambda\mathbf{u}_3$ and $\mathbf{a} = (1-\mu)\mathbf{u}_1 + \mu\mathbf{b}$ for some $\lambda, \mu \in [0, 1]$, where $\mathbf{b}$ = position vector of $B$. But then $\mathbf{a} = \alpha_1\mathbf{u}_1 + \alpha_2\mathbf{u}_2 + \alpha_3\mathbf{u}_3$ where $\alpha_1 = (1-\mu), \alpha_2 = \mu(1-\lambda)$ and $\alpha_3 = \lambda\mu$, is a convex combination of $\mathbf{u}_1, \mathbf{u}_2, \mathbf{u}_3$. For uniqueness, suppose $\mathbf{a}$ also equals $\beta_1\mathbf{u}_1 + \beta_2\mathbf{u}_2 + \beta_3\mathbf{u}_3$. Let $\gamma_i = \alpha_i - \beta_i, i = 1, 2, 3$. Then $\gamma_1\mathbf{u}_1 + \gamma_2\mathbf{u}_2 + \gamma_3\mathbf{u}_3 = 0$. If $\gamma_1 \ne 0$ (say), then $\mathbf{u}_1 = -\dfrac{\gamma_2}{\gamma_1}\mathbf{u}_2 - \dfrac{\gamma_3}{\gamma_1}\mathbf{u}_3$. $\gamma_1 + \gamma_2 + \gamma_3 = \alpha_1 + \alpha_2 + \alpha_3 - (\beta_1 + \beta_2 + \beta_3) = 1 - 1 = 0$ gives $-\dfrac{\gamma_2}{\gamma_1} - \dfrac{\gamma_3}{\gamma_1} = 1$. But this means $P$ lies on the line $QR$ (although not necessarily in the segment $\overline{QR}$), contradicting that $P, Q, R$ are non-collinear. So $\gamma_1, \gamma_2, \gamma_3$ are all $0$. For (iii), if $\alpha_1\mathbf{u}_1 + \alpha_2\mathbf{u}_2 + \alpha_3\mathbf{u}_3$ and $\beta_1\mathbf{u}_1 + \beta_2\mathbf{u}_2 + \beta_3\mathbf{u}_3$ are convex combinations and $\lambda \in [0, 1]$, then write $(1-\lambda)(\alpha_1\mathbf{u}_1 + \alpha_2\mathbf{u}_2 + \alpha_3\mathbf{u}_3) + \lambda(\beta_1\mathbf{u}_1 + \beta_2\mathbf{u}_2 + \beta_3\mathbf{u}_3)$ as a covnex combiantion of $\mathbf{u}_1, \mathbf{u}_2, \mathbf{u}_3$.

**5.** Follow the hint for $\alpha_m < 1$ and note that $\dfrac{\alpha_1}{1 - \alpha_m}, \dfrac{\alpha_2}{1 - \alpha_m}, \dots \dfrac{\alpha_{m-1}}{1 - \alpha_m}$ are all non-negative and their sum is $1$ since $\alpha_1 + \dots + \alpha_{m-1} = 1 - \alpha_m$. If $\alpha_m = 1$, then $\alpha_1 = \alpha_2 = \dots = \alpha_{m-1} = 0$ and $\alpha_1\mathbf{u}_1 + \dots + \alpha_m\mathbf{u}_m$ is simply $\mathbf{u}_m$.

**6.** Let $T$ be the region above the graph of $f$ on $S$, i.e. $T = \{(x, y) : x \in S, y \ge f(x)\}$. Then $T$ is a convex set. The points $(x_1, f(x_1)), \dots, (x_m, f(x_m))$ are in $T$. By the last exercise, $(\sum\limits_{i=1}^{m} \alpha_i x_i, \sum\limits_{i=1}^{n} \alpha_i f(x_i))$ is also in $T$ which implies $f(\sum\limits_{i=1}^{n} \alpha_i x_i) \le \sum\limits_{i=1}^{n} \alpha_i f(x_i)$. Setting each $\alpha_i = \dfrac{1}{m}$ gives $f(\dfrac{x_1 + \dots x_m}{m}) \le$

$\dfrac{f(x_1) + \ldots + f(x_m)}{m}$. If $f$ is strictly convex on $S$ then equality can hold only when all $x_i$'s are equal. For otherwise suppose $x_1 < x_2$ (say). Let $x_1' = x_2' = \dfrac{x_1 + x_2}{2}$ and $x_i' = x_i$ for $i = 3, 4, \ldots, m$. Then $f(x_1') = f(x_2') < \dfrac{f(x_1) + f(x_2)}{2}$ and so

$$\sum_{i=1}^{m} f(x_i') < \sum_{i=1}^{m} f(x_i) = mf(\frac{x_1 + \ldots + x_m}{m}) = mf(\frac{x_1' + x_2' + \ldots + x_m'}{m}).$$

So $f(\dfrac{x_1' + x_2' + \ldots + x_m'}{m}) > \dfrac{1}{m} \sum\limits_{i=1}^{m} f(x_i')$, contradicting the earlier part.

7. Let $f(x)$ be the tax if the annual income is $x$. Then $f$ is a convex function. If $x_1, \ldots, x_m$ are the incomes in $m$ years then $\sum\limits_{i=1}^{m} f(x_i)$ is the total tax to be paid without averaging the income, while $mf(\dfrac{x_1 + \ldots + x_m}{m})$ is the tax due if the income is averaged over the $m$ years.

8. The equation of the tangent at $(c, f(c))$ is $y = L_1(x) = f(c) + f'(c)(x - c)$. By Theorem (3.3), $f(x) - L_1(x) = \frac{1}{2}f''(c)(x - c)^2 + R_2(x)$ where $\dfrac{R_2(x)}{(x-c)^2} \to 0$ as $x \to 0$. Taking $\epsilon = \frac{1}{4}f''(c)$, $\exists\, \delta > 0$ such that $0 < |x - c| < \delta$ implies $|\dfrac{R_2(x)}{(x-c)^2}| < \frac{1}{4}f''(c)$. But then $\frac{1}{2}f''(c)(x - c)^2 + R_2(x) > \frac{1}{4}f''(c)(x - c)^2 > 0$ for $0 < |x - c| < \delta$. So $f(x) - L_1(x) > 0$.

9. The first part follows from the last exercise since $f''(0) = \frac{1}{2} + 0 > 0$. For $x \neq 0, f''(x)$ comes out as $\frac{1}{2} + 12x^2 \sin \frac{1}{x} - 6x \cos \frac{1}{x} - \sin \frac{1}{x}$ (see Exercise (9.8)). As $x \to 0, \frac{1}{2} + 12x^2 \sin \frac{1}{x} - 6x \cos \frac{1}{x} \to \frac{1}{2}$. So $\exists\, \delta > 0$ such that $|\frac{1}{2} + 12x^2 \sin \frac{1}{x} - 6x \cos \frac{1}{x}| < \frac{3}{4}$ for all $x \in (0, \delta)$. But there are intervals arbitrarily close to 0 on which $\sin \frac{1}{x} < -\frac{3}{4}$. (Specifically, these intervals can be chosen to contain the points $\dfrac{2}{(4n + 3)\pi}$ for $n \in \mathbb{N}$.) On these interevals $f$ is (strictly) concave downdards.

10. Let $a < c < d < b$. Since the graph is above the tangent at $c$, we get $f(d) \geq f(c) + f'(c)(d - c)$ or $f'(c) \leq \dfrac{f(d) - f(c)}{d - c}$. Similarly, since the graph is above the tangent at $(d, f(d))$, we get $\dfrac{f(d) - f(c)}{d - c} \leq f'(d)$. In particular this gives $f'(c) \leq f'(d)$ and hence $f'$ is monotonically increasing on $(a, b)$. Apply Theorem (10.5). (Continuity at the end points $a, b$ is not given. But, for concavity we need the theorem only for closed subintervals contained in $(a, b)$.)

11. The definition implies that such a function must be linear, i.e. of the form $Ax + B$ for some constants $A, B$. Every point is a point of inflection.

**12.** Existence of $f''(c)$ means $\exists\, \delta > 0$ such that $f'$ exists in $(c - \delta, c + \delta)$ and also that $f'$ is continuous at $c$. From the hypothesis and Theorem (10.5), $f'$ is increasing on $(c - \delta, c)$ and decreasing on $(c, c + \delta)$ (or vice versa). By continuity of $f$ at $c$, $f'$ is increasing on $(c - \delta, c)$ and decreasing on $[c, c + \delta)$ (or vice versa). So $f'$ has a local maximum (or minimum) at $c$. Apply Theorem (5.2) to $f'$.

**13.** Apply Exercise (3.13).

**14.** $f(x) = x + \sin x$, $f'(x) = 1 + \cos x \geq 0$ for all $x$. So $f$ is increasing throughout $\mathbb{R}$. $f''(x) = -\sin x = 0$, when $x = n\pi$, $n \in \mathbb{N}$. These are the points of inflection. For even $n$, $f'(n\pi) = 2$ while for odd $n$, $f'(n\pi) = 0$. Draw small line segments with these slopes at $(n\pi, n\pi)$. Then draw the curve so as to touch these segments from above or below (to be decided by the sign of $\sin x$ near the points.)

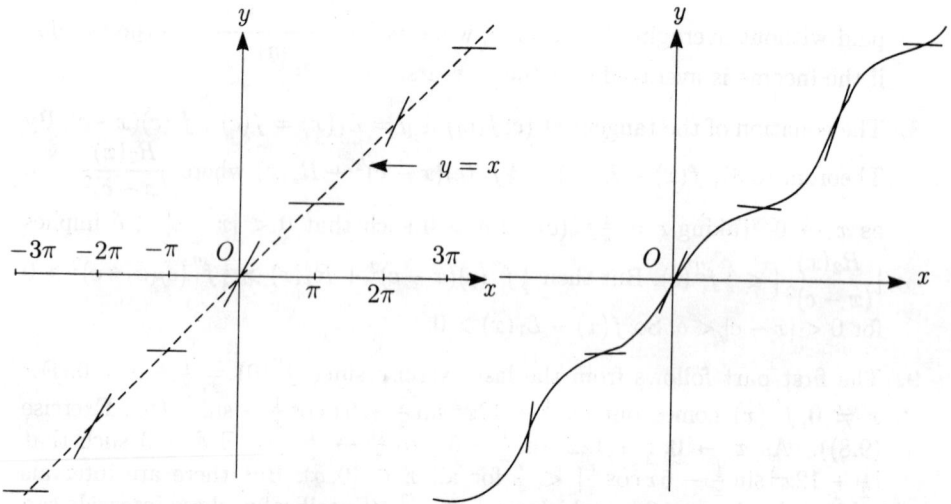

**15.** The first part follows from the 'strict' version of Theorem (10.5) and the fact that $\dfrac{d^2}{dx^2}(e^x) = e^x > 0$ for all $x$. For the second part, let $y_1, \ldots, y_m$ be positive real numbers. Let $x_i = \ln y_i$, i.e., $y_i = e^{x_i}$ for $i = 1, \ldots, n$ and $\bar{x} = \dfrac{x_1 + \ldots + x_m}{m}$. Then $e^{\bar{x}} \leq \dfrac{y_1 + \ldots + y_m}{m}$, with equality holding iff all $x_i$'s are equal (which is equivalent to all $y_i$'s being equal). But $e^{\bar{x}} = e^{\left(\frac{x_1 + \ldots + x_m}{m}\right)} = (e^{x_1} e^{x_2} \ldots x^{x_m})^{1/m} = (y_1 y_2 \ldots y_m)^{1/m}$.

**16.** For the first part note that $\dfrac{d^2 y}{dx^2} = 2 \sec x \tan x > 0$ if $x \in (0, \dfrac{\pi}{2})$. In any triangle $ABC$, $A + B + C = \pi$. So if $ABC$ is acute angled, then by Exercise (10.6), $\dfrac{\tan A + \tan B + \tan C}{3} \geq \tan\left(\dfrac{A + B + C}{3}\right) = \tan\dfrac{\pi}{3} = \sqrt{3}$ with equality holding iff $A = B = C$. The sine and the cosine functions are strictly concave downwards on $[0, \frac{\pi}{2}]$. So by a similar reasoning, an acute an-gled triangle $ABC$ is equilateral iff $\sin A + \sin B + \sin C = \frac{3\sqrt{3}}{2}$ and also iff $\cos A + \cos B + \cos C = \frac{3}{2}$.

## Section 4.11

2.  $f$ vanishes at the end-points of an interval of the form $\left[ \dfrac{1}{(2n+1)\pi}, \dfrac{1}{2n\pi} \right]$, $n \in$ $I\!N$ and is positive at all its interior points. So each such interval contains at least one local maximum of $f$. For local minima consider $\left[ \dfrac{1}{2n\pi}, \dfrac{1}{(2n-1)\pi} \right]$.

3.  Only when the function is locally constant, i.e. constant in a neighbourhood of that point.

4.  Let $a, b$ be strict local maxima of $f$, with $a < b$. Then the minimum of $f$ on $[a, b]$ must occur at an interior point.

5.  $f'(x) = 4x > 0$ if $x > 0$ and $f'(x) = 2x < 0$ if $x < 0$. Non-existence of $f''(0)$, is similar to that in Exercise (3.13).

6.  First part is similar to the last exercise. For the second, $f''(0) = 0$ (since $n \geq 4$).

7.  $f$ is strictly increasing over $I\!R$.

8.  Write $f(x) = \dfrac{1}{n!} f^{(n)}(c)(x-c)^n + R_n(x)$ where $\dfrac{R_n(x)}{(x-c)^n} \to 0$ as $n \to \infty$ by Theorem (3.3). Choose $\delta > 0$ so that $\left| \dfrac{R_n(x)}{|x-c|^n} \right| < \dfrac{1}{2(n!)} |f^{(n)}(c)|$ for $0 < |x - c| < \delta$. Then for $0 < |x - c| < \delta$, $f(x) - f(c)$ has the same sign as that of $f^{(n)}(c)(x-c)^n$. If $n$ is odd, this sign is different for $x > c$ and $x < c$. If $n$ is even then the sign depends only on $f^{(n)}(c)$ for $x$ on either side of $c$. For a point of inflection, the criterion is that $n$ is odd and $\geq 3$.

9.  (a) Consider the function $f(0) = 0$ and $f(x) = e^{-1/x^2}$ for $x \neq 0$ (see Fig. 4.6.1). There is a strict local minimum at 0. But $f^{(n)}(0) = 0$ for all $n$.

    (b) For every $x$, $f(x) \geq 0 = f(0)$. So $f$ has a local minimum at 0. Since $f(\frac{1}{n\pi}) = 0$ for all $n \in I\!N$, this minimum is not strict. Since $f(x) > 0$ for all $x$ in the interval $\left( \dfrac{1}{(n+1)\pi}, \dfrac{1}{n\pi} \right)$ but vanishes at both its end points, $f$ is neither increasing nor decreasing in any interval containing $[\dfrac{1}{(n+1)\pi}, \dfrac{1}{n\pi}]$. The interval $(0, \delta)$ contains such intervals for large $n$. Similarly for $(-\delta, 0)$.

10. Maximum and minimum over $(-\infty, \infty)$ are 5 and -5 respectively. Over $[0, \frac{\pi}{2}]$, the maximum occurs at $\tan^{-1}(\frac{3}{4})$ and equals 5 while the minimum

occurs at the end-point $\frac{\pi}{2}$ and equals 3. (For a trigonometric solution, write $3\sin\theta + 4\cos\theta$ as $5\sin(\theta + \alpha)$ where $\alpha = \tan^{-1}(\frac{4}{3})$.)

**11.** Let $x$ be the side of the square cut off at each corner. The problem amounts to maximising $f(x) = x(a - 2x)(b - 2x)$ for $x \in [0, \alpha]$ where $\alpha = \min\{\frac{a}{2}, \frac{b}{2}\}$. $f'(x)$ is a quadratic with roots $\frac{1}{6}(a + b \pm \sqrt{a^2 - ab + b^2})$. The greater root $\geq \alpha$. Comparing $f(0), f(\alpha)$ and $f(\frac{1}{6}(a + b - \sqrt{a^2 - ab + b^2}))$ the maximum occurs at $x = \frac{1}{6}(a + b - \sqrt{a^2 - ab + b^2})$ while the minimum (which is 0) occurs at both 0 and $\alpha$.

**12.** For the circle, we have to maximise $f(\theta) = (a\cos\theta - a\cos\alpha)^2 + (a\sin\theta - a\sin\alpha)^2$ where $\alpha$ is fixed and $\theta \in [0, 2\pi]$. By an easy calculation, $f(\theta) = 2a^2(1 - \cos(\theta - \alpha))$ which is maximum when $\theta = \alpha + \pi$, i.e. when the points are diametrically opposite. For the ellipse, the function to be maximised is $g(\theta) = (a\cos\theta - a\cos\alpha)^2 + (b\sin\theta - b\sin\alpha)^2$ with $a > b > 0$. $g'(\theta)$ comes out as $(b^2 - a^2)\sin 2\theta + 2a^2\cos\alpha\sin\theta - 2b^2\sin\alpha\cos\theta$. So $g'(\alpha + \pi) = 2(b^2 - a^2)\sin 2\alpha$ which is non-zero (unless $\alpha = 0, \frac{\pi}{2}, \pi$ or $\frac{3\pi}{2}$, which correspond to the ends of the axes of the ellipse). Hence the diametrically opposite point is not the farthest point, except when the point happens to be at one of the ends of the major axis.

**13.** Follow the notation in the figure. Then the time taken $= \dfrac{a\sec\theta}{u} + \dfrac{b - a\tan\theta}{v}$
$= f(\theta)$ (say). By common sense we may suppose $\theta \in [0, \alpha]$ where $\tan\alpha = \frac{b}{a}$.
$f'(\theta) = \dfrac{a\tan\theta\sec\theta}{u} - \dfrac{a\sec^2\theta}{v} = a\sec^2\theta(\dfrac{\sin\theta}{u} - \dfrac{1}{v})$. If $u > v$ then $f'(\theta) < 0$
for all $\theta \in [0, \frac{\pi}{2})$ (and hence for all $\theta \in [0, \alpha]$). So $f$ is decreasing throughout and the quickest path occurs when $\theta = \alpha$, i.e. when the swimmer swims straight from $P$ to $R$. Suppose $u \leq v$. Set $u = v\sin\phi$. Then $\phi$ is the only critical point of $f$. If $\phi \geq \alpha$, then again the quickest path is to swim from $P$ to $R$. (This happens when $au > bv$ i. e. when the distance $OR$ is relatively small.) If $\phi < \alpha$, then $f'(\theta) < 0$ if $\sin\theta < \frac{u}{v}$, i.e. if $\theta < \phi$ and $f'(\theta) > 0$ if $\theta > \phi$. So $f$ attains its minimum when $\theta = \phi$.

**14.** The reasoning would be valid if the times taken by the two parts of the journey were independent of each other. But that is not the case here. (A somewhat similar fallacy is to conclude that since the minimum values of $\sin\theta$ and $\cos\theta$ are $-1$ each, therefore the minimum value of $\sin\theta + \cos\theta$ is $-2$, when it is in fact $-\sqrt{2}$).

**15.** The vertex $A$ must lie on the arc of a circle with $BC$ as a chord. For maximum area the altitude through $A$ is maximum, which happens when $AB = AC$.

**16.** Here $A$ must lie on a line parallel to $BC$. $\angle A$ is maximum when $AB = AC$. This and the last problem are said to be duals of each other. The function specifying the constraint in each is also the funciton to be optimised in the other. In the last problem the area was to maximised subject to the measure of $\angle A$ remaining a constant. In the present one, the measure of $\angle A$ is to be maximised subject to the area being a constant.

**17.** With the notation in the figure, $|AP|+|BP| = \sqrt{a^2 + x^2} + \sqrt{c^2 + (b-x)^2} = f(x)$ (say) is to be minized over $[0, b]$. Setting $f'(x) = 0$ gives $xc = a(b-x)$, or $x = \dfrac{ab}{a+c}$. For this value of $x$, $P$ divides $OD$ in the ratio $a : c$ and hence lies on $A'B$ where $A'$ is the reflection of $A$ in the line along the river. Geometrically, $|AP| = |A'P|$ and $|A'P| + |PB|$ is minimum when $A', P, B$ are collinear.

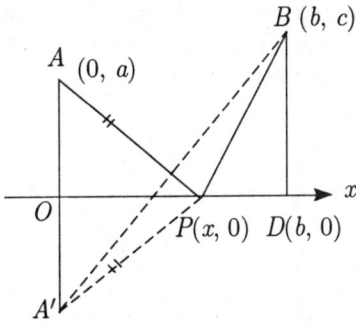

Exercise 11.17                 Exercise 11.18

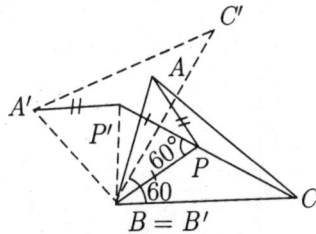

**18.** Here $|PA| + |PB| + |PC|$ is a function of two variables, viz. the two co-ordinates of $P$. So the methods studied in this section do not apply. But a geometric solution can be given as follows. Rotate the triangle $ABC$ around the point $B$ through $60°$ so as to get $\triangle A'BC'$ congruent to $\triangle ABC$. Given a point $P$ inside $ABC$, let $P'$ be its image under this rotation. Then $|PA| = |P'A'|$. Also $\triangle BPP'$ is equilateral and so $|PB| = |PP'|$. Hence $|PA|+|PB|+|PC|$ equals $|CP| + |PP'| + |P'A'|$ which is minimum when $C, P, P', A'$ are collinear. For this, $\angle$'s $BPC, CPA$ and $APB$ must all be $120°$. This argument also applies if $ABC$ is not acute angled provided none of its angles exceeds $120°$. If $\angle B$ (say) $> 120°$, then $|PA| + |PB| + |PC|$ is minimum when $P$ coincides with $B$.

**19.** The problem is equivalent to minimising $f(t) = (ut - a)^2 + (b - vt)^2, t \geq 0$. Either by calculus, or simply by completing squares, the minimum distance is $\dfrac{|av - bu|}{\sqrt{u^2 + v^2}}$. (What happens if $av = bu$?)

**20.** If $r$ is the radius of the variable circle then the area of $QSR$ is $\dfrac{r^2\sqrt{4 - r^2}}{4}$. The maximum area is $\dfrac{4}{3\sqrt{3}}$, occuring when $r = \dfrac{2\sqrt{2}}{\sqrt{3}}$.

**21.** The total yield with $x$ additional trees is $(100+x)(963-5x) = f(x)$. Treating $x$ as a continuous variable, $f$ is increasing for $x < 46.3$ and decreasing for

$x > 46.3$. So the maximum occurs at 46.3. But $x$ takes only non-negative integers as values. So the maximum occurs either at 46 or at 47. A direct comparison shows that it occurs at 46.

**22.** $f'(x) = 3ax^2 + 2bx + c$ has no real zeros if $b^2 < 3ac$. If $b^2 = 3ac$, then $f'(x)$ is a perfect square (or its negative) and hence does not change sign. If $b^2 > 3ac$, then $f'$ has two distinct real roots, say, $\alpha$ and $\beta$ with $\alpha < \beta$. If $a > 0, f$ has a local maximum at $\alpha$ and a local minimum at $\beta$. The reverse holds if $a < 0$. $f''(x) = 6ax + 2b$ exists everywhere and changes sign at $x = \dfrac{-b}{3a}$, which is the only point of inflection. (It is easy to show that $\frac{-b}{3a} = \frac{\alpha+\beta}{2}$, i.e., the point of inflection is midway between the local extrema.) The graphs for $a > 0$ are as shown below. For $a < 0$, take their reflections in the $x$-axis.

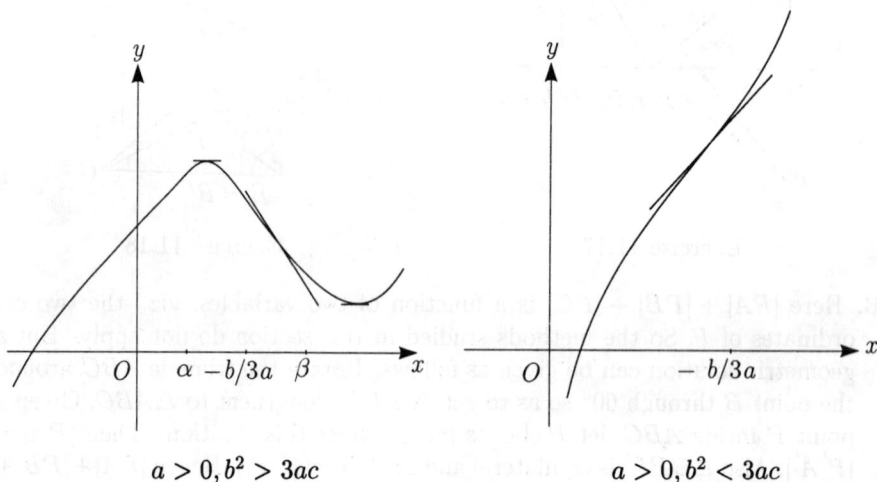

$$a > 0, b^2 > 3ac \qquad\qquad a > 0, b^2 < 3ac$$

**23.** $f(x)$ can be found from the data. But even without it, $f'(x) > 0$ for $x < 0$ and for $x > 2$ and $f'(x) < 0$ if $0 < x < 2$. Also $f''(x) = 2x - 2$ gives 1 as the only point of inflection. A qualitative sketch of the graph of $f$ is similar to the first graph of the last exercise with $\alpha = 0, \beta = 2$.

**24.** (i) Rewriting $f(x) = \dfrac{x}{x-1}$ as $1 + \dfrac{1}{x-1}$ shows $y - 1 = \dfrac{1}{x-1}$ and so if the origin is shifted to $(1,1)$, the graph is that of the more familiar function $y = \dfrac{1}{x}$. But even without observing this special feature, the graph can be drawn by the procedure given. The only discontinuity occurs at $x = 1$. As $x \to 1^+, f(x) \to \infty$ while as $x \to 1^-, f(x) \to -\infty$. So the line $x = 1$ is a vertical asymptote. As $x \to \infty, f(x) \to 1^+$ while as $x \to -\infty, f(x) \to 1^-$. So the line $y = 1$ is also an asymptote. $f'(x) = -\dfrac{1}{(x-1)^2} < 0$ for all $x \neq 1$. So the function is decreasing in $(-\infty, 1)$ and also in $(1, \infty)$. Further, $f''(x) = \dfrac{2}{(x-1)^3}$ shows that the graph is concave downwards on $(-\infty, 0)$

and upwards on $(1, \infty)$. The points $(0,0)$ and $(2,2)$ are on the graph. For (ii) the analysis is similar except that there are two vertical asymptotes, $x = 0$ and $x = 1$ and a local maximum at $x = \frac{1}{2}$. (In the sketch below for (ii), the scale on the $y$-axis is 4 times that on the $x$-axis.)

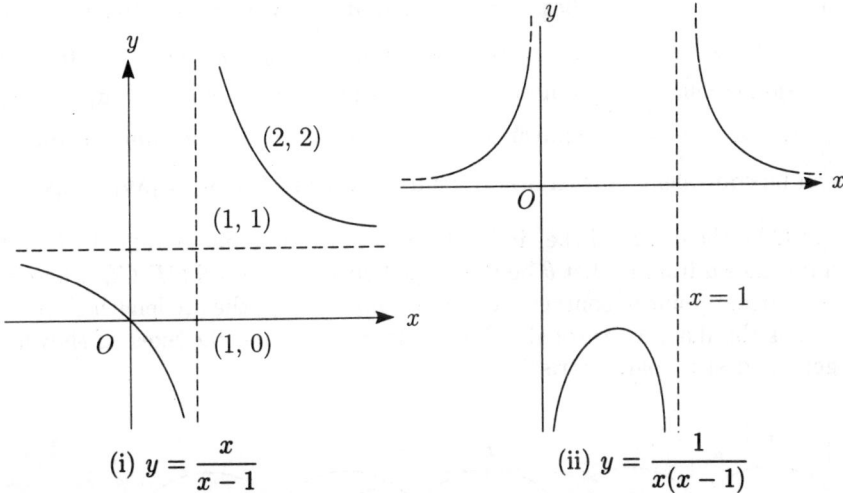

(i) $y = \dfrac{x}{x - 1}$          (ii) $y = \dfrac{1}{x(x - 1)}$

**25.** The problem amounts to finding the maximum and the minimum of the function $f(x) = \dfrac{\sin x}{x}$ on $[0, \frac{\pi}{2}]$ (with the understanding that $f(0) = 1$.) Since $f'(x) = \dfrac{x \cos x - \sin x}{x^2}$ and $x < \tan x$ for $x \in (0, \frac{\pi}{2})$, $f$ is strictly decreasing on $[0, \frac{\pi}{2}]$. So $\alpha = \frac{2}{\pi}$ and $\beta = 1$.

## Section 4.12

**1.** $-\sin(xy)(y + x\dfrac{dy}{dx}) + 2y\dfrac{dy}{dx} - \dfrac{2}{\pi} = 0$ gives $\dfrac{dy}{dx} = \dfrac{2\pi + 4}{4\pi - \pi^2}$ when $x = \dfrac{\pi}{2}, y = 1$.

So the equation of the tangent is $y - 1 = \dfrac{2\pi + 4}{4\pi - \pi^2}(x - \dfrac{\pi}{2})$.

**2.** $5y^4 - 8y^3 + 6x^5 y\dfrac{dx}{dy} + x^6 + \dfrac{dx}{dy} = 0$ gives $\dfrac{dx}{dy} = -\dfrac{17}{13}$ when $x = 1, y = 2$.

**3.** Since $\beta = \alpha \circ f$, the range of $\alpha$ contains that of $\beta$. But $f$ being a bijection, $\alpha = \beta \circ f^{-1}$ which shows that the other way inclusion is also true. So, $\alpha, \beta$ cover the same sets of points in $\mathbb{R}^2$. Since $f$ is an order-preserving bijection, any broken line path along $\alpha$ is also a broken line path along $\beta$ and vice versa. So both have the same lengths. Since $f$ and $f^{-1}$ have non-zero derivatives everywhere, smoothness of $\alpha$ implies that of $\beta$ and vice versa. Finally if $\alpha = (\alpha_1, \alpha_2)$ and $\beta = (\beta_1, \beta_2)$ then $\beta_1 = \alpha_1 \circ f$ and $\beta_2 = \alpha_2 \circ f$. By the chain rule, $(\beta_1'(t), \beta_2'(t)) = (\alpha_1'(f(t))f'(t), \alpha_2'(f(t))f'(t))$. As $f'(t) = b - a > 0$, these two vectors have the same direction (cf. Exercise (12.5) below).

4. If $\alpha = (\alpha_1, \alpha_2)$ and $-\alpha = (\beta_1, \beta_2)$, then $\beta_i(t) = \alpha_i(a + b - t)$ gives $\beta_i'(t) = -\alpha_i'(a + b - t)$ for $i = 1, 2$. So $-\alpha$ is smooth iff $\alpha$ is smooth. Also the tangent to $-\alpha$ at $t$ and that to $\alpha$ at $(a + b - t)$ are oppositely directed.

5. $\mathbf{u} = 1\mathbf{u}, \mathbf{u} = \lambda\mathbf{v}$ implies $\mathbf{v} = \frac{1}{\lambda}\mathbf{u}$ and $\mathbf{u} = \lambda\mathbf{v}, \mathbf{v} = \mu\mathbf{w}$ together imply $\mathbf{u} = (\lambda\mu)\mathbf{w}$. When $\lambda, \mu$ are positive so are $\frac{1}{\lambda}$ and $\lambda\mu$. Also $1 > 0$. So the relation is reflexive, symmetric and transitive. For any non-zero $\mathbf{u}$, the vector $\frac{1}{|\mathbf{u}|}\mathbf{u}$ is a unit vector equivalent to $\mathbf{u}$. Also no two distinct unit vectors $\mathbf{u}, \mathbf{v}$ can be equivalent to each other, for $\mathbf{u} = \lambda\mathbf{v}$ gives $1 = |\mathbf{u}| = |\lambda\mathbf{v}| = \lambda|\mathbf{v}| = \lambda$.

6. Let $C$ be the centre of the circle. Then $C$ moves in a line parallel to $L$ and at a distance $a$ from it. Let $\theta$ be the angle between the radii $CP, CQ$ where $Q$ is the current point of contact. As there is no slipping, the arc length $PQ(= a\theta)$ equals the distance through which $C$ moves. Taking the axes as shown, we get the desired parametrisation.

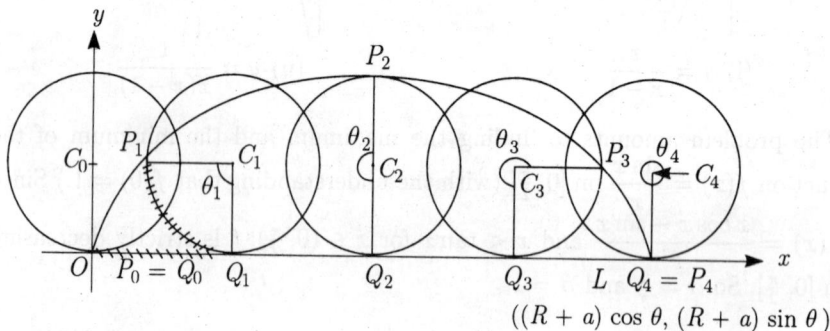

7. Take the centre of the fixed circle as the origin $O$ and the initial point of contact as $(R, 0)$. Let $\theta$ be the angle which the radius $OQ$ makes with the $x$-axis where $Q$ is the current point of contact. Then the loci are :

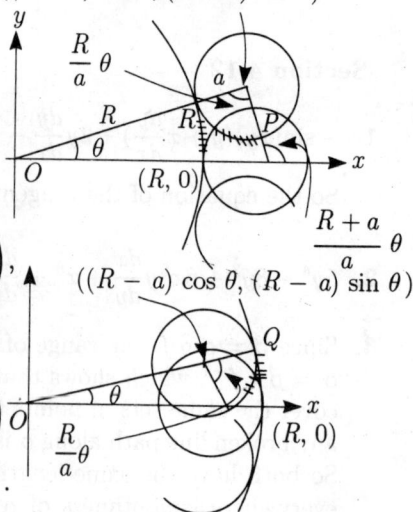

(i) $x(\theta) = (R + a)\cos\theta - a\cos\left(\frac{R + a}{a}\theta\right),$

$y(\theta) = (R + a)\sin\theta - a\sin\left(\frac{R + a}{a}\theta\right)$

(ii) $x(\theta) = (R - a)\cos\theta + a\cos\left(\frac{R - a}{a}\theta\right),$

$y(\theta) = (R - a)\sin\theta - a\sin\left(\frac{R - a}{a}\theta\right).$

If $R/a$ is rational then the functions $x(\theta), y(\theta)$ are periodic (cf. Exercise (8.8)) with a common period and hence the curve is closed in each case. Otherwise the moving point never returns to the same position again.

**8.** Two planes always intersect in a line except when they are either the same plane or parallel to each other. The condition for the former is that $\dfrac{a_1}{a_2} =$ $\dfrac{b_1}{b_2} = \dfrac{c_1}{c_2} = \dfrac{d_1}{d_2}$ while that for the latter is $\dfrac{a_1}{a_2} = \dfrac{b_1}{b_2} = \dfrac{c_1}{c_2} \neq \dfrac{d_1}{d_2}$. The equality of the first three ratios is equivalent to the simultaneous vanishing of $a_1 b_2 - a_2 b_1, b_1 c_2 - b_2 c_1$ and $c_1 a_2 - c_2 a_1$. For the second part, let $x_1 = x_0 + t(b_1 c_2 - b_2 c_1), y_1 = y_0 + t(c_1 a_2 - c_2 a_1), z_1 = z_0 + t(a_1 b_2 - a_2 b_1)$. From $a_1 x_0 + b_1 y_0 + c_1 z_0 = d_1$ it is easily seen that $a_1 x_1 + b_1 y_1 + c_1 z_1 = d_1$. Similarly $a_2 x_1 + b_2 y_1 + c_2 z_1 = d_2$. So $(x_1, y_1, z_1)$ lies on $L$ for all $t \in \mathbb{R}$. Conversely let $(x_1, y_1, z_1)$ be any point on $L$. Then $x_1 - x_0, y_1 - y_0, z_1 - z_0$ satisfy the system of equations: $a_1(x_1 - x_0) + b_1(y_1 - y_0) + c_1(z_1 - z_0) = 0$ and $a_2(x_1 - x_0) + b_2(y_1 - y_0) + c_2(z_1 - z_0) = 0$ whence $\dfrac{x_1 - x_0}{b_1 c_2 - b_2 c_1} = \dfrac{y_1 - y_0}{c_1 a_2 - a_2 a_1} = \dfrac{z_1 - z_0}{a_1 b_2 - a_2 b_1} = t$ (say) which shows that $x_1 = x_0 + t(b_1 c_2 - b_2 c_1)$ etc. Finally, in order that $L$ can be parametrised as $y = g(x), z = h(x), t$ must be expressible in terms of $x$. The condition for this is $b_1 c_2 - b_2 c_1 \neq 0$.

**9.** The first assertion is trivial and so is the second for $r = 0$ since in this case $a, b$ are both $0$ making $S_2$ parallel to the $xy$-plane. For $r > 0$ follow the hint. The unit vectors $\mathbf{u}, \mathbf{v}$ along $O'P, O'Q$ are $\left(\dfrac{a}{r\sqrt{r^2 + 1}}, \dfrac{b}{r\sqrt{r^2 + 1}}, r + c\right)$ and $\left(-\dfrac{b}{r}, \dfrac{a}{r}, 0\right)$ respectively. Equating $(x, y, ax + by + c)$ with $(0, 0, c) + x'\mathbf{u} + y'\mathbf{v}$ and solving the resulting system of equations gives the $(x', y')$-coordinates of a point $(x, y, ax + by + c)$ in $S_2$. For this point to lie on $C$, we must also have $x^2 + y^2 = 1$ which reduces to $x'^2 + y'^2(r^2 + 1) = r^2 + 1$. This gives an ellipse in the plane $S_2$ centred at $O'$ with major axis $2\sqrt{r^2 + 1}$ and minor axis $2$.

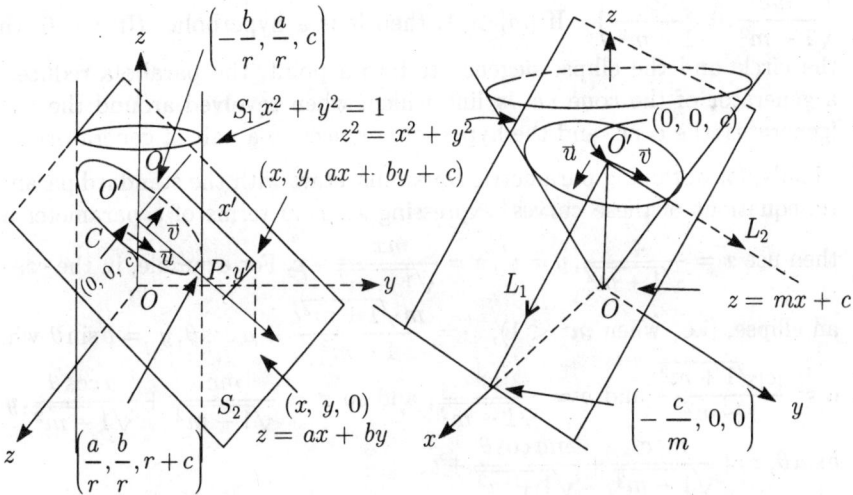

Exercise (12.9)    Exercise (12.10)

**10.** Because of the rotational symmetry of the cone, the $y$ - axis may be taken to be parallel to the plane. Hence the plane may be taken as $z = mx + c$, except when the plane is of the form $x = k = $ a constant. In the latter case, its intersection with the cone is the curve $x = k, z^2 - y^2 = k^2$, which is a rectangular hyperbola, parallel to the hyperbola $z^2 - y^2 = k^2$ in the $y$-$z$ plane. (If $k = 0$, this hyperbola degenerates into a pair of straight lines.)

If the plane is $z = mx + c$, it contains the point $(0, 0, c)$. Call it $O'$. Let $L_1$ be the line of intersection of the given plane with the plane $y = 0$. Then $L_1$ may be parametrised by $x = t, y = 0, z = c + mt$. Let $L_2$ be the line through $(0, 0, c)$ parallel to the $y$ - axis i.e. the line $x = 0, x = c$. Then $L_1, L_2$ are at right angles and the vectors $\mathbf{u} = (\dfrac{1}{\sqrt{1 + m^2}}, 0, \dfrac{m}{\sqrt{1 + m^2}})$ and $\mathbf{v} = (0, 1, 0)$ are, respectively, unit vectors along their directions. Taking $O'$ as the origin and $L_1, L_2$ as axes a co-ordinate system $(x', y')$ can be set up in the cutting plane $z = mx + c$. Following the same method as in the solution to the last exercise, a point $(x, y, mx + c)$ in the plane has co-ordinates $(x', y')$ where $x' = x\sqrt{1 + m^2}$ and $y' = y$. (See the figure last page where only the upper half of the cone is shown. For the case shown there, $c > 0$ and $-1 < m < 0$.)

The equation $x^2 + y^2 = (mx + c)^2$ of the conic then reduces to $(1 - m^2)x'^2 + (1 + m^2)y'^2 - 2mc\sqrt{1 + m^2}x' = c^2(1 + m^2)$. Clearly, $m = 0$ gives a circle of radius $|c|$, while $m = \pm 1$ gives a parabola. If $m \neq \pm 1$, dividing by $1 - m^2$ and completing the squares the equation of the conic becomes $\left(x' - \dfrac{m\sqrt{1 + m^2}c}{1 - m^2}\right)^2 + \left(\dfrac{1 + m^2}{1 - m^2}\right)y'^2 = \dfrac{c^2(1 + m^2)}{(1 - m^2)^2}$. For $|m| < 1, 1 - m^2 >$

0 and this is an ellipse centred at $\left(\dfrac{m\sqrt{1 + m^2}c}{1 - m^2}, 0\right)$ (which corresponds to

$\left(\dfrac{mc}{1 - m^2}, 0, \dfrac{c}{1 - m^2}\right)$). If $|m| > 1$, then it is a hyperbola. (If $c = 0$, then the circle and the ellipse degenerate into a point, the parabola reduces to a generator of the cone i.e. a line which, when revolved around the $z$-axis 'generates' the cone, and the hyperbola reduces to a pair of generators.)

Finally, for obtaining parametric equations, start with the standard parametric equations of these curves, expressing $x', y'$ in terms of a parameter and then use $x = \dfrac{x'}{\sqrt{1 + m^2}}, y = y', z = \dfrac{mx'}{\sqrt{1 + m^2}} + c$. For example, in the case of

an ellipse, (i.e. when $m^2 < 1$), $x' = \dfrac{m\sqrt{1 + m^2}c}{1 - m^2} + a\cos\theta, y' = b\sin\theta$ where

$a = \dfrac{|c|\sqrt{1 + m^2}}{1 - m^2}$ and $b = \dfrac{|c|}{\sqrt{1 - m^2}}$, and so $x = \dfrac{mc}{\sqrt{1 - m^2}} + \dfrac{a\cos\theta}{\sqrt{1 + m^2}}, y = $

$b\sin\theta, z = \dfrac{m^2c}{\sqrt{1 - m^2}} + \dfrac{ma\cos\theta}{\sqrt{1 + m^2}} + c$.

**11.** The first part is a trivial verification. To prametrise $C$ locally at $P$ in terms of $x$, solve the system $z = xy^2$ and $z\sin x + y = \frac{\pi}{12} + 1$ for $y$ and $z$. Substituting the first in the second gives a quadratic in $y$ with roots

$$\frac{-1 \pm \sqrt{1 + 4x(\frac{\pi}{12} + 1)\sin x}}{2x \sin x}.$$ The negative square root is discarded since $y = $

1 when $x = \frac{\pi}{6}$. So $y = \dfrac{-1 + \sqrt{1 + 4x(\frac{\pi}{12} + 1)\sin x}}{2x \sin x} = g(x)$ and $z = xy^2$

$$= \frac{2 + 4x(\frac{\pi}{12} + 1)\sin x - 2\sqrt{1 + 4x(\frac{\pi}{12} + 1)\sin x}}{4x \sin^2 x} = h(x).$$

## Section 4.13

1. The point $\alpha'$ obtained by the binary search at the end of the $n^{\text{th}}$ iterate differs from a true zero $\alpha$ of $f$ by at most $\dfrac{b-a}{2^n}$. But $|\alpha' - \alpha| \leq \dfrac{b-a}{2^n}$ gives $|f(\alpha') - f(\alpha)| \leq \frac{M(b-a)}{2^n} < \epsilon$. $f(\alpha) = 0$ gives the result.

2. Take $f(x) = x^2 - 2, a = 1, b = 2$. The sequence of iterates given by the binary search is 1.5, 1.25, 1.375, 1.4375, 1.40625, 1.421875, 1.4140625 ....
The sequence obtained by the Newton-Raphson method, starting with $x_0 = 1.5$ is $1.5, 1.4166667, 1.4142157, 1.4142136, 1.414236$. So just in 3 iterates (with one more for confirmation) we get the value correct upto 7 places of decimals. Calculation of $5^{1/3}$ is similar to that of $2^{1/3}$ in the text. Starting from $x_0 = 1.5$, the Newton-Raphson method gives 1.70997595 as the value in 4 iterations. Starting from $a = 1, b = 2$, the binary search gives 1.71875 in 5 iterations.

3. Let $\theta = \frac{\pi}{7}$. Then $4\theta + 3\theta = \pi$ and so $\sin 4\theta = \sin 3\theta$ which gives $4\cos^3 \theta \sin \theta - 4\cos \theta \sin^3 \theta = 3\cos^2 \theta \sin \theta - \sin^3 \theta$ and finally $8\cos^3 \theta - 4\cos^2 \theta - 4\cos \theta + 1 = 0$. So $\cos \frac{\pi}{7}$ is a zero of $f(x) = 8x^3 - 4x^2 - 4x + 1$. Starting from $x_0 = \cos \frac{\pi}{6} = \dfrac{\sqrt{3}}{2} = 0.8660254$, the Newton-Raphson method gives $x_1 = 0.9039152, x_2 = 0.9009871, x_3 = 0.9009688, x_4 = 0.9009689$. The first four terms of the power series expansion of $\cos \frac{\pi}{7}$, viz. $1 - \frac{1}{2}(\frac{\pi}{7})^2 + \frac{1}{24}(\frac{\pi}{7})^4$, $-\frac{1}{720}(\frac{\pi}{7})^6$ gives $\cos \frac{\pi}{7} = 0.9009687$. The next term, $\frac{1}{8!}(\frac{\pi}{7})^8$, is less than $10^{-7}$. The second method is easier to implement. The reason is that the cosine function is analytic and has a power series expansion in which the coefficients (viz. $\frac{(-1)^n}{(2n-1)!}$) are easy to calculate and the convergence is so rapid that only a few terms are needed. Newton-Raphson method is a general method which takes no advantage of these special features.

4. Taking $f(x) = \tan x - 1$ and $x_0 = 1$, the Newton iterates are 0.8372778, 0.7881802, 0.7854059, 0.7853981. The last one gives $\pi$ as 3.1415924. Similarly with $f(x) = \sin x - \frac{1}{2}$ and $x_0 = \frac{1}{2}$, the iterates are 0.5234444, 0.5235986, 0.5235986, giving $\pi = 3.1415916$.

5. Since $h'(x) = \frac{1}{3}x^{-2/3}, x_n = x_{n-1} - \dfrac{h(x_{n-1})}{h'(x_{n-1})}$ gives $x_n = -2x_{n-1}$ and hence by induction on $n, x_n = (-2)^n x_0$. This is divergent unless $x_0 = 0$.

**6.** Following the method in Exercise (11.22), the graph of $y = x^3 - 12x^2$ is as shown below. $f'(x) > 0$ if $x < 0$ or $x > 8$ and for $0 < x < 8, f'(x) < 0$.

If $x_0 = 7$, or more generally, if $x_0 \in (6,8)$, then $x_1 = x_0 - \dfrac{x_0^3 - 12x_0^2}{3x_0^2 - 24x_0} =$

$\dfrac{2x_0(x_0 - 6)}{3(x_0 - 8)} < 0$ and the subsequent iterates will all be negative. Also

$|x_n| < \frac{2}{3}|x_{n-1}|$ and so $x_n \to 0$ as $n \to \infty$. The same holds if $x_0 < 0$.

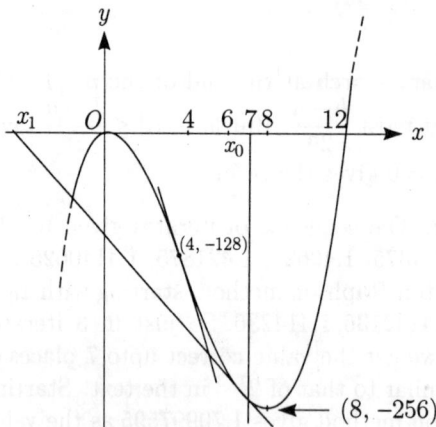

If $x_0 \in (0,6)$ then it can be shown by induction that $0 < x_n \le \frac{1}{2}x_{n-1} < 6$ for all $n$ and so $x_n \to 0$. If $x_0 = 6$ then $x_1 = 0$ and hence $x_n = 0$ for all $n \ge 1$. If $x_0 \in (8,12)$ then $x_1 > 12$ and therefore the iterates are bigger than 12 but get closer to 12. (To see this geometrically, note that for any $x > 8, f'(x) > 0$ and as $f$ is concave upwards on $(8, \infty)$ (in fact on $(4, \infty)$), the tangent at $(x, f(x))$ lies below the graph.) So the sequence of iterates converges to 12 iff $x_0 > 8$.

**7.** (a) is trivial. For (b), take $S = I\!R - \{0\}$ and $f(x) = -x$. Then $f^2(x) = x$ for all $x$. But $f$ has no fixed point. For (c), $f^m(f(x_0)) = f(f^m(x_0)) = f(x_0)$. As $f^m$ has only one fixed point, $f(x_0)$ must equal $x_0$. By (a), $f$ can have no other fixed point.

**8.** $|g(x) - g(y))| = \frac{1}{2}|x - y|$ and so $g$ is a contraction. $g(x) = x$ would mean $x = 10$. So $g$ has no fixed piont in $[0,1]$. Theorem (13.2) is not contradicted since $g$ does not map $[0,1]$ into $[0,1]$.

**9.** As $g'(x) = 1 - \frac{1}{x^2} > 0$ for all $x > 1, g$ is strictly increasing on $[1, \infty)$. If $1 \le x < y$ then $\frac{1}{y} < \frac{1}{x}$ and so $0 < g(y) - g(x) = y - x - (\frac{1}{x} - \frac{1}{y}) < y - x$. Hence $|g(x) - g(y)| < |x - y|$. Given $0 \le \alpha < 1$, choose $x$ so large that $x(x + 1) > \frac{1}{1-\alpha}$ and take $y = x + 1$. Then $|g(y) - g(x)| > \alpha|y - x|$. So $g$ is not a contraction. If $x_0 \in [1, \infty)$ and $x_n = x_{n-1} + \frac{1}{x_{n-1}}$ for $n \ge 1$, then certainly the sequence $\{x_n\}$ is monotonically increasing. So as $n \to \infty$, either $x_n \to \infty$ or $x_n \to L$ for some real $L$ $(> 1)$. But in the latter event since $x_{n+1} \to L$, we get $L = L + \frac{1}{L}$ which is impossible.

**10.** Follow the hint. If $|x_{k+1} - x_k| \le \alpha^k|x_1 - x_0|$, then $|x_{k+2} - x_{k+1}| = |g(x_{k+1}) - g(x_k)| \le \alpha|x_{k+1} - x_k| \le \alpha^{k+1}|x_1 - x_0|$, completing the inductive step. Given $\epsilon > 0$, find $p$ so that $\dfrac{\alpha^p}{(1 - \alpha)} < \dfrac{\epsilon}{|x_1 - x_0|}$. (This is possible since $\alpha^p \to 0$ as $p \to \infty$.) Then for $n > m \ge p, |x_n - x_m| \le$

$$\sum_{i=m}^{n-1} |x_{i+1} - x_i| \leq \sum_{i=m}^{n-1} \alpha^i |x_1 - x_0| < |x_1 - x_0| \sum_{i=p}^{\infty} \alpha^i = |x_1 - x_0| \frac{\alpha^p}{1-\alpha} < \epsilon.$$

The Cauchy sequence must converge to some point $x^*$ in $\mathbb{R}$ by Theorem (3.5.4). But since $a \leq x_n \leq b$ for all $n$, $x^* \in [a, b]$. Finally, by continuity of $g$, $g(x^*) = g(\lim_{n\to\infty} x_n) = \lim_{n\to\infty} g(x_n) = \lim_{n\to\infty} x_{n+1} = x^*$. Uniqueness of the fixed point follows as in the proof of Theorem (13.2).

**11.** The proof above goes through (for any complete metric space).

# CHAPTER 5

## Section 5.1

**1.** The coarsest partition is $(a, b)$, i.e. one in which there is only one subinterval viz. $[a, b]$ itself. There is no finest partition. For the last part, the answer is 'no'. For example, consider the two partitions $(0, \frac{1}{2}, 1)$ and $(0, \frac{1}{3}, 1)$ of $[0, 1]$.

**2.** (i) If $f(x) = k$ for all $x \in [a, b]$, then for every $P \in \mathcal{P}[a, b]$, $U(P; f) = L(P; f) = k(b - a)$. So $\int_a^{\overline{b}} f(x)dx$ and $\int_{\underline{a}}^b f(x)dx$ both equal $k(b - a)$. Hence $f$ is integrable with integral $k(b - a)$.

(ii) If $c > 0$ then for every partition $P = (x_0, x_1, \ldots, x_n)$ of $[a, b]$, we have $\sup_{x\in[x_{i-1}, x_i]} cf(x) = c \sup_{x\in[x_{i-1}, x_i]} f(x)$ for $i = 1, \ldots, n$ and similarly for inf. Hence $U(P; cf) = cU(P; f)$ and $L(P; cf) = cL(P; f)$. So $\sup_{P\in\mathcal{P}[a,b]} U(P; cf) = c \sup_{P\in\mathcal{P}[a,b]} U(P; f)$, i.e. $\int_a^{\overline{b}} cf(x)dx = c \int_a^{\overline{b}} f(x)dx$. A similar equality holds for the lower Riemann integrals. Hence $\int_a^b cf(x)dx$ exists and equals $c \int_a^b f(x)dx$.

For $c < 0$, we have $\sup_{x\in[x_{i-1}, x_i]} cf(x) = c \inf_{x\in[x_{i-1}, x_i]} f(x)$ etc. and so $\int_a^{\overline{b}} cf(x)dx = c \int_{\underline{a}}^b f(x)dx$ and $\int_{\underline{a}}^b cf(x)dx = c \int_a^{\overline{b}} f(x)dx$. Still, equality of $\int_a^{\overline{b}} f(x)dx$ and $\int_{\underline{a}}^b f(x)dx$ implies that of $\int_a^{\overline{b}} cf(x)dx$ and $\int_{\underline{a}}^b cf(x)dx$. The case $c = 0$ is trivial.

3. Follow the hint and apply Exercise (2.6.12) and its analogue to get $x \in A, y \in B$ such that $x > a^* - \epsilon$ and $y < b^* + \epsilon$. But then, by the hypothesis, $x \leq y$ and so $a^* - \epsilon < b^* + \epsilon$ contradicitng $a^* - \epsilon = b^* + \epsilon$.

4. If $a^* = b^*$, and $\epsilon > 0$, then by Exercise (2.6.12) and its analogue, $\exists\, x \in A, y \in B$ such that $x > a^* - \frac{\epsilon}{2}$ and $y < b^* + \frac{\epsilon}{2}$, which implies $y - x < b^* + \frac{\epsilon}{2} - (a^* - \frac{\epsilon}{2}) = \epsilon$. Conversely, if $a^* \neq b^*$, then because of the last exercise, we must have $a^* < b^*$. But then, for every $x \in A, y \in B, y - x \geq b^* - a^*$. So the given condition is violated if $\epsilon = b^* - a^*$ or less (see the figure below).

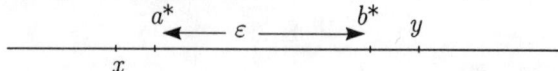

$$a^* \quad\quad \xleftarrow{\quad \epsilon \quad} \quad\quad b^* \qquad y$$

$$\rule{6cm}{0.4pt}$$

$$x$$

5. Let $A, B$ be, respectively, the sets of all lower and upper Riemann sums of $f$. Then by Corollary (1.3) $x \leq y$ for all $x \in A, y \in B$. Integrability of $f$ is equivalent to the equality of $a^*$ and $b^*$. If $f$ is integrable, and $\epsilon > 0$, then by the last exercise $\exists$ partitions $P_1, P_2$ of $[a, b]$ such that $U(P_2, f) - L(P_1; f) < \epsilon$. Let $P = P_1 \cup P_2$ be their common refinement. Then because of Theorem (1.2), $U(P; f) \leq U(P_2; f)$ and $L(P; f) \geq L(P_1; f)$. So $U(P; f) - L(P; f) \leq U(P_2; f) - L(P_1; f) < \epsilon$ as shown in the figure below. Conversely, if the given condition holds then by the last exercise, $a^* = b^*$.

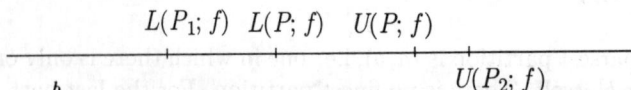

$$L(P_1; f) \quad L(P; f) \quad U(P; f)$$

$$\rule{6cm}{0.4pt}$$

$$\qquad\qquad\qquad\qquad\quad U(P_2; f)$$

6. Suppose $I = \int_a^b f(x)dx$. Given $\epsilon > 0$, the last exercise gives $P \in \mathcal{P}[a, b]$ such that $U(P; f) - L(P; f) < \epsilon$. Since $U(P; f) \geq I \geq L(P; f)$, this means $U(P; f)$ and $L(P; f)$ both lie in $(I - \epsilon, I + \epsilon)$. Conversely suppose the given condition holds. Then applying it with $\frac{\epsilon}{2}$ (instead of $\epsilon$) where $\epsilon > 0$ is given, we get $P \in \mathcal{P}[a, b]$ such that $U(P; f)$ and $L(P; f)$ both lie in $(I - \frac{\epsilon}{2}, I + \frac{\epsilon}{2})$. But then $U(P; f) - L(P; f) < \epsilon$. So by the last exercise $f$ is integrable on $[a, b]$. Let $J = \int_a^b f(x)dx$. If $J \neq I$, choose $\epsilon > 0$ so small that the intervals $(I - \epsilon, I + \epsilon)$ and $(J - \epsilon, J + \epsilon)$ are mutually disjoint. (This time draw your own figures to see this.) From the first part and the given condition, $\exists\, P, Q \in \mathcal{P}[a, b]$ such that $U(P; f), L(P; f)$ are in $(J - \epsilon, J + \epsilon)$ and $U(Q; f), L(Q; f)$ are in $(I - \epsilon, I + \epsilon)$. Let $R = P \cup Q$. Then by Theorem (1.2), $L(P; f) \leq L(R; f) \leq U(R; f) \leq U(P; f)$ and so $L(R; f), U(R; f)$ lie in $(I - \epsilon, I + \epsilon)$. By a similar reasoning they also lie in $(J - \epsilon, J + \epsilon)$, a contradiction. So $J = I$.

7. Let $f$ be integrable on $[a, c]$ and on $[c, b]$ with $I_1 = \int_a^c f(x)dx$ and $I_2 = \int_c^b f(x)dx$. Given $\epsilon > 0$, by the last exercise, $\exists$ partitions $P_1, P_2$ of $[a, c], [c, b]$ respectively such that $I_i - \frac{\epsilon}{2} < L(P_i; f) \leq U(P_i; f) < I_i + \frac{\epsilon}{2}$, for $i = 1, 2$. Let $P$ be the partition of $[a, b]$ obtained by juxtaposing $P_1$ and $P_2$. Then $L(P; f) = L(P_1; f) + L(P_2; f)$ and similarly for $U(P; f)$. So both $L(P; f)$ and $U(P; f)$ lie in $(I_1 + I_2 - \epsilon, I_1 + I_2 + \epsilon)$. By the last exercise,

$f$ is integrable over $[a, b]$ and $\int_a^b f(x)dx = I_1 + I_2$. Conversely, suppose $f$ is integrable on $[a, b]$. Given $\epsilon > 0$, by Exercise (1.5), $\exists P \in \mathcal{P}[a, b]$ such that $0 \leq U(P; f) - L(P; f) < \epsilon$. Obtain $Q$ from $P$ by inserting an additional node $c$. (If $c$ is already a node of $P$ then $Q = P$.) Then $Q$ is a refinement of $P$ and so by Theorem (1.2), $0 \leq U(Q; f) - L(Q; f) \leq U(P; f) - L(P; f) < \epsilon$. Also $Q$ is obtained by juxtaposing partitions, say, $Q_1$ and $Q_2$ of $[a, c]$ and $[c, b]$ respectively. So $\epsilon > U(Q; f) - L(Q; f) = (U(Q_1; f) - L(Q_1; f)) + (U(Q_2; f) - L(Q_2; f))$. As both the summands are non-negative, each is $< \epsilon$. Hence by Exercise (1.5), $f$ is integrable on $[a, b]$ and also on $[c, b]$. By the earlier part, $\int_a^c f(x)dx + \int_c^b f(x)dx = \int_a^b f(x)dx$. More generally if $P = (x_0, x_1, \ldots, x_n)$ is any partition of $[a, b]$ then $f$ is integrable on $[a, b]$ iff it is so on $[x_{i-1}, x_i]$ for each $i = 1, \ldots, n$ and further $\int_a^b f(x) = \sum_{i=1}^n \int_{x_{i-1}}^{x_i} f(x)dx$.

8. Let $P = (x_0, \ldots, x_n)$ be any partition of $[a, b]$. From the hint, for every $i = 1, \ldots, n$ $\exists$ a rational $\xi_i$ and an irrational $\eta_i$ in $(x_{i-1}, x_i)$. So $M_i = 1$ and $m_i = 0$ for each $i$. Hence $U(P; f) = \sum_{i=1}^n M_i \Delta x_i = \sum_{i=1}^n \Delta x_i = 1$ and similarly $L(P; f) = 0$.

9. Boundedness of $h$ is trivial. Let $P = (x_0, \ldots, x_n)$. For $i = 1, \ldots, n$, let $M_i(f), M_i(g)$ and $M_i(h)$ be, respectively, the suprema of $f, g, h$ on $[x_{i-1}, x_i]$. Then for every $x \in [x_{i-1}, x_i], h(x) = f(x) + g(x) \leq M_i(f) + M_i(g)$ and so $M_i(h) = \sup_{x \in [x_{i-1}, x_i]} h(x) \leq M_i(f) + M_i(g)$. Multiply by $\Delta x_i$ and add to get $U(P; h) \leq U(P; f) + U(P; g)$. The proof of the other inequality is similar.

Now suppose $f, g$ are integrable over $[a, b]$ with $I_1 = \int_a^b f(x)dx$ and $I_2 = \int_a^b g(x)dx$. Given $\epsilon > 0$, Exercise (1.6) gives partitions $P, Q$ of $[a, b]$ such that $I_1 - \frac{\epsilon}{2} < L(P; f) \leq U(P; f) \leq I_1 + \frac{\epsilon}{2}$ and $I_2 - \frac{\epsilon}{2} < L(Q; g) \leq U(Q; g) < I_2 + \frac{\epsilon}{2}$. Let $R$ be a common refinement of $P$ and $Q$. Then by Theorem (1.2), $I_1 - \frac{\epsilon}{2} < L(R; f) \leq U(R; f) < I_1 + \frac{\epsilon}{2}$ and $I_2 - \frac{\epsilon}{2} < L(R; g) \leq U(R; g) < I_2 + \frac{\epsilon}{2}$. Adding and using the first part, $I_1 + I_2 - \epsilon < L(R; h) < U(R; h) < I_1 + I_2 + \epsilon$. So by Exercise (1.6) again, $\int_a^b h(x)dx$ exists and equals $I_1 + I_2$. For the more general statement apply what is just proved and Exercise (1.2) (ii).

10. By the last exercise $\int_a^b g(x)dx - \int_a^b f(x)dx = \int_a^b (g(x) - f(x))dx \geq 0$ the last inequality coming the from the fact that when the integrand is non-negative so are all lower Riemann sums.

11. Make three cases (illustrated on the next page) : (i) $0 \leq m$ (ii) $m < 0 < M$ and (iii) $M \leq 0$. In (i) $t = m, T = M$ while in (iii) $T = -m$ and $t = -M$

and so $T - t = M - m$ in both. In (ii), $T = \max\{M, -m\}$ and $t \geq 0$. So, $T - t \leq T \leq M - m$. Strict inequality occurs, for example, when $A = [-1, 2]$.

$$
\begin{array}{ccccccc}
O & m & M & m & O & M & m & OM & -m & m & M O \\
\end{array}
$$

$$
\begin{array}{ccc}
t \quad T & T \quad -T & T \quad -T \quad -t \\
(i) & (ii) & (iii)
\end{array}
$$

12. Certainly, $|f|$ is bounded if $f$ is bounded on $[a, b]$. Given $\epsilon > 0$, by Exercise (1.5), $\exists P \in \mathcal{P}[a, b]$ such that $U(P; f) - L(P; f) < \epsilon$. So $\sum_{i=1}^{n}(M_i - m_i)\Delta x_i < \epsilon$ where $M_i, m_i$ are the supremum and infimum of $f$ on $[x_{i-1}, x_i]$ for $i = 1, \ldots, n$. The last exercise implies $U(P; |f|) - L(P; |f|) \leq U(P; f) - L(P; f)$. Apply Exercise (1.5) again. For the second part apply Exercises (1.10) and (1.2)(ii) to the inequalities (i) $f(x) \leq |f|(x)$ for all $x \in [a, b]$ and (ii) $-f(x) \leq |f|(x)$ for all $x \in [a, b]$. For the last part, merely observe that $\int_{a}^{b} M \, dx = M(b - a)$ by Exercise (1.2) (i).

13. Let $f(x) = 1$ if $x$ is rational and $f(x) = -1$ if $x$ is irrational. By an argument similar to that in Exercise (1.8), $f$ is not integrable on $[0, 1]$. But $|f|$ is the constant function 1 which is integrable by Exercise (1.2)(i).

14. Since $f(x_{i-1}) \leq f(x) \leq f(x_i)$ for all $x \in [x_{i-1}, x_i]$, $f(x_i)$ is the supremum (in fact the maximum) of $f$ on $[x_{i-1}, x_i]$. Similarly, $f(x_{i-1}) = \inf_{x \in [x_{i-1}, x_i]} f(x)$.

15. First assume $f$ is monotonically increasing on $[a, b]$. Follow the hint and note that if $P = (x_0, \ldots, x_n)$ then by the last exercise. $U(P; f) - L(P; f) = \frac{b-a}{n} \sum_{i=1}^{n}(f(x_i) - f(x_{i-1})) = \frac{b-a}{n}(f(b) - f(a)) < \epsilon$. So by Exercise (1.5), $f$ is integrable on $[a, b]$. If $f$ is monotonically decreasing, then $-f$ is monotonically increasing and hence integrable. By Exercise (1.2)(ii), $f$ is integrable on $[a, b]$.

16. Here $f$ is monotonically increasing and so $U(P; f) = \sum_{i=1}^{n} f(x_i)(x_i - x_{i-1})$ where $x_i = a + \frac{i}{n}(b - a)$ for $i = 0, 1, \ldots, n$. A straightforward calculation yields $U(P; f) = (\frac{b-a}{n}) \sum_{i=1}^{n}(a + \frac{i}{n}(b-a)) = (\frac{b-a}{n})(na + \frac{b-a}{n} \sum_{i=1}^{n} i) = (\frac{b-a}{n})(na + \frac{b-a}{n} \cdot \frac{n(n+1)}{2}) = \frac{b-a}{2n}(an + bn + b - a)$. Similarly, the other part follows from $L(P; f) = \sum_{i=1}^{n} f(x_{i-1})(x_i - x_{i-1})$.

17. Let $I = \frac{1}{2}(b^2 - a^2)$. If $P$ is as in the last exercise, then $L(P; f)$ and $U(P; f)$ differ from $I$ by $\frac{(b-a)^2}{2n}$ each. Given $\epsilon > 0$, choose $n$ so that $\frac{(b-a)^2}{2n} < \epsilon$. For $f(x) = x^2$, assume first $0 \leq a < b$. Then $f$ is monotonically increasing. If $P$ is as in the last exercise, then very similar but more involved calculations (requiring $\sum_{i=1}^{n} i^2 = \frac{n(n+1)(2n+1)}{6}$ in addition to $\sum_{i=1}^{n} i = \frac{n(n+1)}{2}$) give

that $U(P; f) - (\dfrac{b^3 - a^3}{3}) = \dfrac{(b-a)(6ab - 6a^2 + 3(b-a)^2)}{6n} + \dfrac{(b-a)^3}{6n^2}$ which

tends to 0 as $n \to \infty$. Similarly $L(P; f) - \frac{b^3 - a^3}{3} \to 0$ as $n \to \infty$. Hence by

Exercise (1.6), $\int\limits_a^b f(x)dx = \frac{b^3 - a^3}{3}$. The calculations are similar if $a < b \le 0$.

If $a < 0 < b$, then $f$ is neither increasing nor decreasing on $[a, b]$. Apply the earlier parts to $[a, 0]$ and $[0, b]$ and use Exercise (1.7).

**18.** Let the lengths of the parallel sides $DE, BC$ be $u, v$ respectively and $a$ be the perpendicular distance between them. Choose the co-rodinates as shown in the figure next page so that $h, k$ are both positive. Then the area of the trapezeum $BCDE = $ area of $OACD - $ area of $OABE = \int\limits_0^a f(x)dx - \int\limits_0^a g(x)dx$

where $f(x), g(x)$ are the functions whose graphs are the lines $DC$ and $EB$

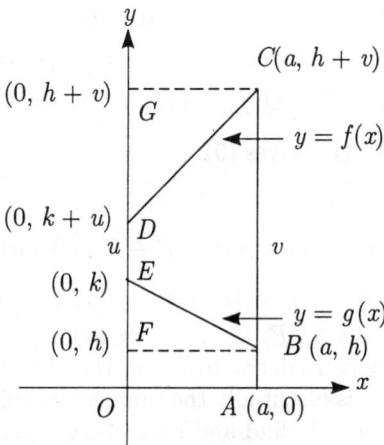

respectively. A little calculation gives $f(x) = k + u + \dfrac{(h + v - k - u)x}{a}$ and $g(x) = k + \dfrac{h - k}{a}x$. Using Exercises (1.9), (1.2)(i) and (1.17), their integrals come out, respectively, as $(k + u)a + \frac{1}{2}(h + v - k - u)a$ and $ka + \frac{1}{2}(h - k)a$. So the area of the trapezeum $BCDE = \frac{1}{2}a(u + v)$. For a geometric proof drop perpendiculars from $B, C$ to the line $DE$ and subtract the areas of the right angled triangles $FBE$ and

$DCG$ from that of the rectangle $CGFB$. (In case either of these triangles is outside the rectangle, its area is to be added rather than subtracted.)

**19.** Depending upon the ordering of $a, b, c$ there are 6 possibilities. The case $a < c < b$ is covered by Exercise (1.7). If $a < b < c$ then $\int\limits_a^c f(x)dx =$

$\int\limits_a^b f(x)dx + \int\limits_b^c f(x)dx = \int\limits_a^b f(x)dx - \int\limits_c^b f(x)dx$. So $\int\limits_a^b f(x)dx = \int\limits_a^c f(x)dx +$

$\int\limits_c^b f(x)dx$. Similarly handle other cases. Exercise (1.9) holds even when $b < a$.

But the inequalities about the integrals in Exercises (1.10) and (1.12) have to be reversed.

## Section 5.2

**1.** For every $\xi$ based on $P, R(P, \xi, f) \le U(P; f)$ holds trivially and so $\sup \mathcal{R}(P) \le U(P; f)$. If strict inequality holds, let $\epsilon = U(P; f) - \sup \mathcal{R}(P)$. Then

$\epsilon > 0$. Let $P = (x_0, \ldots, x_n)$. For each $i$, $\exists \xi_i \in [x_{i-1}, x_i]$ such that $f(\xi_i) > M_i + \dfrac{\epsilon}{b-a}$ where $M_i = \sup\{f(x) : x \in [x_{i-1}, x_i]\}$. But then $R(P, \boldsymbol{\xi}, f) =$

$$\sum_{i=1}^{n} f(\xi_i)(x_i - x_{i-1}) > \sum_{i=1}^{n} M_i(x_i - x_{i-1}) + \sum_{i=1}^{n} \frac{\epsilon(x_i - x_{i-1})}{b-a} = U(P; f) + \epsilon,$$

a contradiction. So $\sup \mathcal{R}(P) = U(P; f)$. Similarly $\inf \mathcal{R}(P) = L(P; f)$. For the last part, let $[a, b] = [0, 1]$, $f(x) = x$ if $x < 1$ and $f(1) = 0$. Let $P = (0, 1)$ be the partition with only one subinterval. Then $U(P; f) = 1$, but there exists no $\xi_1 \in [0, 1]$ for which $f(\xi_1) = 1$ and so for every $\boldsymbol{\xi}$ based on $P$, $R(P, \boldsymbol{\xi}, f) < 1$.

2.  Let $m_i = \inf\{f(x) : x \in [x_{i-1}, x_i]\}$. Then $L(P; f) = \sum_{i=1}^{n} m_i \Delta x_i = L_g + L_b$ where $L_g, L_b$ are respectively the sums of the terms corresponding to the good and the bad indices respectively. The reasoning for (15) also gives $|L_b| < \dfrac{\epsilon}{4}$. Analogous to (16), write $L(P \cup Q; f) = L_g + M_b$ where $M_b$ is the sum of those terms in $L(P \cup Q; f)$ which do not occur in $L_g$. The reasoning for (17) also shows $|M_b| < \dfrac{\epsilon}{4}$. Then $L(P \cup Q; f) - L(P; f) = M_b - L_b \leq |M_b| + |L_b| < \dfrac{\epsilon}{2}$. Combined with (13), this gives (9).

3.  Let $I_1 = \int_a^{\overline{b}} f(x)dx$ and $I_2 = \int_{\underline{a}}^{b} f(x)dx$. Given $\epsilon > 0$, $\exists P \in \mathcal{P}[a, b]$ such that $I_1 \leq U(P; f) < I_1 + \epsilon$. By Theorem (1.2), $I_1 \leq U(Q; f) < I_1 + \epsilon$ whenever $Q$ is a refinement of $P$. In other words, $U(P; f) \to I_1$ (or more precisely, $U(P; f) \to I_1^+$) as $P$ gets finer and finer. To derive from this that $U(P; f) \to I_1^+$ as $\mu(P) \to 0$, the argument needed is essentially the same as that needed for the implication (i) $\Rightarrow$ (ii). Given $\epsilon > 0$, find and fix $Q = (y_0, \ldots, y_k) \in \mathcal{P}[a, b]$ such that whenever $P$ is a refinement of $Q$, $U(P, f) < I_1 + \dfrac{\epsilon}{2}$. Choose $\delta = \dfrac{\epsilon}{4M(k-1)}$ where $M$ is a positive upper bound for $|f(x)|$ on $[a, b]$. Now if $\mu(P) < \delta$ then $I_1 \leq U(P \cup Q; f) < I_1 + \dfrac{\epsilon}{2}$. Analogous to (18) show that $U(P; f) < U(P \cup Q; f) + \dfrac{\epsilon}{2}$. Then $I_1 \leq U(P; f) < I_1 + \epsilon$. So $U(P; f) \to I_1^+$ as $\mu(P) \to 0$. Similarly $L(P; f) \to I_2^-$ as $\mu(P) \to 0$.

4.  Apply the last exercise with the notations in its solution. If $f$ is integrable on $[a, b]$ then $I_1 = I_2$. Since $U(P; f), L(P; f)$ both tend to $I_1(= I_2)$, their difference tends to 0 as $\mu(P) \to 0$. Conversely, if $U(P; f) - L(P; f) \to 0$ as $\mu(P) \to 0$ then $I_1 = \lim_{\mu(P) \to 0} U(P; f) = \lim_{\mu(P) \to 0} [U(P; f) - L(P; f)] + \lim_{\mu(P) \to 0} L(P; f) = 0 + I_2$. So $I_1 = I_2$.

5.  The direct implication follows from Corollary (2.4). Indeed given $\epsilon > 0$, find $\delta > 0$ corresponding to $\frac{\epsilon}{2}$. Then whenever $\mu(P) < \delta$ and $\boldsymbol{\xi}, \boldsymbol{\eta}$ are based on $P$, we have $|R(P, \boldsymbol{\xi}, f) - I| < \frac{\epsilon}{2}$ and $|R(P, \boldsymbol{\eta}, f) - I| < \frac{\epsilon}{2}$ which

together imply $|R(P, \boldsymbol{\xi}, f) - R(P, \boldsymbol{\eta}, f)| < \epsilon$ by the triangle inequality. For the converse, given $\epsilon > 0$ apply the condition with $\frac{\epsilon}{2}$ to get $\delta > 0$ such that whenever $\mu(P) < \delta$ and $\boldsymbol{\xi}, \boldsymbol{\eta}$ are based on $P, |R(P, \boldsymbol{\xi}, f) - R(P, \boldsymbol{\eta}, f)| < \frac{\epsilon}{2}$. For any such $P$, by Exercise (2.1), $\exists \boldsymbol{\xi}, \boldsymbol{\eta}$ based on $P$ such that $U(P; f) < R(P, \boldsymbol{\xi}, f) + \frac{\epsilon}{4}$ and $L(P; f) > R(P, \boldsymbol{\eta}, f) - \frac{\epsilon}{4}$. This implies $U(P; f) - L(P; f) < \frac{\epsilon}{2} + \frac{\epsilon}{4} + \frac{\epsilon}{4} = \epsilon$. So by the last exercises $f$ is integrable on $[a, b]$.

6.  (i) $\int_0^1 x \, dx$ (ii) $\int_0^1 \sin \pi x \, dx$ (iii) $\int_0^2 \dfrac{x \, dx}{\sqrt{1 + x^2}}$ (iv) Write the given sum as

$\dfrac{1}{n} \sum_{r=1}^n \ln(1 + \dfrac{r}{n})$. Its limit is $\int_0^1 \ln(1 + x) dx$.

7.  $\int_0^2 \rho(x) dx$.

8.  (i) $\int_0^2 \sigma(x)\rho(x) dx$ (ii) $\int_0^2 \sigma(x)\rho(x) dx / \int_0^2 \rho(x) dx$. (The contribution to the total income from people living in $[x_{i-1}, x_i]$ is approximately $\sigma(x_i)\rho(x_i)\Delta x_i$ and not $\sigma(x_i)\Delta x_i$.)

9.  $\int_0^L \dfrac{c \, dx}{(\sigma(x))^2}$ where $c$ is some constant.

10. The net displacement over the time interval from $t = a$ to $t = b$.

## Section 5.3

1.  Apply Exercise (1.12) for the first part. The second follows by the Sandwich Theorem.

2.  Let $A = \{f(x) : x < c\}$ and $B = \{f(x) : x > c\}$. Then $f(c)$ is an upper bound for $A$ and a lower bound for $B$. So sup $A$ and inf $B$ exist and sup $A \le$ inf $B$. To finish the proof it suffices to show that sup $A = \lim_{x \to c-} f(x)$ and inf $B = \lim_{x \to c+} f(x)$. For this apply Theorem (2.6.5) (with a minor change of notation).

3.  $g$ is monotonically increasing on $[a, b]$ iff $\lim_{x \to c-} f(x) \le L \le \lim_{x \to c+} f(x)$.

4.  For the first assertion, partition $[0, 1]$ into three parts, $[0, \dfrac{1}{2^n}], [\dfrac{1}{2^n}, \dfrac{1}{2^{n-1}}]$ and $[\dfrac{1}{2^{n-1}}, 1]$ and apply Exercise (1.7) along with Exercise (1.2)(i). For the second assertion, note that for every $x \in [0, 1], f_n(x)$ is eventually zero.

5.  Splitting is necessary if $a < 0 < b$ and $n$ is even. (cf. Exercise (1.17).)

**6.** Let $S_n(x) = \sum\limits_{k=0}^{n} a_k x^k$. By the last exercise and Exercise (1.9) (applied repeatedly), $S_n$ is integrable on $[a,b]$ for all $n$. By Theorem (3.11.5) $S_n(x)$ converges uniformly to $f(x)$ on $[a,b]$. Theorem (3.2) implies the result.

**7.** For the first part, partition $[a,b]$ with nodes at integer points. For every $k \in \mathbb{Z}, [x] = k$ for all $x \in (k, k+1)$. Suppose $f, g$ are two step functions defined on $[a,b]$. Let $P = (x_0, \ldots, x_n)$ and $Q = (y_0, y_1, \ldots, y_m)$ be partitions of $[a,b]$ such that $f$ is constant on $(x_{i-1}, x_i)$ for every $i = 1, \ldots, n$ and $g$ is constant on $(y_{j-1}, y_j)$ for every $j = 1, \ldots, m$. Let $P \cup Q = (z_0, \ldots, z_p)$. Then for every $k = 1, \ldots, p, f, g$ are both constants on $(z_{k-1}, z_k)$ and so are $f + g, f - g$ and $fg$. So these are step functions. For the composite, note that if $f$ is a step function and $g$ is any function (not necessarily a step function) such that $g \circ f$ is defined (i.e. such that the domain of $g$ contains the range of $f$), then $g \circ f$ is a step function.

**8.** Let $f$ be a step function on $[a,b]$. Suppose $P = (x_0, \ldots, x_n)$ is the partition of $[a,b]$ such that $f(x) = c_i$ (say) for all $x \in (x_{i-1}, x_i), i = 1, \ldots, n$. Then $f$ coincides with a constant function on $[x_{i-1}, x_i]$ except possibly at the two end-points $x_{i-1}$ and $x_i$. So by Exercise (1.2) (i) and Theorem (3.1), $f$ is integrable on $[x_{i-1}, x_i]$ and $\int\limits_{x_{i-1}}^{x_i} f(x)dx = \int\limits_{x_{i-1}}^{x_i} c_i dx = c_i(x_i - x_{i-1})$. By Exercise (1.7), $f$ is integrable on $[a,b]$ and $\int\limits_{a}^{b} f(x)dx = \sum\limits_{i=1}^{n} c_i(x_i - x_{i-1})$.

**9.** Let $P = (x_0, \ldots, x_n)$. For $i = 1, \ldots, n$, let $M_i, m_i$ be, respectively, the supremum and the infimum of $f(x)$ on $[x_{i-1}, x_i]$. Define $g(x) = m_i$ and $h(x) = M_i$ if $x \in (x_{i-1}, x_i)$. Define $g(x_i)$ and $h(x_i) = f(x_i)$ for $i = 0, \ldots, n$. Then $g, h$ are step functions with $g(x) \leq f(x) \leq h(x)$ for all $x$. By the last exercise, $\int\limits_{a}^{b} g(x)dx = \sum\limits_{i=1}^{n} m_i(x_1 - x_{i-1})$ and $\int\limits_{a}^{b} h(x)dx = \sum\limits_{i=1}^{n} M_i(x_i - x_{i-1})$.

**10.** By Exercises (1.2) (i), (1.17) and also (1.19) (needed when $x < 0$), $F(x) = \dfrac{x^{n+1}}{n+1}$ for $n = 0, 1, 2$. So $F'(x) = f(x)$ in all cases.

**11.** For $x \in [0,1], F(x) = \int\limits_{0}^{x} 0 dt = 0$. For $x \in [1,2], F(x) = \int\limits_{0}^{1} 0 dt + \int\limits_{1}^{x} 1 dt = x - 1$.

For $x \in [2,3], F(x) = \int\limits_{0}^{1} 0 dt + \int\limits_{1}^{2} 1 dt + \int\limits_{2}^{x} 2 dt = 1 + 2(x - 2) = 2x - 3$. Here $F$ is differentiable except at $x = 1$ and 2 which are also the points of discontinuity of $f$. At every other point $x, F'(x) = f(x)$. So the guess is that $F(x)$ is differentiable wherever $f(x)$ is continuous and $F'(x) = f(x)$.

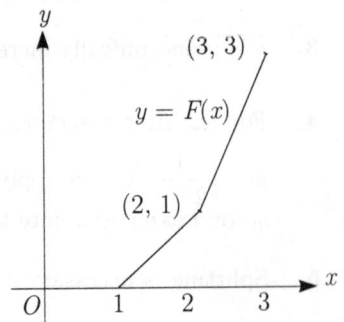

## Section 5.4

**1.** Modify the proof of Theorem (4.1), by choosing $\xi_i, \eta_i$ to be, respectively, points in $[x_{i-1}, x_i]$ where $f$ attains its maximum and minimum. The existence of such points follows from Theorem (4.1.4). Then $U(P; f) - L(P; f) = \sum_{i=1}^{n} (f(\xi_i) - f(\eta_i))(x_i - x_{i-1})$. The rest of the proof goes through. Exercise (1.5) shows $f$ is integrable on $[a, b]$.

**2.** If $c_1 = a$ then while choosing $r > 0$, ensure that $[c_1, c_1 + r] \subset [a, b)$ (instead of $[c_1 - r, c_1 + r] \subset (a, b)$). The other restrictions on $r$ remain the same. In the course of the proof, drop the interval $[b_0, a_1]$ and replace $[a_1, b_1]$ by $[c_1, b_1]$. (Equivalently call $c_1$ as $a_1$.) Similarly if $c_k = b$, then ensure $[c_k - r, c_k] \subset (a, b]$, drop the interval $[b_k, a_{k+1}]$ and replace $[a_k, b_k]$ by $[a_k, c_k]$.

**3.** $(g \circ f)(x) = x^2 = (f \circ g)(x)$ for all $x \in \mathbb{R}$. So $\int_0^2 (g \circ f)(x) dx = \int_0^2 (f \circ g)(x) dx = \int_0^2 x^2 dx = \dfrac{8}{3}$ by Exercise (1.17). But $\int_0^4 g(x) dx \int_0^2 f(x) dx = (\int_0^4 x^2 dx)(\int_0^2 x dx) = \dfrac{64}{3} \cdot \dfrac{1}{2} = \dfrac{32}{3}$ while $\int_0^2 g(x) f(x) dx = \int_0^2 x^3 dx = 4$.

**4 and 5.** Similar to Exercises (1.11) and (1.12) respectively.

**6.** Note that $f(x) \geq \dfrac{1}{n}$ iff $x$ can be written as $\dfrac{p}{q}$ where $p, q$ are positive integers, with no common factor and $q \leq n$. Also since $\dfrac{p}{q} \in [0, 1]$, for each $q \geq 2, p$ can take at most $q - 1$ possible values viz., $1, 2, \ldots, q-1$. So there are only finitely many (in fact, not more than $n(n-1)+1$) rationals of this type in $[0, 1]$. Now given $\epsilon > 0$, choose $n \in \mathbb{N}$ so that $\dfrac{1}{n} < \epsilon$. Let $S = \{x \in [0, 1] : f(x) \geq \dfrac{1}{n}\}$. Then $S$ is a finite set. Let $\delta = \min\{|x - x_0| : x \in S, x \neq x_0\}$. Then $\delta > 0$ and the deleted neighbourhood $(x_0 - \delta, x_0 + \delta) - \{x_0\}$ can contain no points of $S$. Hence $|f(x)| < \dfrac{1}{n} < \epsilon$ for all $x$ in this deleted neighbourhood. The second part follows immediately from the first because, to say $f$ is discontinuous at $c$ means $f(c) \neq 0$ which is the case iff $c > 0$ and is a rational.

**7.** Note that $|f(x)| \leq 1$ for all $x \in [0, 1]$. Let $\epsilon > 0$ be given. Fix $k \in \mathbb{N}$ such that $\dfrac{1}{k} < \dfrac{\epsilon}{2}$. Let $c_1, \ldots, c_r$ be the points in $[0, 1]$ where $f(x) \geq \dfrac{1}{k}$. Assume $0 < c_1 < c_2 < \cdots < c_r = 1$. Choose points $a_1, \ldots, a_r, b_1, \ldots, b_{r-1}$ such that $0 < a_1 < c_1 < b_1 < a_2 < c_2 < b_2 < a_3 \ldots < a_{r-1} < c_{r-1} < b_{r-1} < a_r < c_r = 1$ and $\sum_{i=1}^{r-1} (b_i - a_i) + (1 - a_r) < \dfrac{\epsilon}{2}$. Let $P$ be the partition $(0, a_1, b_1, a_2, b_2, \ldots, a_r, c_r)$ of $[0, 1]$. Because of the denseness of irrationals, $L(P; f) = 0$. As for $U(P; f)$, note that the supremum of $f$ on $[0, a_1], [b_1, a_2], \ldots, [b_{r-1}, a_r]$ is less than $\dfrac{1}{k}$. As their lengths add up to less than 1, their contribution to $U(P; f)$ is at most $\dfrac{1}{k}$. The lengths of the remaining intervals add up to less than $\dfrac{\epsilon}{2}$. So since $|f(x)| \leq 1$ for all $x$, their contribution to $U(P; f)$ is at most $\dfrac{\epsilon}{2}$. Hence $U(P; f) < \epsilon$. So $U(P; f) - L(P; f) < \epsilon$. By Exercise (1.5), $f$ is integrable on $[0, 1]$.

**8.** Merely note that $(g \circ f)(x) = 0$ if $x = 0$ or irrational, while $(g \circ f)(x) = 1$ if $x$ is a positive rational. By the same reasoning as in Exercise (1.8), $g \circ f$ is not integrable over $[0, 1]$.

**9.** By Theorem (4.1.3) and continuity of $f$, the set $f^{-1}(\{y_0\})$ is non-empty. Also it is precisely the set of zeros of the continuous function $g$ defined by $g(x) = f(x) - y_0$. So by Exercise (4.1.2) it is a closed set. As it is also bounded, by Exercise (4.5.7) it has a least and a greatest element. Let $a^*, b^*$ be respectively the least and the greatest elements of $f^{-1}(\{y_0\})$. If $a^* = b^*$, then $f^{-1}(\{y_0\})$ is a single point, otherwise, by monotonicity of $f$, it is the interval $[a^*, b^*]$.

**10.** Let $y_1, \ldots, y_k$ be the points of discontiuity of $g$. By the last exercise, for $i = 1, \ldots, k, f^{-1}(\{y_0\}) = a_i^*$ or an interval $[a_i^*, b_i^*]$ with $a_i^* < b_i^*$. These intervals are mutually disjoint. If $x \in [a, b] - \bigcup\limits_{k=1}^{n} [a_i^*, b_i^*]$, then $g$ is continuous at $f(x)$ and so $g \circ f$ is continuous at $x$ as $f$ is continuous everywhere. If $x \in (a_i^*, b_i^*)$, then $(g \circ f)$ is constant on $(a_i^*, b_i^*)$ (with value $y_i$) and so continuous at $x$. So the only points of discontinuity of $f$ are $a_1^*, \ldots, a_k^*, b_1^*, \ldots, b_k^*$. Since $g$ is bounded, so is $g \circ f$. By Theorem (4.2), $g \circ f$ is integrable on $[a, b]$.

### Section 5.5

**1.** More generally $\sin(\theta) + \sin(\theta + \beta) + \sin(\theta + 2\beta) + \ldots + \sin(\theta + (n-1)\beta) = \dfrac{\sin(\theta + \frac{n-1}{2}\beta) \sin \frac{n\beta}{2}}{\sin \frac{\beta}{2}}$. To prove this, multiply each term on the left by $2 \sin \dfrac{\beta}{2}$ and use the formula $2 \sin A \sin B = \cos(A - B) - \cos(A + B)$. Now, by Exercise (1.7), $\int\limits_a^b \sin x \, dx = \int\limits_0^b \sin x \, dx - \int\limits_0^a \sin x \, dx$. One of the Riemann sums

of $\int\limits_0^b \sin x \, dx$ is $\dfrac{b}{n} \sum\limits_{k=1}^{n} \sin(\dfrac{kb}{n}) = \dfrac{b}{n} \dfrac{\sin(\dfrac{n+1}{2}\dfrac{b}{n}) \sin \dfrac{nb}{2n}}{\sin \dfrac{b}{2n}}$ which tends to $2 \sin^2 \dfrac{b}{2}$

as $n \to \infty$. So $\int\limits_0^b \sin x \, dx = 2 \sin^2 \dfrac{b}{2} = 1 - \cos b$. Similarly $\int\limits_0^a \sin x \, dx = 1 - \cos a$.

So $\int\limits_a^b \sin x \, dx = \cos a - \cos b$. Similarly $\sum\limits_{k=0}^{n-1} \cos(\theta + k\beta) = \dfrac{\cos(\theta + \frac{n-1}{2}\beta) \sin \frac{n\beta}{2}}{\sin \frac{\beta}{2}}$

yields $\int\limits_a^b \cos x \, dx = \sin b - \sin a$.

**2.** $\lim\limits_{x \to \frac{1}{2}} f(x) = 0 \neq f(\dfrac{1}{2})$. So $f$ has a simple discontinuity at $\frac{1}{2}$ and hence cannot be a derived function by Exercise (4.5.8). However since $\frac{1}{2}$ is the only discontinuity of $f$, and it is bounded, it is integrable on $[0, 1]$ by Theorem (4.2).

**3.** It is easily seen that $f(x) = \begin{cases} 1 - 2x & \text{for } x \le 0, \\ 1 & \text{for } x \in [0,1] \\ 2x - 1 & \text{for } x > 1 \end{cases}$ . So an antideriva-

tive $F$ with $F(0) = 0$ is given by $F(x) = \int_0^x 1 - 2t\,dt = x - x^2$ if $x \le 0$, $F(x) =$

$\int_0^x 1\,dt = x$ if $x \in [0,1]$ and $F(x) = \int_0^1 1\,dt + \int_1^x 2t - 1\,dt = 1 + x^2 - x$ if $x \ge 0$.

**4.** After establishing $f$ is integrable on $[a,b]$, the proof of Theorem (5.3) uses continuity of $f$ only at $c$ and hence goes through unchanged. The converse is false. Consider the function $f(x)$ in Exercise (5.2). It is discontinuous at $\frac{1}{2}$. But $F(x) = \int_0^x 0\,dt = 0$ for all $x \in [0,1]$ and hence is differentiable everywhere on $[0,1]$, including at $\frac{1}{2}$.

**5.** Write $G(t) = k - \int_a^x f(t)\,dt$ where $k = \int_a^b f(t)\,dt$ is a constant. By Theorem

(5.3), $G'(x) = -\frac{d}{dx}(\int_a^x f(t)\,dt)) = -f(x)$. For the second part write $H(x)$

as $\int_a^{v(x)} f(t)\,dt - \int_a^{u(x)} f(t)\,dt$. Apply Theorem (5.3) and the chain rule to each term.

**6.** (i) Factorise $x^2 - x_0^2$ as $(x + x_0)(x - x_0)$. By Theorem (5.3), as $x \to x_0$,

$\frac{1}{x - x_0} \int_{x_0}^x \sin(t^2)\,dt \to (\sin x_0^2)$. So for $x_0 \ne 0$ the given limit is $\frac{1}{2x_0} \sin(x_0^2)$.

For $x_0 = 0$, one has to consider the function $F(x) = \int_0^x \sin(t^2)\,dt$ and apply

the L'Hôpital's rule to the ratio $\frac{F(x)}{x^2}$ to reduce the limit to $\lim_{x \to 0} \frac{\sin(x^2)}{2x}$ which comes out as 0.

(ii) Let $F(x) = \int_0^{x^2} \frac{t^2\,dt}{t^6 + 1}$. Then by Exercise (5.5), $F'(x) = \frac{x^4}{x^{12} + 1} \cdot 2x$.

$\lim_{x \to 0} \frac{F(x)}{x^6}$ can now be found using L'Hôpital's rule. It comes out to be $\frac{1}{3}$.

**7.** (i) $\frac{1}{2}$ (ii) 2 (iii) $\sqrt{5} - 1$. (To find an antiderivative of $\frac{x}{\sqrt{1 + x^2}}$, put $t = x^2 + 1$.)

**8.** $\frac{1}{n} \sum_{k=1}^n \sin(\frac{k^2}{n^2})$ is a Riemann sum of $f(x) = \sin(x^2)$ on $[0,1]$. As $n \to \infty$,

it tends to $\int_0^1 \sin(x^2)\,dx$ which is a fixed positive real number. So as $n \to$

$\infty$, $\sum_{k=1}^n \sin(\frac{k^2}{n^2}) = n \cdot (\frac{1}{n} \sum_{b=1}^n \sin(\frac{k^2}{n^2})) \to \infty$ while $\frac{1}{n^2} \sum_{k=1}^n \sin(\frac{k^2}{n^2}) \to 0$.

**9.** Let $S$ be the given sum. Then $\frac{S}{100} = \sum\limits_{k=1}^{10,000} \sqrt{\frac{k}{10,000}} = 10,000R$ where $R =$

$\frac{1}{10,000} \sum\limits_{k=1}^{10,000} \sqrt{\frac{k}{10,000}}$ is one of the Riemann sums of $f(x) = \sqrt{x}$ on $[0,1]$ based on the partition $P$ of $[0,1]$ into 10,000 equal parts. (If we take $10^6 \sqrt{x}$ instead of $\sqrt{x}$, then $S$ itself is a Riemann sum.) As $\mu(P) = \frac{1}{10,000}$ is very small, this Riemann sum is approximately $\int\limits_0^1 \sqrt{x}dx = \frac{2}{3}$. So $S \simeq 100 \cdot 10000 \cdot \frac{2}{3} =$ $6,66,666.666\ldots$. To get an upper bound on the error, note that since $\sqrt{x}$ is a monotonically increasing function, $R$ is actually the upper Riemann sum for the partition $P$. Hence $R \geq \int\limits_0^1 \sqrt{x}dx$. But then, for the same reason, $R$ can also be looked at as the lower Riemann sum for the same function, viz. $f(x) = \sqrt{x}$ over the interval $[0,1.0001]$ divided into 10,001 equal parts. Hence $R \leq \int\limits_0^a \sqrt{x}dx$ where $a = 1.0001$. This integral equals $\frac{2}{3}a^{3/2}$. Taking the first order approximation for the function $(1+x)^{3/2}$ near the point 1, $a^{3/2} \simeq 1.00015$ which gives $R \leq 0.666766666$. Hence the exact value of the given sum lies somewhere between 6,66,666.666 and 6,66,766.666. So the error is at most 100. (The value obtained with a FORTRAN programme comes out to be 6,66,716.459 which is almost the arithmetic mean of the two bounds.)

**10.** Take a partition $P = (x_0, x_1, \ldots, x_n)$ of $[a,b]$. The lines $y = x_i$ chop the desired area into $n$ vertical strips. If $\mu(P)$ is small, then the $i^{\text{th}}$ chip may be regarded approximately as a rectangle with base $[x_{i-1}, x_i]$ and height $|f_1(x_i) - f_2(x_i)|$. So the approximate area is $\sum\limits_{i=1}^n |f_1(x_i) - f_2(x_i)|\Delta x_i$. Take limit as $\mu(P) \to 0$ to get the exact area. (Note that since we are taking $|f_1(x) - f_2(x)|$, it does not matter which graph lies above the other. In fact the two graphs may cross each other any number of times.) The answer to the last part is $\int\limits_c^d |g_1(y) - g_2(y)|dy$, obtained by chopping the desired area horizontally rather than vertically.

**11.** Taking the centre as the origin, the upper and the lower semicircles are the graphs of $y = f_1(x) = \sqrt{a^2 - x^2}$ and $y = f_2(x) = -\sqrt{a^2 - x^2}$; $-a \leq x \leq a$. By the last exercise, the area of the circle = area between these two graphs $= \int\limits_{-a}^a 2\sqrt{a^2 - x^2}dx$. The substitution $x = a\sin t$ converts $\int \sqrt{a^2 - x^2}dx$ to $\int a^2 \cos^2 t\,dt$ which, using $\cos^2 t = \frac{1}{2}(1 + \cos 2t)$ comes out as $\frac{a^2 t}{2} + \frac{a^2}{2}\sin t \cos t$ or $\frac{a^2}{2}\sin^{-1}(\frac{x}{a}) + \frac{1}{2}x\sqrt{a^2 - x^2}$. Hence area of a circle of radius $a = a^2 \sin^{-1}(\frac{x}{a}) + x\sqrt{a^2 - x^2}\Big|_{-a}^a = \pi a^2$. For the second part, take the centres of the two circles at $(0,0)$ and $(R,0)$. By symmetry the common area splits into two equal parts on either side of their common chord and hence equals $4 \int_{R/2}^{a} \sqrt{a^2 - x^2}\,dx = \pi a^2 - 2a^2 \sin^{-1}(\frac{R}{2a}) - \frac{R}{2}\sqrt{4a^2 - R^2}$.

**12.** Identifying the points of intersection of the given curves, taken two at a time, the region bounded by the three curves is the shaded region in the figure below. Its area splits into two (unequal) parts, $PRO$ and $ORQ$. They are respectively, $\int_{-\sqrt{2}}^{0} x + \sqrt{4 - x^2}\,dx$ and $\int_{0}^{\sqrt{2}} -\frac{1}{\sqrt{2}}x^2 + \sqrt{4 - x^2}\,dx$ which come out to be $\frac{\pi}{2}$ and $\frac{\pi}{2} + \frac{1}{3}$. So total area is $\pi + \frac{1}{3}$.

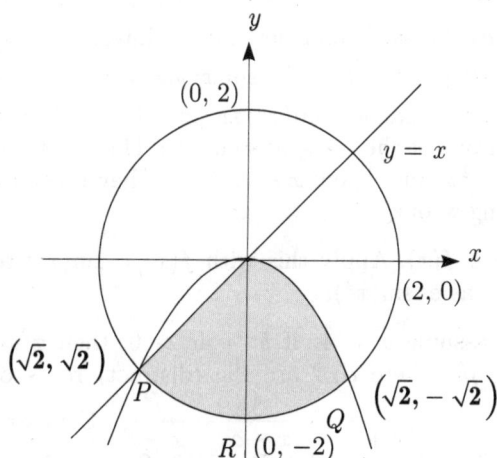

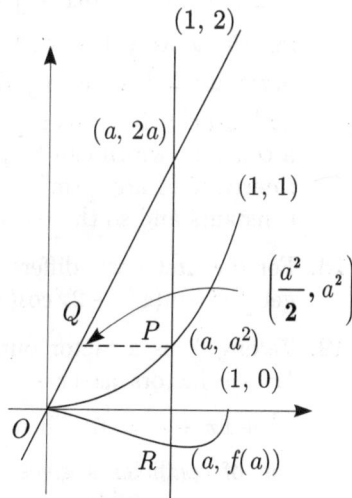

Exercise (5.12)                 Exercise (5.13)

**13.** Referring to the figure above, let $P = (a, a^2), Q = (\frac{a^2}{2}, a^2), R = (a, f(a))$. Then area of $OPQ$ = area $ORP$ gives $\int_{0}^{a^2} \sqrt{y} - \frac{y}{2}\,dy = \int_{0}^{a} x^2 - f(x)\,dx$.

Integrating, $\frac{2}{3}a^3 - \frac{a^4}{4} = \frac{a^3}{3} - \int_{0}^{a} f(x)\,dx$, or with a change of notation,

$\int_{0}^{x} f(t)\,dt = \frac{x^4}{4} - \frac{x^3}{3}$. By Theorem (5.3), $f(x) = x^3 - x^2$.

**14.** Follow the hint. By Theorem (5.3), $H'(x) = f(x)g(x) - f'(x)G(x) - f(x)G'(x) + f'(x)G(x) = 0$ for all $x \in [a,b]$. By Theorem (4.5.1), $H$ is constant on $[a,b]$. In particular, $H(b) = H(a) = -f(a)G(a)$ which gives the result.

**15.** $\int \ln x\,dx = \int 1 \cdot \ln x\,dx = \int \frac{d}{dx}(x)\ln x\,dx = x\ln x - \int x \cdot \frac{d}{dx}(\ln x)\,dx = x\ln x - \int 1\,dx = x\ln x - x$. Similarly $\int \tan^{-1} x\,dx = x\tan^{-1} x - \int \frac{x}{1 + x^2}\,dx =$

$x \tan^{-1} x - \dfrac{1}{2} \ln(1 + x^2)$. Finally, $\int x^2 \sin x dx = -x^2 \cos x + \int 2x \cos x dx = -x^2 \cos x + 2x \sin x - \int 2 \sin x dx = -x^2 \cos x + 2x \sin x + 2 \cos x$.

**16.** $\int \sec^3 \theta d\theta = \int \dfrac{d}{d\theta}(\tan \theta) \sec \theta d\theta = \tan \theta \sec \theta - \int \tan \theta \dfrac{d}{d\theta}(\sec \theta) d\theta = \tan \theta \sec \theta - \int \tan^2 \theta \sec \theta d\theta = \tan \theta \sec \theta + \int \sec \theta d\theta - \int \sec^3 \theta d\theta$. So $2 \int \sec^3 \theta d\theta = \tan \theta \sec \theta + \int \sec \theta d\theta = \tan \theta \sec \theta + \ln(\tan \theta + \sec \theta)$. The substitution $x = a \tan \theta$ gives $\int \sqrt{a^2 + x^2} dx = \dfrac{x}{2}\sqrt{a^2 + x^2} + \dfrac{a^2}{2} \ln(\dfrac{x + \sqrt{a^2 + x^2}}{a})$.

**17.** (i) $\int \sin^2 x dx = \int \dfrac{d}{dx}(-\cos x) \sin x dx = -\cos x \sin x + \int \cos x \cos x dx = -\cos x \sin x + \int 1 dx - \int \sin^2 x dx$. So $\int \sin^2 x dx = \dfrac{x}{2} - \dfrac{\sin x \cos x}{2}$. Writing $\sin^2 x$ as $\dfrac{1}{2}(1 - \cos 2x)$ gives the same anti-derivative. Integration by parts applied twice to $\int \sin(x^2) dx$ gives $\int (1 - \dfrac{4x^4}{3}) \sin(x^2) dx = x \sin(x^2) - \dfrac{2}{3} x^3 \cos(x^2)$. But this is no good for finding $\int \sin(x^2) dx$ since $1 - \dfrac{4x^2}{3}$ is not a constant which can be pulled out of the integral sign. (ii) The three antiderivatives are $\dfrac{1}{2} \sin^2 x$, $-\dfrac{1}{2} \cos^2 x$ and $\dfrac{1}{4}(\sin^2 x - \cos^2 x)$. They differ by constants and so there is nothing wrong.

**18.** For the first part, differentiate $e^x f(x)$. Apply this with $f(x) = \sin(x^2)$ to get $\int e^x (\sin(x^2) + 2x \cos(x^2)) dx$ as $e^x \sin(x^2)$.

**19.** Taking $\dfrac{1}{a}$ as a factor out, we assume $a = 1$. If $b^2 - 4c > 0$, then $x^2 + bx + c$ factors as $(x - \alpha)(x - \beta)$ where $\alpha, \beta$ are the (distinct) roots of $x^2 + bx + c = 0$. Then write $\dfrac{x}{x^2 + bx + c}$ as $\dfrac{A}{x - \alpha} + \dfrac{B}{x - \beta}$. Comparison of coefficients gives $A + B = 1$ and $A\beta + B\alpha = 0$. Solving for $A$ and $B$, $\int \dfrac{x dx}{x^2 + bx + c} = \dfrac{\alpha}{\alpha - \beta} \ln(x - \alpha) - \dfrac{\beta}{\alpha - \beta} \ln(x - \beta)$. By a similar calculation, $\int \dfrac{dx}{x^2 + bx + c} = \dfrac{1}{\beta - \alpha} \ln(\dfrac{x - \alpha}{x - \beta})$. If $b^2 = 4c$ then $x^2 + bx + c = (x + \dfrac{b}{2})^2$ and $\int \dfrac{dx}{x^2 + bx + c}$ equals $\dfrac{-1}{x + \frac{b}{2}}$. To find $\int \dfrac{x dx}{(x + \frac{b}{2})^2}$, write the numerator as $(x + \dfrac{b}{2}) - \dfrac{b}{2}$. Then $\int \dfrac{x dx}{(x + \frac{b}{2})^2}$ equals $\ln(x + \dfrac{b}{2}) + \dfrac{b}{2(x + \frac{b}{2})}$. Finally, if $b^2 < 4c$ then $x^2 + bx + c = (x + \dfrac{b}{2})^2 + \alpha^2$ where $\alpha = \sqrt{4c - b^2}$. Now $\int \dfrac{dx}{(x + \frac{b}{2})^2 + \alpha^2} = \dfrac{1}{\alpha} \tan^{-1}(\dfrac{x + \frac{b}{2}}{\alpha})$, while the other integral $\int \dfrac{x dx}{(x + \frac{b}{2})^2 + \alpha^2}$ comes out as $\int \dfrac{x + \frac{b}{2}}{(x + \frac{b}{2})^2 + \alpha^2} dx - \dfrac{b}{2} \int \dfrac{dx}{(x + \frac{b}{2})^2 + \alpha^2} = \dfrac{1}{2} \ln((x + \dfrac{b}{2})^2 + \alpha^2) - \dfrac{b}{2\alpha} \tan^{-1}(\dfrac{x + \frac{b}{2}}{\alpha})$.

**20.** The denominator factors as $x(x+1)(x^2+x+1)$. Write $\dfrac{3x^2 - 1}{x(x + 1)(x^2 + x + 1)}$

as $\dfrac{A}{x} + \dfrac{B}{x+1} + \dfrac{Cx+D}{x^2+x+1}$. $A, B, C, D$ can be determined by writing a system of four equations. But there are easier ways. Note that if we multiply both the sides by $x$ and take limits as $x \to 0$ then every term on the right except the first tends to 0 while the first term, which is a constant $A$ tends to $A$. This gives $A$ as $\lim\limits_{x\to 0} \dfrac{3x^2-1}{(x+1)(x^2+x+1)} = -1$. Similarly, multiplying both the sides by $x+1$ and taking limits as $x$ tends to $-1$, we get $B = -2$. Substituting these values and expanding, $3x^2 - 1 = -(x^3 + 2x^2 + 2x + 1) - 2(x^3 + x^2 + x) + (x^2+x)(Cx+D)$. Comparing coefficients, we get $C = 3$ and $D = 4$. Writing $\dfrac{3x+4}{x^2+x+1}$ as $\dfrac{\frac{3}{2}(2x+1)}{x^2+x+1} + \dfrac{\frac{5}{2}}{(x+\frac{1}{2})^2 + (\frac{\sqrt{3}}{2})^2}$ gives the antiderivative as $-\ln x - 2\ln(x+1) + \dfrac{3}{2}\ln(x^2+x+1) + \dfrac{5}{\sqrt{3}}\tan^{-1}(\dfrac{2x+1}{\sqrt{3}})$.

21. (i) $-2\sqrt{1+\cos x}$ (ii) $-\sqrt{1-\sin x}$ (put $u^2 = 1 + \sin x$ or rationalise) (iii) $\dfrac{x}{a^2\sqrt{a^2+x^2}}$ (iv) $2\sqrt{1-x}\cos(\sqrt{1-x}) - 2\sin(\sqrt{1-x})$ (v) $\dfrac{1}{8}\ln(1 - \sin x) - \dfrac{1}{12}\ln(1 + \sin x) - \dfrac{1}{24}\ln(5 - \sin x)$ (Put $u = \sin x$ and resolve into partial functions. Note that $1 - \sin x, 1 + \sin x$ and $5 - \sin x$ are all non-negative. See Exercise (7.16).) (vi) $-\dfrac{1}{x}\tan^{-1}x + \ln x - \dfrac{1}{2}\ln(x^2+1)$ (vii) $\dfrac{ae^{ax}\cos(bx+c) + be^{ax}\sin(bx+c)}{a^2+b^2}$ (viii) $\dfrac{ae^{ax}\sin(bx+c) - be^{ax}\cos(bx+c)}{a^2+b^2}$ (ix) $\dfrac{1}{2}\ln(\dfrac{\sqrt{e^{2x}+1}-1}{\sqrt{e^{2x}+1}+1})$ (Put $e^x = u = \tan\theta$. To get the final answer in this form, a little manipulation is needed; cf. Exercise (6.4.3).) (x) $e^x(x\ln x - 1)$.

22. With the substitution $t = \tan\dfrac{x}{2}$, $\int \dfrac{dx}{2+\cos x}$ becomes $\int \dfrac{2dt}{3+t^2}$, which equals $\dfrac{2}{\sqrt{3}}\tan^{-1}(\frac{t}{\sqrt{3}})$, i.e., $\dfrac{2}{\sqrt{3}}\tan^{-1}\left(\dfrac{\tan(\frac{x}{2})}{\sqrt{3}}\right)$. The method also works for $\int \dfrac{dx}{2+\cos^2 x}$, but the computations are rather cumbersome. A slicker way is to rewrite it as $\int \dfrac{\sec^2 x\,dx}{2\sec^2 x + 1}$ which becomes $\int \dfrac{du}{2u^2+3}$ where $u = \tan x$. So an antiderivative is $\dfrac{1}{\sqrt{6}}\tan^{-1}\left(\dfrac{u\sqrt{2}}{\sqrt{3}}\right)$, i.e., $\dfrac{1}{\sqrt{6}}\tan^{-1}\left(\dfrac{\sqrt{2}\tan x}{\sqrt{3}}\right)$.

(Moral : Don't apply any method mechanically just because it works. Look for alternate easier ways!)

## Section 5.6

1. Given an interval of the form $[a, a + T]$, $\exists$ a unique $k \in \mathbb{Z}$ such that $a \le kT < a + T$. Split the interval $[a, a + T]$ into two subintervals $[a, kT]$

and $[kT, a + T]$. We show separately that $f$ is integrable over each of these two subintervals.

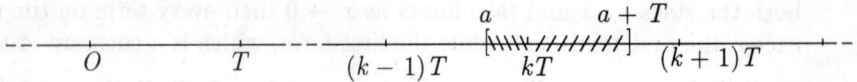

By periodicity, Riemann sums of $f$ over $[(k - 1)T, kT]$ correspond to those of $f$ on $[0, T]$. So $f$ is integrable over $[(k - 1)T, kT]$ and $\int_{(k-1)T}^{kT} f(x)dx = \int_0^T f(x)dx$. Hence by Exercise (1.12), $f$ is integrable over $[(k - 1)T, a]$ and also over $[a, kT]$. Again by periodicity, the Riemann sums over $[(k - 1)T, a]$ correspond to those over $[kT, a + T]$. So the integral $\int_{kT}^{a+T} f(x)dx$ exists and equals $\int_{(k-1)T}^a f(x)dx$. So by Exercise (1.12), $f$ is integrable over $[a, a + T]$ and $\int_a^{a+T} f(x)dx = \int_a^{kT} f(x)dx + \int_{kT}^{a+T} f(x)dx = \int_a^{kT} f(x)dx + \int_{(k-1)T}^a f(x)dx = \int_{(k-1)T}^{kT} f(x)dx = \int_0^T f(x)dx$.

2.  The function $G(x)$ in the hint is differentiable and by Exercise (5.5), $G'(x) = f(x + T) - f(x) = 0$ for all $x$. So $G$ is a constant.

3.  By Exercise (4.5.21), $\exists\ M > 0$ such that $|h(x) - h(y)| \le M|x - y|$ for all $x, y \in [a, b]$. Given $\epsilon > 0$, let $\delta = \dfrac{\epsilon}{M}$. Then $\mu(P) < \delta \Rightarrow t_i - t_{i-1} < \delta$ for all $i \Rightarrow h(t_i) - h(t_{i-1}) < \epsilon$ for all $i \Rightarrow \mu(h(P)) < \epsilon$.

4.  Instead of duplicating the argument, define $g : [c, d] \to [a, b]$ by $g(t) = h(c + d - t)$. Then $g'(t) > 0$ for all $t \in (c, d)$, $g(c) = a$ and $g(d) = b$. Theorem (6.5) applied to $g$, gives $\int_a^b f(x)dx = \int_c^d f(g(t))g'(t)dt$. Since $g(t) = h(c + d - t)$ and $g'(t) = -h'(c + d - t)$, Theorem (6.1) shows that the second integral equals $\int_c^d f(g(t))h'(t)dt$.

5.  First establish the integrability of $u + v$ for (i), of $u^2 + v^2$ for (ii) and of $\sqrt{u^2 + v^2}$ for (iii), using, respectively, Exercise (1.9), Lemma (4.4), and Theorem (4.3) (applied after proving (ii) and taking $g(x) = \sqrt{x}$). As a result, the three sums approach the respective integrals if $\alpha_i = \beta_i$ and so as in the proof of Theorem (6.4), the problem is reduced to showing that each of the following three differences, viz.,

    (a) $\sum_{i=1}^n [(u(\alpha_i) + v(\beta_i)) - (u(\beta_i) + v(\beta_i))]\Delta t_i$ (for (i)),

(b) $\sum_{i=1}^{n} [u^2(\alpha_i) + v^2(\beta_i)) - (u^2(\beta_i) + v^2(\beta_i))]\Delta t_i$ (for (ii)) and

(c) $\sum_{i=1}^{n} [\sqrt{u^2(\alpha_i) + v^2(\beta_i)} - \sqrt{u^2(\beta_i) + v^2(\beta_i)}]\Delta t_i$ (for (iii))

can be made arbitrarily small. For the first two sums, this follows directly from the uniform continuity of $u$ and $u^2$ respectively. For the third sum, use the fact that $|\sqrt{x} - \sqrt{y}| \le \sqrt{|x-y|}$ for $x, y \ge 0$ (cf. Exercise (4.2.9)) and the uniform continuity of $u^2$.

6. Since each $f_n'$ is continuous, $g$ is continuous on $[a, b]$ by Theorem (4.1.5) (applied to $f_n'$ and not to $f_n$). Let $f(x_0) = \lim_{n\to\infty} f_n(x_0)$ (which is given to exist) and for any $x \in [a, b]$, let $f(x) = f(x_0) + \int_{x_0}^{x} g(t)dt$. By Theorem (5.1), $f_n(x) = f_n(x_0) + \int_{x_0}^{x} f_n'(t)dt$ and so by Theorem (3.2), $f_n(x) \to f(x)$ pointwise on $[a, b]$. To show the convergence is, in fact, uniform, note that

$$|f_n(x) - f(x)| \le |f_n(x_0) - f(x_0)| + |\int_{x_0}^{x}(f_n'(t) - g(t))dt|$$
$$\le |f_n(x_0) - f(x_0)| + \int_{x_0}^{x}|f_n'(t) - g(t)|dt$$
$$\le |f_n(x_0) - f(x_0)| + (b-a)\sup_{a \le t \le b}|f_n'(t) - g(t)|$$

for all $x \in [a, b]$. But as $f_n' \rightrightarrows g$, on $[a, b]$, $\sup_{a \le t \le b}|f_n'(t) - g(t)| \to 0$ as $n \to \infty$. Also $f_n(x_0) \to f(x_0)$ as $n \to \infty$. So each term can be made arbitrarily small (independently of $x$) if $n$ is sufficiently large. Finally, from the definition of $f(x)$ and Theorem (5.3), we get $f'(x) = g(x)$ for all $x \in [a, b]$.

7. Split the integral on the left as the integral over $[0, \frac{\pi}{4}]$ and the integral over $[\frac{\pi}{4}, \frac{\pi}{2}]$. By Theorem (6.1), the first integral equals $\int\limits_{0}^{\pi/4} f(\cos 2x)\sin(\frac{\pi}{4} - x)dx$ while the substitution $u = x - \frac{\pi}{4}$ reduces the second integral to $\int_{0}^{\pi/4} f(\cos 2u)\sin(\frac{\pi}{4} + u)du$ which is the same as $\int_{0}^{\pi/4} f(\cos 2x)\sin(\frac{\pi}{4} + x)dx$. Add the two integrals to get the result.

8. In each case, let $I$ denote the integral to be evaluated. In (i), by Theorem (6.1), $I = \int_{0}^{\pi/2} \frac{\pi}{2}\frac{\cos x \sin x}{\sin^4 x + \cos^4 x}dx - I$. So, $I = \int_{0}^{\pi/2}\frac{\pi}{4}\frac{\cos x \sin x}{\sin^4 x + \cos^4 x}dx$. Writing $\sin^4 x + \cos^4 x$ as $(\sin^2 x + \cos^2 x)^2 - 2\sin^2 x \cos^2 x = 1 - \frac{1}{2}\sin^2 2x = \frac{1}{2} + \frac{1}{2}\cos^2 2x$ and putting $\cos 2x = u$, $I$ comes out as $\frac{\pi}{8}\int_{-1}^{1}\frac{du}{1+u^2} = \frac{\pi^2}{16}$. A similar reasoning for (ii), along with the substitution $t = \tan\frac{x}{2}$ (cf. Exercise (5.22)), gives $I = \frac{\pi}{2}\int_{0}^{\pi}\frac{dx}{1 + a\sin x} = \frac{\pi}{\sqrt{1-a^2}}(\frac{\pi}{2} - \tan^{-1}\frac{a}{\sqrt{1-a^2}})$. (Strictly speaking, here the substitution $t = \tan\frac{x}{2}$ converts $I$ into what is called an improper integral, to be studied in Section 5.10. The use of improper integrals may be avoided by using the substitution $t = \tan\frac{x}{2}$ only

for finding an anti-derivative of $\frac{1}{1+a\sin x}$ and not for converting the definite integral $\int_0^\pi \frac{dx}{1+a\sin x}$ to $\int_0^\infty \frac{2dt}{1+t^2+2at}$. That is, after obtaining an anti-derivative in terms of $t$, viz., $\frac{2}{\sqrt{1-a^2}}\tan^{-1}(\frac{t+a}{\sqrt{1-a^2}})$ we convert it back in terms of $x$, viz. $\frac{2}{\sqrt{1-a^2}}\sin^{-1}\left(\frac{\sin\frac{x}{2}+a\cos\frac{x}{2}}{\sqrt{1+a\sin x}}\right)$ and *then* apply Theorem (5.1). Obviously, this is very clumsy.) In (iii), $I = \frac{\pi}{8}\ln 2$. In (iv) use $\tan^{-1}(\frac{1}{1-x+x^2}) = \tan^{-1}x + \tan^{-1}(1-x)$ to get $I = 2\int_0^1 \tan^{-1}x\,dx = \frac{\pi}{2} - \ln 2$ (integrating by parts).

9. Writing $(tx + 1) - x$ as $bx + 1$ where $b = t - 1$ is a constant w.r.t. $x$, we get $\int_0^1 (tx+1-x)^n dx = \frac{1}{(n+1)b}(bx+1)^{n+1}\Big|_0^1 = \frac{1}{n+1}\frac{(b+1)^{n+1}-1}{b} = \frac{1}{n+1}\sum_{k=0}^n (b+1)^k$. (This is valid for $b \neq 0$ i.e. for $t \neq 1$. But if $b = 0$, then also it holds since each side equals 1.) So $\int_0^1 (tx+1-x)^n dx = \frac{1}{n+1}\sum_{k=0}^n t^k$ for all $t$ (including $t = 1$). But we can also evaluate the given integral by first expanding the integrand by the binomial theorem in terms of powers of $t$ and noting that $t$ is a constant in this integration. This gives it is as a polynomial in $t$, viz., $\sum_{k=0}^n a_k t^k$ where $a_k = \binom{n}{k}\int_0^1 x^k(1-x)^{n-k}dx$. We thus have two expressions for the same integral, as polynomials in the parameter $t$. As they are equal for all values of $t$, coefficients of like powers must be equal. This gives the result. (Here the parameter $t$ plays a double role. As far as integration w.r.t. $x$ is concerned, it is a constant. At the same time it is a variable and so we can consider functions of it and use theorems about them. In the present problem, uniqueness of polynomial expansion was used. This is a powerful technique called the **variation of parameter**. It can also be used to evaluate integrals like (ii) in the last exercise (where the parameter is $a$) but the theoretical justifications needed for it are beyond our scope.)

10. The reduction formula is obtained by writing $\sin^n x$ as $\sin^{n-1}x\frac{d}{dx}(-\cos x)$ and integrating by parts. $\int_0^{\pi/2} \sin^0 x\,dx$ and $\int_0^{\pi/2} \sin x\,dx$ are $\frac{\pi}{2}$ and 1 respectively. Let $I_{2m} = \int_0^{\pi/2} \sin^{2m}x\,dx$. Then $I_{2m} = \frac{2m-1}{2m}I_{2m-1} = \frac{2m-1}{2m}\cdot\frac{2m-3}{2m-2}I_{2m-2} = \ldots = \frac{(2m-1)(2m-3)\ldots(5.3)}{(2m)(2m-2)\ldots 6.4.2}I_0 = \frac{(2m)!}{2^m m!2^m m!}\frac{\pi}{2}$. The other result is proved similarly.

**11.** Since $0 \le \sin x \le 1$ for all $x \in [0, \frac{\pi}{2}]$, $\sin^{2m+1}(x) \le \sin^{2m}(x) \le \sin^{2m-1}(x)$ for all $x \in [0, \frac{\pi}{2}]$. So by Exercise (1.10), $I_{2m+1} \le I_{2m} \le I_{2m-1}$.

**12.** Let $a_n = \dfrac{2.2}{1.3} \cdot \dfrac{4.4}{3.5} \cdot \dfrac{6.6}{5.7} \cdots \dfrac{(2n)(2n)}{(2n-1)(2n+1)}$. Then $a_n = \dfrac{(2^n n!)^2}{(2n)!} \dfrac{(2^n n!)^2}{(2n+1)!} = \dfrac{\pi}{2} \dfrac{I_{2n+1}}{I_{2n}}$. By the last exercise and the Sandwich Theorem, $\lim\limits_{n \to \infty} \dfrac{I_{2n}}{I_{2n+1}} = 1$.

So $\dfrac{I_{2n+1}}{I_{2n}} \to 1$ as $n \to \infty$.

**13.** Let $I_n = \int_0^{\pi/2} \cos^n \theta\, d\theta$. Then $I_n = \dfrac{n-1}{n} I_{n-2}$. (Either duplicate the reasoning for Exercise (6.10) or use Theorem (6.1)) Putting $x = \sin\theta$, $\int_0^1 (1-x^2)^n dx$

equals $\int_0^{\pi/2} \cos^{2n+1}\theta\, d\theta$. As this also equals $\int_0^{\pi/2} \sin^{2n+1}\theta\, d\theta$, Exercise (6.10) gives the result. The last assertion follows by expanding $(1-x^2)^n$ by the binomial theorem and integrating term-by-term.

**14.** Integration by parts gives, as in Exercise (6.10), $nI_n = (n-1)I_{n-2} + \int_0^{\pi/2} \sin^{n-1} x \cos x\, dx$. The last integral equals $\int_0^1 u^{n-1} du$ where $u = \sin x$.

**15.** Denote the integral by $I_m$. Apply induction on $m$. The cases $m = 0$ and $m = 1$ are trivial to verify. For $m \ge 2$, note that $I_m + I_{m-2}$
$$= 2\int_0^\pi \frac{1 - \cos(m-1)x \cos x}{1 - \cos x} dx = 2I_{m-1} + 2\int_0^\pi \frac{[\cos(m-1)x](1 - \cos x)}{1 - \cos x} dx =$$
$$2I_{m-1} + 2\int_0^\pi \cos(m-1)x\, dx = 2I_{m-1} + 0 = 2I_{m-1}.$$

**16.** (i) is trivial. For (ii) apply integration by parts. For (iii), go on applying (ii) repeatedly (with different values of $m, n$ every time) till you end up with
$$I_{m,n} = \frac{n(n-1)\dots 1}{(m+1)\dots(m+n)} I_{m+n,0}.$$
Now apply (i).

**17.** Let $S$ be the given sum. Then $S = \sum\limits_{k=0}^{n} \dfrac{(-1)^k}{\binom{n}{k}} = (n+1) \sum\limits_{k=0}^{n} \dfrac{(-1)^k}{(n+1)\binom{k+(n-k)}{k}}$

$= (n+1) \sum\limits_{k=0}^{n} \int_0^1 (-x)^k (1-x)^{n-k} dx = (n+1) \int_0^1 \sum\limits_{k=0}^{n} (-x)^k (1-x)^{n-k} dx.$

The integrand is the sum of a geometric progression with first term $(1-x)^n$ and common ratio $\dfrac{-x}{1-x}$. So $S = (n+1) \int_0^1 \dfrac{(1-x)^n [1 - (\frac{-x}{1-x})^{n+1}]}{1 + \frac{x}{1-x}} dx =$

$(n+1) \int_0^1 (1-x)^{n+1} + (-1)^{n+2} x^{n+1} dx.$ Since $\int_0^1 x^{n+1} dx = \int_0^1 (1-x)^{n+1} dx$

$= \dfrac{1}{n+2}$, the result follows.

**Section 5.7**

1. Modify the solutions to Exercises (2.1.1) and (2.6.3). (The argument is, in fact, slightly easier and gives $e < 3$.)

2. Let $\{x_n\}$ be a Cauchy sequence of rationals. Then $\{x_n\}$ is bounded (see the first part of the proof of Theorem (3.5.4)). Fix a rational $M$ such that $x_n \leq M$ for all $n$. Given $\epsilon > 0$, first find, as in the proof of Theorem (7.2), some $k \in I\!N$ such that $e^{1/k} - 1 < \dfrac{\epsilon}{e^M}$. For any such fixed $k$, there exists $p \in I\!N$ such that $|x_n - x_m| < \frac{1}{k}$ whenever $m \geq n \geq p$. For any such $m, n$, either $x_m \geq x_n$ or $x_n > x_m$. In the first case, $|e^{x_m} - e^{x_n}| = |e^{x_n}(e^{x_m - x_n} - 1)| < e^M(e^{1/k} - 1) < \epsilon$. In the second case, $|e^{x_m} - e^{x_n}| = |e^{x_n} - e^{x_m}| = |e^{x_m}(e^{x_n - x_m} - 1)| < e^M(e^{1/k} - 1) < \epsilon$. So $\{e^{x_n}\}$ is a Cauchy sequence. If $\lim\limits_{n \to \infty} x_n = \lim\limits_{n \to \infty} y_n$ then $|x_n - y_n| \to 0$ as $n \to \infty$ and writing $e^{x_n} - e^{y_n}$ as $e^{x_n}(1 - e^{y_n - x_n})$, it follows by a similar argument that $e^{x_n} - e^{y_n} \to 0$ as $n \to \infty$. So $\lim\limits_{n \to \infty} e^{x_n} = \lim\limits_{n \to \infty} e^{y_n}$.

3. Given $x \in I\!R$, let $\{x_n\}$ be any sequence of rationals converging to $x$. Define $e^x$ as $\lim\limits_{n \to \infty} e^{x_n}$. By the last exercise this is well-defined. If $x \in Q$, we may set $x_n = x$ for all $n$. In that case $\{e^{x_n}\}$ is a constant sequence with limit $e^x$. So the new definition coincides with the old one.

4. $f(0) = f(0 + 0) = f(0) + f(0)$ implies $f(0) = 0$. Let $c = f(1)$. Prove the equality first for $x \in I\!N$ by induction and then for all $x \in Z\!\!\!Z$, by noting $f(-n) + f(n) = f(0) = 0$ for all $n \in I\!N$. Next, if $x = \dfrac{p}{q}$ with $p \in Z\!\!\!Z, q \in I\!N$, then $cp = f(p) = f(qx) = f(x) + \ldots + f(x)(q \text{ times}) = qf(x)$, giving $f(x) = \dfrac{cp}{q} = cx$. Finally continuity of $f(x)$ and of the function $g(x) = cx$, along with Exercise (4.1.4) shows $f(x) = cx$ for all $x \in I\!R$.

5. Indeed, continuity of $f$ at any one point, say $x_0$, implies continuity at any other point, say $x_1$, since $f(x_1 + h) - f(x_1) = f(h) = f(x_0 + h) - f(x_0)$.

6. Define $g : I\!R \to I\!R$ by $g(x) = f(e^x)$. Then $g(x + y) = f(e^x e^y) = f(e^x) + f(e^y) = g(x) + g(y)$ for all $x, y \in I\!R$. Continuity of $f$ at 1 implies that of $g$ at 0 by Exercise (4.1.5). Apply the last two exercises to get $g(x) = cx$ for some $c \in I\!R$. So $f(e^x) = cx$, or $f(y) = c\ln y$ for all $y > 0$.

7. These follow from the corresponding laws for $e^x$. For example, $a^{x+y} = e^{(x+y)\ln a} = e^{x\ln a} \cdot e^{y\ln a} = a^x a^y$.

8. $\dfrac{d}{dx}(x^r) = \dfrac{d}{dx}(e^{r\ln x}) = (\text{by Chain Rule}) \ e^{r\ln x}\dfrac{r}{x} = \dfrac{rx^r}{x} = rx^{r-1}$ by the last exercise.

**9.** The same proof holds since even for an irrational exponent $r(> 0)$, $0 < \dfrac{1}{(n+1)^r} < \dfrac{1}{n^r}$ for all $n \in \mathbb{N}$ and so Theorem (6.2) applies. For $r \leq 0$, $\dfrac{1}{n^r} \nrightarrow 0$ as $n \to \infty$.

**10.** The condition implies $\dfrac{d}{dx}(\ln f(x) - kx) = 0$ for all $x \in \mathbb{R}$. Hence by Theorem (4.5.1), $\ln f(x) - kx = c$ for some constant $c$. This gives $f(x) = e^{kx}e^c$ for all $x$. Put $a = e^k$ and $b = e^c$.

**11.** $f(0) = f(0+0) = f(0)f(0)$ gives $f(0) = 0$ or $f(0) = 1$. In the first case $f(x) = f(x+0) = f(x)f(0) = 0$ for all $x$. Assume $f(0) = 1$. $f(x) = f(\frac{x}{2} + \frac{x}{2}) = (f(\frac{x}{2}))^2$ shows $f(x) \geq 0$ for all $x$. But $f(x) \neq 0$ as otherwise $f(0) = f(x + (-x)) = f(x)f(-x) = 0$. So $f(x) > 0$ for all $x$. Define $g(x) = \ln f(x)$. Then $g(x+y) = g(x) + g(y)$ for all $x$. Also $g$ is continuous at $0$ since $f$ is continuous at $0$ and $\ln$ is continuous everywhere. By Exercises (7.4) and (7.5), $g(x) = cx$ for some $c$. So $f(x) = e^{cx} = a^x$ where $a = e^c$.

**12.** By induction on $n$, prove that $\dfrac{d^n}{dx^n}(\ln(1+x)) = \dfrac{(n-1)!(-1)^{n-1}}{(1+x)^n}$ for $n \geq 1$.

This is valid for all $x > -1$ and hence, in particular for $|x| < 1$. The given series can be shown to be the Maclaurin series of $\ln(1 + x)$ (cf. Exercise (4.6.18)). But proving that $R_n(x) \to 0$ as $n \to \infty$ is rather clumsy. Instead, start with the geometric series $\displaystyle\sum_{n=0}^{\infty} (-1)^n x^n$ which converges to $\dfrac{1}{1+x}$ in $(-1,1)$. For $|x| < 1$, the convergence is uniform on $[0, x]$ (or on $[x, 0]$ if $x < 0$) and term-by-term integration is valid by Theorem (3.2).

**13.** $\ln(1.04) = \displaystyle\sum_{n=1}^{\infty} \dfrac{(-1)^{n+1}(0.04)^n}{n}$. This is an alternating series and so (cf. Exercise (3.8.1)), the partial sum $S_n$ differs from $\ln(1.04)$ by at most $\dfrac{(0.04)^{n+1}}{n+1}$. To get an error less than $10^{-5}$, it suffices to take $n = 3$. So $\ln(1.04) \simeq 0.04 - \dfrac{(0.04)^2}{2} + \dfrac{(0.04)^3}{3} = 0.03922$, upto five places of decimals.

**14.** Let $u = \log_a x$ and $v = \log_a y$. Then $a^u = x$ and $a^v = y$. Then (i) follows from $a^{u+v} = a^u a^v$ while (ii) follows from $(a^u)^y = a^{uy}$, (see Exercie (7.7)). For (iii), let $w = \log_x y$. Then $x^w = y$. So $y = (a^u)^w = a^{uw}$, giving $uw = \log_a y = v$. So $w = \dfrac{v}{u}$.

**15.** This can be proved by generalising the proof of Theorem (7.3), which effectively means re-doing Exercise (2.7.2), but with a valid argument. An alternate approach is to work with logarithms. Since $(\dfrac{d}{dx} \ln x)_{x=1}$ is 1, $\displaystyle\lim_{h \to 0} \dfrac{\ln(1+h) - \ln 1}{h} = 1$. In particular for a fixed non-zero $x$, $\dfrac{\ln(1+\frac{x}{n})}{\frac{x}{n}} \to 1$

as $n \to \infty$; i.e. $\ln(1 + \frac{x}{n})^n \to x$. Since the exponential function is continuous, by Theorem (4.1.2), $(1 + \frac{x}{n})^n \to e^x$ as $n \to \infty$. For $x = 0$, the assertion is trivially true.

**16.** For $x > 0, \ln|x| = \ln x$ and has derivative $\frac{1}{x}$. For $x < 0, \ln|x| = \ln(-x)$ and so $\frac{d}{dx}(\ln|x|) = -\frac{1}{x}(-1) = \frac{1}{x}$.

**17.** It suffices to find the minimum of $\ln(x^{1/x}) = f(x)$ (say) for $0 < x < \infty$. Then $f(x) = \frac{\ln x}{x}$ and $f'(x) = \frac{1 - \ln x}{x^2}$ which vanishes at $x = e$, is positive for $0 < x < e$ and negative for $x > e$. So $f(x)$ has a global maximum (there being no end-points). As $e < 3 < \pi, f(3) > f(\pi)$ which gives $\frac{\ln 3}{3} > \frac{\ln \pi}{\pi}$ i.e. $\pi \ln 3 > 3 \ln \pi$ and hence $3^\pi > \pi^3$.

**18.** $\int_0^1 \ln(1 + x)dx = \int_1^2 \ln u\,du = (u \ln u - u)|_1^2 = 2 \ln 2 - 1$.

**19.** (i) The given sum equals $\frac{1}{n} +$ a Riemann sum of $\frac{1}{1+x}$ on $[0, 1]$. So its limit is $\int_0^1 \frac{dx}{1+x} = \ln(1 + x)\Big|_0^1 = \ln 2$. (ii) $I_n = \int_0^1 \frac{nx^{n-1}}{1+x}dx = \frac{x^n}{1+x}\Big|_0^1 +$ $\int_0^1 \frac{x^n}{(1+x)^2}dx \le \frac{1}{2} + \int_0^1 x^n dx = \frac{1}{2} + \frac{1}{n+1}$. So $I_n \to \frac{1}{2}$ as $n \to \infty$. (Note that the problem asks only for $\lim_{n\to\infty} I_n$ and not for evaluation of $I_n$ per se.)

**20.** The series $\sum_{n=1}^\infty \frac{(-1)^{n+1}}{n}$ is convergent by Leibnitz's test and so to find its value it suffices to evaluate $\lim_{n\to\infty} S_{2n}$. Since $S_{2n+2} - S_{2n} = \frac{1}{2n+1} - \frac{1}{2n+2} = \frac{1}{2n+1} + \frac{1}{2n+2} - \frac{1}{n+1}$, it follows by induction on $n$ that $S_{2n}$ equals $\frac{1}{n+1} + \frac{1}{n+2} + \dots + \frac{1}{2n}$ which tends to $\ln 2$. So $\sum_{n=1}^\infty \frac{(-1)^{n+1}}{n} = \ln 2$.

**21.** Note that the function $f(x) = x^m$ is concave downwards on $(0, \infty)$ for $0 < m < 1$ and concave upwards if $m > 1$. Apply Exercise (4.10.6). If $m = 1$, then equality holds.

**22.** Let $y_i = \ln x_i$ for $i = 1, 2, \dots n$ and use the fact that the exponential function is strictly concave upwards on $(-\infty, \infty)$ along with Exercise (4.10.6) (applied to $y_1, y_2, \dots, y_n$ and not to $x_1, x_2, \dots, x_n$). The A.M.-G.M. inequality results by taking each $\alpha_i$ to be $\frac{1}{n}$.

**23.** For the first part, show that because of the hypothesis and Theorem (6.1), $\int_a^{2a} f(x)dx$ equals $\int_0^a f(x)dx$. For the second part, use the symmetry of the sine function about the point $\frac{\pi}{2}$ and the fact that $\sin \theta \ge \frac{2\theta}{\pi}$ and hence $e^{-R\sin\theta} \le e^{-2R\theta/\pi}$ for all $0 \le \theta \le \pi/2$. Integrating the latter gives the result. (Actually, strict inequality holds, cf. Exercise (6.5.9)(c).)

## Section 5.8

**1.** Rationalising, $\left|\dfrac{b}{a}x - b\sqrt{\dfrac{x^2}{a^2} - 1}\right| = \dfrac{b^2}{\left|\dfrac{b}{a}x + b\sqrt{\dfrac{x^2}{a^2} - 1}\right|} \to 0$ as $x \to \infty$. So,

$$\dfrac{\frac{b}{a}x}{b\sqrt{\frac{x^2}{a^2} - 1}} = 1 + \dfrac{\frac{b}{a}x - b\sqrt{\frac{x^2}{a^2} - 1}}{b\sqrt{\frac{x^2}{a^2} - 1}} \to 1 + 0 = 1 \text{ as } x \to \infty.$$

**2.** Let $b = a^\lambda$ where $\lambda = \dfrac{\ln b}{\ln a}$. Then $\lambda > 1$ since $0 < a < b$. Since $\dfrac{\ln_a x}{\ln_b x} = \dfrac{\ln x/\ln a}{\ln x/\ln b} = \lambda$ (by Exercise (7.14)(iii)), we get the first part. But $\dfrac{b^x}{a^x} = a^{x(\lambda - 1)} \to \infty$ as $x \to \infty$, since $\lambda > 1$.

**3.** $\ln(\ln x), \log_3 x, \ln x, \sqrt{x}, x^e, x^{\ln x}, e^{x/2}, 2^x, e^x, (\ln x)^x, x^x, (ex)^x$ and $x^{2x}$. (To compare $\sqrt{x}, x^e, e^{x/2}, 2^x, e^x, x^{\ln x}$ and $(\ln x)^x$, compare their logarithms. To compare $e^x, (\ln x)^x, x^x, (ex)^x$ and $x^{2x}(= (x^2)^x)$ compare their bases.)

**4.** A trivial example would be to let $f(x) = x$ for all $x$ while let $g(x) = x^2$ if $[x]$ is even and $g(x) = \sqrt{x}$ if $[x]$ is odd. An example where both $f, g$ are continuous and monotonically increasing can be constructed as follows. Let $f(x) = x$ for all $x \geq 0$. For $n \in \mathbb{N}$, let $A_n$ and $B_n$ be the points $((2n-1)! + \frac{1}{2}, (2n)!)$ and $((2n+1)! - \frac{1}{2}, (2n)!)$ respectively. Let $g(x)(x \geq 0)$ be the function whose graph is the (infinite) broken like path $O - A_1 - B_1 - A_2 - B_2 - A_3 - B_3 \ldots - A_n - B_n - A_{n+1} \ldots$. Then $g(x) = (2n)!$ for $(2n-1)! + \frac{1}{2} \leq x \leq (2n+1)! - \frac{1}{2}$. So at $x = (2n-1)! + \frac{1}{2}$, $\dfrac{g(x)}{f(x)} \simeq 2n$ while at $x = (2n+1)! - \frac{1}{2}$, $\dfrac{f(x)}{g(x)} \simeq 2n + 1$.

**5.** Apply Theorem (3.6.2). $\displaystyle\sum_{n=1}^{\infty} 2^n \dfrac{1}{2^n \ln(2^n)} = \dfrac{1}{\ln 2} \sum_{n=1}^{\infty} \dfrac{1}{n}$ is divergent by Theorem (3.6.1). So the given series also diverges. If $\dfrac{\ln n}{n^\alpha} \to A$ for some $\alpha$ and $A > 0$, then the given series would be comparable to $\displaystyle\sum_{n=2}^{\infty} \dfrac{1}{n^{\alpha+1}}$ by Corollary (3.9.2). But the latter is divergent (Exercise (7.9)), a contradiction.

**6.** The first part was established in the proof of Theorem (8.2). For the second part, a comparison between $H_n - \ln n$ and $H_{n+1} - \ln(n+1)$ reduces to the comparison between $\ln(\dfrac{n+1}{n})$ and $\dfrac{1}{n+1}$. Expanding $\ln(1 + \dfrac{1}{n})$ by Exercise (7.12) as $\dfrac{1}{n} - \dfrac{1}{2n^2} + \dfrac{1}{3n^3} - \dfrac{1}{4n^4} + \ldots$, noting that $\dfrac{1}{n} - \dfrac{1}{n+1} = \dfrac{1}{n(n+1)} \geq \dfrac{1}{2n^2} > \dfrac{1}{2n^2} - (\dfrac{1}{3n^3} - \dfrac{1}{4n^4}) - (\dfrac{1}{5n^5} - \dfrac{1}{6n^6}) \ldots$ we get $\ln(\dfrac{n+1}{n}) > \dfrac{1}{n+1}$ and hence $H_n - \ln n > H_{n+1} - \ln(n+1)$. For the remark, $H_{n-1} - \ln n < \gamma < H_n - \ln n$ for all $n$.

7. The equation of the tangent at $M = (\frac{2k+1}{2}, \ln(\frac{2k+1}{2}))$ is $y - \ln(\frac{2k+1}{2}) = \frac{2}{2k+1}(x - \frac{2k+1}{2})$. So $S$ is $(k, \ln(\frac{2k+1}{2}) - \frac{1}{2k+1})$ and hence $PS$ equals $\ln \frac{2k+1}{2k} - \frac{1}{2k+1}$. Similarly $QR$ equals $\frac{1}{2k+1} + \ln(\frac{2k+1}{2k+2})$. The perpendicular distance between the parallel sides $PS$ and $QR$ is 1.

8. Directly substitute and simplify.

9. From Wallis formula (exercise (6.12)), we have

$$\frac{\pi}{2} = \lim_{n \to \infty} \frac{2.2}{1.3} \cdot \frac{4.4}{3.5} \cdots \frac{2n \cdot 2n}{(2n-1)(2n+1)}$$

$$= \lim_{n \to \infty} \frac{4^n (n!)^2}{1^2 \cdot 3^2 \cdot 5^2 \cdots (2n-1)^2 (2n+1)}$$

$$= \lim_{n \to \infty} \frac{4^n (n!)^2 4^n (n!)^2}{((2n)!)^2 (2n+1)} = \lim_{n \to \infty} \frac{1}{2n+1} \left[ \frac{4^n (n!)^2}{(2n)!} \right]^2$$

$$= \lim_{n \to \infty} \frac{n\lambda_n^4}{2\lambda_{2n}^2 (2n+1)}.$$

For the last part note that as $n \to \infty$, $\lambda_n$ as well as $\lambda_{2n}$ tend to $\lambda$ while $\frac{n}{2n+1} \to \frac{1}{2}$.

10. $\binom{100}{30} = \frac{100!}{70! 30!} \simeq \frac{100^{100}}{e^{100}} \cdot \frac{e^{70}}{70^{70}} \cdot \frac{e^{30}}{30^{30}} \cdot \frac{\sqrt{200\pi}}{\sqrt{140\pi}\sqrt{60\pi}} = \frac{10^{100}}{70^{70} 30^{30}} \frac{1}{\sqrt{42\pi}}$.
Taking logarithms and anti-logarithms (with base 10), this comes out to be $2.9464591 \times 10^{25}$.

11. The first part is straight computation. For the second part, let $a_n = \left( \frac{1.3.5 \ldots (2n-1)}{2.4.6 \ldots 2n} \right)^p$ and $b_n = \left( \frac{(2n)^{2n} \sqrt{4\pi n} e^n e^n}{e^{2n} n^n n^n 2\pi n 4^n} \right)^p = \frac{1}{\pi^{p/2} n^{p/2}}$. By Stirling's approximation we get that $\lim_{n \to \infty} a_n / b_n = 1$ and so by Theorem (3.9.2), $\sum_{n=1}^{\infty} a_n$ behaves the same way as $\sum_{n=1}^{\infty} b_n$ and hence as $\sum_{n=1}^{\infty} \frac{1}{n^{p/2}}$. By Exercise (7.9), we see that the series converges for $p > 2$ and diverges for $p \leq 2$. (It is easy to show that $\lim_{n \to \infty} \frac{a_n}{a_{n-1}} = 1$ and $\lim_{n \to \infty} n(1 - \frac{a_n}{a_{n-1}}) = \frac{p}{2}$. So the ratio test fails for all $p > 0$ and Raabe's test works for $p \neq 2$ but fails for $p = 2$. The reason for superiority of the Stirling approximation is, of course, that while the ratio and the Raabe's test are based on the size of $a_n$ *relative to* that of $a_{n-1}$, the Stirling approximation tells you something about its absolute size. Stirling's approximation can, of course, be used for $p \neq 2$ also. But these cases do not need such a powerful tool.)

12. Measure time $t$ (in years) starting from the beginning of 1950. Let $x = f(t), y = g(t)$ be the population of $A, B$ respectively at time $t$. The data is $\dfrac{dx}{dt} = 0.03x, \dfrac{dy}{dt} = 0.04y$ and $f(0) = 2g(0)$. Solving, $x = f(t) = f(0)e^{0.03t}$ and $y = g(t) = g(0)e^{0.04t}$. So $\dfrac{x}{y} = \dfrac{f(t)}{g(t)} = 2e^{-0.01t}$. The populations will be equal when $e^{0.01t} = 2$, i.e. when $t = 100\ln 2 \simeq 69.31471$, i.e. sometime during 2019, (to be more precise, sometime during April, 2019).

14. Solving, $x(t) = x(0)e^{-\lambda t}$. So $x(t) = \dfrac{1}{2}x(0)$ when $t = \dfrac{\ln 2}{\lambda}$. This is independent of $x(0)$.

15. Let $t =$ time since death, $x = f(t) =$ amount of $C_{14}$ at time $t$ and $y = g(t) =$ amount of $C_{12} = k$ (a constant). Then $\dfrac{dx}{dt} = e^{-\lambda x}$ where $\lambda = \dfrac{\ln 2}{T}$ by the last exercise. Also $x(0) = \alpha k$. So $x = f(t) = \alpha k e^{-(t\ln 2)/T} \cdot f(t) = \beta k$ gives $\dfrac{\beta}{\alpha} = e^{-(t\ln 2)/T}$, i.e. $t = \dfrac{\ln \alpha - \ln \beta}{\ln 2}T$.

16. Let $x = x(t) =$ population at time $t$ and $y_n =$ annual food production in the $n^{th}$ year. Choose suitable units of population and food so that $x(0) = 1$ and $y_0 = 0.9$. Then $x(t) = e^{0.03t}$ for $t \geq 0$, while $y_n = 1.04y_{n-1}$ for all integers $n \geq 1$. So, by induction, $y_n = (0.9)(1.04)^n$. For self-sufficiency, $y_n \geq x(n)$, i.e., $(1.04)^n \geq \dfrac{10}{9}e^{.03n}$. Thus $n$ is the smallest integer $\geq \dfrac{\ln 10 - \ln 9}{\ln(1.04) - 0.03} \simeq 11.426504$. So $n = 12$.

17. If one rupee is invested, then the total at the end of the $n^{th}$ year is $(1 + \dfrac{\alpha}{100})^n$. This equals 2 when $n = \dfrac{\ln 2}{\ln(1 + \frac{\alpha}{100})}$. By Exercise (7.12), $\ln(1 + \dfrac{\alpha}{100})$ equals $\dfrac{\alpha}{100} - \dfrac{\alpha^2}{2.100^2} + \dfrac{\alpha^3}{3.100^3} \ldots$ if $\alpha \leq 100$. For small $\alpha$, the terms from the second onwards are very small and hence $\ln(1 + \dfrac{\alpha}{100}) \simeq \dfrac{\alpha}{100}$. So the amount is doubled when $n = \dfrac{\ln 2}{\frac{\alpha}{100}} = \dfrac{100\ln 2}{\alpha}$. The validity of the rule is based on the fact that $100\ln 2 \simeq 69.31471 \simeq 70$.

18. For the first part it suffices to show that $f(x) = x\ln(1 + \dfrac{1}{x})$ is monotonically increasing on $(0, \infty)$. A straight-forward calculation gives, $f'(x) = \ln(1 + \dfrac{1}{x}) - \dfrac{1}{x+1}$ and $f''(x) = \dfrac{-1}{x(x+1)^2}$. So $f'$ is strictly decreasing on $(0, \infty)$. But $\lim\limits_{x \to \infty} f'(x) = 0$. So $f'(x) > 0$ for all $x > 0$. By a similar calculation, $(x + 1)\ln(1 + \dfrac{1}{x})$ and hence $(1 + \dfrac{1}{x})^{x+1}$ is decreasing on $(0, \infty)$.

19. Trivial for $x = 0$. Otherwise, by Lagrange's MVT (applied to the function $\ln(1 + t)$), get $c$ so that $\ln(1 + x) = \dfrac{x}{1 + c}$. Note that $|c| < |x| \leq \dfrac{1}{2}$ implies

$|1 + c| \geq \frac{1}{2}$. For the second part, the inequality $\ln(1 + b_n)| \leq 2|b_n|$ holds for all $n \geq m$, (say), and so by comparison test, the series $\sum_{n=m}^{\infty} \ln(1 + b_n)$ converges, implying the convergence of the product $\prod_{n=m}^{\infty}(1 + b_n)$ and hence also of the product $\prod_{n=1}^{\infty}(1 + b_n)$. The base of the loarithm does not matter since a change of base results in multiplying every term of the series of logarithms by a non-zero constant.

## Section 5.9

1. With the substitution in the hint, the integral in (19) becomes $\int_0^h u(h-u)du$.
   For the second integral, put $x = u + x_i$ to reduce the integral to
   $$\int_{-h}^{h}(u + h)u^2(h - u)du = \int_{-h}^{h} h^2u^2 - u^4 du = 2\left(\frac{h^5}{3} - \frac{h^5}{5}\right) = \frac{4h^5}{15}.$$

2. When $f$ is concave upwards, for each $i$, the chord joining $(x_{i-1}, y_{i-1})$ to $(x_i, y_i)$ lies above the graph of $f$ on $[x_{i-1}, x_{i+1}]$. So $I_i \leq J_i$ and hence $I - T_n = \sum_{i=1}^{n}(I_i - J_i) \leq 0$. Similarly $I - T_n \geq 0$ if $f$ is concave downwards.

3. The determinant of the coefficients of the system of the three equations
   $Ax_k^2 + Bx_k + C = y_k, k = i - 1, i, i + 1$ is $\begin{vmatrix} x_{i-1}^2 & x_{i-1} & 1 \\ x_i^2 & x_i & 1 \\ x_{i+1}^2 & x_{i+1} & 1 \end{vmatrix}$, which (either
   by subtracting rows or by brute force expansion) comes out as
   $(x_i - x_{i-1})(x_{i+1} - x_{i-1})(x_i - x_{i+1})$ and hence is non-zero since the $x_i$'s are distinct. By Cramer's rule, there is a unique solution for $A, B, C$.

4. Here we need to solve a system of $k + 1$ linear equations in $k + 1$ unknowns. The determinant of the coefficients is

   $$\begin{vmatrix} x_0^k & x_0^{k-1} & \cdots & x_0 & 1 \\ x_1^k & x_1^{k-1} & \cdots & x_1 & 1 \\ \vdots & \vdots & \vdots & \vdots & \vdots \\ x_k^k & x_k^{k-1} & \cdots & x_k & 1 \end{vmatrix}$$

   This is a well-known determinant called the **Vandermonde determinant**. By induction on $k$ (along with some clever manipulations with rows and columns) its value can be shown to be $(-1)^{k+1} \prod(x_j - x_i)$ where the product runs over all pairs of indices $i, j$ with $0 \leq i < j \leq k$. In particular, it is non-zero when all $x_i$'s are distinct. This consequence (which is, by far, the most frequent reason why these determinants are used) can also be proved by a slick, inductive argument, without actually computing the determinant. Treat the determinant as a polynomial in the variable $x_0$ (with $x_1, \ldots, x_k$ constants). It is of degree $k$ since the coefficient of $x^k$ is a Vandermonde determinant of order $k - 1$, which is non-zero by the induction hypothesis. This polynomial already has $x_1, \ldots, x_k$ as roots and hence cannot vanish for any other value of $x_0$. The second part of the question follows by observing that $h_i(x_j) = 1$ if $j = 1$ and $= 0$ otherwise. Uniqueness of $h(x)$ follows from the fact if $k(x)$ is another such polynomial then $h(x) - k(x)$ is a polynomial of degree $\leq k$ with more than $k$ roots viz., $x_0, \ldots, x_k$. So $h(x) - k(x) = 0$.

**5.** Let $G_i(x)$ be as in Lemma (9.6). Then both $f, G_i$ are polynomials of degree 3 (or lower) with values $y_k$ at $x_k$ for $k = i - 1, i$ and $i + 1$. So $f(x) = A(x - x_{i-1})(x - x_i)(x - x_{i+1})$ and $G_i(x) = B(x - x_{i-1})(x - x_i)(x - x_{i+1})$ for some constants $A, B$. $f'(x_i) = G_i'(x_i)$ gives $-Ah^2 = -Bh^2$, i.e., $A = B$.

Hence $f(x) = G_i(x)$ for all $x \in [x_{i-1}, x_{i+1}]$. So $\int\limits_{x_{i-1}}^{x_{i+1}} f(x)dx = \int\limits_{x_{i-1}}^{x_{i+1}} G_i(x)dx$

which, after adding over all odd $i$'s from 1 to $n - 1$ gives $\int\limits_a^b f(x)dx = S_n$.

**6.** If $x = x_{i-1}, x_i$ or $x_{i+1}$, then (22) holds with any $c$. For any other $x \in [x_{i-1}, x_{i+1}]$ define a function $K : [x_{i-1}, x_{i+1}] \longrightarrow \mathbb{R}$ by $K(y) = f(y) - g_i(y) - (f(x) - g_i(x)) \dfrac{(y - x_{i-1})(y - x_i)(y - x_{i+1})}{(x - x_{i-1})(x - x_i)(x - x_{i+1})}$. Then $K$ vanishes at $x_{i-1}, x_i, x_{i+1}$ and also at $x$. So $K'''(c) = 0$ for some $c \in (x_{i-1}, x_{i+1})$. But

$K'''(y) = f'''(y) - (f(x) - g_i(x)) \dfrac{6}{(x - x_{i-1})(x - x_i)(x - x_{i+1})}$ which implies

(22). So $|I_i - J_i| \leq \int\limits_{x_{i-1}}^{x_{i+1}} |f(x) - g_i(x)|dx \leq \dfrac{M}{6} \int\limits_{x_{i-1}}^{x_{i+1}} |(x - x_{i-1})(x - x_i)(x -$

$x_{i+1})|dx = \dfrac{M}{6} \int\limits_{-h}^{h} |u(h^2 - u^2)|du$ (where $u = x - x_i$) $= \dfrac{M}{6} 2 \int\limits_0^h h^2 u - u^3 du =$

$\dfrac{M}{3}(\dfrac{h^4}{2} - \dfrac{h^4}{4}) = \dfrac{Mh^4}{12}$. Hence $|I - S_n| \leq \sum\limits_{\substack{i=1 \\ i \text{ odd}}}^{n} |I_i - J_i| \leq \dfrac{Mh^4}{12} \cdot \dfrac{n}{2} =$

$\dfrac{M}{24}(b - a)h^3$.

**7.** Let $h = \dfrac{b - a}{n}$ and divide $[a, b]$ into $n$ equal parts by $x_1, \ldots, x_{n-1}$. Note that here $L, R, M$ are *not* computed w.r.t. the partition $(x_0, x_1, x_2, x_3, \ldots, x_{n-2}, x_{n-1}, x_n)$ but w.r.t. the coarser partition $(x_0, x_2, x_4, \ldots, x_{n-2}, x_n)$. Their values are, respectively, $2h \sum\limits_{\substack{i=0 \\ i \text{ even}}}^{n-2} y_i, 2h \sum\limits_{\substack{i=2 \\ i \text{ even}}}^{n} y_i$ and $2h \sum\limits_{\substack{i=1 \\ i \text{ odd}}}^{n-1} y_i$. So $L + R + 4M$ equals $6S_n$ by direct summation. Each $L, R, M$ is a Riemann sum of $f$ over $[a, b]$ based on the partition $(x_0, x_2, x_4, \ldots, x_{n-2}, x_n)$ whose mesh is $\dfrac{2(b - a)}{n}$.

By Corollary (2.4), each $L, R, M$ tends to $\int\limits_a^b f(x)dx$ as $n \to \infty$ through even

values. Hence $\dfrac{1}{6}L + \dfrac{1}{6}R + \dfrac{2}{3}M \to \int\limits_a^b f(x)dx$ since $\dfrac{1}{6} + \dfrac{1}{6} + \dfrac{2}{3} = 1$.

8.  The values, respectively, are $L_n(=$ left-end point rule$) = 0.718771, R_n = 0.668771, T_n = 0.693771$ and $S_n = 0.69315$.

9.  $f^{(4)}(x) = \dfrac{24(1 - 10x^2 + 5x^4)}{(1 + x^2)^5} = 24h(u)$ where $h(u) = \dfrac{1 - 10u + 5u^2}{(1 + u)^5}$ and
    $u = x^2$. To maximise $|h(u)|$ on $[0, 1]$, note that $|h(u)|$ is not differentiable
    where $h(u) = 0$, but clearly the maximum cannot occur at such points.
    The only other critical point in $[0, 1]$ is $\dfrac{1}{3}$ (obtained by setting $h'(u) =$
    $\dfrac{50u - 15u^2 - 15}{(1 + u)^6}$ equal to 0). Comparison of $|h(0)|, |h(\frac{1}{3})|$ and $|h(1)|$ gives
    $|h(0)|$ i.e. 24 as the maximum of $|h(u)|$ for $0 \le u \le 1$. So $|f^{(4)}(x)| \le 24 \cdot \dfrac{27}{64} \le$
    24, for all $x \in [0, 1]$.

## Section 5.10

1.  (a) follows directly from the definition. (b) is false as it is, but true in a
    modified form viz., $\triangle(a_n b_n) = (\triangle a_n)b_{n+1} + a_n \triangle b_n$. Similarly $\triangle(\dfrac{a_n}{b_n}) =$
    $\dfrac{(\triangle a_n)b_n - a_n \triangle b_n}{b_n b_{n+1}}$.

2.  (i) The case $k = 1$ is just the definition. For $k > 1$, the inductive hy-
    pothesis and the linearity of $\triangle$ (i.e. part (a) of the last exercise) imply,
    $$\triangle^{k+1}a_n = \triangle(\triangle^k a_n) = \sum_{i=0}^{k}(-1)^{k-i}\binom{k}{i}\triangle a_{n+i} = \sum_{i=0}^{k}(-1)^{k-i}\binom{k}{i}(a_{n+i+1} -$$
    $$a_{n+i}) = a_{n+k+1} + \sum_{i=0}^{k-1}(-1)^{k-i}\binom{k}{i}a_{n+i+1} - \sum_{i=1}^{k}(-1)^{k-i}\binom{k}{i}a_{n+i} - (-1)^k a_n.$$
    With a change of index, the second summation is $\sum\limits_{i=0}^{k-1}(-1)^{k-i}\binom{k}{i+1}a_{n+i+1}$.
    Using the identity $\binom{k}{i} + \binom{k}{i+1} = \binom{k+1}{i+1}$ and with a change of index again,
    the two sums add up to $\sum\limits_{i=1}^{k}(-1)^{k+1-i}\binom{k+1}{i}a_{n+i}$. Adding the two remain-
    ing terms, this becomes $\sum\limits_{i=0}^{k+1}(-1)^{k+1-i}\binom{k+1}{i}a_{n+i}$. For (ii) the method is
    similar and the calcuations easier. $a_{n+1}$ is merely $a_n + \triangle a_n$, which by
    the inductive hypothesis equals $\sum\limits_{k=0}^{n}\binom{n}{k}\triangle k a_0 + \sum\limits_{k=0}^{n}\binom{n}{k}\triangle^{k+1}a_0$, where $\triangle^0$
    is the indentity operator i.e. $\triangle^0 a_i = a_i$ for all $i$. The first sum equals
    $a_0 + \sum\limits_{k=1}^{n}\binom{n}{k}\triangle^k a_0$ while the second equals $\sum\limits_{k=0}^{n-1}\binom{n}{k}\triangle^{k+1}a_0 + \triangle^{n+1}a_0$ and

hence $\sum_{k=1}^{n} \binom{n}{k-1} \Delta^k a_0 + \Delta^{n+1} a_0$. Adding and using $\binom{n}{k} + \binom{n}{k-1} = \binom{n+1}{k}$, the

terms give $\sum_{k=0}^{n+1} \binom{n+1}{k} \Delta^k a_0$.

Slightly more sophisticated proofs for both (i) and (ii) can be given using the terminology of operators. Just as differentiation is an operator $D$ acting on the set of functions (see Section 4.4), $\Delta$ can be thought of as an operator acting on the set of sequences. If $\{a_n\}_{n=0}^{\infty}$ is any sequence, then $\{\Delta a_n\}$ is the sequence $\{a_{n+1} - a_n\}_{n=0}^{\infty}$. Let $I$ be the identity operator, i.e., $\{I a_n\}$ is the sequence $\{a_n\}$ itself. In addition let $E$ be operator defined by $\{E a_n\} = \{a_{n+1}\}_{n=0}^{\infty}$. Quite appropriately, $E$ is called the right- shift operator. Note that $E a_0 = a_1, E^2 a_0 = a_2, \ldots, E^n a_0 = a_n$. We also have the operator equation $E = I + \Delta$ (which simply means $a_{n+1} = a_n + (a_{n+1} - a_n)$). Since $I$ and $\Delta$ commute with each other (i.e. $I(\Delta a_n) = \Delta(I a_n)$), $(I + \Delta)^n$ can be expanded by the binomial theorem to give (ii). Also writing $\Delta$ as $E - I$ and noting that $E$ and $I$ commute with each other, the binomial expansion of $(E - I)^n$ gives (i).

3.  The first part is trivial. For the direct implication in the second part, apply induction on $k$. If $f(n)$ is a polynomial of degree $k-1$ (or less) then $\Delta f(n) = f(n+1) - f(n)$ is a polynomial of degree $k-2$ (or less) and so by the induction hypothesis, $\Delta^{k-1}(\Delta f(n)) \equiv 0$, whence $\Delta^k(f(n)) \equiv 0$. For the converse, note that $\Delta^k a_n \equiv 0$ implies $\Delta^r a_0 = 0$ for all $r \geq k$. So by (ii) of the last exercise,

$a_n = \sum_{r=0}^{k-1} \binom{n}{r} \Delta^r a_0$, which is a polynomial in $n$ of degree $k - 1$ (or less).

4.  Modify the proof of Theorem (10.3) by showing that after we get $\left| \sum_{n=p}^{q} a_n b_n \right|$

$\leq M(|b_q| + |b_p| + \sum_{n=p}^{q-1} |b_n - b_{n+1}|)$, the hypothesis implies that if $p, q$ are

sufficiently large then $|b_q|, |b_p|$ and $\sum_{n=p}^{q} |b_n - b_{n+1}|$ can all be made small.

5.  Add and subtract $A_{p-1} b_q$ to the R.H.S. of (1) to get $\left| \sum_{n=p}^{q} a_n b_n \right| \leq$

$|A_q - A_{p-1}||b_q| + |A_{p-1}||b_q - b_p| + \left| \sum_{n=p}^{q-1} A_n(b_{n+1} - b_n) \right|$. By monotonicity

of $\{b_n\}$, $b_{n+1} - b_n$ has the same sign for all $n$. Convergence of $\sum_{n=1}^{\infty} a_n$ and of

$\{b_n\}$ implies $|A_q - A_{p-1}|$ and $|b_q - b_p|$ can be made arbitrarily small. The rest of the proof is similar to that of Theorem (10.3).

6.  (a) The function $f$ (whose graph is shown in the figure on the next page) is a step function on $[0, n]$ with the non-zero value $k$ on the interval $[k - \dfrac{1}{2k^3}, k + \dfrac{1}{2k^3}]$ of length $\dfrac{1}{k^3}$ if $1 \leq k < n$ and value $n$ on the

interval $[n - \dfrac{1}{2k^3}, n]$ whose length is $\dfrac{1}{2n^3}$. Apply Exercise (3.8) to get

the first part. If $b \leq n$, then $F(b) = \int\limits_0^b f(x)dx \leq \sum\limits_{k=1}^n \dfrac{1}{k^2} - \dfrac{1}{2n^2}$, which is

bounded since the series $\sum\limits_{n=1}^\infty \dfrac{1}{n^2}$ is convergent. Also $f(x) \geq 0$ for all $x$

implies $F(b)$ is monotonically increasing. So $\lim\limits_{b \to \infty} \int\limits_0^b f(x)dx$ exists (and

in fact equals the sum of the series $\sum\limits_{n=1}^\infty \dfrac{1}{n^2}$, which, by the non-rigorous

argument in Exercise (1.4.9), equals $\dfrac{\pi^2}{6}$).

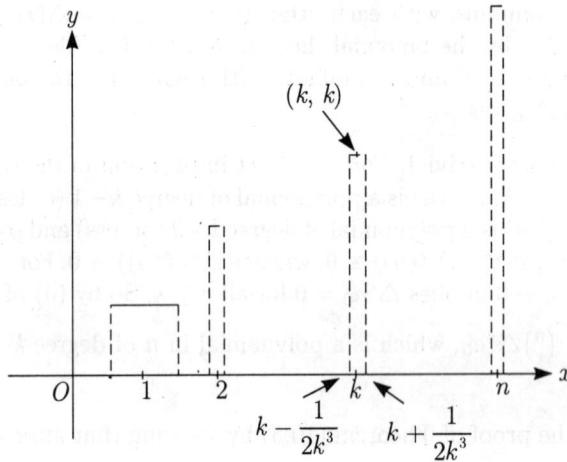

(b) Remove the jumps in the graph of the function $f$ by replacing vertical
segments (which are not really parts of the graph) by steep line segments (which are to become parts of the graph). Thus for example,
join $(n - \dfrac{1}{2n^3}, 0)$ to $(n - \dfrac{1}{4n^3}, n)$ and $(n + \dfrac{1}{4n^3}, n)$ to $(n + \dfrac{1}{2n^3}, 0)$ so
as to get the graph of a function $g$. Note that $0 \leq g(x) \leq f(x)$ for all
$x \geq 0$. So $\int\limits_0^\infty g(x)dx$ exists.

7. Given $\epsilon > 0$, find $R$ as in the hypothesis. Let $p \in \mathbb{N}$ be such that $a + p \geq R$.
Then whenever $p \leq m \leq n, |I_n - I_m| = |\int\limits_{a+m}^{a+n} f(x)dx| \leq \int\limits_{a+m}^{a+n} |f(x)|dx < \epsilon$.
So $\{I_n\}$ is a Cauchy sequence.

8. (i) follows from linearity for integrals over $[a, b]$ for every $b \geq a$ and then
taking limits as $b \to \infty$. For (ii) apply Theorem (10.5) after noting that
$|\int\limits_{b_1}^{b_2} f(x)dx| \leq \int\limits_{b_1}^{b_2} |f(x)|dx$ for all $a \leq b_1 \leq b_2$. For (iii) also, use Theorem

(10.5) after noting that for all $a \le b_1 \le b_2$, $|\int_{b_1}^{b_2} f(x)dx| = \int_{b_1}^{b_2} f(x)dx \le$

$\int_{b_1}^{b_2} g(x)dx = |\int_{b_1}^{b_2} g(x)dx|$. Finally, for (iv), suppose $\lim_{x\to\infty} \dfrac{f(x)}{g(x)} = L > 0$.

Fix $\delta > 0$ and $R_1 \ge a$ such that $L - \delta > 0$ and $L - \delta \le \dfrac{f(x)}{g(x)} \le L + \delta$ for all

$x \ge R_1$. Then $0 \le (L-\delta)g(x) \le f(x) \le (L+\delta)g(x)$ and so by (iii), $\int_{R_1}^{\infty} f(x)dx$

and $\int_{R_1}^{\infty} g(x)dx$ either both converge or both diverge. Since $\int_{a}^{R_1} f(x)dx$ and

$\int_{a}^{R_1} g(x)dx$ are finite real numbers, $\int_{a}^{\infty} f(x)dx$ and $\int_{a}^{\infty} g(x)dx$ both converge or
both diverge.

**9.** By Exercise (3.8), for $n \in \mathbb{N}$, $\int_{1}^{n+1} f(x)dx = S_n$, a partial sum of $\sum\limits_{n=1}^{\infty} a_n$. So

convergence of $\int_{1}^{\infty} f(x)dx$ implies that of $\sum\limits_{n=1}^{\infty} a_n$. For the converse, note that

for $n \in \mathbb{N}$ and $b \in [n, n+1)$, $|\int_{1}^{b} f(x)dx - S_{n-1}| = |(b-n)a_n| \le |a_n| \to 0$ as

$n \to \infty$. So $\int_{1}^{b} f(x)dx \to \sum\limits_{n=1}^{\infty} a_n$ as $b \to \infty$.

**10.** Let $a_n = \dfrac{(-1)^n}{n} + \dfrac{1}{n \ln n}$ and $b_n = \dfrac{(-1)^n}{n}$ for $n \ge 2$. Then $\sum\limits_{n=2}^{\infty} b_n$ is conver-

gent by Exercise (3.8.1), but $\sum\limits_{n=2}^{\infty} \dfrac{1}{n \ln n}$ and hence $\sum\limits_{n=1}^{\infty} a_n$ is divergent (cf.

Exercise (8.5)). Still $\dfrac{a_n}{b_n} = 1 + \dfrac{(-1)^n}{\ln n} \to 1$ as $n \to \infty$.

**11.** For the first part note that $\sin x$ is positive on $(n\pi, (n+1)\pi)$ if $n$ is even

and negative if $n$ is odd. Also $\dfrac{|\sin(\pi + x)|}{\pi + x} = \dfrac{|\sin x|}{\pi + x} \le \dfrac{|\sin x|}{x}$, for all $x \in$

$[n\pi, (n+1)\pi]$ which implies $|I_{n+1}| \le |I_n|$. Further $|\dfrac{\sin x}{x}| \le \dfrac{|\sin x|}{n\pi}$ for all $x \in$

$[n\pi, (n+1)\pi]$ and so $|I_n| = \int_{n\pi}^{(n+1)\pi} |\dfrac{\sin x}{x}|dx \le \dfrac{1}{n\pi} \int_{n\pi}^{(n+1)\pi} |\sin x|dx = \dfrac{2}{n\pi} \to 0$

as $n \to \infty$. By the Leibnitz test (Exercise (3.8.1)), $\sum\limits_{n=1}^{\infty} I_n$ is convergent and

hence $\int_{\pi}^{b} \dfrac{\sin x}{x}dx$ tends to a finite limit as $b \to \infty$ through integral multiples

of $\pi$. To show that it also tends to the same limit as $b \to \infty$ (through real
values) note that if $b \in [n\pi, (n+1)\pi)$, then, because of the inequalities

above, $|\int\limits_{\pi}^{b}\frac{\sin x}{x}dx - \int\limits_{\pi}^{n\pi}\frac{\sin x}{x}dx| = \int\limits_{n\pi}^{b}|\frac{\sin x}{x}|dx \leq \frac{b-n\pi}{n\pi} < \frac{\pi}{n\pi} = \frac{1}{n} \to 0$

as $n \to \infty$. Hence $\int\limits_{\pi}^{\infty}\frac{\sin x dx}{x}$ converges. For the shorter proof, $\int\limits_{\pi}^{b}\frac{\sin x}{x}dx =$

$\frac{-\cos x}{x}|_{\pi}^{b} - \int\limits_{\pi}^{b}\frac{\cos x}{x^2}dx = -\frac{1}{\pi} - \frac{\cos b}{b} - \int\limits_{\pi}^{b}\frac{\cos x}{x^2}dx$. As $b \to \infty$, $\frac{\cos b}{b} \to 0$. Also

$\int\limits_{\pi}^{b}|\frac{\cos x}{x^2}|dx \leq \int\limits_{\pi}^{b}\frac{dx}{x^2} = \frac{1}{\pi} - \frac{1}{b} \to \frac{1}{\pi}$ as $b \to \infty$. So $\int\limits_{1}^{\infty}\frac{\cos x}{x^2}dx$ is absolutely

convergent and hence convergent.

**12.** $\int\limits_{\pi}^{n\pi}|\frac{\sin x}{x}|dx = \sum\limits_{k=1}^{n-1}\int\limits_{k\pi}^{(k+1)\pi}|\frac{\sin x}{x}|dx = \sum\limits_{k=1}^{n-1}|I_k|$. For $x \in [k\pi, (k+1)\pi]$, $|\frac{\sin x}{x}| \geq$

$\frac{|\sin x|}{(k+1)\pi}$ and hence $|I_k| \geq \frac{1}{(k+1)\pi}|\int\limits_{k\pi}^{(k+1)\pi}\sin x dx| = \frac{2}{(k+1)\pi}$. The result

follows from the divergence of the series $\sum\limits_{n=1}^{\infty}\frac{2}{(n+1)\pi}$.

**13.** Write $\int\limits_{0}^{\infty}\frac{\sin x}{x}dx$ as $\int\limits_{0}^{\pi}\frac{\sin x}{x}dx + \int\limits_{\pi}^{\infty}\frac{\sin x}{x}dx$. Since $\frac{\sin x}{x} \to 1$ as $x \to 0^+$

the first integral is not improper and represents a fixed finite number. So $\int\limits_{0}^{\infty}\frac{\sin x}{x}dx$ behaves the same way as $\int\limits_{\pi}^{\infty}\frac{\sin x}{x}dx$, studied in the last two exercises.

**14.** $f(n) = n$ for all $n$ while $f(n + \frac{1}{2}) = 0$ for all $n > 1$.

**15.** $|f(n^2)| = \frac{1}{n^2}$ while $|f(n)| = \frac{1}{n}$ for all $n$. Apply Theorems (3.6.3) and (3.6.1).

**16.** Let $1 \leq a < b \leq 2$, and $M > 0$ be given. Choose $r \in \mathbb{N}$ such that $r > M$. The interval $(a, b)$ contains infinitely many rationals. Of these only finitely many have reduced forms $\frac{p}{q}$ with $q \leq r$. Let $x$ be any rational in $(a, b)$ which is not one of these. Then $x$, in its reduced form is $\frac{p}{q}$ for some $p, q$ with $q > r$. So $f(x) = q > M$.

**17.** The analgoue of Theorem (10.5) would read "$\int\limits_{a}^{b}f(x)dx$ exists if and only if

for every $\epsilon > 0$, $\exists \delta > 0$ such that whenever $0 < \delta_1 < \delta_2 < \delta$, $|\int\limits_{a+\delta_1}^{a+\delta_2}f(x)dx| < \epsilon$". The direct implication is easy. For the converse, begin by showing

that the sequence $\{\int\limits_{a+\frac{1}{n}}^{b}f(x)dx\}$ is a Cauchy sequence. The proofs of the

analogues of the various parts of Theorem (10.6) are the same except for marginal changes. (i) follows from linearity for integrals over $[a + \delta, b]$ by taking limits as $\delta \to 0^+$. For (ii) and (iii) use the analogue of Theorem (10.5) just proved. In (iv) the hypothesis is that $f(x) \geq 0, g(x) \geq 0$ for all $x \in (a, b]$ and $\dfrac{f(x)}{g(x)}$ tends to a finite non-zero limit, say $L$, as $x \to a^+$. Fix $\epsilon > 0$ and $\delta_0 > 0$ such that $L - \epsilon > 0$ and $\dfrac{f(x)}{g(x)} \in (L - \epsilon, L + \epsilon)$ whenever $a < x < a + \delta_0$. Then apply (iii) (i.e. its analogue) to the inequalities $(L - \epsilon)g(x) < f(x)$ and $f(x) < (L + \epsilon)g(x)$.

**18.** Fix any $b >$ all real roots (if any) of $p(x)$. Then for $x > \max\{a, b\}, p(x), q(x)$ maintain their signs and we may suppose both are positive (as otherwise replace $p(x)$ by $-p(x)$ etc.). Let $f(x) = \dfrac{p(x)}{q(x)}$ and $g(x) = \dfrac{1}{x^{n-m}}$. Then $\dfrac{f(x)}{g(x)} \to \dfrac{a_m}{b_n} > 0$ where $a_m, b_n$ are the leading coefficients of $p(x), q(x)$ respectively (cf. Exercise (3.3.1), except that here $x \to \infty$ through all real values). So by Theorem (10.6) (iii), $\int\limits_a^\infty f(x)dx$ converges iff $\int\limits_a^\infty g(x)dx$ does which is the case iff $n - m \geq 2$ by Theorem (10.7). (Note that $m, n$ are integers.)

**19.** The substitution $u = \ln x$ converts $\int\limits_2^\infty \dfrac{dx}{x(\ln x)^p}$ to $\int\limits_{\ln 2}^\infty \dfrac{du}{u^p}$ which is convergent iff $p > 1$. By Theorem (10.8) (or rather, its analgoue for $[2, \infty)$ instead of $[1, \infty)$), the series $\sum\limits_{n=2}^\infty \dfrac{1}{n(\ln n)^p}$ is convergent iff $p > 1$. (We need to verify that $\dfrac{1}{n(\ln n)^p}$ is monotonically decreasing. This is certainly true if $p \geq 0$. If $p < 0$ write $p = -k$ with $k > 0$. Then $\dfrac{1}{n(\ln n)^p} = \dfrac{(\ln n)^k}{n} \geq \dfrac{1}{n}$ for all $n \geq 3$ and so the divergence of $\sum\limits_{n=2}^\infty \dfrac{1}{n}$ implies that of $\sum\limits_{n=2}^\infty \dfrac{(\ln n)^k}{n}$.)

**20.** (i) absolutely convergent ($e^x \geq x^9$ eventually) (ii) divergent (iii) divergent (near $0$, $\dfrac{\cos x}{x}$ is comparable to $\dfrac{1}{x}$) (iv) convergent but not absolutely convergent (similar to $\int\limits_\pi^\infty \dfrac{\sin x}{x}dx$) (v) reducible to (iv) with $u = x^2$ (vi) divergent (vii) divergent (compare with $\int\limits_1^\infty \dfrac{dx}{x}$) (viii) absolutely convergent (put $x = \dfrac{1}{u}$ and compare with $\int\limits_1^\infty \dfrac{du}{u^2}$) (ix) absolutely convergent. (Split as $\int\limits_0^1 \dfrac{dx}{(1+x)\sqrt{x}}$ and $\int\limits_1^\infty \dfrac{dx}{(1+x)\sqrt{x}}$. Compare the first with $\int\limits_0^1 \dfrac{dx}{\sqrt{x}}$ and the second with $\int\limits_1^\infty \dfrac{dx}{x\sqrt{x}}$.)

**21.** For the integral in the hint, using the substitution $\cot x = u$, $\int\limits_0^\pi \dfrac{dx}{1 + A^2 \sin^2 x}$

$= \int\limits_{-\infty}^{\infty} \dfrac{du}{u^2 + 1 + A^2} = \dfrac{1}{\sqrt{1 + A^2}} \tan^{-1}\Big(\dfrac{u}{\sqrt{1 + A^2}}\Big)\Big|_{-\infty}^{\infty} = \dfrac{\pi}{\sqrt{1 + A^2}}$. In the

given integrals, the integrands are non-negative and so they converge or diverge according to the infinite series $\sum\limits_{n=0}^{\infty} I_n$ and $\sum\limits_{n=0}^{\infty} J_n$ respectively converge

or diverge, where $I_n = \int\limits_{n\pi}^{(n+1)\pi} \dfrac{dx}{1 + x^4 \sin^2 x}$ and $J_n = \int\limits_{n\pi}^{(n+1)\pi} \dfrac{dx}{1 + x^2 \sin^2 x}$

for $n = 0, 1, 2, \dots$. Now $I_n = \int\limits_0^\pi \dfrac{du}{1 + (n\pi + u)^4 \sin^2 u} \le \int\limits_0^\pi \dfrac{du}{1 + (n\pi)^4 \sin^2 u} =$

$\dfrac{\pi}{\sqrt{1 + (n\pi)^4}}$ by the result in the hint. Similarly, $J_n \ge \dfrac{\pi}{\sqrt{1 + (n+1)^2\pi^2}}$.

Comparison with the convergent series $\sum\limits_{n=1}^{\infty} \dfrac{1}{n^2}$ shows that $\sum\limits_{n=1}^{\infty} \dfrac{1}{\sqrt{1 + (n\pi)^4}}$

converges while comparison with $\sum\limits_{n=1}^{\infty} \dfrac{1}{n}$ shows that $\sum\limits_{n=1}^{\infty} \dfrac{1}{\sqrt{1 + (n+1)^2\pi^2}}$

diverges.

**22.** Split the integral as $\int\limits_0^1 e^{-x} x^{\alpha-1} dx + \int\limits_1^\infty e^{-x} x^{\alpha-1} dx$. The former compares with

$\int\limits_0^1 \dfrac{dx}{x^{1-\alpha}}$ and converges if $1 - \alpha < 1$, i.e., $\alpha > 0$. The latter converges for all

$\alpha$ (cf. (i) in Exercise (10.20)).

**23.** For (i), integrate by parts. Repeated application of (i) along with the direct

calculation of $\Gamma(1)$ as 1 gives (ii). For (iii) convert $\int\limits_0^\infty e^{-x^2} dx$ to $\int\limits_0^\infty e^{-u} \dfrac{du}{2\sqrt{u}} =$

$\dfrac{1}{2} \int\limits_0^\infty e^{-u} u^{\frac{1}{2}-1} du = \dfrac{1}{2}\Gamma(\dfrac{1}{2})$.

**24.** (i) Follow the hint to reduce the integral to $\int\limits_0^1 \dfrac{du}{\sqrt{u}\sqrt{2 - u^2}}$. $\lim\limits_{u\to 0+} \dfrac{1}{\sqrt{2 - u^2}} =$

$\dfrac{1}{\sqrt{2}} > 0$. So the integral behaves like $\int\limits_0^1 \dfrac{du}{u^{1/2}}$, which is convergent. The

second part follows by observing that $\sqrt{1 - y^2}$ is an even function of $y$.

(ii) By (i), $F(1) = \dfrac{1}{2} \int\limits_{-1}^1 \dfrac{dy}{\sqrt{1 - y^2}} = \dfrac{\pi}{2}$. Also $F(-y) = -F(y)$ for all $y \in [0, 1]$.

So $F(-1) = -\dfrac{\pi}{2}$. $F$ is differentiable on $(-1, 1)$ with $F'(y) = \dfrac{1}{\sqrt{1 - y^2}}$ for

$y \in (-1, 1)$ by Theorem (5.3). Continuity of $F$ at 1 and $-1$ follows from

the very definition of the improper integrals $\int\limits_0^1 \dfrac{dy}{\sqrt{1 - y^2}}$ and $\int\limits_{-1}^0 \dfrac{dy}{\sqrt{1 - y^2}}$.

By Theorem (4.5.5), $F$ is strictly increasing on $[-1, 1]$. By the Intermediate

Value Property, $F$ maps $[-1, 1]$ onto $[F(-1), F(1)]$, i.e. onto $[-\frac{\pi}{2}, \frac{\pi}{2}]$.

(iii) Apply Exercise (4.1.9) and the corresponding properties of $F$.

(iv) By Exercise (4.4.8), differentiablity of $\sin x$ on $(-\frac{\pi}{2}, \frac{\pi}{2})$ follows from that of $F$ on $(-1, 1)$. Moreover, $\frac{d}{dx}(\sin x) = \frac{1}{F'(\sin x)} = \sqrt{1 - \sin^2 x}$.

For $x \in (\frac{\pi}{2}, \frac{3\pi}{2})$, $\frac{d}{dx}(\sin x) = -\sqrt{1 - \sin^2(x - \pi)} = -\sqrt{1 - \sin^2 x}$. Hence by periodicity, $\frac{d}{dx}(\sin x) = \sqrt{1 - \sin^2 x}$ if $x \in (n\pi - \frac{\pi}{2}, n\pi + \frac{\pi}{2})$ for even $n$ while $\frac{d}{dx}(\sin x) = -\sqrt{1 - \sin^2 x}$ if $x \in (n\pi - \frac{\pi}{2}, n\pi + \frac{\pi}{2})$ for odd $n$. For differentiability at $\frac{\pi}{2}$, use Exercise (4.5.7) and its analogue for left-handed derivatives.

Thus $\left[\frac{d}{dx}(\sin x)\right]_{x = \frac{\pi}{2}-} = \lim_{h \to 0+} \sqrt{1 - \sin^2(\frac{\pi}{2} - h)} = \sqrt{1 - 1} = 0$. Similarly,

$\left[\frac{d}{dx}(\sin x)\right]_{x = \frac{\pi}{2}+} = \lim_{h \to 0+} -\sqrt{1 - \sin^2(\frac{\pi}{2} + h)} = -\sqrt{1 - 1} = 0$. By a similar reasoning, $\frac{d}{dx}(\sin x)$ exists and equals 0 at $x = -\frac{\pi}{2}$. By periodicity, this holds for other odd multiples of $\frac{\pi}{2}$.

(v) follows from (iv) (and the continuity of the square-root function).

(vi) Since $\sqrt{x}$ is differentiable w.r.t. $x$ for $x > 0$ with derivative $\frac{1}{2\sqrt{x}}$, the first part follows from (iv) and (v) except when $\sin x = \pm 1$, i.e., when $x$ is an odd multiple of $\frac{\pi}{2}$. At such points, apply Exercise (4.5.7) and continuity of the sine function, the same way as for (iv). For example, $\left[\frac{d}{dx}(\cos x)\right]_{x = \frac{\pi}{2}+} = \lim_{h \to 0} -\sin(\frac{\pi}{2} + h) = -\sin(\frac{\pi}{2})$.

(vii) Following the hint and using (v) and (vi), $f'(x) = -\sin(x + y) + \sin x \cos y + \cos x \sin y$ and $f''(x) = -\cos(x + y) + \cos x \cos y - \sin x \sin y$. So $f''(x) + f(x) = 0$. Hence $\frac{d}{dx}([f'(x)]^2 + [f(x)]^2) = 2f'(x)[f''(x) + f(x)] = 0$. By Theorem (4.5.1), $[f'(x)]^2 + [f(x)]^2 = [f'(0)]^2 + [f(0)]^2 = 0 + 0 = 0$. This forces both $f'(x)$ and $f(x)$ to vanish identically.

(viii) Use (vii) and the facts that $\sin(\frac{\pi}{2}) = 1, \cos(\frac{\pi}{2}) = \sqrt{1 - \sin^2 \frac{\pi}{2}} = 0$.

Similarly, $\sin(\frac{\pi}{2} + x) = \cos x$. Repeated applications give $\sin(x + \pi) = -\sin x$ and $\cos(x + \pi) = -\cos x$ which also follow from (iv).

# CHAPTER 6

## Section 6.1

1. Let $m_1, m_2$ be the number of matches played by $A$, with batting averages $a_1, a_2$ in the two halves. Let $n_1, n_2, b_1, b_2$ be the corresponding figures for $B$. Then their season's averages, say $a$ and $b$, are $\dfrac{m_1 a_2 + m_2 a_2}{m_1 + m_2}$ and $\dfrac{n_1 b_1 + n_2 b_2}{n_1 + n_2}$ respectively. These are points which divide the intervals $[a_1, a_2]$ and $[b_1, b_2]$ in certain ratios. Choose $a_1, a_2, b_1, b_2$ and points $a \in [a_1, a_2], b \in [b_1, b_2]$ so that $b_1 < a_1, b_2 < a_2$ but $b > a$, e.g. $b_1 = 10, b_2 = 50, a_1 = 15, a_2 = 65, m_1 = 4, m_2 = 1, n_1 = 1$ and $n_2 = 1$. Then $a = 25$ and $b = 30$.

2. (i) $\dfrac{a}{2} + b$, attained at $\dfrac{1}{2}$  (ii) $\dfrac{1}{3}$, attained at $\dfrac{1}{\sqrt{3}}$  (iii) $\dfrac{1}{4}$, attained at $\dfrac{1}{\sqrt[3]{4}}$

   (iv) $\dfrac{1}{6}$, attained at $\dfrac{1}{2} \pm \dfrac{1}{2\sqrt{3}}$.

3. The first part follows by the linearity of integrals (Exercise (5.1.9)). For the second part, the last exercise provides a counter-example if we take $f(x) = x$ and $g(x) = x^2$ on $[0, 1]$.

4. For $i = 1, \ldots, n, \int_{x_{i-1}}^{x_i} f(x)dx = M_i \Delta x_i$, where $\Delta x_i = (x_i - x_{i-1})$. By Exercise (5.1.7), $\dfrac{1}{b-a} \int_a^b f(x)dx = \sum\limits_{i=1}^{n} \dfrac{\Delta x_i M_i}{b-a}$ which is a weighted average since $\sum\limits_{i=1}^{n} \Delta x_i = b - a$.

5. Let $F(x) = \int_a^x f(t)dt$. Then by Theorems (5.5.3) and (4.5.4), $\dfrac{1}{b-a} \int_a^b f(x)dx$
   $= \dfrac{1}{b-a}(F(b) - F(a)) = F'(c) = f(c)$ for some $c \in (a, b)$.

6. Take any example in Exercise (1.2) and change the value of $f$ at the point where the average is attained. By Theorem (5.3.1), this does not change the integral $\int_0^1 f(x)dx$ and hence not the average either.

7. This time, let $F(x) = \int_a^x f(t)g(t)dt$ and $G(x) = \int_a^x g(t)dt$. By Theorem (5.5.3) and Exercise (4.5.11), there exists some $c \in (a, b)$ such that
   $$\dfrac{\int_a^b f(x)g(x)dx}{\int_a^b g(x)dx} = \dfrac{F'(c)}{G'(c)} = \dfrac{f(c)g(c)}{g(c)} = f(c).$$

8. $\int_0^3 f(x)g(x)dx = \int_0^2 -xdx + \int_2^3 xdx = -2 + \dfrac{5}{2} = \dfrac{1}{2}$ and $\int_0^3 g(x)dx = 2 - 1 = 1$. So the average of $f$ over $[0, 3]$ w.r.t. the weight function $g$ is $\frac{1}{2}$. But there is no point $c$ in $[0, 3]$ such that $f(c) = \frac{1}{2}$.

**9.** The mean values are (i) $\dfrac{2}{3}$ (ii) $\dfrac{1}{3}$ and (iii) $\dfrac{1}{2}$. As $f(x) = x$, these are also the points where they are attained.

**10.** If $f, g$ are integrable over $[a, b]$, $a = x_0 < x_1 < \ldots < x_n = b$ is a partition of $[a, b]$ and $\int_{x_{i-1}}^{x_i} g(x)dx > 0$ for $i = 1, \ldots n$, then the average of $f$ on $[a, b]$ w.r.t. $g$ is the weighted average of the averages of $f$ over $[x_{i-1}, x_i]$ (w.r.t. $g$) with weights $\int_{x_{i-1}}^{x_i} f(x)dx$, $i = 1, \ldots, n$. The proof is an application of Exercise (5.1.7) to $\int_a^b f(x)g(x)dx$.

**11.** Follow the hint. The function $H$ is continuous by Theorem (5.3.3). Also $H(a) = f(a)\int_a^b g(x)dx$ and $H(b) = f(b)\int_a^b g(x)dx$. Non-negativity of $g$, coupled with $f(a) \le f(x) \le f(b)$ for all $x$ gives $H(a) \le \int_a^b f(x)g(x)dx \le H(b)$. So there is some $c \in [a, b]$ such that $H(c) = \int_a^b f(x)g(x)dx$. For the second assertion, note that $W_1 = \int_a^c g(x)dx$, and $W_2 = \int_a^b g(x)dx$. The result follows by dividing the earlier part throughout by $W_1 + W_2$.

**12.** $\int_0^L \rho(x)\sigma(x)dx / \int_0^L \rho(x)dx$.

**13.** This is a paraphrase of Problem (1.6), if we identify a head with a boy and a tail with a girl.

**14.** Let $a_n$ be the number of winning sequences of length $n$. Then $a_1 = 0, a_2 = 1$. For $n > 2$, classify such sequences according to their first entry. There are $a_{n-1}$ among them which begin with a $T$. If the first entry is $H$ then the second one must be $T$ and the number of such sequences is $a_{n-2}$. So $a_n = a_{n-1} + a_{n-2}$. Hence $a_n =$ the $(n-1)^{\text{th}}$ Fibonacci number $F_{n-1}$, which by (2) in Section (3.10) equals $\dfrac{1}{\sqrt{5}}(\alpha^{n-1} - \beta^{n-1})$ where $\alpha = \dfrac{1 + \sqrt{5}}{2}$ and $\beta = \dfrac{1 - \sqrt{5}}{2}$. Since $p_n = \dfrac{a_n}{2^n}$, the average cost of winning the game is $\displaystyle\sum_{n=1}^{\infty} \dfrac{1}{2\sqrt{5}}((\dfrac{\alpha}{2})^{n-1} - (\dfrac{\beta}{2})^{n-1})$ which can be evaluated using $\displaystyle\sum_{n=1}^{\infty} nx^{n-1} = \dfrac{1}{(1-x)^2}$. (This is also the first solution to Exercise (3.10.6).) For the second solution, let $E$ be the expected cost of winning. The first toss gives a $T$ with probability $\dfrac{1}{2}$ and the first two tosses are $HH$ or $HT$ with probability $\dfrac{1}{4}$ each. So, analogously to (7), $E = \dfrac{1}{2}(E+1) + \dfrac{1}{4}2 + \dfrac{1}{4}(E+2)$ giving $E = 6$.

**15.** To wind up even in the longer run, the amount of the reward should match the average cost of winning, say $E$. Following the second method of solution to Problem (1.6), we get $E = \dfrac{1}{8}3 + \dfrac{1}{8}(E+3) + \dfrac{1}{4}(E+2) + \dfrac{1}{2}(E+1)$, the terms corresponding to the initial tosses $HHH, HHT, HT$ and $T$ respectively. Solving, $E = 14$.

**16.** By the hypothesis, the averages of $f$ over the intervals $[a - h, a], [a, a + h]$ and $[a - h, a + h]$ are $f(a - \frac{h}{2}), f(a + \frac{h}{2})$ and $f(a)$ respectively. Exercise (1.4) gives the equality in the hint, which by symmetry also holds for $h < 0$. Differentiation w.r.t. $h$ gives $\frac{1}{2} f'(a + \frac{h}{2}) - \frac{1}{2} f'(a - \frac{h}{2}) = 0$, or $f'(a + \frac{h}{2}) = f'(a - \frac{h}{2})$. For any two $x, y$ with $x < y$ (say), set $a = \dfrac{x + y}{2}$ and $h = y - x$ to get $f'(x) = f'(y)$. Hence $f'$ is constant, say $A$. Then $f(x) = Ax + B$ for some constant $B$.

**17.** (i) $\frac{1}{1} \int_0^1 y \, dy = \frac{1}{2}$ (ii) $\dfrac{\int_0^{\sqrt{2}} x\sqrt{4 - x^2} \, dx}{2\sqrt{2}} = \dfrac{2\sqrt{2} - 1}{3}$.

**18.** Let $[a, b] = [c, d] = [0, 1], h(t) = t^2$ and $f(x) = 1$. Then the average value of $f$ is 1 while that of $f \circ h$ is $\frac{1}{3}$.

**19.** This is essentially a duplication of Exercise (4.10.4), parts (i) and (iii).

**20.** The first part is a duplication of Exercise (4.10.4), part (ii). The answers to the second part are (i) A (ii) the centroid (i.e. the intersection of medians) (iii) the point which divides $CA$ in the ratio $\alpha : (1 - \alpha)$ (iv) the incentre (v) the orthocentre.

**21.** The number of heads ($= k$, say) can vary from 0 to $n$. $P(k = r)$ is $\binom{n}{r} p^r (1 - p)^{n-r}$. Hence the average value of $k$ is $\sum\limits_{r=0}^{n} \binom{n}{r} p^r (1 - p)^{n-r} r = (1 - p)^n \sum\limits_{r=0}^{n} r \binom{n}{r} \left(\dfrac{p}{1 - p}\right)^r$. Differentiating $(1 + x)^n = \sum\limits_{r=0}^{n} \binom{n}{r} x^r$ and multiplying by $x, nx(1 + x)^{n-1} = \sum\limits_{r=0}^{n} r \binom{n}{r} x^r$. Hence the average value of $k$ is $(1 - p)^n \dfrac{np}{1 - p} \left(1 + \dfrac{p}{1 - p}\right)^{n-1} = np$.

**22.**

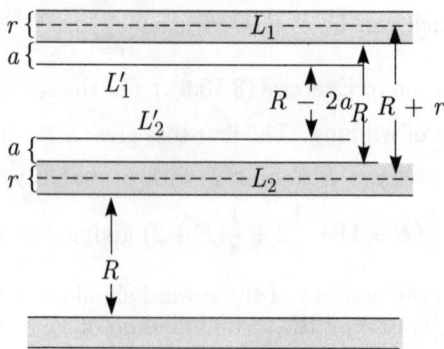

Because of the largeness of the window and the randomness of the throw, we may assume that the centre, say $P$, of the ball lies randomly between the middle lines, say $L_1$ and $L_2$ of any two consecutive strips. The distance between $L_1$ and $L_2$ is $R + r$. The ball will pass through iff $P$ lies between the lines $L_1'$ and $L_2'$ which are at distances $a + \dfrac{r}{2}$ from $L_1, L_2$ respectively.

As the distance between $L_1'$ and $L_2'$ is $R - 2a$, the probability, say $p$, of the ball passing through, is $\dfrac{R - 2a}{R + r}$ (or 0 if $R < 2a$). Using the same reasoning as in the last exercie, the expected number of times the ball will pass through in $n$ throws is $np$, i.e. $\dfrac{n(R - 2a)}{R + r}$.

**23.** Let $P = (x_0, x_1, \ldots, x_n)$ be any partition of $[a, b]$. Set $\alpha_i = \dfrac{x_i - x_{i-1}}{b - a}$ for $i = 1, 2, \ldots, n$. Then for any $\boldsymbol{\xi} = (\xi_1, \xi_2, \ldots, \xi_n)$ based on $P$, Exercise (4.10.6) gives $f(\dfrac{1}{b - a} R(P, \boldsymbol{\xi} ; \phi)) \leq \dfrac{1}{b - a} R(P, \boldsymbol{\xi} ; f \circ \phi)$. Taking limits as $\mu(P) \to 0$, the result follows from continuity of $f$ and of $\phi$.

## Section 6.2

**1.** (i) and (iv) are additive. In (ii) and (v) the quantities involved are proportional, respectively, to $\sqrt{L}$ and $L, L$ being the relevant length. So (ii) is not additive while (v) is. In (iii), the telephone charge is usually not additive as a function of the distance between the points of the call. As a function of the duration of the call, it is additive.

**2.** Parametrise the graph by $x = t, y = f(t), a \leq t \leq b$ and apply Theorem (2.4).

**3.** From Exercise (4.6.9), $\dfrac{d}{dx}(\sinh x) = \cosh x$. It is also easy to see from the definitions (cf. Exercise (3.3.14)) that $\sqrt{1 + \sinh^2 x} = \cosh x$. Let $A = (x_1, y_1), B = (x_2, y_2)$ with $x_1 < x_2$. Then the length of the portion of the catenary $y = a \cosh(\dfrac{x}{a})$ is, by the last exercise, $\displaystyle\int_{x_1}^{x_2} \sqrt{1 + \sinh^2 \dfrac{x}{a}} \, dx$ which comes to $\displaystyle\int_{x_1}^{x_2} \cosh \dfrac{x}{a} \, dx = a \sinh \dfrac{x}{a} \Big|_{x_1}^{x_2} = a \sinh \dfrac{x_2}{a} - a \sinh \dfrac{x_1}{a}$.

**4.** $a \int_0^{2\pi} \sqrt{(1 - \cos\theta)^2 + \sin^2\theta} \, d\theta = a \int_0^{2\pi} 2 \sin\dfrac{\theta}{2} \, d\theta = 8a$.

**5.** $(R + a) \displaystyle\int_0^{2\pi a/R} \sqrt{2 - 2\cos\dfrac{R}{a}\theta} \, d\theta = \dfrac{8a(R + a)}{R}$. As $R \to \infty$, this tends to $8a$ which is consistent with the last exercise, because as $R \to \infty$, the fixed circle approaches a straight line and so an epicycloid tends to a cycloid.

**6.** (i) $\int_0^1 \sqrt{1 + t^2 + 2t} \, dt = \int_0^1 (1 + t) \, dt = \dfrac{3}{2}$. (To find $\dot{x}$, it is easier to put $t = \tan\theta$ and use the chain rule.) (ii) $\int_0^{\pi/4} \sqrt{1 + \cos 2t} \, dt = \int_0^{\pi/4} \sqrt{2} \cos t \, dt = 1$.

**7.**

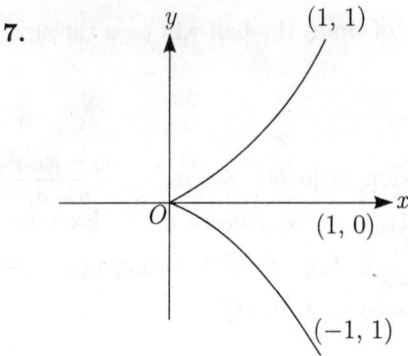

The curve can be prametrised by $x = y^{2/3}, y = y; -1 \le y \le 1$. Its length is $\int\limits_{-1}^{1} \sqrt{\frac{4}{9}y^{-2/3} + 1}\,dy$. This is an improper integral and has to be evaluated by splitting at 0 and then using some substitution such as $\frac{2}{3}y^{-1/3} = \tan\theta$. An easier approach is to note the symmetry of the curve w.r.t. the $x$-axis.

The upper portion is $y = x^{3/2}; 0 \le x \le 1$ and by Exercise (6.2), its length is $\int\limits_{0}^{1} \sqrt{1 + \frac{9}{4}x}\,dx = \frac{8}{27}(1 + \frac{9}{4}x)^{3/2}\Big|_{0}^{1} = \frac{13\sqrt{13} - 8}{27}$.

**8.** To prove the desired analogue of Exercise (5.6.5) (iii), use the fact that $|\sqrt{x} - \sqrt{y}| \le \sqrt{|x - y|}$ (cf. Exercise (4.2.9))and the traingle inequality to get that for each $i = 1, \ldots, n$,

$|\sqrt{[u(\xi_i)]^2 + [v(\eta_i)]^2 + [w(\zeta_i)]^2} - \sqrt{[u(\xi_i)]^2 + [v(\xi_i)]^2 + [w(\xi_i)]^2}|$

$\le |\sqrt{|[v(\eta_i)]^2 - [v(\xi_i)]^2| + |[w(\zeta_i)]^2 - [w(\xi_i)]^2|}|$

$\le \sqrt{2M}\sqrt{|v(\eta_i) - v(\xi_i)| + |w(\zeta_i) - w(\xi_i)|}$ where $M$ is any common upper bound on $|v(t)|, |w(t)|, t \in [a, b]$. By uniform continuity of $v$ and $w$ the difference between the Riemann sum $\sum\limits_{i=1}^{n} \sqrt{[u(\xi_i)]^2 + [v(\xi_i)]^2 + [w(\xi_i)]^2}\Delta t_i$ and the given sum can be made arbitrarily small if the mesh of the partition $(t_0, t_1, \ldots, t_n)$ is sufficiently small.

**9.** $2\pi\sqrt{2}$ ; $\pi\sqrt{4\pi^2 + 2} + \ln(\sqrt{2}\pi + \sqrt{1 + 2\pi^2})$.

**10.** There exist partitions, say, $P = (u_0, \ldots, u_m)$ and $Q = (s_0, s_1, \ldots, s_p)$ of $[a, b]$ such that $f$ is continuously differentiable on $[u_{j-1}, u_j]$ for $j = 1, \ldots, m$ and $g$ is continuously differentiable on $[s_{k-1}, s_k]$ for every $k = 1, \ldots, p$. Obtain $R = (t_0, t_1, \ldots t_n)$ by superimposing $P$ and $Q$. Then Theorem (2.4) applies to $[t_i - 1, t_i]$ for each $i = 1, 2, \ldots, n$ and the additivity of the arc length gives the result.

## Section 6.3

**1.** Take a typical partition of $R = R(a, b; c, d)$ into $mn$ rectangles by lines of the form $x = x_i$ and $y = y_j$ where $a = x_0 < x_1 < x_2 < \ldots < x_{m-1} < x_m = b$ and $c \le y_0 < y_1 < \ldots < y_n = d$. Let $R_{ij} = R(x_{i-1}, x_i; y_{j-1}, y_j)$ for $i = 1, \ldots, m; j = 1, 2, \ldots, n$. Then $\sum\limits_{(i,j)} A(R_{ij}) = \sum\limits_{(i,j)} [(x_i - x_{i-1})(y_j - y_{j-1})] = $

$(\sum\limits_{i=1}^{m}(x_i - x_{i-1}))(\sum\limits_{j=1}^{n}(y_j - y_{j-1})) = (b - a)(d - c)$. Since all such sums equal $A(R)$, their supremum is $A(R)$. In the case of a box $S$, chop it by planes of the form $x = x_i, y = y_j$ and $z = z_k$. The triply indexed sum $\sum\limits_{(i,j,k)} V(S_{ijk})$ equals $V(S)$ for every such chopping.

2. Given mutually disjoint subsets $R_1$ and $R_2$, and a chopping by horizontal and vertical lines, there will in general be many rectangles which are contained neither in $R_1$ nor in $R_2$ but are contained in $R_1 \cup R_2$. These rectangles contribute to $A(R_1 \cup R_2)$ but not to $A(R_1)$ or to $A(R_2)$. So, showing that $A(R_1 \cup R_2) \leq A(R_1) + A(R_2)$ is not easy. It would require a careful estimate of the contribution of such troublesome rectangles and a proof that this contribution can be made arbitrarily small if the chopping is sufficiently fine.

3. The congruence in the hint follows by noting that for $y \in \mathbb{R}, (c, y) \in R$ iff $(c + \alpha, y + \beta) \in R'$. So $l(c) = l'(c + \alpha)$ where $l(c)$ is the length of the intersection of $R$ with $x = c$ and $l'(c + \alpha)$ is the length of the intersection of $R'$ with $x = c + \alpha$. Hence $\int_a^b b(x)dx = \int_{a+\alpha}^{b+\alpha} l'(x)dx$. For the unbounded case chop $R$ by lines of the form $x = m, y = n$ $(m, n \in \mathbb{Z})$ and $R'$ by lines of the form $x = m + \alpha, y = n + \beta$ $(m, n\mathbb{Z})$.

4.

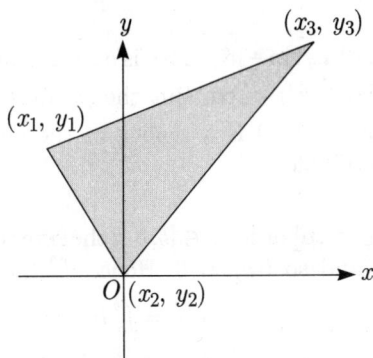

Write down the equations of the sides of the triangle. Then its area is

$\int_{x_1}^0 |y_1 + \dfrac{y_3 - y_1}{x_3 - x_1}(x - x_1) - \dfrac{y_1 x}{x_1}|dx + \int_0^{x_3} |y_1 + \dfrac{y_3 - y_1}{x_3 - x_1}(x - x_1) - \dfrac{y_3 x}{x_3}|dx$ with the understanding that if $x_1 = 0$, the first integral is 0 while if $x_3 = 0$ then the second one is 0. The expressions inside the absolute value signs maintain their signs over the intervals of integration.

Keeping in mind that $x_1 \leq 0$ and $x_3 \geq 0$, the two integrals come out to be, respectively, $\dfrac{-x_1|x_1 y_3 - x_3 y_1|}{2(x_3 - x_1)}$ and $\dfrac{x_3|x_1 y_3 - x_3 y_1|}{2(x_3 - x_1)}$. Hence the area is $\frac{1}{2}|x_1 y_3 - x_3 y_1|$. For a general triangle, this becomes $\frac{1}{2}|(x_1 - x_2)(y_3 - y_2) - (x_3 - x_2)(y_1 - y_2)|$ which equals the desired formula as seen by subtracting the second row of the determinant from the other two.

5. In the last exercise, if the triangle is right angled at $(x_2, y_2)$, which may be taken as $(0, 0)$, then $y_1 y_3 + x_1 x_3 = 0$. With a little manipulation, $\frac{1}{2}|x_1 y_3 - x_3 y_1| = \frac{1}{2}\sqrt{(x_1 y_3 - x_3 y_1)^2} = \frac{1}{2}\sqrt{(x_1^2 + y_1^2)(x_3^2 + y_3^2)}$. Thus the area of a right angled triangle is $\frac{1}{2}$ times the product of its shorter sides. Now split

the given rectangle into two right angled triangles and use the additivity of the areas.

**6.**

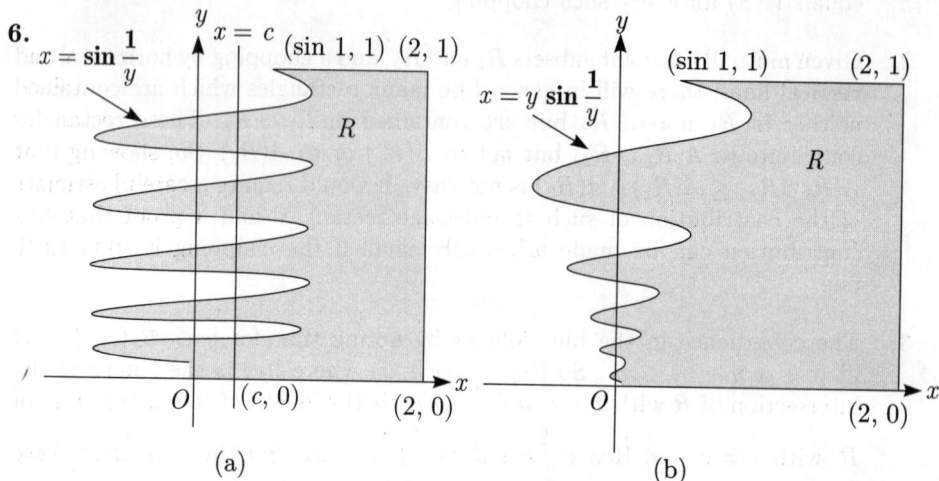

(a)                                                    (b)

(a) This is clear from the sketch, since for every $n \in \mathbb{N}, x = c$ meets the portion of the curve $x = \sin \frac{1}{y}$, $\frac{1}{(n+1)\pi} \leq y \leq \frac{1}{n\pi}$ exactly once. The portions between two successive points of intersections are alternatively inside and outside $R$.

(b) By horizontal slicing, $A(R) = \int_0^1 (2 - y \sin \frac{1}{y}) dy$. This is a proper integral since $y \sin \frac{1}{y} \to 0$ as $y \to 0^+$. So $A(R) < \infty$. But the portion of the boundary of $R$ between $(0,0)$ and $(0, \frac{1}{\pi})$ is a replica of the curve in Exercise (4.7.7) which is not rectifiable.

**7.** Suppose $a' < a < b < b'$. Then for $c \in [a', a]$ or for $c \in [b, b']$, the intersection of $R$ with the line $x = c$ is empty and so $l(c) = 0$. Hence $\int_{a'}^a l(x)dx = \int_b^{b'} l(x)dx = 0$, whence, by Exercise (5.1.7), $\int_{a'}^{b'} l(x)dx = \int_a^b l(x)dx$.

**8.** Merely interchange $x$ with $y$ and $a$ with $b$ in the derivation of (1).

**9.** $R_2$ is the disjoint union of $R_1$ and $(R_2 - R_1)$. So $A(R_2) = A(R_1) + A(R_2 - R_1) \geq A(R_1)$ since $A(R_2 - R_1) \geq 0$. A similar argument applies for volumes.

**10.** Splitting $R$ as in the figure on the next page, $A(R) = A(R_1) + A(R_2) =$
$$\int_0^{r\cos\theta} x \tan\theta dx + \int_{r\cos\theta}^r \sqrt{r^2 - x^2} dx$$
$$= \tan\theta \frac{r^2 \cos^2\theta}{2} + \int_{\frac{\pi}{2}-\theta}^{\pi/2} r^2 \sin^2 u du = \frac{r^2}{2}(\sin\theta\cos\theta + \theta - \sin\theta\cos\theta) = \frac{r^2\theta}{2}.$$
If $\frac{\pi}{2} < \theta \leq 2\pi$, subtract a suitable number of quarters of the circle so that the remaining sector subtends an acute angle at the centre.

**11.** The tangent to $y = \tan x$ at $(\frac{\pi}{4}, 1)$ cuts the $x$-axis at $(\frac{\pi}{4} - \frac{1}{2}, 0)$. The desired
   area is $\displaystyle\int_0^{\pi/4} \tan x \, dx - \int_{\frac{\pi}{4} - \frac{1}{2}}^{\pi/4} (2x + 1 - \frac{\pi}{2}) dx = \frac{1}{2}\ln 2 - \frac{1}{4}$.

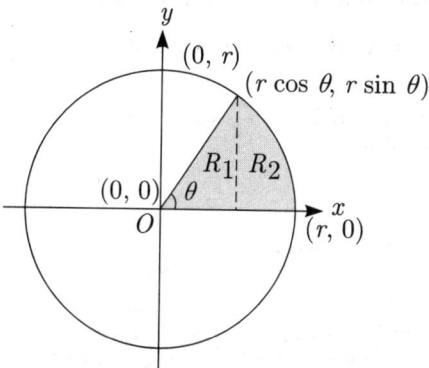

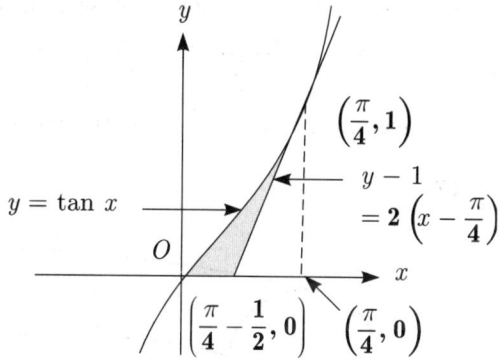

Exercise (3.10)          Exercise (3.11)

**12.** The region $R$ is bounded by four
   parabolas with a common focus
   at $M$, the centre of the square
   and directrices along the sides
   of the square. Let $R_1$ be the
   portion of $R$ bounded by the
   lines $MO, MQ$ and the parabola
   $(x - \frac{1}{2})^2 + (y - \frac{1}{2})^2 = y^2$, i.e. $y =$
   $x^2 - x + \frac{1}{2}$. Then by symmetry,

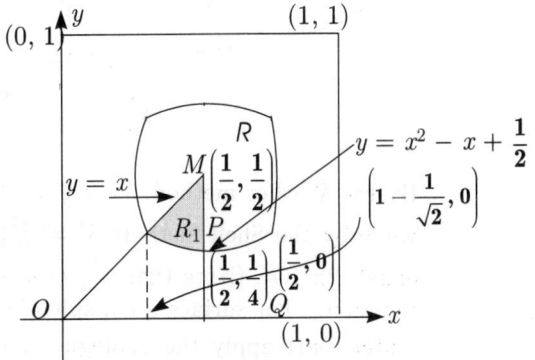

$$A(R) = 8A(R_1) = 8\int_{1 - \frac{1}{\sqrt{2}}}^{1/2} x - (x^2 - x + \frac{1}{2}) dx = \frac{8 - 5\sqrt{2}}{3\sqrt{2}}.$$

**13.** The cross section by a plane of the form $z = k$ (where $0 \le k \le h$) is still a
   square of side $\dfrac{h - k}{h} r$. So (3) and its derivation remain intact. The surface
   area, changes, however and in fact tends to $\infty$ as $|a|$ and/or $|b| \to \infty$.

**14.** Here, regardless of whether the cone is right or not, the cross section by a
   plane parallel to the base is a circle and calculations similar to those leading
   to (5) hold.

**15.** The cross section by a plane parallel to the plane of $R$ (the 'base' plane)
   and at a height $h$ from it $(0 \le k \le h)$, is a figure similar to $R$, the constant of proportionality of the linear dimensions being $\dfrac{h - k}{h}$. So its area
   is $(\dfrac{h - k}{h})^2 A(R)$ and the same proof goes through. (A rigorous calculation
   of the area of the cross section would require that $R$ be sliced. Then the
   cross section is sliced in a corresponding manner, the length as well as the

breadth of each slice being $\dfrac{h-k}{h}$ times that of the corresponding slice of $R$.)

**16.** For $-c \le k \le c$, the cross section of the solid is a region bounded by the ellipse $\dfrac{x^2}{a^2(1-\frac{k^2}{c^2})} + \dfrac{y^2}{b^2(1-\frac{k^2}{c^2})} = 1; z = k$. By (1), the area of this cross section is $\pi ab(1 - \dfrac{k^2}{c^2})$. So the volume is $\displaystyle\int_{-c}^{c} \pi ab(1 - \dfrac{z^2}{c^2})dz = \dfrac{4}{3}\pi abc$.

**17.**

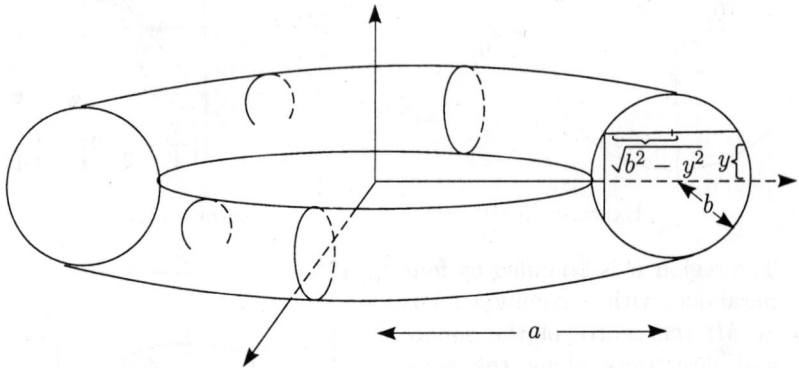

By the Washer method, $V = \int_{-b}^{b} \pi[(a + \sqrt{b^2 - y^2})^2 - (a - \sqrt{b^2 - y^2})^2]dy$ while by the Shell method, $V = \int_{-b}^{b} 2\pi(a + x)2\sqrt{b^2 - x^2}dx$. Each integral equals $2\pi^2 ab^2$. (Note that $\int_{-b}^{b} x\sqrt{b^2 - x^2}dx = 0$ as the integrand is an odd function.) For surface area, split the torus surface into two parts. For the 'outer' part apply the analogue of (14) to get $\int_{-b}^{b} 2\pi f(y)\sqrt{1 + [f'(y)]^2}dy$ where $f(y) = a + \sqrt{b^2 - y^2}$. For the inner part, a similar formula holds with $f(y) = a - \sqrt{b^2 - y^2}$. Since $[f'(y)]^2 = \dfrac{y^2}{b^2 - y^2}$ in both the cases, the total surface area of the torus is $\displaystyle\int_{-b}^{b} 2\pi(2a)\dfrac{b}{\sqrt{b^2 - y^2}}dy = 4\pi^2 ab$.

**18.** Take the base as the disc bounded by $x^2 + y^2 = r^2$ in the $x$-$y$ plane and the fixed diameter along the $x$-axis. Then for $-r \le c \le r$, the cross section of the cake by the plane $x = c$, is an isosceles, right angled triangle of area $\frac{1}{2}(2\sqrt{r^2 - c^2})^2$. So the volume is $\int_{-r}^{r} 2(r^2 - x^2)dx = \frac{8}{3}r^3$.

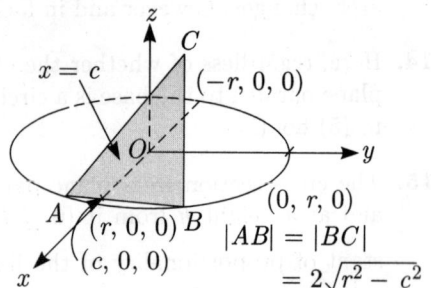

**19.** $R_1$ and $R_1'$ generate the same solid, $S_1$. Similarly $R_2$ and $R_2'$ both generate $S_2$. $R_1$ and $R_2'$ lie on the same side of $L$ and so their intersection, $R_1 \cap R_2'$

generates $S_1 \cap S_2$, which is $S_3$. The solid generated by $R$ is $S_1 \cup S_2$. $S_1 \cup S_2$ is the disjoint union of $(S_1 - S_3), (S_2 - S_3)$ and $S_3$. By additivity of the volume, $V(S_1 \cup S_2) = V(S_1 - S_3) + V(S_2 - S_3) + V(S_3)$. The result follows by noting $V(S_i - S_3) + V(S_3) = V(S_i)$ for $i = 1, 2$. (The result as well as the proof resembles Exercise (3.7.7).)

**20.** $O, P_1, P_2$ are collinear and so are $O, M_1, M_2$. Let $\theta$ be the angle between these two lines. Then $|OP_i| = \sec\theta|OM_i|$ and $|P_iM_i| = \tan\theta|OM_i|$ for $i = 1, 2$. Subtracting the area of the curved surface of the smaller cone from that of the larger one and using (9), the area of the band $= \pi \tan\theta \sec\theta(|OM_1|^2 - |OM_2|^2) = \pi \tan\theta \sec\theta(|OM_1| + |OM_2|)|M_1M_2|$. The R.H.S. of (10) also reduces to this since $|P_iM_i| = \tan\theta|OM_i|$, for $i = 1, 2$ and $|P_1P_2| = |OP_1| - |OP_2| = \sec\theta|OM_1| - \sec\theta|OM_2| = \sec\theta|M_1M_2|$.

**21.** Let $P = (t_0, t_1, \ldots, t_n)$ be a partition of $[a, b]$. Let $P_i = \alpha(t_i)$ be the point on $C$ for $i = 0, i, \ldots, n$. For each $i = 1, \ldots, n$, the area obtained by the portion of $C$ between $P_{i-1}$ and $P_i$ is approximately $2\pi\sigma(t_i)|P_{i-1}P_i|$ which by the Mean Value Theorem equals $2\pi\sigma(t_i)\sqrt{[f'(\xi_i)]^2 + [g'(\eta_i)]^2 + [h'(\zeta_i)]^2}\Delta t_i$ for some $\xi_i, \eta_i, \zeta_i \in [t_{i-1}, t_i]$. By hypothesis, these areas do not overlap and so the total area is approximately $\sum_{i=1}^{n} 2\pi\sigma(t_i)\sqrt{[f'(\xi_i)]^2 + [g'(\xi_i)]^2 + [h'(\eta_i)]^2}\Delta t_i$. By an argument similar to that used in Exercise (2.8), this sum tends to the given integral as $\mu(P) \to 0$.

**22.** The parametrisation of $C$ is $x = a(\theta - \sin\theta), y = a(\cos\theta - 1); 0 \le \theta \le 2\pi$. Because of symmetry about the axis $x = a\pi$, it suffices to revolve only the portion of $C$ for $0 \le \theta \le \pi$. The surface area is $\int_0^\pi 2\pi(a\pi - a\theta + a\sin\theta)\sqrt{(\frac{dx}{d\theta})^2 + (\frac{dy}{d\theta})^2}d\theta$, i.e. $2\pi a^2 \int_0^\pi (\pi - \theta + \sin\theta)2\sin\frac{\theta}{2}d\theta$ which equals $2\pi a^2(4\pi - \frac{16}{3})$. The volume enclosed is obtained by integrating $\pi(a\pi - x)^2$ w.r.t. $y$. In terms of $\theta$, this is $\int_0^\pi \pi(a\pi - a\theta + a\sin\theta)^2 a\sin\theta d\theta$, i.e.

$\pi a^3 \int_0^\pi (\pi - \theta + \sin\theta)^2 \sin\theta d\theta = \pi a^3 \int_0^\pi (\theta + \sin\theta)^2 \sin\theta d\theta$ by Theorem (5.6.1). Using the identities $2\sin^2\theta = 1 - \cos 2\theta$ and $\sin^3\theta = \frac{3}{4}\sin\theta - \frac{1}{4}\sin 3\theta$ and integration by parts this comes out as $\pi a^3(\frac{3\pi^2}{2} - \frac{8}{3})$.

**23.** The hint follows by letting $R$ be the radius of the sphere and solving $(R - h)^2 + r^2 = R^2$. The volume of the cap is $\int_{R-h}^{R} \pi(R^2 - y^2)dy = \pi(Rh^2 - \frac{h^3}{3})$ $= \frac{\pi}{6}(3r^2h + h^3)$. The surface area (of the curved portion) is obtained by the same method as for (15) except that $r$ is to be replaced by $R$ and we integrate w.r.t. $y$, from $y = R - h$ to $y = R$. The answer is $\int_{R-h}^{R} 2\pi R dy = 2\pi Rh$ $= \pi(r^2 + h^2)$.

## Section 6.4

**1.**  $$\iint_R f(x,y)dA = \int_0^2 [\int_0^3 (3x^2y - 2e^x y)dy]dx = \int_0^2 (\frac{3x^2y^2}{2} - e^x y^2)\Big|_{y=0}^{y=3} dx$$
$$= \int_0^2 (\frac{27x^2}{2} - 9e^x)dx = \frac{9}{2}x^3 - 9e^x\Big|_0^2 = 36 - 9e^2 + 9 = 45 - 9e^2.$$

**2.**  $$\int_c^d [\int_a^b y\cos(xy)dx] dy = \int_c^d [\sin(xy)\Big|_{x=a}^{x=b}] dy = \int_c^d \sin(by) - \sin(ay)dy$$
$$= -\frac{\cos(bd)}{b} + \frac{\cos(ad)}{a} + \frac{\cos(bc)}{b} - \frac{\cos(ac)}{a}. \text{ The other iterated integral is}$$
$$\int_a^b \int_c^d y\cos(xy)dydx = \int_a^b [\frac{y}{x}\sin(xy) + \frac{\cos(xy)}{x^2}]\Big|_{y=c}^{y=d} dx \quad \text{which} \quad \text{equals}$$
$$\int_a^b \frac{d\sin(xd)}{x} + \frac{\cos(xd)}{x^2} - \frac{c\sin(xc)}{x} - \frac{\cos(xc)}{x^2}dx. \text{ None of the four terms of}$$
the integrand has an anti-derivative in a closed form. The first two terms
together have  $\dfrac{-\cos(xd)}{x}$  as an antiderivative while the last two terms together have  $\dfrac{\cos(xc)}{x}$  as an antiderivative. So the integral can be evaluated,
but the first method is obviously simpler.

**3.**  For the first part, note that  $\dfrac{1 + \sqrt{1 - y^2}}{1 - \sqrt{1 - y^2}} = \dfrac{(1 + \sqrt{1 - y^2})^2}{1 - (1 - y^2)}$ . The second
part follows from the fact that  $y\ln y \to 0$  as  $y \to 0^+$  by L'Hôpital's rule
(applied to  $\dfrac{\ln y}{1/y}$ ).

**4.**

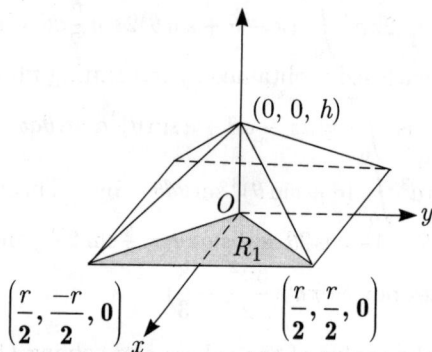

Take the base of the pyramid to be the square bounded by the lines $x = \pm\frac{r}{2}$ and $y = \pm\frac{r}{2}$ in the x-y plane and the vertex to be the point $(0,0,h)$. The four slanting faces of the pyramid project onto four traingles each having $(0,0)$ as a vertex and the other two vertices coming from those of the square. Consider one of these triangles, say $R_1$, with vertices $(0,0)$, $(\frac{r}{2},\frac{r}{2})$ and $(\frac{r}{2},-\frac{r}{2})$.

The equation of the face of the pyramid above it is of the form $ax + by + z = h$ (since $(0,0,h)$ lies on it). Since $(\frac{r}{2},\pm\frac{r}{2},0)$ also lie on this face, $a,b$ can be determined as $a = \frac{2h}{r}, b = 0$. So the portion of the pyramid lying above $R_1$ has volume $\iint_{R_1}(h - \frac{2h}{r}x)dA$. As an iterated in-

tegral, this equals $\int_0^{r/2} [\int_{-x}^x (h - \frac{2h}{r}x)dy]dx = \int_0^{r/2} 2xh - \frac{4hx^2}{r}dx = \frac{hr^2}{12}.$

The other three portions of the pyramid are all congruent to this one and hence (or by a similar calculaiton) also have volume $\dfrac{hr^2}{12}$ each.

**5.** Call the vertices $O, A, B, C$ and suppose the edges meeting at $O$ are mutually perpendicular. Set up a co-ordinate system with $O = (0,0,0)$, $A = (a,0,0)$, $B = (0,b,0)$ and $C = (0,0,c)$. Then the equation of the plane through $A, B, C$ is $\dfrac{x}{a} + \dfrac{y}{b} + \dfrac{z}{c} = 1$, and so the volume of the tetrahedron is $\iint_R c(1 - \dfrac{x}{a} - \dfrac{y}{b})dA$ where $R$ is the triangle in the $x$-$y$ plane with vertices at $(0,0), (a,0)$ and $(0,b)$. This integral equals $\displaystyle\int_0^a [\int_0^{b(1-\frac{x}{a})} c(1 - \dfrac{x}{a} - \dfrac{y}{b})dy]dx$

$= c \displaystyle\int_0^a b(1 - \dfrac{x}{a})^2 - \dfrac{b(1-\frac{x}{a})^2}{2}dx = -\dfrac{bca}{2}\dfrac{1}{3}(1 - \dfrac{x}{a})^3\Big|_0^a = \dfrac{1}{6}abc$. The result follows since the base triangle has area $\frac{1}{2}ab$ and the altitude on it is $c$.

**6.** Clear since $\int_c^d g(x)h(y)dy = g(x)\int_c^d h(y)dy$ for all $x \in [a,b]$.

**7.** $\sin(b+c) - \sin(a+c) - \sin(b+d) + \sin(a+d)$.

**8.** (i) $\displaystyle\int_0^2 [\int_{-y/2}^{1-y} x^2 y \, dx]dy = \int_0^2 \dfrac{y}{3}[(1-y)^3 + \dfrac{y^3}{8}]dy = \dfrac{2}{15}$

(ii) $\displaystyle\int_{-1}^2 [\int_{x^2}^{x+2} (x-y)dy]dx = \int_{-1}^2 x(x+2-x^2) - \dfrac{(x+2)^2}{2} + \dfrac{x^4}{2}dx = -\dfrac{99}{20}$.

**9.** (i) $\displaystyle\int_0^4 [\int_0^{x^2} \sin(x^3)dy]dx = \int_0^4 x^2 \sin(x^3)dx = -\dfrac{1}{3} - \dfrac{\cos(64)}{3}$.

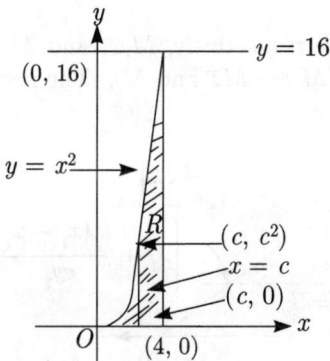

(i)

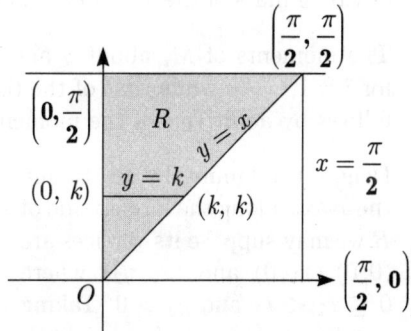

(ii)

(ii) $\displaystyle\int_0^{\pi/2} [\int_0^y \dfrac{\sin y}{y}dx]dy = \int_0^{\pi/2} \dfrac{\sin y}{y} y \, dy = \int_0^{\pi/2} \sin y \, dy = 1$.

**10.** For a discrete mass distribution with masses $m_i$ at $(x_i, y_i), i = 1, \ldots, n$, the moment about $L$ is, because of the hint, $\displaystyle\sum_{i=1}^n \dfrac{m_i(ax_i + by_i + c)}{\sqrt{a^2 + b^2}}$. The assertion follows since $\sum m_i = M, \sum M_i x_i = M_y$ and $\sum M_i y_i = M_x$. For a continuous mass distribution spread out over a region $R$ with mass density $\rho(x,y)$ at $(x,y) \in R$, the moment about $L$ is $\iint_R \dfrac{ax+by+c}{\sqrt{a^2+b^2}} \rho(x,y) dA$ which, by linearity of double integrals, equals the given expression.

11. Let the centre of mass be $(\overline{x}, \overline{y})$ and the equation of the line be $ax+by+c = 0$. Since $M_x = M\overline{y}$ and $M_y = M\overline{x}$, by the last exercise, the moment about the line is $\dfrac{M}{\sqrt{a^2+b^2}}(a\overline{x} + b\overline{y} + c)$ which is 0 iff $(\overline{x}, \overline{y})$ lies on the line.

12.

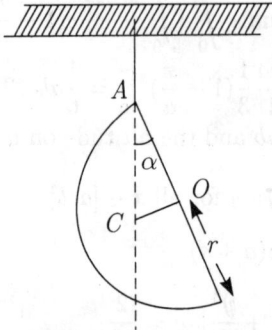

Let $r$ and $O$ be the radius and the centre, $A$ the end of the diameter at which the string is attached and $C$ the centre of mass of the half-disc. The only forces on the half-disc are the tension in the string and its weight acting at $C$. For equilibrium, the moment of mass around the line contianing the string must be 0.

So this line must pass through $C$. By Problem (1.3) (i), $CO = \frac{4r}{3\pi}$. So $\alpha = \tan^{-1}\left(\frac{4}{3\pi}\right)$.

13. Let the co-ordinates of $A, B, C$ be $(x_1, y_1), (x_2, y_2)$ and $(x_3, z_3)$. Then their position vectors (w.r.t. the origin) are also $(x_1, y_1), (x_2, y_2)$ and $(x_3, y_3)$. So $P = (\alpha x_1 + \beta x_2 + \gamma x_3, \alpha y_1 + \beta y_2 + \gamma y_3)$. Since $\alpha + \beta + \gamma = 1$, this is also the centre of mass of the system of masses $\alpha, \beta, \gamma$ placed at $A, B, C$ respectively.

14. The moments of $M_i$ about $y$ and $x$ axis are respectively, $M_i \overline{x_i}$ and $M_i \overline{y_i}$, for $i = 1, \ldots, k$ while that of the total mass $M$ are $M\overline{x}$ and $M\overline{y}$. The result follows by additivity of the moments.

15. Denote the lamina by $R$. Taking the $x$-axis along the largest side of $R$ we may suppose its vertices are $(0,0), (x_1, 0)$ and $(x_2, y_2)$ where $0 \le x_2 < x_1$ and $y_2 > 0$. Taking $\rho \equiv 1$ (without loss of generality), mass of $R$ is $\frac{1}{2} x_1 y_2$. Moreover,

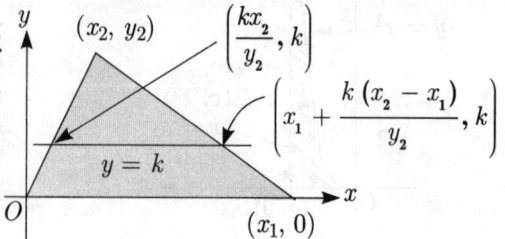

$$M_x = \iint_R y \, dA = \int_0^{y_2} \int_{yx_2/y_2}^{x_1 + y(x_2 - x_1)/y_2} y \, dx \, dy = \int_0^{y_2} \left(x_1 - \frac{yx_1}{y_2}\right) y \, dy = \frac{x_1 y_2^2}{6}.$$

So $\overline{y} = \dfrac{x_1 y_2^2}{6} \dfrac{2}{x_1 y_2} = \dfrac{y_2}{3}$. A similar (but slightly more involved) calculation yields $\overline{x} = \dfrac{x_1 + x_2}{3}$. So $(\overline{x}, \overline{y})$ is the centroid of $R$.

**16.** Let $h$ the height of the triangle. The moment of the semi-disc about the diameter must equal that of the triangle. So, by Problem (4.3) (i) and the last exercise, $\dfrac{\pi r^2}{2}\dfrac{4r}{3\pi} = rh\dfrac{h}{3}$. Hence $h = \sqrt{2}r$.

**17.** All points at a distance $\frac{200}{3}$ km. from the centre of the island. They lie closer to the coast because the average is more influenced by the coastal areas than the central ones. (More precisely, between two portions of the island with equal areas, the one closer to the coast contributes more to $\iint f(x,y)dA$ than the one closer to the centre.

**18.** Similar to the last exercise. In (i), portions of equal areas contribute equally to the mass. But in (ii) those which are farther from the diameter contribute more than those that are nearer. So the centre of mass is pushed away from the diameter.

**19.** $I_x = \displaystyle\iint_R y^2 \rho(x,y)dA$    equals    $\displaystyle\int_0^a c_1 \int_{-\sqrt{a^2-y^2}}^{\sqrt{a^2-y^2}} y^2 dxdy$    for    (i)    and

$\displaystyle\int_0^a c_2 \int_{-\sqrt{a^2-y^2}}^{\sqrt{a^2-y^2}} y^3 dxdy$ for (ii), where $c_1, c_2$ are some constants. These come

out to be $\dfrac{c_1\pi a^4}{8}$ and $\dfrac{4c_2 a^5}{15}$ respectively.   $I_y = \displaystyle\iint_R x^2\rho(x,y)dA$ equals

$c_1 \displaystyle\int_{-a}^a \int_0^{\sqrt{a^2-x^2}} x^2 dydx = \dfrac{c_1\pi a^4}{8}$ for (i) and $c_2 \displaystyle\int_{-a}^a \int_0^{\sqrt{a^2-x^2}} x^2 ydydx =$

$c_2 \displaystyle\int_{-a}^a \dfrac{x^2(a^2-x^2)}{2}dx = \dfrac{2c_2 a^5}{15}$ for (ii).

**20.** Take the plane of the region $R$ as the $x$-$y$ plane and the line $L$ as the $x$-axis. As $L$ does not pass through the interior of $R$, we may suppose $y \geq 0$ for all $(x,y) \in R$. Taking the density function $\rho(x,y) \equiv 2\pi$, the moment of $R$ about $L$ is $\iint_R 2\pi ydA$. When converted to an iterated integral, the moment equals $\int_c^d 2\pi yl(y)dy$ where $R$ lies between $y = c$ and $y = d$ (say) and for $b \in [c,d], l(k)$ is the length of the intersection of $R$ with the line $y = k$. On one hand this equals $M\overline{y} = 2\pi A\overline{y}$. But on the other hand, this is exactly the volume $V$ of $S$, obtained by the shell method.

**21.** $S$ is the ball of radius $a$. So $V = \frac{4\pi}{3}a^3$ and $A = \frac{1}{2}\pi a^2$. By the last exercise
$\overline{y} = \dfrac{V}{2\pi A} = \dfrac{4a}{3\pi}$. By symmetry, $\overline{x} = 0$.

**22.** $S$ is the solid bounded below by the unit disc in the $x$-$y$ plane and above by the 'dome' $z = 1 - x^2 - y^2$. For $k \in [0,1]$, the intersection of $S$ with the plane $z = k$ is the disc $\{(x,y,k) : x^2 + y^2 \leq 1 - k\}$. So $\iiint_S \dots dV$

equals $\displaystyle\int_0^1 \int_{-\sqrt{1-z}}^{\sqrt{1-z}} \int_{-\sqrt{1-z-y^2}}^{\sqrt{1-z-y^2}} \dots dxdydz$ and also $\displaystyle\int_0^1 \int_{-\sqrt{1-z}}^{\sqrt{1-z}} \int_{-\sqrt{1-z-x^2}}^{\sqrt{1-z-x^2}} \dots dydxdz$

by symmetry between $x$ and $y$. For expressing this as an iterated integral in which the outermost variable is $y$, note first that $S$ lies between the planes $y = -1$ and $y = 1$. Also for each $k \in [-1, 1]$, the intersection of $S$ with the plane $y = k$ is the set $\{(x, k, z) : x^2 \leq 1 - k^2, 0 \leq z \leq 1 - x^2 - k^2\}$. It is helpful to sketch the projection of this set onto the $x$-$z$ plane as shown below.

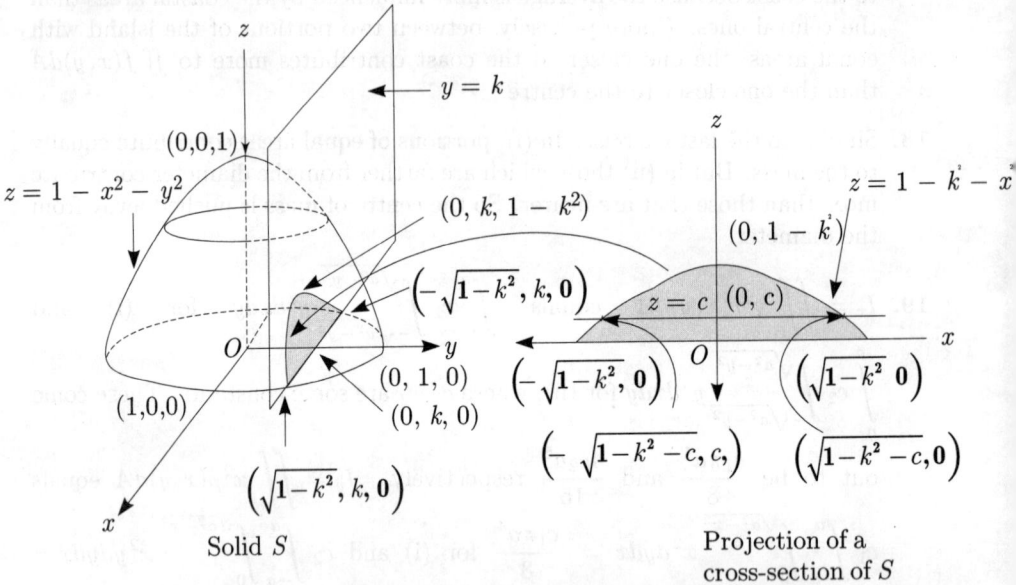

Solid $S$

Projection of a cross-section of $S$

From this projection it is easily seen that $\iiint_S \ldots dV$ equals the iterated

integrals $\int_{-1}^{1} \int_{-\sqrt{1-y^2}}^{\sqrt{1-y^2}} \int_{0}^{1-y^2-x^2} \ldots dz dx dy$ and $\int_{-1}^{1} \int_{0}^{1-y^2} \int_{-\sqrt{1-y^2-z}}^{\sqrt{1-y^2-z}} \ldots dx dz dy$. Be-

cause of the symmetry between $x$ and $y$, the expressions

$\int_{-1}^{1} \int_{-\sqrt{1-x^2}}^{\sqrt{1-x^2}} \int_{0}^{1-x^2-y^2} \ldots dz dy dx$ and $\int_{-1}^{1} \int_{0}^{1-x^2} \int_{-\sqrt{1-x^2-z}}^{\sqrt{1-x^2-z}} \ldots dy dz dx$ represent the

remaining two iterated integrals. In the present problem the integrand is $xy+z$ and taking the first iterated integral, the triple integral $\iiint_S xy + z \, dV$

equals $\int_{0}^{1} \int_{-\sqrt{1-z}}^{\sqrt{1-z}} \int_{-\sqrt{1-z-y^2}}^{\sqrt{1-z-y^2}} xy + z \, dx dy dz$ which comes out as

$$\int_{0}^{1} \int_{-\sqrt{1-z}}^{\sqrt{1-z}} (y\frac{x^2}{2} + zx)\Big|_{x=-\sqrt{1-z-y^2}}^{x=\sqrt{1-z-y^2}} dy dz = \int_{0}^{1} \int_{-\sqrt{1-z}}^{\sqrt{1-z}} 2z\sqrt{1 - z - y^2} dy dz$$

$$= \int_{0}^{1} z\pi(1 - z) dz = \frac{\pi}{6}.$$

**23.** Take the base of the hemi-sphere to be the disc $\{(x, y); x^2 + y^2 \leq a^2\}$. By symmetry, $\bar{x} = \bar{y} = 0$. In (i) $M = c_1 \dfrac{2\pi}{3} a^3$ and $M_{xy} = c_1 \iiint_S z\, dV$ where $c_1$

is a constant. Write $M_{xy}$ as $c_1 \displaystyle\int_0^a \int_{-\sqrt{a^2-z^2}}^{\sqrt{a^2-z^2}} \int_{-\sqrt{a^2-z^2-x^2}}^{\sqrt{a^2-z^2-x^2}} z\, dy\, dx\, dz.$

Note that for $k \in [0,1]$ the integral $\displaystyle\int_{-\sqrt{a^2-k^2}}^{\sqrt{a^2-k^2}} \int_{-\sqrt{a^2-k^2-x^2}}^{\sqrt{a^2-k^2-x^2}} 1\, dy\, dx$ is simply

the area of the cross-section by the plane $z = k$, which is a disc of ra-

dius $\sqrt{a^2 - k^2}$. So $M_{xy} = c_1 \int_0^a z\pi(a^2 - z^2)dz$. Hence $\bar{z} = \dfrac{3}{8}$. By a similar

reasoning, for (ii), $\bar{z} = \dfrac{\int_0^a z^2\pi(a^2 - z^2)dz}{\int_0^a z\pi(a^2 - z^2)dz} = \dfrac{8a}{15}$.

For the cone, take the base to be the disc $\{(x,y) : x^2 + y^2 \le a^2\}$ and the

vertex at $(0,0,h)$ $(h > 0)$. Then $\bar{x} = \bar{y} = 0$ for both (i) and (ii), while $\bar{z} = \dfrac{h}{4}$

for (i) and $\dfrac{2h}{5}$ for (ii).

24. Take the rod as the set $S = \{x,y,z) : 0 \le z \le h, x^2 + y^2 \le r^2\}$ and
$\rho(x,y,z) \equiv 1$. Then $M = \pi r^2 h$. In (i), $I = \displaystyle\iiint_S (x^2 + y^2)dV$ equals

$$\int_0^h \int_{-r}^r \int_{-\sqrt{r^2-x^2}}^{\sqrt{r^2-x^2}} (x^2 + y^2)dy\,dx\,dz = h \int_{-r}^r 2(x^2 + \frac{r^2 - x^2}{3})\sqrt{r^2 - x^2}dx = \frac{\pi r^4 h}{2}.$$

In (ii), take the line $L$ as $y = 0, z = \dfrac{h}{2}$. Then $I = \displaystyle\iiint_S (z - \frac{h}{2})^2 + y^2 dV$

$$= \int_0^h (z - \frac{h}{2})^2 \pi r^2 dz + \frac{\pi}{4} r^4 h = \frac{\pi r^2 h}{12}(h^2 + 3r^2).$$ In (iii), taking $L$ as the $x$-

axis, $I = \displaystyle\int_0^h (z - h)^2 \pi r^2 dz + \frac{\pi}{4} r^4 h = \frac{\pi r^2 h}{12}(4h^2 + 3r^2).$ So the radius of gy-

ration equals $\dfrac{r}{\sqrt{2}}$ for (i), $\sqrt{\dfrac{h^2 + 3r^2}{12}}$ for (ii) and $\sqrt{\dfrac{4h^2 + 3r^2}{12}}$ for (iii). (The

evaluation of some of these integrals can be simplified by the use of polar
co-ordinates, see Section 6.8.)

25. Without loss of generality let $L$ be the $z$-axis. Then $L'$ is the line $x = \bar{x}, y = \bar{y}$. Let $r(x,y,z)$ and $r'(x,y,z)$ be the (perpendicular) distances of $(x,y,z)$ from $L, L'$ respectively. Then $[r(x,y,z)]^2 = x^2 + y^2$ and $[r'(x,y,z)]^2 = (x - \bar{x})^2 + (y - \bar{y})^2 = [r(x,y,z)]^2 + (\bar{x}^2 + \bar{y}^2) - 2x\bar{x} - 2y\bar{y}$. So $I_{L'} - I_L = \iiint_S[(\bar{x}^2 + \bar{y}^2) - 2\bar{x}x - 2\bar{y}y]\rho(x,y,z)dV = M(\bar{x}^2 + \bar{y}^2) - 2\bar{x}M\bar{x} - 2\bar{y}M\bar{y} = -M(\bar{x}^2 + \bar{y}^2) = -Mh^2$.

## Section 6.5

1.  (i) Choose $c_1, c_2$ (for example, $c_1 = -\infty$ and $c_2 = \infty$) so that $P(c_1 \leq x \leq c_2) = 1$. Then $P(a_1 \leq x \leq a_2; b_1 \leq y \leq b_2)$ is the same as $P(a_1 \leq x \leq a_2; b_1 \leq y \leq b_2; c_1 \leq z \leq c_2)$ and so by (26) $x$ and $y$ are mutually independent. The same holds for other pairs. (ii) follows from the fact that the length of any of the sides puts no restriction on those of the others. In (iii), the lengths of any two sides of a triangle can be chosen independently of each other but once these are chosen, the length of the third side is constrained to be less than their sum. Thus, for example, if $x, y, z$ are the lengths and each varies uniformly over, say, $[0, 2]$ so that $P(a_1 \leq x \leq a_2) = \dfrac{a_2 - a_1}{2}$ for $0 \leq a_1 \leq a_2 \leq 2$ and similarly for $y$ and $z$, then $P(0 \leq x \leq \frac{1}{2}, 0 \leq y \leq \frac{1}{2}, 1 \leq z \leq 2) = 0$ but $P(0 \leq x \leq \frac{1}{2})P(0 \leq y \leq \frac{1}{2})P(1 \leq z \leq 2) = \frac{1}{4}\frac{1}{4}\frac{1}{2} = \frac{1}{32}$.

2.  Let $S$ be the set in the hint and $S_1, \ldots, S_6$ the six parts into which it is divided. No two of these parts have an interior point in common and so for any function $f(x, y, z)$ defined on $S$, $\iiint_S f(x, y, z)dV = \sum_{i=1}^{6} \iiint_{S_i} f(x, y, z)dV$. The answers to (i) and (ii) are obtained by taking $f(x, y, z) = 1$ and $f(x, y, z) = 8xyz$ respectively. Because of symmetry (of $S$ as well as of $f$), in each case $\iiint_S f(x, y, z)dV = 6 \iiint_{S_1} f(x, y, z)dV$ where $S_1 = \{(x, y, z) : 0 \leq x \leq y \leq z \leq 1 \text{ and } z - x \leq \frac{1}{2}\}$. If $(x, y, z) \in S_1$ then for $x \in [0, \frac{1}{2}], z \in [x, x + \frac{1}{2}]$, while for $x \in [\frac{1}{2}, 1], z \in [x, 1]$. So $\iiint_{S_1} f(x, y, z)dV$ splits into two parts which equal $\int_0^{1/2} \int_x^{x+1/2} \int_x^z f(x, y, z)dydzdx$ and $\int_{1/2}^1 \int_x^1 \int_x^z f(x, y, z)dydzdx$. With this, the answers to (i) and (ii) are $\frac{1}{2}$ and $\frac{45}{64}$ respectively.

3.  Let $V_n$ be the volume of the unit $n$-ball. Then the volume of an $n$-ball of radius $r$ is $V_n r^n$. For $z \in [-1, 1]$, the cross section of the unit $n$-ball by the hyperplane $x_n = z$ is an $(n - 1)$-ball of radius $\sqrt{1 - z^2}$. So $V_n = \int_{-1}^1 V_{n-1}(\sqrt{1 - z^2})^{n-1}dz = 2V_{n-1}I_n$ where $I_n = \int_0^1 (\sqrt{1 - z^2})^{n-1}dz$. Putting $z = \sin\theta$, $I_n = \int_0^{\pi/2} \cos^n\theta d\theta = \int_0^{\pi/2} \sin^n\theta d\theta$ by Theorem (5.6.1). Repeated applications of this recurrence relation, and the fact $V_1 = 2$ give $V_n = 2^n I_1 I_2 \ldots I_n$. From exercises (5.6.13) and (5.6.10), $I_{2m-1}I_{2m} = \dfrac{\pi}{4m}$ for $m = 1, 2, 3, \ldots$. So $V_n = \dfrac{\pi^k}{k!}$ if $n = 2k$ and $V_n = \dfrac{2^{2k+1}\pi^k}{k!(2k + 1)!}$ if $n = 2k+1$.

4.  Let $p_n$ be the probability that a record occurs at the $n^{\text{th}}$ stage. Since $(x_1, \ldots, x_n)$ varies over the unit $n$-cube with uniform probability density and the unit cube has volume 1, $p_n$ is simply $V(S)$ where $S = \{(x_1, \ldots, x_n) : 0 \leq x_i \leq 1 \text{ for every } i = 1, \ldots, n \text{ and } x_n > x_i \text{ for } i = 1, \ldots, n - 1\}$. For $z \in [0, 1]$, the cross section of $S$ by the hyperplane $x_n = z$ is the set $\{(x_1, \ldots, x_{n-1}, z) : 0 \leq x_i < z \text{ for } i = 1, \ldots, n - 1\}$. This may be identified

with an $(n-1)$-cube of side $z$ and so has $z^{n-1}$ as its $(n-1)$-dimensional volume. So $V(S) = \int_0^1 z^{n-1}dz = \frac{1}{n}$. It is tempting to deduce the last part by merely adding $p_1, p_2, \ldots, p_n$. But some justification is needed. Let $E_n$ be the expected number of records during the first $n$ stages. Then last entry creates a record with probability $p_n(=\frac{1}{n})$. So by the same reasoning as in the second solution to Problem (1.5), $E_n = p_n(E_{n-1}+1)+(1-p_n) = E_{n-1}+p_n$, for $n \geq 2$. It now follows that $E_n = p_1 + p_2 + \ldots + p_n = H_n$.

5. Imitate the procedure in Example 6. The distance of a chord of length $x$ ($0 \leq x \leq 2$) from the centre is $\sqrt{1 - \frac{x^2}{4}}$. For $c \in [0,2]$, $0 \leq x \leq c$ iff $\sqrt{1 - \frac{c^2}{4}} \leq \sqrt{1 - \frac{x^2}{4}} \leq 1$. So $P(x \leq c) = 1 - \sqrt{1 - \frac{c^2}{4}}$. So $p(x) = \frac{d}{dx}\left(1 - \sqrt{1 - \frac{x^2}{4}}\right)$

$= \dfrac{x}{4\sqrt{1 - \frac{x^2}{4}}}$. Therefore, the average chord length comes out to be

$$\int_0^2 \frac{x^2}{4\sqrt{1 - \frac{x^2}{4}}}dx = \int_0^{\pi/2} 2\sin^2\theta d\theta = \frac{\pi}{2}.$$

6. The first moments are $\frac{a+b}{2}, \frac{3}{4}, 1$ and $\frac{1}{a}$ respectively, while the second moments are $\frac{1}{3}(a^2 + ab + b^2), \frac{3}{5}, \frac{11}{10}$ and $\dfrac{2}{a^2}$ respectively.

7. (i) $p(x) = \dfrac{1}{x^2}$ on $[1, \infty)$ (ii) $p(x) = \dfrac{2}{3x^3}$ on $[1, \infty)$.

8. $E(x) = \int_{-\infty}^{\infty} xp(x)dx = \int_{-\infty}^{a} xp(x)dx + \int_a^{\infty} xp(x)dx = \int_0^{\infty}(a-u)p(a-u)du + \int_0^{\infty}(a+u)p(a+u)du = a\int_0^{\infty} p(a-u)du + a\int_0^{\infty} p(a+u)du = a\int_{-\infty}^{a} p(x)dx + a\int_a^{\infty} p(x)dx = a\int_{-\infty}^{\infty} p(x)dx = a1 = a$. This argument is not valid for $p(x) = \dfrac{1}{\pi(1+x^2)}$ (with $a = 0$) because $\int_{-\infty}^{\infty} \dfrac{xdx}{\pi(1+x^2)}$ is undefined.

9. (a) Suppose $f(c) > 0$ for some $c \in (a, b)$. Taking $\epsilon = \frac{1}{2}f(c)$, $\exists \delta > 0$ s.t. $(c-\delta, c+\delta) \subset (a, b)$ and for all $x \in (c-\delta, c+\delta)$, $f(x) \in (f(c) - \frac{1}{2}f(c), f(c) + \frac{1}{2}f(c))$. So $f(x) \geq \frac{1}{2}f(c)$ for all $x \in (c-\delta, c+\delta)$. Hence $\int_{c-\delta}^{c+\delta} f(x)dx \geq 2\delta\frac{1}{2}f(c) = \delta f(c)$. Since $f(x) \geq 0$ everywhere, $\int_a^{c-\delta} f(x)dx \geq 0$ and $\int_{c+\delta}^{b} f(x)dx \geq 0$. So by Exercise (5.1.7), $\int_a^b f(x)dx = \int_a^{c-\delta} f(x)dx + \int_{c-\delta}^{c+\delta} f(x)dx + \int_{c+\delta}^{b} f(x)dx \geq \delta f(c) > 0$ a contradiction. If $c = a$ or $c = b$, modify the argument slightly. If $f$ is only piecewise continuous, then $f$ vanishes except possibly at a finite number of points (viz. the points of discontinuity.)

(b) $(b-a)\sum_{i=1}^{n} f(x_i)$ is simply a Riemann sum of $f(x)$ based on the partition of $[a, b]$ into $n$ equal parts. As $n \to \infty$, the mesh of this partition tends to 0 and so $\sum_{i=1}^{n} f(x_i) \to \frac{1}{b-a}\int_a^b f(x)dx$ which is positive in the present case.

(c) Apply (a) to the function $h - g$.

**10.** $\mathrm{Var}(x) = \int\limits_{-\infty}^{\infty} (x - \mu)^2 p(x)dx = \int\limits_{-\infty}^{\infty} x^2 p(x) - 2\mu \int\limits_{-\infty}^{\infty} xp(x)dx + \mu^2 \int\limits_{-\infty}^{\infty} p(x)dx.$

The result follows from $\int\limits_{-\infty}^{\infty} p(x)dx = 1$ and $\int\limits_{-\infty}^{\infty} xp(x)dx = \mu.$

**11.** $\iint_{T_2} (x - y)dA = \int_0^1 [\int_0^x (x - y)dy]dx = \int_0^1 x^2 - \frac{x^2}{2}dx = \frac{1}{3}.$ Similarly,
$\iint_{T_1} 4xy(y - x)dA = \int_0^1 [\int_x^1 4xy^2 - 4x^2 ydy]dx$
$= \int_0^1 \frac{4x}{3}(1 - x^3) - 2x^2(1 - x^2)dx = \int_0^1 \frac{2}{3}x^4 - 2x^2 + \frac{4}{3}xdx = \frac{2}{15}.$

**12.** $P(\frac{1}{2} \leq x \leq 1) = \int_{1/2}^1 p(x)dx = \int_{1/2}^1 2xdx = \frac{3}{4},$ while $P(0 \leq x \leq \frac{1}{2}) = \int_{1/2}^1 2xdx = \frac{1}{4}.$ So the probability that $x, y$ both lie in the same half of $[0, \frac{1}{2}]$ is $(\frac{3}{4})^2 + (\frac{1}{4})^2 = \frac{5}{8},$ which is higher than the corresponding probability for (i), viz. $\frac{1}{2}.$ As all such pairs contribute to $P(|x - y| \leq \frac{1}{2}),$ this probability for (ii) is higher than that for (i). Also the average distance is smaller because closer pairs of houses are more likely.

**13.** In (i) $x$ and $y$ assume the values $\frac{1}{4}, \frac{1}{2}, \frac{3}{4}$ and 1 with probabilities $\frac{1}{4}$ each. The distance $|x - y|$ takes values $0, \frac{1}{4}, \frac{1}{2}$ and $\frac{3}{4}$ with probabilities $\frac{4}{16}, \frac{6}{16}, \frac{4}{16}$ and $\frac{2}{16}$ respectively. So $P(|x - y| < 1/2) = \frac{5}{8}$ and $E(|x - y|) = \frac{6+8+6}{64} = \frac{5}{16}.$ In (iii), the probabilities of $x$ (or $y$) assuming the values $\frac{1}{4}, \frac{1}{2}, \frac{3}{4}$ and 1 are $\frac{1}{10}, \frac{1}{5}, \frac{3}{10}$ and $\frac{2}{5}$ respectively. $|x - y|$ equals $0, \frac{1}{4}, \frac{1}{2}, \frac{3}{4}$ with probabilities $\frac{3}{10}(= \frac{1}{100} + \frac{1}{25} + \frac{9}{100} + \frac{4}{25}), \frac{2}{5}, \frac{11}{50}$ and $\frac{2}{25}$ respectively. So $P(|x - y| < \frac{1}{2}) = \frac{7}{10}$ and $E(|x - y|) = \frac{27}{100}.$

**14.** For the first part in the alternate approach in the hint, simply expand $\binom{j}{4} = \frac{j(j - 1)(j - 2)(j - 3)}{24}.$ The formulae for $\sum\limits_{j=1}^{n} j^2$ and $\sum\limits_{j=1}^{n} j^3$ are $\frac{n(n + 1)(2n + 1)}{6}$ and $\frac{n^2(n + 1)^2}{4}$ respectively. The identity $\sum\limits_{j=1}^{n} \binom{j}{4} = \binom{n+1}{5}$ follows by summing up $\binom{j}{4} = \binom{j+1}{5} - \binom{j}{5}$ over $j$.

**15.** $\sum\limits_{j=1}^{n} \sum\limits_{i=1}^{n} ij|i - j|$

$= \sum\limits_{j=1}^{n} \sum\limits_{i=1}^{j} ij(j - i) + \sum\limits_{j=1}^{n} \sum\limits_{i=j+1}^{n} ij(i - j)$

$= \sum\limits_{j=1}^{n} [j^2 \sum\limits_{i=1}^{j} i - j \sum\limits_{i=1}^{j} i^2 - j^2 \sum\limits_{i=j+1}^{n} i + j \sum\limits_{i=j+1}^{n} i^2]$

$= \sum\limits_{j=1}^{n} [\frac{j^3(j + 1)}{2} - \frac{j^2(j + 1)(2j + 1)}{6} - \frac{j^2 n(n + 1) - j^3(j + 1)}{2}$

$\quad + \frac{j[n(n + 1)(2n + 1) - j(j + 1)(2j + 1)]}{6}].$

Let $A_n$ be the coefficient of $n^5$ in this sum. Then $A_n = B_n + C_n$ where $B_n$ is the coefficient of $n^5$ coming from $\sum_{j=1}^{n} j^4$ and $C_n$ is the coefficient of $n^5$ coming from the other terms. From the last exercise, $B_n = \frac{1}{5}(\frac{1}{2} - \frac{2}{6} + \frac{1}{2} - \frac{2}{6}) = \frac{1}{15}$, while $C_n = -\frac{1}{6} + \frac{1}{6} = 0$. So $A_n = \frac{1}{15}$ and the limit of the R.H.S. of (26) as $n \to \infty$ equals $\frac{4}{15}$.

**16.**

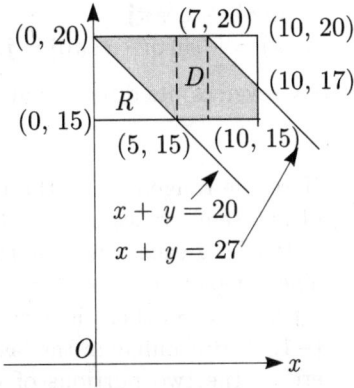

Here we have two independent random variables, say $x$ and $y$, varying uniformly over the intervals $[0, 10]$ and $[15, 20]$ respectively. So the random variable, $(x, y)$, varies uniformly over the rectangle $R = R(0, 10; 15, 20)$ whose area is 50 units. The portion, say $D$, of $R$ lying between the lines $x + y = 20$ and $x + y = 27$ has area 33 units. So the desired probability is $33/50 = 0.66$. The average waiting time is

$$\frac{\iint_D (27 - x - y) \, dxdy}{33}$$ which comes out (after splitting $D$ into three parts by vertical lines as shown) as $\frac{727}{198} \approx 3.6717$ minutes.

## Section 6.6

1.  For $\theta \in [0, 2\pi]$, $s = s(\theta) = \int_0^\theta \sqrt{(\frac{dx}{dt})^2 + (\frac{dy}{dt})^2} dt = 4a - 4a \cos \frac{\theta}{2}$. (cf. Exercise (2.4)). So $\theta = 2 \cos^{-1}(\frac{4a - s}{4a})$, for $0 \le s \le 8a$. Substitution gives the answer.

2.  This time $0 \le s \le 16a$. For $s \in [0, 8a]$, the parametrisation is as in the last exercise. For $s \in [8a, 16a], \theta \in [2\pi, 4\pi]$ and $s - 8a$ is the length of the portion of the second arch. So to get the arc length parametrisation, replace $s$ by $s - 8a$ and add $2\pi$ to the values of $\theta$. Thus

$$x(s) = \begin{cases} 2a \cos^{-1}\left(\dfrac{4a - s}{4a}\right) + \dfrac{(s - 4a)\sqrt{s(8a - s)}}{8a} & \text{if } 0 \le s \le 8a \\[2ex] 2a \cos^{-1}\left(\dfrac{12a - s}{4a}\right) + 2a\pi + \dfrac{(s - 12a)\sqrt{(s - 8a)(16a - s)}}{8a} & \\[1ex] & \text{if } 8a \le s \le 16a \end{cases}$$

and similarly,

$$y(s) = \begin{cases} \dfrac{s(8a - s)}{8a} & \text{if } 0 \le s \le 8a \\[2ex] \dfrac{(s - 8a)(16a - s)}{8a} & \text{if } 8a \le s \le 16a. \end{cases}$$

**3.**

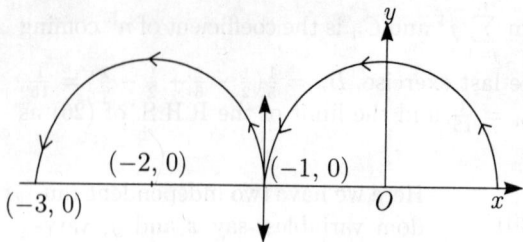

The range of $\alpha$ is the union of two semi-circles or radii 1 each touching each other externally. $\alpha$ is continuously differentiable on $(0, \pi)$ as well as on $(\pi, 2\pi)$ and $|\alpha'(\theta)| = 1$ for all $\theta$ in these intervals. At $\theta = \pi, \alpha'(\pi) = -\sin \pi \mathbf{i} + \cos \pi \mathbf{j} = -\mathbf{j}$, while $\alpha'(\pi) = -\sin 0\mathbf{i} + \cos 0\mathbf{j} = \mathbf{j}$.

Hence the curve changes (in fact completely reverses) its direction at $\theta = \pi$. So there is a cusp at $(-1, 0)$.

**4.**

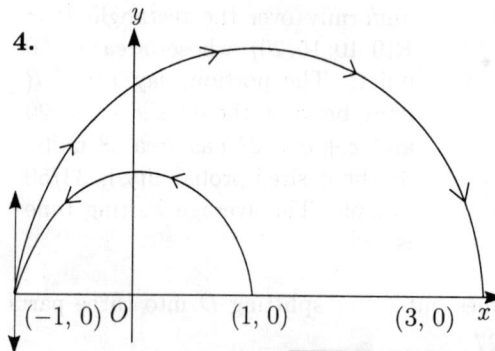

Here the range of $\alpha$ is the union of two semi-circles of radii 1 and 2, touching each other internally. The tangents to $\alpha$ at $\theta = \pi$ are $-\mathbf{j}$ and $2\mathbf{j}$. So there is a cusp at $(-1, 0)$. But unlike in the last exercise, the two portions of $\alpha$ on either side of $\pi$ lie on the same side of the tangent.

**5.** (a) $f'_+(0) = \lim\limits_{t \to 0^+} \dfrac{\sqrt{t^2 + t^3}}{t} = \lim\limits_{t \to 0^+} \sqrt{1 + t} = 1$ and $f'_-(0) = \lim\limits_{t \to 0^-} \dfrac{-\sqrt{t^2 + t^3}}{t}$

$= \lim\limits_{u \to 0^+} \dfrac{-\sqrt{u^2 - u^3}}{-u} = \lim\limits_{u \to 0^+} \sqrt{1 - u} = 1$. So $f'(0)$ exists and equals 1. The same reasoning shows that $g'_+(0) = 1$ and $g'_-(0) = -1$ and so $g$ is not differentiable at 0. As for differentiability of $f$ at $-1$, we need only consider $f'_+(-1) = \lim\limits_{t \to -1^+} \dfrac{-\sqrt{t^2 + t^3} - 0}{t + 1} = \lim\limits_{t \to -1^+} \dfrac{t\sqrt{1 + t}}{t + 1} =$

$-\lim\limits_{t \to -1^+} \dfrac{1}{\sqrt{1 + t}}$, which does not exist. At all other $t \geq -1, f$ is differentiable since $t^2 + t^3 > 0$ and $\sqrt{x}$ is a differentiable function of $x$ for $x > 0$.

(b)

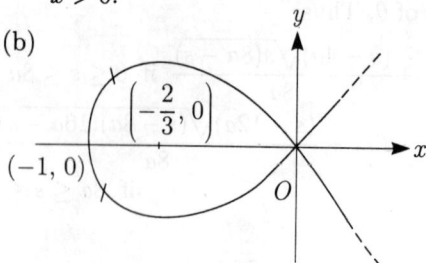

The parts of the curve, $C_1, C_2, C_3, C_4$ correspond to $t \in (-\infty, -2], t \in [-2, -1], t \in [-1, 0]$ and $t \in [0, \infty)$, respectively. (This is easier to see for $C_3$ and $C_4$. For $C_1$ and $C_2$ make a substitution $u = -(t + 2)$.)

(c) At $t = -1, \alpha(t) = -\mathbf{i} = (-1, 0)$ and neither $\alpha'_+(-1)$ nor $\alpha'_-(-1)$ exists. $\alpha(t) = \mathbf{0} = (0, 0)$ for $t = -2$ and also for $t = 0$. But, $\alpha'(-2) = -\mathbf{i} + \mathbf{j}$ while $\alpha'(0) = \mathbf{i} + \mathbf{j}$.

7. $\alpha'(t) = -\sin t \mathbf{i} + \cos t \mathbf{j}$. Although each $-\sin t$ and $\cos t$ vanishes for some values of $t$ in $(0, 2\pi)$, there is no (common) value of $t$ in $(0, 2\pi)$ at which both $\sin t$ and $\cos t$ vanish.

8. Follow the notation of Definition (6.1). If $f : [a, b] \to \mathbb{R}$ is the equivalence between $\alpha$ and $\beta$ then its restriction to $[a, c_1]$ gives the equivalence between $\alpha/[a, c_1]$ and $\beta/[c, f(c_1)]$. Also $f/[c_1, b]$ is an equivalence between $\alpha/[c_1, b]$ and $\beta/[f(c_1), d]$. All that is needed in the proof is to insert the point $c_1$, if necessary, as a node in the partition of $[a, b]$. The converse follows similarly, by putting together partitions of $[a, c_1]$ and of $[c_1, b]$ to get a desired partition of $[a, b]$.

9. Let $\alpha, \beta$ be a pair of two mutually equivalent curves (in the notation of Definition (6.1)) and $\alpha^*, \beta^*$ be another such pair (notations obtained by putting $*$ over corresponding symbols for $\alpha, \beta$, except that the partition, say $P = (t_0, t_1, \ldots, t_n)$ of $[a, b]$ need not correspond to the partition, say, $P^* = (t_0^*, t_1^*, \ldots, t_m^*)$ of $[a^*, b^*]$). Let $X = \alpha(t) = \alpha^*(t^*)$ be a common point lying on $\alpha$ and $\alpha^*$. Then this point also equals $\beta(s)$ and $\beta^*(s^*)$ for (unique) $s \in [c, d]$ and $s^* \in [c^*, d^*]$. The angle between $\alpha$ and $\alpha^*$ at $X$ is the angle between the vectors $\alpha'(t)$ and $\alpha^{*'}(t)$, while that between $\beta$ and $\beta^*$ is the angle between $\beta'(s)$ and $\beta^{*'}(s^*)$. The result says that these angles are equal. To prove this, call these angles as $\theta_1$ and $\theta_2$ respectively. Then

$$\cos\theta_1 = \frac{\alpha'(t) \cdot \alpha^{*'}(t^*)}{|\alpha'(t)||\alpha^{*'}(t^*)|} \text{ and } \cos\theta_2 = \frac{\beta'(s) \cdot \beta^{*'}(s^*)}{|\beta'(s)||\beta^{*'}(s^*)|}. \text{ Since } \alpha = \beta \circ f \text{ and}$$

$\alpha^* = \beta^* \circ f^*$, by Exercise (6.6) (iii), we have $\alpha'(t) = f'(t)\beta'(s)$ and $\alpha^{*'}(t^*) = f^{*'}(t^*)\beta^{*'}(s^*)$. Also $f'(t) > 0$ and $f^{*'}(t^*) > 0$. So $|\alpha'(t)| = f'(t)|\beta'(s)|$ and $|\alpha^{*'}(t^*)| = f^{*'}(t^*)|\beta^{*'}(s^*)|$. Substituting, the result follows.

10. The integrand is $\sqrt{1 + \left(\sin\frac{1}{t} - \frac{1}{t}\cos\frac{1}{t}\right)^2} = g(t)$ (say). Let $n \in \mathbb{N}$. For all

$$t \in \left[\frac{1}{(2n+1)\pi}, \frac{1}{(2n+\frac{2}{3})\pi}\right], \sin\frac{1}{t} \geq 0 \text{ and } \cos\frac{1}{t} \leq -\frac{1}{2} \text{ imply } g(t) \geq \frac{1}{t}\left|\cos\frac{1}{t}\right|$$

$$\geq \left(2n + \frac{2}{3}\right)\frac{\pi}{2}. \text{ So } \int_{1/(2n+1)\pi}^{1/(2n+2/3)\pi} g(t)dt \geq \frac{(2n+\frac{2}{3})\pi}{2}\left(\frac{1}{(2n+\frac{2}{3})\pi} - \frac{1}{(2n+1)\pi}\right)$$

$= \frac{1}{6(2n+1)}$. The intervals $\left[\frac{1}{(2n+1)\pi}, \frac{1}{(2n+\frac{2}{3})\pi}\right]; n \in \mathbb{N}$ are mutaully disjoint and $g$ is non-negative everywhere. So $\int_0^\pi g(t)dt \geq \sum_{n=1}^\infty \frac{1}{12n+2}$ which is divergent.

11. By symmetry about the line $x = \pi, \bar{x} = \pi$ for both (i) and (ii). In (i), $\bar{y} = (\int_0^{2\pi} y\frac{ds}{d\theta}d\theta)/\int_0^{2\pi}\frac{ds}{d\theta}d\theta$. Using the calculations in Exercise (6.2), the denominator is $8a$. The numerator equals $2\int_0^{2\pi} a^2(1 - \cos\theta)\sin\frac{\theta}{2}d\theta = a^2\int_0^{2\pi} 3\sin\frac{\theta}{2} - \sin\frac{3\theta}{2}d\theta = a^2\left[-6\cos\frac{\theta}{2} + \frac{2}{3}\cos\frac{3\theta}{2}\right]\Big|_0^{2\pi} = \frac{32a^2}{3}$. So $\bar{y} = \frac{4a}{3}$. In

(ii), $\bar{y} = (\int_0^{2\pi} a^2(1 - \cos\theta)^2 2a \sin\frac{\theta}{2} d\theta)/ \int_0^{2\pi} a(1 - \cos\theta) 2a \sin\frac{\theta}{2} d\theta = 2aI_5/I_3$
where $I_n = \int_0^{\pi/2} \sin^n \theta d\theta$. Use Exercise (5.6.10) to get $\bar{y} = \frac{8a}{5}$.

12. Parametrise $C$ by $\alpha : [a, b] \to I\!\!R^2$, where $\alpha(t) = f(t)\mathbf{i} + g(t)\mathbf{j}$. Without loss of generality, let $L^*$ be the $x$-axis, and suppose $C$ lies in the upper half-plane. Then $L = \int_a^b \sqrt{[f'(t)]^2 + [g'(t)]^2} dt$. Let $S$ be the surface of revolution. Then $A(S) = \int_a^b 2\pi g(t) \sqrt{f'(t)]^2 + [g'(t)]^2} dt$. The result follows from the fact that if $(\bar{x}, \bar{y})$ is the centre of gravity of $C$ then its distance from $L^*$ is $\bar{y}$ which also equals $\int_a^b g(t) \sqrt{[f'(t)]^2 + [g(t)]^2} dt/L$.

13. The length of a circle of radius $b$ is $2\pi b$ while its centre is also its centre of gravity. Hence the surface area of the torus $= 2\pi b \times 2\pi a = 4\pi^2 ab$.

14. Let the base be the disc $\{(x, y, 0) : x^2 + y^2 \leq r^2\}$ and the apex be at $(0, 0, h)$. By symmetry, the centre of mass is at $(0, 0, \bar{z})$ for some $\bar{z} > 0$.

   For (i), $M = \pi r \sqrt{r^2 + h^2}$ while $M_{xy} = \int_0^h 2\pi \frac{r}{h}(h - z)z \frac{ds}{dz} dz$. Since $\frac{ds}{dz} = \frac{\sqrt{r^2 + h^2}}{h}$, this comes out to be $\frac{\pi r h \sqrt{r^2 + h^2}}{3}$ and so $\bar{z} = \frac{h}{3}$. For (ii)

   $M = \int_0^h 2\pi \frac{r}{h}(h - z)^2 \frac{\sqrt{r^2 + h^2}}{h} dz = \frac{2}{3}\pi r h \sqrt{h^2 + r^2}$, while $M_{xy} = $

   $\int_0^h 2\pi \frac{r}{h}(h - z)^2 z \frac{\sqrt{r^2 + h^2}}{h} dz = \frac{\pi}{6}\pi r h^2 \sqrt{h^2 + r^2}$. So $\bar{z} = \frac{h}{4}$.

15. Let $S$ be the 'upper' half of the torus, obtained by revolving the semi-circle $\{(x, 0, z) : (x - a)^2 + z^2 = b^2, z \geq 0\}$ around the $z$-axis, where $0 < b < a$ (see the figure next page). Then $M = 2\pi^2 ab$ while $M_{xy} = \int_{a-b}^{a+b} 2\pi xz \frac{ds}{dx} dx$

   $= \int_{a-b}^{a+b} 2\pi x \sqrt{b^2 - (x - a)^2} \sqrt{1 + \frac{(x - a)^2}{b^2 - (x - a)^2}} dx = 2\pi b \int_{a-b}^{a+b} x dx = 4\pi ab^2$.

   So $\bar{z} = \frac{M_{xy}}{M} = \frac{2b}{\pi}$. By symmetry $\bar{x} = \bar{y} = 0$.

   Suppose, however, that $S$ is the horizontal half of a torus obtained by revolving every point of the circle $C = \{(x, 0, z) : (x - a)^2 + z^2 = b^2\}$ through a semi-circle around the $z$-axis, so that $x \geq 0$ at all points of $S$. ($S$ looks like a bent tube.) This time $\bar{y} = 0 = \bar{z}$ by symmetry. $M$ is the same viz., $2\pi^2 ab$. To find $M_{yz}$, note that for $c \in (-b, b)$, the plane $z = c$ cuts $S$, into two semi-circles of radii $a - \sqrt{b^2 - c^2}$ and $a + \sqrt{b^2 - c^2}$, each centred at $(0, 0, c)$. The moments about $y$-$z$ plane of the semi-circular bands of width $ds$ each, along these semi-circles are simply $ds$ times the moments of these semi-circles along their diameters. By Problem (6.10), these latter moments are $2(a - \sqrt{b^2 - c^2})^2$ and $2(a + \sqrt{b^2 - c^2})^2$. So $M_{yz}$ equals

   $2\int_{-b}^{b} [(a - \sqrt{b^2 - z^2})^2 + (a + \sqrt{b^2 - z^2})^2] \frac{ds}{dz} dz = 4\int_{-b}^{b} (a^2 + b^2 - z^2) \sqrt{1 + \left(\frac{dx}{dz}\right)^2} dz$

$$= 4\int_{-b}^{b}(a^2 + b^2 - z^2)\sqrt{1 + \frac{z^2}{b^2 - z^2}}\,dz = 4b\int_{-b}^{b}(a^2 + b^2 - z^2)\frac{dz}{\sqrt{b^2 - z^2}}$$

$$= 2\pi b(2a^2 + b^2). \text{ So } \bar{x} = \frac{M_{yz}}{M} = \frac{(2a^2 + b^2)}{\pi a}.$$

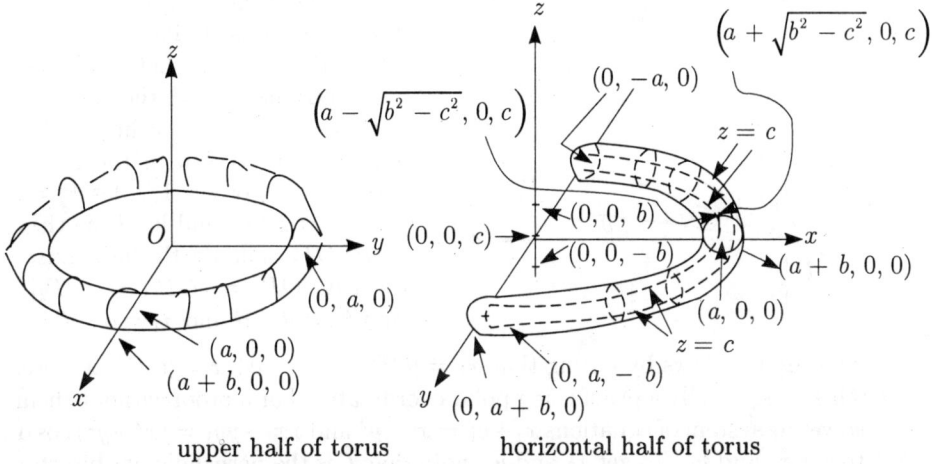

upper half of torus      horizontal half of torus

**16.** (i) Let the torus surface $S$ be as in last exercise. The upper or the lower half of $S$ can be easily parametrised in the form $z = f(x, y)$, where $(x, y)$ varies over the annulus $\{(x, y) : a - b \leq \sqrt{x^2 + y^2} \leq a + b\}$. The parametrisation of the entire $S$ is obtained analogously to that of a sphere shown in Fig.6.6.6. The distance $PM$ equals $a + b\cos v$ for some $v \in [0, 2\pi]$. Hence a parametrisation of $S$ is $\alpha(u, v) = (a + b\cos v)\cos u\mathbf{i} + (a + b\cos v)\sin u\mathbf{j} + b\sin v\mathbf{k}$, where $0 \leq u \leq 2\pi$ and $0 \leq v \leq \pi$. (More generally, the method works for any surface obtained by revolving a parametrised curve in the $x$-$z$ plane around the $z$-axis.)

(ii) Simply let $\alpha(u, v) = a\cos u\mathbf{i} + a\sin u\mathbf{j} + v\mathbf{k}$ for $0 \leq u \leq 2\pi$, $v \in \mathbb{R}$. (If the cylinder has a finite height, restrict $v$ suitably. More generally, this method works for any surface obtained by moving a parametrised curve in the $x$-$y$ plane parallel to the $z$-axis.)

**17.** Suppose $e$ is a point of impropriety of $\int_a^b f(x)dx$ (either because $e = \pm\infty$ or because $f$ is not bounded in a neighbourhood of $e$). Then $f^{-1}(e)$ (which is unique since $h$ is a bijection under the given hypothesis) is also a point of impropriety of $\int_a^b f(h(t))h'(t)dt$ . Assume $e \neq \pm\infty$. By Theorem (5.6.5),

$$\int_{e+\delta}^{b} f(x)dx = \int_{f^{-1}(e)+\epsilon}^{d} f(h(t))h'(t)dt \text{ where } \epsilon = f^{-1}(e + \delta) - f^{-1}(e). \text{ As}$$

$\delta \to 0^+, \epsilon \to 0^+$ and vice versa. So either both $\displaystyle\lim_{\delta \to 0^+} \int_{e+\delta}^{b} f(x)dx$ and

$\displaystyle\lim_{\epsilon \to 0^+} \int_{f^{-1}(e)+\epsilon}^{d} f(h(t))h'(t)dt$ exist and are equal or neither exists. A similar reasoning applies to other cases.

**18.** Merely break the interval $[c, d]$ at points where $h'$ is discontinuous and apply the last exercise to each subinterval.

**19.**

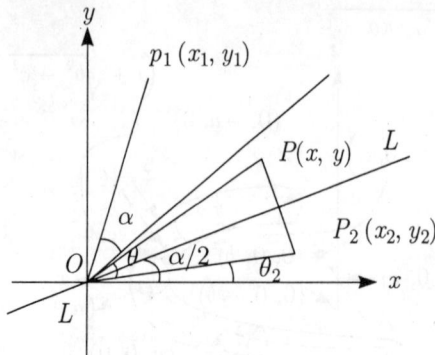

Call $(x, y)$ as $P$ and assume it is not the origin $O$ (as otherwise the assertion is trivial with $P_1 = P_2 = P = (0,0)$). Let $\theta$ be the angle $OP$ makes with the positive $x$-axis. Let $L$ be the line making an angle $\alpha/2$ with the positive $x$-axis. For $i = 1, 2$, let $P_i = (x_i, y_i) = T_i(P)$ and let $\theta_i$ be the angle $OP_i$ makes with the $x$-axis. Clearly $|OP| = |OP_1| = |OP_2|$ and $\theta_1 = \theta + \alpha$ and $\theta_2 = \alpha - \theta$.

The result follows by noting that $x_1 = |OP_1| \cos(\theta + \alpha), x = |OP| \cos \theta$ etc. This is essentially a proof using polar coordinates. For a proof without them solve the system of equations $x_1^2 + y_1^2 = x^2 + y^2$ and $xx_1 + yy_1 = (x^2 + y^2) \cos \alpha$ to get $x_1$ and $y_1$. To get $x_2$ and $y_2$, note that $L$ is the perpendicular bisector of the segment $PP_2$ and write a system of two equations for $x_2$ and $y_2$.

**20.** If $P, Q$ are distinct, then $A, B, C$ would all lie on the prependicular bisector of the line segment $PQ$ and hence be collinear.

**21.** Following the hint, since $T$ preserves all distances, $B, C$ are at unit distance from $A$ (which is $O$). Hence $B = (\cos \alpha, \sin \alpha)$ and $C = (\cos \beta, \sin \beta)$ for some $\alpha, \beta$. The condition $|BC| = \sqrt{2}$ gives $\beta = \alpha \pm \frac{\pi}{2}$. In the first case, let $T_1$ be the rotation around $O$ through the angle $\alpha$. In the second case let $T_1$ be reflection in the $x$-axis followed by a rotation around $O$ through angle $\alpha$. In both the cases show that for any point $P$ in the plane, $T(P)$ and $T_1(P)$ are equidistant from $A, B, C$ to get that they are equal.

For the second assertion, let $T$ be a given congruence of the plane. Let $A$ be the image of the origin and let $R$ be the parallel translation by the vector $\overrightarrow{AO}$. Then the composite $R \circ T$ is a congruence which fixes the origin. So by the first assertion, $R \circ T$ is either a rotation or a reflection followed by a rotation. Now $R^{-1}$ is the translation by the vector $\overrightarrow{OA}$ and $T$ is the composite $R^{-1} \circ (R \circ T)$.

**22.** Because of the symmetry of the sphere and of the population density function about the equator, the answer will be the same if we confine ourselves to the Northern hemisphere. Denote the latitude (in radians) by $\theta$, where $0 \leq \theta \leq \frac{\pi}{2}$. The probability density function $\rho$ is only a function of $\theta$, given by $\rho(\theta) = \frac{2}{\pi}(\frac{\pi}{2} - \theta)$. Let $R$ be the radius of the Earth. Then the portion of Earth between latitudes $\theta$ and $\theta + \Delta\theta$ has (approximate) area $2\pi(R\cos\theta)R\Delta\theta$ and accomodates (approximately) $2\pi R^2(\cos\theta)\frac{2}{\pi}(\frac{\pi}{2} - \theta)\Delta\theta$

people. So the total pulation is $4R^2 \int_0^{\pi/2} (\frac{\pi}{2} - \theta) \cos\theta d\theta = 4R^2$ and the desired probability is $\frac{1}{4R^2} \int_0^{\alpha} 4R^2 (\frac{\pi}{2} - \theta) \cos\theta d\theta$ where $\alpha$ radians $= 23.5°$ i.e. $\alpha = 0.4102$. An easy calculation gives this as $\frac{\pi}{2} \sin\alpha - \alpha \sin\alpha - \cos\alpha + 1$, i.e. $0.5457$. Note that this is independent of $R$.

## Section 6.7

1. For (i) differentiate $|\mathbf{w}|^2 = \mathbf{w} \cdot \mathbf{w}$ and use Exercise (6.6) (i). For (ii), differentiate $\left(\dfrac{1}{|\mathbf{w}|} \mathbf{w}\right)$ using Exercise (6.6) (ii) and use (i) to get $\left(\dfrac{1}{|\mathbf{w}|}\right)'$ as
$$\frac{1}{|\mathbf{w}|^2} \frac{\mathbf{w} \cdot \mathbf{w}'}{|\mathbf{w}|}.$$

2. Apply the last exercise (with $s = t$ and $\mathbf{w} = \dot{\mathbf{r}}$) to get $\dfrac{d}{dt}\left(\dfrac{\dot{\mathbf{r}}}{|\dot{\mathbf{r}}|}\right) =$
$\dfrac{(\dot{\mathbf{r}} \cdot \dot{\mathbf{r}})\ddot{\mathbf{r}} - (\dot{\mathbf{r}} \cdot \ddot{\mathbf{r}})\dot{\mathbf{r}}}{(\dot{\mathbf{r}} \cdot \dot{\mathbf{r}})^{3/2}}$. Now let $s$ be the arc length. Then $\dfrac{\dot{\mathbf{r}}}{|\dot{\mathbf{r}}|} = \mathbf{u}$ and $\mathbf{u}' =$
$\dfrac{d}{dt}\left(\dfrac{\dot{\mathbf{r}}}{|\dot{\mathbf{r}}|}\right) / \dfrac{ds}{dt}$. But $\dfrac{ds}{dt} = |\dot{\mathbf{r}}| = (\dot{\mathbf{r}} \cdot \dot{\mathbf{r}})^{1/2}$. So $\mathbf{u}' = \dfrac{(\dot{\mathbf{r}} \cdot \dot{\mathbf{r}})\ddot{\mathbf{r}} - (\dot{\mathbf{r}} \cdot \ddot{\mathbf{r}})\dot{\mathbf{r}}}{(\dot{\mathbf{r}} \cdot \dot{\mathbf{r}})^2}$. Finally
$\kappa = \sqrt{\mathbf{u}' \cdot \mathbf{u}'}$ gives $\kappa^2 = \mathbf{u}' \cdot \mathbf{u}' = \dfrac{[(\dot{\mathbf{r}} \cdot \dot{\mathbf{r}})\ddot{\mathbf{r}} - (\dot{\mathbf{r}} \cdot \ddot{\mathbf{r}})\dot{\mathbf{r}}] \cdot [(\dot{\mathbf{r}} \cdot \dot{\mathbf{r}})\ddot{\mathbf{r}} - (\dot{\mathbf{r}} \cdot \ddot{\mathbf{r}})\dot{\mathbf{r}}]}{(\dot{\mathbf{r}} \cdot \dot{\mathbf{r}})^4}$. To
finish the proof use, $(\mathbf{a} - \mathbf{b}) \cdot (\mathbf{a} - \mathbf{b}) = |\mathbf{a}|^2 + |\mathbf{b}|^2 - 2\mathbf{a} \cdot \mathbf{b}$.

3. Differentiating, $\dot{\mathbf{r}} = -a\sin t\mathbf{i} + b\cos t\mathbf{j}$ and $\ddot{\mathbf{r}} = -a\cos t\mathbf{i} - b\sin t\mathbf{j}$. So $\dot{\mathbf{r}} \cdot \dot{\mathbf{r}} = a^2\sin^2 t + b^2\cos^2 t = a^2(1 - e^2\cos^2 t)$, $\mathbf{r} \cdot \ddot{\mathbf{r}} = (a^2 - b^2)\sin t\cos t = a^2 e^2\sin t\cos t$ and $\ddot{\mathbf{r}} \cdot \ddot{\mathbf{r}} = a^2\cos^2 t + b^2\sin^2 t = a^2(1 - e^2\sin^2 t)$. The last exercise gives the first part by a straight computation. This formula also shows that the curvature is maximum (minimum) when $x^2$ is maximum (minimum). Hence the points of maximum curvature are $(\pm a, 0)$ and those of minimum curvature are $(0, \pm b)$.

4. The curvature at a point $(at^2, 2at)$ on the parabola $y^2 = 4ax$ is $\dfrac{1}{2a(1 + t^2)^{3/2}}$. For a hyperbola, the parametrisation $x = a\cosh t, y = b\sinh t$ is more convenient than the more customary $x = a\sec\theta, y = b\tan\theta$. Using $\cosh^2 t - \sinh^2 t = 1$, the curvature at the point $(a\cosh t, b\sin ht)$ comes out as
$\dfrac{\sqrt{e^2 - 1}}{a(e^2\cosh^2 t - 1)^{3/2}}$ or as $\dfrac{ab}{(e^2 x^2 - a^2)^{3/2}}$ where $e$, the eccentricity, is given by $a^2 + b^2 = a^2 e^2$. For the conical helix, the curvature at the point $(\theta\cos\theta, \theta\sin\theta, \theta)$ is $\dfrac{\sqrt{\theta^4 + 5\theta^2 + 8}}{(\theta^2 + 2)^{3/2}}$.

5. $\cos\alpha = \dfrac{\dot{\mathbf{r}} \cdot \mathbf{i}}{|\dot{\mathbf{r}}||\mathbf{i}|} = \dfrac{\dot{x}}{|\dot{\mathbf{r}}|}$. Similarly for $\cos\beta$ and $\cos\gamma$. If $t$ is the arc length then $|\dot{\mathbf{r}}| = 1$ and so $\cos\alpha, \cos\beta, \cos\gamma$ are simply $\dot{x}, \dot{y}, \dot{z}$ respectively.

**6.** (i) follows from the fact that the unit vector making an angle $\theta$ with the positive $x$-axis ($\theta \in [0, 2\pi]$) is $\cos\theta \mathbf{i} + \sin\theta \mathbf{j}$ and the formula $\mathbf{u} = \dot{\mathbf{r}}/|\dot{\mathbf{r}}| = \frac{\dot{x}}{|\dot{\mathbf{r}}|}\mathbf{i} + \frac{\dot{y}}{|\dot{\mathbf{r}}|}\mathbf{j}$. For (ii), differentiate (i) and note that $-\sin\phi = \cos(\phi + \frac{\pi}{2})$ and $\cos\phi = \sin(\phi + \frac{\pi}{2})$. For (iii) merely note that $\frac{d\mathbf{u}}{ds} = \frac{d\phi}{ds}\frac{d\mathbf{u}}{d\phi}$ and that $\frac{d\mathbf{u}}{d\phi} = -\sin\phi \mathbf{i} + \cos\phi \mathbf{j}$ is a unit vector. To get (iv), differentiate $\sin\phi = \frac{\dot{y}}{(\dot{x}^2 + \dot{y}^2)^{1/2}}$ to get $\cos\phi \frac{d\phi}{ds}\frac{ds}{dt} = \frac{\dot{x}(\dot{x}\ddot{y} - \dot{y}\ddot{x})}{(\dot{x}^2 + \dot{y}^2)^{3/2}}$ and put $\cos\phi = \frac{\dot{x}}{(\dot{x}^2 + \dot{y}^2)^{1/2}}$ and $\frac{ds}{dt} = |\dot{\mathbf{r}}| = \sqrt{\dot{x}^2 + \dot{y}^2}$.

**7.** Note that $\tan\phi = $ slope of the tangent $= f'(x)$. Differentiation w.r.t. $x$ gives $\sec^2\phi \frac{d\phi}{dx} = f''(x)$. So $\frac{d\phi}{ds} = \frac{f''(x)}{\sec^2\phi \frac{ds}{dx}} = \frac{f''(x)}{(1 + (f'(x))^2)^{3/2}}$. As the denominator is always positive, the signed curvature of a curve is positive (negative) throughout for a curve which is concave upwards (downwards).

**8.** $f''(x) = e^x > 0$ for all $x$. So the curve is concave upwards on $\mathbb{R}$. By the last exercise, $\kappa = \frac{e^x}{(1 + e^{2x})^{3/2}}$. An easy calculation shows that $\kappa$ is maximum when $x = \ln\frac{1}{\sqrt{2}}$. So the radius of curvature is minimum at $(\ln\frac{1}{\sqrt{2}}, \frac{1}{\sqrt{2}})$.

**9.** By Exercise (7.5), the road is steepest when $\gamma$ is minimum, i.e. $\cos\gamma$ is maximum. A direct calculation gives $\dot{s} = (t + 2)$ and $\cos\gamma = \frac{\dot{z}}{\dot{s}} = \frac{2\sqrt{t}}{t + 2}$ which attains its maximum on $[\frac{1}{2}, 3]$ at $t = 2$. Using Exercise (7.2), $\kappa$ comes out as $\frac{1}{\sqrt{t}(t + 2)^2}$ which is maximum at $t = \frac{1}{2}$. So the road is steepest at $(4, 2, \frac{8\sqrt{2}}{3})$ and bends most abruptly at $(1, \frac{1}{8}, \frac{\sqrt{2}}{3})$.

**10.** Denote the height by $z$. Take $x = 0, y = 0$. Then $\ddot{z}\mathbf{k} = \mathbf{a} = -g\mathbf{k}$. So $\ddot{z} = -g$, giving $\dot{z} = -gt + c$, for some constant $c$. $\dot{z}(0) = u$ determines $c$ as $u$. Itegrating $\dot{z} = -gt + u$, we get $z = -\frac{1}{2}gt^2 + ut + \lambda$ for some constant $\lambda$. $z(0) = h$ gives $\lambda = h$.

**11.** The motion is in the vertical plane containing the initial velocity vector. Taking this plane as the $x$-$z$ plane, $\ddot{x}\mathbf{i} + 0\mathbf{j} + \ddot{z}\mathbf{k} = \mathbf{a} = -g\mathbf{k}$. So $\dot{x} = c_1, \dot{z} = -gt + c_2$. The initial velocity $v_0\cos\alpha\mathbf{i} + v_0\sin\alpha\mathbf{k}$ gives $c_1 = v_0\cos a$ and $c_2 = v_0\sin\alpha$. Integrating further and taking the origin at the starting point, $x = x(t) = (v_0\cos\alpha)t$ and $z = z(t) = (v_0\sin\alpha)t - \frac{1}{2}gt^2$, which are parametric equations of a (vertically downward) parabola, with vertex at $\left(\frac{v_0^2\sin\alpha\cos\alpha}{g}, \frac{v_0^2\sin^2\alpha}{g}\right)$. The particle hits the ground again at $t = $

$$\frac{2v_0 \sin \alpha}{g} \text{ at the point } \left( \frac{2v_0 \sin \alpha \cos \alpha}{g}, 0, 0 \right).$$

**12.** Take the centre of the Earth at $(0,0,0)$. The equation of motion is $\ddot{z} = -\frac{GM}{z^2}$. Since $v = $ speed upwards $= \dot{z}$, $\ddot{z}$ equals $\frac{dv}{dt}$ and hence $\frac{dv}{dz}\frac{dz}{dt} = \frac{dv}{dz}v$.

So $\frac{dv}{dz}v = -\frac{GM}{z^2}$. Integrating w.r.t.z, $v^2 = \frac{2GM}{z} + v_0^2 - \frac{2GM}{R}$ where $v_0 = $ initial speed. When the rocket starts returning to the Earth, $v = 0$ but $z > 0$. If $v_0^2 - \frac{2GM}{R} > 0$, this can never happen.

**13.** Apply (22) with $\kappa = \frac{1}{\ell}, v = \ell\omega$. Then the centripetal force on the particle

is $\frac{m}{\ell}\ell^2\omega^2 = m\ell\omega^2$ in magnitude. This must not exceed $T$. So $\omega \leq \sqrt{\frac{T}{m\ell}}$.

**14.** The direct implications are trivial. For the converse of (a), $\kappa \equiv 0$ gives $\mathbf{u} = \mathbf{a}$ a constant vector, say, $a\mathbf{i} + b\mathbf{j} + c\mathbf{k}$. So $\mathbf{r} = \mathbf{r}(s) = (as + a_0)\mathbf{i} + (bs + b_0)\mathbf{j} + (cs + c_0)\mathbf{k}$ which is a straight line. Similarly for (b), $\tau \equiv 0$ gives $\mathbf{b} = $ a constant vector, say, $a\mathbf{i} + b\mathbf{j} + c\mathbf{k}$. But then, $\dot{\mathbf{u}} \cdot \mathbf{b} = 0$ give $a\frac{dx}{ds} + b\frac{dy}{ds} + c\frac{dz}{ds} \equiv 0$, i.e., $\frac{d}{ds}(ax + by + cz) \equiv 0$. So $ax + by + cz = \lambda$, a constant. Hence the curve is planar.

**15.** (i) follows straight from the definition. For (ii) take the dot product of $\mathbf{u} \times \mathbf{v}$ with $\mathbf{u}$ (and with $\mathbf{v}$) and show it is 0. For (iii) write $|\mathbf{w}|^2 = (u_2v_3 - u_3v_2)^2 + (u_3v_1 - u_1v_3)^2 + (u_1v_2 - u_2v_1)^2$ as $(u_1^2 + u_2^2 + u_3^2)(v_1^2 + v_2^2 + v_3^2) - (u_1v_1 + u_2v_2 + u_3v_3)^2$ and use $\cos\theta = u_1v_1 + u_2v_2 + u_3v_3$. (iv) is, strictly speaking, a non-mathematical statement and so only an intuitive argument can be given. Its truth depends upon the assumption that the system $\mathbf{i}, \mathbf{j}, \mathbf{k}$ is right-handed. (Note that here the order of $\mathbf{i}, \mathbf{j}, \mathbf{k}$ matters. The systems $\mathbf{i}, \mathbf{j}, \mathbf{k}$ and $\mathbf{i}, \mathbf{k}, \mathbf{j}$ can never be both right handed or both left handed.) For every pair of non-zero, non-parallel vectors $\mathbf{a}, \mathbf{b}$, there are precisely two vectors, say, $r(\mathbf{a}, \mathbf{b})$ and $l(\mathbf{a}, \mathbf{b})$ which are perpendicular to both $\mathbf{a}$ and $\mathbf{b}$ and have length $|\mathbf{a}||\mathbf{b}|\sin\theta$ each, where $\theta$ is the angle between $\mathbf{a}$ and $\mathbf{b}$. They are oppositely directed and we assume that the labelling is so done that $(\mathbf{a}, \mathbf{b}, r(\mathbf{a}, \mathbf{b}))$ is right-handed while $(\mathbf{a}, \mathbf{b}, l(\mathbf{a}, \mathbf{b}))$ is left-handed. By (ii) and (iii), $\mathbf{a} \times \mathbf{b}$ equals one of the two vectors $r(\mathbf{a}, \mathbf{b})$ and $l(\mathbf{a}, \mathbf{b})$. Now move the pair $(\mathbf{i}, \mathbf{j})$ continuously to the pair $(\mathbf{u}, \mathbf{v})$. This means find continuous vector-valued functions $\mathbf{a}(t)$ and $\mathbf{b}(t); 0 \leq t \leq 1$ so that $\mathbf{a}(0) = \mathbf{i}, \mathbf{b}(0) = \mathbf{j}, \mathbf{a}(1) = \mathbf{u}$ and $\mathbf{b}(1) = \mathbf{u}$. Ensure that for every value of $t, \mathbf{a}(t)$ and $\mathbf{b}(t)$ are non-zero and not multiples of each other. The simplest choice is $\mathbf{a}(t) = (1 - t + tu_1)\mathbf{i} + tu_2\mathbf{j} + tu_3\mathbf{k}$ and $\mathbf{b}(t) = tv_1\mathbf{i} + (1 - t + tv_2)\mathbf{j} + tv_3\mathbf{k}$. This will work in all except a few degenerate cases (e.g. when $\mathbf{u} = -\lambda\mathbf{i}$ or $\mathbf{v} = -\mu\mathbf{j}$ for some positive $\lambda, \mu$) in which slight modifications would suffice. Now $\mathbf{a}(t) \times \mathbf{b}(t)$ is a continuous function of $t$ since its components are continuous functions of $t$. The functions $l(\mathbf{a}(t), \mathbf{b}(t))$ and $r(\mathbf{a}(t), \mathbf{b}(t))$ are also continuous functions of $t$. (For this, there is no mathematical proof. Indeed, this is the intuitive part of the argument.) For every $t \in [0,1]$, $\mathbf{a}(t) \times \mathbf{b}(t)$ coincides with one of $r(\mathbf{a}(t), \mathbf{b}(t))$ and $l(\mathbf{a}(t), \mathbf{b}(t))$ and is opposite to the other. By continuity, either $\mathbf{a}(t) \times \mathbf{b}(t) = r(\mathbf{a}(t), \mathbf{b}(t))$ for all $t \in [0,1]$ or $\mathbf{a}(t) \times \mathbf{b}(t) = l(\mathbf{a}(t), \mathbf{b}(t))$ for all $t \in [0,1]$. The right-handedness of $(\mathbf{i}, \mathbf{j}, \mathbf{k})$ implies the first possibility holds at $t = 0$ and hence everywhere. Put $t = 1$ to get the right-handedness of $(\mathbf{u}, \mathbf{v}, \mathbf{u} \times \mathbf{v})$.

Finally, for (v), first assume $\mathbf{u}, \mathbf{v}$ are non-zero and not parallel. Let $\mathbf{w}$ be the vector defined by the given symbolic determinant. Then duplicating the reasoning in (i) to (iv), $(\mathbf{u}, \mathbf{v}, \mathbf{w})$ is a right handed system with $|\mathbf{w}| = |\mathbf{u}||\mathbf{v}| \sin \theta$, $\theta$ being the angle between $\mathbf{u}$ and $\mathbf{v}$. There cannot be two distinct vectors with these properties. Since $\mathbf{u} \times \mathbf{v}$ also has these properties, $\mathbf{u} \times \mathbf{v}$ must equal $\mathbf{w}$. If $\mathbf{u}$ or $\mathbf{v}$ is $\mathbf{0}$ or they are parallel to each other, then $\mathbf{u} \times \mathbf{v}$ and $\mathbf{w}$ are both $\mathbf{0}$ and hence equal. For a proof not using (iv) (which is based on an intuitive argument), note that because of (i), it suffices to verify (v) in the 9 cases when $\mathbf{u}$ and $\mathbf{v}$ equal $\mathbf{i}, \mathbf{j}$ or $\mathbf{k}$. Three of these (where $\mathbf{u} = \mathbf{v}$) are trivial. The others are mutually similar. So, suppose $\mathbf{u} = \mathbf{i}$ and $\mathbf{v} = \mathbf{j}$. Let $\mathbf{p} = p_1\mathbf{i} + p_2\mathbf{j} + p_3\mathbf{k}$, $q = q_1\mathbf{i} + q_2\mathbf{j} + q_3\mathbf{k}$ and $\mathbf{r} = r_1\mathbf{i} + r_2\mathbf{j} + r_3\mathbf{k}$. Substituting these in $\mathbf{i} = a_1\mathbf{p} + a_2\mathbf{q} + a_3\mathbf{r}$ and equating the coefficients of $\mathbf{i}, \mathbf{j}, \mathbf{k}$ on the two sides gives the following system of equations in the unknowns $a_1, a_2, a_3$

$$\begin{aligned} a_1p_1 + a_2q_1 + a_3r_1 &= 1 \\ a_1p_2 + a_2q_2 + a_3r_2 &= 0 \\ \text{and} \quad a_1p_3 + a_2q_3 + a_3r_3 &= 0. \end{aligned}$$

Since $r_1^2 + r_2^2 + r_3^2 = 1$ and $r_1 = p_2q_3 - p_3q_2$ etc., this system can be solved by the well-known Cramer's rule (or directly), to get $a_1 = q_2r_3 - q_3r_2$, $a_2 = r_2p_3 - r_3p_2$ and $a_3 = p_2q_3 - p_3q_2$. Expressing $r$'s in terms of $p$'s and $q$'s and using $p_1^2 + p_2^2 + p_3^2 = 1 = q_1^2 + q_2^2 + q_3^2$ and $p_1q_1 + p_2q_2 + p_3q_3 = 0$, we get $a_1 = p_1, a_2 = q_1$ and $a_3 = r_1 = p_2q_3 - p_3q_2$. So $\mathbf{i} = p_1\mathbf{p} + q_1\mathbf{q} + r_1\mathbf{r}$. Similarly, $\mathbf{j} = p_2\mathbf{p} + q_2\mathbf{q} + r_2\mathbf{r}$ and $\mathbf{k} = p_3\mathbf{p} + q_3\mathbf{q} + r_3\mathbf{r}$. (There are elegant ways using matrices to get these. But they are beyond our scope.) A straightforward subsitu-

tion now shows $\begin{vmatrix} \mathbf{p} & \mathbf{q} & \mathbf{r} \\ p_1 & q_1 & r_1 \\ p_2 & q_2 & r_2 \end{vmatrix} = (q_1r_2 - q_2r_1)\mathbf{p} + (r_1p_2 - r_2p_1)\mathbf{q}$

$+(p_1q_2 - p_2q_1)\mathbf{r} = p_3\mathbf{p} + q_3\mathbf{q} + r_3\mathbf{r} = \mathbf{k}$ as was to be proved !

16. (i) By the last exercise, $\mathbf{v} \times \mathbf{w} = a_1\mathbf{p} + a_2\mathbf{q} + a_3\mathbf{r}$ where $a_1 = v_2w_3 - v_3w_2$ etc. By the properties of the dot product given in Exercise (4.8.13), and the fact that $\mathbf{p} \cdot \mathbf{q} = \mathbf{q} \cdot \mathbf{r} = \mathbf{r} \cdot \mathbf{q} = 0$ while $\mathbf{p} \cdot \mathbf{p} = 1$ etc., $\mathbf{u} \cdot (\mathbf{v} \times \mathbf{w})$ is precisely the given determinant. (ii) First note that $|\mathbf{v} \times \mathbf{w}|$ is the area of the parallelogram with sides $\mathbf{v}$ and $\mathbf{w}$. Now, $|(\mathbf{u}\ \mathbf{v}\ \mathbf{w})| = |\mathbf{u}|\ |(\mathbf{v} \times \mathbf{w})| \cos \phi$ where $\phi$ is the angle between $\mathbf{v} \times \mathbf{w}$ and $\mathbf{u}$. But $|\mathbf{u}| \cos \phi$ is the perpendicular distance between the two parallel faces of the parallelopiped, each having $\mathbf{v}$ and $\mathbf{w}$ along its edges. Each slice parallel to these faces has the same area, viz., $|\mathbf{v} \times \mathbf{w}|$. So $(|\mathbf{v} \times \mathbf{w}|)(|\mathbf{u}| \cos \phi)$ is the volume of the box.

17. Resolve $\mathbf{u}, \mathbf{v}$ and $\mathbf{u} \times \mathbf{v}$ along $\mathbf{i}, \mathbf{j}$ and $\mathbf{k}$ and differentiate componenetwise.

**18.** Following the hint, $\dfrac{d\mathbf{p}}{ds} = \dfrac{d\mathbf{b}}{ds} \times \mathbf{u} + \mathbf{b} \times \dfrac{d\mathbf{u}}{ds} = -\tau\mathbf{p} \times \mathbf{u} + \mathbf{b} \times \kappa\mathbf{p} = \tau\mathbf{b} - \kappa\mathbf{u}.$

**19.** (8), (25) and (30) serve to resolve $\mathbf{r}', \mathbf{r}'', \mathbf{r}'''$ along the right-handed system $(\mathbf{u}, \mathbf{p}, \mathbf{b})$ of mutually perpendicular unit vectors, viz. $\mathbf{r}' = 1\mathbf{u} + 0\mathbf{p} + 0\mathbf{b}$, $\mathbf{r}'' = \mathbf{u}' = 0\mathbf{u} + \kappa\mathbf{p} + 0\mathbf{b}$ and $\mathbf{r}''' = (\kappa\mathbf{p})' = \kappa\mathbf{p}' + \kappa'\mathbf{p} = -\kappa^2\mathbf{u} + \kappa'\mathbf{p} + \tau\kappa\mathbf{b}$. By

Exercise (7.16), $(\mathbf{r}'\,\mathbf{r}''\,\mathbf{r}''')$ is the determinant $\begin{vmatrix} 1 & 0 & 0 \\ 0 & \kappa & 0 \\ -\kappa^2 & \kappa' & \tau\kappa \end{vmatrix} = \tau\kappa^2.$

**20.** The first part follows from Exercises (7.2) and (7.15) (iii). For the second part, note that $\dot{\mathbf{r}} = \mathbf{r}'\dot{s} = \dot{s}\mathbf{u}$, $\ddot{\mathbf{r}} = \dot{s}(\mathbf{r}'\dot{s})' = \dot{s}(\mathbf{r}''\dot{s} + \mathbf{r}'\frac{\ddot{s}}{\dot{s}}) = \dot{s}^2\kappa\mathbf{p} + \ddot{s}\mathbf{u}$. Now, $(\dot{\mathbf{r}}\,\ddot{\mathbf{r}}\,\dddot{\mathbf{r}})$ also equals $(\dot{\mathbf{r}} \times \ddot{\mathbf{r}}) \cdot \dddot{\mathbf{r}} = (\dot{s}^3\kappa\mathbf{b}) \cdot \dddot{\mathbf{r}}$. To evaluate it we only need the component of $\dddot{\mathbf{r}}$ along $\mathbf{b}$ (and not the components along $\mathbf{u}$ and $\mathbf{p}$). Now, $\dddot{\mathbf{r}} = \dot{s}(\ddot{\mathbf{r}})' = \dot{s}(\dot{s}^2\kappa\mathbf{p} + \ddot{s}\mathbf{u})'$. The coefficient of $\mathbf{b}$ in this comes only from $\mathbf{p}'$ which equals $\tau\mathbf{b} - \kappa\mathbf{u}$ by Exercise (7.19). So, $(\dot{\mathbf{r}}\,\ddot{\mathbf{r}}\,\dddot{\mathbf{r}}) = \dot{s}^6\kappa^2\tau$. Substitution for $\kappa$ gives the result.

**21.** The torsion at a point $(\theta\cos\theta, \theta\sin\theta, \theta)$ is $\dfrac{\theta^2 + 6}{\theta^4 + 5\theta^2 + 8}$.

**22.** Mathematically, 'wrapping' is a parametrisation which preserves lengths of horizontal as well as vertical line elements (as otherwise there would be a stretching or compressing.) Modify the parametrisation in the answer to Exercise (6.16)(ii) so that arc lengths are preserved. Thus, the parametrisation $\alpha(u, v) = a\cos(\frac{u}{a})\mathbf{i} + a\sin(\frac{u}{a})\mathbf{j} + v\mathbf{k}$ transforms the line $u = at$, $v = ct$ i.e. the line $v = \frac{c}{a}u$ in the $u$-$v$ plane into the helix considered in the text.

**23.** Let $\mathbf{e}$ be a unit vector along the fixed direction. Then $\mathbf{u} \cdot \mathbf{e} = $ constant. So $(\mathbf{u} \cdot \mathbf{e})' = 0$, which gives $\kappa\mathbf{p} \cdot \mathbf{e} = 0$ and hence $\mathbf{p} \cdot \mathbf{e} = 0$. (We assume $\kappa \neq 0$ as otherwise $\mathbf{p}$ is undefined.) Differentiating, $\mathbf{p}' \cdot \mathbf{e} = 0$. By Exercise (7.18), this gives $\kappa(\mathbf{u} \cdot \mathbf{e}) = \tau(\mathbf{b} \cdot \mathbf{e})$. Also $(\mathbf{b} \cdot \mathbf{e})' = -\tau\mathbf{p} \cdot \mathbf{e} = 0$. So $\mathbf{b} \cdot \mathbf{e} = $ constant. Since $\mathbf{u} \cdot \mathbf{e}$ is also a constant, $\kappa/\tau$ is a constant. For the converse, suppose $\kappa = \lambda\tau$ where $\lambda$ is a constant. Let $\mathbf{e} = \mathbf{u} + \lambda\mathbf{b}$. Then $\mathbf{e}' = \mathbf{u}' + \lambda\mathbf{b}' = \kappa\mathbf{p} - \lambda\tau\mathbf{p} = 0$, whence $\mathbf{e}$ is a constant vector. Also $\mathbf{e} \neq 0$ since $|\mathbf{e}| = \sqrt{1 + \lambda^2} \geq 1$. Now, $(\mathbf{u} \cdot \mathbf{e})' = \kappa\mathbf{p} \cdot (\mathbf{u} + \lambda\mathbf{b}) = 0$ since $\mathbf{p}, \mathbf{u}, \mathbf{b}$ are mutually perpendicular. So $\mathbf{u} \cdot \mathbf{e} = $ constant, i.e. the tangent makes a fixed angle with $\mathbf{e}$.

**24.** Following the hint, the unit vectors along the axis, along the normal at $P$ and along $\overrightarrow{PF}$ come out to be, respectively, $\mathbf{i}$, $\cos\theta\mathbf{i} - \sin\theta\mathbf{j}$ and $\cos 2\theta\mathbf{i} - \sin 2\theta\mathbf{j}$. The result follows by taking the relevant dot products. For an ellipse with foci at $F_1$ and $F_2$, the normal at any point $P$, divides the angle between $PF_1$ and $PF_2$. So the echo of a sound emanating from $F_1$ can be heard distinctly at $F_2$ as in a whispering gallery. (From the comments after the proof of Theorem (6.8.1), a parabola is a limiting case of an ellipse as one of its foci tends to infinity. The result proved here is consistent with this observation.)

**25.** A unit vector along the direction $\overrightarrow{P_0C_0}$ is $\dfrac{-y_0'\mathbf{i}+\mathbf{j}}{\sqrt{1+y_0'^2}}$ and so the point $C_0$ is $\left(x_0 - \dfrac{\rho y_0'}{\sqrt{1+y_0'^2}},\ y_0 + \dfrac{\rho}{\sqrt{1+y_0'^2}}\right)$ and $\sigma^2 = \left(\Delta x + \dfrac{\rho y_0'}{\sqrt{1+y_0'^2}}\right)^2 +$

$\left(f(x_0+\Delta x) - y_0 - \dfrac{\rho}{\sqrt{1+y_0'^2}}\right)^2.$

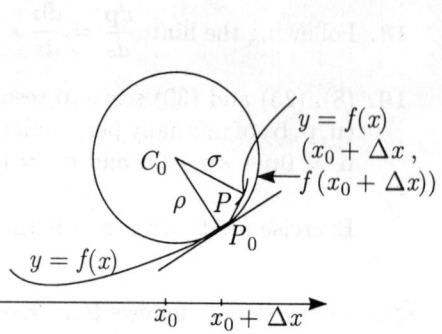

(i), (ii) and (iii) are obtained by respectively putting $\Delta x = 0$, $f(x_0 + \Delta x) - y_0 = y_0'\Delta x$ and $f(x_0+\Delta x) - y_0 = y_0'\Delta x + \frac{1}{2}y_0''(\Delta x)^2$, the last one also requiring $\dfrac{1}{\rho} = \dfrac{y_0''}{(1+y_0'^2)^{3/2}}$. (Here it is tacitly assumed that $y_0'' > 0$. A minor modification is needed if $y_0'' < 0$. Or simply replace $f(x)$ by $-f(x)$.) (ii) is reasonable to expect because upto the first order of approximation, $C_0P_0P$ is a right- angled triangle with $C_0P$ as the hypotenuse. When $y_0' \neq 0$, the term $y_0'y_0''(\Delta x)^3$ is the dominating term in $\sigma^2 - \rho^2$, for sufficiently small $|\Delta x|$. So $\sigma^2 - \rho^2$ has a different sign for $\Delta x > 0$ than for $\Delta x < 0$.

## Section 6.8

**1.** For the first part apply the cosine rule to the triangle formed by the line joining the two points and the lines through them parallel to the axes. The equation of a circle with centre $(a, b)$ and radius $r$ is $Ax^2 + By^2 + 2Hxy + 2Fx + 2Gy + C = 0$ where $A = B = 1, H = \cos\alpha, F = -a - b\cos\alpha, G = -b - a\cos\alpha$ and $C = a^2 + b^2 + 2ab\cos\alpha - r^2$. Clearly the coefficients satisfy the given conditions. Conversely, if $A = B\ (\neq 0)$ and $H = A\cos\alpha$, then dividing throughout by $A$ we may suppose $A = B = 1$ and $H = \cos\alpha$. Determine $a, b$ by solving the system $a + (\cos\alpha)b = -F$ and $(\cos\alpha)a + b = -G$. Then the equation represents a circle with centre $(a, b)$ and radius $\sqrt{a^2 + b^2 + 2ab\cos\alpha - C}$. For the last assertion, a direct argument can be given. But a slicker way is to set up a rectangular cartesian co-ordinate system $(x', y')$ with the same origin and the $x'$-axis along the $x$- axis. Then a point $(x, y)$ in the oblique system corresponds to $(x', y')$ in the rectangular system where $x' = x + y\cos\alpha$ and $y' = y\sin\alpha$. Note that

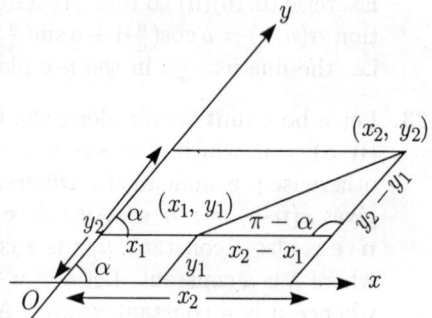

the determinants $\begin{vmatrix} x_1 + y_1\cos\alpha & y_1\sin\alpha & 1 \\ x_2 + y_2\cos\alpha & y_2\sin\alpha & 1 \\ x_3 + y_3\sin\alpha & y_3\sin\alpha & 1 \end{vmatrix}$ and $\begin{vmatrix} x_1 & y_1 & 1 \\ x_2 & y_2 & 1 \\ x_3 & y_3 & 1 \end{vmatrix}$ are either both zero or both non-zero. So the condition for collinearity remains the same. Replacing $x_3$ and $y_3$ in the second determinant by $x$ and $y$ respectively gives the equation of the straight line through $(x_1, y_1)$ and $(x_2, y_2)$ as $(y_1 - y_2)x - (x_1 - x_2)y + (x_1 y_2 - x_2 y_1) = 0$. This is a linear equation.

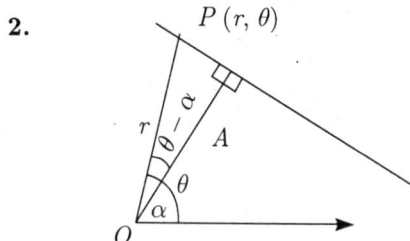

**2.**

$P\,(r,\,\theta)$

$A$ and $\alpha$ are, respectively, the length of and the angle with the axis made by the perpendicular from the pole to the line. If the line passes through the pole then its equation is of the form $(\theta - \alpha)(\theta - \pi - \alpha) = 0$ for some $\alpha$.

**3.** (i) $\dfrac{ed}{1 - e^2}$  (ii) $\left(\dfrac{2e^2 d}{1 - e^2}, \pi\right)$

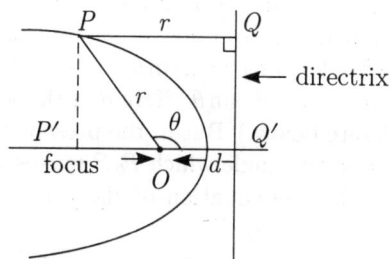

**4.** See the figures below.

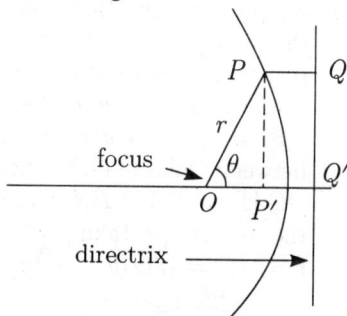

← directrix

focus

directrix

**5.** See the figures below. (c) to (f) suggest that the graphs of $r = \sin n\theta$ or of $r = \cos n\theta$ would be '$n$-petaled roses', with the 'petals' occuring in alternate sectors of angular width $\dfrac{\pi}{n}$ each.

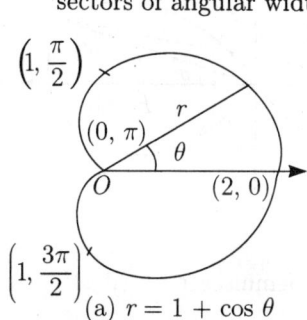

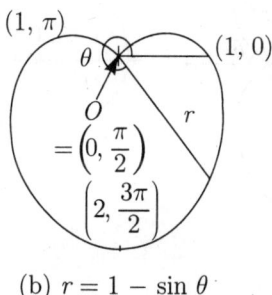

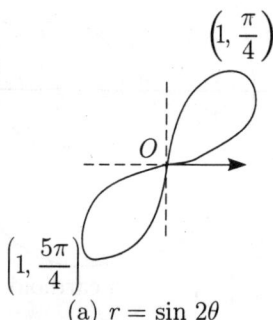

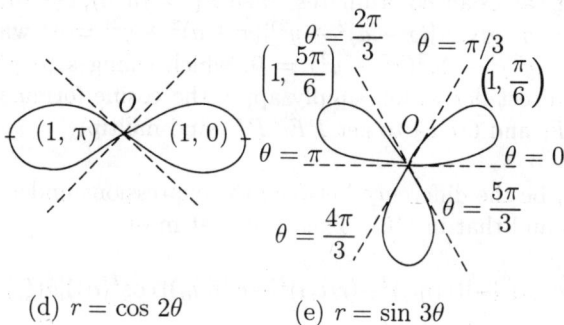

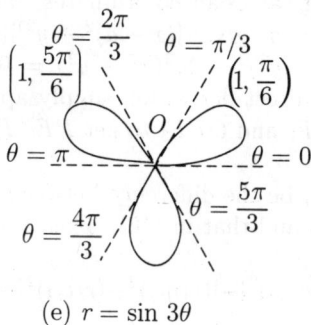

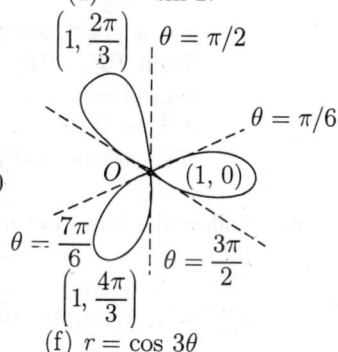

$\left(1, \dfrac{\pi}{2}\right)$

$(0, \pi)$

$(2, 0)$

$\left(1, \dfrac{3\pi}{2}\right)$

(a) $r = 1 + \cos\theta$

$(1, \pi)$

$(1, 0)$

$= \left(0, \dfrac{\pi}{2}\right)$

$\left(2, \dfrac{3\pi}{2}\right)$

(b) $r = 1 - \sin\theta$

$\left(1, \dfrac{\pi}{4}\right)$

$\left(1, \dfrac{5\pi}{4}\right)$

(a) $r = \sin 2\theta$

$(1, \pi)$   $(1, 0)$

(d) $r = \cos 2\theta$

$\theta = \dfrac{2\pi}{3}$   $\theta = \pi/3$

$\left(1, \dfrac{5\pi}{6}\right)$   $\left(1, \dfrac{\pi}{6}\right)$

$\theta = \pi$   $\theta = 0$

$\theta = \dfrac{4\pi}{3}$   $\theta = \dfrac{5\pi}{3}$

(e) $r = \sin 3\theta$

$\left(1, \dfrac{2\pi}{3}\right)$   $\theta = \pi/2$

$\theta = \pi/6$

$O$   $(1, 0)$

$\theta = \dfrac{7\pi}{6}$   $\theta = \dfrac{3\pi}{2}$

$\left(1, \dfrac{4\pi}{3}\right)$

(f) $r = \cos 3\theta$

**6.** The number of petals would be doubled for (c) and (d). For (e) and (f), the same petals would get traversed twice. Thus, (c) and (d) would become, respectively,

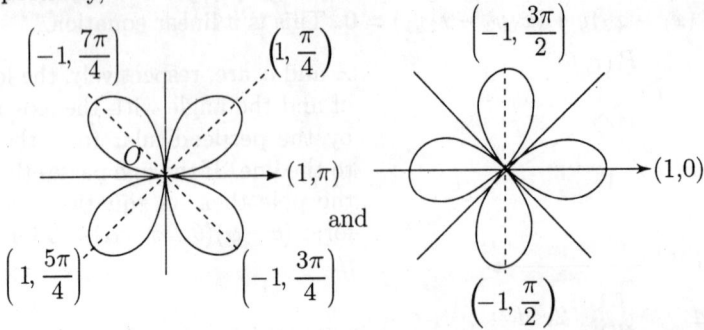

$$\left(-1, \frac{7\pi}{4}\right) \qquad \left(1, \frac{\pi}{4}\right) \qquad \left(-1, \frac{3\pi}{2}\right)$$

$$\qquad\qquad\qquad (1,\pi) \qquad\qquad\qquad (1,0)$$

and

$$\left(1, \frac{5\pi}{4}\right) \qquad \left(-1, \frac{3\pi}{4}\right) \qquad \left(-1, \frac{\pi}{2}\right)$$

**7.**   (a) Keeping in mind that now $a$ is the diameter and not the radius of both the circles and shifting the origin to the starting position of the moving point (i.e. to the point $P_0 = (\frac{a}{2}, 0)$), the equation of the epicycloid becomes $x(\theta) = a\cos\theta - \frac{a}{2}\cos 2\theta - \frac{a}{2}$; $y(\theta) = a\sin\theta - \frac{a}{2}\sin 2\theta$, i.e. $x(\theta) = a(1 - \cos\theta)\cos\theta$; $y(\theta) = a(1 - \cos\theta)\sin\theta$. Here $\theta$ is the angle between $OQ$ and $OP_0$. (See the figure below.) But in the present case, $OQ$ is parallel to $P_0P$ and so $\theta$ is also the angle which $P_0P$ makes with the $x$-axis. So taking $P_0$ as the pole, the equation of the cardioid is $r = r(\theta) = \sqrt{(x(\theta))^2 + (y(\theta))^2} = a(1 - \cos\theta)$.

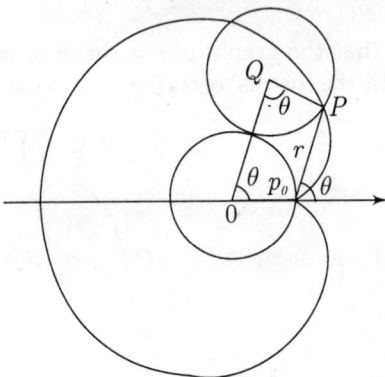

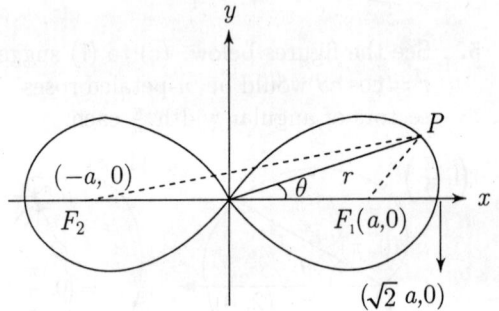

a cardioid                                              a lemniscate

(b) For a proof using cartesian co-ordinates, take $F_1 = (a, 0), F_2(-a, 0)$. Then $PF_1 \cdot PF_2 = a^2$ gives $[(x - a)^2 + y^2][(x + a)^2 + y^2] = a^4$ which simplifies to $(x^2 + y^2)^2 - 2a^2(x^2 - y^2) = 0$, which changes to $r^2 = 2a^2\cos 2\theta$. For a direct derivation simply apply the cosine formula to the triangles $OPF_1$ and $OPF_2$ to get $PF_1, PF_2$ and multiply.

**8.** Follow the hint. Let $u_i$ be the difference between the expressions under the $i^{\text{th}}$ radical sign in (15) and that in (16). Then $|u_i|$ is at most

$$r_i^2|(\cos^2\alpha_i)(\dot\theta(\xi_i))^2 - (\dot\theta(t_i))^2| + |(\dot r(\eta_i))^2 - (\dot r(t_i))^2| + r_i|\dot r(\eta_i)|\cos^2(\alpha_i)|\dot\theta(\xi_i)|^2\Delta t_i$$

$$(*)$$

Taking $A = \cos^2 \alpha_i, B = (\dot\theta(\xi_i))^2, A' = 1$ and $B' = (\dot\theta(t_i))^2$ in the last part of the hint, we get $|(\cos^2 \alpha_i)(\dot\theta(\xi_i))^2 - (\dot\theta(t_i))^2| \leq (\sin^2 \alpha_i)(\dot\theta(\xi_i))^2 + |(\dot\theta(\xi_i))^2 - (\dot\theta(t_i))^2|$. Further, $\sin^2 \alpha_i \leq \alpha_i^2 \leq \dfrac{(\Delta\theta_i)^2}{4} = \dfrac{1}{4}(\dot\theta(\xi_i))^2(\Delta t_i)^2$. Hence the first term of the R.H.S. of (*) is at most $M^2(\dfrac{M^2}{4}(\Delta t_i)^2) + M^2|(\dot\theta(\xi_i))^2 - (\dot\theta(\eta_i))^2|$. Also, the last term is at most $M^4\Delta t_i$ since $\cos^2 \alpha_i \leq 1$. Hence

$$|u_i| \leq \frac{M^4}{4}(\Delta t_i)^2 + M^2|(\dot\theta(\xi_i))^2 - (\dot\theta(t_i))^2| + |(\dot r(\eta_i))^2 - (\dot r(t_i))^2| + M^4\Delta t_i \quad (**)$$

Use uniform continuity of $\dot r^2$ and $\dot\theta^2$ on $[a, b]$ to find $\delta_1 > 0$ such that whenever $0 < \Delta t_i < \delta, |(\dot\theta(\xi_i))^2 - (\dot\theta(t_i))^2| < \dfrac{\epsilon_1}{4M^2}$ and $|\dot r(\eta_i))^2 - (\dot r(t_i))^2| < \dfrac{\epsilon_1}{4}$ where $\epsilon_1 = (\dfrac{\epsilon}{b-a})^2$. Also the first and the last terms of (**) can be made less than $\dfrac{\epsilon_1}{4}$ each by taking $\Delta t_i < \dfrac{\sqrt{\epsilon_1}}{M^2}$ and $\Delta t_i < \dfrac{\epsilon_1}{4M^4}$ respectively. Let $\delta = \min\{\delta_1, \dfrac{\sqrt{\epsilon_1}}{M^2}, \dfrac{\epsilon_1}{4M^4}\}$. Then $\mu(R) < \delta$ ensures $|u_i| < \epsilon_1$ and as noted in the hint, completes the proof.

9. This follows straight from Theorem (8.2) by taking $\theta = t$ and $r = f(t)$ as a parametrisation of $C$. The answers to (i) and (ii) are $8a$ and $3\pi\sqrt{36\pi^2 + 1} + \dfrac{1}{2}\ln(6\pi + \sqrt{36\pi + 1})$ respectively. For the lemniscate the length of each loop is $a\sqrt{2}\int_{-\pi/4}^{\pi/4}\sqrt{\sec 2\theta}d\theta$ which cannot be evaluated in a closed form.

10. The length of a line element is $ds = \sqrt{\dot r^2 + r^2\dot\theta^2}dt$. Its perpendicular distance from the axis is $r \sin\theta$ while that from the $y$-axis is $r \cos\theta$. Hence the surface area is $\int_a^b 2\pi r(t)\sin(\theta(t))\sqrt{(\dot r(t))^2 + (r(t))^2\dot\theta(t))^2}dt$ for (i) and $\int_a^b 2\pi r(t)\cos(\theta(t))\sqrt{(\dot r(t))^2 + (r(t))^2(\dot\theta(t))^2}dt$ for (ii). For the lemniscate $r^2 = 2a^2\cos 2\theta$, these are $8\pi a^2(1 - \dfrac{1}{\sqrt{2}})$ and $\dfrac{8\pi a^2}{\sqrt{2}}$ respectively.

11. $\Delta A \approx$ area of $\triangle OPQ = \dfrac{1}{2}r(r+\Delta r)\sin\Delta\theta$. So, $\dfrac{dA}{d\theta} = \lim\limits_{\Delta\theta \to 0}\dfrac{1}{2}r(r+\Delta r)\dfrac{\sin\Delta\theta}{\Delta\theta} = \dfrac{1}{2}r^2$. Hence the area swept is $\int_{\theta_1}^{\theta_2}\dfrac{1}{2}r^2 d\theta = \dfrac{1}{2}\int_{\theta_1}^{\theta_2}[f(\theta)]^2 d\theta$. If the curve is given parametrically as $r = r(t)$ and $\theta = \theta(t), a \leq t \leq b$, the area is $\dfrac{1}{2}\int_a^b (r(t))^2\dfrac{d\theta}{dt}dt$.

12. (i) $\dfrac{a^2}{2}\int_0^{2\pi}(1 + \cos^2\theta - 2\cos\theta)d\theta = \dfrac{3\pi a^2}{2}$ (ii) $2a^2\int_{-\pi/4}^{\pi/4}\cos 2\theta d\theta = 2a^2$

(iii) $\dfrac{ed^2}{2}\int_0^{2\pi}\dfrac{d\theta}{(1 + e\cos\theta)^2}$ using Theorem (8.1). To find an antiderivative for $\dfrac{1}{(1 + e\cos\theta)^2}$ put $u = \tan\dfrac{\theta}{2}$ and then $u^2 = v$ and finally use partial

fractions. The substitutions $d = a(\dfrac{1}{e} - e)$ and $e^2 = \dfrac{a^2 - b^2}{a^2}$ reduce the answer to $\pi ab$, which is much easier to obtain in cartesian co-ordinates.

13. As $\Delta\theta \to 0, Q \to P$ and so by definition, the direction of the secant tends

to that of the tangent at $P$. Also $| \overrightarrow{PQ} | = 2\sin\dfrac{\Delta\theta}{2}$ and so $\lim\limits_{\Delta\theta\to 0}|\dfrac{\overrightarrow{PQ}}{\Delta\theta}| =$

$\lim\limits_{\Delta\theta\to 0}\dfrac{2\sin\dfrac{\Delta\theta}{2}}{2\dfrac{\Delta\theta}{2}} = 1$. This proves (a). For (b), note that $e_r(P)$ and $e_r(Q)$

are simply the vectors $\overrightarrow{OP}$ and $\overrightarrow{OQ}$. So $\dfrac{d}{d\theta}(e_r)$ at $P = \lim\limits_{\Delta\theta\to 0}\dfrac{\overrightarrow{OQ} - \overrightarrow{OP}}{\Delta\theta} =$

$\lim\limits_{\Delta\theta\to 0}\dfrac{\overrightarrow{PQ}}{\Delta\theta} =$ unit tangent vector at $P = e_\theta$. $e_\theta(Q)$ and $e_\theta(P)$ are obtained by

a $90°$ counterclockwise rotation of $e_r(Q)$ and $e_\theta(P)$ respectively. So $\dfrac{d}{d\theta}(e_\theta)$

at $P$ will be obtained by rotating $\dfrac{d}{d\theta}(e_r)$ at $P$ counterclockwise by $90°$.

Since $\dfrac{d}{d\theta}(e_r)$ at $P$ is $e_\theta($ at $P)$, by rotating it counterclockwise through $90°$, we get $-e_r$, proving (c).

14. (a) follows from linearity of differentiation. For (b) note that $((f(\theta))^2 + (f'(\theta))^2)' \equiv 0$ whence $(f(\theta))^2 + (f'(\theta))^2 = k$, a constant. The initial conditions determine $k$ as $0$ and hence both $f(\theta)$ and $f'(\theta)$ vanish for all $\theta$. Applying (a) and (b) to $f_1(\theta) - f_2(\theta)$ gives (c).

15. If $\dfrac{c^2}{ka_0} - 1 < 0$, then at $\theta = 0, r$ would attain its maximum and not minimum,

contrary to the choice of the initial ray. (The possibility $\dfrac{c^2}{ka_0} - 1 = 0$ is, however, not ruled out. In that case $r = a$ constant and the orbit is a circle, which can be considered as a degenerate ellipse.)

16. By Exercise (8.14), $\dfrac{dA}{dt} = \dfrac{1}{2}r^2\dfrac{d\theta}{dt}$ which is a constant by (27). Conversely,

constancy of $\dfrac{dA}{dt}$ implies $r^2\dot\theta$ is a constant and hence $\dfrac{d}{dt}(r^2\dot\theta) = 0$.

So $r(2\dot{r}\dot\theta + r\ddot\theta) = 0$. As $r \neq 0$, (21) implies that the acceleration is radial.

17. We already have $r^2\dot\theta = c$ (say). If, further, $r = \dfrac{ed}{1 + e\cos\theta}$, then $\dot{r} =$

$\dfrac{e^2 d\sin\theta}{(1 + e\cos\theta)^2}\dot\theta = \dfrac{r^2\sin\theta}{d}\dot\theta = \dfrac{c\sin\theta}{d}$. Hence $\ddot{r} = \dfrac{c\cos\theta}{d}\dot\theta = \dfrac{c^2\cos\theta}{dr^2}$. So the

radial component of the acceleration, viz. $\ddot{r} - r\dot\theta^2$ comes out as $\dfrac{c^2\cos\theta}{dr^2} - \dfrac{c^2}{r^3}$

which reduces to $-\dfrac{c^2}{der^2}$. As the transverse component is $0$, $\mathbf{a}$ and hence the force on the planet is directed towards the Sun and proportional to $\dfrac{1}{r^2}$.

**18.** For an ellipse with semi-major axis $a$ and eccentricity $e, d$ (the distance between a focus and the corresponding directrix) equals $a(\dfrac{1}{e} - e)$. So $de = a(1 - e^2)$. Since $a_0$ is the value of $r(= \dfrac{de}{1 + e\cos\theta})$ at $\theta = 0$, we get $a_0 = a(1-e)$. Also from the last exercise, the radial component of the acceleration is $-\dfrac{c^2}{der^2}$, which also equals $-\dfrac{k}{r^2}$. This gives the formula for $c$ in terms of $k$. Finally, $\dfrac{dA}{dt} = \dfrac{1}{2}r^2\dot\theta = \dfrac{1}{2}c$. The area of the ellipse is $\pi ab = \pi a^2\sqrt{1 - e^2}$ and is swept in time $T, T$ being the period. So $T = \dfrac{\pi a^2\sqrt{1 - e^2}}{\dfrac{1}{2}c} = \dfrac{2\pi a^2\sqrt{1 - e^2}}{\sqrt{k}\sqrt{a}\sqrt{1 - e^2}} = 2\pi a\sqrt{\dfrac{a}{k}}$.

**19.** By the last exercise, $\dfrac{T^2}{a^3} = \dfrac{4\pi^2}{k}$, where $k = GM$. $G$ is a universal constant while $M$ is the mass of the Sun. So $\dfrac{T^2}{a^3}$ is the same for all planets in the same solar system.

**20.**

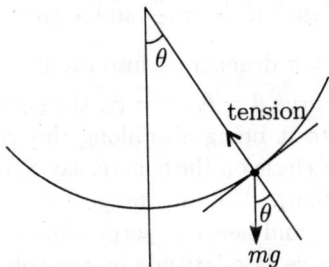

Let $m$ be the mass of the bob of the pendulum. Take the fixed end of the pendulum as the pole and the vertically downward ray from it as the initial ray. The tension in the pendulum is radial. The only other force on the bob of the pendulum is its weight $mg$ acting vertically downwards. Also $r = L$, a constant and so $\dot r = \ddot r = 0$.

The result follows by considering the transverse component of the acceleration. To solve the equation exactly, elliptic integrals are needed. However, if $\theta$ is numerically small then approximating $\sin\theta$ by $\theta$ reduces the equation to $\ddot\theta + \dfrac{g}{L}\theta = 0$ which can be solved analogously to (33) in the text.

**21.** (i) $\int_0^1 \int_0^{2\pi} r^3\cos^2\theta \, d\theta dr = \int_0^1 \pi r^3 dr = \dfrac{\pi}{4}$. (ii) $\int_0^1 \int_0^{2\pi} re^{r^2} d\theta dr = \pi(e - 1)$ (iii) $\pi(1 - \cos 1)$.

**22.** (i) is clear from the figure below. (ii) follows from (i) since the integrand is non-negative everywhere. Write $e^{-(x^2+y^2)}$ as $e^{-x^2}e^{-y^2}$ and use Exercise (4.6) to get (iii). For (iv), write $e^{-(x^2+y^2)}$ as $e^{-r^2}$ and convert the double integral to $\int_0^R \int_0^{\pi/2} re^{-r^2} d\theta dr$. If we let $R \to \infty$, in (iv) we

get $\lim\limits_{R \to \infty} \iint_{D_R} e^{-(x^2+y^2)} dA = \dfrac{\pi}{4}$. Hence by (ii) and the Sandwich Theorem,

$\lim\limits_{R \to \infty} \iint_{S_R} e^{-(x^2+y^2)} dA$ also exists and equals $\dfrac{\pi}{4}$. So by (iii) $\int_0^\infty e^{-x^2} dx =$

$\lim\limits_{R \to \infty} \int_0^R e^{-x^2} dx = \sqrt{\dfrac{\pi}{4}} = \dfrac{\sqrt{\pi}}{2}$.

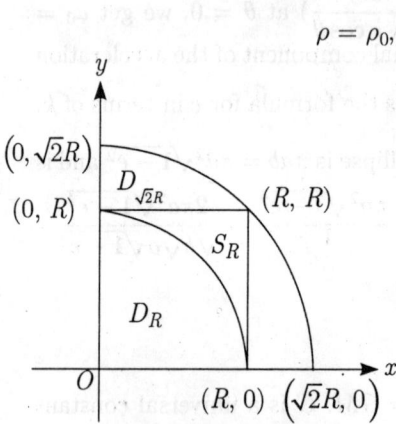

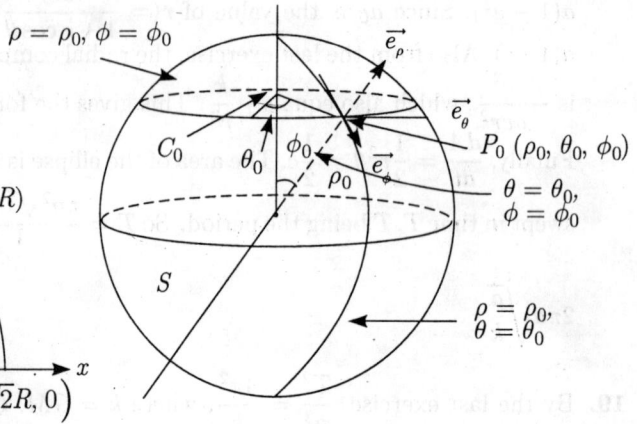

Exercise (8.22)                                        Exercise (8.23)

23. $\theta = \theta_0$ is a vertical half-plane bounded on one side by the $z$-axis and making
    an angle $\theta_0$ with the $x$-$z$ plane. The surface $\phi = \phi_0$ is an inverted cone if
    $0 < \phi_0 < \dfrac{\pi}{2}$, a plane (viz. the $x$-$y$ plane) if $\phi_0 = \dfrac{\pi}{2}$ and a cone again if
    $\dfrac{\pi}{2} < \phi_0 < \pi$. The cases $\phi_0 = 0$ and $\phi_0 = \pi$ degenerate into the positive and
    the negative $z$-axis respectively. The curve $\theta = \theta_0, \phi = \phi_0$ through $P_0$ is a
    half-ray from $O$, the unit tangent $\mathbf{e}_\rho$ to it being also along this ray. The
    curve $\rho = \rho_0, \theta = \theta_0$ is the arc of a great circle on the sphere, say $S$, of radius
    $\rho_0$ centred at $O$ (or popularly, a meridian). The unit tangent to it, $\mathbf{e}_\phi$ lies
    in the tangent plane of the sphere at $P_0$ and hence is perpendicular to $OP_0$
    and hence to $\mathbf{e}_\rho$. The curve $\rho = p_0, \phi = \phi_0$ is a latitude on the sphere $S$. It
    is the intersection of $S$ with a horizontal plane. So it is a circle with centre,
    say $C_0$, on the $z$-axis. $\mathbf{e}_\theta$ is a vector in this plane perpendicular to $C_0P_0$.
    $\mathbf{e}_\theta$ is also lies in the tangent plane of the sphere $S$ at $P_0$. Hence $\mathbf{e}_\theta \perp \mathbf{e}_\rho$.
    Finally, $\mathbf{e}_\phi$ and $\mathbf{e}_\theta$ are orthogonal since $\mathbf{e}_\phi$ lies in a vertical plane through
    $O, P_0$ and $C_0$ while $\mathbf{e}_\theta$ is perpendicular to this plane (since $\theta = \theta_0$ on this
    plane.)

# Index